CARCINOGENICITY

CARCINOGENICITY

Testing, Predicting, and Interpreting Chemical Effects

EDITED BY

Kirk T. Kitchin

Environmental Carcinogenesis Division
National Health and Environmental Effects Research Laboratory
U.S. Environmental Protection Agency
Research Triangle Park, North Carolina

MARCEL DEKKER, INC. NEW YORK • BASEL

To all my teachers. Without having had good teachers,
it is difficult to accomplish anything of value in
scientific endeavor.

Library of Congress Cataloging-in-Publication Data

Carcinogenicity : testing, predicting, and interpreting chemical effects / edited by Kirk T. Kitchin.
p. cm.
Includes bibliographic references and index.
ISBN 0-8247-9893-7 (alk. paper)
1. Carcinogenicity testing. I. Kitchin, Kirk T.
RC268.65.C369 1999
616.99'4071--dc21 98-44484
CIP

This book is printed on acid-free paper.

Headquarters
Marcel Dekker, Inc.
270 Madison Avenue, New York, NY 10016
tel: 212-696-9000; fax: 212-685-4540

Eastern Hemisphere Distribution
Marcel Dekker AG
Hutgasse 4, Postfach 812, CH-4001 Basel, Switzerland
tel: 44-61-261-8482; fax: 44-61-261-8896

World Wide Web
http://www.dekker.com

The publisher offers discounts on this book when ordered in bulk quantities. For more information, write to Special Sales/Professional Marketing at the headquarters address above.

Copyright © 1999 by Marcel Dekker, Inc. All Rights Reserved.

Neither this book nor any part may be reproduced or transmitted in any form or by any means, electronic or mechanical, including photocopying, microfilming, and recording, or by any information storage and retrieval system, without permission in writing from the publisher.

Current printing (last digit):
10 9 8 7 6 5 4 3 2 1

PRINTED IN THE UNITED STATES OF AMERICA

Foreword

There is an adage in cancer research that all investigators working in this field must be absolutely convinced that their approach is the best. The logic of this statement is obvious, for if they did not have such a basis for their work, they would be doing something else, unless economic or other circumstances disallowed their ideal. One may apply this adage to the methods used to identify chemical agents that are carcinogenic in animals, with the goal of extrapolating these findings to an accurate judgment of cancer risk in the human. However, the *accuracy* of the test results may not be directly applicable to the question of the *risk* of cancer *induction* in the human upon exposure to a test agent. It is this quandary with which regulators, investigators, commercial vendors, and the public wrestle. After the demonstration by a number of investigators, primarily in Europe and Japan, that administration of chemicals by direct application or by feeding could induce cancer in a variety of tissues in rodents, logic dictated that such findings may be related to the human situation. It was from such an experimental basis, developed in the first half of the 20th century, that the chronic two-year bioassay "cancer test" originated at the National Cancer Institute and in other laboratories in the world. During the last half of this century, the chronic two-year bioassay has become the "gold standard" for the identification of chemicals that, at a chronic continuous dose that produces "marginal" evidence of toxicity, results in a statistically significant increase in the incidence of neoplasms in the test animal.

Because of the exponential increase in our chemical, biological, and pharmaceutical knowledge, the number of chemicals that have been introduced into the environment of our planet has increased by three or four orders of magnitude during this century. Fortunately, an appreciation of the potential impact of this increase in environmental chemicals on human health and disease has been paralleled by an increasing understanding of health issues by the average individual in relation to the potential dangers of these environmental changes. Unfortunately, the knowledge accumulated by the public comes largely from the media, where scientific information is often tainted by the inaccurate and faulty overemphasis in reporting health issues dealing with environmental concerns. This, coupled with some degree of ignorance of the subject on the part of most individuals, has in turn led to an inordinate fear of any factor even potentially related to an increased risk of cancer. Therefore, in a manner analogous to testing millions of chemicals for their potential therapeutic effect on cancer (as promulgated by the Cancer Chemotherapy National Service Center during its existence earlier this century), the use of the chronic bioassay has expanded enormously. It is safe to say that hundreds of millions—if not billions—of dollars, millions of rats and mice, and untold working hours have been spent on the chronic cancer bioassay to determine the potential cancer risk of numerous chemicals in animals, with a potential for extrapolation to the human. Unlike screens for cancer therapy, however, information from a large number of other methodologies, both in vivo and in vitro, has been used during the last two decades to supplement, reinforce, or abrogate the results seen with the "gold standard."

It is thus appropriate, and one may say long overdue, that a comprehensive consideration of the chronic cancer bioassay, its facts and fallacies, as well as many of the major adjuncts found useful in interpreting the data obtained from the chronic bioassay, is presented in a single, accessible volume. This book, through its many expert authors, presents an excellent compendium of this type of information. No text can be exhaustive, but it can be interpretive, as this volume strives to be. Even in this area, one finds that interpretations may be extreme or "middle of the road." The extremes of extrapolation from no risk on the one hand to an almost certain risk at the other extreme are clearly unwarranted. The problem is whether the middle course is warranted in some or all cases. No one would argue that the chronic bioassay is a perfect system for the extrapolation of carcinogenic risk from animals to humans. The problem is how to deal with the imperfections of the system and to logically, scientifically, and accurately increase the predictability of human cancer risk from animal and other studies. This text is presented as a state-of-the-art treatise on the chronic bioassay and several other methodologies that have been used to predict carcinogenicity but without cancer as an endpoint. Bacterial mutagenicity testing is a rapid, economical approach but always with a nagging percentage of results that do not coincide with the gold standard. SAR models are based almost entirely on the chronic bioassay, while the results of a number of biochemical tests have found varying degrees of usefulness. Identification of the stage(s) at which a chemical acts may be useful in some areas of risk estimation, but might find its greatest effectiveness in the development of biologically based models for extrapolation of risk. The exponential explosion of our knowledge of the molecular mechanisms of neoplasia has led to the possibility that genetically altered animals may serve as a useful adjunct to the gold standard.

The future will determine the best methods for the ideal of extrapolating animal data to human cancer risk. Hopefully, as knowledge of carcinogenesis and cancer continues to expand almost exponentially, methods for testing, predicting, and interpreting the toxic and carcinogenic effects of chemicals as applied to human risk will advance concurrently and be useful. A popular version of the subject as covered in this treatise, which is scientifically accurate and socially neutral, may be a step in the direction of decreasing some members of the population now at risk of cancer development, as well as the fear of cancer, from the environment in which we live.

Henry C. Pitot, M.D., Ph.D.
Professor of Oncology and of
Pathology and Laboratory Medicine
McArdle Laboratory for Cancer Research
Madison, Wisconsin

Preface

The primary reason for producing a book on cancer is obvious: the enormous personal, human, and societal cost of the disease. Few people do not know someone whose life has been adversely affected by this killer.

Cancer is not a monolithic entity as it is sometimes perceived, but a family of related diseases whose manifestations are extremely different from tissue to tissue, animal to animal, and sometimes even between males and females. For instance, cardiac muscle, bone, and adipose tissue have low rates of carcinogenesis, whereas the colon, mammary gland, and prostate have considerably higher rates. Humans rarely develop liver cancer, apart from that caused by aflatoxin ingestion or hepatitis B virus (see Chapter 15). Rodents, however—male mice in particular (see Chapter 4)—frequently develop liver cancer in response to chemical exposure. In another example, male humans rarely get mammary cancer; women, unfortunately, often do. How to interpret the results of cancer tests when mouse liver is the sole site of cancer is a scientific, regulatory, political, and economic debate that continues to rage without any sign of resolution.

In the twentieth century, the causes and mechanisms of cancer have largely been perceived from several standpoints: pathological, DNA-based, mutational, oncogenetic, antioncogenetic, and cellular. These approaches, although certainly worthwhile, cannot begin to embrace the true magnitude of the biological problem. If a typical cell has about 70,000 genes coding for about 70,000 proteins, which proteins or genes should we study first? Which do we leave out, hoping that they will not later be identified as being important? An extensive study of 20 proteins—or the genes that account for these proteins—necessarily omits about 69,980 proteins or genes, any one or combination of which might be an important cause of a particular cancer. This is the problem facing the experimental biologist—there is always so much more that is unstudied than what can be studied. Unfortunately, the ultimate answer (such as the "cause" of cancer) is always more likely to be in the vast pile of what is unstudied.

For scientists, physicians, and those involved in public health, books offer a way to ascertain the undisputed scientific center without falling into the puzzling traps of conflicting data and poorly chosen or executed experiments. We are living in an age of rapid information retrieval and, perhaps, overload. Scientists working in the area of oncogenes cannot possibly read every scientific and clinical paper published in the field. This book attempts to provide in one volume both an overview of our present knowledge and information on how to build on it by compiling and evaluating data.

The volume is divided into three parts, on testing, predicting, and interpreting carcinogenicity. The part on testing addresses the following questions. Why do we test for chemical carcinogenicity? How do we carry out such tests? How did carcinogenicity protocols develop? What are the advantages and disadvantages of using the maximum tolerated dose in the cancer bioassay? What are the principal areas of the rodent cancer bioassay in which scientific interpretations differ? What sense can we make of positive mouse liver data? How can one obtain information from reliable sources of chemical carcinogenicity data?

The part on predicting carcinogenicity illustrates the varied approaches of individuals and organizations that have responded to the challenge of how to more efficiently and effectively achieve goals similar to those of the rodent cancer bioassay. Four whole-animal predictive systems are presented. Chapter 13 explains how rat liver foci systems, as opposed to a full rodent bioassay, offer considerable quality with a reduction in time and expense. Transgenic animal models, as described in Chapter 14, can contribute greater sensitivity and shorter detection times for certain carcinogenic mechanisms, although such transgenic model systems may not perform well for all known mechanisms of carcinogenesis.

The determination of biochemical parameters following acute chemical treatment to rats has been utilized to predict chemical carcinogenicity (Chapter 10). The biochemical parameters system is based on three carcinogenic mechanisms: DNA damage, cell toxicity, and cell proliferation. The fourth, a novel, whole-animal system, requires chemical treatment for one month followed by one enzymatic determination of the hepatic enzyme pyruvate kinase (Chapter 11).

Two factors known to be important in carcinogenesis are proliferation of the peroxisome, a subcellular organelle (Chapter 12), and intracellular gap junction communication (Chapter 9). They are possible sources of short-term tests for chemical carcinogenicity. The mutational ability of chemicals, computer-based systems, and the k_e test for electrophilicity are three well-known ways to predict the carcinogenicity of chemicals (Chapters 6 to 8).

The part on interpreting carcinogenicity presents human, mouse, and rat data for organs with high animal and low human cancer rates (group A), high animal and high human cancer rates (group B), and low animal and high human cancer rates (group C). In group A, the use of animal cancer data obtained in the liver, kidney, forestomach, and thyroid gland are perceived by some as being hyperresponsive, too sensitive, and of limited or no utility for extrapolating to human cancer risks. Others see considerable value and utility in the animal cancer data obtained in group A organs. The liver is such a responsive and important organ in the interpretation of carcinogenesis data that the discussion of this subject area has been broken up into three chapters for human, rat, and mouse data (Chapters 15, 13, and 4).

Group B organs (mammary gland, hematopoietic, urinary bladder, oral cavity, and skin) are less of an interpretive battleground than group A organs. For group B organs, all four major mechanisms of carcinogenesis (electrophile generation, oxidation of DNA, receptor–protein interactions, and cell proliferation) are known to be important. The high cancer rates for group B organs in both experimental animals and humans may at first give us a false sense of security about how well the experimental animal models are working. As we are better able to understand the probable carcinogenic mechanism(s) in the same organ in the three species, we may find that the important differences between the three species are more numerous than we suspect. This is particularly true for receptor-based and for cell-proliferation-based carcinogenic mechanisms.

Animal cancer data of group C organs are the opposite of group A organs. Group C organs have low animal cancer rates and high human cancer rates. In contrast to the continuing clamor and interpretive battleground associated with group A organs, there is little debate over group C organs. Few voices have questioned the adequacy of the present-day animal bioassay to protect the public health from possible cancer risks in these group C organs. Improved efforts must be made toward the development of cancer-predictive systems or short-term tests for cancer of the prostate gland, pancreas, colon/rectum, and cervix/uterus. We must also work harder to develop mechanistically based tests to identify chemicals that act through three of the major mechanisms of carcinogenesis: oxidation of DNA, receptor–protein interactions, and cell proliferation. Only for electrophile generation have we developed a variety of tests that perform well mechanistically and operationally.

In Chapter 29, an integrative attempt is made to synthesize and summarize what we currently know about the four principal carcinogenic mechanisms, interspecies differences, and interorgan differences in carcinogenesis. A sequential mutation/cell proliferation cascade model of carcinogenesis that can fit many organs can be seen in Figure 1 of this chapter. Table 2 presents a data matrix of the four major carcinogenic mechanisms by 14 organs in three species. The many gaps and question marks in the table indicate areas in which future research may be worthwhile.

Books on carcinogenicity can summarize the state of the art. It is hoped that this book will be useful to readers seeking to understand chemical carcinogenicity, planning carcinogenicity experiments, and contributing to future knowledge about carcinogenicity.

The creation of this book depended on the contributions of many people and organizations. I wish to thank the chapter contributors for taking valuable time away from their research programs and other scientific activities to present the state of the art in their research areas. In particular, Dr. David Clayson has synthesized elements of the organ-specific carcinogenesis process in mice, rats, and humans into a thorough and holistic chapter. Janice Brown, of the U.S. Environmental Protection Agency, assisted in many aspects of the book's development, including editing, word processing, and computer graphics. Dr. John Allis contributed computer expertise. The EPA kindly provided the place, time, and computer systems with which to undertake this large project. Russell Dekker, Chief Publishing Officer of Marcel Dekker, Inc., suggested both the original concept for the book and its title. Two other individuals at Marcel Dekker who were invaluable to the production of the book were Kerry Doyle, as a resource person on matters of logistics and management, and Henry Boehm, who supervised its production.

Kirk T. Kitchin

Contents

Predicting Carcinogenicity

Interpreting Carcinogenicity: Organs with High Animal and Low Human Cancer Rates

Contributors

Rajesh Agarwal, Ph.D.* Department of Dermatology, Case Western Reserve University, Cleveland, Ohio

Marko Jan Appel, Ph.D. Division of Toxicology, TNO Nutrition and Food Research Institute, Zeist, The Netherlands

George Bakale, Ph.D. Department of Radiology, Case Western Reserve University School of Medicine, Cleveland, Ohio

Maarten C. Bosland Nelson Institute of Environmental Medicine, New York University Medical Center, New York, New York

Janice L. Brown Biochemistry and Pathobiology Branch, Environmental Carcinogenesis Division, U.S. Environmental Protection Agency, Research Triangle Park, North Carolina

H. B. Bueno-de-Mesquita, M.D., M.P.H., Ph.D. Department of Chronic Diseases and Environmental Epidemiology, National Institute of Public Health and the Environment, Bilthoven, The Netherlands

Wai Nang Choy, Ph.D., D.A.B.T. Safety Evaluation Center, Schering-Plough Research Institute, Lafayette, New Jersey

---

**Current affiliation*: Scientist, Center for Cancer Causation and Prevention, AMC Cancer Research Center, Denver, Colorado.

David B. Clayson* CARP, Ontario, Canada

David M. DeMarini, Ph.D. Environmental Carcinogenesis Division, U.S. Environmental Protection Agency, Research Triangle Park, North Carolina

William C. Eastin, Ph.D. Toxicology Operations Branch, Environmental Toxicology Program, National Institute of Environmental Health Sciences, Research Triangle Park, North Carolina

Clifford R. Elcombe, Ph.D.† Zeneca Central Toxicology Laboratory, Cheshire, United Kingdom

Harald Enzmann American Health Foundation, Valhalla, New York

Tony R. Fox, Ph.D. Cancer Program, Chemical Industry Institute of Toxicology, Research Triangle Park, North Carolina

Shoji Fukushima, M.D., Ph.D. First Department of Pathology, Osaka City University Medical School, Osaka, Japan

Thomas L. Goldsworthy, Ph.D. Cancer Program, Chemical Industry Institute of Toxicology, Research Triangle Park, North Carolina

Paul Grasso, B.Sc., M.D., FRCPath. School of Biological Science, University of Surrey, Guildford, Surrey, United Kingdom

F. F. Hahn Lovelace Respiratory Research Institute, Albuquerque, New Mexico

Gordon C. Hard, B.V.Sc., Ph.D., D.Sc., FRCVS, FRCPath., FATox.Sci. American Health Foundation, Valhalla, New York

Masao Hirose, M.D., Ph.D. First Department of Pathology, Medical School, Nagoya City University, Nagoya, Japan

James Huff, M.D. Intramural Research, National Institute of Environmental Health Sciences, Research Triangle Park, North Carolina

Nobuyuki Ito, M.D., Ph.D. Nagoya City University, Nagoya, Japan

Kirk T. Kitchin, Ph.D. Biochemistry and Pathobiology Branch, Environmental Carcinogenesis Division, National Health and Environmental Effects Research Laboratory, U.S. Environmental Protection Agency, Research Triangle Park, North Carolina

---

*Retired.

†*Current affiliation*: Biomedical Research Centre, University of Dundee, Ninewells Hospital and Medical School, Dundee, Scotland.

Vladimir Krutovskikh, M.D., Ph.D. Unit of Multistage Carcinogenesis, International Agency for Research on Cancer, Lyon, France

Gabriel A. Kune, M.D. University of Melbourne, Melbourne, Victoria, Australia

Moushumi Lahiri, Ph.D. Department of Dermatology, Case Western Reserve University, Cleveland, Ohio

James S. MacDonald, Ph.D. Department of Drug Safety and Metabolism, Schering-Plough Research Institute, Kenilworth, New Jersey

Orest T. Macina, Ph.D. Department of Environmental and Occupational Health, University of Pittsburgh, Pittsburgh, Pennsylvania

Marios Marselos, M.D., Ph.D. Department of Pharmacology, Medical School, University of Ioannina, Ioannina, Greece

Hasan Mukhtar, Ph.D. Department of Dermatology, Case Western Reserve University, Cleveland, Ohio

Minako Nagao, Ph.D. Carcinogenics Division, National Cancer Center Research Institute, Tokyo, Japan

Elizabeth W. Newcomb, Ph.D. Department of Pathology, Kaplan Cancer Center, New York University Medical Center, New York, New York

Shirley Price, B.Sc., M.Sc., Ph.D. Robens Institute of Industrial and Environmental Health and Safety, University of Surrey, Guildford, Surrey, United Kingdom

Elizabeth H. Romach, Ph.D. Cancer Program, Chemical Industry Institute of Toxicology, Research Triangle Park, North Carolina

Herbert S. Rosenkranz, Ph.D. Department of Environmental and Occupational Health, University of Pittsburgh, Pittsburgh, Pennsylvania

Dhananjaya Saranath, Ph.D. Laboratory of Cancer Genes, Cancer Research Institute, Tata Memorial Centre, Mumbai, India

Harvey E. Scribner, Ph.D., D.A.B.T. Toxicology Department, Rohm and Haas Company, Spring House, Pennsylvania

Takashi Sugimura, M.D., Ph.D. Carcinogenesis Division, National Cancer Center Research Institute, Tokyo, Japan

Raymond W. Tennant, Ph.D. Laboratory of Environmental Carcinogenesis/Mutagenesis, National Institute of Environmental Health Sciences, Research Triangle Park, North Carolina

Geraldine Anne Thomas, B.Sc., Ph.D. Department of Histopathology, Addenbrooke's Hospital, University of Cambridge, Cambridge, United Kingdom

Jonathan D. Tugwood, Ph.D. Zeneca Central Toxicology Laboratory, Alderley Park, Macclesfield, Cheshire, United Kingdom

Henk J. van Kranen, Ph.D. Laboratory of Health Effects Research, National Institute of Public Health and the Environment, Bilthoven, The Netherlands

Luis Vitetta, Ph.D. University of Melbourne, Melbourne, Victoria, Australia

Hideki Wanibuchi, M.D., Ph.D. First Department of Pathology, Osaka City University Medical School, Osaka, Japan

Gary M. Williams, M.D. Naylor Dana Institute, American Health Foundation, Valhalla, New York

Ruud A. Woutersen, Ph.D. Division of Toxicology, TNO Nutrition and Food Research Institute, Zeist, The Netherlands

Shinji Yamamoto, M.D., Ph.D. First Department of Pathology, Osaka City University Medical School, Osaka, Japan

Hiroshi Yamasaki, Ph.D. Unit of Multistage Carcinogenesis, International Agency for Research on Cancer, Lyon, France

Susumu Yanagi College of Nursing, Nara Medical University, Nara, Japan

Ying Ping Zhang Department of Environmental and Occupational Health, University of Pittsburgh, Pittsburgh, Pennsylvania

1

The Cancer Bioassay

Paul Grasso and Shirley Price
University of Surrey, Guildford, Surrey, England

I. A HISTORICAL NOTE

Laboratory animals were employed to study biological phenomena long before cancer research was established as a branch of the biological sciences (Rodericks, 1992). Early workers recognized that as animals approached the end of their life span, many developed tumors, some of which were fatal (Rodericks, 1992). The striking similarity to cancer development in the later decades of life in humans did not escape the attention of these workers. Thus, it must have appeared natural to turn to laboratory animals to investigate the observations of Perceival Pott (1775) on the development of scrotal cancer in chimney sweeps and of Rehn on the development of bladder cancer in workers in the aniline dye industry (Rehn, 1895). These two observations could be said to have launched the concept of chemical carcinogenesis.

Early experience provided some justification for the view that animals could reliably detect carcinogenic activity. First Yamagiwa and Ichikawa (1918) and then others (see Clayson, 1962) showed that tar and soot produce skin tumors when painted on the skin of rabbits and mice and that the tumors were of the same type as the scrotal cancers in chimney sweeps. Furthermore, identification of the carcinogenic activity of individual chemicals in soot and coal tar was possible because of the sensitivity of laboratory animals, in particular mice and rabbits, to these chemicals (Cook, 1932).

Attempts to reproduce bladder cancer in laboratory rodents failed, but several years after Rhen's observations, Hueper (1938) was successful using the dog. Over the years it was observed that other internal organs of animals were sensitive to the carcinogenic effect of chemicals, and, in particular, agents known to be carcinogenic in humans (with the exception of arsenic) were also carcinogenic in laboratory animals, although not necessarily producing the same type of tumor. Experience led to the view that animals could be used to investigate the potential carcinogenic effect of chemicals of unknown carcinogenic activity.

At first the tests were conducted on small numbers of animals, but experience showed that while this was successful with potent carcinogenic substances, use of small groups could jeopardize the validity of the test with carcinogenic substances of lower potency, with noncarcinogenic substances, and where animals are accidentally lost (e.g., through an intercurrent infection) or where the incidence of tumors in controls is high. Use of small groups makes statistical analysis virtually impossible. Gradually, the numbers in the groups increased to the numbers employed currently.

After decades of data accumulated from cancer bioassays in rats and mice, these experimental results can be qualitatively compared to the epidemiological data available in humans. On an organ basis, rodents and humans are sometimes similar and sometimes different in the degree of responsiveness to cancer (comparing human cancer rates, which apparently occur spontaneously with rodent cancer rates when treated with various chemicals). The cancer rates are low in humans and high in rodents for liver, kidney, forestomach, and thyroid gland. The cancer rates are high in both humans and rodents for lung, mammary, hematopictic system, urinary bladder, oral cavity, and skin. Finally, the cancer rates are high in humans and low in rodents for prostate, pancreas, colon/rectum, and cervix/uterus.

II. AIMS OF CARCINOGENICITY STUDIES

The aim of a carcinogenicity study is to discover the potential of a particular chemical to increase the incidence of tumors in laboratory animals under strictly controlled exposure conditions. In essence, a group of animals is exposed to a sufficient amount of the chemical for an adequate period of their life span by an acceptable route. Behind this simple statement lies a veritable jungle of differing ideas and opinions on its interpretation. None of these differing views has been satisfactorily resolved, but over the years experience has led to the acceptance of certain requirements and conventions for the conduct of a carcinogenicity study.

Most of these requirements and conventions have been embodied in guidelines for carcinogenicity testing issued by various agencies or official bodies. They reflect the experience gained in the last three decades but also display major weaknesses such as lack of uniformity (despite many attempts to harmonize recommendations for a test procedure) and inadequate scientific validation. In fact, one possible major criticism of the guidelines is that few attempts are made to justify the recommendations by referring to the scientific literature. Those guidelines that do so often fail to provide a citation. To these drawbacks one has to add the difficulties created by the mode of presentation of the guidelines. Some are tersely presented and resemble a series of cookbook recipes. Others are better presented but do not appear to follow any logical order. Yet others deal in depth with some topics and very sketchily with others. It comes as no surprise that scientists entrusted with the responsibility of setting up a carcinogenicity test are often disappointed with these cancer bioassay guidelines.

In this chapter it is hoped to provide a balanced account of the requirements and conventions that are commonly employed in carcinogenicity testing and comments made on their strengths and weaknesses.

III. THE TEST SUBSTANCE

The substance to be tested for carcinogenicity may be chemically "pure" or may consist of complex mixtures of chemicals. Few substances are absolutely pure, and many so designated contain a number of impurities, albeit in minute quantities. In general terms pharmaceutical products are of a high degree of purity. On the other hand, industrial chemicals and agochemicals may contain a substantial amount of impurities and are often referred to as "technical grade." Difficulties often arise when deciding the degree of purity of the substance required for the test. Ideally, the pure chemical and the "technical grade" should be tested simultaneously (IARC, 1980), but cost constraints may not allow this approach to be adopted. Since humans are likely to come into contact with "technical" grade materials, this grade should be tested first, but this carries the risk of noncarcinogenic components diluting carcinogenic components to the extent that amounts of the carcinogen present are too small to exert an observable effect in the animal study. If such a situation is suspected, then the individual components should be tested separately. Sometimes the minor components of a mixture are more carcinogenic than the major components.

The following basic information about a test substance is required before setting up the carcinogenicity study:

1. Chemical structure—this may create an "alert" as to possible carcinogenic activity (Ashby, 1991)
2. Identity and levels of impurities
3. Physicochemical properties—these decide the physical state of the

substance at ambient temperatures and the chemical reactivity with tissues with which it comes into contact. Both are important in deciding the route and mode of testing for carcinogenicity.

4. Stability of the substance in the product likely to come into contact with humans or animals and the solubility in the medium in which it is administered, e.g., injection fluid, diet, or aerosol
5. Probable daily level, route, and pattern of exposure to humans
6. Availability of methods of chemical analysis and suitability of these to the medium in which the test substance is likely to be administered to experimental animals
7. Pharmacokinetic data after a single or multiple exposure—these can assist in selecting the most suitable dosage regimen at the beginning of the experiment and in interpreting final results (Davies, 1992; Usui, 1992; Weissinger, 1992)

In the case of complex mixtures of chemicals (e.g., mineral oils or tobacco smoke), specimens representative of the whole range of materials to which exposure may occur may need to be tested, since analysis of the materials and the testing of individual components may be impossible or impractical.

IV. CHOICE OF TEST SPECIES AND STRAIN

Ideally the species selected should resemble humans as closely as possible in its physiology and in the metabolism of the compound to be tested; this is, however, rarely achievable. For purely practical reasons the choice is restricted to three species: rat, hamster, and mouse. Other species present serious drawbacks—principally large size and long life span. Large animals require more accommodation space, and much larger amounts of the test substance are needed to provide the correct dosage, an important factor in product development. In addition, their life span is usually long (7–10 years for dogs), again a major problem with product development. These drawbacks add considerably to the cost of the test and delay the discovery of any weak carcinogenic activity. In addition, smaller numbers of larger animals are used for purely practical reasons, and this reduces the surety of negative findings. The virtual restriction of choice to the three rodent species is not an undue problem, since experience has shown that they readily develop tumors with a large number of chemicals, including those that are carcinogenic in humans (IARC, 1979).

There has been much debate as to whether to use outbred or inbred strains for carcinogenicity testing. Inbred strains of rats and hamsters are difficult to obtain so that the choice is virtually limited to one of a number of outbred strains. The number of inbred strains of mice available is large, and a wide choice is possible. In theory, inbred strains are preferable because their tumor incidence is usually known with some precision, whereas in outbred mice the tumor incidence may vary markedly from one generation to the next due to dietary changes, genetic drift, or other causes (Tucker, 1979). In practice, it has become conventional to use outbred strains of mice, partly because influential government agencies have from time to time advocated their use (FDA, 1971; Ministry of Health and Welfare, Canada, 1973) and partly because of the concern that a selected strain may not metabolize the test substance to produce the ultimate carcinogenic chemical species (Page, 1977a,b). It is possible to circumvent this pitfall in metabolic capacity by employing three or more inbred strains in any one test, on the lines advocated by Festing (1975) for testing pharmacological or biochemical responses in inbred mice. This advice is sound, but the high cost of doing a conventional carcinogenicity study usually puts strict limits on the design, scope, and size of the test.

It is desirable to obtain as much information as possible on the background incidence of tumors in the strain of animals employed in a study (IARC, 1980). This background "historical control incidence" can be obtained from the breeder or from other long-term tests using the same strain. Historical control data are of value in the interpretation of a borderline result or of an unusual (high or low) incidence of tumors in concurrent controls (Haseman et al., 1984).

Whether to use a strain with a low or high natural cancer incidence requires consideration. A strain with a high incidence is likely to be more sensitive to carcinogens, but a high incidence in controls makes statistical analysis difficult if not impossible, thus placing in jeopardy the validity of the test (Lee, 1993). Besides, such strains may harbor a virus or some genetic proclivity for the production of a high incidence of particular tumors (Bailey et al., 1970), which may make it unclear whether any substance increasing cancer incidence is acting as a primary carcinogen or is merely stimulating the natural process of cancer development in the particular strain.

Occasionally other species are employed in carcinogenicity testing, but there must be good reason for doing so. For example, the dog is the species of choice for testing the carcinogenicity of aromatic amines, a class of chemicals carcinogenic in humans but not readily producing tumors in mice and rats (Parkinson et al., 1993, 1995). Similarly, concern about the significance in humans of the hepatic tumors produced in rodents by peroxisome proliferators (Grasso, 1994) was considerably allayed when it was shown that the marmoset, like humans, was resistant to the production of peroxisome proliferation and did not develop tumors in a 7-year study (Tucker and Orton, 1995).

In selecting the strain of animal, both freedom from or resistance to disease and longevity are of primary importance. A strain free from common infections will ensure that enough animals will survive long enough to develop tumors and allow appropriate statistical analysis. A long life span allows animals to survive beyond the minimum exposure duration of experiments (generally 2 years in rats and 18 months in mice), thus enhancing the sensitivity of the study.

V. DIET

It is common practice to allow animals free (ad libitum) access to food and water throughout a carcinogenicity study. Commercial feeds usually contain adequate amounts of vitamins and minerals but may also be unusually rich in fat and protein. It is a common experience that rodents gain rapidly in weight during the first weeks of the study and as the experiment progresses become obese and lose their agility and level of spontaneous activity and their sleek clean appearance. The rapid onset of obesity indicates that the animals are overeating and oversleeping. There is now well-documented evidence that the incidence of some tumors (e.g., mammary, hepatic, pulmonary, pituitary) in mice and rats is increased by excessive intake of calories (Roe, 1979; Keenan et al., 1992). It is clearly undesirable to have a high background incidence of tumors since this may affect the validity of results. At the moment some attempts are being made to reduce the calorie intake, in the expectation that it will increase longevity and reduce the natural incidence of tumors. Use of calorie restriction in carcinogenicity studies must be made with caution since knowledge about whether reduction of caloric intake will also reduce the sensitivity of tissues to genotoxic carcinogens is sparse and fragmentary.

VI. CAGING

Rodents are usually housed five animals of the same sex to a cage. This communal housing may occasionally jeopardize the validity of a study. The most important problems are loss of animals from fighting, particularly in some strains of mice (I. F. Gaunt, personal communication, 1989), and cannibalism of dead animals (IARC, 1980). Some authorities advise single caging to overcome these difficulties, but this may lead to other problems.

For example, the incidence of liver tumors has been shown to increase in single-caged C3H mice compared with conventionally caged ones (Peraino et al., 1973). Larger species are usually housed singly but occasionally must be housed in groups depending on the availability of resources.

VII. ROUTE OF ADMINISTRATION

There is agreement among toxicologists that the route of administration should parallel as closely as possible human exposure. Departure from this recommendation may be acceptable if the route of administration selected can be shown to provide results that are interpretable as relevant to human exposure. For example, it would not be advisable to carry out inhalation studies for carcinogenicity on the aerosol from a new type of tobacco or on the fumes from a mineral oil obtained by a new refining process when repeated application of the tobacco aerosol condensate or of the mineral oil to the skin of mice is likely to yield acceptable results with a considerable reduction in testing costs.

A. Oral Route

The oral route is appropriate for examining the carcinogenic activity of most substances ingested by humans. The test substance may be administered mixed with the diet or dissolved in drinking water to which the experimental animals are allowed unlimited or limited access. Administration of chemicals may also be by gavage using a chemical dissolved or suspended in a suitable medium. This has the advantage of a fairly accurate control of dosage and simulates the taking of medication at regular daily intervals, but not necessarily the intake of substances in food and drink. Daily gavage results in a high concentration of the test chemical in the stomach and intestines with the possibility of causing serious mucosal damage. In addition, the high concentrations administered may produce blood and tissue levels that, even though of limited duration, are so high that normal metabolic and protective mechanisms are overwhelmed, leading to pathological changes that could not occur with normal dosage. Incorporation of the test substance in food or dissolving it in drinking water allows a much more normal and uniform level of exposure of blood and tissue levels and furthermore allows a larger daily dose to be administered.

B. Inhalation Route

This route is used to test for the carcinogenicity of dusts, mists, aerosols, and gases. The techniques used for generating an atmosphere containing the required concentration of any such test substances are complicated and require considerable expertise and special equipment (WHO, 1978b). The advice of a respiratory physiologist and of an inhalation toxicologist must be sought in the design of the experiment, the type of equipment to be employed, and the precautions necessary for ensuring a valid result. It may also be necessary to consult a physical chemist if there are difficulties in producing a test atmosphere to ensure that it resembles that to which humans will be exposed and to design techniques for checking uniformity and stability.

The following information is helpful when designing a test by the inhalation route:

1. The respiratory rate is, on average, 18 per minute in humans, 120 in rat, and 150 in mouse (WHO, 1978a)
2. The respiratory exchange per minute is approximately 8.51 liters in humans, 150–200 ml in rat, and 25 ml in mouse (Mauderly and Kritchevsky, 1979).
3. Rodents breathe only through the nose, while humans may breathe through the mouth as well.
4. Resistance to air flow may develop in the airways of experimental animals if the inhaled material is too irritating (Mead, 1960).
5. Head-only exposure is recommended to avoid the ingestion of the test substance following self or mutual grooming and to prevent the animals from burying their nose in their neighbour's fur and filtering the inhaled air (IARC, 1980). However, head-only exposure usually involves a degree of restraint and is not normally conducted for periods longer than 6 hours a day 5 days per week. Longer periods of exposure may lead to serious "stress" and endanger the health of the animals (Powell and Hosey, 1965).

C. Topical Application

Substances that would normally come into contact with human skin are tested for carcinogenicity by topical application to the skin of mice. Other species are considerably less sensitive than mice. Topical application to the skin of mice has detected the carcinogenic activity of soot and mineral oils (Grimmer et al., 1982), polycyclic hydrocarbons (Clayson, 1962), alkylating agents, and other proximate carcinogens (Grasso and Crampton, 1972).

The test substance is usually applied in a solvent, which assists its passage across the keratin layer so that it comes into contact with the epidermal layer of cells. The volume applied varies (usually between 0.1 and 0.5 ml). Care should be taken to prevent spread from the site of application, particularly when using solvents of low viscosity such as acetone.

Daily application is not usually recommended; 2 or 3 times weekly suffices. A mild degree of irritation (redness or thickening) at the site of application is permissible, but ulceration of the skin must be avoided. Repeated ulceration may not only destroy the tissue from which tumors might develop but might in itself produce tumors, e.g., as seen with repeated application of caustic soda (Schmahl, 1982).

A preliminary experiment over several weeks to gauge the degree of epidermal reaction to the test substance is usually recommended. Sequential topical application is generally used in the detection of the initiating and promoting activity of a particular chemical (Berenblum, 1955).

D. Parenteral Route

The subcutaneous, intraperitoneal, or intravenous routes of injection of the test substance are sometimes used for testing carcinogenic activity. Subcutaneous injection was popular at one time because of the advantages it offers: accurate control of dosage, prevention of aerial oxidation, and the small amounts needed for testing (an important consideration for a newly synthesized or expensive chemical). The volume administered may vary between 0.5 and 5 ml. Administration may vary in frequency from daily to 2 days per week. The test substance is usually dissolved in water or in vegetable oil, but other vehicles have been employed. Care should be taken to choose a vehicle that is locally nonirritating and systemically nontoxic. Liquid substances have on occasion been administered undiluted, but this practice is not recommended unless the test substance is a nonirritant (Grasso and Golberg, 1966). Solids have been implanted into the subcutaneous tissues of the rat for carcinogenicity testing, but this procedure is no longer practiced.

Repeated intraperitoneal injection has occasionally been used in place of repeated subcutaneous injection. With this route there is a danger of inducing "chemical peritonitis" if the test substance is too irritating or of inducing a bacterial peritonitis if the gut is punctured. The use of this route is to be discouraged, even though it offers the advantage of rapid systemic diffusion of the test substance.

Occasionally, carcinogenicity testing has been attempted by the intravenous route. This route also offers the advantage of rapid systemic diffusion of the test substance. Only a limited number of intravenous injections can be administered to a laboratory rodent, however, because accessible veins soon become thrombosed. The total dose of the chemical that can be given by this route is necessarily much less than by other routes; thus intravenous administration has a limited scope in testing substances for carcinogenicity.

E. Other Routes

Chemicals have been tested for carcinogenic activity by other routes of exposure for specific reasons. For example, some antifertility compounds intended for vaginal application have been tested by the intravaginal route in mice (Boyland et al., 1966). Dusts have been administered intraperitoneally or intrapleurally to avoid the difficulties (and cost) of producing atmospheric suspensions suitable for administration by the inhalation route (Wagner, 1962). Intratracheal application has been used to bypass the nasal passages of the test animal (Grimmer et al., 1982). Results from these routes may not be sufficiently reliable to determine the carcinogenic activity of a chemical, but they often yield valuable information on some specific aspect of the carcinogenic processes involved.

VIII. DOSE SELECTION

The doses selected for carcinogenicity studies, particularly the highest dose, have been and still are the subject of considerable scientific debate. It has been stated that "high test dose selection is the most controversial and perhaps the most important element" in designing the protocol for a carcinogenicity test (Food Safety Council, 1978). It has been and still is the practice to select the highest dose compatible with long-term survival (maximum tolerated dose, or MTD) (Weissinger, 1992). Such doses are essential in order to ensure that the tissues of the test animal are exposed to the highest possible concentration of the chemical. These high tissue concentrations are necessary because the number of animals employed is, for logistical reasons, strictly limited (often 50 animals per dose level per sex per species) and not representative of the total population, which in theory may contain individuals that are particularly susceptible to developing tumors from the test substance. For example, a three-dose cancer bioassay will typically contain 300 rats treated with a test chemical; however, there are over 250 million people to protect just in the United States alone. When one contemplates trying to protect millions of people from a disease with an incidence of just less than one case of lung cancer expected per 1000 American citizens per year with a 300-rat sample size, one quickly understands the requirement that at least one dose level be at the MTD level. The number of experimental animals in an experimental group statistically required to achieve a 95% probability of detecting at least one animal with cancer has been calculated. To detect cancer that occurs at a probability of 50, 10, 1.0, 0.1, 0.001, and 0.0001% requires a group size of 5, 29, 299, 2995,

29956, and 299572 animals, respectively (Uehleke, 1983). The largest individual experimental group ever run involved 2109 mice in the 2-AAF NCTR study.

Administration of the MTD provides a degree of reassurance that absence of tumor development in a carcinogenicity study is not the result of insufficient dosage. The alternative to the MTD is the use of large numbers of animals (milli- or megamouse/rat experiments). In theory, such large numbers would ensure that animals have been included that would develop tumors with relatively low doses of the test chemical, thus reflecting better the population at large; in practice, large experiments of this sort are impracticable. For example, in an experiment carried out on 24,000 mice (the so-called ED01 experiment), there was no clear evidence of increased tumor incidence in liver or urinary bladder at the lowest doses (30–35 ppm in food) of acetylaminofluorene administration despite the fact that over 1000 mice had survived for 24 months at each of these dose levels (Kodell, 1983; Hughes, 1983).

Although there are no hard and fast rules for selecting the MTD for any particular substance, there seems to be some agreement as to what is to be aimed at and what is to be avoided (Food Safety Council, 1978; Munro, 1977). In general, the toxic potential of the chemical should be established by conventional toxicity tests. In addition, some knowledge of the toxicokinetics and metabolism should be available (Weissinger, 1992). Effects to be avoided include depression of growth rate greater than 10% of untreated controls, recognizable injury of one or more of the tissues or organs, excessive stimulation of glandular activity through normal or pathological mechanisms, and excessive pharmacological effects (Grasso, 1979). For substances that have extremely low toxicity e.g., some food colors, it is usual to limit the administration to 5% of the diet to avoid possible problems of nutrition deficiency or diet palatability.

There is less disagreement about setting the intermediate and lowest doses. In general the lowest dose should be a simple multiple of the expected human dose, while the intermediate dose is recommended as the geometric mean of this and the top dose.

There is no widely accepted procedure to select the dose at which novel foods should be administered to rodents for carcinogenicity testing. It is more accurate to talk of "levels" since they usually replace a substantial proportion (up to 50%) of the test animal's diet (de Groot et al., 1970). Nutritional factors mainly govern the selection of the level at which such substances can be administered, e.g., the relative contents of phospholipid and protein, the nature of the protein, and the amount of nucleic acid residues present (de Groot et al., 1970).

IX. DURATION OF STUDY

The rodents selected are usually about 6 weeks old and have been weaned onto an adult diet for at least 2 weeks. After about 2 weeks of acclimatization, they are exposed to the test chemical. Ideally treatment is then continued for the life span of the species: strictly speaking, "life span" means the average life expectancy of the animal. This varies somewhat from strain to strain and must be judged by experience of the longevity of the species to be used. Life span is to be distinguished from lifelong studies, where each animal is allowed to die from natural causes or from the carcinogenic or toxic effect of the test substance. Experience has shown that a lifelong study is impracticable, since the experiment must continue until the last animal dies. In every experiment, there is usually some animal that substantially exceeds the average life expectancy, so that a lifelong experiment in rats could easily go into the fourth year, adding considerably to the cost (see Peto et al., 1991).

There is a more cogent argument against lifelong studies. Normally, the incidence of tumors in most strains of rats is about 25% by the time the rats are 24 months old (approximately the same as that for humans). After this the incidence of tumors rises exponentially. For example, in a study on nitrosamines (see Peto et al., 1991), the tumor rate in untreated controls was 30% at 27 months and became 85% at 3½ years. This means that the longer the animals are allowed to live, the greater will be the background noise and the less sensitive, in statistical terms, the experiment will become. The same is probably true in mice and hamsters.

Some investigators, aware of this difficulty, advocate the termination of the experiment at 2 years in rats and 18 months in mice and hamsters. Such a regime is, however, too inflexible and might mean that the animals are killed at the time chemical-induced tumors are beginning to make an appearance.

The International Agency for Research on Cancer (IARC), in their Supplement 2 (1980), draw attention to the difficulties involved in deciding when to terminate a long-term cancer test and suggest a sensible compromise—that survivors in all groups be killed and the whole experiment terminated when mortality in the control or low-dose group reaches 75%. They also add this important rider: "An experiment is not really satisfactory if the mortality in the control or low dose group is higher than 50% before the end of week 104 of age in rats, week 96 for mice and week 80 for hamsters."

X. NUMBER OF ANIMALS PER GROUP

It is essential to have a sufficient number of animals alive at the end of a carcinogenicity study to carry out a valid statistical analysis of the tumor incidence in the test and control groups. The statistical approach to this problem is complex, and the reader is advised to consult one of the publications on this topic (see Lee, 1993). Nevertheless, experience has shown that, in general, this may be achieved by having group sizes of 50 males and 50 females per dose level at the beginning of the experiment, provided that no excessive deaths occur. This figure should be regarded as a "minimum," and it is not always prudent to adhere to it, particularly with mice and hamsters, where, on occasion, an unexplained higher-than-expected mortality may occasionally occur. Group sizes of about 80–100 of each sex group per dose in rats, mice, and hamsters at the start of the experiment are recommended.

An important issue in the design of carcinogenicity tests is the number of control groups required for a particular experiment. To have a single control group in a study of more than three dose levels would increase the chance of obtaining a false-negative or false-positive result, and the need for an additional control group should be seriously contemplated (Haseman et al., 1986).

Control groups may be completely untreated, as in many experiments where the test chemical is incorporated into the diet or added to the drinking water. Other procedures, such as head-only exposure in inhalation tests where the animals have to be restrained for several hours, repeated instillation in inhalation studies or repeated treatment by gavage produce a degree of stress that might affect the incidence of tumors (Haseman et al., 1985). In these circumstances the controls should be sham, e.g., administered the vehicle alone. In some instances it may be necessary to include an untreated control group as well as a sham-treated group in order to ascertain the effect of stress on tumor incidence.

XI. ANALYSIS OF THE DIET

The diet should be analyzed to ensure that it contains adequate amounts of the principal nutrients (carbohydrates, fats, and proteins) as well as adequate amounts of essential minerals and vitamins (IARC, 1980). Knowledge of the composition of the feed components is of particular importance when the test material itself is a nutrient (e.g., industrially treated protein or starch, protein from bacteria or yeast, or irradiated food). Such knowledge makes it possible to prepare a balanced diet even though the test material may be incorporated at levels as high as 20–60% (de Groot et al., 1970). Analysis of the diet should include a search for common dietary constituents that may influence carcinogenesis, for example, antioxidants, chlorinated hydrocarbons, substances with estrogenlike activity, nitrates and nitrites, nitrosamines, heavy metals, polycyclic aromatic hydrocarbons, and mycotoxins.

Periodic analysis of the feed is also important to ensure that the concentration of the test substance is within the limits set originally to ensure that the intended amount of the substance is delivered to the animal. Analysis may reveal test chemical degradation or the production of substances that may possibly interfere with the conduct of the experiment or with the results obtained.

XII. OBSERVATIONS DURING THE TEST

A regular check needs to be made of the feed and water consumption, body weights, state of health, and general behavior of the animals. These are standard procedures in any modern experimental animal facility. Data from urinalysis, if required, can be obtained without undue discomfort to the animals, but invasive procedures, such as taking blood samples, may cause some stress and are best done on separate or satellite groups receiving the same treatment as the test animals. Satellite groups are also needed when it is necessary to kill animals at intervals during treatment in order to follow the development of any premalignant lesions. These extra groups should be reported on separately, and any cancer findings should not be combined with those in the main experiment without the advice of a statistician.

XIII. MULTIGENERATION STUDIES

There seems to be some evidence that fetal tissues are more sensitive to carcinogens than adult tissues of the same species and strain (Toth, 1968) so that in theory carcinogens stand a better chance of being detected if exposure is commenced in utero (IARC, 1980). Basically, a multigenerational study consists of dosing male and female animals of the F_0 generation with the test substance, mating them, and then expose their offspring (F_1 generation) to the same test substance at predetermined doses, soon after weaning, using the same protocol as that employed in a conventional carcinogenicity study.

In practice there are reasons for suspecting that it may not be possible to obtain meaningful results with a specific test substance. First, the trasplacental passage is known to vary considerably for the same compound within the same litter, and it is likely to differ even more so between different litters (Ruddick et al., 1978). Second, the rate and site of metabolism of the compound may influence considerably the availability of the compound itself or of its active metabolites for transplacental passage. Finally, there is the possibility that the compound may be inactivated by the maternal tissues, by the placenta, or by the mammary glands (during lactation).

Furthermore there is some evidence that nutritional factors may have a role in influencing the tumorigenic response. In a multigeneration study by Olsen (1986) it was found that butylated hydroxytoluene (BHT) produced a high incidence of liver tumors in the highest tested dose in rats of the F_1 generation. BHT had not been found to be carcinogenic in several previous conventional studies in rats (McFarlane et al., 1997). Critical investigations of the effect of BHT on pregnant rats and on the progeny in the immediate postnatal period revealed that at the highest BHT dose, the growth rate of the F_1 generation of rats was much less than in the lower doses and in controls. This slow growth rate was found to be due to poor milk production by the dams maintained on the highest dose of BHT. Slow growth was maintained for several months after weaning, even though the rats were given a nutritionally adequate diet after weaning (McFarlane et al., 1996). The poor nutrition during lactation was thought to imply that there was a shortage of some essential amino acids (e.g., choline), a situation known to produce tumors (Ghoshal et al., 1987). It would thus appear that, in a multigeneration study, there are a number of imponderables that make the interpretation of the results a difficult task.

XIV. CONDUCTING THE NECROPSY

The necropsy is one of the most important procedures in any carcinogenicity study. Unless it is properly conducted, valuable information may be lost, which might make the test results uninterpretable. Every effort should be made to record fully, preferably with diagrams or photos (the Polaroid camera is a useful gadget in this respect), every "lump and bump" and to make a tentative diagnosis. This information can be of great help when examining the tissues microscopically. In fact, some workers consider that thoroughness at the necropsy table is just as important as proficiency at histological examination.

In order to minimize errors at necropsy, the following points are worth considering:

1. Animals in extremis should be killed before they die naturally in order to diminish the risk of autolysis.
2. The necropsy team for a particular study should consist as far as possible of the same individuals (even on weekends) in order to ensure uniformity of reporting.

3. A standard number of tissues should be taken from each major organ (e.g., five from liver, two from kidney). If, for any reason, more than the standard number of specimens are taken from test animals, additional specimens should also be taken from controls. Small organs, such as thyroid and prostate, should be taken whole.
4. It is desirable to form an opinion as to whether any tumors found could be the cause of death.

XV. CONCLUSION

A rodent carcinogenicity study is time consuming and expensive, and its results may well decide whether a particular chemical has any future use in society. Great care must be given to its planning and execution. Particular attention should be given to establishing the highest test dose in order to avoid toxicity, which may lead not only to a considerable waste of animals from toxicity but also to cancer development by a nongenotoxic mechanism. Further time and expense may then be necessary to study this nongenotoxic mechanism in order to determine its relevance to human exposure to the chemical. Carefully conducted toxicity tests prior to the carcinogenicity study are invaluable in this respect.

The guidelines of any regulatory authority to which test results are to be submitted for evaluation should also be consulted at an early stage so that special requirements associated with the chemical type or its potential usage can be included in the experimental test design.

ACKNOWLEDGMENTS

The authors are grateful to Professor M. Sharratt, Dr. J. C. Cohen, and Alan Mann for their invaluable assistance in the preparation of the manuscript.

REFERENCES

Ashby J. Determination of the toxic status of a chemical. Mut Res 1991; 248:221–231.

Bailey PC, Leach WB, Hartley SW. Characteristics of a new inbred strain of mice (PBA) with a high tumor incidence: Preliminary report. J Natl Cancer Inst 1970; 45(1):59–73.

Berenblum I. The significance of the sequence of irritating and promoting actions on the process of skin carcinogenesis in the mouse. Br J Cancer 1955; 9:268–271.

Boyland E, Roe FJC, Mitchley BCV. Tests of certain constituents of spermicide for carcinogenicity in genital tract of female mice. Br J Cancer 1966; 20:184–189.

Clayson DB. The aromatic hydrocarbons and related compounds. In: Clayson DB, ed. Chemical Carcinogenesis. London: Churchill, 1962:135–175.

Committee on Carcinogenicity of Chemicals in Food, Consumer Products and the Environment. Guidelines on Carcinogenicity Testing. Report on Health and Social Subjects No. 42 London, HMSO, 1991.

Cook JW, Hieger I, Kennaway EL, Maynerod WV. The production of cancer by pure hydrocarbons—Part I. Proc Roy Soc B 1932; 111:455–496.

Davies D. Dose levels—how should they be selected and what is the role of pharmacokinetics? An academic/regulatory viewpoint. In: McAuslane JAN, Lumley CE, Walker S, eds. The Carcinogenicity Debate. Lancaster, UK: Quay Publications, 1992:159–164.

de Groot AP, Til HP, Feron VJ. Safety evaluation of yeast grown on hydrocarbons. I. One year feeding study in rats with yeast grown on gas-oil. Food Cosmet Toxicol 1970; 8:267–276.

Festing FW. A case for using inbred strains of laboratory animals in evaluating the safety of drugs. Food Cosmet Toxicol 1975; 13:369–375.

Food Safety Council. Chronic toxicity testing. Proposed system for food safety assessment. Food Cosmet Toxicol 1978; 16(suppl. 2):97–108.

Food and Drug Administration Advisory Committee on Protocols for safety evaluation. Panel on Carcinogenesis. Report on Cancer Testing in the Safety of Food Additives and Pesticides. Toxicol Appl Pharmacol 1971; 20:419–438.

Ghoshal A, Rushmore T, Faber E. Initiation of carcinogenesis by a dietary deficiency of choline in the absence of added carcinogens. Cancer Lett 1987; 37: 289–296.

Grasso P. Hepatic changes associated with peroxisome proliferation. In: Gibson G, Lake B, eds. The Peroxisome. New York: Elsevier, 1994:639–652.

Grasso P. Carcinogenic risk from food—real or imaginary? Chem Ind (London) 1979; 3:73–76.

Grasso P, Crampton RF. The value of the mouse in carcinogenicity testing. Food Cosmet Toxicol 1972; 10:418–426.

Grasso P, Golberg L. Subcutaneous sarcoma as an index of carcinogenic potency. Food Cosmet Toxicol 1966; 4:297–320.

Grimmer G, Dettbarn G, Brune H, Deutsch-Wenzel R, Misfeld J. Quantification of the carcinogenic effect of polycyclic aromatic hydrocarbons in used engine oil by topical application onto the skin of mice. Int Arch Occup Environ Health 1982; 50:95–100.

Haseman JK, Huff JK, Boorman GA. Use of historical control data in carcinogenicity in rodents. Toxicol Pathol 1984; 12:126–135.

Haseman JK, Huff JE, Rao GN, Arnold JE, Boorman GA, McConnel EE. Neoplasms observed in untreated and corn oil gavage control groups of F 344/N rats and (C57Bl/N*C3H/Hen) F1 (B6C3F1) mice. J Natl Cancer Inst 1985; 75(5):975–984.

Haseman JK, Winbush JS, O'Donnell MW. Use of dual control groups to estimate false positive rates in laboratory animal carcinogenicity studies. Fundam App Toxicol 1986; 7:573–584.

Heston WE, Vlahakis G. Factors in the causation of spontaneous hepatoma in mice. J Natl Cancer Inst 1966; 37:839–843.

Hueper WC. "Aniline tumours" of the bladder. Arch Pathol 1938; 25:856–899.

Hughes DH, Bruce RD, Hart RW, Fishbein L, Gaylor GW, Carlton WW. A report on the workshop on biological and statistical implications of the ED01 study and related data-bases. Fundam Applied Toxicol 1983; 3:129–136.

IARC Monographs on the Evaluation of the Carcinogenic Risk of Chemicals to Humans. Suppl. 2. Long Term and Short Term Screening Assays for Carcinogens: A Critical Appraisal. Lyons: International Agency for Research on Cancer, 1980.

IARC Monographs on the Evaluation of the Carcinogenic Risk of Chemicals to Humans. Suppl. I. Lyons: International Agency for Research on Cancer, 1979.

Keenan K, Smith P, Ballam G, Soper K, Bokelman D. The effect of diet and dietary optimisation (caloric restriction) on survival in carcinogenicity studies—an industry viewpoint. In: McAuslane JA, Lumley CE, Walker S, eds. The Carcinogenicity Debate. Lancaster, UK: Quay Publishing, 1992:77–102.

Kodell JL, Gaylor DW, Greenman DL, Littlefield NA, Farmer JH. The Society of Toxicology task-force re-examination of the ED01 study—response. Fundam Appl Toxicol 1983; 3:A3–A8.

Lee PN. Statistics. In: Anderson D, Conning DM, eds. Experimental Toxicology—the basic issues. 2d ed. London: Royal Society of Chemistry, 1993:405–440.

Mauderly JL, Kritchevsky J. Respiration of unsedated F-344 rats and the effect of confinement in exposure tubes. In: ITRI Annual Report 1978–1979, LF-69, pp. 475–478. Available from the National Technical Information Service, Springfield, VA.

McFarlane M, Cottrell S, Price SC, Grasso P, Bremmer J, Bomhard ME, Hinton RH. Hepatic and associated response of rats to pregnancy, lactation and simultaneous treatment with butylated hydroxytoluene. Fd Cosmet Toxicol 1997; 35. In press.

Mead J. Control of respiratory frequency. J Appl Physiol 1960; 15:325.

Ministry of Health and Welfare Canada. The Testing of Chemicals for Carcinogenicity, Mutagenicity and Teratogenicity. 1973.

Monro IC. Considerations in chronic toxicity testing: the chemical, the dose, the design. J Environ Pathol Toxicol 1977; 1:183–197.

Olsen P, Meyer O, Bille N, Wurtzen G. Carcinogenicity study on butylated hydroxytoluene (BHT) in Wistar rats exposed in utero. Food Cosmet Toxicol 1986; 24:1–12.

Page NP. Current concepts in a bioassay program in environmental carcinogenesis. In: Kraybill H, Melhman M, eds. Environmental Carcinogenesis. Washington, DC: Hemisphere Publishers, 1977a:7-717.

Page NP. Current concepts in a bioassay program in environmental carcinogenesis. In: Kraybill MF, Melhman MA, eds. Environmental Cancer. New York: John Wiley and Sons, 1977b:114–129.

Parkinson C, Lumley CE, Walker SR. The value of information generated by long-term toxicity studies in the dog for the nonclinical safety assessment of pharmaceutical compounds. Fundam Appl Toxicol 1995; 25:115–123.

Parkinson C, Grasso P. The use of the dog in toxicity tests on pharmaceutical compounds. Hum Exp Toxicol 1993; 12:99–109.

Peraino C, Fry RDM, Staffedt E. Enhancement of spontaneous hepatic genesis in C3H mice by dietary phenobarbital. J Natl Cancer Inst 1973; 51:1349–1350.

Peto R, Gray R, Brantom P, Grasso P. Effects on 4080 rats of chronic ingestion of N-Nitrosodiethylamine or N-Nitrosodimethylamine: A detailed dose response study. Cancer Res 1991; 51:(23 part 2):6407–6491.

Powell CH, Hosey AD. The Industrial Environment and Its Evaluation and Control. U.S. DHS Publ. 614, 1965.

Rehn L. Überblasentumoren bei Fuchsinarbeiten. Arch Klin Chir 1895; 50:588.

Rodericks JV. Calculated Risks. Cambridge, England: Cambridge University Press, 1992:5–256.

Ruddick JA, Ashanullah M, Craig J, Stavric B. Uptake and distribution of 14C saccharine in the rat foetus. In: Proceedings, Canadian Federation of Biological Sciences, 21st annual meeting, London, Ontario, Abstract 635, p. 159.

Roe FJC. Food and cancer. J Hum Nutr 1979; 33:405–415.

Sanockij IV, ed. Methods for Determining Toxicity and Hazards of Chemicals. Moscow: Medicina,

Schmahl D. Carcinogenic activity of KOH and NaOH by topical application to mice. In: Schmahl D, ed. Maligne Tumoren—Enstethung, Wachtsum, Chemotherapie. Arzneimittel Forschung 1984; Suppl 21:290–291.

Tucker MI. The effects of food restriction on tumours in rodents. Int J Cancer 1979; 23:803–807.

Tucker MJ, Orton TC. Comparative Toxicology of Hypolipidaemic Fibrates. London: Taylor Francis Ltd, 1995:15–18.

Toth B. A critical review of experiments in chemical carcinogenesis using new-born animals. Cancer Res 1968; 28:727–738.

Uehleke H. Thresholds in acute and long term animal studies. In: Cancer and the Environment. New York: Mary Ann Liebert, Inc., 1983.

Usui T. Dose levels—How should they be selected and what is the role of pharmacokinetics? An industry viewpoint. In: McAuslane JAN, Lumley CE, Walker SR, eds. The Carcinogenicity Debate. Lancaster, UK: Quay Publications, 1992:165–179.

Wagner JC. Experimental production of mesothelial tumours of the pleura by implantation of dusts in laboratory animals. Nature 1962; 196:180–183.

Weissinger J. Dose levels—How should they be selected and what is the role of pharmacokinetics—a regulatory viewpoint. In: McAuslane JAN, Lumley CE, Walker SW, eds. The Carcinogenicity Debate. Lancaster, UK: Quay Publications, 1992: 147–152.

WHO. Factors influencing the design of toxicity studies. In: Principles and Methods for Evaluating the Toxicity of Chemicals. Part 1. Geneva: World Health Organisation. 1978b:62–94.

WHO Inhalation exposure. In: Principles and Methods for Evaluating the Toxicity of Chemicals. Part 1. Geneva: World Health Organisation. 1978a:149–235.

Yamagiwa K, Ichikawa K. Experimental study of the pathogenesis of carcinoma. J Cancer Res 1918; 3:1.

2

Value, Validity, and Historical Development of Carcinogenesis Studies for Predicting and Confirming Carcinogenic Risks to Humans

James Huff
National Institute of Environmental Health Sciences, Research Triangle Park, North Carolina

> *As we sit around hoping for a cure for cancer to be discovered, we should focus on preventing it.*
>
> —A. Radwan, letter, *Time*, May 16, 1994

I. OVERVIEW

Chemicals—both natural and synthetic—cause cancer. Not all chemicals cause cancer, so we must be able to identify with reasonable certainty those relatively and proportionately few that do. Carcinogenesis bioassays using laboratory animals have a solid history of identifying the natural and synthetic chemicals, mixtures of chemicals, drugs, and commercial products that are most likely or predictive to be carcinogenic to humans (Fung et al., 1993, 1995; Huff et al., 1988, 1991; Huff 1994; IARC, 1997; Huff, 1998). Nearly 30 agents that cause cancer in humans were first found to induce cancer in animals (Huff, 1993). Unfortunately, however, bioassays have not been—nor can be—used to evaluate or discover industrial processes or occupational-exposure circumstances that cause or might cause cancer in humans. Examples include urinary bladder cancer and leukemia in workers in the rubber industry; nasal cancers in furniture-makers; and lung cancer in furniture-makers and workers in the aluminum production, iron and steel founding, and coke production industries (Tomatis et al., 1989; IARC, 1997). Equally difficult to test for carcinogenesis are environmental mixtures, dietary factors, and actual human exposures to myriad and multiple chemicals, coupled with varied lifestyles and socioeconomic situations (Huff, 1994; Huff and Barrett, 1995; Tomatis et al., 1997). For example, testing a single ingredient of a food (e.g., D-limonene in oranges) without testing all the factors in that food most often leads to incomplete or misleading conclusions.

Despite these obstacles, all chemicals known to cause cancer in humans that could be properly tested in animals are likewise carcinogenic (Huff, 1994a). Does this imply that any chemical that causes cancer in animals will be carcinogenic in humans? No. Carcinogens are not equal, and each must be evaluated regarding strength of the evidence. In fact, using a multifactorial matrix approach including mechanistic information, exposures, and levels and potency of responses, we have predicted that less than 5–10% of all chemicals would eventually be considered reasonably anticipated to cause cancer in humans (Fung et al., 1995). Importantly, most agents tested in animals do not cause cancer under conditions of long-term bioassays (Huff and Hoel, 1992; Huff, 1993a). Further, several experimental anticarcinogens have been identified in routine bioassays (e.g., Douglas and Huff, 1984; Chhabra et al., 1988; Haseman and Johnson, 1996; Chan et al., 1996) that could be pursued for purposes of preventing cancer. Likewise, in foods there are abundant antimutagens and anticarcinogens, especially in fruits and vegetables, that tend to be preventative.

Prudent public health policy obligates us to continue with the strategy of reducing or eliminating exposures to chemical carcinogens, to chemicals in general, and to unhealthy workplace circumstances (Tomatis et al., 1997). Additionally, we need to persevere in our efforts to reduce or eliminate unnecessary industrial emissions and chemical contaminations (more than 2.2 billion pounds per year in the United States) of our air, animal, and plant life; lands and food crops; waters and fishes; and diets. Thus, cancer incidences and mortalities related to and influenced by chemicals, as well as other chemically associated diseases, will be restricted, reduced, or eradicated. We must continue to approach this goal. After all, don't public health and human decency commit us to keeping our individual and corporate trash confined to our own back yards or to disposing of it properly and safely without undue harm to others? More faithfully carrying out these recovery or recycling processes would reduce substantially the amounts of chemicals being unloaded into the envrionment.

II. INTRODUCTION

If a man will begin with certainties, he shall end in doubts; but if he will be content to begin with doubts he will end in certainties.
—F. Bacon, *The Advancement of Learning*, Book 1, iv, 8, 1605

Chemicals cause cancer. One must therefore strive to identify the chemicals—either naturally occurring or synthetically produced—that are or will be most likely to represent a carcinogenic hazard to humans. Relatively few of the universe of chemicals in commerce have been shown or are predicted to cause cancer eventually (Fung et al., 1995); one must pledge to rid the environment of all of them. To do otherwise—for whatever reason—would be anathema to humanity, and especially to future generations. The pro-chemical-at-any-cost cacophony one hears today rests largely on economic gain for the few, while the real health and economic expenses are levied to the public, especially to workers. Simply witness the current harangue about tobacco, and the exploitation of Third World countries with asbestos (Sells, 1994; Giannasi and Thebaud-Mony, 1997). Peto (1980), for example, exposes those on both sides of the environmental-chemical conundrum who would bend the truth for their personal largesse. Granted, we must also evaluate in certain cases the benefits of agents or exposure circumstances shown or strongly suspected to cause cancer in humans; these include cancer chemotherapeutic agents, certain drugs, and some societal necessities such as gasoline (that is, until viable substitutes become available for our mobile society).

In this chapter, the primary emphasis is on the usefulness, value, and relevance of the long-term carcinogenesis bioassay toward avoiding, reducing, and preventing carcinogenic risks to humans. Further, some attention is given to a select overview of the history of the early efforts in chemical carcinogenesis and development of carcinogenesis bioassays. Particular topics covered include: 1) methods for identifying carcinogens, 2) value and validity of experimental carcinogenesis, 3) agreement between laboratory and human evidence, and 4) cumulative importance and public health worthiness of experimental findings.

The six major approaches in current use for identifying existing or predicted cancer causes in humans are detailed in Table 1, and Table 2 lists the key advantages and disadvantages of each. Much as we would like to reduce our reliance on long-term carcinogenesis bioassays, and perhaps use other methods including shorter or mid-term assays (Ward and Ito, 1988), transgenic models (Tennant et al., 1996), and prospective predictive techniques (Bristol et al., 1996; Huff et al., 1996), no proposed paradigm has been proven to be a better indicator of potential human cancer risks than

Table 1 Sources and Strategies for Identifying Human Cancer Hazards and Chemical Carcinogens in Experimental Animals

Human investigations
1. Epidemiological studies
2. Case reports

Basic and applied research and testing
3. Long-term bioassays
4. Mid-term in vivo assays
5. Short-term in vivo and in vitro assays
6. Transgenic models
7. Structure-biological activity relationships and artificial intelligence systems
8. Mechanism-based inference

the long-term carcinogenesis bioassays (Huff et al., 1988, 1991; Huff, 1993, 1998). Several in vivo methods to shorten the duration and costs have been attempted, but none has yet been fully successful, although the Ito medium-term assay has strong promise (Ito et al., 1988; Ito et al., 1992; Shirai, 1997) (Table 3). In fact, most animal studies "rely on non-specific endpoints from chronic studies and where there is little understanding of the mechanism of toxic action: these studies are much more difficult to replace" (Purchase, 1997).

Wishful ideas to utilize a battery of tests (typically "short-term" tests) to better characterize the toxicity and carcinogenicity of chemicals have not proven particularly promising or propitious. Flow-chart configurations and paradigms for "new" carcinogen-identification strategies surface periodically, sometimes with little or no scientific basis. Some deserve our attention (Williams and Weisburger, 1981; Schwetz and Gaylor, 1997; Chapter 3 in this volume), but most represent futile or self-serving efforts. Alternatives or supplemental assays for the long-term bioassay do not seem viable for the near future. Approximately 10 years or more are required to take a new thought-method from the stage of its description to acceptance by regulatory authorities (Purchase, 1997). This appears to be where we are today. Nonetheless, creative and innovative efforts must continue.

This chapter begins with a selective résumé of several significant early events in chemical carcinogenesis, as well as important activities that have been instrumental in preventive carcinogenesis. Also, to illustrate the last two decades of chemical carcinogenesis testing efforts, the initiatives and findings of the National Cancer Institute and the National Toxicology Program are summarized.

Table 2 Methods for Identifying Cancer Risks: Advantages and Disadvantages

Epidemiological studies

Advantages

1. Humans are ultimate indicators of disease
2. Evaluation of sensitive populations
3. Occupational exposure cohorts
4. Environmental sentinel alerts

Disadvantages

1. Generally retrospective (death certificates, recall biases, etc.)
2. Insensitive, costly, lengthy
3. Reliable exposure data not available or difficult to obtain
4. Combined, multiple, and complex exposures
5. Lack of appropriate cohorts
6. Experiments on humans cannot be done for ethical and moral reasons
7. Cancer detection, not prevention, is the achievement

Long-term chemical carcinogenesis bioassays

Advantages

1. Prospective and retrospective (validation) evaluations
2. Excellent correlation with identified human carcinogens
3. Exposure levels and conditions known
4. Identifies chemical toxicity and carcinogenicity effects
5. Results obtained relatively quickly
6. Qualitative comparisons among chemical classes
7. Integrative and interactive biological systems related closely to humans

Disadvantages

1. Rarely replicated
2. Resource-intensive (staff, money, facilities)
3. Shortage of adequate facilities, scientific expertise and experience, or research interest
4. Debate regarding relevancy of results (rodents are not humans, use of "high doses" [MTD], routes of exposure, too sensitive)
5. Exposure levels often exceed those anticipated for humans
6. Single chemical bioassays do not mimic human multiple-exposure circumstances

Mid-term in vivo bioassays and short-term in vivo and in vitro assays

Advantages

1. More rapid and typically less expensive
2. Large samples and most often easily replicated
3. Biologically valuable end-points measured (e.g., mutations, preneoplasia)
4. Assays easily manipulated (e.g., two-stage models; organ systems: liver, urinary bladder)
5. Single or multiple chemicals, simple or complex mixtures, environmental or occupational samples
6. Screening assays to select chemicals for subsequent bioassys and mechanistic studies

(continued)

Table 2 Continued

Disadvantages
1. In vitro not fully predictive for in vivo, especially for negatives
2. In vivo measures purported preneoplastic lesions
3. Not generally predictive for toxicity or carcinogenicity
4. Usually organism- or organ-specific
5. Systems manipulated toward select and sensitive endpoints
6. Not reliable surrogates for whole animals, long-term bioassays, or humans
7. Potencies not comparable to whole animals or humans

Chemical structure–biological–activities associations
Advantages
1. Relatively easy, rapid, and inexpensive
2. Reliable for certain chemical classes (e.g., nitrosamines, antibiotics, anthraquinones, benzidine dyes)
3. Developed using biological information, yet not dependent on animal or human data
4. Computer-adaptable and -compatible

Disadvantages
1. Not "biological"
2. Many exceptions to formulated rules
3. False positives and false negatives
4. Disparate theoretical systems and versions available with conflicting predictions
5. Retrospective, and rarely (but becoming) prospective

Mechanism-based inferences
Advantages
1. Reasonably accurate for classes of chemicals, or known mechanisms for different chemicals
2. Permits evaluations of chemicals without doing long-term bioassays or other in vivo studies
3. Can "test" more chemicals and classes of chemicals
4. Allows refinements and further hypotheses
5. Adjust or orient risk assessments to sensitive populations (genetic susceptibility)
6. Strengthen risk assessments (low doses and species extrapolation)

Disadvantages
1. Mechanisms of chemical carcinogenesis undefined, multiple, and probably chemical- and/or class-specific
2. Exceptions to posed mechanisms: DNA interactions, adduct formation, stones, irritation, cell proliferation, receptor-mediated, toxicity, genotoxic/nongenotoxic, α2u globulin, structural alerts, epoxides, others
3. Nonconfirmation of purported mechanisms, or nonreproducible
4. Flaws lead to faulty risk assessments (under-/overestimates)

Table 3 Possible In Vivo Alternatives to the "Historical" Long-Term Chemical Carcinogenesis Bioassay

---

Alternative systems used in the past as possible replacements for the carcinogenicity bioassay include (in alphabetical order):

1. *A/J mouse (strain A) lung adenoma model*: interpertioneal injection (3/week for 8 weeks); at 24 weeks gross lung tumors
2. *Fish: medaka; guppy*: environmental conditions; evaluate mixtures, chemical combinations; multigenerational; flexible duration
3. *Ito medium-term initiation assay*: a) give initiator (usually liver); chemical in feed; partial hepatectomy; 8 weeks; examine liver (and posibly other organs); b) give initiators (liver, kidney, intestine, others); chemical in feed; partial hepatectomy; 12 to 20 weeks; evaluate liver, kidney, and other specific organs
4. *Liver, two-stage rat model*: partial hepatectomy; administer known target-specific initiator; give test chemical (usually in feed); evaluate from 10 to 26 to 52 weeks; liver (foci and/or tumors)
5. *Local subcutaneous injection (mouse)*: single or multiple injections; observe skin lesions at 13 to 52 weeks; more recently internal organs examined
6. *Neonatal mouse*: single (usually) exposure; interperitoneal injection or oral gavage; examine animals at ≤ 1 yr; typically liver, but often others
7. *Skin painting (typically mouse, and specialty strain models)*: multiple exposures; 26–78 weeks; observe skin tumors
8. *Transgenics (usually mice)*: a) skin application; multiple exposures; 26 weeks; observe skin tumors (internal tumors less typically); b) other routes (oral, feed, s.c., i.p.), same protocol
9. Others
   Toad: s.c.; skin; or gavage; 13 weeks
   Hamster cheek pouch: 3 times/week; painted, and sometimes irritated; 30 weeks; local tumors
   Rat mammary gland: initiator; chemical; 32 weeks
   Urinary bladder: initiator; drinking water 4 weeks; chemical 32 weeks
   Stomach/forestomach: MNNG; chemical; 16–20 weeks
   Organ-specific: kidney; lung; thyroid; pancreas (see item 3 above)

In vivo alternatives: comments and conclusions

Advantages

1. Typically quicker; less expensive; reasonably reliable (especially positives)
2. Fewer animals; smaller facilities; less resources
3. Chemicals: classes; mixtures; combinations; test alternative variables (e.g., diet, duration, single/multiple routes)
4. Screen to select chemicals for long-term bioassays (especially negative chemicals) and to study/explore mechanisms (particularly for potent, positive chemicals)

(continued)

Table 3 Continued

Disadvantages
5. Diminished sensitivity: "single" target site; reduced sex/species correlations, and strength of evidence
6. Nontumor (nonneoplastic) endpoints typically absent: biomarkers, enzymes, foci, and toxic or preneoplastic lesions
7. Screen: notably strong positives; not sensitive for nonconclusive results—noncarcinogenic or "equivocal" and weakly positive evidence (thus, "false negatives"?)
8. Usually not "validated" and often not standardized

III. THE CANCER DILEMMA

Cancers are the second leading cause of death in the United States, with heart diseases being number 1 and cerebrovascular diseases number 3. As cardiovascular diseases are conquered, cancers are predicted to become the primary killer of Americans in the early 21st century (Davis et al., 1994). More than 2.25 million new cancer cases will be diagnosed in 1998 (Huff et al., 1996, chap 1; NCI, 1997; ACS, 1997, 1998), with nearly 565,000 cancer-associated deaths; both figures divide almost equally between females and males (Figure 1). Organs or systems associated with the leading sites of cancer morbidity and mortality include the lung, colon and rectum, breast, and prostate gland.

Even though much is known about certain causes of cancer (e.g., occupational and cancer chemotherapeutic agents), we unfortunately do not know the etiologies of the overwhelming majority of cancers (Doll and Peto, 1981; Schmahl et al., 1989). In the 1980s, for instance, there were more than 4.5 million cancer deaths, 9 million new cancer cases, and 12 million people under medical care for cancer. The numbers keep rising (Bailar and Smith, 1986; Davis and Hoel, 1990; Miller et al., 1993; Kosar et al., 1995; Bailar and Gornik, 1997), and predictions (perhaps conservative) indicate that one in three (about 85 million Americans now living) will eventually develop cancer (ACS, 1994). One widely accepted factor influencing cancer causation is the effect of our modern industrialized and chemically based society. While making our lives longer and better, this revolution has not been without adverse health impact. We are devoting considerable effort to identify and then overcome or change unhealthy practices and habits that lead to or exacerbate diseases (Tomatis, 1990; Hakama et al., 1990). The extent of these changes often depends on outcomes of risk-benefit and cost-benefit analyses. Further, because many

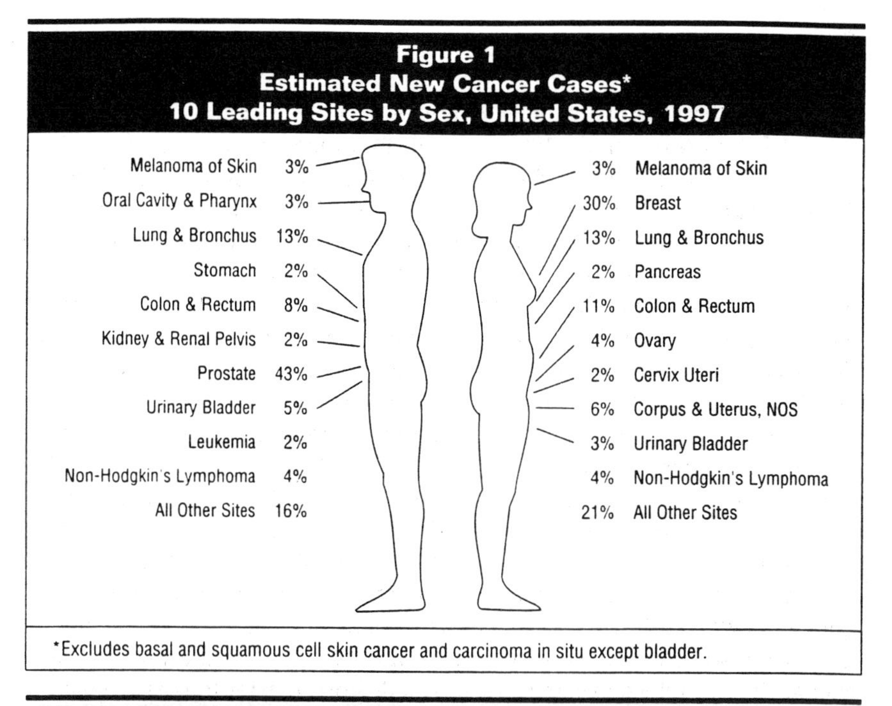

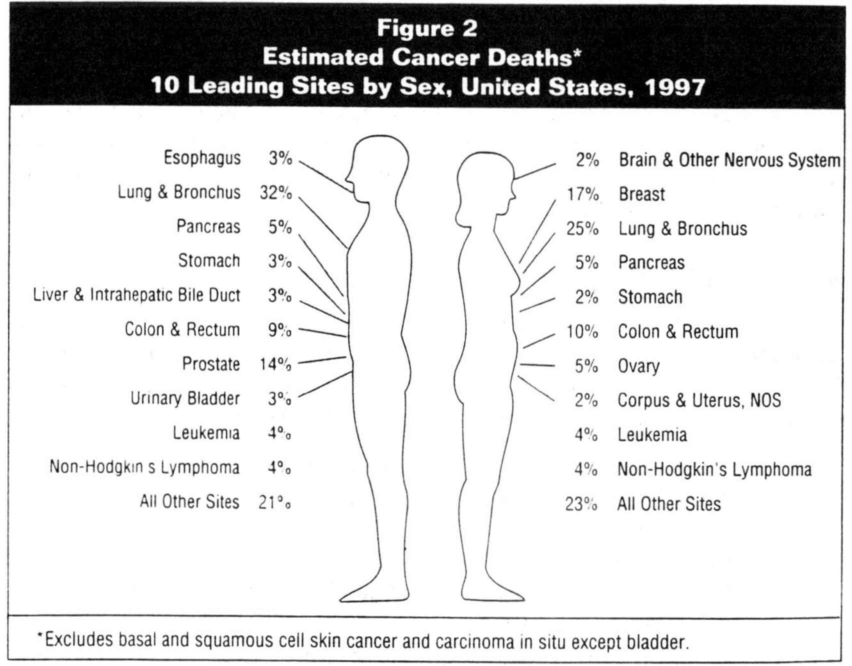

Figure 1 Cancer incidence and mortality: 10 leading sites. (From *CA: A Cancer Journal for Clinicians*.)

cancers are geographically idiosyncratic (e.g., Parkin et al., 1992; Coleman et al., 1993; Geddes, et al., 1993), various investigators have concluded that most cancers (up to 90%) are preventable; this hypothesis hinges greatly on dietary influences (Doll and Peto, 1981; Willet, 1996; DeMarini, 1998).

Cancer—a multi-disease phenomena—remains as much a concern and enigma now as it was when first described as *karkinos* ("new growth," from the Greek word for crab). We have learned much about these diseases in the last century, and the avalanche of information accelerates as we near the 21st century, yet, as a collection of maladies "contrary to nature" (Shimkin, 1977), not nearly enough is understood to stem significantly the devastating tide that has begun to consume more and more lives around the globe as human lifespans increase (Parkin et al., 1993). This is especially true for those of lower socioeconomic status and in undeveloped countries, where both general health care and occupational health and safety are minimal. Nonetheless, as we make advancements to extend our knowledge and understanding of these disease etiologies, many believe that intervention strategies and better treatment regimens are imminent. Meanwhile, implementation of known prevention plans (e.g., aimed at tobacco and lung cancers) may go far to reduce existing cancer trends and societal burdens (Tomatis et al., 1997; Bailar and Gornik, 1997). Reducing or eliminating exposures to cancer-causing agents could have a dramatic effect on cancer morbidity and mortality, and attendant sequelae. Clearly, one need not know a molecular mechanism to eliminate or reduce exposures to known cancer-causing substances. A better understanding of mechanism and interindividual variability can help quantify risk reduction with lowered exposures, especially for sensitive subpopulations.

A. Cancer Rates

The debate continues as to whether cancer incidence and mortality rates are increasing or decreasing, whether the primary causes of the major cancers have been identified, and whether basic and mechanistic knowledge will conquer the "war on cancer" begun 25 years ago. Today, the facts are:

1. Cancer incidence rates are not falling (Kosary et al., 1995; Ries et al., 1997; Parker et al., 1997; Bailar and Gornik, 1997).
2. Cancer mortality rates are declining slightly (Parker et al., 1997; Bailar and Gornik, 1997).
3. Exposures to myriad chemical carcinogens are increasing (EPA, 1997; Huff et al., 1996).
4. Environmental contamination (air and soil; rivers, lakes, streams, aquifers, and drinking water; crops, forests, foods, and natural and agricultural ornamentals) is escalating (EPA, 1997).

5. Cancer-prevention strategies are not now being adequately emphasized (Tomatis et al., 1997; NCI, 1997).

Cancer rates in young children and in adults 65 years and older are continuing to rise (age is certainly one prime carcinogenic factor in the elderly, but not enough to explain the increases in this age category). In children, certain cancers are mounting inexplicably, and we know very little about the role of environmental chemicals in childhood cancers (Goldman, 1998). What we should know is that any additional carcinogens (e.g., pesticide residues in and on foods) should not be permitted and must be avoided. A carcinogenic pesticide, or any other agent known to cause cancer, remains carcinogenic regardless of the amount present on foods; in the biological balance of life, any further burdens of carcinogenic substances, even in low concentrations, do not represent good public health policy.

Some would argue that the statements above are exaggerations and should be ignored, that humans in modern countries are the healthiest ever, that the human lifespan is the longest in history, and that exposures to synthetic (manmade) chemicals are collectively less harmful than indigenous chemicals in plants and edible foods. Yes, on average we are—and gratefully so—healthier and enjoy longer life than earlier generations, yet we might be even healthier and live even longer if exposures to the huge amounts of hazardous chemicals we are involuntarily exposed to were curtailed or eliminated. Unfortunately, good health is not evenly distributed. Sensitive subpopulations, lower socioeconomic groups, workers exposed to hazardous agents, and some Third World and developing countries should not have to suffer excess diseases or any other untoward health effects from avoidable adverse exposures of the consumer, in the workplace, or in the environment.

A major concern that many hold reflects the "exportation" of cancer risks from industrial developed countries to developing countries. Occupational cancers are rising with minimal or less than adequate attention to occupational-exposure standards (Pearce et al., 1994). Thus, driven by economics and not safety, industries settle in less-regulated countries with lower labor costs and fewer occupational and environmental regulations to control exposures to harmful agents. As the scourges of infectious diseases and malnutrition are overcome, populations increase and live longer, exposures to carcinogens are on the increase, resulting in greater risks of cancer.

The fact that cancer dates from antiquity does not mean that we must consider these horrific diseases an inevitable consequence of "old age." Obviously if we were successful in reducing the lifetime cancer burden from the current one in three to something less, cancer morbidity and deaths would not be nullified completely. However, much pain and suffering could

Table 4 Cancer Facts and Figures

In 1997: An estimated 1,382,400 humans were newly diagnosed with cancer—785,800 males and 596,600 females (158 persons an hour or 3792 a day)—and an additional 900,000 persons with skin cancer. An estimated 560,000 humans died from cancer—294,100 males and 265,900 females (64 persons an hour or 1534 individuals a day).

During the 1980s: There were 4.5 million cancer deaths, 9 million new cancer cases, and 12 million individuals with cancer under medical care.

Of living Americans: 1 in 3 will develop cancer during their lifetime; 85 million people will get cancer.

1957–1959 vs. 1987–1989 (each per 100,000 population)
Deaths, males: from 180 to 216 (20% increase)
Deaths, females: from 138 to 140 (1% increase)

1971–1973 vs. 1991–1993 (each per 100,000 population)
Deaths, males: from 205 to 219 (7% increase)
Deaths, females: from 132 to 142 (8% increase)

National cancer death rates (age-adjusted and compared to 1930; per 100,000)
1930: 143, base
1940: 152 (6% increase)
1970: 163 (14% increase)
1990: 175 (22% increase)
1993: 182 (27% increase)

be avoided. And, as incremental gains are made in preventing cancers, eventually we hope to make cancer a rare disease. Table 4 lists both current and some historical rates for cancer in the United States. These diseases continue to inflict devastating consequences, and need our renewed attention to prevention.

IV. THE CHEMICAL RIDDLE

> *If we are going to live so intimately with these chemicals—eating and drinking them, taking them into the very marrow of our bones—we had better know something about their nature and their power.*
>
> —R. Carson, *Silent Spring*, Houghton Mifflin, Cambridge, MA, 1962, 368 pp.

Chemicals cause cancer. Fortunately, only a small percentage of the many tested or yet-to-be tested chemicals are or will be predicted to be carcinogenic to humans (Fung et al., 1995; Huff et al., 1996). However, the chemical armamentarium grows and grows (Table 5). Several environmental and lifestyle risk factors for cancer have been identified (Table 6), but these etiological causative circumstances represent only a small proportion of the total numbers of cancers. Nonetheless, several easily avoidable risks of cancer have been identified (Tables 7 and 8). Thus, most causes of cancer have not been determined with any certainty. After tobacco and sunlight, the major source of human cancers is diet (DeMarini, 1998). Of course, diets vary widely throughout the world, as well as in confined areas of a community. They are typically family-related or individualized within peer groups or ethnic centers of like-minded peoples; geographic food availability is important, especially among those of little means (Tomatis, 1995, 1997; Kogevinas et al., 1997). With increases in global-economy interactive exports, eating habits are becoming more homogenized, especially in industrialized and developed countries. Fast-food exports from the United States are everywhere, and younger generations are shifting their dietary habits away from their parents' customs to those of the world market. Ethnic

Table 5 Chemicals and Cancer: Background to the Problem

1. 13 million synthesized or characterized chemical entities, 65,000–85,000 in common and/or commercial use.
2. 2000 agents (+/−?) have been tested for carcinogenicity.
3. Millions of combinations; mixtures; consumer products.
4. Synthetic organic chemicals produced in the U.S.: 177,800,000,000 kg, or 391,200,000,000 lb.
5. Top 50 U.S. chemical production = 650 million lb (64% inorganics, 36% organics).
6. Uncharacterized, unidentified, and unpredictable exposures
   a. Occupational workplace circumstances
   b. "Pollution" of air, water, soil
   c. Waste sites and facilities
7. "Local" unique exposures and hot spots
   a. Arsenic smelters; refineries
   b. Chemical plants
   c. Accidents/spills: TCDD, methyl isocyanate, pesticides, solvents
8. Individual or personalized multiplicative exposures: workplace, drugs, smoking, alcohol, lifestyle, social, pesticides, chlorinated and fluorinated drinking water, smog, diet, acid rain, "home" (insulation, formaldehyde, solvents, etc.), cosmetics, fuel exhausts, nature, hobbies, pollution burdens, occupation, etc.

Table 6 Environmental and Lifestyle Risk Factors for Cancer

Air, water, soil pollution and contamination
Alcoholic beverages
Anticancer drugs
Biological agents
Chemicals carcinogenic to animals
Chemicals carcinogenic to humans
Chlorination of water and disinfection byproducts (trihalomethanes)
Chronic diseases and chronic infections
Combinatorial exposure circumstances
Diet and dietary factors
Drugs
Engine exhausts (including diesel)
Environmental estrogens
Environmental tobacco smoke
Exercise-deficient (sedentary or inactive)
Familial factors (genetics and susceptibilities)
Hormones
Immunosuppressive agents
Ionizing radiation
Occupations, workplaces, and exposure circumstances
Obesity and lack of fitness
Pesticides and other agricultural agents
Socioeconomic status
Solar radiation and artificial tanning (sunlamps, sunbeds)
Tobacco and tobacco products—cigarette, cigar, pipe, snuff usage
Others, including age

Table 7 Easily Avoidable Risks of Cancer

Certain occupations and workplaces known to cause cancer (not easy to give up livelihood)
Tobacco and tobacco products (not easy if sidestream exposures)
Solar, ionizing, and radon radiations
Prevent, eliminate, and reduce exposures to identified carcinogens (not easy if unknown exposures)
Avoid or temper excessive use of alcoholic beverages
Unhealthy diets and dietary factors (pesticides on foods not easily avoided)
Obesity and lack of fitness
Air, water, and soil pollution (not easily avoided)

Table 8 Primary Cancer Prevention Priorities

- Reduce or eliminate exposures to agents, mixtures, and exposure circumstances known to cause:
  - Cancer in humans
  - Cancer in laboratory animals
  - Mutations in animals, cell cultures, or bacteria
- Avoid exposures to those agents:
  - Mechanistically thought to be carcinogenic
  - Structurally related to identified carcinogens or classes of carcinogens
- Alter or realign unhealthy or hazardous lifestyle and cultural habits
  - Adopt exercise, fitness, and healthy lifestyle
  - Eat balanced diet with daily amounts of fruits and vegetables; limit fat intake, especially from animal sources
  - Stop using tobacco products or don't start, and urge others to quit using tobacco
  - Avoid too much sun and artificial tanning equipment; use sun protection
  - Be physically active; achieve and maintain a healthy body weight
  - Limit consumption of alcoholic beverages, if you drink; urge others not to begin drinking alcoholic beverages

cuisines are likewise worldwide. Whether these shifts in dietary and lifestyle habits will then decrease or eliminate differences in cancer site–specific incidences will only be known decades from now.

A debate continues regarding the disease burden—especially that of cancer—that may be related to chemicals, especially to whether synthetic or natural chemicals play the larger negative role (Ames and Gold, 1997; Huff et al., 1991; Tomatis et al., 1989). Important in this ongoing confrontation of opinions is the definition of a "natural" chemical or agent. In many cases, natural chemicals would not be remotely hazardous because no exposures would occur without human intervention. Asbestos, tobacco, various metals, and other agents are naturally occurring, yet, until their use is called for by a human need or profit motive, these natural agents do no harm. Thus, such agents need to be regarded not as truly natural but as "synthetic" in the sense that, without the evolutionary development of humans, these agents can be considered innocuous. Petroleum oil and all the attendant products—gasoline, plastics, solvents—should likewise be considered synthetic even though oil is simply "there." The key factor is the use humans make of these agents, and the inevitable human exposures to these chemicals that pose the ultimate carcinogenic problems. Thus, "it is important to eliminate or reduce exposure to all carcinogens, be they natural or synthetic, to the extent feasible, since that will reduce the risk of cancer, even if that reduction cannot currently be calculated" (Lijinsky, 1989).

Chemicals and combinations of chemicals, especially pesticides which are designed specifically to kill organisms, are indeed hazardous (Browning, 1965). Much effort has been devoted to identifying, setting standards for, and regulating chemical carcinogens, albeit with little attention directed toward the more frequent and perhaps less obvious or drastic effects of chemicals on other organ systems. Teratogens have received attention as have mutagens, but chemicals that cause other myriad effects have typically gone unnoticed; these effects include an array of toxicities from respiratory insults to gastrointestinal maladies, from immunological compromise to hormone disruption. For example, a relatively "new" disease—multiple chemical sensitivity—has surfaced in certain susceptible individuals, who are considered extraordinarily sensitive to chemicals (Berkson, 1996). Debilitation from this disease is high and usually prolonged; relief is minimal or absent. In another example, tobacco causes more non-cancer disabilities than most people realize, including effects on the cardiovascular, respiratory, gastrointestinal, urinary tract, and other systems. Thus, we need to give more attention to chemical effects other than cancer.

V. BEGINNINGS OF CANCER RESEARCH AND TESTING

A. Experimental Efforts

The second decade of the 1900s, in a historical sense, marked the beginning of the experimental era of chemical carcinogenesis, and this scientific phenomenon has grown exponentially ever since. A short and selective historical overview of chemical carcinogenesis is given below, with selected highlights of signal or "first" discoveries and accomplishments. With respect to long-term carcinogenesis studies to identify potential human carcinogens, the momentum of the 1960s, 1970s, and 1980s has been slowed in the 1990s. Nonetheless, knowledge continues to be sought—and in some cases gained—about etiological causes of cancer (Tomatis, 1990; Huff, 1994, 1998), cancer-prevention strategies (Tomatis et al., 1997), and possible cures of certain cancers (Kelloff et al., 1997). The optimal way of dealing with any disease—especially the multiple-diseases cancer—is prevention. "Prevention is obviously the most practical method of dealing with problems of health. Not having a disease at all is far better than the most effective treatment" (Blumberg, 1992). Certainly, chemoprevention has some potential to provide an important means for cancer prevention (Wattenberg, 1997), especially when more incontestable information becomes available.

Before launching into history, a few comments are in order about the major and relatively rapid advances being made in laboratory techniques that have allowed significant increases in knowledge about 1) basic biological processes of growth control and its aberrations, 2) mechanisms involved in cancer initiation, promotion, progression, and suppression, 3) the elaborateness of the multistage process of carcinogenesis, 4) immunological mechanisms and other fundamental processes, 5) the value and use of molecular-biology techniques in clinical problems, and 6) the toxicology and pathology of the induced cancer process. "The list is unending and often leads the uninitiated clinician to believe that the resolution of the mystery of the cancer cell and its successful control and cure are almost at hand." (Desai, 1994). Unfortunately, this is not the situation.

Despite these and other advances, the gains made in cancer chemotherapeutics and chemoprevention (Kelloff et al.l., 1997) appear to rely more on early diagnosis and more efficient therapeutic procedures in radiotherapy, chemotherapy, and surgery (Desai, 1994). Not to be unduly pessimistic, Janin (1996) opined that "advances in science and technology [are] not equivalent to advances in medicine." Time, evolution, and experience using recent and integrated discoveries in the molecular biology of carcinogenesis will likely lead to further advances in clinical medicine, and especially to better outcomes for detection, treatment, and prevention of cancers. Meanwhile, we must recognize that much of the knowledge about carcinogenesis that we capture today will not be applicable for some or many years in the future.

The key to success in curtailing cancer morbidity and mortality is prevention; prevention or reduction in cancer incidences requires more active and directed efforts. To this end, chemical carcinogenesis bioassays for identifying potential human carcinogens represent amazing tools for planning and instituting prevention strategies. These bioassays were first utilized in the early part of the 20th century; in fact, the first direct evidence of exposing animals to chemicals for the purpose of detecting agents that might be carcinogenic appeared near the end of the 1910s (Shimkin, 1980).

B. History of Carcinogenesis Testing

The revolution and testing age of chemical carcinogenesis began when Yamagiwa and Ichikawa (1918) showed that coal tar experimentally applied to rabbit ears caused skin carcinomas. In these studies the authors were attempting to verify the soundness of Virchow's irritation hypothesis of carcinogenesis, a theory based on clinical experience. Earlier, in 1905, Yamagiwa had expressed a notion that "the repetition or continuation of chronic irritation may cause a precancerous alteration in epithelium previously normal. If the irritant continue its action, carcinoma may be the outcome, even though no specific agent has been interpolated." We now know that this theory that toxicity (with respect to its role in a sequence of irritation, degeneration, regeneration, and cell proliferation) "causes" cancer is incorrect (Huff, 1991, 1993, 1995; Farber, 1996), yet we also recognize that certain toxic effects may have some influence on the carcinogenesis processes. The proof for this is limited, but in some cases the idea does appear to have some biological plausibility. However, the current rush to endorse and adopt generically this theory of toxicity and cancer is fraught with danger (Huff, 1996; Melnick et al., 1997).

Yamagiwa and Ichikawa (1918) divided the course of their experiments—based on microscopic characteristics—into four periods, with detailed and ample descriptions: 1) atypical growth of the epithelium, 2) appearance of folliculoepithelioma (actually papillomatous new growths), 3) production of carcinoma, and 4) metastasis. The authors end by restating (and confirming) their above-quoted hypothesis that irritation leads to cancer; this notion, at the time, was certainly a logical conclusion to draw from these results. In 1944, Berenblum stressed that "the belief that chronic irritation is the cause of cancer has as its counterpart the attitude that irritation has nothing whatever to do with cancer, neither takes fully into account all that is known about the mechanism of carcinogenesis, and both suffer from oversimplification." Since then, others have come to similar conclusions (Weinstein, 1990, 1992; Zito, 1992; Huff, 1995; Farber, 1995, 1996).

In 1925, Murphy and Sturm produced primary lung tumors in mice after application of coal tar to the skin. This finding represents a particularly significant breakthrough, especially at the early developmental stages of chemical carcinogenesis, because few chemicals have been shown to induce systemic cancers resulting from skin or dermal exposures. In fact, the authors held the view that (Murphy and Sturm, 1925):

> a cancer of the skin had occurred as a result of the tar, had given off a metastasis to the lung, and then either healed or sloughed out. But frequent and careful examination of the animals during life failed to show any lesions of the skin suggesting cancer, and at autopsy there were no scars or other evident abnormalities in the painted areas. Furthermore, histological examination of the nodules in the lungs revealed a structure and a type of cell different from the metastases which occur from the tar skin cancers.

Their experiments to prove this eventual theory were quite ingenious: they applied the coal tar at different successive locations on the skin to avoid or prevent the development of application-site skin lesions. In fact, high incidences of lung tumors occurred. "The incidence ranged from 60.0 per cent in one experiment to 78.3 per cent in another. Control mice from the same stock but 3 to 6 months older, and for that reason the more liable to spontaneous lung tumors, failed to show a single instance of such growths."

Another example of a chemical shown experimentally to cause internal cancers (lung) from skin exposures is 1,2-dibromoethane (EDB, ethylene dibromide) (Van Duuren, 1979; Huff, 1983); in this instance, skin tumors were also induced. Urethane is another example of a chemical that may cause lung cancers after skin exposure (Salmon and Zeise, 1991). More recently, both diethanolamine and coconut oil acid diethanolamine condensate applied dermally caused liver and kidney tumors in mice, while none was induced at the site of application (NTP, 1998).

However, the idea that a chemical must induce "distant-site" rather than "local" tumors in order to be considered a "real carcinogen" continues to be held. That is, some currently believe that if tumors are caused at the site of application (by inhalation in the nose or lung; on the skin for dermal; orally for mouth, forestomach, or stomach; or subcutaneously for injection-site tumors), then the chemical is not really carcinogenic but an artifact of the mode of chemical exposure. This is speculative gerrymandering, a view likely thought up in the inner sanctum of some corporate headquarters. Thus, for these and other studies, one should pay little if any attention to these sleight-of-hand configurations; no consistent scientific data have been generated to support these correlations between exposure route and tumor-site. More than 20 years ago, for instance, Tomatis (1977) clearly confirmed the screening value of the subcutaneous route of exposure for identifying chemical carcinogens. Nothing has changed since. Interestingly, as examples, the first evidence of carcinogenicity for ethylene oxide (carcinogenic to humans) and epichlorohydrin (probably carcinogenic to humans) was obtained in subcutaneous tests in mice (IARC, 1976, 1985). However, it is important to understand that route of exposure should not be ignored altogether.

Cook et al. (1932) synthesized and tested a large number of pure polycyclic aromatic hydrocarbons, in particular those related to 1,2-benzanthracene, for skin-cancer induction in mice. Of the many four-, five-, six-, and eight-ring compounds tested, none of the hydrocarbons in the *pure state* produced cancer except 1,2:5,6-dibenzanthracene, This chemical "shows undiminished carcinogenic power when very highly purified; it has been shown to be active in nine different media, and has produced cancer of the skin when applied in a concentration of 0.003 per cent in benzene" (Cook et al., 1932). Regarding solvents for these hydrocarbon experiments, benzene was used for all to neutralize any influence of the solvent. However, in some cases the chemical was poorly soluble in benzene, and thus "these considerations do not lessen the value of positive results, but they do show the need for caution in accepting negative results." In addition, the authors indicate that "in this paper we can do no more than record the observations which we have made; we cannot say what results might be obtained under other conditions." Honest doubts such as these are rarely expressed in today's scientific testing arena, in which generic statements are often made on the basis of one study or flimsy data. Baserga (1990) placed this in perspective by stating that "as so often happens in science, timid suggestions made at the end of a discussion—usually as carefully worded hypotheses—become transformed in the next paper by that author (or others) into scientific fact." Unfortunately, this continues to happen (Huff et al., 1996; Melnick et al., 1997).

Sasaki and Yoshida (1935), observing hepatomas in rats fed o-amidoazotoluene (o-toluene-azo-o-toluidine), described them as having developed "step-by-step via a steadily progressive proliferation of the liver cells." This may have been an early consideration of the stages or steps of the carcinogenesis process we now use, perhaps expressed with too much certainty. The experiments were quite complete and well designed in the sense that some groups of animals were removed, killed, and evaluated histologically every 15 days for up to 300 days. Other groups, instead of being killed at 15-day intervals, were removed from chemical exposure to measure time-dependent regression-progression of lesions at the end of the 300-day experiments. (Today these are typically called "stop studies" and usually include additional groups given control feed after 12, 15, or 18 months of exposure to a chemical. The animals are killed at the 2-year termination, and are compared to continuously exposed groups.) After the exposures were stopped, the tumors continued to develop, with a required period of administration of about 4 months. For experiments in which the exposure levels were reduced, the onset of carcinoma was considerably delayed; for example, when one-half the dose was used, the time required to develop carcinoma was about twice the original time. This important finding should remind us that using exposures below the optimal level or the maximally tolerated exposure level or of short duration often leads to "false-negative" results.

Further, Sasaki and Yoshida also explored a structure-activity issue by exposing animals to p-amidoazotoluene and p-amidoazobenzene; neither of these "demonstrated a discernible organotropy, i.e., no detectable liver changes, apart from some minor regressive changes." One might wonder whether o-amidoazobenzene (not evaluated) would have induced liver tumors. Thus, the orthoconfiguration for these chemicals seems to be necessary for a carcinogenic response to be rendered. This isomeric influence on carcinogenesis occurs commonly, but without consistency regarding which isomers will be carcinogenic. For instance p-dichlorobenzene is carcinogenic whereas the monochlorobenzene and o-dichlorobenzene are not carcinogenic. For phenylenediamines the results are of a degree: 4-chloro-p-phenylenediamine is not carcinogenic in rats or mice; 4-chloro-m-phenylenediamine causes cancer in male rats and female mice but not in female rats or male mice; and 4-chloro-o-phenylenediamine induces cancer in both rats and mice, both male and female. In some cases there is no or little difference between isomeric chemicals and the ability to induce cancer: m- and p-cresidine are both carcinogenic, although the meta-isomer is apparently not carcinogenic in mice.

A final example of isomerism is nitrophenols, of which all three isomers studied caused cancer only in male rats, yet at three different organs sites, with none in female rats or mice: 1-amino-4-nitrophenol in the kidney; 2-amino-5-nitrophenol in the pancreas; 4-amino-2-nitrophenol in the urinary bladder (and possibly in female rats as well). Thus, predictions of carcinogenicity are nearly impossible regarding isomeric carcinogens and noncarcinogens, and—currently at least—the only solid avenue by which to show differences or similarities in carcinogenicity is to do the long-term bioassay studies. Thus, one must remain cautious when attempting to predict carcinogenic activity from structural isomerization.

These seminal findings of Sasaki and Yoshida (1935) with incremental levels of exposure are reminiscent of radiation experiments using fractionated levels of radiation versus single exposures to higher levels of irradiation (total exposures being the same): in either case, tumors result but typically at much later times, and even with different target organs. When half the dose of o-amidoazotoluene was administered, the time required for development of carcinoma was about twice the original time. Obviously, if exposure is too low (as in some carcinogenicity studies), tumor activity may not manifest during the animals' lifetime and the experiment should be considered a "false negative," or uninterpretable, although some consider results at "low exposures" to be adequate negatives. Even exposures for 2 years might not be long enough for some late-stage carcinogens to induce cancer, for example, of certain metals, especially cadmium. Thus, we must continue to use the highest non-life-threatening exposure level when testing unknown chemicals for potential carcinogenic effects (Huff et al., 1988; 1994).

In 1941, Berenblum, considering the significance of cocarcinogenesis, offered the suggestion that the "three phases of carcinogenesis—(a) the development of the preneoplastic phase (latent period) or precarcinogenic action, (b) the conversion of this into the wart stage, or epicarcinogenic action, and (c) the malignant transformation of these warts, or metacarcinogenic action—are probably not simply stages of one single carcinogenic process, but independent processes." This was an early version of what is now believed to be the carcinogenesis paradigm: the initiation, promotion, and progression stages of carcinogenesis, the fourth stage being metastases. "True carcinogens are capable of producing all three actions, though to different degrees in different species."

Berenblum (1941) concludes this paper with some words on the possible influence on the carcinogenic processes of chemicals that cause irritation:

> the demonstration that most of the ordinary skin irritants are not in themselves carcinogenic [has] tended to distract clinical attention from the possibility that irritation might play a part in the development of tumors in man. However, the fact that certain noncarcinogenic irritants are capable of producing cocarcinogenic, epicarcinogenic, and metacarcinogenic effects introduces new conceptions of possible extraneous factors of a noncarcinogenic nature influencing tumor development in man.

Apparently the work of Berenblum (1941) centers on the issue of cocarcinogenicity whereby each chemical may have a specific function to impart on the process rather than being directed specifically on chemicals that have both irritant and carcinogenic activities. Thus, as with most "mechanistic" factors that are frequently chemical-specific, and difficult or near impossible to extend beyond the experimental conditions being reported, irritation per se cannot be assumed a priori to influence the carcinogenesis process; simply stated: sometimes it does and most times it doesn't (Hoel et al., 1988; Huff, 1991, 1993, 1995; Farber, 1995, 1996). Conversely, one cannot assume that irritation may never influence the carcinogenesis process. Another question that remains difficult to answer is whether the inflammation–irritation sequelae come before, during, or after carcinogenesis.

In the middle 1950s, Magee and Barnes (1956) introduced a potent class of hepatocarcinogenic compounds—nitrosamines—by feeding rats dimethylnitrosamine (DMN). Only 10 rats of each gender were fed diets containing 50 ppm DMN, with five controls of each sex. "Between the 26th and 40th week, 19 of the 20 animals developed primary hepatic tumours, metastatic spread being present in 7 cases." An attempt to produce tumors in six rabbits by DMN was unsuccessful: exposures were 20 ppm DMN for 10 weeks, then 30 ppm for 4 weeks, and 50 ppm for 8 weeks. The limited numbers of rabbits, short durations, and low dosing schedule make this an inadequate negative. Prophetically, the authors ended their paper with the understated notion that DMN, "by virtue of its chemical and physical properties, may be of some value in the investigation of hepatic carcinogenesis."

Subsequently, of course, these chemicals have been shown to induce multisite cancers in both genders of various species and strains, and have been used to extend our knowledge of mechanisms of carcinogenesis (Lijinsky, 1992). Very few nitrosamines have been shown not to cause cancer in laboratory animals; this phenomenon may be due to incomplete testing in multiple species, or to uniqueness in this class of typically carcinogenic agents. Interestingly, individual nitrosamines (like other chemicals) frequently induce cancers in different organs in different mammalian species and sexes. Even though consistently carcinogenic among species, this "nonconcordance" observation has been mistakenly used by some as a basis for criticism of the value and usefulness of bioassays for preventing human cancers. The same is true for the observation that certain target organs in animals that are frequent sites of carcinogenicity might not be concordant with human cancers, e.g., liver tumors in rodents (Huff, 1998).

Importantly, "it is important to recognize that the target site for human carcinogens is not always the same as in rodents" (Williams, 1995). This observed nonconcordance in tumor sites should not be used to diminish the value and utility of the findings for hazard extrapolation. In every case, however, when comparing and testing human carcinogens in animals [or vice versa, since nearly one-third of chemicals causing cancers in humans were first discovered to do so in animals (Huff, 1993)], there is at least one target organ common to both species (Wilbourn et al., 1986; Tomatis et al., 1989; Huff, 1994). Weisburger (1988), in fact, asks the cogent question: "Why not differences in species response?" If one simply utilized all the available epidemiological, mechanistic, and animal data to protect public health from threats of cancer-causing chemicals, rather than to continue the debate ad nauseam (with little or no new data) on whether animal data are relevant to the prediction of hazards to humans, then public health and science would be better served. We need to accept the reality that "there are differences in response of different species to many carcinogens" (Linjisky, 1993), and attempt to discover why.

Since these early experiments (e.g., those of Magee and Barnes in 1956), many nitrosamines have been tested for carcinogenicity, shown to induce cancers, utilized as positive experimental controls, and used as tools for mechanistic research (Lijinsky, 1992). Nitrosamines as a possible carcinogen to humans has been the subject of much speculation, but adequate information in humans is lacking. Nearly all the nitrosamines that have been tested are carcinogenic, and these can reasonably be expected to be carcinogenic to humans (IARC, 1987; Lijinsky, 1992). What impact these may have on the incidence and mortality of cancers remains to be discovered. Nitrosamine carcinogenesis gained particular significance when it was found that these chemicals occur in foods and cigarette smoke, and that they were formed in the gastrointestinal tract in the presence of foods and nitrites (Shimkin, 1977). Finally, according to Lijinsky (1992), "while no comprehensive assessment of human exposure to these preformed nitrosamines can be made, nor of the carcinogenic risk they pose, the exceptional potency of these systemic carcinogens mandates that they cannot be dismissed as insignificant, even when the concentration at which they occur is small." On human cancers in general, Schmahl (1988) has noted that:

> the occurrence of cancer at an advanced age can also be ascribed to the summation of effects of low and extremely low doses [of carcinogens] propagated over a lifetime. Cancer occurring in old age may thus be solely attributed to the impact of very low doses of carcinogens, depending on their capacity of summation. In the wake of the increasing mean life expectancy observed worldwide, the effect of such low doses become manifest.

This also strengthens the need to test combinations of chemicals for possible carcinogenicity (Takayama et al., 1987, 1989).

Overall, these early papers may serve to introduce readers to some of the early pioneers and the beginnings of the scientific field of chemical carcinogenesis. Many of these papers can be found in the collections compiled by Shimkin (1977, 1980), and those new to this area of interest might do well to explore the history on which much of our current efforts are based. This not only will yield some historical insights but also help prevent needless repetition of some of these past efforts, and perhaps avert mistakes of ignorance. After all, "history is the witness that testifies to the passing of time; it illumines reality, vitalizes memory, provides guidance in daily life, and brings us tidings of antiquity" (Cicero, 43 B.C.). One might also avoid claiming discoveries of what may already have been discovered.

C. The "Pre-Experimental" Era

In the early days of modern cancer epidemiology, emphasis was placed on occupational and industrial cancers; these were relatively recognizable and hence gathered both rapt attention and vested-interest contention. In more recent years, the relative proportion of other cancer-causing agents seems to have increased compared to industrial chemicals; these agents include viruses, parasites, and retroviruses (combined as "biological agents") and drugs. The latter are comparatively easy to study and to follow up, given that records of dose and treatment duration are usually recorded.

The enormous build-up of environmental contamination from industrial pollution since the 1940s and 1950s represents a difficult challenge to cancer epidemiology, especially because of the relatively low levels of exposures to particular chemicals. However, we have little knowledge about the impact on human cancers from multiple and myriad combinations of relatively small amounts of individual carcinogens. Of course, lower levels of distinct carcinogens would appear to present a lowered risk, but simply reducing the amount of contamination for one carcinogenic agent does not then make that agent a noncarcinogen; if anything, it would seem to render that agent a less potent carcinogen, but certainly the intrinsic carcinogenic activity would still pertain.

Importantly, a major problem is that there are so many carcinogens in the environment that we have difficulty identifying etiological associations for human cancers. This is especially true for environmental "clusters" (Neutra, 1990), wherein geographical groupings of cancers often cannot be explained. Some insist that cancers of old age may in fact be due to or at least influenced by the cumulative impact of lifetime exposures to low levels of many carcinogens (Schmahl, 1988; Habs and Schmahl, 1988). Others believe that these environmental combinations have no bearing on cancer risk (Ames and Gold, 1997). Unfortunately, the real answer probably lies between the two opinions. Thus, one must continue to evaluate this potential environmental syncarcinogenesis, and to begin more earnestly to test combinations of chemicals in laboratory animals (Tomatis et al., 1998).

Some would suggest that the "pre-experimental" era began with the brilliant clinical connection drawn in 1775 by the surgeon Sir Percival Pott, who proposed that chimney sweeps developed scrotal cancer as a direct consequence of exposure to a defined substance in their occupation. His 800-word account (Pott, 1775) documented for the first time this malady, which the sweeps called soot wart, "a disease which always makes its first attack on, and its first appearance in the inferior part of the scrotum; where it produced a superficial, painful, ragged, ill-looking sore, with hard and rising edges." His description concluded, "when the disease has got head, it is rapid in its progress, painful in all its attacks, and most certainly destructive in its event." We now realize that agents such as soot, coal tars and coal-tar pitches, creosotes, and shale oils of various compositions are carcinogenic to humans (IARC, 1987).

Of course, the first reported occupational cancer was not chemically caused but remains monumental. In 1713, Bernardino Ramazzini (the "father of occupational medicine") discovered and reported on the higher risk of breast cancer among single women (nuns) compared to that among married women. This dedicated physician's observation was the first association of cancer causation with occupation (being a nun). The hypothesis was that the greater risk for breast cancer could be related to the absence of an active reproductive life, in particular, pregnancy and lactation. Today breast cancer is the major cause of anxiety, morbidity, and mortality in women, and at long last has become a target for pioneering diagnostic and therapeutic regimens and aggressive prevention strategies (Wolf et al., 1996). Indeed, the age-adjusted breast-cancer mortality rate for white women in the United States dropped 6.8% from 1989 through 1993 (Chu et al., 1996). Incidence rates, however, continue to increase; the percentage increase between 1973 and 1992 for all ages and races was 25.3, with the highest being for women 65 years and older (over 65 = 41.6% increase; under 65 = 16.4 %) (Kosary et al., 1995).

In 1700, Ramazzini published his *De Morbis Artificum Diatriba*. Interestingly, Pott started his 1775 paper on scrotum cancer: "Ramaz[z]ini has written a book de morbis artificum." As reported by Hunter (1978), Ramazzini made a striking addition to the Hippocratic art: "When a doctor visits a working-class home he should be content to sit on a three-legged stool, if there isn't a gilded chair, and he should take time for his examination; and to the questions recommended by Hippocrates, he should add one more—What is your occupation?" Hunter added, admirably, "Did ever a man announce with more point or with less fuss a revolutionary innovation? In an innocent-sounding sentence we find that Ramazzini characterizes and supersedes the medical science and the medical practice of two thousand years." In 1982, an international community of scholars formed an organization in his name, the Collegium Ramazzini, based at his birthplace, Carpi, Italy, to advance the study of occupational and environmental health issues around the world. Founded by Drs. Irving Selikoff and Cesare Maltoni, the Collegium Ramazzini offers a bridge between the world of scientific discovery and the social and political centers that must act on these discoveries to conserve life.

Just after the middle 1700s, John Hill (1761), a British physician, discovered and reported in a treatise the hazards of using snuff:

> Whether or not the polypusses [later defined as "all the frightful symptoms of an open cancer"] which attend Snuff-takers are absolutely caused by that custom, or whether the principles of the disorder were there before and Snuff only irritated the parts and hastened the mischief, I shall not pretend to determine: but even supposing the latter only to be the case, the damage is certainly more than the indulgence is worth. No man should venture upon Snuff who is not sure that he is not liable to cancer: and no man can be sure of that.

This is the first reported instance of cancer caused by lifestyle, culture, or habit. Of main importance, and relevant today, Hill pointed to preventive measures of controlling or reducing snuff-induced cancers while not yet knowing all or even most of the answers to the scientific questions surrounding this issue. We seem to have strayed from that public health initiative to cancer control by insisting on knowing all or most of the answers before implementing appropriate prevention strategies. Tobacco smoke is obviously a modern example of the difficulty of 1) reaching political and social agreement on the degree of association between cancer and an agent or activity and 2) eliminating the agent or activity.

Near the middle of the 19th century, Elmslie (1866) reported an interesting study he had made of the etiology of epithelioma among the people of Kashmir, in northern India. During the severe winters, Kashmiris carry, "whenever they go, earthen-ware pots, which they have denominated kangris. The kangris or portable braziers are made of clay of varying fineness, and are usually covered with wicker-work, more or less ornamented according to the price of the article. Men and women, young and old, rich and poor, Hindu and Musselman (Muslim), all have their kangri, and all consider it indispensable in the cold season." These clay pots, filled with charcoal, generated considerable heat and smoke. Customarily, while outdoors, both men and women "carry the kangri under their loose woolen gowns, and in close proximity of the abdomen. When indoors, or in a sitting posture, the Kashmiris place the kangri between their thighs." Elmslie ends his marvelous commentary with the conclusion that, "to say the least, it seems highly probable that the disease [epithelioma] is caused by the injurious effects of the heat of the kangris on the skin of the abdomen and thighs, the very part with which the utensil comes in contact when used." Obviously, the rough texture of the pots subjected those areas of the skin to continual irritation and sores, which probably contributed to the eventual lesions.

Jonathon Hutchinson, in 1888, in the brief opening paragraph to his 12-page article, described with utmost clarity and completeness the diagnosis, cause, and progression of arsenic-induced cancer in humans exposed to Fowler's solution:

> In the following statement I have two separate propositions to maintain, one of which is, however, of great importance to the other. The first is that by the prolonged internal use of arsenic the nutrition of the skin may be seriously affected, and that, amongst other changes, warty and corn-like indurations may be produced. The second goes much further and asserts that if the drug be continued these "arsenic corns" may assume a tendency to grow downwards and pass into epithelial cancer.

The remainder of his article details the various case reports and correspondence surrounding the long use of Fowler's solution (1% potassium arsenite) and the associated arsenic keratosis and cancer. Inorganic arsenicals were widely used in medicine a few decades ago for the treatment of a large variety of disorders, including arthritis, asthma, psoriasis, and syphilis, and are still used extensively in agriculture and wood treatment (Chan and Huff, 1997). Arsenical keratoses are multiple, discrete, hard hyperkeratotic wartlike lesions on the palms of the hands and soles of the feet. They characteristically occur in persons a decade or more after exposure to arsenic (Wintrobe et al., 1974). The lesions are similar to actinic or solar keratoses in that they are premalignant. Inorganic arsenicals are the only well-recognized carcinogens that cause both skin and visceral malignancies in humans. Certain occupational exposures such as in coal gasification and coke production as well as coal tars, coal-tar pitches, and soot also induce skin and internal cancers (Tomatis et al., 1989). Clearly this involves dermal and inhalation as well as oral exposures to these agents.

Near the close of the 19th century, in 1895, Rehn delivered a lecture on the etiology of urinary bladder tumors in which he proposed that "for a majority of bladder tumors one can only imagine that substances were present in solution in the urine discharged from the kidneys and that these caused a tumor to form by means of chemical stimulus." Even though Rehn said that "these substances completely escape our knowledge," he went on to tell the audience "about a number of illnesses which, with a high degree of probability, can be attributed to chemical stimulus." Involved in the treatment of three workers from an aniline factory who were suffering from bladder tumors, Rehn thoroughly investigated the situation and formed three significant conclusions:

1. The gases created by the production of fuchsin "lead to disorders in the urinary apparatus."
2. "Many years of working in the fuchsin operation can lead to the development of bladder tumors as a result of the constant stimulus."
3. The harmful effect is "essentially due to the inhalation of aniline vapors."

Since these astute observations and causative assignment, many other substances and occupations have been associated with the formation of cancer of the urinary bladder and urinary system (Tomatis et al., 1989; Huff, 1994, 1998a, 1998b). For several of these human carcinogens, the evidence was first determined in experimental animals (Huff, 1993). Agents known to cause cancer of the urinary bladder in humans include 4-aminobiphenyl; analgesic mixtures containing phenacetin; aviation fuels; benzidine; chlornaphazine; coal tars and coal-tar pitches; cyclophosphamide; 2-naphthylamine; *Schistosomia hematobium*; tobacco; trichloroethylene; untreated mineral oils; and agents used in boot and shoe manufacture and repair, painting, coal gasification, iron and steel founding, the rubber industry, and the manufacture of auramine. Other chemicals that cause tumors of the urinary bladder in animals have been identified in experimental systems (Huff et al., 1991; Huff, 1998, 1998a); studies from the National Toxicology Program have identified 24 unique chemicals in this category (http://ntp-server.niehs.nih.gov/htdocs/Sites/Site_Cnt.html). Some of these experimental findings might serve as leads for epidemiological investigations.

In summary, Ramazzini (1713) and Pott (1775) identified and published the first accounts of occupational cancers; Hill (1761) and, a century later, Elmslie (1866) recorded the first correlations between lifestyle and cancer; Hutchinson (1888) published what were perhaps the first case reports of therapeutically induced cancer; and Rehn (1895) made the first association of an industrial process, chemicals, and cancer. Much has happened since these historic and distinguished events, and much more will be accomplished in the future. Discovering the absolute causes of cancers and preventing cancers from occurring are not easy, yet we will continue to pursue these objectives.

VI. RELEVANT COMPILATIONS AND EVALUATIONS

A. PHS 149: Agents Tested for Carcinogenic Activity

In the late 1940s, Jonathan Hartwell sought out the available published papers on testing chemicals for carcinogenic activity, and compiled and printed the first volume of the U.S. Public Health Service's excellent continuing series *Survey of Compounds Which Have Been Tested for Carcinogenic Activity* (PHS 149, 1951–1997). Catalogued within these volumes through 1996 is carcinogenic information on 9151 individual chemicals (i.e., unique CAS numbers) and 21,137 synonyms. Virtually all literature published on chemical carcinogenesis testing through 1996 is available via PHS 149 (V. Fung, NCI, personal communication).

The contents of these volumes are not critical reviews or assessments of the carcinogenic activity of tested substances, and users must draw their own conclusions from the data presented. In addition, neither positive nor negative findings should be taken to imply proof of the existence or absence of carcinogenic activity. Thus, users may have to search out the original articles to get a more complete indication of the carcinogenicity of a particular agent.

Articles are selected for inclusion if they meet three criteria: 1) the study involved testing a compound for carcinogenic activity, 2) the test was performed on vertebrates or invertebrates (excluding microorganisms), and 3) a positive result was obtained or a negative result was confirmed by at least 60 days of observation following exposure in utero or at birth, or 120 days of observation following exposure of weanling or older animals.

Unfortunately, these criteria do not include the duration of an experiment or the appropriateness of the exposure levels used. This evaluation is left to the user.

For each entry that meets the above criteria, these data are reported: 1) name of the agent, 2) full reference citation of the source of information, 3) common name and numbers of animals, 4) species, strain, sex, and age at start of test, 5) preparation and dose of chemical, 6) route of administration or site of application, 7) pathology examination level (gross or external without dissection, autopsy, and/or microscopic examination of tissue sections), 8) animals with tumors (number of tumor-bearing animals, specific tumor topography divided into eight predesignated areas), 9) details on tumors found, 10) survival, and 11) duration of experiment.

Nine useful indexes are available: 1) bibliography, 2) authors, 3) chemical name, 4) CAS Registry Number Index, 5) route of administration, 6) site of application, 7) animal species/strain, 8) tumor site, and 9) vehicle. PHS 149 is often the first source one should search to determine if an agent has ever been tested for carcinogenic activity.

B. IARC Monographs, Bulletins, and Directories

In 1965, the International Agency for Research on Cancer (IARC) was established in Lyons, France. Six years later it implemented a program to evaluate critically the published information on chemicals and carcinogenic risks, and to periodically make available their findings as the *IARC Monographs on the Evaluation of Carcinogenic Risks to Humans* (IARC, 1997). The specific objectives are to review data on the carcinogenicity of agents to which humans are known to be exposed and on specific exposure situations, to evaluate these data in terms of human risk with the help of international working groups of experts in chemical carcinogenesis and related fields, and to indicate where additional research efforts are needed. As of 1997, 69 volumes had been published, containing evaluations of 836 chemicals, groups of chemicals, industrial processes, occupations, lifestyle factors, biological agents, complex mixtures, and other exposure circumstances (Tables 9 and 9a).

The IARC also prepares an ideal companion to the *Monographs*, the PHS 149 series, and the NTP technical reports: the *Directory of Agents Being Tested for Carcinogenicity*. A recent edition (IARC, 1996) contained information from 65 institutes in 21 countries on 533 chemicals or agents being tested for carcinogenicity, as well as a listing of 186 published references on 53 chemicals or agents. It provides a reasonably comprehensive catalog of chemicals at various stages of testing that should be consulted

Table 9 Numbers of Chemicals, Groups of Chemicals, and Exposure Circumstances Evaluated for Carcinogenicity by the IARC and the NTP (about 2000 chemicals studied for carcinogenicity)

International Agency for Research on Cancer (IARC) evaluations: 836

IARC category of evidence	Total	Biological			Exposure circumstances
		Chemicals	Agents	Mixtures	
Carcinogenic to humans Group 1	74	40	9	12	13
Probably carcinogenic to humans Group 2A	56	44	3	5	4
Possibly carcinogenic to humans Group 2B	225	203	5	13	4
Not classifiable as to its carcinogenicity to humans Group 3	480	455	6	12	7
Probably not carcinogenic to humans Group 4	1	1	—	—	—
Totals	836	743	23	42	28

National Toxicology Program (NTP) evaluations: about 800

Known to be carcinogenic to humans	29
Reasonably anticipated to be carcinogenic to humans	169
Agents or exposure circumstances not listed[a]	~600

[a]Considered not meeting guideline criteria for listing as being carcinogenic in two species.

before the undertaking of time-consuming and costly long-term carcinogenesis experiments.

The *Directory of On-going Research in Cancer Epidemiology* contains descriptions of 1101 projects in 80 countries. A joint project between the IARC and the German Cancer Research Centre (DKFZ), the directory is a valuable source of information about most of the studies being undertaken on cancer epidemiology (IARC, 1996a).

Additionally, the IARC convenes and sponsors scientific symposia and publishes the proceedings in their Scientific Publications Series (since 1971), all typically centered on cancer. IARC Monographs and Scientific Publications are listed in Table 10.

Table 9a IARC Monographs on the Evaluation of Carcinogenic Risks to Humans

Categories of evidence used by the IARC

Group 1: The agent (mixture) is carcinogenic to humans. The exposure circumstance entails exposures that are carcinogenic to humans.

Group 2 (two classifications)

Group 2A: The agent (mixture) is probably carcinogenic to humans. The exposure circumstance entails exposures that are probably carcinogenic to humans.

Group 2B: The agent (mixture) is possibly carcinogenic to humans. The exposure circumstance entails exposures that are possibly carcinogenic to humans.

Group 3: The agent (mixture, exposure circumstance) is unclassifiable as to carcinogenicity to humans.

Group 4: The agent (mixture, exposure circumstance) is probably not carcinogenic to humans.

Group 1: List of agents, mixtures, and exposures carcinogenic to humans (74 entries)

Agents

1. Aflatoxins, naturally occurring
2. 4-Aminobiphenyl
3. Arsenic and arsenic compounds (evaluation applies to the group of compounds as a whole and not necessarily to all in the group)
4. Asbestos
5. Azathioprine
6. Benzene
7. Benzidine
8. Beryllium and beryllium compounds (evaluated as a group)
9. N,N-Bis(2-chloroethyl)-2-naphthylamine (Chlornaphazine)
10. Bis(chloromethyl)ether and chloromethyl methyl ether (tech grade)
11. 1,4-Butanediol dimethanesulfonate (Busulphan; Myleran)
12. Cadmium and cadmium compounds (evaluated as a group)
13. Chlorambucil
14. l-(2-Chloroethyl)-3-(4-methylcyclohexyl)-1-nitrosourea (Methyl-CCNU)
15. Chromium (VI) compounds (evaluated as a group)
16. Cyclosporin
17. Cyclophosphamide
18. Diethylstilboestrol
19. Erionite
20. Ethylene oxide[a]
21. *Helicobacter pylori* (infection with)
22. Hepatitis B virus (chronic infection with)
23. Hepatitis C virus (chronic infection with)
24. Human immunodeficiency virus type 1 (infection with)
25. Human papillomavirus type 16
26. Human papillomavirus type 18
27. Human T-cell lymphotropic virus type I

Table 9a Continued

28. Melphalan
29. 8-Methoxypsoralen (Methoxsalen) plus ultraviolet A radiation
30. MOPP and other combined chemotherapy including alkylating agents
31. Mustard gas (sulfur mustard)
32. 2-Naphthylamine
33. Nickel compounds (evaluated as a group)
34. Oestrogen replacement therapy
35. Oestrogens, nonsteroidal
36. Oestrogens, steroidal (evaluated as a group)
37. *Opisthorchis viverrini* (infection with)
38. Oral contraceptives, combined (conclusive evidence that these agents protect against ovary and endometrium cancers)
39. Oral contraceptives, sequential
40. Radon and its decay products
41. *Schistosoma haematobium* (infection with)
42. Silica, crystalline (inhaled in the form of quartz or cristobalite from occupational sources)
43. Solar radiation
44. Talc containing asbestiform fibres
45. Tamoxifen (conclusive evidence that this agent reduces the risk of contralateral breast cancer)
46. 2,3,7,8-Tetrachlorodibenzo-p-dioxin[a]
47. Thiotepa
48. Treosulfan
49. Vinyl chloride

Mixtures

50. Alcoholic beverages
51. Analgesic mixtures containing phenacetin
52. Betel quid with tobacco
53. Coal-tar pitches
54. Coal tars
55. Mineral oils, untreated and mildly treated
56. Salted fish (Chinese-style)
57. Shale oils
58. Soots
59. Tobacco products, smokeless
60. Tobacco smoke
61. Wood dust

Exposure circumstances

62. Aluminium production
63. Auramine, manufacture
64. Boot and shoe manufacture and repair

Table 9a Continued

65. Coal gasification
66. Coke production
67. Furniture and cabinet making
68. Haematite mining (underground) with exposure to radon
69. Iron and steel founding
70. Isopropanol manufacture (strong-acid process)
71. Magenta, manufacture of
72. Painter (occupational exposure as a)
73. Rubber industry
74. Strong-inorganic-acid mists containing sulfuric acid (occupational exposure to)

Group2A: List of agents, mixtures, and exposures probably carcinogenic to humans (56 entries)

Agents

1. Acrylamide[b]
2. Acrylonitrile
3. Adriamycin[b]
4. Androgenic (anabolic) steroids
5. Azacitidine[b]
6. Benz(a)anthracene[b]
7. Benzidine-based dyes[b]
8. Benzo(a)pyrene[b]
9. Bischloroethyl nitrosourea (BCNU)
10. 1,3-Butadiene
11. Captafol[b]
12. Chloramphenicol[b]
13. (2-Chloroethyl)-3-cyclohexyl-1-nitrosourea (CCNU)[b]
14. para-Chloro-ortho-toluidine and its strong acid salts
15. Chlorozotocin[b]
16. Cisplatin[b]
17. Clonorchis sinensis (infection with)[b]
18. Dibenz(a,h)anthracene[b]
19. Diethyl sulfate
20. Dimethylcarbamoyl chloride[b]
21. Dimethyl sulfate[b]
22. Epichlorohydrin[b]
23. Ethylene dibromide[b]
24. N-Ethyl-N-nitrosourea[b]
25. Formaldehyde
26. Human papillomavirus type 31
27. Human papillomavirus type 33
28. IQ (2-amino-3-methylimidazo(4,5-f)quinoline)[b]
29. 5-Methoxypsoralen[b]

Table 9a Continued

30. 4,4-Methylene bis(2-chloroaniline) (MOCA)[b]
31. N-Methyl-N-nitro-N-nitrosoguanidine (MNNG)[b]
32. N-Methyl-N-nitrosourea[b]
33. Nitrogen mustard
34. N-Nitrosodiethylamine[b]
35. N-Nitrosodimethylamine[b]
36. Phenacetin
37. Procarbazine hydrochloride[b]
38. Styrene-7,8-oxide[b]
39. Tetrachloroethylene
40. Trichloroethylene
41. 1,2,3-Trichloropropane
42. Tris(2,3-dibromopropyl)phosphate[b]
43. Ultraviolet radiation A[b]
44. Ultraviolet radiation B[b]
45. Ultraviolet radiation C[b]
46. Vinyl bromide[b]
47. Vinyl fluoride

Mixtures

48. Creosotes
49. Diesel engine exhaust
50. Hot mate
51. Nonarsenical insecticides (occupational exposures in spraying and application of)
52. Polychlorinated biphenyls

Exposure circumstances

53. Art glass, glass containers, and pressed ware (manufacture of)
54. Hairdresser or barber (occupational exposure)
55. Petroleum refining (occupational exposure)
56. Sunlamps and sunbeds

[a]Overall evaluation upgraded from Group 2A to Group 1 with supporting evidence from other data relevant to the evaluation of carcinogenicity and its mechanisms.
[b]Overall evaluation upgraded from Group 2B to 2A with supporting evidence from other data relevant to the evaluation of carcinogenicity and its mechanisms).

C. NTP Bioassay Technical Reports

In the late 1960s and early 1970s, having the advantage of the growing collection of the PHS 149 series and the IARC Monographs, the National Cancer Institute (NCI) pioneered an extensive carcinogenesis bioassay testing program to evaluate the potential carcinogenicity of chemicals (Sontag et al., 1976; Weisberger, 1983). These studies typically involve exposing

(*text continues on p. 63*)

Table 10 IARC Publications Relevant to Cancer Risk Identification, Evaluation, and Assessment

IARC Monographs on the Evaluation of Carcinogenic Risks to Humans

Volume 1
Some Inorganic Substances, Chlorinated Hydrocarbons, Aromatic Amines, N-Nitroso Compounds, and Natural Products
1972; 184 pages; ISBN 92 832 1201 0
(out of print)

Volume 2
Some Inorganic and Organometallic Compounds
1973; 181 pages; ISBN 92 832 1202 9
(out of print)

Volume 3
Certain Polycyclic Aromatic Hydrocarbons and Heterocyclic Compounds
1973; 271 pages; ISBN 92 832 1203 7
(out of print)

Volume 4
Some Aromatic Amines, Hydrazine and Related Substances, N-Nitroso Compounds and Miscellaneous Alkylating Agents
1974; 286 pages; ISBN 92 832 1204 5

Volume 5
Some Organochlorine Pesticides
1974; 241 pages; ISBN 92 832 1205 3
(out of print)

Volume 6
Sex Hormones
1974; 243 pages; ISBN 92 832 1206 1
(out of print)

Volume 7
Some Anti-Thyroid and Related Substances, Nitrofurans and Industrial Chemicals
1974; 326 pages; ISBN 92 832 1207 X
(out of print)

Volume 8
Some Aromatic Azo Compounds
1975; 357 pages; ISBN 92 832 1208 8

Volume 9
Some Aziridines, N-, S- and O-Mustards and Selenium
1975; 268 pages; ISBN 92 832 1209 6

Volume 10
Some Naturally Occurring Substances
1976; 353 pages; ISBN 92 832 1210 X
(out of print)

Volume 11
Cadmium, Nickel, Some Epoxides, Miscellaneous Industrial Chemicals and General Considerations on Volatile Anaesthetics
1976; 306 pages; ISBN 92 832 1211 8
(out of print)

Volume 12
Some Carbamates, Thiocarbamates and Carbazides
1976; 282 pages; ISBN 92 832 1212 6

Volume 13
Some Miscellaneous Pharmaceutical Substances
1977; 255 pages; ISBN 92 832 1213 4

Volume 14
Asbestos
1977; 106 pages; ISBN 92 832 1214 2
(out of print)

Volume 15
Some Fumigants, the Herbicides 2,4-D and 2,4,5-T, Chlorinated Dibenzodioxins and Miscellaneous Industrial Chemicals
1977; 354 pages; ISBN 92 832 1215 0
(out of print)

Volume 16
Some Aromatic Amines and Related Nitro Compounds – Hair Dyes, Colouring Agents and Miscellaneous Industrial Chemicals
1978; 400 pages; ISBN 92 832 1216 9

Volume 17
Some N-Nitroso Compounds
1978; 365 pages; ISBN 92 832 1217 7

Volume 18
Polychlorinated Biphenyls and Polybrominated Biphenyls
1978; 140 pages; ISBN 92 832 1218 5

Volume 19
Some Monomers, Plastics and Synthetic Elastomers, and Acrolein
1979; 513 pages; ISBN 92 832 1219 3
(out of print)

Volume 20
Some Halogenated Hydrocarbons
1979; 609 pages; ISBN 92 832 1220 7
(out of print)

Volume 21
Sex Hormones (II)
1979; 583 pages; ISBN 92 832 1521 4

Volume 22
Some Non-Nutritive Sweetening Agents
1980; 208 pages; ISBN 92 832 1522 2

Volume 23
Some Metals and Metallic Compounds
1980; 438 pages; ISBN 92 832 1523 0
(out of print)

Volume 24
Some Pharmaceutical Drugs
1980; 337 pages; ISBN 92 832 1524 9

Volume 25
Wood, Leather and Some Associated Industries
1981; 412 pages; ISBN 92 832 1525 7

Volume 26
Some Antineoplastic and Immunosuppressive Agents
1981; 411 pages; ISBN 92 832 1526 5

Volume 27
Some Aromatic Amines, Anthraquinones and Nitroso Compounds, and Inorganic Fluorides Used in Drinking Water and Dental Preparations
1982; 341 pages; ISBN 92 832 1527 3

Volume 28
The Rubber Industry
1982; 486 pages; ISBN 92 832 1528 1

Volume 29
Some Industrial Chemicals and Dyestuffs
1982; 416 pages; ISBN 92 832 1529 X

Volume 30
Miscellaneous Pesticides
1983; 424 pages; ISBN 92 832 1530 3

Volume 31
Some Food Additives, Feed Additives and Naturally Occurring Substances
1983; 314 pages; ISBN 92 832 1531 1

Volume 32
Polynuclear Aromatic Compounds, Part 1: Chemical, Environmental and Experimental Data
1983; 477 pages; ISBN 92 832 1532 X

Volume 33
Polynuclear Aromatic Compounds, Part 2: Carbon Blacks, Mineral Oils and Some Nitroarenes
1984; 245 pages; ISBN 92 832 1533 8
(out of print)

Volume 34
Polynuclear Aromatic Compounds, Part 3: Industrial Exposures in Aluminium Production, Coal Gasification, Coke Production, and Iron and Steel Founding
1984; 219 pages; ISBN 92 832 1534 6

Table 10 Continued

Volume 35
Polynuclear Aromatic Compounds: Part 4: Bitumens, Coal-Tars and Derived Products, Shale-Oils and Soots
1985; 271 pages; ISBN 92 832 1535 4

Volume 36
Allyl Compounds, Aldehydes, Epoxides and Peroxides
1985; 369 pages; ISBN 92 832 1536 2

Volume 37
Tobacco Habits Other than Smoking; Betel-Quid and Areca-Nut Chewing; and Some Related Nitrosamines
1985; 291 pages; ISBN 92 832 1537 0

Volume 38
Tobacco Smoking
1986; 421 pages; ISBN 92 832 1538 9

Volume 39
Some Chemicals Used in Plastics and Elastomers
1986; 403 pages; ISBN 92 832 1239 8

Volume 40
Some Naturally Occurring and Synthetic Food Components, Furocoumarins and Ultraviolet Radiation
1986; 444 pages; ISBN 92 832 1240 1

Volume 41
Some Halogenated Hydrocarbons and Pesticide Exposures
1986; 434 pages; ISBN 92 832 1241 X

Volume 42
Silica and Some Silicates
1987; 289 pages; ISBN 92 832 1242 8

Volume 43
Man-Made Mineral Fibres and Radon
1988; 300 pages; ISBN 92 832 1243 6

Volume 44
Alcohol Drinking
1988; 416 pages; ISBN 92 832 1244 4

Volume 45
Occupational Exposures in Petroleum Refining; Crude Oil and Major Petroleum Fuels
1989; 322 pages; ISBN 92 832 1245 2

Volume 46
Diesel and Gasoline Engine Exhausts and Some Nitroarenes
1989; 458 pages; ISBN 92 832 1246 0

Volume 47
Some Organic Solvents, Resin Monomers and Related Compounds, Pigments and Occupational Exposures in Paint Manufacture and Painting
1989; 535 pages; ISBN 92 832 1247 9

Volume 48
Some Flame Retardants and Textile Chemicals, and Exposures in the Textile Manufacturing Industry
1990; 345 pages; ISBN: 92 832 1248 7

Volume 49
Chromium, Nickel and Welding
1990; 677 pages; ISBN: 92 832 1249 5

Volume 50
Some Pharmaceutical Drugs
1990; 415 pages; ISBN: 92 832 1259 9

Volume 51
Coffee, Tea, Mate, Methylxanthines and Methylglyoxal
1991; 513 pages; ISBN: 92 832 1251 7

Volume 52
Chlorinated Drinking-Water; Chlorination By-products; Some other Halogenated Compounds; Cobalt and Cobalt Compounds
1991; 544 pages; ISBN: 92 832 1252 5

Volume 53
Occupational Exposures in Insecticide Application, and Some Pesticides
1991; 612 pages; ISBN 92 832 1253 3

Volume 54
Occupational Exposures to Mists and Vapours from Strong Inorganic Acids; and other Industrial Chemicals
1992; 336 pages; ISBN 92 832 1254 1

Volume 55
Solar and Ultraviolet Radiation
1992; 316 pages; ISBN 92 832 1255 X

Volume 56
Some Naturally Occurring Substances: Food Items and Constituents, Heterocyclic Aromatic Amines and Mycotoxins
1993; 600 pages; ISBN 92 832 1256 8

Volume 57
Occupational Exposures of Hairdressers and Barbers and Personal Use of Hair Colourants; Some Hair Dyes, Cosmetic Colourants, Industrial Dyestuffs and Aromatic Amines
1993; 428 pages; ISBN 92 832 1257 6

Volume 58
Beryllium, Cadmium, Mercury and Exposures in the Glass Manufacturing Industry
1994; 444 pages; ISBN 92 832 1258 4

Volume 59
Hepatitis Viruses
1994; 286 pages; ISBN 92 832 1259 2

Volume 60
Some Industrial Chemicals
1994; 560 pages; ISBN 92 832 1260 6

Volume 61
Schistosomes, Liver Flukes and *Helicobacter pylori*
1994; 280 pages; ISBN 92 832 1261 4

Volume 62
Wood Dusts and Formaldehyde
1995; 405 pages; ISBN 92 832 1262 2

Volume 63
Dry cleaning, Some Chlorinated Solvents and Other Industrial Chemicals
1995; 558 pages; ISBN 92 832 1263 0

Volume 64
Human Papillomaviruses
1995; 409 pages; ISBN 92 832 1264 9

Volume 65
Printing Processes, Printing Inks, Carbon Blacks and Some Nitro Compounds
1996; 578 pages; ISBN 92 832 1265 7

Volume 66
Some Pharmaceutical Drug
1996; 514 pages; ISBN 92 832 1266 5

Volume 67
Human Immunodeficiency Viruses and Human T-cell Lymphotropic Viruses
1996; 424 pages; ISBN 92 832 1267 3

Volume 68
Silica, Some Silicates, Coal Dust and para-Aramid Fibrils
1997; 506 pages; ISBN 92 832 1268 1

Volume 69
Polychlorinated Dibenzo-dioxins and Dibenzofurans
1997; 666 pages; ISBN 92 832 1269 X

Table 10 Continued

IARC Scientific Publications

No. 1
Liver Cancer
1971; 176 pages; ISBN 0 19 723000 8

No. 2
Oncogenesis and Herpesviruses
Edited by P.M. Biggs, G. de Thé and L.N. Payne
1972; 515 pages; ISBN 0 19 723001 6

No. 3
N-Nitroso Compounds: Analysis and Formation
Edited by P. Bogovski, R. Preussman and E.A. Walker
1972; 140 pages; ISBN 0 19 723002 4

No. 4
Transplacental Carcinogenesis
Edited by L. Tomatis and U. Mohr
1973; 181 pages; ISBN 0 19 723003 2

No. 5/6
Pathology of Tumours in Laboratory Animals. Volume 1: Tumours of the Rat
Edited by V.S. Turusov
1973/1976; 533 pages; ISBN 92 832 1410 2

No. 7
Host Environment Interactions in the Etiology of Cancer in Man
Edited by R. Doll and I. Vodopija
1973; 464 pages; ISBN 0 19 723006 7

No. 8
Biological Effects of Asbestos
Edited by P. Bogovski, J.C. Gilson, V. Timbrell and J.C. Wagner
1973; 346 pages; ISBN 0 19 723007 5

No. 9
N-Nitroso Compounds in the Environment
Edited by P. Bogovski and E.A. Walker
1974; 243 pages; ISBN 0 19 723008 3

No. 10
Chemical Carcinogenesis Essays
Edited by R. Montesano and L. Tomatis
1974; 230 pages; ISBN 0 19 723009 1

No. 11
Oncogenesis and Herpes-viruses II
Edited by G. de-Thé, M.A. Epstein and H. zur Hausen
1975; Two volumes, 511 pages and 403 pages; ISBN 0 19 723010 5

No. 12
Screening Tests in Chemical Carcinogenesis
Edited by R. Montesano, H. Bartsch and L. Tomatis
1976; 666 pages; ISBN 0 19 723051 2

No. 13
Environmental Pollution and Carcinogenic Risks
Edited by C. Rosenfeld and W. Davis
1975; 441 pages; ISBN 0 19 723012 1

No. 14
Environmental N-Nitroso Compounds. Analysis and Formation
Edited by E.A. Walker, P. Bogovski and L. Griciute
1976; 512 pages; ISBN 0 19 723013 X

No. 15
Cancer Incidence in Five Continents, Volume III
Edited by J.A.H. Waterhouse, C. Muir, P. Correa and J. Powell
1976; 584 pages; ISBN 0 19 723014 8

No. 16
Air Pollution and Cancer in Man
Edited by U. Mohr, D. Schmähl and L. Tomatis
1977; 328 pages; ISBN 0 19 723015 6

No. 17
Directory of On-Going Research in Cancer Epidemiology 1977
Edited by C.S. Muir and G. Wagner
1977; 599 pages; ISBN 92 832 1117 0 (out of print)

No. 18
Environmental Carcinogens. Selected Methods of Analysis. Volume 1: Analysis of Volatile Nitrosamines in Food
Editor-in-Chief: H. Egan
1978; 212 pages; ISBN 0 19 723017 2

No. 19
Environmental Aspects of N-Nitroso Compounds
Edited by E.A. Walker, M. Castegnaro, L. Griciute and R.E. Lyle
1978; 561 pages; ISBN 0 19 723018 0

No. 20
Nasopharyngeal Carcinoma: Etiology and Control
Edited by G. de Thé and Y. Ito
1978; 606 pages; ISBN 0 19 723019 9

No. 21
Cancer Registration and its Techniques
Edited by R. MacLennan, C. Muir, R. Steinitz and A. Winkler
1978; 235 pages; ISBN 0 19 723020 2

No. 22
Environmental Carcinogens: Selected Methods of Analysis. Volume 2: Methods for the Measurement of Vinyl Chloride in Poly(vinyl chloride), Air, Water and Foodstuffs
Editor-in-Chief: H. Egan
1978; 142 pages; ISBN 0 19 723021 0

No. 23
Pathology of Tumours in Laboratory Animals. Volume II: Tumours of the Mouse
Editor-in-Chief: V.S. Turusov
1979; 669 pages; ISBN 0 19 723022 9

No. 24
Oncogenesis and Herpesviruses III
Edited by G. de-Thé, W. Henle and F. Rapp
1978; Part I: 580 pages, Part II: 512 pages; ISBN 0 19 723023 7

Table 10 Continued

No. 25
Carcinogenic Risk: Strategies for Intervention
Edited by W. Davis and C. Rosenfeld
1979; 280 pages; ISBN 0 19 723025 3

No. 26
Directory of On-going Research in Cancer Epidemiology 1978
Edited by C.S. Muir and G. Wagner
1978; 550 pages; ISBN 0 19 723026 1
(out of print)

No. 27
Molecular and Cellular Aspects of Carcinogen Screening Tests
Edited by R. Montesano, H. Bartsch and L. Tomatis
1980; 372 pages; ISBN 0 19 723027 X

No. 28
Directory of On-going Research in Cancer Epidemiology 1979
Edited by C.S. Muir and G. Wagner
1979; 672 pages; ISBN 92 832 1128 6
(out of print)

No. 29
Environmental Carcinogens. Selected Methods of Analysis. Volume 3: Analysis of Polycyclic Aromatic Hydrocarbons in Environmental Samples
Editor-in-Chief: H. Egan
1979; 240 pages; ISBN 0 19 723028 8

No. 30
Biological Effects of Mineral Fibres
Editor-in-Chief: J.C. Wagner
1980; Two volumes, 494 pages & 513 pages; ISBN 0 19 723030 X

No. 31
N-Nitroso Compounds: Analysis, Formation and Occurrence
Edited by E.A. Walker, L. Griciute, M. Castegnaro and M. Börzsönyi
1980; 835 pages; ISBN 0 19 723031 8

No. 32
Statistical Methods in Cancer Research.Volume 1: The Analysis of Case-control Studies
By N.E. Breslow and N.E. Day
1980; 338 pages; ISBN 92 832 0132 9

No. 33
Handling Chemical Carcinogens in the Laboratory
Edited by R. Montesano, H. Bartsch, E. Boyland, G. Della Porta, L. Fishbein, R.A. Griesemer, A.B. Swan and L. Tomatis
1979; 32 pages; ISBN 0 19 723033 4
(out of print)

No. 34
Pathology of Tumours in Laboratory Animals. Volume III: Tumours of the Hamster
Editor-in-Chief: V.S. Turusov
1982; 461 pages; ISBN 0 19 723034 2

No. 35
Directory of On-going Research in Cancer Epidemiology 1980
Edited by C.S. Muir and G. Wagner
1980; 660 pages; ISBN 0 19 723035 0
(out of print)

No. 36
Cancer Mortality by Occupation and Social Class 1851–1971
Edited by W.P.D. Logan
1982; 253 pages; ISBN 0 19 723036 9

No. 37
Laboratory Decontamination and Destruction of Aflatoxins B1, B2, G1, G2 in Laboratory Wastes
Edited by M. Castegnaro, D.C. Hunt, E.B. Sansone, P.L. Schuller, M.G. Siriwardana, G.M. Telling, H.P. van Egmond and E.A. Walker
1980; 56 pages; ISBN 0 19 723037 7

No. 38
Directory of On-going Research in Cancer Epidemiology 1981
Edited by C.S. Muir and G. Wagner
1981; 696 pages; ISBN 0 19 723038 5
(out of print)

No. 39
Host Factors in Human Carcinogenesis
Edited by H. Bartsch and B. Armstrong
1982; 583 pages;
ISBN 0 19 723039 3

No. 40
Environmental Carcinogens: Selected Methods of Analysis. Volume 4: Some Aromatic Amines and Azo Dyes in the General and Industrial Environment
Edited by L. Fishbein, M. Castegnaro, I.K. O'Neill and H. Bartsch
1981; 347 pages; ISBN 0 19 723040 7

No. 41
N-Nitroso Compounds: Occurrence and Biological Effects
Edited by H. Bartsch, I.K. O'Neill, M. Castegnaro and M. Okada
982; 755 pages; ISBN 0 19 723041 5

No. 42
Cancer Incidence in Five Continents Volume IV
Edited by J. Waterhouse, C. Muir, K. Shanmugaratnam and J. Powell
1982; 811 pages; ISBN 0 19 723042 3

No. 43
Laboratory Decontamination and Destruction of Carcinogens in Laboratory Wastes: Some N-Nitrosamines
Edited by M. Castegnaro, G. Eisenbrand, G. Ellen, L. Keefer, D. Klein, E.B. Sansone, D. Spincer, G. Telling and K. Webb
1982; 73 pages; ISBN 0 19 723043 1

No. 44
Environmental Carcinogens: Selected Methods of Analysis.
Volume 5: Some Mycotoxins
Edited by L. Stoloff, M. Castegnaro, P. Scott, I.K. O'Neill and H. Bartsch
1983; 455 pages; ISBN 0 19 723044 X

No. 45
Environmental Carcinogens: Selected Methods of Analysis. Volume 6: N-Nitroso Compounds
Edited by R. Preussmann, I.K. O'Neill, G. Eisenbrand, B. Spiegelhalder and H. Bartsch
1983; 508 pages; ISBN 0 19 723045 8

No. 46
Directory of On-going Research in Cancer Epidemiology 1982
Edited by C.S. Muir and G. Wagner
1982; 722 pages; ISBN 0 19 723046 6
(out of print)

No. 47
Cancer Incidence in Singapore 1968–1977
Edited by K. Shanmugaratnam, H.P. Lee and N.E. Day
1983; 171 pages; ISBN 0 19 723047 4

No. 48
Cancer Incidence in the USSR (2nd Revised Edition)
Edited by N.P. Napalkov, G.F. Tserkovny, V.M. Merabishvili, D.M. Parkin, M. Smans and C.S. Muir
1983; 75 pages; ISBN 0 19 723048 2

No. 49
Laboratory Decontamination and Destruction of Carcinogens in Laboratory Wastes: Some Polycyclic Aromatic Hydrocarbons
Edited by M. Castegnaro, G. Grimmer, O. Hutzinger, W. Karcher, H. Kunte, M. Lafontaine, H.C. Van der Plas, E.B. Sansone and S.P. Tucker
1983; 87 pages; ISBN 0 19 723049 0

Table 10 Continued

No. 50
Directory of On-going Research in Cancer Epidemiology 1983
Edited by C.S. Muir and G. Wagner
1983; 731 pages; ISBN 0 19 723050 4
(out of print)

No. 51
Modulators of Experimental Carcinogenesis
Edited by V. Turusov and R. Montesano
1983; 307 pages; ISBN 0 19 723060 1

No. 52
Second Cancers in Relation to Radiation Treatment for Cervical Cancer: Results of a Cancer Registry Collaboration
Edited by N.E. Day and J.C. Boice, Jr
1984; 207 pages; ISBN 0 19 723052 0

No. 53
Nickel in the Human Environment
Editor-in-Chief: F.W. Sunderman, Jr
1984; 529 pages; ISBN 0 19 723059 8

No. 54
Laboratory Decontamination and Destruction of Carcinogens in Laboratory Wastes: Some Hydrazines
Edited by M. Castegnaro, G. Ellen, M. Lafontaine, H.C. van der Plas, E.B. Sansone and S.P. Tucker
1983; 87 pages; ISBN 0 19 723053

No. 55
Laboratory Decontamination and Destruction of Carcinogens in Laboratory Wastes: Some N-Nitrosamides
Edited by M. Castegnaro, M. Bernard, L.W. van Broekhoven, D. Fine, R. Massey, E.B. Sansone, P.L.R. Smith, B. Spiegelhalder, A. Stacchini, G. Telling and J.J. Vallon
1984; 66 pages; ISBN 0 19 723054 7

No. 56
Models, Mechanisms and Etiology of Tumour Promotion
Edited by M. Börzsönyi, N.E. Day, K. Lapis and H. Yamasaki
1984; 532 pages; ISBN 0 19 723058 X

No. 57
N-Nitroso Compounds: Occurrence, Biological Effects and Relevance to Human Cancer
Edited by I.K. O'Neill, R.C. von Borstel, C.T. Miller, J. Long and H. Bartsch
1984; 1013 pages; ISBN 0 19 723055 5

No 58
Age-related Factors in Carcinogenesis
Edited by A. Likhachev, V. Anisimov and R. Montesano
1985; 288 pages; ISBN 92 832 1158 8

No. 59
Monitoring Human Exposure to Carcinogenic and Mutagenic Agents
Edited by A. Berlin, M. Draper, K. Hemminki and H. Vainio
1984; 457 pages; ISBN 0 19 723056 3

No. 60
Burkitt's Lymphoma: A Human Cancer Model
Edited by G. Lenoir, G. O'Conor and C.L.M. Olweny
1985; 484 pages; ISBN 0 19 723057 1

No. 61
Laboratory Decontamination and Destruction of Carcinogens in Laboratory Wastes: Some Haloethers
Edited by M. Castegnaro, M. Alvarez, M. Iovu, E.B. Sansone, G.M. Telling and D.T. Williams
1985; 55 pages; ISBN 0 19 723061 X

No. 62
Directory of On-going Research in Cancer Epidemiology 1984
Edited by C.S. Muir and G. Wagner
1984; 717 pages; ISBN 0 19 723062 8
(out of print)

No. 63
Virus-associated Cancers in Africa
Edited by A.O. Williams, G.T. O'Conor, G.B. de Thé and C.A. Johnson
1984; 773 pages; ISBN 0 19 723063 6

No. 64
Laboratory Decontamination and Destruction of Carcinogens in Laboratory Wastes: Some Aromatic Amines and 4-Nitrobiphenyl
Edited by M. Castegnaro, J. Barek, J. Dennis, G. Ellen, M. Klibanov, M. Lafontaine, R. Mitchum, P. van Roosmalen, E.B. Sansone, L.A. Sternson and M. Vahl
1985; 84 pages; ISBN: 92 832 1164 2

No. 65
Interpretation of Negative Epidemiological Evidence for Carcinogenicity
Edited by N.J. Wald and R. Doll
1985; 232 pages; ISBN 92 832 1165 0

No. 66
The Role of the Registry in Cancer Control
Edited by D.M. Parkin, G. Wagner and C.S. Muir
1985; 152 pages; ISBN 92 832 0166 3

No. 67
Transformation Assay of Established Cell Lines: Mechanisms and Application
Edited by T. Kakunaga and H. Yamasaki
1985; 225 pages; ISBN 92 832 1167 7

No. 68
Environmental Carcinogens: Selected Methods of Analysis. Volume 7: Some Volatile Halogenated Hydrocarbons
Edited by L. Fishbein and I.K. O'Neill
1985; 479 pages; ISBN 92 832 1168 5

No. 69
Directory of On-going Research in Cancer Epidemiology 1985
Edited by C.S. Muir and G. Wagner
1985; 745 pages; ISBN 92 823 1169 3
(out of print)

No. 70
The Role of Cyclic Nucleic Acid Adducts in Carcinogenesis and Mutagenesis
Edited by B. Singer and H. Bartsch
1986; 467 pages; ISBN 92 832 1170 7

No. 71
Environmental Carcinogens: Selected Methods of Analysis. Volume 8: Some Metals: As, Be, Cd, Cr, Ni, Pb, Se, Zn
Edited by I.K. O'Neill, P. Schuller and L. Fishbein
1986; 485 pages; ISBN 92 832 1171 5

No. 72
Atlas of Cancer in Scotland, 1975–1980: Incidence and Epidemiological Perspective
Edited by I. Kemp, P. Boyle, M. Smans and C.S. Muir
1985; 285 pages; ISBN 92 832 1172 3

No. 73
Laboratory Decontamination and Destruction of Carcinogens in Laboratory Wastes: Some Antineoplastic Agents
Edited by M. Castegnaro, J. Adams, M.A. Armour, J. Barek, J. Benvenuto, C. Confalonieri, U. Goff, G. Telling
1985; 163 pages; ISBN 92 832 1173 1

No. 74
Tobacco: A Major International Health Hazard
Edited by D. Zaridze and R. Peto
1986; 324 pages; ISBN 92 832 1174 X

No. 75
Cancer Occurrence in Developing Countries
Edited by D.M. Parkin
1986; 339 pages; ISBN 92 832 1175 8

Table 10 Continued

No. 76
Screening for Cancer of the Uterine Cervix
Edited by M. Hakama, A.B. Miller and N.E. Day
1986; 315 pages; ISBN 92 832 1176 6

No. 77
Hexachlorobenzene: Proceedings of an International Symposium
Edited by C.R. Morris and J.R.P. Cabral
1986; 668 pages; ISBN 92 832 1177 4

No. 78
Carcinogenicity of Alkylating Cytostatic Drugs
Edited by D. Schmähl and J.M. Kaldor
1986; 337 pages; ISBN 92 832 1178 2

No. 79
Statistical Methods in Cancer Research. Volume III: The Design and Analysis of Long-term Animal Experiments
By J.J. Gart, D. Krewski, P.N. Lee, R.E. Tarone and J. Wahrendorf
1986; 213 pages; ISBN 92 832 1179 0

No. 80
Directory of On-going Research in Cancer Epidemiology 1986
Edited by C.S. Muir and G. Wagner
1986; 805 pages; ISBN 92 832 1180 4
(out of print)

No. 81
Environmental Carcinogens: Methods of Analysis and Exposure Measurement. Volume 9: Passive Smoking
Edited by I.K. O'Neill, K.D. Brunnemann, B. Dodet and D. Hoffmann
1987; 383 pages; ISBN 92 832 1181 2

No. 82
Statistical Methods in Cancer Research. Volume II: The Design and Analysis of Cohort Studies
By N.E. Breslow and N.E. Day
1987; 404 pages; ISBN 92 832 0182 5

No. 83
Long-term and Short-term Assays for Carcinogens: A Critical Appraisal
Edited by R. Montesano, H. Bartsch, H. Vainio, J. Wilbourn and H. Yamasaki
1986; 575 pages; ISBN 92 832 1183 9

No. 84
The Relevance of *N*-Nitroso Compounds to Human Cancer: Exposure and Mechanisms
Edited by H. Bartsch, I.K. O'Neill and R. Schulte-Hermann
1987; 671 pages; ISBN 92 832 1184 7

No. 85
Environmental Carcinogens: Methods of Analysis and Exposure Measurement. Volume 10: Benzene and Alkylated Benzenes
Edited by L. Fishbein and I.K. O'Neill
1988; 327 pages; ISBN 92 832 1185 5

No. 86
Directory of On-going Research in Cancer Epidemiology 1987
Edited by D.M. Parkin and J. Wahrendorf
1987; 685 pages; ISBN: 92 832 1186 3
(out of print)

No. 87
International Incidence of Childhood Cancer
Edited by D.M. Parkin, C.A. Stiller, C.A. Bieber, G.J. Draper. B. Terracini and J.L. Young
1988; 401 page; ISBN 92 832 1187 1
(out of print)

No. 88
Cancer Incidence in Five Continents, Volume V
Edited by C. Muir, J. Waterhouse, T. Mack, J. Powell and S. Whelan
1987; 1004 pages; ISBN 92 832 1188 X

No. 89
Methods for Detecting DNA Damaging Agents in Humans: Applications in Cancer Epidemiology and Prevention
Edited by H. Bartsch, K. Hemminki and I.K. O'Neill
1988; 518 pages; ISBN 92 832 1189 8
(out of print)

No. 90
Non-occupational Exposure to Mineral Fibres
Edited by J. Bignon, J. Peto and R. Saracci
1989; 500 pages; ISBN 92 832 1190 1

No. 91
Trends in Cancer Incidence in Singapore 1968–1982
Edited by H.P. Lee, N.E. Day and K. Shanmugaratnam
1988; 160 pages; ISBN 92 832 1191 X

No. 92
Cell Differentiation, Genes and Cancer
Edited by T. Kakunaga, T. Sugimura, L. Tomatis and H. Yamasaki
1988; 204 pages; ISBN 92 832 1192 8

No. 93
Directory of On-going Research in Cancer Epidemiology 1988
Edited by M. Coleman and J. Wahrendorf
1988; 662 pages; ISBN 92 832 1193 6
(out of print)

No. 94
Human Papillomavirus and Cervical Cancer
Edited by N. Muñoz, F.X. Bosch and O.M. Jensen
1989; 154 pages; ISBN 92 832 1194 4

No. 95
Cancer Registration: Principles and Methods
Edited by O.M. Jensen, D.M. Parkin, R. MacLennan, C.S. Muir and R. Skeet
1991; 296 pages; ISBN 92 832 1195 2

No. 96
Perinatal and Multigeneration Carcinogenesis
Edited by N.P. Napalkov, J.M. Rice, L. Tomatis and H. Yamasaki
1989; 436 pages; ISBN 92 832 1196 0

No. 97
Occupational Exposure to Silica and Cancer Risk
Edited by L. Simonato, A.C. Fletcher, R. Saracci and T. Thomas
1990; 124 pages; ISBN 92 832 1197 9

No. 98
Cancer Incidence in Jewish Migrants to Israel, 1961-1981
Edited by R. Steinitz, D.M. Parkin, J.L. Young, C.A. Bieber and L. Katz
1989; 320 pages; ISBN 92 832 1198 7

No. 99
Pathology of Tumours in Laboratory Animals, Second Edition, Volume 1, Tumours of the Rat
Edited by V.S. Turusov and U. Mohr
1990; 740 pages; ISBN 92 832 1199 5
For Volumes 2 and 3 (Tumours of the Mouse and Tumours of the Hamster), see IARC Scientific Publications Nos. 111 and 126.

No. 100
Cancer: Causes, Occurrence and Control
Editor-in-Chief: L. Tomatis
1990; 352 pages; ISBN 92 832 0110 8

No. 101
Directory of On-going Research in Cancer Epidemiology 1989–1990
Edited by M. Coleman and J. Wahrendorf
1989; 828 pages; ISBN 92 832 2101 X

No. 102
Patterns of Cancer in Five Continents
Edited by S.L. Whelan, D.M. Parkin and E. Masuyer
1990; 160 pages; ISBN 92 832 2102 8

Table 10 Continued

No. 103
Evaluating Effectiveness of Primary Prevention of Cancer
Edited by M. Hakama, V. Beral, J.W. Cullen and D.M. Parkin
1990; 206 pages; ISBN 92 832 2103 6

No. 104
Complex Mixtures and Cancer Risk
Edited by H. Vainio, M. Sorsa and A.J. McMichael
1990; 441 pages; ISBN 92 832 2104 4

No. 105
Relevance to Human Cancer of N-Nitroso Compounds, Tobacco Smoke and Mycotoxins
Edited by I.K. O'Neill, J. Chen and H. Bartsch
1991; 614 pages; ISBN 92 832 2105 2

No. 106
Atlas of Cancer Incidence in the Former German Democratic Republic
Edited by W.H. Mehnert, M. Smans, C.S. Muir, M. Möhner and D. Schön
1992; 384 pages; ISBN 92 832 2106 0

No. 107
Atlas of Cancer Mortality in the European Economic Community
Edited by M. Smans, C. Muir and P. Boyle
1992; 213 pages + 44 coloured maps; ISBN 92 832 2107 9

No. 108
Environmental Carcinogens: Methods of Analysis and Exposure Measurement. Volume 11: Polychlorinated Dioxins and Dibenzofurans
Edited by C. Rappe, H.R. Buser, B. Dodet and I.K. O'Neill
1991; 400 pages; ISBN 92 832 2108 7

No. 109
Environmental Carcinogens: Methods of Analysis and Exposure Measurement. Volume 12: Indoor Air
Edited by B. Seifert, H. van de Wiel, B. Dodet and I.K. O'Neill
1993; 385 pages; ISBN 92 832 2109 5

No. 110
Directory of On-going Research in Cancer Epidemiology 1991
Edited by M.P. Coleman and J. Wahrendorf
1991; 753 pages; ISBN 92 832 2110 9

No. 111
Pathology of Tumours in Laboratory Animals, Second Edition. Volume 2: Tumours of the Mouse
Edited by V. Turusov and U. Mohr
1994; 800 pages; ISBN 92 832 2111 1

No. 112
Autopsy in Epidemiology and Medical Research
Edited by E. Riboli and M. Delendi
1991; 288 pages; ISBN 92 832 2112 5

No. 113
Laboratory Decontamination and Destruction of Carcinogens in Laboratory Wastes: Some Mycotoxins
Edited by M. Castegnaro, J. Barek, J.M. Frémy, M. Lafontaine, M. Miraglia, E.B. Sansone and G.M. Telling
1991; 63 pages; ISBN 92 832 2113 3

No. 114
Laboratory Decontamination and Destruction of Carcinogens in Laboratory Wastes: Some Polycyclic Heterocyclic Hydrocarbons
Edited by M. Castegnaro, J. Barek, J. Jacob, U. Kirso, M. Lafontaine, E.B. Sansone, G.M. Telling and T. Vu Duc
1991; 50 pages; ISBN 92 832 2114 1

No. 115
Mycotoxins, Endemic Nephropathy and Urinary Tract Tumours
Edited by M. Castegnaro, R. Plestina, G. Dirheimer, I.N. Chernozemsky and H. Bartsch
1991; 340 pages; ISBN 92 832 2115 X

No. 116
Mechanisms of Carcinogenesis in Risk Identification
Edited by H. Vainio, P. Magee, D. McGregor and A.J. McMichael
1992; 615 pages; ISBN 92 832 2116 8

No. 117
Directory of On-going Research in Cancer Epidemiology 1992
Edited by M. Coleman, E. Demaret and J. Wahrendorf
1992; 773 pages; ISBN 92 832 2117 6

No. 118
Cadmium in the Human Environment: Toxicity and Carcinogenicity
Edited by G.F. Nordberg, R.F.M. Herber and L. Alessio
1992; 470 pages; ISBN 92 832 2118 4

No. 119
The Epidemiology of Cervical Cancer and Human Papillomavirus
Edited by N. Muñoz, F.X. Bosch, K.V. Shah and A. Meheus
1992; 288 pages; ISBN 92 832 2119 2

No. 120
Cancer Incidence in Five Continents, Vol. VI
Edited by D.M. Parkin, C.S. Muir, S.L. Whelan, Y.T. Gao, J. Ferlay and J. Powell
1992; 1020 pages; ISBN 92 832 2120 6

No. 121
Time Trends in Cancer Incidence and Mortality
By M. Coleman, J. Estéve, P. Damiecki, A. Arslan and H. Renard
1993; 820 pages; ISBN 92 832 2121 4

No. 122
International Classification of Rodent Tumours.
Part I. The Rat
Editor-in-Chief: U. Mohr
1992–1996; 10 fascicles of 60–100 pages; ISBN 92 832 2122 2

No. 123
Cancer in Italian Migrant Populations
Edited by M. Geddes, D.M. Parkin, M. Khlat, D. Balzi and E. Buiatti
1993; 292 pages; ISBN 92 832 2123 0

No. 124
Postlabelling Methods for the Detection of DNA Damage
Edited by D.H. Phillips, M. Castegnaro and H. Bartsch
1993; 392 pages; ISBN 92 832 2124 9

No. 125
DNA Adducts: Identification and Biological Significance
Edited by K. Hemminki, A. Dipple, D.E.G. Shuker, F.F. Kadlubar, D. Segerbäck and H. Bartsch
1994; 478 pages; ISBN 92 832 2125 7

No. 126
Pathology of Tumours in Laboratory Animals, Second Edition. Volume 3: Tumours of the Hamster
Edited by V. Turosov and U. Mohr
1996; 464 pages; ISBN 92 832 2126 5

No. 127
Butadiene and Styrene: Assessment of Health Hazards
Edited by M. Sorsa, K. Peltonen, H. Vainio and K. Hemminki
1993; 412 pages; ISBN 92 832 2127 3

No. 128
Statistical Methods in Cancer Research. Volume IV. Descriptive Epidemiology
By J. Estève, E. Benhamou and L. Raymond
1994; 302 pages; ISBN 92 832 2128 1

No. 129
Occupational Cancer in Developing Countries
Edited by N. Pearce, E. Matos, H. Vainio, P. Boffetta and M. Kogevinas
1994; 191 pages; ISBN 92 832 2129 X

Table 10 Continued

No. 130
Directory of On-going Research in Cancer Epidemiology 1994
Edited by R. Sankaranarayanan, J. Wahrendorf and E. Démaret
1994; 800 pages; ISBN 92 832 2130 3

No. 132
Survival of Cancer Patients in Europe: The EUROCARE Study
Edited by F. Berrino, M. Sant, A. Verdecchia, R. Capocaccia, T. Hakulinen and J. Estève
1995; 463 pages; ISBN 92 832 2132 X

No. 134
Atlas of Cancer Mortality in Central Europe
W. Zatonski, J. Estéve, M. Smans, J. Tyczynski and P. Boyle
1996; 300 pages; ISBN 92 832 2134 6

No. 135
Methods for Investigating Localized Clustering of Disease
Edited by F.E. Alexander and P. Boyle
1996; 235 pages; ISBN 92 832 2135 4

No. 136
Chemoprevention in Cancer Control
Edited by M. Hakama, V. Beral, E. Buiatti, J. Faivre and D.M. Parkin
1996; 160 pages; ISBN 92 832 2136 2

No. 137
Directory of On-going Research in Cancer Epidemiology 1996
Edited by R. Sankaranarayan, J. Warendorf and E. Démaret
1996; 810 pages; ISBN 92 832 2137 0

No. 139
Principles of Chemoprevention
Edited by B.W. Stewart, D. McGregor and P. Kleihues
1996; 358 pages; ISBN 92 832 2139 7

No. 140
Mechanisms of Fibre Carcinogenesis
Edited by A.B. Kane, P. Boffetta, R. Saracci and J.D. Wilbourn
1996; 135 pages; ISBN 92 832 2140 0

both genders of two species of laboratory animals to chemicals for 2 years by various routes of exposure (Chhabra et al., 1990). In 1976, the NCI published the first of nearly 200 technical reports (NCI, 1976; Chu et al., 1981).

Since 1982, the National Toxicology Program (NTP) has continued these efforts, and about 300 more technical reports have been issued (Haseman et al., 1987; Huff and Soward, 1998). Table 11 lists the chemicals tested, evaluated, and printed in the NTP Technical Reports series. Typical design protocols of 2-year bioassays are given in Table 12, while possible design alternatives are narrated in Table 13.

D. NTP Reports on Carcinogens

When the NTP was created in 1978, Congress also decided that a report on carcinogens should be compiled, printed, and distributed (Huff, 1998). Two categories of carcinogens have been promulgated: 1) those "known to be carcinogenic to humans," based almost exclusively on epidemiological findings, and 2) those "reasonably anticipated to be carcinogenic to humans," based most often on experimental observations from long-term carcinogenesis bioassays in animals (usually rodents). The eighth edition (NTP, 1998) contains listings of 29 agents, mixtures of chemicals, and

(*text continues on p. 82*)

Table 11 Summary Results of Long-Term Chemical Carcinogenesis Studies Reported by the National Cancer Institute (NCI) and the National Toxicology Program (NTP) (sorted by chemical name; revised February 1998)

Chemical name	CAS no.	TR no.	MR/FR/MM/FM	Overall
Acetaminophen (4-Hydroxyacetanilide)	103-90-2	TR-394	– ± – –	E
Acetohexamide	968-81-0	TR-050	– – – –	N
Acetonitrile	75-05-8	TR-447	± – – –	E
Acronycine	7008-42-6	TR-049	+ + is is	P
Agar	9002-18-0	TR-230	– – – –	N
Aldicarb	116-06-3	TR-136	– – – –	N
Aldrin	309-00-2	TR-021	± ± + –	P
Allyl Chloride	107-05-1	TR-073	– – ± ±	E
Allyl Glycidyl Ether	106-92-3	TR-376	± – + ±	P
Allyl Isothiocyanate	57-06-7	TR-234	+ ± – –	P
Allyl Isovalerate	2835-39-4	TR-253	+ – – +	P
2-Aminoanthraquinone	117-79-3	TR-144	+ is + +	P
1-Amino-2,4-Dibromoanthraquinone	81-49-2	TR-383	+ + + +	P
3-Amino-4-Ethoxyacetanilide	17026-81-2	TR-112	– – + –	P
3-Amino-9-Ethylcarbazole HCl	6109-97-3	TR-093	+ + + +	P
1-Amino-2-Methylanthraquinone	82-28-0	TR-111	+ + – +	P
2-Amino-4-Nitrophenol	99-57-0	TR-339	+ – – –	P
2-Amino-5-Nitrophenol	121-88-0	TR-334	+ – – –	P
4-Amino-2-Nitrophenol	119-34-6	TR-094	+ ± – –	P
2-Amino-5-Nitrothiazole	121-66-4	TR-053	+ – – –	P
11-Aminoundecanoic Acid	2432-99-7	TR-216	+ – ± –	P
dl-Amphetamine Sulfate	60-13-9	TR-387	– – – –	N
Ampicillin Trihydrate	7177-48-2	TR-318	± – – –	E
Anilazine	101-05-3	TR-104	– – – –	N
Aniline Hydrochloride	142-04-1	TR-130	+ + – –	P
o-Anisidine Hydrochloride	134-29-2	TR-089	+ + + +	P
p-Anisidine Hydrochloride	20265-97-8	TR-116	± – – –	E
Anthranilic Acid	118-92-3	TR-036	– – – –	N
Aroclor 1254	11097-69-1	TR-038	± ± nd nd	E
Asbestos, Amosite	12172-73-5	TR-249	Hamster= – –	N
Asbestos, Amosite	12172-73-5	TR-279	– – nd nd	N
Asbestos, Amosite + Dimethyl Hydrazine	12172-73-5	TR-279	is is nd nd	I

Chemical name	CAS no.	TR no.	MR/FR/MM/FM	Overall
Asbestos, Chrysotile(IR)	12001-29-5	TR-246	Hamster= – –	N
Asbestos, Chrysotile(IR) + Dimethyl Hydrazine	12001-29-5	TR-246	Hamster=is is	I
Asbestos, Chrysotile(SR)	12001-29-5	TR-246	Hamster= – –	N
Asbestos, Chrysotile(IR)	12001-29-5	TR-295	+ – nd nd	P
Asbestos, Chrysotile(IR) + Dimethyl Hydrazine	12001-29-5	TR-295	is is nd nd	I
Asbestos, Chrysotile(SR)	12001-29-5	TR-295	– – nd nd	N
Asbestos, Crocidolite	12001-28-4	TR-280	– – nd nd	N
Asbestos, Tremolite	14567-73-8	TR-277	– – nd nd	N
L-Ascorbic Acid	50-81-7	TR-247	– – – –	N
Aspirin, Phenacetin, and Caffeine	8003-03-0	TR-067	– ± – –	E
5-Azacytidine	320-67-2	TR-042	is is is +	P
AZT + Interferon AD	30516-87-1	TR-469	nd nd ± +	P
Azinphosmethyl	86-50-0	TR-069	± – – –	E
Azobenzene	103-33-3	TR-154	+ + – –	P
Barium Chloride Dihydrate	10326-27-9	TR-432	– – – –	N
Benzaldehyde	100-52-7	TR-378	– – + +	P
Benzene	71-43-2	TR-289	+ + + +	P
Benzethonium Chloride	121-54-0	TR-438	– – – –	N
Benzofuran	271-89-6	TR-370	– + + +	P
Benzoin	119-53-9	TR-204	– – – –	N
p-Benzoquinone Dioxime	105-11-3	TR-179	– + – –	P
1,2,3-Benzotriazole	95-14-7	TR-088	± ± – ±	E
Benzyl Acetate	140-11-4	TR-250	± – + +	P
Benzyl Acetate	140-11-4	TR-431	– – – –	N
Benzyl Alcohol	100-51-6	TR-343	– – – –	N
o-Benzyl-p-Chlorophenol	120-32-1	TR-424	– ± + –	P
o-Benzyl-p-Chlorophenol	120-32-1	TR-444	2	
2-Biphenylamine Hydrochloride	2185-92-4	TR-233	– – ± +	P
2,2-Bis(Bromomethyl)-1,3-Propanediol	3296-90-0	TR-452	+ + + +	P
Bis(2-Chloro-1-Methylethyl) Ether	108-60-1	TR-191	– – nd nd	N
Bis(2-Chloro-1-Methylethyl) Ether	108-60-1	TR-239	nd nd + +	P
Bisphenol A	80-05-7	TR-215	± – – –	E
Boric Acid	10043-35-3	TR-324	nd nd – –	N
Bromodichloromethane	75-27-4	TR-321	+ + + +	P

Table 11 Continued.

Chemical name	CAS no.	TR no.	MR/FR/MM/FM	Overall
Bromoethane (Ethyl Bromide)	74-96-4	TR-363	+ ± ± +	P
1,3-Butadiene	106-99-0	TR-288	nd nd + +	P
1,3-Butadiene	106-99-0	TR-434	nd nd + +	P
Tert-Butyl Alcohol	75-65-0	TR-436	+ – ± +	P
Butylated Hydroxytoluene	128-37-0	TR-150	– – – –	N
Butyl Benzyl Phthalate	85-68-7	TR-213	is + – –	P
Butyl Benzyl Phthalate	85-68-7	TR-458	+ ± nd nd	P
N-Butyl Chloride	109-69-3	TR-312	– – – –	N
t-Butylhydroquinone	1948-33-0	TR-459	– – – –	N
Gamma-Butyrolactone	96-48-0	TR-406	– – ± –	E
Calcium Cyanamide	156-62-7	TR-163	– – – –	N
Caprolactam	105-60-2	TR-214	– – – –	N
Captan	133-06-2	TR-015	– – + +	P
Carbromal	77-65-6	TR-173	– – – –	N
d-Carvone	2244-16-8	TR-381	nd nd – –	N
Chloramben	133-90-4	TR-025	– – ± +	P
Chloraminated Water	10599-90-3	TR-392	– ± – –	E
Chlordane (Analytical Grade)	57-74-9	TR-008	– – + +	P
Chlordecone (Kepone)	143-50-0	No TR No.	+ + + +	P
<u>Chlorendic Acid</u>	115-28-6	TR-304	+ + + –	P
Chlorinated Paraffins: C12, 60% Chlorine	108171-26-2	TR-308	+ + + +	P
Chlorinated Paraffins: C23, 43% Chlorine	108171-27-3	TR-305	– ± + ±	P
Chlorinated Trisodium Phosphate	56802-99-4	TR-294	is is – –	N
Chlorinated Water	7782-50-5/ 7681-52-9	TR-392	– ± – –	E
2-Chloroacetophenone (CN)	532-27-4	TR-379	– ± – –	E
4-(Chloroacetyl) Acetanilide	140-49-8	TR-177	– – – –	N
p-Chloroaniline	106-47-8	TR-189	± – ± ±	E
p-Chloroaniline Hydrochloride	20265-96-7	TR-351	+ ± + –	E
o-Chlorobenzal-malononitrile (CS)	2698-41-1	TR-377	– – – –	N
Chlorobenzene	108-90-7	TR-261	± – – –	E
Chlorobenzilate	510-15-6	TR-075	± ± + +	P
Chlorodibromomethane	124-48-1	TR-282	– – ± +	P

Chemical name	CAS no.	TR no.	MR/FR/MM/FM	Overall
Chloroethane	75-00-3	TR-346	± ± is +	P
2-Chloroethanol (Ethylene Chlorohydrin)	107-07-3	TR-275	– – – –	N
2-Chloroethyltrimethyl–ammonium Chloride	999-81-5	TR-158	– – – –	N
Chloroform	67-66-3	No TR No	+ – + +	P
3-Chloro-2-Methylpropene	563-47-3	TR-300	+ + + +	P
2-Chloromethylpyridine Hydrochloride	6959-47-3	TR-178	– – – –	N
3-Chloromethylpyridine Hydrochloride	6959-48-4	TR-095	+ ± + +	P
4-Chloro-m-Phenylenediamine	5131-60-2	TR-085	+ – – +	P
4-Chloro-o-Phenylenediamine	95-83-0	TR-063	+ + + +	P
2-Chloro-p-Phenylenediamine Sulfate	61702-44-1	TR-113	– – – –	N
Chloropicrin	76-06-2	TR-065	is is – –	N
Chloroprene	126-99-8	TR-467	+ + + +	P
1-Chloro-2-propanol	127-00-4	TR-477	– – – –	N
Chlorothalonil	1897-45-6	TR-041	+ + – –	P
3-Chloro-p-Toluidine	95-74-9	TR-145	– – – –	N
5-Chloro-o-Toluidine	95-79-4	TR-187	– – + +	P
4-Chloro-o-Toluidine Hydrochloride	3165-93-3	TR-165	– – + +	P
Chlorpheniramine Maleate	113-92-8	TR-317	– – – –	N
Chlorpropamide	94-20-2	TR-045	– – – –	N
C.I. Acid Orange 3	6373-74-6	TR-335	– + – –	P
C.I. Acid Orange 10	1936-15-8	TR-211	– – – –	N
C.I. Acid Red 14	3567-69-9	TR-220	– – – –	N
C.I. Acid Red 114	6459-94-5	TR-405	+ + nd nd	P
C.I. Basic Red 9 Monohydrochloride	569-61-9	TR-285	+ + + +	P
C.I. Direct Black 38	1937-37-7	TR-108	+ + nd nd	P
C.I. Direct Blue 6	2602-46-2	TR-108	+ + nd nd	P
C.I. Direct Blue 15	2429-74-5	TR-397	+ + nd nd	P
C.I. Direct Blue 218	28407-37-6	TR-430	+ – + +	P
C.I. Direct Brown 95	16071-86-6	TR-108	– + nd nd	P
C.I. Disperse Blue 1	2475-45-8	TR-299	+ + ± –	P
C.I. Disperse Yellow 3	2832-40-8	TR-222	+ – – +	P
C.I. Pigment Red 3	2425-85-6	TR-407	+ + + –	P
C.I. Pigment Red 23	6471-49-4	TR-411	± – – –	E
C.I. Solvent Yellow 14	842-07-9	TR-226	+ + – –	P

Table 11 Continued.

Chemical name	CAS no.	TR no.	MR/FR/MM/FM	Overall
C.I. Vat Yellow 4	128-66-5	TR-134	– – + –	P
Cinnamyl Anthranilate	87-29-6	TR-196	+ – + +	P
Clonitralid	1420-04-8	TR-091	– ± is –	E
Cobalt Sulfate Heptahydrate	10026-24-1	TR-471	+ + + +	P
Coconut Oil Acid Diethanolamine Condensate	68603-42-9	TR-479	– ± + +	P
Codeine	76-57-3	TR-455	– – – –	N
Corn Oil	8001-30-7	TR-426	2	
Coumaphos	56-72-4	TR-096	– – – –	N
Coumarin	91-64-5	TR-422	+ ± + +	P
m-Cresidine	102-50-1	TR-105	+ + is –	P
p-Cresidine	120-71-8	TR-142	+ + + +	P
Cupferron	135-20-6	TR-100	+ + + +	P
Cytembena	21739-91-3	TR-207	+ + – –	P
Daminozide	1596-84-5	TR-083	– + ± –	P
D & C Red No. 9	5160-02-1	TR-225	+ ± – –	P
D & C Yellow No. 11	8003-22-3	TR-463	+ + nd nd	P
Decabromodiphenyl Oxide	1163-19-5	TR-309	+ + ± –	P
Diallyl Phthalate	131-17-9	TR-242	nd nd ± ±	E
Diallyl Phthalate	131-17-9	TR-284	– ± nd nd	E
4,4'-Diamino-2,2'-Stilbenedisulfonic Acid, Disodium	7336-20-1	TR-412	– – – –	N
2,4-Diaminoanisole Sulfate	39156-41-7	TR-084	+ + + +	P
2,4-Diaminophenol Dihydrochloride	137-09-7	TR-401	– – + –	P
2,4-Diaminotoluene (2,4-Toluene Diamine)	95-80-7	TR-162	+ + – +	P
Diarylanilide Yellow	6358-85-6	TR-030	– – – –	N
Diazinon	333-41-5	TR-137	– – – –	N
Dibenzo-p-Dioxin	262-12-4	TR-122	– – – –	N
1,2-Dibromo-3-Chloropropane	96-12-8	TR-206	+ + + +	P
1,2-Dibromo-3-Chloropropane	96-12-8	TR-028	+ + + +	P
1,2-Dibromoethane	106-93-4	TR-210	+ + + +	P
1,2-Dibromoethane	106-93-4	TR-086	+ + + +	P
2,3-Dibromo-1-Propanol	96-13-9	TR-400	+ + + +	P
Dibutyltin Diacetate	1067-33-0	TR-183	– is – –	N
1,2-Dichlorobenzene (o-Dichlorobenzene)	95-50-1	TR-255	– – – –	N

Chemical name	CAS no.	TR no.	MR/FR/MM/FM	Overall
1,4-Dichlorobenzene (p-Dichlorobenzene)	106-46-7	TR-319	+ – + +	P
2,7-Dichlorodibenzo-p-Dioxin	33857-26-0	TR-123	– – ± –	E
p,p'-Dichlorodiphenol-dichloroethylene	72-55-9	TR-131	– – + +	P
Dichlorodiphenyl-trichloroethane (DDT)	50-29-3	TR-131	– – – –	N
1,1-Dichloroethane	75-34-3	TR-066	– ± – ±	E
1,2-Dichloroethane	107-06-2	TR-055	+ + + +	P
2,4-Dichlorophenol	120-83-2	TR-353	– – – –	N
2,6-Dichloro-p-Phenylenediamine	609-20-1	TR-219	– – + +	P
1,2-Dichloropropane (Propylene Dichloride)	78-87-5	TR-263	– ± + +	P
1,3-Dichloropropene (Telone II)	542-75-6	TR-269	+ + is +	P
Dichlorvos	62-73-7	TR-342	+ ± + +	P
Dichlorvos	62-73-7	TR-010	– – – –	N
Dicofol	115-32-2	TR-090	– – + –	P
N,N'-Dicyclohexylthiourea	1212-29-9	TR-056	– – – –	N
Dieldrin	60-57-1	TR-021	– – ± –	E
Dieldrin	60-57-1	TR-022	– – nd nd	N
Diesel Fuel, Marine	No CAS#	TR-310	nd nd ± ±	E
Diethanolamine	111-42-2	TR-478	– – + +	P
Di(2-Ethylhexyl) Adipate	103-23-1	TR-212	– – + +	P
Di(2-Ethylhexyl) Phthalate	117-81-7	TR-217	+ + + +	P
Di(p-Ethylphenyl) Dichloroethane	72-56-0	TR-156	– – – ±	E
Diethyl Phthalate	84-66-2	TR-429	– – ± ±	E
Diethyl Phthalate/ Dimethyl Phthalate	84-66-2/ 131-11-3	TR-429	3	
N,N'-Diethylthiourea	105-55-5	TR-149	+ + – –	P
Diglycidyl Resorcinol Ether (DGRE)	101-90-6	TR-257	+ + + +	P
3,4-Dihydrocoumarin	119-84-6	TR-423	+ – – +	P
1,2-Dihydro-2,2,4-Trimethylquinoline (Monomer)	147-47-7	TR-456	+ – – –	P
Dimethoate	60-51-5	TR-004	– – – –	N
Dimethoxane	828-00-2	TR-354	– – ± –	E
2,4-Dimethoxyaniline Hydrochloride	54150-69-5	TR-171	– – – –	N
3,3'-Dimethoxybenzidine Dihydrochloride	20325-40-0	TR-372	+ + nd nd	P
3,3'-Dimethoxybenzidine-4,4'-Diisocyanate	91-93-0	TR-128	+ + – –	P
N,N-Dimethylaniline	121-69-7	TR-360	+ – – ±	P

Table 11 Continued.

Chemical name	CAS no.	TR no.	MR/FR/MM/FM	Overall
3,3'-Dimethylbenzidine Dihydrochloride	612-82-8	TR-390	+ + nd nd	P
Dimethyl Hydrogen Phosphite	868-85-9	TR-287	+ ± – –	P
Dimethyl Methylphosphonate	756-79-6	TR-323	+ – is –	P
Dimethyl Morpholino-phosphoramidate	597-25-1	TR-298	+ + – –	P
Dimethyl Terephthalate	120-61-6	TR-121	– – ± –	E
Dimethylvinyl Chloride (DMVC)	513-37-1	TR-316	+ + + +	P
2,4-Dinitrotoluene	121-14-2	TR-054	+ + – –	P
1,4-Dioxane	123-91-1	TR-080	+ + + +	P
Dioxathion	78-34-2	TR-125	– – – –	N
Diphenhydramine Hydrochloride	147-24-0	TR-355	± ± – –	E
5,5-Diphenylhydantoin (Phenytoin)	57-41-0	TR-404	± – – +	P
2,5-Dithiobiurea	142-46-1	TR-132	– – – ±	E
Emetine Hydrochloride	316-42-7	TR-043	is is is is	I
Endosulfan	115-29-7	TR-062	is – is –	N
Endrin	72-20-8	TR-012	– – – –	N
Ephedrine Sulfate	134-72-5	TR-307	– – – –	N
Epinephrine Hydrochloride	55-31-2	TR-380	is is is is	I
1,2-Epoxybutane	106-88-7	TR-329	+ ± – –	P
Erythromycin Stearate	643-22-1	TR-338	– – – –	N
Estradiol Mustard	22966-79-6	TR-059	– – + +	P
Ethionamide	536-33-4	TR-046	– – – –	N
Ethyl Acrylate	140-88-5	TR-259	+ + + +	P
Ethylbenzene	100-41-4	TR-466	+ + + +	P
Ethylene Glycol	107-21-1	TR-413	nd nd – –	N
Ethylene Oxide	75-21-8	TR-326	nd nd + +	P
Ethylene Thiourea (ETU)	96-45-7	TR-388	+ + + +	P
Ethyl Tellurac	20941-65-5	TR-152	± – ± ±	E
Eugenol	97-53-0	TR-223	– – ± ±	E
FD & C Yellow No. 6	2783-94-0	TR-208	– – – –	N
Formulated Fenaminosulf	140-56-7	TR-101	– – – –	N
Fenthion	55-38-9	TR-103	– – ± –	E
Fluometuron	2164-17-2	TR-195	– – ± –	E
Furan	110-00-9	TR-402	+ + + +	P

Chemical name	CAS no.	TR no.	MR/FR/MM/FM	Overall
Furfural	98-01-1	TR-382	+ – + +	P
Furfural Alcohol	98-00-0	TR-482	+ ± + –	P
Furosemide	54-31-9	TR-356	± – – +	P
Geranyl Acetate	105-87-3	TR-252	– – – –	N
Glycidol	556-52-5	TR-374	+ + + +	P
Guar Gum	9000-30-0	TR-229	– – – –	N
Gum Arabic	9000-01-5	TR-227	– – – –	N
HC Blue 1	2784-94-3	TR-271	± + + +	P
HC Blue 2	33229-34-4	TR-293	– – – –	N
HC Red 3	2871-01-4	TR-281	– – ± is	E
HC Yellow 4	59820-43-8	TR-419	± – – –	E
Heptachlor	76-44-8	TR-009	– ± + +	P
Hexachlorocyclo-pentadiene	77-47-4	TR-437	– – – –	N
1,2,3,6,7,8-Hexachlorodibenzo-p-Dioxin	57653-85-7	TR-198	± + + +	P
1,2,3,6,7,8-Hexachlorodibenzo-p-Dioxin	57653-85-7	TR-202	nd nd – –	N
Hexachloroethane	67-72-1	TR-361	+ – nd nd	P
Hexachloroethane	67-72-1	TR-068	– – + +	P
Hexachlorophene	70-30-4	TR-040	– – nd nd	N
4-Hexylresorcinol	136-77-6	TR-330	– – ± –	E
Hydrazobenzene	122-66-7	TR-092	+ + – +	P
Hydrochlorothiazide	58-93-5	TR-357	– – ± –	E
Hydroquinone	123-31-9	TR-366	+ + – +	P
8-Hydroxyquinoline	148-24-3	TR-276	– – – –	N
ICRF-159	21416-87-5	TR-078	– + – +	P
IPD (3,3'-Iminobis-1-Propanol Dimethanesulfonate (Ester) Hydrochloride)	3458-22-8	TR-018	± ± ± ±	E
Iodinated Glycerol	5634-39-9	TR-340	+ – – +	P
Iodoform	75-47-8	TR-110	– – – –	N
Isobutene	115-11-7	TR-487	+ – – –	P
Isobutyl Nitrite	542-56-3	TR-448	+ + + +	P
Isobutyraldehyde	78-84-2	TR-472	– – – –	N
Isophorone	78-59-1	TR-291	+ – ± –	P
Isophosphamide	3778-73-2	TR-032	– + – +	P
Isoprene	78-79-5	TR-486	+ + nd nd	P

Table 11 Continued.

Chemical name	CAS no.	TR no.	MR/FR/MM/FM	Overall
Lauric Acid Diethanolamine Condensate	120-40-1	TR-480	– – – +	P
Lasiocarpine	303-34-4	TR-039	+ + nd nd	P
Lead Dimethyldithiocarbamate	19010-66-3	TR-151	– – – –	N
d-Limonene	5989-27-5	TR-347	+ – – –	P
Lindane	58-89-9	TR-014	– – – –	N
Lithocholic Acid	434-13-9	TR-175	– – – –	N
Locust Bean Gum	9000-40-2	TR-221	– – – –	N
Malaoxon	1634-78-2	TR-135	– – – –	N
Malathion	121-75-5	TR-024	– – – –	N
Malathion	121-75-5	TR-192	– – nd nd	N
Malonaldehyde, Sodium Salt	24382-04-5	TR-331	+ + – –	P
Manganese Sulfate Monohydrate	10034-96-5	TR-428	– – ± ±	E
d-Mannitol	69-65-8	TR-236	– – – –	N
Melamine	108-78-1	TR-245	+ – – –	P
dl-Menthol	15356-70-4	TR-098	– – – –	N
2-Mercaptobenzothiazole	149-30-4	TR-332	+ + – ±	P
Mercuric Chloride	7487-94-7	TR-408	+ ± ± –	P
Methoxychlor	72-43-5	TR-035	– – – –	N
8-Methoxypsoralen	298-81-7	TR-359	+ – nd nd	P
Alpha-Methylbenzyl Alcohol	98-85-1	TR-369	+ – – –	P
Methyl Bromide	74-83-9	TR-385	nd nd – –	N
Methyl Carbamate	598-55-0	TR-328	+ + – –	P
Methyldopa Sesquihydrate	41372-08-1	TR-348	– – ± –	E
4,4'-Methylenebis(N,N-Dimethyl)Benzenamine	101-61-1	TR-186	+ + ± +	P
Methylene Chloride	75-09-2	TR-306	+ + + +	P
4,4'-Methylenedianiline Dihydrochloride	13552-44-8	TR-248	+ + + +	P
Methyl Methacrylate	80-62-6	TR-314	– – – –	N
2-Methyl-1-Nitroanthraquinone	129-15-7	TR-029	+ + + +	P
N-Methylolacrylamide	924-42-5	TR-352	– – + +	P
Methyl Parathion	298-00-0	TR-157	– – – –	N
Methylphenidate Hydrochloride	298-59-9	TR-439	– – + +	P
Mexacarbate	315-18-4	TR-147	– – – –	N
Michler's Ketone	90-94-8	TR-181	+ + + +	P

Chemical name	CAS no.	TR no.	MR/FR/MM/FM	Overall
Mirex	2385-85-5	TR-313	+ + nd nd	P
Molybdenum Trioxide	1313-27-5	TR-462	± – + +	P
Monochloroacetic Acid	79-11-8	TR-396	– – – –	N
Monuron	150-68-5	TR-266	+ – – –	P
Nalidixic Acid	389-08-2	TR-368	+ + ± –	P
Naphthalene	91-20-3	TR-410	nd nd – +	P
<u>1,5-Naphthalenediamine</u>	2243-62-1	TR-143	– + + +	P
N-(1-Naphthyl) Ethylenediamine Dihydrochloride	1465-25-4	TR-168	– – – –	N
Navy Fuels JP-5	8008-20-6	TR-310	nd nd – –	N
Nickel (II) Oxide	1313-99-1	TR-451	+ + – ±	P
Nickel Sulfate Hexahydrate	10101-97-0	TR-454	– – – –	N
Nickel Subsulfide	12035-72-2	TR-453	+ + – –	P
Nithiazide	139-94-6	TR-146	– + + ±	P
Nitrilotriacetic Acid (NTA)	139-13-9	TR-006	+ + + +	P
Nitrilotriacetic Acid Trisodium Monohydrate	18662-53-8	TR-006	+ + nd nd	P
Nitrilotriacetic Acid Trisodium Monohydrate	18662-53-8	TR-006	± ± – –	E
<u>5-Nitroacenaphthene</u>	602-87-9	TR-118	+ + – +	P
3-Nitro-p-Acetophenetide	1777-84-0	TR-133	– – + –	P
p-Nitroaniline	100-01-6	TR-418	nd nd ± –	E
<u>5-Nitro-o-Anisidine</u>	99-59-2	TR-127	+ + ± +	P
o-Nitroanisole	91-23-6	TR-416	+ + + +	P
4-Nitroanthranilic Acid	619-17-0	TR-109	– – – –	N
6-Nitrobenzimidazole	94-52-0	TR-117	– – + +	P
p-Nitrobenzoic Acid	62-23-7	TR-442	– + – –	P
<u>Nitrofen</u>	1836-75-5	TR-026	is + + +	P
Nitrofen	1836-75-5	TR-184	– – + +	P
Nitrofurantoin	67-20-9	TR-341	+ – – +	P
Nitrofurazone	59-87-0	TR-337	± + – +	P
<u>Nitromethane</u>	75-52-5	TR-461	– + + +	P
1-Nitronaphthalene	86-57-7	TR-064	– – – –	N
p-Nitrophenol	100-02-7	TR-417	nd nd – –	N
2-Nitro-p-Phenylenediamine	5307-14-2	TR-169	– – – +	P
4-Nitro-o-Phenylenediamine	99-56-9	TR-180	– – – –	N
3-Nitropropionic Acid	504-88-1	TR-052	± – – –	E

Table 11 Continued.

Chemical name	CAS no.	TR no.	MR/FR/MM/FM	Overall
N-Nitrosodiphenylamine	86-30-6	TR-164	+ + – –	P
p-Nitrosodiphenylamine	156-10-5	TR-190	+ – + –	P
Beta-Nitrostyrene	102-96-5	TR-170	– – – –	N
5-Nitro-o-Toluidine	99-55-8	TR-107	– – + +	P
Ochratoxin A	303-47-9	TR-358	+ + nd nd	P
Oleic Acid Diethanolamine Condensate	93-83-4	TR-481	– – – –	N
Oxazepam	604-75-1	TR-443	nd nd + +	P
Oxazepam	604-75-1	TR-468	± – nd nd	E
4,4'-Oxydianiline	101-80-4	TR-205	+ + + +	P
Oxytetracycline Hydrochloride	2058-46-0	TR-315	± ± – –	E
Ozone	10028-15-6	TR-440	– – ± +	P
Ozone (130 weeks)	10028-15-6	TR-440	– – ± +	P
Parathion	56-38-2	TR-070	± ± – –	E
Penicillin VK	132-98-9	TR-336	– – – –	N
Pentachloroanisole	1825-21-4	TR-414	+ ± + –	P
Pentachloroethane	76-01-7	TR-232	± – + +	P
Pentachloronitrobenzene	82-68-8	TR-061	– – – –	N
Pentachloronitrobenzene	82-68-8	TR-325	nd nd – –	N
Pentachlorophenol, Dowicide EC-7	87-86-5	TR-349	nd nd + +	P
Pentachlorophenol, Purified	87-86-5	TR-483	– – nd nd	N
Pentachlorophenol, Purified	87-86-5	TR-483	– + nd nd	P
Pentachlorophenol, Technical	87-86-5	TR-349	nd nd + +	P
Pentaerythritol Tetranitrate	78-11-5	TR-365	± ± – –	E
Phenazopyridine Hydrochloride	136-40-3	TR-099	+ + – +	P
Phenesterin	3546-10-9	TR-060	– + + +	P
Phenformin Hydrochloride	834-28-6	TR-007	– – – –	N
Phenol	108-95-2	TR-203	– – – –	N
Phenolphthalein	77-09-8	TR-465	+ + + +	P
Phenoxybenzamine Hydrochloride	63-92-3	TR-072	+ + + +	P
Phenylbutazone	50-33-9	TR-367	± + + –	P
p-Phenylenediamine Dihydrochloride	624-18-0	TR-174	– – – –	N
Phenylephrine Hydrochloride	61-76-7	TR-322	– – – –	N
1-Phenyl-3-Methyl-5-Pyrazolone	89-25-8	TR-141	– – – –	N

Chemical name	CAS no.	TR no.	MR/FR/MM/FM	Overall
N-Phenyl-2-Naphthylamine	135-88-6	TR-333	– – – ±	E
o-Phenylphenol	90-43-7	TR-301	nd nd – –	N
N-Phenyl-p-Phenylenediamine	101-54-2	TR-082	– – – –	N
1-Phenyl-2-Thiourea	103-85-5	TR-148	– – – –	N
Phosphamidon	13171-21-6	TR-016	± ± – –	E
Photodieldrin	13366-73-9	TR-017	– – – –	N
Phthalamide	88-96-0	TR-161	– – – –	N
Phthalic Anhydride	85-44-9	TR-159	– – – –	N
Picloram	1918-02-1	TR-023	– ± – –	E
Piperonyl Butoxide	51-03-6	TR-120	– – – –	N
Piperonyl Sulfoxide	120-62-7	TR-124	– – + –	P
Pivalolactone	1955-45-9	TR-140	+ + – –	P
Polybrominated Biphenyl Mixture (Firemaster FF-1)	67774-32-7	TR-244	+ + + +	P
Polybrominated Biphenyl Mixture (Firemaster FF-1)	67774-32-7	TR-398	+ + + +	P
Polysorbate 80 (Glycol)	9005-65-6	TR-415	± – – –	E
Polyvinyl Alcohol	9002-89-5	TR-474	nd nd nd –	N
Primidone	125-33-7	TR-476	± – + +	P
Probenecid	57-66-9	TR-395	– – – +	P
Procarbazine Hydrochloride	366-70-1	TR-019	+ + + +	P
Profavin Hydrochloride	952-23-8	TR-005	± – ± ±	E
Promethazine Hydrochloride	58-33-3	TR-425	– – – –	N
Propylene	115-07-1	TR-272	– – – –	N
1,2-Propylene Oxide	75-56-9	TR-267	+ + + +	P
Propyl Gallate	121-79-9	TR-240	± – ± –	E
Pyrazinamide	98-96-4	TR-048	– – – is	N
Pyridine	110-86-1	TR-470	+ ± + +	P
Pyridine	110-86-1	TR-470	± nd nd nd	
Pyrimethamine	58-14-0	TR-077	– – is –	N
Quercetin	117-39-5	TR-409	+ – nd nd	P
Reserpine	50-55-5	TR-193	+ – + +	P
Resorcinol	108-46-3	TR-403	– – – –	N
Rhodamine 6G	989-38-8	TR-364	± ± – –	E
Rotenone	83-79-4	TR-320	± – – –	E
Roxarsone	121-19-7	TR-345	± – – –	E

Table 11 Continued.

Chemical name	CAS no.	TR no.	MR/FR/MM/FM	Overall
Safflower Oil	8001-23-8	TR-426	2	
Salicylazosulfapyridine	599-79-1	TR-457	+ + + +	P
Scopolamine Hydrobromide Trihydrate	6533-68-2	TR-445	– – – –	N
Selenium Sulfide	7446-34-6	TR-197	nd nd – –	N
Selenium Sulfide	7446-34-6	TR-194	+ + – +	P
Selsun	EMTDP-74	TR-199	nd nd – –	N
Sodium Azide	26628-22-8	TR-389	– – nd nd	N
Sodium Diethyldithiocarbamate	148-18-5	TR-172	– – – –	N
Sodium Fluoride	7681-49-4	TR-393	± – – –	E
Sodium Xylenesulfonate	1300-72-7	TR-464	– – – –	N
Stannous Chloride	7772-99-8	TR-231	± – – –	E
Styrene	100-42-5	TR-185	– – ± –	E
Succinic Anhydride	108-30-5	TR-373	– – – –	N
Sulfallate	95-06-7	TR-115	+ + + +	P
Sulfisoxazole	127-69-5	TR-138	– – – –	N
3-Sulfolene	77-79-2	TR-102	– – – –	N
4,4'-Sulfonyldianiline (Dapsone)	80-08-0	TR-020	+ – – –	P
Talc	14807-96-6	TR-421	+ + – –	P
Tara Gum	39300-88-4	TR-224	– – – –	N
2,3,7,8-Tetrachlorodibenzo-p-Dioxin	1746-01-6	TR-201	nd nd ± +	P
2,3,7,8-Tetrachlorodibenzo-p-Dioxin	1746-01-6	TR-209	+ + + +	P
Tetrachlorodiphenyl-ethane	72-54-8	TR-131	± – – –	E
1,1,1,2-Tetrachloroethane	630-20-6	TR-237	± – + +	P
1,1,2,2-Tetrachloroethane	79-34-5	TR-027	± – + +	P
Tetrachloroethylene	127-18-4	TR-311	+ + + +	P
Tetrachloroethylene	127-18-4	TR-013	is is + +	P
2,3,5,6-Tetrachloro-4-Nitroanisole	2438-88-2	TR-114	– – – –	N
Tetrachlorvinphos	961-11-5	TR-033	– + + +	P
Tetracycline Hydrochloride	64-75-5	TR-344	– – – –	N
Tetraethylthiuram Disulfide	97-77-8	TR-166	– – – –	N
Tetrafluoroethylene	116-14-3	TR-450	+ + + +	P
1-Trans-Delta-9-Tetrahydrocannabinol	1972-08-3	TR-446	– – ± ±	E
Tetrahydrofuran	109-99-9	TR-475	+ – – +	P

Chemical name	CAS no.	TR no.	MR/FR/MM/FM	Overall
Tetrakis(Hydroxymethyl) Phosphonium Chloride (THPC)	124-64-1	TR-296	– – – –	N
Tetrakis(Hydroxymethyl) Phosphonium Sulfate (THPS)	55566-30-8	TR-296	– – – –	N
Tetranitromethane	509-14-8	TR-386	+ + + +	P
Theophylline	58-55-9	TR-473	– – – –	N
4,4-Thiobis(6-Tert-Butyl-M-Cresol)	96-69-5	TR-435	– – – –	N
4,4'-Thiodianiline	139-65-1	TR-047	+ + + +	P
β-Thioguanidine Deoxyriboside	789-61-7	TR-057	± + is is	P
Titanium Dioxide	13463-67-7	TR-097	– – – –	N
Titanocene Dichloride	1271-19-8	TR-399	± ± nd nd	E
Tolazamide	1156-19-0	TR-051	– – – –	N
Tolbutamide	64-77-7	TR-031	– – – –	N
Toluene	108-88-3	TR-371	– – – –	N
2,6-Toluenediamine Dihydrochloride (2,6-Diamino-toluene Dihydrochloride)	15481-70-6	TR-200	– – – –	N
2,5-Toluenediamine Sulfate	6369-59-1	TR-126	– – – –	N
2,4- & 2,6-Toluene Diisocyanate	26471-62-5	TR-251	+ + – +	P
o-Toluidine Hydrochloride	636-21-5	TR-153	+ + + +	P
Toxaphene	8001-35-2	TR-037	± ± + +	P
Triamterene	396-01-0	TR-420	± – + +	P
Tribromomethane	75-25-2	TR-350	+ + – –	P
Tricaprylin	538-23-8	TR-426	2	
1,1,1-Trichloroethane	71-55-6	TR-003	is is is is	I
1,1,2-Trichloroethane	79-00-5	TR-074	– – + +	P
Trichloroethylene	79-01-6	TR-273	is is nd nd	I
Trichloroethylene	79-01-6	TR-243	is – + +	P
Trichloroethylene	79-01-6	TR-273	is is nd nd	I
Trichloroethylene	79-01-6	TR-002	– – + +	P
Trichlorofluoromethane	75-69-4	TR-106	is is – –	N
2,4,6-Trichlorophenol	88-06-2	TR-155	+ – + +	P
1,2,3-Trichloropropane	96-18-4	TR-384	+ + + +	P
Tricresyl Phosphate	1330-78-5	TR-433	– – – –	N
Triethanolamine	102-71-6	TR-449	± – ± +	P

Table 11 Continued.

Chemical name	CAS no.	TR no.	MR/FR/MM/FM	Overall
Trifluralin	1582-09-8	TR-034	– – – +	P
2,4,5-Trimethylaniline	137-17-7	TR-160	+ + ± +	P
Trimethylphosphate	512-56-1	TR-081	+ – – +	P
Trimethylthiourea	2489-77-2	TR-129	– + – –	P
Triphenyltin Hydroxide	76-87-9	TR-139	– – – –	N
Tris(Aziridinyl)-Phosphine Sulfide(Thiotepa)	52-24-4	TR-058	+ + + +	P
Tris(2-Chloroethyl) Phosphate	115-96-8	TR-391	+ + ± ±	P
Tris(2,3-Dibromopropyl) Phosphate	126-72-7	TR-076	+ + + +	P
Tris(2-Ethylhexyl)Phosphate	78-42-2	TR-274	± – – +	P
Trisodium Ethylenediamine-tetraacetate Trihydrate (EDTA)	150-38-9	TR-011	– – – –	N
L-Tryptophan	73-22-3	TR-071	– – – –	N
Turmeric, Oleoresin (Curcumin)	8024-37-1	TR-427	– ± ± ±	E
4-Vinylcyclohexene	100-40-3	TR-303	is is is +	P
4-Vinyl-1-Cyclohexene Diepoxide	106-87-6	TR-362	+ + + +	P
Vinylidene Chloride	75-35-4	TR-228	– – – –	N
Vinyl Toluene	25013-15-4	TR-375	– – – –	N
Xylenes (Mixed)	1330-20-7	TR-327	– – – –	N
2,6-Xylidine	87-62-7	TR-278	+ + nd nd	P
Zearalenone	17924-92-4	TR-235	– – + +	P
Ziram	137-30-4	TR-238	+ – – ±	P

Positive Studies	252
Negative Studies	153
Equivocal Studies	65
Inadequate Studies	8
TOTAL	478

Abbreviations: P or + = positive evidence of carcinogenicity; E and ± = equivocal evidence of carcinogenicity; N and – = negative or no evidence of carcinogenicity; I and IS = inadequate carcinogenesis study; ND = not tested in this species.

Chemical names bolded, underlined, and italicized would be *candidates* for being classified as "reasonably anticipated to be carcinogenic to humans" by the National Toxicology Program, and as "sufficient evidence of carcinogenicity" by the International Agency for Research on Cancer. **Boldface** = positive cancer results in each of the four experiments; <u>underline</u> = positive cancer results in three of the four experiments, and *italics* = positive cancer results in two of the four experiments (but only those involving both species).

1. Under the conditions of this 1-year mouse skin initiation/promotion study in Swiss (CD-1®) mice, o-benzyl-p-chlorophenol was a cutaneous irritant and a weak skin tumor promoter relative to strong promoters such as TPA. o-Benzyl-p-chlorophenol had no activity as an initiator or as a complete carcinogen.

2. These studies demonstrate that safflower oil and tricaprylin do not offer significant advantages over corn oil as a gavage vehicle in long-term rodent studies. Corn oil, safflower oil, and tricaprylin each caused hyperplasia and adenoma of the exocrine pancreas, decreased incidences of mononuclear cell leukemia, and reduced incidences or severity of nephropathy in male F344/N rats. There was an increased incidence of squamous cell papillomas of the forestomach in F344/N rats receiving 10 mL tricaprylin/kg. Further, the use of corn oil as a gavage vehicle may have a confounding effect on the interpretation of chemical-induced proliferative lesions of the exocrine pancreas and mononuclear cell leukemia in male F344/N rats.

3. In an initiation/promotion model of skin carcinogenesis, there was no evidence of initiating activity of diethylphthalate or dimethylphthalate in male Swiss (CD-1®) mice. Further, there was no evidence of promotion activity of diethylphthalate or dimethylphthalate in male Swiss (CD-1®) mice. The promoting activity of TPA following DMBA initiation was confirmed in these studies.

Table 12 Basic NTP Design Protocol for Long-Term Chemical Carcinogenesis Studies

1. **Purpose:** Identify chronic toxicity (nonneoplastic) and carcinogenic (neoplastic) effects of chemicals.
2. **Animals:** The following are the overwhelming choices of the NTP for toxicity and carcinogenicity experiments, although other species and strains are occasionally used when appropriate
   a. Fischer 344 hybrid rats
   b. B6C3F1 inbred mice (C3H × C57Bl/6)
3. **Groups and group sizes**
   a. Both males and females of each species and strain
   b. 50 to 60 or more animals in each sex, species, and treatment group
   c. Concurrent controls and two to four experimental groups
4. **Exposure levels:** Chosen to show some minimal yet obvious chemical-associated effects of a degree not to compromise well-being or growth and survival
5. **Exposure duration:** 24 months (approximately two-thirds of the lifespan of these strains)
6. **Routes of exposure:** Priority typically given to mimic human exposures. Historically, via feed (44%), oral intubation (27%), inhalation (10%), skin (9%), drinking water (6%), i.p. injection (2%)
7. **Pathology:** All animals, controls and exposed; complete gross and histological examination. Special target organs may be step-sectioned.

Table 13 Alternative Protocol Design Considerations for Long-Term Chemical Carcinogenesis Studies

1. **Animals:** Chemical- and target-site-specific. For example, for a potential leukemogen or testicular carcinogen, Fischer 344 rats should *not* be used, because this strain has very high control background incidences. Also, for a possible mammary carcinogen, do not use Sprague-Dawley rats, which have a high backgound incidence. For skin carcinogens, do not use insensitive B6C3F1 mice or, in general, rats. For potential lung chemical carcinogens, do not use hamsters. Thus, it is essential to know the tumor incidences of "control" animals before selection.
2. **Sex and species:** Instead of using both sexes, consider a single sex of two species, e.g., male Fischer rats and female B6C3F1 mice. For short-chain aliphatic halogenated solvents, use two strains of mice (rats appear to be "resistant" or generally not responsive to these solvents).

Table 13 Continued

3. **Groups and groups sizes:** At least 100 per group, with up to 1000 per group for a predictably "weak" environmental carcinogen (e.g., electromagnetic fields) or a chemical with extensive human population exposures (e.g., fluoride or aspartame).
4. **Exposures:** Highest concentration, or "dose," at the minimal toxic exposure (MTE) level, observed from shorter-term toxicology studies; lower exposure levels could be selected at a "metabolic saturation point" or at a "pharmacokinetic inflection point." Lowest levels may be a few or several multiples of human exposures or environmental levels. Consider intermittent regimens of exposures rather than continuous (feed or drinking water) or 6 hours per day for 5 days per week (inhalation). This is consistent with more realistic exposure patterns, except for drugs.
5. **Routes:** Mimic main route of human exposure when possible. However, because inherent carcinogenic potential is typically irrelevant to route of exposure, perhaps the easiest and cheapest route should be utilized, i.e., in food. Use multiple routes when appropriate; e.g., chemical(s) could be given by inhalation and in drinking water and by skin painting and gavage to better simulate actual human exposures (e.g., to pesticides, solvents, water-supply contaminants as trihalomethanes, and added chemicals such as fluoride).
6. **Duration:** Flexible. a) At least 2 years for "unknowns"; b) longer periods for metals (e.g., cadmium carcinogenesis was not apparent before 30 months) and most other chemicals; c) supplemental routine shorter-exposure groups (e.g., 13, 25, or 52 weeks' exposure and then "stop" exposures for remaining duration) to determine tumor progression/regression; d) preconception, gestation, lactation, and F1 generation exposures for 30–36 months if large human populations are exposed (e.g., to fluoride, food additives, EMF, drugs, ozone); intermittent and random exposures for solvents.
7. **Chemicals:** Use multiple exposures and mixtures closer to human exposure patterns, e.g., oxazapam by gavage, fluoride by drinking water, pesticides by feed, solvents by skin. Combinations need to be evaluated versus single chemicals with high purity.
8. **Pathology:** More selective; in general, "standard" histopathology is costly, excessive, and unproductive. More effort needs to be directed toward lesions observed grossly. Species-specific high background tumor organs should be avoided, e.g., testicular interstitial cell tumors in Fischer rats. More efforts should be given to multiple sections for predicted or grossly observed target organs, e.g., solvents and kidney; hormones and glands; aniline dyes and spleen; halogenated hydrocarbons and liver; "site of application" such as nose and lung via inhalation, stomach, and forestomach by gavage, oral via drinking water.

exposure circumstances "known to be carcinogenic to humans" and 169 agents, mixtures, and exposure situations "reasonably anticipated to be carcinogenic to humans" (Table 14). Agents recommended for the ninth edition are listed in Table 15.

VII. DEFINITIONS OF CHEMICAL CARCINOGENS

Over the years, definitions of chemical carcinogens have changed somewhat. Most modifications have attempted to include more details—often restrictive and didactic—and the latest trends or discoveries in mechanistic understanding. However, simplicity coupled with scientific judgment often leads to better and more broad-based individual, cooperative, and harmonic evaluation criteria that are useful and fundamental for cancer testing and epidemiological data. Three are described below.

In 1959, Zwickey and Davis outlined in fewer than 20 sentences a carcinogenicity screening design that apparently prefigured the one eventually and essentially adopted by the National Cancer Institute (Weisberger, 1983) and others (Montesano et al., 1986; Huff and Moore, 1984; Huff et al., 1988; Chhabra et al., 1990). These pathologists begin their four-page article (Zwickey and Davis, 1959) by stating that "carcinogens are those substances which produce a significant increase in tumor incidence when administered at any dosage level by any route of administration in any species of animal as compared to controls." Today, we seem to use more extensive and detailed definitions, yet in its simplicity this definition remains quite relevant.

A more comprehensive definition of carcinogen was conceived by Hueper and Conway in their wonderful book *Chemical Carcinogenesis and Cancers* (1964)

> Carcinogens may be defined as chemical, physical and parasitic agents of natural and man-made origin which are capable under proper conditions of exposure of producing cancers in animals, including man, in one or several organs and tissues, regardless of the route of exposure and the dose and physical state of the agent used. Such cancers would not have occurred without the intervention of these agents. . . . Carcinogens thus do not merely produce a significant increase in cancer incidence when administered at any dose level, by any route of administration and to any species and strain, but elicit cancers located at sites definitely related to the particular carcinogen and route of its introduction employed.

(*text continues on p. 87*)

Table 14 NTP Reports on Carcinogens

A. Categories of evidence used by the NTP
   1. Known to be carcinogenic to humans
   2. Reasonably anticipated to be carcinogenic to humans

B. List of agents, mixtures, and exposures known to be carcinogenic to humans (29 entries)
   1. Aflatoxins
   2. 4-Aminobiphenyl
   3. Analgesic mixtures containing phenacetin
   4. Arsenic and certain arsenic compounds
   5. Asbestos
   6. Azathioprine
   7. Benzene
   8. Benzidine
   9. Bis(chloromethyl) ether (BCME)
   10. Technical grade chloro-methyl methyl ether (CMME)
   11. 1,4-Butanediol dimethyl sulfonate (Myleran®)
   12. Chlorambucil
   13. (2-Chloroethyl)-3-(4-methylcyclo-hexyl)-1-nitrosourea (MECCNU)
   14. Chromium and certain chromium compounds
   15. Coke oven emissions
   16. Conjugated estrogens
   17. Cyclophosphamide
   18. Cyclosporin A (ciclosporin)
   19. Diethylstilbestrol
   20. Erionite
   21. Melphalan
   22. Methoxsalen with ultraviolet A therapy (PUVA)
   23. Mustard gas
   24. 2-Naphthylamine
   25. Radon
   26. Soots, tars, and mineral oils
   27. Thorium dioxide
   28. Thiotepa
   29. Vinyl chloride

C. List of agents, mixtures, and exposures reasonably anticipated to be carcinogenic to humans (152 entries)
   1. Acetaldehyde
   2. 2-Acetylaminofluorene
   3. Acrylamide
   4. Acrylonitrile
   5. Adriamycin
   6. 2-Aminoanthraquinone
   7. o-Aminoazotoluene
   8. 1-Amino-2-methylanthraquinone

(continued)

Table 14 Continued

9. Amitrole
10. o-Anisidine hydrochloride
11. Azacitidine (5-azacytidine)
12. Benzotrichloride
13. Beryllium and certain beryllium compounds
14. Bischloroethyl nitrosourea
15. Bromodichloromethane
16. 1,3-Butadiene
17. Butylated hydroxyanisole
18. Cadmium and certain cadmium compounds
19. Carbon tetrachloride
20. Ceramic fibers (respirable size)
21. Chlorendic acid
22. Chlorinated paraffins (C12, 60% chlorine)
23. (2-Chloroethyl)-3-cyclo-hexyl-1-nitrosourea (CCNU)
24. Chloroform
25. 3-Chloro-2-methylpropene
26. 4-Chloro-o-phenylene-diamine
27. p-Chloro-o-toluidine and p-chloro-o-toluidine HCl
28. Chlorozotocin
29. Basic red 9 monoHCl
30. Cisplatin
31. p-Cresidine
32. Cupferron
33. Dacarbazine
34. Danthrone (1,8-Dihydroxy-anthraquinone)
35. DDT (Dichlorodiphenyltri-chloroethane)
36. 2,4-Diaminoanisole sulfate
37. 2,4-Diaminotoluene
38. 1,2-Dibromo-3-chloropro-pane
39. 1,2-Dibromoethane (ethylene dibromide)
40. 1,4-Dichlorobenzene
41. 3,3′-Dichlorobenzidine and 3,3′-dichlorobenzidine DiHCl
42. 1,2-Dichloroethane
43. Dichloromethane (methylene chloride)
44. 1,3-Dichloropropene (technical grade)
45. Diepoxybutane
46. Di(2-ethylhexyl) phthalate
47. Diethyl sulfate
48. Diglycidyl resorcinol ether
49. 3,3′-Dimethoxybenzidine and 3,3′-dimethoxybenzidine HCl
50. 4-Dimethylaminoazo-benzene
51. 3,3′-Dimethylbenzidine
52. Dimethylcarbamoyl chloride

Table 14 Continued

53. 1,1-Dimethylhydrazine
54. Dimethyl sulfate
55. Dimethylvinyl chloride
56. 1,6-Dinitropyrene
57. 1,8-Dinitropyrene
58. 1,4-Dioxane
59. Direct black 38
60. Direct blue 6
61. Disperse blue 1 (1,4,5,8-tetraaminoanthraquinone)
62. Epichlorohydrin
63. Estrogens (not conjugated): estradiol-17β
64. Estrogens (not conjugated): estrone
65. Estrogens (not conjugated): ethinylestradiol
66. Estrogens (not conjugated): mestranol
67. Ethyl acrylate
68. Ethylene oxide
69. Ethylene thiourea
70. Ethyl methanesulfonate
71. Formaldehyde (gas)
72. Furan
73. Glasswool
74. Glycidol
75. Hexachlorobenzene
76. Hexachloroethane
77. Hexamethylphosphoramide
78. Hydrazine and hydrazine sulfate
79. Hydrazobenzene
80. Iron dextran complex
81. Kepone® (chlordecone)
82. Lead acetate and lead phosphate
83. Lindane and other hexa-chlorocyclohexane isomers
84. 2-Methylaziridine (propyleneimine)
85. 4,4′-Methylenebis(2-chloroaniline) (MBOCA)
86. 4,4′-Methylenebis(N,N-dimethylbenzenamine)
87. 4,4′-Methylenedianiline and MDA hydrochloride
88. Methyl methanesulfonate
89. N-Methyl-N′-nitro-N-nitro-soguanidine
90. Metronidazole
91. Michler's ketone
92. Mirex
93. Nickel and certain nickel compounds
94. Nitrilotriacetic acid (NTA) and nitrilotriacetic acid Na
95. o-Nitroanisole
96. 6-Nitrochrysene

(continued)

Table 14 Continued

97. Nitrofen
98. Nitrogen mustard HCl
99. 2-Nitropropane
100. 1-Nitropyrene
101. 4-Nitorpyrene
102. N-Nitrosodi-N-butylamine
103. N-Nitrosodiethanolamine
104. N-Nitrosodiethylamine
105. N-Nitrosodimethylamine
106. N-Nitroso-n-ethylurea
107. (N-Nitrosomethylamino)-1-(3-pyridyl)-1-butanone (NNK)
108. N-Nitroso-N-methylurea
109. N-Nitrosomethylvinyl-amine
110. N-Nitrosomorpholine
111. N-Nitrosonornicotine
112. N-Nitrosopiperidine
113. N-Nitrosopyrrolidine
114. N-Nitrososarcosine
115. Norethisterone
116. Ochratoxin A
117. 4,4′-Oxydianiline
118. Oxymetholone
119. Phenacetin
120. Phenazopyridine HCl
121. Phenoxybenzamine HCl
122. Phenytoin (diphenylhydantoin)
123. Polybrominated biphenyls
124. Polychlorinated biphenyls
125. Polycyclic aromatic hydrocarbons (15 listings)
126. Procarbazine HCl
127. Progesterone
128. 1,3-Propane sultone
129. β-Propiolactone
130. Propylene oxide
131. Propylthiouracil
132. Reserpine
133. Saccharin
134. Safrole
135. Selenium sulfide
136. Silica, crystalline (respirable size)
137. Streptozotocin
138. Sulfallate
139. 2,3,7,8-Tetrachlorodiben-zo-p-dioxin (TCDD)
140. Tetrachloroethylene (perchloroethylene; PERC)

Table 14 Continued

141.	Tetranitromethane
142.	Thioacetamide
143.	Thiourea
144.	Toluene diisocyanate
145.	o-Toluidine and o-toluidine hydrochloride
146.	Toxaphene
147.	2,4,6-Trichlorophenol
148.	1,2,3-Trichloropropane
149.	Tris(1-aziridinyl)phosphine sulfide (Thiotepa)
150.	Tris(2,3-dibromopropyl) phosphate
151.	Urethane (ethyl carbam-ate)
152.	4-Vinyl-1-cyclohexene diepoxide

This latter comment appears to endorse what some have labeled as irrelevant and immaterial, that is, application- or exposure-site carcinogenesis. These sites are indeed relevant. Human cancers likewise appear at exposure sites—as a few examples, with tobacco smoke in the lung and oral cavity; with asbestos in the lung; with solar radiation on the skin; and with alcoholic beverages in the oral cavity and on the pharynx, larynx, and esophagus. Thus, exposure-site carcinogenesis in animals must be considered as

Table 15 Agents Recommended for the Ninth NTP Report on Carcinogens

Agent	Recommendation for change
1,3-Butadiene	Known—upgrade
Benzidine-based dyes	Known—class
Cadmium	Known—upgrade
Chloroprene	Reasonably—1st
Phenolphthalein	Reasonably—1st
Saccharin	Delist—mechanism
Smokeless tobacco	Known—1st
Strong inorganic acid mists	Known—1st
Tamoxifen	Known—1st
TCDD	Known—upgrade mechanism
Tetrafluoroethylene	Reasonably—1st
Tobacco smoking	Known—1st
Trichloroethylene	Reasonably—1st
UV radiation	Known—1st

important as distant-site tumor responses; there is no scientific evidence to do otherwise.

In their wonderful book, Hueper and Conway (1964) also point out that the term "carcinogen" characterizes chemicals by their most important property. Practically all known carcinogens may also induce the development of benign tumors, such as papillomas and adenomas. In fact, weak carcinogens may sometimes, when inadequately tested, produce only benign neoplasms, and under such conditions display merely a "tumorigenic" effect. However, this has been documented for only a few chemicals (Huff et al., 1989).

In 1977, the National Cancer Advisory Board stated that "benign neoplasms may endanger the life of the host by a variety of mechanisms including hemorrhage, encroachment on a vital organ, or unregulated hormone production" and that "benign neoplasms may represent a stage in the evolution of a malignant neoplasm and in other cases may be 'end points' which do not undergo transition to malignant neoplasms." This view was endorsed in 1984 by the American Industrial Health Council, and in 1985 the Office of Science and Technology Policy reported that "truly benign tumors in rodents are rare and most tumors diagnosed as benign really represent a stage in the progression to malignancy." Furthermore, it is not yet known whether benign neoplasia in rodents correspond to benign or malignant neoplasia in other species, including humans. Accordingly, we consider chemically induced benign neoplasia an important indicator of a chemical's carcinogenic activity in rodents, and believe it should continue to be made an integral part of the overall weight-of-the-evidence evaluation process for identifying potential human carcinogenic health hazards (Huff et al., 1989).

As a final comment on definitions of a carcinogen, most research and commercial organizations and regulatory agencies worldwide endorse some distinction between agents and exposure circumstances that cause cancer in humans and/or in animals and those agents or exposure circumstances that don't. Conversely, some individuals seem to think the term "carcinogen" is not only unnecessary but misleading (Flamm and Hughes, 1997). The Interdisciplinary Panel on Carcinogenicity (1984) declared that "the carcinogenicity of a substance in animals is established when administration in adequately designed and conducted experiments results in an increase in the incidence of one or more types of malignant (or, where appropriate, a combination of benign and malignant) neoplasms in treated animals as compared to untreated animals maintained under identical conditions except for exposure to the test compound." NTP (1998), NRC/NAS (1993), and OTA (1987) are examples of reports that endorse categories of carcinogens. The key, of course, is to recognize that not all carcinogens are equal with respect to potential human cancer hazards.

VIII. EXPERIMENTAL RELEVANCY OF ANIMAL CANCER DATA TO HUMAN CANCER HAZARDS PREDICTION

Throughout the history of chemical carcinogenesis, experimental animal bioassay results have been viewed both as being beneficial to public health for identifying chemical carcinogens and as predicting carcinogenic risks to humans. Based on mammalogy and evolution, as well as on pharmacological extrapolation from animals to humans, this validated practice of using scientific knowledge gained from one mammalian species (e.g., laboratory animals) to predict effects in another mammalian species (i.e., humans) remains the cornerstone of medical practice and scientific research. Similarly, animal toxicologic data—including carcinogenesis—have historically been interpreted as being the primary means of securing and protecting human health from both toxic and carcinogenic effects. Experimental animal data have recently been instrumental in establishing occupational workplace standards of exposures as well as in planning strategies to clean up contaminated areas of the environment and to reduce the overall pollution of the global environment (Lubchenco, 1998).

Over the last 10–15 years, in particular, arguments have been raised that challenge this belief. Much of this debate centers on science and posed mechanisms of carcinogenic activity; also, however, considerable discourse seems to originate from self-interest, personal or corporate advantage, and economics. In 1980, Peto insisted that "historically, the most powerful pressure groups in U.S. society have been large financial interests, which have almost always put financial advantage before human health." Peto continues, "Where other industries (besides tobacco) have been found to cause cancer (or dust-induced diseases) in their workers or in the consumers of their products, their immediate response has usually been to delay acceptance of the findings, to minimize their relevance to current practice, and in general to delay or obstruct any hygienic measures which will cost money." He emphasized that the "politics of cancer are becoming increasingly polarized with the environmentalist taking one extreme view and industrialists the other." This is unfortunate, largely because the major issue is health, and health policies are being thwarted by endless delays and bickerings. Fagin and Lavelle (1996) pointedly detail how, in their opinion, "the chemical industry manipulates science, bends the law, and endangers your health."

One corporate executive spoke out recently about this issue of denial (Sells, 1994):

> As a manager with Johns-Manville . . . for more than 30 years, I witnessed one of the most colossal corporate blunders of the twentieth century. This blunder was not about the manufacture and sale of a dangerous product. Manville's blunder was not even its frequently cited failures to warn workers and customers of what it knew to be the dangers of asbestos during the 1940s, when so much of the damage to workers' health was done. The blunder was denial. What took thousands of lives and destroyed an industry was management's failure to insist on its own responsibilities.

Too often these divulgences are not made until someone leaves an industry, as with recent revelations from scientists formerly in the tobacco industry who have come forth with long-concealed information on the harm caused by tobacco smoke.

According to Levin (1992), "the question remains whether we can learn to use scientific information more expeditiously to prevent disease and protect the public's health or whether, for each hazard, we must await accumulation of a sufficient weight of data to crush attempts at delay, concealment, and deceit." This is a harsh statement, admittedly, yet it is sometimes necessary to take such a critical stance. Witness the 10-year delay (from 1977 to 1987) in reducing occupational exposure to benzene (OSHA, 1987; Huff, 1992a). Millions of dollars were spent on this delay by both industry and the government. Millions more, however, were made by industry while negotiations for reduced standards took place over that 10-year period; meanwhile, people continued to be exposed to unnecessarily high levels of benzene.

The quest and reality of hazard-identification efforts—and cancer research in general—are cancer cessation, prevention, or reduction. These goals can be met to some extent by reducing direct (occupational and consumer products) and indirect (environmental, food contaminants, and nondisclosure) exposures to chemicals that cause cancer and other diseases. Current consensus information indicates that the incidence and mortality rates of cancer continue to rise (Bailar and Smith, 1986; Miller et al., 1992, 1993; Ries et al., 1994; ACS, 1995; Wingo et al., 1995; Huff et al., 1996; Kosary et al., 1995; Bailar and Gornik, 1997; Parker et al., 1997). The good news is that between 1991 and 1995, the national cancer death rate fell 2.6%. Most of the decline can be attributed to decreases in mortality from cancers of the lung, colon–rectum, and prostate in men, and the breast, colon–rectum, and gynecological sites in women (ACS, 1998). Perhaps more reliance on animal-study results, overall weight of evidence, and mechanisms of carcinogenesis will lead to better protection from carcinogenic agents and help strengthen our efforts at prevention. However, as Wynder (1994) reminds us, "Students of the history of medicine have long known that diseases can be prevented or effectively treated long before causative mechanisms and therapeutic activity are understood." Unfortunately, what we hear from vested interests is that we must wait for more mechanistic information before we declare a particular carcinogen to be carcinogenic, the argument being that we should not make any hasty decisions regarding carcinogenic potential based only on animal data without mechanistic information or epidemiological findings. This delay tactic is replete with self-serving interests and for the most part should be uniformly ignored. Of course, when new relevant findings do become available, then new decisions can be made.

Identified risk factors for human cancer include age, dietary fat and diet (calories?) per se, genetic susceptibility or inheritance, education level and socioeconomic factors, hormonal perturbations, environmental and commercial chemicals, occupations, workplace exposure circumstances, lifestyle and cultural habits, solar and ionizing radiations, and biological agents. In many cases, resultant cancers are thought to involve combinations of these risk situations. Unfortunately, most cancers have not been associated unequivocally with any etiological factor.

IX. METHODS FOR IDENTIFYING CARCINOGENS

Methods of choice for identifying carcinogens center on epidemiological findings and experimental research and testing results. For each there are advantages and limitations (Tables 1 and 2; Huff and Hoel, 1992). Identified obstacles to the use of these two methods are being pursued, and, for some obstacles, the issues have been overcome or explained; nonetheless, the task remains clouded with difficulties, uncertainties, ambiguities, and a multitude of opinions. Perhaps the major point of contention, and where others originate as well, hinges on whether findings from experimental carcinogenesis studies relate or predict similar qualitative effects in humans. This is, of course, the decided basis for doing chemical-exposure studies in laboratory animals in the first place.

This historical and time-accepted premise comes from the obvious biological fact that mammals are mammals, and thus there are more similarities than differences among members of the class. Since laboratory animals are mammals and humans are mammals, one can assume with reasonable certainty that findings from one species (e.g., rats or mice) can be used to predict findings in the other (e.g., humans). Often quantitative differences are discovered—sometimes qualitative ones as well—but rarely are major incongruities found between mammalian species. However, because inconsistencies are more exciting than similarities, the former seem more likely to be published, resulting in exaggerated attention being given to the stated disparities. Nonetheless, even when species differences have been promoted as reasons for not using animal data to protect human health, one learns that an agent previously known to be carcinogenic to animals is later found to cause cancer in humans as well, e.g., 4-aminobiphenyl, 1,3-butadiene, trichloroethylene, solar radiation, and vinyl chloride (Huff, 1993, IARC, 1997).

Much of the debate about animal-to-human relevancy seems to have attained momentum over the last decade or two, at the maturity of chemical carcinogenesis bioassays. Not so much clamor was raised while model chemicals or laboratory curiosities were being tested and declared carcinogenic. But when large-volume commerce chemicals were identified as being carcinogenic to laboratory animals and hence likely to be carcinogenic to humans as well, those in industry and others mounted intense efforts to discredit the carcinogenesis testing paradigm, or to conjure pointed exceptions and objections to the rodent-to-human connection. Interestingly, this took place only when the chemical showed carcinogenic activity, never when the chemical exhibited no evidence of carcinogenicity. The basic argument at that time, and even today, centers on the declaration that "such-and-such a chemical has been used successfully for *x* numbers of years and no epidemics of cancers have been found among workers." Fascinatingly, when this statement was made in public sessions, almost without exception, no epidemiological studies had in fact been done or were even planned for the chemical in jeopardy.

Unfortunately, we now know that an all too sizable number of agents have been shown to cause cancer in humans *after* being first discovered to cause cancer in laboratory animals (Table 16) (Tomatis, 1979; Huff, 1993; Vainio et al., 1995; IARC, 1997). This may or may not be easily avoidable, since humans may be exposed to many agents before any testing or epidemiological studies have been initiated or completed. In fact, for most—if not all—industrial chemical-exposure circumstances, bioassays are typically done after many years of production and use. Drugs or pharmaceuticals intended for use by humans are usually an exception in that the newer drugs are generally tested during development; older drugs and over-the-counter medications are generally exempt from such requirements.

One of the latest examples is 1,3-butadiene. We first discovered this to be a particularly potent carcinogen in the early 1980s (Huff et al., 1985), and subsequently refined the dose-response-tumor relationship (Melnick et al., 1990a; Melnick and Huff, 1992, 1993). Because we believed that the

Table 16 Chemicals and Cancer in Humans: Evidence of Carcinogenicity First Observed in Experimental Animals and Subsequently by Epidemiological Evidence: Chemicals Causally or Probably Associated with Cancer in Humans

1. Acrylonitrile
2. Aflatoxins
3. 4-Aminobiphenyl
4. Analgesic mixtures with phenacetin
5. Asbestos
6. Azathioprine
7. Betel quid with tobacco
8. Beryllium and beryllium compounds
9. bis(Chloromethyl)ether and chloromethylmethylether (tech)
10. 1,3-Butadiene
11. Cadmium and cadmium compounds
12. Chlorambucil
13. Chlorinated water and by-products
14. Chlornaphazine
15. Cyclosporin
16. Coal-tar pitches
17. Coal-tars
18. Cyclophosphamide
19. DDT and related compounds
20. Diethylstillbestrol
21. Dibromoethane
22. Estrogens, nonsteroidal
23. Estrogen replacement therapy/steroid estrogens
24. Ethyl acrylate
25. Ethylene oxide
26. Formaldehyde
27. Gasoline
28. Glass wool
29. Lead and lead compounds
30. Melphalan
31. 8-Methoxysoralen plus UVA radiation
32. 4,4′-Methylene bis(2-chloro-aniline)
33. 4,4′-Methylene dianiline diHCI
34. Mustard gas
35. Myleran
36. 2-Naphthylamine
37. Ochratoxin A
38. Oral contraceptives, combined
39. Phenacetin
40. Radon gas
41. Silica, crystalline

(*continued*)

Table 16 Continued

42.	Solar radiation
43.	Tetrachlorodibenzo-p-dioxin
44.	Thiotepa
45.	Trichloroethylene
46.	Vinyl chloride

Sources: Huff, 1993; IARC, 1997.

"current workplace standards for exposure to butadiene should be reexamined in view of these findings" (Huff et al., 1985), we convened an international conference with the affected union and rubber industry, together with regulators and scientists engaged in broad-based butadiene research (Melnick et al., 1990; EHP, 1990). At the time of our initial findings, the OSHA occupational exposure standard was 1000 ppm; more than 10 years later it remained at 1000 ppm, 160 times greater than the lowest exposure level tested of 6.25 ppm, which was found to be carcinogenic. No exposure concentration of 1,3-butadiene has yet been studied that is not carcinogenic to animals.

For more then 10 years, OSHA has attempted to lower the occupational standard for 1,3-butadiene to 2 ppm, but these regulatory efforts have been consistently delayed or blocked by industrial interests. Even before the conference, evidence was available that workplace exposures to 1,3-butadiene were associated with human cancers. Since these early experimental (Huff et al., 1985) and first epidemiological findings (Mantanoski and Schwartz, 1987; Santos-Burgoa, 1988), unequivocal evidence now exists that 1,3-butadiene causes cancers of the hematopoietic and lymphatic systems in humans (Mantanoski et al., 1990, 1993). One would have hoped that the OSHA-proposed exposure rules for 1,3-butadiene, now set officially at 1 ppm, had been promulgated more expeditiously (OSHA, 1996): 11 years elapsed between the animal findings and reducing the OSHA standard from 1000 to 1 ppm (8 hours time-weighted average; 5 ppm for 15-minute excursions).

A. Optimal Exposure Levels for Carcinogenesis Experiments: Concepts, Principles, Guidelines, and Experience

Tables 17–23 review the dose-selection issues. Brief comments on some of the major concerns are offered below.

Table 17 The Dose-Selection Issue

1. Among the most important—and most argued about—issues surrounding the entire area of long-term chemical carcinogenesis studies are the scientific criteria behind the selection of exposure concentrations and the toxicological philosophy.
2. A key point in doing toxicology studies (or experiments in any other discipline) is that some chemical-associated effects must be observed; otherwise the investigation will be considered a waste of time and scientific resources and therefore of questionable value.
3. A major challenge in designing long-term (i.e., for a duration of 2 years) toxicology experiments is setting the exposure levels "just right," to allow a reasonably normal health status (appearance, body weight, etc.) for the animals while "guaranteeing" obvious evidence of chronic toxicity over and above that seen typically in aged animals.
4. Pragmatically, this objective is exceedingly difficult to attain entirely, especially if one considers that the selection of exposure concentrations is based on results obtained from relatively short exposure periods (ordinarily, 13-week experiments) in young, robust animals.
5. Rarely in toxicological practice does one "hit" precisely on the optimal exposure level; and routinely this awareness remains unknown until the studies are fully completed.
6. This retrospective issue becomes particularly important from a hazard-identification point of view, if in a positive carcinogenicity study the exposure levels are considered extremely excessive for maintaining proper health and longevity of the animals and carcinogenic effects are observed. Even in this case, however, one should not discount a positive carcinogenic result without repeating the experiments with different protocols, or developing alternative explanations with strong scientific support. To do otherwise would not be in the best interests of public health. Of course, a no-evidence-of-carcinogenicity result in a scenario of low body-weight gain and poor survival, or when the exposure levels had minimal or no observable effect (exposures too low), would probably be considered an inadequate experiment that might have to be repeated.
7. Any conclusion of a study, whether adequate or indadequate or whether the carcinogenic results are positive or negative, depends on the overall completeness and strength of the experimental evidence; these underlying provisions must be stated clearly to support any conclusions.

Most national and international research, testing, regulatory, and chemical standards-setting organizations endorse the underlying concept of exposure-level selection based on the prediction that some chemical-associated toxic responses should be seen in the long-term carcinogenicity studies. These organizations include:

Table 18 Selection of Exposure Concentrations for Long-Term Chemical Carcinogenesis Studies: Background, Data, Facts, and Conclusions

1. Since 1975, the National Cancer Institute and, subsequently, the National Toxicology Program have designed, conducted, and evaluated nearly 500 chemical carcinogenesis studies.
2. Protocols typically include both sexes of two species of rodents (rats and mice), with concurrent controls and two to three exposure groups per sex/species group, and exposures for 2 years.
3. Using 90-day experimental results, top exposure levels are chosen *prospectively* to elicit some toxicological responses.
4. This upper exposure level is the minimally toxic exposure (MTE) level, and should be predicted not to adversely affect "normal" well-being or longevity.
5. Selecting optimal MTE levels for identifying potential carcinogens involves a gamut of multidisciplinary factors utilizing a holistic team approach.
6. Key *MTE* selection criteria should include
   Chemically related clinical signs
   Water and food consumption
   Organ and body weights
   Hematology and urinalyses
   Pharmacological and toxicological effects
   Dose-response findings
   Metabolism, disposition, and excretion of the chemical
   Histopathology
   Biochemical and molecular events
   Morbidity and mortality
7. Often overlooked is the concept that these MTE levels must be chosen *in advance* based on shorter-term results, and, importantly, one does not know until the end of the long-term experiments whether optimal, adequate, and appropriate exposure concentrations for protecting public health were in fact achieved.
8. Nearly one-half of the chemicals studied were associated with a carcinogenic response in at least one organ of one sex of one species.
9. In perspective, however:
   14% caused cancer in *each* of the four sex/species experimental groups
   8% caused cancer in *three of four* experimental groups
   18% caused cancer in *two of four* experimental groups
   12% caused cancer in *one of four* experimental groups
   48% caused cancer in *none of four* experimental groups
10. Significantly, for 94% of chemicals, carcinogenic responses were observed at two of three exposure levels, with just 6% of chemicals being carcinogenic at the MTE "top dose only."
11. Thus, based on available biological and toxicological data, use of MTE levels for identifying potential human carcinogens is valid and scientifically sound, and should continue.
12. Further, the term *maximum tolerated dose* (MTD) should no longer be used because it is scientifically incorrect and misleading.

Table 19 Dose-Selection Highlights and Comments

1. Most chemical carcinogens do not cause cancer only at the highest exposures used.
2. For the few chemicals that do, there are no scientific reasons for these to be considered any less relevant for human hazard identification.
3. Most critics of the exposure selection criteria do not seem to fully realize the basic issue: long-term carcinogenesis studies are carried out to identify long-term toxicological effects, including carcinogenic potential, and therefore the exposure levels chosen must elicit some measurable biological effects.
4. Based mainly on short-term experiments, exposure levels are chosen primarily so as not to compromise reasonable health and longevity and to show real toxicological indications of long-term exposure.
5. Otherwise—if no chemical-effects were observed—the entire paradigm of sufficient chemical exposure levels in toxicological research and testing would collapse. This would be a waste of laboratory and financial resources, time, and scientific expertise.
6. Most prospective "dose" selections, and subsequently most completed experiments, do not indicate excessive exposure levels or come close to "exceeding" the toxicological concept of MTE.
7. Relatively few individuals dealing with these complicated and controversial issues have been involved intimately with:
   a. Nominating and selecting chemicals for evaluation
   b. Designing, conducting, and evaluating these important and complex studies
   c. Interpreting and presenting the findings and conclusions in public peer review sessions
   d. Publishing the results in scientific journals

ATSDR (Agency for Toxic Substance Disease Registry)
The Collegium Ramazzini
EPA (Environmental Protection Agency)
FDA (Food and Drug Administration)
IARC (International Agency for Research on Cancer)
NCI (National Cancer Institute)
NIEHS (National Institute of Environmental Health Sciences)
NIOSH (National Institute of Occupational Safety and Health)
NRC/NAS (National Research Council/National Academy of Sciences)
NTP (National Toxicology Program)
OSHA (Occupational Safety and Health Administration)
OSTP (Office of Science and Technology Policy)
OTA (Office of Technology Assessment)

Table 20 Uncertainties and Definition of Experimental MTE

1. Perhaps the most important, and surely one of the most debatable, issues in identifying chemicals considered likely to be harmful to humans centers on the scientific and pragmatic rationale for selecting exposure levels (e.g., "dose").
2. Part of the argument and misunderstanding comes from the apparent discrepancies of the various definitions used to describe the dose-selection process or, in particular, the "top dose."
3. Another major conundrum arises from what seems to simply be a lack of acute and detailed knowledge and awareness of the multidisciplinary approach used to actually choose long-term exposures.
4. Rarely in current practice does one select and utilize exposures that could possibly jeopardize the long-term experiments.
5. Thus, a conservative and safe tendency has evolved in dose selection, often using exposure levels below historical precedents.
6. One of the earliest and most thoughtful contributions—and perhaps the origin of much current misunderstanding—defines the maximum tolerated dose (MTD) as follows (Sontag et al., 1976; emphasis added):
   a. The highest dose of the test agent given during the chronic study that can be *predicted not to alter* the animals' normal longevity from effects other than carcinogenicity.
   b. The MTD is *estimated* after a review of the subchronic data. Since these data may not always be easily interpretable, a *degree of judgment* is often necessary in *estimating* the MTD.
   c. The MTD should be the highest dose that causes no more than a *10% weight decrement* . . . (although a depressed weight gain is a clinical sign of toxicity, this particular effect is acceptable when estimating the MTD) . . . as compared to the appropriate control groups . . . and *does not produce* mortality, clinical signs of toxicity, or pathological lesions (*other than* those that may be related to a neoplastic response) that would be *predicted* to shorten the animal's natural life span.
   d. Other measurements (e.g., organ function, body burden, absorption and excretion) also may be used to aid in *predicting* the MTD.

NRC/NAS (1993), for example, recommends that the maximum tolderated dose (MTD) "continue to be one of the doses used in carcinogenicity bioassays. Other doses, ranging downward from MTD/2 possibly to MTD/10 or less, should also be used. The capacity of the test animal to absorb and metabolize the test chemical should be taken into account in selection of doses below the MTD."

Public health prudence, cancer-preventive strategies, and good sense should demand no less when attempting to identify potential carcinogenic hazards to humans. The eventual risk-assessment and -management pro-

Table 21 Utilization and Endorsement of the Use of Minimally Toxic Exposure Levels in Long-Term Carcinogenicity Bioassays

1. Most national and international research, testing, regulatory, and chemical standards-setting organizations endorse the underlying concept of exposure-level selection based on the basic toxicological edict that some chemical-associated toxic responses should be seen in the long-term carcinogenicity studies:
   ATSDR (Agency for Toxic Substances and Disease Registry)
   EPA (Environmental Protection Agency)
   FDA (Food and Drug Administration)
   IARC (International Agency for Research on Cancer)
   NCI (National Cancer Institute)
   NIEHS (National Institute of Environmental Health Sciences)
   NIOSH (National Institute of Occupational Safety and Health)
   NRC/NAS (National Research Council/National Academy of Sciences)
   NTP (National Toxicology Program; participating members include the research and regulatory agencies within the Department of Health and Humans Services)
   OSHA (Occupational Safety and Health Administration)
   Research and regulatory agencies in other countries
   As an example, the NRC/NAS (1993) recommends that "the MTD should continue to be one of the doses used in carcinogenicity bioassays. Other doses, ranging downward from MTD/2 possibly to MTD/10 or less, should also be used. The capacity of the test animal to absorb and metabolize the test chemical should be taken into account in selection of doses below the MTD."
2. Public health prudence, preventive strategies, and good sense require use of the MTD when attempting to identify potential carcinogenic hazards to humans.
3. This initial hazard-identification step is not the appropriate stage or forum in which to debate and argue the societal benefits and political realities versus safe levels of human exposure and possible risks of cancer. The proper place for these debates is at the risk-assessment and -management phases of the process.
4. Meanwhile, research on mechanisms of carcinogenic activity should proceed to investigate, advance, and shed further biological knowledge on chemical carcinogenesis processes that will be helpful in continuing to protect and promote better health for workers, the environment, and the general population.

cesses—not the hazard-identification step in the research and testing phases—are the appropriate places in which to debate and argue the societal benefits and political realities versus levels of human exposure and possible risks. Meanwhile, research on mechanisms of carcinogenic activity should proceed in order to shed further biological information on the chemical-carcinogenesis processes that will be helpful in continuing to protect and

Table 22 Do Chemicals Cause Cancer *Only* at the MTD?: Experimental Results I

1. To investigate the claim that "many chemicals cause cancer in animals only at the MTD," we evaluated results of 99 chemicals tested for carcinogenic potential:
   47 chemicals did not cause any carcinogenic effect
   52 chemicals induced a carcinogenic response in at least one organ of one sex of one species. Of these 52:
      34 produced confirming, statistically significant carcinogenic effects at lower exposure levels
      18 were viewed as being "statistically" carcinogenic only at the "highest exposure level," importantly, and often ignored. (15 of these 18 showed biologically, but not statistically, elevated tumor incidences at the lower exposures in the same organs. In most cases, the dose-response trend was statistically significant.)
2. Thus, in fact, only three chemicals of this 99-chemical dataset (3%) showed carcinogenic responses "restricted" to the MTD, unsupported by increased tumor incidences at lower doses. Further, two of the three chemicals were single-organ, single-sex carcinogens, and the third chemical induced the same single-organ responses (nasal tumors) in each of the four sex/species experiments.
3. Therefore, only one of the 99 chemicals fits the sufficient-evidence-of-carcinogenicity category based on the carcinogenic responses seen singularly at the MTD, while not being supported statistically or biologically by any evidence of carcinogenicity in the lower-exposure groups.

Source: Hoel et al., 1988.

promote better health for workers, the environment, and the general population.

Selection criteria for selecting appropriate dose levels or exposure concentrations for long-term chemical carcinogenesis experiments are reasonably straightforward (Huff et al., 1991; Huff et al., 1994; Toth, 1997; Bucher et al., 1996). Short-term toxicology experiments are done to establish concentrations that cause or induce minimal toxic effects, and exposure levels are chosen that are predicted to not cause any debilitating effects over the course of the long-term experiments, other than those due to cancer and associated effects. The basic tenet in selecting and using optimal exposures in long-term studies is to protect the public by making as certain as feasible that the study results are obtained using the most sensitive experimental conditions and that these results are the most reliable for predicting either human health hazards or lack of any hazard. Thus, some toxic effect must be seen to help ensure that any observed noncarcinogenicity will be valid. Therefore, the highest exposures that can be used without compromising reasonable health and longevity of the experimental animals should be used

Table 23 Do Chemicals Cause Cancer *Only* at the MTD?: Experimental Results II

1. Expanding our retrospective research survey to update, expand, and better clarify this MTD use and clarification issue, we evaluated 450 long-term chemical carcinogenesis studies for unique carcinogenesis at the "highest dose used":
   234 chemicals did not cause any carcinogenic effect
   Of the 216 chemicals considered to be carcinogenic in at least one of the four sex/species experiments:
   - Only 13 (6%) produced increased site-specific tumor rates limited to the top exposure concentration.
   - Additionally, most of the affected organs were uncommon tumor sites, and the carcinogenic responses at the top exposure levels were unequivocal positive carcinogenic effects.
   - Of these 13 chemicals
     - Four chemicals caused uncommon kidney tumors
     - Three chemicals caused uncommon urinary bladder tumors
     - Two chemicals caused both nasal cavity (uncommon) and liver tumors
     - One chemical each caused tumors either in spleen, lung, or thyroid gland, or leukemia (total of four chemicals)
2. These findings should further dispel the interpretation and belief that the majority of chemicals tested in long-term bioassays are "high-dose-only" carcinogens. These data do not support this notion.

to guarantee the maximum reliability of regulatory safety and health decisions.

Arguments regarding "dose selection" typically surface with respect to the often changing or different viewpoints on the concepts underlying the rationale for choosing and following established guidelines in dose selection. Most of these arguments tend to revolve around the notion that the relatively high exposures used in these experiments are in themselves carcinogenic, and that the human population is only rarely if ever exposed environmentally to such concentrations. Hence, and to strengthen the claim that exposures are too high, one hears about theoretical mechanistic considerations: metabolic overload, pharmacokinetic plateau, altered distribution, metabolism, excretion pathways, and high-dose toxicity (e.g., inflammation, regenerative cell proliferation). These mechanistic considerations are claimed to lead to carcinogenicity. However, these claims have yet to be validated.

Most of the available evidence does not support the contention that "toxicity" or cell proliferation per se leads to or provokes the carcinogenesis processes (Hoel et al., 1988; Huff et al., 1991; Melnick, 1992; Zito, 1992; Weinstein, 1992; Huff, 1993; Ward et al., 1993; Huff, 1995 a,b; Farber,

1995, 1996). That is not to imply that cell proliferation is not important in the carcinogenesis process; obviously without cell proliferation cancer would not exist. The point is that chemicals that induce cell proliferation are not necessarily carcinogens. In fact, as reported by Haddow (1938), "there are at least two groups of data that question the validity of the cell proliferation speculation: 1) a negative association between cell proliferation and cancer in several organs and tissues and 2) the well-known observations that the majority of chemical carcinogens are inhibitors, not stimulators, of cell proliferation" (quoted in Farber, 1976, 1996).

Historically, individuals and committees have debated the use and usefulness of the MTD; this term, although useful shorthand, is inaccurate and has led to misunderstanding or strained debates. The main differing views regarding the MTD concept have been set forth by Toth (1997) and Foran (1997). More meaningful and descriptive terms have been suggested. *Minimally toxic exposure* (MTE) (Huff et al., 1991), for example, is more accurate because the shorter-term experiments are designed to elicit a spectrum of toxic responses. It is this degree or totality of toxic effects that establishes the base for selecting exposure levels for the longer-term experiments. Toth (1997) supports the use of *estimated maximum tolerated dose* (EMTD), simply because one can never be sure that the top exposure level is indeed optimal until the 2-year experiments are well underway or even have been completed. Although this term is certainly more accurate and relevant than MTD, it does not convey the actual conditions as MTE does.

In any event, the underlying principles of exposure selection need to be defined for clarity. In fact, the definition of dose selection developed in 1976 by the National Cancer Institute (Sontag et al., 1976; Table 20) can be still used with impunity today. However, one must read the entire definition, which has often been misquoted with respect to the actual parameters of importance. Individuals opposed to the concept of animal cancer testing, and in particular to the use of the MTD, typically cite only body-weight alterations and chemically related mortality. Interestingly, some individuals would even argue that a study should be discounted if the body-weight differences are only 10.1%, or 0.1% over what has become a more strict interpretation of the original intention of using body-weight differences between the control and exposed groups. This is obviously a slight exaggeration, but some persons have in fact urged us to discount long-term animal studies if the body-weight differences exceed 10%. Further, the use of differences in body-weight *gain* should be replaced with total body-weight comparisons; and differences of up to 15 to 20% should not be used arbitrarily to negate a valid study. Of course, the totality of the experimental findings must be considered and evaluated before a judgment is rendered on a single parameter. Further, lighter and less obese animals are more healthy, live longer, and in fact have fewer tumors than do heavier or obese control groups, which simplifies evaluations and conclusions.

In many instances, both body-weight decreases and mortality increases in chemically exposed groups are due to induced carcinogenic effects. This fact is frequently overlooked or ignored. An interesting phenomenon arises, with a "negative" study having significantly lowered body weights and perhaps increased mortality. Curiously, few "negative" experimental results have been criticized as being flawed and thus irrelevant. In fact, only rarely does a "negative" study ever get questioned as being a "false negative," while "positive" studies are often construed as being "false positives." From a public health view, one should be more comfortable with false-positive findings than with false-negative results.

Those who insist that any nonneoplastic toxic response leads to or confounds any eventual carcinogenic response need to be reminded that these long-term experiments are in fact toxicology studies, and thus toxic effects are essential, with one of these endpoints being cancer. Hence, to use exposure levels such as those often seen environmentally would be wasteful of time, resources, and professional efforts, because invariably no toxic or carcinogenic effects would be observed given the limitations of animal experiments. Conversely, one needs to evaluate for toxic and carcinogenic activity combinations of the many agents released into the environment.

B. New Arguments Devised to Discount Experimental Findings

A "mechanistic trend" seems to be developing to discount or modulate certain chemically induced carcinogenic responses in laboratory animals as being "rodent-specific" and thus not relevant to humans. This growing din is premature at best, and, at worse, antithetical to public health (Weinsten, 1996):

> In the euphoria surrounding recent advances in the molecular biology of cancer, we must remain sober about the limitations in translating our knowledge and accomplishments into more effective strategies for cancer prevention and treatment. We still lack hard facts about the exogenous factors that cause cancers of the breast, prostate and colon, even though today they account for about half of new cancer cases in the United States and Western Europe and their incidence is increasing in other countries. Although dietary factors have been implicated, experts in the field disagree on the precise components and the underlying mechanisms.
>
> Within the mechanistic framework of oncogenes and tumour-

> suppressor genes, the multistep evolution of cancer cells seems to be much more complex than was originally thought when the first oncogenes were discovered. We now know that a given tumour often carries several mutated genes as well as gross chromosomal abnormalities, that more than a hundred different genes have been found to be mutated and/or abnormally expressed in various types of human cancer and that each year this list gets longer.

So far, the list of agent-induced tumors with their assigned "rodent-specific mechanism" that some consider irrelevant to humans include the following (Omen et al., 1997; IARC, 1998).

1. Male rat kidney tumors if or when associated with alpha-2u-globulin induced nephropathy. The evidence for this speculation is sketchy and inconsistent, with many exceptions to this posed mechanism (Huff, 1992, 1996). Alpha-2u-globulin is more likely a carrier molecule for a particular chemical, or even a biomarker of cellular damage or exposure (Huff, 1992b, 1993b; Melnick, 1992a, 1993), and perhaps a coincidental and simultaneous occurrence (Melnick et al., 1997; Melnick and Kohl, 1998). At least one agent—gasoline—known to induce alpha-2u-globulin in rodents has also been strongly associated with kidney cancer in humans (Lynge et al., 1997).

2. Forestomach tumors with local hyperplasia. The argument centers on the fact that humans do not have a forestomach. Humans, in common with other nonrodent mammals, do have an esophagus of the same cell type as the rodent forestomach. The squamous epithelium at the squamocolumnar junction of the cardiac portion of the stomach can be a site for either squamous cell carcinoma or adenocarcinoma (NTP, 1986; Huff, 1992a). Further, the debate continues, it is the local—rather than the diffuse—hyperplasia that is due to chemical irritation, leads to preneoplasia and then to neoplasia; and thus the lesions are not relevant to humans. This speculation is unsupported by the overwhelming available information (Hoel et al., 1988; Huff, 1992). At least two known human carcinogens induce tumors of the forestomach in rodents: benzene and vinyl chloride.

3. Male rat urinary bladder tumors with cytotoxicity and reactive hyperplasia from precipitated chemicals. Again this is not a mechanism, simply a collection of observations. Animal studies have repeatedly shown that stones often exist with or without tumors. Sometimes stones are present in both sexes and tumors occur only in one sex, or calculi are found in two species and tumors are seen only in one species. Thus, many exceptions exist, and no actual mechanistic studies have proved this observational correlation. Moreover, data in humans link the occurrences of stones or precipitates with cancers of the urinary bladder (Chow et al., 1997).

4. Lung tumors with overwhelmed pulmonary clearance mecha-

nisms. As with the other tumors, there is little scientific evidence that lung-burden overload has any relationship to lung tumors. In a study exploring this pulmonary-overload hypothesis, inhaled talc caused pheochromocytomas of the adrenal gland in male and female rats, tumors of the lung in female rats, and no carcinogenic activity in male or female mice (Abdo, 1993). The principal toxic lesions associated with inhalation exposure to the same concentrations of talc in rats included chronic granulomatous inflammation, alveolar epithelial hyperplasia, squamous metaplasia and squamous cysts, and interstitial fibrosis of the lung. These lesions were accompanied by impaired pulmonary function. Interestingly, there is some evidence in humans that talc is associated with pulmonary fibrosis, lung cancer, and ovarian neoplasms (Thomas and Stewart, 1987; Thomas, 1990; Harlow and Weiss, 1989; NTP, 1993). Obviously each of the human effects may be considered associated with local exposures.

5. Thyroid gland tumors with sustained excessive hormonal stimulation. This situation seems to stem from the notion that goitrogenic chemicals will inevitably cause tumors as well. In certain cases, this is applicable and in others it is not. For example, salicylazosulfapyridine (SASP) is widely used for the treatment of ulcerative colitis and Crohn's disease and has been beneficial in the treatment of psoriasis and rheumatoid arthritis. SASP was nominated for testing because of its widespread use in humans and because it is a representative chemical from a class of aryl sulfonamides. It was considered a suspect carcinogen because reductive cleavage of the azo linkage yields a p-amino aryl sulfonamide (sulfapyridine), and a related p-amino aryl sulfonamide (sulfamethoxazole) has been shown to produce thyroid neoplasms in rats. In 16-day studies, SASP caused hypothyroidism, evinced by decreased serum triiodothyronine and thyroxine concentrations, increased concentrations of thyroid-stimulating hormone (TSH), enlarged thyroid glands, thyroid-gland follicular cell hyperplasia, and an increase in TSH-producing cells in the pars distalis of the pituitary gland in male and female rats. In 13-week SASP studies, male and female rats had minimal but consistent changes in thyroid-gland follicular cells, and had red, enlarged thyroid glands (at about 1/10 the dose in the 16-day studies). Also, decreased serum triiodothyronine and thyroxine concentrations and increased TSH concentration, similar to differences observed in the 16-day study, occurred in male rats.

Despite these "predictive" findings in the shorter-term SASP studies that would indicate the thyroid as a target organ, thyroid-gland hyperplasia seen in the 13-week study was not observed in the 2-year study, and there was no evidence of chemically related thyroid-gland follicular cell adenomas or carcinomas. However, in the 2-year gavage studies, there was carcinogenic activity of SASP in male and female rats based on increased incidences of neoplasms in the urinary tract: transitional epithelial papilloma of the urinary bladder in males and uncommon transitional epithelial papillomas of the kidney and the urinary bladder in females. In male and female mice, there were increased incidences of hepatocellular neoplasms.

The point here is that the rationale for discounting the thyroid gland as a useful target organ—"sustained excessive hormonal stimulation"—did not result in the predicted outcome (i.e., cancer of the thyroid glands). Thus, one needs to remain careful when either predicting certain carcinogenic effects or discounting a priori the value of certain carcinogenic effects observed in animals.

First, these tumors and the posed tumor "mechanisms" are not mechanisms in the classic sense because they are simply histological observations made after a bioassay has been completed. That is, if a male rat has a tumor of the urinary bladder, and that same bladder contains calculi or a stone, then a claim is made that there is an association. Likewise for the other four "mechanisms" mentioned above. These limited observations do not represent scientific proof (Huff, 1992, 1993a, 1995). Most, if not all, are basically phenomenological or observational modalities, not mechanisms in the true molecular sense. Besides, human tumors are also preceded by local or diffuse hyperplasia, impaired airways, hormonal disruption, or many other concomitant organ disturbances.

Second, these associations have rarely if ever been proven experimentally. Again, for example, there are select data to show that male rats with kidney tumors induced by certain chemicals also have alpha-2u-globulin-associated nephropathy, and those with tumors of the urinary bladder have calculi. However, there are chemicals that also induce this same nephropathy that do not cause male rat kidney tumors, and there are chemicals that cause stone formation that do not result in tumors of the urinary bladder (Huff, 1996). How then does one "selectively" assign a mechanism, and claim that that mechanism is unique to rodents and thus irrelevant to humans? We obviously know that chemicals seem to exhibit different mechanisms for different tumor types, and chemicals induce different tumors in different species; therefore, why should we insist that a chemical causing tumors of the liver in rodents necessarily causes that same tumor in humans?

4-Aminobiphenyl causes tumors of the liver, mammary glands, and urinary bladder in rodents and tumors of the urinary bladder in humans. Yes, there is site concordance, but the first experimental results showed only liver tumors. Benzidine causes liver (and other) tumors in mice, rats, and hamsters, and tumors of the urinary bladder in humans; subsequently dogs were found to develop tumors of the urinary bladder from exposure to benzidine. The point is that we must not be too quick to discount certain tumor types simply because we believe they are or may be rodent-specific. For the human leukemogen benzene, the first tumors observed in rodents occurred in the Zymbal gland, an organ of which humans appear to have only a vestigial counterpart. A leukemia model for benzene has yet to be found, even though benzene causes lymphoma in mice. Thus, one needs to be careful when making public health decisions with limited or dogmatic information.

Third, these "mechanistic associations" are inconsistently allied with a cause-and-effect paradigm. In many long-term bioassays, animals frequently exhibit these "mechanisms" (be they local inflammation or regenerative hyperplasia and cell proliferation, goiter, urinary-bladder stones, or alpha-2u) and yet tumors of those same organs do not develop. Thus, without due consistency in cause and effect, one begins to embrace a mechanism that only works some of the time, and this only leads to skepticism and lack of confidence or acceptance.

Fourth, in almost all bioassays, these tumors or mechanisms are not the only carcinogenic responses induced by a particular chemical. Does this mean that chemicals elicit multiple mechanisms? Probably. Does this mean that tumor A is not relevant to humans but tumor B or C is or may be relevant? What would one decide if each of the above five tumor sites were induced by a single chemical? Are these effects considered not relevant to humans only if the chemicals are nongenotoxic? Clearly more scientific thought needs to be devoted to issues of discounting chemically induced tumors in animals as being irrelevant to humans.

Large collections of evaluated empirical data do not support a relationship between "mechanistic associations" and cancers (Hoel et al., 1986; Tennant et al., 1990; Huff 1992, 1993, 1996). That is not to say these "associative mechanisms" never or cannot influence a particular site-specific tumor pathogenesis by some cocarcinogenic consequence. However, before altering public health policy by using mechanistic evidence, the mechanism should meet exhaustive scientific scrutiny, and should be operative all the time (Huff, 1995). For example, if alpha-2u-globulin-induced nephropathy is observed in male rats, and this is posed as a mechanism for inducing kidney tumors, then tumors of the kidney should be induced by *all* chemicals that incite this nephrotic syndrome. This does not happen consistently, and thus I contend that alpha-2u-globulin-induced nephropathy cannot be considered a mechanism of tumorigenicity or as being irrelevant to human cancer risk (Huff, 1996; Melnick et al., 1997).

Similar inconsistent experimental findings are available for each of the other four posed rodent-specific mechanisms. These findings are discussed elsewhere (Hoel et al., 1988; Huff, 1992, 1993, 1996; Farber, 1995, 1996).

In contrast, mechanisms have been used to declare chemicals as being human carcinogens in the absence of overwhelming or scientifically convincing epidemiological evidence (e.g., ethylene oxide, tetrachlorodibenzo-p-dioxin, and trichloroethylene); that is, associations with cancer in humans have been made, but were not strong enough for that agent or exposure circumstance to be declared a human carcinogen based solely on epidemiological information. Using experimental data and mechanistic information, the International Agency for Research on Cancer has upgraded two chemicals from the category "probably carcinogenic to humans" (category 2A) to "carcinogenic to humans" (category 1) based on solid mechanistic and bioassay information (ethylene oxide, tetrachlorodibenzo-p-dioxin). Likewise, because benzidine is a "human carcinogen" and benzidine-based dyes are all carcinogenic to laboratory animals, are usually more potent carcinogens than the parent compound, and metabolically form benzidine in vivo, one could propose that all benzidine-based dyes should be upgraded to the category "carcinogenic to humans." Other, similar classes of chemicals would also fit this situation.

C. Key Findings from Long-Term Chemical Carcinogenesis Studies

Table 24 presents narrative findings gleaned from my experience and findings over the last two decades doing long-term chemical carcinogenesis studies. Individual comments are not specifically referenced, but most of the topics are covered in the references cited in the Bibliography. Many of the comments are simply statements of facts as we have come to realize them. Key findings that pertain to the value and validity of long-term bioassays for identifying potential carcinogenic risks to humans are listed in Table 24. These 27 factors are not listed in any particular order, and some are covered in the text. For others the reader is referred to the Bibliography. Publications by me are available upon request.

He who knows and knows he knows,
He is wise—follow him;
He who knows not and knows he knows not,
He is a child—teach him;
He who knows and knows not he knows,
He is asleep—wake him;
He who knows not and knows not he knows not,
He is a fool—shun him.

—Ancient Arabic proverb

Table 24 Select Key Findings and Conclusions from Long-Term Chemical Carcinogenicity Studies (listed randomly; see text and references for more details)

1. Carcinogenesis findings from studies in laboratory animals are scientifically reasonable for identifying potential carcinogenic hazards to humans.
2. Most chemicals are not considered potentially carcinogenic to humans.
3. Approximately 5 to 10% of all chemicals might be predicted to be potentially carcinogenic to humans.
4. Chemicals can be grouped by qualitative strength of evidence based on empirical indicators of "potency."
5. Malignant and benign tumors (same site or cell type) can be combined for evaluating and interpreting carcinogenicity.
6. Benign tumors are relevant for judging carcinogenicity; few chemicals induce only benign tumors.
7. Site-specific tumor analysis is useful to determine chemically induced carcinogenesis.
8. Liver neoplasia is valid for identifying potential cancer hazard to humans.
9. Rodent-specific organs are useful for identifying potential cancer risks to humans: Zymbal glands, forestomach, Harderian gland, preputial gland.
10. Chemical-induced cellular toxicity and chemical carcinogenesis are not reliably correlated.
11. Cellular proliferation per se does not cause cancer.
12. Chemical-induced cell replication per se cannot be used to predict carcinogenesis.
13. Most chemical carcinogens do not cause cancer only at the highest doses or exposure levels used.
14. Purported "mechanisms" are not yet understood well enough for generic utilization.
15. "Mechanisms of carcinogenesis" have not been sufficiently developed to discount carcinogenic effects observed in rodents from predicting potential cancer hazards to humans.
16. Route of exposure has little or no effect on the innate carcinogenic potential of chemicals.
17. Corn oil used in oral studies has little or no influence on chemical carcinogenesis.
18. Rodent interspecies concordance in carcinogenic response is good.
19. Known human carcinogens induce carcinogenesis in animals, with at least one concordant tumor site.
20. Concordance between in vitro genetic toxicity and in vivo carcinogenesis is relatively low.
21. Caging and cage location exhibit no impact on chemical carcinogenesis.
22. Chemical structure alone does not allow cancer prediction.
23. Rodent carcinogenicity false-positive rates are relatively low; little is known about false-negative rates.
24. Nearly 30% of chemicals shown to cause cancer in humans were first observed to be carcinogenic in laboratory animals.
25. Basing regulatory decisions on animal data is prudent public health practice.
26. Epidemiological investigations should be considered for chemicals considered carcinogenic in multiple species and those that cause cancers at multiple sites.
27. Cancer prevention is strengthened by avoiding or eliminating exposures to agents causing cancers in laboratory animals and/or in humans.

X. CONCLUSIONS AND COMMENTS

From the beginnings of chemical carcinogenesis early in this century, we have come a long way to better understanding how and which agents are likely to cause cancer. However, we may also have regressed to a certain extent by either neglecting or being ignorant of the past. Much has been learned since those early investigators served as pioneers in the development of this field. We have struggled collectively, especially over the last 25 years, in our quest to conquer cancer. Much has been learned and considerable progress made. Nonetheless, in many aspects the war remains far from being won (Bailar and Gornik, 1997). No magic bullets have been discovered. No universal, generic cancer chemotherapeutic treatment breakthroughs have been made. No common mechanism of cancer has been attained, or likely ever will be. No etiological consensus has been reached regarding most of the common cancers afflicting and killing humans.

Because of these discouraging facts, one wonders if we have failed miserably in our crusade to elimate cancers, or if we are simply gaining ground at a slight or imperceptible pace? In my opinion, we are gaining. And the future of cancer prevention appears encouraging. However, we could and should do more, and we need to consider revising current strategies of devoting most of our efforts toward basic and mechanistic research and treatment regimens. These are valuable and necessary, of course, but we must objectively re-evaluate the progress we think we have made in reducing cancer incidence and mortality.

One area that demands more, or renewed, attention is prevention (Tomatis et al., 1997). And a key to prevention of cancers—or at least to a reduction in incidences and mortalities—rests on the idea of discovering or even rediscovering the "causes" of cancer and then taking aggressive action to eliminate or greatly curtail them. *Causes* is in quotation marks because one need not know the actual *etiological* causes or the mechanisms of cancer to effectively introduce active and functional prevention conditions—a colossal task, one that has been ongoing with varying degrees of effort and success for decades and even centuries. More needs to be done, since the numbers of people with cancer and the numbers of people dying with and by cancer have not fallen. The rates of increase may have slowed somewhat but not nearly enough for an overall assessment to be made of any particular gains in our collective war on cancer. Thus, prevention seems to be a major direction to pursue for reducing the incidence of and mortality from cancer—a single word denoting more than 200 individual diseases.

The theme of this chapter and one of the mainstays for promoting prevention strategies is the identification of potential chemical human carcinogens by using long-term bioassays in laboratory animals. Of the six main avenues for detecting carcinogens or potential carcinogenic agents (see Tables 1 and 2), only the long-term bioassay using rodents or other laboratory animals has shown repeatedly, or possesses still, the greatest potential for public health success. All agents shown so far by epidemiological methods to cause cancer in humans have likewise been shown to cause cancer in laboratory animals, with nearly a third being first identified in animals (Huff, 1993). Note that one human carcinogenic metal, arsenic, has yet to be adequately tested in animals, and some individuals use this single "nonconcordance" to deny the value of animal bioassays (Chan and Huff, 1997).

These 2- to 3-year bioassays are not perfect (Huff, 1991), but they are the best of what is currently available to detect with reasonable scientific assuredness the agents that will potentially be carcinogenic to humans (Tables 9a, 11, 14, and 24). Of course, one must take other vital information into account while extrapolating carcinogenesis findings from rodents to humans and for initiating prevention action. Are humans exposed? At what exposure concentrations? What are the durations and frequencies of exposure? Are exposures to other carcinogens or cocarcinogens occurring? Are susceptible populations such as children being exposed disproportionately? Answers to these and other questions may yield different schemes of exposure prevention.

The key to prevention success could reliably be stated thus: discover and identify the agents, mixtures of agents, and particular or peculiar exposure circumstances (including occupations and lifestyle habits) that are associated with, or even thought to be associated with, cancer causation, and eliminate or substantially diminish those exposures (Table 6–8, 9a, 11, and 14–16). Obviously this can be done without knowing the exact causal agent (witness the knowledge that tobacco was associated with cancer in the 1700s and declared a carcinogen by the Surgeon General of the United States in the middle 1960s) or without knowing the mechanism or mechanisms of carcinogenesis of the cancer-associated exposure (this includes every identified carcinogen, since no complete mechanism has yet been fully described). Moreover, prudent public health activities in prevention can be, and historically have been, implemented with little more knowledge than that an agent or exposure circumstance is carcinogenic to animals and/or to humans. One need not know or understand the mechanisms of cancer but simply recognize the conditions under which cancer may arise and has arisen. Unfortunately, this prime yet simple public health strategy has become more difficult in recent years, especially if the agent is economically important (Peto, 1980). Regulatory and prevention actions have been delayed or stymied by debate and the search for biological mechanisms or for specific etiologically defined causes of cancer. This is not the right strategy for reducing cancer.

A newer trend of delay involves allegations that certain "rodent-specific tumors" (Huff, 1992; Karstadt and Haseman, 1997) or rodent-unique mechanisms (Huff, 1996, 1998; Melnick and Kohn, 1998; IARC, 1998) have little or nothing to do with humans. These delaying tactics by those wishing to reduce reliance on experimental data obtained from animals has gained acceptance in certain quarters. To some, "for a moment the lie becomes truth" (Dostoyevsky), but the truth tends not to be buried for long. However, as long as such obfuscatory and self-serving polemics persist, public health suffers. Perhaps the time has come to worry about protecting ourselves from cancer rather than pretend that carcinogenic agents and myriad suspect carcinogenic exposure circumstances are harmless. We have that obligation.

ACKNOWLEDGMENTS

I appreciate the helpful comments and valued suggestions from Kamal Abdo, Po Chan, and Matthew Longnecker. For help in preparing tables, I thank Sharon Soward. I am also grateful to Kirk Kitchin for inviting me to prepare this chapter, for his editing skills, and especially for his patience while this chapter was being written. Kerry Doyle was most helpful and thoughtful in coordinating and copyediting this chapter.

Do the right thing because it is right.

— Buddha

REFERENCES

Abdo K (1993). Toxicology and carcinogenesis studies of talc (CAS 14807-96-6) in F344/N rats and B6C3F1 mice (inhalation studies). NTP Tech Report Series 421:1-286. National Toxicology Program, Research Triangle Park, NC.

ACS (1994). Cancer Facts and Figures—1994. American Cancer Society, Atlanta, 30 pp.

ACS (1995). Cancer Facts and Figures—1995. American Cancer Society, Atlanta, 32 pp.

ACS (1997). Cancer Facts and Figures—1997. American Cancer Society, Atlanta, 32 pp.

ACS (1998). Cancer Facts and Figures—1998. American Cancer Society, Atlanta, 36 pp.

Ames BN, Gold LS (1997). Environmental pollution, pesticides, and the prevention of cancer: misconseptions. FFASED J 11:1041–1052.
Bailar JC, Gornik MHS (1997). Cancer undefeated. N Engl J Med 336:1569–1574.
Bailar JC, Smith EM (1986): Progress against cancer? N Engl J Med314:1226–1232.
Baserga R (1990). The cell cycle: myths and realities. Cancer Res 50:6769–6771.
Berenblum I (1941). The mechanism of carcinogenesis: a study of the significance of cocarcinogenic action and related phenomena. Cancer Res 1:807–814.
Berenblum I (1944). Irritation and carcinogenesis. AMA Arch Pathol 38:233–244.
Berkson JB (1996). A Canary's Tale. The Final Battle. Politics, poisons, and pollutions the environment and the public health. Vol I The Oyssey (1988–1996). 302 pp.
Browning E (1965). Toxicity and Metabolism of Industrial Solvents. Elsevier, New York, 739 pp.
Bucher JR, Portier CJ, Goodman JI, Faustman EM, Lucier GW (1996). National Toxicology Program studies: principles of dose selection and applications to mechanistic based risk assessment.
Fund Appl Toxicol 31:1–8.
Bucher JR, Portier CJ, Chhabra RS, Lucier, GW (1998). National Toxicology Program Principles and Procedures. I. Dose Selection. Environ Health Perspect. In press.
Chan PC, Huff JE (1997). Arsenic carcinogenesis in animals and in humans: mechanistic, experimental, and epidemiologic evidence. Environ Carcino Ecotox Revs C15(2):83–122.
Chan PC, Sills RC, Braun AG, Haseman JK, Bucher JR (1996). Toxicity and carcinogenicity of D9-tetrahydrobcannabinol in Fischer rats and B6C3F1 mice. Fund Appl Toxicol 30:109–117.
Chhabra RS, Huff JE, Haseman JK, Hall A, Baskin G, Cowan M (1988). Inhibition of some spontaneous tumors by 4-hexylresorcinol in F344/N rats and B6C3F1 mice. Fund Appl Toxicol 11:685–690.
Chhabra RS, Huff JE, Schwetz BS, Selkirk J (1990). An overview of prechronic and chronic toxicity/carcinogenicity experimental study designs and criteria used by the National Toxicology Program. Environ. Health Perspect. 86:313–321.
Chow W-H, Lindblad P, Gridley G, Nyren O, McLaughlin JK, Linet MS, Pennello GA, Adami H-O, Fraumeni JF (1997). Risk of urinary tract cancers following kidney or ureter stones. J Natl Cancer Inst 89:1453–1457.
Chu KC, Cueto C, Ward JM (1981). Factors in the evaluation of 200 National Cancer Institute carcinogen bioassays. J Toxicol Environ Health 8:251–280.
Chu KC, Tarone RE, Kessler LG, Ries LA, Hankey BF, Miller BA, Edwards BK (1996). Recent trends in U.S. breast cancer incidence, survival, and mortality rates. JNCI 88:1571–1579.
Cook JW, Hieger I, Kennaway EL, Mayneord WV (1932). The production of cancer by pure hydrocarbons. Part 1. Roy Soc Proced, Part B, 111:455–484.
DeMarini DM, Huff JE (1998). Genetic toxicity assessment: toxicology test methods. In: Stellman JM, ed. ILO Encyclopaedia of Occupational Health and Safety. Vol 33. 4th ed. Geneva: International Labour Office (ILO), pp 43–45.

Desai PB (1994). Understanding the biology of cancer: has this any impact on treatment? J Cancer Res Clin Oncol 120:193–199.

Douglas JF, Huff JE (1984). No evidence of carcinogenicity of l-ascorbic acid in rodents. J Toxicol Environ Health 14:605–609.

Dunnick JK, Elwell MR, Huff JE, Barrett JC (1995). Chemically Induced mammary gland cancer in the National Toxicology Program's carcinogenesis bioassy. Carcinogenesis 16:173–179.

EHP (1990). Proceedings: Symposium on the toxicology, carcinogenesis, and human health aspects of 1,3-butadiene. Environ Health Perspect 86:1–171.

Elmslie WJ (1866). Etiology of epithelioma among the Kashmiris. Indian Med Gaz 1:324–326.

Fagin D, Lavelle M (1996). Toxic Deception. How the chemical industry manipulates science, bends the law, and endangers your health. Birch Lane Press, Carol Pub Group, Secaucus, NJ.

Farber E (1976). The pathology of experimental liver cell cancer. In: Cameron HM, Linsell DA, Warwick GP, eds. Liver Cell Cancer. Elsevier, Amsterdam, p 243.

Farber E (1995). Cell proliferation as not a major risk factor for cancer: a concept of doubtful validity. Cancer Res 55:3759–3762.

Farber E (1996). Cell proliferation is not a major risk factor for cancer. Mod Pathol 9:606.

Flamm WG, Hughes D (1997). Does the term carcinogen send the wrong message? Cancer Lett 117:189–194.

Foran JA (1997). Principles for the selection of doses in chronic rodent bioassays. ILSI Risk Science Working Group on Dose Selection. Environ Health Perspect 105:18-20.

Fung VA, Huff JE, Weisburger E, Hoel DG (1993). Predictive strategies for selecting 379 NCI/NTP chemicals evaluated for carcinogenic potential: scientific and public health impact. Fund Appl Toxicol 20:413–436.

Fung VA, Barrett JC, Huff JE (1995). The carcinogenesis bioassay in perspective: application in identifying human cancer hazards. Environ Health Perspect 103:680–683.

Giannasi F, Thebaud-Mony A (1997). Occupational exposures to asbestos in Brazil. Int J Occup Environ Health 3:150–157.

Goldman L (1998). Chemicals and children's environment: what we don't know about risks. Environ Health Perspect. In press.

Habs M, Schmahl D (1988). Combination effects in different organs in small rodents. In: Schmahl D, ed. Combination Effects in Chemicals Carcinogenesis, pp 75–92.

Haddow A (1938). Cellular inhibition and the origin of cancer. Acta Umo Int Contra Cancrum 3:342.

Harlow BL, Weiss NS (1989). A case-control study of borderline obarian tumors: the influence of perineal exposure to talc. Am J Epidemiol 130:390–394.

Haseman JK, Huff JE. (1987). Species correlation in long-term carcinogenicity studies. Cancer Lett 37:125–132.

Haseman JK, Huff JE (1991). Arguments that discredit animal studies lack scientific support. Chem. Engineer News 69:49–51.

Haseman JK, Johnson FM (1996). Analysis of National Toxicology Program rodent bioassay data for anticarcinogenic effects. Mut Res 350:131–141.
Haseman JK, Lockhart A (1994). Relationship between use of the maximum tolerated dose and study sensitivity for detecting rodent carcinogenicity. Fund Appl Toxicol 22:382–391.
Haseman JK, Huff JE, Rao GN, Arnold JE. Boorman GA, McConnell EE (1985). Neoplasms observed in untreated and corn oil gavage control groups of F344/N rats and (C57Bl/6N × C3H/HeN)F1 (B6C3F1) mice. JNCI 75: 975–984.
Haseman JK, Huff JE, Zeiger E, McConnell EE (1987). Comparative results of 327 chemical carcinogenicity studies. Environ Health Perspect 74:229–235.
Haseman JK, Huff JE, Rao GN, Eustis SL. (1989). Sources of variability in rodent carcinogenicity studies. Fund Appl Toxicol 12:793–804.
Hill J (1761). Cautions Against the Immoderate Use of Snuff. Founded on the known qualities of Tobacco Plant; And the effects it must produce when this way taken into the body: and enforced by instances of persons who have perished miserably of diseases, occasioned, or rendered incurable by its use. Baldwin and Jackson, London, 57 pp.
Hoel DG, Haseman JK, Hogan MD, Huff JE, McConnell EE (1988). The impact of toxicity on carcinogenicity studies: implications for risk assessment. Carcinogenesis 9:2045–2052.
Hueper WC, Conway WD (1964). Chemical Carcinogenesis and Cancers. Charles C Thomas, Springfield, IL.
Huff JE (1986). The value of in-life and retrospective data audits, 99–104. In: Hoover BK, Baldwin JK, Uelner AF, Whitmire E, Davies CL, Bristol DW, eds. Managing Conduct and Data Quality of Toxicology Studies. Princeton, NJ: Scientific, 352 pp.
Huff JE (1992). A historical perspective of the classification developed and used for chemical carcinogens by the National Toxicology Program during 1983–1992. Scand J Work Environ Health 18(suppl 1):74–82.
Huff JE (1992). Design strategies, results, and evaluations of long-term chemical carcinogenesis studies. Scand J Work Environ Health 18(suppl 1):31–37.
Huff JE (1992a). Applicability to humans of rodent-specific sites of chemical carcinogenicity: tumors of the forestomach and of the harderian, preputial, and zymbal glands induced by benzene. J Occup Med Toxicol 1:109–141.
Huff JE (1992b). Chemical toxicity and chemical carcinogenesis: is there a causal connection? A comparative morphological evaluation of 1500 experiments. In: Vainio, Magee, McGregor, McMichael,eds. Mechanisms of Carcinogenesis in Risk Identification. IARC Sci Pub 116. Lyons, France, pp 437–475.
Huff JE (1993). Absence of morphologic correlation between chemical toxicity and chemical carcinogenesis. Environ Health Perspect 101(suppl 5):45–54.
Huff JE (1993). Chemicals and cancer in humans: first evidence in experimental animals. Environ Health Perspect 100:201–210.
Huff JE (1993a). Issues and controversies surrounding qualitative strategies for identifying and forecasting cancer causing agents in the human environment. Pharmacol Toxicol 72(suppl 1):12–27.

Huff JE (1993b). Absence of morphologic correlation between chemical toxicity and chemical carcinogenesis. Environ Health Perspect 101(suppl 5):45–54.

Huff JE (1994). Carcinogenic hazards from eating fish and shellfish contaminated with disparate and complex chemical mixtures. In: Yang RSH, ed. Toxicology of chemical mixtures: from real life examples to mechanisms of toxicological interactions. New York: Academic Press, pp 157–194.

Huff JE (1994a). Chemicals causally associated with cancers in humans and in laboratory animals: a perfect concordance. In: Waalkes MP, Ward JM, eds. Carcinogenesis. New York: Raven Press, pp 25–37.

Huff JE (1995). Mechanisms, chemical carcinogenesis, and risk assessment: cell proliferation and cancer. Am J Indust Med 27:293–300.

Huff JE (1996). α2u-Globulin nephropathy, posed mechanisms, and white ravens. Environ Health Perspect 104:1264–1267.

Huff JE (1996). Chemically induced cancers in hormonal organs of laboratory animals and of humans. In: Huff JE, Boyd JA, Barrett JC, eds. Cellular and Molecular Mechanisms of Hormonal Carcinogenesis: Environmental Influences. New York: Wiley-Liss, pp 77–102.

Huff JE (1998). NTP Report on Carcinogens: history, concepts, procedures, progress. Eur J Oncol In press.

Huff JE (1998a). Chemicals associated with tumours of the kidney, urinary bladder, and thyroid glands from 2,000 NTP/NCI long-term chemical carcinogenesis experiments in laboratory rodents. In: Species Differences in Thyroid, Kidney and Urinary Bladder Carcinogenesis. IARC Sci Pub. In press.

Huff JE (1998b). Carcinogenesis results in animals predict cancer risks to humans. In: Wallace RB, ed. Maxcy-Rosenau-Last's Public Health & Preventive Medicine. 14th ed. Appleton & Lange, Norwalk, CT. In press.

Huff JE, Barrett JC (1998). Breast cancer and associated environmental risk factors. Environ Health Perspect In press.

Huff JE, Haseman JK (1991). Exposure to certain pesticides may pose real carcinogenic risk. Chem Engineer News 69:33–37. (Reprinted in J Pest Reform (1991) 11:10–14; Pesticide News (1991) 12:7–10.

Huff JE, Haseman JK (1991). Long-term chemical carcinogenesis experiments for identifying potential human cancer hazards: collective data base of the National Cancer Institute and National Toxicology Program (1976–1991). Environ Health Perspect 96:23–31.

Huff JE, Hoel DG (1992). Perspective and overview of the concepts and value of hazard identification as the initial phase of risk assessment for cancer and human health. Scand J Work Environ Health 18 (suppl 1):83–89.

Huff JE, Moore JA (1984). Carcinogenesis studies design and experimental data interpretation/evaluation at the National Toxicology Program. In: Jarvisalo J, Pfaffli P, Vainio H, eds. Industrial Hazards of Plastics and Synthetic Elastomers. Alan R Liss, New York, pp 43–64.

Huff JE, Rall DP (1992). Relevance to humans of carcinogenesis results from laboratory animal toxicology studies. In: Last JM, Wallace RB, eds. Maxcy-Rosenau-Last's Public Health and Preventive Medicine. 13th ed. Norwalk, CT: Appleton and Lange, pp 433–440, 453–457.

Huff JE, Soward S (1998). Carcinogenesis bioassays of 500 chemicals: results, evaluations, tumor site specificity, prevalence categories, and human risk potential. Submitted.

Huff JE, Melnick RL, Solleveld HA, Haseman JK, Powers M, Miller RA (1985). Multiple organ carcinogneicity of 1,3-butadiene in B6C3F1 mice after 60 weeks of inhalation exposure. Science 227:548–549.

Huff JE, McConnell EE, Haseman JK, Boorman GA, Eustis SL, Schwetz BA, Rao GN, Jameson CW, Hart LG, Rall DP (1988). Carcinogenesis studies: results from 398 experiments on 104 chemicals from the US National Toxicology Program. Ann NY Acad Sci 534:1–30.

Huff JE, Eustis SE, Haseman JK. (1989). Occurrence and relevance of chemically induced benign neoplasms in long-term carcinogenicity studies. Cancer Metast Rev 8:1–21.

Huff JE, Bucher JR, Yang RSH (1991). Carcinogenesis studies in rodents for evaluating risks associated with chemical carcinogens in aquatic food animals. Environ Health Perspect 90:127–132.

Huff JE, Cirvello J, Haseman JK, Bucher JR (1991). Chemicals associated with site-specific neoplasia in 1394 long-term carcinogenesis experiments in laboratory rodents. Environ Health Perspect 93:247–271.

Huff JE, Haseman JK, Rall DP (1991). Scientific concepts, value, and significance of chemical carcinogenesis studies. Ann Rev Pharmacol Toxicol 31:621–652.

Huff JE, Bucher JR, Schwetz BA, Barrett JC (1994) Optimum exposure levels for chemical carcinogenesis experiments: concepts, principles, guidelines, and experience. SOT Toxicologist 14:139 (abstr 475). Ann Meeting, Society of Toxicology, Dallas, March 1994.

Huff JE, Boyd JA, Barrett JC (1996). Environmental influences on hormonal carcinogenesis. In: Huff JE, Boyd JA, Barrett JC, eds. Cellular and Molecular Mechanisms of Hormonal Carcinogenesis: Environmental Influences. New York: Wiley-Liss, pp xiii-xix.

Huff JE, Boyd JA, Barrett JC (1996). Hormonal carcinogenesis and environmental influences: background and overview. In: Huff JE, Boyd JA, Barrett JC, eds. Cellular and Molecular Mechanisms of Hormonal Carcinogenesis: Environmental Influences. New York: Wiley-Liss.

Huff JE, Weisburger E, Fung VA (1996). Multicomponent criteria for predicting carcinogenicity: 30 chemicals data set. Environ Health Perspect 104(suppl 5): 1105–1112.

Hunter D. (1978). The father of occupational medicine. In: The Diseases of Occupations. 6th ed. Hodder and Stoughton, London, pp 33–37.

Hutchinson J (1888). On some examples of arsenic-keratosis of the skin and arsenic-cancer. Trans Pathol Soc London 39:352–363.

IARC (1985). Ethylene Oxide 189–226. In: Allyl Compounds, Aldehydes, Epoxides, and Peroxides. Vol 36. IARC Monographs on the Evaluation of Carcinogenic Risk of Chemicals to Humans. International Agency for Research on Cancer, Lyons, France.

IARC (1987). Overall evaluations of carcinogenicity: an updating of IARC Monographs 1–42. Suppl 7:1–440. IARC Monographs on the Evaluation of Carcinogenic Risks to Humans. International Agency for Research on Cancer, Lyons, France.

IARC (1996). Directory of agents being tested for carcinogenicity. IARC Sci Pub 17. Lyons, France, 247 pp.

IARC (1996a). Directory of on-going research in cancer epidemiology, 1996. IARC Sci Pub 137. International Agency for Research on Cancer, Lyons, France, 814 pp.

IARC (1976). Epichlorohydrin, 131–139. In: Cadmium, nickel, some epoxides, miscellaneous industrial chemicals, and general considerations on volatile anaesthetics. Vol 11. IARC Monographs on the Evaluation of Carcinogenic Risk of Chemicals to Man. International Agency for Research on Cancer, Lyons, France.

IARC (1997). IARC Monographs on the Evaluation of Carcinogenic Risks to Humans. Vols 1–69. International Agency for Research on Cancer, Lyons, France.

Infante PF (1993). Use of rodent carcinogenicity test results for determining potential cancer risk to humans. Environ Health Perspect 101(suppl 5):143–148.

Interdisciplinary Panel on Carcinogenicity (1984). Criteria for evidence of chemical carcinogenicity. Science 225:682–687.

Ito N, Imaida K, Tsuda H, Shibata MA, Aoki T, de Camargo JLV, Fukushima S (1988). Wide-spectrum initiation models: possible application to medium-term multiple organ bioassays for carcinogenesis modifiers. Jpn J Cancer Res 79:413–417.

Ito N, Shirai T, Hasegawa R (1992). Medium-term bioassays for carcinogens. In: Vainio H, Magee P, McGregor D, McMichael A, eds. Mechanisms of Carcinogenesis in Risk Identification. IARC Sci. Pub 116. International Agency for Research on Cancer, Lyons, France, pp 353–388.

Janin N (1996). Vingt ans de progress dans la dissection des mecanismes de la cancerogenese: rapports entre magie de la science et exigences de la medecine. Presse Med 25:1669–1671.

Kelloff GJ, Hawk ET, Karp JE, Crowell JA, Boone CW, Steele VE, Lubet RA, Sigman CC (1997). Progress in clinical chemoprevention. Semin Oncol 24:1–13.

Kogevinas M, Pearce N, Susser M, Boffetta P, eds (1997). Social inequalities and cancer. IARC Sci Pub 138. Lyons, France, 397 pp.

Kosary CL, Ries LAG, Miller BA, Hankey BF, Harras A, Edwards BK, eds. (1995). SEER Cancer statistics review, 1973–1992. NIH Pub 96–2789. National Cancer Institute, Bethesda, MD.

Levin SM (1992). Prevention delayed is prevention denied. Amer J Indust Med 22: 435–6.

Lijinsky W (1989). Environmental cancer risks—real and unreal. Environ Res 50: 207–209.

Lijinsky W (1990). In vivo testing for carcinogenicity. In: Cooper CS, Grover PL, eds. Handbook of Experimental Pharmacology. Vol 94/I. New York: Springer-Verlag, pp 179–209.

Lijinsky W (1992). Chemistry and biology of N-nitroso compounds. Cambridge Monographs on Cancer Research. Cambridge, 464 pp.

Lijinsky W (1993). Species differences in carcinogenesis. In Vivo 7:65–72.
Lubchenco J (1998). Entering the century of the environment: a new social contract for science. Science 279:491–497.
MacGregor JT, Shane BS, Spalding J, Huff JE (1995). Carcinogenicity and genotoxicity assays for cancer risk to humans. In: Screening and Testing Chemicals in Commerce. OTA-BP-ENV-166. Workshop Proceedings on Genotoxic and Carcinogenic Assays for Identifying Carcinogens. Washington, DC: Office of Technology Assessment, pp 11–28.
Magee PN, Barnes JM (1956). The production of malignant primary hepatic tumors in the rat by feeding dimethylnitrosamine. Br J Cancer 10:114–122.
Mantanoski GM, Schwartz L (1987). Mortality of workers in the syrene-butadiene polymer production. J Occup Med 29:675–680.
Mantanoski GM, Santos-Burgoa, C, Schwartz L (1990). Mortality of a cohort of workers in the syrene-butadiene manufacturing industry (1943–1982). Environ Health Perspect 86:107–117.
Mantanoski GM, Francis M, Correa-Villasenor A, Elliott E, Santos-Burgoa C, Schwartz L (1993). Cancer epidemiology among styrene-butadiene rubber workers. IARC Sci Publ 127. Lyons, France, pp 363:374.
Maronpot RR, Haseman JK, Boorman G, Eustis S, Rao GN, Huff JE (1987). Liver lesions in B6C3F1 mice: the National Toxicology Program experience and position. Arch Toxicol Suppl 10:10–26.
Melnick RL (1992). Does chemically induced hepatocyte proliferation predict liver carcinogenesis? FASEB J 6:2698–2706.
Melnick RL (1992a). An alternative hypothesis on the role of chemically induced protein droplet (α2u-globulin) nephropathy in renal carcinogenesis. Reg Toxicol Pharmacol 16:111–125.
Melnick RL (1993). Critique does not validate assumptions in the model on α2u-globulin and renal carcinogenesis. Reg Toxicol Pharmacol 18:365–368.
Melnick RL, Huff JE (1992). 1,3-Butadiene: toxicity and carcinogenicity in laboratory animals and in humans. Rev Environ Contam Toxicol 124:111–144.
Melnick RL, Huff JE (1993). 1,3-Butadiene induces cancer in experimental animals at all concentration from 6.25 to 8000 parts per million. In: Sorsa M, Peltonen K, Vainio H, Hemminiki K, eds. Butadiene and styrene: assessment of health hazards. IARC Sci Pub 127. Lyons, France, pp 309–322.
Melnick RL, Huff JE (1993). Liver carcinogenesis is not a predicted outcome of chemically induced hepatocyte proliferation. Toxicol Ind Health 9:415–438.
Melnick RL, Kohn MC (1998) Possible mechanisms of induction of renal tubular cell neoplasms in rats associated with α2u-globulin: role of protein accumulation versus ligand delivery to the kidney. In: Species Differences in Thyroid, Kidney and Urinary Bladder Carcinogenesis. IARC Sci Pub. Lyons, France. In press.
Melnick RL, Huff JE, Bird MG, Acquavella JF (1990). Symposium overview: toxicology, carcinogenesis, and human health aspects of 1,3-butadiene. Environ Health Perspect 86:3–5.
Melnick RL, Huff JE, Chou BJ, Miller RA (1990). Carcinogenicity of 1,3-butadiene

in C57Bl/6 × C3H F1 mice at low exposure concentrations. Cancer Res 50: 6592–6599.

Melnick RL, Barrett JC, Huff JE, eds (1993). Cell proliferation and chemical carcinogenesis: proceedings. Environ Health Perspect 101(suppl 5):1–285.

Melnick RL, Kohn MC, Huff JE (1997). Weight of evidence versus weight of speculation to evaluate the α2u-globulin hypothesis [letter]. Environ Health Perspect 105:904–906.

Miller BA, Ries LAG, Hankey BF, Kosary CL, Edwards BK, eds (1992). SEER Cancer Statistics Review: 1973–1989. NIH Pub 92-2789. Bethesda, MD: National Cancer Institute.

Miller BA, Ries LAG, Hankey BF, Kosary CL, Harras A, Devesa SS, Edwards BK, eds (1993). SEER Cancer Statistics Review: 1973–1990. NIG Pub 93-2789. Bethesda, MD: National Cancer Institute.

Montesano R, Bartsch H, Vainio H, Wilbourn J, Yamasaki H, eds (1986). Long-term and Short-term Assays for Carcinogens: A Critical Appraisal. IARC Sci Pub 83. International Agency for Research on Cancer. Lyons, France, 564 pp.

Murphy JB, Strum E (1925). Primary lung tumors in mice following the cutaneous application of coal tar. J Exptl Med 42:693–700.

NCI (1976). Carcinogenesis bioassay of trichloroethylene (CAS 79-01-6). NCI Carcinogenesis Tech Rept Ser 2, NCI-CG-TR-1. National Cancer Institute, Bethesda, MD, 197 pp.

Neutra R (1990). Counterpoint from a cluster buster. Am J Epidemiol 132:1–8.

NRC/NAS (1993). Advantages and disadvantages of bioassays that use the MTD. In: Issues in Risk Assessment. Washington, DC: National Research Council, National Academy of Sciences, pp 43- 51.

NTP (1986). Toxicology and Carcingoenesis Studies of Diglycidyl Resorcinol Ether (Technical Grade) (CAS 101-90-6) in F344/N Rats and B6C3F1 Mice (Gavage Studies). NTP Tech Rept Ser 257. National Toxicology Program, Research Triangle Park, NC.

NTP (1998). Toxicology and Carcinogenesis Studies of Diethanolamine (CAS 111-42-2) in F344/N Rats and B6C3F1 Mice (Dermal Studies). National Toxicology Program, Research Triangle Park, NC. In press.

NTP (1998a). Toxicology and Carcinogenesis Studies of Coconut Oil Acid Diethanolamine Condensate (CAS 68603-42-9) in F344/N Rats and B6C3F1 Mice (Dermal Studies). National Toxicology Program, Research Triangle Park, NC. In press.

OSHA (1987). Occupational exposure to benzene: final rule. Fed Register 52(176): 34460–34578.

OTA (1987). Identifying and regulating carcinogens: background paper. Office of Technology Assessment, Congress of the United States, 251 pp.

Parker SL, Tong T, Bolden S, Wingo PA (1997). Cancer statistics, 1997. CA Cancer J Clin 47:5–27.

Peto R (1980). Distorting the epidemiology of cancer: the need for a more balanced overview. Nature 284:297–300.

Pott P (1775). Cancer scroti. In: Chirurgical Observations Relative to the Cataract, the Polypus of the Nose, the Cancer of the Scrotum, the Different Kinds of Ruptures, and the Mortification of the Toes and Feet. London, pp 63-68.

Purchase IFH (1997). Prospects for reduction and replacement alternatves in regulatory toxcicology. Toxicol In Vitro 11:313-319.

Rall DP, Hogan MD, Huff JE, Schwetz B A, Tennant TW (1987). Alternatives to using human experience in assessing health risks. Ann Rev Public Health 8: 355-385.

Ramazzini B (1713). De Morbis Artificum Diatriba. Revised with translations and notes. University of Chicago Press, Chicago, 1940.

Rao G, Huff JE (1990). Refinement of long-term toxicity and carcinogenesis studies. FAAT 15:33-43.

Rehn L (1895). Bladder tumors in fuchsin workers [from German]. Archiv Klin Chirur 50:588-600.

Ries LAG, Miller BA, Hankey BF, Kosary CL, Harras A, Edwards BK, eds (1994). SEER Cancer Statistics Review: 1973 – 1991, Tables and graphs. NIH Pub 94-2789. Bethesda, MD: National Cancer Institute, 449 pp.

Ries LAG, Kosary CL, Hankey BF, Miller BA, Harras A, Edwards BK, eds (1997). SEER Cancer Statistics Review: 1973 – 1994, Tables and graphs. NIH Pub 94-2789. Bethesda, MD: National Cancer Institute, 479 pp.

Salmon AG, Zeise L (1991). Risks of carcinogenesis from urethane exposure. Boca Raton, FL: CRC Press, 231 pp.

Santos-Burgoa, C (1988). Case-control study of lympho-hematopoietic malignant neoplams within a cohort of styrene-butadiene polymerization workers. Doctoral thesis, Johns Hopkins School of Hygiene and Public Health, Baltimore, MD.

Sasaki T, Yoshida T (1935). Liver carcinoma induced by feeding o-amidoazotoluene [from German]. Virch Arch Patholog Anat 295:175-220.

Schmahl D (1988). Combination Effects in Chemical Carcinogenesis. New York: VCH Pub, 279 pp.

Schwetz B, Gaylor D (1997). New directions for predicting carcinogenesis. Mol Carcinog 20:275-279.

Sells B (1994). What asbestos taught me about managing risk. Harvard Business Review, March-April, pp 76-90.

Shiari T (1997). A medium-term rat liver bioassay as a rapid in vivo test for carcinogenic potential: a historical review of model development and summary of results from 291 tests. Toxicol Pathol 25:453-460.

Shimkin MB (1977). Contrary to nature. DHEW Pub (NIH) 79-720. Bethesda, MD: National Institutes of Health, 498 pp.

Shimkin MB (1980). Some classics of excperimental oncology: 50 selections, 1775-1965. DHEW Pub (NIH) 80-2150. Bethesda, MD: National Institutes of Health, 739 pp.

Sontag JM, Page NP, Saffiotti U (1976). Guidelines fo Carcinogen Bioassay in Small Rodents. NCI Carcinogenesis Tech Rept Ser 1, NCI-CG-TR-1. National Cancer Institute, Bethesda, MD, 65 pp.

Takayama S, Nakatsuru Y, Sato S (1987). Carcinogenic effect of the simultaneous administration of five heterocyclic amines to F344 rats. Jpn J Cancer Res 78: 1068-1072.

Takayama S, Hasegawa H, Ohgaki H (1989). Combination effects of forty carcinogens administered at low doses to male rats. Jpn J Cancer Res 80:732–736.

Tennant RW, Elwell MR, Spaulding JW, Greisemer RA (1991). Evidence that toxic injury is not always associated with induction of chemical carcinogenesis. Mol Carcinog 4:420–440.

Tennant RW, Spaulding J (1996). Predictions for the outcome of rodent carcinogenicity bioassays: identification of trans-species carcinogens and noncarcinogens. Environ Health Perspect 104(suppl 5):1095–1100.

Tennant RW, Spalding J, French JE (1996). Evaluation of trangenic mouse bioassays for identifying carcinogens and noncarcinogens. Mut Res 365:119–127.

Thomas TL (1990). Lung cancer mortality among pottery workers in the United States. IARC Sci Pub 97:75–81. Lyons, France.

Thomas TL, Stewart PA (1987). Mortality from lung cancer and respiratory disease among pottery workers exposed to silica and talc. Am J Epidemiol 125:35–43.

Tomatis L (1995). Socioeconomic factors and human cancers. Int J Cancer 62:121–125.

Tomatis L (1977). Comment on metholodogy and interpretation of results. JNCI 59:1341.

Tomatis L (1979). The predictive value of rodent carcinogenicity tests in the evaluation of human risks. Ann Rev Pharmacol Toxicol 19:511–530.

Tomatis L (1997). Poverty and cancer. IARC Sci Pub 138. Lyons, France, pp 25–39.

Tomatis L, Aitio A, Wilbourn J, Shuker L (1989). Human carcinogens identified so far. Jpn J Cancer Res 80:795–807.

Tomatis L, Huff JE, Hertz-Picciotto I, Sandler D, Bucher J, Boffetta P, Axelson O, Blair A, Taylor J, Stayner L, Barrett JC (1997). Avoided and avoidable risks in cancer. Carcinogenesis, pp 95–105.

Tomatis L, Melnick RL, Haseman JK, Barrett JC, Huff JE (1998). Alleged misconceptions belittle environmental cancer risks. In press.

Toth B (1997). Facts, myths, and reflections on the use of maximum tolerated dose in chemical carcinognesis [review]. Int J Oncol 10:529–534.

Vainio H, Wilbourn JD, Sasco AJ, Partensky C, Gaudin N, Haseltine E, Eragne I (1995). Identification of human carcinogenic risk in IARC Monographs [in French]. Bull Cancer 82:339–348.

Waalkes MP, Infante P, Huff JE (1994). The scientific fallacy of route specificity of carcinogenesis with particular reference to cadmium. Regul Toxicol Pharmacol 20:119–121.

Ward JM, Ito N (1988). Development of new medium-term bioassays for carcinogens. Cancer Res 48:5051–5054.

Ward JM, Uno H, Kurata Y, Weghorst CM, Jang J-J (1993). Cell proliferation not associated with carcinogenesis in rodents and humans. Environ Health Perspect 101(suppl 5):125–135.

Weinstein IB (1991). Mitogenesis is only one factor in carcinogenesis. Science 251: 387–388.

Weinstein IB (1992). Toxicity, cell proliferation, and carcinogenesis. Mol Carcinog 5:2–3.

Weinstein IB (1996). Tunnel vision. Review of Weinberg RA, Racing to the Beginning of the Road: The Search for the Origin of Cancer (Harmony Books). Nature 383:777–778.

Weisburger EK (1983). History of the bioassay program of the National Cancer Institute. Prog Exp Tumor Res 26:187–201.

Weisburger EK (1988). Why not differences in species response? Comm Toxicol 2: 279–288.

Williams GM (1995). PSEBM 208:141–143.

Wilbourn J, Haroun L, Heseltine E, Kaldor J, Partensky C, Vainio H. (1986). Response of experimental animals to human carcinogens: an analysis based upon the IARC Monographs Programme, Carcinogenesis 7:1853–1863.

Williams GM, Weisburger JH (1981). Systematic carcinogen testing through the decision point approach. Ann Rev Pharmacol Toxicol 21:393–416.

Wingo PA, Tong T, Bolden S (1995): Cancer Statistics, 1995. CA Cancer J Clin 45: 8–30.

Wintrobe MM, editor in chief (1974). Harrison's Principles of Internal Medicine. 7th ed, pp 2029–2030.

Wolff M, Coleman G, Barrett JC, Huff JE (1996). Breast cancer and environmental risk factors: epidemiological and experimental findings. Ann Rev Pharmacol Toxicol 36:573–596.

Wynder EL (1994). Studies in mechanism and prevention: striking a proper balance. Am J Epidemiol 139:547–549.

Yamagiwa K, Ichikawa K (1918). Experimental study of the pathogenesis of carcinoma. Cancer Res 3:1–29.

Zito R (1992). Cell proliferation in experimental carcinogenesis. J Exp Clin Cancer Res 11:3–6.

Zwickey RE, Davis KJ (1959). Carcinogenicity screening. In: Appraisal of the Safety of Chemicals in Foods, Drugs, and Cosmetics. Editorial Committee, Association of Food and Drug Officials of the United States, Baltimore, MD, pp 79–82.

3

The Maximum Tolerated Dose and Secondary Mechanisms of Carcinogenesis

James S. MacDonald
Schering-Plough Research Institute, Kenilworth, New Jersey

Harvey E. Scribner
Rohm and Haas Company, Spring House, Pennsylvania

The assessment of potential human carcinogenic risk from chemical exposure is a challenging task. In addition to consideration of chemical structure, results of short-term genetic toxicity studies and results from in vivo subchronic and chronic animal studies, the principal tool used to assess potential human risk has been chronic rat and mouse carcinogenicity studies. The use of data from these studies has been based on the assumption that exposure of a mammalian species for most of its life span to as much chemical as it can tolerate will identify chemicals hazardous to humans, who may be exposed to much lower doses for shorter periods of time. A key assumption in this approach is that very high exposures are necessary to enhance the sensitivity of an assay that for pragmatic reasons can use only relatively few numbers of animals. In rodents we use 50 animals per treatment group; in humans we wish to protect the health of approximately 250,000,000 individual citizens of the United States. The determination of the maximum tolerated dose has been the subject of debate for many years; the development and consequences of the use and subsequent interpretation of data obtained using high doses in rodent carcinogenicity assays will be discussed in this chapter.

The use of long-term studies in rodents to assess carcinogenic potential (referred to hereafter as rodent bioassays) was conceived over 30 years ago (1). The observation in the early 1960s that tumors could be produced in mice after exposure to alkylating chemicals led to the suggestion that such rodent bioassays could predict potential carcinogenic responses in humans. This resulted in the rapid expansion of the use of rodent bioassays to screen chemicals for carcinogenic potential under the direction of the National Cancer Institute (NCI) in the early 1970s (1,2). In these early studies with known human carcinogens, the tumor response was observed to be dose-dependent, and it was shown that doses near those maximally tolerated were required to produce tumors in rodents (2,3). In fact, some recent reviews of the U.S. National Toxicology Program (NTP) database have shown a general correlation between the magnitude of the maximally tolerated dose and the carcinogenic response in rodents (4). This requirement to use toxic doses evolved over the early years of the bioassay into what is now commonly referred to as the maximal tolerated dose (also referred to by some groups as the minimum toxic dose), or MTD. MTD has been best described in an often quoted statement as "the highest dose of the test agent during the chronic study that can be predicted to not alter the animals' longevity from effects other than carcinogenicity" (5).

This original MTD concept has been revised and debated extensively over the ensuing years, with most discussion centering on how to put this concept into practice in the bioassay (3,6,7). At the center of these discussions has been Sontag's (5) further expansion on the MTD definition as a dose that "causes no more than a 10% weight decrement, as compared to the appropriate control groups, and does not produce mortality, clinical signs of toxicity, or pathologic lesions (other than those that may be related to a neoplastic response) that would be predicted to shorten an animal's natural life span."

The determination of the MTD is based on results of subchronic studies in mice and rats and takes into consideration all the data available from antemortem and postmortem studies. Different groups have defined the MTD differently, placing more emphasis on one or another of the possible endpoints, but the basic approach has not changed for more than 20 years (8–12).

A major difficulty with this approach is that data from 90-day studies must be used to predict the response to the test chemical over 2 years of exposure (8 times longer than 90 days). A dose that produces a 10% decrement in rate of body weight gain in 90 days may result in unacceptable toxicity and mortality over the course of 2 years, leading to the loss of that dose treatment group for risk-assessment purposes. Conversely, rodents may adapt to such a dose and show no difference from the concurrent control group in any parameter. This inevitably leads to a discussion of the adequacy of the bioassay because the dose used may not have truly been the maximally tolerated dose.

Since the original definition of the MTD, several alternative approaches to selection of the top dose in the rodent bioassay have been proposed. Initially, these were based on pharmacokinetic endpoints measuring saturation of metabolic processes of the test agent (13,14). More recently, a method of establishing the top dose based on multiples of human plasma levels of drug and/or metabolites has been accepted by the international regulatory community (15). Even though these endpoints are not truly based on the maximum tolerated dose, they drive the dose to a high level relative to human exposure levels. Much evidence is becoming available that it is precisely this approach that leads to difficulties in interpretation of the results of these studies and the determination of the significance of the findings for assessment of human carcinogenic risk. It is this concept of the use of high doses and the problems they may cause that will be explored in the following section.

The current practice of the NTP is to use doses in the rodent bioassay that produce a wide variety of toxic effects in 90-day studies (16,17). A review of these procedures by the Advisory Committee to the NTP (18), however, suggested that a more comprehensive approach to the design and interpretation of the rodent bioassays was indicated by recent advances in understanding the carcinogenic process as well as a growing appreciation of the role of secondary mechanisms of carcinogenesis in rodent species. A secondary mechanism here refers to a mechanism that generally acts at high doses but does not operate at low doses.

While this more comprehensive approach is an appropriate direction to pursue simply because of the significant advances made in the area of cancer biology, it is also dictated by the bioassay results from a broad range of chemicals. The data in Table 1 show the incidence of positive findings in rodent carcinogenicity studies in several recent surveys. As there was no attempt to identify chemicals in these surveys and as many of the surveys access similar databases, there is certainly much overlap in these data. Nevertheless, the striking finding is that no matter which database is used, approximately 50% of chemicals tested in rodent carcinogenicity studies are found to be positive in the rat or mouse. More recently, the percentage of NTP chemicals testing positive has been lower (approximately 22%) because fewer suspected carcinogens are now being selected for study by the NTP (19). Nevertheless, the high percentage of positive assays is striking because if one examines the available databases for compounds presently known to be carcinogenic in humans, the number is quite small. According to the IARC database, there are only 17 known carcinogenic human phar-

Table 1 Incidence of Positive Findings in Rodent Carcinogenicity Studies

Database	Number of chemicals	Incidence		
		Rat only	Mouse only	One or both
NTP	301	–	–	54
FDA	256	18	10	51
CPMP	175	22	7	48
CMR	140	18	11	56
JPMA	99	14	12	41
PDR	241	21	11	54
CPB	955	–	–	51

Numbers represent % of total number of chemicals showing a positive response in chronic rodent carcinogenicity bioassays (NTP, National Toxicology Program; FDA, U.S. Food and Drug Administration (82); CPMP, Committee on Proprietary Medicinal Products (83); CMR, Centre for Medicine Research (82); JPMA, Japanese Pharmaceutical Manufacturers Association (82); PDR, U.S. Physician's Desk Reference (20); CPB, Carcinogenic Potency Database (84).

maceuticals and 69 known human carcinogens overall (21). As shown in Table 2, these carcinogenic pharmaceuticals fall into a small subset of categories (predominantly antineoplastics), and most are clearly genotoxic. If one expands the area of concern to all chemicals, IARC recognizes only 69 (21) agents as human carcinogens, and if one excludes hormones, immunosuppressants, and inorganic agents (metals, asbestos), most of these are clearly genotoxic as well. This discrepancy between the number of rodent carcinogens and the number of human carcinogens provokes one to ask why so many rodent bioassays are positive. Either there are many more human carcinogens waiting to be found with higher human exposures,

Table 2 Pharmaceutical Agents Listed as Human Carcinogens by IARC

Class (No.)	Genotoxicity
Antineoplastics (9)	+
Immunosuppressants (2)	+/–
Dermatologicals (3)	+
Hormones (4)	+/–
Miscellaneous (1)	+

Source: Ref. 20.

longer human exposure, or more careful and sensitive epidemiologic studies, or the design of the rodent bioassay has rendered it too sensitive at the expense of specificity (percentage of noncarcinogens that test negative).

Another important observation can be made from the results of rodent bioassays conducted over the years. If one excludes chemicals that are clearly genotoxic, the positive responses in rodent bioassays are limited to relatively few tissues (20,22,23). A good example of this is shown in Table 3. As these data are restricted to marketed human pharmaceuticals and most of these are not genotoxic, this list should be fairly representative of the tumor profile of nongenotoxic agents in rodent bioassays. Over 90% of tumors in rats occurred in liver and endocrine organs. Tumors in endocrine organs accounted for over 70% of all positive findings in the rat. The situation in the mouse is similar, with lung and liver tumors predominant.

Reviewing the NTP database in 1985, Haseman noted that approximately two thirds of the agents determined to be positive in rodent bioassays would not have been called positive if one-half the MTD had been used instead of the full MTD (7). A later examination of this database altered this conclusion somewhat (i.e., one third of the chemicals would not have been judged positive rather than two thirds). The authors concluded that restriction of the dose to one-half the MTD would reduce the sensitivity of the bioassay (24). With few exceptions, similar tissue site effects were seen at doses lower than the MTD (24). Although the incidence was lower at lower doses, there were, in most instances, no unique effects observed at the MTD (4). Thus, while sensitivity may have been reduced by reducing the dose, the organ specificity of the tumor response was essentially unchanged (4).

Table 3 Target Organs Found Positive in Rodent Carcinogenicity Studies

Rats		Mice	
Thyroid	15	Liver	18
Liver	13	Lung	17
Testis	13	Mammary gland	6
Mammary gland	10	Blood	6
Adrenal	9	Ovary	5
Pituitary	8		
All other organs or tissues reported	≤4	All other organs or tissues reported	≤3
No. of positive studies	76	Number of positive studies	62

Data extracted from *U.S. Physicians' Desk Reference*. Numbers indicate total number of carcinogenicity studies in which a tumorigenic response was seen in the indicated tissues.
Source: Ref. 20.

Williams and Weisburger pioneered the classification of carcinogens based on mechanism of action into genotoxic and epigenetic agents (25). While it is understood today that such a clear distinction is too simplistic, it is appropriate to examine agents that are not positive in well-designed conventional genetic toxicology assays. There is a growing body of evidence that rodent tumors detected at high doses of these nongenotoxic agents may result from processes that do not suggest human carcinogenic risk. It is important to emphasize here that simply determining that rodent tumors result from secondary mechanisms is not adequate to conclude that the rodent tumors have no relevance for human risk. It is imperative that sufficient data be generated to demonstrate that the tumorigenic process does not operate in humans at the expected exposure levels. With this caveat, what does the use of the MTD for nongenotoxic agents mean for the prediction of human risk? Let us examine the specifics of five examples of nongenotoxic rodent carcinogenesis: urinary bladder, renal tumors, forestomach, endocrine organs, and receptor-mediated tumors.

I. URINARY BLADDER TUMORS SECONDARY TO CYTOTOXIC EFFECTS

In the early 1970s it was shown that exposure of male rats to high doses of sodium saccharin for 2 years resulted in urinary bladder tumors (26). Similar results have since been reported for other sodium salts of weak acids (26). These findings led to the identification and regulation of this artificial sweetener as a possible human carcinogen and widespread product labeling of sodium saccharin.

Subsequently, much evidence has accumulated to suggest that the conclusion regarding potential human risk is not warranted. The most critical piece of evidence relative to the current discussion was that this response in the urinary bladder is dependent on exposure to extremely high doses. Rodent doses of 5% in the diet are required to produce an increase in urinary bladder tumors, and doses of 1% or less do not produce any evidence of a chemical-related effect (26–28).

The available evidence indicates that the tumor response with these agents is secondary to a proliferative response in the urothelium elicited by damage to superficial urothelial cells by microprecipitates (26,29). These precipitates are amorphous silicates formed in the presence of high concentrations of specific urinary proteins, require the presence of large amounts of sodium, and are dependent on high urinary pH (>6.5) (26,29,30). Another important observation was that the increase in tumor yield was dependent on administration of the agent during the neonatal period. If administration of sodium saccharin was delayed until 6 weeks of age in the rats, no tumors were seen (26). During the neonatal period active cell proliferation occurs in the urothelium. Later in life the rate of cell proliferation in the urothelium is much lower. Thus, the neonatal period is more sensitive to the further induction of cell proliferation and subsequent tumor formation (26).

It is clear that the use of very high doses in the early studies of sodium saccharin produced a series of events that are now known to be rat-specific, high-dose–dependent effects not seen in other species or at lower doses. Extrapolation from these high-dose effects to the lower human exposure conditions where the necessary preconditions for the bladder tumor response do not exist is not appropriate (31).

II. RENAL CELL TUMORS SECONDARY TO SPECIES-SPECIFIC CYTOTOXIC EFFECT

By a mechanism analogous to that described for the rat urinary bladder, exposure of male rats to high doses of several hydrocarbons (e.g., trimethylpentane, d-limonene) induces a nephropathy in subchronic studies (32,33). With continued administration, this cytotoxic response induces a responsive hyperplasia, which progresses through micronodule formation to renal cell adenoma and adenocarcinoma (34–38) (see Chapter 16).

The mechanism of this tumor formation has been extensively studied and is now well characterized (37). It is clear that tumor development is not directly attributable to the chemicals but rather to a chemical-protein complex. This class of chemicals is capable of binding to a specific rat protein, α_{2u}-globulin, which is secreted into the urine in large amounts. When this complex is reabsorbed by the proximal tubules, it is resistant to lysosomal degradation and thus accumulates in the tubular epithelial cells. This complex is cytotoxic at high concentrations and leads to cellular necrosis with a regenerative hyperplasia. It is this sustained hyperplastic effect in the presence of continued high-level exposure that is the direct tumorigenic stimulus to the kidney. Exposure to high levels of these chemicals in other species or strains of rat that do not produce this specific urinary protein do not result in the cytotoxic response in the tubular epithelial cell, which is the necessary precursor to tumorigenesis.

In addition to the species-specific nature of this response, it is important to emphasize for our current discussion that the cytotoxic response is dependent on exposure to high levels of the chemicals. As with most secondary mechanisms, there is a threshold below which the cytotoxic response is not seen and thus no tumors are produced in the kidney. The scientific weight of evidence for agents that induce male rat hydrocarbon nephropathy was adequate for the EPA to exclude tumors found by this mechanism from the risk-assessment process (38).

III. FORESTOMACH TUMORS SECONDARY TO A CYTOTOXIC RESPONSE

The forestomach of rodents is an anatomical structure not found in most other mammals. It is lined by squamous epithelium and, in this regard, resembles the histology of the esophagus most closely. Tumors in this organ have been the cause for concern for human safety with such agents as the antioxidant butylated hydroxyanisole (BHA) (39).

Rodents will tolerate relatively high doses of BHA with little evidence of adverse effect on conventional measures of toxicity (e.g., body weight gain). At the high doses required to demonstrate an MTD, however, this agent produces a cytotoxic effect in the squamous epithelial cells of the forestomach leading to acanthosis and, with continued BHA administration, papillomas and squamous cell carcinomas. In an important series of experiments, Ito and colleagues were able to clearly demonstrate the dose-response nature of this effect. Doses of 2% in the diet yielded squamous cell carcinomas. Intermediate doses of 1% BHA yielded acanthosis and papillomas but no carcinomas. At 0.5% BHA in the diet no tumors were seen (40).

As with the examples above, BHA is not the proximate tumorigenic stimulus. At these very high concentrations in the diet, the chemical induces a cytotoxic response. The resultant responsive hyperplasia leads ultimately to tumor formation (41,42). In this case, the use of the MTD has resulted in the production of a biological effect that will not occur at much lower human exposure levels. Without the mechanistic data relating the observed tumors to the cytotoxic response, the findings at the high dose would lead to a misinterpretation of potential human risk.

A similar example comes from the findings in this organ with ethyl acrylate (EA), which is a commercially important monomer used in latex paints, textiles, paper coatings, leather finishing, adhesives, and plastics. Although EA did not induce a dose-related increased in tumors from chronic inhalation, dermal, or drinking water dosing (43–45), chronic gavage administration of 100 or 200 mg/kg EA in corn oil resulted in a dose-related increase in the incidence of both benign and malignant epithelial tumors in the forestomach of rats and mice (46). These lesions were accompanied by gross signs of irritation (inflammation, edema, hyperplasia, and hyperkeratosis) in the target tissue. No indications of toxicity or increases in tumor incidence were observed in other organs. When the same dosing regimen was employed for 13 weeks and the animals allowed to recover for 19 months, neoplasia was not observed (47). In addition to this observation, mechanistic studies from the same laboratory have suggested that chronic irritation induced by gavage dosing of EA is critical for the induction of the forestomach tumors (48–51).

A single dose of 200 mg/kg EA in corn oil has been shown to induce vesicle formation, erosion, and ulceration of the squamous epithelium without corresponding mucosal damage in the glandular stomach (48). Two weeks of gavage dosing with 2–200 mg/kg EA in corn oil caused a dose-dependent increase in the incidence and severity of forestomach lesions in male F344 rats, ranging from minimal to mild epithelial hyperplasia at 20 mg/kg through moderate to marked hyperplasia at 200 mg/kg (52). Equivalent doses of EA in the drinking water for 2 weeks were much less effective in inducing forestomach hyperplasia. The glandular stomach remained unaffected in either case. Similar results, but with fewer dose levels, were observed in a 13-week study (53). After 2 weeks of EA dosing at the maximum tolerated dose, 200 mg/kg, the acute forestomach toxicity correlated with significant depletion of nonprotein sulfhydryl content (52).

Carcinogenic risk of EA is clearly associated with toxicity: irritation, inflammation, and subsequent hyperplasia. Dose levels that do not induce histopathological indications of toxicity do not yield tumors in rodents and would not be expected to represent a human carcinogenic risk. Further information on forestomach and glandular stomach carcinogenesis is presented in Chapter 16.

IV. TUMORS IN ENDOCRINE ORGANS SECONDARY TO ALTERATION OF ENDOCRINE HOMEOSTASIS

As indicated above, tumors in endocrine organs account for approximately 70% of all positive findings in rat carcinogenicity studies. An extensive amount of work has been done in an effort to understand the mechanism behind these endocrine tumors. This work has led to the conclusion that these tumors do not arise directly as a result of the chemical agent. Rather these tumors are secondary to a biochemical or pharmacological alteration of normal cellular endocrine homeostasis resulting from exposure to high doses of the test agents. It is the resulting trophic stimulus of specific hormones that results in the hyperplastic effect leading ultimately to tumorigenesis. This general mechanism has been established for the tumor types listed in Table 4.

While it is not possible within the scope of this chapter to elaborate

Table 4 Hormonal Mechanisms of Endocrine Tumors

Endocrine tissue	*Tumor type*	*Trophic hormone*	*Examples*
Thyroid	Follicular cell adenoma and adenocarcinoma	TSH	Sulfonamides, thioureas, phenobarbital
Testes	Leydig cell adenomas and adenocarcinomas	LH	Flutamide, finasteride, ammonium perflouroctanoate
Ovary	Tubular adenomas	FSH, LH	Nitrofurantoin
Mammary gland	Fibroadenomas and fibrosarcomas	Prolactin	Haloperidol, reserpine
Uterus	Endometrial tumors	FSH, estrogen	Bromociptine (and other dopamine agonists)
Adrenal	Pheochromocytomas	Catecholamines	Reserpine
Pancreas	Acinar cell adenomas	CCK	Trypsin inhibitors
	Islet cell adenomas	?	Corticosteriods
Glandular stomach	ECL cell carcinoids	Gastrin	Omeprazole, ranitidine
Mesovarial ligament	Leiomyoma, leiomyosarcoma	Cholinergic agents	β-Adrenergic receptor agonists (e.g., salbutamol)

TSH, Thyroid-stimulating hormone; LH, luteinizing hormone; FSH, follicle-stimulating hormone; CCK, cholecystokinin; ECL, enterochromaffinlike.

fully on all of these secondary mechanisms, several observations common to all of these examples are appropriate to consider in the context of dose selection in the bioassay. The tumorigenic response observed with agents that elicit endocrine tumors in rodents is generally dose responsive. As the dose of the test agent increases, the extent of perturbation of endocrine homeostasis increases, resulting in a stronger or more prolonged trophic stimulus to the target cell. The dose of the test agent required to do this may be quite high, as in the case of finasteride-induced alterations of LH and the resultant Leydig cell adenomas in mice (54). Conversely, the required tumorigenic dose of the test agent may be low relative to human exposure levels [as in the case of omeprazole and gastrin-induced ECL cell carcinoids in the glandular stomach of rats (55)]. In each of these cases, however, an experimental dose can be defined below which no significant endocrine disruption is seen and, consequently, no tumors are produced. As mentioned earlier, however, simply identifying that the tumors in rodents are secondary to altered endocrine homeostasis is not sufficient to conclude that these findings have no significance for human safety assessment. In these instances a comprehensive evaluation of the rodent bioassay data is essential to show that the necessary tumorigenic stimulus does not operate in humans at the expected exposure levels. This dissociation may be shown to be due simply to marked differences in doses required to produce the observed effect in rodents and that expected in humans [e.g., finasteride and Leydig cell tumors in mice (54)].

A marked difference in sensitivity of the target cell population between rodents and humans or a difference in the endocrine physiology between the species may also be important in understanding the significance of the rodent bioassay findings for human risk [e.g., omeprazole and ranitidine and gastrin-induced ECL cell carcinoids (55,56,86), various agents, and induction of TSH-induced thyroid follicular cell adenomas (57)]. It is in this area that the use of high doses in the rodent bioassay creates the most difficulty. In the early years of the use of the bioassay, endocrine tumors in rodents caused much concern. With the extensive work that has been conducted over the years, these tumors are not now generally regarded as important for human safety if they fall into these widely studied and characterized classes and the appropriate data are available to support a secondary endocrinological mechanism. Yet the use of the MTD in the bioassay will continue to yield experimental findings that are widely viewed as unimportant for human risk assessment. It is also important to point out here that the means of determining that the rodent findings are not predictive of human risk are much easier to obtain relatively speaking than for environmental chemicals. The absence of definitive human exposure response data for environmental chemicals may make interpretation of high-dose rodent studies quite difficult.

V. RECEPTOR-BASED MECHANISMS OF TUMORIGENESIS

A final area that should be discussed is that of receptor-mediated tumorigenic responses in rodents and the use of high doses. This is an important area that we are only beginning to understand. Data are rapidly becoming available, the significance of which is not yet clearly understood.

High doses of a variety of classes of agents (fibrates, phthalate esters) have been shown to produce hepatocellular adenomas and carcinomas in rats associated with an increase in the appearance of peroxisomes, subcellular organelles (58). The association between this hepatic peroxisome proliferation and hepatocellular tumors has been the subject of much conjecture.

While human epidemiology data suggest that chemicals that induce peroxisomal proliferation as well as hepatocellular tumors in rodents are not human tumorigens (59,60; see also Chapter 13), the data on the mechanism of tumor induction in rodents are not clear. There are a number of hypotheses to explain the association between peroxisome proliferation and tumor development and why these tumors are not important for human safety (59,60; see also Chapter 13). The discovery of a peroxisome proliferator-activated receptor (PPAR) (61) has made clear and convincing risk assessment more difficult at the present time. Information on questions such as the relationship between binding to the receptor and subsequent induction of peroxisomal enzymes and the difference between receptor subtypes in rodents and humans is necessary before a clear understanding of the dose-response relationship can be achieved. This discovery has, however, made possible an ultimately better understanding of the relationship between the positive findings in the rodent bioassay and what may happen with human exposure. The use of the MTD for peroxisome proliferating chemicals without a clear understanding of the relative interactions with the PPAR in the test species and humans will make risk assessment using bioassay data difficult at best.

A second area of current intensive investigation in receptor-based mechanisms of tumorigenesis that exemplifies difficulties associated with a default approach to the use of the MTD in rodents is the pleiotrophic responses to chemical binding to the aromatic hydrocarbon (Ah) receptor. Many years ago, it was shown that a wide variety of cellular responses was initiated by binding of aromatic hydrocarbons to a specific protein identified as the Ah receptor (62,63). A series of molecular events initiated by this binding and subsequent binding of the Ah receptor–chemical complex to DNA recognition sites leads to the expression of specific genes and the consequent cellular effects of their protein products (62). Although the specific nature of the cascade of events is not yet fully defined, it is generally believed that the diverse effects of compounds such as dioxin (2,3,7,8,-tetrachlorodibenzo-*p*-dioxin) as well as the broad species toxicity differences observed are related to the diverse pathways that are downstream of the initial binding to the Ah receptor (62,64–66). In fact, it can be shown that these responses are not due to a direct toxic effect of the chemical per se but, rather, result directly from this receptor binding (62).

Rats have been shown to be quite sensitive to dioxin, showing tumors at very low dose levels (67,68). Dietary exposure to levels as low as 0.01 mg/kg/day resulted in an increase in hepatocellular tumor incidence (68,69). Importantly, tumors were only detected in animals showing significant hepatotoxicity (69); in the absence of hepatoxicity, no liver tumors were seen.

Superficial evaluation of these data would suggest a significant potential risk for humans. Epidemiological data are available for the worst accidental exposure to this agent in Seveso, Italy, in 1976. Although initial follow-up reports from the decade following this accident did not reveal an increase in the incidence of human cancers (70), subsequent studies have suggested slight increases in certain types of cancers among the most heavily exposed people (71). Similar epidemiological studies in exposed workers have demonstrated increased cancer risks, although the magnitude of the increased risk is also small in those studies (72). It is clear, however, from these human epidemiological studies that the significant tumorigenic response predicted by the rodent studies has not been seen in humans. These data indicate that important quantitative differences exist between species in response to receptor binding or receptor binding–induced effects.

A greater understanding of the molecular biology and the species differences in the responses to binding to the Ah receptor will result in a more accurate risk assessment (73,74).

VI. DISCUSSION AND CONCLUSIONS

During the early years of the rodent bioassay, the concept evolved that to detect subtle carcinogenic effects with adequate statistical confidence, it was necessary to use the highest doses compatible with long-term survival of the test animal. Thus, the cancer bioassay was created to serve a hazard-identification function in overall risk assessment. Over the quarter century during which rodent cancer bioassays have been employed, this approach has led to some elegant descriptions of rodent responses to toxic doses of chemicals and a much more complete understanding of the biology of these responses.

With these data now in hand, however, it is appropriate to ask whether high doses are necessary to evaluate human tumorigenic risk. A review of the NTP database reveals "relatively few instances in which the MTD produced unique effects which are not also observed (with reduced incidence) at lower dose levels" (4). In fact, these and other authors (4,75,76) pointed out a weak correlation between low MTD value (toxic chemicals) and higher carcinogenic potential in rodents. Subsequently, Monro and Davies demonstrated that known human carcinogens are positive in the rodent bioassay at relatively low doses (77). What then is the value of high doses in light of the ultimate objective of our efforts—evaluation of human carcinogenic risk? It is clear that we can increase the sensitivity of the bioassay by increasing the dose. Does the use of the MTD enhance or detract from the process of human risk assessment? As suggested by Rozman and colleagues (78), the very use of the MTD makes one of the most important functions of experimental toxicology—the definition of the dose-response relationship and threshold doses—very difficult. Using MTD-generated data for quantitative risk assessment does not take into account the qualitative nature of the response and forces the arbitrary distinction between "carcinogens" and "noncarcinogens." The purpose of the bioassay should be to assess potential human risk under the conditions of human exposure.

This chapter demonstrates that high doses of chemicals produce effects that range from cytotoxicity and the resultant regenerative hyperplasia through disturbances of endocrine homeostasis to modulation of gene expression through binding to cellular receptors or other regulatory elements. Tumors that result from these secondary mechanisms that are operative only at high doses in rodents may not operate at low doses and may not have relevance to human safety. It is important, however, not to oversimplify this point. In their critical review of this area, Melnick and coauthors argue that we currently have insufficient data to dismiss tumors produced in rodents by these mechanisms as unimportant for human safety (79). They argue that more research into the underlying mechanisms and how these mechanisms operate at expected exposure levels in humans is needed to permit firmer conclusions. A good example of this approach has been used to understand the differences in tumorigenic responses in mice to chloroform administered either in corn oil or by gavage. In the absence of the marked and sustained cell proliferation seen when chloroform is administered in corn oil, comparable exposures to chloroform in drinking water produces no increase in liver tumors (80).

In the examples cited above, the induction of tumors at doses at or above the MTD are not relevant to human risk at reasonably expected human exposure levels. The body of evidence we have collected clearly indicates that it is imperative to take a global approach to characterize the biological response to a test agent. These data can then be used to select dose levels of bioassays that provide useful and relevant information for risk assessment. This approach is becoming more generally accepted (15,81). If, for example, test agents are nongenotoxic in a well-validated battery of short-term genetic tests and the 90-day rodent subchronic studies and early mechanistic work indicates a similar mode of action to the agents cited above, then chronic dose levels at or above the MTD will be of little use in the safety evaluation of that agent. While it is difficult to absolutely predict tumor response in specific tissues, methodology does exist to design studies to detect biologically relevant tumors and understand those tumors that have little or no relevance to humans.

For more than 20 years we have adhered to a methodology that requires, in some cases, dose levels that clearly overwhelm the biology of the test animal and are of no relevance to human exposure scenarios. With the resultant database in hand and our current technology, we can design, perform, interpret, and evaluate valid human risk assessments without the simplistic reliance on a maximally tolerated dose.

REFERENCES

1. Weisburger EK. History of the bioassay program of the National Cancer Institute. Prog Exp Tumor Res 1983; 26:187–201.
2. Page N, ed. Concept of a bioassay program in environmental carcinogenesis. In: Kraybill HF, Mehlman HA, eds. Advances in Medical Toxicology. New York: Wiley & Sons, 1977:87–171.
3. McConnell EE. The maximum tolerated dose. The debate. J Am Coll Toxicol 1989; 8(6):1115–1120.
4. Haseman JK, Seilkop SK. An examination of the association between maximum-tolerated dose and carcinogenicity in 326 long-term studies in rats and mice. Fundam Appl Toxicol 1992; 19:207–213.
5. Sontag JM, Page NP, Saffiotti U. Guidelines for carcinogen bioassay in small rodents. Natl Cancer Inst 1976:76–801.
6. Apostolou A. Relevance of maximum tolerated dose to human carcinogenic risk. Regul Toxicol Pharmacol 1990; 11:68–80.
7. Haseman JK. Issues in carcinogenicity testing: dose selection. Fundam Appl Toxicol 1985; 5:66–78.
8. International Life Sciences Institute. The selection of doses in chronic toxicity/carcinogenicity studies. In: Grice HC, ed. Current Issues in Toxicology. New York: Springer Verlag, 1984:9–49.
9. IARC. Long-term and short-term screening assays for carcinogens: a critical appraisal. IARC Monograph Evaluating Carcinogenic Risk of Chemicals to Man International Agency for Research on Cancer 1980; (suppl.2):21–83.
10. Organization for Economic Cooperation and Development (OECD). Guidelines for Testing of Chemicals, Section 4, No. 451. Paris: OECD, 1981:8.
11. U.S. Environmental Protection Agency. Pesticide Assessment Guidelines. Subdivision F Hazard Evaluation. Human & Domestic Animals, rev. ed. Springfield, VA: U.S. Dept. Of Commerce. National Technical Information Service. PB86-108958, 1984:120.
12. Clayson DB, Iverson F, Mueller R. An appreciation of the maximum tolerated dose: an inadequately precise decision point in designing a carcinogenesis bioassay? Teratogenesis Carcinog Mutagenesis 1991; 11:279–296.
13. Gehring PJ, Wantanabe PG, Blau GE. Pharmacokinetic studies in evaluation

of the toxicological and environmental hazard of chemicals. In: Mehlman MA, Shapiro RE, Blumenthal H, eds. New Concepts in Safety Evaluation. New York: Wiley & Sons, 1976:195–270.
14. Levy G. Dose dependent effects in pharmacokinetics. In: Tedeschi DH, Tedeschi RE, eds. Importance of Fundamental Principles in Drug Evaluation. New York: Raven Press, 1968:141–172.
15. Contrera JF, Jacobs AC, Prasanna HR, Mehta M, Schmidt WJ, DeGeorge J. A systemic exposure-based alternative to the maximum tolerated dose for carcinogenicity studies of human therapeutics. J Am Coll Toxicol 1995; 14(1): 1–10.
16. Chhabra RS, Huff JE, Schwetz BA, Selkirk J. An overview of prechronic and chronic toxicity/carcinogenicity experimental study designs and criteria used by the National Toxicology Program. Environ Health Perspect 1990; 86:313–321.
17. Bucher JR, Portier CJ, Goodman JI, Faustman EM, Lucier GW. Workshop Overview. National Toxicology Program Studies: principles of dose selection and applications to mechanistic based risk assessment. Fundam Appl Toxicol 1996; 31:1–8.
18. National Toxicology Program Board of Scientific Counselors. Final Report of the Advisory Review. Fed Reg 1992; 57:31721–31730.
19. Fung VA, Barrett JC, Huff J. The carcinogenesis bioassay in perspective: Application in identifying human cancer hazards. Environ Health Perspect 1995; 103:7–8.
20. Davies TS, Monro A. Marketed human pharmaceuticals reported to be tumorigenic in rodents. J Am Coll Toxicol 1995; 14(2):90–107.
21. World Health Organization. International Agency for Research on Cancer. IARC Monographs of the Evaluation of Carcinogenic Risks to Humans. List of Evaluations, 1995:1–7.
22. MacDonald JS, Lankas GR, Morrissey RE. Toxicokinetic and mechanistic considerations in the interpretation of the rodent bioassay. Toxicol Pathol 1994; 22(2):124–140.
23. McClain RM. Mechanistic considerations in the regulation and classification of chemical carcinogens. In: Kotsonis FN, Mackey M, Hjelle J, eds. Nutritional Toxicology. New York: Raven Press, Ltd., 1994:273–303.
24. Haseman JK, Lockhart A. The relationship between use of the maximum tolerated dose and study sensitivity for detecting rodent carcinogenicity. Fundam Appl Toxicol 1994; 22:382–391.
25. Williams GM, Weisburger JH. Chemical carcinogenesis. In: Amdur MO, Doull J, Klassen CD, eds. Casarett & Doull's Toxicology, The Basic Science of Poisons. (4th ed.). New York: McGraw-Hill, 1993:127–200.
26. Ellwein LB, Cohen SM. The health risks of saccharin revisited. Crit Rev Toxicol 1990; 20:311–326.
27. Fukushima S, Arai M, Nakanowatari J, Hibino T, Okuda M, Ito N. Differences in susceptibility to sodium saccharin among various strains of rats and other animal species. Gann 1983; 74:8–20.
28. Schoenig GP, Goldenthal EI, Geil RG, Frith CH, Richter WR, Carlborg FW.

Evaluation of the dose response and in utero exposure to saccharin in the rat. Food Chem Toxicol 1985; 23:475–490.
29. Cohen SM, Cano M, Earl RA, Carson SD, Garland EM. A proposed role for silicates and protein in the proliferative effects of saccharin on the male rat urothelium. Carcinogenesis 1991; 12(9):1551–1555.
30. Cohen SM. Human relevance of animal carcinogenicity studies. Reg Toxicol Pharmacol 1995; 21:75–80.
31. Cohen SM, Ellwein LB. Risk assessment based on high-dose animal exposure experiments. Chem Res Toxicol 1992; 5:742–748.
32. Alden CL, Ridder G, Stone L. The pathogenesis of the nephrotoxity of volatile hydrocarbons in the male rat. In: Mehlman MA, Hemstreet GP, Thorpe JJ, Weaver NK, eds. Renal Effects of Petroleum Hydrocarbons. Princeton, NJ: Princeton Scientific Publishers 1984:107–120.
33. Alden CL. A review of unique male rat hydrocarbon nephropathy. Toxicol Pathol 1986; 14:109–111.
34. Swenberg JA, Short B, Borghoff S, Strasser J, Charbonneau M. The comparative patho-biology of α_{2u}-globulin nephropathology. Toxicol Appl Pharmacol 1989; 97:35–46.
35. Short BG, Burnett VL, Swenberg JA. Histopathology and cell proliferation induced by 2,2,4-trimethylpentane in the male rat kidney. Toxicol Pathol 1986; 14:194–203.
36. Short BG, Burnett VL, Cox MG, Bus JS, Swenberg JA. Site-specific renal cytotoxicity and cell proliferation in male rats exposed to petroleum hydrocarbons. Lab Invest 1987; 57:564–577.
37. Hard GC, Rodgers IS, Baetcke KP, Richards WL, McGaughy RE, Valcovic LR. Hazard evaluation of chemicals that cause accumulation of I_{2u}-globulin, hyaline droplet nephropathy, and tubule neophasia in the kidneys of male rats. Environ Health Perspect 1993; 99:313–349.
38. Alpha$_{2u}$-Globulin: Association with Chemically Induced Renal Toxicity and Neoplasia in the Male Rat. EPA/62a5/3-91/019F. Washington, DC U.S. Environmental Protection Agency, 1991.
39. Flamm WG, Lehman-McKeeman LD. The human relevance of the renal tumor-inducing potential of *d*-limonene in male rats: implications for risk assessment. Reg Toxicol Pharmacol 1991; 13:70–86.
40. Ito N, Fukushima S, Tamano S, Hirose M, Hagiwara A. Dose response in butylated hydroxyanisole induction of forestomach carcinogenesis in F344 rats. J Natl Cancer Inst 1986; 77:1261–1265.
41. Clayson DB, Iverson F, Nera EA, Lok E. The significance of induced forestomach tumors. Ann Rev Pharmacol Toxicol 1990; 30:441–463.
42. Grice HC. Safety evaluation of butylated hydroxyanisole from the perspective of effects on forestomach and oesophageal squamous epithelium. Food Chem Toxicol 1988; 26:717–723.
43. Miller RR, Young JT, Kociba RJ, Keyes DG, Bodner KM, Calhoun LL, Ayres JA. Chronic toxicity and oncogenicity bioassay of inhaled ethyl actylate in Fischer 344 rats and B6C3F1 mice. Drug Chem Toxicol 1985; 8:1–42.
44. DePass IR, Fowler EH, Meckley DR, Weil CS. Derman oncogenicity bioas-

says of acrylic acid, ethyl acrylate, and butyl acrylate. J Toxicol Environ Health 1984; 14:115–120.

45. Borzelleca JF, Larson PS, Hennigar GR, Iluf EG, Crawford EM, Smith RB. Studies on the chronic oral toxicity of monomeric ethyl acrylate and methyl methacrylate. Toxicol Appl Pharmacol 1964; 6:29–36.
46. National Toxicology Program. Carcinogenesis Bioassay of Ethyl Acrylate. Technical Report Series 259, Publication (NIH) 82-2515. Research Triangle Park, NC: U.S. Department of Health and Human Services, Public Health Service, National Institute of Health, 1986.
47. Ghanayem BI, Matthews HB, Maronpot RR. Sustainability of forestomach hyperplasia in rats treated with ethyl acrylate for thirteen weeks and regression after cessation of dosing. Toxicol Pathol 1991; 19:273–297.
48. Ghanayem BI, Maronpot RR, Matthews HB. Ethyl acrylate induced gastric toxicity. 1. Effect of single and repetitive dosing. Toxicol Appl Pharmacol 1985; 80:323–335.
49. Ghanayem BI, Maronpot RR, Matthews HB. Ethyl acrylate-induced gastric toxicity. II. Structure-toxicity relationships and mechanism. Toxicol Appl Pharmacol 1985; 80:336–344.
50. Ghanayem BI, Maronpot RR, Matthews HB. Ethyl acrylate-induced gastric toxicity. III. Development and recovery of lesions. Toxicol Appl Pharmacol 1986; 83:576–583.
51. Ghanayem BI, Maronpot RR, Matthews HB. Association of chemically induced forestomach cell proliferation and carcinogenesis. Cancer Lett 1986; 32: 271–278.
52. Frederick CB, Hazelton GA, Frantz JD. Histopathologic and biochemical response of the stomach of male F 344/N rats following two weeks of oral dosing with ethyl acrylate. Toxicol Pathol 1990; 18:247–256.
53. Frederick CB, Chang-Mateu IM. Contact site carcinogenicity: estimation of an upper limit for risk of dermal dosing site tumors based on oral dosing site carcinogenicity. In: Gerrity TR, Henry CJ, eds. Principles of Route-to-Route Extrapolation for Risk Assessment. New York: Elsevier, 1990:237–270.
54. Prahalada S, Majka JA, Soper KA, Nett TM, Bagdon WJ, Peter CP, Burek JD, MacDonald JS, van Zwieten MJ. Leydig cell hyperplasia and adenomas in mice treated with finasteride, a 5α-reductase inhibitor: possible mechanism. Fundam Appl Toxicol 1994; 22:211–219.
55. Havu N, Mattsson H, Ekman L, Carlsson E. Enterochromaffin-like cell carcinoids in the rat gastric mucosa following long-term administration of ranitidine. Digestion 1990; 45:189–195.
56. Larsson H, Carlsson E, Hakanson R, Mattsson H, Nilsson G, Seensalu R, Wallmark B, Sundler F. Time-course of development and reversal of gastric endocrine cell hyperplasia after inhibition of acid secretion. Studies with omeprazole and ranitidine in intact and antrectomized rats. Gastroenterology 1988; 95:1477–1486.
57. Hill RN, Erdreich LS, Paynter OE, Roberts PA, Rosenthal SL, Wilkinson CF. Review thyroid follicular cell carcinogenesis. Fundam Appl Toxicol 1989; 12:629–697.

58. Reddy JK. Carcinogenicity of peroxisome proliferators: evaluation and mechanisms. Biochem Soc Trans 1990; 18:92–94.
59. Stott WT. Chemically induced proliferation of peroxisomes: implications for risk assessment. Reg Toxicol Pharmacol 1988; 8:125–159.
60. Ashby J, Brady A, Elcombe CR, Elliott BM, Ishmael J, Odum J, Tugwood JD, Kettle S, Pruchase IFH. Mechanistically-based human hazard assessment of peroxisome proliferator-induced hepatocarcinogenesis. Human Exp Toxicol 1994; 13(suppl 2):S1–117.
61. Isseman I, Green S. Activation of a membrane of the steroid hormone receptor super family by peroxisome proliferation. Nature (London) 1990; 347:645–650.
62. Landers JP, Bunce NJ. The Ah receptor and the mechanisms of dioxin toxicity. Biochem J 1991; 276:273–287.
63. Whitlock JP, Jr. Mechanism of dioxin action: relevance to risk assessment. In: Gallo MA, Scheuplein RJ. Banbury Report 35: Biological Basis for Risk Assessment of Dioxins and Related Compounds, 1991:351–359.
64. Whitlock JP, Jr. Genetic and molecular aspects of 2,3,7,8-tetrachlorodibenzo-p-dioxin action. Annu Rev Pharmacol Toxicol 1990; 30:251–277.
65. Poland A, Knutson JC. 2,3,7,8-tetrachloro-dibenzo-p-dioxin and related aromatic hydrocarbons: examination of the mechanism of toxicity. Annu Rev Pharmacol Toxicol 1982; 22:517–554.
66. Safe SH. Comparative toxicology and mechanism of action of polychlorinated dibenzo-p-dioxins and dibenzofurans. Annu Rev Pharmacol Toxicol 1986; 26: 371–399.
67. Huff JE, Salmon AG, Hopper NK, Zeise L. Long-term carcinogenesis studies on 2,3,7,8-tetrachlorodibenzo-p-dioxin and hexachlorodibenzo-p-dioxins. Cell Biol Toxicol 1991; 7:67–94.
68. Kociba RJ, Keyes DG, Beyer JE, Carreon RM, Wade CE, Dittenber DA, Kalnins RP, Frauson LE, Park CN, Barnard SD, Hummel RA, Humiston CG. Results of a two-year chronic toxicity and oncogenicity study of 2,3,7,8-tetrachlorodibenzo-p-dioxin in rats. Toxicol Appl Pharmacol 1978; 46:279–303.
69. Goodman DG, Sauer RM. Hepatotoxicity and carcinogenicity in female Sprague-Dawley rats treated with 2,3,7,8-tetrachlorodibenzo-p-dioxin (TCDD): a pathology working group reevaluation. Reg Toxicol Pharmacol 1992; 15:245–252.
70. Johnson ES. Human exposure to 2,3,7,8-TCDD and risk of cancer. Crit Rev Toxicol 1992; 21:451–463.
71. Bertazzi PA, Pesatori AC, Consonni D, Tironi A, Landi MT, Zocchetti C. Cancer incidence in a population accidentally exposed to 2,3,7,8-tetrachlorodibenzo-para-dioxin. Epidemiology 1993; 4(5):398–406.
72. Fingerhut MA, Halpern WE, Mallow DA, Piacitelli LA, Honchar PA, Sweeny MH, Greife AL, Dill PA, Streenland K, Suruda AJ. Cancer mortality of workers exposed to 2,3,7,8-tetrachlorodibenzo-*p*-dioxin. N Engl J Med 1991; 324:212–218.
73. DeVito MJ, Birnbaum LS. Dioxins: model chemicals for assessing receptor-mediated toxicity. Toxicology 1995; 102:115–123.

74. Demby KB, Lucier GW. Receptor-mediated carcinogenesis: the role of biological effect modeling for risk assessment of dioxin and tamoxifen. Prog Clin Biol Res 1996; 394:113–129.
75. Hoel DG, Haseman JK, Hogan MD, Huff J, McConnell EE. The impact of toxicity on carcinogenicity studies: Implications for risk assessment. Carcinogenesis 1988; 9(11):2045–2052.
76. Haseman JK, Lockhart A. The relationship between use of the maximum tolerated dose and study sensitivity for detecting rodent carcinogenicity. Fundam Appl Toxicol 1994; 22:382–391.
77. Monro A, Davies TS. High dose levels are not necessary in rodent studies to detect human carcinogens. Cancer Lett 1993; 75:183–194.
78. Rozman KK, Kerecsen L, Viluksela MK, Osterle D, Demi E, Viluksela M, Stahl BU, Grein H, Doull J. A toxicologist's view of cancer risk assessment. Drug Metab Rev 1996; 28(1&2):29–52.
79. Melnick RL, Kohn MC, Portier CJ. Implications for risk assessment of suggested nongenotoxic mechanisms of chemical carcinogenesis. Environ Health Perspect 1996; 104(suppl 1):123–134.
80. Larson JL, Wolf DC, Butterworth BE. Induced cytotoxicity and cell proliferation in the hepatocarcinogenicity of chloroform in female $B6C3F_1$ mice: comparison of administration by gavage in corn oil vs ad libitum in drinking water. Fundam Appl Toxicol 1994; 22:90–102.
81. National Research Council. Use of the maximum tolerated dose in animal bioassays for carcinogenicity. In: Goldstein et al., eds. Issues in Risk Assessment. Washington, DC: National Academy Press, 1993:15–83.
82. Monro A. Rapporteurs' Report. Testing for carcinogenic potential. In: D'arcy PF, Harron DWG, eds. Proceedings of Third International Conference on Harmonization. Greystone Books, Ltd. 1996:261-268.
83. Van der Laan JW. Regulatory viewpoint, testing for carcinogenic potential. In: D'arcy PF, Harron DWG, eds. Proceedings of Third International Conference on Harmonization. Greystone Books, Ltd. 1996:269–273.
84. Gold LS, Sawyer CB, Magaw R, Backman GM, De Veciana M, Levinson R, Hopper NK, Havender WR, Bernstein L, Peto R, Pike MC, Ames BN. Carcinogenic potency database of the standardized results of animal bioassays. Environ Health Perspect 1984; 58:9–319.
85. Marselos M, Vaino H. Carcinogenic properties of pharmaceutical agents evaluated in the IARC monographs programme. Carcinogenesis 1991; 12(10): 1751–1766.
86. Havu N. Enterochromaffin-like cell carcinoids of gastric mucosa in rats after lifelong inhibition of gastric secretion. Digestion 1986; 35(suppl 1):42–55.

4
Mouse Liver Carcinogenesis

Thomas L. Goldsworthy, Elizabeth H. Romach, and Tony R. Fox
Chemical Industry Institute of Toxicology, Research Triangle Park, North Carolina

I. INTRODUCTION

The assignment of potential human carcinogenicity to a chemical relies heavily on the outcome of the rodent bioassay. The rodent organ site displaying the highest incidence of chemically induced tumors is the liver. In a recent survey of 533 chemicals tested in the National Toxicology Program (NTP) bioassay, 299 were identified as mouse carcinogens. Of these, 171 (57%) induced mouse liver tumors (1,2). Mouse hepatocarcinogens are represented by multiple chemical classes and include chemicals used as food additives, drugs, pesticides, and industrial and environmental agents and their intermediates. The relevance of the induction of mouse liver tumors to human risk has been an unresolved, contentious issue within the scientific and regulatory communities for over two decades. A number of factors have contributed to this controversy, but a factor that has been at the forefront is the inherently high spontaneous liver tumor incidence found in mouse strains used in the bioassay. A resolution of this issue is widely believed to require a mechanistic understanding of how these tumors arise spontaneously and how both genotoxic and nongenotoxic chemicals enhance this response.

In this chapter we discuss some of the issues that have an impact on the mouse liver tumor controversy. Our discussions include an examination of factors influencing the mouse liver tumor response, strain and species considerations, and the relationship of mouse liver tumor development to the hazard identification and risk assessment processes. This chapter emphasizes the role of cell and molecular growth alterations in mouse hepatocarcinogenesis. It is not the intent of our review to debate the value of the rodent bioassay; this issue is addressed in the chapters on validity, bioassay, and risk assessment in this volume. Instead, we wish to point out the scientific issues that are sources of concern regarding the mouse liver tumor controversy. Because we strongly believe that the ultimate solution to this controversy rests on a mechanistic understanding of the process of mouse liver tumor development, the bulk of this chapter will focus on a discussion of what is currently known and where research needs still exist.

II. ISSUES AND CONTROVERSIES

Although results from the rodent bioassay for carcinogenicity have a significant influence on the carcinogenic risk-assessment process, the development of tumors in rats and mice is not definitive evidence that a chemical is a human carcinogen. A number of factors complicate the extrapolation of rodent bioassay results in establishing potential human risk. Among these factors are differences in metabolism and pharmacokinetics between the test species and humans; inherent problems with the bioassay design such as the use of relatively low numbers of animals exposed to a high and limited dose range of the test chemical; lack of mechanistic understanding of the cancer process; and the existence of human and rodent cancer susceptibility genes, resulting in the occurrence of high incidences of tissue-specific spontaneous tumors.

The high susceptibility to tumor formation has centered primarily on the liver tumors induced in the B6C3F1 mouse, which is used in the NTP carcinogenicity bioassay. Historical tumor data indicate that the incidence of control liver tumors in male B6C3F1 hybrid mice averages around 31%, with a range of 7–55%. Female B6C3F1 mice have a spontaneous liver tumor incidence of approximately 6%, with a range of 2–8% (3). This background cancer incidence is frequently increased after chemical exposure. The controversy generated from this response is based on the relevance of the increase in tumors that occurs in an animal already predisposed to this type of tumor. The relevance of the response takes on even more uncertainty when (a) the only positive response in the bioassay is the mouse liver tumor, (b) it is induced by a chemical that is considered nongenotoxic, and (c) the chemical was administered at high exposure levels. For these reasons, the scientific community remains divided over the relevance of the induction of these tumors and the utility of the bioassay as a valid test system for the evaluation of human chemical carcinogenic risk.

Those opposed to using mouse liver tumor data for human risk assessment state that this endpoint has little value because of the high spontaneous background, e.g., these tumors are a peculiar anomaly for a genetically susceptible strain that bears little relevance to the total human population. The mouse is often stated to be an overly sensitive model test system that is not reflective of the normal human response. Thus, the high frequency of chemicals that induce liver tumors in mice is of questionable significance. This is particularly true when tumor induction occurs following exposure to high doses of the test agent, which may compromise the normal physiology of the animal or induce liver toxicity and regenerative cell proliferation. Also, because the liver in this strain of mouse contains endogenous retrovirus proviral vectors integrated into its genome, it is possible that these play a role in the tumorigenic response and thus limit the utility of the mouse for risk extrapolations. And finally, it is asserted that the mouse liver tumor model is irrelevant to human risk because liver cancer is relatively uncommon in humans in the United States.

On the other hand, those supporting the use of mouse liver tumor data in risk-assessment decisions state that a sensitive model is necessary to detect a potential positive response because the number of animals typically used in the bioassay is relatively small. The fact that so many chemicals induce liver tumors should not be surprising, since the liver is where the test chemical will, in most cases, have the highest concentrations and because the liver is the primary organ to activate proximate carcinogens. Tissue-specific cancer susceptibility genes have been identified in both humans and rodents. Thus one cannot discount the B6C3F1 mouse as irrelevant because there may also be individuals that are equally susceptible in heterogeneous human populations. Human liver cancer is a common disease worldwide. Furthermore, there is no experimental evidence to suggest that the basic processes of tumor development are qualitatively different between mice and humans.

Groups who are either for or against the use of mouse liver tumor data in human risk assessment both present convincing cases for their particular position. Our current understanding of how these tumors develop, however, is insufficient to substantiate the validity of either position. For this reason, we advocate a more complete mechanistic description of these processes. Only then can this scientific debate and regulatory impasse be resolved.

III. CANCER BIOASSAYS

In the 1960s, the National Cancer Institute (NCI) began testing chemicals for carcinogenicity due to increasing public pressure to perform routine animal testing of industrial chemicals. The National Cancer Act of 1971 provided legislative authority for NCI to plan and develop an expanded, intensified, and coordinated cancer research program (4). In response to this legislation, NCI produced a document providing guidelines for the bioassay of chemicals for carcinogenic potential in small rodents. Because of the increased concern about the exposure of human populations to harmful chemicals, the Department of Health, Education, and Welfare launched the NTP in 1978 to test chemicals of public health concern as well as to develop and validate new test methods. The NTP hazard identification is largely based on animal studies. Since its beginning in 1978, NTP has tested approximately 500 chemicals for carcinogenicity. In more recent years, other institutions and private organizations have begun the routine testing of chemicals for carcinogenicity using the NTP bioassay as a guide.

The current NTP carcinogen bioassay was based on a document produced by the NCI in 1976. Its basic core has remained essentially unchanged over the past 20 years despite recent additions to improve information on mechanisms. A 90-day subchronic study is done prior to the chronic bioassay to help determine the maximum tolerated dose (MTD), which is defined as the highest dose of the test chemical that does not cause death or chronic toxicity. Originally, the ideal MTD should cause no more than a 10% reduction in body weight in the 90-day study and not exceed 5% of the diet to avoid affecting the nutritive balance of the diet (5). For the cancer bioassay, a test chemical is administered to groups of 50 mice or rats of each sex per dose group. Each bioassay includes a control group, a high-dose group given the maximum tolerated dose, and at least one more group that usually receives 50% or less of the MTD. The NTP has recently reviewed the factors important in the selection of doses for cancer bioassays and the application of NTP studies to mechanistic-based risk assessments (6). The strain of mouse most frequently used in the NTP chronic bioassay is the B6C3F1 hybrid, a cross between the male C3H and the female C57BL mouse strains. The in-life portion of the bioassay is followed by a histopathological study to identify and classify lesions that result from treatment with the test compound. (For more details, see Chapter 1 of this volume.)

There is a strong correlation between mutagenicity and rodent carcinogenicity. However, we now recognize that many chemicals that were not identified as mutagens by in vitro or in vivo tests are carcinogenic to rodents. A study analyzing 114 chemicals assayed by NTP shows that approximately 50% of the chemicals that were carcinogenic in rodent bioassays were not mutagenic (7,8). A genotoxic carcinogen is defined as one that is capable of producing cancer by *directly* altering the genetic material of a cell. A nongenotoxic carcinogen is capable of inducing cancer by some other mechanism(s) *not* involving *direct* gene damage (9). Many of the chemicals that test positive in rodent carcinogen bioassays are nongenotoxic, and thus scientists and regulators need to understand and employ the mechanisms of carcinogenesis to appropriately assess the human risk resulting from exposure to these chemicals. The division of agents into genotoxic or nongenotoxic has proven helpful in understanding modes of action, but the two categories are clearly not mutually exclusive. The multistep nature of cancer strongly implies that multiple genes, mechanisms, and causes exist in carcinogenesis. Understanding the mechanisms of carcinogenesis requires having enough information about the biology of the cancer process to recognize what biological processes are disrupted by the chemical exposure and how this disruption contributes to tumor formation. This information is needed to reasonably assess human risk resulting from human exposure to carcinogenic agents.

IV. MORPHOLOGY AND PATHOGENESIS OF MOUSE LIVER TUMORS

Understanding the morphological and biological behavior of neoplastic lesions in rodents is important for our understanding of the pathogenesis of neoplasia and in the toxicological evaluation of chemicals. Extrapolation of animal data to humans is based on evaluation of the information available and in part relies on the accurate diagnosis of the proliferative lesions in the rat and mouse liver. Both hepatocellular and nonhepatocellular primary tumors are found in the livers of mice. Nonhepatocellular tumors include hepatoblastomas and tumors of vascular, stromal and ductular origin; these lesions are found in mice, rats, monkeys, and humans (10). A limited number of mouse liver carcinogens induce nonhepatocellular tumors. Given that the vast majority of spontaneous and chemically induced mouse liver tumors are hepatocellular tumors, these lesions are the focus of the following discussion.

A. Nomenclature and Morphology

Numerous reports have been devoted to establishing histological criteria for diagnosing mouse hepatocellular tumors. The terminology (e.g., hepatoma, type A and type B nodules, nodular hyperplasia, and mouse liver tumor types I, II, III, and IV) used to diagnose these lesions has been variable and controversial and has changed over time. Although the importance of an acceptable and uniform classification of mouse liver tumors cannot be understated, numerous conflicting reports on the nomenclature and morphological criteria of these lesions currently exist. The Society of Toxicologic Pathologists (STP) and the International Life Sciences Institute (ILSI) have established a joint committee to aid in the harmonization of nomenclature and diagnostic criteria used in toxicological pathology. The classification guidelines for proliferative lesions in mice are expected to be finalized in 1997.

Current classifications of mouse hepatic lesions are based on similar histogenic and morphological features found in rats and other species. Many of the current problems associated with diagnosis and significance of mouse liver tumors stem from the lack of understanding of the various stages in the carcinogenic process. Similar controversies concerning the pathogenesis of mouse tumors exist for liver lesions in rats and other species.

To improve consistency and study-to-study comparability, the NTP has adopted uniform and conventional nomenclature for hepatoproliferative lesions in mice. Terminology, based on morphological appearance, includes *foci of cellular alteration, hepatocellular adenoma*, and *hepatocellular carcinoma*. Descriptions of these terms can be found in a review (11). These terms associate biological behavior with a specific morphological appearance and are believed to represent a spectrum of changes in the pathogenesis of mouse liver neoplasia.

B. Pathogenesis Features

In the past, the neoplastic classification of hepatoproliferative lesions in mice has been questioned. Scientists now widely accept that a typical hepatocellular carcinoma in the mouse exhibits morphological abnormalities similar to carcinomas in other species and displays functional and biological behavior indicative of malignancy. Metastatic potential of mouse liver tumors is related to tumor size, type, and degree of differentiation. These tumors metastasize with a frequency similar to liver tumors found in other species. Carcinomas from mouse liver can be readily transplanted. Although morphological differences between spontaneous and chemically induced neoplasms have been reported, there is no overall relationship between the class of carcinogen, morphology, and pathogenesis of tumor development. These findings suggest that liver tumor classification cannot by itself identify the type of carcinogen or mechanism of action.

Compared with hepatocellular carcinomas, the neoplastic classification and behavior of hepatocellular adenomas and foci are more controversial. The bulk of evidence, including morphology, transplantability, and progressive growth, suggests that mouse hepatocellular adenomas are appropriately designated as benign neoplasms. The evidence of reversibility is more controversial and often case specific and cannot be discerned by morphological criteria alone. Foci of cellular alteration are localized lesions that are usually distinguished from surrounding hepatic parenchymal tissue by altered morphological, biochemical, or enzymatic parameters. These lesions are believed to be reversible cell populations that may progress to neoplasia. Limited evidence suggests that some foci may progress directly to carcinoma without intermediate adenoma formation. Foci of cellular alterations are much less common in mice than in rats (for rat liver foci, see Chapter 12). The potential of focal phenotypic subtypes to progress to neoplasia can differ under the influence of certain external and internal stimuli. Thus, the pathology and etiology in the pathogenesis of liver tumors in mice are apparently quite similar to lesions in rats, monkeys, and humans. Neoplastic development in tissues other than liver also exhibit similar lesions and pathogenesis.

In rodents and humans, males almost invariably exhibit a higher incidence and quicker latency of hepatic tumors than females (11,12). This finding is very apparent in liver tumor induction in most strains of mice and is also observed in experimental models using chemical initiation. Both castration and ovariectomy can alter spontaneous and chemically induced liver tumor induction (13,14). Castration of male mice decreases growth rate and hepatic tumor multiplicity, whereas ovariectomy of female mice increases hepatic growth and tumor development above that for intact females. Testosterone administration to ovariectomized female mice results in a further increase in tumor development, demonstrating that testosterone acts as a tumor promoter in mice. Strain-dependent effects of sex hormones on mouse hepatocarcinogenesis has also been demonstrated (15). Estrogen administration to diethylnitrosamine (DEN)-initiated female mice decreased the number and growth rate of hepatic foci (16), demonstrating the inhibitory actions of estrogen on tumor promotion. We are investigating the hypothesis that some chemicals may selectively induce liver tumors in cancer bioassays in female but not male mice through modification of estrogen and testosterone levels and subsequent altered tumor promotion. Unleaded gasoline, a female mouse liver carcinogen and tumor promoter, lacked tumor-promoting activity in ovariectomized female mice (17). Collectively, these studies support a role for sex hormones in modifying mouse hepatocarcinogenesis, probably through carcinogen metabolism, tumor promotion and progression, or both.

C. Markers

Researchers continue to search for distinctive and critical biological differences between normal and neoplastic hepatic lesions. Although many "biological" markers have been identified in hepatocarcinogenesis, no single marker or set of markers has been shown to be ubiquitous to the process of neoplasia or to a specific histogenic stage or type of lesion (18). Marker or phenotypic heterogeneity is the rule rather than the exception in hepatocarcinogenesis as well as in other nonhepatic neoplasms in general. Heterogeneity of marker appearance is observed in the staining patterns of many tumors. In our experience, the most consistent markers in mouse liver carcinogenesis are the altered hematoxylin and eosin (H&E) staining of foci on routine paraffin sections. The four common types of mouse liver foci identified with H&E analysis are basophilic, clear, acidophilic, and mixed (11). The most common immunohistochemical marker of mouse hepatic foci is glucose-6-phosphatase (19).

Differences in histochemical markers between benign and malignant mouse neoplasms have been reported, suggesting that transformation events may be associated with changes in the biochemical or histochemical markers. Certain treatment regimens and chemicals can alter and induce specific markers, which can differ from those seen in spontaneous lesions. Markers of preneoplastic and neoplastic lesions (e.g., glutathione-S-transferase II) may show sex-specific alterations and are believed to be under hormonal control. Markers are useful because they allow the researcher to identify and characterize the altered cell population of interest. However, we emphasize that markers may only reflect, but not define, the fundamental or critical nature of the alteration. The application of molecular techniques and the emphasis on the oncogenes, tumor-suppressor genes, and signals that affect cell cycle regulation will undoubtedly lead to the identification of new markers for describing the pathogenesis process.

Our group has been interested in identifying markers that may provide the early precancerous cell populations a selective advantage for subsequent growth and progression (20). We observed alterations in the expression of the hepatocyte growth regulators transforming growth factor alpha (TGF-α) and transforming growth factor beta (TGF-β) in mouse altered hepatic foci. A large proportion of eosinophilic foci display an increase in histochemical staining for TGF-α. In contrast, the vast majority of basophilic foci display no change in TGF-α immunostaining but do exhibit decreased immunostaining for TGF-β and the activating mannose-6-phosphate (M-6-P) receptor (20). The lack of TGF-β–induced inhibitory growth signals and apoptosis signals in lesions as compared with surrounding hepatic parenchyma is expected to provide these lesions with a survival and growth advantage. Interestingly, TGF-α immunostaining is increasingly observed in basophilic mouse lesions as they progress from foci to adenoma to carcinoma, suggesting that TGF-α may be one marker of progression (G. J. Moser and T. L. Goldsworthy, unpublished observations). Similar conclusions have been made in transgenic TGF-α mice (21) and in rat hepatocarcinogenesis (22,23).

D. Implications

In summary, hepatocellular tumors of mice have morphological and pathogenic characteristics similar to those of other species, including rats and humans. Whereas differences do exist for some characteristics, the similarities in pathogenesis are just as striking. Characteristics similar to other epithelial tumors of mice and other species also exist. These comparative findings do not support the notion of dismissing positive mouse liver tumors in safety evaluations because of its biological nature. However, other, yet-to-be-defined biological characteristics of mouse liver carcinogenesis may do so. A number of different nonhepatic tumors are found in laboratory rodents in frequencies that exceed the natural incidences of the same tumor observed in humans. Instead of discarding these tumors as biologically irrelevant, scientists must focus on understanding the mechanisms underlying mouse liver tumor induction and its relevance to potential hazard of the chemical to humans. Identification of the genes and factors that influence the susceptibility for developing spontaneous and chemically induced mouse liver tumors would help determine how these hepatic lesions might be more appropriately used in human cancer risk assessments.

V. CELLULAR AND MOLECULAR ALTERATIONS REGULATING GROWTH IN MOUSE CARCINOGENESIS

Cancer is a multistep process involving in part the activation of oncogenes and the inactivation of tumor suppressor genes. Alterations in multiple genes are believed to be required for the formation of malignancy. The multistep nature of carcinogenesis has begun to be defined for some human tumors such as those found in the colon (24,25). In contrast, the molecular mechanisms of mouse liver tumorigenesis remain largely unknown.

Studies are needed to define the molecular targets involved in the formation of mouse liver tumors. Only after the molecular targets disrupted during the neoplastic process are identified can the mechanisms of mouse liver carcinogenicity begin to be elucidated. Identifying gene targets within the mouse liver will allow researchers to establish the dose at which a target is disrupted. This in turn will make it possible to extrapolate the process to humans. Defining the molecular targets of carcinogenesis in mouse liver is a critical first step toward resolving the controversy surrounding the use of in rodent cancer bioassay data for assessing human risk to environmental and occupational chemicals. In the following section, we outline the current knowledge of molecular data involved during mouse liver tumor formation.

A. The *ras* Oncogene

There are three mammalian *ras* oncogenes, which are designed as H, K, and N-*ras*. All three genes code for a 21-kDa protein that is involved in a complex series of biochemical reactions mediating the transfer of extracellular information to the nucleus for use in the control of cellular growth and proliferation. The *ras* genes are activated in tumors by the induction of specific point mutations clustered within specific codons in exons 1, 2, and 3. These point mutations most commonly occur at codons 12, 13, 61, and 117. Mutations in the *ras* genes have frequently been detected in human tumors as well as in spontaneous and chemically induced animal tumors.

The first identified molecular alteration in mouse liver tumors was the activation of the H-*ras* oncogene (26,27). Since this early finding, numerous reports have documented the mutational frequencies and mutational spectrum found in this gene for spontaneous and chemically induced tumors of the B6C3F1 mouse and other mice of differing susceptibilities of liver tumorigenesis. The mutational activation of this gene is frequently found in mouse liver tumors, and data concerning its activation have potential use in chemical carcinogenic risk assessment. A detailed review of the *ras* oncogene and its involvement in mouse hepatocarcinogenesis is beyond the scope of this chapter. We refer those interested in a more extensive overview to a recently published review (28).

Research on *ras* gene activation in mouse liver tumor development has focused primarily on three areas: a comparison in the frequency of H-*ras* activation between spontaneous liver tumors and those induced by various genotoxic and nongenotoxic chemicals; an evaluation of whether various types of chemical carcinogens induce specific H-*ras* mutational spectra; and a determination of whether differences in H-*ras* mutation frequency occur between strains of mice that vary in liver tumor susceptibility.

*Comparison of Frequency of H-*ras* Activation Between Spontaneous Liver Tumors and Those Induced by Various Genotoxic and Nongenotoxic Chemicals*

The initial identification of activated H-*ras* proto-oncogenes in spontaneous and chemically induced mouse liver tumors suggested that the frequencies and mutational spectrum could allow a distinction between these two types of tumors. If an increase or decrease in H-*ras* activation is observed in chemically induced liver tumors compared with spontaneous neoplasms, this would imply that the chemical exposure preferentially favored a *ras* or non-*ras* pathway, respectively. This concept has had some usefulness towards making this distinction. However, when one examines the large database that has been generated over the past 10 years, the analysis of *ras* activation is clearly not as straightforward as originally hoped. This is, in part, due to the range of the response observed in both spontaneous and chemically induced tumors. The complexity of *ras* activation is best illustrated by examining the data from several selected studies.

Most investigations that analyze H-*ras* mutations in mouse spontaneous liver tumors have been conducted using the B6C3F1 mouse. Slightly over half of the analyzed tumors contain mutations in the *ras* gene (56%; 183/333) (Table 1). This frequency can vary from as low as 42% to as high as 82% depending on the study.

A wide range of responses in H-*ras* activation has been reported for chemically induced liver tumors in the B6C3F1 mouse (Table 2). This frequency varies from as high as 100% in tumors induced with the genotoxic agent N-hydroxy-2-acetylaminofluorene (N-OH-AAF) or 1-hydroxy 2,3-dihydroestragole (OH-DHE) to as low as 0% with the nongenotoxic rodent hepatocarcinogen chlordane. This comparative data has led to the idea that analysis of H-*ras* mutation frequencies in chemically induced tumors may be used to assess whether the increase in tumor response was a result of a particular chemical exposure. In addition, these data could be used to determine whether a chemical was acting via a genotoxic or a nongenotoxic mechanism. This is based on the premise that if a chemical is genotoxic, there should be an increase in the number of H-*ras* mutated cells within the liver, which would subsequently lead to an increase in the number of tumors harboring this alteration. To some extent, the above hypothesis holds true. However, when *ras* analysis is expanded to include a broad spectrum of chemicals, the concept is obviously too simplistic. This can be exemplified by an examination of data from tumors induced with diethylnitrosamine, a potent genotoxic carcinogen, and methylene chloride, a chemical generally considered nongenotoxic. The frequency of H-*ras* activation in these tumors is 26% and 76%, respectively (Table 2), as compared to the frequency of 56% observed in spontaneous tumors. There were fewer tumors with H-*ras* mutations after treatment with the genotoxic agent and more after exposure to the nongenotoxic chemical in this particular case. This points

Table 1 Spectrum of H-*ras* Codon 61 Mutations in Spontaneous Liver Tumors of B6C3F1 Mice

H-*ras* codon 61 mutations	Codon 61 CAA → AAA	Codon 61 CAA → CGA	Codon 61 CAA → CTA
183/333 (56%)	106/177 (60%)	50/177 (28%)	21/177 (12%)

Table 2 Examples of H-*ras* Activation in Chemically Induced Liver Tumors in B6C3F1 Mice

Treatment	No. of tumors analyzed	Tumors with any H-*ras* mutations (%)	Distribution of H-*ras* mutations: H-ras codon 61 (CAA) mutations (%)			Ref.
			AAA (%)	CGA (%)	CTA (%)	
N-OH-AAF	7	100	100	0	0	27
OH-DHE	11	100	0	50	50	27
Urethan	29	59	6	24	71	153
Vinyl carbamate	55	62	15	18	68	27,154
Methylene chloride	50	76	42	42	16	155
Furan[a]	29	35	40	10	0	156
Tetrachloroethylene[b]	53	28	33	20	33	157
Ciprofibrate[c]	39	36	50	7	0	72,158
Diethylnitrosamine	240	26	16	32	15	30,31,153,159,160
Chloroform	24	21	80	0	20	158
Phenobarbital	15	7	100	0	0	158
Chlordane	30	0	0	0	0	162

[a] 2 codon 117 AAG → AAC; 2 codon 117 AAG → AAT; 1 unknown.
[b] 1 codon 117 AAG → AAT; 1 exon 2 insert.
[c] 1 codon 117 AAG → AAC; 3 codon 117 AAG → AAT.

to the complexity of the chemically induced carcinogenic process and the fact that a specific classification of a chemical based on a single gene alteration may not provide sufficient information to define the carcinogenic mode of action of the chemical.

The observation that the frequency of chemically induced *ras* activation in hepatic tumors may be dose dependent adds a second needed dimension to the interpretation and utilization of *ras* activation data. Differing frequencies of *ras* mutation have been reported for vinyl carbamate (29) and DEN-induced liver tumors (30,31). Dose-dependent differences with respect to *ras* mutations (32) and mutational spectra of other target genes (30,33) have also been observed in nonliver organs in other experimental model systems. These data support the idea that other factors play a role in tumor induction at high doses and may include dose-dependent differences in DNA binding specificity, DNA repair, and cell death. Identification of dose-dependent differences in *ras* mutational activation, a mechanistic component of tumor induction, clearly challenges a basic tenet of carcinogen risk assessment. This tenent is that chemical carcinogens produce cancer by the same mechanism at either high or low doses.

*H-*ras *Mutational Spectrum and Its Implications for Risk Assessment*

For a variety of mutagenic agents, a unique chemical-specific mutation spectrum has been detected within a defined DNA sequence. In other words, each individual chemical has its own mutational fingerprint. This concept has led to the hypothesis that an analysis of the spectrum of mutations within the H-*ras* gene in the B6C3F1 mouse could be used to assess whether a chemical has in vivo genotoxic activity. Such an analysis is based on a comparison between the mutational spectrum observed in spontaneous tumors with that observed in tumors induced by a specific chemical agent. A difference between the two could be interpreted as being the result of the in vivo mutational activity of the chemical. If a chemical showed the same frequency and mutation spectrum in induced and spontaneous tumors, then it might be concluded that the chemical was promoting spontaneous lesions.

Most of the point mutations identified in hepatocellular neoplasms are located in the 61st codon of the *ras* gene, which has the normal nucleotide sequence of CAA. The three commonly occurring alterations at this position in spontaneous tumors are AAA, CGA, and CTA, with proportions of approximately 5 : 2 : 1, respectively (Table 1). On rare occasions, mutations are detected in the 117th codon of H-*ras* and in codon 13 of the K-*ras* gene.

Analysis of codon 61 mutations has been conducted for a number of chemicals that induce liver tumors in the B6C3F1 mouse. Tumors induced with vinyl carbamate (VC) and 1-hydroxy-2,3-dihydroestragole (OH-DHE) exhibit a codon 61 mutational profile, suggesting that these agents are inducing specific H-*ras* mutations and are therefore acting as in vivo mutagens. The mutational spectrum reveals the preferential occurrence of CAA to CGA or CTA mutations for OH-DHE and CAA to CTA for VC (Table 1). These particular mutations are relatively uncommon in spontaneous tumors and suggest a chemical-specific effect. In contrast, tumors induced with methylene chloride have a spectrum of mutations similar to that observed in spontaneous tumors. Such a result may have been expected given that methylene chloride is generally considered to be nongenotoxic. On the other hand, these tumors have a relatively high frequency of H-*ras* mutations, suggesting that methylene chloride has either direct or indirect genotoxic activity. So, as was found with the mutation frequency analysis, mutational spectral analysis can aid in the evaluation of the action of a chemical but does not provide definitive mechanism of action in and of itself.

*Effect of Strain on H-*ras *Mutation Frequency*

In addition to the influence of the individual carcinogenic agent, the frequency and spectrum of H-*ras* mutations is also dependent on the mouse strain used. In tumors induced by diethylnitrosamine (DEN) and VC, the frequency of H-*ras* activation ranges from 3 to 48% depending on the mouse strain (Table 3). There is some suggestion that the spectrum of mutations also shows a differential strain response. For example, the most predominant DEN-induced mutation in the B6C3F1 mouse is a CAA to CGA transition, whereas the CAA to AAA mutation is seen most frequently in the CD1 strain. However, when considering the spectrum in VC-induced tumors, the most frequent mutation is the CAA to CTA transversion in all the strains examined even though the frequency of H-*ras* activation varies in these strains from 10 to 100%. In contrast, the frequency of H-*ras*-containing tumors is similar, with 2,3,7,8-tetrachlorodibenzo-*p*-dioxin (TCDD) and methylclofenapate in both the hepatocarcinogenesis-susceptible B6C3F1 and less susceptible C57BL strains.

The data discussed above emphasize the complexity of H-*ras* mutational activation in murine hepatocarcinogenesis. H-*ras* activation is dependent on several factors, which include the specific chemical and strain of mouse used. Tumors induced with both genotoxic and nongenotoxic chemicals harbor these mutations, although there is a trend for their frequency to be lower with the nongenotoxic agents. For chemicals where exposure induces liver tumors that contain novel mutations in H-*ras*, the enhanced tumor burden may be attributed to the treatment regimen. Strains with

Table 3 Effect of Strain Differences on H-*ras* Activation and Mutational Spectrum

Group	H-*ras* mutations	H-*ras* codon 61 CAA → AAA	H-*ras* codon 61 CAA → CGA	H-*ras* codon 61 CAA → CTA	Ref.
Diethylnitrosamine (DEN)					
B6C3F1	63/240 (26%)	16	32	15	See Table 1
C3H	57/122 (47%)	29	25	3	31,163–165
C57BL	2/59[a] (3%)	0	1	0	31,153,165,166
B6CF1	1/18 (6%)	1	0	0	153
CD1	8/25 (32%)	7	1	0	167
Vinyl Carbamate (VC)					
B6C3F1	34/55 (62%)	5	6	23	27,154
B6D2F1	23/33 (70%)	4	9	10	29
B6CF1	1/10 (10%)	0	0	1	29
C57BL/6J	18/73 (25%)	0	7	11	29

[a]1Codon 61 CAA → CCA.

both high and low susceptibility to liver tumor development contain H-*ras* mutations, indicating that this gene does not by itself determine susceptibility to hepatocarcinogenesis. Studies discussing the nature of strain susceptibility follow in a later section of this chapter. Although *ras* activation is thought to be an important step in the multistep process of liver tumor development in mice, it is clearly not obligatory since many spontaneous and chemically induced tumors do not contain *ras* mutations.

B. Oncogenes and Growth Factors

Genes and proteins involved in growth signaling and cell cycle control are commonly altered in cancer and are believed to represent molecular targets for carcinogenesis. Much of what is known about these genetic alterations and their effects on mechanisms of tumorigenesis in mouse liver come from studies utilizing transgenic animals. Transgenic mice have been genetically engineered to integrate foreign DNA. Transgenic mice can contain activated oncogenes or genes encoding growth factors under the control of liver-specific enhancer/promoters such as albumin. Such transgenic mice have been used to study the effects of these genes on liver tumor formation. Transgenic animals can be made by transfecting the DNA of a gene of interest into the germline of an animal. Generally, a plasmid containing the foreign DNA is injected into the nucleus of an egg. The egg is subsequently implanted into a pseudopregnant mouse, which then gives birth to offspring expressing the foreign gene of interest (34). The generation of transgenic animals is a very powerful research tool that has been used to elucidate some of the processes and genes involved in neoplastic development (35) as well as an experimental chemical carcinogen–predicting system (see Chapter 7). Several lines of transgenic mice have been generated that result in the spontaneous formation of mouse liver tumors, suggesting that the transgene may play a role in hepatic tumor formation.

Liver Tumor Induction in Transgenic Mice

TGF-α Transgenic Mice. Studies utilizing mice carrying the TGF-α transgene have been useful in understanding the role that this gene plays in hepatocarcinogenesis. TGF-α is a growth factor that belongs to the epidermal growth factor superfamily of proteins. TGF-α exerts its mitogenic effect in the liver by binding to the epidermal growth factor receptor (EGFR). A number of studies implicate TGF-α as playing a role in carcinogenesis. Enhanced production of TGF-α has been reported in human cancers, including liver cancer (36,37). Of TGF-α transgenic mice, 75% develop hepatocellular carcinomas spontaneously (38). Chemical treatment of TGF-α transgenic mice affects their liver tumor response (39). When TGF-α transgenic mice are initiated with 5 mg/kg body weight DEN followed by promotion with 0.05% phenobarbital, they develop tumors with a shorter latency and an increased incidence as compared to mice without the transgene given the same chemical treatment. Thus the carcinogenic effect of the initiator (DEN) and the promoting effect of phenobarbital were accelerated in transgenic mice. Therefore, TGF-α transgenic mice have an enhanced sensitivity to development of hepatocellular carcinoma with DEN initiation and phenobarbital promotion. These data suggest that TGF-α is a target for carcinogenesis in rodent liver (40,41). Work done with TGF-α transgenic mice has helped to identify other genes that may cooperate with TGF-α in tumor formation. For example, one third of the liver tumors from TGF-α transgenic mice also had elevated levels of the RNA transcript for the c-*myc* oncogene, and 75% of these tumors had elevated insulinlike growth factor II (IGF-II) (42).

c-*myc* Transgenic Mice. The c-*myc* gene encodes for a nuclear phosphoprotein that binds to specific DNA sequences and activates transcription. Overexpression and amplification of c-*myc* is associated with a variety of human and rodent cancers (43). This gene is known to be activated in rodent liver following partial hepatectomy and in developing fetal liver, suggesting that c-*myc* normally plays a role in control of liver growth. Elevated levels of the *myc* protein disrupt control of cell proliferation (44,45). Therefore, inappropriate expression of c-*myc* may contribute to carcinogenesis by affecting pathways involved in growth control.

Transgenic mice expressing a c-*myc* transgene in the liver under control of an albumin enhancer/promoter develop liver tumors (46). However, the incidence is low and the latent period long. Deregulation of the *myc* oncogene alone is hypothesized not to be sufficient to lead to neoplasia but that some other alteration must cooperate with *myc* in tumor development. Double transgenic mice expressing c-*myc* and TGF-α were used to investigate the interaction of these two oncogenes in the tumorigenic process. Coexpression of c-*myc* and TGF-α transgenes in mouse liver resulted in a tremendous acceleration of neoplastic development as compared with expression of either transgene alone (47). This work supports the hypothesis that these two genes cooperate in the formation of liver neoplasia.

Another transgenic line that develops liver tumors are animals engineered to express the simian virus 40 (SV40) T antigen under control of a liver-specific promoter/enhancer. Although T antigens have no known cellular counterpart in mammals, they have been shown to transform a variety of cell types, including hepatocytes (47–49). SV40 T antigen transgenic mice develop liver tumors by 3–7 months of age. SV40 T antigen binds the retinoblastoma (RB) and p53 tumor suppressor proteins (50). SV40 T antigen may be tumorigenic, in part, through its ability to deregulate RB and p53.

Because of the link between activated *ras* and mouse liver tumors Sandgren et al. (46) developed a transgenic mouse containing a human mutant H-*ras* gene. Transgenic mice engineered with a liver-specific albumin promoter/enhancer fused to the *ras* oncogene coding sequence die of lung neoplasms between 1 and 5 months. At 5 months of age, 2 of 24 surviving mice had developed HCC. Dual transgenic mice generated with both *myc* and *ras* oncogenes developed large focal adenomas, hepatocellular carcinomas, and/or cholangiocarcinomas in approximately one third of the mice examined between 1 and 4 months of age. The remaining two thirds of these mice were sacrificed between 3 and 4 months of age due to the development of *ras*-induced lung tumors.

Dual transgenic mice containing the SV40 T antigen and *myc* developed hepatic tumors that resembled those induced by SV40 alone, although they had a shorter latent period. These dual transgenic mice died at 3–6 weeks of age from tumor burden as opposed to 3–5 months of age in mice with only the SV40 T antigen transgene. Work done with dual transgenic mice suggests that oncogenes cooperate during hepatic transformation in vivo. This work supports similar studies done in vitro.

IGF-II Transgenic Mice. Rogler and coworkers generated a transgenic mouse that expresses IGF-II in the adult tissues (51). IFG-II is a growth factor important in cellular proliferation and differentiation in the developing mouse embryo. This gene is not expressed in adult rodent liver. However, IGF-II expression has been reported to be reactivated in a number of different tumor types. The IGF-II gene is not expressed in the normal adult liver even in liver regenerating following partial hepatectomy. Reactivation of IGF-II transcription has been observed in hepatocellular carcinomas in rats (52), in hepatitis virus–induced carcinogenesis in woodchuck liver tumors (53,54), and in human liver cancers (55). Of IGF-II transgenic mice, 75% overexpress IGF-II relative to surrounding nontumor tissue. Transgenic mice expressing human IGF-II develop a spectrum of tumors at a higher frequency than controls, the most prevalent tumor type being HCC (56).

Growth-Related Gene Alterations in Spontaneous Liver Tumors

We are investigating the role of gene alterations involved in cell cycle control during B6C3F1 liver tumor development. Initial analysis has focused on gene alterations in tumor tissue as compared to surrounding nontumor tissue (Table 4). Once we have identified gene alterations in tumor tissue, studies will focus on the point during tumor development when these molecular alterations occur, what biological responses are associated with the specific gene alterations (e.g., apoptosis, proliferation, other phenotypic

Table 4 Alterations in Gene Expression

Gene	% Spontaneous HCC with alterations
IGF-II	61
c-*myc*	20
Cyclin D1	27
EGFR	91
TGF-α	0

changes), and the ways in which chemicals may perturb or affect these observed specific molecular alteration and their associated responses. We anticipate that this research approach will aid in identifying both critical genes and associated pathways involved in the neoplastic process in B6C3F1 liver.

Our results to date indicate that approximately 20% of spontaneous HCC analyzed from male B6C3F1 mice overexpressed c-*myc* mRNA relative to surrounding nontumor tissue (57). We also looked at expression of epidermal growth factor receptor mRNA and found that 91% of the tumors analyzed had a decrease in expression as compared to controls. TGF-α expression was unchanged in the tumors analyzed to date as determined by Northern blot analysis. This result may be due to the lack of sensitivity of Northern analysis since immunohistochemistry for the TGF-α protein revealed that cellular subpopulations within some of the tumors were expressing the TGF-α protein. Recent data generated using TGF-α null mice suggest that TGF-α is not required for early events in chemically induced hepatocarcinogenesis but suggest that it may be important in progression from foci to large tumors (21). It is possible that some other alteration has occurred in this pathway that substitutes for alterations in the TGF-α gene itself.

Recently, proteins involved in control of the cell cycle have been implicated as playing a role in the cancer process. One such protein is cyclin D1, which is involved in G1 progression of the cell cycle along with its associated cyclin-dependent kinase (cdk 4/5). Deregulated expression of cyclin D1 has been reported in several types of human cancer, including breast cancer, squamous cell carcinomas (58), and hepatocellular carcinoma (59). Additionally, cyclin D1 has been shown to be altered in mouse squamous cell cancers (60). Human HCC analyzed for expression of cyclin D1 RNA revealed that 22% of the tumors analyzed had an increase in expression of this transcript when compared to normal liver (61). In our evaluation of spontaneous HCC, 27% analyzed exhibited an increase in expression of cyclin D_1 mRNA and protein as compared to nontumor tissue. Our current research focus is the identification of alterations in other genes and proteins involved in the control of the cell cycle.

Interestingly, 61% of spontaneous HCC from male B6C3F1 mice have reactivated IGF-II gene expression. The IGF-II gene is paternally imprinted; therefore only the paternal allele is expressed. Some of the tumors analyzed in this study produced a large amount of IGF-II transcript. A loss of imprinting in the tumor tissue could lead to increases in the amount of transcript produced above that normally found in fetal liver. An increase in expression of IGF-II could result from a dysregulation of another gene that regulates the expression of IGF-II. For example, the DNA-binding protein encoded by the Wilm's tumor 1 gene (WT1) acts as a repressor of IGF-II expression (62). It has also been suggested that reexpression of IGF-II in tumors may reflect an undifferentiated state of the tumor cells. More work is needed to determine the mechanism of the reexpression of IGF-II transcripts in this and other systems. The biological significance of a reexpression of IGF-II in HCC in mice remains to be determined. It is, however, reasonable to speculate that an abnormal expression of this growth factor may give tumor cells a growth advantage.

Our investigations and those of others involving the molecular alterations in spontaneous and chemically induced mouse liver tumors clearly demonstrate that individual tumors, even within the same animal, exhibit different combinations of gene alterations. For example, we have observed (Table 4) all possible single or multiple combinations of myc, *ras*, cyclin D_1, and IGF-II in the analysis of about 30 individual tumors.

C. Tumor-Suppressor Genes

In addition to cellular oncogenes, the other family of cancer genes that has been shown to play an important role in carcinogenesis are the tumor-suppressor genes. To date, over a dozen tumor-suppressor genes have been identified and found to be inactivated in a wide spectrum of human tumors. These genes are involved in the regulation and control of such diverse biological processes as apoptosis, cell cycle control, and the regulation of gene transcription.

The most frequently inactivated tumor-suppressor gene in human tumors, including the liver, is the p53 gene (63). Investigations aimed at understanding the role that tumor-suppressor genes play in rodent carcinogenesis have been aided by the development of knockout strains of mice. With in vitro homologous recombination and selective breeding, a gene of interest can be made homozygously nonfunctional in mice. Several investigators have used this technology to create mice deficient in the p53 gene

(64–66). These mice are developmentally normal but have a high frequency of spontaneous tumors, the most prevalent being lymphomas. None of these mice developed liver tumors as a result of p53 inactivation; this result may be, in part, due to a shortened life span. Mice both homozygous and heterozygous for p53 inactivation were exposed to the potent liver carcinogen dimethylnitrosamine or to diethylnitrosamine. These animals showed an increased susceptibility to tumors including hemangiosarcomas of the liver, but there was no induction of hepatocellular carcinomas (65,67). These data suggest that p53 may not play a role in the induction of mouse liver tumors.

Several studies have been undertaken to directly determine whether p53 inactivation occurs within mouse liver tumors. Goodrow et al. (68) analyzed 9 spontaneous tumors and 34 chemically induced (8 DEN, 8 DMBA, 8 4-aminoazobenzene and 10 N-OH-2AAF) CD-1 liver tumors for mutations in the p53 gene. No mutations were detected in 43 tumors. Also, in two different studies, no p53 mutations were present when a combined 107 DEN-induced liver tumors from C3H/He, B6C3F1, and C57BL/6J mice were examined (69,70). Likewise, no p53 mutations were found in 6 spontaneous and 8 phenobarbital tumors from the C3H/He strain of mouse (70).

Because p53 inactivation appears to be a rather late event in the progression of human neoplasms, it has been suggested that mouse liver tumors were not suitably advanced to incur p53 aberrations. To investigate this hypothesis, p53 mutations were evaluated in hepatoblastomas, a highly malignant embryonal mouse liver tumor. No aberrations of the p53 gene were detected within exons 5–8 in any of the 16 tumors examined (71). These data support the idea that p53 does not play a major role in the development of hepatocellular tumors in mice.

There are very limited data implicating the involvement of other tumor suppressor genes in mouse liver tumor development. Besides p53, the other most frequently studied tumor-suppressor gene is the retinoblastoma susceptibility gene, Rb. The use of knockout mice for this gene has been somewhat limited because homozygous null mice exhibit embryonic lethality. Heterozygous Rb+/−mice as well as dual knockout Rb+/−; p53-/-mice have, however, been used to evaluate the role of Rb in mouse tumor development (66). Both types of mice displayed an increased sensitivity towards the development of certain tumor types; however, there was no impact on liver tumorigenesis. No studies have reported Rb inactivation from direct analysis of liver tumors, although 1 of 20 tumors induced by methylene chloride did show loss of heterozygosity in the vicinity of the Rb locus (72).

A number of investigators have analyzed mouse liver tumors for loss of heterozygosity in an attempt to identify other candidate tumor suppressor genes. This analysis has not resulted in a definitive conclusion. A study by Davis et al. (73) looked for the loss of heterozygosity in 142 spontaneous and chemically induced tumors from the B6C3F1 mouse at 78 different loci. Approximately 30% of the tumors exhibited loss of heterozygosity. Most of the loss of heterozygosity was observed at seven loci, and some of these occurred at regions that were syntenic to areas in the human genome where other known tumor suppressor genes are found, such as Wilm's, retinoblastoma, APC, MCC, and DCC. In contrast to these results, several other studies did not detect loss of heterozygosity in mouse liver tumors. In one study, 24 B6C3F1 mouse liver tumors were examined at 13 different loci, and 76 loci were examined in 58 tumors derived from HSB mice [(C3H/He × *Mus spretus*) × C57BL/6JBy]. Minisatellite fingerprint analysis detected one case of LOH and less than 1% genomic rearrangements in polymorphic and nonpolymorphic bands (74). In addition, no changes were observed at loci where hepatocellular susceptibility genes or markers homologous to loci were frequently lost in human hepatocellular carcinomas. This study suggests that the inactivation of tumor-suppressor genes may be an uncommon event in the generation of mouse liver tumor development.

Recent studies by Sargent et al. (75) investigated ploidy and karyotypic alterations associated with early events in mouse hepatocarcinogenesis. Hepatic foci and dysplasias were correlated to increase in chromosomal breakage and aneuploidy. The relatively specific and high frequency of breakage observed on chromosomes 1, 4, 7, and 12 suggests that these genetic regions are important to the hepatocarcinogenesis process. These regions correspond to tumor susceptibility genes in the mouse as well as displaying loss of heterozygosity in mouse liver tumors. Importantly, the breakpoints in mouse chromosomes are at equivalent human chromosomal regions that are rearranged in human tumors (76). Because genetic linkage groups are highly conserved between mice and humans, this information may provide clues for the identification of tumor genes that are important in both murine and human hepatocellular carcinomas.

D. Critical Pathways

Molecular components, e.g., oncogenes and tumor-suppressor genes, regulate the cell cycle through cell-signaling pathways (Fig. 1). The number of known molecular targets of carcinogenesis in mouse liver continues to increase and includes growth factors (IGF-II, TGF-α, TGF-β), growth factor receptors (IGF-II/M-6-P), signal transduction proteins (*ras*), cell cycle proteins (cyclin D1), and transcription factors (c-myc). These molecular com-

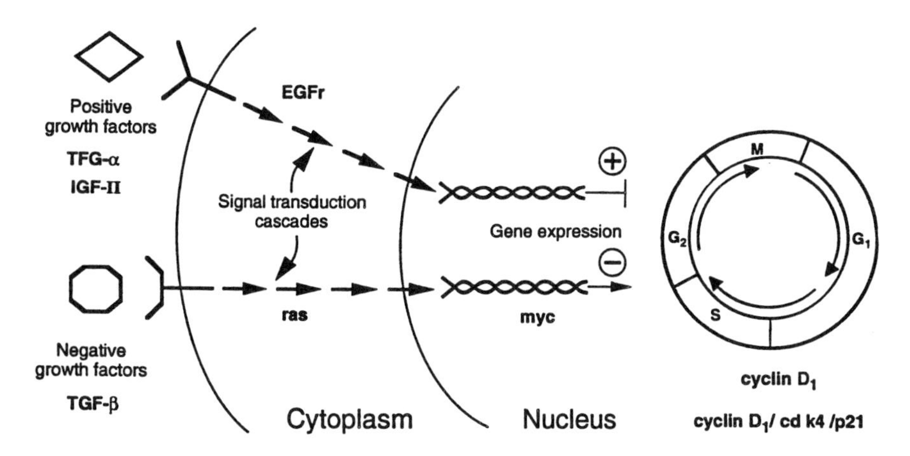

Figure 1 Molecular components regulating cell proliferation: a graphic model illustrating various components involved in cell cycle control. Pathways consisting of both negative and positive signals work in coordination with one another to regulate cell growth. Molecules involved in these processes can be functionally grouped into one of five different categories consisting of external growth factors, membrane-bound receptors that interact with the external ligands, components of signal transduction cascade pathways located within the cytoplasm, transcription factors within the nucleus regulating gene expression, and components involved in controlling the cell cycle itself. Numerous oncogenes and tumor-suppressor genes have been identified within these five areas that, when activated, can lead to a disregulation of cell cycle controls. The text describes genes altered in mouse liver carcinogenesis, including TGF-α and TGF-β (growth factors), M-6-P_R and EGF_R (membrane-bound growth factor receptors), *ras* (signal transduction pathways), *myc* (transcription factor), and cyclin D1 (cell cycle component).

ponents regulate the cell cycle through extracellular and intracellular cell-signaling pathways (Fig. 1). The molecular components so far identified in mouse liver carcinogenicity are, not surprisingly, genes that affect cell cycle control pathways. In all likelihood, these growth-related genes and associated pathways are the targets for chemical carcinogens. Genes involved in other cellular pathways also need to be explored, including those involved in repair, differentiation, apoptosis, senescence, immortalization, and metastasis. These are all pathways that, if altered, could contribute to development of neoplasia.

In summary, some of the gene targets involved in mouse liver tumor formation are now being identified. By identifying the targets of carcinogenesis and analyzing the combinations of alterations that occur in an individual tumor, investigators should be able to draw some conclusions about which genes are involved in growth-related pathways and therefore which pathways are altered in carcinogenesis. Disruption of multiple pathways is necessary for formation of tumors in animals (Fig. 2). More targets for disruption by carcinogens need to be identified before researchers can begin to identify the mechanisms whereby chemicals cause liver tumors in mice used in cancer bioassays. This information is needed so that data collected in rodent bioassays can be better used to accurately assess exposure–internal dose-response relationships.

E. Growth Control in Liver Tumor Development

Carcinogenesis in a complex, multistage process in which normal cellular proliferation becomes altered (77,78). Chemical carcinogens may act by modulating expression of or inducing mutations in genes that control nor-

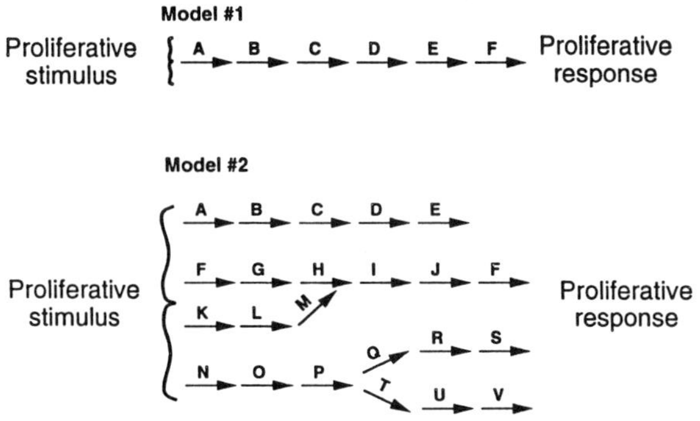

Figure 2 Molecular pathways regulating cell proliferation: models illustrating potential molecular pathways regulating cell proliferation. Model #1 represents a single linear pathway where a proliferative stimulus, such as a growth factor, would bind to its receptor (A) to initiate a sequential biochemical cascade (B–F), ultimately resulting in a cell entering the cell cycle and going through cellular division. The letters above the arrows represent genes (some of which may be oncogenes or tumor-suppressor genes) coding for the proteins participating in the series of biochemical reactions involved in the regulatory process. In this model, a single gene alteration theoretically could lead to the disruption in the normal control of cellular proliferation. Model #2 is representative of a more complex scenario where multiple biochemical pathways are involved in regulating cellular proliferation. This may consist of a combination of several single linear pathways or pathways that initially begin independently but then either converge or diverge downstream of the initiation point. The key element of this model is that multiple pathways may be required to cause complete disruption in the control of cell proliferation.

mal proliferation with subsequent clonal expansion of the resulting precancerous or cancerous cells. Emphasis has been placed on the utilization of cell proliferation data in the toxicological evaluations of chemicals. Research on apoptosis, a programmed form of cell death, has recently converged with the study of cell proliferation in assessing the altered growth observed in cancer development.

Chemicals that induce cancer can be subdivided into two broad groups: mutagenic and nongenotoxic carcinogens. Mutagenic chemicals or their metabolites interact directly with DNA to induce mutations or chromosomal alterations. Replication of the chemical-induced DNA damage by DNA synthesis before it is repaired is needed for the induction of mutation. Thus, mutagens are more effective carcinogens at doses that induce a proliferative response or in normally proliferating cells (78).

Nongenotoxic carcinogens do not interact directly with DNA and are generally negative in in vitro and in vivo tests of genotoxicity. Increases in cell proliferation induced by various nongenotoxic carcinogens have been implicated in the formation of liver tumors in rodents. Apoptosis, a genetically programmed process of cell death, has recently been shown to be as essential as cell proliferation in multistage hepatocarcinogenesis (79). Alterations of cell proliferation or apoptosis in target tissue by nongenotoxic agents may play a key role in the pathogenesis of tumors by increasing the survival of mutated cells or affording cancer cells a preferential growth advantage. Thus, chemically induced cell proliferation or loss of apoptosis or both may potentiate the effect of nongenotoxic agents and contribute to their neoplastic potential (80–84).

Proliferation

The type of proliferative responses to chemical exposures in the liver is often classified as being either cytotoxic or mitogenic (81,85). Cytotoxicants first produce hepatocyte death and then subsequent regenerative growth. This sustained hepatocyte proliferation can increase both background and induced mutations through DNA-replication errors. In contrast, hepatic mitogens produce a transient increase in hepatocyte proliferation in the absence of hepatotoxicity. Mitogens generally produce a sustained increase in liver weight for the duration of mitogen exposure. However, cell proliferation levels return to normal control rates or lower after the liver has reached its new size. Mitogenic agents can provide a selective growth advantage to initiated cells by decreasing apoptosis, inhibiting normal hepatocyte proliferation, or increasing cell proliferation in preneoplastic cell populations.

Various procedures have been used to determine the extent and duration of chemically induced cell proliferation and its role in carcinogenesis (80,81,85–87). To demonstrate an association between extended cell proliferation and tumorigenicity, experimental conditions for measuring proliferation should mimic those used in the cancer bioassay (85,86). Dose, route of exposure, strain, species, sex, animal age, diet, and environmental conditions should be reproduced in the cell proliferation–testing regimen. The demonstration of quantitative dose-response equivalence between cell proliferation and hepatocarcinogenic response is important. In the ideal situation, the evaluation of cell proliferation is conducted in parallel with the evaluation of carcinogenicity.

Cell proliferation over the extended periods of times required for nongenotoxic agents to increase the incidence of hepatic tumors has been evaluated in a limited number of quantitative studies. Chemically induced cell proliferation studies have focused primarily on a comparison of early proliferative responses between chemically treated and untreated animals after a single dose or subacute exposures. Strain-dependent differences in cell proliferation and gene expression have been noted (88). Most nongenotoxic agents require continued exposure over extended periods of time for tumor development. Proliferative responses seen after acute exposure do not necessarily reflect the proliferative response observed after chronic exposure. For example, continued exposure to the nongenotoxic mitogens phenobarbital (PB) or unleaded gasoline (UG) produces a short-term increase in mouse hepatocyte proliferation that is not sustained (89). In contrast, the cytotoxic agents furan (90) and chloroform (91) increase nonfocal hepatocyte proliferation after both acute and chronic treatment. Proliferative responses clearly differ for different hepatocarcinogens, thus limiting generalizations and the potential to predict responses from one agent to another.

Proliferation and Tumor Promotion

Foci of altered hepatocytes are putative preneoplastic lesions capable of progressing to hepatocellular carcinomas. Proliferative changes in nonfocal hepatocytes affect the initiation of a tumor, whereas proliferative changes in hepatic foci relate to tumor development and progression. Mouse hepatic foci generally have an increased hepatic labeling index (LI) relative to normal nonfocal hepatocytes (92) and may reflect clonal expansion of a single, presumably initiated hepatocyte (78). Focal growth may be due to preferential inhibition of proliferation in nonfocal hepatocytes, inhibition of focal hepatocyte apoptosis, selective proliferation of focal hepatocytes, or a combination of these three factors. As in the case of nonfocal hepatocyte proliferation, focal hepatocyte LI can vary greatly depending upon the hepatocarcinogenic agent as well as the phenotypic classes of hepatic foci. Furthermore, proliferation after acute or subchronic exposures in nonfocal hepatocytes does not predict the chemically induced proliferative responses observed in specific focal cell populations. The predictive capability of proliferative responses is further limited by substantial variations observed in hepatic foci within the same animal and in the LI in hepatic foci within animals of the same treatment group (92). For some nongenotoxic agents (phenobarbital, peroxisome proliferators), an improved understanding of long-term effects on cell loss as well as cell proliferation in focal and nonfocal hepatocytes is needed to elucidate the mechanisms and role of these carcinogens in induced growth alterations. For other nongenotoxic carcinogens such as chloroform, correlations with dose response and tumor outcome are very good (9).

To further investigate the role of cell proliferation in mouse hepatocarcinogenesis, we performed an initiation-promotion experiment (16). Female B6C3F1 mice were initiated with a single dose of the mutagenic agent DEN (5 mg/kg body weight). DNA synthesis rates in focal and nonfocal hepatocytes were assessed following 3.5 days of exposure to BrdU. BrdU is incorporated into nuclear DNA during the S phase of DNA synthesis in the cell cycle. Sixteen weeks after DEN initiation, the mean hepatic focal LI was approximately 20%; the mean nonfocal hepatic LI was only approximately 0.5%. The increased cell proliferation of focal hepatocytes is presumed to be critical to the net growth rate and progression of these tumor precursor populations. A wide range of labeling indices (4–35%) was observed in the hepatic foci, suggesting that individual lesions exhibit unique growth characteristics. This complicates generalizations regarding lesion growth and cancer potential and necessitates studies on individual foci to link gene alterations to focal growth and progression.

Tumor promoters are one class of nongenotoxic agents that accelerate tumor development by enhancing clonal expansion (78). Liver tumor promoters or inhibitors can reversibly modify the net growth rate of hepatic foci by altering focal or nonfocal cell proliferation, apoptosis, or both. In the above-mentioned tumor initiation-promotion studies, UG, a hepatic mitogen and tumor promoter, increased the size of DEN-initiated hepatic foci (Table 2). In contrast, treatment with ethinyl estradiol (EE) decreased the size of foci induced by DEN (16). To understand these treatment-induced differences in the size of foci, cell proliferation was assessed in and out of precancerous lesions. UG promotion significantly increased mean focal LI compared with DEN alone by approximately 60%, while EE treatment significantly decreased mean focal LI approximately 60%. Thus, UG-induced increases in focal size correlated with stimulation of focal hepatic LI and EE-induced decreases in focal size correlated with inhibition of focal hepatic LI. This supports the concept that tumor promoter- or inhibitor-induced focal hepatocyte proliferative alterations may be critical for treatment-related changes in clonal hepatocyte growth. Nonfocal hepatocyte proliferation was similar in all treatment groups (DEN/Control, DEN/UG, and DEN/EE) and was similar to noninitiated controls. These data further emphasize that cell proliferation in nonfocal hepatocytes after either acute or subchronic treatment does not necessarily correlate with or predict hepatic foci growth and that the promoters may exert a selective proliferative stimulus to preneoplastic cell populations compared with nonfocal cells.

Growth Factors and Cell Proliferation

Although the mechanisms of proliferative alterations in chemically induced hepatic foci are unknown, hepatocyte growth factors are known to play a pivotal role in liver growth. The regulation of cell proliferation is a highly complex process requiring the interplay and balance of multiple positive and negative growth factors. Positive growth factors such as TGF-α stimulate liver cell division. Negative growth factors, which include TGF-β, inhibit cell division in hepatocytes. Growth alterations that occur in tumor development could arise from the uncoupling of the normal regulatory constraints imposed by the balance of growth factors and their signaling mechanisms. For example, a preneoplastic liver cell may have altered growth factors concentrations, modulations in the quantity of growth factor receptors, and modifications in postreceptor pathways or responses.

TGF-β is a potent mitoinhibitory protein believed to play a role in the regulation of hepatocyte proliferation and apoptosis (93–96). The production of this growth factor is increased following a wave of proliferation in regenerating liver. TGF-β is believed to function by dampening the proliferative response, thereby maintaining homeostasis in the liver. There is evidence that deregulation of this growth factor may play a role in liver tumor formation. Studies in rats and humans have shown that proliferative lesions and tumors in the liver have decreased TGF-β receptors as demonstrated by immunohistochemical studies (97,98).

TGF-β is synthesized as a preprotein that must be activated before binding the TGF-β receptor to elicit a biological response (99). The inactive form of TGF-β contains two M-6-P residues. The latent form of TGF-β is activated by binding to the IGF-II/M-6-P receptor (100–103). Extracellular TGF-β latent complex binds to the IGF-II/M-6-P receptor through its M-6-P residues and is then cleaved to the active form by the proteolytic enzyme plasmin. The M-6-P receptor also binds IGF-II, which can be then subsequently internalized and degraded. Therefore, the M-6-P receptor is important, not only in activation of a negative liver growth factor, but also in degrading a positive growth factor. Loss of this receptor could result in uncontrolled liver growth. Some liver tumors in humans have lost or have mutated the IGF-II/M-6-P receptor (104,105).

To determine if increased LI in DEN-induced hepatic foci was related to modulations of hepatic growth regulators, we immunohistochemically stained DEN/control (Ct), DEN/UG, and DEN/EE liver sections for TGF-β and M-6-P/IGF-II receptor (20). Over 95% of DEN/Ct and DEN/UG foci, which were predominantly basophilic, were negative for TGF-β and M-6-P/IGF-II receptor immunostaining relative to nonfocal hepatocytes. In contrast, only about one half of DEN/EE foci were negative for TGF-β and M-6-P/IGF-II receptor. The remaining DEN/EE foci were either positive for TGF-β or elicited a staining intensity similar to that of background nonfocal hepatocytes.

These data demonstrate that DEN/Ct and DEN/UG hepatic foci elicited decreased TGF-β and M-6-P/IGF-II receptor relative to nonfocal hepatocytes. The concomitant modulation of TGF-β and M-6-P/IGF-II supports the hypothesis that decreased mature TGF-β was the result of a reduction in the activation of the secreted latent form of TGF-β by M-6-P/IGF-II receptor. This is consistent with work done in the rat liver by Jirtle et al. (106,107). Since TGF-β is a negative regulator of hepatocyte proliferation, the selective growth advantage of foci may be due to the negative growth effects that TGF-β exerts via M-6-P/IGF-II receptor activation on normal nonfocal but not focal hepatocytes. This could be highly advantageous to focal hepatocytes, allowing them to survive by blocking apoptosis, increasing proliferation, or both under in vivo conditions. The increased hepatic LI of focal hepatocytes relative to nonfocal hepatocytes is consistent with the selective growth advantage of foci.

Apoptosis

During recent years, it has become evident that programmed cell death, termed apoptosis, influences the multiple stages of hepatocarcinogenesis (79,84). The number of DNA-damaged cells or initiated cells may be increased by mitosis or diminished by apoptosis. Treatments leading to an increase in apoptosis may eliminate DNA-damaged cells and initiated cells and counter clonal expansion and suppress tumor growth. Similarly, clonal expansion of initiated cell populations may be stimulated by increased cell division or through decreased cell loss. Thus, the delicate balance between cell proliferation and apoptosis not only has an effect on organ growth but also may have an impact on the survival and growth of initiated cells and preneoplastic and tumor cell populations.

A number of chemical agents, including liver carcinogens and tumor promoters, has been shown to affect hepatocyte growth processes. Although most of this work have been done in rat liver, similar effects would be expected to occur during mouse liver carcinogenesis. Treatments that prevent preneoplastic growth and tumor development by enhancing lesion-specific apoptosis include food restriction and monoterpene perillyl alcohol administration (108,109). Inhibition or suppression of apoptosis has been proposed as an important mechanism of action for some types of nongenotoxic carcinogens. Disruption of apoptotic processes results in the survival and outgrowth of damaged or initiated cells. The altered hepatocytes would thus remain as targets for mitogenic stimulation and neoplastic alteration. Agents known to induce transient mitogenic activity such as phenobarbital, cyprolerone acetate, and peroxisome proliferators also inhibit apoptotic processes (110,111). Withdrawal of such agents may reverse tumor growth by stimulating apoptosis. Moreover, tumor promoters such as TCDD may promote foci growth, not by increasing cell proliferation, but by specifically inhibiting apoptosis in the cells of altered hepatic foci (112). Both decreases and increases in nonfocal hepatocyte apoptosis have been observed following treatment with the nongenotoxic mouse carcinogens dichloroacetic acid and furan, respectively (83,113). Although the precise mechanisms of these chemically induced cell growth alterations are unknown, nongenotoxic chemical-induced changes in growth factor signaling and cell communication may be critical in influencing the balance of hepatocyte growth processes (11,79,84,114–117).

Death or viability signals in the liver are produced via a wide variety of stimuli. Activators of hepatocyte cell death include DNA damage, food deprivation, differentiation, cytotoxic agents, TGF-β, fas ligand, tumor necrosis factor, and growth factor withdrawal. Inhibitors of hepatocyte cell death include growth factors and tumor promoters. Extracellular signals for apoptosis are recognized by receptors that trigger intracellular signaling. Death signals are believed to stimulate protease activity, which leads to increased endonuclease activity, alteration of the cell surface, and cytoskeleton reorganization. These alterations lead to fragmentation of the cell into apoptotic bodies that are subsequently phagocytosed by surrounding hepatocytes or macrophages. Although many different signals are capable of affecting apoptotic pathways, several evolutionarily conserved genes seem to be involved in the regulation of this process (119–122). All molecules involved in controlling apoptosis are believed to be constitutively expressed and, hence, always present in most mammalian cells. Cell survival would therefore be dependent on extrinsic and intrinsic signals that control the apoptotic pathways.

A number of cell cycle signaling genes, including myc, fos, cycA, E2F-1, p53, p21, and cdc-2, are involved in the induction of apoptosis.

Thus, proliferative signals can also function as death signals. Given that the same signals can trigger opposing responses, identifying the factors that determine which response occurs becomes critical. Cell type, context, and nature of the stress are at least part of the response determinants. TGF-β1 and structural analogs can trigger apoptosis in hepatocytes (95,123). Hepatocytes containing high levels of TGF-β1 are considered to be preapoptotic cells. In addition, the transmembrane protein fas, which belongs to the tumor necrosis factor and nerve growth factor receptor family, is a potent inducer of apoptosis in various tissues, including the liver (124). The number of genes and gene products found to be capable of affecting the regulation of apoptosis is growing at a rapid rate and continues to demonstrate the complexity of how genes may interact to influence the development of neoplasia (for review of liver regulation of apoptosis, see Refs. 79,84,115).

Various techniques have been used to identify apoptotis in rodent liver (83,125). The most commonly used staining methods for detection and quantitation of apoptotic cells and bodies in paraffin-embedded liver sections are H&E and nucleotide incorporation assays with in situ end-labeling. More recent methods are based on eosin fluorescence and TGF-β immunohistochemistry. The choice of method depends on many factors, including the scientific issue being addressed and the availability and type of cell population under study. Since apoptosis is highly regulated, apoptotic cells may be identified by the genes and proteins that mediate the apoptotic processes. Examples of such proteins include the vitronectin receptor, p53, c-*myc*, fas, bcl-2, and interleukin-1β-converting enzyme (126,127). Studies of these genes are essential for elucidating their specific roles in apoptotic processes and to determine potential chemical interaction. However, their use as general markers of apoptosis may be limited since these genes are only transiently expressed and active during the apoptotic process.

In summary, whether a net increase and growth of initiated cells, preneoplastic lesions, and hepatic tumors occurs is highly dependent on how a chemical affects the rates of both hepatic cell proliferation and cell death. Responses of cells to apoptotic signals change throughout neoplastic progression. Experimental information about hepatic cell proliferation and cell death rates induced by different types and doses of chemical treatments must therefore be examined throughout tumor development. This information is important for understanding the underlying mechanisms of chemically induced liver growth control and hepatocarcinogenesis. Experimental data such as liver weight, cell proliferation index, and apoptotic index are useful in that they can be applied to biologically based cancer models that mathematically describe transition rates, birth rates, and death rates for cells at different stages of carcinogenesis. Biologically based models incorporating cell birth and death processes (82) have been developed for mouse liver carcinogenesis and should be helpful in describing treatment- and dose-related growth effects by mouse carcinogens.

The induction of mouse liver tumors, both spontaneously and as the result of chemical exposure, is likely to occur by multiple mechanisms. This review has focused mainly on the cellular and gene alterations involved in growth control since altered growth is a hallmark of neoplastic cells. Other factors and mechanisms known to be involved with mouse liver hepatocarcinogenesis include but are not limited to modulations in receptors, growth signaling, and methylation, cell communication, oxidative stress, and ploidy. Genetic instability, altered repair, and cell cycle checkpoint controls have also recently been implicated in hepatocarcinogenesis. A description of these mechanisms in mouse liver carcinogenesis is given in volumes by Popp et al. (128) and Stevenson et al. (129). However, it is noteworthy that these mechanisms are not specific to the induction of mouse liver tumors. Similar mechanisms are operative and implicated in most if not all chemically induced rodent tumor responses as well as in human cancers.

VI. GENETIC SUSCEPTIBILITY IN MOUSE LIVER TUMOR DEVELOPMENT

Differences in the susceptibility of various mouse strains to develop liver tumors have been well documented (130). Strains such as C3H/HeJ, DBA/2J, and B6C3F1 are highly sensitive to the development of liver tumors, both spontaneously and as a consequence of chemical exposure. In contrast, strains like SWR/J, AJ, and C57BL/6J are comparatively resistant. A mechanistic understanding of this susceptibility or lack thereof is currently very rudimentary. However, efforts by several groups over the past 10 years have made important progress towards establishing the genetic basis for this susceptibility.

The importance of a genetic component to liver tumor susceptibility in these mice was clearly established in an elegant study conducted by Kitagawa and colleagues (131). They demonstrated that strain-specific liver tumor susceptibilities were controlled by intrinsic genetic differences in the hepatocytes, not by factors in the cellular environment. Chimeric (C3H/He)/(C57BL/6J) mice were created and foci induced by exposure to diethylnitrosamine. Six to nine months after treatment, the numbers and type of hepatic foci were scored. Using antibodies specific to the C3H/He antigens, immunohistochemical discrimination of C57BL/6J and C3H/He foci could be made. Quantitative analysis revealed that both the number and size of C3H/He foci in the chimeric livers were larger than the C57BL/6J foci. A greater tendency for malignant progression was also associated with the C3H/He cells. These results indicate that strain differences depend on the genetic composition of the cells themselves and not on the cellular milieu.

Studies in several laboratories have established that this genetic susceptibility lies not with a single gene, but with allelic differences at multiple loci. This was first established by Drinkwater and colleagues when they used segregation analysis of tumor susceptibility in recombinant inbred mice to identify the existence of at least two specific loci controlling the sensitivity to liver tumor induction (132). They designated these genes as *Hcs* (hepatocarcinogen sensitivity) and concluded that approximately 85% of the susceptibility in the C3H/HeJ mouse was due to a single gene located on chromosome 1 (133). The alleles at this locus are codominant, resulting in an intermediate sensitivity in heterozygotes such as the B6C3F1 mouse.

Since these early efforts, several other laboratories have used genetic linkage analysis to identified other putative *Hcs* genes in the C3H/HeJ mouse. Data from Dragani and coworkers suggest that three *Hcs* genes are located on chromosomes 7, 8, and 12 (134). The locations of these genes are in chromosomal regions homologous to regions frequently reported to be deleted in human hepatocellular carcinomas (32,135–137). Using a cross between (C3H/He × *Mus spretus*)F_1 mice and the C57BL6J strain to increase the number of informative polymorphic markers, Manenti and colleagues identified three additional regions on chromosomes 2, 5, and 19 that also showed linkage with hepatocellular tumor development (74). These studies suggest that as many as six or seven different genes may be involved in the genetic susceptibility to develop liver tumors in the C3H/HeJ mouse.

The story is complicated even more when this type of analysis is extended to other strains. Analysis of male C57BR/cdJ mice indicates that, for this sensitive strain, a major susceptibility gene resides on chromosome 17, a distinctly different gene from that which imparts susceptibility in the C3H/HeJ mouse (138). The complexity is further enhanced by the recent report describing the genetic factors influencing the susceptibility of male DBA/2J mice to develop liver tumors. When exposed to DEN perinatally, DBA/2J mice are 20-fold more sensitive to hepatocarcinogenesis than are C57BL/6J strain. Linkage analysis has revealed that the D2 mice carry multiple cancer-sensitivity loci and two major cancer-resistant genes (139). These studies suggest that the nature of the genes involved in dictating the susceptibility of mice to develop liver tumors is dependent on multiple factors involving combinations and potential interactions between both cancer-sensitivity and -resistance genes.

Since these genes have yet to be isolated, their biological function is unknown. Functionally these genes could exert their influence on tumor susceptibility at the level of initiation, promotion, or progression. Independent studies by several different investigators have been conducted to determine at what stage the *Hcs* genes in the C3H/HeJ mouse affect susceptibility. In one study liver tumors were induced with diethylnitrosamine in male C57BL/6J, B6C3F1, and C3H/HeJ mice. Although the incidence of tumors in these mice was similar (76, 95, and 100%, respectively) the tumor multiplicity was 20 times greater in the C3H/HeJ mice (39/animal) than in the C57BL/6J mice (2.3/animal), with an intermediate number in the B6C3F1 mice (22/animal) (132). These data suggest that the difference in susceptibility may be at the level of initiation (the greater number of tumors in the susceptible mice is due to a greater number of initiated cells in the susceptible strains). However, adduct measurements indicate that levels of N-7 and O-6 ethyl guanine in the liver are relatively equivalent in the C3H/HeJ and C57BL/6J mice, which suggest that the liver tumor susceptibility may not be due to differences in initiation. Analysis of the developing preneoplastic lesions revealed that number and size of the foci increase over time in both strains but that the growth rate was 1.7 times greater in the C3H/HeJ mice compared with the C57BL/6J strain (140). Based on these data, it was concluded that the *Hcs* genes exert their influence on tumor susceptibility at the level of promotion. Further work will be needed to determine whether the *Hcs* genes present in the other strains of mice influence liver tumor development (similarly) and how the interplay of susceptibility and resistance genes affect this process. This effort will be aided significantly by the isolation and characterization of this specific class of cancer genes. Discovery of mouse liver tumor susceptibility genes could lead to the identification of human homologs that confer risk for liver cancer in humans.

Some genetically predisposed forms of human cancers have been attributed to heritable mutations in genes involved with DNA mismatch repair processes (141–143). Mutations in these genes are in part responsible for maintaining the sequence integrity of the genome. Changes in the sequence length of DNA microsatellites, which are markers of genomic instability, have been used as potential indicators for mutations in the DNA mismatch repair genes (144–146). Because the B6C3F1 mouse is genetically susceptible to develop liver tumors, we were interested in determining whether this predisposition also involved potential mutations in these genes. To this end, we analyzed whether tumors derived in this strain of mouse contained alterations in microsatellite sequences (147). Our analysis of 48 liver tumors at 24 different microsatellite sequences revealed that the microsatellite sequence alterations observed were the result of spontaneous mutations within these microsatellite sequences and not the result of mutations in mismatch repair genes. Consequently, from these results we conclude that the type of genetic susceptibility found in certain heritable human cancers is probably not operational in the development of liver tumors in the genetically predisposed B6C3F1 mouse.

VII. PERSPECTIVES

It has been nearly 30 years since the inception of the NCI (NCI/NTP) NTP cancer bioassay program. Over this period, there have been much rhetoric and numerous volumes written debating both sides of the issue regarding the relevance of animal bioassay data for human carcinogenic risk assessment. In this book, for example, there are chapters on the cancer bioassay (see Chapter 1), the maximum tolerated dose (see Chapter 3), and the validity of the cancer bioassay (see Chapter 2). In spite of much research and discussion, the controversy surrounding this important social and scientific problem remains unresolved, and the question still remains: Can chemically induced mouse liver tumors be used to establish safe exposure levels in humans? To address this question, one must first consider several other questions and issues that have an impact on the process of mouse liver tumor development.

A. Mouse and Human Biology

One issue for determining the relevance of mouse liver tumors is whether the process of mouse liver tumorigenesis is biologically different from carcinogenesis in humans. Over the past 10 years, significant progress has been made in elucidating the specific cellular and molecular alterations involved in the neoplastic transformation of human cells. Much of this information has been utilized to define the carcinogenic process involved in murine hepatocarcinogenesis. From this comparative analysis, we have seen that, in general, the biology of murine hepatocarcinogenesis is similar to the biology that is operational in human tumors. Many of the same molecular alterations found in human tumors are detected in mouse liver tumors. The biological processes controlling growth, apoptosis, immortalization, and differentiation have been exquisitely conserved throughout evolution; and the disregulation of one or more of these processes is the characteristic hallmark of cancer in all species examined. Not surprisingly, very few fundamental differences therefore exist between mouse and human tumors at the molecular, biochemical, and cellular levels. This fact clearly supports the conduct of experimental studies of mouse liver carcinogenesis to help elucidate critical molecular and cellular processes involved in cancer development in humans.

B. Genetic Susceptibility

Another closely related issue is that concerning the genetic predisposition of the B6C3F1 mouse to develop liver tumors. Again, this issue raises the possibility that this genetic susceptibility implies some unique biological difference between mice and humans. Consequently, chemically induced tumors within the mouse may not be relevant to human chemical exposures. If logic similar to that in the above discussion is used, the genetic alterations leading to increased susceptibility in the B6C3F1 mouse are likely not unique to this species. Genetic alterations that result in a heightened sensitivity towards tumor development clearly exist in other rodents and may exist within subsets of the human population. Therefore, the tumor response observed in control and chemically treated mice cannot be dismissed as a genetic anomaly specific to mice. What could be argued is that the tumor response observed in the bioassay is not a response that would be observed in the general human population. The mice exposed in the bioassay are nearly identical genetically. So the bioassay may more likely represent a human exposure scenario involving a subset population of genetically similar or sensitive human individuals. Similar chemical exposure levels to the genetically diverse human population (or genetically diverse mouse population) may not elicit the same response as that observed in the B6C3F1 mouse cancer bioassay.

C. Mouse-Liver-Only Carcinogens

Rodent carcinogens that induce tumors in only a single anatomical site in one sex and one species (e.g., male mouse liver only) are often the most difficult carcinogens to employ for assessing potential human risk. Such chemically induced specific tumor responses, which include but are not limited to mouse liver carcinogens, are likely to be a consequence of strain- or species-specific factors. Chemicals that induce multiple site, multiple species, or multiple sex tumor responses clearly suggest a greater carcinogen risk (148). Use of multiple strains of each test species without increasing the total number of animals used has been proposed as one way to improve the cancer bioassays (149). This multistrain factorial design will be useful in obtaining test results that are specific to one animal strain and aiding in data extrapolation across genotypes.

D. Central Importance of Bioassay Dose and Dose-Response Relationships

A strong argument remains for questioning the relevance of the mouse liver tumor data based on issues involving the basic design of the bioassay itself. One issue having the greatest impact on the outcome of the bioassay relates to the high doses of test chemical used. As was discussed above, the animals are exposed to a very limited range of high doses (e.g., MTD, MTD/2). The maximum tolerated dose frequently compromises the normal physiology of the animal and is often at exposure levels several orders of magnitude higher than the typical human exposure. Determining potential human carcinogenic risks of agents from current testing protocols is a difficult if not an impossible task. The current risk-assessment extrapolation procedures can lead to costly regulation that has minimal benefit to the protection of human health. It can be argued that the single most important factor determining the outcome of the rodent bioassay and contributing to risk extrapolation difficulties is the dose used in the bioassay study. Most, if not all, chemically induced biological responses and pathways are dose dependent; in all probability, so is tumor development. Dose-response relationships in carcinogenesis for initiators (150) and promoters (151) have been published. However, only a limited number of studies have critically examined rodent tumor responses or proposed mechanisms of action at the low exposures that are typical of human exposures.

E. Modes of Action

Given the number of mouse liver carcinogens and their potential environmental and economic impact, research must be focused to achieve a resolution of key issues as quickly as possible. The field of carcinogenesis is progressing to the point where the mode of action for a chemical might be used to evaluate potential human cancer risks. Mode of action is defined as relevant and fundamental steps in the carcinogenic process (152). For a direct DNA-reactive mutagenic agent, not all the complex pathways of activation, binding specificity, DNA repair, resulting DNA damage and adducts, subsequent mutations, and consequences of oncogenes and suppressor genes are fully understood. However, the general concept of DNA reactivity resulting in gene mutation as a key mode of action has been and continues to be important in the assessment of potential cancer risk. Similarly, a number of other biochemical and biological endpoints relevant to the process of carcinogenesis provide important information in helping assess potential carcinogenic activity. These endpoints include but are not limited to alterations in cell proliferation and apoptosis, cell toxicity, inhibition of gap junctions, receptor activation and signaling, cell cycle checkpoint perturbations, genomic instability, and alterations in gene methylation.

Nongenotoxic mouse liver carcinogens are known to have different modes of action. The induction of mouse liver tumors caused by a nongenotoxic mode of action exhibiting a threshold dose-response relationship should be treated differently in risk assessment from DNA-reactive, mutagenic mouse hepatocarcinogens. The second U.S. EPA guidelines on cancer risk assessment allow this distinction. We emphasize that modes of action may not be mutually exclusive and that modes of action will likely have differing dose-response activities. The complexities and potential interaction of modes of action imply a general need to generate and evaluate information on a chemical-by-chemical basis. The incorporation of this new knowledge into the risk-assessment process should provide research incentives for generating the data needed to reduce the use of default assumptions in risk assessments.

F. Research Strategies

To help resolve the mouse liver tumor controversy, we developed a mouse hepatocarcinogenesis research strategy based on the consideration of data needs for carcinogen risk assessments. Our research strategy focuses on the cellular and molecular growth alterations in mouse liver tumor induction. Phase I involves the identification of the critical cellular targets for the chemical interaction that are responsible for inducing the neoplastic response. These molecular targets include oncogenes, tumor-suppressor genes, and components of cell cycle regulation, cell-to-cell communication, immortalization, DNA repair, DNA replication fidelity, differentiation, and metastasis. Once the targets for chemically induced alterations have been identified, the effect of dose on the induction of the target response can be evaluated (Phase II). This should allow the determination of whether the same biological response observed at high doses can also occur at low-dose exposures. Studies of this type should make it more feasible to investigate the validity of concepts such as linear, supralinear, sublinear, and threshold responses and their applicability for risk assessment. Information obtained in Phase I and II can be utilized to extrapolate to the potential effects in humans once it is determined whether the targets disrupted in the mouse are the same as those affected in humans and whether the same response is elicited in humans at doses comparable to those used in the bioassay (Phase III).

G. Weight of Evidence

Evaluation of potential human hazards should make use of all biological information. No single test, including the cancer bioassay itself, can establish potential human hazards and predict safe levels of exposure for humans. Such decisions need to be based on the consideration of all information, particularly the toxicological and biological features of the agent in question. The weight of the evidence will hopefully include the mode of action of a carcinogenic agent, the likelihood of human carcinogenic effects, and the conditions under which effects of the agent may be expressed.

H. Summary

The induction of mouse liver tumors by a number of agents is an indisputable fact. Much of the controversy regarding mouse liver tumor induction stems from the use of the bioassay data in risk assessment and not from the validity of the data generated in the bioassay study. Tumor data from cancer bioassays are clearly a useful tool for hazard identification and assessing potential carcinogenic activity of environmental chemicals. The question remains as how best to utilize the data generated in rodent bioassays in human risk assessment. It can be argued that the single most important factor affecting the outcome of the rodent bioassay and contributing to extrapolation difficulties is the exposure concentration (or dose). Nongenotoxic agents that induce a tumor response only at high doses and only in the mouse liver clearly need to be viewed differently from mutagenic chemicals. Improved methodology and ability to determine the mode of action of an agent is an important step in risk characterization. Significant advances in our understanding of mouse hepatocarcinogenesis have resulted from the use of molecular, biochemical, and genetic approaches. The full use of all biological information and the emphasis on the mode of action should greatly affect risk characterizations for mouse liver carcinogens. This review focused on the currently available information on cell and molecular alterations in mouse hepatocarcinogenesis. The challenge for the future is to develop a mechanistic understanding of the relationship between route of exposure, tissue dose, and biological response and then to integrate all the information to improve our ability to assess human health risks.

ACKNOWLEDGMENTS

We would like to express our appreciation for the contributions of all members of the CIIT research program. We thank Drs. Julian Preston, Byron Butterworth, Robert Maronpot, Glenda Moser, Ronny Fransson-

Steen, Patricia Fernandez, and Russell Cattley for scientific discussion and/ or editorial review; Linda Smith and Sadie Leak for typing of the manuscript; and Dr. Barbara Kuyper for editorial assistance. We are especially grateful to Robert Maronpot and Julian Preston for years of valuable intellectual input and encouragement, and collaborative efforts. This work was supported in part by a National Research Service Award (E.H.R.; Grant Number 1 F32 ES05692-01A1).

REFERENCES

1. Gold LS, Slone TH, Manley NB, Bernstein L. Target organs in chronic bioassays of 533 chemical carcinogens. Environ Health Perspect 1991; 93: 233-246.
2. Gold LS, Manley NB, Slone TH, Garfinkel GB, Rohrbach L, Ames BN. The fifth plot of the carcinogenic potency database: results of animal bioassay published in the general literature through 1988 and by the National Toxicology Program through 1989. Environ Health Perspect 1993; 100:65-135.
3. International Expert Advisory Committee to the Nutrition Foundation. The Relevance of Mouse Liver Hepatomas to Human Carcinogenic Risk. Washington, DC: Nutrition Foundation Inc., 1983.
4. Sontag JM, Page NP, Safioti V. Guidelines for Carcinogen Bioassay in Small Rodents. NCI-CG-TR-1USDHEW, Stock Number 017-042-00118-8. Washington, DC: U.S. Government Printing Office, 1976.
5. Chu KC, Cuento C Jr, Ward JM. Factors in the evaluation of 200 National Cancer Institute bioassays. J Toxicol Environ Health 1981; 8:251-280.
6. Bucher JR, Portier CJ, Goodman JI, Gaustman EM, Lucier GW. National Toxicology Program studies: principles of dose selection and applications to mechanistic based risk assessment. Fundam Appl Toxicol 1996; 31:1-8.
7. Tennant RW, Margolin BH, Shelby MD, Zeiger E, Haseman JK, Spaulding J, Caspary W, Resnick M, Stasiewicz S, Anderson B, Minor R. Prediction of chemical carcinogenicity in rodents from in vitro genetic toxicity assays. Science 1987; 236:933-941.
8. Tennant RW, Zeiger E. Genetic toxicology: Current status of methods of carcinogen identification. Environ Health Perspect 1993; 100:307-315.
9. Butterworth BE, Conolly RB, Morgan KT. A strategy for establishing mode of action of chemical carcinogens as a guide for approaches to risk assessments. Cancer Lett 1995; 129-146.
10. Ward JM. Morphology of potential preneoplastic hepatocyte lesions and liver tumors in mice and a comparison with other species. In: Popp JA, ed. Mouse Liver Neoplasia. Washington, DC: Hemisphere, 1984:1-26.
11. Maronpot RR, Haseman JJ, Boorman GA, Eustis SE, Rao GN, Huff JR. Liver lesions in B6C3F1 mice: The National Toxicology Program experience and position. Arch Toxicol 1987; 10:10-26.
12. Williams GM. Sex hormones and liver cancer. Lab Invest 1982; 46:352-354.

13. Vesselinovitch SD, Itze L, Mihailovich N, Rao KVN. Modifying the role of partial hepatectomy and gonadectomy in ethylnitrosourea-induced hepatocarcinogenesis. Cancer Res 1980; 40:1538-1542.
14. Kemp CJ, Drinkwater NR. The androgen receptor and liver tumor development in mice. In: Stevenson DE, McClain RM, Popp JA, Slaga TJ, Ward JM, Pitot HC, eds. Mouse Liver Carcinogenesis: Mechanisms and Species Comparisons. New York: Alan R. Liss, Inc., 1990:203-214.
15. Poole TM, Drinkwater NR. Strain dependent effects of sex hormones on hepatocarcinogenesis in mice. Carcinogenesis 1996; 17(2):191-196.
16. Standeven AM, Wolf DC, Goldsworthy TL. Interactive effects of unleaded gasoline and estrogen on liver tumor promotion in female B6C3F1 mice. Cancer Res 1994; 54:1198-1204.
17. Goldsworthy TL, Wolf DC, Wong BA, Moss OR, Roberts K, Moser GM. Lack of hepatic tumor-promoting activity of unleaded gasoline in ovariectomized mice. Toxicologist 1996; 30:163.
18. Dragan YP, Sargetnn L, Xu Y-D, Xu Y-H, Pitot HC. The initiation-progression model of rat hepatocarcinogenesis. Proc Soc Exp Biol Med 1993; 202:16-24.
19. Drinkwater NR, Hanigan MH, Kemp CJ. Genetic and epigenetic promotion of hepatocarcinogenesis. In: Stevenson DE, McClain RM, Popp JA, Slaga TJ, Ward JM, Pitot HC, eds. Mouse Liver Carcinogenesis: Mechanisms and Species Comparisons. New York: Alan R. Liss, Inc., 1990:163-176.
20. Moser GJ, Wolf DC, Harden R, Standeven AM, Mills J, Jirtle RL, Goldsworthy TL. Cell proliferation and regulation of negative growth factors in mouse liver foci. Carcinogenesis 1996; 17(9):1835–1840.
21. Russell WE, Kaufmann WK, Sitaric S, Luetteke NC, Lee DC. Liver regeneration and hepatocarcinogenesis in transforming growth factor-α-targeted mice. Mol Carcinog 1996; 15:183-189.
22. Kaufman WR, Zhang Y, Kaufman DG. Association between expression of transforming growth factor-alpha and progression of hepatocellular foci to neoplasms. Carcinogenesis 1992; 13:1481-1483.
23. Dragan Y, Teegvarden J, Campbell H, Hsia S, Pitot H. The quantitation of altered hepatic foci during multistage hepatocarcinogenesis in the rat. transforming growth factor α expression as a marker for the stage of progression. Cancer Lett 1995; 93:73-83.
24. Vogelstein B, Fearon ER, Hamilton SR, Kern SE, Preisinger AC, Leppert M, Nakamura Y, White R, Smits AM, Bos JL. Genetic alterations during colorectal-tumor development. N Engl J Med 1988; 319:525-532.
25. Fearon ER, Vogelestein B. A genetic model for colorectal tumorigenesis. Cell 1990; 61:759-767.
26. Fox T, Watanabe P. Detection of a cellular oncogene in spontaneous liver tumors of B6C3F1 mice. Science 1985; 228:546-597.
27. Wiseman RW, Stowers SJ, Miller EC, Anderson MW, Miller JA. Activating mutations of the c-Ha-ras protooncogene in chemically induced hepatomas of the male B6C3F1 mouse. Proc Natl Acad Sci USA 1986; 83:5825-5829.

28. Maronpot RR, Fox T, Malarkey DE, Goldsworthy TL. Mutations in the ras proto-oncogene: clues to etiology and molecular pathogenesis of mouse liver tumors. Toxicology 1995; 101:125-156.
29. Stanley LA, Devereux TR, Foley J, Lord PG, Maronpot RR, Orton TC, Anderson MW. Proto-oncogene activation in liver tumors of hepatocarcinogenesis-resistant strains of mice. Carcinogenesis 1992; 13:2427-2433.
30. Chen B, Liu L, Castonguay A, Maronpot RR, Anderson MW, You M. Dose-dependent ras mutation spectra in N-nitrosodiethylamine-induced mouse liver tumors and 4-(methylnitrosamino)-1-(3-pyridyl)-1-butanone induced mouse lung tumors. Carcinogenesis 1993; 14:1603-1608.
31. Buchmann A, Bauer-Hofmann R, Mahr J, Drinkwater NR, Luz A, Schwarz M. Mutational activation of the c-Ha-ras genein liver tumors of different rodent strains: correlation with susceptibity to hepatocarcinogenesis. Proc Natl Acad Sci USA 1991; 88:911-915.
32. Zhang W, Hirohashi S, Tsudu H, Shimosato Y, Yokota S, Terada M, Sugimura T. Frequent loss of heterozygosity on chromosomes 16 and 4 in human hepatocellular carcinoma. Jpn J Cancer Res 1990; 81:108-111.
33. Wei C, Chang R, Wong C, Bhacheck N, Cui X, Henning E, Yagi H, Sayer J, Jerina D, Preston B, Conney A. Dose-dependent differences in the profile of mutations induced by an ultimate carcinogen from benzo[a]pyrene. Proc Natl Acad Sci USA 1991; 88:11227-11230.
34. Camper SA. Research applications of transgenic mice. Biotechniques 1987; 5:638-649.
35. Goldsworthy TL, Recio L, Brown K, Donehower LA, Mirsalis JC, Tennant RW, Purchase IFH. Transgenic animals in toxicology. Fundam Appl Toxicol 1994; 22:8-19.
36. Liu C, Tsao MS, Grisham JW. Transforming growth factors produced by normal and neoplastically transformed rat liver epithelial cells in culture. Cancer Res 1988; 48:850-855.
37. Raymond VW, Lee DC, Grisham JW, Earp HS. Regulation of transforming growth factor alpha messenger RNA expression in a chemically transformed rat hepatic epithelial cell line by phorbal ester and hormones. Cancer Res 1989; 49:3608-3612.
38. Lee GH, Merlino G, Fausto N. Development of liver tumors in transforming growth factor alpha transgenic mice. Cancer Res 1992; 52:5162-5170.
39. Tamano S, Merlino GT, Ward JM. Rapid development of hepatic tumors in transforming growth factor alpha transgenic mice associated with increased cell proliferation in precancerous hepatocellular lesions initiated by N-nitrosodiethylamine and promoted by phenobarbital. Carcinogenesis 1994; 15:1791-1798.
40. Luetteke NC, Michalopoulos GK, Teixido J, Gilmore R, Massague J, Lee DC. Characterization of high molecular weight transforming growth factor alpha produced by rat hepatocellular carcinoma cells. Biochemistry 1988; 27: 6487-6494.
41. DiMarco E, Pierce JH, Fleming TP, Kraus MH, Molloy CJ, Aaronson SA, diFiore PP. Autocrine interaction between TGF-α and the EGF-receptor:

quantitative requirements for induction of the malignant phenotype. Oncogene 1989; 4:831-838.

42. Takagi H, Sharp R, Hammermeister C, Goodrow T, Bradley MO, Fausto N, Merlino G, Molecular and genetic analysis of liver oncogenesis in transforming growth factor α transgenic mice. Cancer Res 1992; 52:5171-5177.
43. Garte SJ. The c-myc oncogene in tumor progression. Crit Rev Oncogen 1993; 4(4):435-449.
44. Campisi J, Gray HE, Pardee AB, Dean M, Sonenshein GE. Cell-cycle control of c-myc but not c-ras expression is lost following chemical transformation. Cell 1984; 36(2):241-247.
45. Zimmerman K, Alt WE. Expression and function of myc family genes. Crit Rev Oncogen 1990: 2(1);75-95.
46. Sandgren EP, Quaife CJ, Pinkert CA, Palmiter RD, Brinster RL. Oncogene-induced liver neoplasia in transgenic mice. Oncogene 1989; 4:715-724.
47. Murakami H, Sanderson ND, Nagy P, Marino P, Merlino G, Thorgeirsson RS. Transgenic mouse model for synergistic effects of nuclear oncogenes and growth factors in tumorigenesis: interaction of c-myc and transforming growth factor a in hepatic oncogenesis. Cancer Res 1993; 53:1719-1723.
48. Messing A, Chen HY, Palmiter RD, Brinster RL. Peripheral neuropathies, hepatocellular carcinomas and islet cell adenomas in transgenic mice. Nature 1985; 316:461-463.
49. Held WA, Mullins JJ, Kuhn NJ, Gallagher JF, Gu GD, Gross KW. T antigen expression and tumorigenesis in transgenic mice containing a mouse major urinary protein SV40 T antigen hybrid gene. EMBO J 1989; 8:183-191.
50. Ludlow JW. Interactions between SV40 large-tumor antigen and the growth suppressor proteins pRB and p53. FASEB J 1993; 7:866-871.
51. Rogler CE. Altered body composition and increased frequency of diverse malignancies in insulin-like growth factor-II transgenic mice. J Biol Chem 1994; 169(1):13779-13784.
52. Blatt J, White C, Dienes S, Friedman H, Foley TP. Production of an insulin-like growth factor by osteosarcoma. Biochem Biophys Res Commun 1984; 123(1):373-376.
53. Fu XX, Su CY, Lee Y, Hintz R, Biempica L, Snyder R, Rogler CE. Insulin-like growth factor II expression and oval cell proliferation associated with hepatocarcinogenesis in woodchuck hepatitis virus carriers. J Virol 1988; 62(9):3422-3430.
54. Yang DY, Rogler CE. Analysis of insulin-like growth factor II (IGF-II) expression in neoplastic nodules and hepatocellular carcinomas of woodchucks utilizing in situ hybridization and immunocytochemistry. Carcinogenesis 1991; 12:1893-1901.
55. Cariani E, Lasserre C, Seurin D, Hamelin B, Kemeny F, Franco D, Czech MP, Ullrich A, Brechot C. Differential expression of insulin like growth factor II mRNA in human primary liver cancers, benign liver tumors and liver cirrhosis. Cancer Res 1988; 48:6844-6849.
56. Schirmacher P, Held WA, Yang D, Chisari FV, Rustum Y, Rogler CE. Reactivation of insulin-like growth factor II during hepatocarcinogenesis in

transgenic mice suggests a role in malignant growth. Cancer Res 1992; 52: 2549-2556.
57. Romach EH, Goldsworthy TL, Maronpot RR, Fox TR. Altered gene expression in spontaneous hepatocellular carcinoma in male B6C3F1 mice. Molec Carcinogen 1997. In press.
58. Lammie GA, Fantl V, Smith R, Schuuring E, Brookes S, Michaelides R, Dickson C, Arnold A, Perers G. D11S287, a putative oncogene on chromosome 11q13, is amplified and expressed in squamous cell and mammary carcinomas and linked to BCL-1. Oncogene 1991; 6:439-444.
59. Zhang YJ, Jiang W, Chen CJ, Lee CS, Kahn SM, Santella RM, Weinstein BI. Amplification and overexpression of cyclin D1 in human hepatocellular carcinoma. Biochem Biophy Res Commun 1993; 196(2):1010-1016.
60. Bianchi AB, Fischer SM, Robles AI, Rinchik EM, Conti CJ. Overexpression of cyclin D1 in mouse skin carcinogenesis. Oncogene 1993; 8:1127-1133.
61. Nishida N, Fukuda Y, Komeda T, Kita R, Sando T, Furukawa M, Amenomori M, Shibagaki I, Nakao K, Ikenaga M, Ishizaki K. Amplification and overexpression of the cyclin D1 gene in aggressive human hepatocellular carcinoma. Cancer Res 1994; 54:3107-3110.
62. Drummond IA, Madden SL, Rohvernutter P, Bell GI, Sukhatve VP, Rauscher FJ. Repression of the insulin-like growth factor II gene by the Wilms tumor suppressor WT1. Science 1992; 257:674-678.
63. Hollstein M, Sidransky D, Vogelstein B, Harris CC. p53 Mutations in human cancers. Science 1991; 253:49-53.
64. Donehower LA, Harvey M, Slagle BL, McArthur MJ, Montgomery CA, Butel JS, Bradley A. Mice deficient for p53 are developmentally normal but susceptible to spontaneous tumours. Nature 1992; 356:215-221.
65. Harvey M, McArthur MJ, Montgomery CA, Butel JS, Bradley A, Donehower LA. Spontaneous and carcinogen-induced tumorigenesis in p53-deficient mice. Nature Genetics 1993; 5:225-229.
66. Williams BO, Remington L, Albert DM, Mukai S, Brunson RT, Jacks T. Cooperative tumorigenic effects of germlike mutations in Rb and p53. Nature Genetics 1994; 4:480-484.
67. Kemp C. Hepatocarcinogenesis in p53-deficient mice. Mol Carcinogen 1995; 12:132-136.
68. Goodrow TL, Stoner RD, Leander KR, Prahalada SR, Van Zwieten MJ, Bradley MO. Murine p53 intron sequences 5-8 and their use in polymerase chain reaction/direct sequencing analysis of p53 mutations in CD-1 mouse liver and lung tumors. Mol Carcinogen 1992; 5:9-15.
69. Kress S, Konig J, Schweizer J, Lohrke H, Bauer-Hofmann R, Schwarz M. p53 Mutations are absent from carcinogen-induced mouse liver tumors but occur in cell lines established from these tumors. Mol Carcinogen 1992; 6: 148-158.
70. Rumsby PC, Davies MJ, Evans JG. Screening for p53 mutations in C3H/He mouse liver tumors derived spontaneously or induced with diethylnitrosamine or phenobarbitone. Mol Carcinogen 1994; 9:71-75.

71. Calvert RJ, Tashiro Y, Buzard GS, Diwan BA, Weghorst GM. Lack of p53 point mutations in chemically induced mouse hepatoblastomas: an end-stage, highly malignant hepatocellular tumor. Cancer Lett 1995; 95:175-180.
72. Hegi ME, Söderkvist P, Foley JF, Schoonhoven R, Swenberg JA, Kari F, Maronpot R, Anderson MW, Wiseman RW. Characterization of mutations in methylene chloride-induced lung tumors from B6C3F1 mice. Carcinogenesis 1993; 14:803-810.
73. Davis LM, Caspary WJ, Sakallah AS, Maronpot RR, Wiseman R, Barret JC, Elliot R, Hozier JC. Loss of heterozygosity in spontaneous and chemically induced tumors of the B6C3F1 mouse. Carcinogenesis 1994; 15:1637-1645.
74. Manenti G, Binelli G, Gariboldi M, Canzian F, DeGregorio L, Falvella FS, Dragani TA, Pierotti MA. Multiple loci affect genetic predisposition to hepatocarcinogenesis in mice. Genomics 1994; 23:118-124.
75. Sargent LM, Sanderson ND, Thorgiersson SS. Ploidy and karyotypic alterations associated with early events in the development of hepatocarcinogenesis in transgenic mice harboring c-myc and transforming growth factor a transgenes. Cancer Res 1996; 56:2137-2142.
76. Searle AG, Peter L, Lyon MF, Evans EP, Edwards JH, Buckel VJ. Chromosome maps of man and mouse. III. Genomics 1987; 1:3-18.
77. Peto RM, Roe JFC, Lee PN, Levy L, Clark J. Cancer and aging, mice and man. Br J Cancer 1975; 32:411-426.
78. Pitot HE, Sirica AE. The stages of initiation and promotion in hepatocarcinogenesis. Biochim Biophys Acta 1980; 605:191-205.
79. Schulte-Hermann R, Grasl-Kraupp B, Bursch W. Apoptosis and hepatocarcinogenesis. In: Jirtle RJ, ed. Liver Regeneration and Carcinogenesis. San Diego: Academic Press, 1995:141-178.
80. Butterworth BE, Goldsworthy TL. The role of cell proliferation in multistage carcinogenesis. Proc Soc Exp Biol Med 1991; 198:683-687.
81. Butterworth BE, Popp JA, Conolly RB, Goldsworthy TL. Chemically induced cell proliferation in carcinogenesis. IARC Sci Publ 1992; 116:279-305.
82. Goldsworthy TL, Conolly RB, Fransson-Steen R. Apoptosis and cancer risk assessment. Mut Res 1996; 365:71–90.
83. Goldsworthy TL, Fransson-Steen R, Maronpot RR. Quantitation of hepatocyte apoptosis: relevance for mechanistic studies of liver growth and cancer. Toxicol Pathol 1996; 24:24-35.
84. Schulte-Hermann R, Bursch W, Grasl-Kraupp B. Active cell death (apoptosis) in liver biology and disease. In: Boyer JL, Ockner RK, eds. Progress in Liver Disease. Vol. XIII. Philadelphia: W.B. Saunders Company, 1995:1-35.
85. Goldsworthy TL, Morgan KT, Popp JA, Butterworth BE. Guidelines for measuring chemically-induced cell proliferation in specific rodent target organs. In: Butterworth BE, Slaga TJ, eds. Chemically Induced Cell Proliferation: Implications for Risk Assessment. New York: Wiley-Liss, 1991:253-284.
86. Goldsworthy TL, Butterworth BE, Maronpot RR. Concepts, labeling procedures and design of cell proliferation studies relating to carcinogenesis. Environ Health Perspect 1993; 101(5):59-66.

87. Foley J, Thai T, Maronpot RR, Butterworth BE, Goldsworthy TL. Comparison of proliferating cell nuclear antigen to tritiated thymidine as a marker of proliferating hepatocytes in rats. Environ Health Perspect 1993; 101(suppl 5):199-206.
88. Bennett LM, Farnham PJ, Drinkwater RR. Strain-dependent differences in DNA synthesis and gene expression in the regenerating livers of C57BL/6J and C3H/HeJ mice. Mol Carcinogen 1995; 14:46-52.
89. Standeven AM, Goldsworthy TL. Promotion of hepatic preneoplastic lesions and induction of CYP2B by unleaded gasoline vapor in female B6C3F1 mouse liver. Carcinogenesis 1993; 14:2137-2141.
90. Wilson DM, Goldsworthy TL, Popp JA, Butterworth BE. Evaluation of genotoxicity, pathological lesions and cell proliferation of livers and rats and mice treated with furan. Environ Mol Mutagen 1992; 19:209-222.
91. Larson JL, Wolf DC, Butterworth BE. Induced cytotoxicity and cell proliferation in the hepatocarcinogenicity of chloroform in female B6C3F1 mice: comparison of administration by gavage in corn oil vs. ad libitum in drinking water. Fundam Appl Toxicol 1994; 22:90-102.
92. Standeven AM, Goldsworthy TL. Promotion of hepatic preneoplastic lesions in male B6C3F1 mice by unleaded gasoline. Environ Health Perspect 1995; 103:2-6.
93. Thoresen GH, Refsnes M, Christofferson T. Inhibition of hepatocyte DNA synthesis by transforming growth factor β1 and cyclic AMP: effect immediately before G1/S border. Cancer Res 1992; 52:3598-3603.
94. Oberhammer F, Bursch W, Parzefall W, Breit P, Erber E, Stadler M, Shulte-Hermann R. Effect of transforming growth factor β on cell death of cultured rat hepatocytes. Cancer Res 1991; 51:2478-2485.
95. Oberhammer FA, Pavelka M, Sharma A, Tiefenbacher R, Purchio A, Bursch W, Schulte-Hermann R. Induction of apoptosis in cultured hepatocytes and in regression liver by transforming growth factor b1. Proc Natl Acad Sci USA 1992; 89:5408-5412.
96. Russell WE, Coffey RJ, Ouellette AJ, Moses HL. Transforming growth factor beta reversibly inhibits the early proliferative response to partial hepatectomy in the rat. Proc Natl Acad Sci USA 1988; 85:5126-5130.
97. Jirtle RL, Hankins GR, Reisenbichler H, Boyer IJ. Regulation of mannose 6-phosphate/insulin-like growth factor-II receptors and transforming growth factor beta during liver tumor promotion with phenobarbital. Carcinogenesis 1994; 15:1473-1478.
98. Sue SR, Chari RS, Kong FM, Mills JJ, Fine RL, Jirtle RL, Meyers WC. Transforming growth factor-beta receptors and mannose 6-phosphate/insulin-like growth factor-II receptor expression in human hepatocellular carcinoma. Ann Surg 1995; 222:171-178.
99. Sha X, Brunner AM, Purchio AF, Gentry LE. Transforming growth factor beta 1: importance of glycosylation and acidic proteases for processing and secretion. Mol Endocrinol 1989; 3:1090-1098.
100. Kovacina KS, Steele-Perkins G, Purchio AF, Lioubon M, Miyazono K, Hel-

din C-H, Roth RA. Interactions of recombinant and platelet transforming growth factor-β1 precursor with the insulin-like growth factor II/mannose 6-phosphate growth factor type II receptor. Biochem Biophy Res Comm 1989; 160:393-403.
101. Kornfield S. Structure and function of the mannose 6-phosphate/insulin-like growth factor II receptors. Annu Rev Biochem 1992; 61:307-330.
102. Dennis PA, Rifkin DB. Cellular activation of latent transforming growth factor beta requires binding to the cation-independent mannose 6-phosphate/ insulin-like growth factor type II receptor. Proc Natl Acad Sci USA 1991; 88(2):580-584.
103. Harpel JG, Metz CN, Kojima S, Rifkin DB. Control of transforming growth factor-beta activity: latency vs. activation. Prog Growth Factor Res 1992; (4): 321-325.
104. DeSouza AT, Hankins GR, Washington MK, Orton TC, Jirtle RL. M6P/ IGF2R gene is mutated in human hepatocellular carcinomas with loss of heterozygosity. Nature Genetics 1995; 11:447-449.
105. DeSouza AT, Hankins GR, Washington MK, Fine RL, Orton TC, Jirtle RL. Frequent loss of heterozygosity on 6q at the mannose 6-phosphate/insulin-like growth factor II receptor locus in human hepatocellular tumors. Oncogene 1995; 10:1725-1729.
106. Jirtle RL, Carr BI, Scott CD. Modulation of insulin-like growth factor-II/ mannose 6-phosphate receptors and transforming growth factor-β1 during liver regeneration. J Biol Chem 1991; 266(33):22444-22450.
107. Jirtle RL, Meyer S, Brackbrough SS. Liver tumor promoter phenobarbital: a biphasic modulator of hepatocyte proliferation. Implications for risk assessment. In: Progress in Clinical and Biological Research. New York: Wiley-Liss, 1991:209-216.
108. Grasl-Kraupp B, Bursch W, Ruttkay-Nedecky B, Wagner A, Lauer B, Schulte-Hermann R. Food reduction eliminates preneoplastic cells through apoptosis and antagonizes carcinogenesis in rat liver. Proc Natl Acad Sci USA 1994; 91:9995-9999.
109. Mills JJ, Chari RS, Boyer IJ, Gould MN, Jirtle RL. Induction of apoptosis in liver tumors by the monoterpene perilly alcohol. Cancer Res 1995; 55: 979-983.
110. Bursch W, Paffe S, Putz B, Barthel G, Schulte-Hermann R. Determination of the length of the histological stages of apoptosis in normal and in altered hepatic foci of rats. Carcinogenesis 1990; 11:847-853.
111. Roberts RA, Soames AR, Gill JH, James NH, Wheeldon EB. Non-genotoxic hepatocarcinogens stimulate DNA synthesis and their withdrawal induces apoptosis, but in different hepatocyte populations. Carcinogenesis 1995; 16: 1693-1698.
112. Stinchcombe S, Buchmann A, Boch KW, Schwarz M. Inhibition of apoptosis during 2,3,7,8-tetrachlorodibenzo-p-dioxin-mediated tumor promotion in rat liver. Carcinogenesis 1995; 16:1271-1275.
113. Snyder RD, Pullman J, Carter JH, Carter HW, DeAngelo AB. In vivo ad-

ministration of dichloroacetic acid suppresses spontaneous apoptosis in murine hepatocytes. Cancer Res 1995; 55(17):3702-3705.
114. Bursch W, Oberhammer F, Jirtle RL, Askari M, Sedivy R, Grasl-Kraupp B, Purchio AF, Schulte-Hermann R. Transforming growth factor-β1 as a signal for induction of cell death by apoptosis. Br J Cancer 1993; 67:531-536.
115. Oberhammer FA, Roberts RA. Apoptosis: a widespread process involved in liver adaptation and carcinogenesis. In: Arias IM, Boyer JL, Fausto N, Jakoby WB, Schachter DA, Schafritz, eds. The Liver: Biology and Pathobiology. 3d ed. New York: Raven Press, Ltd, 1994:1547-1556.
116. James NH, Roberts RA. The peroxisome proliferator class of non-genotoxic hepatocarcinogens synergize with epidermal growth factor to promote clonal expansion of initiated rat hepatocytes. Carcinogenesis 1994; 15:2687-2694.
117. Gill JH, Molloy CA, Shoesmith KJ, Bayly AC, Roberts RA. The rodent non-genotoxic hepatocarcinogen Nafenopin and EGF alter the mitosis/apoptosis balance promoting hepatoma cell clonal growth. Cell Death Diff 1995; 2:211-217.
118. Trosko JE, Goodman JI. Intercellular communication may facilitate apoptosis: implications for tumor promotion. Mol Carcinogen 1994; 11:8-12.
119. Alison MR, Sarraf CE. Apoptosis: regulation and relevance to toxicology. Human Exp Toxicol 1995; 14:234-247.
120. Canman CE, Kastan MB. Induction of apoptosis by tumor suppressor genes and oncogenes. In: Eastman A, ed. Seminars in Cancer Biology. San Diego: Academic Press, 1995; 6:17-26.
121. Craig RW. The bcl-1 gene family. Sem Cancer Biol 1995; 6:35-44.
122. Steller H. Mechanisms and genes of cellular suicide. Science 1995; 267:1445-1449.
123. Hully JR, Chang L, Schwall RH, Widmer HR, Terrell TG, Gillett NA. Induction of apoptosis in the murine level with recombinant human activin A. Hepatology 1994; 20:854-862.
124. Nagata S, Goldstein P. The fas death factor. Science 1995; 267:1449-1456.
125. Bursch W, Taper HS, Lauer B, Schulte-Hermann R. Quantitative histological and histochemical studies on the occurrence and stages of controlled cell death (apoptosis) during regression of rat liver hyperplasia. Virchows Arch Cell Pathol 1985; 50:153-166.
126. Harrington EA, Fanidi A, Evan GI. Oncogenes and cell death. Curr Opin Gene Devel 1994; 4:120-129.
127. Savill J, Dransfield I, Hogg N, Haslett C. Vitronectin receptor-mediated phagocytosis of cells undergoing apoptosis. Nature 1990; 343:170-173.
128. Popp JA. Mouse Liver Neoplasia. Washington, DC: Hemisphere, 1984.
129. Stevenson DE, McClain RM, Popp JA, Slaga TJ, Ward JM, Pitot HC. Mouse Liver Carcinogenesis: Mechanisms and Species Comparisons. Vol. 331. New York: Wiley-Liss, 1990.
130. Drinkwater NR, Bennett LM. Genetic control of carcinogenesis in experimental animals. In: Ito N, Sugano H, eds. Modification of Tumor Development in Rodents. Vol. 33. Basel: Karger, 1991:1-20.

131. Lee GH, Monura K, Kanda H, Kusakabe M, Yoshibi A, Sakakura T, Kitagawa T. Strain specific sensitivity to diethylnitrosamine-induced carcinogenesis is maintained in hepatocytes of C3H/H3N ↔ C55BL/6N chimeric mice. Cancer Res 1991; 51:3257-3260.
132. Drinkwater NR, Ginsler JJ. Genetic control of hepatocarcinogenesis in C57BL/65 and C3H/HeJ inbred mice. Carcinogenesis 1986; 7(10):1701-1707.
133. Bennett LM, Winkler ML, Drinkwater NR. A gene that determines the high susceptibility of the C3H/HeJ strain of mouse to liver tumor induction is located on chromosome one. Proc Am Assoc Cancer Res 1993; 33:144.
134. Gariboldi M, Manenti G, Canzian F, Falvella FS, Pierotti MA, Della Porta G, Binelli G, Dragani TA. Chromosome mapping of murine susceptibility loci to liver carcinogenesis. Cancer Res 1993; 53:209-211.
135. Wang HP, Rogler CE. Deletions in human chromosome arms 11p and 13g in primary hepatocellular carcinomas. Cytogenet Cell Genet 1988; 48:72-78.
136. Fujimori M, Tokino T, Hino O, Kitaguwi T, Imamura T, Okamoto E, Mitsunoba M, Ishikawa T, Nakagama H, Harada H, Yagura M, Matsubara K, Nakamura Y. Allotype study of primary hepatocellular carcinoma. Cancer Res 1991; 51:89-93.
137. Tsuda H, Zhang W, Shimosato Y, Yokota J, Terada M, Sugimura T, Miyamura T, Hirohashi S. Allele loss on chromosome 16 associated with progression human hepatocellular carcinoma. Proc Natl Acad Sci USA 1990; 87:6791-6794.
138. Poole TM, Winkler ML, Kuehler DA, Drinkwater NR. Two loci account for the sensitivity of male C57BR/cdJ mice to hepatocarcinogenesis. Proc Am Assoc Cancer Res 1994; 35:122.
139. Lee GH, Drinkwater NR. The Hcv (hepatocarcinogen resistance) loci of DBA/2J mice partially suppress phenotypic expression of the Hcs (hepatocarcinogen sensitivity) loci of C3H/H3J mice. Carcinogenesis 1995; 16:1993-1996.
140. Hanigan MH, Kemp CJ, Ginsler JJ, Drinkwater NR. Rapid growth of preneoplastic lesions in hepatocarcinogen-sensitive C3H/H3J male mice relative to C57/BL/6J male mice. Carcinogenesis 1988; 9(6):885.
141. Fishel R, Lesco MK, Rao MRS, Copeland NG, Jenkins NA, Garber J, Kane M, Kolodner R. The human mutator gene homolog MSH2 and its association with hereditary non-polyposis colon cancer. Cell 1993; 75:1027-1038.
142. Papadopoulos N, Nicolaides NC, Wei Y-F, Ruben SM, Carter KC, Rosen CA, Haseltine WA, Fleischmann RD, Fraser CM, Adams MD, Venter JC, Hamilton SR, Petersen GM, Watson P, Lynch HT, Peltomäki P, Mecklin J-P, de la Chapelle A, Kinzler KW, Vogelstein B. Mutation of a mut L homolog in hereditary colon cancer. Science 1994; 263:1625-1629.
143. Bronner CE, Baker SM, Morrison PT, Warren G, Smith LG, Lescoe MK, Kane M, Earabino C, Lipford J, Lindblom A, Tannergard P, Bollag RJ, Godwin AR, Ward DC, Nordenskjelk M, Fishel R, Kolodner R, Liskay RM. Mutation in the DNA mismatch repair gene homologue hmLH1 is associated with hereditary non-polyposis colon cancer. Nature 1994; 368:258-261.

144. Ionov Y, Peinado MA, Malkhosyan S, Shibata D, Perucho M. Ubiquitous somatic mutations in simple repeated sequences reveal a new mechanism for colonic carcinogenesis. Nature 1993; 260:558-561.
145. Peltomaki P, Lothe RA, Aaltonen LA, Pylkkänen L, Nyström-Lahti M, Seruca R, David LM, Holm R, Ryberg D, Haugen A, Brogger A, Borresen AL, de al Chapelle A. Microsatellite instability is associated with tumors that characterize the hereditary non-polyposis colorectal carcinoma syndrome. Cancer Res 1993; 53:5853-5855.
146. Thibodeau SN, Bren G, Schaid D. Microsatellite instability in cancer of the proximal colon. Science 1993; 260:816–819.
147. Fox TR, McMillen PJ, Maronpot RR, Goldsworthy TL. Genomic instability, as measured by microsatellite alterations, is not associated with liver tumor development in the genetically susceptible B6C3F1 mouse. Tox Appl Pharm 1997; 143:167–172.
148. Ashby J, Tennant RW. Definitive relationships among chemical structure, carcinogenicity and mutagenicity for 301 chemicals tested by the U.S. NTP. Mutat Res 1991; 257:229-306.
149. Festing MFW. Use of a multistrain assay could improve the NTP carcinogenesis bioassay. Environ Health Perspect 1995; 103:44-52.
150. Zeise L, Wilson R, Crouch EA. Dose-response relationships for carcinogens: a review. Environ Health Perspect 1987; 73:259-308.
151. Kitchin KT, Brown JL, Setzer RW. Dose-response relationship in multistage carcinogenesis: promoters. Environ Health Perspect Suppl 1994; 102(1):255-264.
152. Butterworth BE, Eldridge SR. A decision tree approach for carcinogen risk assessment. In: McClain RM, Slaga TJ, LeBoeuf R, Pitot H, eds. Growth Factors and Tumor Promotion: Implications for Risk Assessment. New York: Wiley-Liss, Inc., 1995:49-70.
153. Dragani TA, Manenti G, Colombo BM, Falvella FS, Gariboldi M, Pierotti MA, Della Porta G. Incidence of mutations at codon 61 of the Ha-*ras* gene in liver tumors of mice genetically susceptible and resistant to hepatocarcinogenesis. Oncogene 1991; 6:333-338.
154. Watson MA, Devereux TR, Anderson MW, Maronpot RR. H-*ras* oncogene mutation spectra in B6C3F1 and C57BL/6 mouse liver tumors provide evidence for TCDD promotion of spontaneous and vinyl carbamate-initiated liver cells. Carcinogenesis 1995; 16:1705-1710.
155. Devereux TR, Foley JF, Maronpot RR, Kari F, Anderson MW. *Ras* proto-oncogene activation in liver and lung tumors from B6C3F1 mice exposed chronically to methylene chloride. Carcinogenesis 1993; 14:795-801.
156. Reynolds SH, Stowers SJ, Patterson RM, Maronpot RR, Aaronson SA, Anderson MW. Activated oncogenes in B6C3F1 mouse liver tumors: implications for risk assessment. Science 1987; 237:1309-1316.
157. Anna CH, Maronpot RR, Pereira MA, Foley JF, Malarkey DE, Anderson MW. *Ras* proto-oncogene activation in dichloroacetic acid-, trichloroethylene-, and tetrachloroethylene-induced liver tumors in B6C3F1 mice. Carcinogenesis 1994; 15:2255-2261.

158. Fox TR, Schumann AM, Watanabe PG, Yano BL, Maher VM, McCormick JJ. Mutational analysis of the H-*ras* oncogene in spontaneous C57BL/6 × C3H/He mouse liver tumors and tumors induced with genotoxic and nongenotoxic hepatocarcinogens. Cancer Res 1990; 50:4014-4019.
159. Stowers SJ, Wiseman RW, Ward JM, Miller EC, Miller JA, Anderson MW, Eva A. Detection of activated proto-oncogenes in *N*-nitrosodiethylamine-induced liver tumors: a comparison between B6C3F1 mice and Fischer 344 rats. Carcinogenesis 1988; 9:271-276.
160. Buchmann A, Mahr J, Bauer-Hofmann R, Schwarz M. Mutations at codon 61 of the Ha-*ras* proto-oncogene in precancerous liver lesions of the B6C3F1 mouse. Mol Carc 1989; 2:121-125.
161. Richardson KK, Helvering LM, Copple DM, Rexroat MA, Linville DW, Engelhardt JA, Todd GC, Richardson FC. Genetic alterations in the 61st codon of the H-*ras* oncogene isolated from archival sections of hepatic hyperplasias, adenomas and carcinomas in control groups of B6C3F1 mouse bioassay studies conducted from 1979 to 1986. Carcinogenesis 1992; 13:935-941.
162. Malarkey DE, Devereux TR, Dinse GE, Mann PC, Maronpot RR. Hepatocarcinogenicity of chlordane in B6C3F1 and B6D2F1 male mice: evidence for regression in B6C3F1 mice and carcinogenesis independent of *ras* protooncogene activation. Carcinogenesis 1995; 16:2617-2625.
163. Rumsby PC, Barrass NC, Phillimore HE, Evans JG. Analysis of Ha-*ras* oncogene in C3H/He mouse liver tumors derived spontaneously or induced with diethylnitrosamine or phenobarbitone. Carcinogenesis 1991; 12:2331-2336.
164. Bauer-Hofmann R, Buchmann A, Mahr J, Kress S, Scharz M. The tumour promoters dieldrin and phenobarbital increase the frequency of c-Ha-*ras* wild-type, but not of c-Ha-*ras* mutated focal liver lesions in male C3H/He mice. Carcinogenesis 1992; 13:477-481.
165. Muller O, Kress S, Schwarz M. Absence of mutations in the functional parts of the p120-GAP gene in carcinogen-induced mouse liver tumors. Carcinogenesis 1992; 13:1903-1905.
166. Lord PG, Hardaker KJ, Loughlin JM, Marsden AM, Orton TC. Point mutation analysis of *ras* genes in spontaneous and chemically induced C57Bl/10J mouse liver tumors. Carcinogenesis 1992; 13:1383-1387.
167. Manam S, Storer RD, Prahalada S, Leander KR, Kraynak AR, Ledwith BJ, van Zwieten MJ, Bradley MO, Nichols WW. Activation of Ha-, Ki-, and N-*ras* genes in chemically induced liver tumors from CD-1 mice. Cancer Res 1992; 52:3347-3352.

5

Information Sources on the Carcinogenicity of Chemicals

Kirk T. Kitchin
U.S. Environmental Protection Agency, Research Triangle Park, North Carolina

I. INTRODUCTION

Once an animal carcinogenicity test is completed and summarized, the final report is issued and often the publication of the research results occurs. Individual scientists, organizations, and societies must then come to grips with the following questions: "How high quality is this particular carcinogenicity study?", "How should we interpret these results?", and "What risk assessments and regulatory decisions should be made on the basis of this new information?"

The amount, complexity, and different types of information available on chemical carcinogenicity can pose problems of informational overload. For example, a Medline search on the keywords of cancer, rat cancer, rat liver cancer, and rat lung cancer produced 111,875, 3,412, 1,101 and 299 entries, respectively, for the 5 years 1990–1994. Unless you work full time in the area of chemical carcinogenicity, using the primary scientific literature on chemical carcinogenicity may pose a challenge. This chapter's purpose is to present to the reader 11 information sources on chemical carcinogenicity data which often provide a great deal of quality information with less effort and time required than the primary scientific literature. A book

---

This document has been reviewed in accordance with U.S. Environmental Protection Agency policy and approved for publication. Mention of trade names or commercial products does not constitute endorsement or recommendation for use.

chapter of this type cannot hope to be comprehensive; due to space limitations many excellent information sources of chemical carcinogenicity data are not mentioned here. Depending on the reason for obtaining chemical carcinogenicity information, time and money resources available, scientific training, and library skills, appropriate search procedures can vary greatly. If several of the information sources recommended here do not provide the information on chemical carcinogenicity for which you are searching, it does not mean such information is not available; the information may simply exist somewhere else.

II. INFORMATION SOURCES

Any of the first seven compilations and information sources are often an excellent place to start. They may provide you with much of what you seek and/or direct you to primary scientific literature of high quality and appropriateness to your purpose. Depending on your purpose, any particular information source may best meet your needs; thus, the following cancer information sources are arranged in no particular order.

A. International Agency for Research on Cancer (IARC) Monographs

The IARC, part of the World Health Organization, is headquartered in Lyon, France. Based on both human and animal data, IARC reviews, evaluates and publishes *Monographs on the Evaluation of Carcinogenic Risks to Humans* and also the *List of IARC Evaluations* (1). IARC is continually producing new monographs on different chemicals, mixtures, exposure circumstances, and physical or biological agents (124 volumes by 1993, 800 evaluations of carcinogenicity to humans by 1995). In 1987 IARC published a summary which reviewed the data available from monographs 1-42 and listed the agent, the degree of evidence for human and animal carcinogenicity, the overall IARC evaluation, and the monograph volume and year on which the carcinogenicity evaluation was based (2). The IARC monographs are written by international experts committed to providing detailed, balanced reviews of scientific information necessary to answer the question: "Does this chemical cause cancer in either animals or man and thus does it currently pose a risk to human health?" This excellent information source is often the first place to look for information on chemical carcinogens.

B. Ashby and Coworkers' Compilations

Three compilations of chemical carcinogenicity have been published by Ashby and colleagues in the journal *Mutation Research* (3-5). In the first publication in 1988 (3), data on the chemical structure, *Salmonella* mutagenicity, and the extent of animal carcinogenicity by the National Cancer Institute (NCI) or the National Toxicology Program (NTP) bioassay test results of 222 chemicals were presented. An alphabetical list of the 222 test chemicals (115 rodent carcinogens, 24 equivocal chemicals, and 83 noncarcinogens) directs you to the chemical identification number, which then locates the desired data on a particular chemical. A great deal of useful information (chemical abstract service number, structural formula, Salmonella assay data, structural alert data, carcinogenic status in male rats, female rats, male mice, and female mice, level of carcinogenic effect across animal species, sex and tissue, route of chemical administration, maximum dose for rats and mice, duration of the study, tumor site, and percentage of tumor-bearing animals) is presented in a remarkably compact space. The main body of this paper (3) is a 62-page-long table. In this 99-page study, the authors' main purpose was to explore the connection between structural alerts, *Salmonella* mutagenicity, and animal carcinogenicity data.

The next *Mutation Research* article, in 1991 (4), presented similar results for 301 chemicals (162 rodent carcinogens, 39 equivocal chemicals, and 100 noncarcinogens) tested by the U.S. National Toxicology Program. This 79-page 1991 compilation contains some data on compounds not present in the first compilation of 222 chemicals. An alphabetical index of chemicals leads one to the chemical identification number and the table in which the desired chemical carcinogenicity data is contained. The main purpose of this paper was to examine the data within certain chemical functional groups and determine the relationship between structural alerts, mutagenicity in *Salmonella*, and rodent carcinogenicity. Because of this purpose, the easier-to-use single-table format of the 1988 article could not be used. Thus, the 1988 paper is easier to use; the 1991 paper has more compounds in it. In both of these papers, the NCI/NTP carcinogenic evaluations of chemical carcinogenicity are used as the source of the animal carcinogenicity data. This has both advantages and disadvantages. The main advantage is the similarity of the protocols over the years during which the U.S. government has been running this cancer bioassay program in either the NCI or the NTP section of NIEHS. Disadvantages include the fact that no one single cancer bioassay protocol is always the best experimental design for any particular individual chemical.

The third major chemical carcinogenicity compilation was done by Ashby and Paton (5), published in *Mutation Research* in 1993. This 72-page study utilizes the animal cancer database developed by Gold and colleagues (522 rodent carcinogens) (6) and the human cancer database of IARC (55 chemicals and exposures) (7). Although structural alerts and genotoxicity are again the main theme of the paper, this study can be utilized as a good information source on the chemical carcinogenicity of many chemicals. Table 7 of this paper (5) presents 35 different tissues of rats and mice (adrenal, bone, central nervous system, clitoral gland, esophagus, gall bladder, harderian gland, hematopoietic system, kidney, large intestine, liver, lung, mammary gland, mesovarium, myocardium, nasal cavity, oral cavity, ovary, pancreas, peritoneal cavity, pituitary gland, preputial gland, prostate, skin, small intestine, spleen, stomach, subcutaneous testes, thyroid gland, urinary bladder, uterus, vagina, vascular system, and Zymbal's gland) and the number of chemical carcinogens (as well as structural alert positive carcinogens) that cause cancer in either mice alone, rats alone, or rats and mice. The cancer data are not further broken down by the sex of the experimental animals. Table 1 of this paper (5) presents the rodent organ in which cancer occurred for the chemical carcinogens compiled by Gold et al. (6). In this database 510 chemicals were tested for carcinogenicity in either rats or mice. Of the 250 carcinogens tested in both rats and mice, 64 were carcinogenic in mice only, 43 were carcinogenic in rats only and 143 were carcinogenic in both rats and mice (5).

C. Biennial Report on Carcinogens

The *Biennial Report on Carcinogens* (previously known as the *Annual Report on Carcinogens*) is prepared by the Department of Health and Human Services of the U.S. government. This report contains a list of all substances either known to be human carcinogens or reasonably anticipated to be human carcinogens or to which a significant number of persons residing in the United States are exposed. The agencies participating in the preparation of the report are the Agency for Toxic Substances and Disease Registry, the National Institute for Occupational Safety and Health, the Consumer Product Safety Commission, the U.S. Environmental Protection Agency, the Food and Drug Administration, the National Cancer Institute, the National Institute of Environmental Health Sciences, the National Library of Medicine, and the Occupational Safety and Health Administration. By statutory requirement the report must use a particular two-category carcinogen-classification scheme: chemicals "known to be carcinogenic" and chemicals or exposures that "may reasonably be anticipated to be carcinogens." The 1994 *Seventh Annual Report on Carcinogens* contains 180 chemicals including 24 listed as known to be human carcinogens and 156 listed as reasonably anticipated to be a human carcinogen (8). The report on each of the chemicals uses the subheadings carcinogenicity, properties, use, production, exposure, and regulations. No information on noncarcinogenic chemicals is included in this report. The 473 page *Annual Report on Carcin-*

ogens Summary contains only text. The two-volume, 1063-page full report contains text and tables of related federal regulations for each entry.

D. Integrated Risk Information System (IRIS)

The IRIS database was created by the U.S. EPA and is publicly available from the National Technical Information Service (NTIS) and also from the National Library of Medicine via the Toxicology Data Network (TOXNET). IRIS contains information on over 500 chemicals for health effects and over 200 chemicals for cancer effects. For chemicals evaluated for their carcinogenicity, the carcinogenicity assessment for lifetime exposure, the unit risk estimate, and present regulatory limits are presented. For noncarcinogenic chemicals, the main sections of an IRIS file are chronic health hazard assessments for noncarcinogenic effects (oral RfD and inhalation RfC information) and health hazard assessments for varied exposure durations. The IRIS database is continually being updated and modified. Of the chemical carcinogen information sources mentioned in this chapter, the IRIS database is the most oriented towards risk assessment and regulation of chemicals.

E. The Gene-Tox Carcinogen Data Base

The Gene-Tox Carcinogen Data Base was published in *Mutation Research* in 1987 (9). Five hundred and six selected chemicals were included in this compilation and then placed in one of eight categories. The categories of degrees of evidence of chemical carcinogenicity were sufficient positive (252 chemicals), limited positive (99 chemicals), inadequate (with a positive indication) (13 chemicals), inadequate (48 chemicals), inadequate (with a negative indication) (32 chemicals), equivocal (1 chemical), limited negative (21 chemicals), and sufficient negative (40 chemicals). The chemicals' carcinogenic potentials were judged by the International Agency for Research on Cancer (185 chemicals), by the NCI/NTP (28 chemicals), or by an expert panel (293 chemicals selected because of the available genetic toxicology data as well as carcinogenicity data). Chemicals can be located in this 195-page paper by alphabetical index, by CAS registry number, or by chemical classification by functional groups. A strong point of this particular report by Nesnow et al. (9) is that a substantial number of chemicals were classified as noncarcinogenic to experimental animals at either the limited or the sufficient degree of evidence. The Gene-Tox Carcinogen Data Base has been analyzed by chemical class, strain, species, gender, route of administration, tumor site, and tumor type (10). This database is online in TOXNET at the National Library of Medicine or as a searchable DBASE file for personal computer use (from Dr. Nesnow, MD-68, US EPA, RTP, NC, 27711).

F. Carcinogen Potency Database

The Carcinogenicity Potency Database (CPDB) is a widely used resource on the results of chronic, long-term animal cancer tests. It provides a single, standardized, and easily accessible database that includes sufficient information on each experiment to permit investigations into many research areas of carcinogenesis. The CPDB is a guide to the literature of animal cancer tests and includes references to published experimental results. Both qualitative and quantitative information are reported on 5000 positive and negative experiments on 1230 chemicals, including all technical reports of the NCI/NTP and results from the general literature that meet a set of inclusion criteria. There is great diversity in the testing of chemicals reported in the database; while most chemicals have been tested in rats or mice, some have been tested in hamsters, dogs, prosimians, or monkeys. For each experiment a user-friendly format includes data on the species, strain, and sex of test animal; features of experimental protocol such as route of administration, duration of dosing, dose level(s) in mg/kg body weight/day, and duration of experiment; histopathology and tumor incidence; shape of the dose-response curve; published author's opinion as to carcinogenicity; and literature citation. A measure of carcinogenic potency, the TD_{50}, its statistical significance, and confidence limits, are given for each site in the CPDB. TD_{50} is the chronic dose rate (mg/kg/day) that would halve the probability of an animal remaining tumor-free by the end of a standard life span for the species. The range of statistically significant TD_{50} values for chemicals in the CPDB that are carcinogenic in rodents is more than 10 million-fold.

The CPDB has been published in plot form in six papers since 1984 (11–16). A combined plot that represents results from the six papers alphabetically by chemical name is available both in printed form and on computer tape or diskette. A SAS database is also available. A form to order these is available through the following World Wide Web URL address: http://potency.berkeley.edu/cpdb.html. Also provided on the Web are a description of the plot of the CPDB, the inclusion rules, a list of publications and abstracts, a HERP table that ranks possible carcinogenic hazards to rodent carcinogens, a summary of TD_{50} values and carcinogenicity in rats and mice for all chemicals in CPDB, and an example of the plot format.

Gold and colleagues have used their CPDB to examine many issues in carcinogenesis, including the proportion of chemicals that are positive for several datasets, the role of cell division in rodent bioassays, tautologous aspects of the interspecies correlation in carcinogenic potency, concordance of carcinogenic response between species, reproducibility of results in "near-replicate" experiments, comparison of target organs of mutagenic and nonmutagenic rodent carcinogens, carcinogenic identification on the basis of two versus four sex-species groups, ranking possible carcinogenic hazards of naturally occurring and synthetic chemicals (HERP), comparison of results for heterocyclic amines with other chemicals in the CPDB, setting priorities among possible carcinogenic hazards in the workplace, quick estimation of the regulatory, virtually safe dose based on the MTD, and a review of causes and prevention of human cancer.

G. Chemical Carcinogenesis Research Information System (CCRIS)

The CCRIS database is part of TOXNET in the National Library of Medicine system. The National Cancer Institute has arranged the CCRIS data by species, strain and sex of animal, route, dose, tumor site and type of lesion, results of the carcinogenicity test, and literature reference. The CCRIS is well organized, brief, and quite clear in several important aspects such as strain, sex, route, and dose of the experimental study. This information is not always fully provided by all carcinogenicity information sources. The cancer references may be to NCI, NTP, IARC, or journal articles. In CCRIS the cancer data is presented but not discussed. So for discussion and interpretation of the data, you must go to the cited reference.

H. Specialty Compilations of Chemical Carcinogens

Depending on the type of chemical that interests you, good compilations, summaries, or review articles of just this specialty research area may be available. For example, these types of articles are available for halogenated hydrocarbons (17) and N-nitroso (18,19) compounds. These specialty type of data compilations often cover their chosen compounds well but do not contain information on other, diverse chemicals. Finding this type of article or compilation may take some ingenuity or experience.

I. Chemicals Carcinogenic to Humans

Sixty-nine agents, mixtures, and exposure circumstances which were classified by IARC as group 1 (carcinogenic to humans) are given in the List of IARC Evaluations (1). The IARC monographs classified 57 agents and exposures as group 2A (probably carcinogenic to humans) and 215 agents and exposures as group 2B (possibly carcinogenic to humans) (1). Group 3 (cannot be classified as to its carcinogenicity to humans) contained 458 agents and exposures, while group 4 (probably not carcinogenic to humans) contained 1 chemical (1).

In a five-page 1991 paper, Vainio et al. (7) presented the overall structure of the IARC classification system and listed 55 human carcinogens and their target organs. Another article on human carcinogens was published in 1988 by Shelby (20). Shelby's 13-page article presented the genetic toxicology, rodent bone marrow cytogenetics test data, *Salmonella* mutagenicity data, structural electrophilic centers, and primary sites of cancer induction in test animals and in humans of the 23 human carcinogens known at that time.

J. NCI/NTP Technical Reports

For chemicals that have been bioassayed for possible carcinogenicity by the NCI or NTP, a technical report (often called a blue book) is prepared for each bioassayed chemical. These reports contain a wealth of details on the bioassay experiment and experimental results that can never be summarized in any compilation of chemical carcinogens. These books can be 150–400 pages long and normally cover only one chemical per blue book. To find the technical report number and the year of publication of these NCI/NTP technical reports, other animal cancer compilations (3,4,8) are often useful.

K. Survey of Compounds That Have Been Tested for Carcinogenic Activity

This multiple volume compilation (tens of thousands of pages which are available in hard copy only) is published and updated by the NCI (21). This compilation is organized by chemical compound. Within each compound, data are presented on the literature reference, study animal, preparation and dose of the compound, route and site of compound administration, pathology examination level, types of animals with tumors, animal survival, and duration of experiment. Indexes for the chemical name, CAS Registry number, route of administration, site of application, animal species/strain, tumor site, vehicle, and primary and secondary authors of the cancer study allow the reader to find information in many different ways. This survey does not critically review the experimental studies, assess the reliability of the cancer bioassay, or exclude studies of lessor quality. It aims to be comprehensive; the best interpretation of the data is left to the readers.

L. Medline Searches of the Scientific Literature

If the preceding 11 approaches to obtaining chemical carcinogenicity information do not meet your needs, then you can try searching Medline directly. Searches can be organized by one, two, or more keywords over chosen time periods. Good keywords may include chemical name, species of animal, gender of animal, tissue of interest, type of tumor, certain journal names (e.g., *Mutation Research, Journal of the National Cancer Institute*), or even the word review itself. This type of journal article search will not find chemical carcinogenicity information contained in book chapters or books.

III. DISCUSSION

The criteria used in evaluating chemical carcinogenicity evolves over time and also varies between different groups (e.g., IARC, U.S. EPA, and NTP). Thus, direct comparisons between different information sources for cancer data can cause confusion. For example, DDT is classified as noncarcinogenic by Ashby and Tennant (3) based on only the carcinogenicity data from the NTP bioassays, but the same chemical is classified as a carcinogen by IARC (1), by the *Seventh Annual Report on Carcinogens* (8), and also by Nesnow et al. (9).

The number of chemicals classified as noncarcinogens by different information sources varies greatly. For example, the number of noncarcinogens is one according to IARC (1) (only one group 4 chemical out of 800 chemicals total), 100 according to Ashby and Tennant (4) (100 out of 301 chemicals evaluated), and 61 according to Nesnow et al. (9) (61 out of 506 chemicals). Thus the percentage of chemicals judged to be "noncarcinogenic" by different groups using different criteria ranges from a low of 0.14% to a high of 33.2%, a difference of over 200-fold.

Some information sources are continually updated and data on new chemicals added (e.g., IARC, Biennial Report on Carcinogens, CPDB, CCRIS, and IRIS). Other information sources are either the work of individuals or are the outcome of particular research projects. This type of information source may become dated and of less importance and utility over time.

The perfect information source on chemical carcinogenicity data has not yet been created. To move towards that goal requires decisions on issues such as (a) how high quality a carcinogenicity experiment must be to be included in a database, (b) the criteria for chemical carcinogenicity in animals and humans, and (c) whether the goal should be either comprehensiveness or brevity and clarity. These and similar issues cannot be resolved to the satisfaction of all interested users of chemical carcinogenicity information. Nonetheless, incorporating several positive features of three to six of the stronger presently available databases of information on chemical carcinogenicity into a larger chemical carcinogenicity database is both desirable and well within the capacity of present-day computer storage and retrieval systems.

ACKNOWLEDGMENT

I thank Janice Brown for her help in preparing this manuscript.

REFERENCES

1. International Agency for Research on Cancer. List of IARC Evaluations. IARC Monographs of the Evaluation of Carcinogenic Risks to Humans. Lyon, France: IARC, June 1995.
2. International Agency for Research on Cancer. IARC Monographs on the Evaluation of Carcinogenic Risks to Humans, Overall Evaluations of Carcinogenicity: An Updating of IARC Monographs Volumes 1 to 42. Lyon, France: IARC, Supplement 7, 1987.
3. Ashby J, Tennant RW. Chemical structure, Salmonella mutagenicity and extent of carcinogenicity as indicators of genotoxic carcinogenesis among 222 chemicals tested in rodents by the U.S. NCI/NTP. Mut Res 1988; 204:17–115.
4. Ashby J, Tennant RW. Definitive relationships among chemical structure, carcinogenicity and mutagenicity for 301 chemicals tested by the U.S. NTP. Mut Res 1991; 257:229–311.
5. Ashby J, Paton D. The influence of chemical structure on the extent and sites of carcinogenesis for 522 rodent carcinogens and 55 different human carcinogen exposures. Mut Res 1993; 286:3–74.
6. Gold LS, Slone TH, Manley NB, Bernstein L. Target organs in chronic bioassays of 533 chemical carcinogens. Environ Health Perspect 1991; 93:233–246.
7. Vainio H, Coleman M, Wilbourn J. Carcinogenicity evaluations and ongoing studies: The IARC Databases. Environ Health Perspect 1991; 96:5–9.
8. National Toxicology Program. Seventh Annual Report on Carcinogens. Research Triangle Park, NC: U.S. Department of Health and Human Services, 1995.
9. Nesnow S, Argus M, Bergman H, Chu K, Frith C, Helmes T, McGaughy R, Ray V, Slaga TJ, Tennant R, Weisburger E. Chemical carcinogens: a review and analysis of the literature of selected chemicals and the establishment of the Gene-Tox Carcinogen Data Base. Mut Res 1986; 185:1–195.
10. Nesnow S, Bergman H. An analysis of the Gene-Tox Carcinogen Data Base. Mut Res 1988; 205:237–253.

11. Gold LS, Sawyer CB, Magaw R, Backman GM, de Veciana M, Levinson R, Hopper NK, Havender WR, Bernstein L, Petro R, Pike MC, Ames BN. A carcinogenic potency database of the standardized results of animal bioassays. Environ Health Perspect 1984; 58:9–319.
12. Gold LS, de Veciana M, Backman GM, Magaw R, Lopipero P, Smith M, Blumenthal M, Levinson R, Bernstein L, Ames BN. Chronological supplement to the carcinogenic potency database: standardized results of animal bioassays published through December 1982. Environ Health Perspect 1986; 67:161–200.
13. Gold LS, Slone TH, Backman GW, Magaw R, Da Costa M, Lopipero P, Blumenthal M, Ames BN. Second chronological supplement to the carcinogenic potency database: standardized results of animal bioassays published through December 1984 and by the National Toxicology Program through May 1986. Environ Health Perspect 1987; 74:237–329.
14. Gold LS, Slone TH, Backman GM, Eisenberg S, Da Costa M, Wong M, Manley NB, Rohrbach L, Ames BN. Third chronological supplement to the carcinogenic potency database: standardized results of animal bioassays published through December 1986 and by the National Toxicology Program through June 1987. Environ Health Perspect 1990; 84:215–286.
15. Gold LS, Manley NB, Slone TH, Garfinkel GB, Rohrbach L, Ames BN. The fifth plot of the carcinogen potency database: results of animal bioassays published in the general literature through 1988 and by the National Toxicology Program through 1989. Environ Health Perspect 1993; 100:65–168.
16. Gold LS, Manley NB, Slone TH, Garfinkle GB, Ames BN, Rohrbach L, Stern BR, Chow K. Sixth plot of the carcinogenic potency database: results of animal bioassays published in the general literature 1989–1990 and by the National Toxicology Program 1990–1993. Environ Health Perspect 1995; 103(suppl. 8):3–122.
17. Greim H, Wolff T. Carcinogenicity of organic halogenated compounds. In: Searle CE, ed. Chemical Carcinogens. 2d ed. Washington, DC: ACS Monograph 182, 1984:525–575.
18. Magee PN, Montesano R, Preussman R. N-Nitroso compounds and related carcinogens. In: Searle CE, ed. Chemical Carcinogens. Washington, DC: ACS, 1976:491–625.
19. Lijinsky W. Structure-activity relations in carcinogenesis by N-nitroso compounds. In: Rao TK, Linjinsky W, Epler JL, eds. Genotoxicity of N-nitroso Compounds. New York: Plenum, 1984:189–231.
20. Shelby MD. The genetic toxicity of human carcinogens and its implications. Mut Res 1988; 204:3–15.
21. National Cancer Institute. Survey of Compounds Which Have Been Tested for Carcinogenic Activity. NIH Publication No. 94-3765. Washington, DC: U.S. Department of Health and Human Services, 1994.

6

Use of Mutagenicity for Predicting Carcinogenicity

David M. DeMarini
U.S. Environmental Protection Agency, Research Triangle Park, North Carolina

I. INTRODUCTION

The evaluation of agents for their ability to induce mutations has evolved considerably since the development by Muller in 1927 (1) of the first assay to detect mutagens. Since then, more than 200 assays have been developed that measure mutations in DNA; however, only a few assays (< 10) are in use today for routine genetic toxicity assessment (2). Muller recognized early on that mutagenesis was valuable as an endpoint in its own right—as a predictor of potential genetic damage to the germ cells and, thus, as a means of detecting agents that might cause birth defects and harm to future generations (3). However, in the 1940s and 1950s, systemic efforts were initiated to screen chemicals for mutagenic activity with the notion that such data might also be predictive of potential genetic damage to somatic cells and, thus, could be used to detect agents that might cause cancer (4,5). This chapter reviews some of the mutagenicity assays used currently, what they measure, how the data can be used for predicting carcinogenicity, and recent developments in the field of genetic toxicology.

This document has been reviewed in accordance with U.S. Environmental Protection Agency policy and approved for publication. Mention of trade names or commercial products does not constitute endorsement or recommendation for use.

II. RATIONALE AND CONCEPTUAL BASIS FOR MUTAGENICITY ASSAYS

The history of the field of genetic toxicology and the use of mutagenicity as a surrogate measure of carcinogenicity have been described previously and will not be recounted here (6,7). However, the central organizing principle of this discipline is rooted in the long-standing observation that cancer cells beget cancer cells, implying that genetic changes (mutations) underlie the conversion of a normal cell to a tumor cell. Thus, an agent that can induce mutations might have the potential to induce cancer. In recent years, an enhanced understanding of the molecular workings of normal vs. tumor cells has reinforced the idea that mutations in key regulatory genes involved in basic aspects of metabolism, cell division, and DNA repair conspire to produce a tumor cell (8–10). The myriad of enzymatic and biochemical changes that can be measured in tumor cells reflect and are a consequence of this underlying genetic change.

A second motivating force for using mutagenicity as a predictor of carcinogenicity is the fact that most mutagenicity assays can be performed in less time and for less money than any version of a rodent carcinogenicity assay. Thus, there have been both scientific and financial reasons for trying to predict an agent's carcinogenicity based on its mutagenicity. The following discussion presents an overview of the genetic toxicity assays that are in use currently and how predictive they are for carcinogenicity.

III. GENETIC TOXICITY ASSAYS CURRENTLY IN USE

There are three general classes of mutations—gene, chromosomal, and genomic—all of which have been associated with various types of tumor cells (8–10). Assays that measure gene mutation are those that detect the substitution or addition/deletion of nucleotides within a gene. Assays that measure chromosomal mutation are those that detect breaks or chromosomal rearrangements involving one or more chromosomes. Assays that measure genomic mutation are those that detect changes in the number of chromosomes, a condition called aneuploidy. Although the exact types and numbers of assays used for genetic toxicity assessment are constantly evolving and vary from country to country, the most common ones include assays for (a) gene mutation in bacteria and/or cultured mammalian cells and (b) chromosomal mutation in cultured mammalian cells and/or bone marrow within living mice. Some of the assays within this second category can also detect aneuploidy.

A. Bacterial Assays

The primary system in most widespread use for mutagenicity testing is the Salmonella (Ames) mutagenicity assay and, to a much lesser extent, a similar bacterial mutagenicity assay in strain WP2 of *Escherichia coli* (11). Additional bacterial assays that measure DNA damage based on induction or extent of DNA repair, such as the SOS Chromotest or UMU assay, have smaller data bases than the Salmonella assay but give results that are ~90% similar to those of Salmonella (12,13). Although there are many strains of Salmonella available for mutagenesis testing, an extensive study showed that the use of only two strains (TA98 and TA100) was sufficient to detect ~90% of the mutagens detectable using four or five strains (14). Thus, these two strains are now used for most screening purposes; however, other strains are available for more extensive testing.

Recent molecular studies have identified the types of mutations that are recovered by strains TA98 and TA100 and the range of mutations (mutation spectrum) produced by various mutagens in these strains (15–23). Although a review of these studies is beyond the scope of this chapter, some general conclusions have emerged from these molecular studies in Salmonella. In strain TA100, the primary target for reversion to histidine independence involves two CCs (or two GGs on the other strand) in which most (but not all) types of base substitutions can be recovered. Most aromatic mutagens typically found in the environment, such as benzo[a]pyrene (BAP), 1-nitropyrene (1NP), and aromatic amines such as 4-aminobiphenyl (4AB) and 2-acetylaminofluorene (2AAF), induce primarily C-to-A transversions and, secondarily, C-to-T transitions. Most alkylating agents, such as ethyl methanesulfonate, induce primarily C-to-T transitions and, secondarily C-to-A transversions. The base-substitution mutation spectrum produced by agents in Salmonella is similar to the spectrum produced by the agents in other organisms, including mammalian cells and mice in vivo (15).

In strain TA98, the primary mutation induced by most agents is a two-base deletion (a frameshift) within the sequence CGCGCGCG (16). Some agents, such as the heterocyclic amine cooked-food mutagen Glu-P-1, 2AAF, and 1NP, induce only this mutation in TA98 (and TA1538), and they are equally potent in TA98, which contains the pKM101 plasmid, as they are in TA1538, which does not contain the plasmid (17–19). In addition, these mutagens are more potent as frameshift mutagens than as base-substitution mutagens, i.e., they are more potent in TA98 than in TA100. These mutagens tend to form only one planar DNA adduct, which promotes slippage, leading to the single type of frameshift.

In contrast, mutagens such as BAP, 4AB, and the chlorinated furanone drinking-water mutagen MX, induce primarily the two-base deletion within the CGCGCGCG sequence, but they also produce a second class of mutation called complex frameshifts, which are frameshifts with an adjacent base substitution (20–22). These mutagens are more potent in TA98 than in TA1538, and they are more potent as base-substitution mutagens than as frameshift mutagens, i.e., they are more potent in TA100 than in

TA98. These mutagens tend to form several types of DNA adducts or one type of adduct that can assume at least two conformations. One adduct or conformation is planar, which promotes the production of the two-base deletion, and the other adduct is nonplanar, which promotes misincorporation (i.e., base substitution), followed by slippage, which leads to the frameshift.

Another important feature of the Salmonella assay is its use for studying complex mixtures. As discussed previously (23), it is the most used bioassay for evaluating the mutagenicity of environmental samples (air, food, water, etc.), mixtures resulting from industrial processes, and body fluids such as urine (24). Molecular analysis of revertants of Salmonella induced by complex mixtures has shown that the mutation spectrum of a complex mixture reflects the dominance of one or a few chemical classes within the mixture. Thus, the mutation spectrum of urban air reflects the dominance of polycyclic aromatic hydrocarbons (PAHs) (20), that of chlorinated drinking water reflects the dominance of certain chlorinated organics (22), and that of cigarette smoke reflects the presence of PAHs and aromatic amines (25).

Agents that produce primarily DNA strand breakage, such as oxidative mutagens, are not detected well or at all by TA98 and TA100, because DNA strand breakage will not revert the alleles contained by these strains. Oxidative mutagens can be detected in microbial systems, however, by the use of Salmonella strains TA102 or TA104 or by the various DNA damage or SOS assays, such as the umu-test, Chromotest, or prophage-induction assay (26–28).

B. Mammaliam Cell Mutagenicity Assays

Less than one tenth of the chemicals that are tested routinely in contract testing laboratories for mutagenicity in Salmonella are also tested for mutagenicity in mammalian cells. This is due partly to regulatory standards, which generally do not require mammalian cell mutagenicity data, as well as to the expense and time required to perform such assays. Mammalian cell mutagenicity assays usually require 2–3 weeks, rather than the 2–3 days required for bacterial mutagenicity assays, and they are generally 10 times more costly than bacterial assays. Nonetheless, they can serve a useful purpose and are used when additional information about an agent is desired (29,30).

Routine testing of agents for mutagenicity in mammalian cells is performed primarily in one of two genes: *hprt* and *tk*. The assays used for this purpose include the CHO/HPRT, the TK6, and mouse lymphoma L5178Y/TK$^{+/-}$ assays. Thus, Chinese hamster ovary (CHO) cells, a human lymphoblastoid cell line (TK6), and a mouse cell line (L5178Y) are most used for routine mutagenesis screening in mammalian cells. An additional assay that is also used is performed in AS52 cells, which are derived from CHO cells and contain the *gpt* gene, which is the bacterial homolog of the *hprt* gene. Mutations in several other genes are also measured, shuttle vector systems are also available, and other cell lines are used that contain various DNA repair mutations as well as some human or rodent genes involved in metabolism, such as P450 gene (30,31). The mammalian cell mutagenicity assays permit the recovery of mutations within the gene being studied (gene mutations) as well as mutations involving regions of the chromosome flanking the gene (chromosomal mutations). However, this latter type of mutation is recovered to a much greater extent in the mouse lymphoma and AS52 assays, apparently due to the location of the gene in these assays (32).

Molecular analyses of mutations in mammalian cell systems have indicated that these systems recover the types of mutations associated with cancer cells—both gene and chromosomal mutations (31–38). By far the vast majority of molecular studies have been performed at the *hprt* locus, with over 1000 mutant sequences identified thus far (38). The advantage of this database is that the results can be compared to those obtained in vivo in rodents and humans (39,40). In general, mutagens that have been tested in several mammalian cell systems have produced similar mutation spectra. Thus, for example, BAP produces primarily C-to-A transversions in these systems, as it does also in bacterial systems.

One of the many insights that have emerged from mammalian cell studies is the observation that some deletions of the gene (usually of whole exons, or coding portions of the gene) are due to the production of a base substitution at a splice-site junction (a site at which introns are removed prior to formation of the functional mRNA). In addition, large deletions covering many thousands of bases have been detected, and complex molecular rearrangements have also been noted (31). Because mammalian cells are more complex than bacterial cells, a wider array of mutations involving more complex changes has been detected in these systems. Not surprisingly, many of the mutations parallel those detected in human disease genes, such as cystic fibrosis or Lysh-Nyhan syndrome, and have provided the opportunity to explore mechanisms of mutation that are especially relevant to humans.

C. Cytogenetic Assays

Although chromosomal mutations can be detected or inferred using some of the mammalian cell mutagenicity assays described above, chromosomal mutations are generally identified by cytogenetic assays. These involve exposing mammalian cells (rodent or human) or rodents to an agent, staining the chromosomes in the target cells, and then visually examining the chromosomes through a microscope to detect alterations in the structure or number of chromosomes. Although a variety of endpoints can be examined, the two that are currently accepted by regulatory agencies are chromosomal aberrations and micronuclei; however, sister chromatid exchanges (SCEs) are also measured in some studies (41).

Considerable training and expertise are required to score cells for the presence of chromosomal aberrations, making this a costly procedure in terms of time and money. In contrast, little training is required to score micronuclei, and their detection can be automated. Micronuclei appear as small dots within the cell that are distinct from the nucleus. Micronuclei result from either chromosome breakage or from aneuploidy, and molecular methods can distinguish between the two (42). Because of the ease of scoring micronuclei compared to chromosomal aberrations, and because recent studies indicate that agents that induce chromosomal aberrations in the bone marrow of mice also induce micronuclei in this tissue (43), micronuclei are now commonly measured as an indication of the ability of an agent to induce chromosomal mutations.

Nearly all tumor cells contain cytogenetically detectable chromosomal aberrations, and specific chromosomal abnormalities are mechanistically linked with various types of cancers (44). Thus, chromosomal mutations are highly relevant to the cancer process and are, therefore, an integral feature of genetic toxicology. Currently, the most common types of cytogenetic analyses are performed in CHO cells because of the large database in these cells and because of the relative ease of scoring aberrations in this cell line. Because in vivo studies are considered more relevant than in vitro studies, cytogenetic analyses (chromosomal aberrations and micronuclei) are also performed in mouse bone marrow. As discussed at the end of this chapter, new techniques are being introduced and are under development that will complement and/or replace these methods.

IV. PREDICTIVITY OF MUTAGENICITY FOR CARCINOGENICITY

Systematic studies over a 25-year period (1970–1995), especially at the U.S. National Toxicology Program (NTP) in North Carolina, have resulted in the routine use of the limited number of mutagenicity assays described above. At the NTP, various mutagenicity assays were evaluated for their ability to detect as mutagenic those agents that caused cancer in rodents and that were suspected of causing cancer in humans—as well as for their ability to detect as nonmutagenic those agents that were not carcinogenic in rodents. The results of this effort and of other similar analyses were that (a) the carcinogenicity of a set of agents in a mouse is ~70% concordant with the carcinogenicity of the agents in a rat, (b) the mutagenicity of a set of agents in any of several mutagenicity assays is also ~60–70% concordant with the carcinogenicity of the agents in rodents, (c) combinations of mutagenicity assays did not significantly improve the predictivity of the single mutagenicity assays for rodent carcinogenicity, and (d) 84% of human carcinogens as defined by the International Agency for Research on Cancer (IARC) are carcinogenic in rodents (45–51). In all of these analyses, improvements in sensitivity generally came at the expense of specificity.

The genetic toxicity assays in use today reflect these strengths and limitations. Thus, the current assays rest on a large database, are relatively inexpensive and easy to perform, and detect the main classes of genetic changes that are associated with cancer cells. Positive results in these assays have only slightly less predictivity for rodent carcinogens than positive results in rodent carcinogenicity assays have for predicting the ICRC-identified human carcinogens. Some important observations have emerged from this massive study that support the current use of these mutagenicity assays and that offer insights into the value of these assays as predictors of rodent and human carcinogenicity.

Most of the IARC Group I human carcinogens (a) are carcinogenic in at least two species of laboratory animals (usually rats and mice), (b) induce tumors at multiple sites in these animals, and (c) are mutagenic in one of more of the standard mutagenicity assays (usually Salmonella and a cytogenetic assay). This has led to the conclusion that the identification of an agent as a trans-species carcinogen indicates that the agent is likely to be both a human carcinogen and mutagenic (49,50). Also, 90% of the carcinogens detected by the standard NTP rodent carcinogenicity assay could be detected by using only male rats and female mice (52).

The value of mutagenicity assays for detecting potential human carcinogens is also emphasized by the finding that essentially all of the IARC Group I human carcinogens (>90%) are detected as mutagenic by a combination of the Salmonella and mouse bone marrow micronucleus assays (49,53,54). Additional confidence in using the micronucleus assay instead of the more complicated and expensive chromosomal aberration assay has been provided by the finding that nearly all of the agents that induce chromosomal aberrations in mouse bone marrow also induce micronuclei in mouse bone marrow (43).

By far the most impressive use of mutagenicity data has been in a unique exercise sponsored by the NTP in which various approaches were used to predict the a priori carcinogenicity of 44 chemicals while the chemicals were being tested in the NTP rodent carcinogenicity assay. The most successful approach permitted 85% of the rodent carcinogens to be predicted by using only the data from a 90-day rodent toxicity assay, data from a set of Salmonella mutagenicity assays, and knowledge of the chemical structure of the agent (55). This impressive result supports the view that mutagenicity data alone can have some, but only limited, predictivity for carcinogenicity. However, in combination with just a few additional pieces of information, mutagenicity data can provide highly useful data for carcinogenicity prediction. Overall this approach had 85% sensitivity, 65% specificity, and 75% concordance (55).

As mentioned at the beginning of this chapter, the original purpose of mutagenicity assays was to detect germ cell mutagens. Today, only a few germ cell mutagens have been identified: of the 29 chemicals that have been tested for germ cell mutation in the mouse specific-locus assay, only 13 are positive (56). However, the extensive mutagenicity testing that has occurred over the years has shown that most of these germ cell mutagens are detectable in the mouse bone marrow cytogenetic assay (57,58). Thus, the mouse bone marrow cytogenetic assay may have the ancillary value of identifying rodent germ cell mutagens. Studies of survivors of the atomic bombs in Japan have not identified ionizing radiation as a germ cell mutagen (59). However, data on the ability of various cancer chemotherapeutic agents (mostly alkylating agents) to induce mutations in the germ cells and bone marrow of mice suggest that many of the cancer chemotherapeutic agents are likely human germ cell mutagens (58). The use of mutagenicity assays to identify potential human germ cell mutagens is another important use to which the assays can be put.

Because of the lack of 100% concordance between mutagenicity and carcinogenicity, much consideration has been given to the existence, prevalence, and mechanisms of so-called nongenotoxic carcinogens (60,61). Recognizing the limitations of both the carcinogenicity and mutagenicity assays, a comprehensive analysis of a set of putative nongenotoxic carcinogens found that < 10% of these were, in fact, not mutagenic when they had been tested adequately for the three main categories of mutagenicity (gene, chromosomal, and genomic) (62). Many of the putative nongenotoxic carcinogens have been tested for mutagenicity only in the Salmonella assay; until such agents are also tested for their ability to induce chromosomal or genomic mutations, they cannot be considered as validated nongenotoxic carcinogens. Although nongenotoxic carcinogens may exist, they appear to comprise just a few percent of the known rodent carcinogens (62).

A final observation that has resulted from the extensive mutagenicity evaluation of thousands of agents is that, among the "universe" of identified mutagens, most are capable of inducing both gene and chromosomal mutation. Most mutagens appear to induce one category better than they induce the other, but most mutagens appear to induce both. For example, benzene may be one of the few mutagens that induce only chromosomal mutation and not gene mutation (63,64). I am unaware of a mutagen that induces only gene mutation and not chromosomal mutation. This observation may help explain why several different mutagenicity assays (e.g., Salmonella plus a cytogenetic assay) do not appear to complement each other to improve the ability to detect mutagens (45). This also has implications for the mechanism by which mutagens induce these two main categories of mutation—perhaps these categories can be viewed as a continuum or a range of the types of mutations produced by cells when exposed to a mutagen (65).

V. FUTURE ROLE FOR MUTAGENICITY AS A PREDICTOR OF CARCINOGENICITY

With advances in DNA technology, the human genome project, and an improved understanding of the role of mutation in cancer, new carcinogenicity as well as mutagenicity assays are being developed that will be incorporated into standard screening procedures. Foremost among these developments is the use of transgenic cells and rodents. For example, the NTP is currently validating various transgenic lines of mice that capitalize on the role of the *p53* gene or *ras* genes in tumor induction (66,67). The animals were constructed to have an alteration in the expression of these genes, and an additional mutation may be all that is necessary to initiate tumor development. In fact, the time to tumor in these transgenic mice is much faster than in standard strains of mice. All of these systems are based on the role that mutations play in tumor etiology, and because of their speed and potential reliability, they may replace, for all practical purposes, the standard 2-year rodent carcinogenicity assay for screening purposes. The development of these systems will greatly accelerate carcinogenicity screening and will permit rapid molecular analyses of the mutations present in the tumors. Such results can be compared to mutagenesis data generated in vivo and in vitro.

As mentioned earlier, transgenic cell lines are already in use, such as the AS52 cell line for molecular analysis of mutagenesis in mammalian cells (36) and the various cell lines containing human P450 genes to characterize the metabolism of promutagens to mutagens (68). In addition, transgenic strains of Salmonella have been constructed that contain rodent or human genes, such as the rat glutathione *S*-transferase theta gene, which has permitted the detection as mutagenic some of the carcinogenic chlorinated organics that were not previously detectable as mutagenic in Salmonella (69). Further developments will improve the ability of bacterial and mammalian cell systems to detect these and other chemical classes of carcinogens.

Transgenic rodents for mutagenesis testing are currently being developed and evaluated (70–74). The potential advantage of these systems is that they will permit the detection of mutations in vivo in the rodent target organs and that such information might be predictive of the rodent target organ carcinogenicity of an agent. In addition, the mutations in the transgene can be identified by DNA sequence analysis, which provides information about the mutation spectrum of a chemical mutagen in vivo.

The first generation of transgenic rodents for mutagenesis testing have used as the transgene the *lacI* or *lacZ* genes from *E. coli* or a mutant allele of the *trp* gene from bacteriophage lambda (70–74). The *lacI* system in *E. coli* has been used extensively for basic research on the use of mutation spectra to deduce mutational mechanisms (75). Thus, a considerable database for this gene exists in bacteria. Comparison of the mutation spectra of certain agents at this gene in bacteria and the transgenic mouse has revealed differences in the mutation spectra that reflect differences in the DNA metabolism of bacteria and mice (76). Initial comparative studies, however, also indicate that the transgenic systems may be as responsive to mutagens as endogenous genes, such as *hprt*, for somatic cell mutation (39) and potentially as responsive as the mouse specific locus assay for germ cell mutation (77). The cost and time required to use these systems makes them impractical for routine testing on a large scale; however, second-generation systems may be more cost-effective, more sensitive, and easier to use.

Although molecular analysis of mutations in bacterial systems, such as the Salmonella assay, can now be performed as described earlier using probe hybridization and PCR/DNA sequence analysis (15–22,25), new systems have been developed that permit mutation identification without the need for molecular techniques. In particular, a set of strains of Salmonella (78) and *E. coli* (79) are now available that revert by specific mutations. These strains permit the detection of all 6 types of base substitutions and several types of +1 or −1 frameshifts or +2 or −2 frameshifts within repeating G or GC sequences. Thus, the mutation spectrum of an agent is indicated by which strains are reverted by the agent. Although these systems are easy to use, they are unlikely to gain any more widespread use than molecular analysis of the standard bacterial strains. For the near future, it appears that for general screening purposes, simply having knowledge that an agent is mutagenic may supersede the need to know precisely what mutations are produced by the agent.

Molecular advances in cytogenetics now permit highly detailed evaluations of chromosomal mutations. These include the use of probes that hybridize to specific genes to permit the detection of deletions and rearrangements, especially translocations (80). Fluorescent probes are easily visualized as colored sectors on the chromosome, and less training is required to identify chromosomal mutations by these molecular methods than by classical cytogenetic techniques. PCR-based methods are also valuable to identify specific translocations at frequencies that would not be detectable by cytogenetic methods (81). As mentioned previously, molecular methods are now available to distinguish micronuclei that result from chromosome breakage vs. those that result from nondisjunction (42). This can be useful because agents that cause the former would be considered likely to cause chromosomal aberrations, whereas the latter would cause aneuploidy through disruption of, for example, the spindle apparatus. Applications of flow cytometry are also available to speed the analysis of chromosomal changes (82). Although these methods are not yet in routine use in genetic toxicology or for screening large numbers of agents for mutagenic activity, their selective use for characterizing specific chemicals will likely increase in the coming years.

One method that is gaining use in a routine fashion for detection of DNA breaks is the single-cell gel electrophoresis (SCGE) or comet assay (83,84). This simple but powerful technique is inexpensive and rapid and can be used for cells treated in vitro or in vivo. Although this assay detects DNA breaks, the presence of broken DNA following treatment with an agent may be due to DNA breaks induced by the agent or to lesions that are in the midst of being repaired by DNA repair processes. This second category can be distinguished from the first by the inclusion of an agent that inhibits DNA repair. An agent that produced DNA strand breaks only in the presence of such an inhibitor would not be considered a clastogen, i.e., an agent that breaks DNA and/or causes chromosomal aberrations.

One goal of genetic toxicology is to identify potential human as well as rodent carcinogens. In this regard, the human genome project is rapidly identifying many of the genes and mutant forms (alleles) of the genes involved in phase I and phase II metabolism (85). The ability to perform PCR-based genotyping and phenotyping on individuals who are exposed to an agent in vivo (86) or whose blood cells are exposed in vitro (87) followed by the measurement of a genotoxic endpoint may permit the identification of susceptible populations. Although such human-based mutagenicity studies will not be a part of initial screening programs for identifying potential carcinogens, such research may play an important role in clarifying the potential risk of rodent carcinogens to humans. PCR-based genotyping methods combined with mutation-detection assays for changes in certain oncogenes or tumor-suppressor genes are also becoming important components of molecular epidemiology (88,89).

VI. CONCLUSIONS

The past quarter century of research in environmental mutagenesis has revealed, perhaps unexpectedly, that we live in a virtual sea of mutagens. Urban air, chlorinated drinking water, combustion emissions, and much food are mutagenic and/or carcinogenic (20,22,90–92). Fortunately, the environment, especially certain foods, contains many antimutagens and anticarcinogens (93). The presence of so many naturally occurring mutagens and carcinogens, coupled with the presence of many antimutagens and anticarcinogens as well as highly refined DNA repair mechanisms, has raised the issue of whether or not industrial chemicals or newly introduced agents contribute much to the "mutagenic burden" that we already encounter (94). Nonetheless, prudence suggests that the ability to screen agents (either natural or synthetic) for mutagenic activity is an important capability in the toxicologist's and public health official's armamentarium.

After many years of use and the generation of a large and systematically developed database, mutagenicity assessment of agents can now be performed with just a few assays at a relatively small cost in a short period of time. The data produced can be combined with several other kinds of information to provide useful predictions of the ability of an agent to be a carcinogen in rodents and, presumptively, in humans. Such an ability limits the introduction into the environment of mutagenic agents and permits the development of alternative, nonmutagenic agents. Future developments should lead to even better methods with greater predictivity than are currently available.

REFERENCES

1. Muller HJ. Artificial transmutation of the gene. Science 1927; 66:84–87.
2. Li AP, Heflich RH. Genetic Toxicology. Boca Raton, FL: CRC Press, 1991.
3. Carlson EA. Genes, Radiation, and Society. The Life and Work of H.J. Muller. Ithaca, NY: Cornell University Press, 1981.
4. Tatum EL. Chemically induced mutations and their bearing on carcinogenesis. Ann NY Acad Sci 1947; 49:87–97.
5. Barratt RW, Tatum EL. Carcinogenic mutagens. Ann NY Acad Sci 1958; 71: 1072–1084.

6. Brockman HE, de Serres FJ. Short-term tests for genetic toxicity. In: Woodhead AD, Waters MD, eds. Short-Term Tests for Environmentally Induced Chronic Health Effects. Washington, DC: US EPA-600/8-83-002, 1983:11-37.
7. Casciano DA. Introduction: historical perspectives of genetic toxicology. In: Li AP, Heflich RH, eds. Genetic Toxicology. Boca Raton, FL: CRC Press, 1991:1-12.
8. Greenblatt MS, Bennett WP, Hollstein M, Harris CC. Mutations in the p53 tumor suppressor gene: clues to cancer etiology and molecular pathogenesis. Cancer Res 1994; 54:4855-4878.
9. Weinberg RA. Oncogenes, antioncogenes, and the molecular bases of multistep carcinogenesis. Cancer Res 1989; 49:3713-3721.
10. MacPhee DG. Mismatch repair, somatic mutations, and the origins of cancer. Cancer Res 1995; 55:5489-5492.
11. Gatehouse D, Haworth S, Cebula T, Gocke E, Kier L, Matsushima T, Melcion C, Nohmi T, Ohta T, Venitt S, Zeiger E. Recommendations for the performance of bacterial mutation assays. Mutat Res 1994; 312:217-233.
12. Quillardet P, Hofnung M. The SOS Chromotest: a review. Mutat Res 1993; 297:235-279.
13. Oda Y, Nakamura S, Oki I, Kato T, Shinagawa H. Evaluation of the new system (*umu*-test) for the detection of environmental mutagens and carcinogens. Mutat Res 1985; 147:219-229.
14. Zeiger E, Risko KJ, Margolin BH. Strategies to reduce the cost of mutagenicity screening with the Salmonella assay. Environ Mutagen 1985; 7:901-911.
15. Koch WH, Henrikson EN, Kupchella E, Cebula TA. *Salmonella typhimurium* strain TA100 differentiates several classes of carcinogens and mutagens by base substitution specificity. Carcinogenesis 1994; 15:79-88.
16. DeMarini DM, Bell DA, Levine JG, Shelton ML, Abu-Shakra A. Molecular analysis of mutations induced at the *hisD3052* allele of Salmonella by single chemicals and complex mixtures. Environ Health Perspect 1993; 101(suppl 3): 207-212.
17. Levine JG, Knasmuller S, Shelton ML, DeMarini DM. Mutation spectra of Glu-P-1 in Salmonella: Induction to hotspot frameshifts and site-specific base substitutions. Environ Mol Mutagen 1994; 24:11-22.
18. Shelton ML, DeMarini DM. Mutagenicity and mutation spectra of 2-acetylaminofluorene at frameshift and base-substitution alleles in four DNA repair backgrounds of Salmonella. Mutat Res 1995; 327:75-86.
19. DeMarini DM, Shelton ML, Bell DA. Mutation spectra of chemical fractions of a complex mixture: Role of nitroarenes in the mutagenicity of municipal waste incinerator emissions. Mutat Res 1996; 349:1-20.
20. DeMarini DM, Shelton ML, Bell DA. Mutation spectra in Salmonella of complex mixtures: comparison of urban air to benzo[a]pyrene. Environ Mol Mutagen 1994; 24:262-275.
21. Levine JG, Schaaper RM, DeMarini DW. Complex frameshift mutations mediated by plasmid pKM101: Mutational mechanisms deduced from 4-

aminobiphenyl-induced mutation spectra in Salmonella. Genetics 1994; 136: 731–746.

22. DeMarini DM, Abu-Shakra A, Felton CF, Patterson KS, Shelton ML. Mutation spectra in Salmonella of chlorinated, chloraminated, or ozonated drinking water extracts: comparison to MX. Environ Mol Mutagen 1995; 26:270–285.
23. DeMarini DM. Environmental mutagens/complex mixtures. In: Li AP, Heflich RH, eds. Genetic Toxicology. Boca Raton, FL: CRC Press, 1991:285–302.
24. Choi BCK, Connolly JG, Zhou RH. Application of urinary mutagen testing to detect workplace hazardous exposure and bladder cancer. Mutat Res 1995; 341:207–216.
25. DeMarini DM, Shelton ML, Levine JG. Mutation spectrum of cigarette smoke condensate in Salmonella: comparison to mutations in smoking-associated tumors. Carcinogenesis 1995; 16:2535–2542.
26. Urios A, Herrera G, Blanco M. Detection of oxidative mutagens in strains of *Escherichia coli* deficient in the OxyR or MutY functions: dependence on SOS mutagenesis. Mutat Res 1995; 332:9–15.
27. Quillardet P, de Bellecombe C, Hofnung M. The SOS Chromotest, a colorimetric bacterial assay for genotoxins: validation study with 83 compounds. Mutat Res 1985; 147:79–95.
28. DeMarini DM, Lawrence BK. Prophage induction by DNA topisomerase II poisons and reactive-oxygen species: role of DNA breaks. Mutat Res 1992; 267:1–17.
29. Moore MM, DeMarini DM, de Serres FJ, Tindall KR. Mammalian Cell Mutagenesis, Banbury Report No. 28. Cold Spring Harbor, NY: Cold Spring Harbor Laboratory, 1987.
30. Aaron CS, Bolcsfoldi G, Glatt HR, Moore M, Nishi Y, Stankowski L, Theiss J, Thompson E. Mammalian cell gene mutation assays working group report. Mutat Res 1994; 312:235–239.
31. Glatt H. Comparison of common gene mutation tests in mammalian cells in culture: a position paper of the GUM Commission for the Development of Guidelines for Genotoxicity Testing. Mutat Res 1994; 313:7–20.
32. DeMarini DM, Brockman HE, de Serres FJ, Evans HH, Stankowski Jr LF, Hsie AW. Specific-locus mutations induced in eukaryotes (especially mammalian cells) by radiation and chemicals: a perspective. Mutat Res 1989; 220:11–29.
33. Applegate ML, Moore MM, Broder CB, Burrell A, Juhn G, Kasweck KL, Lin Wadhams A, Hozier JC. Molecular dissection of mutations at the heterozygous thymidine kinase locus in mouse lymphoma cells. Proc Natl Acad Sci USA 1990; 87:51–55.
34. Cariello NF, Skopek TR. Analysis of mutations occurring at the human *hprt* locus. J Mol Biol 1993; 231:41–57.
35. Nelson SL, Jones IM, Fuscoe JC, Burkhart-Schultz K, Grosovsky AJ. Mapping the end points of large deletions affecting the *hprt* locus in peripheral blood cells and cell lines. Radiat Res 1995; 141:2–10.

36. Tindall KR, Stankowski Jr LF. Molecular analysis of spontaneous mutations at the *gpt* locus in Chinese hamster ovary (AS52) cells. Mutat Res 1989; 220: 241–253.
37. Scheerer JB, Adair GM. Homology dependence of targeted recombination at the Chinese hamster *aprt* locus. Mol Cell Biol 1994; 14:6663–6673.
38. Cariello NF. Data base and software for the analysis of mutations at the human *hprt* gene. Nucleic Acids Res 1994; 22:3547–3548.
39. Skopek TR, Kort KR, Marino DR. Relative sensitivity of the endogenous *hprt* gene and *lacI* transgene in ENU-treated Big Blue™ B6C3F1 mice. Environ Mol Mutagen 1995; 26:9–15.
40. Albertini RJ, Nicklas JA, O'Neill JP. Somatic cell gene mutations in humans: biomarkers for genotoxicity. Environ Health Perspect 1993; 101(suppl 3):193–201.
41. Preston RJ. Mechanisms of induction of chromosomal alterations and sister chromatid exchanges: Presentation of a generalized hypothesis. In: Li AP, Heflich RH, eds. Genetic Toxicology. Boca Raton, FL: CRC Press, 1991:41–65.
42. Afshari AJ, McGregor PW, Allen JW, Fuscoe JC. Centromere analysis of micronuclei induced by 2-aminoanthraquinone in cultured mouse splenocytes using both a gamma-satellite DNA probe and anti-kinetochore antibody. Environ Mol Mutagen 1994; 24:96–102.
43. Shelby MD, Witt KL. Comparison of results from mouse bone marrow chromosome aberration and micronucleus tests. Environ Mol Mutagen 1995; 25: 302–313.
44. Hartwell L. Defects in a cell cycle checkpoint may be responsible for the genomic instability of cancer cells. Cell 1992; 71:543–546.
45. Tennant RW, Margolin BH, Shelby MD, Zeiger E, Haseman JK, Spalding J, Caspary W, Resnick M, Stasiewicz S, Anderson B, Minor R. Prediction of chemical carcinogenicity in rodents from in vitro genetic toxicity assays. Science 1987; 236:933–941.
46. Ashby J, Tennant RW. Chemical structure, Salmonella mutagenicity and extent of carcinogenicity as indicators of genotoxic carcinogenesis among 222 chemicals tested in rodents by the U.S. NCI/NTP. Mutat Res 1988; 204:17–115.
47. Ashby J, Tennant RW, Zeiger E, Stasiewicz S. Classification according to chemical structure, mutagenicity to Salmonella and level of carcinogenicity of a further 42 chemicals tested for carcinogenicity by the U.S. National Toxicology Program. Mutat Res 1989; 223:73–103.
48. Tennant RW, Ashby J. Classification according to chemical structure, mutagenicity to Salmonella and level of carcinogenicity of a further 39 chemicals tested for carcinogenicity by the U.S. National Toxicology Program. Mutat Res 1991; 257:209–227.
49. Ashby J, Paton D. The influence of chemical structure on the extent and sites of carcinogenesis for 522 rodent carcinogens and 55 different human carcinogen exposures. Mutat Res 1993; 286:3–74.

50. Tennant RW. Stratification of rodent carcinogenicity bioassay results to reflect relative human hazard. Mutat Res 1993; 286:111–118.
51. Wilbourn J, Haroun L, Heseltine E, Kaldor J, Partensky C, Vainio H. Response of experimental animals to human carcinogens: an analysis based upon the IARC Monographs programme. Carcinogenesis 1986; 7:1853–1863.
52. Ashby J, Purchase IFH. Will all chemicals be carcinogenic to rodents when adequately evaluated? Mutagenesis 1993; 8:489–493.
53. Shelby MD. The genetic toxicity of human carcinogens and its implications. Mutat Res 1988; 204:3–15.
54. Shelby MD, Zeiger E. Activity of human carcinogens in the Salmonella and rodent bone-marrow cytogenetics tests. Mutat Res 1990; 234:257–261.
55. Ashby J, Tennant RW. Prediction of rodent carcinogenicity for 44 chemicals: results. Mutagenesis 1994; 9:7–15.
56. Bentley KS, Sarrif AM, Cimino MC, Auletta AE. Assessing the risk of heritable gene mutation in mammals: Drosophila sex-linked recessive lethal test and tests measuring DNA damage and repair in mammalian germ cells. Environ Mol Mutagen 1994; 23:3–11.
57. Shelby MD, Erexson GL, Hook GJ, Tice RR. Evaluation of a three-exposure mouse bone marrow micronucleus protocol: results with 49 chemicals. Environ Mol Mutagen 1993; 21:160–179.
58. Shelby MD. Human germ cell mutagens. Environ Mol Mutagen 1994; 23(suppl 24):30–34.
59. Neel JV, Lewis SE. The comparative radiation genetics of humans and mice. Annu Rev Genet 1990; 24:327–362.
60. Butterworth BE. Consideration of both genotoxic and nongenotoxic mechanisms in predicting carcinogenic potential. Mutat Res 1990; 239:117–132.
61. Butterworth BE, Popp JA, Conolly RB, Goldsworthy TL. Chemically induced cell proliferation in carcinogenesis. IARC Sci Publ 1992; 116:279–305.
62. Jackson MA, Stack HF, Waters MD. The genetic toxicology of putative nongenotoxic carcinogens. Mutat Res 1993; 296:241–277.
63. Huff JE, Haseman JK, DeMarini DM, Eustus S, Maronpot RR, Peters AC, Persing RL, Crisp CE, Jacobs AC. Multiple-site carcinogenicity of benzene in Fisher 344 rats and B6C3F1 mice. Environ Health Perspect 1989; 82:125–163.
64. Rothman N, Haas R, Hayes RB, Li G-L, Wiemels J, Campleman S, Quintana PJE, Xi L-J, Dosemeci M, Titenko-Holland N, Meyer KB, Lu W, Zhang LP, Bechtold W, Wang Y-Z, Kolachana P, Yin S-N, Blot W, Smith MT. Benzene induces gene-duplicating but not gene-inactivating mutations in the glycophorin A locus in exposed humans. Proc Natl Acad Sci USA 1995; 92:4069–4073.
65. Galloway SM, Greenwood SK, Hill RB, Bradt CI, Bean CL. A role for mismatch repair in production of chromosome aberrations by methylating agents in human cells. Mutat Res 1995; 346:231–245.
66. Tennant RW, French JE, Spalding JW. Identifying chemical carcinogens and assessing potential risk in short-term bioassays using transgenic mouse models. Environ Health Perspect 1995; 103:942–950.
67. Goldsworthy TL, Recio L, Brown K, Donehower LA, Mirsalis JC, Tennant

RW, Purchase IF. Transgenic animals in toxicology. Fundam Appl Toxicol 1994; 22:8–19.

68. Penman BW, Chen L, Gelboin HV, Gonzalez FJ, Crespi CL. Development of a human lymphoblastoid cell line constitutively expressing human CYP1A1 cDNA: substrate specificity with model substrates and promutagens. Carcinogenesis 1994; 15:1931–1937.
69. Thier R, Taylor JB, Pemble SE, Humphreys WG, Persmark M, Ketterer B, Guengerich FP. Expression of mammalian glutathione *S*-transferase 5-5 in *Salmonella typhimurium* TA1535 leads to base-pair mutations upon exposure to dihalomethanes. Proc Natl Acad Sci USA 1993; 90:8576–8580.
70. Short JM, ed. Transgenic systems in mutagenesis and carcinogenesis. Mutat Res 1994; 307:427–595.
71. Morrison V, Ashby J. A preliminary evaluation of the performance of the Muta™Mouse (*lacZ*) and BigBlue™ (*lacI*) transgenic mouse mutation assays. Mutagenesis 1994; 9:367–375.
72. Provost GS, Kretz PL, Hamner RT, Matthews CD, Rogers BJ, Lundberg KS, Dycaico MJ, Short JM. Transgenic systems for in vivo mutation analysis. Mutat Res 1993; 288:133–150.
73. Douglas GR, Gingerich JD, Gossen JA, Bartlett SA. Sequence spectra of spontaneous *lacZ* gene mutations in transgenic mouse somatic and germline tissues. Mutagenesis 1994; 9:451–458.
74. Burkhart JG, Burkhart BA, Sampson KS, Malling HV. ENU-induced mutagenesis at a single A:T base pair in transgenic mice containing phi X174. Mutat Res 1993; 292:69–81.
75. Halliday JA, Glickman BW. Mechanisms of spontaneous mutation in DNA repair-proficient *Escherichia coli*. Mutat Res 1991; 250:55–71.
76. Burkhart JG, Malling HV. Mutagenesis and transgenic systems: perspective from the mutagen, *N*-ethyl-*N*-nitrosourea. Environ Mol Mutagen 1993; 22:1–6.
77. Burkhart JG, Malling HV. Mutations among the living and the undead. Mutat Res 1994; 304:315–320.
78. Gee P, Maron DM, Ames BN. Detection and classification of mutagens: a test of base-specific *Salmonella* tester strains. Proc Natl Acad Sci USA 1994; 91:11606–11610.
79. Cupples CG, Miller JH. A set of *lacZ* mutations in *Escherichia coli* that allow rapid detection of each of the six base substitutions. Proc Natl Acad Sci USA 1989; 86:5345–5349.
80. Tucker JD, Lee DA, Ramsey MJ, Briner J, Olsen L, Moore DH II. On the frequency of chromosome exchanges in a control population measured by chromosome painting. Mutat Res 1994; 311:193–202.
81. Baffa R, Negrini M, Schichman SA, Huebner K, Croce CM. Involvement of the ALL-1 gene in a solid tumor. Proc Natl Acad Sci USA 1995; 92:4922–4926.
82. Cher ML, MacGrogan D, Bookstein R, Brown JA, Jenkins RB, Jensen RH. Comparative genomic hybridization, allelic imbalance and fluorescence in situ

hybridization on chromosome 8 in prostate cancer. Genes Chromosom Cancer 1994; 11:153–162.
83. Tice RR, Andrews PW, Hirai O, Singh NP. The single cell gel (SCG) assay: an electrophoretic technique for the detection of DNA damage in individual cells. Adv Exp Med Biol 1991; 283:157–164.
84. McKelvey-Martin VJ, Green MHL, Schmezer P, Pool-Zobel BL, De Meo MP, Collins A. The single cell gel electrophoresis assay (comet assay): a European review. Mutat Res 1993; 288:47–63.
85. Guengerich FP. Catalytic selectivity of human cytochrome P450 enzymes: relevance to drug metabolism and toxicity. Toxicol Lett 1994; 70:133–138.
86. Hirvonen A, Nylund L, Kociba P, Husgafvel-Pursiainen K, Vainio H. Modulation of urinary mutagenicity by genetically determined carcinogen metabolism in smokers. Carcinogenesis 1994; 15:813–815.
87. Norppa H, Hirvonen A, Järventaus H, Uusküla M, Tasa G, Ojajärvi A, Sorsa M. Role of *GSTT1* and *GSTM1* genotypes in determining individual sensitivity to sister chromatid exchange induction by diepoxybutane in cultured human lymphocytes. Carcinogenesis 1995; 16:1261–1264.
88. Mao L, Lee DJ, Tockman MS, Erozan YS, Askin F, Sidransky D. Microsatellite alterations as clonal markers for the detection of human cancer. Proc Natl Acad Sci USA 1994; 91:9871–9875.
89. Mao L, Hruban RH, Boyle JO, Tockman M, Sidransky D. Detection of oncogene mutations in sputum precedes diagnosis of lung cancer. Cancer Res 1994; 54:1634–1637.
90. Wakabayashi K, Sugimura T, Nagao M. Mutagens in food. In: Li AP, Heflich RH, eds. Genetic Toxicology. Boca Raton, FL: CRC Press, 1991:303–338.
91. DeMarini DM, Lewtas J. Mutagenicity and carcinogenicity of complex combustion emissions: emerging molecular data to improve risk assessment. Toxicol Environ Chem 1995; 49:157–166.
92. Ames BN, Profet M, Gold LS. Dietary pesticides (99.99% all natural). Proc Natl Acad Sci USA 1990; 87:7777–7781.
93. Bronzetti G, Hayatsu H, De Flora S, Waters MD, Shankel DM. Antimutagenesis and Anticarcinogenesis Mechanisms III. New York: Plenum, 1993.
94. Ames BN. The causes and prevention of cancer. Proc Natl Acad Sci USA 1995; 92:5258–5265.

7

Improved Predictivity of Chemical Carcinogens: The Use of a Battery of SAR Models

Orest T. Macina, Ying Ping Zhang, and Herbert S. Rosenkranz
University of Pittsburgh, Pittsburgh, Pennsylvania

I. INTRODUCTION

The application and reliance on structure-activity relationship (SAR) approaches to predict and understand mechanisms of toxicity is increasing. In turn, this has led to the realization that the development, validity, and predictivity of SAR models is greatly dependent upon the inclusion criteria used for data bases (i.e., learning sets), the complexity of the biological phenomenon under investigation, the criteria used for acceptance of a model, the methods used to measure predictivity, and the approach taken to interpret the predictions.

The methods described herein, although they are illustrated with CASE/MULTICASE, are generic and applicable to other SAR methods as well as to the combination of methods. The overriding concern of the present approach is to extract the maximum amount of useful information from SAR models so as to reflect the complexity of the biological phenomenon under investigation.

In previous studies using the CASE/MULTICASE methodologies, we described methods for assessing the informational content of SAR models (1), the optimal size of the data base (2), the effect of varying the ratios of actives to inactives (2,3), the procedures for validating models (4), and methods for expressing predictivity (5). Thus in a recent exercise to predict the activity of 100 chemicals, TOPKAT (TOxicity Prediction by Komputer

Aided Tools) (6) and CASE/MULTICASE methods were compared (7). They both showed concordances of ~78% between predicted and experimental results. However, TOPKAT was able to make predictions on only 61 chemicals, while CASE/MULTICASE made predictions on all of the chemicals. Using the methods developed to evaluate predictions (5), this translated to χ^2 values of 13.6 and 25.2, respectively (7), which indicate a significant difference in performance.

While CASE/MULTICASE predictions actually consist of four separate predictions (see below), in the past we have used only one of these for analytical purposes. In fact, we observed for many "simple" data bases that there was little variability among the four. However, a recent detailed analysis of rodent carcinogenicity SAR models using an inductive learning machine (RL4) (8) revealed that the results of the four predictions did not parallel one another (9). This caused us to reevaluate the analytical procedures used to make predictions. The present investigation represents the resulting approach, which is applicable to other SAR models as well. Moreover, since there is a relationship between the results of *Salmonella* mutagenicity assays and carcinogenicity (10), in the present study we also investigated whether coupling the results of the *Salmonella* assay to the predictions of SAR models improved overall predictive performance.

II. CARCINOGENICITY DATA BASES

The extent of a biological activity (i.e., the potency) assigned to compounds comprising a CASE/MULTICASE data base (i.e., learning set) range from 10 to 99 on the CASE/MULTICASE unit scale. These arbitrary internal CASE/MULTICASE potency units are obtained from a logarithmic transformation of the raw biological data, e.g., log $1/TD_{50}$ in which TD_{50} values are expressed in mmol/kg/day. Alternatively, CASE/MULTICASE units can also be assigned in consideration of the biological activity category to which a compound belongs (e.g., active/inactive; or no evidence, equivocal evidence, and clear evidence of carcinogenicity, etc.). By convention, units of 10–19 represent inactive or very-low-potency compounds; units of 20–29 indicate marginally active compounds; and units of 30–99 represent biologically active chemicals of increasing potency.

The carcinogenicity data utilized in our present series of investigations consists of results for compounds obtained under the aegis of the U.S. National Toxicology Program (NTP) (10–12) and the Carcinogenicity Potency Data Base (CPDB) assembled by Gold and associates (13–17). Data for both the rat and the mouse are available in both compilations. Furthermore, the rat and mouse data have been combined into a generalized rodent data base, wherein a compound carcinogenic to either one of the species is classified as a rodent carcinogen. For construction of our data bases, we consider only organic compounds.

The NTP data is classified according to a compound's ability to exhibit carcinogenicity in more than one species and/or sex and/or organ sites (11,12). We have utilized these carcinogenicity classifications as a basis for assigning CASE/MULTICASE units. The units and the corresponding classification are as follows:

- 60: Agents carcinogenic to both rats and mice at one or more sites
- 50: Agents carcinogenic only to the rat or the mouse at two or more sites
- 40: Agents carcinogenic to the rat or the mouse at a single site in both sexes
- 30: Agents carcinogenic at only a single site in a single sex of a single species
- 20: Agents adequately evaluated but for which equivocal evidence of carcinogenicity was obtained
- 10: Agents adequately evaluated and concluded to be noncarcinogenic

Presently, the NTP rodent data base contains 313 compounds, 173 of which are active (CASE/MULTICASE units of 60, 50, 40, or 30), 39 marginal (units of 20), and 101 inactive (units of 10). The mouse data base consists of 319 compounds (131 actives, 20 marginals, and 168 inactives), while the rat data base contains 316 compounds (131 actives, 28 marginals, and 157 inactives).

The CPDB data bases express carcinogenicity in terms of TD_{50} values; i.e., the dose at which 50% of the animals remain tumor-free. We have first transformed the CPDB TD_{50} information into the log of the reciprocal of the TD_{50} expressed in mmol/kg. Then we transformed those units into CASE/MULTICASE units ranging from 10 to 99. The rodent compilation consists of 433 compounds, 261 of which are classified as active (units of 30–99), 10 as marginal (units of 20–29), and 162 as inactive (units of 10–19). The mouse data base contains 636 compounds (291 actives, 11 marginals, and 334 inactives). The corresponding rat data base consists of 745 compounds, 383 of which are active, 14 marginal, and 348 inactive.

III. CASE

A detailed description of the algorithm employed within the Computer Automated Structure Evaluation (CASE) program has been presented previously (18). Input is in the form of a data base composed of a set of chemical structures of interest and their respective experimental biological activities. The CASE program proceeds to identify descriptors consisting of molecular fragments, ranging from 2 to 10 heavy atoms along with their associated hydrogens, which account for the biological activity of the compounds under study. The molecular fragments are generated as a result of decomposing each individual chemical structure into its constituent parts. Each fragment is "labeled" with respect to its origin within active or inactive compounds. A binomial distribution is assumed, and fragments exhibiting a statistically significant nonrandom distribution among the active and inactive classes of compounds are retained. These fragments serve as indicators of biological activity.

At this point within the analysis, CASE utilizes the statistically significant fragments in order to classify compounds as active in inactive. Predictions for submitted compounds are expressed as percent probabilities (0–100%) of being active or inactive. The probabilities of multiple fragments that may occur in an individual compound are combined using Bayes' theorem (19) in order to arrive at an overall probability of activity. In addition to the fragments identified as being responsible for activity and inactivity, the CASE program also utilizes molecular fragments to derive a quantitative structure-activity relationship (QSAR). Descriptors related to transport properties (e.g., the logarithm of the octanol/water partition coefficient) are screened as potential variables for the QSAR equation. The derived QSAR model is global in nature, i.e., all of the compounds within a data base are accounted for by the equation. The CASE QSAR model is utilized in order to predict the potencies (extent of biological activity) of compounds submitted for analysis. These predictions are expressed in CASE units for the CPDB data bases and may be transformed back into TD_{50} values using the original transformation equation.

IV. MULTICASE

The Multiple Computer Automated Structure Evaluation (MULTICASE) program evolved from the CASE program. A detailed description of the algorithm employed within MULTICASE has been published (20). Like CASE, MULTICASE utilizes molecular fragments as descriptors of biological activity. MULTICASE also identifies relevant two-dimensional distances between atoms within a chemical structure as potential descriptors.

Unlike CASE, MULTICASE utilizes the set of statistically significant descriptors (fragment and/or distance) in order to find a descriptor (biophore) that has the highest probability of being responsible for the observed biological activity. Compounds within the data base containing the primary biophore are removed from the analysis, and subsequent biophores are selected that explain the activity of the remaining compounds. This iterative process is continued until either all of the active compounds are accounted for or no statistically significant descriptors remain. Thus, MULTICASE breaks the total data base into logical subsets of compounds based on the commonality of structural features (biophores). This is unlike CASE, which attempts to derive a model for the entire data set. It is assumed that the MULTICASE biophores may be representative of different mechanisms of action. The presence of a biophore within a compound indicates a potential for exhibiting the biological activity under study. Predictions made on compounds submitted as unknowns consist of the identification of the biophore(s) and the percent probability of the compound being biologically active due to this biophore. A chemical is presumed to be inactive if it contains no biophores.

After identifying the primary biophore, MULTICASE attempts to derive a QSAR within each group of compounds containing a particular biophore in order to identify molecular features that modulate the activity; i.e. to explain the extent of activity. These features, termed modulators, are selected from the pool of molecular fragments, distance descriptors, calculated electronic indices (molecular orbital energies, charge densities), and calculated transport parameters (octanol/water partition coefficient, water solubility). These localized QSAR models are only valid within the group of compounds sharing the same biophore.

VII. CASE/MULTICASE PREDICTIONS

Submission of an unknown compound to the CASE/MULTICASE system results in predictions based on both CASE and MULTICASE SAR models. The predictions generate MULTICASE potency units (MCu: 10–99), MULTICASE probabilities (MC%: 0–100%), CASE potency units (Cu: 10–99), and CASE probabilities (C%: 0–100%). The potencies and probabilities resulting from CASE may not necessarily correspond to those produced from MULTICASE. Discrepancies between the two models are due to differences in the approach towards selecting fragments responsible for activity and performing QSAR analyses (i.e., global vs. local). It may be beneficial, however, to combine all four indicators of activity (MCu, MC%, Cu, C%) as this may overcome limitations associated with an individual CASE or MULTICASE SAR model.

A typical example of a CASE/MULTICASE prediction is shown in Figure 1. CASE/MULTICASE predictions are listed for the chemical acronycine (CAS # 7008-42-6) tested against the CPDB mouse carcinogen data

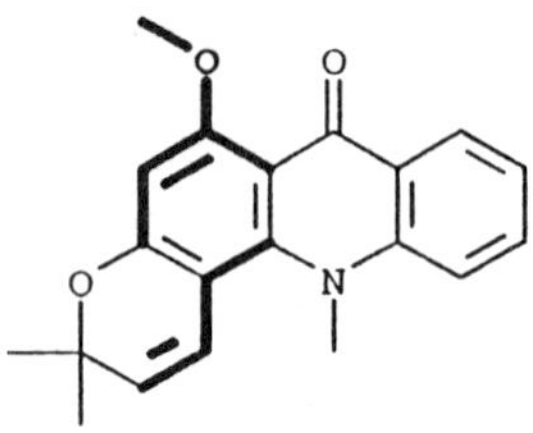

Acronycine

```
*******************************************************************************
A0E- Mouse carcinogens                                          - CPDB     2.80
-------------------------------------------------------------------------------

*** Predictions based on MULTICASE :

The molecule contains the Biophore   (nr.occ.= 1):
                    CH  =CH
                           \
                            C.   =C.
                           /
   *** 15 out of the known  16 molecules ( 94%) containing such Biophore
       are mouse carcinogens with an average activity of  61. (conf.level=100%)
                                                       Constant is     4.3
   ** The following modulator(s) is/are also present:
  ( 1) CH =C. -CO -C. =                              Inactivating  -8.8
       Ln Nr.Bi/Mol.Wt.    = -5.77 ;          Nr.Bioph/MW contrib.is  69.3
   ** The probability that this molecule is a mouse carcinogen is  94.1%  **

   ** The compound is predicted to be EXTREMELY active  ( 65) **
   ** The projected mouse carcinogen activity is   65.0    CASE units **

 *** Predictions based on CASE :

  80 % chance of being   ACTIVE due to substructure (Conf.level=  87%) :
        CH3-O  -C  =C. -
  94 % chance of being   ACTIVE due to substructure (Conf.level= 100%) :
        CH =CH -C. =C. -
  80 % chance of being   ACTIVE due to substructure (Conf.level=  87%) :
        C. =CH -C  =C. -

 *** OVERALL, the probability of being a mouse carcinogen is  99.6% ***
     ** The compound is predicted to be MARGINALLY active ( 20) **
       ** The predicted activity is   20.0     CASE units **
```

Figure 1 Example of a CASE/MULTICASE prediction. The MULTICASE biophore and CASE activating fragments are shown in bold within the structure.

base. The first part of the output deals with MULTICASE predictions. Acronycine contains the biophore CH=CH−C.=C. (where C. refers to a carbon common to two rings) present in 15 carcinogenic compounds within the learning data base of 16 chemicals of similar structure. There are two modulators associated with the biophore. The overall conclusion from MULTICASE is that acronycine has a 94.1% chance of being a carcinogen with an activity calculated to be 65 MULTICASE units (corresponding to a TD_{50} value of 0.033 mmol/kg/day). The CASE results are listed in the second half of the output. CASE finds several activating (carcinogenic) fragments within the structure of acronycine. The overall conclusion from CASE is a probability of 99.6% chance of being a carcinogen with a calculated CASE unit of 20 (marginal activity; TD_{50} of 51.0 mmol/kg/day). For acronycine the potency predictions of CASE and MULTICASE differ considerably.

In order to combine all four predictions into a single overall evaluation, we used a sequential application of Bayes' theorem (19,21,22). This approach is based on the fact that the joint probability of two events can be written as the product of the probability of one of the events and the conditional probability of the second event, given the first event. Designating the two events as "A" (or "not A") and "+" (or "−"), Bayes' theorem takes the form:

$$\text{Prob}(A/+) = \frac{\text{Prob}(+/A)*\text{Prob}(A)}{\text{Prob}(+/A)*\text{Prob}(A) + \text{Prob}(+/\text{not } A)*\text{Prob}(\text{not } A)}$$

Prob(A/+) represents the probability that we are dealing with the state A, given that we have observed the data +. Prob(+/A) represents the probability of the observed data +, given that we are dealing with A. Prob(A) is our current belief concerning the probability of A. In the above form, Prob(A/+) is termed the posterior probability, i.e., the updated probability given that we have observed a +. Prob(A) and Prob(not A) are termed prior probabilities since they are decided upon before any new data are introduced.

Bayes' theorem takes the following form within a predictive setting utilizing CASE/MULTICASE results:

$$\text{Prob}(\text{Active}/*) = \frac{\text{Sensitivity}*\text{Prob}(\text{Active})}{\text{Sensitivity}*\text{Prob}(\text{Act}) + (1 - \text{Specificity})*\text{Prob}(\text{not active})}$$

Using this form of Bayes' theorem, together with estimates of sensitivity and specificity associated with our respective prediction SAR models (i.e., Cu, C%, MCu, MC%), and the prior estimate of the probability of activity of the chemical, we can estimate the new probability that we have an active chemical, given that a positive prediction is obtained when the model is applied. For our purposes we have assumed an initial prior probability of 0.50, which reflects the prevalence of carcinogens in the NTP data base. However, any other prior probability value may be appropriate as the purpose of the exercise is to determine to what extent the new probabilities differ from the prior ones. Each additional application of Bayes' theorem (i.e., models pertaining to Cu, C%, MCu, MC%) uses the previous posterior probability of activity as the new prior probability and the relevant estimates of sensitivity and specificity pertaining to the SAR model being utilized. By sequentially assimilating the data (pertaining to the predicted Cu, C%, MCu, MC%) through the procedure outlined above, we obtain an overall estimate of the probability of activity for the chemical under consideration. This probability is based on consideration of all four metrics (probabilities and units) obtained from the CASE/MULTICASE system. The overall strategy is to determine the effect of additional knowledge (i.e., a specific prediction) on the initial probability.

The Bayesian approach can be expanded further in situations where more than one SAR model (data base) is available for particular biological phenomenon. This is germane to the prediction of carcinogenicity, wherein several SAR models (NTP: rodent, mouse, and rat; CPDB: rodent, mouse, and rat) can be used in making assessments of carcinogenic potential. The quality of individual predictions is dependent upon the learning set used to derive the respective SAR models, as well as on the complexity of the biological phenomenon being modeled. Given the multiple mechanisms operating in the multiple sequential stages of carcinogenesis, we do not expect an individual carcinogenicity SAR model to be as good as models for simpler endpoints (e.g., mutagenicity). It is anticipated, therefore, that the utilization of a battery of carcinogenic SAR models for an overall prediction will overcome some of the limitations associated with an individual SAR model.

VIII. VALIDATION OF CARCINOGENICITY DATA BASES

Obviously, application of Bayes' approach is dependent upon estimates of sensitivity and specificity of individual SAR models. We have explored a variety of models to determine their parameters. The exact strategy used depends upon the size and nature of the data base. This is due to the fact that the predictivity of an SAR model is dependent upon size until an optimal data base size has been reached. Thus, for large data bases (e.g., the NTP-based *Salmonella* mutagenicity data base) we may simply not include in the learning set a randomly chosen selection of 100 chemicals and then use the SAR model derived from the remaining chemicals to predict the 100 chemicals that were left out and thus determine sensitivity and specificity (however, see below on how their values may be predicted).

We have previously determined that the optimal size of data bases may be approximately 350 chemicals (2). Many of the available data bases, including the NTP carcinogenicity data bases, are in that range. Thus, it is possible that the deletion of 100 chemicals from the learning set would result in a deteriorization of the predictive performance of the SAR model. We have investigated other approaches including iterative removal of randomly selected, nonoverlapping sets of 10% of the learning set and subsequent predictions of activity. The procedure is performed 10 times. The sensitivity and specificity is then calculated from the aggregated predictions. The most rigorous approach is to perform a "leave-out one" type of validation. This tedious exercise consists of removing one compound at a time from the complete data base and rederiving a CASE/MULTICASE SAR model for the reduced set and submitting the chemical originally removed to the CASE/MULTICASE SAR model for prediction of Cu, C%, MCu, and MC%. These predictions can then be compared to the experimental results of the cancer bioassay. After this exercise is complete (315 times for each NTP data base), sensitivities and specificities can be calculated as described below. For larger data bases (CPDB), for which a complete "leave-out one" exercise would be impractical, it is possible to define the size of an acceptable subset of chemicals for which a "leave-out one" approach can be performed.

Once having performed the above exercise (complete leave-out one for NTP, and subset of 57 chemicals for CPDB), in order to determine the sensitivity and specificity of our SAR model, cut-off values for the individual MCu, MC%, Cu, and C% must be established, i.e., the demarcation between indicators of activity and inactivity. Thus, for example, for C% we might define that predictions of $\geq 50\%$ and $< 50\%$ indicate carcinogenicity and lack of carcinogenicity, respectively. Obviously rather than assign such cut-off values arbitrarily, the values should be chosen for optimal predictivity.

The establishment of optimal demarcation values was based upon investigation of the individual values for MCu, MC%, Cu, and C% obtained in the validation studies described above. Ranges were selected within which the cut-off values were considered for selection. The range for Cu and MCu was between 17 and 45 units, while the range for both C% and MC% was between 45 and 75%. These ranges are considered reasonable with respect to the biological activity represented. Considering cut-off values at higher values would lead to unrealistic demarcations between active and inactive chemicals.

Each of the values within the acceptable ranges were considered sequentially as potential cut-off values with respect to Cu, C%, MCu, and MC%. The chemicals employed as test cases (for which predictions were made) during the cross-validation studies were used as the calibration set for determining predictions of activity/inactivity on the basis of the cut-off value being considered. The predicted activities based on independent consideration of Cu, C%, MCu, and MC% were used to determine carcinogenicity. The respective concordances, sensitivities, and specificities were determined for each value being considered. The individual optimal cut-off values for Cu, C%, MCu, and MC% were determined by sequential application of the following ad hoc rules:

1. The concordance between calculated and experimental values must be greater than 0.50, otherwise the predictive index is discarded.
2. Starting with the highest concordance available, search for the first value of sensitivity greater than 0.50. When more than one cut-off value satisfies this condition, the median value is chosen.
3. If no sensitivity greater than 0.50 is available, search for a sensitivity of greater than or equal to 0.40 with the additional condition that the corresponding specificity achieves a value greater than 0.80.
4. If rules 2 and 3 remain unsatisfied, search for a sensitivity greater than or equal to 0.30 with the additional condition that the sum of the sensitivity and specificity for the respective cut-off value is greater than 1.00. Predictive indices not satisfying this last rule are discarded.

The optimal cut-off values as well as the corresponding concordances, sensitivities, and specificities are listed in Table 1 for the NTP and CPDB rodent and mouse data bases. The NTP and CPDB rat data bases were found not to satisfy any of the above four conditions. Based upon these considerations, the utilization of the rat carcinogenicity SAR model for predictions is therefore unwarranted. The rat SAR model may, however, be useful for mechanistic studies as well as cross-species comparisons. In the present context, the NTP rodent and mouse SAR models employ only the MCu, MC%, and Cu% as predictive indices to arrive at a conclusion of activity. The exclusion of Cu is presumably derived from the fact that in the NTP models Cu values were assigned arbitrarily leased upon spectrum of carcinogenicity, i.e., multiple species, multiple sites, single species, etc.

Table 1 Optimal Cut-Off Values

Data base	MULTICASE unit				MULTICASE probability				CASE unit				CASE probability			
	Cut-off.	Conc.	Sens.	Spec.	Cut-off	Conc.	Sens.	Spec.	Cut-off	Conc.	Sens.	Spec.	Cut-off	Conc.	Sens.	Spec.
N__Rod.	22	0.61	0.61	0.62	72%	0.61	0.55	0.67	Unusable				58%	0.56	0.65	0.45
N__Mice	30	0.64	0.33	0.85	57%	0.65	0.40	0.82	Unusable				53%	0.58	0.46	0.66
N__Rat	Unusable															
G__Rod.	30	0.54	0.53	0.56	68%	0.58	0.69	0.56	20	0.60	0.66	0.52	58%	0.58	0.59	0.56
G__Mice	29	0.67	0.42	0.85	50%	0.68	0.46	0.85	20	0.74	0.54	0.88	66%	0.70	0.67	0.73
G__Rat	Unusable															

N__Rod. = NTP Rodent carcinogenicity data base; N__Mice = NTP Mouse carcinogenicity data base; N__Rat = NTP Rat carcinogenicity data base; G__Rod. = CPDB Rodent carcinogenicity data base; G__Mice = CPDB Mouse carcinogenicity data base; G__Rat = CPDB Rat carcinogenicity data base.

The CPDB rodent and mouse SAR models, on the other hand, utilize all four indices. In the latter data bases Cu and MCu values were derived from actual TD_{50} values.

The lack of predictivity of the rat SAR model is surprising. It is noteworthy as both the NTP and CPDB rat carcinogenicity data bases resulted in the same finding. Since approximately the same chemicals result in a predictive mouse carcinogenicity model, it would appear that the problem is intrinsic to the rat. If the assumption is made that carcinogenicity in the rat is also encoded within the structure of the chemicals, then one of the possible explanations of the lack of predictivity is that a greater variety of mechanisms exist leading to carcinogenesis in the rat relative to the mouse. Since the presence of multiple mechanisms would require a larger number of representative chemicals, it may be that in the rat carcinogenicity data set there is an underrepresentation of chemicals that induce some of the rat-specific mechanisms.

IX. DETERMINATION OF BAYESIAN PROBABILITIES

It is to be noted that the differences in MCu, MC%, Cu, and Cu% within a single carcinogenicity data base presumably reflect the complexity of the biological phenomenon being modeled. Thus, when modeling *Salmonella* mutagenicity, all of these parameters are approximately the same (unpublished results).

Once the optimal cut-off values are determined, they can be utilized for classifying a prediction as active (+) or inactive (−). Compounds whose predicted values of Cu, C%, MCu, and MC% fall above the respective cut-off values are classified as active (+). The individual activity classifications (+ or −) corresponding to the four individual prediction indices (CASE/MULTICASE units/probabilities) can then be combined according to Bayes' theorem to provide an overall probability of activity. The number of possible patterns increases according to 2^N, where N is the number of indices utilized. For the NTP rodent and mouse data bases there are 8 possible patterns (the Cu index is discarded), while for the CPDB rodent and mouse SAR models there are 16 possible outcomes.

Table 2 lists the calculated probabilities resulting from combining the activity classifications (+/−) with respect to MCu, MC%, Cu, and C%. The overall probabilities for each possible activity pattern were calculated by application of Bayes' theorem with the appropriate values of the previously determined sensitivity and specificity (Table 1) and a prior probability of 0.50. The listed probabilities pertain to the probability of being a carcinogen based on the predicted CASE/MULTICASE activity pattern.

Table 2 Probabilities of Carcinogenicity: Possible Combinations of SAR Predictions

NTP Rodent		NTP Mouse		CPDB Rodent				CPDB Mouse			
Pattern	Prob.	Pattern	Prob.	Pattern	Prob.	Pattern	Prob.	Pattern	Prob.	Pattern	Prob.
+ + * +	0.7597	+ + * +	0.8687	+ + + +	0.7769	− − + +	0.4614	+ + + +	0.9897	− + − +	0.7308
+ + * −	0.6754	+ + * −	0.8000	− + + +	0.7082	+ − + −	0.4017	− + + +	0.9590	+ − − +	0.6976
+ − * +	0.5603	− + * +	0.7033	+ + + −	0.6554	− + − −	0.3865	+ − + +	0.9521	+ + − −	0.6699
− + * +	0.5534	+ − * +	0.6853	+ + − +	0.6235	+ − − +	0.3689	+ + + −	0.9459	− − + −	0.4686
+ − * −	0.4561	− + * −	0.5890	− + + −	0.5699	− − + −	0.3187	+ + − +	0.9176	− − − +	0.3599
− + * −	0.4492	+ − * −	0.5684	+ − + +	0.5515	− − − +	0.2894	− − + +	0.8288	− + − −	0.3309
− − * +	0.3330	− − * +	0.4383	− + − +	0.5357	+ − − −	0.2420	− + + −	0.8098	+ − − −	0.2959
− − * −	0.2473	− − * −	0.3206	+ + − −	0.4749	− − − −	0.1819	+ − + −	0.7835	− − − −	0.0929

The pattern is ordered as MCu, MC%, Cu, C%. *represents a discarded index. Probability $>$ 0.5 is scored as +. Probability $<$ 0.5 is scored as −.

For our purposes, we have arbitrarily chosen a probability value of 0.50 as a demarcation between classifying a compound as active (carcinogenic) or inactive (noncarcinogenic).

The predicted activities of the test set of chemicals previously utilized for data base validation studies were reclassified according to the activity patterns listed in Table 2. Compounds with probabilities greater than 0.50 were classified as carcinogens. Comparison between projected activity and the results of cancer bioassays resulted in the observed concordances, sensitivities, and specificities shown in Table 3. Both the NTP and CPDB mouse SAR models have the highest concordances and specificities. The highest sensitivities were observed for the NTP rodent and CPDB mouse models.

X. PREDICTIONS BASED ON A BATTERY OF SAR MODELS

Each individual carcinogenicity SAR model may be limited by the type and quality of data utilized for model construction and development. Thus, as shown above by these criteria, the rat SAR models based upon either NTP or CPDB data are not useful for predictive purposes. A strategy to overcome limitations of individual SAR models may entail combining the results of different carcinogenicity SAR models to arrive at an overall conclusion of carcinogenicity. Application of a battery of such SAR models may yield higher prediction success than utilization of a single model.

We sought to investigate this hypothesis by comparing the predictions from a battery of carcinogenicity SAR models with predictions from individual models. The SAR models utilized involve the NTP and CPDB rodent and mouse SAR models. Utilizing the respective sensitivities and specificities (from Table 3) and applying Bayes' theorem results in the calculated probabilities of carcinogenic activity shown in Table 4 for all possible activity patterns. These probabilities result from combining all the possible predictions for the four carcinogenic SAR models (NTP rodent, NTP mouse, CPDB rodent, and CPDB mouse). As previously, we arbitrarily chose a probability value of 0.50 or greater as indicative of carcinogenic activity.

Table 3 Performance Characteristics of Four SAR Carcinogenicity Models

Data Base	Concordance	Sensitivity	Specificity
NTP Rodent	0.619	0.605	0.636
NTP Mouse	0.649	0.397	0.822
CPDB Rodent	0.561	0.500	0.640
CPDB Mouse	0.719	0.583	0.818

Table 4 Predictions Based Upon Combining Carcinogenicity SAR Models

Pattern	Probability	Pattern	Probability	Pattern	Probability	Pattern	Probability
+ + + +	0.9429	− + − +	0.7764	+ + − −	0.5906	− + − −	0.3553
+ + − +	0.9029	+ − − +	0.7536	− − − +	0.5332	+ − − −	0.3268
− + + +	0.8606	+ + + −	0.7204	− + + −	0.4949	− − + −	0.2238
+ − + +	0.8447	− − + +	0.6701	+ − + −	0.4633	− − − −	0.1535

The order within the pattern is NTP rodent, NTP mouse, CPDB rodent, and CPDB mouse. Probability > 0.5 is scored as +.

An "independent" test set of compounds not included in any of our carcinogenicity data bases with known carcinogenic activities was selected to compare the predictive ability of individual SAR models as well as batteries thereof. The compounds chosen consist of 30 chemicals, which have data available from the NTP. Twenty-seven of the compounds had been shown to be carcinogenic in at least one sex of one species. In order to supplement the three noncarcinogenic compounds evaluated experimentally, we included a randomly chosen set of 19 physiological compounds. For the present analysis, we assumed that these chemicals are noncarcinogenic in nature.

This independent test set consisting of 49 compounds (27 actives/22 inactives) was submitted to the NTP rodent, NTP mouse, CPDB rodent, and CPDB mouse SAR models. The individual compounds and the respective activity patterns generated are listed in Table 5. The assigned probability values correspond to the values previously determined by applying Bayes' theorem to combinations of predictions (see Table 4). By classification of compounds with probabilities greater than 0.50 as carcinogens, we obtain a concordance of 0.735 (sensitivity of 0.704 and specificity of 0.773) for predictions based on the battery of carcinogenic SAR models. By comparison, the concordances, sensitivities, and specificities for utilizing individual SAR model predictions are listed in Table 6. The concordances for individual model predictions range from 0.580 to 0.700. Sensitivities range from 0.370 to 0.667, while the specificities range form 0.652 to 1.000. The observed concordance (0.735) and sensitivity (0.704) for the battery of carcinogenic SAR models lends support to the strategy of combining predictions from different SAR models to arrive at an overall conclusion of carcinogenic activity. Thus, the combined SAR battery performed better than any of its four components.

XI. SAR MODELS COMBINED WITH *SALMONELLA* MUTAGENICITY RESULTS

Results from the *Salmonella* assay are widely accepted as indicative of a compound's potential to exhibit DNA reactivity. Furthermore, it has been proposed that the *Salmonella* assay can be used for identifying potentially carcinogenic compounds (10–12). Although this is true for "genotoxic" carcinogens, the *Salmonella* mutagenicity system fails to detect nongenotoxic carcinogens, which are presently thought to represent a substantial portion of carcinogens (23).

We sought to investigate the possibility of utilizing the experimental results of *Salmonella* mutagenicity in combination with the battery of car-

Table 5 Prediction of the Carcinogenicity of Independent Test Set

No.[a]	Chemical name	NTP carcinogenicity	1234	Probability
*1	ACRONYCINE	+	+ + − +	0.9029
2	ALLYL CHLORIDE	+	− − − −	0.1535
*3	5-AZACYTIDINE	+	+ − + +	0.8447
*4	1,3-BUTADIENE	+	− − − +	0.5332
5	*TERT*-BUTYL ALCOHOL	+	− − − −	0.1535
*6	CHLORDECONE	+	+ + + +	0.9429
*7	CHLOROFORM	+	− − − +	0.5332
8	C.I. RED 114	+	− − + −	0.2238
9	CI DIRECT BLUE 15	+	+ − + −	0.4633
*10	CYTEMBENA	+	+ + + −	0.7240
*11	FURAN	+	− + + +	0.8606
*12	ICRF 159	+	− − + +	0.6701
*13	ISOPHOSPHAMIDE	+	+ − + +	0.8447
*14	LASIOCARPINE	+	+ + − −	0.5960
*15	8-METHOXYPSORALEN	+	+ + + +	0.9429
*16	MIREX	+	+ + + +	0.9429
17	NAPHTHALENE	+	− − − −	0.1535
−18	β-NITROSTYRENE	−	− − − −	0.1535
*19	OCHRATOXIN A	+	+ − + +	0.8447
20	OXAZEPAM	+	− − − −	0.1535
*21	PHENOXYBENZAMINE HYDROCHLORIDE	+	+ + + +	0.9429
−22	o-PHENYLPHENOL	−	− − + −	0.2238
*23	PROBENECID	+	− + + +	0.8606
*24	PROCARBAZINE	+	− − + +	0.6701
25	QUERCETIN	+	+ − + −	0.4633
26	β-THIOGUANINE DEOXYRIBOSIDE	+	− − + −	0.2238
−27	TRICRESYL PHOSPHATE	−	+ − + −	0.4633
*28	THIO-TEPA	+	− − + +	0.6701
*29	4-VINYLCYCLOHEXENE	+	− − − +	0.5332
*30	2,6-XYLIDINE	+	+ + + +	0.9429
31	Thiamine	−	+ − + +	0.8447
−32	Niacinamide	−	− − − −	0.1535
−33	Isoleucine	−	− − − −	0.1535
−34	Stearic acid	−	+ − + −	0.4633
35	β-Carotene	−	− − − +	0.5332
36	Methionine	−	− − − +	0.5332
−37	Fructose	−	− − − −	0.1535
−38	Spermine	−	+ − − −	0.3268
−39	Pyruvic acid	−	− − − −	0.1535
−40	Adenine	−	+ − + −	0.4633

(continued)

Table 5 Continued

No.[a]	Chemical name	NTP carcinogenicity	1234	Probability
−41	Glycine	−	− − − −	0.1535
−42	Ribose	−	− − − −	0.1535
−43	*p*-Aminobenzoic acid	−	+ − + −	0.4633
−44	Nicotinamide	−	− − − −	0.1535
−45	Valine	−	− − − −	0.1535
46	Riboflavine	−	− − − +	0.5332
−47	Niacin	−	− − − −	0.1535
48	Cholesterol	−	− − + +	0.6701
−49	Lactose	−	− − − −	0.1535

1: NTP Rodent; 2: NTP Mouse; 3: CPDB Rodent; 4: CPDB Mouse. Lowercase physiological chemicals are of assumed but not demonstrated noncarcinogenicity.
[a]*: Positive compound predicted as active; −: negative compound predicted as inactive.

cinogenic SAR models, thereby improving overall predictivity. The same independent test set of compounds used to determine the predictive ability of a battery of carcinogenic SAR models (see Table 5) was utilized for the exercise. All of the physiological compounds (assumed to be noncarcinogenic) were also assumed to be *Salmonella* negative (nonmutagenic). Indeed, glycine and *p*-aminobenzoic acid have been experimentally determined to be nonmutagenic (24). The probabilities of carcinogenicity previously derived from the application of the carcinogenic SAR model battery (see Table 5) were combined with the available *Salmonella* results in our Bayesian approach. A sensitivity of 0.767 and a specificity of 0.730 were assigned to the *Salmonella* mutagenicity results. These values were calculated from the *Salmonella* mutagenicity for the chemicals tested for carcinogenicity by the NTP and/or present in the CPDB data bases. Table

Table 6 Individual SAR Model Predictions for the Independent Test Set

Data Base	Concordance	Sensitivity	Specificity
NTP Rodent	0.580	0.481	0.696
NTP Mouse	0.660	0.370	1.000
CPDB Rodent	0.660	0.667	0.652
CPDB Mouse	0.700	0.630	0.783
SAR Battery	0.735	0.704	0.773

7 lists the results of combining the *Salmonella* data with the battery of carcinogenic SAR models. Once again, using a probability value of 0.50 as the demarcation between carcinogens and noncarcinogens, we obtain a concordance of 0.778 with a sensitivity and specificity of 0.609 and 0.955, respectively.

By comparison, the battery of SAR models (without *Salmonella*; see Table 5) resulted in a concordance of 0.735 (sensitivity of 0.704 and specificity of 0.773). Inclusion of the *Salmonella* results appears to degrade the sensitivity (from 0.704 to 0.609), while the specificity increases (from 0.773 to 0.955). Table 8 lists the comparison of utilizing three strategies for detecting carcinogens. Consideration of only the 26 compounds for which there are experimentally derived data on carcinogenicity and *Salmonella* mutagenicity data further substantiates the problem of relying solely on mutagenicity as an indicator of carcinogenicity. Using *Salmonella* results alone to predict the activity of the 23 carcinogens yields a sensitivity of only 0.478.

The above results confirm that the results obtained from the *Salmonella* assay are not necessarily predictive of carcinogenicity (25). Indeed it would appear that a battery consisting solely of SAR models related to carcinogenicity offers a better estimate of a compound's potential to exhibit carcinogenicity than the inclusion of the *Salmonella* results.

XII. CONCLUSIONS

The ability of an SAR model to predict carcinogenicity is hampered by the complexity of the biological phenomenon as well as by the fact that since rodent carcinogenicity assays are very costly and time-consuming, repeat carcinogenicity assays are seldomly performed. Thus the reproducibility of the standard used (i.e., carcinogenicity bioassays) to calibrate the SAR models is unknown. This is unlike, for example, the situation with the *Salmonella* mutagenicity assay, the interlaboratory reproducibility of which is 82% (26); thus we can evaluate to the significance of a concordance of 77% between the experimental and predicted results. The absolute predictivity of SAR models of rodent carcinogenicity cannot be determined due to lack of information regarding experimental reproducibility of the cancer bioassays. In the absence of this vital piece of information, we have attempted to improve predictive accuracy by using all of the information provided by the SAR models, including using batteries of such assays in the same fashion that earlier batteries of short-term assays were used to predict carcinogenicity (22,24). Based upon a test set of 49 chemicals for which carcinogenicity has been reported and noncarcinogenicity has been as-

Table 7 Prediction of the Carcinogenicity of Independent Test Set with Salmonella Data Included

No.[a]	Chemical name	NTP carcinogenicity	1234	1234 Probability	SAL	1234 + SAL Probability
1	ACRONYCINE	+	+ + − +	0.9029	n	n/a
2	ALLYL CHLORIDE	+	− − − −	0.1535	+	0.3400
*3	5-AZACYTIDINE	+	+ − + +	0.8447	+	0.9392
*4	1,3-BUTADIENE	+	− − − +	0.5332	+	0.7644
5	*TERT*-BUTYL ALCOHOL	+	− − − −	0.1535	−	0.0547
*6	CHLORDECONE	+	+ + + +	0.9429	−	0.8406
7	CHLOROFORM	+	− − − +	0.5332	−	0.2672
8	C.I. RED 114	+	− − + −	0.2238	+	0.4780
9	CI DIRECT BLUE 15	+	+ − + −	0.4633	−	0.2160
*10	CYTEMBENA	+	+ + + −	0.7240	+	0.8817
*11	FURAN	+	− + + +	0.8606	−	0.6633
12	ICRF 159	+	− − + +	0.6701	n	n/a
13	ISOPHOSPHAMIDE	+	+ − + +	0.8447	n	n/a
*14	LASIOCARPINE	+	+ + − −	0.5960	+	0.8074
*15	8-METHOXYPSORALEN	+	+ + + +	0.9429	+	0.9791
*16	MIREX	+	+ + + +	0.9429	−	0.8406
17	NAPHTHALENE	+	− − − −	0.1535	−	0.0547
−18	β-NITROSTYRENE	−	− − − −	0.1535	+	0.3400
*19	OCHRATOXIN A	+	+ − + +	0.8447	−	0.6433
20	OXAZEPAM	+	− − − −	0.1535	−	0.0547
*21	PHENOXYBENZAMINE	+	+ + + +	0.9429	+	0.9791
−22	o-PHENYLPHENOL	−	− − + −	0.2238	−	0.0933
*23	PROBENECID	+	− + + +	0.8606	−	0.6633
24	PROCARBAZINE	+	− − + +	0.6701	−	0.3933

*25	QUERCETIN	+	+ − + −	0.4633	+	0.7103
26	β-THIOGUANINE DEOXYRIBOSIDE	+	− − + −	0.2238	n	n/a
−27	TRICRESYL PHOSPHATE	−	+ − + −	0.4633	−	0.2106
*28	THIO-TEPA	+	− − + +	0.6701	+	0.8523
29	4-VINYLCYCLOHEXENE	+	− − − +	0.5332	−	0.2672
*30	2,6-XYLIDINE	+	+ + + +	0.9429	+	0.9791
31	Thiamine	−	+ − + +	0.8447	−	0.6344
−32	Niacinamide	−	− − − −	0.1535	−	0.0547
−33	Isoleucine	−	− − − −	0.1535	−	0.0547
−34	Stearic acid	−	+ − + −	0.4633	−	0.2160
−35	β-Carotene	−	− − − +	0.5332	−	0.2672
−36	Methionine	−	− − − +	0.5332	−	0.2672
−37	Fructose	−	− − − −	0.1535	−	0.0547
−38	Spermine	−	+ − − −	0.3268	−	0.1342
−39	Pyruvic acid	−	− − − −	0.1535	−	0.0547
−40	Adenine	−	+ − + −	0.4633	−	0.2160
−41	Glycine	−	− − − −	0.1535	−	0.0547
−42	Ribose	−	− − − −	0.1535	−	0.0547
−43	*p*-Aminobenzoic acid	−	+ − + −	0.4633	−	0.2160
−44	Nicotinamide	−	− − − −	0.1535	−	0.054 7
−45	Valine	−	− − − −	0.1535	−	0.0547
−46	Riboflavine	−	− − − +	0.5332	−	0.2672
−47	Niacin	−	− − − −	0.1535	−	0.0547
−48	Cholesterol	−	− − + +	0.6701	−	0.3933
−49	Lactose	−	− − − −	0.1535	−	0.0547

1: NTP Rodent; 2: NTP Mouse; 3: CPDB Rodent; 4: CPDB Mouse. Lowercase physiological chemicals are of assumed but not demonstrated noncarcinogenicity.

n: No *Salmonella* data available.

[a]*: Positive compound predicted as active; −: negative compound predicted as inactive.

Table 8 Sensitivities of Strategies to Detect Carcinogens

Battery	Concordance	Sensitivity	Specificity
Salmonella	0.500	0.478	0.667
Carcinogenic SAR Model Battery	0.735	0.704	0.773
Salmonella + Carcinogenic SAR Model Battery	0.778	0.609	0.955

sumed, it would appear that the approach described herein improves carcinogen predictions significantly. In view of that apparent success and the fact that the method described is generic, it may be appropriate to use this approach for other SAR models or to couple different SAR models to improve overall predictivity.

In summary, over the last 3–5 years, SAR methods applicable to carcinogenicity have matured and been refined. The quality of the mechanistic and predictive information has been improved considerably. The present report summarizes only one aspect of the possible refinement. It is our belief that the methodologies for formal SAR model development using the currently available methodologies have reached their maximum. Additional improvements will derive from a better understanding of the nature of data bases, improved validation procedures, and the multiple SAR type of approach described herein.

ACKNOWLEDGMENTS

This study was supported by the U.S. Department of Defense (Contract No. DAAA21-93-C-0046), the Center for Alternatives to Animal Testing, and the University of Pittsburgh Cancer Institute.

REFERENCES

1. Takihi N, Zhang YP, Klopman G, Rosenkranz HS. Development of a method to assess the informational content of structure-activity data bases. Quality Assurance: Good Practice, Regulation, and Law 1993; 2:255–264.
2. Liu M, Sussman N, Klopman G, Rosenkranz HS. Estimation of the optimal data base size for structure-activity analyses: the *Salmonella* mutagenicity data base. Mutat Res. 1996; 358:63–72.
3. Liu M, Sussman N, Klopman G, Rosenkranz HS. Structure-activity and mechanistic relationships: the effects of chemical overlap on the structural overlap in databases of varying size and composition. Mutat Res. 1996; 372:79–85.

4. Zhang YP, Sussman N, Macina TO, Rosenkranz HS, Klopman G. Prediction of the carcinogenicity of a second group of chemicals undergoing carcinogenicity testing. Environ Health Perspect 1996; 104(5):1045–1050.
5. Klopman G, Rosenkranz HS. Quantification of the predictivity of some short-term assays for carcinogenicity in rodents. Mutat Res 1991; 253:237–240.
6. Enslein K. An overview of structure-activity relationships as an alternative for carcinogenicity, mutagenicity, dermal and eye irritation, and acute oral toxicity. Toxicol Ind Health 1988; 4:479–498.
7. Zeiger E, Ashby J, Bakale G, Enslein K, Klopman G, Rosenkranz HS. Prediction of *Salmonella* mutagenicity. Mutagenesis. 1996; 11:471–484.
8. Lee Y, Rosenkranz HS, Buchanan BG, Mattison DR, Klopman G. Learning rules to predict rodent carcinogenicity of non-genotoxic chemicals. Mutat Res 1995; 328:127–149.
9. Lee Y, Zhang YP, Sussman N, Rosenkranz HS. Evaluation of predicted probabilities and potencies generated by the SAR expert systems MULTICASE and CASE: application to rodent carcinogenicity databases. In preparation.
10. Ashby J, Tennant RW. Definitive relationships among chemical structure, carcinogenicity and mutagenicity for 301 chemicals tested by the U.S. NTP. Mutat Res 1991; 257:229–306.
11. Ashby J, Tennant RW. Chemical structure, *Salmonella* mutagenicity and extent of carcinogenicity as indicators of genotoxic carcinogenesis among 222 chemicals tested in rodents by the U.S. NCI/NTP. Mutat Res 1988; 204:17–115.
12. Ashby J, Tennant RW, Zeiger E, Stasiewicz S. Classification according to chemical structure, mutagenicity to *Salmonella* and level of carcinogenicity of a further 42 chemicals tested for carcinogenicity by the U.S. National Toxicology Program. Mutat Res 1989; 223:73–103.
13. Gold LS, Sawyer CB, Magaw R, Backman GM, deVeciana M, Levinson R, Hooper NK, Havender WR, Berstein L, Peto R, Pike MC, Ames BA. A carcinogenic potency database of the standardized results of animal bioassays. Environ Health Perspect 1984; 58:9–319.
14. Gold LS, deVeciana M, Backman GM, Lopipero M, Smith M, Blumenthal R, Levinson R, Berstein L, Ames BN. Chronological supplement to the carcinogenic potency database: standardized results of animal bioassays published through December 1982. Environ Health Perspect 1986; 67:161–200.
15. Gold LS, Slone THE, Backman GM, Magaw R, DaCosta M, Lopipero P, Blumental M, Ames BN. Second chronological supplement to the carcinogenic potency database: standardized results of animal bioassays published through December 1984 and by the National Toxicology Program through May 1986. Environ Health Perspect 1987; 74:237–329.
16. Gold LS, Slone THE, Backman GM, Eisenberg S, DaCosta M, Wong M, Manley NB, Rohrbach L, Ames BN. Third chronological supplement to the carcinogenic potency database; standardized results of animal bioassays published through December 1986 and by the National Toxicology Program through June 1987. Environ Health Perspect. 1990; 84:215–286.
17. Gold LS, Manley NB, Slone THE, Garfinkle THE, Rohrbach L, Ames BN.

Fifth plot of the carcinogenic potency database: Results of animal bioassays published in the general literature through 1986 and by the National Toxicology Program through 1989. Environ Health Perspect 1993; 100:86–135.
18. Klopman G. Computer automated structure evaluation of organic molecules. J Am Chem Soc 1984; 106:7315–7324.
19. Iversen GR. Bayesian Statistical Inference. Newbury Park, CA: Sage Publications, 1984.
20. Klopman G. MULTICASE 1. A hierarchical computer automated structure evaluation program. Quant Struct Act Rel 1992; 11:176–184.
21. Pet-Edwards J, Haimes YY, Chankong V, Rosenkranz HS, Ennever FK. Risk Assessment and Decision Making Using Test Results: The Carcinogenicity Prediction and Battery Selection Approach. New York: Plenum Press, 1989.
22. Chankong V, Haimes YY, Rosenkranz HS, Pet-Edwards J. The Carcinogenicity Prediction and Battery Selection (CPBS) method: a Bayesian approach. Mutat Res 1985; 153:135–166.
23. Rosenkranz HS, Klopman G. Structural basis of carcinogenicity in rodents of genotoxicants and nongenotoxicants. Mutat Res 1990; 228:105–124.
24. Rosenkranz HS, Ennever FK, Chankong V, Ped-Edwards J, Haimes YY. An objective approach to the deployment of short-term tests predictive of carcinogenicity. Cell Biol Toxicol 1986; 2:425–440.
25. Zeiger E. Carcinogenicity of mutagens: predictive capability of the *Salmonella* mutagenesis assay for rodent carcinogenicity. Cancer Res 1987; 47:1287–1296.
26. Piegorsch WW, Zeiger E. Measuring intra-assay agreement for the Ames *Salmonella* assay. In: Rienhoff O, Linberg DAB, eds. Lecture Notes in Medical Information. Berlin: Springer, 1991:35–41.

8
The k_e Test

George Bakale
Case Western Reserve University School of Medicine, Cleveland, Ohio

I. INTRODUCTION

Minimization of the exposure of human populations to biohazardous chemicals is possible only if cost-effective and efficacious means are available to identify and monitor potentially biohazardous components of our global environment. The overall socioeconomic impact of having such means available to identify potential carcinogens in the environment was evaluated by Lave and Omenn, who concluded that "the development and validation of short-term tests for screening new and existing chemicals should be accelerated" (1). A decade later this need is still critical and is far greater for an efficacious test for screening potentially hazardous complex mixtures. This was discussed by Anderson, who noted that 35 of the 52 agents first evaluated by the IARC as being human carcinogens were mixtures of chemicals (2). In addition to these clear needs for a short-term screening test that can be applied to mixtures, numerous questions have arisen in the last decade pertaining to rodent bioassays, one of the classical means used to identify carcinogens. These questions include the cost-effectiveness and information value of such tests (3), the relevance of the results of rodent bioassays to human disease (4), and the question of ethics, which is reflected in the objections of animal rights activists (5).

This clear-cut need to identify carcinogens appeared to be largely solved more than two decades ago when the Ames *Salmonella* test was introduced as a screening tool to identify potential carcinogens in 1971, which prompted Ames et al. to proclaim 2 years later that "carcinogens are mutagens" (6). This assertion appeared to be verified in subsequent validation studies in which the *Salmonella* test, hereafter referred to as the Ames test, was found to identify carcinogens with an accuracy of ~90% (e.g., Ref. 7). This strong correlation between mammalian carcinogenicity and bacterial mutagenicity was the basis for Bridges' noting in his 1976 review of short-term tests that the DNA of the Ames tester strains could be viewed as a sensitive probe of a chemical's electrophilicity (8), which strongly influenced our subsequent research in developing the k_e test. During the intervening two decades, more than 200 carcinogen-screening tests (CSTs) based on genotoxicity and/or mutagenicity have been developed (9). However, during the last decade extensive studies of these CSTs have established that a significant fraction (~1/3) of all carcinogens tested are nongenotoxic. For mutagenicity tests like the Ames test, nongenotoxic carcinogens pose the problem of false negatives, and there is also the problem of false positives, which is generally of less concern (1). Recognition that CSTs are fallible and do not have an accuracy of 90% in identifying presently known carcinogens has prompted the development of batteries of tests in which several tests are combined to enhance the sensitivity of the battery relative to its test components (10). This goal is attainable, however, only if the tests of a battery are complementary, i.e., if one component test yields a correct positive response to a carcinogen that yields false-negative responses in other tests in the battery (11). Very few contemporary CSTs are complementary to the Ames test, which is by far the most widely used CST. The k_e test, however, has been found to complement the Ames test, and it is this physicochemical screening test that is now discussed along with its electron-based endpoint. Because this endpoint is electronically accessible, the k_e test could spawn a new generation of electronic electrophile-monitoring devices capable of reducing human exposure to biohazards from a myriad of environmental sources.

II. RATIONALE: CARCINOGENS ARE ELECTROPHILES

A. The k_e Test: A Physicochemical Ames Test

The k_e and Ames tests have common roots in the somatic mutation theory of carcinogenesis, which states that the one physicochemical property shared by most carcinogens is electrophilicity and that electrophiles react with electron-rich biotargets such as DNA in the initiation step of carcinogenesis. This principle was initially proposed by Boveri more than 80 years ago and was the crux of decades of cancer research conducted by E. and J. Miller, who verified that electrophilic chemicals react with electron-rich cellular DNA (12), often at oxygen and nitrogen atoms. The Millers' definitive studies on electrophile-DNA interaction being the initiating step in the multistep process of chemical carcinogenesis are directly pertinent to the Ames test, since the electron-rich DNA of the Ames tester strains serves as a sensitive probe of a test chemical's electrophilicity (8). Moreover, the Millers' recognizing that many classes of chemicals require metabolic activation to be sufficiently electrophilic for the initiating step of carcinogenesis to occur is also pertinent to the Ames test, since mammalian metabolic activation enzymes are a component of the Ames test protocol.

Freely diffusing excess electrons in cyclohexane are used in the k_e test to measure electrophilicity analogously to the DNA of the Ames tester strains that serve as a sensitive probe of a chemical's electrophilicity. This common thread shared by the two tests unravels, however, when the role of metabolic activation is considered. Many of the chemicals that must be activated to the ultimate reactive form that induces mutagenesis in bacteria or carcinogenesis in eukaryotes yield positive responses in the k_e test without metabolic activation.

In the early stages of developing the k_e test, we attempted to resolve this apparent paradox by using reversed micelles to encapsulate microsomes and the test chemical in an effort to effect activation of the chemical that would increase k_e (G. Bakale and R. D. McCreary, unpublished results). We found, however, that electron attachment to reversed micellar systems was itself too complex (13,14) to detect any changes in k_e that might be induced by microsomal activation. In lieu of direct experimental evidence, a rationale was offered in which we proposed that freely drifting excess electrons in cyclohexane are more nucleophilic than are bound electrons in the target DNA of *Salmonella*, rodents, or humans (15). Consequently, electrons in cyclohexane attach to some carcinogens at every encounter, whereas the same procarcinogens are unreactive with a biotarget unless the k_e-positive electrophile is converted to a stronger electrophile that reacts with an electron-rich but less nucleophilic biotarget. An implication of this rationale is that a "prechemical" electron transfer occurs prior to the initiating step of carcinogen-biotarget interaction, which can culminate in adduct formation, intercalation, alkylation, etc., each of which disrupts subsequent transcription of the target if not excised by the host's repair facilities. The proposed prechemical electron transfer step may in some cases involve "single-electron transfer" or SET reactions, which during the last two decades have been found by physical organic chemists to be more ubiquitous than had been previously assumed (16). This rationale is also consistent with the k_e test complementing the Ames test, which is discussed in detail following consideration of a key assumption in the k_e-carcinogenicity rationale, which involves the dependence of electron transport and reactions on the reaction media.

B. Not All Electrons Are Equal: Dependence of Electron Transport and Reactions on Environment

In order to explain the intimate relationship between electron reactivity and the medium in which reaction occurs, it is first necessary to consider the physical parameter that is used as a measure of electron drift in an electric field E, which is denoted as the electron mobility u_e and is expressed in units of cm^2/Vs. This direct measure of the degree to which an electron is free to diffuse in searching for an electrophile can be viewed simply as the electron's drift velocity, v_d (cm/s), while diffusing in an electric field of E volts/cm or V/cm:

$$u_e(cm^2/Vs) = \frac{v_d\ (cm/s)}{E(V/cm)}$$

The dependence of u_e on the electron's environment is demonstrated by u_e values at $-162°C$ ranging from 400 to $\sim 0.001\ cm^2/Vs$ in methane and ethane, respectively (17). These two simple alkanes are obviously similar chemically but interact very differently with excess electrons drifting through each alkane due to the markedly different structure of spherical methane relative to that of cylindrical ethane. Our more recent studies involved mixtures of spherical tetramethylsilane (TMS) and planar cyclohexane, the solvent used in most of our studies of the k_es of carcinogens. In these studies conducted at room temperature the u_e values ranged from 100 to 0.22 cm^2/Vs in pure TMS and cyclohexane, respectively (18). These examples of the structure of the media that surrounds electrons influencing u_e extends to solvent structure influencing k_e, which one obviously expects for electron reactions that approach being diffusion-controlled, which has been demonstrated in a myriad of solute/solvent systems (e.g., Refs. 19,20). The next step in developing the rationale for the k_e-carcinogenicity correlation relates to the structure of the microenviroment where a potent electrophile approaches the site of interaction with a potential electron-rich biotarget such as mitochondrial DNA or more highly structured nuclear chromatin. Although u_e in the vicinity of DNA has been measured (21) and numerous studies of facile electron-transfer reactions in DNA have been reported for which the term "DNA π-way" was recently coined (22), details of the dynamics of electrophile-DNA interaction are not yet available. It should be clear, however, that an electron diffusing in cyclohexane is more free to attach to an unactivated procarcinogen than is an electron that must "exit" the DNA π-way in order to initiate carcinogen-DNA interaction. This obvious difference in reaction environments is the basis for the k_e test yielding positive responses both to procarcinogens as well as to many nongenotoxic carcinogens. The k_e test, however, can be regarded as the "empirical 'litmus paper' test, with no known theoretical basis" for which Bridges cited a need 20 years ago (8). One of the objectives of this chapter is to persuade the reader that the k_e-carcinogenicity empirical correlation has a theoretical basis that is consistent with current physicochemical knowledge of electron-transfer processes in biological systems.

Before commencing with experimental measurements of k_e, it is noted that one group has sufficient regard for the predictive value of k_e that they have applied a QSAR (quantitative structure-activity relationship) technique to calculating values of k_e. This is the work of Benigni's group, who have developed a method to attempt to calculate values of k_e from several physicochemical properties (23). More recently, this group made a comparison of the performance of the k_e test with the Ames *Salmonella* test and Ashby's structural-alert (SA) method of predicting carcinogenicity (24); for 205 chemicals studied, Benigni et al. concluded that "the k_e system performed better than the other systems. . . . The main role of k_e in risk assessment consists in producing a probabilistic estimate of rodent carcinogenicity . . . (that) . . . can be used to rank the chemicals in a priority scale for subsequent and more detailed studies" (25). Benigni later used calculated values of k_e in combination with Ashby's SAs to predict the carcinogenicity of chemicals that were then being screened in long-term rodent bioassays under the aegis of the National Toxicology Program (NTP) (26) and more recently has made a further modification in applying calculated values of k_e to predict the carcinogenicity of chemicals currently being tested by the NTP (27).

This background section of the k_e test is concluded by referring again to the conclusion drawn by Benigni et al., namely, the value of using k_e to prioritize the probable carcinogenicity of pure chemicals (see above). This statement is equally applicable to using k_e to prioritize the relative carcinogenicity of complex mixtures of chemicals; however, Benigni's method of calculating k_e, Ashby's SAs, and all other structure-activity methods cannot be easily applied to the problem of mixtures. In contrast to the latter methods, the electron-based nature of the k_e test makes it an ideal basis for the development of an instrument that can be used to rank the relative electrophilic potencies of chemical mixtures in a variety of matrices that can range from toxic effluents in air, water, or waste to traces of carcinogens in human urine or blood in personnel-monitoring applications.

III. EXPERIMENTAL: MEASUREMENT OF k_e

Measuring the reaction rate constant of a species as reactive as an electron requires that the species be generated in a nonreactive medium in which the test chemical is soluble and that a means of monitoring the rate of interaction of the two species be available. The first criterion eliminates such common solvents as water, in which electrons are solvated in <1 ps, and carbon tetrachloride, which attaches electrons at a diffusion-controlled rate (28). Although most hydrocarbons are nonreactive with electrons and are therefore candidate solvents, cyclohexane was used in most of our k_e studies for practical as well as theoretical reasons. The practical reasons include cyclohexane being inexpensive and available in high purity as well as having low toxicity and convenient liquid and vapor pressure ranges that facilitate purification and measurement of k_es at room temperature. The theoretical basis for using cyclohexane in our initial k_e-screening studies (29) was that the value of u_e in cyclohexane is similar to the value that was estimated for electrons in the sheath of water that surrounds DNA (30), which was subsequently found to be underestimated (21). A k_e-screening study in which isooctane was used as the solvent yielded k_e responses that were not significantly different than those measured in cyclohexane despite the 25-fold greater value of u_e in isooctane (31). We therefore continued to use cyclohexane as the primary solvent in subsequent k_e-screening studies due to the aforementioned experimental advantages that it provides.

The second requirement for measuring k_es is a means of monitoring the attachment rate of electrons to the test chemical which was available in the pulse-conductivity (PC) technique that we had developed to study the transport and reaction properties of electrons in nonpolar liquids (28). A block diagram of the PC method is shown in Figure 1. A Van de Graaff generator is the source of pulsed ionizing radiation that is used to produce a single 15-ns pulse of 1 MeV electrons that irradiates a parallel-plate ion chamber containing a solution of the test chemical. These high-energy electrons pass through the cyclohexane solvent and produce secondary electrons that are thermalized in much less than a nanosecond and then drift in the electric field between the electrodes of the ion chamber. Only a small fraction of the electrons reach the anode, which can be illustrated by considering that the time t_d required to drift the interelectrode distance d of 0.6 mm in an electric field of 30 kV/cm in cyclohexane for which $u_e = 0.22$ cm^2/Vs; t_d is given by:

$$t_d = \frac{d}{Eu_e} = \frac{0.06\ \text{cm}}{[30{,}000\ \text{V/cm}][0.22\ \text{cm}^2/\text{Vs}]} = 9.1\ \mu\text{s}$$

This value of t_d is more than 10-fold longer than the $t_{1/2}$ of 700 ns typically obtained in "pure" cyclohexane with the purification procedures that we routinely use to conduct k_e measurements. The 700 ns half-life reflects attachment of electrons to adventitious electron-attaching impurities, which are present in sufficient concentration to result in a rate of attachment that is >10-fold faster than the rate of electron drift to the anode. Conse-

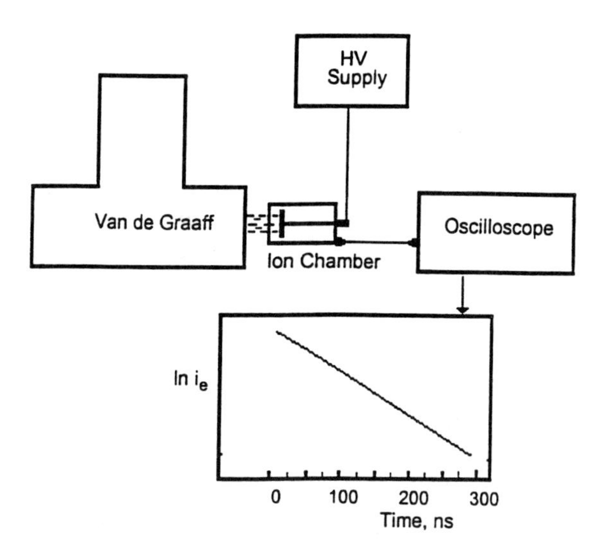

Figure 1 Block diagram of pulse-conductivity system used to measure k_e. A 15 ns pulse of 1 MeV electrons from a Van de Graaff generator produces ions and electrons in a cyclohexane solution of the test chemical that is contained in an ion chamber. Electrons that escape geminate recombination drift in an electric field maintained by 1800 V from the HV supply, which are applied across parallel electrodes separated by 0.6 mm. Electrons drifting in this field of 30 kV/cm induce a current that is monitored and recorded on a 100 MHz storage oscilloscope. The half-life of the exponentially decaying electrons is obtained from a first-order plot of the logarithm of the electron current i_e in the nanosecond time regime. An efficient electron-attaching solute at a 2 μM concentration reduces the electron half-life to ~100 ns and yields a k_e of $3.5 \times 10^{12}\ M^{-1}s^{-1}$.

quently, <1/10 of the thermal electrons reach the anode in "pure" cyclohexane, and this fraction is further reduced if an electrophilic test solute is present in the solvent at a concentration greater than that of the impurities.

The preceding comparison was presented to illustrate that the number of electrons that reach the electrodes is not critical to observing the electron current, i_e. Instead, it is during the electrons' drift in the electric field that a current is induced in the circuit shown in Figure 1, and it is this current that is readily monitored with the oscilloscope. Attachment of the high-mobility electrons to electrophiles converts the electrons to sluggish anions having a mobility of $<0.001\ cm^2/Vs$, and it is this conversion that effects a concomitant decrease of i_e that is recorded on the oscilloscope (28). Typical values of the experimental conditions used to obtain k_e are provided in the figure legend; for these conditions the electron current decays exponentially and k_e is evaluated from the equation:

$$k_e = \frac{\ln 2}{\{[S]t_{1/2}\}}$$

in which [S] is the solute concentration and $t_{1/2}$ is the electron half-life. The logarithm term in the above equation is a reflection of the pseudo-first-order kinetics that apply to our irradiation conditions, which indicates that electron attachment predominates over other modes of decay available to the electrons, such as neutralization at the anode (see above) or recombination with cations, both of which have been discussed in detail (e.g., Refs. 28–33).

IV. PREDICTIVITY OF k_e: VALIDATION STUDIES

The best indicator of the value of the k_e test as a predictive tool for carcinogens is provided in the results of four major validation studies in which the k_e test was used to screen chemicals having known carcinogenic properties. These four screening studies include (a) a study of 34 carcinogens and 51 noncarcinogens (32) for which the rodent carcinogenicity and Ames *Salmonella* mutagenicity were reported by Kawachi et al. (34), (b) a study of 46 nonmutagenic carcinogens and 19 mutagenic noncarcinogens tested by the National Cancer Institute (NCI) or the NTP (35) and reported by Zeiger (36), (c) a study of 61 mutagenic carcinogens and 44 nonmutagenic noncarcinogens (37) also tested by the NCI/NTP and also included in the survey of Zeiger (36), and (d) a study of 23 chemicals for which k_es were screened while NTP rodent bioassays to establish carcinogenicity were in process (38). Of the 85 chemicals studied by Kawachi et al., 14 were also included in the subsequent NTP study, and only the NTP results obtained for these chemicals are included in the analysis that follows.

The responses of the k_e and Ames tests for these 264 chemicals (150 carcinogens, 114 noncarcinogens) are compared in Table 1 using the three measures of the predictivity of short-term tests most frequently used. Sensitivity and specificity are the percentages of carcinogens and of noncarcinogens, respectively, that a test correctly identifies. Accuracy or concordance is a combination of these two criteria. From the results presented in Table 1 it is apparent that the k_e test compares favorably with the Ames test for all three measures of predictivity. The k_e test sensitivity, specificity, and accuracy are 63, 72, and 67%, respectively, and for the Ames test the same respective criteria are 58, 75, and 66%. These minor differences in the predictive responses of the two tests suggest that either test could be used in most screening applications with almost equal probabilities of identifying a chemical as a carcinogen or noncarcinogen, but analysis of the same re-

Table 1 Comparison of Three Measures of Predictivity for the Ames *Salmonella* and k_e Tests from Screening Results of Four Validation Studies

Measure of predictivity	Ames *Salmonella* test	k_e test
Sensitivity $= \dfrac{\text{Correct + responses}}{\text{Carcinogens tested}} \times 100\%$	$\dfrac{87}{150} = 58\%$	$\dfrac{95}{150} = 63\%$
Specificity $= \dfrac{\text{Correct − responses}}{\text{Noncarcinogens tested}} \times 100\%$	$\dfrac{86}{114} = 75\%$	$\dfrac{82}{114} = 72\%$
Accuracy $= \dfrac{\text{All correct responses}}{\text{All chemicals tested}} \times 100\%$	$\dfrac{173}{264} = 66\%$	$\dfrac{177}{264} = 67\%$

Source: Refs. 32, 35, 37, 38.

sponses in greater depth reveals the advantage of using both tests in a two-arm battery in order to exploit their independent endpoints.

V. COMPLEMENTARY VALUE OF k_e

In view of the markedly different probes of electrophilicity which the Ames tester strains and the k_e test use, namely, DNA reactivity versus electron attachment, the responses of the k_e test are expected not to be highly concordant with *Salmonella* mutagenicity, which was confirmed in our analysis of the k_e responses to the NCI/NTP-tested chemicals included in the 1987 Zeiger survey (39). The high degree of independence between the Ames and k_e tests reported therein was verified in a recent interlaboratory screening of coded *Salmonella* mutagens and nonmutagens (40), which underscored the potential value of k_e in complementing the Ames test that we had reported earlier (39). The need for CSTs that complement those based on DNA reactivity such as the Ames test and Ashby's structural alerts (24) was cited by the U.S. Interagency Staff Group on Carcinogens in 1986 (41) and was reiterated 3 years later by Ashby, who noted "a major need for new assays with which to predict non-genotoxic carcinogens" (9). In the intervening 3 years Zeiger (36) and Tennant et al. (42) reported, respectively, a significant decline in the accuracy of the Ames test to known rodent carcinogens and the failure of three in vitro STTs to complement the Ames test. Both of these studies were based on results obtained in long-term rodent bioassays and in short-term in vitro tests conducted under the aegis of the NCI and NTP.

As stated in the introduction, few tests have been found to complement the Ames test, the most widely used CST. In our earlier study of k_e complementing the Ames test and SAs, 107 chemicals classified as rodent carcinogens by the NCI/NTP were considered and the Ames and k_e test sensitivities were again comparable, i.e., 57 and 59%, respectively. If the Ames and k_e tests are used together in a screening battery for these 107 NTP carcinogens, the sensitivity of the battery is 82%. A two-element battery of k_e combined with Ashby's SAs correctly identifies 91% of the 107 carcinogens correctly. This study prompted our concluding that "k_e provides valuable information for identifying carcinogens and should be considered in future carcinogen-screening strategies" (39).

This conclusion can be reiterated for the results presented in Figure 2A, which again illustrates the increased sensitivity of the Ames-k_e battery relative to that of the Ames test alone. For the 150 carcinogens included in the extended database of Table 1, the sensitivity of the Ames-k_e battery is 83%, whereas the sensitivity of the Ames test alone is 58% and of the k_e test is 62%. The 25% enhancement of the sensitivity of the k_e-Ames battery relative to that of the Ames test provides a screening tool having predictivity at the level for which Bridges cited a need two decades ago (8). Also pertinent to carcinogen-screening strategy are the results illustrated in Figure 2B, which are based on the 114 noncarcinogens in the same database. The k_e test yields only 19 false-positive responses to the 85 nonmutagenic noncarcinogens screened, which decreases the specificity by 17% or significantly less than the sensitivity is increased in using the k_e-Ames battery. Whether this trade-off in the increased sensitivity being partially offset by the decreased specificity in using the Ames-k_e battery is tolerable would depend upon the screening application, which was discussed in our earlier study of the complementary value of k_e (39). Our more recent study of the responses of the k_e test and several structure-based CSTs to 100 coded chemicals for which the *Salmonella* mutagenicity responses were determined should shed additional light on those applications where the k_e test can be used most efficaciously (40). In concluding, it is noted that a study of the complementary value of the k_e test and several other CSTs has been conducted (43) and is discussed in detail in Chapter 10 in this volume.

VI. CONCLUSIONS AND FUTURE DIRECTIONS

The first step in environmental risk assessment is hazard identification, which consists of evaluating various types of studies directed at determining if potentially biologically hazardous agents are present in the environment being assessed (e.g., Ref. 44). The k_e test provides a reliable and unique

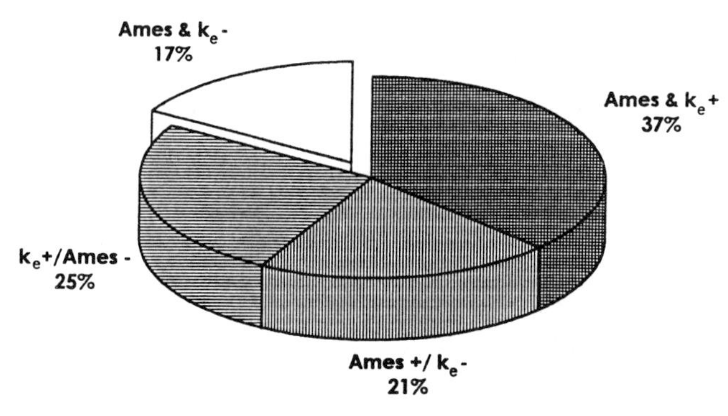

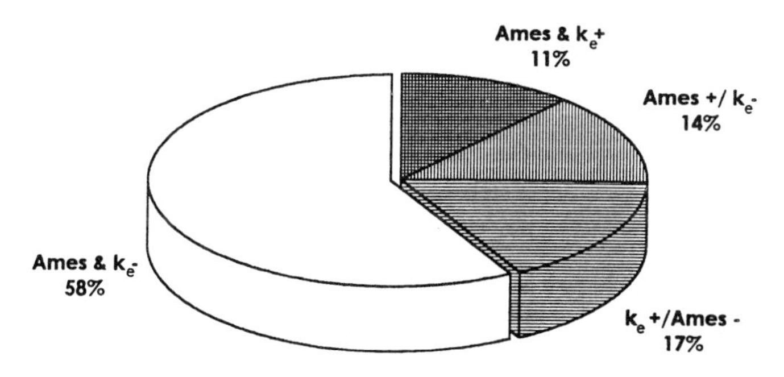

Figure 2 (A) Increased sensitivity of an Ames-k_e battery relative to using each test alone in screening 150 carcinogens. (B) Decreased specificity of this battery for 114 putative noncarcinogens. In both A and B, the cross-hatched areas denote both tests yielding positive responses, whereas the vertical and horizontal hatchings indicate positive Ames and k_e responses, respectively, and the unhatched slices denote both tests yielding negative responses. The increased sensitivity of the battery more than offsets its decreased specificity, which is an attribute that would be sought in most screening applications (39).

means of conducting this determination more cost-effectively than methods that are currently used. Although various analytical instruments have been developed with which one or several potential biohazards in the environment can be identified, these chemical-specific devices provide no indication of the biohazard potential of the sample. On the other hand, bioassay-directed analyses that indicate the overall genotoxicity of complex mixtures have been developed (e.g., Ref. 45), but such assays are insensitive to nongenotoxic biohazards, are time-consuming, and generally cannot be used in applications that require on-stream monitoring. A third approach to identifying biohazards is the use of structure-activity relationships, but these computer-based methods are inapplicable to many "real-world" sampling problems and, in addition, rely on CST results in their "learning" databases that force these methods to be noncomplementary to CSTs. In contrast to these screening methods, the k_e test has an electron-based endpoint that complements the most widely used in vitro carcinogen-screening tests and is also compatible with the multitude of sampling protocols that have been developed for both chemical and biological analyses of the gas, liquid, and solid phases. The use of electrons as the "titrant" with which the electrophilicity of samples is measured permits the latest electronic technology to be directly linked to assessing biohazard potential. Thus, the k_e test can serve as a physicochemical bridge to unite the fields of risk assessment, genetic toxicology, and analytical chemistry, which would ultimately permit the biohazardous potential of chemicals to be monitored from their point of introduction into the global environment to their eventual site of disposal.

ACKNOWLEDGMENT

The author thanks R. D. McCreary for years of dedicated service in measuring the k_es upon which this review is based. The author also thanks the U.S. Department of Energy and the National Institute for Occupational Safety and Health for support of our carcinogen-screening studies.

REFERENCES

1. Lave LB, Omenn GS. Cost-effectiveness of short-term tests for carcinogenicity. Nature 1986; 324:9–34.
2. Anderson D. The use of short-term tests in detecting carcinogenicity of complex mixtures. In: Vainio H, Sorsa M, McMichael AJ, eds. Complex Mixtures and Cancer Risk. Lyon: IARC, 1990:89–100
3. Lave LB, Ennever FK, Rosenkranz HS. Omenn GS. Information value of rodent bioassays. Nature 1988; 366:631–634.
4. Ames BN, Gold LS. Too many rodent carcinogens: mitogenesis increases mutagenesis. Science 1990; 24:970–971.
5. Rack L, Spira H. Animal rights and modern toxicology. In: Mehlman MA, ed. Benchmarks: Alternative Methods in Toxicology. Princeton, NJ: Princeton Scientific, 1989:193–203.
6. Ames BN, Durston WE, Yamasaki E, Lee FD. Carcinogens are mutagens: a simple test system combining liver homogenates for activation and bacteria for detection. Proc Natl Acad Sci USA 1973; 2281–2285.

7. McCann J, Ames BN. Detection of carcinogens as mutagens in the Salmonella/microsome test. Proc Natl Acad Sci USA 1976; 73:950–954.
8. Bridges BA. Short-term screening tests for carcinogens. Nature 1976; 261:195–200.
9. Ashby J. Origins of common uncertainties in carcinogen/mutagen screening. Environ Mol Mutagen 1989; 14(suppl 16):51–59.
10. Pet-Edwards J, Haimes YY, Chankong V, Rosenkranz HS, Ennever FK. Risk Assessment and Decision Making Using Test Results. The Carcinogenicity Prediction and Battery Selection Approach. New York: Plenum Press, 1989.
11. Ashby J, de Serres FJ, Draper MH, Ishidate M, Margolin B, Matter BE, Shelby MD. Evaluation of short-term tests for carcinogens. Prog Mutat Res 1985; 5:117–174.
12. Miller JA, Miller EC. Ultimate chemical carcinogens as reactive mutagenic electrophiles. In: Hiatt HH, Watson JD, Winsten JA, eds. Origins of Human Cancer, Book B. Cold Spring Harbor, NY: Cold Spring Harbor Lab, 1977: 605–628.
13. Bakale G, Beck G, Thomas JK. Electron capture in water pools of reversed micelles. J Phys Chem 1981; 85:1062–1064.
14. Bakale G, Beck G, Thomas JK. Dynamics of electron attachment to AOT/H_2O reversed micelles. J Phys Chem 1992; 96:2328–2334.
15. Bakale G. Theoretical implications of the k_e test. In: Politzer P, Martin FJ, eds. Chemical Carcinogens. Amsterdam: Elsevier, 1988:320–344.
16. Ashby EC. Single-electron transfer, a major reaction pathway in organic chemistry. An answer to recent criticisms. Acc Chem Res 1988; 414–421.
17. Bakale G, Tauchert W, Schmidt WF. Electron transport in mixtures of liquid methane and ethane. J Chem Phys 1975; 63:4470–4473.
18. Bakale G, Lacmann K, Schmidt WF. Electron mobility in mixtures of tetramethylsilane with isooctane and cyclohexane. Physics Lett 1992; A168:209–212.
19. Warman JM. The dynamics of electrons and ions in non-polar liquids. In: Baxendale JH, Busi F, eds. The Study of Fast Processes and Transient Species by Electron Pulse Radiolysis. Dordrecht: Reidel, 1982:433–533.
20. Holroyd RH. The electron: its properties and reactions. In: Farhataziz, Rodgers MAJ, eds. Radiation Chemistry. Principles and Applications. New York: VCH, 1987:201–235.
21. Van Lith D, de Haas MP, Warman JM, Hummel A. Highly mobile charge carriers in hydrated DNA and collagen formed by pulsed ionization. Biopolymers 1983; 22:807–810.
22. Murphy CJ, Arkin MR, Ghatlia ND, Bossmann S, Turro NJ, Barton JK. Fast photoinduced electron transfer through DNA intercalation. Proc Natl Acad Sci USA 1994; 91:5315–5319.
23. Benigni R, Andreoli C, Giuliani A. Structure-activity studies of chemical carcinogens: use of an electrophilic reactivity parameter in a new QSAR model. Carcinogenesis 1989; 10:55–61.
24. Ashby J. Fundamental structural alerts to potential carcinogenicity or noncarcinogenicity. Environ Mol Mutagen 1985; 7:919–921.

25. Benigni R, Cotta-Ramusino M, Andreoli C, Giuliani A. Electrophilicity as measured by k_e: molecular determinants, relationship with other physical chemical and quantum mechanical parameters, and ability to predict rodent carcinogenicity. Carcinogenesis 1992; 13:547-553.
26. Benigni R. QSAR prediction of rodent carcinogenicity for a set of chemicals currently bioassayed by the US National Toxicology Program. Mutagenesis 1991; 6:423-425.
27. Benigni R, Andreoli C, Zito R. Prediction of rodent carcinogenicity of further 30 chemicals bioassayed by the US National Toxicology Program. Environ Health Persp 1996; 104(suppl 5):1041-1044.
28. Bakale G, Gregg EC, McCreary RD. Decay of quasifree electrons in pulse-irradiated liquid hydrocarbons. J Chem Phys 1972; 57:4246-4254.
29. Bakale G, McCreary RD, Gregg EC. Quasifree electron attachment to carcinogens in liquid cyclohexane. Cancer Biochem Biophys 1981; 5:103-109.
30. Bakale G, Gregg EC. Conjecture on the role of dry charges in radiosesitization. Br J Cancer 1978; 37(suppl III):24-28.
31. Bakale G, McCreary RD, Gregg EC. Quasifree electron attachment to carcinogens. Int J Quantum Chem Quantum Biol Symp 1982; 9:15-25.
32. Bakale G, McCreary RD. A physico-chemical screening test for chemical carcinogens: the k_e test. Carcinogenesis 1987; 8:253-264.
33. Bakale G. Detection of mutagens and carcinogens by physiochemical techniques. In: Saxena J, ed. Hazard Assessment of Chemicals. Vol. 6. Current Developments. Washington, DC: Hemisphere, 1989:85-124.
34. Kawachi T, Yahagi T, Kada T, Tazima Y, Ishidate M, Sasaki M, Sugiyamura T. Cooperative programme on short-term assays for carcinogenicity in Japan. In: Montesano R, Bartsch H, Tomatis L, Davis W, eds. Molecular and Cellular Aspects of Carcinogen Screening Tests. Lyon: IARC, 1980:323-330.
35. Bakale G, McCreary RD. Response of the k_e test to NCI/NTP-screened chemicals I: non-genotoxic carcinogens and genotoxic noncarcinogens. Carcinogenesis 1990; 11:1811-1818.
36. Zeiger E. Carcinogenicity of mutagens: predictive capability of the Salmonella mutagenesis assay for rodent carcinogenicity. Cancer Res 1987; 47:1287-1296.
37. Bakale G, McCreary RD. Response of the k_e test to NCI/NTP-screened chemicals II: genotoxic carcinogens and nongenotoxic noncarcinogens. Carcinogenesis 1992; 13:1437-1445.
38. Bakale G, McCreary RD. Prospective k_e screening of potential carcinogens being tested in rodent bioassays by the NTP. Mutagenesis 1992; 7:91-94.
39. Ennever FK, Bakale G. Response of the k_e test to NCI/NTP-screened chemicals III: complementary value of k_e in screening carcinogens. Carcinogenesis 1992; 13:2059-2065.
40. Zeiger E, Ashby J, Bakale G, Enslein K, Klopman G, Rosenkranz HS. Prediction of Salmonella mutagenicity. Mutagenesis 1996; 11:471-484.
41. U.S. Interagency Staff Group on Carcinogens. Chemical carcinogens: a review of the science and its associated principles. Environ Health Perspect 1986; 67: 201-282.
42. Tennant RW, Margolin BH, Shelby MD, Zeiger E, Haseman JK, Spalding J,

Caspary W, Resnick M, Stasiewicz S, Anderson B, Minor R. Prediction of chemical carcinogenicity in rodents from in vitro genetic toxicity assays. Science 1987; 236:933–941.
43. Kitchin KT, Brown JL, Kulkarni AP. Complementarity of genotoxic and nongenotoxic predictors of rodent carcinogenicity. Teratogen Carcinogen Mutagen 1994; 13:83–100.
44. Vainio H, Sorsa M, McMichael AJ, eds. Complex Mixtures and Cancer Risk. Lyon: IARC, 1990.
45. Schuetzle D, Lewtas J. Bioassay directed chemical analysis in environmental research. Anal Chem 1986; 58:1060A–1067A.

9

Gap Junctional Intercellular Communication as a Method to Detect and Predict Carcinogenicity

Vladimir Krutovskikh and Hiroshi Yamasaki
International Agency for Research on Cancer, Lyon, France

In testing, predicting, and assessing the risk of carcinogenicity, it is becoming more and more recognized that mechanistic information on actions of agents should be taken into consideration. During the past two decades, genotoxicity has been assumed to be the principal mechanism of carcinogenesis, and, thus, various short-term tests based on mutations have been developed. However, it has become clear that not all carcinogens (or their metabolites) are DNA-reactive and therefore, such non–DNA-reactive carcinogens should not be detected in conventional mutation tests. The solution may be found in creation of alternative short-term tests based upon nonmutagenic mechanisms operating in carcinogenesis.

One such nonmutagenic mechanism of carcinogenesis that is supported by abundant experimental evidence is the inhibition of gap junctional intercellular communication (GJIC). The idea to develop a short-term screening test to detect or predict carcinogenicity based upon inhibition of GJIC was formed when it was found that tumor promoters could specifically impair intercellular communication in cell culture (Trosko et al., 1993). There are a number of different versions of this assay, but none have been perfected so far (reviewed by Barrett et al., 1980; Swierenga and Yamasaki, 1992); further attempts to improve these methods are being made in several laboratories. A large body of data obtained during the last decade in this field strongly confirmed that the inhibition of GJIC by different carcinogens, particularly nongenotoxic ones, is rather a specific event. About 300 substances have been tested by GJIC assay so far; nearly 60% of nongenotoxic, but only 30% of genotoxic carcinogens possessed the ability to uncouple cells (Budunova and Williams, 1994).

Any reliable carcinogen-detection method, including GJIC, should be based on firm scientific support. In fact, substantial progress has been made during the last few years in understanding the molecular mechanisms regulating GJIC machinery and their impairment by carcinogens. In this chapter, we review the current knowledge of structure, function, regulation and study methods of GJIC, as well as the role of GJIC disorders in carcinogenesis with particular regard to its possible application for carcinogen identification and characterization.

I. STRUCTURE AND PHYSIOLOGY OF GAP JUNCTIONS

A. Structural Components—Connexins

The structure and function of gap junctions have been extensively reviewed in the scientific literature (Beyer et al., 1990; Paul, 1995; Kumar and Gilula, 1996). Therefore, we will give here just a brief overview of relevant information.

Gap junctions are plaquelike clusters of intercellular aqueous channels that mediate communication between the cytoplasm of contiguous cells. They are formed by two semi-channels, called connexons, each coming from one of two adjacent cells, making the gap junction a unique cellular structure resulting from a joint effort of two cells. Each connexon consists of six identical or similar transmembrane protein molecules (connexins) oligomerized within a plasma membrane. In each cell, a pool of connexons exists as a potential source for assembling of gap junctions, but the process of gap junction formation from connexons is not yet fully understood. It seems that gap junction assembly is a rather delicate process and may be easily affected by a variety of factors, including different biologically harmful chemicals (see below).

The principal molecular element of the gap junction is a transmembrane protein, connexin. Connexins represent a rather big protein family; at least 12 different connexin genes are found in mammals. All connexins have a common membrane topology, namely, two terminal tails located in the cytoplasm, four membrane-spanning domains separated by two extracellular loops and one intracellular loop (Hertzberg et al., 1988). On the other hand, there is a great difference between individual connexins in terms of size and amino acid sequences of different parts of the proteins. If the N-terminal tail is rather short for all connexins, the length of the C-terminal tail varies greatly from connexin to connexin. Thus, as has been shown experimentally, the phosphorylation of the C-terminal tail of certain connexins (c×43) plays an important functional role (Musil et al., 1990). The heavier the molecular weight of the connexin, the longer its C-tail and the greater the possibility of its phosphorylation. Despite the great diversity in amino acid sequences between individual connexins, basically their topologically similar domains perform similar functions. Thus, it has been shown that two extracellular loops are responsible for interaction with opposed counterparts to establish a gap junction. Those parts of the protein molecule are the most conservative of the individual connexins. The high homology of extracellular loops of different connexins suggested their heterologous compatibility, which may result in the assembly of individual gap junction from different species of connexin proteins. It has been shown, however, that due to the second extracellular domain, such heterotypic compatibility between individual connexins is rather selective (White et al., 1994; Kumar and Gilula, 1996). Much less is known about the functional role of the intracellular loops of connexins, except its possible implication into gating control by pH. However, recently a functional domain of six residues, responsible for gap junction permeability, was located in this part of connexin 43 (Backer et al., 1995). The fact that different individual connexins possess the greatest diversity of amino acid composition in the intracytoplasmic loop is widely used for obtaining highly specific anti-individual connexin antibodies.

Of the four membrane-spanning domains of connexin, the third one from the N-terminal end is of particular interest functionally. This third domain may face the lumen of the gap junctional channel (Milks et al., 1988). Therefore, any allosteric modifications in this part of the connexin could be crucial for gap junction permeability.

More recently, the function of another membrane-spanning domain of connexin has been studied: the first from N-terminal end. It appears to be involved in the process of connexin targeting into the lateral plasma membrane (Leube, 1995).

B. Physiological Roles of GJIC

Gap junctions are the only means by which cells, being organized in tissue as a sort of syncytium, can share different messengers of low molecular weight and, therefore, maintain tissue homeostasis. The nature of these messengers could be diverse, namely, ions, nutrients, nucleotides, and metabolites (Saez et al., 1989), both of endogenous origin and xenobiotics, but their molecular weight cannot exceed 1000 daltons. The passage of messengers through gap junctions is a passive process requiring no energy.

Besides a vast and vaguely defined role of GJIC in tissue homeostasis, several other specific physiological roles have been postulated for GJIC.

Electrical Coupling

Gap junctions serve as electrical synapses in excitable cells, such as cardiac myocytes and central nervous system neurons (Furshpan and Potter, 1959).

Growth Regulation

Gap junctions are considered an important pathway of distribution in tissue signals keeping cell proliferation under a negative control. There is a strong correlation between the level of GJIC functional efficiency and growth rate of cells. Thus, the bursting of proliferation in tissue by different mitogenic stimuli drastically inhibits GJIC, and conversely enhancement of GJIC by certain chemicals (e.g., retinoids) inhibits the rate of cell proliferation (Yamasaki et al., 1993). Since cancer could be considered in a broad sense a disease of cell proliferation, this particular function of GJIC is the most relevant to carcinogenesis. A possible implication of GJIC in apoptosis has also been proposed (Trosko and Goodman, 1994).

Tissue Response to Hormones

Hormonally activated cells may release waves of second messengers such as cyclic nucleotides, calcium, and inositol phosphates through gap junctions into nonactivated cells and stimulate the latter, increasing the number of hormone-responsive cells in tissue (Murani-Silem et al., 1991).

Regulation of Embryonic Development

Gap junctions mediate the spreading of morphogenic signals in embryos and play a role in defending the boundaries of developmental compartments. As shown by the use of the knock-out gene technique in mice, the impairment of GJIC during embryogenesis leads to developmental anomalies (Reaume et al., 1995). Gap junctions also play a principal role in diffusional feeding and signaling in avascular tissues as well as synchronizing the labor process during childbirth.

C. Regulation of GJIC

Knowledge of basic mechanisms of regulation of GJIC function is important in increasing the specificity of carcinogen detection by GJIC tests. The function of GJIC could be regulated, and therefore, affected by chemicals at various stages of gap junction formation, i.e., connexin gene transcription, stability of their mRNA, translation, following transportation of connexin protein from the endoplasmatic reticulum and Golgi apparatus into the lateral plasma membrane, oligomerization into connexons, and establishment of functional gap junctions. Permeability of already established gap junctions also could be modulated "in place" by "gating" mechanisms. At least three types of GJIC function modulation are recognized (Holder et al., 1993): (a) fast control, or gating within milliseconds, by closing/opening of the gap junction channel; (b) intermediate control, requiring minutes or hours, when redistribution of connexin proteins between lateral plasma membrane and cytoplasm depot take place; and (c) long-term control, occurring on the level of mRNA synthesis and stability. Among a few known molecular mechanisms regulating GJIC are the involvement of cell-adhesion molecules (CAM) in the establishment of intercellular interactions and the phosphorylation of the C-terminal domain of certain connexins. It has been shown that targeted abrogation of CAM function disrupts functions of established GJ (Hertig et al., 1996). Conversely, reestablishment of intercellular adhesion in CAM-deficient tumorigenic cells by CAM gene transfection resulted in increased GJIC (Jongen et al., 1991).

It is becoming obvious that the function of individual connexins could be regulated differentially, sometimes by unique mechanisms or biochemical pathways, e.g., functional modulation by phosphorylation of cx43 (Musil et al., 1990). Phosphorylation sites for cAMP-dependent kinase, protein kinase C, and MAP kinase have been identified in its C-terminal tail. The pattern of connexins expressed in established cell lines is often different from that seen in the normal tissue from which these cells originated; usually in vitro cells express cx43. Therefore, in the case of identification of a chemical as positive as GJIC inhibition by an in vitro assay, one may suggest that mechanisms responsible for observed inhibition could be specific for cultured cells and do not necessarily reflect in vivo situations. This circumstance reinforces the importance of studying GJIC in vivo on the one hand and limits the credibility of in vitro GJIC test systems on carcinogenicity on the other.

As discussed below, various regulatory points of GJIC are impaired during carcinogenesis and by carcinogenic insults. So far, posttranslational steps appear to be more vulnerable than those preceding them.

II. ROLE OF GJIC DISORDERS IN CARCINOGENESIS

The putative role of GJIC disorders in tumor development was postulated long ago (Kanno Matsui, 1968; Loewenstein and Kanno, 1966). There are already several comprehensive reviews on this topic (Yamasaki, 1990; Holder et al, 1993; Hotz-Wagenblatt and Shalloway, 1993; Trosko et al., 1993; Budunova, 1994; Ruch, 1994; Yamasaki et al., 1995). Initially, it was found that the communication capacity of cultivated tumor cells varies depending on the degree of their malignancy; the more malignant they are, the less they communicate (Loewenstein and Kanno, 1966). Then it was shown that tumor promoters strongly and specifically inhibited GJIC in cultured cells. Both of these findings suggested that GJIC disorders are causally related to tumor formation, presumably by means of abrogation of negative growth control in tissue. It was then found that certain tumorigenic cells may show high levels of communication with each other (homologous-type GJIC), but not with surrounding normal counterparts (heterologous-type GJIC), suggesting that the defect of heterologous GJIC may facilitate rapid selective clonal propagation and malignant progression of transformed cells (Yamasaki, 1990).

Later, the relevance of this in vitro finding to real carcinogenesis was confirmed by in vivo observation of the rat hepatocarcinogenic model; direct dye-transfer measurement of cell coupling revealed the selective lack of heterologous communication between some, but not all, GST-p–positive putative preneoplastic foci and surrounding liver parenchyma (Krutovskikh et al., 1991).

Basically, two principal mechanisms of aberrant connexin functions are considered to contribute to carcinogenesis at the genetic and functional levels. Genetic alteration of GJIC may occur at the initiation stage of carcinogenesis, when genotoxic carcinogens induce mutations in connexin genes or other genes related to cell-cell interaction machinery. GJIC may also occur at later stages of tumor progression as a result of progressive accumulation of secondary genetic alterations due to general genetic instability. Mutational inactivation of connexin genes may occur in either their regulatory or their coding region, or both.

To be established as a functional gap junction, connexin proteins need to be oligomerized into hexamers to form a connexon. As has recently been shown, even if only one allele of a connexin gene is mutated in its coding region, a partial presence of mutated protein within the pool of connexin proteins could be enough to abolish overall intercellular communication by dominant-negative mechanism (Bruzzone et al., 1994; Omori et al., 1996). There have been few studies in which the possibility of connexin gene mutations have been examined in an effort to explain the GJIC failure observed in tumors. Our laboratory has found one cx32 gene mutation in 7 rat liver tumors induced by a nitrosamine (Omori et al., 1996). However, no cx32 gene mutations were found in 20 liver and 22 gastric human tumors (Krutovskikh et al., 1994; Mironov et al., 1994), nor was any mutation of human cx37 found in 8 lung adenocarcinomas and 18 sporadic breast carcinomas (Krutovskikh et al., 1996). Meanwhile, our preliminary results suggest that mutations of cx43 occur in some malignant meningiomas and mutations of cx37 in hepatic angiosarcomas. Thus, it is possible that connexin gene mutations play a certain, albeit limited role in GJIC disorders associated with carcinogenesis.

Certain species differences probably exist in terms of mechanisms of GJIC impairment during carcinogenesis. Thus, while cx32 gene expression in rat liver tumors was decreased, its expression usually remains high in human liver tumors; however, the GJIC in both benign and malignant human liver neoplasms was always very low. Immunohistochemical analysis of human hepatocellular carcinomas revealed abundant presence of intracytoplasmic connexin proteins in tumor cells, suggesting that GJIC abrogation in these tumors occurred by means of disturbance of connexin proteins transportation from its intracytoplasmic pool into lateral plasma membrane (Krutovskikh et al., 1994).

It has been demonstrated that the expression of liver-specific cx32 is inhibited in experimental rat hepatocarcinogenic models at initial stages of tumor formation, and this could be considered as a good marker of early preneoplastic liver lesions (Neveu et al., 1990; Krutovskikh et al., 1991). Moreover, it was shown that altered expression of cx32 in liver preneoplastic focal lesions is different from other commonly used histochemical markers, such as GST-p or γGT; only some of GST-p–positive foci induced in rat liver by carcinogens possess cx32-negative phenotype. Decreased or fully abrogated expression of cx32 as a marker of preneoplastic lesions biologically appears to be more relevant to hepatocarcinogenesis, compared to usual histochemical enzyme markers. This gap junction protein was also found to be a good marker of liver lesions developed in rats carrying a sequence of SV40 large T-antigen oncogene under albumin gene promoter as a transgene, while none of the other above-mentioned markers were informative in this model of hepatocarcinogenesis (Hully et al., 1994).

Another widely observed GJIC disorder in carcinogenesis is found in those cells exposed to promoting agents. The role of functional disorders of GJIC in tumor promotion was postulated and demonstrated in different studies first in vitro (Murray and Fitzgerald, 1979; Yotti et al., 1979; Enmoto et al., 1981) and then in vivo (Tateno et al., 1994; Krutovskikh et al., 1995a,b).

Surprisingly little is known concerning the precise mechanisms of GJIC inhibition in cell culture by tumor promoters. Being nonmutagenic, most tumor promoters exert reversible effects on cell coupling, mostly on the level of connexin protein assembly into gap junctions, without essential changes of its expression (Asamoto et al., 1991). Thus, it was shown that TPA-induced inhibition of communication is due to redistribution of connexin proteins between cytoplasmic and lateral membrane pools, either by withdrawing already functionally established gap junctions from lateral membrane inside cytoplasm (Berthoud et al., 1992) or by abrogation of assembling of nascent connexin proteins into new gap junctions (Lampe, 1994). However, the total amount of connexin proteins in cells usually remains unchanged after TPA exposure. It is worth mentioning that a similar process of intensive intracytoplasmic accumulation of connexin proteins is found in tumor promoter–treated hepatocytes in vivo (Krutovskikh et al., 1995) as well as in human and some rat liver tumors (Krutovskikh et al., 1994; Omori et al., 1996), suggesting the importance of this particular mechanism of GJIC disorder during tumor progression. However, due to the general lack of knowledge concerning basic mechanisms of connexin trafficking into lateral membrane, as well as the limited number of observations, it is not possible to draw a general conclusion about the role of these phenomena in carcinogenesis.

While 60% of nongenotoxic carcinogens suppress cell coupling, only 30% of genotoxic carcinogens inhibit GJIC in short-term dye-coupling assays. Genotoxic carcinogens' inhibition of cell communication cannot be explained by their direct DNA reactivity with connexin genes, since the effect appears to be too rapid to be accounted for by genetic changes. To conclude, GJIC is specifically inhibited during different stages of carcinogenesis, and nongenotoxic mechanisms are mainly responsible for this.

III. EVIDENCE OF TUMOR SUPPRESSION BY CONNEXIN GENES

The putative tumor-suppressive role of connexins was suggested most convincingly on the basis of studies with successful conversion of malignant cell phenotypes into nontumorigenic ones by transfection of connexin genes. Thus, rat glioma cells (Naus et al., 1992) and chemically transformed mouse fibroblasts (Rose et al., 1993) transfected with the cx43 gene, unlike their parental cells, did not produce tumors in nude mice. Similarly, human liver tumor cells transfected with the cx32 gene showed reduced tumor growth (Eghbali et al., 1991). The subtractive hybridization approach also revealed the connexin26 gene as functionally silent in human breast cancer, suggesting its putative tumor-suppressive role (Lee et al., 1991). Apparently different connexins are playing differing roles in growth regulation and tumor suppression. Thus, when HeLa cells (which do not express detectable levels of any connexin genes examined) are transfected with cx26, cx40, cx43, and cx32 genes, the transfected HeLa cells showed an increased level of GJIC. However, only those transfectants with the cx26 gene (which is a cervix-specific gene) grew more slowly in vitro and lost their tumorigenicity (Mesnil et al., 1995).

While it is becoming clearer that connexins play an important role in cell growth control, how connexins or GJIC regulate the cell cycle is not known. The earliest hypothesis concerning the role of GJIC in growth control suggested that gap junctions provide a conduit between cells for the dispersion of a diffusible intracellular factor (i.e., negative growth regulator) to control cell growth (Loewenstein and Kanno, 1966). The GJIC field has not advanced significantly from this original hypothesis so far. However, in transformed cells transfected with cx43, changes in the expression of genes involved in regulating the cell cycle, namely, a decrease in cyclins A, D1, and D2 and cyclin-dependent kinases CDK5 and CDK6, were recently reported (Chen et al., 1995). These findings support the notion that GJIC can affect cell cycle-related gene expression, but again, nothing is known about the mechanisms of that phenomenon.

IV. METHODS TO DETECT AND QUANTIFY GJIC

There are a number of ways to estimate the degree of GJIC of cultured cells, of which the dye-transfer (DT) and metabolic-cooperation (MC) methods are the most common.

A. Dye-Transfer Assay

The simplest and most widespread method used to study GJIC involves visualizing intercellular communication by means of introduction into the cytoplasm artificial tracer molecules, usually a fluorescent dye, followed by observation on the dye as it spreads through a cellular monolayer. There are two principal ways to deliver trace dye molecules into the cell: by microinjection into a single cell or via mechanical damage of the cellular membrane of cell groups. The first method is more precise and suitable for further quantitative estimation of communication, but it requires expensive microinjection equipment (Fitzgerald et al., 1983). The other method, called "scrape loading," in which the dye is introduced into a group of cells through a freshly made scrape of the cell monolayer with a blade (El-Fouly et al., 1987), is less quantitative, but much simpler and easier. No particular equipment is needed to scrape load. Even an epifluorescent attachment for the microscope is not needed for scrape loading, since the presence of fluorescent dye in cells can be revealed by photo-oxidation of diaminodenzidine to form a permanent stain visible by light microscopy (Lübbe and Albus, 1989). The extent of GJIC studied by this method could be estimated by different means: manually counting the number of fluorescent cells, measuring the distance of dye travel from the scrape line, or quantifying the amount of cellular dye uptake (McKarns and Doolittle, 1992). Alternatively, flow cytometry of scrape-loaded cells could be used for subsequent quantitative measure of intercellular communication (Kavanagh et al., 1987).

The disadvantages of both microinjection and scrape-loading DT assays are (a) the traumatization of cells during initial intracytoplasmic introduction of dye into donor cells and (b) the difficulty of measuring dye-transfer kinetics. Both of these factors may lead to underestimation of the effect of chemicals. Of the two techniques, the microinjection dye-transfer technique is more widely used; more than 70 substances have been studied in this assay (Budunova and Williams, 1994).

Another type of fluorescent dye-transfer assay of cell coupling is the fluorescence-recovery/redistribution after photobleaching (gap FRAP analysis). In this procedure, cells are exposed to 6-carboxyfluorescein diacetate in culture media. All cells in the population are internally labeled by this stain, which is hydrolyzed intracellularly to 6-carboxyfluorescein, a hydrophilic fluorescein derivative that is not membrane-diffusible and is therefore retained in the cell. Any labeled cell may then be photobleached by a laser beam whose width is approximately equal to the cell diameter or by a series of laser pulses with a diameter of about 1 μm. After photobleaching, the bleached dye molecules from one cell and the unbleached dye molecules from an adjacent, contacting cell may be redistributed through gap junctions. Monitoring the redistribution of these labeled reporter molecules as a function of time results in a single exponential recovery curve that yields a rate constant for dye transport. The principal advantages of this method are the atraumatic delivery of the measurable partitioning agent, the possibility of reexperimentation on the same series of cells, and the continuous monitoring of the communication pattern. A disadvantage is the expensive special equipment required to perform this assay (Wade et al., 1986).

The tracer molecules used for any dye-transfer assay of GJIC measurement must be of low molecular weight. Usually, Lucifer Yellow molecules produce satisfactory results, but in certain cases, other dyes, such as neurobiotine, have been successfully used. Any artificial dye used for measuring GJIC is a foreign material for gap junctions with properties often rather different from the natural physiological messengers circulating through gap junctions. Therefore, artificial tracers may give only approximately correct data about physiological messengers in GJIC. Because of this limitation, some natural tracers for gap junctions—e.g., radioactive labeled uridine, other nucleotides, or their metabolites—could be a better choice.

B. Metabolic-Cooperation Assay

Until recently, the metabolic-cooperation assay to determine GJIC ability was used intensively. The principle of this assay is based on cocultivation of genetically deficient cells with their metabolically competent counterparts. The latter ones can produce toxic metabolites from nontoxic chemical precursors present in the culture media. Such toxic metabolites may pass through gap junctions into metabolically incompetent cells which are unable to produce this toxin by themselves. Therefore, these metabolically incompetent cells eventually could be killed as is expected for metabolically competent ones ("kiss of death"). The most common pair of cells used in metabolic-cooperation studies are Chinese hamster V79 lung fibroblasts (more than 200 out of 250 substances tested by metabolic cooperation were examined in the V79 assays). If the hypoxantine phosphoribosyl transferase–deficient ($HPRT^-$) clones of these cells are cocultivated with their wild-type ($HPRT^+$) counterparts in the presence of 6-thioguanine (the nontoxic precursor), the $HPRT^+$ cells phosphorylate the synthetic purine to a toxic metabolite, which is then transferred, presumably through gap junctions, to $HPRT^-$ cells, resulting in cell deaths (Pitts and Simms, 1977). An interesting version of the Chinese hamster V79 cell assay to detect GJIC is to use endpoints other than cytotoxicity, i.e., a sister-chromatid exchange and gene mutation at the HPRT locus, and has been successfully used to detect the inhibitory effect of TPA and cigarette smoking condensate (Jongen et al., 1986).

Since metabolic cooperation is naturally occurring in tissues so that some metabolically incompetent cells can be compensated for by neighboring cells, this assay may be a natural way to measure GJIC in cell culture. Results from screening of chemicals with Chinese hamster V79 cell metabolic-cooperation assay provide satisfactory results—up to 75% of correlation with in vivo carcinogenicity, according to some studies (Elmore et al., 1987).

Efforts to standardize the V79 MC system have been done. Thus, in an interlaboratory validation study to evaluate the usefulness of this assay to predict the tumor-promoting activity of 15 selected chemicals of different structures, 9 were detected as positive in all three participating laboratories, and for 23 chemicals tested in two labs, there was agreement on 16 chemicals. With the exception of the peroxides and alkalis, the metabolic-cooperation data were in general agreement with the in vivo data (Bohrman et al., 1988).

Another metabolic cooperation system was described for two human fibroblast cell lines, each deficient in one of two consecutive enzymes involved in citrulline incorporation into proteins, i.e., arigninosuccinate synthetase (ASS^-) and argininosuccinate lyase (ASL^-). Normal human fibroblasts can convert [^{14}C]citrulline into arginine, which will then be incorporated into protein. Both of the deficient cell lines show rates of incorporation of [^{14}C]citrulline of $<1\%$ that in normal cells, whereas cocultures show the level of incorporation approaching that of normal cells. The essential component of the coculture system is the intercellular transfer of argininosuccinate from ASL^- cells to ASS^- cells (Davidson et al., 1980).

The principal drawback of this version of the method is that an incorporation of citrulline could be blocked at any point during the following processes: from its transport into the cells, conversion to argininosuccinate, passage through gap junctions, conversion into arginine, and incorporation of arginine into protein. Therefore, in order to demonstrate that test agents specifically block GJIC, it is necessary to show that none of the other steps in the process of citrulline incorporation are affected. The advantage of this version of the metabolic-cooperation technique is that its endpoint is not based on cell toxicity, and therefore a wider range of doses could be applied for testing chemicals.

Despite the generally satisfactory results obtained by metabolic-cooperation assays, they have several drawbacks. The method is rather sensitive to many parameters of cell cultivation and, therefore, requires thorough controlling of experimental conditions. Since the endpoint of some versions of the MC assay is cell toxicity, compounds must be used at nontoxic concentrations. The effect of chemicals on cell coupling in this assay can be detected only by statistical techniques. The metabolic-cooperation method is unsuitable for examining the onset, duration, or reversal inhibition, because the final results of the assay are not available until 7 days after its initiation. And, most importantly, the unique pair of cells used in this method are common target cells for the majority of chemical carcinogens.

Despite the great differences between the DT and MC methods of studying GJIC, both yielded very similar data; up to 74% concordance between the methods was reported (Budunova and Williams, 1994). In certain cases the DT and MC methods are complementary. Recently DT is becoming more widely preferred, mostly due to its flexibility in being applied to a wide variety of target cells and to in vivo situations.

V. CHOICE OF TARGET CELLS FOR GJIC ASSAY

The sensitivity and specificity of the GJIC assay in detecting carcinogens depends to a great extent on the type of cells used. They must preferably fulfill the following criteria: to be established as a stable cell line, to communicate well, and to contain metabolic and phenotypic capabilities of the cells of their original tissue. The latter is the most important and at the same time the most difficult to achieve. Initial GJ studies by dye-transfer assay have been based on the use of murine fibroblasts, which were suitable cells to see the GJIC inhibitory effect of many well-known tumor-promoting agents. However, it soon became evident that in order to detect nongenotoxic carcinogenic activity of unknown agents by a dye-transfer assay, the cells most resembling carcinogen-specific target tissue need to be used. The fact that about 80% of already known nongenotoxic carcinogens are rodent liver-specific carcinogens made rat hepatocytes the cells most appropriate to use as a model for this assay. Experimental studies carried out with different rat liver cell lines confirmed this choice (Williams, 1980; Tailing et al., 1982; Mesnil et al., 1993).

However, advances in gap junction molecular biology revealed an essential difference between hepatocytes in vivo and their corresponding established cell lines in terms of the pattern of their connexin expression as well as the mechanisms responsible for their functional regulation. If in vivo hepatocytes co-express cx32 and cx26, almost all available rat liver cell lines express in vitro cx43 (Stutenkemper et al., 1992). It is well established that cx26, cx32, and cx43 are regulated differently, notably through specific phosphorylation pathways (Musil et al., 1990). Thus, TPA, a strong tumor promoter in skin and a strong protein kinase C inducer, strongly inhibits GJIC in many different cell lines that express cx43, but not in primary rat hepatocyte culture, which continue to express cx32 and cx26 (Saez et al., 1991).

Thus, primary cultures of rodent hepatocytes has advantages, since primary hepatocytes usually continue to express the same set of in vivo connexins for a certain period of time after initiation of culture. The strain, species, and target cell specificity of the inhibitory effects on GJIC of liver tumor promoters on primary culture of rodent hepatocytes was found to correlate with their in vivo tumor-promotion activity (Klauning et al., 1987).

Another example of the importance of target cell specificity in detecting the GJIC inhibitory effect is 2,3,7,8,-tetrachlordibenzo-*p*-dioxin (TCDD). This strong in vivo hepatospecific tumor promoter was repeatedly reported as negative in in vitro GJIC tests, when fibroblasts were used as model cells (Lincoln et al., 1987). However, when tested on primary rat hepatocyte culture, TCDD did show a downregulation effect on GJIC (Baker et al., 1995). Recent in vivo GJIC functional studies carried out on rat liver treated by different hepatospecific tumor promoters essentially corroborated an earlier finding of inhibitory effects of these compounds on cultured hepatocytes (Sugie et al., 1987; Tateno et al., 1994; Krutovskikh et al., 1995b).

Subsequent in vivo study of putative mechanisms of GJIC inhibition revealed differential, sometimes even opposite effects of several tested promoters on different connexin species expressed in the liver. Thus, if all four tested hepatopromoters (PB, PCBs, DDT, and CF) demonstrated intensive disactivation of the principal rat liver cx32 function, mostly by means of its massive translocation from lateral membrane of hepatocyte into cytoplasm, the local induction of another rat liver connexin (i.e., cx26) was often observed in those parts of the liver lobule where cx32 localization was most affected. Nevertheless, overall evaluation of GJIC function by dye microinjection into the liver indeed revealed a decreased overall level of communication (Krutovskikh et al., 1995a).

While improved culture methods have indeed helped cultured cells mimic in vivo cells, the maintenance of the expression of specific connexin gene species for long culture times appears to be difficult. The stabilization of particular connexin expression in in vitro conditions by means of stable gene transfection may be one way; such an approach has been performed with HeLa cells transfected with various types of conexin genes (Mazzoleni et al., 1996). It is, however, still not possible to mimic connexin gene regulation, since the cells can be transfected only with coding, but not yet regulatory, regions of the connexin genes.

VI. PERSPECTIVES

Since all short-term tests for carcinogens are based on only one particular mechanism among many associated with different aspects of tumorigenesis, none is expected to possess 100% concordance in detecting and predicting chemical carcinogenicity. In other words, short-term tests cannot fully replace long-term carcinogenic bioassays. There are simply too many presently known different mechanisms of carcinogenesis. Therefore, the observed positive correlation between the carcinogenicity of substances and their capacity to inhibit in vitro cell coupling (~70%) is a rather satisfactory outcome.

There are several limitations of GJIC assays as potential routine screening tests for chemical carcinogenicity. Thus, positive correlation between inhibition of communication and carcinogenicity is rather artificial, since in an overwhelming majority of cases this positive correlation has been found for compounds with already known carcinogenicity. At the same time, the animal carcinogenicity of many substances with known GJIC inhibitory activity in vitro remains to be determined. Conversely, very few noncarcinogens have been examined by the GJIC assay. There are few cases so far in which the compound was first detected as positive in terms of GJIC inhibition in vitro and then confirmed as potential animal carcinogen in long-term study in vivo. Among such examples are polybrominated biphenyls (Trosko et al., 1980; Jensen et al., 1982), chlordane and heptachlor (Tailing et al., 1982; Williams and Numoto, 1984), and pyretroid insecticide fenvalerate (Flodstrom et al., 1988).

The intercellular communication network can only be studied properly if target cells are in a communication network. From this point of view, one of the promising recent advances in the field is a tissue slice technique, which has already been successfully applied in several laboratories to detect the GJIC modulating effect of tumor-promoting and -preventing compounds (Tateno et al., 1994; Krutovskikh et al., 1995a,b). The principal advantage of this technique is its nearly full metabolic and biological matching with the in vivo situation. Another unique feature of the tissue slice method is the possibility of studying the effect of chemicals on intercellular communication in human tissue, since surgically removed human tissue could be exposed to test compounds in culture medium. The only disadvantage of this version of the GJIC assay is the limited number of tissues that can be used (i.e., mostly liver).

Another potentially powerful alternative to GJIC in vitro methods to detect and predict carcinogenicity is to use transgenic animals with null mutation or dominant-negative mutation of connexin genes. Such transgenic animals are predicted to be intrinsically highly susceptible to carcinogenesis, and, therefore, even very weak carcinogenic chemicals might cause cancer in such transgenic animals. Such approaches are being taken by several laboratories, including ours.

As discussed in this chapter, the mechanism-based information on chemicals is an important determination in predicting their carcinogenicity. Many carcinogens, especially those with nongenotoxic activities, induce tumors by various cell- and tissue-specific mechanisms. In addition, some may even be species-specific. Therefore, we suggest that fixed protocols for mechanism-based short-term tests be selected with a large degree of flexibility. Even if the experimental results are generated from nonstandardized protocols, reliable mechanistic information should be considered for predicting and characterizing carcinogenic hazard. When a given chemical is studied based on a working hypothesis with flexible experimental designs, results derived from such a study should be taken as seriously as, or sometimes even more seriously than results obtained from a test carried out under a fixed protocol.

REFERENCES

Asamoto M, Oyamada M, Ei Aoumari, Gros D, Yamasaki H. Molecular mechanisms of TPA-mediated inhibition of gap junctional intercellular communication: evidence for action on the assembly or function but not the expression of connexin 43 in rat liver epithelial cells. Mol Carcinogenesis 1991; 4:322–327.

Backer DL, Evans WH, Green CR, Warner A. Functional analysis of amino acid sequences in connexin43 involved in intercellular communication through gap junctions. J Cell Sci 1995; 108:1455–1467.

Baker TK, Kwaitkowski AP, Madhukar BV, Klaunig JE. Inhibition of gap junction intercellular communication by 2,3,7,9,-tetrachlordibenzo-p-dioxin (TCDD) in rat hepatocytes. Carcinogenesis 1995; 16:2321–2326.

Barrett JC, Kakunaga T, Kuroki T, Neubert D, Trosko JE, Vasiliev JM, Williams GM, Yamasaki H. In vitro assays that may be predictive of tumor promoting agents. In: Long-term and short-term assays for carcinogens: a critical appraisal. Lyon: IARC, 1989:287–303.

Berthoud VM, Ledbetter MLS, Hertzberg EL, Saez JC. Connexin43 in MDCK cells: regulation by tumor promoting phorbol ester and calcium. Eur J Cell Biol 1992; 57:40–50.

Beyer EC, Paul DL, Goodenough DA. Connexin family of gap junction proteins. J Membrane Biol 1990; 116:187–194.

Bohrman JS, Burg JR, Elmore E, Gulati DK, Barfknecht TR, Niemeier RW, Dames BL, Toraason M, Lamgenbach R. Interlaboratory studies with the Chinese hamster V79 cell metabolic cooperation assay to detect tumor-promoting agents. Environ Mol Mutagenesis 1988; 12:33–51.

Bruzzone R, White TW, Scherer SS, Fischbeck KH, Paul DL. Null mutations of connexin32 in patients with X-linked Charcot-Marie-Tooth disease. Neuron 1994; 13:1253–1260.

Budunova IV. Alteration of gap junctional intercellular communication during carcinogenesis. Cancer J 1994; 7:228–237.

Budunova IV, Williams GM. Cell culture assays for chemicals with tumor-promoting or tumor inhibiting activity based on the modulation of intercellular communication. Cell Biol Toxicol 1994; 10:71–116.

Chen SC, Pelletier DB, Ao P, Boynton AL. Connexin43 reverses the phenotype of transformed cells and alter their expression of cyclin/cyclin dependent kinases. Cell Growth Diff 1995; 6:681–690.

Davidson JS, Baumgarten I, Harley EH. Metabolic cooperation between argininosuccinate lyase-deficient fibroblasts. Exp Cell Res 1980; 150:367–378.

Eghbali B, Kessler JA, Reid LM, Roy C, Spray DC. Involvement of gap junctions in tumorigenesis: transfection of tumor cells with connexin 32 c DNA retards growth in vivo. Proc Natl Acad Sci USA 1991; 88:10701–10705.

El-Fouly MH, Trosko JE, Chang CC. Scrape loading and dye-transfer. A rapid and simple technique to study gap junctional intercellular communication. Exp Cell Res 1987; 168:422–430.

Elmore E, Milman HA, Wyatt GP. Applications of the Chinese hamster V79 metabolic cooperation assay in toxicology. In: Milman HA, Elmore E, eds. Advances in Modern Environmental Toxicology, Vol. 14: Biochemical Mechanisms and Regulation of Intercellular Communication. New York: Princeton Scientific Publishing, 1987:165–292.

Enomoto T, Sasaki Y, Shiba Y, Kanno Y, Yamasaki H. Tumor promoters case a rapid and reversible inhibition of the formation and maintenance of electrical cell coupling in culture. Proc Natl Acad Sci USA 1981; 78:5628–5632.

Fitzgerald DJ, Knowles SE, Ballard FJ, Murray AW. Rapid and reversible inhibition of junctional communication by tumor promoters in mouse cell line. Cancer Res 1983; 43:3614–3618.

Flodstrom S, Warngard L, Ljungquist S, Ahlborg UG. Inhibition of metabolic cooperation in vitro and enhancement of enzyme altered foci incidence in rat liver by pyrethroid insecticide fenvalerate. Arch Toxicol 1988; 61:218–223.

Furshpan EJ, Potter DD. Transmission at the giant moto synapses of the crayfish. J Physiol 1959; 145:289–325.

Hertig CM, Eppenberger-Eberhardt M, Koch S, Eppenberger HM. N-Cadherin in adult rat cardiomyocytes in culture. I. Functional role of N-cadherin and impairment of cell-cell contact by a truncated N-cadherin mutant. J Cell Sci 1996; 109:12–10.

Hertzberg EL, Disher RM, Tiller AA, Zhou Y, Cook RG. Topology of the Mr 27,000 liver gap junction protein. J Biol Chem 1988; 263:19105–19111.

Holder JW, Elmore E, Barrett JC. Gap junction function and cancer. Cancer Res 1993; 53:3475–3485.

Hotz-Wagenblatt A, Shalloway D. Gap junctional communication and neoplastic transformation. Crit Rev Oncogen 1993; 4:541–558.

Hully GR, Su Y, Lohse JK, Griep AE, Sattler CA, Haas MJ, Dragan Y, Peterson J, Neveu M, Pitot H. Transgenic hepatocarcinogenesis in the rat. Am J Pathol 1994; 145:384–397.

Jensen RK, Sleight SD, Goodman JI, Aust AD, Trosko JE. Polybrominated byphenyls as promoters in experimental hepatocarcinogenesis. Carcinogenesis 1982; 3:1183–1186.

Jongen WMF, Fitzgerald M, Asamoto M, Piccoli C, Slaga TJ, Gros D, Takeichi M, Yamasaki H. Regulation of connexin43-mediated gap junctional intercellular communication by Ca++ in mouse epidermal cells is controlled by E-cadherin. J Cell Biol 1991; 114:545–555.

Jongen WM, Hakkert BC, van de Poll ML. Inhibitory effect of the phorbol ester TPA and cigarette smoking condensate on the mutagenicity of benzo[a]pyrene in a co-cultivation system. Mutat Res 1986; 159:133–138.

Kanno Y, Matsui Y. Cellular uncoupling in cancerous stomach epithelium. Nature 1968; 218:775–776.

Kavanagh TJ, Martin GM, El-Fopuly MH, Trosko JE, Chang Ch-Ch, Rabinovich PS. Flow cytometry and scrape loading/dye transfer as a rapid quantitative measure of intercellular communication in vitro. Cancer Res 1987; 47:6046–6051.

Klaunig JE, Ruch RJ. Role of inhibition of intercellular communication in carcinogenesis. Lab Invest 1990; 62:135–144.

Klaunig JE, Ruch RJ. Stain and species effects on the inhibition of hepatocyte intracellular communication by liver tumor promoters. Cancer Lett 1987; 36: 161–168.

Krutovskikh VA, Mazzoleni G, Mironov N, Omori Y, Aguelon A-M, Mesnil M,

Berger F, Partensky C, Yamasaki H. Altered homologous and heterologous gap junctional intercellular communication in primary human liver tumors associated with aberrant protein localization but not gene mutation of connexin 32. Int J Cancer 1994; 56:87–94.
Krutovskikh VA, Mesnil M, Mazzoleni G, Yamasaki H. Inhibition of rat gap junction intercellular communication by tumor promoting agents in vivo; association with aberrant localization of connexin proteins. Lab Invest 1995a; 72(5): 571–577.
Krutovskikh VA, Mironov NM, Yamasaki H. Human connexin37 is polymorphic but not mutated in tumors. Carcinogenesis (in press).
Krutovskikh VA, Oyamada M, Yamasaki H. Sequential changes of gap junctional intercellular communications during multistage rat liver carcinogenesis: direct measurement of communication in vivo. Carcinogenesis 1991; 12:1701–1706.
Krutovskikh VA, Yamasaki H. Ex-vivo dye transfer assay as an approach to study gap junctional intercellular communication disorders in hepatocarcinogenesis. In: Kanno Y, et al., eds. Progress in Cell Research. Vol. 4. Elsevier Science B.V., 1995b:93–97.
Kumar NM, Gilula NB. The gap junction communication channel. Cell 1996; 84: 381–388.
Lampe PD. Analyzing phorbol ester effects on gap junctional communication: a dramatic inhibition of assembly. J Cell Biol 1994; 127:1895–1905.
Lee WS, Tomasetto C, Sager R. Positive selection of candidate tumor-suppressor genes by subtractive hybridization. Proc Natl Acad Sci USA 1991; 88:2825–2829.
Leube RE. The topogenic fate of the polytopic transmembrane proteins, synaptophysin and connexin, is determined by their membrane-spanning domains. J Cell Sci 1995; 108:883–894.
Lincoln DW, Kampcik SJ, Gierthy JF. 2,3,7,8,-Tetrachlordibenzo-p-dioxin (TCDD) does not inhibit intercellular communication in Chinese hamster V79 cells. Carcinogenesis 1987; 8:1817–1820.
Loewenstein WR, Kanno Y. Intercellular communication and the control of the tissue growth; lack of communication between cancer cells. Nature 1966; 209: 1248–1249.
Lübbe J, Albus K. The postnatal development of layer Vl pyramidal neurons in the cat's striate cortex, as visualized by intracellular Lucifer Yellow injections in aldehyde-fixed tissue. Dev Brain Res 1989; 45:29–38.
McKarns SC, Doolittle DJ. Limitations of scrape loading/dye transfer technique to qualify inhibition of gap junctional intercellular communication. Cell Biol Toxicol 1992; 8:89–103.
Malcolm AR, Mills LJ, Schulz KM, Rupp HL, Madhukar BV, Trosko JE. Effects of gap junctional communication between V79 cells measured by metabolic cooperation and fluorescerecovery after photobleaching. The Toxicologist. Abstracts of the 31th Annual Meeting. 1992; 12:377.
Mazzoleni G, Camplani A, Telo P, Pozzi A, Tanganelli S, Elfgang CM, Willecke K, Ragnotti C. Effect of tumor-promoting and anti-promoting chemicals on

the viability and junctional coupling of human HeLa cells transfected with DNAs coding for various murine connexin proteins. Comp Biochem Physiol 1996; 113:247–256.

Mesnil M, Krutovskikh VA, Piccoli C, Elfgang C, Traub O, Willecke K, Yamasaki H. Negative growth control of HeLa cells by connexin genes: connexin species specificity. Cancer Res 1995; 55:629–639.

Mesnil M, Piccoli C, Yamasaki H. An improved long-term culture of rat hepatocytes to detect liver tumor-promoting agents: results with phenobarbital. Eur J Pharmacol 1993; 248:59–66.

Mesnil M, Yamasaki H. Cell-cell communication and growth control of normal and cancer cells: evidence and hypothesis. Mol Carcinogenesis 1993; 7:14–17.

Milks LC, Kumar NM, Houghten R, Unwin N, Gilula NB. Topology of the 32-kd liver gap junction protein determined by site-directed antibody localizations. EMBO J 1988; 7:2967–2975.

Mironov NM, Aguelon AM, Potapova GI, Omori Y, Gorbunov OV, Klimenkov AA, Yamasaki H. Alterations of $(CA)_n$ DNA repeats and tumor suppressor genes in human gastric cancer. Cancer Res 1994; 54:41–44.

Munari-Silem Y, Audebet C, Rouset B. Hormonal control of cell to cell communication: regulation by thyrotropin of the gap junction-mediated dye transfer between thyroid cells. Endocrinology 1991; 128:3299–3309.

Musil LS, Cunningham BA, Edelman GM, Goodenough DA. Differential phosphorylation of the gap junction protein connexin43 in junctional communication-competent and -difficient cell lines. J Cell Biol 1990; 111:2077–2088.

Musil LS, Goodenough DA. Biochemical analysis of connexin43 intercellular transport, phosphorylation and assembly into gap junctional plaques. J Cell Biol 1991; 115:1357–1374.

Murray AW, Fitzgerald DJ. Tumor promoters inhibit metabolic cooperation in cocultures of epidermal and 3T3 cells. Biochem Biophys Res Commun 1979; 91:395–401.

Naus CC, Elisevich K, Zhu D, Belliveau DJ, Del Maestro RF. In vitro growth of C6 glioma cells transfected with connexin43 cDNA. Cancer Res 1992; 52:4208–4213.

Neveu M, Hully J, Paul D, Pitot H. Reversible alterations in the expression of the gap junctional protein connexin32 during tumor promotion in rat liver and its role during cell proliferation. Cancer Commun 1990; 2:21–31.

Omori Y, Krutovskikh VA, Tsuda H, Yamasaki H. Connexin32 mutation in a chemically-induced rat liver tumor. Carcinogenesis 1996; 17:2077–2080.

Omori Y, Mesnil M, Yamasaki H. Connexin 32 mutations from X-linked Charcot-Marie-Tooth disease patients; functional defects and dominant-negative effects. Mol Biol Cell 1996; 7:907–916.

Paul DL. New functions of gap junctions. Curr Opin Cell Biol 1995; 7:665–672.

Pitts JD, Simms JW. Permeability of junctions between animal cells. Exp Cell Res 1977; 104:153–163.

Reaume AG, De Sousa PA, Kulkarni S, Langille BL, Zhu D, Davies TC, Juneja SC, Kidder GM, Rossant J. Cardiac malformation in neonatal mice lacking connexin43. Science 1995; 267:1831–1834.

Rose B, Mehta PP, Loewenstein WR. Gap junction protein gene suppresses tumorigenicity. Carcinogenesis 1993; 14:1073–1075.

Ruch RJ. The role of gap junctional intercellular communication in neoplasia. Anal Clin Lab Sci 1994; 24:216–231.

Saez JC, Connor JA, Spray DC, Bannett MVL. Hepatocyte gap junctions are permeable to the second messenger, inositol 1,4,5-triphosphate, and to calcium ions. Proc Natl Acad Sci USA 1989; 86:2708–2712.

Saez JC, Spray DC, Herzburg EL. Gap junctions: biochemical properties and functional regulation under physiological and toxicological conditions. In vitro toxicology. J Mol Cell Toxicol 1990; 3:69–86.

Stutenkemper R, Geisse S, Schwarz HJ, Look J, Traub O, Nicholson BJ, Willecke K. The hepatocyte-specific phenotype of murine liver cells correlates with high expression of connexin32 and connexin26 but very low expression of connexin43. Exp Cell Res 1992; 201:43–54.

Sugie S, Mori H, Takahashi M. Effect of in vivo exposure to the liver tumor promoters phenobarbital or DDT on the gap junctions of rat hepatocytes: a quantitative freeze-fracture analysis. Carcinogenesis 1987; 8:45–51.

Sweirenga SH, Yamasaki H. Performance of tests for cell transformation and gap junction intercellular communication for detecting nongenotoxic carcinogenic activity. In: Vainio H, Magee P, McGregor D, McMikchel A, eds. Mechanisms of Carcinogenesis in Risk Identification. Lyon: IARC, 1992:165–193.

Tateno C, Ito S, Tanaka M, Oyamada M, Yoshitake A. Effect of DDT on hepatic gap junctional intercellular communication in rats. Carcinogenesis 1994; 15: 517–521.

Telang S, Tong C, Williams GM. Epigenetic membrane effects of possible tumor promoting type on cultured liver cells by the non-genotoxic organochlorine pesticides chlordane and heptachlor. Carcinogenesis 1982; 3:1175–1178.

Trosko JE, Dawson B, Chang C-C. PBB inhibits metabolic cooperation in Chinese hamster cells in vitro: its potential as tumor promoter. Environ Health Perspect 1981; 37:179–182.

Trosko JE, Goodman JI. Intercellular communication may facilitate apoptosis: implication for tumor promotion. Mol Carcinogen 1994; 11:8–12.

Trosko JE, Madhukar BV, Chang C-C. Endogenous and exogenous modulation of gap junctional intercellular communication: toxicological and pharmacological implications. Life Sci 1993; 53:1–19.

Wade MH, Trosko JE, Shindler M. A fluorescence photobleaching assay of gap junction-mediated communication between human cells. Science 1986; 232: 429–432.

Williams GM. Classification of genotoxic and epigenetic hepatocarcinogens using liver culture assays. Ann NY Acad Sci 1980; 349:273–282.

Williams GM, Numoto S. Promotion of mouse liver neoplasms by the organochlorine pesticides chlordane and heptachlor in comparison to dichlordiphenyltrichlorethane. Carcinogenesis 1984; 5:1689–1696.

Yamasaki H. Gap junctional intercellular communication and carcinogenesis. Carcinogenesis 1990; 7:1051–1058.

Yamasaki H. Role of disrupted gap junctional intercellular communication in detection and characterization of carcinogens. Mutat Res 1996; 365:91–105.

Yamasaki H, Krutovskikh VA, Mesnil M, Columbano A, Tsuda H, Ito N. Gap junctional intercellular communication and cell proliferation during rat liver carcinogenesis. Environ Health Perspect 1993; 101(suppl 5):191–198.

Yamasaki H, Mesnil M, Omori Y, Mironov NM, Krutovskikh VA. Intercellular communication and carcinogenesis. Mutat Res 1995; 333:181–188.

Yotti LP, Chang CC, Trosko JE. Elimination of metabolic cooperation in Chinese hamster cells by tumor promoter. Science 1979; 206:1089–1091.

10

Predicting Chemical Carcinogenicity by In Vivo Biochemical Parameters

Kirk T. Kitchin and Janice L. Brown
U.S. Environmental Protection Agency, Research Triangle Park, North Carolina

I. INTRODUCTION

A. Predicting Chemical Carcinogenicity

Predicting chemical carcinogenicity is difficult because of the multiple mechanisms of carcinogenesis, the many stages through which carcinogenesis proceeds, and the remarkable species, sex, and organ specificity of the carcinogenic process. Initiation (the first mutational event), promotion (clonal expansion of the number of initiated cells), and progression (subsequent mutational events) are the three major temporal stages of carcinogenesis.

To date, most of the research in predicting chemical carcinogenicity has been limited to the initiation (mutational) stage of carcinogenesis (1–6). Since the scientific breakthrough of the development of the Ames test in 1975 (7), virtually hundreds of short-term tests have been developed to varying extents. Surprisingly, no individual short-term test or composite battery of genotoxic tests has emerged as superior to the original *Salmonella* mutation test (1).

This document has been reviewed in accordance with U.S. Environmental Protection Agency policy and approved for publication. Mention of trade names or commercial products does not constitute endorsement or recommendation for use.

This chapter is based on our published work (43,44). Reproduction permission has been granted.

In this chapter, some studies our laboratory conducted for the purpose of better predicting chemical carcinogenicity are presented. Our study design centered on the promotion stage of carcinogenesis, rather than the initiation stage. The two best studied experimental systems of promotion of carcinogenesis are mouse skin and rat liver.

Only in vivo tests offer the complete pharmacokinetic and pharmacodynamic response of the intact mammalian organism. Since pharmacokinetic factors such as absorption, distribution, metabolism, and excretion cannot be satisfactorily included in short-term in vitro tests, the results of these tests are difficult to extrapolate to whole animals or humans. Regulatory agencies such as the U.S. EPA and FDA require extrapolation of in vitro data before it is utilized for human risk assessment. Therefore, whole animals (adult female rats) were used as the test system. Rat liver was selected as the primary organ of study because of the knowledge base available in this model system of cancer promotion, the high blood flow the liver receives from the gastrointestinal tract after oral dosing, the enzymatic capabilities the liver possesses, and the large numbers of hepatocarcinogenic chemicals known (both with and without mutagenicity in the Ames test).

B. Biochemical Parameters

The materials and methods used to produce the data in Table 1 have been described previously (10). Briefly, female Sprague-Dawley rats (CD strain) were first orally dosed with various treatment chemicals 21 hours before sacrifice of the rats; the second dose was given 4 hours before sacrifice by decapitation. Liver tissue and blood was obtained from treated and control animals, and five biochemical assays were performed.

The five selected biochemical parameters were markers for promotion of carcinogenesis [hepatic ornithine decarboxylase activity (ODC) and hepatic cytochrome P-450 content (P450)], cell toxicity [serum alanine aminotransferase activity (ALT)], cell electrophile defense [hepatic reduced glutathione content (GSH)], and hepatic DNA damage (DD). The alkaline elution technique for measuring DNA damage detects both alkali-labile sites (e.g., base-free sites and phosphotriesters) as well as existing single-strand breaks in DNA. Only DD has been previously evaluated as a cancer predictor (8,9). In the DD assay alkaline pH conditions cause the DNA double helix to become single stranded and alkali-labile sites to be converted into DNA strand breaks. Polycarbonate filters physically retain the long undamaged DNA strands while the shorter damaged DNA strands flow through the 2 μm pores in the inert polycarbonate filter. The amount of DNA eluted resulting from these strand breaks as well as the amount of

DNA remaining on top of the filter is measured fluorometrically using a DNA fluorescent probe (Hoechst 33258).

Promotion of carcinogenesis has fewer biochemical or morphological markers than does carcinogenic initiation. Most promoters of carcinogenesis induce ODC activity in vivo. ODC is the first and rate-limiting enzyme of polyamine biosynthesis. ODC is involved with cell mitosis and proliferation. Many promoters of carcinogenesis induce hepatic cytochrome P-450 as well. Microsomal cytochrome P-450 is a mixed-function oxygenase which is known to metabolize many substrates and also to activate certain carcinogens in vivo. Alanine aminotransferase (ALT) catalyzes the transfer of the amino group of alanine to α-ketoglutarate to produce pyruvic and glutamic acids. The release of the hepatic enzyme ALT into the serum is indicative of liver cell damage and death. Reduced glutathione (GSH) is a cellular defense against electrophilic proximate carcinogens, which might damage tissue macromolecules such as DNA. Glutathione may detoxify electrophiles by direct nonenzymatic conjugation or via an enzymatic process involving the glutathione S-transferases. When the cellular content of glutathione is depleted, electrophiles may damage cellular macromolecules such as DNA at a greatly increased rate.

The work with biochemical parameters represented in this chapter contrasts with most previous work to identify cancer predictors in that it has (a) favored an in vivo approach, (b) used a whole animal system, (c) included four nongenotoxic as well as one genotoxic endpoints, (d) assumed carcinogenesis was caused by multiple causes, and (e) aimed at high-specificity cancer predictors. On the other hand, nearly all other cancer predictor research has (a) favored in vitro approaches, (b) used single-cell systems, (c) selected only genotoxic endpoints, (d) assumed carcinogenesis was caused only by mutation, and (e) aimed at high-sensitivity cancer predictors. Aiming at high sensitivity attempts to maximize the number of chemical carcinogens that test positive; aiming at high specificity attempts to maximize the number of noncarcinogens that test negative. Some high-specificity cancer predictors have the advantage of being able to be combined into composite cancer predictors with higher, improved predictive characteristics.

C. Interpretation of Results of the Biochemical Parameters

For genotoxic parameters such as DD, the basis for an interpretation of increased cancer risk is the somatic mutation theory of carcinogenesis. However, for the rest of the biochemical parameters the interpretation is less obvious. With this data set of 111 chemicals, only increases were observed with DD and ODC. For P450, ALT, and GSH both statistically significant increases and decreases in the biochemical parameters were observed. The interpretation of an increase in cancer risk was made for significant changes ($p < 0.05$) in the biochemical parameters only in certain directions (i.e., increases in DD, ODC, P450, or ALT or decreases in GSH). Any decreases in P450 or ALT or increase in GSH were not interpreted as indicating an increased risk of rodent cancer. This interpretation of the data is consistent with what is now known of the causes and mechanisms of chemical carcinogenesis.

These five individual biochemical parameters can be used either individually or in various combinations using logical "and" or "or" linkage to create composite cancer predictors of varying sensitivity and specificity. For a two-element composite cancer predictor connected with a logical "and" (e.g., ODC and P450), both ODC and P450 must be increased in order for the composite cancer predictor to be increased. For a two-element composite cancer predictor connected with a logical "or" (e.g., DD or P450 at low dose), the composite cancer predictor registers a positive signal if either of the two elements is positive. By grouping individual cancer predictors into composite cancer predictors in certain combinations, a better prediction may be obtained.

D. Comparision with Other Published Results

Rodent cancer bioassay results were principally from six sources and are individually cited in Table 1 (1–3,8,11,12). Structural alert (SA) data came from the work of Ashby et al. (3,4,13). Mutation in *Salmonella tymphimurium* (AMES) data were obtained from several sources (1–3,7,14–16) and are individually referenced in Table 1. The rate at which excess electrons attach to the test chemical dissolved in liquid cyclohexane (k_e) has been used as a cancer predictor. The k_e test is a measure of electrophilicity. In this chapter the k_e test and SA are considered to be genotoxic parameters. The k_e data were compiled from the Bakale group (17–21). The data on cell mutation in mouse lymphoma cells (MOLY), chromosome aberrations in Chinese hamster ovary cells (ABS), and the sister-chromitid exchange in Chinese hamster ovary cells (SCE) are from the National Toxicology Program (NTP) studies (1–5).

For determining operational characteristics presented in Tables 2 and 4–7, only complete data sets were used. Thus, the experimental sample size, or N, of the studies vary; for the biochemical parameters (DD, ODC, P450, ALT, GSH, CP, CT, and TS), N is 111 (49 carcinogens and 62 noncarcinogens), for AMES 109 (49 carcinogens and 60 noncarcinogens), for SA 83 (43 carcinogens and 40 noncarcinogens), for k_e 72 (34 carcinogens and 38 noncarcinogens), for mutation in *Salmonella* TA1537 (abbreviated

Table 1 Activities of Carcinogens and Noncarcinogens in Various Cancer Predictor Tests

CHEMICAL	DD	OCD	ALT	P450	GSH	AMES	Ames Ref.	TA1537	SA	Cancer Ref.
Carcinogens										
1,1,1,2-TETRACHLOROETHANE	–	–	–	–	–	neg	3	–	–	3
1,1,2,2,-TETRACHLOROETHANE	–	inc	–	–	–	neg	3	–	–	3
1,1-DICHLOROETHYLENE	–	inc	–	–	–	positive	7	inc		30
1,2-DIBROMOETHANE	inc	–	–	–	–	positive	2	–	inc	12
1,2-DIBROMO-3-CHLOROPROPANE	inc	inc	–	–	–	positive	3		inc	12
1,2-DICHLOROETHANE	inc	–	–	–	dec	positive	2	–	inc	12
1,2-DIMETHYLHYDRAZINE	inc	inc	–	–	–	neg	7		inc	8
1,3-DICHLOROPROPENE	inc	inc	inc	–	dec	positive	2	–		2
1,4-DIOXANE	inc	inc	–	inc	–	neg	3	–	–	12
11,-AMINOUNDECANOIC ACID	–	inc	–	inc	–	neg	3		–	3
2,3,7,8-TETRACHLORODIBENZO-*p*-DIOXIN	–	–	–	inc	–	neg	1	–	–	12
2,4,5-TRIMETHYLANILINE	–	inc	–	–	–	positive	3		inc	3
2,4,6-TRICHLOROPHENOL	–	inc	–	inc	–	neg	3	–	–	12
2,4-DIAMINOTOLUENE	–	inc	–	–	–	positive	3		inc	12
4,4′-OXYDIANILINE	–	inc	–	inc	dec	positive	3		inc	3
ACTINOMYCIN D	–	–	–	–	–	neg	24		-	11
AFLATOXIN B1	–	inc	inc	inc	inc	positive	7	inc	inc	12
AURAMINE O	inc	–	–	–	–	positive	16	–	inc	11
A-HEXACHLOROCYCLOHEXANE	–	inc	–	inc	–	neg	24		–	38
BENZO(A)PYRENE	–	–	–	–	inc	positive	7	inc	inc	12
BUTYLATED HYDROXYTOLUENE	inc	inc	–	inc	–	neg	3		–	22
CARBON TETRACHLORIDE	–	inc	inc	–	–	neg	23		inc	12

(*continued*)

Table 1 Continued

CHEMICAL	DD	OCD	ALT	P450	GSH	AMES	Ames Ref.	TA1537	SA	Cancer Ref.
CHLORENDIC ACID	–	inc	–	–	–	neg	2	–	–	4
CHLOROFORM	–	inc	inc	–	–	neg	27	–	inc	12
CLOFIBRATE	–	–	inc	–	–	neg	41		–	39
CUPFERRON	–	–	–	–	–	positive	3		inc	3
C.I. DISPERSE YELLOW 3	–	–	–	–	–	positive	2		inc	11
C.I. SOLVENT YELLOW 14	–	inc	inc	inc	–	positive	2		inc	2
DECABROMODIPHENYL OXIDE	–	–	–	–	–	neg	3	–	–	3
DICHLOROVOS	–	–	–	–	–	positive	3	–	inc	5
DIRECT BLACK 38	–	inc	–	–	–	positive	2	inc		11
DIRECT BLUE 6	–	–	–	–	–	pos/red	2	–		11
DIRECT BROWN 95	–	inc	–	–	–	pos/red	2	–		31
DISPERSE ORANGE 11	–	–	–	inc	–	positive	3		inc	11
H C BLUE 1	–	inc	–	–	–	positive	3		inc	3
HEXACHLOROBENZENE	–	inc	–	–	–	neg	32	–	–	12
HYDRAZOBENZENE	–	inc	–	–	–	positive	3		inc	3
KEPONE	–	inc	–	inc	–	neg	25		–	12
METHYLENE CHLORIDE	inc	inc	–	–	–	pos/neg	34	–	inc	34
MICHLER'S KETONE	inc	–	–	–	–	positive	3		inc	12
MIREX	–	inc	inc	–	–	neg	25		–	12
MONURON	–	inc	–	inc	-	neg	1	–	–	11
NAFENOPIN	–	inc	inc	inc	inc	neg	28			39
N-NITROSOPIPERIDINE	inc	inc	–	–	–	positive	37		inc	36
p,p'-DDE	–	inc	–	inc	dec	neg	3	–	–	38
p,p-DDT	–	–	–	inc	–	neg	3	–	–	12
PHENOBARBITAL	–	inc	–	inc	–	neg	16	–	–	30

SELENIUM SULFIDE	inc	inc	–	–	–	positive	3		–	3
TOXAPHENE	–	inc	–	inc	dec	positive	3	–	inc	12
Noncarcinogens										
1-NITRONAPHTHALENE	–	inc	–	–	dec	positive	3	–	inc	3
1-PHENYL-2-THIOUREA	–	–	–	–	–	neg	3		–	3
2,4,5-TRICHLOROPHENOL	–	inc	–	inc	–	neg	41	–		35
2,4-DICHLOROPHENOXYACETIC ACID	–	–	–	inc	–	neg	25	–		35
2,4-DIMETHOXYANILINE	–	inc	–	inc	–	positive	2		inc	11
2,5-DIAMINOTOLUENE	–	–	inc	–	–	positive	15			2
2,6-DIAMINOTOLUENE	–	inc	–	–	dec	positive	1		inc	1
2,6-DIMETHYL-*N*-NITROSOPIPERIDINE	–	–	–	–	–	neg	37			37
2-CHLOROETHANOL	–	inc	inc	–	dec	positive	3		inc	3
2-CHLOROMETHYLPYRIDINE	–	inc	–	–	dec	positive	2		inc	3
2-CHLORO-*p*-PHENYLENE-DIAMINE	–	inc	–	inc	–	positive	2		inc	3
3-NITROPROPRIONIC ACID	–	inc	–	–	dec	positive	3		inc	11
4-CARBOXY-*N*-NITROSO-PIPERIDINE	–	–	–	–	–	neg	37			37
4-CHLOROACETYLACETANILIDE	–	inc	–	–	–	positive	2		inc	3
4-CYCLOHEXYL-*N*-NITROSOPIPERIDINE	–	–	–	–	–	neg	37			37
4-NITROANTHRANILIC ACID	–	–	–	–	–	positive	2		inc	3
4-NITRO-o-PHENYLENEDIAMINE	–	inc	–	–	–	positive	3		inc	3
8-HYDROXYQUINOLINE	–	inc	–	inc	–	positive	3		inc	11
ALDICARB	–	–	–	–	–	neg	3	–	–	3
AMMONIUM,(2-CHLOROETHYL)-TRIMETHYLCHLORIDE	–	–	–	–	–	neg	3		–	3

(continued)

Table 1 Continued

CHEMICAL	DD	OCD	ALT	P450	GSH	AMES	Ames Ref.	TA1537	SA	Cancer Ref.
ANISOLE,2,3,5,6-TETRA-CHLORO-4-NITRO-	–	–	–	–	–	neg	3		–	3
ANTABUSE (TETRAETHYL-THIURAM DISULFIDE)	–	inc	–	–	–	neg	2			11
a-NAPHTHYLAMINE	–	inc	–	–	dec	positive	7	–		11
BARBITURIC ACID	–	inc	–	–	–	na	42			40
BENZO(E)PYRENE	–	–	–	–	–	positive	7			8
CALCIUM CYANAMIDE	–	–	–	–	–	positive	3		inc	3
CAPROLACTAM	–	–	inc	–	–	neg	3	–	–	11
CARBROMAL	–	inc	–	–	–	neg	3		–	3
COUMAPHOS	–	–	–	–	–	neg	3		inc	11
CYCLOHEXYLAMINE	–	–	–	–	–	neg	16	–		3 3
DEXON (FENAMINOSULF)	–	–	–	–	–	positive	3		inc	3
DIETHYLDITHIOCARBAMATE, SODIUM	–	inc	inc	–	–	neg	3		–	11
DIMETHOATE	–	inc	–	–	–	positive	3	–	inc	11
DIOXATHION	–	–	–	–	–	positive	2		inc	11
EDTA, TRISODIUM SALT	–	–	–	–	–	neg	3		–	11
ENDRIN	–	–	–	inc	–	neg	3	–	–	3
FLUORENE	–	inc	inc	inc	dec	neg	7			8
HC BLUE 2	–	inc	–	–	–	positive	2		inc	3
HYCANTHONE METHANE-SULFONATE	–	–	–	–	–	positive	7			11
IODOFORM	–	inc	inc	–	–	positive	2		–	3
LEAD DIMETHYLDITHIO-CARBAMATE	–	–	–	–	–	positive	2		–	3

MALAOXON	–	–	–	–	–	neg	3		inc	3
MALATHION	–	inc	–	–	–	neg	3	–	inc	3
MALEIC HYDRAZINE	–	–	–	–	–	neg	7			11
METHOTREXATE	–	–	–	–	–	neg	16	–		11
METHYLPARATHION	–	inc	–	–	–	positive	3	–	inc	3
NAPTHALENE	–	inc	–	–	dec	neg	7	–		8
n-BUTYLCHLORIDE	–	inc	–	–	dec	neg	3		inc	3
N-PHENYL-*p*-PHENYLENE-DIAMINE	–	inc	–	inc	–	neg	3		inc	3
N-(1-NAPHTHYL)ETHYLENE-DIAMINE	–	inc	–	–	–	positive	2			11
o-DICHLOROBENZENE	–	inc	–	–	dec	neg	1	–	–	3
PENTACHLORONITROBENZENE	–	–	–	–	dec	neg	3		–	3
PHENOL	–	–	–	–	–	neg	1	–	–	3
PHOTODIELDRIN	–	–	–	inc	–	positive	3		–	3
PIPERIDINE	–	–	–	–	–	positive	14			8
PYRENE	–	inc	–	–	–	neg	7			8
p-PHENYLENEDIAMINE	–	inc	–	–	–	positive	3		inc	3
SODIUM BARBITAL	–	inc	–	–	–	na	42			40
SUCROSE	–	–	–	–	–	neg	26			42
TITANIUM(IV)OXIDE	–	–	–	–	–	neg	1	–	–	11
TOLAZAMIDE	–	–	–	–	dec	positive	2		–	3
TPA (12-O-TETRADECANOYL-PHORBOL-13-ACETATE)	–	–	–	–	–	neg	7			8

Rodent carcinogens are listed first followed by the noncarcinogens. Chemicals are listed by increasing number of their substituent groups (e.g., 1,2-dichloroethane, 1,2-dimethylhydrazine, 1,3-dichloropropene, 1,4-dioxane, etc.) and then in alphabetical order.
For SA, inc means a prediction of carcinogenic activity, while – is a prediction of no carcinogenic activity.
–, No change; inc, increase (at $p < 0.05$ level); dec, decrease (at $p < 0.05$ level); neg, negative; na, not available; pos/neg, positive only with a sealed container; pos/red, positive only with reductive metabolism.
Source Ref. 44

TA1537) 42 (26 carcinogens and 16 noncarcinogens), for chromosome aberrations in Chinese hamster ovary cells (ABS) 31 (21 carcinogens and 10 noncarcinogens), for cell mutation in mouse lymphoma L5178Y (MOLY) 27 (17 carcinogens and 10 noncarcinogens), for sister-chromatid exchange in Chinese hamster ovary cells (SCE) 25 (15 carcinogens and 10 noncarcinogens), and for the composite predictor (AMES and k_e and SA and ABS) 21 (13 carcinogens and 8 noncarcinogens). The lack of completeness of data sets obtained from different published sources is a widespread problem in the cancer-prediction area. Until centrally funded research programs generate large, complete data sets of several different cancer predictors, comparative studies must continue to rely on partially complete data sets.

II. RESULTS

A. Individual Cancer Predictors

Table 1 lists the chemical assay results of the five in vivo biochemical parameters (DD, ODC, ALT, P450, and GSH), short-term test results for AMES and TA1537, SA data, and cancer references for the 49 carcinogens and 62 noncarcinogens tested. The resulting data set, along with compiled k_e, ABS, MOLY, and SCE data, was used to determine the operational characteristics (e.g., sensitivity, specificity, concordance) presented in Table 2 and Tables 4–7. Table 1 is useful for making comparisons of individual test results for a single chemical. For example, the chemical carcinogen 1,4-dioxane significantly increased three biochemical parameters (DD, ODC, and P450), did not change ALT or GSH, and was found not mutagenic in *Salmonella* (Table 1).

Table 2 presents the operational characteristics of 12 individual cancer predictors. As complete data were not available for all 111 chemicals, the number of data available (N) for each individual predictor is given. Both DD and TA1537 had a specificity and positive predictivity of 100%. ALT and P450 also did well with specificities of 90 and 86%, respectively. By sacrificing specificity, high sensitivity can be achieved with several individual cancer predictors (SCE 87%, MOLY 77%, k_e 74%, and ODC 67%).

In the individual cancer predictor section, Table 1 shows that DD is a good, new individual cancer predictor with the desirable characteristics of 100% specificity as well as 100% positive predictivity. The overall sensitivity for DD is 24.5% for rodent carcinogens as a whole and 37.5% for the AMES-positive carcinogens (the 24 presumed mutagenic carcinogens) of this data set. The sensitivity of AMES was 53% for rodent carcinogens. The lower sensitivities observed for DD and AMES is because many of the 49 rodent carcinogens included in this study are not presently thought to act

Table 2 Individual Cancer Predictors

Predictor	Sensitivity	Specificity	Positive predictivity	Negative predictivity	Concor-dance	N
DD	24.5	100.0	100.0	62.6	66.7	111
ODC	67.4	51.6	52.4	66.7	58.6	111
P450	34.7	85.7	65.4	62.4	63.1	111
ALT	18.4	90.3	60.0	58.3	58.6	111
GSH	10.2	80.7	29.4	53.2	49.6	111
AMES	53.1	53.3	48.2	58.2	53.2	109
TA1537	15.4	100.0	100.0	42.1	47.6	42
SA	53.5	42.5	50.0	46.0	48.2	83
k_e	73.5	44.7	54.4	65.4	58.3	72
ABS	61.9	60.0	76.5	42.9	61.3	31
MOLY	76.5	30.0	65.0	42.9	59.3	27
SCE	86.7	30.0	65.0	60.0	64.0	25

ODC, P450, ALT, and GSH are nongenotoxic cancer predictors. All the other cancer predictors listed depend on genotoxicity either directly (e.g., AMES or ABS) or indirectly (SA and k_e). The first five predictors are biochemical parameters, and the next seven are various cancer predictors developed by others. For genotoxic predictors the source of data is Refs. 17–21, 36–40 (for the k_e data), and 1 and 5 (for the ABS, MOLY, and SCE data).
Source: Ref. 44.

via genotoxic or mutational mechanisms of action. Thus, neither AMES nor DD should detect these nonmutagenic carcinogens.

B. Combined Cancer Predictors

As a single predictor, GSH did not find any chemical carcinogens not already detected by the other four biochemical parameters, nor did GSH improve the operational characteristics when combined with other predictors. Therefore, GSH is not used as a predictor in the subsequent tables when looking for optimal operational characteristics or complementarity to other cancer predictors.

Omitting GSH reduces the number of possible patterns of biochemical parameter results from 32 to 16 patterns. Table 3 shows each of these 16 possible patterns, the number of chemicals in each pattern, and the rodent carcinogenicity data for these chemicals. All 8 patterns in which DD was increased (12 chemicals) agreed perfectly with the carcinogenicity data. From the patterns of biochemical parameter results given in Table 3, one can see two useful combined composite cancer predictors that utilize a logical "and". The patterns representing CP (14 chemicals) and CT (7 chem-

Table 3 Patterns of the Four Biochemical Parameters and Carcinogenicity Results for 111 Chemicals

	Patterns				Total Chemicals	Rodent cancer	
	DD	ALT	ODC	P450		Positive	Negative
1	INC	INC	INC	INC	0	0	0
2	INC	INC	INC	–	1	1	0
3	INC	INC	–	INC	0	0	0
4	INC	INC	–	–	0	0	0
5	INC	–	INC	INC	2	2	0
6	INC	–	INC	–	5	5	0
7	INC	–	–	INC	0	0	0
8	INC	–	–	–	4	4	0
9	–	INC	INC	INC	4	3	1
10 (CT)	–	INC	INC	–	7	3	4
11	–	INC	–	INC	0	0	0
12	–	INC	–	–	3	1	2
13 (CP)	–	–	INC	INC	14	9	5
14	–	–	INC	–	30	10	20
15	–	–	–	INC	6	3	3
16	–	–	–	–	35	8	27
Total					111	49	62

–, No increase or decrease; INC, increase (at $p < 0.05$).
Only P450 showed decreases as well as increases. Patterns 1–8 are collectively termed DD. Pattern 10 is CT, while pattern 13 is CP.
Source: Ref. 43.

icals) contained 21 chemicals, of which 12 were rodent carcinogens. CP, for example, was increased by 9 carcinogens and only 5 noncarcinogens. CT was increased by 3 carcinogens and 4 noncarcinogens. The two most frequently occurring patterns of biochemical parameter results were no increase in any of the 4 biochemical assays (35 chemicals) and an increase in ODC alone (30 chemicals). These two patterns contained 58% of the total number of chemicals. No chemicals were present in 5 of the 16 possible patterns of biochemical assay results (Table 3).

Table 4 presents individual and composite cancer predictors of high concordance, high sensitivity, or high specificity. The three composite predictors that give a high concordance of 71–74% [(TA1537 or TS), TS, and ([AMES and k_e and SA and ABS] or CP)] are composed of both genotoxic and nongenotoxic parameters. The resulting concordance values of these composite cancer predictors are higher than those obtainable by any one

Table 4 Cancer Predictors of High Concordance, High Sensitivity, or High Specificity

Predictor	Sensitivity	Specificity	Positive predictivity	Negative Predictivity	Concordance	N
High concordance						
[TA1537 or TS]	61.5	93.8	94.1	60.0	73.8	42
TS	55.1	87.1	77.1	71.1	73.0	111
([AMES and k_c and SA and ABS] or CP)	61.5	87.5	88.9	58.3	71.4	21
High sensitivity						
[ODC or k_c or AMES]	100.0	15.8	51.5	100.0	55.6	72
[P450 or k_c or AMES]	97.1	26.3	54.1	90.9	59.7	72
[DD or ODC or ALT or P450]	83.7	43.6	54.0	77.1	61.3	111
High specificity						
[DD or P450 at low dose]	40.8	100.0	100.0	68.1	73.9	111
DD	24.5	100.0	100.0	62.6	66.7	111
TA1537	15.4	100.0	100.0	42.1	47.6	42
CT	16.3	93.6	66.7	58.6	59.5	111
CP	28.6	91.9	73.7	62.0	64.0	111

CP, [ODC and P450]; CT, [ALT and ODC]; TS, {DD or [ODC and P450] or [ALT and ODC]}; [DD or P450 at the low dose only] is the high specificity mode of the in vivo biochemical parameters.

Source: Ref. 44.

individual cancer predictor from which they are made. These high-concordance predictors also do well with respect to specificity and positive predictivity (77–94%) and have sensitivities of 55% or more, making them good overall predictors of carcinogenicity.

By combining one or more nongenotoxic cancer predictors with one or more genotoxic cancer predictors, very high sensitivity and negative predictivity can be obtained. For example, the composite predictor of (ODC or k_e or AMES) is 100% for both sensitivity and negative predictivity. For the predictor (P450 or k_e or AMES), sensitivity and negative predictivity are 97 and 91%, respectively. Perfect specificity and positive predictivity can be gained from three of the cancer predictors [(DD or P450 at low dose), DD alone, and TA1537 alone]. It is interesting to note that although TA1537 has a specificity and positive predictivity of 100%, the overall Ames test (actually a composite of four or five *Salmonella* strains both with and without exogenous activation) has a specificity of only 53% and positive predictivity of only 48%.

Table 5 presents the data on operational characteristics from a biological mechanism of action, rather than a mathematical, point of view. Five individual genotoxic cancer predictors are given in the first section of Table 5. The concordances range between 47 and 66%. For nongenotoxic cancer predictors, two individual and three composite cancer predictors are presented. All of these nongenotoxic cancer predictors are from our biochemical parameter studies. The concordances of these nongenotoxic cancer predictors range from 59 to 66%. However, when genotoxic and nongenotoxic cancer predictors are used together, synergy or complementarity results. The improved concordances of these genotoxic and nongenotoxic cancer predictors range from 71 to 74%. Thus substantial increases in concordance have resulted from combining genotoxic and nongenotoxic cancer predictors (Table 5). For individual genotoxic and nongenotoxic cancer predictors, the maximum concordances were both 66% (Table 5).

The three cancer predictors (DD, CP, and CT) individually detected 12, 14, and 7 carcinogens, respectively, and this is graphically illustrated in Figure 1. CP is (ODC and P450) while CT is (ODC and ALT). Two chemicals were detected by both DD and CP; one chemical was detected by both DD and CT, and three chemicals were detected by CP and CT (Fig. 1). Because of the complementarity or synergy obtained when DD, CP, and CT are combined, 27 different rodent carcinogens are detected. Linking CP to DD detects 12 new chemical carcinogens (e.g., kepone, *p,p'*-DDE) not already detected by DD. When CT is linked to (DD or CP), three new chemical carcinogens (carbon tetrachloride, chloroform, and mirex) are detected. The cancer predictor DD responds to chemicals that initiate the carcinogenesis process via genotoxic events. 1,2-Dibromoethane, *N*-nitroso-

Table 5 Several Genotoxic, Nongenotoxic, and Composite Cancer Predictors

Predictor	Sensitivity	Specificity	Positive predictivity	Negative predictivity	Concordance	N
Genotoxic						
DD	24.5	100.0	100.0	62.6	66.7	111
AMES	53.1	53.3	48.2	58.2	53.2	109
TA1537	15.4	100.0	100.0	42.1	47.6	42
k_c	73.5	44.7	54.4	65.4	58.3	72
SA	53.5	42.5	50.0	46.0	48.2	83
ABS	61.9	60.0	76.5	42.9	61.3	31
Nongenotoxic						
ODC	67.4	51.6	52.4	66.7	58.6	111
P450	34.7	85.7	65.4	62.4	63.1	111
CP	28.6	91.9	73.7	62.0	64.0	111
CT	16.3	93.6	66.7	58.6	59.5	111
[CP or CT]	38.8	87.1	70.4	64.3	65.8	111
Composite (genotoxic and nongenotoxic)						
TS	55.1	87.1	77.1	71.1	73.0	111
[TA1537 or TS]	61.5	93.8	94.1	60.0	73.8	42
([AMES and k_c and SA and ABS] or CP)	61.5	87.5	88.9	58.3	71.4	21

Source: Ref. 44.

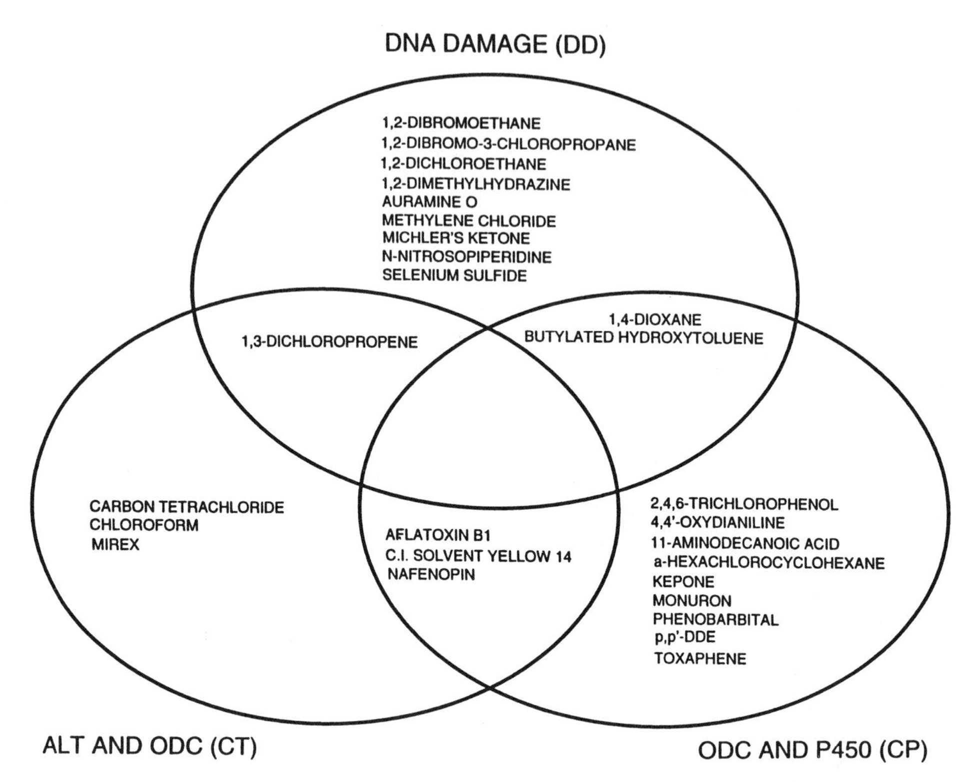

Figure 1 The 27 rodent carcinogens identified by three biochemical parameters: DD, CP, and CT. Each of the 27 rodent carcinogens was identified by one or two, but never all three of the biochemical parameters. For example, nafenopin was identified by both CP and CT. (From Ref. 43.)

piperidine, and 1,2-dimethylhydrazine are examples of chemicals that increased DD. The cancer predictor CP responds to chemicals that cause cancer by promotion of carcinogenesis or cell proliferation. Phenobarbital, *p,p'*-DDE, and α-hexachlorocyclohexane are examples of the carcinogens that increased CP. The cancer predictor CT responds to chemicals that cause cancer by cell toxicity, damage, and death. Carbon tetrachloride, chloroform, and aflatoxin B_1 are examples of carcinogens that increased CT (Fig. 1). Aflatoxin B_1 came the closest to increasing all three cancer predictors—DD, CP, and CT. At 1 mg/kg, aflatoxin B_1 significantly increased DD in some but not all statistical tests.

From the data presented in the combined cancer predictors section, four major interpretations emerge. The first is that CP has been discovered to be a new useful cancer predictor that responds well to the positive influences (e.g., trophic effects or cell proliferation) that a cell may experience. Second, CT is a new cancer predictor that responds well to the negative influences (cell damage and death, regenerative hyperplasia) that a cell may experience. Third, combining genotoxic and nongenotoxic cancer predictors can make better composite cancer predictors that possess demonstrated synergy and complementarity. Fourth, combining cancer predictors that monitor different mechanisms of carcinogenic action is a good basis for making progress. TS is an example of a combined cancer predictor that includes three different mechanisms of action. For different cancer predictors to be combined, they must first be of high specificity or one will be numerically overwhelmed with false positives. CP, CT, and TS had specificities of 92, 94, and 87%, respectively.

C. Predicting Rodent Hepatic Cancer

Rodent liver, particularly mouse liver, is responsive to many genotoxic and nongenotoxic carcinogens (3). The term rodent means rat or mouse. Substantial data and research experience show that predicting rodent liver cancer is difficult for any cancer-predicting test. The 49 rodent liver carcinogens in this data set include such difficult-to-predict hepatocarcinogens as carbon tetrachloride, nafenopin, 1,4-dioxane, *p,p'*-DDE, kepone, mirex, and α-hexachlorocyclohexane. The concordance for predicting liver adenoma and/or carcinoma in rodents was 73, 51, and 47% for TS, the Ames test, and SA, respectively (Table 6). Compared to either the Ames test or

Table 6 Operational Characteristics of the Predictions of Rodent Liver Cancer and/or Adenoma by TS, the Ames Test and SA.

	TS	Ames test	TS	SA
Sensitivity	56	51	58	53
Specificity	83***	52	80***	42
Positive predictivity	69**	41	71*	43
Negative predictivity	74*	62	69	51
Concordance	73***	51	70**	47
Number of chemicals	109	109	83	83

A chemical is considered positive if it caused hepatic adenoma only, adenoma and carcinoma, or carcinoma only. Statistical significance of TS versus either the Ames test (for 109 chemicals; data appear in two columns at the left of the table) or SA (for 83 chemicals, data appear in two columns at the right of the Table) is given by asterisks (*, $p < 0.05$; **, $p < 0.01$; ***, $p < 0.001$).
Source: Ref. 43.

SA, the specificity and positive predictivity of TS was 27–38 percentage units higher (Table 6). Thus, a cancer prediction system largely based on rat liver better predicted rodent liver cancer results than did two other predictive systems which are not as biologically differentiated and integrated.

An even harder problem is predicting the carcinogenicity of chemicals that cause only liver tumors in either mice, rats, or rodents but do not cause tumors at extrahepatic sites. Table 7 shows the sensitivity, specificity, and concordance of TS, SA, and the Ames test for these difficult chemicals, which cause cancer only in the liver.

For chemicals that produced only rodent liver cancer, the sensitivity was 53, 37, and 33% for TS, Ames, and SA, respectively (Table 7). For predicting mouse liver cancer only, the sensitivity was 58, 32, and 39% for TS, Ames, and SA, respectively. But with respect to specificity, TS was superior to both Ames and SA. For chemicals that produced only rodent liver cancer, the specificity was 72, 48, and 39% for TS, Ames, and SA, respectively (Table 7). For predicting mouse liver cancer only, the specificity was 73, 47, and 40% for TS, Ames, and SA, respectively. With respect to concordance, TS was also superior to Ames and SA. For chemicals that produced only rodent liver cancer, the concordance was 69, 46, and 39 for TS, Ames, and SA, respectively (Table 7). For predicting mouse liver cancer

Table 7 Sensitivity, Specificity, and Concordance for TS, the Ames Test, and SA in Predicting Chemicals That Cause Only Liver Cancer.

	Sensitivity	Specificity	Concordance
TS			
Rodent	53	72	69
Rat	58	73	71
Mouse	58	73	71
Ames			
Rodent	37	48**	46**
Rat	47	50**	50**
Mouse	32	47**	44**
SA			
Rodent	33	39***	39***
Rat	40	41***	41***
Mouse	39	40***	40***

Statistical significance of TS versus either the Ames test (for 109 chemicals) or structural alerts (for 83 chemicals) is given by asterisks (**, $p < 0.01$; ***, $p < 0.001$).
Source: Ref. 43.

only, the concordance was 71, 44, and 40% for TS, Ames, and SA, respectively.

TS clearly has major advantages compared to either Ames or SA in predicting the carcinogenicity of chemicals which were active only in rodent, rat, or mouse liver. The advantages arise from two principal sources. First, TS detects both nongenotoxic and genotoxic carcinogens, and second, TS avoids a high rate of false positives with mutagenic noncarcinogens (see Fig. 2). As risk assessments of chemical carcinogens are often based on rodent hepatic tumors, it is encouraging that TS can perform well in predicting the carcinogenicity of chemicals active in only rodent, rat, and mouse liver.

D. False Positives and False Negatives

Short-term mutagenicity tests, such as the Ames test, work well for some chemicals, but not for others. The Ames mutagenicity test and rodent cancer bioassay results can be used to operationally classify chemicals into the four categories of mutagenic carcinogen, mutagenic noncarcinogen, nonmutagenic carcinogen, and nonmutagenic noncarcinogen (Fig. 2). Two of these predictive classifications (mutagenic carcinogens and nonmutagenic noncarcinogens) are perfect, and the other two incorrect classifications (mutagenic noncarcinogens and nonmutagenic carcinogens) offer us the opportunity for improvement.

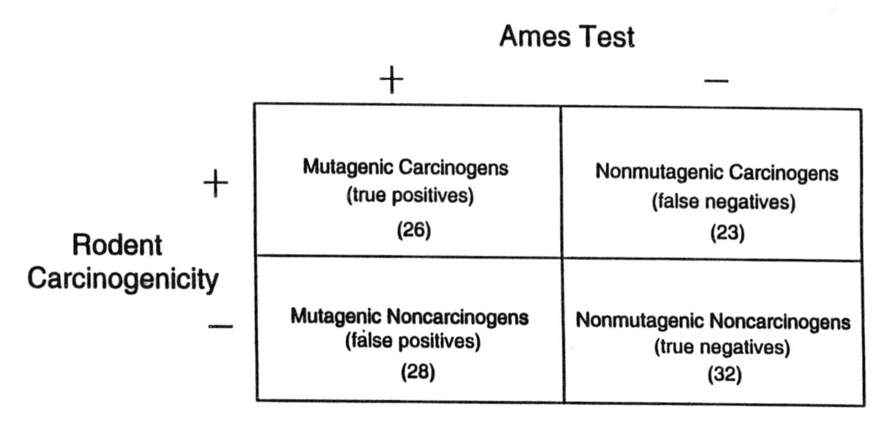

Figure 2 Contingency table obtained by operationally using results from the Ames test and rodent carcinogenicity bioassays. In each of the boxes of the contingency table is given the number of chemicals of that type (e.g., there are 23 nonmutagenic carcinogens in this study). (From Ref. 44.)

Table 8 presents a list of carcinogens that are positive for TS but negative in the Ames test as well as noncarcinogens that are negative for TS but positive in the Ames test. Table 8 shows that TS often did better than the Ames *Salmonella* mutagenicity test for chemical carcinogens that were DNA unreactive but possessed cell-killing, cell-proliferating, or cell-tropic properties. The length of the second part of Table 8 shows that TS generally avoided large numbers of false positives, which is a large problem for many single cell–based mutagenicity systems. Because of the false-positive problem of many short-term tests, it is difficult to connect them with a logical "or." One is soon overwhelmed with large numbers of false positives.

To think and compare by chemicals is an alternative to thinking and comparing by the percentages of the various operational characteristics. The biochemical parameters correctly predicted several carcinogens, which were nonmutagenic in an Ames test (e.g., DDE, mirex, and 2,4,6-trichlorophenol). The biochemical parameters were particularly successful in avoiding false positives (e.g., hycanthone methanesulfonate and dioxathion) (31). Thus the biochemical parameters were able to contribute in both of the two possible areas of improvement (false positives, false negatives) relative to an Ames test alone. It is in the area of nonmutagenic carcinogens (false negatives) that nongenotoxic cancer predictors are likely to make the most important impact.

For mutagenic noncarcinogens, TS (86% agreement with rodent carcinogenicity), DD (100%), CP (93%), and CT (93%) predict rodent cancer bioassay results much better than the short-term mutagenicity tests SA (18%), MOLY (0%), ABS (20%), and SCE (0%) (Fig. 3). For nonmutagenic carcinogens, TS (61%) is a better predictor of rodent bioassay results than all mutagenicity assays except SCE (67%) (which performed poorly with genotoxic noncarcinogens, 0% agreement) (Fig. 3).

It is in the area of nonmutagenic carcinogens (false negatives) that nonmutagenic cancer predictors are likely to make the most important impact. When the two categories of mutagenic noncarcinogens and nonmutagenic carcinogens are combined, the biochemical indicators TS (74% agreement with rodent carcinogenicity), DD (61%), CP (71%), and CT (59%) predict rodent cancer bioassay results substantially better than any of the four short-term mutagenicity tests examined [SA (16%), MOLY (18%), ABS (18%), and SCE (36%)] (Fig. 4).

Thus, the biochemical parameters in general and TS in particular are able to take advantage of the opportunity for improvement presented by the mutagenic noncarcinogens and nonmutagenic carcinogens of Figure 2. Because the selection criteria for the chemicals of Figures 3 and 4 was an incorrect cancer prediction by the Ames assay by definition, the 0% of the

Table 8 Ames Test False Positives and Negatives

Carcinogens positive in TS, but negative in the Ames test (false negatives)
1,2-Dimethylhydrazine
1,4-Dioxane
11-Aminoundecanoic acid
2,4,6-Trichlorophenol
α-Hexachlorocyclohexane
Butylated hydroxytoluene
Carbon tetrachloride
Chloroform
Kepone
Mirex
Monuron
Nafenopin
p,p'-DDE
Phenobarbital
Noncarcinogens negative in TS, but positive in the Ames test (false positives)
1-Nitronaphthalene
2,5-Diaminotoluene
2,6-Diaminotoluene
2-Chloromethylpyridine
2-Chloro-*p*-phenylenediamine
3-Nitroproprionic acid
4-Nitro-o-phenylenediamine
α-Naphthylamine
Benzo (ϵ) pyrene
Calcium cyanamide
Dexon (fenaminosulf)
Dimethoate
Dioxathion
HC blue 2
Hycanthone methanesulfonate
Lead dimethlydithiocarbamate
Methylparathion
N-(1-Naphthyl)ethylenediamine
Photodieldrin
Piperidine
p-Phenylenediamine
Tolazamide

Source: Ref. 43.

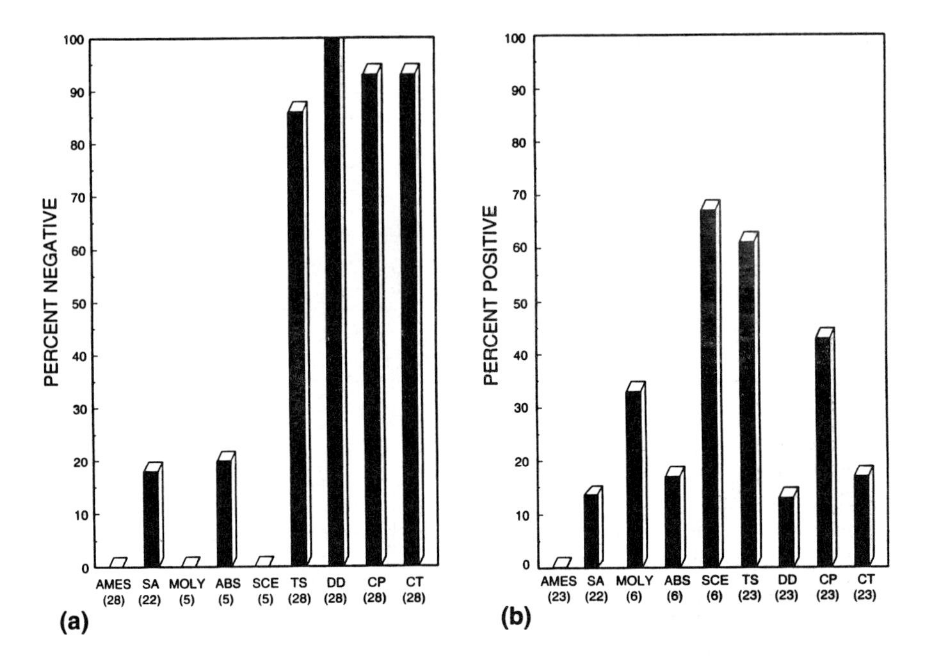

Figure 3 Percentage of correct predictions for (a) 28 genotoxic noncarcinogens (a negative prediction is correct) or (b) 23 nongenotoxic carcinogens (a positive prediction is correct) by SA, MOLY, ABS, SCE, TS, DD, CP, and CT. As the Ames test was used to select these two categories of false positives and false negatives, the percentage of the chemicals correctly identified by the Ames test must be 0% by definition. The Ames data are included only to show if the results of other predictors are similar or dissimilar to the Ames test. The number of chemicals is given in parentheses below the predictor assay name. (From Ref. 43.)

chemicals correctly identified in *Salmonella* does not reflect negatively on the utility of the Ames assay. However, the SA, MOLY, ABS, and SCE assays were not preselected for disconcordance, but still gave results similar to those of the Ames test, repeating the same predictive errors.

III. CONCLUSION

A. Presently Available Data

The reasons why TS worked so well in predicting whole animal cancer bioassay results include the following: (a) high specificity of the three individual biochemical parameters employed, (b) whole animal pharmacokinetics (absorption, distribution, metabolism, and excretion) present in this study and absent in most, if not all, short-term mutagenicity tests, (c) the parameter CP can respond to chemicals that cause cancer by promotion of carcinogenesis, cell proliferation, or mitogenesis, (d) the parameter CT can respond to chemicals that cause cancer by cell toxicity, damage, and death with the resulting compensatory cell proliferation, (e) taken together, DD, CP, and CT detect many more different types of carcinogenesis than DD does alone. As with all initial reports, experimental studies with additional chemicals are desirable to further test the predictivity of these in vivo biochemical parameters.

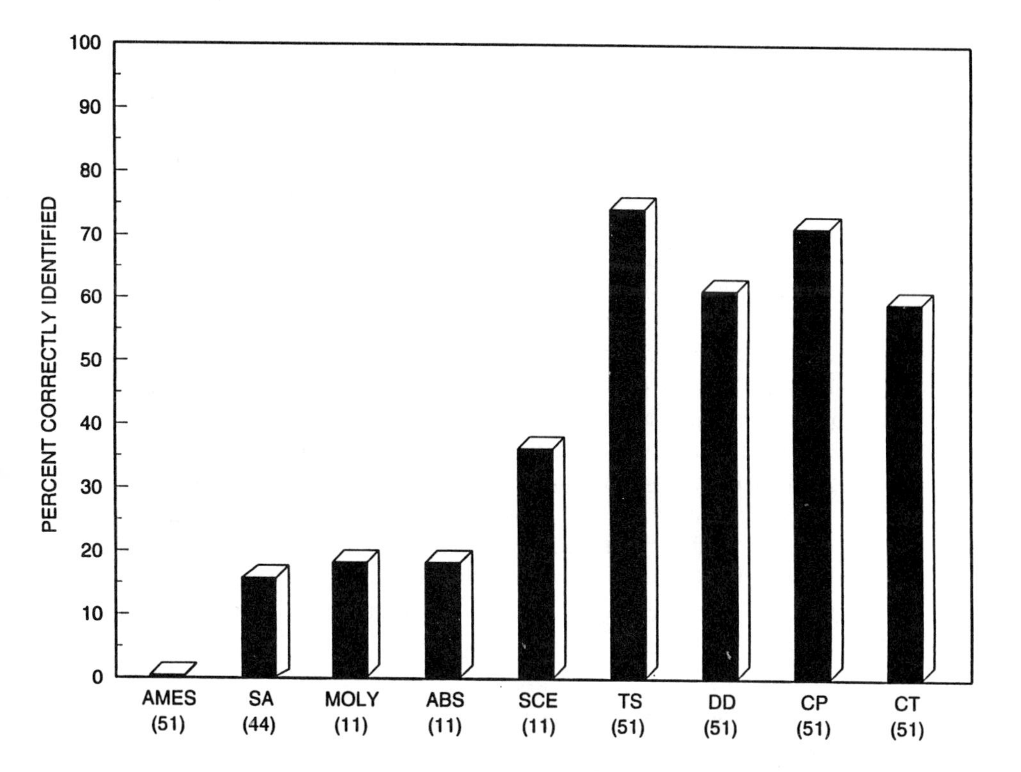

Figure 4 Percentage of correct predictions for genotoxic noncarcinogens and nongenotoxic carcinogens by SA, MOLY, ABS, SCE, TS, DD, CP, and CT. As the Ames test was used to select these two categories of false positives and false negatives, the percentage of the chemicals correctly identified by the Ames test must be 0% by definition. The Ames data is included only to show if the results of other predictors are similar or dissimilar to the Ames test. There are 51 chemicals in this comparison; the *n* for each predictor is given in parentheses below the assay name. (From Ref. 43.)

At present we have large amounts of data from many different cell-based systems which utilize mutation or genotoxicity as an endpoint. However, this area of research seems to have reached an impasse, as discussed by Tennant et al. in 1987 (1). General mutation-based systems may not be able to reach higher levels of predictive performance than might be achievable with specific mutation (e.g., oncogenes, tumor-suppressor genes)-based systems. This area—mutation of specific oncogenes and tumor-suppressor genes—offers opportunities for development of improved systems for predicting chemical carcinogenesis. The somatic mutation theory once energized the development of general mutational systems in an effort to better predict chemical carcinogenicity. Subsequently, specific, much more detailed theories of oncogene and tumor-suppressor gene involvement in carcinogenesis were developed. These specific oncogene and tumor-suppressor gene theories may now be ready to fuel a further advance in developing predictive systems of chemical carcinogenicity.

Parameters that do not depend on DNA, mutagenicity, genotoxicity, or the somatic mutation theory of cancer have been underexplored and hence underutilized in rodent cancer prediction. Multiple predictors of carcinogenesis which are based on several different mechanisms may outperform predictors based only on genotoxic mechanisms (43–45). Nonmutagenic predictors of carcinogenesis should receive more attention than currently given because (a) the initial results using biochemical parameters are promising (43) and (b) effective batteries of mutagenic or genotoxic cancer predictors have not been developed after many years of effort (1). DNA is a cellular macromolecule that both controls cellular processes and is controlled by other cellular constituents and processes (e.g., DNA polymerases, DNA repair enzymes, protein transcription factors, signal transduction systems, and the cell proliferation machinery). It is wise to remember that about 99.8% of the mass of a typical cell is not DNA. Thus, nongenotoxic cancer predictors are an important area of opportunity for future research.

B. Future Research Prospects in Nongenotoxic Cancer Predictors

As nongenotoxic cancer predictors are still a relatively unexplored area, many excellent research opportunities are available. High-specificity cancer predictors could be developed in areas such as cell proliferation (e.g., incorporation of ^{3}H-thymidine or bromodeoxyuridine into DNA), cell death and regenerative hyperplasia, apoptosis, peroxisomal proliferators, and receptor-mediated carcinogenesis for chemicals such as hormones, halogenated dibenzodioxins, and possibly peroxisomal proliferating chemicals. Signal transduction pathways also offer the potential for the development of systems for predicting chemical carcinogenesis. In the future an excellent paradigm for progress in the area of predicting rodent carcinogenicity bioassay results will be the use of multiple cancer predictors for the multiple mechanisms of multistage carcinogenesis (Fig. 5).

There is also a need for fairly early cancer predictors, which might be obtained between 1 and 30 days after an in vivo chemical exposure is begun. Multiple cancer predictors obtained from different organs of the same exposed animal can save animal expense and possibly predict the target organ for carcinogenesis. Two longer-term proposals in the hepatic area are the γ-glutamlytranspeptidase system (46) and the 8-week system determining the placental form of glutathione S-transferase (47).

Finally, there is a need for organ-specific indicators of carcinogenesis in organs in which high human tumor rates occur (e.g., prostate, breast, gastrointestinal tract, lung, etc.). This type of cancer predictor obtained at much later time points (e.g., 3–18 months) ought to have excellent predictive characteristics for the organs with high human cancer rates.

Recently, a more holistic view of the multiple causes of cancer has come into prominence. Instead of focusing on DNA and mutations in DNA to the exclusion of almost all other biology, more modern views of cancer

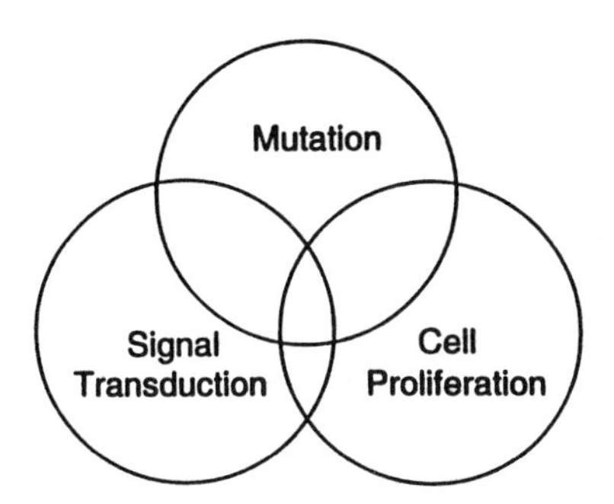

Figure 5 Three causes of cancer. A chemical may work through one, two, or three mechanisms at the same time. Alternatively, a chemical may act sequentially through two or three or more mechanisms in the multiple stages of carcinogenesis. Notice the similarity in the Venn diagrams for Figure 5 (causes of cancer) and Figure 1 (predictors of cancer). The top circles are mutation (Fig. 5) and cancer predictors that depend on mutation or genotoxicity (DD of Fig. 1). The right-hand circles are cell proliferation (Fig. 5) and cancer predictors that depend on cell cycle and cell phenotypic changes (CP of Fig. 1). The left-hand circles differ: for Figure 5, it is signal transduction pathways; for Figure 1, it is cell toxicity or regenerative hyperplasia (CT).

causality have stressed other cellular processes, particularly that of cell proliferation and cellular signal transduction pathways. Figure 5 presents a Venn diagram of these concepts and their interrelationships. It is particularly noteworthy that mutations in signal transduction pathways or mutations in the control of cell proliferation are more likely to predispose a cell to the carcinogenic pathway than a mutation in some other part of the genome. Chemicals that first and primarily act via cell proliferation may cause mutational or signal transduction changes as a secondary effect. Similarly, chemicals that act first and primarily via signal transduction pathways may cause cell proliferative or even mutational effects as a secondary effect. Examples of chemical carcinogens that may act through signal transduction pathways include 2,3,7,8-tetrachlorodibenzo-*p*-dioxin, estrogen, and diethylstilbesterol. Each of these concepts offers us opportunities to develop better systems for predicting chemical carcinogenicity.

ACKNOWLEDGMENTS

We thank Kay Rigsbee and Fay Poythress of ROW Sciences, Inc. for their animal dosing and sacrificing. Dr. Rory Conolly contributed to early phases of this research project. Drs. Ashoke Mitra and Ira Richards assisted Dr. Arun Kulkarni's work on 10 of the chemicals of Table 1.

ABBREVIATIONS AND DEFINITIONS

ABS, chromosome aberrations in Chinese hamster ovary cells.
ALT, rat serum alanine aminotransferase activity.
AMES, mutation in *Salmonella typhimurium*.
Concordance, % of agreement between test system and rodent carcinogenicity.
CP, a composite parameter positive if both the biochemical assays ODC and P450 are individually increased. CP was chosen to stand for Cell Proliferation.
CT, a composite parameter positive if both the biochemical assays ALT and ODC are individually increased. CT was chosen to stand for Cell Toxicity.
DD, rat hepatic DNA damage as determined by alkaline elution.
GSH, rat hepatic reduced glutathione content.
k_e, the rate at which excess electrons attach to the test chemical dissolved in liquid cyclohexane.
MOLY, cell mutation in mouse lymphoma L5178Y cells.

Negative predictivity, % of test negatives that are noncarcinogens.
ODC, rat hepatic ornithine decarboxylase activity.
P450, rat hepatic cytochrome P-450 content.
Positive predictivity, % of test positives that are carcinogens.
SA, structural alerts.
SCE, sister chromatid exchange in Chinese hamster ovary cells.
Sensitivity, % of carcinogens that test positive.
Specificity, % of noncarcinogens that test negative.
TA1537, mutation in *Salmonella typhimurium* strain TA1537.
TS, a composite parameter positive if any of the parameters DD, CP, or CT are increased. Defined in terms of the individual biochemical assays, TS = [DD or (ODC and P450) or (ALT and ODC)]. TS stands for This Study.

REFERENCES

1. Tennant RW, Margolin BH, Shelby MD, Zeiger E, Haseman JK, Spalding J, Caspary W, Resnick M, Stasiewicz S, Anderson B, Minor R. Prediction of chemical carcinogenicity in rodents from in vitro genetic toxicity assays. Science 1987; 236:933–941.
2. Zeiger E. Carcinogenicity of mutagens: predictive capability of the Salmonella mutagenesis assay for rodent carcinogenicity. Cancer Res 1987; 47:1287–1296.
3. Ashby J, Tennant RW. Chemical structure, Salmonella mutagenicity and extent of carcinogenicity as indicators of genotoxic carcinogenesis among 222 chemicals tested in rodents by the U.S. NCI/NTP. Mutat Res 1988; 204:17–115.
4. Ashby J, Tennant RW, Zeiger E, Stasiewicz S. Classification according to the chemical structure, mutagenicity to Salmonella and level of carcinogenicity of a further 42 chemicals tested for carcinogenicity by the U.S. National Toxicology Program. Mutat Res 1989; 223:73–103.
5. Zeiger E, Haseman JK, Shelby MD, Margolin BH, Tennant RW. Evaluation of four in vitro genetic toxicity tests for predicting rodent carcinogenicity: confirmation of earlier results with 41 additional chemicals. Environ Molecular Mutagen 1990; 16(suppl. 18):1–14.
6. Ashby J, Tennant RW. Definitive relationships among chemical structure, carcinogenicity and mutagenicity for 301 chemicals tested by the U.S. NTP. Mutat Res 1991; 257:229–311.
7. McCann J, Choi E, Yamasaki E, Ames B. Detection of carcinogens as mutagens in the Salmonella microsome test: assay of 300 chemicals. Proc Natl Acad Sci USA 1975; 72(12):5135–5139.
8. Sina JF, Bean CL, Dysart GR, Taylor VI, Bradley MO. Evaluation of the alkaline elution/rat hepatocyte assay as a predictor of carcinogenic/mutagenic potential. Mutat Res 1983; 113:357–391.
9. Swenberg JA, Petzold GL. The usefulness of DNA damage and repair assays

for predicting carcinogenic potential of chemicals. In: Butterworth C, ed. Strategies for Short-Term Testing for Mutagens/Carcinogens. West Palm Beach, FL: CRC Press, 1979:77–86.
10. Kitchin KT, Brown JL. Biochemcial effects of three carcinogenic chlorinated methanes in rat liver. Teratogen Carcinogen Mutagen 1989; 9:61–69.
11. Nesnow S, Argus M. Bergman H, Chu K, Frith C, Helmes T, McGaughy R, Ray V, Slaga TJ, Tennant R, Weisburger E. Chemical carcinogens: a review and analysis of the literature of selected chemicals and the establishment of the Gene-Tox Carcinogen Data Base. Mutat Res 1986; 185:1–195.
12. National Toxicology Program. Fourth Annual Report on Carcinogens. Research Triangle Park, NTP 85-002, 1985.
13. Ashby J, Paton D. The influence of chemical structure on the extent and sites of carcinogenesis for 522 rodent carcinogens and 55 different human carcinogen exposures. Mutat Res 1993; 286(1):3–74.
14. Bempong MA, Scull FE. Mutagenic activity of N-chloropiperidine. J Environ Pathol Toxicol 1980; 4:345–354.
15. Dunkel VC, Zeiger E, Brusnick D, McCoy E, McGregor D, Mortelmans K, Rosenkranz HS, Simmon VF. Reproducibility of microbial mutagenicity assays. II. Testing of carcinogens and noncarcinogens in Salmonella typhimurium and Escherichia coli. Environ Mutagen 1985; 7(suppl 5):1–248.
16. IARC Working Group. IARC Monographs on the Evaluation of Carcinogenic Risks to Humans, Genetic and Related Effects: an Updating of Selected IARC Monographs from Volumes 1 to 42. Supp. 6. Lyon: International Agency for Research on Cancer, 1987.
17. Bakale G, McCreary RD. A physico-chemical screening test for chemical carcinogens: the k_e test. Carcinogenesis 1987; 8:253–264.
18. Bakale G. Theoretical implications of the k_e carcinogen-screening test. In: Politzer P, Martin J Jr, eds. Chemical Carcinogens—Activation Mechanisms, Structural and Electronic Factors, and Reactivity. New York: Elsevier, 1988: 322–344.
19. Bakale G, McCreary RD. Response of the k_e test to NCI/NTP-screened chemicals. I. Nongenotoxic carcinogens and genotoxic noncarcinogens. Carcinogenesis 1990; 11:1811–1818.
20. Bakale G, McCreary RD. Response of the k_e test to NCI/NTP-screened chemicals. II. Genotoxic carcinogens and nongenotoxic noncarcinogens. Carcinogenesis 1992; 13(8):1437–1445.
21. Bakale G, McCreary RD, Gregg EC. Quasifree electron attachment to carcinogens and liquid cyclohexane. Cancer Biochem Biophys 1981; 5:103–109.
22. Clapp NK, Tydall RB, Cumming RB, Otten JA. Effects of butylated hydroxytoluene alone or with diethylnitrosamine in mice. Food Cosmet Toxicol 1974; 12:367–371.
23. U.S. EPA. Health assessment document for carbon tetrachloride. Washington, DC: EPA/600/8-82001F, 1984.
24. Rinkus SJ, Legator MS. Chemical characterization of 465 known or suspected carcinogens and their correlation with mutagenic activity in the Salmonella typhimurium system. Cancer Res 1979; 39:3289–3318.

25. Probst GS, McMahon RE, Hill LE, Thompson CZ, Epp JK, Neal SB. Chemically-induced unscheduled DNA synthesis in rat primary hepatocyte cultures: a comparison with bacterial mutagenicity using 218 compounds. Environ Mutagen 1981; 3:11-32.
26. Trueman RW. Activity of 42 coded compounds in the Salmonella reverse mutation test. Progr Mutat Res 1981; 1:343-350.
27. U.S. EPA. Health assessment document for chloroform. Washington, DC: EPA/600/8-84-004A, 1984.
28. Reddy JK, Scarpelli DG, Subbarao V, Lalwani N. Chemical carcinogens without mutagenicity: peroxisome proliferators as a prototype. Toxicol Pathol 1983; 11:172-178.
29. Soderman J, ed. CRC Handbook of Identified Carcinogens and Noncarcinogens: Carcinogenicity-Mutagenicity Database. Vol. 1. Boca Raton, FL: CRC Press, 1982:1-655.
30. Gold LS, Sawyer CB, Magaw R, Backman GM, de Veciana M, Levinson R, Hooper NK, Havender WR, Bernstein L, Petro R, Pike MC, Ames BN. A carcinogenic potency database of the standardized results of animal bioassays. Environ Health Perspect 1984; 58:9-319.
31. National Cancer Institute. Thirteen-week Subcronic Toxicity Studies of Direct Blue 6, Direct Black 38, and Direct Brown 95 Dyes. Tech. Rep. Ser. No. 108, DHEW Publication No. (NIH) 78-1358. Bethesda, MD: 1978.
32. U.S. EPA. Health assessment document for chlorinated benzene. EPA/600/8-84-015F, 1985.
33. Bopp BA, Sonders RC, Kesterson JW. Toxicological aspects of cyclamate and cyclohexylamine. Crit Rev Toxicol 1986; 16:213-306.
34. U.S. EPA. Health assessment document for dichloromethane (methylene chloride). Washington, DC: EPA/600/8-82/004F, 1985.
35. Vainio H, Hemminki K, Wilbourn J. Data on the carcinogenicity of chemicals in the IARC Monographs. Carcinogenesis 1985; 6:1653-1665.
36. Magee PN, Montesano R, Preussman R. Nitroso compounds and related carcinogens. In: Searle CE, ed. Chemical Carcinogens. Washington, DC: ACS, 1976:491-625.
37. Lijinsky W. Structure-activity relations in carcinogenesis by N-nitroso compounds. In: Rao, TK, Linjinsky W, Epler JL, eds. Genotoxicity of N-Nitroso Compounds. New York: Plenum, 1984:189-231.
38. Greim H, Wolff T. Carcinogenicity of organic halogenated compounds. In: Searle CE, ed. Chemical Carcinogens. 2nd ed. Washington, DC: ACS Monograph 182, 1984:525-575.
39. Reddy JK, Lalwani ND. Carcinogenesis by hepatic peroxisome proliferators: evaluation of the risk of hypolipidemic drugs and industrial plasticizers to humans. Crit Rev Toxicol 1983; 12:1-58.
40. Nims RW, Devor DE, Henneman JR, Lubet RA. Induction of alkoxyresorufin O-dealkylases, epoxide hydrolase, and liver weight gain: correlation with liver tumor-promoting potential in a series of barbiturates. Carcinogenesis 1987; 8: 67-71.
41. Rasanen L, Hattula ML, Arstila AU. The mutagenicity of MCPA and its soil

metabolites, chlorinated phenols, catechols and some widely used slimicides in Finland. Bull Environ Contam Toxicol 1977; 18:565–571.
42. NIOSH. Registry of Toxic Effects of Chemical Substances. Stock No. 017-033-00166-9. Rockville, MD: U.S. Printing Office, HEW No. 76–191, 1976.
43. Kitchin KT, Brown JL, Kulkarni AP. Predictive assay for rodent carcinogenicity using in vivo biochemical parameters: operational characteristics and complementarity. Mutat Res 1992; 266:253–272.
44. Kitchin KT, Brown JL, Kulkarni AP. Predicting rodent carcinogenicity by in vivo biochemical parameters. Environ Carcinogen Ecotoxicol Rev 1994; C12(1):63–88.
45. Ashby J, Tennent RW. Prediction of rodent carcinogenicity for 44 chemicals: results. Mutagenesis 1994; 9:7–15.
46. Pitot HC, Goldsworthy TL, Moran S, Keenan W, Glauert HP, Maronpot RR, Campbell HA. A method to quantitate the relative initiating and promoting potencies of hepatocarcinogenic agents in their dose-response relationships to altered hepatic foci. Carcinogenesis 1987; 8:1491–1499.
47. Hasegawa R, Ito N. Liver medium-term bioassay in rats for screening of carcinogens and modifying factors in hepatocarcinogenesis. Food Chem Toxicol 1992; 30(11):979–992.

11
Liver Pyruvate Kinase as a Predictor of Promotion of Hepatocarcinogenesis

Susumu Yanagi
College of Nursing, Nara Medical University, Nara, Japan

I. INTRODUCTION

Since the discovery that a variety of promoters of skin carcinogenesis can induce dermal ornithine decarboxylase (ODC) (1), it has been postulated that the elevation of the ODC enzyme by promoters may be specific for the skin-tumor system (2). The hypothesis has been applied to other promoters of gastric (3), mammary (4), colon (5), and bladder (6) carcinogenesis. Two research groups studied the effect of hepatocarcinogenesis promotion on ODC induction using the potent promoter phenobarbital (PB), but no evidence of ODC induction in rodent liver was detected (7,8). Subsequently, we showed that PB in the diet induced a slight but significant increase in ODC activity in rat liver that had been previously treated with an initiating dose of diethylnitrosamine (DEN). The increase was observed only after using a controlled feeding schedule that minimizes the secondary effect of food intake (9). However, the elevation of hepatic ODC was minimal, rendering it not useful for the development of a screening test for hepatocarcinogenesis promoters. Kitchin and Brown (10) observed a large increase in hepatic ODC activity after acute oral administration of PB. However, it is unknown whether or not acute PB administration promotes hepatic carcinogenesis.

During the course of the ODC study we also assayed pyruvate kinase

(PK) activity in the same liver samples since we had previously been studying PK isozyme patterns in the liver during hepatocarcinogenesis (11,12), in tumor-bearing animals (12,13), and in hepatic tumors of various degrees of differentiation (14). We had expected that the activity of PK, the last key enzyme in glycolysis, might be elevated in the livers of rats fed PB because the glycolytic activity of liver tumors is always higher than that of normal liver (15,16). Contrary to our expectation, PK activity in the livers of rats fed PB and a number of other hepatocarcinogenesis promoters was significantly and consistently lower than in those fed the basal diet (17).

This surprising correlation between known promoters of hepatic carcinogenesis and the observed decrease in total rat liver PK activity became the basis of a major research project. Thus we investigated the utility of the observed decrease in hepatic PK assay as a potential screening test for hepatocarcinogenesis promoters (18). In this review we summarize our findings and discuss the strength and weakness of this assay.

II. MATERIALS AND METHODS

A. Administration of Test Chemicals

All test chemicals were added in a commercially available basal powder diet (MF, Oriental Yeast, Tokyo). For known promoters, doses administered were the same as the effective doses for hepatocarcinogenesis promoters. For chemicals with unknown promoting activity, the doses were determined by a small-scale preliminary test, with the maximal dose being the highest dose that did not cause a significant decrease in body weight of the animals. Feeding of at least 4 weeks is generally needed to avoid spurious results due to nonspecific acute toxic effects (17).

B. Animals

Four-week-old male Wistar rats were obtained from Kiwa Laboratory Animals, Wakayama, Japan. After a week of adaptation to our animal room, groups of 12 rats were fed basal diet or diets containing the test chemicals for up to 4 weeks. Six rats of each group were sacrificed by decapitation at the end of the second and fourth weeks. It is important to avoid the use of ether for euthanasia because the anesthetic agent caused a significant decrease in liver PK activity.

Assay of Pyruvate Kinase Activity

The livers were taken out and homogenized using a Potter-Elvehjem homogenizer. The homogenizing buffer consisted of 20 mM Tris-HCl (pH 7.5), 100 mM KCl, 5 mM $MgSO_4$, and 1 mM EDTA. The homogenates were centrifuged at 20,000 × *g* for 1 hour. The supernatant fraction was used for PK assay using the 2,4-dinitrophenylhydrazine method (19,20) with a slight modification (17). The assay mixture contained 50 mM potassium phosphate buffer (pH 6.7), 10 mM $MgSO_4$, 100 mM KCl, 4 mM ADP, 2.5 mM phosphoenol pyruvate, and 0.5 mM fructose 1,6-diphosphate (FDP) in a total volume of 0.5 ml. The homogenizing buffer and the assay mixture have been selected to give the highest possible PK activity. The reaction was started by an addition of 0.5 ml assay mixture into 0.1 ml enzyme solution, which had been appropriately diluted with the homogenizing buffer containing 0.5 mM FDP. After 3 minutes of the enzyme reaction at 37°C, 1.5 ml of 2,4-dinitrophenylhydrazine (0.0125% in 2 N HCl) was added into the reaction mixture and the mixture was left in the same incubator for another 10 minutes. Then 4 ml of 2.4 N NaOH containing 1 mM EDTA was added to the mixture, and the absorbance at 510 nm was determined within 10 minutes at room temperature. PK activity was expressed as μmol of pyruvate/min/g liver tissue. Using this method we could assay more samples in a shorter time with inexpensive reagents as compared with the lactate dehydrogenase coupling method reaction (21).

Evaluation of Promoting Activity

In order to compare the promoting activity of a variety of promoters, a relative promoting activity (RPA) index was calculated for each tested chemicals using our experimental data or published data from the literature on promotion of hepatic cancer. For studies in which hyperplastic nodules (HN) or γ-glutamyltranspeptidase (GGT)-positive or placental glutathione S-transferase-positive foci were used as the indicator of tumor promotion, the RPA was expressed as the ratio of number of HN or foci per cm^2 in the experimental group over the corresponding number in the control group. In these studies the experimental group was initiated and fed an experimental diet containing a test chemical and the control group was initiated but fed the basal diet. For studies in which tumor formation (TF) was followed, RPA was expressed as the ratio of percentages of tumor-bearing animals in the experimental group vs. the control groups. Hepatic promoters show a RPA >1.0. Compounds with RPA below or close to 1.0 are considered inactive promoters of hepatocarcinogenesis.

Interpretation of the Results

Since the hepatic PK activity in rats is regulated by dietary conditions, hormones, and injuries (20,22), it is necessary to have two control groups, namely, the basal diet group (negative control) and the PB diet group (positive control). The nonspecific decrease in PK activity, caused by toxic substance or injuries observed at the end of the second week, was returned to control values by the end of the fourth week (18). As a result of further comparative studies on PK activity-reducing action of various promoters of hepatocarcinogenesis, we suggest that the following criteria be adopted for the screening test of hepatic promoters:

1. A compound should cause a persistent decrease in PK activity in rat liver.
2. The decrease should be significantly different from the control group.
3. The extent of the decrease at week 4 should ideally be at least 20% (i.e., less than 80% of the activity of the control group).

III. RESULTS

A. Comparison with Other Published Results

In order to develop a screening test for promoters of hepatocarcinogenesis, one must first have a good database of such promoters. Different endpoints (e.g., GGT-positive foci, hyperplastic nodules, or tumors) have been used in experiments to determine a chemical's ability to promote hepatocarcinogenesis. Since GGT-positive foci have been used frequently for short-tern in vivo tests for carcinogens (36) and promoters (37), we compared our test data for PK activity to RPAs calculated by either HN, TF, or GGT-foci methods (Table 1). Figure 1 shows a relationship between RPA calculated from GGT-positive foci and RPA from the development of hyperplastic nodules (HN) or tumors (TF) in the liver. Generally speaking, there was a degree of positive correlation between either the RPA based on GGT (the y value) and the RPA based on either HN or TF (Fig. 1). Based on the limited database we examined, it appears that chemical compounds with RPA of less than about 2.5 do not enhance the formation of HN or tumors. It is interesting to note that all these chemical compounds that have an RPA of less than 2.5 failed to decrease the PK activity (see Table 1).

B. Effects of Hepatocarcinogens and Known Promoters

Table 1 shows the effects of a number of known hepatocarcinogens and hepatic promoters on the PK activity and compares this data to the RPA of these chemicals. Most of the hepatocarcinogens and potent hepatic promoters have been shown to decrease the liver PK activity. Among the hepatic carcinogens, the extent of the decrease varied greatly depending on the carcinogen and the time of the determination of PK activity during and

Table 1 Effects on Rat Liver PK Activity of Hepatocarcinogens, Reported Hepatic Promoters, and Some Other Chemicals

Chemicals in diet (concentration)	PK activity [a] Week 2	PK activity [a] Week 4	RPA[b] (markers)	Ref.
Hepatocarcinogens and related chemicals				
3′-MeDAB (0.06%)	58.6*	85.7*	107 (HN)	43
	66.7[c]	40.3[c]		
2-MeDAB (0.06%)	74.7	96.5	–	44
2-FAA (0.025%)	84.9[d]	(week 3)	78 (HN)	43
DEN (0.2 g/kg BW)[e]	80.8	60.0	4 (HN)	43
CD diet		34.0[f]	10 (HN)	35
Chemicals that have been tested for promotion of cancer				
CD diet		34.0[f]	2.6 (GGT)	34
PB (0.05%)	42.3*	54.8*	2.7 (TF)	45
DDT (0.05%)	42.2*	48.3*	2.6 (TF)	45
DH (0.05%)	59.1*	82.5*	0.9 (TF)	45
AB (0.05%)	79.4*	114.5*	1.1 (TF)	45
			3.3 (GGT)	46
EED (10 ppm)	74.0*	77.8*	3.5 (GGT)	37
			3.1 (HN)	37
			3.3 (TF)	47
			4.8 (TF)	48
Testosterone (0.05%)	81.5*	120.7*	2.4 (GGT)	37
			0.8 (TF)	37
DEXA (0.5 ppm)	112.3	105.5	2.5 (GGT)	37
			0 (HN)	37
Cortisone (10 ppm)	68.1*	85.9	1.7 (GGT)	37
Deoxycholate (0.5%)	73.0*	53.3*	10.8 (GGT)	37
	70.4	68.7	6.0 (HN)	37
Cholic acid (0.3%)	104.8	112.0	1.7 (GGT)	49
(0.5%)	91.8	97.6		
CDOCA (0.5%)	96.5	101.9	2.7 (GST)	50
Taurine (0.3%)	103.7	115.6	1.3 (GGT)	49
TBA (0.05%)	41.5*	50.0*	3.8 (GGT)	51
Orotic acid (1%)	77.1*	59.8*	3.0 (GGT)	52
			2.7 (TF)	53
SorFAE (10%)	75.9*	53.5*	2.0 (TF)	41
Suxibuzone (1%)	46.4*	40.5*	4.99 (TF)	42
Aspirin (0.5%)	60.0*	50.5*	0.95 (TF)	38
			0.44 (TF)	39
Saccharin (5%)	–	101.3	0 (HN)	40
Ethanol (5%)	118.3*	104.8	1.00 (TF)	54
(10%)	133.7*	132.7*	0.76 (TF)	54
Lard (15%)	66.0*	68.9*	5.0 (TF)	55
VA-pal (ip 25 mg/rat)	136.5*	120.8*	1.2 (TF)	56

(*continued*)

Table 1 Continued

Chemicals in diet (concentration)	PK activity [a]		RPA[b] (markers)	Ref.
	Week 2	Week 4		
Chemicals that have not been tested for promotion of cancer				
LCA (0.5%)	77.3	105.7		
Nicotinamide (1%)	71.0*	71.2*		
Nicotinic acid (1%)	79.4*	84.5*		
3-ABA (1%)	79.6*	82.3*		
4-A0BA (0.5%)	117.2*	86.5		
o-AAP (1%)	107.2	99.9		
m-AAP (1%)	96.4	90.6		
p-AAP (1%)	82.1*	73.4*		
Phenacetin (1%)	69.0*	47.8*		
IM (50 ppm)	101.4	101.1		
Lecithin (10%)	77.6*	92.2		
SucFAE (10%)	83.7	81.2		
GFAE (10%)	122.1*	69.1*		
PFAE (15%)	106.2	107.8		
VC (2.5%)	110.6	149.2		
VE (0.72%)	121.8*	110.5		

AAP, acetoamidophenol (=acetaminophen); AB, amobarbital; 3-ABA, 3-aminobenzamide; 4-AOBA, 4-acetoxybenzoic acid; CD, choline deficient; CDOCA, chenodeoxycholic acid; DDT, dichlorodiphenyltrichloroethane; DEN, diethylnitrosamine; DEXA, dexamethasone; DH, diphenylhydantoin; EED, ethynylestradiol; 2-FAA, 2-fluorenylacetamide; GFAE, glycerine fatty acid ester; 3′-MeDAB, 3′-methyl-4-dimethylaminoazobenzene; 2-MeDAB, 2-methyl-4-dimethylaminoazobenzene; IM, indomethacin; LCA, lithocholic acid; PB, phenobarbital; PFAE, propyleneglycol fatty acid ester; SorFAE, sorbitan fatty acid ester; SucFAE, sucrose fatty acid ester; TBA, thiobenzamide; Va-pal, retynylpalmitate; VC, vitamin C; VE, α-tocopherol.

[a]PK activity was expressed as percentage of the control group fed the basal diet. Asterisks (*) show a significant difference from the control group ($p < 0.05$).

[b]RPA was calculated from the increase or decrease in the formation of hepatic tumors (TF), hyperplastic nodules (HN), γ-glutamyltranspeptidase-positive foci/cm^2 (GGT) or placental glutathione S-transferase (GST) by the chemicals: RPA > 1, promotion; RPA = 1, no promotion, no inhibition; RPA < 0.1, inhibition.

[c]Calculated from Ref. 27.

[d]Calculated from Ref. 28. The activity was assayed between 2 and 4 months.

[e]A single intraperitoneal injection. The activity was assayed 1 and 7 days after i.p. injection.

[f]The activity was assayed in 12th week.

Source: Slightly modified from Ref. 57.

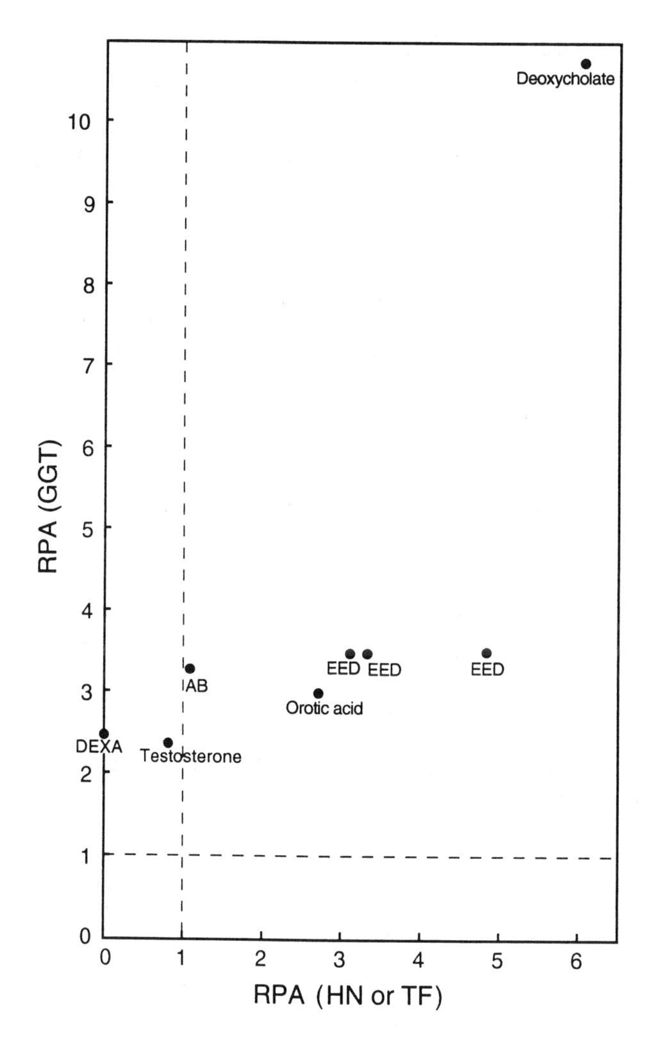

Figure 1 Relationship between relative promotion activities (RPA) calculated from γ-glutamyltranspeptidase-positive foci (GGT) and from hyperplastic nodules (HN) or tumor formation (TF). Abbreviations as in Table . (From Ref. 57.)

after carcinogen administration. 2-Methyl-4-dimethylaminoazobenzene (2-MeDAB), which initiates but does not promote hepatocarcinogenesis (23), caused only a slight, insignificant decrease in PK activity after 2 weeks and had no effect after 4 weeks. In contrast, the isomeric 3′-methyl-4-dimethylaminoazobenzene (3′-MeDAB), which is both an initiator and promoter of hepatocarcinogenesis, caused substantial reduction in the PK activity at the

Table 2 Conditions That Cause a Change in the Activities of Rodent PK Isozymes

Conditons/Chemicals	Days after	No. of rats	Relative pyruvate kinase activities[d] Total	L-type	M2-type	M2 (%)	Ref.
Chemicals							
2.7-Fluorenylenebisacetamide (0.025%) in diet)							
Control (mice)	30–60	8	100	100	100	21.8	12
2,7-FAA (mice)	30	8	104.7	76.2*	191.5*	45.5*	
	60	9	96.7	89.5	118.9	29.9	
Azo dyes (0.06% in diet)							
Control	30–60	9	100	100	100	6.4	11
2-MeDAB	30	4	101.9	101.3	110.2	7.4	
	60	5	81.8	79.8*	100.0	6.6	
3′-MeDAB	30	5	99.1	44.7	851.3*	57.9*	
	60	5	93.8	87.4	182.1	14.2*	
Phenobarbital (0.05% in diet), azo dyes							
Control	30	5	100	100	100	1.7	17
PB	30	3	41.2*	37.6*	206.0	8.7	
Control	60	4	100	100	100	7.5	
PB	60	4	41.0*	38.2	75.2	13.7	
2-MeDAB	30	3	97.3	96.5	142.4	2.5	
3′-MeDAB	30	4	81.0*	77.2*	487.8*	10.5*	
Diethylnitrosamine (ip, 200 mg/kg body weight)							
Control	15	100	100	100	7.0	9	
DEN	1	4	80.8	63.6*	308.4*	26.8*	
Control	7	5	100	100	100	2.5	
DEN	7	4	41.2	26.4*	622.2*	37.3*	
Dietary conditions							
Carbohydrate							
53%	4	4	100	100	100	8.7	11
85%	4	4	258.7*	274.8*	90.0	3.0	
Fasting and alloxane diabetes							
Control	2	6	100	100	100	12.7	11
Fasted	2	4	50.9*	44.6*	94.7	23.5	
Diabetes		4	63.7*	58.9*	96.8	19.2	
Choline deficiency							
Control	84	5	100	100	100	5.0	original
CD	84	5	34.0*	14.5*	396*	59.5*	

(continued)

Table 2 Continued

Conditons/Chemicals	No. of rat tumors	Relative pyruvate kinase activities[d] Total	L-type	M2-type	M2 (%)	Ref.
Liver tumors						
Control	16	100	100	100	20.3	14
HN	14	95.7	69.1	212.6	42.2	
Highly diff.[a]	3	102.9	54.8	312.6	56.3	
Well diff.[b]	5	48.2	17.8	180.2	70.0	
Poorly diff.[c]	2	300.8	52.5	1470.9	91.3	

[a-c]Highly, well, and poorly differentiated hepatocarcinomas, respectively.
[d]Relative values for experimental groups when control activites = 0.
All animals are rats except where indicated (e.g., 2,7-FAA).
For other explanations and abbreviations, see Table 1.

hepatocarcinogenesis, caused substantial reduction in the PK activity at the end of both the second and fourth weeks. The effects of 2-fluorenyl-acetamide (2-FAA) and DEN are evaluated differently because of the use of different experimental protocols (in 2-FAA study, assays for PK were carried out between 2 and 4 months, whereas in the DEN study, the chemical was given in a single i.p. injection). Although most of these hepatic carcinogens tend to cause decrease in total PK activity, the effect was not consistent enough to be useful for a screening test (see below).

In contrast, virtually all the known hepatic promoters (e.g., PB, DDT, EED) have been shown to cause persistent decreases in PK activity (see discussion below).

For chemicals that lack present-day data on their promotional effects, the PK test system found nicotinamide, p-AAP, phenacetin, and GFAE to reduce rat liver PK activity at the important 4-week time point.

C. Conditions That Cause PK Isozyme Changes

It should be noted that the PK assay measures total PK activity. There are at least two types of PK isozymes in rat liver. One is dominant in fetal liver (K- or M_2-type), the other in adult liver (L-type). [The nomenclature used for PK isozymes is quite complicated; readers are referred to the references by Ibsen (24) and Noguchi et al. (25,26) for details.] Table 2 shows the experimental conditions that cause a change in the activities of the PK isozymes. The hepatocarcinogens tend to cause an increase in the fetal type and a decrease in L-type. The extent of the activity changes varied greatly depending on chemicals and time of activity determination (17,27,28). Since extensive changes in hepatic cell types occur during the process of hepatocarcinogenesis (29) and since the increase in the fetal isozyme always occurs in proliferating tissues such as fetal (30) and regenerating livers (20,31) and hepatomas (14,32), the increase in the fetal-type isozyme activity may reflect the change in the liver cell types caused by the administration of hepatic carcinogens (33). In contrast to hepatocarcinogens, most typical hepatic promoters had no appreciable effect on the fetal isozyme but mainly cause a decrease in the L-type isozyme (17).

Recently we completed some studies on the effect of a choline-deficient (CD) diet on liver PK activity in the rat. It is quite an interesting phenomenon that CD alone promotes (34) and induces hepatocarcinogenesis (35) without any additional chemicals. Liver PK activity assayed at the twelfth week of a CD diet was the lowest thus far (Table 1). In addition, the CD diet caused not only a decrease in total PK activity, but also an extensive increase in the fetal isozyme activity (Table 2). This is the same PK-isozyme pattern as observed in the liver of rats administered a complete hepatocarcinogen (17,27,28). The effect of CD diet on PK-isozyme are being further studied in our laboratory.

Figure 2 shows the relationship between the RPA calculated from HN or TF versus the decrease in PK activity. All the potent hepatic promoters with RPA exceeding 2.0 have been shown to cause a reduction in PK activity of at least 20% (i.e., the PK activity in the experimental group is less than 80% of that in the control group). With the exception of aspirin, none of the chemicals with RPA below 1.2 had this type of PK-reducing activity. In hepatocarcinogenesis studies, aspirin was found to have no promoting or enhancing activity in one study (38) and was inhibitory in another (39).

D. Organ Specificity of the PK-Screening Assay

Since the liver is the only organ in which the L-type PK isozyme is major isozyme (30), applicability of the PK-reducing assay for screening promoters seems to be limited to hepatic promoters. For instance, saccharin, which promotes urinary bladder but not liver carcinogenesis (40), did not cause a decrease in liver PK activity (17).

E. Previous Unknown Hepatic Promoters Detected by This PK-Reducing Screening Test

Using the PK-reducing screening test, so far we have detected two previously unknown hepatic promoters. Sorbitan fatty acid ester (SorFAE), a

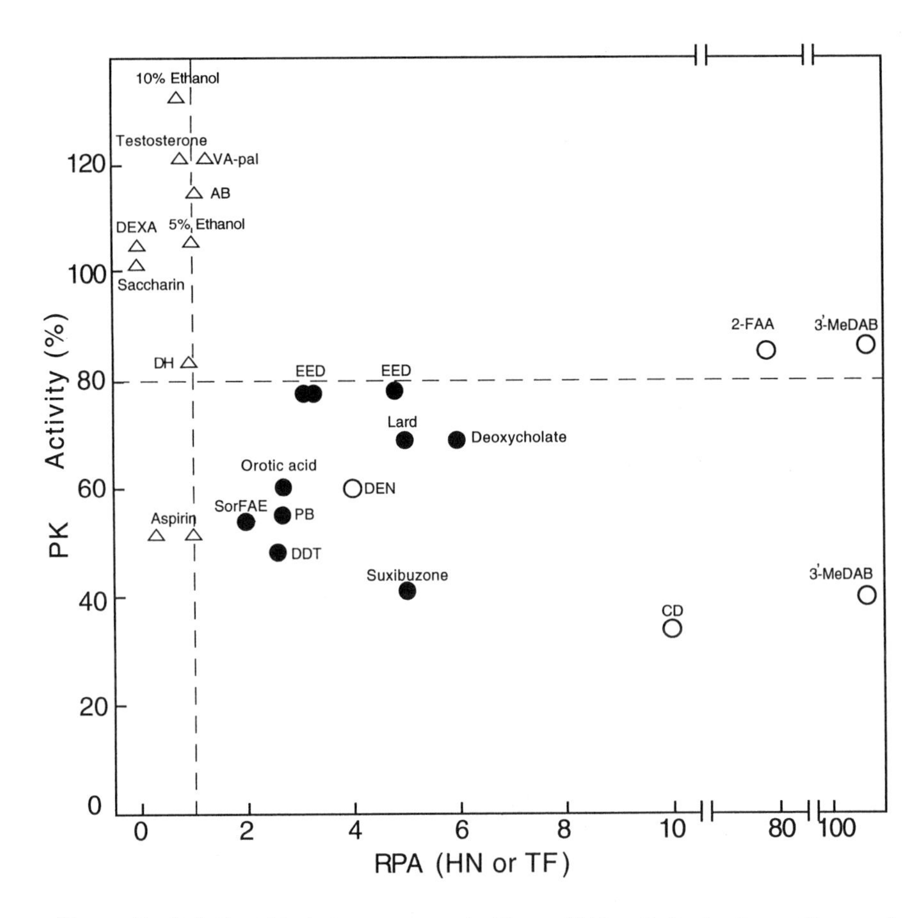

Figure 2 Relationship between pyruvate kinase (PK) activity in the rat liver and RPA calculated from HN or TF. (○), Hepatocarcinogenic conditions; (●), hepatocarcinogenesis promoters; (△), chemicals without carcinogenic and promoting activity. For abbreviations as in Table 1. (Adapted from Ref. 57.)

commonly used emulsifier in the processing of margarine and other food products, has been shown to decrease the PK activity in treated rats (given a diet containing 5% SorFAE) to 59% of the control rats. At the same dietary level, SorFAE has subsequently been found to increase the incidence of liver tumors in 3′-MeDAB–initiated rats from 35% in control to 71% (41). Suxibuzone, a potent antiinflammatory agent, decreased PK activity to 40.5% of control and increased liver tumor incidence from 14 to 86% (42).

IV. DISCUSSION

It was quite unexpected that most hepatic promoters cause a decrease in rat liver PK activity. Since it is well known that in hepatocarcinomas glycolysis is always higher than in normal liver (15,16), it would be expected that the activity of PK, the last regulatory enzyme of glycolysis, would be elevated by hepatic promoters. It is also interesting to note that total PK activity in well-differentiated hepatomas was significantly lower compared with PK activities not only in the normal liver but also in both poorly and highly differentiated hepatomas (Table 2) (14,32). The decrease was due to a decrease in L-type PK activity. PK activity in poorly differentiated hepatomas was higher than in any other liver tissues. However, the isozyme pattern in the poorly differentiated hepatomas was completely different from normal liver and highly differentiated hepatomas. The main isozyme was the fetal type in the former and the L-type in the latter (Table 2). Therefore, for the progression of liver tumors it might be necessary to replace L-type with fetal type. At present, we cannot explain the physiological meaning of the decrease in PK activity during the promoting process.

V. FUTURE PROSPECTS OF THIS ASSAY

A. Advantages

The most important advantage of the PK method for screening potential hepatic promoters may be its simplicity and low cost. Rats are fed a basal diet containing a test compound for 4 weeks without any other treatment. PK assays can be carried out using ordinary laboratory equipments and inexpensive chemical reagents. In contrast, in the GGT-positive foci method, rats must be partially hepatectomized and injected with a potent initiator such as DEN and then histochemical staining of microscope slides of liver sections must be performed (36,37).

B. Disadvantages and Limitations

There are some disadvantages and limitations of the PK method: (a) So far only a limited number of chemical compounds have been testing using this assay. Thus, testing of a broader range of chemicals and confirmation by other laboratories would be desirable to firmly establish the utility of this method. (b) We have no theoretical or mechanistic explanation for the experimentally observed decrease in total PK activity caused by hepatic promoters. (c) Thus far, at least one false-positive—aspirin—has been identified. The sensitivity and specificity of the assay remains to be investigated with a larger number of chemicals.

C. Future Research Prospects

As described above, this method can be relatively easily carried out without highly trained personnel or special equipment. Without additional research and validation, this method can now be utilized as a standard screening test for hepatic promoters. Thus far, SorFAE and suxibuzone have been prospectively detected and subsequently proven to be hepatic promoters based on this PK method. A number of other chemicals cause a persistent decrease in PK activity (Table 1), and it would be interesting to test whether or not any of these compounds are indeed hepatic promoters.

In contrast to promoters, complete hepatocarcinogens cause both a decrease in L-type and an increase in fetal-type PK isozymes. No exceptions to this rule have been observed thus far. For example, the CD diet, which induces hepatic tumors without any additional chemicals, also caused the same change in PK isozyme pattern (Table 2). Thus it may be possible to use the PK isozyme changes for a screening test for hepatocarcinogens. However, since the extent and time of the PK isozyme changes fluctuate greatly, we have not yet been able to establish an isozyme-dependent PK screening test.

ACKNOWLEDGMENTS

The author thanks Drs. D. Nakae and Y. Konishi, Department of Oncological Pathology, for providing the choline-deficient rat livers.

REFERENCES

1. O'Brien TG, Simsiman RC, Boutwell RK. Induction of the polyamine-biosynthetic enzymes in mouse epidermis by tumor-promoting agents. Cancer Res 1975; 35:1662–1670.
2. O'Brien TG, Simsiman RC, Boutwell RK. Induction of the polyamine-biosynthetic enzymes in mouse epidermis and their specificity for tumor promotion. Cancer Res 1975; 35:2426–2433.
3. Furihata C, Yoshida S, Sato Y, Matsushima T. Inductions of ornithine decarboxylase and DNA synthesis in rat stomach mucosa by glandular stomach carcinogens. Jpn J Cancer Res 1987; 78:1363–1369.
4. Abou-El-Ela SH, Prasse KW, Farrell RL, Carroll RW, Wade AE, Bunce OR. Effects of D,L-2-difluoromethylornithine and indomethacin on mammary tu-

mor promotion in rats fed high n-3 and/or n-6 fat diets. Cancer Res 1989; 49: 1434–1440.
5. Takano S, Matsushima M, Erturk E, Bryan GT. Early induction of rat colonic ornithine and S-adenosyl-L-methionine decarboxylase activities by N-methyl-N′-nitro-N-nitrosoguanine and bile salts. Cancer Res 1981; 41:624–628.
6. Verma AK, Erturk E, Bryan GT. Specific binding, stimulation of rodent urinary bladder epithelial ornithine decarboxylase, and induction of transitional cell hyperplasia by the skin tumor promoter 12-0-tetradecanoylphorbol-13-acetate. Cancer Res 1983; 43:5964–5971.
7. Farwell DC, Nolan CE, Herbst EJ. Liver ornithine decarboxylase during phenobarbital promotion of nitrosamine carcinogenesis. Cancer Lett 1978; 5:139–144.
8. Saccone GTP, Pariza MW. Effects of dietary butylated hydroxytoluene and phenobarbital on the activities of ornithine decarboxylase and thymidine kinase in rat liver and lung. Cancer Lett 1978; 5:145–152.
9. Yanagi S, Sasaki K, Yamamoto N. Induction by phenobarbital of ornithine decarboxylase activity in rat liver after initiation with diethylnitrosamine. Cancer Lett 1981; 12:87–91.
10. Kitchin KT, Brown JB. Biochemical effects of two promoters of hepatocarcinogenesis in rats. Food Chem Toxicol 1987; 25:603–605.
11. Endo H, Eguchi M, Yanagi S. Irreversible fixation of increased level of muscle type aldolase activity appearing in rat liver in the early stage of hepatocarcinogenesis. Cancer Res 1970; 30:743–752.
12. Yanagi S, Kamiya T, Ikehara Y, Endo H. Isozymes in the liver from mice given hepatocarcinogen and from tumor-bearing mice. GANN 1971; 62:283–291.
13. Tanaka T, Yanagi S, Miyahara M, Kaku R, Imamura K, Taniuchi K, Suda M. A factor responsible for the metabolic deviations in liver of tumor-bearing animals. GANN 1972; 63:555–562.
14. Yanagi S, Makiura S, Arai M, Matsumura K, Hirao K, Ito N, Tanaka T. Isozyme patterns of pyruvate kinase in various primary liver tumors induced during the process of hepatocarcinogenesis. Cancer Res 1974; 34:2283–2289.
15. Sweeney MJ, Ashmore J, Morris HP, Weber G. Comparative biochemistry of hepatomas. IV. Isotope studies of glucose and fructose metabolism in liver tumors of different growth rates. Cancer Res 1963; 23:995–1002.
16. Burk D, Woods M, Hunter J. On the significance of glucolysis for cancer growth, with special reference to Morris rat hepatomas. J Natl Cancer Res 1967; 38:839–863.
17. Yanagi S, Sakamoto M, Ninomiya Y, Kamiya T. Decrease in L-type pyruvate kinase activity in rat liver by some promoters of hepatocarcinogenesis. J Natl Cancer Inst 1984; 73:887–894.
18. Yanagi S, Sakamoto M, Nakano T. Comparative study on pyruvate kinase activity-reducing action of various promoters of hepatocarcinogenesis. Int J Cancer 1986; 37:459–464.
19. Kimberg DV, Yielding KL. Pyruvate kinase: structural and functional changes induced by diethylstilbestrol and certain steroid hormones. J Biol Chem 1962; 237:3233–3239.

20. Tanaka T, Harano Y, Sue F, Morimura H. Crystallization, characterization and metabolic regulation of two types of pyruvate kinase isolated from rat tissues. J Biochem 1967; 62:71-91.
21. Bucher T, Pfleiderer G. Pyruvate kinase from muscle. In: Colowick SP, Kaplan NO, eds. Methods in Enzymology. New York: Academic Press, 1955: 435-440.
22. Kayne FJ. Pyruvate kinase. In: Boyer PD, ed. The Enzymes. New York: Academic Press, 1973:353-382.
23. Kitagawa T, Pitot HC, Miller EC, Miller JA. Promotion by dietary phenobarbital of hepatocarcinogenesis by 2-methyl-N,N-dimethyl-4-aminoazobenzene in the rat. Cancer Res 1979; 38:112-115.
24. Ibsen KH. Interrelationships and functions of pyruvate kinase isozymes and their variant forms: a review. Cancer Res 1977; 37:341-353.
25. Noguchi T, Inoue H, Tanaka T. The M1- and M2-type isozymes of rat pyruvate kinase are produced from the same gene by alternating RNA splicing. J Biol Chem 1986; 261:13807-13812.
26. Noguchi T, Yamada K, Inoue H, Matsuda T, Tanaka T. The L- and R-type isozymes of rat pyruvate kinase are produced from a single gene by use of different promoters. J Biol Chem 1987; 262:14366-14371.
27. Potter VR, Walker PR, Goodman JI. Survey of current studies on oncogeny as blocked ontogeny: isozyme changes in livers of rats fed 31-methyl-4-dimethylaminoazobenzene with collateral studies on DNA stability. In: Weinhouse S, Ono T, eds. GANN Monographs on Cancer Research. Tokyo: University of Tokyo Press, 1972:121-134.
28. Silber D, Checinska E, Rabczynski J, Kasprzak AA, Kochman M. Isozyme pattern of pyruvate kinase during hepatocarcinogenesis induced by 2-acetylaminofluorene in rat liver. Eur J Cancer 1978; 14:729-739.
29. Farber E. Chemical carcinogenesis: a biologic perspective. Am J Pathol 1982; 106:271-296.
30. Saheki S, Harada K, Sanno Y, Tanaka T. Hybrid isozymes of rat pyruvate kinase: their subunit structure and developmental changes in the liver. Biochim Biophys Acta 1978; 526:116-128.
31. Walker PR, Potter VR. Isozyme studies on adult, regenerating precancerous and developing liver in relation to findings in hepatomas. Adv Enzyme Reg 1972; 10:339-364.
32. Farina FA, Shatton JB, Morris HP, Weinhouse S. Isozymes of pyruvate kinase in liver and hepatomas of the rat. Cancer Res 1974; 34:1439-1446.
33. Reinacher M, Eigenbrodt E, Gerbracht U, Zenk G, Timmerman-Trosiener I, Bentley P, Waechter F, Schulte-Hermann R. Pyruvate kinase isoenzymes in altered foci and carcinoma of rat liver. Carcinogenesis 1986; 7:1351-1357.
34. Takahashi S, Lombardi B, Shinozuka H. Progression of carcinogen-induced foci of γ-glutamyltranspeptidase-positive hepatocytes to hepatomas in rats fed a choline-deficient diet. Int J Cancer 1982; 29:445-450.
35. Mikol YB, Hoover KL, Creasia D, Poirier LA. Hepatocarcinogenesis in rats fed methyl-deficient, amino-acid-defined diets. Carcinogenesis 1983; 4:1619-1629.

36. Tsuda H, Lee G, Farber E. Induction of resistant hepatocytes as a new principle for a possible short-term in vivo test for carcinogens. Cancer Res 1980; 40: 1157–1164.
37. Cameron RG, Imaida K, Tsuda H, Ito N. Promotive effects of steroids and bile acids on hepatocarcinogenesis initiated by diethylnitrosamine. Cancer Res 1982; 42:2426–2428.
38. Yanagi S, Yamashita M, Hiasa Y. Aspirin shares a short-term effect, inhibition of pyruvate kinase activity with tumor-promoting agents, but fails to promote rat livercarcinogenesis. Oncology 1993; 50:275–278.
39. Tang Q, Denda A, Tsujiuchi T, Tsutsumi M, Amanuma T, Murata Y, Maruyama H, Konishi Y. Inhibitory effects of inhibitors of arachidonic acid metabolism on the evolution of rat liver preneoplastic foci into nodules and hepatocellular carcinomas with or without phenobarbital exposure. Jpn J Cancer Res 1993; 84:120–127.
40. Nakanishi K, Fukushima S, Hagiwara A, Tamano S, Ito N. Organ-specific promoting effects of phenobarbital sodium and sodium saccharin in the induction of liver and urinary bladder tumors in male F344 rats. J Natl Cancer Inst 1982; 68:497–500.
41. Yanagi S, Sakamoto M, Takahashi S, Hasuike A, Konishi Y, Kumazawa K, Nakano T. Enhancement of hepatocarcinogenesis by sorbitan fatty acid ester, a liver pyruvate kinase activity-reducing substance. J Natl Cancer Inst 1985; 75:381–384.
42. Yanagi S, Sakamoto M, Takahashi S, Tsutsumi M, Konishi Y, Shibata K, Kamiya T. Promotion of hepatocarcinogenesis by suxibuzone in rats initiated with 3′-methyl-4-dimethylaminobenzene. Cancer Lett 1987; 36:11–18.
43. Tatematsu M, Shirai T, Tsuda H, Miyata Y, Shinohara Y, Ito N. Rapid production of hyperplastic liver nodules in rats treated with carcinogenic chemicals: a new approach for in vivo short-term screening test for hepatocarcinogens. Gann 1977; 68:499–507.
44. Miller JA, Miller EC. The carcinogenic aminoazo dyes. Adv Cancer Res 1953; 1:339–396.
45. Peraino C, Fry RJM, Staffeldt E, Christopher JP. Comparative enhancing effects of phenobarbital, amobarbital, diphenylhydantoin, and dichlorodiphenyltrichloroethane on 2-acetylaminofluorene-induced hepatic tumorigenesis in the rat. Cancer Res 1975; 35:2884–2890.
46. Shinozuka H, Lombardi B, Abanobi SE. A comparative study of the efficacy of four barbiturates as promoters of the development of γ-glutamyltranspeptidase-positive foci in the liver of carcinogen treated rats. Carcinogenesis 1982; 3:1017–1020.
47. Yager JD, Campbell HA, Longnecker DS, Roebuck BD, Benoit MC. Enhancement of hepatocatrcinogenesis in female rats by ethinyl estradiol and mestranol but not estradiol. Cancer Res 1984; 44:3862–3869.
48. Mayol X, Perez-Tomas R, Cullere X, Romero A, Estadella MD, Domingo J. Cell proliferation and tumour promotion by ethinyl estradiol in rat hepatocarcinogenesis. Carcinogenesis 1991; 12:1133–1136.

49. Tsuda H, Nakanishi K, Sakata T, Imaida K, Kitaori M, Ikawa E, Katagiri K, Inaguma H. Promotion effects of primary bile acids and taurine on hepatocarcinogenesis initiated by diethylnitrosamine (in Japanese). Proc Jpn Cancer Assoc 1983; 42:62.
50. Blair PC, Popp JA, Bryant-Varela BJ, Thompson MB. Promotion of hepatocellular foci in female rats by chenodeoxycholic acid. Carcinogenesis 1991; 12: 59–63.
51. Malvaldi G, Chieli E, Saviozzi M. Promotive effects of thiobenzamine on liver carcinogenesis. Gann 1983; 74:469–471.
52. Columbano A, Ledda GM, Rao PM, Rajalakshmi S, Sarma DSR. Dietary orotic acid, a new selective growth stimulus for carcinogen altered hepatocytes in rat. Cancer Lett 1982; 16:191–196.
53. Laurier C, Tatematsu M, Rao PM, Rajalakshimi S, Sarma DSR. Promotion by orotic acid of liver carcinogenesis in rats initiated by 1,2-dimethylhydrazine. Cancer Res 1984; 44:2186–2191.
54. Yanagi S, Yamashita M, Hiasa Y, Kamiya T. Effects of ethanol on hepatocarcinogenesis initiated in rats with 3′-methyl-4-diemthylaminoazobenzene in the absence of liver injuries. Int J Cancer 1989; 44:681–684.
55. McCay PB, King M, Rikans LE, Pitha JV. J Environ Pathol Toxicol 1980; 3: 451.
56. Ohkawa K, Abe T, Hatano T, Takizawa N, Yamada K, Takada K. The facilitated effect of retinol on rat hepatocarcinogenesis induced by 3′-methyl-4-dimethylaminoazobenzene. Carcinogenesis 1991; 12:2357–2360.
57. Yanagi S. Liver pyruvate kinase assay as a predicting screening test for promoters of hepatocarcinogenesis in the rat. Environ Carcin Ecotox Rev 1994; C12:89–104.

12
Predicting Carcinogenicity: Peroxisome Proliferators

Jonathan D. Tugwood and Clifford R. Elcombe*
Zeneca Central Toxicology Laboratory, Alderley Park, Macclesfield, Cheshire, United Kingdom

Peroxisome proliferators (PPs) are a diverse class of chemicals that share the common ability to induce the proliferation of hepatic peroxisomes in rodents and which constitute a discrete class of rodent hepatocarcinogens (1). Many pharmaceutically and industrially important compounds are PPs, including fibrate hypolipidemic drugs for the treatment of ischemic heart disease, industrial plasticizers, and a number of agrochemicals including pesticides and fertilizers. As a consequence, a large amount of research effort has been directed towards a more complete understanding of the peroxisome proliferation phenomenon, with particular emphasis on two main questions:

1. Do the early phenomena of peroxisome proliferation and other hepatic changes have any predictive value for, and mechanistic association with, rodent hepatocarcinogenesis?
2. Do PPs pose a hepatocarcinogenic risk to humans?

The accumulation of a large amount of in vivo and in vitro data from multiple species, compounds, and doses, combined with the application of the latest molecular biology technology, have made possible significant progress towards answering these questions (see Ref. 2 for review). How-

**Current affiliation*: University of Dundee, Ninewells Hospital and Medical School, Dundee, Scotland

ever, a complete understanding of mechanisms, particularly with regard to PP-induced liver growth and cancer, is still some way off.

I. MODELS OF PEROXISOME PROLIFERATOR–INDUCED HEPATOCARCINOGENESIS

Obviously, mechanistic information is crucial both to the development and evaluation of predictive tests for PP-induced hepatocarcinogenesis in the rodent and for the extrapolation to humans for risk-assessment purposes. However, at present the precise mechanism(s) by which PPs cause rodent liver cancer remain to be defined. What is clear is that nearly all hepatocarcinogenic PPs are nongenotoxic, as determined by a variety of in vitro and in vivo assays. This lack of genotoxicity, coupled with a good understanding of the biochemical effects of PPs on rodent hepatocytes, has led to the proposal of two models for the carcinogenicity of PPs. These are depicted schematically in Figure 1.

A. Oxidative Damage

First, it has been proposed that the action of PPs, via the generation of hydrogen peroxide, results in the formation of reactive oxygen species,

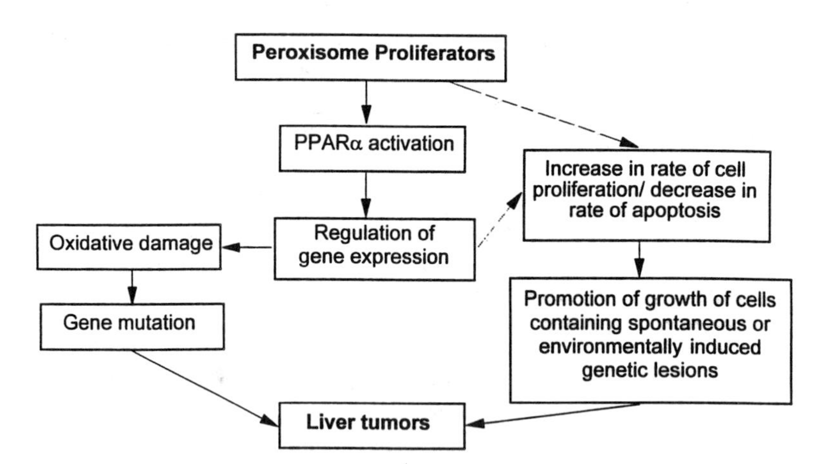

Figure 1 Model depicting alternative mechanisms by which peroxisome proliferators may cause liver tumors in rodents. Broken arrows indicate the possibility that PPs may regulate liver cell growth by a mechanism other than regulation of gene transcription via PPARα activation.

which have the capacity to damage DNA (3). This model suggests that PPs are indirect-acting genotoxic agents and follows from the observation that PPs strongly induce the activity of peroxisomal fatty acyl CoA oxidase (the action of which generates hydrogen peroxide), without a significant concomitant increase in the enzymes that remove hydrogen peroxide, e.g., catalase. Increases in numbers of altered DNA bases such as 8-OH deoxyguanosine have been detected in the livers of animals treated with PPs, although such increases are relatively modest (approximately 1.5–2.5 times control level). Perhaps more significant is the observation that the majority of these altered bases are to be found in the mitochondrial DNA (4), which is consistent with the observation that the mitochondrial DNA is generally more susceptible to oxidative damage than nuclear DNA (5). This calls into question the relevance of low levels of oxidative damage to the carcinogenic action of PPs, particularly since there is no correlation between the carcinogenic potency of PP compounds and the degree to which they apparently cause oxidative damage in the livers of treated animals. This lack of correlation is exemplified by the compounds di(2-ethylhexyl)phthalate and di-(2-ethylhexyl)adipate, both of which cause increases in 8-OH deoxyguanosine levels in rat liver (6), and yet only di(2-ethylhexyl)phthalate is a rat liver carcinogen. Therefore, while certain PPs may indirectly induce a degree of DNA damage, this damage is not detected by standard genotoxicity assays, and oxidative damage per se is insufficient to account for the hepatocarcinogenicity of PPs in rodents.

B. Liver Growth Regulation

The second hypothesis invokes PP-induced liver growth as a mechanism for tumorigenesis. The increase in liver size and weight that follows PP administration to rats and mice is a consequence both of increases in cell numbers due to enhanced cell replication and a decrease in cell death due to a lowering of the rate of apoptosis (see Ref. 7 for review). This is a characteristic not just of PPs, but also of other classes of nongenotoxic carcinogens. One possibility is that the process of DNA replication itself causes mutation, since it "fixes" genetic lesions that in a quiescent cell would be inconsequential (8). Alternatively, PPs may promote the growth of cells or groups of cells ("altered foci") that already have a genetic lesion, acquired either spontaneously or through environmental damage. Indeed, some evidence suggests that nongenotoxic carcinogens act preferentially on these altered cell populations, thus providing them with a growth advantage (9). This concept is supported by data showing that aged animals, which have higher levels of altered hepatic foci as determined by morphological criteria, are more susceptible to PP-induced liver cancer (10). Ashby et al. (2) have examined the relationship between peroxisome proliferation, stimulation of DNA synthesis, and hepatocellular carcinogenesis. Reviewing the available data in the open literature, these authors identified 71 chemicals shown to be peroxisome proliferators; chronic carcinogenicity data were available for 28 of these chemicals. Following careful review only 18 of these studies were considered to be adequate by modern standards (Table 1; see Ref. 2 for a full explanation of the selection criteria). The correlation between peroxisome proliferation and hepatocellular tumors was about 80% in both rats and mice. In general, the more potent the peroxisome proliferator, the higher the tumor incidence, although some exceptions were noted. The only consistent anomaly is observed in studies using di(2-ethylhexyl)adipate, which causes hepatic peroxisome proliferation in both male and female rats and mice, but liver tumors have only been detected in bioassays with female mice. However, none of the 71 compounds show tumorigenic activity in the absence of peroxisome proliferation in either sex of any species.

A similarly strong relationship also was found between stimulation of replicative DNA synthesis and liver tumors, although only 8 out of the 18 chemicals described above were included in this correlation, as insufficient data were available for the remainder. Interestingly, there seemed to be a similar predictivity with early (< 14 days) and subchronic (> 50 days) replicative DNA synthesis.

The other side of the liver growth "equation," programmed cell death or apoptosis, is now recognized as an important factor in xenobiotic-induced liver carcinogenesis (see Ref. 11 for review). Recent data have demonstrated that PPs are able to lower the rate of apoptosis in in vitro cultures of rat hepatocytes (12) and suppress basal levels of apoptosis in rat liver in vivo (13). Since apoptosis is the principal mechanism by which aberrant or damaged cells are deleted, it is conceivable that the suppression of this process by PPs may allow the persistence of altered cells that would otherwise be eliminated. Indeed, the growth of these altered cells may then be preferentially enhanced by PPs, as mentioned above. This coupled enhancement of survival and growth of altered hepatocytes by PPs provides a cogent hypothesis of how these compounds promote liver carcinogenesis. It is clearly not as straightforward as this, however, since it has been shown recently that PPs are able to suppress in vitro apoptotic death of hepatocytes from several species, including guinea pig (14). The guinea pig is unresponsive to PPs (at present there is no carcinogenicity data available in the guinea pig; see below, and Table 2). Further data on the mechanism(s) by which PPs regulate hepatocyte replication and apoptosis will help clarify this situation.

Table 1 Chemicals Reported to Induce Peroxisome Proliferation in the Liver

Acetylsalicylic acid	**Di(2-ethylhexyl)adipate (DEHA)**	2,2,4,4,6,8,8-Heptylmethyl-nonane	Perfluorobutyric acid
Beclobric acid	Di(2-ethylhexyl)phosphate	Hexadecanedioic acid	Perfluorodecanoic acid
Benzbromarone	**Di(2-ethylhexyl)phthalate (DEHP)**	Hexanoic acid	Perfluorooctanoic acid
Butyl Benzyl phthalate	Di(2-ethylhexyl)sebacate	**Lactofen**	Perfluoroocytyl sulphuric acid
Bezafibrate	Dihydroepiandrosterone (DHEA)	LK903	RMI-14514
Bis(carboxymethylthio)-1,10-decane	Diisodecyl phthalate	LS2265	S-8527
BR-931	**Diisononyl phthalate (DINP)**	**LY 171883**	SaH-442348
Chlorinated paraffins ($C_{12\text{-}14}$)	Dimethrin	Medica 16	Simfibrate
Cinnamyl anthranilate	Diundecyl phthalate	Methyl chlorophenoxyacetic acid (MCPA)	Tiadenol
Ciprofibrate	DL-040	**Methyl clofenapate**	**Tibric acid**
Citral	Ethyl 4-(4-chlorophenoxy)butanoate	1-mono(carboxyethylthio) tetradecone	**Trichloroacetic acid**
Clobuzarit (ICI-55897)	2-Ethylhexanoic acid	1-mono(carboxymethylthio) tetradecone	**Trichloroethylene**
Clofibrate	2-Ethylhexanol	Mono-(2-ethylhexyl)phthalate	2,4,5-Trichlorophenoxyacetic acid (2,4,5-T)
Clofibric acid	2-Ethylhexylaldehyde	Mono-n-octyl phthalate	2,2,4-Trimethylpentane
DG5685	Fenofibrate	**Nafenopin**	Tridiphane
DH6463	Gemcadiol	Niadenate	Valproic acid
Dibutyl phthalate	**Gemfibrozil**	OKY-1581	**Wy-14,643**
2,4-Dichlorophenyloxyacetic acid (2,4-D)	Halofenate	**Perchloroethylene**	

Underlined chemicals are those for which well-conducted carcinogenicity bioassays have been performed. Chemicals in bold type are those for which well-conducted carcinogenicity bioassays and relevant quantitative peroxisome proliferation studies have been carried out.
Source: Ref. 2.

Table 2 Species Differences in Responses to Peroxisome Proliferators According to In Vivo and In Vitro Criteria

	In vivo peroxisome proliferation	In vitro peroxisome proliferation	Increase in hepatic S-phase	PPARα activation by PPs	Liver tumors
Mouse	Yes	Yes	Yes	Yes	Yes
Rat	Yes	Yes	Yes	Yes	Yes
Guinea pig	No[a]	No	No	ND	ND
Dog	No	No	ND	ND	ND
Nonhuman primate	No[b]	No	ND	ND	ND
Human	No	No	ND	Yes[c]	No[d]

[a]Two studies have reported marginal (<threefold) increases in peroxisome proliferation in guinea pigs following administration of bezafibrate (69) and tiadenol (70).
[b]One group has reported increased peroxisome proliferation in rhesus and cynomolgus monkeys following ciprofibrate treatment (71).
[c]Of the human PPARαs described to date, two were activated by PPs in in vitro assays (28,49). However, recent data describe a human PPARα variant not similarly activated (51).
[d]Human tumor data obtained from follow-up studies of patients treated with hypolipidemic drugs (see Ref. 40).
ND = Not determined.

II. A RECEPTOR-MEDIATED MECHANISM OF PEROXISOME PROLIFERATOR ACTION

Qualitatively, induction of both peroxisome proliferation and hepatic S-phase by PPs are good predictors of rodent hepatocarcinogenesis, and even if there were no mechanistic link between these parameters and tumor development, they would still be of value. However, determination of the precise molecular events involved in PP-induced liver tumorigenesis would allow both the refinement of existing predictive methodologies and the development and evaluation of new predictive methods.

Given that PPs are not direct-acting mutagens, it is clear that the cell must "sense" the presence of the compounds by some other means in order to elaborate a biochemical response. The most obvious candidates for such a sensory mechanism are receptors, since a wide variety of other chemical signals exert their effects in this way, including hormones, growth factors, and neurotransmitters. Indeed, there is a very good precedent for other nongenotoxic carcinogens mediating their effects via receptors. These compounds are collectively known as dioxins, a term that describes a diverse group of halogenated chemicals, the most potent of which is 2,3,7,8-

tetrachlorodibenzo-*p*-dioxin (TCDD). It had been clear for some time that a particular genetic locus, the *Ah* locus (for *A*romatic *h*ydrocarbon), is important in dioxin-related toxicity (15), and a protein product of this locus was shown to be a nuclear receptor of the 'helix-loop-helix" class (16,17). This molecule forms a heterodimeric complex with another helix-loop-helix protein, the Ah receptor nuclear translocator (Arnt) (18). This receptor complex is activated by binding with TCDD, and the series of molecular events associated with this activation process confers on the receptor complex the ability to bind specifically to DNA and regulate gene transcription. This transcriptional activation process has been best characterized for the cytochrome P4501A1 (CYP1A1) gene, which contains "xenobiotic response elements" (XREs) in the promoter region, which are DNA sequences recognized by the Ahr-Arnt complex and which facilitate TCDD-mediated transcriptional regulation (19–21). It is now generally accepted that all the toxic effects of TCDD and related compounds are mediated by the Ah receptor, supported by the observation that transgenic mice lacking the Ah receptor are unresponsive to dioxins (22). The isolation of the Ah receptor and its subsequent characterization has allowed rapid progress in understanding the mechanisms of dioxin action, although admittedly the picture is still far from complete. Significantly, a number of parallels can be drawn between the toxic effects of TCDD and those of PPs, in addition to their nongenotoxic mode of action. In the liver, both classes of compounds regulate gene expression, cause hyperplasia and fatty changes, and show a degree of species-specificity in their effects: even at occupational exposure levels, there appears to be no adverse effect on liver function in humans (23).

Some time ago it was demonstrated that the regulation of peroxisomal enzymes in the liver by PPs occurs at the transcriptional level (24), and this implies that PPs act via a transcription factor or factors. Shortly afterwards, a novel member of the steroid hormone receptor superfamily was isolated from mouse liver which could be activated by PPs in an in vitro transactivation assay (the peroxisome proliferator-activated receptor, or PPAR) (25). It became rapidly evident that rather than being a unique receptor, the PPAR was one of a complex receptor family, present in various species and expressed in multiple tissues. PPAR "subtypes" have been discovered not only in mouse and rat, but also in *Xenopus* (26), hamster (27), and, significantly, human (28–30). To date, four subtypes have been identified (α, β, γ, and δ), the "original" PPAR from mouse liver being referred to hereafter as PPARα.

The existence of a PPAR "superfamily" raises the question whether any or all of these receptors are involved in PP-induced rodent liver peroxisome proliferation and carcinogenesis. Several lines of evidence implicate PPARα as the important subtype in this respect. First, the degree to which

PPARα can be transcriptionally activated by PPs reflects the potency of the compounds as carcinogens in rodent bioassays (31). Other nongenotoxic rodent liver carcinogens, such as phenobarbital, fail to activate PPARα (our unpublished results). Second, like a number of other ligand-activated transcription factors, the PPARα forms heterodimers with another nuclear receptor, the retinoid X receptor (RXR) (32). The PPARα/RXR heterodimers bind to specific DNA sequences in the promoters of genes that are induced by PPs, thereby regulating their rate of transcription. These DNA sequences, peroxisome proliferator response elements (PPREs), have been identified in the promoters of a number of genes, significantly in the peroxisomal fatty acyl CoA oxidase (33) and bifunctional hydratase/dehydrogenase (34) and in the microsomal cytochrome P450IVA1 (35). Third, and in contrast to the other PPAR subtypes, PPARα transcripts are abundant in rat and mouse liver (36,37). Finally, and most convincingly, homozygous deletion of PPARα from the mouse genome renders the animal incapable of early responses to PPs, including peroxisome proliferation, peroxisomal enzyme induction, and hepatomegaly (38). If these "knockout" animals fail also to develop liver tumors in a bioassay, this will provide a compelling argument for PPARα mediating the pleiotropic rodent responses to PPs, including rodent hepatocarcinogenesis.

III. SPECIES DIFFERENCES IN RESPONSES TO PEROXISOME PROLIFERATORS

Both in vivo and in vitro, responses to PPs are highly variable between species (reviewed in Ref. 1; see Table 2). For example, dogs, nonhuman primates, and guinea pigs appeared to be completely unresponsive to PP administration in that no peroxisome proliferation occurs at dose levels that induce significant hepatic changes in the mouse and rat. In vitro analyses employing primary cell cultures have contributed the significant finding that human hepatocytes also show no response to PPs (39). This is consistent with limited epidemiological data, suggesting that patients receiving fibrate drugs do not show an increased tumor incidence (40).

Therefore, available data indicate that following exposure to PPs, humans do not undergo all of the hepatic changes that, in the rat and mouse, are strongly correlated with hepatocarcinogenesis. The following sections deal in the main with rodent-based systems for assaying the toxicity and carcinogenic potential of PPs, and the relevence of these for predicting human carcinogenicity is discussed.

IV. IN VIVO TESTING FOR THE PEROXISOME PROLIFERATION PHENOMENON

The peroxisome proliferation phenomenon in rodents is characterized by marked hepatomegaly due largely to hepatocellular hypertrophy (peroxisome proliferation and smooth endoplasmic reticulum proliferation) and hyperplasia (increased replicative DNA synthesis and cell division).

A. Peroxisome Proliferation

Screens for peroxisome proliferators have normally involved the administration of chemicals to rats or mice for between 14 and 28 days. After this time period livers are examined in a variety of ways to assess peroxisome proliferation. The definitive measure of peroxisome proliferation is obviously the determination by electron microscopy and quantitative morphometrics of organelle number or, more correctly, organelle volume (or fractional) density. This latter measure determines the volume of the cytoplasm occupied by the peroxisomes (for experimental principles, see Ref. 41).

Because morphometric procedures are time-consuming and expensive, peroxisome proliferation is more frequently estimated by the measurement of an appropriate peroxisomal marker enzyme activity. Perhaps the best enzymatic marker of peroxisomal proliferation is acyl CoA oxidase. This is the first and rate-limiting enzyme in the peroxisomal β-oxidation of fatty acyl CoA esters. Usually, the overall activity of the peroxisomal β-oxidation pathway is determined by observing the reduction of NAD^+ during the metabolism of palmitoyl CoA in the presence of cyanide to inhibit mitochondrial oxidation. This is referred to as cyanide-insensitive palmitoyl CoA oxidation (PCO). This activity may be measured in whole liver homogenates or in cellular fractions obtained from the resuspended 15,000 g (av) pellets. These enzymatic assays are relatively inexpensive and take only a few minutes per sample. Several studies have shown that the determination of a relevant (e.g., peroxisomal β-oxidation) peroxisomal marker enzyme is predictive of the degree of peroxisome proliferation determined by quantitative morphometric procedures. Alternatively, this enzyme activity may be measured directly by the production of H_2O_2 during the oxidation of a fatty acyl CoA (e.g., palmitoyl CoA). Many peroxisomally localized enzymes (e.g., catalase, D-amino acid oxidase, urate oxidase) are not appropriate markers because they are either only weakly induced or not induced during the chemically stimulated process of peroxisome proliferation.

B. Replicative DNA Synthesis (S-Phase)

As described earlier, stimulation of hepatocellular DNA synthesis is a characteristic of peroxisome proliferator administration to rats and mice. This normally occurs early in the study (1–14 days, depending on the chemical and its dose), returns to normal values, and then may increase at a later time. The determination of chemically stimulated replicative DNA synthesis is best performed using surgically implanted osmotic pumps containing a DNA precursor, such as ^{3}H-thymidine or bromodeoxyuridine (BrdU). Hepatocyte nuclei which have incorporated these materials during the labeling period can then be counted under the microscope following autoradiography or immunohistochemistry. The pros and cons of these different methodologies are discussed in depth elsewhere (2,42,43). Microscopic enumeration of the proportion of cells undergoing DNA synthesis during the treatment time is moderately expensive; however, at the present time it is the most reliable marker for hepatocellular DNA synthesis. Others procedures, such as extraction of hepatic DNA and quantitation of radioactivity incorporated by scintillation counting, do not eliminate other cell types within the liver.

C. A Typical Study Designed to Assess a Chemical as a Peroxisome Proliferator

In vivo toxicity studies are normally employed to identify a chemical as a potential peroxisome proliferator. The chemical is administered to animals by an appropriate route for either 28 or 90 days. Osmotic pumps containing BrdU or ^{3}H-thymidine are surgically implanted on day 0 and day 21 or 83 of dosing (7 days prior to sacrifice of the animals). At post mortem samples of liver are taken for (a) the measurement of CN^{-}-insensitive palmitoyl CoA oxidation, (b) the determination of the number of hepatocyte nuclei incorporating the DNA precursor, and, sometimes, (c) electron microscopy and quantitative morphometrics.

This assay is based on an assay of Bronfman et al. (44) for the measurement of CN^{-}-insensitive palmitoyl CoA oxidation. Liver homogenates, 15,000-g pellets, or hepatocyte sonicates may be used as the enzyme source.

Reagents	Concentrations
1. Tris-HCl	60 mM, pH 8.3
2. Reagent A:	
Coenzyme A	75 μM
NAD^{+}	555 μM

Nicotinamide	141 mM
Dithiothreitol	4.2 mM
KCN	3 mM
Bovine serum albumin	0.225 mg/ml (fatty acid free)
3. FAD$^+$	14.9 mg/ml in 60 mM Tris-HCl
4. Palmitoyl CoA	7.54 mg/ml
5. Triton X-100	1% (v/v) in 60 mM Tris-HCl

The procedure is as follows:

1. Samples are diluted 1 : 1 (v/v) with 1% Triton X-100 solution and incubated at 37 °C for 5 minutes prior to assay.
2. Reagent A (2 ml), FAD (0.02 ml), and an aliquot of the resuspended 15,000-g pellet (final protein concentration <0.2 mg/ml) or hepatocyte sonicate (<0.1 mg) are added to each cuvette.
3. Volumes are adjusted to 3 ml with 60 mM Tris-Hcl.
4. Baseline absorbance at 340 nm using a kinetic spectrophotometer is recorded.
5. 0.02 ml palmitoyl CoA solution is added and ΔA_{340} recorded until a linear gradient is obtained.
6. Oxidation is expressed as nmol NAD$^+$ reduced per minute per mg protein using an extinction coefficient of 6.22 $mM^{-1}cm^{-1}$ for NADH.

The above assay has been extensively validated in a large number of laboratories over a long period of time and is generally accepted as being a reliable biochemical indicator of peroxisome proliferation.

V. RECEPTOR-BASED "TRANSACTIVATION" ASSAYS OF PEROXISOME PROLIFERATOR ACTIVITY

Following the discovery of the PPARα and the demonstration of its involvement in rodent peroxisome proliferation, it became possible to use molecular biology techniques to "screen" chemicals for PP activity by determining the ability of the compound to activate PPARα. Experiments such as these have corroborated and extended peroxisome proliferation data obtained from more conventional sources and provided data unconfounded by pharmacokinetics and rate of compound uptake.

The introduction of cloned DNAs into eukaryotic cells (transfection) has proved a valuable technique with a number of applications, including analysis of gene function and gene regulatory elements and the overproduction of proteins for structural study and antibody generation. Perhaps the simplest and most widespread application is the study of gene function by transient transfection analysis, which can rapidly and efficiently generate large amounts of quantitative data.

There are a number of protocols available to introduce DNA into eukaryotic cells efficiently (reviewed in Refs. 45,46), but the following method is the one favored in our own laboratory, represented diagrammatically in Figure 2. The cloned DNAs used are as follows.

1. Expression vector. This contains the full-length complementary DNA (cDNA) encoding the protein of interest, under the control of a suitable promoter to obtain high level expression in eukaryotic cells, e.g., human cytomegalovirus immediate-early promoter (47). In this case the cDNA would specify mouse PPARα.

2. Reporter plasmid. One of the most widely-used "reporter" genes is the bacterial chloramphenicol acetyl transferase (CAT) gene (48), which can be easily and sensitively detected. This is placed under the control of a promoter responsive to the protein/factor encoded by the expression plasmid. In this case the rat acyl CoA oxidase promoter would be suitable, since this is transcriptionally regulated by PPARα (33).

3. β-Galactosidase expression plasmid. This also encodes an easily assayable enzyme, this time under the control of a constitutive promoter. This is included to allow the efficiency with which the cells are transfected

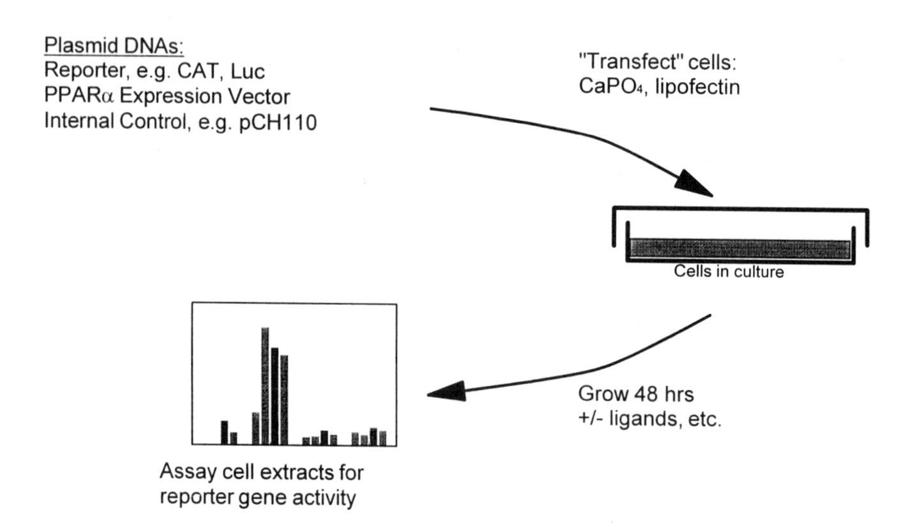

to be determined. A suitable plasmid is pCH110 (Pharmacia Biotech.), which expresses β-galactosidase from the SV40 virus early promoter.

4. An inert "carrier" plasmid, e.g., pBR322 or pBluescript (Stratagene, Inc.) In a typical experiment, 3 μg plasmid DNAs would be transfected onto a monolayer of approximately 2×10^6 cultured hepatoma cells (e.g., Hepa1c1c7). This would comprise 0.2 μg reporter plasmid, 0.3 μg expression vector, 0.5 μg β-gal plasmid, and 2.0 μg carrier (as indicted above). The most commonly employed transfection protocol uses a lipophilic medium (e.g., DOTAP, Boehringer Mannheim) to introduce the DNA into the cells, although alternative protocols use DEAE dextran or calcium phosphate or electroporation (45). Following the introduction of the DNA into the cells, the culture is maintained for 48 hours in the presence of a range of concentrations of the test chemical. Cell extracts are then prepared and assayed for β-galactosidase and CAT activity. The activity of the extract is expressed as percentage acetylated chloramphenicol per unit β-galactosidase, which gives a quantitative measure of the compound's potency as a PPARα activator.

Table 3 shows the results of a typical analysis conducted with a range of compounds that have peroxisome proliferating activity in vivo. There is a good "dose-response" with most of the compounds, although some were toxic to the transfected cells at the highest concentration used. There was also a good concordance between the ability to activate mouse PPARα in the assay and potency in vivo. There are few other reports in the literature where a range of compounds have been analyzed simultaneously in a transactivation assay. The literature does contain many studies in which a few PPs have been tested. In these investigations, often the aim is not to assay potency of the compounds but to characterize a novel receptor or promoter, etc. Nonetheless, in experiments where PPARα activation by Wy-14,643 and clofibrate has been compared, the more potent PP (Wy-14,643) is the most efficient receptor activator. This is a consistent finding despite the differences in methodology and irrespective of whether the receptor origin is mouse (25,31), rat (49), *Xenopus* (26), or human (28,49,50).

Although this in vitro receptor-based assay has a sound theoretical basis and has generated some useful data, it still falls short of being a robust indicator of human carcinogenic risk from peroxisome proliferators. A number of issues need to be resolved before the in vitro assay could be said to have any predictive value.

First, the assay needs establishment of a standardized protocol, then extensive validation by the study of large numbers of compounds, both "positive" and "negative." Although comparatively little data have been generated so far, the present data set was obtained using a bewildering range of reagents, protocols, and systems. One approach would be to select

Table 3 Peroxisome Proliferators Activate Mouse PPARα in an In Vitro Transactivation Assay

	CAT activity (arbitrary units)		
Compound	1 μM	10 μM	100 μM
Dimethylsulfoxide (vehicle)	7.0	7.0	7.0
Wy-14,643	63.6	100*	85.0
Nafenopin	12.7	77.6	104.7
Ciprofibrate	14.3	44.8	81.8
Methylclofenapate	7.1	32.0	Toxic
Bezafibrate	7.7	25.2	67.1
Gemfibrozil	81.4	18.4	6.1
Clofibric acid	5.5	12.4	71.4
Mono-(2-ethylhexyl) phthalate	9.0	32.1	Toxic
Trichloroacetic acid	ND	ND	10.8
Dehydroepiandrosterone	ND	5.6	Toxic
Dehydroepiandrosterone sulfate	ND	ND	4.1

Hepa1c1c7 cells were transfected with a plasmid expressing mouse PPARα, a chloramphenicol acetyl transferase (CAT) reporter plasmid containing part of the rat acyl CoA oxidase promoter, and a control plasmid expressing β-galactosidase (see Sec. V). Cells were maintained for 48 hours in medium containing peroxisome proliferators at the indicated concentrations. CAT activities were determined and normalized to the β-galactosidase standard. The degree of CAT activity generated in this assay using 10 μM Wy-14,643 is arbitrarily assigned a value of 100 units (*); this is equivalent to acetylation of approximately 35% of the chloramphenicol. Each compound was tested in triplicate, and the results shown are an average of 6–12 experiments.
Source: Ref. 31.

an appropriate cell line, e.g., a rodent liver-derived line, and to integrate stably a suitable reporter construct, perhaps a large promoter fragment from a rat PP-responsive gene driving CAT or luciferase expression. Such a cell line could then be distributed for independent evaluation by a number of laboratories. A good starting point would be the evaluation of the 18 chemicals for which adequate, quantitative in vivo data exist for peroxisome proliferation and carcinogenesis (see Table 1), thereby determining the degree of concordance between receptor activation and in vivo potency.

Second, the relevance of the in vitro assay requires an improved understanding of human responses to PPs and the role of human PPARα in mediating such responses. When responses of human and mouse PPARαs to Wy-14,643, nafenopin, and clofibrate are assayed in parallel, both receptors show “dose-dependent” increases in activity (28). Whereas PPARα is highly expressed in rodent liver, preliminary data suggest that human PPARαs are expressed at low, albeit variable, levels in human liver and may also be polymorphic (51).Therefore, while the assay may conceivably predict *rodent* carcinogenesis accurately, the interindividual variation seen in humans would make the prediction of *human* risk from such an assay considerably more complicated.

Therefore, the usefulness of a receptor-based assay may be limited to the prediction of rodent peroxisome proliferation and, probably, rodent carcinogenicity. It is necessary to await further mechanistic research to determine definitively that PPs pose no hepatocarcinogenic hazard to humans.

VI. IN VITRO SCREENING FOR PEROXISOME PROLIFERATORS

The in vivo effects of peroxisome proliferators are readily reproduced in primary hepatocyte cultures. Several laboratories have used such cultures extensively to study peroxisome proliferation, especially aspects of species differences and structure-activity relationships (52–56). Typically cells are exposed to the test chemical for 3 days, with daily changes of the culture media and the test chemical. After this time the cells are harvested and peroxisomal β-oxidation determined as described above. Similarly, hepatocyte replicative DNA synthesis can be determined by measuring the incorporation of labeled DNA precursors included in the culture medium. A typical experimental design is shown in Figure 3. Details of the hepatocyte culture conditions can be found in Ref. 53.

Some investigators have used liver- or liver tumor–derived cell lines to investigate peroxisome proliferation; however, this cannot generally be recommended, since such cell lines typically exhibit much less response than primary hepatocytes to peroxisome proliferators and are additionally compromised with lessened biotransformation capabilities.

In vitro experiments using primary cultures of hepatocytes usually faithfully mimic in vivo experiments in respect to potency rankings and species differences. Exceptions may occur at times because of limitations in the biotransformation of a chemical to the proximate peroxisome proliferator in vitro.

Species differences in response, a characteristic of the peroxisome proliferation phenomenon (Table 2), are readily demonstrable in cultured

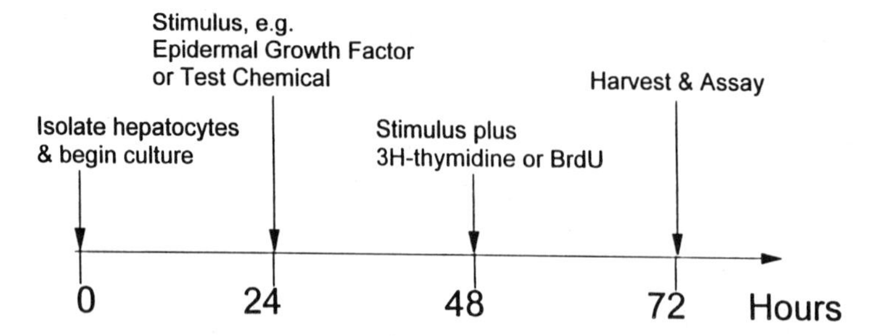

Figure 3 Schematic representation of method used to determine replicative DNA synthesis in primary hepatocyte cultures. In each series of experiments, an inducer of replicative DNA synthesis such as epidermal growth factor is used in parallel with test chemicals in order to determine the responsivity of the hepatocyte preparation. (From Ref. 53.)

hepatocytes. In culture, confounding factors such as absorption and distribution of the chemical may be avoided (57).

Human hepatocytes may be utilized to estimate potential human hazard. To date, peroxisomal β-oxidation and replicative DNA synthesis have not been induced in human hepatocytes exposed to peroxisome proliferators in primary culture. Stimulation of DNA synthesis by epidermal growth factor is used as a surrogate positive control and viability marker in the absence of a "true" positive control peroxisome proliferator for human hepatocytes.

Peroxisome proliferators shown to be ineffective in human hepatocytes include beclobric acid, benzbromarone, ciprofibrate, clofibric acid, 2-ethylhexanoic acid, fenofibric acid, fomesafen, monoethylhexylphthalate, methylcofenapate, nafenopin, trichloroacetic acid, and Wy-14,643 (39,58–62). These 12 compounds are all effective peroxisome proliferators in rat and/or mouse hepatocyte cultures.

Typical species differences in peroxisome proliferation in cultured hepatocytes are shown in Figure 4.

VII. FUTURE PERSPECTIVES ON ASSESSING HUMAN CARCINOGENIC RISK FROM PEROXISOME PROLIFERATORS

In the preceding sections, available systems for determining toxic and carcinogenic potential of PPs have been discussed. It will be apparent that an important gap is the difficulty in extrapolating the data generated from

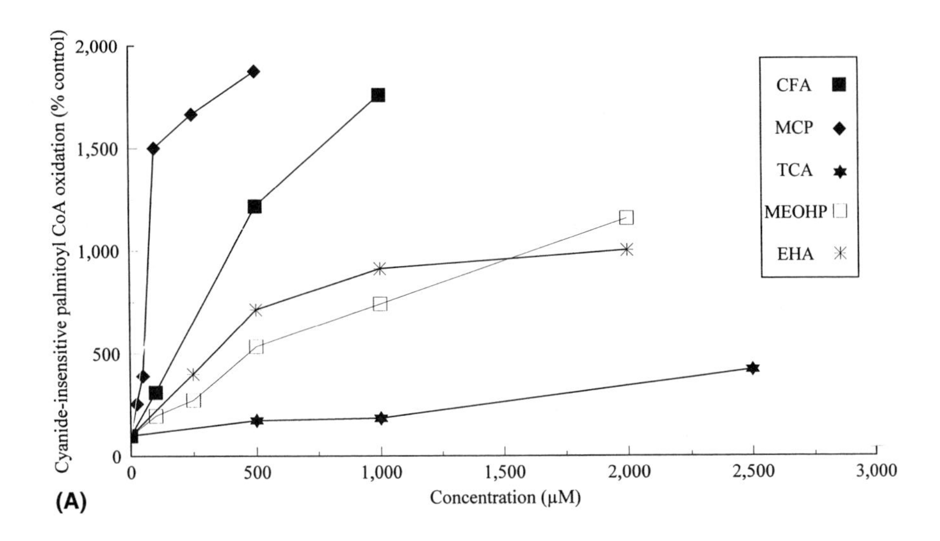

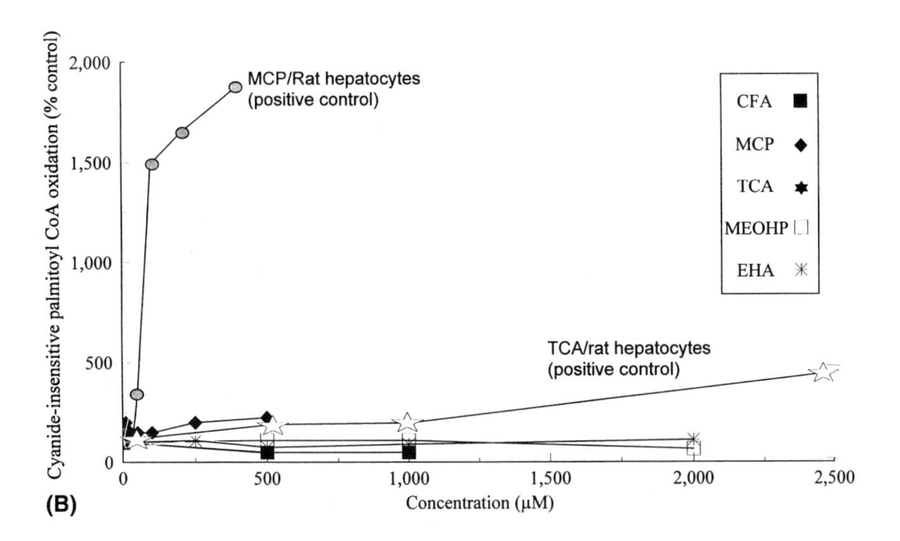

Figure 4 Peroxisome proliferation in hepatocyte cultures. (For description of methodology, see Sec. IV.C.) (A) Rat, (B) guinea pig, and (C) human hepatocyte cultures. CFA = Clofibric acid; MCP = methyl clofenapate; TCA = trichloroacetic acid; MEOHP = mono-(2-ethyl-5-oxyhexyl)phthalate (a metabolite of mono-(2-ethylhexyl)phthalate); EHA = 2-ethylhexanoic acid.

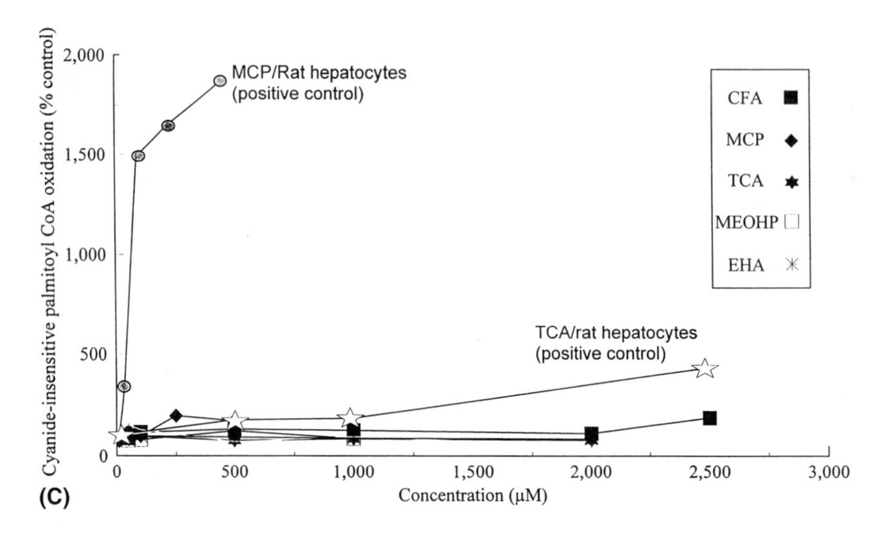

Figure 4 Continued.

predominantly rodent systems to assessing human risk for these chemicals. What is required is a mechanistic understanding of how humans interact with and respond to PPs, and to this end some recent effort has focused on the biology of PPARα in humans.

Given the lack of evidence for human responsiveness to PPs, it is perhaps perplexing that human livers contain PPARα at detectable albeit low levels and that the human receptors are apparently capable of activation by PPs in transactivation assays (28,49). Human hepatocytes certainly have the capacity to respond "beneficially" to fibrate hypolipidemic drugs since these are useful therapeutically. But, in vitro, human hepatocytes show no evidence of peroxisome proliferation elicited by the fibrates in rodent hepatocytes (Fig. 4C). Clearly there are two aspects to the response to PPs: the "hypolipidemic" response, which occurs in all species, and the "peroxisome proliferation" response, which is apparently rodent-specific. The hypolipidemic response is characterized by the reduction of plasma cholesterol levels, achieved partly through the modulation of a number of genes, including those encoding certain lipoproteins and lipases (63,64). Recent evidence suggests that PPARα is involved in regulating the transcription of at least some of these genes, both in rodents and humans (65–67). This differential transcriptional regulation, in rodents versus humans, of the two PP-responsive pathways is likely to be a consequence of complex interactions between liver transcription factors, including PPARα and RXR.

An important component of understanding carcinogenic mechanisms of PPs is knowledge of how these compounds regulate liver growth, since this is a good predictor of the carcinogenic potential of PPs (see above, and Ref. 2). PPs are able to regulate both DNA synthesis and apoptosis in rodent liver (12,68); the challenge is to determine whether these processes, and thereby liver growth, are regulated by PPARα. Coupled with an improved knowledge of the role of PPARα in humans, this would allow a much more realistic evaluation of human carcinogenic risk from PPs and provide an important paradigm for other nongenotoxic carcinogens.

REFERENCES

1. Stringer DA. ECETOC Monograph No. 17 Hepatic Peroxisome Proliferation. Brussels: European Center for Ecotoxicology and Toxicology of Chemicals, 1992.
2. Ashby J, Brady A, Elcombe CR, Elliott BM, Ishmael J, Odum J, Tugwood JD, Kettle S, Purchase IFH. Mechanistically-based human hazard assessment of peroxisome proliferator-induced hepatocarcinogenesis. Hum Exp Tox 1994; 13:S1-S117.
3. Reddy JK, Rao MS. Oxidative DNA damage caused by persistent peroxisome proliferation: its role in hepatocarcinogenesis. Mutat Res 1989; 214:63–68.
4. Cattley RC, Glover SE. Elevated 8-hydroxydeoxyguanosine in hepatic DNA of rats following exposure to peroxisome proliferators: relationship to carcinogenesis and nuclear localisation. Carcinogenesis 1993; 14:2495–2499.
5. Richter C, Park J-W, Ames BN. Normal oxidative damage to mitochondrial and nuclear DNA is extensive. Proc Natl Acad Sci USA 1988; 85:6465–6467.
6. Takagi A, Sai K, Umemura T, Hasegawa R, Kurokawa Y. Significant increase in 8-hydroxyguanosine in liver DNA of rats following short term exposure to peroxisome proliferators di(2-ethylhexyl)phthalate and di(2-ethylhexyl)-adipate. Jpn J Cancer Res 1990; 81:213–215.
7. Schulte-Hermann R. Adaptive liver growth induced by xenobiotic compounds: its nature and mechanism. Arch Toxicol 1979; 2(Suppl):113–124.
8. Ames BN, Gold LS. Too many rodent carcinogens: mitogenesis increases mutagenesis. Science 1990; 249:970–971.
9. Farber E. Clonal adaptation during carcinogenesis. Biochem Pharmacol 1990; 39:1837–1846.
10. Kraupp-Grasl B, Huber W, Taper H, Schulte-Hermann R. Increased susceptibility of aged rats to hepatocarcinogenesis by the peroxisome proliferator nafenopin and the possible involvement of altered liver foci occurring spontaneously. Can Res 1991; 51:666–671.
11. Oberhammer FA, Roberts RA. Apoptosis: a widespread process involved in liver adaptation and carcinogenesis. In: Arias IM, Boyer JL, Fausto N, Jakoby WB, Schachter D, Schafritz DA, eds. The Liver: Biology and Pathobiology. New York: Raven Press, 1994:1547–1556.

12. Bayly AC, Roberts RA, Dive C. Suppression of liver cell apoptosis in vitro by th non-genotoxic hepatocarcinogen and peroxisome proliferator, nafenopin. J Cell Biol 1994; 125:197–203.
13. Roberts RA, Soames AR, Gill JH, James NH, Wheeldon EB. Non-genotoxic hepatocarcinogens stimulate DNA synthesis and their withdrawal induces apoptosis, but in different discrete hepatocyte populations. Carcinogenesis 1995; 16:1693–1698.
14. James NH, Roberts RA. Species differences in response to peroxisome proliferators correlate in vitro with induction of DNA synthesis rather than suppression of apoptosis. Carcinogenesis 1996; 17:1623–1632.
15. Nebert DW, Masahike N, Lang MA, Hjelmeland LM, Eisen HJ. The *Ah* locus, a multigene family necessary for survival in a chemically adverse environment: comparison with the immune system. In: Demerec M, ed. Advances in Genetics. New York: Academic Press, 1982:1–52.
16. Burbach KM, Poland A, Bradfield CA. Cloning of the Ah-receptor cDNA reveals a distinctive ligand-activated transcription factor. Proc Natl Acad Sci USA 1992; 89:8185–8189.
17. Ema M, Sogawa K, Watanabe N, Chujoh Y, Matsushita N, Gotoh O, Funae Y, Fujii-Kuriyama Y. cDNA cloning and structure of the mouse putative *Ah* receptor. Biochem Biophys Res Comm 1992; 184:246–253.
18. Reyes H, Reisz-Porszasz S, Hankinson O. Identification of the *Ah* receptor nuclear translocator protein (Arnt) as a component of the DNA binding form of the *Ah* receptor. Science 1992; 256:1193–1194.
19. Jones PBC, Durrin LK, Galeazzi DR, Whitlock JP Jr. Control of cytochrome P450 gene expression: analysis of a dioxin-responsive enhancer system. Proc Natl Acad Sci USA 1986; 83:2802–2806.
20. Hapgood J, Cuthill S, Denis M, Poellinger L, Gustafsson JA. Specific protein-DNA interactions at a xenobiotic-response element: copurification of dioxin receptor and DNA binding activity. Proc Natl Acad Sci USA 1989; 86:60–64.
21. Wu L, Whitlock JPJ. Mechanism of dioxin action: *Ah* receptor mediated increase in promoter accessibility in vivo. Proc Natl Acad Sci USA 1992; 89: 4811–4815.
22. Fernandez-Salguero P, Pineau T, Hilbert DM, McPhail T, Lee SST, Kimura S, Nebert DW, Rudikoff S, Ward JM, Gonzalez FJ. Immune system impairment and hepatic fibrosis in mice lacking the dioxin-binding Ah receptor. Science 1995; 268:722–726.
23. Calvert JM, Hornung RW, Sweeney MH, Fingerhut MA, Halperin WE. Hepatic and gastrointestinal effects in an occupational cohort exposed to 2,3,7,8-tetrachlorodibenzo-*p*-dioxin. JAMA 1992; 267:2209–2214.
24. Reddy JK, Goel SK, Nemali MR, Carrino JJ, Laffler TG, Reddy MK, Sperbeck SJ, Osumi T, Hashimoto T, Lalwani ND, Rao MS. Transcriptional regulation of peroxisomal fatty acyl-CoA oxidase and enoyl-CoA hydratase/ 3-hydroxyacyl-CoA dehydrogenase in rat liver by peroxisome proliferators. Proc Natl Acad Sci USA 1986; 83:1747–1751.
25. Issemann I, Green S. Activation of a member of the steroid hormone receptor superfamily by peroxisome proliferators. Nature 1990; 347:645–649.

26. Dreyer C, Krey G, Keller H, Givel F, Helftenbein G, Wahli W. Control of the peroxisomal β-oxidation pathway by a novel family of nuclear hormone receptors. Cell 1992; 68:879–887.
27. Aperlo C, Pognonec P, Saladin R, Auwerx J, Boulukos KE. cDNA cloning and characterization of the transcriptional activities of the hamster peroxisome proliferator-activated receptor haPPARgamma. Gene 1995; 162:297–302.
28. Sher T, Yi H-F, McBride OW, Gonzalez FJ. cDNA cloning, chromosomal mapping, and functional characterisation of the human peroxisome proliferator activated receptor. Biochemistry 1993; 32:5598–5604.
29. Greene ME, Blumberg B, McBride OW, Yi HF, Kronquist K, Kwan K, Hsieh L, Greene G, Nimer SD. Isolation of the human peroxisome proliferator activated receptor gamma cDNA: expression in hematopoietic cells and chromosomal mapping. Gene Expr 1995; 4:281–299.
30. Lambe KG, Tugwood JD. A human peroxisome-proliferator-activated receptor-gamma is activated by inducers of adipogenesis, including thiazolidinedione drugs. Eur J Biochem 1996; 239:1–7.
31. Issemann I, Prince RA, Tugwood JD, Green S. The PPAR:RXR heterodimer is activated by fatty acids and fibrate hypolipidaemic drugs. J Mol Endocrinol 1993; 11:37–47.
32. Kliewer SA, Umesono K, Noonan DJ, Heyman RA, Evans RM. Convergence of 9-cis retinoic acid and peroxisome proliferator signalling pathways through heterodimer formation of their receptors. Nature 1992; 358:771–774.
33. Tugwood JD, Issemann I, Anderson RG, Bundell KR, McPheat WL, Green S. The mouse peroxisome proliferator activated receptor recognizes a response element in the 5′ flanking region of the rat acyl CoA oxidase gene. EMBO J 1992; 11:433–439.
34. Zhang B, Marcus SL, Fereydoun GS, Alvares K, Reddy JK, Subramani S, Rachubinski RA, Capone JP. Identification of a peroxisome proliferator responsive element upstream of the gene encoding rat peroxisomal enoyl-CoA hydratase/3-hydroxyacyl-CoA dehydrogenase. Proc Natl Acad Sci USA 1992; 89:7541–7545.
35. Aldridge TC, Tugwood JD, Green S. Identification and characterisation of DNA elements implicated in the regulation of CYP4A1 transcription. Biochem J 1995; 306:473–479.
36. Kliewer SA, Forman BM, Blumberg B, Ong ES, Borgmeyer U, Mangelsdorf DJ, Umesono K, Evans RM. Differential expression and activation of a family of murine peroxisome proliferator-activated receptors. Proc Natl Acad Sci USA 1994; 91:7355–7359.
37. Jones PS, Savory R, Barratt P, Bell AR, Gray TJB, Jenkins NA, Gilbert DJ, Copeland NG, Bell DR. Chromosomal localisation, inducibility, tissue-specific expression and strain differences in three murine peroxisome proliferator-activated receptor genes. Eur J Biochem 1995; 233:219–226.
38. Lee SST, Pineau T, Drago J, Lee EJ, Owens JW, Kroetz DL, Fernandez-Salguero PM, Westphal H, Gonzalez FJ. Targeted disruption of the α isoform of the peroxisome proliferator-activated receptor gene in mice results in abol-

ishment of the pleiotropic effects of peroxisome proliferators. Mol Cell Biol 1995; 15:3012–3022.

39. Blaauboer BJ, van Holstein CW, Bleumink R, Mennes WC, van Peit FN, Yap SH, van Pelt JF, van Iersel AA, Timmerman A, Schmidt BP. The effects of beclobric acid and clofibric acid on peroxisomal β-oxidation and peroxisome proliferation in primary cultures of rat, monkey and human hepatocytes. Biochem Pharmacol 1990; 40:521–528.
40. Frick H, Elo, Haapa K, Heinonen OP, et al. Helsinki heart study: primary-prevention trial with gemfibrozil in middle-aged men with dislipidemia. N Eng J Med 1987; 317:1235–1247.
41. Weibel ER. Stereological principles for morphometry in electron microscope cytology. Int Rev Cytol 1969; 26:235–238.
42. Goldsworthy TL, Morgan KT, Popp JA, Butterworth BE. Guidelines for measuring chemically-induced proliferation in specific rodent target organs. In: Butterworth BE, Slaga TJ, Farland W, McClain M, eds. Chemically Induced Cell Proliferation: Implications for Risk Assessment. New York: Wiley Liss, 1991:253–284.
43. Alison MR. Assessing cellular proliferation; What's worth measuring? Hum Exp Toxicol 1995; 14:935–944.
44. Bronfman M, Ingestrosa NC, Leighton F. Fatty acid oxidation by human liver peroxisomes. Biochem Biophys Res Comm 1979; 88:1030–1036.
45. Ausubel FM, Brent R, Kingston RE, Moore DD, Seidman JG, Smith JA, Struhl K. Current Protocols in Molecular Biology. New York: J. Wiley & Sons Inc., 1995.
46. Tontonoz P, Hu E, Graves RA, Budavari AL, Spiegelman BM. mPPAR-gamma2: tissue-specific regulator of an adipocyte enhancer. Genes Devel 1994; 8:1224–1234.
47. MacGregor A, Caskey CT. Construction of plasmids that express *E. coli* β-galactosidase in mammalian cells. Nucl Acid Res 1989; 17:2365.
48. Gorman CM, Moffat LF, Howard BH. Recombinant genomes which express chloramphenicol acetyl transferase in mammalian cells. Mol Cell Biol 1982; 2: 1044–1051.
49. Mukherjee R, Jow L, Noonan D, McDonnell DP. Human and rat peroxisome proliferator activated receptors (PPARs) demonstrate similar tissue distribution but different responsiveness to PPAR activators. J Ster Biochem Mol Biol 1994; 51:157–166.
50. Yu K, Bayona W, Kallen CB, Harding HP, Ravera CP, McMahon G, Brown M, Lazar MA. Differential activation of peroxisome proliferator-activated receptors by eicosanoids. J Biol Chem 1995; 270:23975–23983.
51. Tugwood JD, Aldridge TC, Lambe KG, Macdonald N, Woodyatt NJ. Peroxisome proliferator-activated receptors: structures and function. In: Reddy JK, Suga T, Mannaerts GP, Lazerow PB, Subramani S, eds. Peroxisomes: Biology and Role in Toxicology and Disease. New York: Annals of the New York Academy of Sciences, 1996; 804:252–265.
52. Lake BG, Pels Rijcken WR, Gray TJB, Foster JR, Gangolli SD. Comparative

studies of the hepatic effects of di- and mono-Octyl Phthalates, di-(2-ethylhexyl)phathalate and clofibrate in the rat. Acta Pharmacol Toxicol 1984: 54:167–176.

53. Mitchell AM, Bridges JW, Elcombe CR. Factors influencing peroxisome proliferation in cultured rat hepatocytes. Arch Toxicol 1985; 55:239–246.
54. Elcombe CR, Bell DR, Elias E, Hasmall SC, Plant NJ. Peroxisome proliferators: species differences in response. In: Reddy JK, Suga T, Mannaerts GP, Lazerow PB, Subramani S, eds. Peroxisomes: Biology and Role in Toxicology and Disease. New York: Annals of the New York Academy of Sciences, 1996; 804:628–635.
55. Lake BG, Lewis DVF, Gray TJB, Beamand JA, Hodder KD, Purchase R, Gangolli SD. Structure activity studies on the induction of peroxisomal enzyme activities by phthalate mono esters in primary rat hepatocyte cultures. Arch Toxicol 1986; 12(Suppl):217–224.
56. Eacho PI, Foxworthy PS, Dillard RD, Whitesitt CA, Herron KD, Marshall WS. Induction of peroxisomal β-oxidation in the rat liver in vivo and in vitro by tetrazole-substituted acetophenones: structure-activity relationships. Tox Appl Pharmacol 1989; 100:177–184.
57. Bentley P, Calder I, Elcombe CR, Grasso P, Stringer D, Weigand H-J. Hepatic peroxisome proliferation in rodents and its significance for humans. Food Chem Toxicol 1993; 31:857–907.
58. Allen KL, Green CE, Tyson CA. Comparative studies on peroxisomal enzyme induction in hepatocytes from rat, cynomolgus monkey and human by hypolipidemic drugs. Toxicologist 1987; 7:253.
59. Bichet N, Cahard D, Fabre D, Remandet B, Gouy D, Cano J-P. Toxicological studies on a benzofuran derivative III. Comparison of peroxisome proliferation in rat and human hepatocytes in primary culture. Toxicol Appl Pharmacol 1990; 106:509–517.
60. Elcombe CR. Species differences in carcinogenicity and peroxisome proliferation due to trichloroethylene; a biochemical human hazard assessment. Arch Toxicol 1985; 8(Suppl):6–17.
61. Hwang JJ, Hsia MTS, Jirtle RL. Induction of sister chromatid exchange and micronuclei in primary cultures of rat and human lymphocytes by the peroxisome proliferator Wy-14,643. Mutat Res 1993; 286:123–133.
62. Staels B, Vu-Dac N, Kosykh VA, Saladin R, Fruchart J-C, Dallongeville J, Auwerx J. Fibrates downregulate apolipoprotein C-III expression independent of induction of peroxisomal acyl coenzyme A oxidase. J Clin Invest 1995; 95: 705–712.
63. Staels B, Van Tol A, Andreu T, Auwerx J. Fibrates influence the expression of genes involved in lipoprotein metabolism in a tissue selective manner in the rat. Arterioscler Thromb 1992; 12:286–294.
64. Staels B, Peinado-Onsurbe J, Auwerx J. Down-regulation of hepatic lipase gene expression and activity by fenofibrate. Biochem Biophys Acta 1992; 1123: 227–230.
65. Hertz R, Bishara-Shieban J, Bar-Tana J. Mode of action of peroxisome proliferators as hypolipidemic drugs. J Biol Chem 1995; 270:13470–13475.

66. Vu-Dac N, Schoonjans K, Laine B, Fruchart J, Auwerx J, Staels B. Negative regulation of the human apolipoprotein A-I promoter by fibrates can be attenuated by the interaction of the peroxisome proliferator-activated receptor with its response element. J Biol Chem 1994; 269:31012–31018.
67. Vu-Dac N, Schoonjans K, Kosyth V, Dallongeville J, Fruchart JC, Staels B, Auwerx J. Fibrates increase human apolipoprotein A-II expression through activation of the peroxisome proliferator-activated receptor. J Clin Invest 1995; 96:741–750.
68. Eacho PI, Lanier PL, Brodhecker CA. Hepatocellular DNA synthesis in rats given peroxisome proliferating agents: comparison of Wy-14,643 to clofibric acid, nafenopin and LY171883. Carcinogenesis 1991; 12:1557–1561.
69. Watanabe T, Horie S, Yamada J, Isaji M, Nishigaki T, Naito J, Suga T. Species differences in the effects of bezafibrate, a hypolipidaemic agent, on hepatic peroxisome-associated enzymes. Biochem Pharmacol 1989; 38:430–437.
70. Oesch F, Schladt L, Steinberg P, Thomas H. Concomitant induction of cytosolic epoxide hydrolase and peroxisomal β-oxidation by hypolipidaemic compounds in rat and guinea pig liver. Arch Toxicol 1988; 12(Suppl):248–255.
71. Reddy JK, Lalwani ND, Qureshi SA, Reddy MK, Moehle CM. Induction of hepatic peroxisome proliferation in non-rodent species, including primates. Am J Pathol 1984; 114:171–183.

13

The Rat Liver Hepatocellular-Altered, Focus-Limited Bioassay for Chemicals with Carcinogenic Activity

Gary M. Williams and Harald Enzmann
American Health Foundation, Valhalla, New York

I. INTRODUCTION

Bioassays using rodent liver foci are widely used for the rapid detection of carcinogenic activity of chemicals (Williams, 1982; Bannasch, 1986; Ito et al., 1989). The utility of this approach rests on two features of the rodent liver: the liver has the broadest capability for bioactivation of carcinogens (Weisburger and Williams, 1982), and preneoplastic liver lesions as indicators of neoplastic development are readily quantified by a variety of techniques (Williams, 1980; Moore and Kitagawa, 1986).

A. Preneoplastic Foci of Altered Hepatocytes

Sasaki and Yoshida (1935) were the first to identify foci of abnormal hepatocytes preceding the occurrence of chemical-induced liver tumors. They described clear and basophilic cell foci that were considered to be precursors to benign and malignant liver cell tumors. Extensive subsequent research established that there is an ordered sequence of phenotypically different lesions leading from clear to basophilic cell foci and finally to tumors (Bannasch, 1968; Williams, 1980; Enzmann and Bannasch, 1987; Pitot et al., 1987). A number of features of hepatocellular altered foci (HAF) indicate that they are preneoplastic (Table 1).

In addition to the clear and basophilic cell foci described by Sasaki and Yoshida (1935), a number of histochemically detectable phenotypic alterations have been used for the quantification of the lesions (Table 2).

In rodents exposed to carcinogens, HAF may precede liver tumors for months or years and outnumber liver tumors hundred- or thousandfold (Watanabe and Williams, 1978; Pitot et al., 1985). Therefore, the quantification of HAF instead of fully developed hepatocellular tumors allows both a more rapid and more sensitive detection of hepatocarcinogenic effects.

The detection of foci of altered hepatocytes after exposure to a test agent for a limited time, e.g., several weeks, or in stop-experiments including a recovery period was proposed nearly 40 years ago (Druckrey et al., 1958) and has frequently been used for the quantification of the dose dependence of carcinogenic effects.

B. The Two-Stage Concept of Carcinogenesis

The pioneering studies on skin carcinogenesis of Friedwald and Rous (1944) and Berenblum and Shubik (1947) introduced the concept of two stages of carcinogenesis: initiation and promotion. Initiation was conceived of as the formation of a neoplastic cell and promotion as the facilitation of the growth of the neoplastic cell into a tumor (Berenblum, 1974). The operational approach originally developed for the skin model was extended to liver carcinogenesis when Peraino and coworkers (1971) discovered the hepatopromoting effect of phenobarbital, and Farber and coworkers (Solt and Farber, 1976; Williams et al., 1976) developed the resistant hepatocyte model. The demonstration by Kitagawa et al. (1978), Pitot et al. (1978), and Williams and Watanabe (1978) that phenobarbital enhanced develop-

Table 1 Evidence for the Preneoplastic Nature of Altered Hepatic Foci

1. Similarity of many morphological and biochemical changes with hepatic tumors
2. Genetic alterations
3. Appearance prior to experimental hepatic tumors induced by:
   - chemicals
   - radiation
   - woodchuck hepatitis virus
4. Persistence in stop experiments
5. Transitions into hepatic adenomas and carcinomas
6. Statistical correlation with hepatic tumors

Table 2 Phenotypic Alterations of Preneoplastic Rat Liver Foci Frequently Used for Quantification

Marker	Alteration	Ref.
Glucose-6-phosphatase	Decrease	Gössner and Friedrich-Freksa, 1964
Adenosinetriphosphatase	Decrease	Schauer and Kunze, 1968
γ-Glutamyl-transpeptidase	Increase	Kalengayi and Desmet, 1975
Iron storage	Decrease	Williams et al., 1976
Glucose-6-phosphate dehydrogenase	Increase	Hacker et al., 1982
Placental glutathione-S-transferase	Increase	Sato et al., 1984

ment of HAF established the possibility to measure promoting effects in short-term studies.

Hepatic regenerative proliferation may be useful to enhance the sensitivity of these test systems (de Gerlache et al., 1982; Tsuda et al., 1987) because an increase in cell replication enhances all steps of neoplastic transformation and neoplastic development (Williams et al., 1996). Other experimental models are designed to take advantage of the high level of proliferation in growing young (Deml et al., 1981) or newborn animals (Vesselinovitch et al., 1985).

To further accelerate the carcinogenic process, relatively complex experimental protocols have been proposed. For example, one approach is to administer the test substance as a single dose followed by an enhancement by a two-third partial hepatectomy followed by selection pressure treatment by 2-acetylaminofluorene including a necrogenic dose of CCl_4 and finally to add a promotion treatment with phenobarbital (de Gerlache et al., 1982). The number and size of foci and nodules induced by such exposures are to be compared with the number and size of foci and nocules in the liver of animals that were not exposed to the test chemical but underwent the other manipulations. The interpretation of these complex experimental protocols is complicated by the fact that the majority of the induced focal and nodular lesions are reversible after the end of the treatment. These reversible nodular lesions have even been regarded as a "physiological" adaptation to the cytotoxic effect of the carcinogen (Farber, 1984). Findings in vitro (Tsukada et al., 1986) and in vivo (Harris et al., 1989) seem to substantiate the assumption that this "physiological reaction" might be beneficial for the survival of the cells or the animals. Hence, it may be doubtful whether these lesions are truly neoplastic and whether they can be used as quantitative indicators of carcinogenic activity.

In order to avoid the problems linked to the evaluation and interpretaion of these complex models, simpler experimental models may be more advantageous. Thus, the following discussion will be limited to classical two-stage protocols and the medium-term assay proposed by Ito et al. (1989, 1994), which requires the induction of regenerative proliferation during the promotion stage.

II. ASSAY DESCRIPTION

A. Testing for Initiating Activity

Inition is regarded as the first stage of carcinogenesis (Berenblum, 1974). The initiation stage is irreversible as judged by extended separation between the point of initiation and the beginning of promotion, and repeated initiation applications are clearly additive (Peraino et al., 1977; Loehrke et al., 1983). Initiation is generally assumed to be the result of a somatic mutation. Although many initiating agents are overtly DNA reactive, apparent "genotoxic" endpoints (mutations, chromosome aberrations) can also be induced by "nongenotoxic" agents (Yamasaki et al., 1992). DNA alterations such as hypomethylation can permanently affect gene expression (Ray et al., 1994), and substances with indirect gene toxicity (Appel et al., 1990) may also exhibit initiating activity. Initiation requires replicative DNA synthesis (Cayama et al., 1978; Ishikawa et al., 1980) in order to convert DNA lesions into permanent alterations. In the liver, initiation is manifested by the rapid induction of HAF. In the absence of promotion, the number of identifiable clones derived from spontaneously initiated cells increases with the age of the animals (Schulte-Hermann et al., 1983; Harada et al., 1989). However, when exogenous promoting agents are administered the number of such foci increases during the first 2 months of life but remains unchanged thereafter (Pitot et al., 1985). This finding has been interpreted to indicate that spontaneous initiation is dependent on the fixation by cell division which occurs frequently in the growing liver early in life but is relatively rare thereafter.

For the detection of initiating activity of a test substance, the substance is given to the animal alone or prior to a subsequent treatment with a promoter. For known strong hepatocarcinogens that may be used as a positive control, a single dose may be sufficient. In order to increase the sensitivity of the assay for less carcinogenic test substances, however, an extended initiation with exposure for several weeks is advantageous (Williams, 1982; Parnell et al., 1988).

After the initiation exposure, the induced effects can be enhanced by exposure of the animals to a promoting chemical for several weeks. The most widely used promoting agent is phenobarbital. In addition, a wide range of different chemicals has been used for promotion including potent carcinogens (Solt et al., 1977), mixtures of polychlorinated biphenyls (Oesterle and Deml, 1988), or putatively harmless substances such as sucrose (Hei and Sudilovsky, 1985) or orotic acid (Laurier et al., 1984). For the evaluation, the animals that received the test chemical and promotion treatment are compared with the animals that received promotion treatment only in respect to the number and/or size of HAF and/or the percentage of liver occupied by focal lesions.

A quantitative description of initiating effects in the liver has been suggested (Pitot et al., 1987). Based on stereological calculations, parameters for the estimation of the relative potency of chemicals as initiating agents are defined as:

$$\text{Initiation index} = \text{no. of foci induced} \times \text{liver}^{-1} \times (\text{mmol/kg body wt})^{-1}$$

Since this initiation index formula requires the estimated number of preneoplastic foci per liver, the morphometric quantification of the observed preneoplastic liver lesions with subsequent stereological calculation is indispensable. Usually, the size and the number of the preneoplastic foci of phenotypically altered hepatocytes and the area of the evaluated section or part of section are measured planimetrically. From these planimetric data, the number of foci per cm^2 and the percentage of liver occupied by focal lesions are calculated. For the stereological calculation of the number of foci per cm^3 or the number of foci per liver, various simple methods are available (Scherer, 1981; Campbell et al., 1982; Enzmann et al., 1987).

B. Testing for Promoting Activity

Considerable evidence indicates that the promotion stage of carcinogenesis is characterized by the clonal expansion of the initiated cell population. The mathematical two-stage model of Moolgavkar and colleagues (1986) has successfully been applied to rat hepatocarcinogenesis (Moolgavkar et al., 1990; Luebeck et al., 1991). Biologically, promoters may stimulate the proliferation of initiated cells or permit their clonal expansion by inhibition of cell-cell communication (Williams, 1981; Trosko et al., 1983; Yamasaki, 1986). Also, promoters may inhibit the proliferation of the surrounding putatively normal cells (Solt et al., 1977; Farber, 1984) or may inhibit apoptosis of previously initiated cells (Schulte-Hermann et al., 1995). The promotion stage is frequently reversible, and some authors postulate a measurable threshold for the action of a promoting agent on initiated cells (Goldsworthy et al., 1984; Maeura and Williams, 1984). It is important to distinguish between an experimental threshold defined as the dose at which no observable promoting effect is obtained in a particular experiment and the more absolute biological or nonexperimental threshold (Kitchin et al., 1994). The finding of an experimental no-effect level does not necessarily prove or imply the existence of a biological threshold for the promoting effect of a given chemical. However, the shape of the dose-response curve or the potential mechanism of promotion may provide evidence for the existence of biological thresholds. For example, the demonstration of an antipromotional effect at low doses of known promoters (Pitot et al., 1987; Whysner et al., 1994) strongly excludes any promotional effect at these doses.

In most experimental systems, the induction of proliferation is a very effective promoting stimulus. Huge variations in the molar potencies of promoters suggest profound differences between postnecrotic regenerative proliferation and mitogen-induced cell proliferation (Kitchin et al., 1994). Whereas the promoting effect of chemically or surgically induced regenerative liver cell proliferation has been widely accepted, the relevance of mitogen-stimulated liver cell proliferation is still a matter of controversy. Although Kitchin and colleagues suggested a linkage between the promoting effect and the mitogenic effects of phenobarbital, hexachlorcyclohexane, and polychlorinated biphenyls, other groups have shown that in some cases mitogen-induced hyperplasia may lack a promoting effect in rat liver (Ledda-Columbano et al., 1993, 1994). Distinct strain differences have also been reported both for the promoting effect of surgical induction of liver cell proliferation (Hanigan et al., 1990) and for the promoting effect of phenobarbital (Lin et al., 1989).

For the detection of promoting activity of a test substance, the animals receive an initiation exposure with a subcarcinogenic dose of a strong hepatocarcinogen followed by exposure to the test chemical for several weeks. The evaluation consists of measurement of the number and/or size and/or the percentage of liver occupied by HAF in the animals that received the initiation followed by exposure to the test chemical compared with the number and/or size and/or the percentage of liver occupied by focal lesions in the animals that received the initiation treatment only.

In contrast to the test for initiating activity, the selection of the appropriate marker reaction for the detection of the induced foci of altered hepatocytes may be crucial for the test for promoting activity, because the phenotype of preneoplastic foci may be modulated by the treatment of the animals (Williams et al., 1980; Maruyama et al., 1990). In the test for initiating activity, samples of liver tissue are taken while the animals are still exposed to the promoter. Therefore, the phenotype of the induced altered hepatic foci is determined by the promoting agent, e.g., phenobarbital, that is the same in all groups. The immunohistochemical demonstration of the placental form of glutathione S-transferase has been shown to be a superior marker for this experimental approach (Ito et al., 1992). However, in the test for promoting activity, samples of liver tissue are usually taken while the animals are still exposed to the test chemical, although a recovery period can be allowed. Accordingly, the phenotype of the induced altered hepatic foci can be influenced by the test chemical. The immunohistochemical demonstration of the placental form of glutathione S-transferase is the most sensitive marker for HAF induced by genotoxic hepatocarcinogens, but some types of preneoplastic lesions are not detectable by this staining. Foci in animals exposed to ciprofibrate (Glauert et al., 1986), tigroid cell foci (Bannasch et al., 1985), enzymatically hyperactive foci (Enzmann et al., 1989), foci in animals exposed to the dietary antioxidant vitamin E (Glauert et al., 1990), and glycogen phosphorylase increased foci (Enzmann et al., 1991a) are negative for the placental form of glutatione S-transferase. As a consequence, the use of more than one histochemical marker (Dragan et al., 1991, 1992) in a test for promoting activity in the rat liver bioassay may be advantageous.

The promoting activity of a chemical can be expressed by a promotion index similar to the initiation index (Rizvi et al., 1988). Based on planimetric measurements, parameters for the estimation of the relative potency of chemicals as promoting agents are defined as:

$$\text{Promotion index} = \frac{Vf}{Vc} \times \text{mmol}^{-1} \times \text{weeks}^{-1}$$

where Vf is the total volume fraction (%) occupied by preneoplastic foci in the livers of rats exposed to the test agent and Vc is the total volume of preneoplastic foci in control animals, which have only been initiated.

Presently, the largest data set in rats is available for the experimental protocol of Ito and coworkers (Ito et al., 1994). This approach is somewhat complex and can be regarded as involving syncarcinogenic promoting activities. The animals are injected with a single intraperitoneal dose of 200 mg/kg diethylnitrosamine as an initiation treatment. After a 2-week recovery period, the animals are exposed to the test compound for 8 weeks. After one week of exposure to the test substance, the animals are subjected to a two-third partial hepatectomy to provoke regenerative cell proliferation. The control group receives the same treatment but is not exposed to the test chemical. For evaluation, the numbers and areas of glutathione S-transferase–positive foci are measured and the results are assessed by comparing the values of foci between the groups. Since, according to this protocol, promoting substances like phenobarbital as well as potent carcinogens like aflatoxins are recognized as positive, a third group has been added to discriminate between predominantly promoting substances like phenobarbital and complete carcinogens that induce foci even without previous initiation treatment. The animals of this group receive the test substance and the partial hepatectomy but an intraperitoneal injection with saline instead of diethylnitrosamine (Ito et al., 1992).

A total of 237 compounds have so far been tested in this system and the results compared with long-term carcinogenicity test findings (Ito et al., 1992). The sensitivity was high for genotoxic hepatocarcinogens (97%) and was satisfactory for nongenotoxic hepatocarcinogens (86%). Carcinogens without hepatocarcinogenic effects are only occasionally (24%) recognized. Until now, only two false-positive results (<1%) have been described: malathion and vinclozolin proved positive, although both have been reported to be noncarcinogenic in rats and mice. Five of the six false-negative hepatocarcinogens are peroxisome proliferators. Although the immunohistochemical demonstration of the placental form of glutathione S-transferase has been shown to be the most sensitive marker for enyzme altered foci in this experimental model (Ito et al., 1992), foci in animals exposed to ciprofibrate (Glauert et al., 1986) or to the dietary antioxidant vitamin E (Glauert et al., 1990) have been shown to be negative for the placental form of glutathione S-transferase. The failure to detect the hepatocarcinogenic effect of peroxisome proliferaters may be due to the selection of the immunohistochemical staining method rather than to the experimental protocol for the in-life part of the assay. The placental form of glutatione S-transferase is inhibited by peroxisome proliferators (Numoto et al., 1984; Furukawa et al., 1985). For these chemicals the value of the use of several complementary marker reactions is obvious. In addition, a number of inhibitory agents for the development of HAF have been detected, including several antioxidants (Williams et al., 1981; Ito et al., 1994), showing that the experimental protocol may be used to detect anticarcinogenic effects, too. The rat has great utility in the detection of liver-promoting activity, but responds to some chemicals that do not promote cancer in other species (Tukomo et al., 1991; Stenbäck et al., 1994).

C. The Evolution of Liver Foci Tests into Multiorgan Models of Promotion

Experimental protocols in rodents that allow the simultaneous detection of carcinogenic effects in several potential target tissues have been developed independently by Ito and coworkers (1992) and Williams and coworkers (Iatropoulos, 1992; Williams et al., 1996). The accelerated bioassay (ABA) introduced as part of the decision point approach for carcinogenicity testing

(Williams et al., 1996) effectively combines the tests for initiating and promoting activity in liver, lung, kidney, bladder, and forestomach in both male and female rats and mammary gland in females. For each of these organs, a tissue-specific initiation exposure of 14 weeks duration is used followed by a promotion exposure (organ-specific) for 24 weeks. The compound under investigation is given at the maximal tolerated dose since this does not usually produce significant organ toxicity with the limited duration of exposure to young animals. Frank cytotoxicity may result in carcinogenic activity by stimulating regenerative cell replication.

The experimental design of the accelerated bioassay contains multiple groups with positive controls. The duration of the entire assay is 38 weeks with an interim sacrifice at 14 weeks and a second at 26 weeks. Measurement of cell proliferation is easily included (Iatropoulos, 1996). The ABA eliminates age-related pathology by examining animals before the evolution of "spontaneous" neoplasms obscures those resulting from activity of the test compound without compromising the essential elements factors of chronicity and latency in carcinogenesis. To date, no false-positive results have been reported for the ABA.

In the medium-term multiorgan carcinogenesis bioassay (Ito et al., 1992), rats are exposed to different carcinogens to achieve initiation in a wide spectrum of organs for 4 weeks prior to the exposure to the test substance for 16 weeks. The animals are sequentially treated with diethylnitrosamine, methylnitrosourea, *N*-butyl-*N*-(4-hydroxybutyl)nitrosamine, 1,2-dimethylhydrazine, and dihydroxy-di-*N*-propylnitrosamine for multiorgan initiation (Hagiwara et al., 1993). Of the known target tissues of these chemicals, initiation is expected in thyroid, lung, liver, pancreas, esophagus, forestomach, small intestine, kidney, and urinary bladder. Positive test results for a series of test chemicals were found in the thyroid, liver, esophagus, forestomach, urinary bladder (Ito et al., 1992), and kidney (Yamamoto et al., 1995). The test reveals enhancing as well as inhibiting effects of a test substance on carcinogenesis (Hirose et al., 1991, 1993).

The medium-term multiorgan carcinogenesis bioassay is partially similar to the ABA in that (a) it does not investigate for an initiating activity of the test chemical, (b) it does not use different groups for the investigation of the effects on different target tissues, and (c) the target organs in which the effects are specifically observed differ somewhat. The medium-term multiorgan carcinogenesis bioassay and the ABA have five target organs in common; liver, kidney, lung, forestomach, and urinary bladder. ABA is designed to include the mammary gland, while the medium-term multiorgan carcinogenesis bioassay includes the thyroid, pancreas, esophagus, and small intestine as target organs.

III. EVALUATION AND INTERPRETATION OF THE LIVER FOCUS TEST

A. The Operational Discrimination of Initiation and Promotion

The discrimination of initiating and promoting activities of a test chemical is crucial for the interpretation of the liver focus test (LFT). Using the operational model described above, a substance with predominantly initiating activity will exhibit an inducing effect on the occurrence of foci when exposure to the test substance is followed by promotion (test for initiating activity). A substance with predominantly promoting activity will exhibit an enhancing effect on the occurrence of foci when given subsequent to initiation (test for promoting activity). A complete or solitary carcinogen (Appel et al., 1990) has initiating activity as well as promoting activity and consequently will be positive in both the test for initiating and promoting activity and will result in the induction of foci after exposure for a limited time without a requirement of either initiation or promotion from a second chemical.

The interpretation of these studies must always take into consideration "spontaneous" or cryptogenic carcinogenesis in the rat liver. Since hepatic foci and even tumors may occur in control animals without deliberate exposure to any agent it must be assumed that a low background level of initiation and promotion is always present in the liver. Consequently, a positive result of a carcinogenicity test with the two-stage model will be essentially an obvious quantitative difference between the groups rather than a qualitative difference (Table 3).

This is especially important to consider when animals are exposed only to the test chemical without additional initiation or promotion. Even a test substance with exclusively promoting activity will enhance the develop-

Table 3 Liver Foci Test: The Two-Stage Concept

	Theoretical Results	Realistic results
Initiation and promotion	+	+ + +
Initiation only	−	+
Promotion only	−	+
Promotion and initiation	−	+
No treatment	−	−/+

There are occasionally tumors in the control animals reflecting a certain degree of "spontanous" initiation and promotion.

ment of foci compared to an untreated control group, because such a test substance can promote the development of foci from preexisting initiated cells (Fig. 1).

Depending on the selected marker for preneoplastic foci and on the evaluation method, detection sensitivity is limited for very small lesions. Usually, only groups of three or more phenotypically altered hepatocytes are regarded as foci. Figure 2 shows the same situation as Figure 1, except that small lesions are regarded as below the detection limit and are omitted from Figure 2. It is obvious that both the initiation-only and the promotion-only groups exhibit more foci than the control group. Moreover, the promotion-only group has larger lesions and seems more affected by the treatment than the initiation-only group.

Therefore, the purely operational distinction between a promoting and initiating chemical may not always be possible. However, a stereological analysis of the data may be helpful to distinguish an effect on the number of lesions indicating initiation from an effect on the size of lesions indicating promotion.

B. Use of Stereological Methods for the Evaluation of the Liver Focus Test

Stereology tries to reconstruct three-dimensional objects from two-dimensional data and to estimate the number of these objects per tissue volume and the volume size distribution of these objects. The larger the object, the higher the probability that the object will be observed in a random section. Various simple methods for the stereological calculation

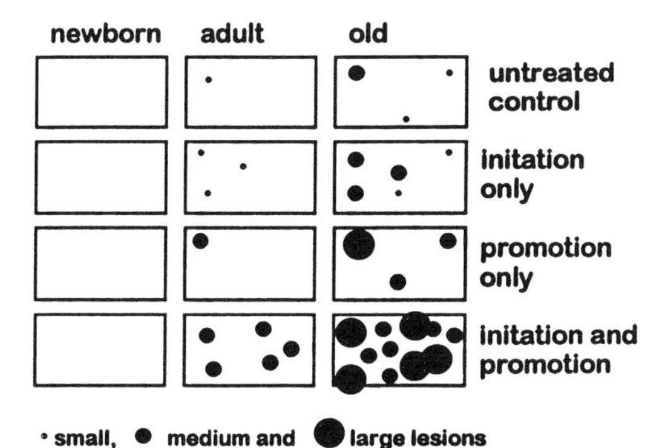

Figure 1 Liver foci test: the two-stage concept with a background of spontanous carcinogenesis.

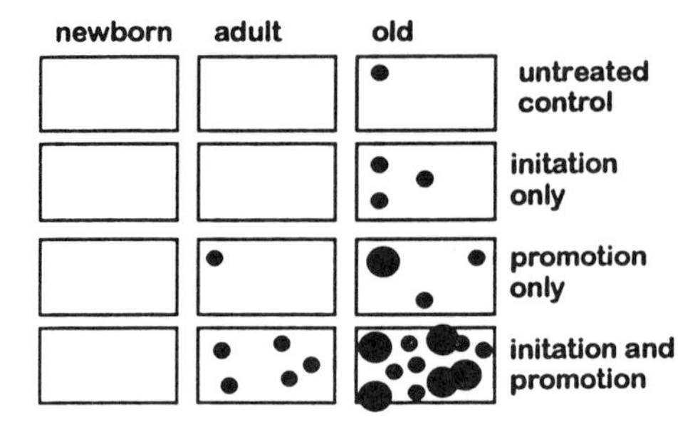

Figure 2 Liver foci test: the two-stage concept with a background of spontanous carcinogenesis and a limited detection sensitivity.

of liver foci have been published (Scherer, 1981; Campbell et al., 1982; Enzmann et al., 1987). All of these methods postulate randomly distributed spherical foci, but this assumption cannot be ideally fulfilled in any biological sample. Especially in tissue samples with very large lesions, foci may be of a more complex shape, including concave bodies (Imaida et al., 1989). In this case, planimetric two-dimensional data may give more reliable results than stereological calculations. In many experiments, however, the shape of the foci in the sections suggests a less complex and exclusively convex shape of the three-dimensional foci. For convex, ellipsoid rather than spherical foci, the bias caused by the nonspherical shape was estimated by Weibel (1979) for prolate (egg-shaped, drawn-out spheres) or oblate (compressed, pill-shaped spheres) ellipsoids. Depending on the axial ratio of the foci, the expected bias is given in Table 4. As long as the axial ratio is less than 2 : 1, the resulting small overestimation of a few percent can usually be neglected in comparison with the treatment-induced effects.

Table 4 Effect of the Shape of Foci on the Estimation of Foci/cm^3

Shape of ellipsoids	Axial ratio	Overestimation of foci/cm^3 (%)
Oblate	1 : 1.5	<3
	1 : 3	12
	1 : 5	45
Prolate	1 : 1.5	<3
	1 : 3	27
	1 : 5	60

It is crucial to recheck the validity and plausibility of the results of a stereological calculation with the originally measured data. A comparison of the *planimetrically measured* versus the *stereologically calculated* percentage of liver occupied by foci is recommended and can also been used for the selection of the most appropriate stereological method (Enzmann et al., 1987; Imaida et al., 1989).

C. The Role of Cell Proliferation

It is widely known that an increase in cell replication enhances all steps of neoplastic transformation and neoplastic development; in contrast, a decrease in cell proliferation inhibits rat hepatocarcinogenesis during the initiation and postinitiation stages (Williams et al., 1992; Kojima et al., 1993; Iatropoulos and Williams, 1996). Therefore, hepatotoxicity, toxic cell loss, and the resulting regenerative proliferation is an important potential confounder in the LFT for carcinogenic effects.

If the test substance given after initiation in the test for promoting activity induces pronounced hepatocyte replication, a promoting effect may be totally dependent on this cell proliferation and be absent at nonhepatotoxic doses that do not induce cell loss and regenerative proliferation. Therefore, we recommend the quantification of cell proliferation as part of the LFT. The discrimination of hepatotoxicity and promoting activity may be impossible without investigations of additional groups exposed to lower doses of the test chemical. In this context, the Ito model or the in ovo carcinogenesis assay, discussed later in this chapter, may be advantageous. In these models, the liver cell proliferation is maximally stimulated either by the two-third hepatectomy or by the rapid growth of the embryonic tissue and consequently a stimulatory effect of the test chemical on cell proliferation will have little if any effect.

It is less recognized that cell proliferation may also mimic an initiating effect. In the mouse skin model of carcinogenesis, the stage one of promotion is defined as a conversion of epidermis into a stage of increased promotability. Remarkably, the stage one of promotion may precede initiation (Fürstenberger et al., 1985). Similar to initiation, the first stage of tumor promotion or conversion seems to be linked to replicative DNA synthesis (Kinzel et al., 1984, 1986) and genotoxic events (Fürstenberger et al., 1989). In contrast to the initiated stage, which is virtually irreversible, the converted stage is slowly reversible, exhibiting a biological half-life of 10–12 weeks (Fürstenberger et al., 1985). A similar mechanism may exist in the liver. Regenerative liver cell proliferation preceding exposure to carcinogens has also been reported to enhance carcinogenesis (Warwick, 1971; Craddock, 1975; Tanaka et al., 1986; Bartsch et al., 1988) and the biological half-life of this effect was similar to the half life of the conversion effect in mouse skin. However, other groups found that prior organ injury by toxic agents does not always act to enhance sensitivity to carcinogenesis (Hagiwara et al., 1993).

Under at least some experimental conditions, hepatocyte replication may mimic both initiating and promoting activity of a test substance. If the enhancing effect is observed exclusively in the presence of a significantly increased liver cell proliferation, a positive LFT is not conclusive proof of carcinogenic potential. In this case, the effects on carcinogenesis may only reflect the hepatotoxic potential of the test substance, and the investigation of lower, not hepatotoxic, doses is recommended.

IV. VALIDITY AND SIGNIFICANCE OF THE LIVER FOCUS TEST

A. Sensitivity and Specificity

Since preneoplastic foci are much more frequently observed than liver tumors, studies based on the quantification of liver cell foci detect effects on hepatocarcinogenesis with higher sensitivity than the conventional life time tumor studies in rodents. For example, using the usual group size of 50 animals and assuming a spontaneous tumor incidence of about 1%, there is little or no chance to detect a substance-induced 100% increase in the incidence of liver tumors in a long-term bioassay. However, a substance-induced 100% increase in foci in a LFT will usually be statistically significant even with only 10 animals per group. However, this high sensitivity of the LFT may occasionally result in an oversensitivity. For example, both sucrose (Hei and Sudilowsky, 1985) and tryptophan (Sidransky et al., 1985) have been reported to enhance the development of liver foci in rats, although both substances are not regarded as liver carcinogens.

The impact of these false-positive results on the specificity of the LFT must be seen in the context of the limited specificity of the chronic rodent bioassay. The generally used independent comparison of tumor incidences for different organs results in an increased probability of false-positive findings. Haseman (1983) calculated that even the use of a statistically significant level of 1% for common tumors and for 5% for rare tumors instead of the widely used significance level of 5% for all tumors corresponds to a 7–8% significance level for a typical animal bioassay testing both sexes of two species. Moreoever, similar to the sucrose effect on liver foci, in the chronic rodent bioassay lactose induced neoplasia of the adrenal medulla in rats and urinary bladder tumors in mice (Roe, 1984). Consequently, lactose was discussed as a potential occupational carcinogenic risk

(Hickey et al., 1979). Exposure of newborn babies to high doses of lactose (in the mothers' milk) is commonplace and cannot be regarded as a cancer risk. These examples clearly illustrate that the specificity of the LFT is at least comparable to the specificity of the chronic rodent bioassay. Obviously, the chronic rodent bioassay and the LFT may result in experimental findings that must be regarded as a false-positive result with regard to the relevance to humans, underlining the importance of including mechanistic considerations in the interpretation of any bioassay.

B. Extrapolation to Humans

Although foci of phenotypically altered hepatocytes were found in the livers of patients with an increased risk for liver tumors (Fischer et al., 1986), the significance of chemical-induced liver tumors in rodents for the occurrence of hepatocellular tumors in humans remains a mtter of controversial debate (McClain, 1990). Whereas the liver is the clearly predominant target organ in chemical carcinogenesis in rats, hepatocellular tumors in humans are much less frequent than other tumor types, and apart from aflatoxin B_1, the chemical induction of human liver tumors seems to be of minor importance (Williams, 1997).

There has been an abundance of reports on the induction of hepatocellular carcinomas in animals by natural or man-made chemicals for the last 50 years. In contrast, their impact on the epidemiology of hepatocellular carcinomas is doubtful or very small (Simonetti et al., 1991). A few recent reports described an increased incidence of liver tumors in chemical workers (Bond et al., 1990; Teta et al., 1991; Chow et al., 1993) and in solvent-exposed workers in the Nordic countries (Hernberg et al., 1988; Lynge et al., 1990). Epidemiological evidence for the induction of hepatocellular carcinomas by the low-dose exposure of environmental pollution has not been reported except for in one report from China. In addition to an increase in stomach cancer, Tao and coworkers (1991/92) found an association of liver cancer incidence and the contamination of river water by mutagens.

Viral hepatitis has been proposed to be responsible for 75–90% (Beasly, 1988), 80% (Kew, 1989), and 90% (Lisker-Melman et al., 1989) of all hepatocellular carcinomas worldwide. In areas with a high incidence of hepatitis B, food is frequently contaminated with aflatoxin, however, and both may act synergistically in hepatocarcinogenesis. Some epidemiological studies provided evidence for an important independent role of aflatoxin (IARC, 1976; Yeh, 1989; Hatch, 1993; Qian et al., 1994), whereas others failed to detect an association between aflatoxin-contaminated food and occurrence of hepatocellular carcinomas (Campbell et al., 1990, Hsing et al., 1991; Yamaguchi et al., 1993). In low-incidence areas, cirrhosis itself is the major etiological association of hepatocellular carcinomas (Kew, 1989), and ethyl alcohol-related cirrhosis is probably the most important cause of hepatocellular carcinomas (Lisker-Melman et al., 1989).

In summary, the LFT allows an excellent prediction of hepatocarcinogenic effects in rodents. The relevance of rodent liver carcinogenesis per se to a potential human exposure may still remain a matter of uncertainty, especially with regard to the effects of low doses. In this context, the high sensitivity of the LFT is advantageous. Instead of extrapolating from the high experimentally used bioassay doses to the low doses of a potential human exposure, the LFT allows dose-response studies over wide dose ranges that may show nonlinearity even for genotoxic carcinogens (Williams et al., 1993; Umemura et al., 1993; Enzmann et al., 1995a).

V. FUTURE PERSPECTIVES OF THE LIVER FOCUS TEST

A. Extension of the LFT to Different Species

The use of preneoplastic liver lesions as indicators of carcinogenicity has been successfully adapted for several species. The comparison of different species may help to demonstrate the species specificity of effects on hepatocarcinogenesis. For example, phenobarbital is a promoter for liver carcinogenesis in rat and mouse, but not in hamster (Stenbäck et al., 1986; Tanaka et al., 1987); butylated hydroxytoluene promotes in the rat, but not in the mouse (Tokumo et al., 1991). There is no evidence for an enhancing effect of phenobarbital on hepatocarcinogenesis in humans at therapeutic doses even after long-term exposure (Olsen et al., 1995), although the dose levels used in humans are clearly promoting in rats (Williams and Whysner, 1996). Consequently, for phenobarbital the hamster seems to be the most humanlike species. This finding underlines the importance of the selection of the *appropriate* species depending on the goal of the experiment.

Similar to the experimental protocols in rat liver, initiation-promotion protocols have been described for hepatocarcinogenesis in the mouse (Tokumo et al., 1991; Ward et al., 1990). However, if the induction of tumors in mice and especially in the mouse liver is not relevant to human safety assessment (Monro, 1993), initiation-promotion tests in the mouse liver will seem of limited or questionable relevance, too. However, there is an increasing interest in experimental models with nonrodent species.

The in ovo carcinogenicity assay (IOCA)

The induction of preneoplastic lesions in embryonic bird liver has been suggested as a rapid and inexpensive test model for hepatocarcinogenic effects. In this model, fertilized eggs are injected with the test chemical prior to incubation.

In turkey embryos, foci of altered hepatocytes can be demonstrated (Enzmann et al., 1992b, 1995b) that were similar to preneoplastic liver cell foci that have been observed in the liver of rodents after exposure to carcinogens. In HE-stained sections, in ovo exposure to carcinogens results in the occurrence of mostly basophilic, occasionally clear and acidophilic hepatocellular foci. The demonstration of decreased activity of glycogen phosphorylase is the most reliable enzyme histochemical marker in ovo (Enzmann et al., 1995b).

In the quail embryos, the in ovo exposure to the nitrosamines induced hyperplastic adenomatous lesions in the embryonic livers, which were composed of cells that morphologically appeared more similar to cholangiocytes than to hepatocytes and were arranged in distinct glandular patterns. These were the predominant type of lesion in embryonic quail liver, whereas HAF were observed less frequently and only after high doses of diethylnitrosamine (DEN), or *N*-nitrosomorpholine (N-MOR) (Enzmann et al., 1996).

The species turkey and quail were preferred over chicken eggs because of the longer incubation period of turkeys (28 days) and quail (24 days) compared with chicken (21 days). Assuming that the development of phenotypically altered preneoplastic hepatic foci requires some time, a longer hatching period may be advantageous.

In addition to the induction of focal preneoplastic lesions, hepatocytes with unusually large nuclei were induced in a dose-dependent manner in both species, even with lower doses that were insufficient for the induction of HAF or hyperplastic adenomatous lesions (Enzmann et al., 1995d, 1996). Clawson and coworkers (1992) showed that enlarged nuclei occur in rats after low doses of carcinogens that fail to induce other signs of unspecific toxic effects. Therefore, this parameter may be a useful complementary marker for carcinogenic effects in the IOCA of carcinogenesis.

The exposure to hepatocarcinogens resulted in damage to the mitochondrial DNA (Enzmann et al., 1995c). Electrophoresis on agarose gels with native mitochondrial DNA and with ribonuclease-treated mitochondrial DNA revealed a DEN-induced effect on the molecular size of the mtDNA. The content of 16 kilobase mtDNA (the normal size) decreased in a dose-dependent manner, whereas the amount of mtDNA fragments of various size increased. Fluorescent staining of the electrophoresis gels allowed the densitometric quantification of the mitochondrial DNA of the regular band at 16 kilobase and the amount of fragments of irregular size (smear). The DEN-induced effect was dose-dependent over the whole dose range from 1.24 to 6.2 mmol/kg. Even the lowest dose (10 mg DEN per egg) showed clear-cut effects. The sensitivity of the IOCA appears to be comparable or superior to the sensitivity of lifetime studies in rodents, even when only a single dose of the test chemical is given prior to incubation

(Table 5). In a lifetime study in rats exposed to diethylnitrosamine for 130 weeks, 1.35 mg of diethylnitrosamine is required for the induction of hyperplastic nodules in the liver and 2.7 mg of diethylnitrosamine per rat is needed for a statistically significant increase in hepatocellular carcinomas (Lijinsky et al., 1981). In the IOCA, 2 mg of DEN per egg induced various types of foci within 24 days (Enzmann et al., 1995b). Similarly, for the induction of hepatocellular carcinomas by NNM in as few as 2% of the exposed rats in a 2-year study, 11 mg of NNM per rat are required (Lijinsky et al., 1998). In the IOCA, 1 mg of NNM per egg is sufficient for the induction of foci within 24 days (Enzmann et al., 1992b). Using the tobacco-specific nitrosamine (4-(n-methyl-*N*-nitrosamino)-1-(3-pyridyl)-1-1-butanone (NNK), we were able to show carcinogenic effects in the IOCA at doses that were in the same range as the human exposure (Table 6). The advantages of bird embryos for caricnogenicity testing are considerable:

The IOCA is rapid—focal and nodular preneoplastic liver lesions can be induced in 24 days and damage to mitochondrial DNA in 4 days.

It does not require highly sophisticated methods or expensive equipment, since routine histology is sufficient for the detection of the induced foci.

Proliferation-mediated effects on carcinogenesis are unlikely to effect the induction of foci in the IOCA. Similar to the partial hepatectomy in the Ito model, the rapid growth of the embryonic liver provides such a high cell proliferation rate that a stimulating effect of the test substance is of little or no added impact.

An additional nonrodent species strengthens the extrapolation of animal studies to humans. If a test chemical enhances hepatocarcinogenesis not only in rodents, but also in embryonic turkey liver, it appears more likely that this effect is not rodent specific but that basic cell mechanisms are affected and, consequently, the test chemical probably is a carcinogenic hazard for exposed humans.

Table 5 Comparison of the Detection Limit for Carcinogenicity of Diethylnitrosamine in Different Test Systems

Test System	Duration	Biological endpoint	Cumulated dose	
IOCA in quail	21 days	Hyperplastic adenomous lesions	5 mg/kg	0.05 mg/egg
IOCA in turkey	24 days	Foci of altered hepatocytes	15 mg/kg	1 mg/egg
Chronic rodent bioassay	104 weeks	Liver tumors	47 mg/kg	11 mg/rat

Table 6 Exposure Dose for NNK in Humans and in the IOCA

A. Lifetime exposure in humans: 100 mg	Single dose turkey eggs: 0.5 mg	Single dose quail eggs: 0.05 mg
B. Lifetime exposure in humans: 1.4 mg/kg b.w.	Turkey eggs: 6 mg/kg egg	Quail eggs: 5 mg/kg egg

The IOCA is considerably less expensive than whole animal experiments. Since the experiments are terminated several days before hatching and fertile eggs are commercially available, no facilities for animal housing are required.

Finally, there is less potential for human exposure during the experiment: there is no excretion of the injected carcinogens from the eggs, the egg shell may be regarded as a barrier at least for nonvolatile chemicals, and the amounts of the carcinogenic chemicals needed for the IOCA are relatively small compared to animal experiments.

In summary, the IOCA may be a valuable approach for short-term carcinogenicity screening, bridging the gap between the in vitro assays for genotoxic effects on hepatocytes (Williams et al., 1989) and whole animal assays like the LFT or the ABA discussed above.

Fish models

Studies on the carcinogenicity of chemicals in fish have been proposed for two indications. The effects of carcinogens in fish may be useful bioindicators for environmental pollution and a possible indicator of cancer risk to humans (Weisburger and Williams, 1991). Second, the exposure of fish to carcinogens has been suggested as an inexpensive model for chemically induced carcinogenesis. The occurrence of neoplasia of hepatocellular, cholangiocellular, epidermal, and oral epithelial origin in freshwater fish (Black and Baumann, 1991), as well as the detection of neoplastic and preneoplastic liver lesions in various wild sea fish species (Myers et al., 1991, 1994) and the demonstration of hepatic DNA damage (Malins and Gunselman, 1994), have been suggested as indicators for environmental pollution with carcinogens. In contrast to lymphoma and leukemia of fish, which are no longer thought to result from exposure to environmental pollutants (Mulcahy, 1992), the hepatomas in fish and mammals are sufficiently similar (Machotka, 1992) to justify the conclusion that carcinogen-induced alterations in the liver of fish may reflect a potential human hazard

(Black and Baumann, 1991). Experimental exposure of fish to carcinogens has been proposed as an interesting approach to demonstration of chemical-induced tumors of various tissues, including biliary and hepatic neoplasms (Belitsky et al., 1994; Goodwin and Grizzle, 1994a, 1994b) and preneoplastic liver foci (Stein et al., 1993), stomach adenomas and nephroblastomas (Fong et al., 1993), melanomas (Anders, 1991; Halaban and Moellmann, 1991), and thyroid tumors (Park et al., 1993).

Bailey and coworkers (1992) emphasize three major advantages for the use of fish for carcinogenesis studies: high sensitivity, low costs, and the value of an additional nonrodent species for confident extrapolation of animal study data to humans. If a test chemical induces tumors not only in rodents, but also in fish, then the carcinogenic activity is not rodent specific. These additional positive fish carcinogenicity data suggest that basic cell mechanisms are affected and that the test chemical probably poses a carcinogenic risk to humans.

The possibility of using huge numbers of animals at moderate expense makes the fish models especially interesting for dose-response studies (Hendricks et al., 1994; Bailey et al., 1994). Several groups have demonstrated that in rodents that the dose-effect curve, even for genotoxic carcinogens, may not be linear (Lijinsky et al., 1988; Umemura et al., 1993; Williams et al., 1993). Similar nonlinear responses were found in trouts for aflatoxin B_1 and aflatoxicol (Bailey et al., 1994). It has been suggested that an estimation of a potential risk for exposed humans should rather be based upon the quantification of the biological effects of low doses of carcinogens instead of simple linear extrapolation from high experimental doses to the low doses of a potential human exposure (Enzmann et al., 1995a). In this context, fish studies may be an interesting and relatively inexpensive approach for investigations on biological effects of low-dose carcinogens.

B. Role of the Liver Focus Test in Safety Assessment of Chemicals

Use of the LFT is likely to increase for several reasons. The liver is the most frequent site of carcinogenicity in rodents and will therefore be of primary interest in toxicology. Especially with regard to the ongoing discussion on the relevance of chemical-induced rodent liver tumors to exposed humans, the sensitive LFT will be an indispensable tool for exposure effect studies in the range of potential human exposure and for studies on the combination effects of low doses of different carcinogens. The chronic rodent bioassay cannot effectively address these questions due to the large numbers of animals and the long time required. In addition, the advanced age of the animals at the end of a chronic rodent bioassay may complicate a sound interpretation of the experiments, because several mechanisms contribute to the aging process as well as to carcinogenesis (Williams and Baker, 1992).

An increasing use of the LFT can also be expected since, unlike the chronic rodent bioassay, inhibiting effects on hepatocarcinogenesis can also be demonstrated. This is important for investigations on combination effects of putatively carcinogenic chemicals as well as for the rapidly growing field of cancer prevention (Williams, 1994). The various types of liver foci tests (in ovo carcinogenicity assay, limited rat liver foci bioassay, accelerated bioassay) cover the various phases of the decision point approach to carcinogenicity testing proposed by Williams and Weisburger (1981, 1988).

In this approach, a test substance is not automatically subject to a chronic rodent bioassay. Instead, in four different stages the question is asked whether the expensive chronic rodent bioassay is necessary or justified (Fig. 3). At each of the levels A through E, the appropriate experimental approaches are selected on a case by case basis.

At stage A, the consideration of structure-activity relationships may lead to the suspicion of carcinogenicity. At stage B, in vitro short-term tests involving primarily DNA damage and mutagenesis provide information on genotoxicity. Most types of DNA damage can be detected by the induction of unscheduled DNA synthesis in rat hepatocytes in vitro (Williams et al., 1989). Cell transformation assays may be performed with a variety of cell preparations or cell lines (Isfort et al., 1994) and may include the in ovo transformation of embryonic avian hepatocytes in ovo. These tests may detect genotoxic as well as epigenetic carcinogens and therefore may also be used at stage C for the detection of epigenetic carcinogens (McClain et al., 1995). At stage C, investigation of epigenetic effects assays of intercellular communication (Budunova and Williams, 1994) and mitogenic effects (Ames and Gold, 1990) are valuable tools. At stage D, several short-term models for the demonstration of carcinogenicity in vivo are available. In rats, limited bioassays for liver, urinary bladder, colon, mammary gland, and kidney are available. In the mouse, skin and lung carcinogenesis are well characterized models for carcinogenicity testing (Weisburger and Williams, 1984). In other experimental approaches, such as transgenetic animals (Merlino, 1994), the fish models may also be used, and the induction of tumors in avian embryos or invertebrates has been suggested for rapid and inexpensive test models.

The definitive safety assessment ultimately depends on stage E and can be based upon the accelerated bioassay, which in many cases may substitute for the conventional chronic rodent carcinogenicity bioassay. It is obvious that even the conclusion that a test substance is a potential

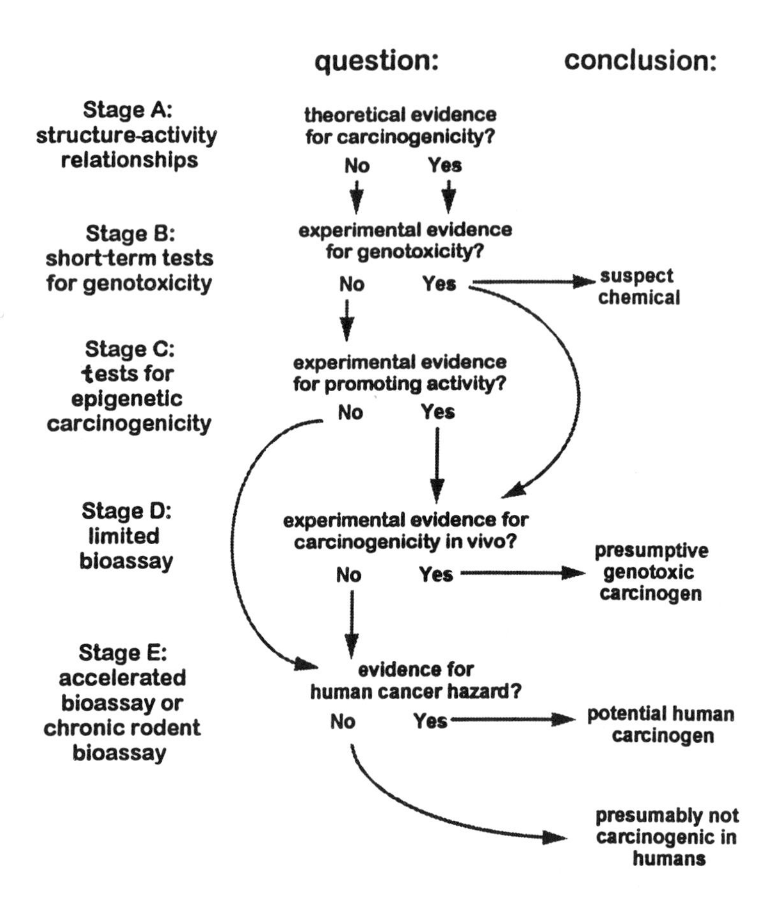

Figure 3 Decision point approach to carcinogen testing.

human carcinogen is not necessarily prohibitive for the use of such a substance if the benefits justify the carcinogenic risk. If possible, a quantitative risk-benefit analysis should be initiated as early as stage D. If the benefits justify the carcinogenic risk, e.g., for a pharmaceutical with a vital indication, humans may be exposed to such a substance whatever the outcome of a chronic bioassay might be and thus actually performing the expensive in vivo test of stage E does not make sense. If additional mechanistic information on the carcinogenic properties of the chemical is desired, the accelerated rodent bioassay appears more appropriate than the chronic rodent bioassay.

VI. CONCLUSIONS

In the context of the increasingly recognized limitations of the chronic rodent bioassay and the ongoing discussion of the maximum tolerated dose, the LFT offers several important advantages:

The high sensitivity allows exposure effect studies, including the low range of potential human exposure.
Experiments provide information on the likely mechanism(s) of action, according to the two-stage concept.
Interference between age-related degenerative alterations and carcinogenesis is avoided, because the experiments are terminated before the animals reach an advanced age.
Enhancing as well as inhibiting effects on different stages of carcinogenesis can be detected.

The evaluation and the interpretation of the LFT, however, may require more sophisticated methods than the standard chronic rodent bioassay, including immunohistochemistry, morphometry, and stereology and a case-by-case adaptation of the experimental protocol. In summary, the LFT is a valuable tool both for toxicological screening and for mechanistic investigations on carcinogenesis and should not only help to detect potential carcinogenic hazards, but should also provide mechanistically based understanding of the relevance of experimental findings.

REFERENCES

Ames BN, Gold LS. Too many rodent carcinogens: mitogenesis increases mutagenesis. Science 1990; 249:970.

Anders A, Zechel C, Schlatterer B, Groger H, Schmidt D, Smith A, Anders F. Genetic and molecular approach for breeding and use of laboratory fish for the detection of agents with carcinogenic and/or promoting activity. Bull Cancer 1991; 78:415–433.

Appel KE, Fürstenberger G, Hapke HJ, Hecker E, Hildebrandt AG, Koransky W, Marks F, Neumann HG, Ohnesorge FK, Schulte-Hermann R. Chemical carcinogenesis: definition of frequently used terms. Cancer Res Clin Oncol 1990; 116:232–236.

Bailey GS, Hendricks J, Dashwood R. Anticarcinogenesis in fish [review]. Mutat Res 1992; 267:243–250.

Bailey GS, Loveland PM, Pereira C, Pierce D, Hendricks JD, Groopman JD. Quantitative carcinogenesis and dosimetry in rainbow trout for aflatoxin B1 and aflatoxicol, two aflatoxins that form the same DNA adduct. Mutat Res 1994; 313:25–38.

Bannasch P. The cytoplasm of hepatocytes during carcinogenesis. Electron and light microscopical investigations of the nitrosomorpholine-intoxicated rat liver. Rec Res Cancer Res 1968; 19:1–100.

Bannasch P. Preneoplastic lesions as end points in carcinogenicity testing. I. Hepatic preneoplasia. Carcinogenesis 1986a; 7:689–695.

Bannasch P. Preneoplastic lesions as end points in carcinogenicity testing. II. Preneoplasia in various non-hepatic tissues [review]. Carcinogenesis 1986b; 7: 849–852.

Bannasch P, Benner U, Enzmann H, Hacker HJ. Tigroid cell foci and neoplastic nodules in the liver of rats treated with a single dose of aflatoxin B1. Carcinogenesis 1985; 6:1641–1648.

Bannasch P, Enzmann H, Klimek F, Weber E, Zerban H. Significance of sequential cellular changes inside and outside foci of altered hepatocytes during hepatocarcinogenesis. Toxicol Pathol 1989; 17:617–629.

Bartsch H, Preat V, Aitio A, Cabral JRP, Roberfroid M. Partial hepatectomy of rats ten weeks before carcinogen administration can enhance liver carcinogenesis. Carcinogenesis 1988; 9:2315–2317.

Beasley RP. Hepatitis B virus. The major etiology of hepatocellular carcinoma. Cancer 1988; 61:10.

Belitsky GA, Lytcheva TA, Khitrovo IA, Safaev RD, Zhurkov VS, Vyskubenko IF, Sytshova LP, Salamatova OG, Feldt EG, Khudoley VV, Mizgirev IV, Khovanova EM, Ugnivenko EG, Tanirbergenov TB, Malinovskaya KI, Revazova YA, Ingel FI, Bratslavsky VA, Terentyev AB, Shapiro AA, Williams GM. Genotoxicity of 1,2-dibromopropane and 1,1,3-tribromopropane in comparison to 1,2-dibromo-3-chloropropane. Cell Biol Toxicol 1994; 10:265–279.

Berenblum I. Frontiers of Biology: Carcinogenesis as a Biological Problem. Amsterdam: North Holland Publishing Company, 1974.

Berenblum J, Shubik P. A new quantative approach to the study of the stages of chemical carcinogenesis in the mouse's skin. Br J Cancer 1947; 1:389–391.

Black JJ, Baumann PC. Carcinogens and cancers in freshwater fishes [review]. Environ Health Perspect 1991; 90:27–33.

Bond GG, McLaren EA, Sabel FL, Bodner KM, Lipps TE, Cook RR. Liver and biliary tract cancer among chemical workers. Am J Indust Med 1990; 18:19–24.

Budunova IV, Williams GM. Cell culture assays for chemicals with tumor-promoting or tumor-inhibiting activity based on the modulation of intercellular communication. Cell Biol Toxicol 1994; 10:71.

Campbell HA, Pitot HC, Potter VR, Laishes BA. Application of quantitative stereology to the evaluation of enzyme-altered foci in rat liver. Cancer Res 1982; 42:465–472.

Campbell TC, Chen JS, Liu CB, Li JY, Parpia B. Nonassociation of aflatoxin with primary liver cancer in a cross-sectional ecological survey in the People's Republic of China. Cancer Res 1990; 50:6882–6893.

Cayama E, Tsuda H, Sarma DSR, Farber E. Initiation of chemical carcinogenesis requires cell proliferation. Nature 1978; 275:60–62.

Chow WH, McLaughlin JK, Zheng W, Blot WJ, Gao YT. Occupational risks for primary liver cancer in Shanghai, China. Am J Indust Med 1993; 24(1):93–100.

Clawson GA, Blankenship LJ, Rhame JG, Wilkinson DS. Nuclear enlargement induced by hepatocarcinogens alters ploidy. Cancer Res 1992; 52:1304–1308.

Craddock VM. Effect of a single treatment with alkylating carcinogens dimethylnityrosamine, diethylnitrosamine and methyl methanesulphonate, on liver regeneration after partial hepatectomy. I. Test for induction of liver carcinomas. Chem Biol Interact 1975; 10:313.

de Gerlache J, Lans M, Taper H, Préat V, Roberfroid M. Promotion of chemically initiated hepatocarcinogenesis. In: Sorsa M, Vaina H, eds. Mutagens in Our Environment. New York: Liss, 1982:35–46.

Deml E, Oesterle D, Wolff T, Greim H. Age-, sex-, and strain-dependent differences in the induction of enzyme-altered islands in rat liver by diethylnitrosamine. J Cancer Res Clin Oncol 1981; 100:125–134.

Dragan YP, Rizvi T, Xu YH, Hully JR, Bawa N, Campbell HA, Maronpot RR, Pitot HC. An initiation-promotion assay in rat liver as a potential complement to the 2-year carcinogenesis bioassay. Fund Appl Toxicol 1991; 16:525–547.

Dragan YP, Xu XH, Goldsworthy TL, Campbell HA, Maronpot RR, Pitot HC. Characterization of the promotion of altered hepatic foci by 2,3,7,8-tetrachlorodibenzo-p-dioxin in the female rat. Carcinogenesis 1992; 13:1389–1395.

Druckrey H, Bresciani F, Schmeider H. Atmung und Glukolyse der Rattenleber waehrend der Behandlung mit 4-Diethylaminoazobenzol. Z Naturforsch 1958; 13b:514.

Enzmann H, Bannasch P. Potential significance of phenotypic heterogeneity of focal lesions at different stages in hepatocarcinogenesis. Carcinogenesis 1987a; 8:1607–1612.

Enzmann H, Edler L, Bannasch P. Simple elementary method for the quantification of focal liver lesions induced by carcinogens. Carcinogenesis 1987b; 8:231–235.

Enzmann H, Ohlhauser D, Dettler T, Benner U, Hacker HJ, Bannasch P. Unusual histochemical pattern in preneoplastic hepatic foci characterized by hyperactivity of several enzymes. Virchows Arch B 1989; 57:99–108.

Enzmann H, Zerban H, Löser E, Bannasch P. Glycogen phosphorylase hyperactive foci of altered hepatocytes in aged rats. Virchows Arch B 1992a; 62:3–8.

Enzmann H, Kaliner G, Watta-Gebert B, Löser E. Foci of altered hepatocytes induced in embryonal turkey liver. Carcinogenesis 1992b; 13:943–946.

Enzmann H, Zerban H, Kopp-Schneider A, Löser E, Bannasch P. Effects of low doses of N-nitrosomorpholine on the development of early stages of hepatocarcinogesis. Carcinogenesis 1995a; 16:1513–1518.

Enzmann H, Kühlem C, Löser E, Bannasch P. Damage to mitochondrial DNA induced by the hepatocarcinogen diethylnitrosamine in ovo. Mutat Res 1995b; 329:113–120.

Enzmann H, Kühlem C, Kaliner G, Löser E, Bannasch P. Rapid induction of preneoplastic liver foci in embryonal turkey liver by diethylnitrosamine. Toxicol Pathol 1995c; 23:560–569.
Enzmann H, Kühlem C, Löser E, Bannasch P. Dose dependence of diethylnitrosamine-induced nuclear enlargement in embryonal turkey liver. Carcinogenesis 1995d; 16:1351–1355.
Enzmann H, Brunnemann KD, Iatropoulos M, Williams GM. Induction of hyperplastic lesions in embryonic quail liver in ovo. Proc Am Assoc Cancer Res 1996; 37:112.
Farber E. Chemical carcinogenesis: a current biological perspective. Carcinogenesis 1984; 5:1–5.
Fischer G, Hartmann H, Droese M, Schauer A, Bock KW. Histochemical and immunohistochemical detection of putative preneoplastic liver foci in woman after long-term use of oral contraceptives. Virchows Arch B 1986; 50:321–337.
Fong AT, Dashwood RH, Cheng R, Mathews C, Ford B, Hendricks JD, Bailey GS. Carcinogenicity metabolism and Ki-ras proto-oncogene activation by 7,12-dimethylbenz[a]anthracene in rainbow trout embryos. Carcinogenesis 1993; 14:629–635.
Friedwald WF, Rous P. Initiating and promoting elements in tumor production; analysis of effects of tar, benzpyrene and methylcholanthrene on rabbit skin. J Exp Med 1944; 80:101–126.
Fürstenberger G, Marks F, Fusenig NE. Prostaglandins, epidermal hyperplasia and skin tumor promotion. In: Fischer SM, Slaga TJ, eds. Arachidonic Acid Metabolism and Tumor Promotion. Boston: Nijoff Publ. 1985:49–72.
Fürstenberger G, Schurich B, Kaina B, Petrusevska RT, Fusenig NE, Marks F. Tumor induction in initiated mouse skin by phorbol esters and methyl methanesulfonate: correlation between chromosomal damage and conversion in vivo. Carcinogenesis 1989; 10:719–752.
Furukawa K, Numoto S, Furuya K, Furukawa NT, Williams GM. Effect of the hepatocarcinogen nafenopin, a peroxisome proliferator, on the activities of rat liver glutathione-requiring enzymes and catalase in comparison to the action of phenobarbital. Cancer Res 1985; 45:5011–5019.
Glauert HP, Beer D, Rao MS, Schwarz M, Xu YD, Goldsworthy TL, Coloma J, Pitot HC. Induction of altered hepatic foci in rats by the administration of hypolipidemic peroxisome proliferators alone or following a single dose of diethylnitrosamine. Cancer Res 1986; 46:4601–4606.
Glauert HP, Beaty MM, Clark TD, Greenwell WS, Tatum V, Chen LC, Borges T, Clark TL, Srinivasan SR, Chow CK. Effect of dietary vitamin E, on the development of altered hepatic foci and hepatic tumors induced by the peroxisome proliferator ciprofibrate. J Cancer Res Clin Oncol 1990; 116:351–356.
Goldsworthy TL, Campbell HA, Pitot C. The natural history and dose-response characteristics of enzyme-altered foci in rat liver following phenobarbital and diethylnitrosamine administration. Carcinogenesis 1984; 5:67–71.
Goodman JI, Ward JM, Popp JA, Klaunig JE, Fox TR. Mouse liver carcinogenesis: mechanisms and relevance. Fundam Appl Toxicol 1991; 17:651.

Goodwin AE, Grizzle JM. Oncogene expression in hepatocytes of the fish Rivulus ocellatus marmoratus during the necrotic and regenerative phases of diethylnitrosamine toxicity. Carcinogenesis 1994a; 15:1985–1992.

Goodwin, AE, Grizzle JM. Oncogene expression in hepatic and biliary neoplasms of the fish Rivulus ocellatus marmoratus: correlation with histologic changes. Carcinogenesis 1994b; 15:1993–2002.

Hagiwara A, Hikarru T, Katsumi I, Seiko T, Shoji F, Nobuyki I. From the known target tissues of these chemicals. Jpn J Cancer Res 1993; 84:237–245.

Halaban R, Moellmann G. Proliferation and malignant transformation of melanocytes [review]. Crit Rev Oncogen 1991; 2:247–258.

Hanigan MH, Winkler ML, Drinkwater NR. Partial hepatectomy is a promoter of hepatocarcinogenesis in C57BL/6J male mice but not in C3H/HeJ male mice. Carcinogenesis 1990; 11:589–594.

Harada T, Maronpot RR, Morris RW, Stizel KA, Boorman GA. Morphological and stereological characterization of hepatic foci of cellular alteration in control Fischer 344 rats. Toxicol Pathol 1989; 17(4 PT 1):579–593.

Harris L, Morris LE, Farber E. Protective value of a liver initiation-promotion regimen against the lethal effect of carbon tetrachloride in rats. Lab Invest 1989; 61:467–470.

Haseman JK. Issues: a reexamination of false-positive rates for carcinogenesis studies. Fundam Appl Toxicol 1983; 3:334–339.

Hatch MC, Chen CJ, Levin B, Ji BT, Yang GY, Hsu SW, Wang LW, Hsieh LL, Santella RM. Urinary aflatoxin levels, hepatitis-B virus infection and hepatocellular carcinoma in Taiwan. Int J Cancer 1993; 54(6):931–934.

Hei TK, Sudilovsky O. Effects of a high sucrose diet on the development of enzyme-altered foci in chemical hepatocarcinogenesis in rats. Cancer Res 1985; 45: 2700–2705.

Hendricks JD, Cheng R, Shelton DW, Pereira CB, Bailey GS. Dose-dependent carcinogenicity and frequent Ki-ras proto-oncogene activation by dietary N-nitrosodiethylamine in rainbow trout. Fundam Appl Toxicol 1994; 23:53–62.

Hernberg S, Kauppinen T, Riala R, Korkala ML, Asikainen U. Increased risk for primary liver cancer among women exposed to solvents. Scand J Work, Environ Health 1988; 14(4):356–365.

Hickey RJ, Clelland RC, Bowers EJ. Essential hormones as carcinogenic hazards [review]. Occup Med 1979; 21:265–268.

Hirose M, Ozaki K, Takaba K, Fukushima S, Shirai T, Ito N. Modifying effects of the naturally occurring antioxidants gamma-oryzanol, phytic acid tannic acid and n-tritriacontane-16,18-dione in a rat wide-spectrum organ carcinogenesis model. Carcinogenesis 1991; 12:1917–1921.

Hirose M, Hoshiya T, Akagi K, Takahashi S, Hara Y, Ito N. Effects of green tea catechins in a rat multi-organ carcinogenesis model. Carcinogenesis 1993; 14: 1549–1953.

Hsing AW, Guo W, Chen J, Li JY, Sone BJ, Blot WJ, Fraumeni JF Jr. Correlates of liver cancer mortality in China. Int J Epidemiol 1991; 20(1):54–59.

Iatropoulos MJ. Accelerated rodent bioassay predictive of chemical carcinogenesis. Exp Toxicol Pathol 1992; 44:481–487.

Iatropoulos MJ, Williams GM. Proliferation markers. Exp Toxicol Pathol 1996; 48:175–181.

Imaida K, Tatematsu M, Kato T, Tsuda H, Ito N. Advantages and limitations of the stereological estimation of placental glutathione S transferase positive rat liver cell foci by computerized three-dimensional reconstruction. Jpn J Cancer Res 1989; 80:326–330.

International Agency for Research on Cancer. IARC Monographs on the Evaluation of Carcinogenic Risk of Chemicals to Man of Some Naturally Occurring Substances. Vol. 10. Lyon, France: IARC, 1976:265–342.

Isfort RJ, Cody DB, Doersen CJ, Kerckaert GA, Leboeuf RA. Alterations in cellular differentiation, mitogenesis, cytoskeleton and growth characteristics during Syrian hamster embryo cell multistep in vitro transformation. Int J Cancer 1994; 59(1):114–125.

Ishikawa T, Takayama S, Kitagawa T. Correlation between time of partial hepatectomy after a single treatment with diethylnitrosamine and induction of adenosinetriphosphatase-deficient islands in rat liver. Cancer Res 1980; 40:4261–4264.

Ito N, Imaida K, Hasegawa R, Tsuda H. Rapid bioassay methods for carcinogens and modifiers of hepatocarcinogenesis CRC Crit Rev Toxicol 1989; 19:386–415.

Ito N, Hasegawa R, Imaida K, Masui T, Takahashi S, Shirai T. Pathological markers for non-genotoxic agent-associated carcinogenesis. Toxicol Lett 1992a; 64–65:613–620.

Ito N, Shirai T, Hasegawa R. Medium-term bioassays for carcinogens [review]. IARC Sci Pub 1992b; 116:353–388.

Ito N, Hasegawa R, Imaida K, Takahashi S, Shirai T. Medium-term rat liver bioassay for rapid detection of carcinogens and modifiers of hepatocarcinogenesis. Drug Metab Rev 1994; 26:431–442.

Kew MC. Role of cirrhosis in hepatocarcinogenesis. In: Bannasch P, Keppler D, Weber G, eds. Liver Cell Carcinoma. Lancaster, UK: Kluwer Academic Publishers, 1989:37–46.

Kinzel V, Loehrke H, Goerttler K, Fürstenberger G, Marks F. Suppression of the first stage of phorbol 12-tetradecanoate 13-acetate-effected tumor promotion in mouse skin by nontoxic inhibition of DNA synthesis. Proc Natl Acad Sci USA 1984; 81:5858–5862.

Kinzel V, Fürstenberger G, Loehrke H, Marks F. Three-stage tumorigenesis in mouse skin: DNA synthesis as a prerequisite for the conversion stage induced by TPA prior to initiation. Carcinogenesis 1986; 7:779–782.

Kitchin KT, Brown JL, Setzer RW. Dose-response relationship in multistage carcinogenesis: promoter. Environ Health Perspect 1994; 102 (suppl 1):255–264.

Kitagawa T, Sugano H. Enhancing effect of phenobarbital on the development of enzyme-altered islands and hepatocellular carcinomas initiated by 3′-methyl-4-(dimethylamino)azobenzene or diethylnitrosamine. Gann 1978; 69(5):679–687.

Kojima T, Tanaka T, Kawamori T, Hara A, Mori H. Chemopreventive effects of dietary D,L-α-difluoromethylornithine, an ornithine decarboxylase inhibitor, on initiation and postinitiation stages of diethylnitrosamine-induced rat hepatocarcinogenesis. Cancer Res 1993; 53:3903–3907.

Laurier C, Tatematsu M, Rao PM, Rajalakshmi S, Sarma DS. Promotion by orotic acid of liver carcinogenesis in rats initiated by 1,2-dimethylhydrazine. Cancer Res 1984; 44(5):2186–2191.

Ledda-Columbano GM, Coni P, Simbula G, Zedda I, Columbano A. Compensatory regeneration mitogen-induced liver growth and multistage chemical carcinogenesis [review]. Environ Health Perspect 1993; 101 (suppl 5):163–168.

Ledda-Columbano GM, Coni P, Columbano A. Cell proliferation and cell death in rat liver carcinogenesis by chemicals [review]. Arch Toxicol 1994; 16 (suppl): 271–280.

Lijinsky W, Reuber MD, Riggs CW. Dose response studies of carcinogenesis in rats by nitrosodiethylamine. Cancer Res 1981; 41:4997–5003.

Lijinsky W, Kovatch RM, Riggs CW, Walters PT. Dose response study with N-nitrosomorpholine in drinking water of F-344 rats. Cancer Res 1988; 48:2089–2095.

Lin ELC, Klaunig JE, Mattox JK, Weghorst CM, McFarland BH, Pereira MA. Comparison of the effects of acute and subacute treatment with phenobarbital in different strains of mice. Cancer Lett 1989; 48:43–51.

Lisker-Melman M, Martin P, Hoofnagle JH. Conditions associated with hepatocellular carcinoma. Med Clin North Am 1989; 73:999–1009.

Loehrke H, Schweizer J, Dederer E, Hesse B, Rosenkranz G, Goerttler H. On the persistence of tumor initiation in the two stage carcinogenesis on mouse skin Carcinogenesis 1983; 4:771–775.

Luebeck EG, Moolgavkar SH, Buchmann A, Schwarz M. Effects of polychlorinated biphenyls in rat liver: quantitative analysis of enzyme-altered foci. Toxicol Appl Pharmacol 1991; 111:469–484.

Lynge E, Thygesen L. Primary liver cancer among women in laundry and dry-cleaning work in Denmark. Scand J Work Environ Health 1990; 16(2):108–112.

Machotka SV. Hepatocellular neoplasia in fish rats and man: a selected comparative review [review]. In Vivo 1992; 6:339–347.

Maeura Y, Williams GM. Enhancing effect of butylated hydroxytoluene on the development of liver altered foci and neoplasms induced by N-2-fluorenylacetamide in rats. Food Chem Toxicol 1984; 22:191.

Malins DC, Gunselman SJ. Fourier-transform infrared spectroscopy and gas chromatography-mass spectrometry reveal a remarkable degree of structural damage in the DNA of wild fish exposed to toxic chemicals. Proc Natl Acad Sci 1994; 91:13038–13041.

Maruyama H, Tanaka T, Williams GM. Effects of the peroxisome proliferator di(2-ethylhexyl)phthalate on enzymes in rat liver and on carcinogen-induced liver altered foci in comparison to the promoter phenobarbital. Toxicol Pathol 1990; 18:257–267.

McClain RM. Mouse liver tumors and microsomal enzyme-inducing drugs: experimental and clinical perspectives with phenobarbital. In: Stevenson DE, McClain RM, Popp JA, Slaga TJ, Ward JM, Pitot HC, eds. Mouse Liver Carcinogenesis Mechanisms and Species Comparisons. New York: Wiley-Liss, 1990:345–365.

McClain RM, Slaga TJ, Leboeuf R, Pitot H. Progress in Clinical And Biological Research: Growth Factors and Tumor Promotion Implications for Risk Assessment. Vol. 391. New York: John Wiley 1995.

Merlino G. Transgenic mice as model for tumorgenesis. Cancer Invest 1994; 12(2): 203–213.

Monro A. How useful are chronic (life-span) toxicology studies in rodents in identifying pharmaceuticals that pose a carcinogenic risk to humans? [review]. Adverse Drug Reactions Toxicol Rev 1993; 12:5–34.

Moolgavkar SH. Carcinogenesis modeling: from molecular biology to epidemiology. Annu Rev Public Health 1986; 7:151–169.

Moolgavkar SH, Luebeck EG, de Gunst M, Port RE, Schwarz M. Quantitative analysis of enzyme-altered foci in rat hepatocarcinogenesis experiments I: single agent regimen. Carcinogenesis 1990; 11:1271–1278.

Moore MA, Kitagawa T. Hepatocarcinogenesis in the rat: the effect of promoters and carcinogens in vivo and in vitro. Int Rev Cytol 1986; 101:125–173.

Mulcahy MF. The current position of poikilotherms [review]. Leukemia 1992; 6 (suppl 3):147S–149S.

Myers MS, Landahl JT, Krahn MM, McCain BB. Relationships between hepatic neoplasms and related lesions and exposure to toxic chemicals in marine fish from the U.S. West Coast [review]. Environ Health Perspect 1991; 90:7–15.

Myers, MS, Stehr CM, Olson OP, Johnson LL, McCain BB, Chan SL, Varanasi U. Relationships between toxicopathic hepatic lesions and exposure to chemical contaminants in English sole (Pleuronectes vetulus), starry flounder (Platichthys stellatus), and white croaker (Genyonemus lineatus) from selected marine sites on the Pacific Coast USA. Environ Health Perspect 1994; 102:200–215.

Numoto S, Furukawa K, Furuya K, Williams GM. Effects of the hepatocarcinogenic peroxisome-proliferating hypolipidemic agents clofibrate and nafenopin on the rat liver cell membrane enzymes gamma-glutamyltranspeptidase and alkaline phosphatase and on the early stages of liver carcinogenesis. Carcinogenesis 1984; 5:1603.

Oesterle D, Deml E. Lack of initiating and promoting activity of thiourea in rat liver foci bioassay. Cancer Lett 1988; 41:245–249.

Olsen JH, Schulgen G, Boice JD Jr, Whysner J, Travis LB, Williams GM, Johnson FB, McGee JO. Antiepileptic treatment and risk for hepatobiliary cancer and malignant lymphoma. Cancer Res 1995; 55:294–297.

Park EH, Chang HH, Lee KC, Kweon HS, Heo OS, Ha KW. High frequency of thyroid tumor induction by N-methyl-N′-nitro-N-nitrosoguanidine in the hermaphroditic fish Rivulus marmoratus. Jpn J Cancer Res 1993; 84:608–615.

Parnell MJ, Exon JH, Koller LD. Assessment of hepatic initiation-promotion properties of trichloroacetic. Arch Environ Contamin Toxicol 1988; 17:429–436.

Peraino C, Fry RJM, Staffeldt E. Reduction and enhancement by phenobarbital of hepatocarcinogenesis induced in the rat by 2-acetylaminofluorene. Cancer Res 1971; 31:1506–1512.

Peraino C, Fry RJM, Staffeldt E. Effects of varying the onset and duration of exposure to phenobarbital on its enhancement of 2-acetylaminofluorene-induced hepatic tumorigenesis. Cancer Res 1977; 37:3623–3628.

Pitot HC, Barness L, Goldsworthy T, Kitagawa T. Biochemical characterization of stages of hepatocarcinogenesis after a single dose of diethylnitrosamine. Nature 1978; 217:456–458.

Pitot HC, Gross LE, Goldsworthy T. Genetics and epigenetics of neoplasia: facts and theories. Carcinogenesis 1985; 10:65–79.

Pitot HC, Goldsworthy TL, Moran S, Kennan W, Glauert HP, Maronpot RR, Campbell HA. A method to quantitate the relative initiating and promoting potencies of hepatocarcinogenic agents in their dose-response relationship to altered hepatic foci. Carcinogenesis 1987; 8:1491–1499.

Qian GS, Ross RK, Yu MC, Yuan JM, Gao YT, Henderson BE, Wogan GN, Groopman JD. A follow-up study of urinary markers of aflatoxin exposure and liver cancer risk in Shanghai, People's Republic of China. Cancer Epidemiol Biomarkers Prev 1994; 3:3–10.

Ray JS, Harbison ML, McClain RM, Goodman JI. Alterations in the methylation status and expression of the raf oncogene in phenobarbital-induced and spontaneous B6C3F1 mouse liver tumors. Mol. Carcinogen 1994; 9:155.

Rizvi TA, Kennan W, Xu YH, Pitot HC. The effects of dose and duration of administration on the promotion index of phenobarbital in multistage hepatocarcinogenesis in the rat. Apmis 1988; 2 (suppl):262–274.

Roe FJ. Perspectives in carbohydrate toxicology with special reference to carcinogenicity. Swedish Dental J 1984; 8:99–111.

Sasaki T, Yoshida T. Experimentelle Erzeugung des Lebercarcinoms durch Fütterung mit o-Amidoazotoluol. Virchows Arch Pathol Anat 1935; 295:175–200.

Scherer E. Use of a programmable pocket calculater for the quantification of precancerous foci. Carcinogenesis 1981; 2:805–807.

Schulte-Hermann R, Timmermann-Trosiener I, Schuppler J. Promotion of spontaneous preneoplastic cells in rat liver as a possible explanation of tumor production by nonmutagenic compounds. Cancer Res 1983; 43:839–844.

Schulte-Hermann R, Bursch W, Kraupp-Grasl B. Active cell dealth (apoptosis) in liver biology and disease. In: Boyer JL, Ockner RK, eds. Progress in Liver Disease. Vol. XIII. Philadelphia: W.B. Saunders Company, 1995:1–35.

Sidransky H, Garrett CT, Murty CN, Verney E, Robinson ES. Influence of dietary trypthophane on the induction of γ-glutamyltranspeptidase-positive foci in the livers of rats treated with hepatocarcinogen. Cancer Res 1985; 45:4844–4847.

Simonetti RG, Cammá C, Fiorello F, Politi F, D'Amico G, Pagliaro L. Hepatocellular carcinoma: a worldwide problem and the major risk factors. Digest Dis Sci 1991; 36(7):962–972.

Solt D, Farber E. New principle for the analysis of chemical carcinogenesis. Nature 1976; 263:701–703.

Solt DB, Medline A, Farber E. Rapid emergence of carcinogen-induced hyperplastic lesions in a new model for the sequential analysis of liver carcinogenesis. Am J Pathol 1977; 88:595–618.

Stein JE, Reichert WL, French B, Varanasi U. 32P-postlabeling analysis of DNA adduct formation and persistence in English sole (Pleuronectes vetulus) exposed to benzo[a]pyrene and 7H-dibenzo[c,g]carbazole. Chem Biol Interact 1993; 88:55–69.

Stenbäck F, Mori H, Furuya K, Williams GM. Pathogenesis of dimethylnitrosamine-induced hepatocellular cancer in hamster liver and lack of enhancement by phenobarbital. J Natl Cancer Inst 1986; 76:327–333.

Stenbäck F, Gebhardt R, Sirma H, Garbay JM, Williams GM. Sequential functional and morphological alterations during hepatocarcinogenesis induced in rats by feeding of a low dose of 2-acetylaminofluorene. Toxicol Pathol 1994; 22:620–632.

Tanaka T, Mori H, Hirota N, Furuya K, Williams GM. Effect of DNA synthesis on induction of preneoplastic and neoplastic lesions in rat liver by a single dose of methylazoxymethanol acetate. Chem Biol Interac 1986; 58:13–27.

Tanaka T, Mori H, Williams GM. Enhancement of dimethylnitrosamine-initiated hepatocarcinogenesis in hamsters by subsequent administration of carbon tetrachloride but not phenobarbital or p,p′-dichlorodiphenyltrichloroethane. Carcinogenesis 1987; 8:1171–1178.

Tao X, Zhu H, Yu S, Zhao Q, Wang J, Wu G, You W, Li C, Zhi W, Bao J, Sun Z. Pilot study on the relationship between male stomach and liver cancer death and mutagenicity of drinking water in the Huangpu River area. Public Health Rev 1991–92; 19:1–4.

Teta MJ, Ott MG, Schnatter AR. An update of mortality due to brain neoplasms and other causes among employees of a petrochemical facility. J Occup Med 1991; 33(1):45–51.

Tokumo K, Iatropoulos MJ, Williams GM. Butylated hydroxytoluene lacks the activity of phenobarbital in enhancing diethylnitrosamine-induced mouse liver carcinogenesis. Cancer Lett 1991; 59193–59199.

Trosko JE, Jone C, Chang CC. Oncogenes, inhibited intercellular communication and tumor promotion [review]. Princess Takamatsu Symposia 1983; 14:101–131.

Tsuda H, Masui T, Ikawa E, Imaida K, Ito N. Compared promoting potential of D-Galactosamine Carbon tetrachloride and partial hepatectomy in rapid induction of preneoplastic liver lesions in the rat. Cancer Lett 1987; 37:163–171.

Tsukada H, Sawada N, Mitaka T, Gotoh M. Effects of liver-tumor promoters on phalloidin sensitivity of rat hepatocytes. Carcinogenesis 1986; 7:335–337.

Umemura T, Tokumo K, Sirma H, Gebhardt R, Poirier MC, Williams GM. Dose response effects of 2-acetylaminofluorene on DNA damage cytotoxicity cell proliferation and neoplastic conversion in rat liver. Cancer Lett 1993; 73:1–10.

Vesselinovitch SD, Hacker HJ, Bannasch P. Histochemical characterization of focal hepatic lesions induced by single diethylnitrosamine treatment in infant mice. Cancer Res 1985; 45:2774–2780.

Ward JM, et al. Liver tumor promoters and other mouse liver carcinogens. In: Stevenson DE, et al., eds. Mouse Liver Carcinogenesis: Mechanisms and Species Comparisons. New York: Wiley-Liss, 1990; 85–108.

Watanabe K, Williams GM. Enhancement of rat hepatocellular-altered foci by the liver tumor promoter phenobarbital: evidence that foci are precursors of neoplasms and that the promoter acts on carcinogen-induced lesions. J Natl Cancer Inst 1978; 61:1311–1314.

Warwick GP. Effect of the cell cycle on carcinogenesis. Fed Proc 1971; 30:1760.

Weibel ER. Stereological Methods. Vol. 1. Practical Methods for Biological Morphometry, London: Academic Press, 1979.

Weisburger JH, Williams GM. Carcinogen testing: current problems and new approaches. Science 1981; 214:401–407.

Weisburger JH, Williams GM. Classification of carcinogens as genotoxic and epigenetic as basis for improved toxicologic bioassay methods. In: Sugimura T, Kondo S, Takebe H, eds. Environmental Mutagens and Carcinogens. New York: Alan R. Liss, Inc., 1982:283–294.

Weisburger JH, Williams GM. Bioassay of carcinogens: in vitro and in vivo tests. In: Searle CE, ed. Chemical Carcinogens. 2nd ed. Washington, DC: American Chemical Society, 1984:1323–1373.

Weisburger JH, Williams GM. Critical effective methods to detect genotoxic carcinogens and neoplasm promoting agents. Environ Health Perspect 1991; 90: 121–126.

Whysner J, Wang CX, Zang E, Iatropoulos MJ, Williams GM. Dose response of promotion by butylated hydroxyanisole in chemically initiated tumours of the rat forestomach. Food Chem Toxicol 1994; 32:215–222.

Williams GM, Klaiber M, Parker SE, Farber E. Nature of early appearing, carcinogen-induced liver lesions resistant to iron accumulation. J Natl Cancer Inst 1976; 57:157–165.

Williams GM. The pathogenesis of rat liver cancer caused by chemical carcinogens. Biochem Biophys Acta 1980; 605:167–189.

Williams GM. Liver carcinogenesis: the role for some chemicals of an epigenetic mechanism of liver tumour promotion involving modification of the cell membrane. Food Cosmet Toxicol 1981; 19:577–583.

Williams GM. Phenotypic properties of preneoplastic rat liver lesions and applications to detection of carcinogens and tumor promoters. Toxicol Pathol 1982; 10:3–10.

Williams GM. Inhibition of chemical-induced experimental cancer by synthetic phenolic antioxidants. In: Williams GM, Sies H, Baker III GT, Erdman JW Jr, Henry CJ, eds. Antioxidants; Chemical, Physiological, Nutritional and Toxicological Aspects. Princeton, NJ: Princeton Scientific Press, 1993:303–308.

Williams GM. Interventive prophylaxis of liver cancer. Eur J Cancer Prev 1994; 3: 89–99.

Williams GM. Chemicals with carcinogenic activity in rodent liver: mechanistic evaluation of human risk. Cancer Lett 1997; 118:1–14.

Williams GM, Weisburger JH. Application of a cellular test battery in the decision

point approach to carcinogen identification [review]. Mutat Res 1988; 205: 79-90.
Williams GM, Baker GT. The potential relationships between aging and cancer [review]. Exp Gerontol 1992; 27:469-476.
Williams GM, Whysner J. Epigenetic carcinogenesis: evaluation and risk assessment. Exp Toxicol Pathol 1996; 48:189-195.
Williams GM, Ohmori T, Katayama S, Rice JM. Alteration by phenobarbital of membrane-associated enzymes including gamma glutamyl transpeptidase in mouse liver neoplasms. Carcinogenesis 1980; 1:813-818.
Williams GM, Maeura Y, Weisburger JH. Simultaneous inhibition of liver carcinogenicity and enhancement of bladder carcinogenicity of N-2-fluorenylacetamide by butylated hydroxytoluene. Cancer Lett 1983; 19:55-60.
Williams GM, Tanaka T, Maeura Y. Dose-dependent inhibition of aflatoxin B1 induced hepatocarcinogenesis by the phenolic antioxidants, butylated hydroxyanisole and butylated hydroxytoluene. Carcinogenesis 1986; 7:1043-1050.
Williams GM, Mori H, McQueen CA. Structure-activity relationships in the rat hepatocyte DNA-repair test for 300 chemicals. Mutat Res 1989; 221:263-286.
Williams GM, Gebhardt R, Sirma H, Stenback F. Non-linearity of neoplastic conversion induced in rat liver by low exposures to diethylnitrosamine. Carcinogenesis 1993; 14:2149-2156.
Williams GM, Iatropoulos MJ, Weisburger JH. Chemical carcinogen mechanisms of action implications for testing methodology. Exp Toxicol Pathol 1996; 48: 101-111.
Yamaguchi G. Hepatocellular carcinoma and its risk factors—their annual changes and effects on the age at onset. Kurume Med J 1993; 40(1):33-40.
Yamamoto S, Konishi Y, Matsuda T, Murai T, Shibata MA, Matsui-Yuasa I, Otani S, Kuroda K, Endo G, Fukushima S. Cancer induction by an organic arsenic compound, dimethylarsenic acid (cacodylic acid) in F344/DuCrj rats after pretreatment with five carcinogens. Cancer Res 1995; 55:1271-1276.
Yamasaki H. Role of cell-cell interaction in tumor promotion [japanese]. Gan to Kagaku Ryoho 1986; 13(3 Pt 2):773-781.
Yamasaki H, Mesnil M, Nakazawa H. Interaction and distinction of genotoxic and non-genotoxic events in carcinogenesis [review]. Toxicol Lett 1992; 64-65: 597-604.
Yeh FS, Yu MC, Mo CC, Luo S, Tong MJ, Henderson BE. Hepatitis B virus, aflatoxins, and hepatocellular carcinoma in Southern Guangxi, China. Cancer Res 1989; 49(9):2506-2509.

14

Transgenic Animals as Predictive Models for Identifying Carcinogens

William C. Eastin and Raymond W. Tennant
National Institute of Environmental Health Sciences, Research Triangle Park, North Carolina

I. BACKGROUND

With more than 70,000 chemicals in commerce (Walker, 1995), the human population is exposed to many chemicals through their use in a wide variety of industrial and consumer products as well as to many chemicals naturally occurring in food and as contaminants in the environment. While it is generally assumed that relatively few of these chemicals are likely to pose a significant risk to human health at the levels of exposure that exist, the health effects of most of these chemicals are generally unknown. As a part of their mandate to protect human health, federal regulatory agencies, e.g., the U.S. Environmental Protection Agency (EPA) and the Food and Drug Administration (FDA), evaluate all available data on each chemical to determine toxicity and to establish potential human risk from exposure. For these agencies and other groups with human health interests, specific data are needed to best provide an adequate toxicological chemical evaluation. Two sources of the more critical data are epidemiological studies and the long-term rodent bioassays for carcinogenicity.

Epidemiology studies provide the most relevant data because they show when human health effects are associated with environmental conditions. When a single environmental agent can be shown to be directly related to a specific human health problem in an exposed population, the risk for that agent is clear. However, the fundamental limitation of epidemiology is that it is a retrospective evaluation of hazards. In addition, it is difficult to relate a health problem to one environmental agent since most exposures are to mixtures of agents and lifestyles within the exposed population vary (Doll, 1996).

The long-term rodent cancer bioassay, usually employing both sexes of two rodent species, is the current protocol used to produce toxicity data on single agents under well-controlled conditions (e.g., National Toxicology Program, 1992). In the last decade, refinements have been made to the protocols and accepted by regulatory agencies to ensure that the data will be collected under established standardized conditions (Organisation for Economic Co-operation and Development, 1981; U.S. Environmental Protection Agency, 1982; U.S. Food and Drug Administration, 1993). While still considered the best test model system for obtaining carcinogenicity data, the 2-year rodent bioassay has several drawbacks. First, these studies require large numbers of animals; second, it normally takes 5–8 years to complete and evaluate the studies before the results are reported; and third, the results must be extrapolated from rodents to humans. Additionally, given the thousands of chemicals currently in commerce without appropriate toxicity data—and new ones are continuously added—and the number of bioassays that can be completed per year, it would take decades to provide the data needed to establish human risk on all of the chemicals using the bioassay alone.

Other methodologies, principally utilizing in vitro systems, have been developed which provide data about genotoxicity, which are used to supplement the process of recognizing potential carcinogens (Zeiger et al., 1990). Genotoxicity data from in vitro studies are collected in a shorter time and therefore many more chemicals have been studied using these systems. However, it is not possible to adequately assess the carcinogenic potential of chemicals without data from in vivo studies. The combined data from these in vivo rodent and in vitro genotoxicity studies have been used to help us understand the relationship between genetic effects and carcinogenesis (Ashby and Tennant, 1991). These studies indicate a greater likelihood that chemicals testing positive for mutagenicity in *Salmonella* will produce an increased incidence of tumors when tested in the rodent bioassay than for those chemicals that were negative in *Salmonella*. It is clear that the development of new testing protocols that would allow more chemicals to be screened for carcinogenic potential in a shorter time frame and still provide data to allow for extrapolation to other species would be extremely beneficial in the quest to protect human health. In this regard, an expanding body of knowledge regarding the role of oncogenes and tumor-suppressor genes in neoplastic events is leading to a better understanding of carcinogenic mechanisms. This knowledge points to the use of transgenic animal species in carcinogenicity bioassays (Adams and Cory, 1991; Bishop, 1991; Christofori and Hanahan, 1994; Weinberg, 1995).

The potential for using transgenic models to screen for chemical carcinogens is based on the premise that genetic factors are the principal determinants of interspecies differences in response to chemicals. That is, the uncertainty and controversies arising from the use of rodent bioassays and the extrapolation of bioassay results to potential human health risks can, in large measure, be explained by specific genetic differences. In the rodent bioassay, the carcinogenic effects of the tested chemical are reported separately by species and by sex. Thus, testing outcomes may be negative, i.e., they may produce no increase in tumor incidence or may range from an increase in the incidence of tumors in one sex of one species at one tissue site to an increase of tumors in both sexes of both species at multiple tissue sites. The results may also include one of the possible combinations of responses between these two extremes.

Chemicals causing an increased incidence at only one site in one sex of one species may be acting through a different mechanism than chemicals producing an increased incidence in both species and more than one tissue site. Approximately 50% of all bioassays have shown strain-specific responses (Ashby and Tennant, 1991). The most extreme specificity is represented among those chemicals that induced tumors at only one site in one of the two sex/species (Tennant, 1993). Therefore, the testing outcomes must be taken into account when extrapolating the carcinogenic potential for human risk assessment.

Strain specificity may be explained by the pattern of inheritance in inbred lines (Silver, 1995). Genes that code for intrinsic functions of growth, development, and maintenance of essential cellular functions occur in common between virtually all species. Such highly conserved genes demonstrate relatively little variation in sequence homology and function similarly even in very diverse species; these genes were often identified or recognized in nonmammalian species before they were identified in mammals. The highly conserved genes thus have relatively few variants and can be contrasted with other genes which are polymorphic or for which there may be many variant alleles.

Standard chemical carcinogenesis studies use inbred rodent strains, an experimental paradigm that has become widely accepted as necessary in order to minimize experimental variation. What is not often recognized is that there are consequences derived from the use of inbred rodents where, in the process of inbreeding, either selectively or randomly, the individual strains come to possess various specific alleles of polymorphic genes in all individuals in the inbred line. In outbred populations such as humans and feral rodents, however, the polymorphisms are variably distributed among individuals and expressed as dominant or recessive.

Many of the polymorphic genes influence interactions of the individual, or of the individual's cells, with the environment or with drugs and chemicals, e.g., cytochrome P-450 enzymes. When rodents are highly inbred, the resultant strains possess only specific alleles of the polymorphic gene families, and their response to chemical exposure may be representative of only a portion of individuals in a randomly breeding population. This consequence of allelic enrichment may be the basis of the strain-specific responses to chemicals (Tennant, 1993; Festing, 1995).

Extensive data collected by the National Toxicology Program for the Fisher (F344) rat and B6C3F1 mouse lines reveal that while the incidence of tumors at specific sites may vary within and between studies, the pattern of spontaneous tumors is repeated generation after generation in the respective lines (Haseman, 1995) and therefore is clearly of genetic origin. For example, in the case of the B6C3F1 mouse, the most well-studied spontaneous tumor has been the hepatocellular carcinoma. The development of this tumor has been linked to a locus designated *hcs*, which is transmitted as a dominant or semi-trait from the C3H male parent (Poole and Drinkwater, 1995). The existence of these spontaneous liver tumors demonstrates that there are specific heritable determinants of neoplastic disease that can be profoundly manifested when such alleles are enriched by inbreeding. The occurrence of these spontaneous tumors presents one of the more complicated problems in interpreting the long-term effects of chemicals and in extrapolating between species. It is difficult to judge which biological effects observed after long-term chemical exposure of rodents would be most likely to occur in humans. One attempts to extrapolate from effects of the chemical in inbred species to potential effects in an outbred human population in which the alleles of cancer susceptibility or determinants of cancer (spontaneous tumor) genes are distributed in a much different frequency. In addition, in extrapolating such strain-specific effects in bioassays, uncertainties arise as to whether the alleles that govern the susceptibility of the test species are even represented in humans and, if so, the actual frequency with which such loci might be represented in the randomly breeding human population.

The induction of strain- and site-specific tumors in bioassays poses one of the most problematic aspects of interpretation. If such tumors are a principal consequence of strain-specific inheritance, then the chemicals that induce only such responses need to be distinguished from those that are tumorigenic in both bioassay species. The rationale for such a distinction is that the transpecies carcinogens most probably interact with more highly conserved genes, such as oncogenes or tumor-suppressor genes. Since tumors are induced in both rodent species, the chemicals are also less subject to the modifying effects of the highly polymorphic genes that influence strain-specific responses. Therefore, it is also probable that transpecies carcinogens are also more likely to be carcinogenic to humans than are the chemicals inducing strain-specific effects.

Another aspect of this rationale is that many chemicals have been identified as carcinogens based upon increases in tumors that occur spontaneously (i.e., background tumors). In some studies, the chemicals concomitantly decreased the incidence of spontaneous tumors at specific sites (Haseman and Johnson, 1996). These results suggest that the action of the chemicals in both cases are indirect in the sense that they modulate the expression of genes associated with spontaneous tumors. The modulation can be either to increase or to decrease the incidence of tumors in those animals. Whichever way the effects are seen, they are most likely related to strain-specific inheritance; thus, these chemicals may be unlikely to be tumorigenic to other species and they are also unlikely to demonstrate anticarcinogenic effects except in a strain-specific manner.

The preceding statements present the rationale for the use of transgenic models as short-term bioassays for identifying carcinogens. Three desirable characteristics of a carcinogen detection system are now presented. First, the carcinogen-detection system should be risk adverse. This means that very few chemicals with the capacity to induce carcinogenic effects in humans should be missed by such systems (few false negatives). Second, alternatives for the 2-year bioassays should not reproduce the highly specific effects of chemicals that are the consequence of strain-specific inheritance (few false positives in the sense of rodent-to-human extrapolation). Third, transgenic bioassays should detect chemicals that induce transpecies carcinogenic effects. These transpecies carcinogenic chemicals carry the greatest probability of being able to induce cancer in exposed human populations (Tennant et al., 1995).

II. IDENTIFICATION OF USEFUL TRANSGENIC MODELS

As the number of transgenic mouse lines developed for studying mutagenic events began to increase, workers at the National Institute of Environmental Health Sciences (NIEHS) recognized that some of these models might also be useful as adjuncts to a bioassay, i.e., to screen for potential carcinogens (Griesemer and Tennant, 1992). For example, transgenic mice overexpressing *pim*-1 are predisposed to T-cell lymphogenesis and appear to develop significant increases in lymphomas when exposed to agents such as *N*-ethyl-*N*-nitrosourea, compared with untreated mice (Breuer et al., 1989).

Transgenic mice containing one copy per cell of the hepatitis B virus (HBV) without the core gene may be a useful model for determining hepatocarcinogens (Sell et al., 1991). A transgenic mouse line carrying a v-Ha-ras oncogene with an embryonic ζ globin promoter has shown a propensity to form dermal papillomas at sites of skin wounding or chemical exposure (Leder et al., 1990). As more transgenic strains become available, use of some of these models may identify carcinogens and lead to a greater understanding of the mechanisms of action of these agents.

Transgenic modifications to the genome can be grouped into three general categories: regulated transgenes, nonexpressed reporter genes, or animals in which the specific genes have been knocked out or whose function is not expressed. These characteristics make transgenic models appropriate adjuncts to toxicology studies and important tools with which to identify potential carcinogens. A large number of lines in all three categories have been developed over the past decade (Leder et al., 1990; Adams and Cory, 1991; Donehower et al., 1992; Cardiff and Muller, 1993; Greenhalgh et al., 1993; Merlino, 1994; Mirsalis et al., 1994; Zinkel and Fuchs, 1994; Viney, 1995). Transgenic models with many different carcinogen-responsive tissues would be the most appropriate systems with which to complement rodent bioassays (Tennant et al., 1995). The p53-deficient ($p53^{def}$) model developed by Donehower et al. (1992) and the Tg.AC (zetaglobin-v-Ha-ras) transgenic line developed by Leder et al. (1990) are being evaluated as adjuncts to the rodent bioassay by the NIEHS and the National Toxicology Program (NTP). Study results from these two prototype lines of transgenic models for carcinogen bioassay will provide the basis for determining the most informative characteristics of transgenic models and for establishing a database with which these and other transgenic models can be readily evaluated for their ability to discriminate between carcinogens and noncarcinogens.

III. THE p53 KNOCKOUT MODEL

The $p53^{def}$ mouse is perhaps one of the most appropriate models for potential carcinogen identification. The p53 gene was one of the first tumor-suppressor genes identified whose function could regulate neoplastic transformation (Finlay et al., 1989; Levine, 1989). The mechanism has been defined by which loss of expression or function of both p53 alleles can result in emergence of the neoplastic phenotype. The p53 gene has been shown to serve as a regulator of the entry of DNA damaged cells into specific stages of the cell cycle (Kastan, 1992). A high proportion of cancers in both humans and in mice have been found to carry mutations in the p53 gene. p53 appears to be an essential component of the pathway by which neoplasias develop in several different tissues (Zambetti, 1993). The concept for the use of the heterozygous $p53^{def}$ (i.e., $p53^{+/-}$) model for detecting potential carcinogens is that the loss of the one remaining functional allele would render animals more specifically sensitive to the effects of genotoxic carcinogens. The phenotype of the knockout mice support this assumption since they remain free of tumors for over a year but have been sensitive to some mutagenic carcinogens (Harvey et al., 1993a,b; Kemp et al., 1994). Studies by Kemp et al. (1993) using the two-stage initiation-promotion protocol (DMBA initiation and TPA promotion) showed that while the incidence and onset of papillomas was similar to that of the wild type, there was an earlier onset and higher frequency of malignant tumors in the $p53^{def}$ mice. Animals that have lost function of both alleles ($p53^{-/-}$) develop a high incidence of predominantly hematopoietic tumors within the first 6 months of life. The homozygous p53 genotype therefore is not useful for potential carcinogen detection, but the phenotype of the heterozygous ($p53^{def}$) animal is very appropriate. The heterozygosity of the p53 transgene makes it a more sensitive model system because it apparently requires only one mutagenic event. Heterozygous mice remain tumor-free for up to one year, and, in NIEHS studies, spontaneous tumors have not been found in animals held for over a year in the absence of any treatment (Tennant, 1995).

A series of studies were conducted by workers at the NIEHS to evaluate the $p53^{def}$ mouse as a model for chemical carcinogen detection (Tennant et al., 1995). The operating hypothesis was that these animals would be preferentially responsive to mutagenic carcinogens since loss of the functional p53 allele would more likely be accomplished via specific mutation. Two chemicals, *p*-cresidine and vinylcyclohexene diepoxide, were selected for pilot $p53^{def}$ model study evaluation because they were shown to be transpecies mutagenic carcinogens at low doses (NCI/NTP, 1979; National Toxicology Program, 1989a) and, therefore, expected to be positive in the $p53^{def}$ line. To examine the specificity of the carcinogenic response, *p*-anisidine (which is structurally similar to *p*-cresidine) was selected (as a negative control) for study. *p*-Anisidine is a mutagen in *Salmonella* assays but not a carcinogen in the 2-year bioassay at doses similar to those at which *p*-cresidine produced a carcinogenic response in rodents. The second pair of chemicals chosen were nonmutagenic carcinogens (reserpine and *n*-methylol-*o*-acrylamide). Reserpine has induced tumors at a single but different tissue sites in male rats (adrenal), male mice (seminal vesicle), and female mice (mammary). *N*-Methylol-*o*-acrylamide induced Harderian gland, liver, and lung tumors only in mice in 2-year bioassays.

The results of those studies in $p53^{def}$ mice showed a clear difference between effects of the mutagenic and nonmutagenic carcinogens. Within 6 months of exposure, *p*-cresidine induced transitional cell carcinomas of the bladder epithelium as in the 2-year bioassay. There was evidence of a dose response, with the highest incidence occurring in male mice, and, most importantly, the tumors were of the same histological type as was diagnosed in the 2-year bioassay in B6C3F1 mice (NTP 1979, TR-142). Comparable induction of bladder tumors were observed when the *p*-cresidine experiments were repeated. Vinylcyclohexene diepoxide induced squamous cell tumors of the skin, which were the same type of tumors that had been induced in the 2-year bioassay. The absence of a response to the two non-mutagenic carcinogens (reserpine and *n*-methylol-*o*-acrylamide) and the mutagenic noncarcinogen (*p*-anisidine) suggests that this model may be specifically responsive to the actions of mutagenic carcinogens.

Tennant et al. (unpublished data) have also evaluated benzene, which was a transpecies multisite carcinogen in the 2-year bioassay but which was not mutagenic in *S. typhimurium*. Administration of benzene via gavage (oral intubation) to $p53^{def}$ mice for 6 months resulted in a high incidence of subcutaneous sarcomas. Based on the previous hypothesis that the $p53^{def}$ line would respond to mutagenic carcinogens, the benzene study results may at first appear to be paradoxical. However, the genotoxicity of benzene can be demonstrated in vivo as a chromosomal aberration. These chromosomal aberrations are dependent on in vivo metabolism that has not been adequately achieved in in vitro mutagenicity assays. Subcutaneous sarcomas were not diagnosed in the 2-year benzene bioassay in B6C3F1 mice. This suggests the possibility of a strain-dependent effect or that the principal effects of benzene, in relationship to expression of the p53 gene, may occur primarily in mesodermal tissue.

The above experimental study results are encouraging, in that it appears that this transgenic model is preferentially responsive to mutagenic carcinogens; target organs are similar to those identified in the 2-year bioassays and the effects were observed after 26 weeks rather than 104 weeks of chemical treatment. Chemical specificity, similar target organ site, and shortened exposure time are features of the $p53^{def}$ line that would make it a good candidate as an adjunct to the traditional bioassay. Although the tumorigenic response to benzene occurred at an unanticipated site, the effect was clearly related to the exposure to benzene and will require additional studies to be understood.

IV. THE Tg.AC LINE

In contrast to the $p53^{def}$ line, the Tg.AC mouse line was created by the germline insertion of a mutated v-Ha-ras under the regulation of a fetal zetaglobin promoter sequence (Leder et al., 1990) and manifests a unique phenotype of wound and chemical-induced skin papillomagenesis. Thus, Tg.AC mice exhibit the characteristic of genetically initiated skin. In Tg.AC mice topical application of tumor promoters, other chemicals or full thickness wounding of the skin (without a previous application of a chemical initiator) results in papillomas induced with a latent period as short as 3–5 weeks (Spalding et al., 1993). The transgene is carried on the FVB/N strain background, a strain not widely used in carcinogenesis studies, but one used frequently in the development of transgenic lines (Hennings et al., 1993). With the exception of lung tumors, FVB/N mice show low incidence of spontaneous tumors (Mahler et al., 1995) and reproduce with good litter sizes. The Tg.AC line has no background frequency of spontaneous skin tumors and often has a general uniformity of response and a high incidence of tumors. Use of skin as the study organ permits a noninvasive way to determine the time to first appearance of tumors and to observe the development of tumors. Following four sequential treatments with 12-O-tetradecanoyl-phorbol-13-acetate (TPA), there was a clear additive relationship between the onset and incidence of papillomas to increases in dose with a no effect level apparent at below 1.25 μg and a maximum response at 5 μg of TPA. Early stages of papilloma development can be seen within 3 weeks of the first treatment with TPA (Spalding et al., 1993). In order to characterize the response of the Tg.AC mice, the well-studied chemical TPA was used as a prototype papilloma inducer. An important observation is an age-related increase in responsiveness to TPA treatment; animals more than 20 weeks of age at first treatment are significantly more responsive than animals 5–10 weeks old. These kinds of observations may provide clues to mechanisms of chemical-genome interactions.

Other observations make the Tg.AC line worthy of further consideration as a model to identify carcinogens and examine mechanisms of tumorigenesis. Both *p*-cresidine and benzene were tested in Tg.AC and $p53^{def}$ mice (Tennant et al., 1995). In the case of the Tg.AC line, both chemicals were administered by skin painting and both chemicals were papillomagenic in the Tg.AC line. *p*-Cresidine also produced gross and microscopic effects in the urinary bladders of some of the Tg.AC mice, including hyperplasia and a carcinoma in situ. The bladder was also the target organ in *p*-cresidine–treated $p53^{def}$ mice in 26-week studies and B6C3F1 mice in the 2-year studies. The studies in the Tg.AC line suggest that (a) the induction of skin papillomas in this mouse model could be used as a reporter phenotype for carcinogens with target organs other than skin in 2-year bioassays (based on the *p*-cresidine and benzene results) and (b) the model is responsive to genotoxic as well as nongenotoxic transpecies carcinogens (based on all studies) (Table 1).

The absence of a response to ethylacrylate in Tg.AC is noteworthy.

Table 1 Comparison of Results of NCI/NTP 2-Year Bioassay in $B6C3F_1$ Mice and 24-Week Subchronic Studies[a] in C57BL/6 p53-Deficient Mice

			$B6C3F_1$			p53[def] (+/−);
Chemical	TR No.	SAL[b]	Sex	Target organ	Route	M,F
p-Anisidine	115	+	M,F	None	Feed	−
Benzene	289	−[c]; +[d]	M,F	Multiple sites	Gavage	+
p-Cresidine	142	+	M,F	Bladder, liver	Feed	+
VCD[e]	362	+	M,F	Skin, ovary	Skin paint	+
NMOA[f]	352	−	M,F	Kidney, liver	Gavage	−
Reserpine	193	−	M,F	Mammary	Feed	−

[a]Replication of the NCI/NTP 2-year bioassay using the same route of administration and dose range.
[b]In vitro Salmonella mutagenesis assay.
[c]In vitro Salmonella mutagenesis assay negative.
[d]In vivo micronucleus assay positive.
[e]4-Vinyl-1-cyclohexene diepoxide.
[f]*N*-Methylol-o-acrylamide.

Ethylacrylate, considered a nonmutagen, induced forestomach tumors in mice and rats via gavage exposure in 2-year bioassays (Table 2) but did not induce skin tumors in Tg.AC. In C3H/HeJ male mice, there was no increased incidence of tumors at the site of application or at distant sites in a lifetime skin-painting study with undiluted ethylacrylate (Hengler and DePass, 1982). Although ethylacrylate produced tumors in both species in the NTP 2-year bioassay, the forestomach was the only target site, and other, nonchemical specific mechanisms may explain this response. One interpretation of the positive 2-year ethylacrylate study results is that oral administration of a bolus dose of a high concentration of an irritant combined with the mechanical effect of repeated gavaging to the same site (forestomach) for 2 years played a role in the increased tumor incidence (National Toxicology Program, 1986). Although the Tg.AC line responds rapidly to topical chemical exposure and full thickness wounding, there appears to be a response specificity since neither occasional superficial wounding by hair clipping nor skin exposure to some irritants including acetic acid and phenol produced skin papillomas. These results suggest that the Tg.AC model may discriminate between tissue and/or exposure specific responses and chemical-related tumorigenesis. Likewise, chemical-specific effects that are strain or species dependent are unlikely to be produced in the transgenic models.

Table 2 Comparison of Results of NTP 2-Year Bioassays in Both Sexes of B6C3F$_1$ Mice Except Where Noted and 20-Week Bioassays in Transgenic Tg.AC Female Mice

		B6C3F$_1$			Tg.AC	
Chemical	SAL[a]	Route[b] (NTP TR No.)	Cancer results	Target sites	Route	Tumor results
Benzene	–[c]; +[d]	Gav (289)	+	Multiple sites	SP	+
Dimethyl vinylchloride[d]	+	Gav (316)	+	Forestomach	Gav	+
p-Cresidine	+	DF (142)	+	Bladder, liver	SP	+
DMBA[f]	+	SP (441)	+	Skin	SP	+
o-Benzyl-*p*-chlorophenol	–	Gav (424)	+[g]	Kidney		
		SP[h] (444)	+	Skin	SP	+
Ethylacrylate	–	Gav (259)	+	Forestomach	SP	–
Mirex[i]	–	DF (313)	+	Liver[i]	SP	+
Triethanolamine	–	SP (449)	+[j]	Liver	SP	–
2-Chloroethanol	+	SP (275)	–		SP	–
Benzethonium chloride	–	SP (438)	–		SP	–
Phenol	–	DW (203)	–		SP	–
Diethanolamine	–	SP	?[k]		SP	–
LADA[l]	–	SP	?[k]		SP	+

[a]Salmonella mutagenesis assay.
[b]Gav = Gavage; DF = dosed feed; SP = skin paint; DW = dosed water.
[c]In vitro Salmonella mutagenesis assay negative.
[d]In vivo micronucleus assay positive.
[e]1-Chloro-2-methylpropene.
[f]7,12-Dimethyl-benzanthracene.
[g]Male only.
[h]Swiss mice.
[i]Rat only NTP 2-year bioassay.
[j]Female only.
[k]Unreported 2-year NTP bioassay.
[l]Lauric acid diethanolamine condensate.

V. SUMMARY

The use of transgenic lines as adjuncts to traditional rodent bioassays to identify chemical carcinogens and noncarcinogens is based on results of a limited number of chemical studies for which there are also data from carcinogenicity bioassays. The current hypothesis, based on results from six chemicals tested in the $p53^{def}$ mice and 13 chemicals tested in the Tg.AC line, is that $p53^{def}$ responds to mutagenic carcinogens and the Tg.AC responds to both mutagenic and nonmutagenic carcinogens. Chemicals showing no evidence of carcinogenesis in the NTP 2-year studies were also negative in both the $p53^{def}$ and the Tg.AC models. The $p53^{def}$ line failed to respond to two nonmutagenic carcinogens (*n*-methylol-acrylamide and reserpine). The Tg.AC line responded to mutagenic and nonmutagenic carcinogens except for two nonmutagenic chemicals that only induced tumors at one site; both of the latter chemicals responses in the 2-year studies may be related to either route of administration or to strain susceptibility.

The results of studies described above provided impetus for the NTP to develop plans to further evaluate the utility of the $p53^{def}$ and the Tg.AC models as adjuncts to the 2-year standard carcinogenesis studies. Fifteen agents have been selected, most from the NTP historical database but also from the human carcinogen literature, to represent a range of chemicals with known mutagenic and carcinogenic effects. Thus, mutagenic and nonmutagenic noncarcinogens, mutagenic and nonmutagenic carcinogens (based on the results of the NTP 2-year studies), and several chemicals established as human carcinogens are being evaluated in several different laboratories under defined testing protocol conditions similar to those used by the NTP for their 2-year studies. These studies were designed simply to evaluate the predictive capacity of these two transgenic models to identify known transpecies and/or multisite rodent and human carcinogens. Should further evaluation support the hypothesis then it is expected that more complete protocols will address other questions, e.g., dose-response and no-effect levels.

The use of transgenic models to identify chemical carcinogens have generated a lot of interest because the studies require fewer animals and shorter exposure times; these changes allow more chemicals to be screened and reduce costs. Under the mission objectives of the International Congress on Harmonization to examine whether the need for long-term studies in two species might be reduced, FDA is proposing revisions to their testing guidelines for carcinogenicity of pharmaceuticals (U.S. Food and Drug Administration, 1996). One of the options under consideration as a replacement for a second rodent species in the long-term bioassay is the use of transgenic mouse models to assay for carcinogenic potential. Members of the pharmaceutical industry have expressed interest in forming partnerships to exchange study information on transgenic mouse model evaluations with NIEHS and the NTP.

REFERENCES

Adams JM, Cory S. Transgenic models of tumor development. Science 1991; 254: 1161–1167.

Ashby J, Tennant RW. Definitive relationships among chemical structure, carcinogenicity and mutagenicity for 301 chemicals tested by the US NTP. Mutat Res 1991; 257:229–306.

Bishop JM. Molecular themes in oncogenesis. Cell 1991; 64:235–248.

Breuer M, Slebos R, Verbeek S, van Lohuizen M, Wientjens E, Berns A. Very high frequency of lymphoma induction by a chemical carcinogen in pim-1 transgenic mice. Nature 1989; 340:61–63.

Cardiff RD, Muller WJ. Transgenic mouse models of mammary tumorigenesis cancer surveys. Mol Pathol Cancer 1993; 16:97–113.

Christofori G, Hanahan D. Molecular dissection of multi-stage tumorigenesis in transgenic mice. Semin Cancer Biol 1994; 5:3–12.

Doll R. Nature and nurture: possibilities for cancer control. Carcinogenesis 1996; 17:177–184.

Donehower LA, Harvey M, Slagle BL, McArthur MJ, Montgomery CAJ, Butel JS, Bradley A. Mice deficient for p53 are developmentally normal but susceptible to spontaneous tumors. Nature 1992; 356:215–221.

Drinkwater NR. The interaction of genes and hormones in hepatocarcinogenesis. In: Cockburn A, Smith L, eds. Nongenotoxic Carcinogenesis. Berlin: Springer Verlag, 1994:219–230.

Festing FW. Properties of inbred strains and outbred stocks, with special reference to toxicity testing. J Toxicol Environ Health 1979; 5:53–68.

Festing MFW. Use of a multistrain assay could improve the NTP carcinogenesis bioassay. Environ Health Perspect 1995; 103:44–52.

Finlay CA, Hinds PW, Levine AJ. The p53 proto-oncogene can act as a suppressor of transformation. Cell 1989; 57:1083–1093.

Greenhalgh DA, Rothnagel JA, Quintanilla MI, Orengo CC, Gagne TA, Bundman DS, Longley MA, Roop DR. Induction of epidermal hyperplasia, hyperkeratosis and papillomas in transgenic mice by a targeted v-Ha-*ras* oncogene. Mol Carcinogen 1993; 7:99–110.

Griesemer R, Tennant R. Transgenic mice in carcinogenicity testing. IARC-Sci-Publ. 1992; 116:429–436.

Hansen L, Tennant RW. Focal transgene expression associated with papilloma development in v-Ha-*ras* transgenic TG.AC mice. Mol Carcinogen 1994; 9: 143–156.

Hartwell L. Defects in a cell cycle checkpoint may be responsible for the genomic instability of cancer cells. Cell 1992; 71:543–546.

Harvey M, McArthur MJ, Montgomery CAJ, Bradley A, Donehower LA. Genetic background alters the spectrum of tumors that develop in p53-deficient mice. FASEB J 1993a; 7:938–943.

Harvey M, McArthur MJ, Montgomery CAJ, Butel JS, Bradley A, Donehower LA. Spontaneous and carcinogen-induced tumorigenesis in p53-deficient mice. Nat Genet 1993b; 5:225–229.

Haseman JK. Data analysis: statistical analysis and use of historical control data. Reg Toxicol Pharmacol 1995; 21:52–59.

Haseman JK, Johnson FM. Analysis of National Toxicology Program rodent bioassay data for anticarcinogenic effects. Mutat Res 1996; 350:131–141.

Hengler WC, DePass LR. Ethyl acrylate: lifetime dermal carcinogenesis study in C3H/HeJ male mice. Report No. 45-513. Pittsburgh: Union Carbide Co., 1982.

Hennings, H, Glick AB, Lowry DT, Krsmanovic LS, Sly LM, Yuspa SH. FVB/N mice: an inbred strain sensitive to the chemical induction of squamous cell carcinomas of the skin. Carcinogenesis 1993; 2353–2358.

Huff JE, Haseman JK, Demarini DM, Eustis S, Maronpot RR, Peters AC, Persing RL, Crisp CE, Jacobs AC. Multiple-site carcinogenicity of benzene in Fischer 344 rats and B6C3F1 mice. Environ Health Perspect 1989; 82:125–163.

Kastan MB, Zhan Q, el-Deiry WS, Carrier F, Jacks T, Walsh WV, Plunkett BS, Vogelstein B, Fornace AJJ. A mammalian cell cycle checkpoint pathway utilizing p53 and GADD45 is defective in ataxia-telangiectasia. Cell 1992; 71: 587–597.

Kemp CJ, Donehower LA, Bradley A, Balmain A. Reduction of p53 gene dosage does not increase initiation or promotion but enhances malignant progression of chemically induced skin tumors. Cell 1993; 74:813–822.

Kemp CJ, Wheldon T, Balmain A. p53-deficient mice are extremely susceptible to radiation-induced tumorigenesis. Nature Genet 1994; 8:66–69.

Leder A, Kuo A, Cardiff RD, Sinn E, Leder P. v-Ha-*ras* transgene abrogates the initiation step in mouse skin tumorigenesis: effects of phorbol esters and retinoic acid. Proc Natl Acad Sci USA 1990; 87:9178–9182.

Levine AJ. The p53 tumor suppressor gene and gene product. Princess Takamatsu Symp 1989; 20:221–230.

Mahler JF, Mann P, Takaoka M, Maronpot RM. Spontaneous lesions of the FVB/N mouse. Toxicol Pathol 1995; 23:744–745.

Merlino G. Regulatory imbalances in cell proliferation and cell death during oncogenesis in transgenic mice. Semin Cancer Biol 1994; 5:13–20.

Mirsalis JC, Monforte JA, Winegar RA. Transgenic animal models for measuring mutations in vivo. Crit Rev Toxicol 1994; 24:255–280.

National Toxicology Program. Carcinogenesis Studies of Ethyl Acrylate (CAS No. 140-88-5) in F344/N Rats and B6C3F1 Mice (gavage studies). Research Triangle Park, NC: National Toxicology Program, 1986.

National Toxicology Program. Toxicology and Carcinogenesis Studies of 4-Vinyl-1-Cyclohexene Diepoxide in F344/N Rats and B6C3F1 Mice. Research Triangle Park, NC: National Toxicology Program, 1989a.

National Toxicology Program. Toxicology and Carcinogenesis Studies of N-

Methyloacrylamide in F344/N Rats and B6C3F1 Mice. Research Triangle Park, NC: National Toxicology Program, 1989b.

National Toxicology Program. Specification for the Conduct of Studies to Evaluate the Toxic and Carcinogenic Potential of Chemical, Biological and Physical Agents in Laboratory Animals. Research Triangle Park, NC: National Toxicology Program, 1992.

NCI/NTP. Bioassay of p-Cresidine for Possible Carcinogenicity. Bethesda: NCI, 1979.

Organisation for Economic Co-operation and Development. OECD Guidelines for Testing of Chemicals. Paris: OECD, 1981.

Poole TM, Drinkwater NR. Hormonal and genetic interactions in murine hepatocarcinogenesis. In: McClain RM, Slaga TJ, Leboeuf R, Pitot H, eds. Growth Factors and Tumor Promotion. New York: Wiley-Liss, 1995:185–194.

Sell S, Hunt JM, Dunsford HA, Chisari FV. Synergy between hepatitis B virus expression and chemical hepatocarcinogens in transgenic mice. Cancer Res 1991; 51:1278–1285.

Silver L. Mouse Genetics. New York: Oxford University Press, 1995:34–42.

Spalding JW, Momma J, Elwell MR, Tennant RW. Chemical induced skin carcinogenesis in a transgenic mouse line (TG.AC) carrying a v-Ha-*ras* gene. Carcinogenesis 1993; 14:1335–1341.

Storer RD, Cartwright ME, Cook WO, Soper KA, Nichols WW. Short-term carcinogenesis bioassay of genotoxic procarcinogens in PIM transgenic mice. Carcinogenesis 1995; 16:285–293.

Tennant RW. Stratification of rodent carcinogenicity bioassay results to reflect relative human hazard. Mutat Res 1993; 286:111–118.

Tennant RW, Ashby J. Classification according to chemical structure, mutagenicity to Salmonella and level carcinogenicity of a future 39 chemicals tested for carcinogenicity by the U.S. National Toxicology Program. Mutat Res 1991; 257:209–227.

Tennant RW, Stasiewicz S, Spalding J. Comparison of multiple parameters of rodent carcinogenicity and in vitro genetic toxicity. Environ Mutagen 1986; 8: 205–227.

Tennant RW, Rao GN, Russfield A, Seilkop S, Braun AG. Chemical effects in transgenic mice bearing oncogenes expressed in mammary tissue. Carcinogenesis 1993; 14:29–35.

Tennant RW, French JE, Spalding JW. Identification of chemical carcinogens and assessing potential risk in short term bioassays using transgenic mouse models. Environ Health Perspect 1995; 103:942–950.

U.S. Environmental Protection Agency. Health effects test guidelines. EPA 560/ 6-82-001. Washington, DC: EPA, 1982.

U.S. Food and Drug Administration. Toxicological principles for the safety assessment of direct food additives and color additives used in food (Draft: Redbook II). Fed Reg 1993; 58 (58):16536–16537.

U.S. Food and Drug Administration. International Conference on Harmonisation; Draft Guideline on Testing for Carcinogenicity of Pharmaceuticals; Notice. Fed Reg 1996; 61 (163):43298–43300.

Viney JL. Transgenic and gene knockout mice in cancer research. Cancer Metastasis Rev 1995; 14:77–90.

Walker J. Toxic substances control act (TSCA) Interagency Testing Committee: data developed under section 4 of TSCA. In: Rand GM, ed. Fundamentals of Aquatic Toxicology. Washington, DC: Taylor & Francis Publishers, 1995.

Weinberg RA. Oncogenes and tumor suppressor genes. Cancer J Clin 1994; 44:160–170.

Weinberg RA. The molecular basis of oncogenes and tumor suppressor genes. Ann NY Acad Sci 1995; 758:331–338.

Zambetti GP, Levine AJ. A comparison of the biological activities of wild-type and mutant p53. FASEB J 1993; 7:855–865.

Zeiger E, Haseman JK, Shelby MD, Hargolin BH, Tennant RW. Evaluation of four in vitro genetic toxicity tests for predicting rodent carcinogenicity: confirmation of earlier results with 41 additional chemicals. Environ Mol Mutagen 1990; 16 (suppl. 18):1–14.

Zinkel S, Fuchs E. Skin cancer and transgenic mice. Sensors Cancer Biol 1994; 5: 77–90.

15

Human Liver Carcinogenesis

Wai Nang Choy
Schering-Plough Research Institute, Lafayette, New Jersey

I. INTRODUCTION

Primary hepatocellular carcinoma (HCC) is one of the most common human cancers worldwide, but it is a relatively rare disease in the western world (1). The annual HCC incidence has been estimated to be about 500 cases per 100,000 person worldwide (2) and less than 4 cases per 100,000 person in the United States (3). High incidences of HCC are reported in Southeast Asia and sub-Saharan Africa. This uneven geographic distribution of HCC is attributed to environmental carcinogens and differences in lifestyles. HCC in Asia and Africa is often associated with aflatoxin exposures and hepatitis virus infections. HCC in the western world, however, is mostly associated with alcoholic cirrhosis and, less frequently, with steroid intake (4). The identification of human carcinogens by regulatory agencies is based on epidemiological studies. Only three human liver carcinogens are included in the IARC human carcinogen list (Group 1; carcinogenic to humans): aflatoxins, oral contraceptives (combined), and vinyl chloride (5). All these chemicals were also found to induce liver tumors in rodents (5,6). Aflatoxins, especially aflatoxin B_1 (AFB1), is the most common liver carcinogen in Asia and Africa (7). Much less frequently, HCC or its precancerous states are induced by continued use of oral contraceptives or androgenic and anabolic steroids (8). Liver tumors are also induced by Thorotrast (^{232}thorium dioxide) in clinical uses (9) and by vinyl chloride in occupational exposures (10). More importantly, a large number of HCC in Asia and Africa is associated with hepatitis B and/or C viruses (HBV and/or HCV) infections (11), which are synergistic factors for aflatoxin liver carcinogenesis. Liver cirrhosis, induced by alcohol or hepatitis virus infection, is also a risk factor for HCC (12).

II. HUMAN LIVER CARCINOGENS

A. Chemical Carcinogens

Aflatoxins

Aflatoxins are a group of mycotoxins produced by *Aspergillus flavus* and *Aspergillus parasiticus*. Aflatoxins are found in peanuts, corn, cottonseed, and food products made from these commodities. AFB1 is the most potent carcinogenic congener. In animal studies, AFB1 induces tumors in mice, rats, hamsters, monkeys, ducks, and fish by various routes of administrations. Most tumors are liver tumors, and kidney, lung, and colon tumors occur less frequently (13–15). The rat is the most sensitive species and liver is the most sensitive site. AFB1 is a genotoxic chemical, and genotoxicity is mediated through its epoxidation to AFB1-8,9-oxide, which reacts with DNA to form the mutagenic adduct, AFB1-N^7-guanine (AFB1-N^7-Gua) (13–15).

Numerous epidemiology studies in Africa and Asia provided strong evidence that AFB1 induces HCC in humans (15). Quantitative risk assessments of early studies are difficult because data on AFB1 exposure are often unreliable and HBV infection history of the HCC patients is incomplete. In a recent regulatory AFB1 risk assessment in which approximately 10 epidemiological studies were evaluated (15), the Yeh et al. (16) study was considered to be the most appropriate for quantitative cancer risk assessment (3). In this study, AFB1-induced human HCC incidences were analyzed separately based on hepatitis virus infection, as monitored by hepatitis B surface antigen (HBsAG). Three mathematical models were used to explore the effect of HBV infection to AFB1 carcinogenesis. Cancer incidences fit the additive model poorly, and the interactive and multiplicity models were used for risk characterization. The cancer potency of AFB1 in HBV carriers was found to be about one order of magnitude higher than in HBV noncarriers. When converted to the American situation with low HBV infection rates, the cancer potency was estimated to be 5.7–45.6 $(mg/kg/day)^{-1}$, a range dependent on the risk assessment model (3).

With recent developments of molecular epidemiology methodologies, numerous attempts have been made to use biomarkers to estimate AFB1 exposure in humans. These biomarkers are mostly AFB1 metabolites or DNA adducts or protein adducts (17). Human urine was monitored for AFB1 metabolites, aflatoxin M_1 and P_1, and AFB1-N^7-Gua DNA adducts

(18,19), and peripheral blood was assayed for aflatoxin-albumin adducts (20). The amount of biomarkers found are used as indices of AFB1 exposures.

Vinyl Chloride

Vinyl chloride is also a rodent carcinogen inducing liver tumors in rats, mice, and hamsters. Other tumors such as lung tumors in mice and rats and gastrointestinal tumors in hamsters were also reported (21,22). Vinyl chloride induces tumors in the liver, brain, lung, and hematolymphopoietic systems in humans (21). The most distinct tumor induced by vinyl chloride in humans is angiosarcoma in the liver (23–26). Angiosarcoma is of endothelial origin, which is different from HCC. Tumor induction was observed in industrial workers exposed to the vinyl chloride monomers in the manufacturing of polyvinyl chloride. Most tumors occurred between 1950 and 1970 before statutory occupational exposure limits for vinyl chloride were established. The association of angiosarcoma and vinyl chloride has been demonstrated in epidemiology studies (21,27), and in case reports (28–31). Epidemiological studies in the United Kingdom (32–34), the United States (35,36), Canada (37), Germany (38), Australia (39), and in a collaborative study of Italy, Norway, Sweden, and the United Kingdom (40) also demonstrated that vinyl chloride induces liver angiosarcoma. In addition, vinyl chloride also was reported to induce HCC (26,28).

Thorotrast (Thorium Dioxide)

Thorotrast is a colloidal preparation of 20–25% thorium dioxide. It was commonly used as a contrast agent to facilitate visualization of roentgenography from 1930 to 1940, but its use was discontinued in the 1950s because of its toxicity to the liver (41). ^{232}Thorium is a radioactive metal with a long biological half-life of about 500 years. It emits predominantly alpha particles but also, to a much lesser degree, beta and gamma radiation. Liver is the target organ of Thorotrast toxicity and carcinogenicity because of Thorotrast's preferential distribution to the liver. For nonmalignant lesions, Thorotrast induces peliosis hepatis (42), periportal fibrossi, veno-occlusive lesions, and cirrhosis (41). Thorotrast is best know to induce hemangiosarcoma (41,43–48), but its induction of hepatic leiomyosarcoma (49), cholangiocarcinoma (49–51), and HCC (51–54) were also reported.

A histopathological characteristic of Thorotrast-induced tumors is the presence of Thorotrast particles in the tumor. This raised the possibility that mechanical irritation may be the cause of Thorotrast tumorigenicity. A rat cancer study was conducted with zirconium dioxide (an analog of thorium dioxide), a chemical that does not emit alpha particles. The negative results of this study indicated that liver tumors induced by Thorotrast were caused by ionizing radiation (55).

Oral Contraceptives

Epidemiology studies have indicated that hepatocellular adenoma is associated with the use of oral contraceptives in women (56–60). Oral contraceptives have many formulations, but they typically contain a synthetic estrogen (usually ethinyl estradiol) and progestin, or progestin alone (61). A combination of estrogen and progestin inhibits gonadotropin secretion and prevents ovulation. The occurrence of hepatocellular adenoma was dependent on the dose, the duration of use, and the age of the user. In studies in U.S. hospitals from 1970 to 1975, high incidences of hepatocellular adenoma or focal nodular hyperplasia were observed in young women aged 26–30 when contraceptives were commonly used (62,63). In case-control studies of women with hepatocellular adenoma, the risk ratios were reported to increase with the duration of contraceptive use, from 1.3 for under 3 months of use to 25 for over 109 months of use (64) and to 503 for over 85 months of use (65). Improved formulations with lower doses being used nowadays have substantially lowered the risk of liver tumors (66).

As for the association of oral contraceptives with HCC, a few case studies conducted in the United States did not show sufficient associations (67–69). In contrast, additional case-control studies conducted in the United States (70–73), Canada (74), Europe (75), the United Kingdom (76,77), and Italy (78,79) all showed significant increases of HCC with relative risks ranging from 4 to 20. These increases were also dependent on the duration of contraceptive use. These positive findings, however, were not confirmed by a multinational World Health Organization (WHO) study conducted in several countries where HBV infections are prevalent (Chile, China, Colombia, Israel, Kenya, Nigeria, Philippines and Thailand) (80) and a case-control study in South African black women (81). It was postulated that the high incidences of HCC in these regions may have masked the small increases of HCC induced by oral contraceptives. Further studies, with separate evaluations of HBV status in patients, are required to clarify these findings.

The above findings indicate that hepatocellular adenomas may develop after continued use of oral contraceptives for longer than approximately 5 years. Such tumors may either regress if the use of oral contraceptive is discontinued or otherwise progress to liver cell dysplasia, which may subsequently develop to HCC (74).

Androgenic and Anabolic Steroids

Hepatocellular adenoma has also been reported to associate with the use of androgenic or anabolic steroids, but tumor incidences are relatively low (82–86). All tumorigenic steroids contain a 17-alkyl substitution.

Androgenic or anabolic steroids are used mostly for treatments of impotence, Fanconi's anemia, refractory anemia, bone marrow aplasia, sex change, and body building. The common steroids involved are methyltestosterone, oxymetholone, and norethandrolone. In athletes and body-builders using such steroids, hepatocellular adenomas have been infrequently reported (87,88). Almost no steroid-induced hepatocellular adenomas, in contrast to HCC, produce α-fetoprotein or metastasize. Similar to tumors induced by oral contraceptives, hepatocellular adenomas induced by adrogenic and anabolic steroids may regress after discontinuation of steroid use (89,90) or may progress to HCC (91–94). HCC was reported in several case studies as related to androgen and corticosteroid treatments of Fanconi's anemia (95) or androgen treatment alone (96,97).

B. Viral Carcinogens

Hepatitis viruses have long been associated with human liver cancer, both as oncogenic viruses or as co-carcinogens (99). There are five hepatitis viruses named A, B, C, D, and E (HAV, HBV, HCV, HDV, and HEV). Only HBV and HCV are known to establish chronic infection in the host resulting in chronic hepatitis, cirrhosis, and liver tumors. HAV and HEV do not induce chronic liver diseases, and HDV is pathogenic only in the presence of HBV. The biology of HBV and HCV, as related to liver carcinogenicity, is described below.

Hepatitis Virus B

Hepatitis virus B (HBV) is a DNA virus (Hepadnavirus) containing a partially double-stranded circular DNA. Its genome of 3.2 kb has been completely sequenced (100,101). Several viral components are immunogenic, and they are used as markers of HBV infection in humans. The HBV surface antigen (HBsAg) is a glycosylated surface protein antigen coded by three genes: S, preS2, and preS1. The viral core protein antigen (HBcAG) is coded by gene C. The viral envelope antigen (HBeAg) is coded by preC, and the transactivating gene antigen (HBxAg) is coded by gene X (100,101). After HBV infection, HBsAg, HBeAg, and viral DNA polymerase are first detected in the blood. Subsequently, anti-HBc and anti-HBe antibodies appear. HBeAg gradually disappears and anti-HBs antibody develops. The presence of HBsAg is commonly used as an indicator for chronic HBV infection.

HBV infection can be replicative or nonreplicative. Neither mode of infection is cytopathic, which allows continued survival of the cell. The HBV DNA is replicated through a reverse transcription process (DNA → RNA → ssDNA → partially dsDNA). Actively replicating HBV produces large amount of HBeAg and single-stranded DNA in the tissues (102–104). The DNA can be episomal or be integrated into the host genome by random nonhomologous recombination (105–108). Viral integration results in the production of anti-HBe antibody accompanied by the absence of replication markers in the serum and viral DNA in the tissues. Accordingly, an increase of anti-HBe antibody and decrease of HBcAg is indicative of a reduction in virus replication (109).

Recent studies on HBV infection are focused on characterization of HBV mutants. Mutations in the preS and S genes resulted in the loss of HBsAg and anti-HBs production (110,111). Mutants in the preC and C genes are often nonsense mutations, which abolish the production of HBeAg (112). Mutations in the X gene (113) and in the DNA polymerase genes (114) were also reported, but their significance in liver carcinogenesis is not clear.

The most commonly used method to screen for HBV infection is the detection of HBsAg in the serum. Serum antibodies, such as anti-HBs, anti-HBc, anti-HBe, and anti-HBx, have also been monitored, and they are assayed by radioimmunoassays or by hemagglutination of erythrocytes coated with their respective antigens. Although HBcAg is not always detectable in serum, anti-HBc is persistent and it is useful to reveal current or past infection. HBV DNA in serum can also be detected by Southern blot hybridization (115) or by polymerase chain reaction (PCR) assays (115).

The prevalence of HBV infection is low in North America, Western Europe, Australia, and South America, with 0–2% HBV carriers in the population. Slightly higher frequencies of carriers to 2–8% were reported in Eastern and Southern Europe, Middle East, and Japan. High frequencies of 8% and above were reported in China, Southeast Asia, and sub-Saharan Africa (116).

Epidemiology studies of HBV and HCC have recently been reviewed and summarized by IARC (99). In cohort studies, HBV infection was estimated to increase the relative risk of HCC to 5.3–148. In case-control studies, the relative risks of HCC were estimated to increase to 5–30 (99). No increases of cancer risks were observed in other organs.

Hepatitis Virus C

Hepatitis virus C (HCV) is a single-stranded RNA virus with a genome of about 9.4 kb (117,118). Several variants with different genomic sequences (cDNA sequences) have been identified (119–121). The original virus was designated as HCV-1 (119). Based on sequence homology, HCV were classified into four types (type I to type IV), and viruses with extensive homologous sequences are referred to as quasispecies (121). The genome codes for a large precursor protein of about 3010 amino acids, which contains the core protein. The glycoprotein envelop 1 is coded by gene E1, the protein 1/envelop 2 is coded by NS1/E2, and proteins 2–5 are coded by NS2, NS3, NS4, NS5, respectively. The HCV genome does not integrate into the host DNA (118). Although viral particles are not detectable, viral RNA (122) and viral antigens are detectable in the serum (123–125). Two generations of anti-HCV antibody assays were developed. The first-generation assay uses the C100-3 antigen (derived from the NS3–NS4 protein region) to detect serum antibody (126). The second-generation assays combines several antigens—5-1-1, C100-3, C33c, and C22-3 (also derived from the NS3–NS4 region)—to enhance the sensitive of antibody detection (127). In addition, a recombinant immunoblot assay (RIBA) using four viral antigens has been developed as a confirmatory assay (128). A combination of these assays has been used to detect HCV infection in epidemiological studies.

The prevalence of HCV infection is low in North America and Western Europe with 0–1% carriers in the population. Slightly higher frequencies of 1–3% were found in the Middle East and Asia (99).

IARC has reviewed and summarized epidemiology studies of HCV and HCC (99). In 17 studies in which HCV was monitored by the first-generation assay, the odds ratio of HCV-induced HCC ranged from 1.3 to 134. In six studies in which HCV was monitored by the second-generation assays, the odds ratios were 1.1–52 (99).

C. Cirrhosis-Inducing Agents

Liver cirrhosis is a possible precancerous condition to HCC. Cirrhosis is the loss of normal lobular architecture accompanied by fibrosis and nodular regeneration. Cirrhosis results in matrix degradation and abnormal matrix formation, followed by cell death and liver regeneration (129). Liver cirrhosis has been found mostly in liver tumors associated with HBV and/or HCV infections or alcohol consumption (130–132). Cirrhosis is commonly observed in the western world, where both liver cancer incidences and hepatitis virus infections are low. Only 5–15% of patients with cirrhosis develop

HCC. The primary cause of cirrhosis in the western world is believed to be excessive alcohol consumption (133). Since many HCC do not arise from cirrhosis, the role of chronic cirrhosis in HCC development still remains speculative. Only chronic cirrhosis with sufficient liver damage may result in this neoplasm (134).

In contrast, cirrhosis is less frequent in Africa and in Southeast Asia despite their prevalence of HBV infections and high HCC incidences. The causes of cirrhosis in these regions are less clear, but cirrhosis has been associated with HBV infections. The duration of cirrhosis in African and Asian patients is relative short, as compared to chronic cirrhosis in western countries. But in Africa and Asia a higher proportion, about 40–50% of cirrhosis patients, eventually develop HCC (135). Cirrhosis is a significant factor for liver carcinogenesis in the western world but not in high-risk areas where the predominant risk factors are HBV infections and AFB1 exposures (136,137).

D. Cofactors

Epidemiology studies show that the most prominent co-factors for HCC are aflatoxins and HBV infection. There is a synergistic effect of AFB1 exposure and HBV infection, which increases the risk of HCC in humans (138–140). HBV infection alone (as measured by the presence of HBsAg) and AFB1 exposure alone associated with a 7.3-fold and 3.4-fold increase in HCC risk, respectively. Co-exposure to HBV and AFB1 produced a 59.4-fold increase of HCC risk as compared to individuals not exposed to either agent (141). A similar finding was observed when epidemiology data (16) were analyzed with mathematical risk assessment models. The HCC data were found to fit better with the multiplicative or interactive model but poorly with the additive model (3).

Alcohol consumption also interacts synergistically with AFB1. Alcohol was reported to induce a twofold increase in the relative risk of HCC (142). This effect has been postulated as due to the ability of ethanol to induce cytochrome P450 isoenzymes, which activate AFB1 to genotoxic forms. Tobacco smoking also induces activation enzymes and enhances some toxification/detoxification reactions. As expected, the combination of alcohol consumption, tobacco smoking, and HBV infection increases the risk of HCC (143).

E. Predisposition Diseases for HCC

Although not related to environmental carcinogenesis, several human genetic diseases are predisposed to HCC. Patients inflicted with hereditary chronic tyrosinaemia (excessive tyrosine) (144), α_1-antitrypsin deficiency

(145), hemochromatosis (146), and Wilson's disease (copper accumulation) (147) all have high incidences of HCC. In addition, physical obstruction of the inferior vena cava has also been reported to induce HCC (148).

III. MECHANISMS OF CHEMICAL CARCINOGENESIS

A. Genotoxic Mechanisms

Chemical Mutagenesis

AFB1 and vinyl chloride are potent genotoxic agents (13,22). Mutagenicity of both chemicals requires metabolic activation. AFB1 and vinyl chloride are metabolized similarly to their respective epoxides, which interact with DNA and form DNA adducts. The mutagenic DNA adduct of AFB1 is AFB1-N^7-Gua, which is found in rodent and human HCC (17,149). The induction of AFB1 adduct is dose-dependent, and its linear dose-response has been used for cancer risk assessment (149). Vinyl chloride induces the DNA adduct N^2-3-ethenoguanine (EG) (150–152). DNA adduct mediated mutagenesis is the putative mechanism of carcinogenesis (153,154).

Recent genetic studies have focused on the genotoxicity of these two liver carcinogens to cancer genes, specifically the activation of oncogenes and inactivation of tumor suppressor genes. As expected, both AFB1 and vinyl chloride induce mutations in cancer genes, often at a few specific hot spots.

The most widely studied tumor-suppressor gene is p53. Approximately 50% of human cancers contain mutations in the p53 gene. Loss of p53 function by sequential mutations at the diploid alleles is associated with tumor progression (155–157). Although not all cancers are defective in the p53 gene, mutations in the p53 gene are interpreted as a highly significant risk for carcinogenesis. In a study of angiosarcomas and HCC from factory workers exposed to vinyl chloride (158), missense mutations in the p53 gene at a highly conserved domain of the coding sequence were identified. Mutations were observed at codon 249 with base-pair substitutions from AGG to TGG (an amino acid change from arginine to tryptophan) and at codon 255 from ATC to TTC (a change from isoleucine to phenylalanine). These specific mutations were similar to those previously observed in HCC (159,160). As for oncogene activation, the MDM2 gene (a murine double min-2 proto-oncogene) was monitored but no activation was observed in vinyl chloride–induced HCC (158).

Mutations in the p53 gene have also been repeatedly observed in AFB1-induced human tumors. p53 mutations were found in about 53% of HCC from areas with high AFB1 exposures and in about 26% in areas of low AFB1 exposures (161). Most mutations, about 50%, in high-aflatoxin-exposure areas occurred at a hot spot in codon 249 (from AGG to AGT, arginine to serine) (161). Codon 249 has been proposed to associate with the binding region of p53 to DNA, which is critical for the regulation of gene expression (162). Mutations in codon 249 have been reported in liver cancers in China (163), Mexico (164), South Africa (165), Taiwan (166), and Thailand (167). In support of this specific base-pair mutation, a recent in vitro mutagenesis study showed that AFB1-N^7-Gua produced predominantly G to T mutations (168). This specific mutation was also observed in normal liver samples from HCC patients indicating that p53 mutations are associated with early tumor development (169). However, these specific mutations do not occur in all HCC. At low AFB1 exposure, mutations in HCC spread over the conserved domains of the p53 gene without specific mutational hot spots (170,171). A study using liver cancers from Taiwan and Japan also failed to demonstrate specific mutations at codon 249 (172). Thus the degree of significance of mutations in codon 249 for HCC is still being resolved. As for oncogene activation, activations of the ras (173,174) and myc oncogenes (175) by AFB1 have been demonstrated in rodent liver carcinomas, but similar observations have not yet been demonstrated in humans so far.

Radiation

Thorotrast induces liver injury both by its accumulation in the liver and the ensuing radioactivity. ^{232}Thorium emits predominantly alpha particles (approx. 90%) and small amounts of beta (10%) and gamma radiation (176). Its biological half-life is estimated to be 500 years. Thus, after exposure to Thorotrast the radiation persists throughout the lifetime of the individual (177). Alpha radiation is capable of inducing double DNA strand breaks which are not repairable, and thus permanent DNA damage results. Thorotrast predominantly accumulates in the liver (approximately 60%), the target of carcinogenesis.

B. Nongenotoxic Mechanisms

Cirrhosis

Cirrhosis is known to associate with HCC, especially in tumors due to alcohol use and HCV infections. Although the mechanism of progression of cirrhosis to HCC is unclear, increased cell proliferation during liver regeneration probably plays an important role. Nevertheless, the progression of cirrhosis to HCC has been described in steps of histopathological changes. The initial change of cirrhosis is liver cell dysplasia (cell enlargement with nuclear pleomorphism and multinucleation, which occurred in about 60% HCC). This is followed by the growth of regenerative nodules, which are later transformed to adenomatous hyperplasic nodules. Adenomatous hyperplastic nodules will develop to atypical adenomatous hyperplastic nodules (containing both hepatocellular and adenomatous hyperplasia), which eventually develop into HCC (178–180). The transition of atypical adenomatous hyperplastic nodules to HCC has been demonstrated by HBV DNA patterns, which showed that multiple nodules were derived from the same origin (181).

Hormonal Carcinogenesis

The estrogen and progestin of oral contraceptives are metabolized in the liver. These steroids are known to alter liver functions and induce cholestatic jaundice, cholecystitis, and cholangitis (61). The development of general liver toxicity to hepatic adenomas or HCC, however, is relatively rare. Although the mechanisms of hormonal hepatocarcinogenesis are unclear, receptor-mediated promotion and cell proliferation are presumed to be involved.

IV. MECHANISM OF VIRAL CARCINOGENESIS

The mechanism of viral carcinogenesis is fundamentally different from chemical carcinogenesis. Viral carcinogenesis is an active process in which the virus transfects its transforming genes into the host cell and direct neoplastic transformation. In addition, for integrating viruses, genome integration may result in insertional mutation, which is similar to chemical mutagenesis.

Since HBV is an integrating virus, both types of causal mechanisms (viral and mutational) may apply. Integration of HBV DNA has been observed in chronic hepatitis (182), and the number of integrated copies increases with age (183). HBV integration occurred in about 80% of HCC patients positive in HBsAg (184,185), but integration was also detected in patients negative in HBsAg (186,187).

Examinations of viral DNA in HCC revealed that the integrated HBV genome is fragmentary and defective in viral replication (188). These fragmented DNA, however, usually retain the S gene (HBsAg) and the X gene (transactivator, often truncated), with their respective enhancer and promoter regions (189–191). The X gene has been demonstrated to transcribe into the host genome to produce fused viral and host transcripts (192–194).

Fused transcripts of the X gene with host genes have been detected in chronic hepatitis (192) and in HCC (193). This finding indicated that the integrated X gene is capable of activating cellular genes leading to carcinogenesis. In fact, transgenic mice carrying the HBV X gene develop tumors early in life (195).

Carcinogenesis may also be induced by insertional mutations. Random viral DNA integration may disrupt host genes and induce mutations. In the studies of HCC, HBV insertions have been reported in the retinoic acid receptor β gene (RARβ) (196), the cyclin A gene (197), and the epidermal growth factor receptor (c-erbB) gene (198). The biological significance of mutations in these genes as related to liver carcinogenesis remains speculative.

Much less information is known about HCV carcinogenesis in humans. The HCV genome does not integrate into host DNA, but actively replicating HCV RNA has been reported in HCC (199). Since almost all HCV-associated HCC arose in patients with cirrhosis or chronic hepatitis, cell proliferation may play an important role in tumor development.

Alterations in oncogene expression in HCC have also been observed. Transfection studies showed that only the N-ras in the ras oncogene family (N-ras, Ha-ras, Ki-ras) was activated (200,201) and overexpressed (202) in about 75% of HCC examined. In addition, oncogenes c-myc, ets-2, and growth factor IGF-II were expressed in almost all HCC examined (201). The activation of these four genes enhances liver cell proliferation and was postulated to result in neoplasm (201).

V. INTERSPECIES EXTRAPOLATION OF HUMAN LIVER CANCER RISK

Out of about 740 agents, mixtures, or exposure circumstances evaluated by IARC in the early 1990s, approximately 350 carcinogens (IARC Group 1, 2A, and 2B) were identified based on the results of rodent cancer bioassays. Many of these chemicals, especially chlorinated hydrocarbons, are also rodent liver carcinogens (204,205). Only 57 of the exposures/chemicals were classified as human carcinogens (IARC Group 1). The three liver carcinogens in IARC's Group 1—aflatoxins, vinyl chloride, and combined oral contraceptives—have been discussed in this chapter (203). Androgenic and anabolic steroids, Thorotrast, and hepatitis viruses are not on IARC's Group 1 list, but existing epidemiology data warrant their discussion as human carcinogens.

Two genotoxic human liver carcinogens, AFB1 and vinyl chloride, are also liver carcinogens in rodents. This finding is consistent with the observations that all human carcinogens except arsenic are known to cause cancer in laboratory animals, with at least one common organ site (205). In fact, the same types of liver tumors were induced in humans and animals, HCC induced by AFB1 (206) and angiosarcomas by vinyl chloride, at least at early stages (207). For Thorotrast, carcinogenicity studies also showed liver tumors in mice, rats, and rabbits (208). Rodent bioassays have been good indicators for these three known human liver carcinogens.

HCC were also induced in rats treated with steroids (209). The mechanism of hormonal carcinogenesis is less clear, but hormones are believed to function as promoters (210). As for hepatitis virus carcinogenesis, animals infected with species-specific hepadnaviruses also developed chronic hepatitis and liver tumors. These viruses, similar to the human viruses, preferentially infect hepatocytes in animals (211). The woodchuck hepatitis virus (WHV) has been intensively studied, and its mechanism of carcinogenesis appears to be similar to HBV in humans (212,213). WHV was found to integrate to woodchuck DNA in DNA fragments at random sites similar to the pattern observed in HBV in human HCC. WHV also induces metabolic enzymes that may enhance carcinogenicity of liver carcinogens (214).

Although the human liver carcinogens discussed above are predictable by rodent cancer bioassays, a very large number of rodent liver carcinogens are not presently considered to be human liver carcinogens or even carcinogens at other sites due to the lack of epidemiological evidence. Many human liver carcinogens may never be detected by epidemiological studies because of their usually low exposures and the presence of confounding factors.

One group of rodent liver carcinogens, however, is not expected to be human carcinogens. This class of compounds, identified by their mechanism of action, is called peroxisome proliferators (see Chapter 12 for an in-depth treatment of peroxisomal proliferators). Peroxisome proliferators are not genotoxic chemicals. They may initiate carcinogenesis by inducing peroxisome proliferation in the livers of mice and rats (215), which generates hydrogen peroxide and produces hydroxyl radicals capable of reacting with DNA (216,217). In fact, chemically modified DNA, 8-hydroxydeoxyguanosine, has been reported in rats after prolonged exposure to peroxisome proliferator chemicals (218). Such perturbation induces cell division in the liver and sustained cell proliferation presumably generates and accumulates spontaneous mutations. The fact that human liver is not susceptible to peroxisome proliferators indicates that this class of rodent liver carcinogens should not pose a significant risk to humans (217,219).

Another important consideration of interspecies risk extrapolation is pharmacokinetics. The liver is the site of metabolism of many xenobiotics, and liver cells, because of their proximity to these toxification/detoxification reactions, are often exposed to carcinogenic metabolites. Both AFB1 and vinyl chloride are activated by cytochrome P450 isoenzymes to generate reactive epoxides, which, if not detoxified by glutathione or hydrolysis in time, will react with DNA to form adducts. The activation of aflatoxin B_1 to AFB1-8,9-epoxide has been demonstrated to be similar in mouse and rat (220), but deactivation of the epoxide is about threefold more efficient in the mouse than in the rat (220). This difference in deactivation may account for the higher AFB1-induced cancer incidence in rats as compared to mice (221,222). In humans, AFB1 is similarly activated by cytochrome P450 3A4 and AFB1-8,9-epoxide is deactivated by glutathione S-transferase and epoxide hydrolase (222,223). Direct comparison of human and rodent detoxification is, however, not available. Humans defective in the detoxification gene, EPHX (epoxide hydrolase), and glutathione S-transferase M1 (GSTM1) are more susceptible to chemical carcinogenesis (224).

VI. SUMMARY

Although primary HCC is one of the most common human cancers in the world, only a few human liver carcinogens have been identified by epidemiology studies so far. Most HCCs occur in Southeast Asia and sub-Saharan Africa, and many of these carcinomas are associated with aflatoxin (AFB1) exposures or hepatitis virus (HBV or HCV) infections. HCC in the western world is rare and often associated with alcoholic cirrhosis or steroid administration. All human liver carcinogens identified also induce liver tumors in at least one rodent species. Rodent cancer bioassays, however, are not specific in identifying human liver carcinogens. The main reasons for these human-rodent differences in hepatocarcinogenicity are (a) low chemical exposures of environmental carcinogens to humans, (b) interspecies variations in chemical metabolism, (c) lack of response in humans to certain chemicals, such as peroxisome proliferators, and (d) the high degree of susceptibility of rodent liver to chemical carcinogenesis. Chapters 4 and 13 of this book further explore hepatocarcinogenesis in mice and the development of rat hepatic foci, respectively.

The mechanism of liver carcinogenesis is chemical or viral specific. For AFB1 and vinyl chloride, genotoxic mechanisms have been demonstrated by the formation of DNA adducts and specific gene mutations, especially in the p53 tumor-suppressor gene. The mechanism of hepatitis virus carcinogenesis is less clear. Integration of HBV DNA fragments into the liver cell genome may induce insertional mutations or transactivation of gene expression in host cells. HCV does not integrate into the host genome, and its carcinogenicity appears similar to that of liver cirrhosis. Furthermore, synergistic carcinogenic effects were observed in co-exposures of

AFB1 and HBV. Oral contraceptives and steroids are not genotoxic, and they induce tumors through epigenetic mechanisms such as promotion of carcinogenesis or enhanced cell proliferation.

REFERENCES

1. Rust VK. Epidemiology of hepatocellular carcinoma. Gastroenterol Clin North Am 1987; 16:545–551.
2. Wands JR, Blum HE. Primary hepatocellular carcinoma. N Engl J Med 1991; 325:729–731.
3. Wu-Williams, Zeise L, Thomas D. Risk assessment for aflatoxin B1: a model approach. Risk Anal 1992; 12:559–567.
4. DiBisceglie AM, Rustgi VK, Hoofnagle JH, Dusheiko GM, Lotze MT. Hepatocellular carcinoma. Ann Intern Med 1988; 108:391–401.
5. Vanio H, Wilbourn J. Cancer etiology: agents causally associated with human cancer. Pharmacol Toxicol 1993; 72:4–11.
6. Huff J. Chemicals causally associated with cancers in humans and in laboratory animals. A perfect concordance. In: Waalkes MP, Wand JM, eds. Carcinogenesis. New York: Raven Press, 1994:25–37.
7. Wogan GN. Aflatoxins as risk factor for hepatocellular carcinoma in humans. Cancer Res 1992; 52 (suppl.):2114s–2118s.
8. Gleeson D, Newbould MJ, Taylor P, McMahon RF, Leachy BC, Warnes TW. Androgen associated hepatocellular carcinoma with an aggressive course. Gut 1991; 32:1084–1086.
9. Baserga R, Yokoo H, Henengar GC. Thorotrast-induced cancer in man. Cancer 1960; 13:1021–1031.
10. Evans DM, Williams WJ, King IT. Angiosarcoma and hepatocellular carcinomas in vinyl chloride workers. Histopathology 1983; 7:377–388.
11. Sallie R, Di Bisceglie AM. Viral hepatitis and hepatocellular carcinoma. Gastroenterol Clin North Am 1994; 23:567–579.
12. Kew MC, Popper H. Relationship between hepatocellular carcinoma and cirrhosis. Semin Liver Dis 1984; 4:136–146.
13. Busby WF Jr, Wogan GN. Aflatoxins. In: Searle CE, ed. Chemical Carcinogens. 2d ed. Vol. 2. Washington, DC: American Chemical Society, 1984:945–1136.
14. IARC (International Agency for Research on Cancer). Aflatoxins. IARC monograph on the evaluation of the carcinogenic risk of chemicals to humans 1987; Supplement 7:83–87.
15. CDHS (California Department of Health Services). Risk Specific Intake Levels for the Proposition 65 Carcinogen: Aflatoxin. Berkeley: Reproductive and Cancer Hazard Assessment Section, CDHS, 1991.
16. Yeh FS, Yu MC, Mo CC, Luo S, Tong MJ, Henderson BE. Hepatitis B virus, aflatoxins and hepatocellular carcinoma in southern Guangxi, China. Cancer Res 1989; 49:2506–2509.

17. Groopman JD, Sabbioni G, Wild CP. Molecular dosimetry of human aflatoxin exposures. In: Groopman JD, Skipper PL, eds. Molecular Dosimetry and Human Cancer. Boca Raton, FL: CRC Press, 1991:303–324.
18. Ross RK, Yuan JM, Yu MC, Wogan GN, Qian GS, Tu JT, Groopman JD, Gao YT, Henderson BE. Urinary aflatoxin biomarkers and risk of hepatocellular carcinoma. Lancet 1992; 339:943–946.
19. Qian GS, Ross RK, Yu MC, Yuan JM, Gao YT, Henderson BE, Wogan GN, Groopman JD. A follow-up study of urinary markers of aflatoxin exposure and liver cancer risk in Shanghai, People's Republic of China. Cancer Epi Biomarkers Prevent 1994; 3:3–10.
20. Wild CP, Hudson GJ, Sabbioni G, Chapot B, Hall AJ, Wogan GN, Whittle H, Montesano R, Groopman JD. Dietary intake of aflatoxins and the level of albumin-bound aflatoxin in peripheral blood in the Gambia, West Africa. Cancer Epi Biomarkers Prevent 1992; 1:229–234.
21. IARC (International Agency for Research on Cancer). Vinyl Chloride, Polyvinyl Chloride and Vinyl Chloride-Vinylacetate Copolymers. IACR monograph on the evaluation of the carcinogenic risk of chemicals to humans 1979; 19:337–348.
22. IARC (International Agency for Research on Cancer). Vinyl Chloride. IACR monograph on the evaluation of the carcinogenic risk of chemicals to humans 1987; Supplement 7:373–375.
23. Tamburro CH. Relationship of vinyl monomers and liver cancers: angiosarcoma and hepatocellular carcinoma. Semin Liver Dis 1984; 4:158–169.
24. Forman D, Bennet B, Stafford J, Doll R. Exposure to vinyl chloride and angiosarcoma of the liver: a report of the register of cases. Br J Ind Med 1985; 42:750–753.
25. Smulevich VB, Fedotova IV, Filatova VS. Increasing evidence of the rise of cancer in workers exposed to vinyl chloride. Br J Ind Med 1988; 45:93–97.
26. Evens DM, Williams WJ, King IT. Angiosarcoma and hepatocellular carcinoma in vinyl chloride workers. Histopathology 1983; 7:377–388.
27. Spirtas R, Kaminski R. Angiosarcoma of the liver in vinyl chloride/polyvinyl chloride workers. 1977 Update of the NIOSH Register. J Occup Med 1978; 20:427–429.
28. Gokel JM, Liebezeit E, Eder M. Hemangiosarcoma and hepatocellular carcinoma of the liver following vinyl chloride exposure. A report of two cases. Virchows Arch Pathol Anat Histol 1976; 372:195–203.
29. Vianna NJ, Brady J, Harper P. Angiosarcoma of the liver: a signal lesion of vinyl chloride exposure. Environ Health Perspect 1981; 41:207–210.
30. Maltoni C, Clini C, Vicini F, Masina A. Two cases of liver angiosarcoma among polyvinyl chloride (PVC) extruders of an Italian factory producing PVC bags and other containers. Am J Ind Med 1984; 5:297–302.
31. Louagie YA, Gianello P, Kestens PJ, Bonbled F, Haot JG. Vinyl chloride induced hepatic angiosarcoma. Br J Surg 1984; 71:322–323.
32. Baxter PJ, Anthony PP, MacSween RNM, Cheuer PJ. Angiosarcoma of the liver in Great Britain 1963–1973. Br Med J 1977; 2:919–921.

33. Baxter PJ, Anthony PP, MacSween RNM, Scheuer PJ. Angiosarcoma of the liver: annual occurrence and etiology in Great Britain. Br J Ind Med 1980; 37:213–221.
34. Baxter PJ. The British hepatic angiosarcoma register. Environ Health Perspect 1981; 41:115–116.
35. Falk H, Herbert J, Crowley S, Ishak KG, Thomas LB, Popper H, Caldwell GG. Epidemiology of hepatic angiosarcoma in the United States, 1964–1974. Environ Health Perspect 1981; 41:107–113.
36. Theriault G, Allard P. Cancer mortality of a group of Canadian workers exposed to vinyl chloride monomer. J Occup Med 1981; 23:671–676.
37. Vianna NJ, Brady JA, Cardamone AT. Epidemiology of angiosarcoma of liver in New York State. NY State J Med 1981; 6:895–899.
38. Weber H, Reinl W, Greiser E. German investigations on morbidity and mortality of workers exposed to vinyl chloride. Environ Health Perspect 1981; 41:95–99.
39. Riordan SM, Lee CK, Harber RW, Thomas MC. Vinyl chloride related hepatic angiosarcoma in a polyvinyl chloride autoclave cleaner in Australia. Med J Aust 1991; 155:125–128.
40. Hagmar L, Langard S, Lundberg I. A collaborative study of cancer incidences and mortality among vinyl chloride workers. Scand J Work Environ Health 1991; 17:159–169.
41. Selinger M, Koff RS. Thorotrast and the liver: a reminder. Gastroenterology 1975; 68:799–803.
42. Okuda K, Omata M, Itoh Y, Ikezaki H, Nakashima T. Peliosis hepatis as a late and fetal complication of Thorotrast liver disease: report of five cases. Liver 1981; 1:110–122.
43. Ishak KG. Mesenchymal tumors of the liver. In: Okuda K, Peters, RL, eds. Hepatocellular Carcinoma. New York: John Wiley, 1976:247–307.
44. De Motta CL, Da Silva Horta J, Taveres MH. Prospective epidemiological study of Thorotrast exposed patients in Portugal. Environ Res 1979; 18:152–153.
45. Baxter PJ, Langlands AO, Anthony PP, MacSween RNM, Scheuer PJ. Angiosarcoma of the liver: a marker tumour for the late effects of Thorotrast in Great Britain. Br J Cancer 1980; 41:446–453.
46. Falk H, Herbert J, Crowley S, Ishak KG, Thomas LB, Popper H, Caldwell GG. Epidemiology of hepatic angiosarcoma in the United States: 1964–1974. Environ Health Perspect 1981; 41:107–113.
47. Yamada S, Hosoda S, Tateno H, Kido C, Takahashi S. Survey of Thorotrast-associated liver cancers in Japan. J Natl Cancer Inst 1983; 70:31–35.
48. Azodo MV, Gutierrez OH, Greer T. Thorotrast-induced ruptured hepatic agiosarcoma. Abdom Imaging 1993; 18:78–81.
49. Shurbaji MS, Olson LJ, Kuhajda P. Thorotrast-associated hepatic leiomyosarcoma and cholangiocarcinoma. Human Pathol 1987; 18:524–526.
50. Rubel LR, Ishak KG. Thorotrast associated cholangiocarcinoma. Cancer 1982; 50:1408–1415.

51. Ito Y, Kojiro M, Nakashima T, Mori T. Pathomorpholgic characteristics of 102 cases of thorotrast-related hepatocellular carcinoma, cholangiocarcinoma and hepatic angiosarcoma. Cancer 1988; 62:1153–1162.
52. Khan AA. Thorotrast-associated liver cancer. Am J Gastroenterol 1985; 80: 699–703.
53. Anderson M, Storm HH. Cancer incidence among Danish Thorotrast-exposed patients. J Natl Cancer Inst 1992; 84:1318–1325.
54. Olsen JH, Schulgen G, Boice JD, Whysner J, Travis LB, Williams GM, Johnson FB, McGee JO. Antiepileptic treatment and risk of hepatobiliary cancer and malignant lymphoma. Cancer Res 1995; 55:294–297.
55. Spiethoff A, Wesch H, Hover KH, Wegener K. The combined and separate action of neutron radiation and zirconium dioxide on the liver of rats. Health Phys 1992; 63:111–118.
56. Prentice RL. Epidemiologic data on exogenous hormones and hepatocellular carcinoma and selected other cancers. Prevent Med 1991; 20:38–46.
57. Rosenberg L. The risk of liver neoplasia in relation to combined oral contraceptive use. Contraception 1991; 43:643–652.
58. WHO (World Health Organization). Steroid Contraception and Risk of Neoplasia. Report of WHO scientific group. Technical report series 619. Geneva: World Health Organization, 1978.
59. Baum JK, Holtz F, Bookstein JJ, Klein EW. Possible association between benign hepatomas and oral contraceptives. Lancet 1973; 2:926–929.
60. WHO (World Health Organization). Oral contraceptives and neoplasia. WHO Tech. Rep Ser 1992; 812:22–25.
61. Greenspan FS, Baxter JD. Basic and Clinical Endocrinology. 4th ed. Norwalk, CT: Appleton & Lange, 1994.
62. Vena J, Murphy GP, Arronoff BL, Baker HW. Primary liver tumors and oral contraceptives: results of a survey. JAMA 1977; 238:2154–2158.
63. Nime F, Pickren JW, Vana J, Aronoff BL, Baker HW, Murphy GP. The histology of liver tumors in oral contraceptive users observed during a national survey of the American College of Surgeons Commission on Cancer. Cancer 1979; 44:1481–1489.
64. Edmondson HA, Henderson B, Benton B. Liver cell adenomas associated with the use of oral contraceptives. N Engl J Med 1976; 294:470–472.
65. Rooks JB, Ory HWI, Ishak KG. Epidemiology of hepatocellular adenoma. The role of oral contraceptive use. JAMA 1979; 242:644–648.
66. Lindgren A, Olsson R. Liver damage from low-dose oral contraceptive. J Intern Med 1993; 234:287–292.
67. Klatskin G. Hepatic tumors: possible relationship to use of oral contraceptives. Gastroenterology 1977; 73:386–394.
68. Prentice RL, Thomas DB. On the epidemiology of oral contraceptives and disease. Adv Cancer Res 1987; 49:285–401.
69. Goodman ZO, Ishak KG. Hepatocellular carcinoma in women: probable lack of etiologic association with oral contraceptive steroids. Hepatology 1982; 2: 440–444.

70. Handerson BE, Preston-Martin S, Edmondson HA, Peters RL, Pike MC. Hepatocellular carcinoma and oral contraceptives. Br J Cancer 1983; 48:437–440.
71. Palmer JR, Rosenberg L, Kaufman DW, Warshauer ME, Stolley PD, Shapiro S. Oral contraceptive use and liver cancer. Am J Epidemiol 1989; 130: 878–882.
72. Yu MC, Tong MJ, Govindarajan S, Henderson BE. Nonviral risk factors for hepatocellular carcinoma in a low-risk population. The non-Asian for Los Angeles County, California. J Natl Cancer Inst 1991; 83:1820–1826.
73. Hsing WA, Hoover RN, McLaughlin JK, Co-Chien HT, Wacholder S, Blot WJ, Fraumeni JF Jr. Oral contraceptives and primary liver cancer among young women. Cancer Causes Control 1992; 3:43–48.
74. Tao LC. Oral contraceptive-associated liver cell adenoma and hepatocellular carcinoma. Cancer 1991; 68:341–347.
75. Trichopoulos D. Etiology of primary liver cancer and the role of steroidal hormones. Cancer Causes Control 1992; 3:3–5.
76. Neuberger J, Forman D, Doll R, Williams R. Oral contraceptives and hepatocellular carcinoma. Br Med J 1986; 292:1355–1357.
77. Forman D, Vincent TJ, Doll R. Cancer of the liver and the use of oral contraceptives. Br Med J 1986; 292:1357–1361.
78. Tavani A, Negri E, Parazzini F, Franceschi S, La Vecchia C. Female hormone utilization and risk of hepatocellular carcinoma. Br J Cancer 1993; 67: 635–637.
79. La Vecchia C, Negri E, Parazzini F. Oral contraceptives and primary liver cancer. Br J Cancer 1989; 59:460–461.
80. WHO (World Health Organization). Combined oral contraceptives and liver cancer. The WHO collaborative study of neoplasia and steroid contraceptives. Int J Cancer 1989; 43:254–259.
81. Kew MC, Song E, Mohammed A, Hodkinson J. Contraceptive steroids as a risk factor for hepatocellular carcinoma: a case-control study in South African black women. Hepatology 1990; 11:298–302.
82. Anthony PP. Hepatoma associated with androgenic steroids. Lancet 1975; I: 685–686.
83. Paradinas F, Bull TB, Westaby D, Murray-Lyon IM. The development of peliosis hepatis and liver tumours during long-term methyltestosterone therapy: a light and electron microscopical study. Histopathology 1977; 1:225–246.
84. Ishak KG. Hepatic lesions caused by anabolic and contraceptive steroids. Semin Liver Dis 1981; 1:116–128.
85. Westaby D, Portmann B, Williams R. Androgen related primary tumours in non-Fanconi patients. Cancer 1983; 51:1947–1952.
86. Ishak KG, Zimmerman HJ. Hepatotoxic effects of the anabolic/androgenic steroids. Semin Liver Dis 1987; 7:230–236.
87. Creagh TM, Rubin A, Evans DJ. Hepatic tumours induced by anabolic steroids in an athlete. J Clin Pathol 1988; 41:441–443.

88. Klava A, Super P, Aldridge M, Horner J, Guillou P. Body builder's liver. J Roy Soc Med 1994; 87:43–44.
89. Steinbrecher UP, Lisbona R, Huang SN, Mishkin S. Complete regression of hepatocellular adenoma after withdrawal of oral contraceptives. Digest Dis Sci 1981; 26:1045–1050.
90. Edmondson HA, Reynolds TB, Henderson B, Benton B. Regression of liver cell adenomas associated with oral contraceptives. Ann Intern Med 1977; 86: 180–182.
91. Tesluk H, Lawrie J. Hepatocellular adenoma. Its transformation to carcinoma in a user of oral contraceptives. Arch Pathol Lab Med 1981; 105:296–299.
92. Gordon SC, Reddy KR, Liverstone AS, Jeffers LJ, Schiff ER. Resolution of a contraceptive-steroid induced hepatic adenoma with subsequent evolution into hepatocellular carcinoma. Ann Intern Med 1986; 105:547–549.
93. Gyorffy EJ, Brefeldt JE, Black WC. Transformation of hepatic cell adenoma to hepatocellular carcinoma due to oral contraceptive use. Ann Intern Med 1989; 110:489–490.
94. Foster JH, Berman MM. The malignant transformation of liver cell adenomas. Arch Surg 1994; 129:712–717.
95. Moldvay J, Schaff Z, Lapis K. Hepatocellular carcinoma in Fanconi's anemia treated with androgen and corticosteroid. Zentralbl Pathol 1991; 137:167–170.
96. Gleeson D, Newbould MJ, Taylor P, McMahon RF, Leachy BC, Warnes TW. Androgen associated hepatocellular carcinoma with an aggressive course. Gut 1991; 32:1084–1086.
97. Balazs M. Primary hepatocellular tumors during long-term androgenic steroid therapy. A light and electron microscopic study of 11 cases with emphasis on microvasculature of the tumours. Acta Morphol Hung 1991; 39:201–216.
98. IARC (International Agency for Research on Cancer). Hepatitis viruses. IARC monographs on the evaluation of carcinogenic risks to humans, 1994; volume 59.
99. Tiollais P, Pourcel C, Dejean A. The hepatitis B virus. Nature 1985; 317: 489–495.
100. Carman WF, Thomas HC. Genetic variation in hepatitis B virus. Gastroenterology 1992; 102:711–719.
101. Tozuka S, Uchida T, Suzuki K, Esumi M, Shikata T. State of hepatitis B virus DNA in hepatocytes of patients with noncarcinomatous liver disease. Arch Pathol Lab Med 1989; 113:20–25.
102. Burrell CJ, Gowans EJ, Rowland R, Hall P, Jilbert AR, Marmion BP. Correlation between liver histology and markers of hepatitis B virus replication in infected patients: a study by in situ hybridization. Hepatology 1984; 4:20–24.
103. Gowans E, Burrell CJ, Jibert AR, Marmion BP. Patterns of single- and double-stranded hepatitis B virus DNA and viral antigen accumulation in infected liver cells. J Gen Virol 1983; 64:1229–1239.

104. Marion PL, Salazar FH, Alexander JJ, Robinson WS. State of hepatitis B viral DNA in a human hepatoma cell line. J Virol 1980; 33:795–806.
105. Chakraborty PR, Ruiz-Opazo N, Shouval D, Shafritz DA. Identification of integrated hepatitis B virus DNA and expression of viral RNA in a HBsAg-producing human hepatocellular carcinoma cell line. Nature 1980; 286:531–533.
106. Brechot C, Pourcel C, Louise A, Rain B, Tiollais P. Presence of integrated hepatitis B virus DNA sequences in cellular DNA of human hepatocellular carcinoma. Nature 1980; 286:533–535.
107. Hadziyannis SJ, Lieberman HM, Karvountzis GG, Shafritz DA. Analysis of liver disease, nuclear HBcAg, viral replication and hepatitis B virus DNA in liver and serum of HBeAg vs. anti-Hbe positive carriers of hepatitis B virus. Hepatology 1983; 3:656–662.
108. Tanaka Y, Esumi M, Shikata T. Persistence of hepatitis B virus DNA after serological clearance of hepatitis B virus. Liver 1990; 10:6–10.
109. Carman WF, Thomas H, Domingo E. Viral genetic variation: hepatitis B virus as a clinical example. Lancet 1993; 341:349–353.
110. Waters JA, Kennedy M, Voet P, Hauser P, Petre J, Carman W, Thomas HC. Loss of the common 'a' determinant of hepatitis B surface antigen by a vaccine-induced escape mutant. J Clin Invest 1992; 90:2543–2547.
111. Carman WF, Jacyna MR, Hadziyannis S, Karayiannis P, McGarvey MJ, Makris A, Thomas HC. Mutation preventing formation of hepatitis B e antigen in patients with chronic hepatitis B infection. Lancet 1989; 2:588–591.
112. Santatonio T, Jung MC, Schneider R, Pastore G, Pape GR, Will H. Selection for a pre-C stop codon mutation during interferon treatment. J Hepatol 1991; 13:368–371.
113. Kim SH, Hong SP, Kim SK, Lee WS, Rho HM. Replication of a mutant hepatitis B virus with a fused X-C reading frame in hepatoma cells. J Gen Virol 1992; 73:2421–2424.
114. Blum HE, Galun E, Liang TJ, von Weizsacker F, Wands JR. Naturally occurring missense mutation in the polymerase gene terminating hepatitis B virus replication. J Virol 1991; 65:1836–1842.
115. Dusheiko G, Xu J, Zuckermann AJ. Clinical diagnosis of hepatitis B infection: applications of the polymerase chain reaction. In: Becker Y, Darai G, eds. Frontiers of Virology. Berlin: Springer Verlag, 1992:67–85.
116. Sobeslavsky O. Prevalence of markers of hepatitis B virus infection in various countries, a WHO collaborative study. Bull World Health Organ 1980; 58: 621–628.
117. Choo QL, Kuo G, Weiner AJ, Overby LR, Bradley DW, Houghton M. Isolation of a cDNA clone derived from a blood-borne non-A, non-B viral hepatitis genome. Science 1989; 244:359–362.
118. Choo QL, Richman KH, Han JH, Berger K, Lee C, Dong C, Gallegos C, Coit D, Medina-Selby A, Barr PJ, Weiner AJ, Bradley DW, Kuo G, Houghton M. Genetic organization and diversity of the hepatitis C virus. Proc Natl Acad Sci USA 1991; 88:2451–2455.

119. Houghton M, Weiner A, Han J, Kuo G, Choo QL. Molecular biology of the hepatitis C viruses: implications for diagnosis, development and control of viral disease. Hepatology 1991; 14:381-388.
120. Takada N, Kakase S, Enomoto N, Takada A, Date T. Clinical backgrounds of the patients having different types of hepatitis C virus genomes. J Hepatol 1991; 14:35-40.
121. Martell M, Esteban JI, Quer J, Genesca J, Weiner A, Esteban R, Guardia J, Gomez J. Hepatitis C virus (HCV) circulates as a population of different but closely related genomes: quasispecies nature of HCV genome distribution. J Virol 1992; 66:3225-3299.
122. Fong TL, Shindo M, Feinstoner SM, Hoofnagle JH, Di Bisceglie AM. Detection of replicative intermediates of hepatitis C viral RNA in liver and serum of patients with chronic hepatitis C. J Clin Invest 1991; 88:1058-1060.
123. Hiramatsu, Hayashi N, Haruna Y, Kasahara A, Fusomoto H, Mori C, Fuke L, Okayama H, Kamada T. Immunohistochemical detection of hepatitis C virus-infected hepatocytes in chronic liver diseased with monoclonal antibodies to core, envelop and NS3 region of the hepatitis C virus genome. Hepatology 1992; 16:306-311.
124. Krawczynski K, Beath MJ, Bradley DW, Kuo G, Di Bisceglie AM, Houghton M, Reyes GR, Kim JP, Choo QL, Alter MJ. Hepatitis C virus antigen in hepatocytes: immunomorphologic detection and identification. Gastroenterology 1992; 102:622-629.
125. Mita E, Hayashi N, Ueda K, Kasahara A, Fusamoto H, Takamizawa A, Matsurbara K, Okayama H, Kamada T. Expression of MPB-HCV NS1/E2 fusion protein in E. coli and detection of anti-NS1/E2 antibody in type C chronic liver disease. Biochem Biophy Res Comm 1992; 183:925-930.
126. Kao G, Choo QL, Alter HJ, Gitnick GL, Redeker AG, Purcell RH, Miyamura T, Dienstag JL, Alter MJ, Stevens CE, Tegtmeier GE, Bonino F, Colombo M, Lee WS, Kuo C, Berger K, Shuster JR, Overby LR, Bradley DW, Houghton M. An assay for circulating antibodies to a major etiologic virus of human non-A, non-B hepatitis. Science 1989; 244:362-364.
127. McHutchison JG, Person JL, Govindarajan S, Valinluck B, Gore T, Lee SR, Nelles M, Polito A, Chien D, DiNello R, Quan S, Kuo G, Redeker AG. Improved detection of hepatitis C virus antibodies in high-risk populations. Hepatology 1992; 15:19-25.
128. van der Poel CL, Cuypers HTM, Reesink HW. Confirmation of hepatitis C virus infection by new four-antigen recombinant immunoblot assay. Lancet 1991; 337:317-319.
129. Anthony PP, Ishak KG, Nayak NC, Poulsen HE, Scheuer PJ, Sobin LH. The morphology of cirrhosis. J Clin Path 1978; 31:395-414.
130. Johnson PJ, Williams R. Cirrhosis and the etiology of hepatocellular carcinoma. J Hepatol 1987; 4:140-147.
131. Colombo M. Hepatocellular carcinoma in cirrhotic. Semin Liv Dis 1993; 13: 374-383.
132. Craig JR, Klatt EC, Yu M. Role of cirrhosis and the development of HCC:

evidence from histologic studies and large population studies. In: Tabor E, Di Bisceglie AM, Purcell RH, eds. Etiology, Pathology, and Treatment of Hepatocellular Carcinoma in North America. Houston: Gulf Publishing, 1991:177–190.
133. Lee FI. Cirrhosis and hepatoma in alcoholics. Gut 1966; 7:77–85.
134. Kew MC, Popper H. Relationship between hepatocellular carcinoma and cirrhosis. Semin Liver Dis 1984; 4:136–146.
135. Omata M. Current perspectives on hepatocellular carcinoma in oriental and African countries compared to development Western countries. Digest Dis 1987; 5:97–115.
136. Tiribelli C, Melato M, Croce LS, Giarelli L, Okuda K, Ohnishi K. Prevalence of hepatocellular carcinoma and relation to cirrhosis: comparison of two different cities in the world: Trieste, Italy and Chiba, Japan. Hepatology 1989; 10:998–1002.
137. Hadengue A, N'Dri N, Benhamou JP. Relative risk of hepatocellular carcinoma in HBsAg positive vs alcoholic cirrhosis. A cross section study. Liver 1990; 10:147–151.
138. Ross RK, Yuen JM, Yu MC, Wogan GN, Qian GS, Tu JT, Groopman JD, Gao YT, Henderson BE. Urinary aflatoxin biomarkers and risk of hepatocellular carcinoma. Lancet 1992; 339:943–946.
139. Ross RK, Yu MC, Henderson BE, Yuan JM, Qian GS, Tu JT, Gao YT, Wogan GN, Groopman JD. Aflatoxin biomarkers. Lancet 1992; 340:119.
140. Qian GS, Ross RK, Yu MC, Yuan JM, Gao YT, Henderson BE, Wogan GN, Groopman JD. A follow-up study of urinary markers of aflatoxin exposure and liver cancer risk in Shanghai, People's Republic of China. Cancer Epidemiol Biomarkers Prev 1994; 3:3–10.
141. Wand JR, Blum HE. Primary hepatocellular carcinoma. N Engl J Med 1991; 325:729–731.
142. Bulatao-Jayme J, Almero EM, Castro CA, Jardeleza TH, Salamat LA. A case-control dietary study of primary liver cancer risk from aflatoxin exposure. Int J Epidemiol 1982; 11:112–119.
143. Yu H, Harris RE, Kabat GC, Wynder EL. Cigarette smoking alcohol consumption and primary liver cancer: a case-control study in the USA. Int J Cancer 1988; 42:325–328.
144. Weinberg AG, Mize CE, Worthan HG. The occurrence of hepatoma in the chronic form of hereditary tyrosinaemia. J Pediatrics 1976; 88:434–438.
145. Eriksson S, Carlson J, Velez R. Risk of cirrhosis and primary liver cancer in alpha-1-antitrypsin deficiency. N Engl J Med 1986; 314:736–739.
146. Adams PC. Hepatocellular carcinoma in hereditary hemochromatosis. Can J Gastroenterol 1993; 7:37–41.
147. Polio J, Enriquez RE, Chow A, Wood WM, Atterbury CE. Hepatocellular carcinoma in Wilson's disease. J Clin Gastroenterol 1989; 11:220–224.
148. Simson IW. Membraneous obstruction of the inferior vena and hepatocellular carcinoma in South Africa. Gastroenterology 1982; 82:171–178.
149. Choy WN. A review of the dose-response induction of DNA adducts by

aflatoxin B1 and its implications to quantitative cancer-risk assessment. Mutat Res 1993; 298:181–198.
150. Fedtke N, Boucheron JA, Turner MJ, Swenberg JA. Vinyl chloride inducted DNA adducts. I: quantitative determination of N^2,3-ethenoguanine based on electrophore labeling. Carcinogenesis 1990; 11:1279–1285.
151. Laib J, Doerjer G, Bolt HM. Detection of N^2,3-ethenoguanine in liver DNA hydrolysates after exposure of the animals to ^{14}C-vinyl chloride. J Cancer Clin Oncol 1985; 109:A7.
152. Fedtke N, Boucheron JA, Walker VE, Swenberg JA. Vinyl chloride induced DNA adducts. II: formation and persistence of 7-(2′-oxoethyl-guanine and N^2,3-ethenoguanine in rat tissue DNA. Carcinogenesis 1990; 11:1287–1292.
153. Eaton DL, Gallagher EP. Mechanisms of aflatoxin carcinogenesis. Ann Rev Pharmacol Toxicol 1994; 34:135–172.
154. Eaton DL, Groopman JD. The Toxicology of Aflatoxins-Human Health, Veterinary and Agricultural Significance. San Diego: Academic Press, 1994.
155. Hollstein M, Sidransky D, Vogelstein B, Harris CC. p53 mutations in human cancer. Science 1991; 253:49–53.
156. Baker SJ, Fearson ER, Nigro JM, Hamilton SR, Preisinger AC, Jessup JM, vanTuinen P, Ledbetter DJ, Baker DF, Nakamura Y, White R, Vogelstein B. Chromosome 17 deletion and p53 gene mutations in colorectal carcinomas. Science 1989; 224:217–221.
157. Hollstein M, Sidransky, D, Vogelstein B, Harris CC. p53 mutations in human cancers. Science 1991; 253:49–53.
158. Hollstein M, Marion MJ, Lehman T, Welsh J, Harris CC, Martel-Planche G, Kusters I, Montesano R. p53 mutations at A:T base pair in angiosarcomas of vinyl chloride-exposed factory workers. Carcinogenesis 1994; 15:1–3.
159. Li D, Cao Y, He L, Wang NJ, Gu J. Aberration of p53 gene in human hepatocellular carcinoma from China. Carcinogenesis 1991; 14:169–173.
160. Oda T, Tsuda H, Scarpe A, Sakamoto M, Hirohashi S. Mutation pattern of the p53 gene as a diagnostic marker for multiple hepatocellular carcinoma. Cancer Res 1992; 53:3674–3679.
161. Greenblatt MS, Bennet WP, Hollstein M, Harris CC. Mutations in the p53 tumor suppressor gene: clues to cancer etiology and molecular pathogenesis. Cancer Res 1994; 55:4855–4878.
162. Cho Y, Gorina S, Jeffrey P, Pavletich NP. Crystal structure of a p53 tumor suppressor-DNA couples: understanding tumorigenic mutations. Science 1994; 364:346–355.
163. Li D, Cao Yuqing, He L, Wang NJ, Gu JR. Aberrations of p53 gene in human hepatocellular carcinoma from China. Carcinogenesis 1993; 14:169–173.
164. Soini Y, Chia SC, Bennet WP, Groopman JD, Wang JS, DeBenedetti VMG, Cawley H, Welsh JA, Hansen C, Bergasa NV, Jones EA, DiBisceglie AM, Trivers GE, Sandoval CA, Calderon IE, Munoz Espinosa LE, Harris CC. An aflatoxin-associated mutational hotspot at codon 249 in the p53 tumor suppressor gene occurs in hepatocellular carcinomas from Mexico. Carcinogenesis 1996; 17:1007–1012.

165. Bressac B, Kew M, Wands J, Ozturk M. Selective G to T mutations of p53 gene in hepatocellular carcinoma from southern Africa. Nature 1991; 350: 429-431.
166. Sheu JC, Huang GT, Lee PH, Chung JC, Chou HC, Lai MY, Wang JT, Lee HS, Shih LN, Yang PM, Wang TH, Chen DS. Mutation of p53 gene in hepatocellular carcinoma in Taiwan. Cancer Res 1992; 52:6098-6100.
167. Hollstein MC, Wild CP, Bleicher F, Chutimataewin S, Harris CC, Srivatanakul P, Montesano R. p53 mutations and aflatoxin B1 exposure in hepatocellular carcinoma patients from Thailand. Int J Cancer 1993; 53:51-55.
168. Bailey EA, Lyer RS, Stone MP, Harris TM, Essigmann JM. Mutational properties of the primary aflatoxin B1-DNA adduct. Proc Natl Acad Sci USA 1996; 93:1535-1539.
169. Aguilar F, Harris CC, Sun T, Hollstein M, Cerutti P. Geographic variation of p53 mutational profile in nonmalignant human liver. Science 1994; 264: 1317-1319.
170. Oda T, Tsuda H, Scarpe A, Sakamoto M, Hirohashi S. p53 gene mutation spectrum in hepatocellular carcinoma. Cancer Res 1992; 52:6358-6364.
171. Kress S, Jahn UR, Buchmann A, Bannasch P, Schwartz M. p53 mutation in human hepatocellular carcinomas from Germany. Cancer Res 1992; 52:3220-3223.
172. Hsieh DPH, Atkinson DN. Recent aflatoxin exposure and mutation at codon 249 of the human p53 gene: lack of association. Food Add Contam 1995; 12: 421-424.
173. McMahon G, Hanson L, Lee J, Wogan GN. Identification of an activated c-Ki-ras oncogene in rat liver tumors induced by aflatoxin B1. Proc Natl Acad Sci USA 1986; 83:9418-9422.
174. Soman NR, Wogan GN. Activation of the c-Ki-ras oncogene in aflatoxin B1-induced hepatocellular carcinoma and adenoma in the rat: detection by denaturing gradient gel electrophoresis. Proc Natl Acad Sci USA 1993; 90: 2045-2049.
175. Tashiro F, Morimura S, Hayashi K, Makino R, Kawamura H. Expression of the C-Ha-ras and C-myc genes in aflatoxin B1-induced hepatocellular carcinomas. Biochem Biophy Res Commun 1986; 138:858-864.
176. Kaul A, Noffz W. Tissue dose in thorotrast patients. Health Phys 1978; 35: 113-121.
177. Kaul A. Dose in liver and spleen after injection of thorotrast into blood. Isotopenpraxis 1969; 5:85-94.
178. Ho JCI, Wu PC, Mak TK. Liver cell dysplasia in association with hepatocellular carcinoma, cirrhosis and hepatitis B surface antigen in Hong Kong. Int J Cancer 1981; 28:571-574.
179. Arakawa M, Kage M, Sugihara S, Nakashima T, Suenaga M, Okuda K. Emergence of malignant lesions within an adenomatous hyperplastic nodule in cirrhotic liver. Observations in five cases. Gastroenterology 1986; 91:198-208.
180. Takayama T, Makuuchi M, Hirohashi S, Sakamoto M, Okazaki N, Takayasu K, Kosuge T, Motoo Y, Yamazaki S, Hasegawa H. Malignant transfor-

mation of adenomatous hyperplasia to hepatocellular carcinoma. Lancet 1990; 336:1150–1153.

181. Hsu HC, Chiou TJ, Chen JY, Lee CS, Lee PH, Peng SY. Clonality and clonal evolution of hepatocellular carcinoma with multiple nodules. Hepatology 1991; 13:923–928.
182. Brechot C, Hadchouel M, Scotto J, Degos F, Charnay P, Trepo C, Tiollais P. Detection of hepatitis B virus DNA in liver and serum: a direct appraisal of the chronic carrier state. Lancet 1981; 2:765–768.
183. Chang MH, Chen PJ, Chen JY, Lai MY, Hsu HC, Lian DC, Liu YG, Chen DS. Hepatitis B virus integration in hepatitis B virus-related hepatocellular carcinoma in childhood. Hepatology 1991; 13:316–320.
184. Shafritz DA, Shouval D, Sherman HI, Hadziyannis SJ, Kew MC. Integration of hepatitis B virus DNA into the genome of liver cells in chronic liver disease and hepatocellular carcinoma. Studies in precutaneous liver biopsies and post-morte, tissue specimens. N Engl J Med 1982; 305:1067–1073.
185. Sakamoto M, Hirohashi S, Tsuda H, Shimosato Y, Makuuchi M, Hosoda Y. Multicentric independent development of hepatocellular carcinoma revealed by analysis of hepatitis B virus integration pattern. Am J Surg Pathol 1989; 13:1064–1067.
186. Lai MY, Chen PJ, Yang PM, Sheu JC, Sung JL, Chen DS. Identification and characterization of intrahepatic hepatitis B virus DNA in HbsAg-seronegative patients with chronic liver disease and hepatocellular carcinoma in Taiwan. Hepatology 1990; 12:575–581.
187. Paterlini P, Driss F, Nalpas B, Pisi E, Franco D, Berthelot P, Brechot C. Persistence of hepatitis B and hepatitis C viral genomes in primary liver cancers from HBsAg-negative patients: a study of low-endemic area. Hepatology 1993; 17:20–29.
188. Tokino T, Matsubara K. Chromosome sites for hepatitis B virus integration in human hepatocellular carcinoma. J Virol 1991; 65:6761–6764.
189. Buendia MA. Hepatitis B virus and hepatocellular carcinoma. Adv Cancer Res 1992; 59:167–226.
190. Caselmann WH, Meyer M, Kekule AS, Lauer U, Hofschneider PH, Koshy R. A trans-activator function is generated by integration of hepatitis B virus preS/S sequences in human hepatocellular carcinoma DNA. Proc Natl Acad Sci USA 1990; 87:2970–2974.
191. Zhou YZ, Butel JS, Li PJ, Finegold MJ, Melnick JL. Integrated state of subgenomic fragments of hepatitis B virus DNA in hepatocellular carcinoma from mainland China. J Natl Cancer Inst 1987; 79:223–231.
192. Takada S, Koike K. Trans-activation function of a 3′ truncated X gene-cell fusion product from integrated hepatitis B virus DNA in chronic hepatitis tissues. Proc Natl Acad Sci USA 1990; 87:5628–5632.
193. Wollersheim M, Debelka U, Hofschneider PH. A transactivating function encoded in the hepatitis B virus X gene is conserved in the integrated state. Oncogene 1988; 3:545–552.
194. Miyaki M, Sato Gotanda T, Matsui T, Mishiro S, Lmai M, Mayumi M. Integration of region X of hepatitis B virus genome in human primary hepa-

tocellular carcinomas propagated in nude mice. J Gen Virol 1986; 67:1449–1454.

195. Kim CM, Koike K, Saito I, Miyamura T, Jay G. Hbx gene of hepatitis B virus induces liver cancer in transgenic mice. Nature 1991; 351:317–320.
196. Dejean A, Bougueleret L, Grzeschik KH, Tiollaris P. Hepatitis B virus DNA integration in a sequence homologous to v-erb-A and steroid receptor gene in a hepatocellular carcinoma. Nature 1986; 322:70–72.
197. Wang J, Zindy F, Chenivesse X, Lamas E, Henglein B, Brechot C. Modification of cyclin A expression by hepatitis B virus DNA integration in a hepatocellular carcinoma. Oncogene 1992; 7:1653–1656.
198. Zhang XK, Egan JO, Hung DP, Sun ZL, Chien VKY, Chiu JF. Hepatitis B virus DNA integration and expression of an erb B-like gene in human hepatocellular carcinoma. Biochem Biophy Res Commun 1992; 188:344–351.
199. Gerber MA, Shieh YSC, Shim KS, Thung SN, Demetris AJ, Schwartz M, Akyol G, Dash S. Short communication. Detection of replicative hepatitis C virus sequences in hepatocellular carcinoma. Am J Pathol 1992; 141:1271–1277.
200. Gu JR, Hu L, Wan DF, Hong JX. Oncogene in primary hepatic cancer. In: Wagner G, Zhang YH, eds. Cancer of the Liver. Heidelberg: Springer-Verlag, 1986:50–57.
201. Gu JR. Molecular aspects of human hepatic carcinogenesis. Carcinogenesis 1988; 9:697–703.
202. Young D, Waiches G, Birchmeier C, Fasano O, Wigler M. Isolation and characterization of a new cellular oncogene encoding a protein with multiple potential transmembranous domains. Cell 1986; 45:711–719.
203. IARC (International Agency for Research on Cancer). IARC monograph on the evaluation of the carcinogenic risk of chemicals to humans. Lyon: IARC, 1987. Supplement 7.
204. Vainio H, Wilbourn J. Cancer etiology: agent causally associated with human cancer. Pharmacol Toxicol 1993; 72:4–11.
205. Huff J. Chemicals causally associated with cancer in humans and in laboratory animals. A perfect concordance. In: Waalkes MP, Ward JM, eds. Carcinogenesis. New York: Raven Press, 1994:25–37.
206. IARC (International Agency for Research on Cancer). IARC monograph on the evaluation of the carcinogenic risk of chemicals to humans. Some naturally occurring carcinogens. Lyon: IARC, 1993.
207. Popper H, Maltoni C, Selikoff IJ, Squire RA, Thomas LB. Comparison of neoplastic hepatic lesions in man and experimental animals. In: Hiatt HH, Watson JD, Winston JA, eds. Origins of Human Cancer. Book C, Human Risk Assessment. Cold Spring Harbor, NY: Cold Spring Harbor Laboratory, 1977:1359–1382.
208. RTECS (Registry of Toxic Effects of Chemical Substances). NIOSH, USDHHS Publication #87-114. Rockville, MD: USDHHS, 1985–1986:4749.
209. Higashi S, Tomita T, Mizumoto R, Nakakuki K. Development of hepatoma in rats following oral administration of synthetic oestrogen and progesterone. Gann 1980; 71:576–577.

210. Yager JD, Yager R. Oral contraceptive steroids as promoters of hepatocarcinogenesis in female Sprague-Dawley rats. Cancer Res 1980; 40:3680–3685.
211. Mason WS, Seeger C, eds. Hepadnaviruses. Berlin: Springer, 1991.
212. Popper H, Roth L, Purcell RH, Tennant BC, Gerin JL. Hepatocarcinogenicity of the woodchuck hepatitis virus. Proc Natl Acad Sci USA 1987; 83: 2994–2997.
213. Rogler CE. Cellular and molecular mechanisms of hepatocarcinogenesis associated with hepadnavirus infection. Curr Top Microbiol Immunol 1991; 168: 103–140.
214. DeFlora S, Hietanen E, Bartsch H. Enhanced metabolic activation of chemical hepatocarcinogens in woodchucks infected with hepatitis B virus. Carcinogenesis 1989; 10:1099–1106.
215. Ashby J, Brady CR, Elcombe CR, Elliott BM, Ishmael J, Odum J, Tugwood JD, Kettle S, Purchase IFH. Mechanistically-based human hazard assessment of peroxisome proliferator-induced hepatocarcinogenesis. Human Exp Toxicol 1994; 13:S1–S117.
216. Reddy JK, Lalwani ND. Carcinogenesis by hepatic peroxisome proliferators: evaluation of the risk of hypolipidemic drugs and industrial plasticizers to humans. CRC Crit Rev Toxicol 1983; 12:1–58.
217. Lake BG. Mechanisms of hepatocarcinogenicity of peroxisome-proliferating drugs and chemicals. Annul Rev Pharmacol Toxicol 1995; 35:483–507.
218. Cattley RC, Glover SE. Elevated 8-hydroxydeoxyguanosine in hepatic DNA of rats following exposure to peroxisome proliferators: relationship to carcinogenesis and nuclear localization. Carcinogenesis 1993; 14:2495–2499.
219. Bentley P, Calder I, Elcombe C, Grasso P, Stringer D, Wiegand HJ. Hepatic peroxisome proliferation in rodents and its significance for humans. Food Chem Toxicol 1993; 11:857–907.
220. Degan GH, Newmann HG. Differences in aflatoxin B1-susceptibility of rat and mouse are correlated with the capability in in vitro to inactivate aflatoxin B1-epoxide. Carcinogenesis 1981; 2:229–306.
221. Wogan GN, Palinanlunga S, Newberne PM. Carcinogenic effects of low dietary levels of aflatoxin B1. Food Cosmet Toxicol 1974; 12:681–685.
222. Moss EJ, Neal GE. The metabolism of aflatoxin B1 by human liver. Biochem Pharmacol 1985; 34:3193–3197.
223. Guengerich FP, Johnson WW, Ueng YF, Yamazaki H, Shimada T. Involvement of cytochrome P450, glutathione S-transferase and epoxide hydrolase in the metabolism of aflatoxin B1 and relevance to risk of human liver cancer. Environ Health Perspect 1996; 104:557–562.
224. McGlynn KA, Rosvold EA, Lustbader ED, Hu Y, Clapper M, Zhou T, Wild CP, Xia XL, Baffoe-Bonnie A, Ofori-Adjei D, Chen GC, London WT, Shen FM, Buetow KH. Susceptibility to hepatocellular carcinoma is associated with genetic variation in the enzymatic detoxification of aflatoxin B1. Proc Natl Acad Sci USA 1995; 92:2384–2387.

16

Comparative Kidney Carcinogenesis in Laboratory Rodents and Humans

Gordon C. Hard
American Health Foundation, Valhalla, New York

I. INTRODUCTION

In rodent bioassays, the kidney is one of the most frequent sites for the induction of cancer by chemicals. This is borne out by the carcinogenicity data base, involving hundreds of chemical carcinogenesis studies (1), originated in the late 1960s by the National Cancer Institute (NCI) and continued after 1978 by the National Toxicology Program (NTP). The high ranking of kidney in this respect is caused by a preponderance of studies showing a positive response in male rats; renal tumor induction is a much less frequent response in mice (1,2). Comparing this high ranking among organ systems in animal tests with the frequency of human renal cancer in the world population (3,4), there is a distinct lack of concordance. In the United States, for example, kidney cancer is number 11 in a frequency ranking of human cancer by organ site (5), accounting for approximately 2.5% of male cancers and approximately 1.5% of female cancers (6). An implication of this discrepancy is that, in some cases, the cancer risks for animals cannot be directly extrapolated to humans. This does not mean that rodent bioassays are not a useful strategy in determining risks for humans. To the contrary, there is very good concordance between the more than 50 chemical agents that are known human carcinogens and their positive animal bioassay results and tumor site equivalence in rodent experimentation (7). Furthermore, the rodent bioassay has provided much of the stimulus for investigative research aimed at elucidating biochemical and molecular mechanisms underlying chemical carcinogenesis. This research data base provides a means whereby, in many cases, interspecies extrapolations can be based on a mode of action. Recent years have witnessed a substantial expansion of our knowledge concerning mechanisms of cancer induction in the kidney, which has had a great impact on improving the risk-assessment process in general for chemical carcinogens. Although the kidney is not one of the major organ sites for cancer in humans, kidney cancer remains an important human health problem because it is unpredictable and carries a high mortality rate. Accordingly, this chapter examines the relationship between rodent and human renal tumors, the groups of chemicals that have induced different types of renal tumors in rodent chronic bioassays, the current understanding of modes of action of known renal carcinogens, the mechanisms underlying renal carcinogenesis, and the predictability of that information for human risk.

II. FACTORS PREDISPOSING THE KIDNEY TO CHEMICAL INJURY

The kidney possesses a number of properties that render it susceptible to chemically induced toxicity. It logically follows that some of these features have implications also for chemical carcinogens. Representing less than 1% of the total body weight, the kidneys are perfused with 25% of the cardiac output of blood. The kidneys are therefore preferentially exposed to large quantities of foreign chemicals distributed by the circulatory system. The tubules of the cortex are particularly vulnerable because this region receives about 80% of the total renal blood flow (8). In addition, the proximal tubule cells have a high capacity for transporting organic ions and certain other substrates directionally from the peritubular circulation into the tubular fluid and vice versa. This process can lead to significant levels of nephrotoxicants within the tubule cell (9). Next to the liver, the kidney is a critical site for biotransformation of many different xenobiotics, being well endowed with enzymatic pathways for carrying out extensive oxidative, reductive, hydrolytic, and conjugative processes. It also possesses metabolic pathways that are not well represented in other organs, such as the mercapturic acid pathway (10). The countercurrent mechanism that concentrates the urine in the tubules can result in concentrations of xenobiotics in the lumen many times higher than in the plasma. This effect is most evident in the medulla and papilla.

Although there may be interspecies differences at the level of specific enzymes, these various predisposing factors apply equally to humans and to rodents. However, one physiological peculiarity that would predispose the male rat kidney, but not the human kidney, to the carcinogenic action of certain chemicals, is the unique presence in male rats of high urinary levels of a specific protein, namely, alpha$_{2u}$-globulin (α_{2u}-g). This protein belongs to the lipocalin family and is synthesized in high levels in the male rat liver (11). The involvement of the protein in male rat renal carcinogenesis is considered in more detail in a subsequent section.

III. RENAL TUMOR TYPES

Renal tumors in rodents can be classified into three broad categories according to their origin and cellular composition identifying either with mature epithelium (tubules or urothelium), connective tissue, or primordial embryonic tissue (12,13). The main tumor types in rats are renal tubule adenoma and carcinoma, taking origin in tubules of the renal parenchyma; transitional or squamous cell carcinomas, arising from the urothelial lining of the renal pelvis; renal mesenchymal tumor and lipomatous tumors, originating from interstitial connective tissue; and nephroblastoma, representing neoplasia of the primordial embryonic tissue. In the mouse, the renal tumors are predominantly renal adenoma/carcinoma and transitional cell carcinoma, with renal sarcoma and nephroblastoma being exceedingly rare.

Renal tumors in humans comprise a similar spectrum, the main types including renal cell adenoma and carcinoma, equivalent to renal tubule tumors in rodents; renal pelvic cancer of transitional or squamous cell types; and nephroblastoma, most frequently referred to as Wilms' tumor (14). The epithelial tumors affect mainly older adults, whereas Wilms' tumor is primarily a neoplasm of childhood. There are also two distinct sets of uncommon or rare mesenchymal tumors occurring in children, on the one hand, and adults, on the other. The connective tissue tumors of childhood or infancy are represented by congenital mesoblastic nephroma (15), clear cell sarcoma (16), also known as bone-metastasizing renal tumor (17), and malignant rhabdoid tumor (18,19), all of which were once considered to be variants of Wilms' tumor (20). The connective tissue/vascular tumors occurring in adulthood include such entities as angiomyolipoma (21,22), leiomyosarcoma (14), juxtaglomerular cell tumor (23), and the nodules frequently encountered at autopsy known as renomedullary interstitial cell tumors (24).

Renal tubule cell tumors of rat and mouse and the human renal cell tumors are counterparts of each other, displaying a rather similar range of morphology. Cytologically, the neoplastic cells identify with renal tubule parenchyma and may have a granular (basophilic or eosinophilic), clear, or mixed staining pattern with architectural organization into tubular, papillary, solid, or cystic patterns. In rats, basophilic tumors are the most frequent staining pattern (12), but in humans clear cell tumors represent the predominant form (about 70%) of all renal cell carcinomas (25). Papillary renal cell neoplasms, a subtype with distinct cytogenetic abnormalities, comprise approximately 14% of human renal cell tumors, chromophobe carcinoma about 4%, and basophilic or eosinophilic tumors about 14% (25,26). Chromophobe tumors have been described in the rat in basic investigations using *N*-nitrosomorpholine as a model of renal cancer induction (27), but in safety evaluation studies they are usually not distinguished from clear cell variants (13). Another uncommon but clinically distinct form, the oncocytoma, characterized ultrastructurally by numerous mitochondria, is encountered in both human and rat. There is also a rare sarcomatoid variant of renal cell carcinoma in humans and a collecting duct carcinoma, also known as Bellini duct carcinoma (26), which has not been described in rodents.

Nephroblastoma in the rat is an epithelial neoplasm of similar morphology to the predominant form of Wilms' tumor, that is, having a composition of blastema, primitive tubule formation, and a benign stroma. The closest human counterpart of rat renal mesenchymal tumor appears to be congenital mesoblastic nephroma of infancy (28).

IV. TUMOR FREQUENCY

By far the most frequently occurring kidney tumor in both humans and laboratory rodents is the renal tubule tumor (adenoma/carcinoma). In humans, data from the National Cancer Institute's Surveillance, Epidemiology, and End Results (SEER) program indicate that the incidence of this tumor type is approximately 12 per 100,000 in the U.S. population (5,29). Spontaneous renal tubule tumor incidence in conventional laboratory rat strains used in the United States and in Europe for toxicology testing is similar between strains. Amalgamating information derived from reports that deal with large population samples for the Fischer 344 (F344) (30–32), Sprague-Dawley (33,34), Osborne-Mendel (35), and Wistar (36,37) strains produces an approximate spontaneous incidence for renal tubule tumor across these strains of 0.3% for males and 0.2% for females (12). Spontaneous renal tubule tumor frequency in the mouse is reported to be similar to that of the rat (38,39).

Thus, spontaneously occurring renal tubule tumor incidence is some 25 times higher in the commonly used laboratory rat strains than in the human population in North America. However, if physiological scaling factors such as life span are taken into account, as suggested in Chapter 1, the rat and human incidences are very close to equal. Where spontaneous frequency has been determined in small population samples of inbred rats, the difference between human and rat becomes even greater. In most reports on spontaneous renal tubule tumor frequency, the incidence in males exceeds that of females for both laboratory species of rodents. This parallels the situation in humans, where renal cell carcinoma occurs at least twice as often in men than in women (29). It might be speculated that the male preponderance of renal tumors may reflect an influence by hormonal factors.

In certain strains of rat, tumor types other than the tubule adenoma/carcinoma represent the predominant spontaneous neoplasm in the kidney. For example, lipomatous tumors have been recorded as the most frequent spontaneous renal neoplasm in Osborne-Mendel rats (35), nephroblastoma the most frequent in the Noble hooded (Nb) strain (40), the WAB/Not strain (41), and in a subline of the Upjohn Sprague-Dawley (42), and transitional cell tumor of the renal pelvis the most frequent in the DA/Han (43) and CHbb:Thom (44) strains. The main relevance of these strain-specific predispositions is in the potential for development of high-incidence animal models for experimental investigation of the cellular and molecular pathogenesis of specific tumor types, as exemplified by the transplacental induction with *N*-ethylnitrosourea (ENU) in the Nb rat as a useful model of nephroblastoma (45). An important animal model of heritable renal tubule carcinoma is represented by the Eker rat (46). Tumor development in Eker rats is inherited as a Mendelian dominant trait with virtually complete penetrance. Heterozygous animals begin to develop kidney tumors by 3–4 months of age, and by one year the incidence is 100%. The homozygous mutant condition is lethal for the fetus at around 10 days of gestation (47).

A further point of variance between humans and laboratory animals is the frequency with which the kidney represents a cancer site for chemical action in rodents. Of the chemicals evaluated in carcinogenicity bioassays by the NCI/NTP up to July 1996, 12% have been found to give a positive renal tumor response in at least one sex of one of the two species (J. E. Huff, personal communication). This stands in contrast to the general paucity of knowledge on the identity of chemicals that might cause renal cancer in humans.

V. RODENT RENAL CARCINOGENS

All told, well over 100 different chemicals of diverse structure have been associated with some level of renal tumor induction in laboratory rodents (12). Up to 1991, 39 of nearly 400 chemicals tested were reported as producing neoplasms of the rodent kidney in the NCI/NTP chronic bioassay data base (48). As of July 1996, the NTP data base had expanded to cover 457 chemicals, of which 56 tested positive for the rodent kidney (J. E. Huff, personal communication). These chemicals, listed in Table 1, include those for which the results were rated as equivocal evidence of carcinogenic activity. Forty-nine of the chemicals were administered orally in the feed or by gavage. For the remainder, the mode of administration was inhalation (3 chemicals), dermal (3 chemicals), or via the drinking water (1 chemical). Of the 56 positive responses, 42 involved the male rat (as opposed to just 17 for the female rat), and for 30 of these the renal tumor response was in the male rat only. In contrast, just 4 chemicals showed a renal carcinogenic response limited to the female rat. In the mouse, 13 chemicals elicited a positive response in the kidney, 11 of these involving the male and 3 the female. Only one chemical, nitrilotriacetic acid, was carcinogenic in both sexes of both species. Only two types of renal neoplasm were encountered in these bioassays. The predominant type by far was the renal tubule tumor, being associated with 51 of the chemicals. The minor tumor type was the transitional cell neoplasm of the renal pelvis, occurring with 8 chemicals. In 3 cases, the chemical was associated with both types of tumor. Interestingly, 5 of the 8 chemicals inducing renal pelvic tumors did so in the female rat, with 2 of them involving the female rat only.

Thus far, bioassays of just over 450 chemicals have not revealed that any were inducers of renal mesenchymal tumor, nephroblastoma, or lipomatous tumors. In experimental carcinogenesis studies, induction of the mesenchymal tumor and nephroblastoma have been with potent genotoxic chemicals usually employing an invasive route of administration. Moreover, the conventional 2-year bioassays are unlikely to detect tumors of embryonal derivation, such as nephroblastoma, because the standard starting age of rats used in the tests is approximately 6 weeks. It also appears that lipomatous tumors have never been induced by chemical agency. The one report claiming an association of these neoplasms with administration of certain "natural" products including two pyrrolizidine alkaloids (49) did not take account of historical control information indicating an unusual occurrence of lipomatous tumors in the rat colony at that time.

A number of potent genotoxic carcinogens, notably nitroso compounds, have been used as experimental models to study the pathogenesis of various rodent renal neoplasms. These include dimethylnitrosamine (DMN), an inducer of both renal tubule tumors and the mesenchymal tumor, diethylnitrosamine, *N*-ethyl-*N*-hydroxyethyl nitrosamine (EHEN), and *N*-nitrosomorpholine, which induce tubule tumors (12). ENU by the transplacental route is a model for nephroblastoma induction (45), while bis-(2-oxopropyl)nitrosamine (BOPN) produces transitional cell tumors of the renal pelvis (50). All of these experimental carcinogens produce high tumor incidences in an appropriate rat strain following abbreviated regimens of exposure, with a relatively rapid time-to-tumor latency period. Another group of potent renal carcinogens are certain cancer chemotherapeutic agents. Of the latter, streptozotocin provides one of the very few high-incidence experimental models of renal tubule tumor induction in the mouse (51). Clearly these potent genotoxins, usually characterized by rapid tumor development, must be regarded as a cancer risk for humans.

Among pharmaceuticals, it is notable that renal pelvic cancer has been produced in rats with either phenacetin or phenazone alone, or in combination with caffeine (52,53). In the diverse group of pesticides, the fungicides captofol (54) and merpafol (55) have induced renal tubule tumors in male rats. Representing a class of naturally occurring compounds, secondary fungal metabolites appear to have a special predilection for rodent kidney carcinogenesis. Mycotoxins that are renal carcinogens include aflatoxin B_1 (56), aflatoxin G_1 (57), ochratoxin A in both rats and mice (58,59), and citrinin (60). Each of these represents an environmental contaminant to which humans may be exposed in certain geographical regions.

There is a diverse group of chemicals, mainly light hydrocarbons, that are known to induce a renal syndrome termed hyaline droplet or α_{2u}-g nephropathy in male rats, but not in female rats or in male or female mice. The nephropathy consists of accumulation of microscopically visible eosinophilic droplets representing enlarged phagolysosomes containing the low molecular weight protein α_u-g in the second (P2) segment of the proximal convoluted tubule. At later stages the pathology is also typified by granular casts at the junction between the pars recta and thin descending limb of Henle and by linear calcification in tubules of the papilla (61). A number of these chemicals that have undergone 2-year carcinogenicity testing have been found to produce a low incidence of renal tubule tumors with a long latency, but again, in male rats only. Some of these carcinogens are listed in Table 2. The associated mechanism and implications for humans are discussed below in a subsequent section.

VI. ETIOLOGY OF HUMAN RENAL CANCER

Although there have been numerous epidemiological investigations into the causative role of environmental factors in human kidney cancer, knowledge of etiology remains exceedingly limited. On the positive side, studies consistently show that cigarette smoking increases the risk of renal cell carcinoma, and it has been suggested that this exposure contributes about one third of all cases (62–64). Which agent(s) within tobacco smoke are responsible for the carcinogenic effect has not been determined. Cigarette smoking is also

Table 1 Chemicals Inducing Renal Tumors in NCI/NTP Rodent Bioassays

Chemical	Mode of administration	Positive species and sex	Tumor type	TR series number[a]
1-Amino-2,4-dibromoanthraquinone	Feed	Male, female rat	RTT	383
1-Amino-2-methylanthraquinone	Feed	Male rat	RTT	111
2-Amino-4-nitrophenol	Gavage	Male rat	RTT	339
o-Anisidine hydrochloride	Feed	Male rat	TCT	089
Aspirin, phenacetin, and caffeine	Feed	Female rat	RTT	067
Benzofuran	Gavage	Female rat	RTT	370
o-Benzyl-*p*-chlorophenol	Gavage	Female rat	TCT	424
		Male mouse	RTT	
2,2-Bis(bromomethyl)-1,3-propanediol	Feed	Male mouse	RTT	452
Bromodichloromethane	Gavage	Male, female rat; male mouse	RTT	321
1,3-Butadiene	Inhalation	Female mouse	RTT	434
t-Butyl alcohol	Water	Male rat	RTT	436
Chlorinated paraffins: C12, 60% chlorine	Gavage	Male rat	RTT	308
Chloroform	Gavage	Male rat	RTT	000
Chlorothalonil	Feed	Male, female rat	RTT	041
C.I. Acid Orange 3	Gavage	Female rat	TCT	335
C.I. Direct Blue 218	Feed	Male mouse	RTT	430
C.I. Pigment Red 3	Feed	Male mouse	RTT	407
C.I. Pigment Red 23	Feed	Male rat	RTT	411
Cinnamyl anthranilate	Feed	Male rat	RTT	196
Coumarin	Gavage	Male, female rat	RTT	422
D&C Yellow No. 11	Feed	Male rat	RTT	463
2,4-Diaminophenol dihydrochloride	Gavage	Male mouse	RTT	401
2,3-Dibromo-1-propanol	Dermal	Male, female rat	RTT	400
1,4-Dichlorobenzene	Gavage	Male rat	RTT	319
3,4-Dihydrocoumarin	Gavage	Male rat	RTT,TCT	423
1,2-Dihydro-2,2,4-trimethylquinoline	Dermal	Male rat	RTT	456

Dimethyl methylphosphonate	Gavage	Male rat	RTT	323
Furfural	Gavage	Male mouse	RTT	382
Furosemide	Feed	Male rat	RTT	356
Geranyl acetate	Gavage	Male rat	RTT	252
Hexachloroethane	Gavage	Male rat	RTT	361
Hydroquinone	Gavage	Male rat	RTT	366
Isophorone	Gavage	Male rat	RTT	291
d-Limonene	Gavage	Male rat	RTT	347
Mercuric chloride	Gavage	Male mouse	RTT	408
8-Methoxypsoralen	Gavage	Male rat	RTT	359
α-Methylbenzyl alcohol	Gavage	Male rat	RTT	369
Methyldopa sesquihydrate	Feed	Male mouse	RTT	348
Mirex	Feed	Male rat	TCT	313
Monuron	Feed	Male rat	RTT	266
Nitrilotriacetic acid (NTA) and NTA trisodium monohydrate	Feed	Male, female rat; male, female mouse	RTT	006
o-Nitroanisole	Feed	Male, female rat	TCT	416
Nitrofurantoin	Feed	Male rat	RTT	341
Ochratoxin A	Gavage	Male, female rat	RTT	358
Pentachloroethane	Gavage	Male rat	RTT	232
Phenolphthalein	Feed	Male rat	RTT	465
Phenylbutazone	Gavage	Male, female rat	RTT,TCT	367
N-Phenyl-2-naphthylamine	Feed	Female mouse	RTT	333
Quercetin	Feed	Male rat	RTT	409
Salicylazosulfapyridine	Gavage	Female rat	TCT	457
Tetrachloroethylene	Inhalation	Male rat	RTT	311
Tetrafluoroethylene	Inhalation	Male, female rat	RTT	450
1,2,3-Trichloropropane	Gavage	Male rat	RTT	384
Triethanolamine	Dermal	Male rat	RTT	449
Tris(2-chloroethyl)phosphate	Gavage	Male, female rat	RTT	391
Tris(2,3-dibromopropyl)phosphate	Feed	Male, female rat; male mouse	RTT	076

[a]Refers to National Toxicology Program Technical Report Series.
RTT = Renal tubule tumor; TCT = transitional cell tumor.
Source: Ref. 48; J. E. Huff, personal communication.

Table 2 Some Chemicals or Mixtures Inducing Male Rat Renal Tubule Tumors Associated with Hyaline Droplet Accumulation

Chemical	Occurrence or use
d-Limonene	Food constituent
RJ-5	Fuel
JP-10	Fuel
Unleaded gasoline	Fuel
1,4-Dichlorobenzene	Insect repellent
Dimethyl methylphosphonate	Plasticizer
Hexachloroethane	Solvent
Isophorone	Solvent
Pentachloroethane	Solvent
t-Butyl alcohol	Solvent
Tetrachloroethylene	Solvent

recognized as the major cause of cancer of the renal pelvis in developed countries (65). The only other chemical-related factor accepted as being a cause of human kidney neoplasia is analgesic-induced renal pelvis cancer. In particular, phenacetin-containing pain killer combinations have been implicated and may act in concert with the renal papillary necrosis also induced by analgesic abuse (66,67). Asbestos is the only occupational exposure that has been consistently linked to renal cell carcinoma (68,69), although asbestos has not been accepted as a rodent kidney carcinogen. Conversely, there is no firm evidence from occupational studies that the heavy metal lead, a very effective carcinogen for the rodent proximal tubule (70,71), is associated with renal cancer in humans.

A number of still-to-be-proven associations between renal cancer and occupational and environmental exposures have been reported and sometimes disputed. For example, based on the experimental evidence linking renal tubule tumor with light hydrocarbons in male rats, the association of kidney cancer and hydrocarbon exposure in the petroleum industry has received much attention. Overall, most of the studies find no clear-cut exposure-response relationship (72–74). In 1972, an association with renal cell cancer was first reported for workers exposed to coke ovens in the iron and steel industry (75). Subsequently, a 30-year follow-up of this same population erased that correlation (76). The risk associated with employment in the laundry and dry-cleaning industries, where there is exposure to some of the same solvents that have induced renal tubule tumors in male rats, remains a controversial issue. Some well-conducted studies have found no association with renal cancer (77), while others continue to do so (69). Recently, an association between the use of diuretic drugs as antihypertensive medications and renal cell cancer has been revealed (78,79). As yet the respective contributions of the hypertension itself and the antihypertensive medication have not been unraveled.

Two hereditary conditions predispose humans to renal cell cancer. Von Hippel-Lindau disease is a dominantly inherited familial cancer syndrome involving multiple organs in which one third of patients die of metastatic renal cell cancer (80–82). The tuberous sclerosis complex is a multisystem autosomal dominant disorder characterized by seizures, mental retardation, and hamartomas. Patients with this disease develop renal cell carcinomas with increased frequency, although the actual incidence is not known. At least 30 cases of kidney cancer have been recorded in tuberous sclerosis patients since 1980 (83).

A number of exposure associations have been investigated in relation to the occurrence of Wilms' tumor (84), including exposures during pregnancy. Intrauterine exposure is particularly relevant because of the embryonal origin of this neoplasm. However, the etiology or any unequivocal risk factors for Wilms' tumor remain to be determined. One exploratory investigation in Brazil indicated a very strong association of Wilms' tumor in offspring of mothers who had a history of repeated exposure to pesticides in agricultural work (85).

VII. MODES OF ACTION OF RENAL CARCINOGENS

Knowledge of the biochemical and molecular mechanisms of renal carcinogenesis in rodents is far from complete, but continues to expand. Certainly the animal data available exceed the information known concerning mechanisms of renal cancer induction in humans. In this respect, animal data represent the foundation on which our understanding of kidney cancer in humans must be guided.

Some rodent renal carcinogens such as *N*-methylnitrosourea (MNU), ENU, and ethyl methane sulfonate are alkylating agents capable of acting directly on cells without biochemical modification. However, most chemicals with tumorigenic activity in rodent kidney require metabolic activation to a reactive species termed the ultimate carcinogen. It is clear that the kidney possesses a number of enzymatic functions, which not only represent detoxification or synthetic mechanisms, but are also capable of metabolic activation of xenobiotics (86). One of the metabolic pathways involved in the proximal tubules is the cytochrome P450 monooxygenase system. Specifically, CYP IIE1 plays a major role in the activation of the potent renal carcinogen DMN in liver (87,88). This isoenzyme is also known to be present in rodent proximal tubules (89,90), and it is presumed that DMN activation at this site is mediated by P450 activity (91,92). CYP IIE1 is also responsible for the metabolism of chloroform in proximal tubules and explains the male mouse vulnerability to chloroform-induced renal tumors in contrast to the female mouse, which possesses considerably less activity for this isoenzyme (89,93). It has been argued that sustained tissue damage is a common predisposing factor for cancer development in rodents (94). In this context, prolonged regenerative cell proliferation, which occurs after the renal tubule damage that is dependent on enzymatic conversion of chloroform to a reactive species, is considered to be the epigenetic mode of action underlying chloroform's carcinogenic activity (95). There are species, strain, and sex differences in the amount of CYP IIE1 present in kidney tubules, and some evidence indicates that humans are at the lower end of the scale (96).

The cytosolic enzyme β-lyase is also located in the proximal tubules, but mainly in the pars recta in rats (97). The rodent renal carcinogen hexachloro-1,3-butadiene (98) and other halogenated alkenes (99) are biotransformed to nephrotoxic metabolites via a multistep pathway involving this β-lyase. The selective renal action of these compounds commences with glutathione S-conjugate formation in the liver, transport of the S-conjugate to the kidney where it is metabolized to a cysteine S-conjugate by one of several brush border enzymes, mainly γ-glutamyltranspeptidase, and finally renal bioactivation by cysteine conjugate β-lyase to yield cytotoxic or mutagenic metabolites (100). Thus, the kidney can be the main or sometimes the exclusive target organ for compounds acting via glutathione-dependent toxicity. One environmental chemical that may lead to rat renal tumors via this complex pathway is the widely used broad-spectrum fungicide chlorothalonil (101). Because the various components of the bioactivation process exist in humans, it could be anticipated that halogenated hydrocarbons metabolized by this pathway would carry the same potential for nephrotoxicity and carcinogenicity in humans as in rodents. However, quantitatively, humans might be less susceptible than rats because interspecies differences exist for two of the enzymes that represent rate-limiting steps, namely, γ-glutamyltranspeptidase and β-lyase. Laboratory rats have extremely high levels of renal γ-glutamyltranspeptidase, 6- to 9-fold greater than in humans (102). In addition, human kidney β-lyase shares properties with the β-lyase enzyme found in other mammalian species, but expressed on a per gram of kidney tissue basis, the rat has about 10 times more of this renal enzyme than humans (103). Such differences could lead to much higher production in rat tubule cells of the nephrotoxic metabolites generated from these chemicals than in humans.

In the medulla, which lacks the cytochrome P-450 monooxygenase system, prostaglandin H synthase of the arachidonic acid cascade represents a possible mechanism for the activation of carcinogens such as *N*-[4-(5-nitro-2-furyl)-2-thiazolyl]formamide (FANFT) and phenacetin, which exert a specific tumorigenic effect on the renal pelvic urothelium (104,105). Prostaglandin H synthase is present mainly in the specialized interstitial and collecting duct cells of the renal medulla (10). The enzyme has two distinct activities, as fatty acid cyclooxygenase and as prostaglandin hydroperoxidase. Through a series of catalytic and electron donor steps, reduction of prostaglandin hydroperoxide results in cooxidation of certain xenobiotics into reactive metabolites. Compounds found to undergo prostaglandin H synthase–dependent cooxidation include acetaminophen, a major metabolite of phenacetin, and nitrofurans (10).

There is as yet little information on the molecular events that follow biotransformation of a renal carcinogen and the events that initiate the conversion of a normal cell to a neoplastic one. It is now widely accepted that the active metabolites of genotoxic carcinogens are electrophilic molecules which can interact with nucleophilic sites on macromolecules such as DNA, RNA, proteins, and lipids (106). In particular, interaction with the informational molecule DNA is regarded as the most relevant event to chemical carcinogenesis involving genotoxic carcinogens (107). The interaction of DMN with rat kidney DNA has been studied, but few data are available for other chemicals. DMN alkylates rat kidney DNA with the formation of both 7-methylguanine and 0^6-methylguanine as the major adducts (108,109). Alkylation at the 0^6 site, a promutagenic lesion, correlates best with organ specificity of carcinogenic action (110,111). Longer biological persistence of the 0^6 residue typifies the DMN-treated rat kidney, whereas 7-methylguanine disappears much faster (108).

Clearly the mechanism of direct DNA reactivity by electrophile-generating chemicals is of relevance to and predictive of human carcinogenicity. Ochratoxin A is an example of a chemical in this class, to which humans are environmentally exposed through food contamination in various parts of the world. This fungal metabolite, produced by certain *Aspergillus* and *Penicillium* genera, has proved to be a potent renal carcinogen in rodent bioassays (59,112). In fact, in the NTP study, ochratoxin produced the highest incidence of renal tubule carcinomas (60% in male rats) found among the 56 chemicals showing a positive carcinogenic response for rodent kidney. Furthermore, the malignant potential of the tumor was underscored by an unusually high rate of metastasis (59). Although some of the short-term genotoxicity test data with ochratoxin is inconclusive, studies have revealed DNA reactivity in the form of single-strand breaks in various tissues (114,115) and, more recently, DNA adduct formation in the mouse and in nonhuman primate kidney cell (115,116). Thus, the bioassay data signal a clear risk for humans associated with ochratoxin A exposure.

However, not all rodent renal carcinogens (or their reactive metabolites) are genotoxic, acting by way of direct adduct formation with kidney DNA. The now classic example of a secondary mechanism of chemical carcinogenesis in rat kidney is the association of a low incidence of renal tubule tumors in conventional strains of male rats with various chemicals that can induce α_{2u}-g nephropathy (117). Representatives of this group of chemicals (see Table 2) include the food constituent *d*-limonene, certain solvents, and hydrocarbon mixtures used as fuels (61). Alpha $_{2u}$-g is a low molecular weight protein synthesized in large amounts in liver of the mature male rat (but not in the female rat), which is partly reabsorbed by the P2 proximal tubules and catabolized and partly excreted in the urine (118). The quantity of α_{2u}-g protein excreted makes the mature male rat physiologically proteinuric and therefore dissimilar in urinary profile to humans. As demonstrated by studies with model chemicals, there is noncovalent binding between the inducing chemical and α_{2u}-g (119,120), which retards the lysosomal degradation of this protein in proximal tubule cells (121), leading to cellular protein overload and single cell degeneration specifically involving the P2 segment. In keeping with the prevalent view that sustained regenerative cell turnover is a risk factor for carcinogenesis (122–125), the compensatory tubule cell proliferation that accompanies the cell loss throughout the chronic period of chemical exposure is believed to be the basis for the increased tumor development (61,117). The induction of renal tubule tumors by this diverse group of chemicals only in male rats correlates with the absence of high circulating levels of α_{2u}-g in female rats and the complete absence of α_{2u}-g in mice. This paradigm proved to be a watershed for the development of contemporary risk assessment policy. Based on the ample biological evidence available, the U.S. Environmental Protection Agency (EPA) has determined that renal tubule tumors produced as a result of the α_{2u}-g accumulation mechanism are not an appropriate endpoint for human hazard identification (126). The points implicated in the EPA scientific policy judgment are that the tumors develop via a secondary or indirect mechanism, and not by direct interaction of the chemical with renal cell DNA, and that the high physiological levels of α_{2u}-g necessary for the effect represent a male rat–specific phenomenon involving a protein that has no homolog in humans.

Another factor that may predispose the male rate to renal tumor formation is the spontaneous age-related condition encountered in virtually all carcinogenicity bioassays, chronic progressive nephropathy (CPN). CPN is a degenerative condition which affects male rats more severely than females (127). There is also a sustained regenerative component to CPN because many affected tubules are characterized by a marked increase in cell proliferative activity (128), and atypical tubule hyperplasia (a preneoplastic lesion) has been reported to occur in very advanced stages of the disease (129,130). A number of chemicals have been observed to exacerbate CPN during the course of 2-year bioassays. One such chemical is hydroquinone, which also produced a minimal incidence of small renal adenomas in male, but not female, rats (131). Based on histopathological evidence related to foci of atypical hyperplasia and other proliferative tubule manifestations, it has been proposed that the hydroquinone-induced renal tumors in male rats result from an exacerbation of CPN to very advanced stages, coupled with a stimulating effect exerted by hydroquinone on the proliferative potential of CPN (132). As CPN is a rodent-specific disease, this particular response may have little relevance for humans.

Some chemicals are believed to induce renal tubule tumors through the mechanism of oxidative damage. These include the transition metal, iron, and potassium bromate. The administration of ferric ion chelated with nitrilotriacetate (Fe-NTA) causes a state of iron overload in rat kidney which is associated with a high incidence of renal tubule adenocarcinoma (133). There is both in vitro and in vivo evidence suggesting that the carcinogenic activity of the complex is mediated through DNA damage (134,135) induced by oxygen radicals (136). This evidence includes significantly increased levels of the marker of oxidative damage, 8-hydroxydeoxyguanosine (8-OH-dG), and of the associated repair activity (137). Lipid peroxidation is also implicated by production of 4-hydroxy-2-nonenal in rat kidney exposed to Fe-NTA, which is a reliable index of free radical–induced lipid peroxidation (138). Potassium bromate (a food additive with oxidizing properties used widely throughout the world for the treatment of wheat flour in the breadmaking process) also induces a high incidence of renal tubule tumors in both sexes of rats following daily oral administration (139). Recent studies have shown that potassium bromate specifically increases the levels of 8-OH-dG in rat kidney DNA (140), which may be attributed to increased levels of lipid peroxidation (141). As with Fe-NTA, such data suggest that oxygen radical formation by way of lipid peroxidation may underlie the renal carcinogenicity of potassium bromate. Virtually nothing is known about the role of oxidative processes in human renal carcinogenesis. Nevertheless, the presumption must be made that this mechanism would be applicable equally to rodents and humans, particularly as the iron overload condition known as hereditary hemochromatosis carries with it a significant late complication of hepatocellular carcinoma in at least 10% of hemochromatosis patients, a risk some 200 times greater than in normal subjects (142).

VIII. CHANGES IN GENE EXPRESSION IN RENAL CARCINOGENESIS

Neoplasms in general are believed to arise as a result of an accumulation of mutations, inherited or somatic, in critical genes involved in the regulation of cell proliferation and/or differentiation. Proliferation of normal cells is controlled by growth-promoting nuclear proto-oncogenes and by growth-constraining suppressor genes (143). According to the contemporary view, tumors develop clonally from a single precursor cell through multistep acquisition of adverse genetic and epigenetic events involving certain nuclear proto-oncogenes and tumor-suppressor genes, gradually conferring an increasing growth advantage.

Little research has been conducted on growth control genes in rodent kidney tumors. Studies by Ohgaki et al. (144,145) on DMN-induced renal tubule carcinomas and renal mesenchymal tumors and ENU-induced and spontaneous nephroblastomas in rats have found a high incidence of point mutation in Ki-*ras* codon 12 and inactivation of *p53* in each of these tumor types. Research by other groups has not confirmed all of these findings. Whereas activation of Ki-*ras* was found to occur commonly in renal mesenchymal tumors induced by methyl(methoxymethyl)nitrosamine (146,147), there was no evidence supporting the involvement of *ras* point mutation in the development of rat renal tubule tumors (148). In addition, only a very low incidence of point mutations was detected in the *p53* tumor suppressor gene from methyl(methoxymethyl)nitrosamine-induced renal mesenchymal tumor (149).

Studies with human renal cell carcinoma demonstrate that neither *ras* family mutations (150,151) nor *p53* gene mutations (152–155) appears to contribute significantly to the genesis of this tumor type, as both are rarely or infrequently involved in the early stage of tumor growth. In this respect, human renal cell carcinoma parallels the findings for rat epithelial kidney tumors. On the other hand, as with a variety of other human tumor types, *p53* gene alteration may be associated with malignant progression in some renal cell carcinomas (156,157).

In contrast to rodents, there have been numerous investigations on human renal tumors at the cytogenetic level identifying in particular, regions of loss of heterozygosity (LOH) on chromosomes. Cytogenetic and molecular genetic studies have identified four genetically distinct groups of renal cell tumors which correlate with histopathological classification. These are nonpapillary (mainly clear cell), papillary, and chromophobe renal cell carcinomas, and oncocytomas, each having a specific combination of chromosomal and mitochondrial DNA alterations (82). The most frequent genetic changes observed in human and renal cell cancer are deletions and translocations involving the short arm of chromosome 3, often resulting in LOH of 3p sequences. LOH on 3p appears to occur only in clear cell carcinomas, and its frequency in this cytological subcategory is at least 60% (82,158). One gene that has been proposed as a tumor suppressor from this region is the von Hippel-Lindau (VHL) disease gene, mutations of which occur exclusively in the nonpapillary subtype of renal cell carcinoma (159).

The only rodent renal tumor to have received in-depth cytogenetic and molecular genetic analysis has been the Eker rat hereditary renal tubule tumor. Studies have excluded involvement of the VHL tumor-suppressor gene locus (160). Instead, the predisposing gene has been mapped to the proximal portion of rat chromosome 10 (161), determined to be the rat homolog of the tuberous sclerosis 2 (*Tsc2*) gene (162). In this model, *Tsc2* is thought to function as a tumor-suppressor gene and its genetic alteration to be a critical event in the development of the Eker rat renal tumors. In humans, kidney tumors occur at an increased frequency in tuberous sclerosis patients underscoring some comparability in these heritable conditions of rat and human. However, the true significance and interrelationship of these molecular genetic events in renal carcinogenesis between species is not known.

IX. SUMMARY AND CONCLUSIONS

In general, humans and rodents develop a similar spectrum of tumor types in the kidney. Renal tubule tumor is the most important, being the type most frequently induced in rodents as well as representing the majority of kidney tumors in humans. Many chemicals have been associated with renal tumors in rodents, but the chemical-related causes of human renal tumors are known in only a few instances. Much more needs to be learned about molecular mechanisms involved in the chemical causation of kidney tumors in laboratory rodents and in their pathogenesis in humans before a correlation of animal results with predictability in humans can be clearly established. At this time, mutations of proto-oncogenes, such as the *ras* gene, appear unlikely to represent potential markers for in vivo genotoxicity related to the kidney.

Nevertheless, renal carcinogens acting through direct genotoxic mechanisms as well as those causing secondary DNA effects through oxidative damage can be assumed to represent equivalent risks for humans. The risk-assessment model for DNA-reactive chemicals should be based on the traditional low-dose linearity default. Renal carcinogens acting via a secondary or indirect mechanism involving threshold phenomena for their tumor-inducing or nephrotoxic effect are best dealt with by the nonlinear approach, providing that sufficient biological data on mode of action has been established. The kidney is one of the few cases where the risk assessment of a significant number of renal carcinogens involves only the hazard-assessment step before reaching a conclusion that interspecies extrapolation of the data to humans is not relevant. These renal carcinogens include a diverse group of hydrocarbons that induce renal tubule tumors in male rats exclusively through a process commencing with hyaline droplet nephropathy involving accumulation of a male rat-specific urinary protein.

Chemically induced kidney neoplasms are a relatively frequent finding in bioassays, contrasting with the situation in humans where the kidney is an infrequent site for primary cancer development. Consideration of the 56 chemicals associated with rodent renal tumor induction to date reveals that only a small proportion of these chemicals represent hyaline droplet inducers acting through a mechanism not relevant to humans. It is possible that other low-incidence male rat renal carcinogens in the NTP list may be associated with CPN, the specific spontaneous aging process that characterizes rodents, as hydroquinone appears to be. However, at the present stage of our limited knowledge concerning the modes of action of renal cancer-inducing chemicals, the majority of the positive chemicals must be assumed to have some predictiveness for humans. In this regard, the chronic bioassay for carcinogenicity in rodents remains one of the useful means for generating relevant data. In many cases, the bioassay results can act as a stimulus for additional scientific investigation to generate information explaining mechanism of action. That conductive process will inevitably help to place interspecies extrapolation and predictiveness for humans on a firmer base.

REFERENCES

1. Huff J, Cirvello J, Haseman J, Bucher J. Chemicals associated with site-specific neoplasia in 1394 long-term carcinogenesis experiments in laboratory rodents. Environ Health Perspect 1991; 93:247–270.
2. Haseman JK, Crawford DD, Huff JE, Boorman GA, McConnell EE. Results from 86 two-year carcinogenicity studies conducted by the National Toxicology Program. J Toxicol Environ Health 1984; 14:621–639.
3. Parkin DM, Laara E, Muir CS. Estimates of the worldwide frequency of sixteen major cancers in 1980. Int J Cancer 1988; 41:184–197.
4. Levi F, Lucchini F, La Vecchia C. Worldwide patterns of cancer mortality, 1985–1989. Eur J Cancer Prev 1994; 3:109–143.
5. Parker SL, Tong T, Bolden S, Wingo PA. Cancer statistics, 1996. CA Cancer J Clin 1996; 46:5–27.
6. Silverberg E. Statistical and epidemiologic data on urologic cancer. Cancer 1987; 60 (suppl):692–717.

7. Tomatis L, Aitio A, Wilbourn J, Shuker L. Human carcinogens so far identified. Jpn J Cancer Res 1989; 80:795–807.
8. Lock EA. Renal drug-metabolizing enzymes in experimental animals and humans. In: Goldstein RS, ed. Mechanisms of Injury in Renal Disease and Toxicity. Boca Raton, FL: CRC Press, 1994:173–206.
9. Berndt WO. Role of transport in chemically-induced nephrotoxicity. In: Goldstein RS, ed. Mechanisms of Injury in Renal Disease and Toxicity. Boca Raton, FL: CRC Press, 1994:235–246.
10. Lash LH. Role of metabolism in chemically induced nephrotoxicity. In: Goldstein RS ed. Mechanisms of Injury in Renal Disease and Toxicity. Boca Raton, FL: CRC Press, 1994:207–234.
11. Roy AK, Chatterjee B. Sexual dimorphism in the liver. Annu Rev Physiol 1983; 45:37–50.
12. Hard GC. Tumours of the kidney, renal pelvis and ureter. In: Turusov VS, Mohr U, eds. Pathology of Tumours in Laboratory Animals. Vol. 1—Tumours of the Rat. 2d ed. Lyon: IARC Scientific Publications No 99, 1990: 301–344.
13. Hard GC, Alden CL, Stula EF, Trump BF. Proliferative lesions of the kidney in rats. In: Guides for Toxicologic Pathology. Washington, DC: STP/ARP/AFIP, 1995:1–19.
14. Bennington JL, Beckwith JB. Tumors of the Kidney, Renal Pelvis, and Ureter. Atlas of Tumor Pathology, Second Series, Fascicle 12. Washington, DC: Armed Forces Institute of Pathology, 1975.
15. Bolande RP, Brough AJ, Izant RJ. Congenital mesoblastic nephroma of infancy. A report of eight cases and the relationship to Wilms' tumor. Pediatrics 1967; 40:272–278.
16. Haas JE, Bonadio JF, Beckwith JB. Clear cell sarcoma of the kidney with emphasis on ultrastructural studies. Cancer 1984; 54:2978–2987
17. Marsden HB, Lawler W. Bone metastasizing renal tumour of childhood. Histopathological and clinical review of 38 cases. Virchows Arch A Pathol Anat Histol 1980; 387:341–351.
18. Haas JE, Palmer NF, Weinberg AG, Beckwith JB. Ultrastructure of malignant rhabdoid tumor of the kidney. A distinctive renal tumor of children. Human Pathol 1981; 12:646–657.
19. Schmidt D, Harms D, Zieger G. Malignant rhabdoid tumor of the kidney. Histopathology, ultrastructure and comments on differential diagnosis. Virchows Arch (Pathol Anat) 1982; 398:101–108.
20. Gonzalez-Crussi F, Baum ES. Renal sarcomas of childhood. A clinicopathologic and ultrastructural study. Cancer 1983; 51:898–912.
21. Hajdu SI, Foote FW. Angiomyolipoma of the kidney: report of 27 cases and review of the literature. J Urol 1969; 102:396.
22. Steiner MS, Goldman SM, Fishman EK, Marshall FF. The natural history of renal angiomyolipoma. J Urol 1993; 150:1782–1786.
23. Kihara I, Kitamura S, Hoshino T, Seida H, Watanabe T. A hitherto unreported vascular tumor of the kidney: a proposal of 'juxtaglomerular cell tumor.' Acta Pathol Jpn 1968; 18:197–206.
24. Lerman RJ, Pitcock JA, Stephenson P, Muirhead EE. Renomedullary inter-

stitial cell tumor (formerly fibroma of renal medulla). Human Pathol 1972; 3:559–568.
25. Thoenes W, Störkel S, Rumpelt HJ. Histopathology and classification of renal cell tumors (adenomas, oncocytomas and carcinomas). The basic cytological and histopathological elements and their use for diagnostics. Pathol Res Pract 1986; 181:125–143.
26. Weiss LM, Gelb AB, Medeiros LJ. Adult renal epithelial neoplasms. Am J Clin Pathol 1995; 103:624–635.
27. Bannasch P, Schacht U, Storch E. Morphogenese und Mikromorphologie epithelialer Nierentumoren bei Nitrosomorpholinvergifteten Ratten I. Induktion und Histologie der Tumoren. Zeits Krebsforsch 1974; 81:311–331.
28. Hard GC. The nature of experimentally induced renal tumors of the rat, and possible implications for human renal cancer. In: Nieburgs HE, ed. Prevention and Detection of Cancer, Part I, Prevention, Vol. 2, Etiology; Prevention Methods. New York: Marcel Dekker, 1978:1435–1442.
29. Motzer RJ, Bander NH, Nanus DM. Renal-cell carcinoma. N Engl J Med 1996; 335:865–875.
30. Goodman DG, Ward JM, Squire RA, Chu KC, Linhart MS. Neoplastic and noneoplastic lesions in aging F344 rats. Toxicol Appl Pharmacol 1979; 48: 237–248.
31. Maekawa A, Kurokawa Y, Takahashi M, Kokubo T, Ogiu T, Onodera H, Tanigawa H, Ohno Y, Furukawa F, Hayashi Y. Spontaneous tumors in F-344/DuCrj rats. Gann 1983; 74:365–372.
32. Solleveld HA, Haseman JK, McConnell EE. Natural history of body weight gain, survival, and neoplasia in the F344 rat. J Natl Cancer Inst 1984; 72: 929–940.
33. Zwicker GM, Eyster RC, Sells DM, Gass JH. Spontaneous renal neoplasms in aged Crl:CD®BR rats. Toxicol Pathol 1992; 20:125–130.
34. Chandra M, Riley MGI, Johnson DE. Spontaneous renal neoplasms in rats. J Appl Toxicol 1993; 13:109–116.
35. Goodman DG, Ward JM, Squire RA, Paxton MB, Reichardt WD, Chu KC, Linhart MS. Neoplastic and nonneoplastic lesions in aging Osborne-Mendel rats. Toxicol Appl Pharmacol 1980; 55:433–447.
36. Crain RC. Spontaneous tumors in the Rochester strain of the Wistar rat. Am J Pathol 1958; 34:311–335.
37. Bomhard E, Rinke M. Frequency of spontaneous tumors in Wistar rats in 2-year studies. Exp Toxicol Pathol 1994; 46:17–29.
38. Percy DH, Jonas AM. Incidence of spontaneous tumors in CD®1 HaM/ICR mice. J Natl Cancer Inst 1971; 46:1045–1065.
39. Bomhard E. Frequency of spontaneous tumors in NMRI mice in 21-month studies. Exp Toxicol Pathol 1993; 45:269–289.
40. Hard GC, Noble RL. Occurrence, transplantation, and histologic characteristics of nephroblastoma in the Nb hooded rat. Invest Urol 1981; 18:371–376.
41. Middle JG, Robinson G, Embleton MJ. Naturally arising tumors of the inbred WAB/Not rat strain. I. Classification, age and sex distribution, and transplantation behavior. J Natl Cancer Inst 1981; 67:629–636.

42. Mesfin GM, Breech KT. Heritable nephroblastoma (Wilms' tumor) in the Upjohn Sprague Dawley rat. Lab Animal Sci 1996; 46:321–326.
43. Deerberg F, Rehm S. Spontaneous renal pelvic carcinoma in DA/Han rats. Z Versuchstierkd 1985; 27:33–38.
44. Tilov T, Köllmer H, Weisse I, Stötzer H. Spontan aufretende Tumoren des Ratten-stammes CHbb:Thom (SPF). Arzneimittelforschung 1976; 26: 45–50.
45. Hard GC. Differential renal tumor response to N-ethylnitrosourea and dimethylnitrosamine in the Nb rat; basis for a new rodent model of nephroblastoma. Carcinogenesis 1985; 6:1551–1558.
46. Eker R, Mossige J. A dominant gene for renal adenomas in the rat. Nature 1961; 189:858–859.
47. Everitt JI, Goldsworthy TL, Wolf DC, Walker CL. Hereditary renal cell carcinoma in the Eker rat: a unique animal model for the study of cancer susceptibility. Toxicol Lett 1995; 82/83:621–625.
48. Barrett JC, Huff J. Cellular and molecular mechanisms of chemically induced renal carcinogenesis. Renal Failure 1991; 13:211–225.
49. Schoental R, Hard GC, Gibbard S. Histopathology of renal lipomatous tumors in rats treated with "natural" products, pyrrolizidine alkaloids and α,β-unsaturated aldehydes. J Natl Cancer Inst 1971; 47:1037–1044.
50. Solé M, Cardesa A, Domingo J, Mohr U. The carcinogenic effect of 2,2-dioxoproplynitrosamine on the renal pelvic epithelium of Sprague-Dawley rats, after chronic subcutaneous injections. J Cancer Res Oncol 1992; 118: 222–227.
51. Hard GC, Identification of a high-frequency model for renal carcinoma by the induction of renal tumors in the mouse with a single dose of streptozotocin. Cancer Res 1985; 45:703–708.
52. Johansson SL. Carcinogenicity of analgesics: long-term treatment of Sprague-Dawley rats with phenacetin, phenazone, caffeine and paracetamol (acetamidophen). Int J Cancer 1981; 27:521–529.
53. Murai T, Mori S, Machino A, Hosono M, Takeuchi Y, Ohara T, Makino S, Takeda R, Hayashi Y, Iwata H, Yamamoto S, Ito H, Fukushima S. Induction of renal pelvic carcinoma by phenacetin in hydronephrosis-bearing rats of the SD/cShi strain. Cancer Res 1993; 53:4218–4223.
54. Tamano S, Kurata Y, Kawabe M, Yamamoto A, Hagiwara A, Cabral R, Ito N. Carcinogenicity of captafol in F344/DuCrj rats. Jpn J Cancer Res 1990; 81:1222–1231.
55. Nyska A, Waner T, Pirak M, Gordon E, Bracha P, Klein B. The renal carcinogenic effect of merpafol in the Fischer 344 rat. Israel J Med Sci 1989; 25:428–432.
56. Epstein SM, Bartus B, Farber E. Renal epithelial neoplasms induced in male Wistar rats by oral aflatoxin B_1. Cancer Res 1969; 29:1045–1050.
57. Butler WH, Greenblat M, Lijinsky W. Carcinogenesis in rats by aflatoxin B_1, G_1 and B_2. Cancer Res 1969; 29:2206–2211.
58. Kanisawa M, Suzuki S. Induction of renal and hepatic tumors in mice by ochratoxin A, a mycotoxin. Gann 1978; 69:599–600.

59. Boorman GA, McDonald MR, Imoto S, Persing R. Renal lesions induced by ochratoxin A exposure in the F344 rat. Toxicol Pathol 1992; 20:236–245.
60. Arai M, Hibino T. Tumorigenicity of citrinin in male F344 rats. Cancer Lett 1983; 17:281–287.
61. Hard GC, Rodgers IS, Baetcke KP, Richards WL, McGaughy RE, Valcovic LR. Hazard evaluation of chemicals that cause accumulation of α_{2u}-globulin, hyaline droplet nephropathy, and tubule neoplasia in the kidneys of male rats. Environ Health Perspect 1993; 99:313–349.
62. McLaughlin JK, Mandel JS, Blot WJ, Schuman LM, Mehl ES, Fraumeni JF. A population-based case-control study of renal cell carcinoma. J Natl Cancer Inst 1984; 72:275–284.
63. Yu MC, Mack TM, Hanisch R, Cicioni C, Henderson BE. Cigarette smoking, obesity, diuretic use, and coffee consumption as risk factors for renal cell carcinoma. J Natl Cancer Inst 1986; 77:351–356.
64. McLaughlin JK, Linblad P, Mellemgaard A, McCredie M, Mandel JS, Schlehofer B, Pommer W, Adami H-O. International renal-cell cancer study. I. Tobacco use. Int J Cancer 1995; 60:194–198.
65. McLaughlin JK, Silverman DT, Hsing AW, Ross RK, Schoenberg JB, Yu MC, Stemhagen A, Lynch CF, Blot WJ, Fraumeni JF. Cigarette smoking and cancers of the renal pelvis and ureter. Cancer Res 1992; 52:254–257.
66. Jensen OM, Knudsen JB, Tomasson H, Sorensen BL. The Copenhagen case-control study of renal pelvis and ureter cancer: role of analgesics. Int J Cancer 1989; 44:965–968.
67. McCredie M, Stewart JH, Day NE. Different roles for phenacetin and paracetamol in cancer of the kidney and renal pelvis. Int J Cancer 1993; 53:245–249.
68. Smith AH, Shearn VI, Wood R. Asbestos and kidney cancer: the evidence supports a causal association. Am J Ind Med 1989; 16:159–166.
69. Mandel JS, McLaughlin JK, Schlehofer B, Mellemgaard A, Helmert U, Linblad P, McCredie M, Adami H-O. International renal-cell cancer study. IV. Occupation. Int J Cancer 1995; 61:601–605.
70. Van Esch GJ, Van Genderen H, Vink HH. The induction of renal tumours by feeding basic lead acetate to rats. Br J Cancer 1962; 16:289–297.
71. Van Esch GJ, Kroes R. The induction of renal tumours by feeding basic lead acetate to mice and hamsters. Br J Cancer 1969; 23:765–771.
72. McLaughlin JK. Renal cell cancer and exposure to gasoline: a review. Environ Health Perspect 1993; 101(suppl 6):111–114.
73. Gamble JF, Pearlman ED, Nicolich MJ. A nested case-control study of kidney cancer among refinery/petrochemical workers. Environ Health Perspect 1996; 104:642–650.
74. Satin KP, Wong O, Yuan LA, Bailey WJ, Newton KL, Wen C-P, Swencicki RE. A 50-year mortality follow-up of a large cohort of oil refinery workers in Texas. J Occup Environ Med 1996; 38:492–506.
75. Redmond CK, Ciocco A, Lloyd JW, Rush HW. Long-term mortality study of steelworkers. VI. Mortality from malignant neoplasms among coke oven workers. J Occup Med 1972; 14:621–629.

76. Constantino JP, Redmond CK, Bearden A. Occupationally related cancer risk among coke oven workers: 30 years of follow-up. J Occup Environ Med 1995; 37:597–604.
77. Blair A, Stewart PA, Tolbert PE, Grauman D, Moran FX, Vaught J, Rayner J. Cancer and other causes of death among a cohort of dry cleaners. Br J Ind Med 1990; 47:162–168.
78. Weinmann S, Glass AG, Weiss NS, Psaty BM, Siscovick DS, White E. Use of diuretics and other antihypertensive medications in relation to the risk of renal cell cancer. Am J Epidemiol 1994; 140:792–804.
79. McLaughlin JK, Chow WH, Mandel JS, Mellemgaard A, McCredie M, Linblad P, Schlehofer B, Pommer W, Niwa S, Adami H-O. International renal-cell cancer study. VIII. Role of diuretics, other anti-hypertensive medications and hypertension. Int J Cancer 1995; 63:216–221.
80. Horton WA, Wong V, Eldridge R. Von Hippel-Lindau disease. Clinical and pathological manifestations in nine families with 50 affected members. Arch Intern Med 1976; 136:769–777.
81. Lamiell JM, Salazar FG, Hsia YE. Von Hippel-Lindau disease affecting 43 members of a single kindred. Medicine (Baltimore) 1989; 68:1–29.
82. Kovacs G. Molecular cytogenetics of renal cell tumors. Adv Cancer Res 1993; 62:89–124.
83. Bjornsson J, Short MP, Kwiatkowski DJ, Henske EP. Tuberous sclerosis-associated renal cell carcinoma. Clinical pathological, and genetic features. Am J Pathol 1996; 149:1201–1208.
84. Sharpe CR, Franco EL. Etiology of Wilms' tumor. Epidemiol Rev 1995; 17: 415–432.
85. Sharpe CR, Franco EL, de Camargo B, Lopes LF, Barreto JH, Johnsson RR, Mauad MA. Parental exposure to pesticides and risk of Wilms' tumor in Brazil. Am J Epidemiol 1995; 141:210–217.
86. Ford SM, Hook JB. Biochemical mechanisms of toxic nephropathies. Semin Nephrol 1984; 4:88–106.
87. Yoo J-SH, Ishizaki H, Yang CS. Roles of cytochrome P450IIE1 in the dealkylation and denitrosation of N-nitrosodimethylamine and N-nitrosodiethylamine in rat liver microsomes. Carcinogenesis 1990; 11:2239–2243.
88. Yamazaki H, Oda Y, Funae Y, Imaoka S, Inui Y, Guengerich FP, Shimada T. Participation of rat liver cytochrome P450 2E1 in the activation of N-nitrosodimethylamine and N-nitrosodiethylamine to products genotoxic in an acetyltransferase-overexpressing Salmonella typhimurium strain (NM 2009). Carcinogenesis 1992; 13:979–985.
89. Henderson CJ, Scott AR, Yang CS, Wolf RC. Testosterone mediated regulation of mouse renal cytochrome P-450 isoenzymes. Biochem J 1989; 266:675–681.
90. Wilke AV, Clay AS, Borghoff SJ. Cytochrome P450 2E1 (P450 2E1) levels in male and female rat kidney: implications for xenobiotic metabolism. Toxicologist 1994; 14:73(abstract).
91. Weekes UY. Metabolism of dimethylnitrosamine to mutagenic intermediates

by kidney microsomal enzymes and correlation with reported host susceptibility to kidney tumors. J Natl Cancer Inst 1975; 55:1199–1201.
92. Hard GC, Mackay RL, Kochhar OS. Electron microscopic determination of the sequence of acute tubular and vascular injury induced in the rat kidney by a carcinogenic dose of dimethylnitrosamine. Lab Invest 1984; 50:659–672.
93. Smith JH, Hook JB. Mechanism of chloroform nephrotoxicity. III. Renal and hepatic microsomal metabolism of chloroform in mice. Toxicol Appl Pharmacol 1984; 73:511–524.
94. Grasso P. Persistent organ damage and cancer production in rats and mice. Arch Toxicol 1987; suppl 11:75–83.
95. Butterworth BE, Templin MV, Borghoff SJ, Conolly RB, Kedderis GL, Wolf DC. The role of regenerative cell proliferation in chloroform-induced cancer. Toxicol Lett 1995; 82/83:23–26.
96. Waziers I, de Cugnene PH, Yang CS, Leroux J-P, Beaune PH. Cytochrome P-450 isoenzymes, epoxide hydrolase and glutathione transferases in rat and human hepatic and extrahepatic tissues. J Pharmacol Exp Ther 1990; 253: 387–394.
97. MacFarlane M, Foster JR, Gibson GG, King LJ, Locke EA. Cysteine conjugate β-lyase of rat kidney cytosol: characterization, immunocytochemical localization, and correlation with hexachlorobutadiene nephrotoxicity. Toxicol Appl Pharmacol 1989; 98:185–197.
98. Nash JA, King LJ, Lock EA, Green T. The metabolism and disposition of hexachloro-1:3-butadiene in the rat and its relevance to nephrotoxicity. Toxicol Appl Pharmacol 1984; 73:124–137.
99. Dekant W, Vamvakas S, Koob M, Köchling A, Kanhai W, Müller D, Henschler D. A mechanism of haloalkene-induced renal carcinogenesis. Environ Health Perspect 1990; 88:107–110.
100. Anders MW, Dekant W, Vamvakas S. Glutathione-dependent toxicity. Xenobiotica 1992; 22:1135–1145.
101. Wilkinson CF, Killeen JC. A mechanistic interpretation of the oncogenicity of chlorothalonil in rodents and an assessment of human relevance. Regul Toxicol Pharmacol 1996; 24:69–84.
102. Lau SS, Jones TW, Sioco R, Hill BA, Pinon RK, Monks TJ. Species differences in renal γ-glutamyltranspeptidase activity do not correlate with susceptibility to 2,bromo-(diglutathion-S-yl)-hydroquinone nephrotoxicity. Toxicology 1990; 64:291–311.
103. Lash LH, Nelson RM, Van Dyke RA, Anders MW. Purification and characterization of human kidney cytosolic cysteine conjugate β-lyase activity. Drug Metab Dispos 1990; 18:50–54.
104. Zenser TV, Davis BB. Enzyme systems involved in the formation of reactive metabolites in the renal medulla: cooxidation via prostaglandin H synthase. Fundam Appl Toxicol 1984; 4:922–929.
105. Bach PH, Bridges JW. The role of metabolic activation of analgesics and nonsteroidal anti-inflammatory drugs in the development of renal papillary necrosis and upper urothelial carcinoma. Prostaglandins Leukotrienes Med 1984; 15:251–274.

106. Miller EC, Miller JA. Searches for ultimate chemical carcinogens and their reactions with cellular macromolecules. Cancer 1981; 47:2327–2345.
107. Dipple A. DNA adducts of chemical carcinogens. Carcinogenesis 1995; 16: 437–441.
108. Nicoll JW, Swann PF, Pegg AE. Effect of dimethylnitrosamine on persistence of methylated guanines in rat liver and kidney DNA. Nature 1975; 254: 261–262.
109. Driver HE, White INH, Butler WH. Dose-response relationships in chemical carcinogenesis: renal mesenchymal tumours induced in the rat by single dose dimethylnitrosamine. Br J Exp Path 1987; 68:133–143.
110. Singer B. N-Nitroso alkylating agents: formation and persistence of alkyl 7 derivatives in mammalian nucleic acids as contributing factors in carcinogenesis. J Natl Cancer Inst 1979; 62:1329–1339.
111. Singer B. DNA damage: chemistry, repair, and mutagenic potential. Reg Toxicol Pharmacol 1996; 23:2–13.
112. Huff JE. Carcinogenicity of ochratoxin A in experimental animals. In: Castegnaro M, Pleština R, Dirheimer G, Chernozemsky IN, Bartsch H, eds. Mycotoxins, Endemic Nephropathy and Urinary Tract Tumours. Lyon: International Agency for Research on Cancer, 1991:229–244.
113. Creppy EE, Kane A, Dirheimer G, Lafarge-Frayssinet C, Mousset S, Frayssinet C. Genotoxicity of ochratoxin in mice: DNA single-strand break evaluation in spleen, liver and kidney. Toxicol Lett 1985; 28:29–35.
114. Kane A, Creppy EE, Roth A, Röschenthaler R, Dirheimer G. Distribution of the [^{3}H]-label from low doses of radioactive ochratoxin A ingested by rats, and evidence for single-strand breaks caused in liver and kidneys. Arch Toxicol 1986; 58:219–224.
115. Pfohl-Leszkowicz A, Chakor K, Creppy EE, Dirheimer G. DNA adduct formation in mice treated with ochratoxin A. In: Castegnaro M, Pleština R, Dirheimer G, Chernozemsky IN, Bartsch H, eds. Mycotoxins, Endemic Nephropathy and Urinary Tract Tumours. Lyon: International Agency for Research on Cancer, 1991:245–253.
116. Grosse Y, Baudrimont I, Castegnaro M, Betbeder AM, Creppy EE, Dirheimer G, Pfohl-Leszkowicz A. Formation of ochratoxin A metabolites and DNA-adducts in monkey kidney cells. Chem-Biol Interact 1995; 95:175–187.
117. Swenberg JA, Short B, Borghoff S, Strasser J, Charbonneau M. The comparative pathobiology of α_{2u}-globulin nephropathy. Toxicol Appl Pharmacol 1989; 97:35–46.
118. Neuhaus OW, Flory W, Biswas N, Hollerman CE. Urinary excretion of α_{2u}-globulin and aluminum by adult male rats following treatment with nephrotoxic agents. Nephron 1981; 28:133–140.
119. Lock EA, Charbonneau M, Strasser J, Swenberg JA, Bus JS. 2,2,4-Trimethylpentane-induced nephrotoxicity. II. The reversible binding of a TMP metabolite to a renal protein fraction containing α_{2u}-globulin. Toxicol Appl Pharmacol 1987; 91:182–192.
120. Lehman-McKeeman LD, Rodriguez PA, Takigiku R, Caudill D, Fey ML. d-Limonene-induced male rat-specific nephrotoxicity: evaluation of the asso-

ciation between d-limonene and α_{2u}-globulin. Toxicol Appl Pharmacol 1989; 99:250–259.

121. Lehman-McKeeman LD, Rivera-Torres MI, Caudill D. Lysosomal degradation of α_{2u}-globulin and α_{2u}-globulin-xenobiotic conjugates. Toxicol Appl Pharmacol 1990; 103:539–548.
122. Ames BN, Shigenaga MK, Gold LS. DNA lesions, inducible DNA repair, and cell division: three factors in mutagenesis and carcinogenesis. Environ Health Perspect 1993; 101(suppl. 5):35–44.
123. Cohen SM. Role of cell proliferation in regenerative and neoplastic disease. Toxicol Lett 1995; 82/83:15–21.
124. Cunningham ML, Matthews HB. Cell proliferation as a determining factor for the carcinogenicity of chemicals: studies with mutagenic carcinogens and mutagenic noncarcinogens. Toxicol Lett 1995; 82/83:9–14.
125. Preston Martin S, Pike MC, Ross RK, Jones PA, Henderson BE. Increased cell division as a cause of human cancer. Cancer Res 1990; 50:7415–7421.
126. Baetcke KP, Hard GC, Rodgers IS, McGaughy RE, Tahan LM. Alpha$_{2u}$-Globulin: Association with Chemically Induced Renal Toxicity and Neoplasia in the Male Rat. Risk Assessment Forum. Washington, DC: U.S. Environmental Protection Agency, 1991.
127. Barthold SW. Chronic progressive nephropathy in aging rats. Toxicol Pathol 1979; 7:1–6.
128. Konishi N, Ward JM. Increased levels of DNA synthesis in hyperplastic renal tubules of aging nephropathy in female F344/NCr rats. Vet Pathol 1989; 26: 6–10.
129. Foley WA, Jones DCL, Osborn GK, Kimeldorf DJ. A renal lesion associated with diuresis in the aging Sprague-Dawley rat. Lab Invest 1964; 13:439–449.
130. Hard GC, Neal GE. Sequential study of the chronic nephrotoxicity induced by dietary administration of ethoxyquin in Fischer 344 rats. Fundam Appl Toxicol 1992; 18:278–287.
131. Kari FW, Bucher J, Eustis SL, Haseman JK, Huff JE. Toxicity and carcinogenicity of hydroquinone in F344/N rats and B6C3F$_1$ mice. Food Chem Toxicol 1992; 30:737–747.
132. Hard GC, Whysner J, English JC, Zang E, Williams GM. Relationship of hydroquinone-associated rat renal tumors with spontaneous chronic progressive nephropathy. Toxicol Pathol, 1997; 25:132–143.
133. Ebina Y, Okada S, Hamazaka S, Ogino F, Li J, Midorikawa O. Nephrotoxicity and renal cell carcinoma after use of iron- and aluminum-nitrilotriacetate complexes in rat. J Natl Cancer Inst 1986; 76:107–113.
134. Toyokuni S, Sagripanti JL. DNA single- and double-strand breaks produced by ferric nitrilotriacetate in relation to renal tubular carcinogenesis. Carcinogenesis 1993; 14:223–227.
135. Toyokuni S, Mori T, Dizdaroglu M. DNA base modifications in renal chromatin of Wistar rats treated with a renal carcinogen, ferric nitrilotriacetate. Int J Cancer 1994; 57:123–128.
136. Umemura T, Sai K, Takagi A, Hasegawa R, Kurokawa Y. Oxidative DNA

damage, lipid peroxidation and nephrotoxicity induced in the rat kidney after ferric nitrilotriacetate administration. Cancer Lett 1990; 54:95–100.

137. Yamaguchi R, Hirano T, Asami S, Sugita A, Kasai H. Increase in the 8-hydroxyguanine repair activity in the rat kidney after the administration of a renal carcinogen, ferric nitrilotriacetate. Environ Health Perspect 1996; 104(suppl 3):651–653.
138. Toyokuni S, Uchida K, Okamoto K, Hattori-Nakakuki Y, Hiai H, Stadtman ER. Formation of 4-hydroxy-2-nonenal-modified proteins in the renal proximal tubules of rats treated with a renal carcinogen, ferric nitrilotriacetate. Proc Natl Acad Sci USA 1994; 91:2616–2620.
139. Kurokawa Y, Hayashi Y, Maekawa A, Takahashi M, Kokubo T, Odashima S. Carcinogenicity of potassium bromate administered orally to F344 rats. J Natl Cancer Inst 1983; 71:965–972.
140. Kasai H, Nishimura S, Kurokawa Y, Hayashi Y. Oral administration of the renal carcinogen, potassium bromate, specifically produces 8-hydroxydeoxyguanosine in rat target organ DNA. Carcinogenesis 1987; 8:1959–1961.
141. Sai K, Takagi A, Umemura T, Hasegawa R, Kurokawa Y. Relation of 8-hydroxydeoxyguanosine formation in rat kidney to lipid peroxidation, glutathione level and relative organ weight after a single administration of potassium bromate. Jpn J Cancer Res 1991; 82:165–169.
142. Weinberg ED. The role of iron in cancer. Eur J Cancer Prev 1996; 5:19–36.
143. Weinberg RA. Oncogenes, antioncogenes, and the molecular bases of multistep carcinogenesis. Cancer Res 1989; 49:3713–3721.
144. Ohgaki H, Kleihues P, Hard GC. Ki-ras mutations in spontaneous and chemically induced renal tumors of the rat. Mol Carcinog 1991; 4:455–459.
145. Ohgaki H, Hard GC, Hirota N, Maekawa A, Takahashi M, Kleihues P. Selective mutation of codons 204 and 213 of the p53 gene in rat tumors induced by alkylating N-nitroso compounds. Cancer Res 1992; 52:2995–2998.
146. Sukumar S, Perantoni A, Reed C, Rice JM, Wenk ML. Activated K-ras oncogenes in renal mesenchymal tumors induced in F344 rats by methyl(methoxymethyl)-nitrosamine. Mol Cell Biol 1986; 6:2716–2720.
147. Higinbotham KG, Rice JM, Perantoni AO. Activating point mutation in Ki-ras codon 63 in a chemically induced rat renal tumor. Mol Carcinog 1992; 5:136–139.
148. Matsumoto K, Tsuda H, Iwase T, Ito M, Nishida Y, Oyama F, Titani K, Ushijima T, Nagao M, Hirono I. Absence of ras family point mutations at codons 12, 13 and 61 in N-ethyl-N-hydroxyethylnitrosamine- or N-nitrosomorpholine-induced renal cell tumors in rats. Jpn J Cancer Res 1992; 83:933–936.
149. Weghorst CM, Dragnev KH, Buzard GS, Thorne KL, Vandeborne GF, Vincent KA, Rice JM. Low incidence of point mutations detected in the p53 tumor suppressor gene from chemically induced rat renal mesenchymal tumors. Cancer Res 1994; 54:215–219.
150. Peter S. Oncogenes and growth factors in renal cell carcinomas. Urol Int 1991; 47:199–202.

151. Rochlitz CF, Peter S, Willroth G, de Kant E, Lobeck H, Huhn D, Herrmann K. Mutations in the ras protooncogenes are rare events in renal cell cancer. Eur J Cancer 1992; 28:333–336.
152. Torigoe S, Shuin T, Kubota Y, Horikoshi T, Danenberg K, Danenberg PV. p53 Gene mutation in primary human renal cell carcinoma. Oncol Res 1992; 4:467–472.
153. Suzuki Y, Tamura G, Satodate R, Fujioka T. Infrequent mutation of p53 gene in human renal cell carcinoma detected by polymerase chain reaction single-strand conformation polymorphism analysis. Jpn J Cancer Res 1992; 83:233–235.
154. Brooks JD, Bova GS, Marshall FF, Isaacs WB. Tumor suppressor gene allelic loss in human renal cancers. J Urol 1993; 150:1278–1283.
155. Uchida T, Wada C, Shitara T, Egawa S, Mashimo S, Koshiba K. Infrequent involvement of p53 mutations and loss of heterozygosity of 17p in the tumorigenesis of renal cell carcinoma. J Urol 1993; 150:1298–1301.
156. Reiter RE, Anglard P, Liu S, Gnarra JR, Linehan WM. Chromosome 17p deletions and p53 mutations in renal cell carcinoma. Cancer Res 1993; 53: 3092–3097.
157. Imai Y, Strohmeyer TG, Fleischhacker M, Slamon DJ, Koeffler HP. p53 Mutations and MDM-2 amplification in renal cell cancers. Modern Pathol 1994; 7:766–770.
158. Van der Hout AH, van den Berg E, van der Vlies P, Dijkhuizen T, Störkel S, Oosterhuis JW, de Jong B, Buys CHCM. Loss of heterozygosity at the short arm of chromosome 3 in renal-cell cancer correlates with the cytological tumour type. Int J Cancer 1993; 53:353–357.
159. Kenck C, Wilhelm M, Bugert P, Staehler G, Kovacs G. Mutation of the VHL gene is associated exclusively with the development of non-papillary renal cell carcinomas. J Pathol 1996; 179:157–161.
160. Walker C, Ahn Y-T, Everitt J, Yuan X. Renal cell carcinoma development in the rat independent of alterations at the VHL gene locus. Mol Carcinog 1996; 15:154–161.
161. Hino O, Mitani H, Nishizawa M, Katsuyama H, Kobayashi E, Hirayama Y. A novel renal cell carcinoma susceptibility gene maps on chromosome 10 in the Eker rat. Jpn J Cancer Res 1993; 84:1106–1109.
162. Hino O, Kobayashi T, Tsuchiya H, Kikuchi Y, Kobayashi E, Mitani H, Hirayama Y. The predisposing gene of the Eker rat inherited cancer syndrome is tightly linked to the tuberous sclerosis (TSC2) gene. Biochem Biophys Res Commun 1994; 203:1302–1308.

17

Forestomach and Glandular Stomach Carcinogenesis

Masao Hirose and Nobuyuki Ito
Nagoya City University, Nagoya, Japan

I. NORMAL STRUCTURE OF THE FORESTOMACH AND GLANDULAR STOMACH

Rodents such as rats, mice, and hamsters have a forestomach located between the esophagus and the glandular stomach. The epithelium of the forestomach consists of three or four layers of stratified squamous epithelium, which is similar to the esophageal epithelium. The forestomach has no well-developed granular layer, and the horny layer is thin. The forestomach and glandular stomach are separated by a limiting ridge, where the mucosa is thicker than in other parts of the forestomach, and the basal, prickle-cell, granular, and horny layers can be easily recognized. The main function of the forestomach is storage of ingested diet prior to digestion in the glandular stomach. Humans do not have a forestomach. On the other hand, the rodent glandular stomach is structurally and functionally similar to human stomach, comprising fundic and pyloric regions. Fundic mucosa consists of surface mucous cells (foveolar or surface epithelium), neck mucous cells, parietal or oxyntic cells, chief or peptic cells, and argentaffin or enterochromaffin cells. The surface mucous cells have mucin in their cytoplasm, which stains red-purple with the alcian blue–periodic acid–Schiff reaction (AB-PAS). Chief cells are basophilic in routine H&E sections and secrete pepsinogen. Parietal cells are eosinophilic and secrete hydrochloric acid. Neck mucous cells are located in the neck region of the gastric pits and stain light blue by AB-PAS. The surface mucous cells and mucous neck cells also stain brown with paradoxical concanavalin A (ConA). Argentaffin cells secrete neurotransmitters and hormones such as serotonin and gastrin and can be immunohistochemically demonstrated with antichromogranine A antibodies. The pyloric mucosa consists of superficial mucous cells and pyloric glands, the latter being stained light blue by AB-PAS and brown by ConA staining. The pyloric glands also contain enterochromaffin cells. Cardiac glands are compound tubular glands that open directly into the gastric pits and are composed of mucous cells that are histologically quite similar to the mucous cells of the pyloric glands or the neck mucous cells. They are not well developed in rodents.

II. FORESTOMACH CARCINOGENS

There are a number of forestomach carcinogens known to be active in rats and mice (1–9) (Table 1). The majority are *N*-nitroso compounds, nitro compounds, aromatic hydrocarbons, heterocyclic amines, and halogenated hydrocarbons which are genotoxic in nature. There are also several nongenotoxic forestomach carcinogens such as phenolic compounds, propionic acid, diallyl phthalate, allyl chrolide, and ethyl acrylate. When phenolic compounds (i.e., catechol, hydroquinone, pyrogallol, propyl gallate, etc.) or ascorbic acid in combination with sodium nitrite are administered to rats, strong cytotoxicity and cell proliferation and/or papillomas are induced in the forestomach within a short period, whereas individual chemicals do not induce such lesions, indicating that long-term treatment induces carcinomas (10,11). Most forestomach carcinogens cause forestomach tumors after intragastric administration or by admixture to the diet or drinking water, although inhaled epichlorohydrin and 1,3-butadiene can induce forestomach tumors (12,13). Combined administration of sodium nitrite and secondary amines can also induce forestomach tumors by formation of *N*-nitroso compounds in the stomach under acidic conditions (14).

III. EARLY CHANGES OF THE FORESTOMACH CAUSED BY FORESTOMACH CARCINOGENS

Most genotoxic as well as nongenotoxic forestomach carcinogens exert strong cytotoxicity as evidenced by necrosis, inflammation, erosion, and ulceration, as well as cell proliferation in the forestomach epithelium. Twelve hours after a single intragastric injection of a carcinogenic dose of the genotoxic agent *N*-methyl-*N'*-nitro-*N*-nitrosoguanidine (MNNG), hydropic degeneration of the forestomach epithelium appears, followed by severe necrosis and inflammation at 24 hours. Almost the entire epithelium except near the ridge disappears within 72 hours of the injection and is then

Table 1 Forestomach Carcinogens

Chemicals	Species
Nitroso compounds	
N-Methylnitrosourea	Rat, hamster
N-Ethylnitrosourea	Rat
N-Propyonitrosourea	Rat
N-Diethylmethylnitrosourea	Hamster
N-Butylnitrosourea	Rat
N-Amylnitrosourea	Rat
N-Methylnitrosourethane	Rat, hamster
N-Ethylnitrosourethane	Rat
N-Propylnitrosourethane	Rat
N-Butylnitrosourethane	Rat
N-Amylnitrosourethane	Rat
N-Nitrososarcosine ethyl ester	Rat
N-Nitrosomorpholine	Rat
N-Hydroxy-*N*-nitrosobenzamine ammonium salt (Cupferon)	Rat
N-Methyl-*N'*-nitro-*N*-nitrosoguanidine	Rat, mouse, hamster
N-Ethyl-*N'*-nitro-*N*-nitrosoguanidine	Rat
N-Nitrosoeffedrine	Rat
Nitro compounds	
8-Nitroquinoline	Rat
4-Nitroquinoline 1-oxide	Rat
Polyaromatic hydrocarbons	
Benzo[*a*]pyrene	Mouse, hamster
Dibenz[*a*,*h*]anthracene	Mouse
7,12-Dimethylbenz[*a*]anthracene	Mouse
3-Methylcholanthrene (20-Methylcholanthrene)	Mouse, hamster
3,4,5,6-Dibenzcarbazol	Mouse
2-Acetylaminofluorene	Hamster
N-Hydroxyaminofluorene	Hamster
Halogenated hydrocarbons	
Methyl bromide	Rat
Chlorofluoromethane	Rat
1,2-Dichloroethane	Rat
1,2-Dibromoethane (Ethylene dibromide)	Rat, mouse
Allyl chrolide	Mouse
Epichlorohydrine	Rat
1,3-Dichloropropene (Telone II)	Rat
1,2-Dibromo-3-chloropropane	Rat, mouse
3-Chloro-2-methylpropene	Rat, mouse
Bis-(2-chloro-1-methylethyl)ether	Mouse

(*Continued*)

Table 1 Continued

Chemicals	Species
Tris-(2,3-dibromopropyl)phosphate	Mouse
3-(Chloromethyl)pyridine hydrochloride	Rat, mouse
4-Chloro-*o*-phenylenediamine	Rat
Aliphatic/Aromatic hydrocarbons	
Ethylene oxide	Rat
1,2-Propylene oxide	Rat
Propionic acid	Rat
β-Propiolactone	Rat
Ethyl acrylate	Rat, mouse
1,3-Butadiene	Mouse
Pivalolactone	Rat
Diallyl phthalate	Mouse
Styrene oxide	Rat, mouse
Heterocyclic amines	
2-Amino-3-methylimidazo[4,5-*f*]quinoline (IQ)	Mouse
2-Amino-3,4-dimethylimidazo[4,5-*f*]quinoline (MeIQ)	Mouse
Phenolic compounds	
Butylated hydroxyanisole (BHA)	Rat, hamster
Caffeic acid	Rat, mouse
Sesamol	Rat, mouse
4-Methoxyphenol	Rat
4-Methylcatechol	Rat
Miscellaneous	
Urethane	Mouse, hamster
Diglycidyl resorcinol ether	Rat, mouse
Sulfallate	Rat
Captafol	Mouse
Aristolochic acid	Rat
Estradiol mustard	Mouse
N-[4-(5-Nitro-2-furyl)-2-thiazolyl]formamide (FANFT)	Mouse
N-[4-(5-Nitro-2-furyl)-2-thiazolyl]acetamide	Mouse
2-(2-Furyl)-3-(5-nitro-2-furyl)acrylamide (AF-2)	Mouse
trans-5-Amino-3-[2-(5-nitro-2-furyl)vinyl]-1,2,4-oxadiazole	Mouse
7-Methoxy-2-nitro-naphtho[2,1-b]furan (R-7000)	Rat
2-(2-Formylhydrazino)-4-(5-nitro-2-furyl)thiazole	Mouse, hamster
Diazoaminobenzene	Mouse
Betel quid	Rat
Snuff	Rat

Source: Modified from Ref. 1.

replaced by cellular regeneration. Continuous oral treatment with *N*-methylnitrosourethane (MNUR) slightly enhances DNA synthesis and results in marked cell necrosis at day 7. Such toxicity, however, has not been reported with genotoxic epichlorohydrin-induced forestomach carcinogenesis (15).

The putative pathway of forestomach carcinogenesis induced by nongenotoxic phenolic compounds is presented in Figure 1. BHA, caffeic acid, and 4-methoxyphenol have been demonstrated to induce overexpression of the protooncogenes *c*-myc, *c*-fos, and *c*-jun only 15 minutes after oral treatment with these carcinogens with no evidence of cytotoxicity or cell proliferation. DNA synthesis as evaluated by anti-bromodeoxyuridine (BrdU) immunohistochemical staining was found to be significantly elevated starting 12 hours after treatment with these carcinogens, followed by cell proliferation and/or cytotoxicity (16,17). BHA is thought to enter into cellular membranes and disrupt mitochondrial function, producing a decrease in ATP and subsequently cell death (18). A continuous process of cellular regeneration leads to development of hyperplasia. In such a state where cell turnover is strongly enhanced, the mucosa would be expected to be very sensitive to carcinogenic insult. In fact, when the genotoxic carcinogens *N*-methylnitrosourea (MNU) or 3,2′-dimethyl-4-aminobiphenyl (DMAB) are administered at the time of marked cell proliferation induced by 2% BHA, high incidences of forestomach tumors result. No lesions are seen when the chemicals are given alone (19). The potential to induce necro-

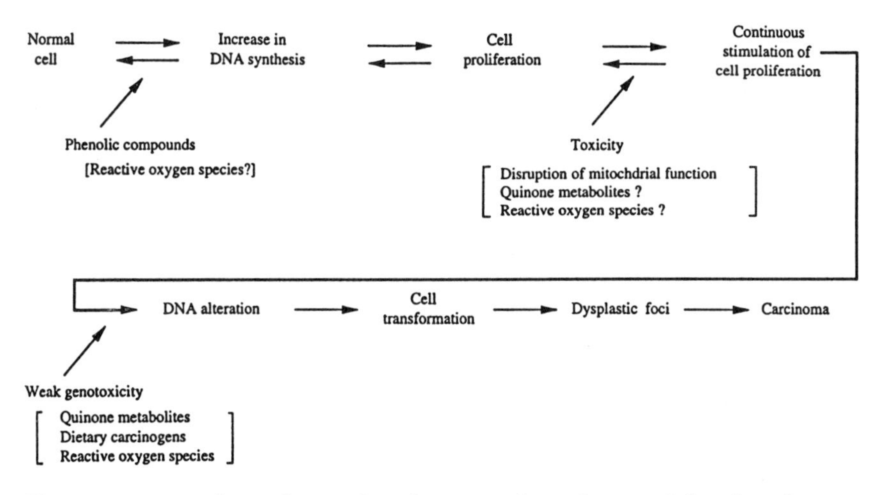

Figure 1 A putative pathway of rat forestomach carcinogenesis by phenolic compounds.

sis, erosion, and ulceration differs between chemicals, and positive association appears to exist between cytotoxicity and carcinogenicity. Thus, among the phenolic compounds tested for cytotoxicity and carcinogenicity, 4-methoxyphenol is the strongest followed by caffeic acid and BHA. Carcinomas preferentially develop in the epithelium at the edges of chronic ulcers with 4-methoxyphenol, and near the limiting ridge, close to the esophageal orifice with BHA, both in areas where hyperplasia is most pronounced. It has been shown that free radicals are generated during prostaglandin H synthase–mediated biotransformation of BHA in vitro and in some organs other than the forestomach (20,21), and very small amounts of genotoxic quinone metabolites are produced during oxidative metabolism of BHA (22). Free radicals, quinone metabolites, and also unknown carcinogens produced in the stomach by reaction of secondary amines and nitrite might thus be responsible for transformation of epithelial cells, which then develop into forestomach tumors under the continued influence of these so-called nongenotoxic carcinogens.

Most forestomach hyperplasia as well as papillomas induced by nongenotoxic carcinogens rapidly regress after cessation of chemical treatment (23–28). However, focal hyperplastic and dysplastic lesions, in which the level of DNA synthesis remains higher than in normal epithelium, do persist after withdrawal of chemical treatment, and such lesions can develop into carcinomas (26–28). Hyperplasias induced by genotoxic carcinogens such as MNNG and MNUR do not regress but rather continuously grow to form papillomas and carcinomas after cessation of carcinogen treatment (27). However, forestomach hyperplastic lesions observed after daily intragastric administration of the genotoxin methyl bromide for 90 days did regress when methyl bromide treatment was discontinued (29), indicating that irritation rather than genotoxicity was largely responsible for the induction of hyperplastic lesions.

At an early stage prior to cancer development, strong toxicity and/or cell proliferation are caused in rat and/or mouse forestomach epithelium by many genotoxic forestomach carcinogens such as MNNG, MNUR, 3-chloro-2-methylpropene, diglycidyl resorcinol ethel, ethylene dibromide, styrene oxide, methyl bromide, 1,2-dibromo-3-chloropropane, 1-chloro-2-methylpropane, 3-chloro-2-methylpropane, and aristolochic acid (29–33), as well as nongenotoxic forestomach carcinogens such as ethyl acrylate (34), propionic acid, BHA, caffeic acid, sesamol, 4-methoxyphenol, and 4-methylcatechol (35–40). A dose-response study of BHA demonstrated that the degree of hyperplasia strongly correlated with the degree of carcinogenicity (41). Dietary treatment with a carcinogenic dose (2%) of BHA for 1 week induces hyperplasia, whereas at 1.0%, 0.5% or lower dose levels (which are generally noncarcinogenic), only an increase in DNA synthesis is found. It should be remembered, however, that even at dose levels lower than the threshold for complete carcinogenicity, chemicals that induce only a weak response in cell proliferation can still cause tumor development after appropriate initiation (42). Therefore, cytotoxicity and/or cell proliferation are important markers for both carcinogenicity and/or promotion of carcinogenesis in the forestomach. Some weak genotoxic carcinogens such as 6-nitroquinoline, epichlorohydrin, and AF-2 induce only a slight increase in cell proliferation (15,27).

IV. PATHOLOGY OF FORESTOMACH PROLIFERATIVE LESIONS AND TUMORS

Proliferative lesions of the rodent forestomach can be classified into hyperplasia, papillomas, papillomatosis, fibroepitheliomas, atypical hyperplasia or dysplasia, carcinomas, benign tumors of connective tissue origin, and sarcomas (43–45).

Hyperplasia is further divided into simple, papillary or nodular, and cystic types. Simple hyperplasia is a lesion in which the mucosa becomes diffusely thickened to five or more layers with hyperkeratosis or parakeratosis but without any increase in connective tissue stroma. Papillary or nodular hyperplasia is a focal upward or downward, endophytic growth of mucosa supported by a fibrovascular core of branching single or secondary stalks. Basal cells are sometimes predominant in downward hyperplasia. Cystic hyperplasia, which is encountered in the lamina propria or in submucosa, is similar to the human epidermal cyst, but the epithelium is often hyperplastic. Mitoses are frequently observed except with basal cell hyperplasia. Various degrees of structural and nuclear atypia are found within hyperplasias after long-term treatment, these then being termed atypical hyperplasia or dysplasia.

The papilloma is a focal polypoid or nodular lesion with complex fibrovascular branching of more than tertiary stalks. Some structural and nuclear atypia can be observed and the tumor cells may be mature squamous cells, basal cells, or mixtures of the two. The term papillomatosis is applied when the entire or broad areas of forestomach epithelium are covered with papillary hyperplasia or papillomatous lesions.

Carcinomas can be classified into well-, moderately, and poorly differentiated squamous cell carcinomas, basal cell carcinomas, and adenosquamous cell carcinomas. The well-differentiated squamous cell carcinoma, characterized by clear differentiation of tumor cells from basal to horny layers and prominent cornification, develops from hyperplastic epithelium with slight dysplasia and infiltrates downward to the submucosa or deeper layers. It may be difficult to recognize preneoplastic lesions. Poorly differentiated squamous cell carcinomas are often associated with ulceration in areas where small atypical cell nests with minimum intracytoplasmic cornification infiltrate down into the deeper layers. Basal cell carcinomas are rare and show endophytic invasive growth of thin strands of small basaloid cells without cornification. Adenosquamous cell carcinomas arise from near the limiting ridge and show both squamous and glandular differentiation. In general, squamous components are moderately or poorly differentiated, while the glandular components show tubular or papillary patterns with or without mucin. Carcinoma in situ is defined as a lesion where carcinoma cells exist only in lamina propria. It can only be recognized when tumor cells are less mature. Carcinomas induced in rats and mice often infiltrate into submucosa and deeper layers and sometimes invade to adjacent tissues such as the glandular stomach, pancreas, liver, and spleen and/or metastasize to liver, kidney, lung, heart, adrenal glands, brain, abdominal cavity, or regional lymph nodes.

Fibroepitheliomas exhibit polypoid growth and are composed of both surface hyperplastic squamous epithelium and excess connective tissue stroma with prominent fibroblastic cells. Mitotic figures are sometimes encountered. Fibrosarcomas as well as squamous cell carcinomas can arise from these lesions.

In addition to the lesions described above, benign or malignant nonepithelial tumors such as leiomyomas, leiomyosarcomas, fibromas, fibrosarcomas, malignant schwannomas, and hemangiosarcomas are sometimes encountered.

V. MUTATIONS IN DNA OF FORESTOMACH TUMORS

A small number of p53 mutations in exon 8 have been demonstrated in rat forestomach squamous cell carcinomas induced by caffeic acid or MNUR, and Ha-ras point mutations were detected in two of three mouse forestomach tumors induced by 2-amino-3,4-dimethylimidazo[4,5-*f*]quinoline (MeIQ) (46). In another study, one papilloma and one carcinoma out of four mouse forestomach tumors were found to have p53 mutations (a G → A transition at the second position of codon 171 and a G → T transversion at the second position of codon 113 with loss of the wild-type allele, respectively) (47).

VI. SPECIES, STRAIN, AND SEX DEPENDENCE OF FORESTOMACH CARCINOGENESIS

While there are species, strain, and sex differences in forestomach carcinogenesis, these differences depend to some extent on the inducing chemical (Table 2). Rats are more susceptible than mice to caffeic acid (5), while

Table 2 Species and Sex Dependence of Forestomach Carcinogenesis Due to Chemical Carcinogens

			Rat				Mouse			
			Male		Female		Male		Female	
Carcinogens	Dose	Route	PAP	SCC	PAP	SCC	PAP	SCC	PAP	SCC
IQ	0.03%	diet	0/40	0/40	0/40	0/40	11/39	5/39	8/36	3/36
MeIQ	0.04%	diet	1/20	0/20	0/20	0/20	5/38	30/38	3/98	24/38
3-Chloro-2-methylpropene	150 mg/kg bw	ig	30/48	2/48	10/50	0/50	—	—	—	—
	200 mg/kg bw	ig	—	—	—	—	30/49	7/49	29/44	2/44
Diglycidyl resorcinol ethel	50 mg/kg bw	ig	6/50	4/50	1/50	3/50	4/50	14/50	5/50	12/100
Styrene oxide	275 mg/kg bw	ig	23/52	35/52	21/52	32/52	—	—	—	—
	375 mg/kg bw	ig	—	—	—	—	22/51	16/51	14/50	10/50
Ethylene dibromide	40 mg/kg bw	ig		45/50		40/50	—	—	—	—
	60 mg/kg bw	ig	—	—	—	—		45/50		46/49
Ethyl acrylate	200 mg/kg bw	ig		12/50				5/50		
BHA	2%	diet	52/52	18/52	49/51	15/51	—	—	—	—
	1%	diet	71/94	0/94	—	—	5/43	2/43	—	—
Caffeic acid	2%	diet	23/30	17/30	24/30	15/30	4/30	3/30	0/30	1/30
4-Methoxyphenol	2%	diet	15/30	23/30	7/30	6/30	—	—	—	—
Sesamol	2%	diet	10/29	9/29	3/30	0/30	0/29	11/29	0/35	5/30

PAP: Papilloma; SCC: squamous cell carcinoma.

mice are much more sensitive to heterocyclic amines. In mice fed MeIQ, invasive forestomach carcinomas with frequent metastases were found in 79% of males and 63% of females, whereas in rats only one (5% incidence) papilloma was found in a male and none in females (9). A similar tendency was observed in animals fed 2-amino-3-methylimidazo[4,5-*f*]quinoline (IQ) (8). Although the reasons for such marked species variation are presently not known, several factors such as differences in microsome- or cytosol-mediated metabolic activation and covalent binding of carcinogenic chemicals to tissue macromolecules in the forestomach as well as in detoxification enzymes and DNA repair systems could be involved (48).

A clear strain difference exists for rat forestomach carcinogenesis induced by the nongenotoxic carcinogen BHA. When male Sprague-Dawley, SHR, and Lewis strains of rats were maintained on a diet containing 2% BHA for up to 2 years, the incidences of forestomach squamous cell carcinomas were 37, 77, and 7%, respectively (49). The observed difference in the incidences of carcinomas was well correlated with the degrees of cytotoxicity and cell proliferation in the forestomach epithelium and might reflect variation in formation of toxic metabolites due to metabolic activation in the target tissue.

Generally, males are more susceptible than females, with clear differences being found in rats treated with the nongenotoxic agents 4-methoxyphenol and sesamol (6,7). Epithelial damage induced by these two chemicals is morphologically very similar, and free radicals are suspected to be involved in their mechanisms of action (50). Thus, it is speculated that differences between the antioxidant defense systems of males and females may account for some of the sex difference.

VII. EVALUATION OF THE HUMAN HAZARD POTENTIAL OF FORESTOMACH CARCINOGENS

Genotoxic carcinogens, even if they are found to be carcinogenic only in rat or mouse forestomach, cannot be regarded as safe for humans because they might initiate cells in other human tissues. Evidence for this scientific interpretation includes the finding that whereas the heterocyclic amine 2-amino-1-methyl-6-phenylimidazo[4,5-*b*]pyridine (PhIP) is usually only carcinogenic for the colon in male rats, PhIP-associated adducts are also formed in nontarget organs such as heart, lung, and pancreas even at low noncarcinogenic doses (51). Furthermore, a single intragastric administration of 7,12-dimethylbenz(a)anthracene (DMBA) induces tumors in the mammary gland, Zymbal's gland, and leukemia in female rats, but forestomach tumors also develop with subsequent administration of a forestomach promoter (52).

With most nongenotoxic chemicals, the major mode of action appears to be cytotoxicity and primary or secondary strong cell proliferation, and generally it requires high dose levels and a long time (more than one year) for development of carcinomas (25,53,54). Moreover, the fact that the induced hyperplastic or early neoplastic lesions regress after discontinuation of chemical treatment (23–27) indicates that the human hazard of nongenotoxic carcinogens active only in the forestomach may be negligible (1,15,55). Strong cell proliferation generally correlates with carcinogenicity, whereas a slight increase in the rate of cell division tends to be associated with promotion. For example, catechol induces tumors in the glandular stomach but not the tongue, esophagus, and forestomach in F344 male rats (56). However, catechol enhances cell proliferation in the forestomach, esophagus, and tongue and potently promotes carcinogenesis in these organs when administered to rats in the promotion stage (57,58). BHA is a forestomach carcinogen, but enhances cell proliferation and rat urinary bladder carcinogenesis in the promotion stage (59). Therefore, it should be born in mind that forestomach carcinogens could exert effects in other organs as well, and tissues with a high level of exposure should be examined carefully for possible promotional activity. The forestomach is histologically similar to the esophagus, but the two should be regarded as different organs from the view of (a) contact time of foods, (b) acidity, and (c) presence of bacteria and swallowed hairs. These factors might influence both exposure and metabolic activation of chemicals and stimulation of cell proliferation.

VIII. EPIDEMIOLOGICAL DATA FOR STOMACH CANCER

The mortality rate for stomach cancer continues to decline substantially in nearly the entire industrial world. However, the mortality rates (per 100,000) still differ widely between countries. In Japan, it exceeds 500 in males and 200 in females, but in the United States it is only 70 in males and 35 in females. The mortality rate is also high in Italy, about 300 in males and 180 in females (60). In all countries, males appear more affected than females. From a large-scale cohort study in Japan, the standardized mortality rates for stomach cancer in males and females were found to be significantly lowered by frequent intake of green and yellow vegetables, soybean paste soup, milk, and green tea (61). A reduced gastric cancer risk has been observed in women with late menopause and a longer duration of fertility; multiparous women have an elevated risk (62). Consumption of salty foods or nitrite, low socioeconomic status, and cigarette smoking are also reported as risk factors (63–65). There are several nested case-control studies in which preexisting infection with *Helicobacter pylori* was shown to markedly increase the risk for gastric carcinoma (66–68). A linear relationship between the years of documented infection with *H. pylori* and the magnitude of gastric cancer risk has been reported (69).

IX. GLANDULAR STOMACH CARCINOGENS

In contrast to the very large number and variety of chemicals known to cause tumors in the forestomach, only a small number of glandular stomach carcinogens have been identified (7,56,70–79) (Table 3). With the exception of two phenolic compounds, all glandular stomach carcinogens are genotoxic. Experimentally, the most commonly used carcinogens are *N*-methyl-*N'*-nitro-*N*-nitrosoguanidine (MNNG), the related *N*-ethyl-*N'*-nitro-*N*-nitroso-guanidine (ENNG), and *N*-methylnitrosourea (MNU). These carcinogens induce higher yields of glandular stomach tumors in rats with continuous oral administration in the drinking water than with intragastric administration. Repeated intragastric administration of MNNG in fact causes a high incidence of forestomach tumors, but the glandular stomach response is much lower (27,70). Repeated intraperitoneal injections of MNU induce malignant lymphomas and other miscellaneous tumors, but not the glandular stomach lesions (80). One or two intragastric

Table 3 Glandular Stomach Carcinogens

Chemical	Species
N-Methyl-*N'*-nitro-*N*-nitrosoguanidine (MNNG)	Rat, mouse
N-Ethyl-*N'*-nitro-*N*-nitrosoguanidine (ENNG)	Rat
N-Propyl-*N'*-nitro-*N*-nitrosoguanidine (PNNG)	Rat
N-Methylnitrosourea (MNU)	Rat
1-Methyl-3-acetyl-1-nitrosourea	Rat
4-Nitroquinoline 1-oxide (4-NQO)	Rat
4-Hydroxyaminoquinoline 1-oxide (4-HAQO)	Mouse
N-(β-Chloroethyl)-*N*-nitrosourethane	Rat
4-(Hydroxyethyl)benzenediazonium ion	Mouse
2,7-Fluorenylenebisacetamide (2,7-FAA)	Rat, mouse
Catechol	Rat, mouse
4-Methylcatechol	Rat
Gyromitra esculenta	Mouse

administrations of *N*-(β-chloroethyl)-*N*-nitrosourethane, a closely related chemical to MNU, caused glandular stomach tumors in 3 of 11 Porton strain rats (76). Continuous oral administration of the nitrosourea derivative 1-methyl-3-acetyl-1-nitrosourea produces high incidences of glandular stomach as well as nervous system lesions in male and female ACI/N rats and BD rats (75). A single intragastric instillation of 4-(hydroxymethyl)-benzenediazonium ion, an ingredient of *Agaricus bisporus*, caused 30–32% incidences of glandular stomach tumors in male and female Swiss mice (77). *N*-Propyl-*N'*-nitro-*N*-nitrosoguanidine (PNNG) is less effective than MNNG for inducing glandular stomach and forestomach tumors (79). Although their genotoxic properties are equivocal (negative in the Ames assay but positive in some other mutation assays), catechol and 4-methylcatechol have recently been identified as new glandular stomach carcinogens. Catechol and 4-methyl catechol are contained in cigarette smoke, wood smoke, hair dye, onions, and black-and-white photo developers and thus can be called environmental carcinogens. Continuous dietary administration of 0.8% catechol for up to 104 weeks induced adenocarcinomas in about 50% of male and female F344 rats, and 96% and 72% incidences of adenomas but no carcinomas in male and female B6C3F$_1$ mice (56). Similar tumor induction was observed in F344 rats of both sexes treated with 0.8% 4-methylcatechol (7). 4-Methylcatechol also induces forestomach carcinomas.

So far, no gastric chemical carcinogens which might be responsible for the development of human gastric cancer have been recognized. Although carcinogens or mutagens have not been identified in humans infected with *H. pylori*, this bacterium has been shown to produce several toxins and other agents such as vacuolating cytotoxin and several proinflammatory cytokines (81–83). Infection with *H. pylori*, especially with strains possessing the cagA gene, is associated with strong acute inflammation (84,85), cell proliferation (86,87), and intestinal metaplasia (88) as well as atrophic gastritis in human gastric mucosa (88,89). During inflammation, genotoxic reactive oxygen species or reactive nitrogen species can be produced by inflammatory cells (89–91). Genotoxic compounds could also be formed in the human gastric contents with atrophic gastritis, possibly by the reaction of sodium nitrite and secondary amines in foods (92,93). In addition, *H. pylori* has potent urease activity and produces ammonia (94), which has experimentally been shown to induce cell proliferation of glandular stomach epithelium and promote MNNG-induced glandular stomach carcinogenesis in rats (95).

An Epstein-Barr virus (EBV) association has been demonstrated in about 7% of human gastric cancer cases in Japan (96) and 16% of cases in North American (97). The presence of the EBV genome was first demonstrated in poorly differentiated gastric cancers with marked stromal lymphocytic infiltration, histologically similar to that seen in nasopharyngeal lymphoepitheliomas. Subsequently, EBV involvement was also suggested for cases of well-differentiated gastric adenocarcinomas, as evidenced by presence of the EBV genome, EBV-encoded small RNA (EBER) in the carcinoma cells, monoclonal proliferation of EBV-infected cells, and elevated antibody titers. EBV-related gastric carcinomas are characterized by a male predominance, preferential occurrence of certain histological types (featuring infiltration of lymphocytes and a lace-pattern appearance of cancer cells), preferential location in the upper and middle parts of the stomach (96,98), and a high prevalence in gastric remnants (99). EBV-infected cells can be demonstrated in normal (100) and atrophic or metaplastic epithelium adjacent to gastric cancers as well as in early gastric cancers (98,100). Therefore, EBV is now believed to be a causative agent for some proportion of human gastric cancers, with transmission through infected lymphocytes. However, whether EBV acts as an initiating or promoting agent and its relation to *H. pylori* are not well understood.

X. GASTROJEJUNOSTOMY AND GASTRIC CANCER

Glandular stomach tumors have been found in rats subjected to gastroenterostomy alone without carcinogen treatment. When male Wistar rats underwent gastrojejunostomy to divert the duodenal contents into the resected stomach through afferent and efferent loops, carcinomas developed more frequently in the gastric mucosa around the anastomosis of the afferent loop (101). Pancreaticoduodenal reflux and bile reflux are both associated with induction of gastric neoplasia (102,103), and carcinomas develop near the pylorus in rats with reflux through the pylorus, whereas the adjacent fundic mucosa is involved in those with reflux through the stoma. Carcinomas do not develop in animals with anastomosis alone without bile reflux, and the induced tumors are well-differentiated and mucinous types. Polyploid or downward proliferative lesions similar to human gastritis cystica polyposa are frequently found (104). Therefore, it has been suggested that initial mucosal damage due to anastomosis and continuous cell proliferation caused by the duodenal contents may be factors responsible for the tumor development observed after gastrojejunostomy (101).

In humans, gastritis cystica polyposa develops in gastric mucosa adjacent to gastrojejunal anastomoses made by the Billroth II procedure (105). A number of gastric stump carcinomas have been reported in patients undergoing this type of stomach resection with carcinomas preferentially found close to the gastroenteric anastomosis (106,107). Therefore it is possible that duodenal secretion might play an important role in neoplastic development in gastric remnants in humans.

XI. ENHANCING FACTORS FOR GLANDULAR STOMACH CARCINOGENESIS

Taurocholic acid, sodium taurocholate, and sodium chloride weakly promote rat glandular stomach carcinogenesis when administered after MNNG initiation (108–110). Coadministration of sodium chloride or surfactants also enhances MNNG-induced rat glandular stomach carcinogenesis (110,111). Sodium chloride is known to induce mucosal damage and increase the permeability of the mucosa to carcinogens (112). In humans, excess consumption of sodium chloride has been implicated as a risk factor for gastric cancer (63–65,113). Long-term administration of H_2-receptor antagonists or inhibitors of gastric H^+,K^+-ATPase, which are commonly used for the management of peptic ulcer disease, results in enterochromaffinlike cell hyperplasia and/or gastric carcinoids via inhibition of acid secretion in rats (114,115). These clinically used drugs are therefore suspected as having promoting potential. However, it was found that administration of cimetidine after MNNG initiation did not enhance, but rather significantly reduced the development of gastric cancers in rats (116). The presence of gastric ulceration is an enhancing factor for rat gastric carcinogenesis. Thus, when animals were treated with MNNG after induction of gastric ulcers in the fundic region by freezing or by administration of iodoacetamide, tumors were induced within metaplastic mucosa (pyloric metaplasia) at the ulcer sites (117,118). The results indicate that regenerating mucosa is more sensitive than normal mucosa and that metaplastic pylori mucosa is more sensitive than normal fundic mucosa. The fact that carcinogens penetrate more easily into pyloric than into fundic mucosa might account for the observed differences between metaplastic pyloric and normal fundic glands (119).

XII. SPECIES, STRAIN, AND SEX DEPENDENCE OF GLANDULAR STOMACH CARCINOGENESIS

Generally male rats are more susceptible than females to glandular stomach carcinogens, including MNNG (120). The incidence of glandular stomach adenocarcinomas induced by oral application of this carcinogen was found to be clearly decreased after castration in males, but not in females (121). In castrated males the incidence increased with progesterone treatment (122).

Estrogen, testosterone, and progesterone receptors have been identified in human gastric tumors (123), which often show progression with pregnancy or delivery (124). Therefore, sex differences in gastric neoplasia in rats and humans can partly be explained by the hormonal status.

Since mice generally demonstrate a low sensitivity to glandular stomach carcinogens, most experimental studies have been carried out using rats. Pronounced variation exists among rat strains, with some dependence on the carcinogens used. Analbuminemic rats (NAR, established from SD stock) are highly susceptible, ACI, SD, WKY, BD, IX, and BN strains are susceptible, and Wistar, F344, and Wistar/Furth are less susceptible, while the Buffalo strain is resistant to the induction of glandular stomach tumors by MNNG (120,125–128). On the other hand, the SD, Lewis, Wistar, and F344 strains are more susceptible than the WKY strain to induction of glandular stomach tumors by catechol (129). The degree of cytotoxicity does not necessarily parallel the incidence of gastric carcinoma induced by MNNG or catechol (120,129). Strain differences appear to be genetically controlled (120) and can be partly accounted for by variation in the metabolic activation system.

XIII. HISTOGENESIS OF GLANDULAR STOMACH TUMORS—RAT VERSUS HUMAN

Glandular stomach tumors in rats develop mostly in the pyloric region or in areas of metaplastic epithelium within the fundic region. The sequential changes observed in rat glandular stomach with carcinogen treatment are shown in Figure 2. The first lesion that becomes apparent on continuous oral treatment with MNNG is diffuse superficial erosion of the pyloric mucosa with inflammatory cell infiltration, followed by regeneration. There may be at least three different outcomes after regeneration. One is focal irregular arrangement of glands with stromal fibrosis, followed by the appearance of atypical glands (atypical epithelium), which may eventually infiltrate down to the submucosal layer or deeper layers (adenocarcinomas, well- or moderately differentiated type). The second is downward growth of a few mature surface epithelia and/or pyloric glands (submucosal hyperplasia), which sometimes form submucosal cysts. Such glands gradually grow larger to form adenomas (or adenomatous hyperplasia). While this lesion sometimes infiltrates into the muscular layer or even into subserosa, there are no structural and cytological atypia and thus it should not be diagnosed as a carcinoma. In some cases, however, atypia do occur and well-differentiated adenocarcinomas develop. The third possible outcome is focal upward growth of surface epithelium and/or pyloric glands (upward

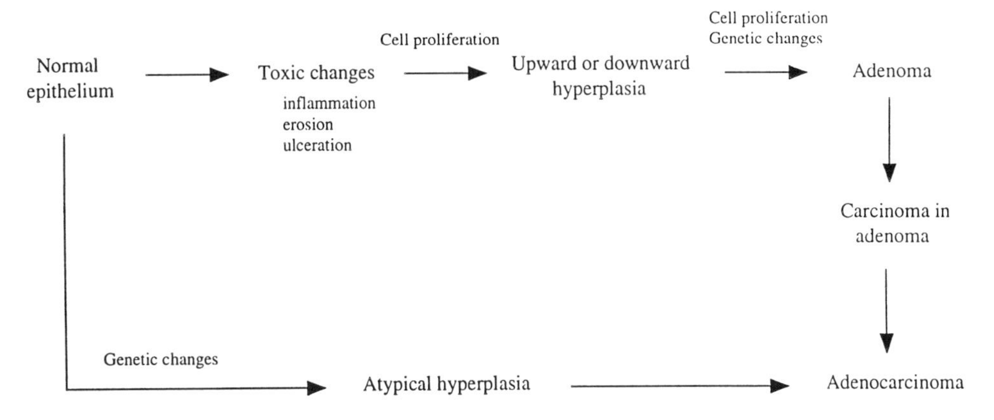

Figure 2 Sequential changes of well- or moderately differentiated rat glandular stomach carcinogenesis.

hyperplasia) resulting in polypoid lesions (adenoma or adenomatous hyperplasia), which develop into well-differentiated adenocarcinomas with increases in structural and cytological atypia (45,130–133). The histogenesis of signet ring cell carcinomas in rats is not well understood. Similar sequential changes are observed in catechol-induced glandular stomach carcinogenesis. However, catechol is exceedingly toxic and causes necrosis, erosion, and severe ulceration with infiltration of inflammatory cell such as neutrophils, eosinophils, and histiocytes in the pyloric region as well as superficial erosion in the fundus at carcinogenic doses. We failed to demonstrate any catechol-induced increase in lipid peroxidation or 8-OH-dG levels in the affected gastric epithelia (134). Submucosal hyperplasia and adenomas induced by dietary administration of catechol for 24 weeks did not regress after cessation of insult, and these lesions appear to have the potential to develop carcinomas without further stimulus (135). Intestinal metaplasia is not commonly observed during MNNG- or catechol-induced rat glandular stomach carcinogenesis, and it is not necessarily preneoplastic in nature (136). Most carcinomas induced in rats are well-differentiated adenocarcinomas with abundant connective tissue; other types are rare. Human gastric atypical epithelium or gastric adenomas are histopathologically quite different from those induced in rats.

The histopathological characteristics of rat submucosal hyperplasia and downward growing adenomas, as well as upward hyperplasia and protruding adenomas, are quite similar to mucosal or submucosal cysts and gastritis cystica polyposa, respectively, in humans (105,137). Such lesions sometimes develop in the stomal areas of remnant stomachs following gastric resection by the Billroth II method, along with carcinomas in some cases. It has been postulated that erosion might give rise to heterotopic cysts, whose development makes the mucosa more prone to chronic erosion, leading to carcinoma development (138). Therefore the experimental glandular stomach carcinogenesis model may be a good model of human gastric carcinogenesis via mucosal or submucosal cysts and gastritis cystica polyposa.

In humans, gastric carcinomas are sometimes classified as intestinal and diffuse categories (139). The former type is thought to be preceded by a sequential chain of events involving chronic gastritis, atrophy, intestinal metaplasia, dysplasia, intramucosal carcinomas, and invasive neoplasia, although well-recognized precursor lesions for the latter are lacking. Infection with *H. pylori* and environmental factors may play roles in the development of intestinal-type cancers (140). It has been postulated that DNA damage due to food-derived carcinogens or mediated by oxidants or nitric oxide occurs in gastric mucosa infected with *H. pylori*. Damaged cells are lost by apoptosis and atrophic gastritis follows. Some cells with DNA damage become transformed, and these develop into dysplasias and then carcinomas, especially of intestinal type (141,142).

XIV. GENE ALTERATIONS IN GASTRIC CARCINOGENESIS

Genetic events during human gastric carcinogenesis have been well summarized by several researchers (140,143). Differences between intestinal and diffuse-type gastric carcinomas have been established. The earliest gene alteration in human gastric mucosa preceding development of intestinal type carcinoma concerns the tpr-*met* rearrangement. Abnormalities in this tpr-*met* rearrangement involving fusion of the translocated promoter region on chromosome 1 to the 5′ region of *met* gene on chromosome 7 have been found in superficial gastritis, mainly due to *H. pylori*, and in gastric carcinomas (144). The *ras* gene product p21 has been found to be overexpressed in nearly all carcinomas of intestinal and diffuse type and in dysplastic and/or metaplastic mucosal regions associated with intestinal-type gastric cancers (145). Point mutations of the c-Ki-*ras* oncogene have been found in a few adenomas and differentiated adenocarcinomas (146). There are also many reports of p53 gene abnormalities in gastric cancer and precancerous lesions. Generally, such alteration first appear at the stage of high-grade dysplasia, before development of early intestinal-type carcinomas, and they are suspected to be linked to nitric oxide (140) or dietary carcinogens (147). Expression of the bcl-2 protooncogene, which encodes a 26 kDa protein effective at inhibiting programmed cell death, has been shown in dysplastic epithelium and intestinal-type gastric carcinomas (148). Allele losses on chromosomes 1q and 7p may also be involved in progression of intestinal-type carcinomas (149). The histogenesis of human intestinal-type gastric carcinoma and possible related genetic changes are shown in Figure 3. In diffuse-type carcinomas, p53 abnormalities are less frequent and appear linked to a more advanced stage (150). In addition to reduction or loss of cadherin and catenins (151), amplification of the k-*sam* gene, which is identical with a gene coding for the receptor tyrosine kinase bek/FGFR2, was demonstrated to be specific to diffuse-type carcinomas (152,153). Amplification of the c-*met* gene which encodes the hepatocyte growth factor receptor has been detected in advanced carcinomas, particularly in scirrhous type, suggesting that amplification of this gene might participate in progression of gastric cancer, especially scirrhous-type carcinomas (154). Enhanced expression of c-*erb* B2 occurs more frequently in carcinomas demonstrating lymph node metastases (155) and in those with microsatellite instability, irrespective of histological type (156), suggesting that it may be responsible for some aspect of tumor progression in both types of tumors. The available data suggest that alternative genetic pathways may exist for intestinal and diffuse-type gastric carcinomas.

XV. CONCLUSIONS

The rodent forestomach, located between the esophagus and glandular stomach, has the function of storing food prior to digestion in the glandular stomach. A number of both genotoxic and nongenotoxic carcinogens target

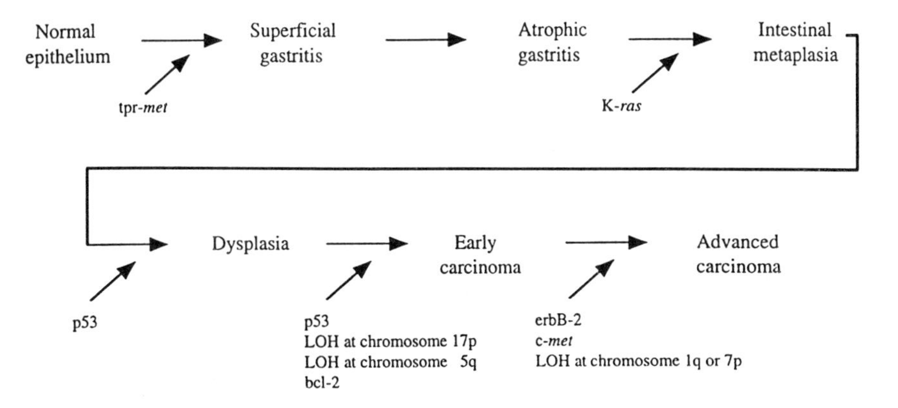

Figure 3 Histogenesis of human intestinal-type gastric cancer and related genetic changes.

the forestomach epithelium. Genetic alteration, cell proliferation, and cytotoxicity all play a role in tumor induction. Cell proliferation appears to be the most important causative factor for nongenotoxic carcinogens. Whether it is directly or indirectly caused, cell proliferation may be a useful marker for the prediction of forestomach promoting or carcinogenicity activity. Since humans have no forestomach, carcinogens that target this organ only may be irrelevant to human health. However, rodent forestomach carcinogens may promote human carcinogenesis in other, different anatomical sites.

The glandular stomach is equivalent to the human stomach functionally and morphologically. Despite the high mortality rate from gastric cancer in some Asian countries, only a few environmental glandular stomach carcinogens have been recognized. Rodent experimentation has identified some promoters such as bile acids and cocarcinogens such as sodium chloride. Recently, *H. pylori* has been shown to be largely responsible for the development of human gastric carcinomas. Inflammation, cell proliferation, intestinal metaplasia, atrophic gastritis, and genotoxicity are all associated with *H. pylori*. The *H. pylori* link appears particularly strong for intestinal type carcinomas. Although some EBV-positive gastric carcinomas have been described, the contribution of EBV to stomach carcinogenesis is not well understood. Since they demonstrate different types of genetic alterations, different causative mechanisms may exist for intestinal and diffuse-type carcinomas. The future elimination of *H. pylori* (and possibly EBV) could considerably reduce the incidence of human gastric carcinomas.

REFERENCES

1. Kroes R, Wester W. Forestomach carcinogens: possible mechanisms of action. Food Chem Toxicol 1986; 24:1083–1089.
2. Hirose M, Maekawa A, Kamiya S, Odashima S. Carcinogenic effect of N-ethyl- and N-amyl-N-nitrosourethanes on female Donryu rats. Gann 1979; 70:653–662.
3. Krishna Murthy AS, McConnell EE, Huff JE, Russfield AB, Good AE. Forestomach neoplasms in Fischer F344/N rats and B6C3F$_1$ mice exposed to diglycidyl resorcinol ether—an epoxy resin. Food Chem Toxicol 1990; 28: 723–729.
4. Salmon RJ, Buisson JP, Zafrani B, Aussepe L, Royer R. Carcinogenic effect of 7-methoxy-2-nitro-naphtho[2,1-b]furan (R7000) in the forestomach of rats. Carcinogenesis 1986; 7:1447–1450.
5. Hagiwara A, Hirose M, Takahashi S, Ogawa K, Shirai T, Ito N. Forestomach and kidney carcinogenicity of caffeic acid in F344 and C57BL/6N × C3H/HeN F$_1$ mice. Cancer Res 1991; 51:5655–5660.

6. Tamano S, Hirose M, Tanaka H, Asakawa E, Ogawa K, Ito N. Forestomach neoplasm induction in F344/DuCrj rats and $B6C3F_1$ mice exposed to sesamol. Jpn J Cancer Res 1992; 83:1279–1285.
7. Asakawa E, Hirose M, Hagiwara A, Takahashi S, Ito N. Carcinogenicity of 4-methoxyphenol and 4-methylcatechol in F344 rats. Int J Cancer 1994; 56: 146–152.
8. Ohgaki H, Kusama K, Matsukura N, Morino K, Hasegawa H, Sato S, Takayama S, Sugimura T. Carcinogenicity in mice of a mutagenic compound, 2-amino-3-methylimidazo[4,5-f]quinoline, from broiled sardine, cooked beef and beef extract. Carcinogenesis 1984; 5:921–924.
9. Ohgaki H, Hasegawa H. Suenaga M, Kato T, Sato S, Takayama S, Sugimura T. Induction of hepatocellular carcinomas and highly metastatic squamous cell carcinomas in the forestomach of mice by feeding 2-amino-3,4-dimethylimidazo[4,5-f]quinoline. Carcinogenesis 1986; 7:1889–1893.
10. Kawabe M, Takaba K, Yoshida Y, Hirose M. Effects of combined treatment with phenolic compounds and sodium nitrite on two-stage carcinogenesis and cell proliferation in the rat stomach. Jpn J Cancer Res 1994; 85:17–25.
11. Yoshida Y, Hirose M, Takaba K, Kimura J, Ito N. Induction and promotion of forestomach tumors by sodium nitrite in combination with asorbic acid or sodium ascorbate in rats with or without N-methyl-N′-nitro-N-nitrosoguanidine pre-treatment. Int J Cancer 1994; 56:124–128.
12. Laskin S, Sellahumar AR, Kushner M, Nelson N, La Medola S, Rush GM, Gatz GV, Dulak NC, Albert RE. Inhalation carcinogenicity of epichlorohydrin in noninbred Sprague-Dawley rats. J Natl Cancer Inst 1980; 65:751–757.
13. Melnick RL, Huff JE, Roycroft JH, Chou BJ, Miller RA. Inhalation toxicology and carcinogenicity of 1,3-butadiene in $B6C3F_1$ mice following 65 weeks of exposure. Environ Health Perspect 1990; 86:27–36.
14. Mirvish SS, Salmasi S, Cohen SM, Patil K, Mahboubi E. Liver and forestomach tumors and other forestomach lesions in rats treated with morpholine and sodium nitrite, with and without sodium ascorbate. J Natl Cancer Inst 1983; 71:81–85.
15. Wester PW, Kroes R. Forestomach carcinogens: pathology and relevance to man. Toxicol Pathol 1988; 16:165–171.
16. Ito N, Hirose M, Takahashi S. Cell proliferation and forestomach carcinogenesis. Environ Health Perspect 1993; 101(suppl 5):107–110.
17. Hirose M, Takahashi S, Shirai T. Characteristics of forestomach carcinogenesis by non-genotoxic phenolic compounds. J Toxicol Pathol 1995; 8:277–284.
18. Thompson D, Moldeus P. Cytotoxicity of butylated hydroxyanisole and butylated hydroxytoluene in isolated rat hepatocytes. Biochem Pharmacol 1988; 37:2201–2207.
19. Ito N, Hirose M, Shibata MA, Tanaka H, Shirai T. Modifying effects of simultaneous treatment with butylated hydroxyanisole (BHA) on rat tumor induction by 3,2′-dimethyl-4-aminobiphenyl, 2,2′-dihydroxy-di-n-propylnitrosamine and N-methylnitrosourea. Carcinogenesis 1989; 10:2255–2259.
20. Schilderman PAEJ, van Maanen JMS, Smeets EJ, ten Hoor F, Kleinjans JCS. Oxygen radical formation during prostaglandin H synthase-mediated

biotransformation of butylated hydroxyanisole. Carcinogenesis 1993; 14: 347–353.

21. Schilderman PAEL, ten Vaarwerk FJ, Lutgerink JT, van der Wurff A, ten Hoor F, Kleinjans JCS. Induction of oxidative DNA damage and early lesions in rat gastro-intestinal epithelium in relation to prostaglandin H synthase-mediated metabolism of butylated hydroxyanisole. Food Chem Toxicol 1995; 33:99–109.
22. Morimoto K, Tsuji K, Iio T, Miyata N, Uchida A, Osawa R, Kitsutaka H, Takahashi A. DNA damage in forestomach epithelium from male F344 rats following oral administration of tert-butylquinone, one of the forestomach metabolites of 3-BHA. Carcinogenesis 1991; 12:703–708.
23. Nera EA, Iverson F, Lok E, Armstrong CL, Karpinski K, Clayson DB. A carcinogenesis reversibility study of the effects of butylated hydroxyanisole on the forestomach and urinary bladder in male Fischer 344 rats. Toxicology 1988; 53:251–268.
24. Masui T, Asamoto M, Hirose M, Fukushima S, Ito N. Disappearance of upward proliferation and persistence of downward basal cell proliferation in rat forestomach papillomas induced by butylated hydroxyanisole. Jpn J Cancer Res 1986; 77:854–857.
25. Ghanayem BI, Matthews HB, Maronpot RR. Sustainability of forestomach hyperplasia in rats treated with ethyl acrylate for 13 weeks and regression after cessation of dosing. Toxicol Pathol 1991; 19:273–279.
26. Hirose M, Masuda A, Hasegawa R, Wada S, Ito N. Regression of butylated hydroxyanisole (BHA)-induced hyperplasia but not dysplasia in the forestomach of hamsters. Carcinogenesis 1990; 11:239–244.
27. Kagawa M, Hakoi K, Yamamoto A, Futakuchi M, Hirose M. Comparison of reversibility of rat forestomach lesions induced by genotoxic and nongenotoxic carcinogens. Jpn J Cancer Res 1993; 84:1120–1129.
28. Ghanayem BL, Sanchez IM, Matthews HB, Elwell MR. Demonstration of a temporal relationship between ethyl acrylate-induced forestomach cell proliferation and carcinogenicity. Toxicol Pathol 1994; 22:497–509.
29. Boorman GA, Hong HL, Jameson CW, Yoshitomi K, Maronpot RR. Regression of methyl bromide-induced forestomach lesions in the rat. Toxicol Appl Pharmacol 1986; 86:131–139.
30. Danse LHJC, van Velsen FL, van der Heijden CA. Methylbromide: carcinogenic effects in the rat forestomach. Toxicol Appl Pharmacol 1984; 72:262–271.
31. Ghanayem BI, Maronpot RR, Matthews HB. Association of chemically induced forestomach cell proliferation and carcinogenesis. Cancer Lett 1986; 32:271–278.
32. Mengs U. On the histopathogenesis of rat forestomach carcinoma caused by aristolochic acid. Arch Toxicol 1983; 52:209–220.
33. Lijinsky W. Rat and mouse forestomach tumors induced by chronic oral administration of styrene oxide. J Natl Cancer Inst 1986; 77:471–476.
34. Ghanayem BI, Maronpot RR, Matthews HB. Ethyl acrylate-induced gastric

toxicity. III. Development and recovery of lesions. Toxicol Appl Pharmacol 1986; 83:576–583.
35. Altmann HJ, Grunow W, Mohr U, Wester PW. Effects of BHA and related phenols on the forestomach of rats. Food Chem Toxicol 1986; 24:1183–1188.
36. Nera EA, Lok E, Iverson F, Ormsby E, Karpinski KF, Clayson DB. Short-term pathological and proliferative effects of butylated hydroxyanisole and other phenolic antioxidants in the forestomach of Fischer 344 rats. Toxicology 1984; 32:197–213.
37. Altmann HJ, Wester PW, Matthiaschk G, Grunow W, van der Heijden CA. Induction of early lesions in the forestomach of rats by 3-tert-butyl-4-hydroxyanisole (BHA). Food Chem Toxicol 1985; 8:723–731.
38. Rodrigues C, Lok E, Nera E, Iverson F, Page D, Karpinski K, Clayson DB. Short-term effects of various phenols and acids on the Fischer 344 male rat forestomach epithelium. Toxicology 1986; 38:103–117.
39. Hirose M, Masuda A, Imaida K, Kagawa M, Tsuda H, Ito N. Induction of forestomach lesions in rats by oral administrations of naturally occurring antioxidants for 4 weeks. Jpn J Cancer Res (Gann) 1987; 78:317–321.
40. Harrison PTC, Grasso P, Badescu V. Early changes in the forestomach of rats, mice and hamsters exposed to dietary propionic and butyric acid. Food Chem Toxicol 1991; 29:367–371.
41. Ito N, Fukushima S, Tamano S, Hirose M, Hagiwara A. Dose response in butylated hydroxyanisole induction of forestomach carcinogenesis in F344 rats. J Natl Cancer Inst 1986; 77:1261–1265.
42. Williams GM. Epigenetic promoting effects of butylated hydroxyanisole. Food Chem Toxicol 1986; 24:1163–1166.
43. Fukushima S, Ito N. Squamous cell carcinoma, forestomach, rat. In: Jones TC, Mohr U, Hunt RD, eds. Digestive System (Monographs on Pathology of Laboratory Animals). New York: Springer-Verlag, 1985:292–295.
44. Fukushima S, Ito N. Papilloma, forestomach, rat. In: Jones TC, Mohr U, Hunt RD, eds. Digestive System (Monographs on Pathology of Laboratory Animals). New York: Springer-Verlag, 1985:925–930.
45. Takahashi M, Hasegawa R. Tumors of the stomach. In: Turusov VS, Mohr U, eds. Pathology of Tumours in Laboratory Animals. Vol. I. Tumours of the Rat. Lyon: International Agency for Research on Cancer, 1990:129–157.
46. Makino H, Ochiai M, Caignard A, Ishizawa Y, Onda M, Sugimura T, Nagao M. Detection of a Ha-ras point mutation by polymerase chain reaction-single strand conformation polymorphism analysis in 2-amino-3, 4-dimethylimidazo[4,5-f]quinoline-induced mouse forestomach tumors. Cancer Lett 1992; 62:115–121.
47. Ushijima T, Makino H, Okonogi H, Hosoya Y, Sugimura T, Nagao M. Mutation, loss of heterozygosity, and recombination of the p53 gene in mouse forestomach tumors induced by 2-amino-3, 4-dimethylimidazo[4,5-f]quinoline. Mol Carcinog 1995; 12:23–30.
48. Övervik E, Hellmold H, Branting C, Gustafsson JÅ. Activation and effects of the food-derived heterocyclic amines in extrahepatic tissues. In: Adamson

RH, Gustafsson JÅ, Ito N, Nagao M, Sugimura T, Wakabayashi K, Yamazoe Y, eds. Heterocyclic Amines in Cooked Food: Possible Human Carcinogens. Princeton, NJ: Princeton Scientific Publishing Co., Inc., 1995: 123–133.

49. Tamano S, Hagiwara A, Shibata M, Tanaka H, Asakawa E, Fukushima S, Ito N. High sensitivity of SHR strain rats to the induction of rat forestomach tumors by BHA (abstr). Proc Jpn Cancer Assoc 1989; 48:38.
50. Futakuchi M, Mizoguchi Y, Hoshiya T, Hirose M, Hasagawa R, Sano M, Shirai T. Effects of aspirin, indomethacin, superoxide dismutase, and catalase on rat forestomach lesion development induced by antioxidants (abstr). Proc Jpn Cancer Assoc 1995; 54:93.
51. Kaderlik KR, Minchin RF, Mulder GJ, Ilett KF, Daugaard-Jenson M, Teitel CH, Kadlubar FF. Metabolic activation pathway for the formation of DNA adducts of the carcinogen 2-amino-1-methyl-6-phenylimidazo[4,5-b]pyridine (PhIP) in rat extrahepatic tissues. Carcinogenesis 1994; 15:1703–1709.
52. Hirose M, Masuda A, Fukushima S, Ito N. Effects of subsequent antioxidant treatment on 7,12-dimethylbenz[a]anthracene-initiated carcinogenesis of the mammary gland, ear duct and forestomach in Sprague-Dawley rats. Carcinogenesis 1988; 9:101–104.
53. Masui T, Asamoto M, Hirose M, Fukushima S, Ito N. Regression of simple hyperplasia and papillomas and persistence of basal cell hyperplasia in the forestomach of F344 rats treated with butylated hydroxyanisole. Cancer Res 1987; 47:5171–5174.
54. Masui T, Hirose M, Imaida K, Fukushima S, Tamano S, Ito N. Sequential changes of the forestomach of F344 rats, Syrian golden hamsters, and $B6C3F_1$ mice treated with butylated hydroxyanisole. Jpn J Cancer Res 1986; 77:1083–1090.
55. Clayson DB, Iverson F, Nera EA, Lok E. The significance of induced forestomach tumors. Annu Rev Pharmacol Toxicol 1990; 30:441–463.
56. Hirose M, Fukushima S, Tannaka H, Asakawa E, Takahashi S, Ito N. Carcinogenicity of catechol in F344 rats and $B6C3F_1$ mice. Carcinogenesis 1993; 14:525–529.
57. Hirose M, Fukushima S, Kurata Y, Tsuda H, Tatematsu M, Ito N. Modification of N-methyl-N′-nitro-N-nitrosoguanidine-induced forestomach and glandular stomach carcinogenesis by phenolic antioxidants in rats. Cancer Res 1988; 48:5310–5315.
58. Yamaguchi S, Hirose M, Fukushima S, Hasegawa R, Ito N. Modification by catechol and resorcinol of upper digestive tract carcinogenesis in rats treated with methyl-N-amylnitrosamine. Cancer Res 1989; 49:6015–6018.
59. Imaida K, Fukushima S, Shirai T, Ohtani M, Nakanishi K, Ito N. Promoting activities butylated hydroxyanisole and butylated hydroxytoluene on 2-stage urinary bladder carcinogenesis and inhibition of γ-glutamyl transpeptidase-positive foci development in the liver of rats. Carcinogenesis 1983; 4:895–899.
60. Davis DL, Hoel D, Fox J, Lopez AD. International trends in cancer mortality

in France, West Germany, Italy, Japan, England and Wales, and the United States. In: Davis DL, Hoel D, eds. Trends in Cancer Mortality in Industrial Countries. New York: The New York Academy of Sciences, 1990:5–48.
61. Hirayama T. A large scale cohort study on cancer risks by diet—with special reference to the risk reducing effects of green-yellow vegetable consumption. In: Hayashi Y, Nagao M, Sugimura T, Takayama S, Tomatis L, Wattenberg LW, Wogan G, eds. Diet, Nutrition and Cancer. Utrecht: Jpn Sci Soc Press, Tokyo/VNU Sci Press, 1986:41–53.
62. LaVecchina C, D'Avanzo B, Franceschi S, Negri E, Parazzini F, Decarli A. Menstrual and reproductive factors and gastric-cancer risk in women. Int J Cancer 1994; 59:761–764.
63. Hirayama T. Epidemiology of stomach cancer. Gann Monogr Cancer Res 1971; 11:3–19.
64. Cipriani F, Buiatti E, Palli D. Gastric cancer in Italy. Ital J Gastroenterol 1991; 23:429–435.
65. Lee JK, Park BJ, Yoo KY, Ahn YO. Dietary factors and stomach cancer: a case-control study in Korea. Int J Epidemiol 1995; 24:33–41.
66. Forman D, Newell DG, Fullerton F, Yarnell JW, Stacey AR, Wald N, Sitas F. Association between infection with Helicobacter pylori and risk of gastric carcinoma. Br Med J 1991; 302:1302–1305.
67. Parsonnet J, Friedman GD, Vandersteen DP, Chang Y, Vogelman JH, Orentreich N, Sibley RK. Helicobacter pylori infection and the risk of gastric carcinoma. N Engl J Med 1991; 325:1170–1171.
68. Lin JT, Wang LY, Wang JT, Wang TH, Yang CS, Chen CJ. A nested case-control study on the association between Helicobacter pylori infection and gastric cancer risk in a cohort of 9775 men in Taiwan. Anticancer Res 1995; 15:603–606.
69. Forman D, Webb P, Personnet J. Helicobacter pylori and gastric cancer. Lancet 1994; 343:243–244.
70. Sugimura T, Kawachi T. Experimental stomach cancer. In: Busch H, ed. Methods in Cancer Research. New York: Academic Press, 1973:245–308.
71. Tatematsu M, Katsuyama T, Fukushima S, Takahashi M, Shirai T, Ito N, Nasu T. Mucin histochemistry by paradoxical concanavalin A staining in experimental gastric cancers induced in Wistar rats by N-methyl-N′-nitro-N-nitrosoguanidine or 4-nitroquinoline 1-oxide. J Natl Cancer Inst 1980; 64: 835–843.
72. Snell KC, Stewart HL, Morris HP. The sequential development of atrophy, precancerous lesions and cancer of the glandular stomach and other organs and tissues of rats ingesting N, N′-2,7-fluorenylene-bisacetamide. In: Morris HP, Yoshida T, eds. Experimental Carcinoma of the Glandular Stomach (Gann Monograph No. 8). Tokyo: Maruzen, 1969:157–196.
73. Mori K, Ota A, Murakami T, Tamura M, Kondo M, Ichimura H. Carcinomas of the glandular stomach and other organs of rats induced by 4-hydroxyaminoquinoline-1-oxide hydrochloride. Gann 1969; 60:627–630.
74. Hirota H, Aonuma T, Yamada S, Kawai T, Saito K, Yokoyama T. Selective

induction of glandular stomach carcinoma in F344 rats by N-methyl-N-nitrosourea. Jpn J Cancer Res (Gann) 1987; 78:634–638.
75. Maekawa A, Odashima S, Nakadate M. Induction of tumors in the stomach and nervous system of the ACI/N rat by continuous oral administration of 1-methyl-3-acetyl-1-nitrosourea. Z Krebsforsch 1976; 86:195–207.
76. Schoental R, Bensted JPM. Gastric tumors and lung lesions in the rat following the intragastric or intraperitoneal administration of N-(β-chloroethyl)-N-nitrosourethan. Cancer Res 1971; 31:573–576.
77. Toth B, Nagel D, Ross A. Gastric tumorigenesis by a single dose of 4-(hydroxymethyl)-benzenediazonium ion of Agaricus bisporus. Br J Cancer 1982; 46:417–422.
78. Toth B, Patil K, Pyysalo H, Stessman C, Gannett P. Cancer induction in mice by feeding the raw false morel mushroom Gyromitra esculenta. Cancer Res 1992; 52:2279–2284.
79. Wang CX, Williams GM. Comparison of stomach cancer induced in rats by N-methyl-N′-nitro-N-nitrosoguanidine or N-propyl-N′-nitro-N-nitrosoguanidine. Cancer Lett 1987; 34:173–185.
80. Tsuda H, Sakata T, Shirai T, Kurata Y, Tamano S, Ito N. Modification of N-methyl-N-nitrosourea initiated carcinogenesis in the rat by subsequent treatment with antioxidants, phenobarbital and ethinyl estradiol. Cancer Lett 1984; 24:19–27.
81. Kuipers EJ, Pérez-Pérez GI, Meuwissen SGM, Blaser MJ. Helicobacter pylori and atrophic gastritis: importance of the cagA status. J Natl Cancer Inst 1995; 87:1731–1732.
82. Tummuru MK, Cover TL, Blaser MJ. Cloning and expression of a high-molecular-mass major antigen of Helicobacter pylori: evidence of linkage to cytotoxin production. Infect Immun 1993; 61:1799–1809.
83. Covacci A, Censini S, Bugnoli M, Petracca R, Burroni D, Macchia G, Massone A, Papini E, Xiang Z, Figura N, Rappuoli R. Molecular characterization at the 128-kDa immunodominant antigen of Helicobacter pylori associated with cytotoxicity and duodenal ulcer. Proc Natl Acad Sci USA 1993; 90:5791–5795.
84. Crabtree JE, Taylor JD, Wyatt JI, Heatley RV, Shallcross TM, Tompkins OS, Rathbone BJ. Mucosal IgA recognition of Helicobacter pylori 120 kDa protein, peptic ulceration, and gastric pathology. Lancet 1991; 338:332–335.
85. Oderda G, Figura N, Bayeli PF, Basagni C, Bugnoli M, Armellini D. Serologic IgG recognition of Helicobacter pylori cytotoxin-associated protein, peptic ulcer and gastroduodenal pathology in childhood. Eur J Gastroenterol Hepatol 1993; 5:695–699.
86. Lynch DA, Mapstone NP, Clarke AM, Sobala GM, Jackson P, Morrison L, Dixon MF, Quirke P, Axon AT. Cell proliferation in Helicobacter pylori associated gastritis and the effect of eradication theory. Gut 1995; 36:346–350.
87. DeKoster E, Buset M, Fernandes E, Deltenre M. Helicobacter pylori: the link with gastric cancer. Eur J Cancer Prev 1994; 3:247–257.
88. Kuipers EJ, Uyterlinde AM, Pena AS, Roosendaal R, Pals G, Nelis GF,

Festen HP, Meuwissen SG. Long-term sequelae of Helicobacter pylori gastritis. Lancet 1995; 345:1525–1528.
89. Correa P. Helicobacter pylori and gastric carcinogenesis. Am J Surg Pathol 1995; 19 (suppl 1):S37–S43.
90. Barvo LE, Mannick EE, Zhang XJ, Ruiz B, Correa P, Miller MJ. H. pylori infection is associated with inducible nitric oxide synthase expression, nitrotyrosine and DNA damage (abstr). Gastroenterology 1995; 108 (suppl):abstr 63.
91. Baik SC, Youn HS, Chung MH, Lee WK, Cho MJ, Ko GH, Park CK, Kasai H, Rhee KH. Increased oxidative DNA damage in Helicobacter pylori-infected human gastric mucosa. Cancer Res 1996; 56:1279–1282.
92. Farinati F, Lima V, Naccarato R, Garro AJ. Mutagenic activity in gastric juice and urine of subjects with chronic atrophic gastritis, gastric epithelial dysplasia and gastric cancer. Cancer Lett 1989; 48:169–175.
93. Mirvish SS. Role of N-nitroso compounds (NOC) and N-nitrosation in etiology of gastric, esophageal, nasopharyngeal and bladder cancer and contribution to cancer of known exposures to NOC. Cancer Lett 1995; 93:17–48.
94. Megraud F, Neman-Shimha V, Brugmann D. Further evidence of the toxic effect of ammonia produced by Helicobacter pylori urease on human epithelial cells. Infect Immun 1992; 60:1858–1863.
95. Tsuji M, Kawano S, Tsuji S, Takei Y, Tamura K, Fusamoto H, Kamada T. Mechanism for ammonia-induced promotion of gastric carcinogenesis in rats. Carcinogenesis 1995; 16:563–566.
96. Tokunaga M, Uemura Y, Tokudome T, Ishidate T, Masuda H, Okazaki E, Kaneko K, Naoe S, Ito M, Okamura A, Shimada A, Sato E, Land CE. Epstein-Barr virus related gastric cancer in Japan: a molecular pathoepidemiological study. Acta Pathol Jpn 1993; 43:574–581.
97. Shibata D, Weiss LM. Epstein-Barr virus associated gastric adenocarcinoma. Am J Pathol 1992; 140:769–774.
98. Uemura Y, Tokunaga M, Arikawa J, Yamomoto N, Hamasaki Y, Tanaka S, Sato E, Land CE. A unique morphology of Epstein-Barr virus-related early gastric carcinoma. Cancer Epidemiol Biomarkers Prev 1994; 3:607–611.
99. Yamamoto N, Tokunaga M, Uemura Y, Tanaka S, Shirahama H, Nakamura T, Land CE, Sato E. Epstein-Barr virus and gastric remnant cancer. Cancer 1994; 74:805–809.
100. Shousha S, Luqmani YA. Epstein-Barr virus in gastric carcinoma and adjacent normal gastric and duodenal mucosa. J Clin Pathol 1994; 47:695–698.
101. Kondo K, Kojima H, Akiyama S, Ito K, Takagi H. Pathogenesis of adenocarcinoma induced by gastrojejunostomy in Wistar rats: role of duodenogastric reflux. Carcinogenesis 1995; 16:1747–1751.
102. Mason RC, Filipe I. The aetiology of gastric stump carcinoma in the rat. Scand J Gastroenterol 1990; 25:961–965.
103. Kuwahara A, Saito T, Kobayashi M. Bile acids promote carcinogenesis in the remnant stomach of rats. J Cancer Res Clin Oncol 1989; 115:423–428.
104. Miwa K, Hasegawa H, Fujimura T, Matsumoto H, Miyata R, Kosaka T, Miyazaki I, Hattori T. Duodenal reflux through the pylorus induces gastric adenocarcinoma in the rat. Carcinogenesis 1992; 13:2313–2316.

105. Litter ER, Gleibermann E. Gastritis cystica polyposa. Cancer 1972; 29:205–209.
106. Toftgaard C. Gastric cancer after peptic ulcer surgery. Annu Surg 1989; 210: 159–164.
107. Osnes M. Early gastric carcinoma in patients with a Billroth II partial gastrectomy. Endoscopy 1977; 9:45–49.
108. Salmon RJ, Laurent M, Thierry JP. Effect of taurocholic acid feeding on methyl-nitro-N-nitroso-guanidine induced gastric tumors. Cancer Lett 1984; 22:315–320.
109. Kobori O, Shimizu T, Maeda M, Atomi Y, Watanabe J, Shoji M, Morioka Y. Enhancing effect of bile and bile acid on stomach tumorigenesis induced by N-methyl-N′-nitro-N-nitrosoguanidine in Wistar rats. J Natl Cancer Inst 1984; 73:853–861.
110. Takahashi M, Hasegawa R. Enhancing effects of dietary salt on both initiation and promotion stages of rat gastric carcinogenesis. In: Hayashi Y, Nagao M, Sugimura T, Takayama S, Tomatis L, Wattenberg LW, Wogan GN, eds. Diet, Nutrition and Cancer. Utrecht: Japan Sci Soc Press, Tokyo/vnu Sci Press, 1986:169–182.
111. Fukushima S, Tatematsu M, Takahashi M. Combined effect of various surfactants on gastric carcinogenesis in rats treated with N-methyl-N′-nitro-N-nitrosoguanidine. Gann 1974; 65:371–376.
112. Sørby H, Kvinnsland S, Svanes K. Effect of salt-induced mucosal damage and healing on penetration of N-methyl-N′-nitro-N-nitrosoguanidine to proliferative cells in the gastric mucosa of rats. Carcinogenesis 1994; 15:673–679.
113. Chyou PH, Nomura AMY, Hankin JH, Stemmermann GN. A case-cohort study of diet and stomach cancer. Cancer Res 1990; 50:7501–7504.
114. Poynter D, Pick CR, Harcourt RA, Selway SAM, Ainge G, Harman IW, Spurling NW, Fluck PA, Cook JL. Association of long lasting unsurmountable histamine H_2 blockade and gastric carcinoid tumors in the rat. Gut 1985; 26:1284–1295.
115. Betton GR, Dormer CS, Wells T, Pert P, Price CA, Buckley P. Gastric ECL-cell hyperplasia and carcinoids in rodents following chronic administration of H_2-antagonists SK&F 93479 and oxmetidine and omeprazole. Toxicol Pathol 1988; 16:288–298.
116. Tatsuta M, Iishi H, Yamamura H, Baba M, Yamamoto R, Taniguchi H. Effect of cimetidine on inhibition by tetragastrin of carcinogenesis induced by N-methyl-N′-nitro-N-nitrosoguanidine in Wistar rats. Cancer Res 1988; 48:1591–1595.
117. Shirai T, Takahashi M, Fukushima S, Tatematsu M, Hirose M, Ito N. Induction of preneoplastic hyperplasia and carcinoma by N-methyl-N′-nitro-N-nitrosoguanidine from regenerated mucosa of ulcers induced by iodoacetamide in fundus of rat stomach. Gann 1978; 69:361–366.
118. Takahashi M, Shirai T, Fukushima S, Ito N, Kokubo T, Furukawa F, Kurata Y. Ulcer formation and associated tumor production in multiple sites within stomach and duodenum of rats treated with N-methyl-N′-nitro-N-nitrosoguanidine. J Natl Cancer Inst 1981; 67:473–479.

119. Sørbye H, Kvinnsland S, Svanes K. Penetration of N-methyl-N′-nitrosoguanidine to proliferative cells in gastric mucosa of rats is different in pylorus and fundus and depends on exposure time and solvent. Carcinogenesis 1993; 14:887–892.
120. Ohgaki H, Kawachi T, Matsukura N, Morino K, Miyamoto M, Sugimura T. Genetic control of susceptibility of rats to gastric carcinoma. Cancer Res 1983; 43:3663–3667.
121. Furukawa H, Iwanaga T, Koyama H, Taniguchi H. Effect of sex hormones on carcinogenesis in the stomachs of rats. Cancer Res 1982; 42:5181–5182.
122. Ando Y, Watanabe H, Fujimoto N, Ito A, Toge T. Progesterone enhancement of stomach tumor development in SD rats treated with N-methyl-N′-nitro-N-nitrosoguanidine. Jpn J Cancer Res 1995; 86:924–928.
123. Wu CW, Chi CW, Chang TJ, Lui WY, P'eng FK. Sex hormone receptors in gastric cancer. Cancer 1990; 65:1396–1400.
124. Furukawa H, Iwanaga T, Hiratsuka M, Imaoka S, Ishikawa O, Kabuto T, Sasaki Y, Kameyama M, Ohigashi H, Nakamori S. Gastric cancer in young adults: growth accelerating effect of pregnancy and delivery. J Surg Oncol 1994; 55:3–6.
125. Sugiyama K, Nagase S, Maekawa A, Onodera H, Hayashi Y. High susceptibility of analbuminemic rats to gastric tumor induction by N-methyl-N′-nitro-N-nitrosoguanidine. Jpn J Cancer Res (Gann) 1986; 77:219–221.
126. Tatematsu M, Aoki T, Inoue T, Mutai M, Furihata C, Ito N. Coefficient induction of pepsinogen 1-decreased pylori glands and gastric cancers in five different strains of rats treated with N-methyl-N′-nitro-N-nitrosoguanidine. Carcinogenesis 1988; 9:495–498.
127. Martin MS, Martin F, Justrabo E, Michiels R, Bastien H, Knobel S. Susceptibility of inbred rats to gastric and duodenal carcinomas induced by N-methyl-N′-nitro-N-nitrosoguanidine. J Natl Cancer Inst 1974; 53:837–840.
128. Watanabe H, Nakagawa Y, Takahashi T, Ito A. Effect of age and strain on N-methyl-N′-nitro-N-nitrosoguanidine tumorigenesis in ACI and Wistar Furth rats. Carcinogenesis 1988; 9:1317–1318.
129. Tanaka H, Hirose M, Hagiwara A, Imaida K, Shirai T, Ito N. Rat strain differences in catechol carcinogenicity to the stomach. Food Chem Toxicol 1995; 33:93–98.
130. Saito T, Inokuchi K, Takayama S, Sugimura T. Sequential morphological changes in N-methyl-N′-nitro-N-nitrosoguanidine carcinogenesis in the glandular stomach of rats. J Natl Cancer Inst 1970; 44:769–783.
131. Kobori O. Analytical study of precancerous lesions in rat stomach mucosa induced by N-methyl-N′-nitro-N-nitrosoguanidine. Cancer Res 1980; 25: 141–150.
132. Ohgaki H, Kusama K, Hasegawa H, Sato S, Takayama S, Sugimura T. Sequential histologic changes during gastric carcinogenesis induced by N-methyl-N′-nitro-N-nitrosoguanidine in susceptible ACI and resistant BUF rats. J Natl Cancer Inst 1986; 77:747–755.
133. Szentirmay Z, Sugar J. Adenocarcinoma, glandular stomach, rat. In: Jones

TC, Mohr U, Hunt RD, eds. Digestive System (Monographs on Pathology of Laboratory Animals). New York: Springer-Verlag, 1985:301–309.
134. Ogiso T, Hirose M, Takahashi S, Hakoi K, Imaida K, Kaneko H, Shirai T. Histogenesis and mechanism of catechol-induced rat glandular stomach tumors (abstr). Proc Jpn Cancer Assoc 1995; 54:71.
135. Hirose M, Wada S, Yamaguchi S, Masuda A, Okazaki S, Ito N. Reversibility of catechol-induced rat glandular stomach lesions. Cancer Res 1992; 52:787–790.
136. Tatematsu M, Furihata C, Katsuyama T, Hasegawa R, Nakanowatari J, Saito D, Takahashi M, Matsushima T, Ito N. Independent induction of intestinal metaplasia and gastric cancer in rats treated with N-methyl-N′-nitro-N-nitrosoguanidine. Cancer Res 1983; 43:1335–1341.
137. Sung ME. Histopathological study of mucosal and submucosal cysts of stomach. Acta Pathol Jpn 1991; 41:31–40.
138. Iwanaga T, Koyama H, Takahashi Y, Taniguchi H, Wada A. Diffuse submucosal cysts and carcinoma of the stomach. Cancer 1975; 36:606–614.
139. Lauren P. The two histologic main types of gastric carcinoma: diffuse and so-called intestinal type carcinoma. An attempt at a histoclinical classification. Acta Pathol Microbiol Scand 1965; 64:31–49.
140. Correa P, Shiao YH. Phenotypic and genotypic events in gastric carcinogenesis. Cancer Res (suppl) 1994; 54:1941s–1943s.
141. Correa P, Miller MJS. Helicobacter pylori and gastric atrophy—Cancer paradoxes. J Natl Cancer Inst 1995; 87:1731–1732.
142. Endo S, Ohkusa T, Saito Y, Fujiki K, Okayasu I, Sato C. Detection of Helicobacter pylori infection in early stage gastric cancer. A comparison between intestinal- and diffuse-type gastric adenocarcinomas. Cancer 1995; 75:2203–2208.
143. Wright PA, Quirke P, Attanoos R, Williams GT. Molecular pathology of gastric carcinomas: progress and prospects. Hum Pathol 1992; 23:848–859.
144. Soman NR, Correa P, Ruiz BA, Wogan GN. The TPR-MET oncogenic rearrangement is present and expressed in human gastric carcinoma and precursor lesions. Proc Natl Acad Sci USA 1991; 88:4892–4896.
145. Czerniak B, Herz F, Gorczyca W, Koss LG. Expression of ras oncogene p21 protein in early gastric carcinoma and adjacent gastric epithelia. Cancer 1989; 64:1467–1473.
146. Kihana T, Tsuda H, Hirota T, Shimosato Y, Sakamoto H, Terada M, Hirohashi S. Point mutation of c-Ki-ras oncogene in gastric adenoma and adenocarcinoma with tubular differentiation. Jpn J Cancer Res 1991; 82:308–314.
147. Poremba C, Yandell DW, Huang Q, Little JB, Mellin W, Schimid KW, Bocker W, Dockhorn-Dworniczak B. Frequency and spectrum of p53 mutation in gastric cancer—a molecular genetic and immunohistochemical study. Virchows Arch 1995; 426:447–455.
148. Lauwers GY, Scott GV, Karpeh MS. Immunohistochemical evaluation of bcl-2 protein expression in gastric adenocarcinomas. Cancer 1995; 75:2209–2213.

149. Sano T, Tsujino T, Yosida K, Nakayama H, Haruma K, Ito H, Nakamura Y, Kajiyama G, Tahara E. Frequent loss of heterozygosity on chromosomes 1q, 5q, and 17p in human gastric carcinomas. Cancer Res 1991; 51:2926-2931.
150. Ranzani GN, Luinetti O, Padovan LS, Calistri D, Renault B, Burrel M, Amadori D, Fiocca R, Solcia E. P53 gene mutations and protein nuclear accumulation are early events in intestinal type gastric cancer but late events in diffuse type. Cancer Epidemiol Biomarkers Prev 1995; 4:223-231.
151. Shimoyama Y, Hirohashi S. Expression of E- and P-cadherin in gastric carcinomas. Cancer Res 1991; 51:2185-2192.
152. Nakatani H, Sakamoto H, Yoshida T, Yokota J, Tahara E, Sugimura T, Terada M. Isolation of an amplified DNA sequence in stomach cancer. Jpn J Cancer Res 1990; 81:707-710.
153. Tahara E. Molecular mechanism of stomach carcinogenesis. J Cancer Res Clin Oncol 1993; 119:265-272.
154. Kuniyasu H, Yasui W, Kitadai Y, Yokozaki H, Ito H, Tahara E. Frequent amplification of the c-met gene in scirrhous type stomach cancer. Biochem Biophys Res Commun 1992; 30:227-232.
155. Yonemura Y, Ninomiya I, Yamaguchi A, Fushida S, Kimura H, Ohoyama S, Miyazaki I, Endou Y, Tanaka M, Sasaki T. Evaluation of immunoreactivity for erbB-2 protein as a marker of poor short term prognosis in gastric cancer. Cancer Res 1991; 51:1034-1038.
156. Lin JT, Wu MS, Shun CT, Lee WJ, Sheu JC, Wang TH. Occurrence of microsatellite instability in gastric carcinoma is associated with enhanced expression of erbB-2 oncoprotein. Cancer Res 1995; 55:1428-1430.

18
Thyroid Carcinogenesis

Geraldine Anne Thomas
Addenbrooke's Hospital, University of Cambridge,
Cambridge, United Kingdom

I. INTRODUCTION

The mammalian thyroid gland is composed of two types of epithelial cell: the follicular cells that line the lumen of the colloid filled follicles and are responsible for the production of thyroid hormones and the C cells, which are derived from the neuroectoderm, are parafollicular in position in most mammals, and are responsible for the production of calcitonin. In humans, C cells are relatively infrequent and tumors of the C cell are rare; in up to 30% of cases, tumors of the C cells will be associated with familial syndromes such as multiple endocrine neoplasia type 2A (MEN2A) and multiple endocrine neoplasia type 2B (MEN2B), which are both linked to an inherited mutation of the ret oncogene (1,2). In laboratory rodents, particularly rats, C cells are relatively much more common than in humans. In the rat several studies have shown that there is an age-related increase in C-cell number (3,4), with a variable frequency of tumor development, depending on the sex and strain of rat (5). C-cell tumors are morphologically distinct from tumors derived from the follicular cell and in immunocytochemistry stain positively for calcitonin. The induction of C-cell tumors is not a common finding in toxicological studies, but it is important to be aware of the age-related increase in C-cell number and spontaneous incidence of C-cell tumors in aged rats in order to separate them from tumors of the follicular cell, which have a much lower spontaneous incidence in old rats. C-cell tumors occur very rarely in aged mice (6).

In contrast, tumors of the thyroid follicular cell are a common finding in humans, occurring in about 4% of the general population, and their incidence increases with age (7–9). In many cases, these nodules occur as part of a multinodular goiter, and the great majority are benign. The finding of a truly solitary nodule (which account for only 25% of clinically detectable tumors) is clinically more important because although in the majority of cases these too will be benign tumors, 10–30% may harbor a malignant neoplasm (10–12). The risk of malignancy is greater at both ends of the age spectrum (i.e., under 20 and over 60 years of age) (11,13).

Spontaneous tumors of the follicular cell are rare in rats with an incidence of 1% or less in Fischer 344 and Sprague-Dawley rats and 2.9% in the Osborne-Mendel strain (14). A similar frequency is observed in mice (15–18); the majority of tumors are benign in both species, with only occasional reports of carcinomas in large studies (15,18). However, an increased incidence of thyroid follicular tumors is a common finding in many toxicological studies, using a variety of chemical agents, and there has been a large increase in carcinomas of the thyroid follicular cell in the populations exposed to fallout from the Chernobyl nuclear accident (19–21), particularly those who were young children at the time of exposure (22).

II. PATHOLOGY OF THYROID FOLLICULAR TUMORS

The majority of thyroid follicular tumors found in humans and animals are benign. Two types of benign tumors may be distinguished: the nodule and the adenoma. Adenomas are defined as being solitary, encapsulated, and having a uniform internal architecture, different from the surrounding thyroid. They may also compress the adjacent gland. Nodules are typically less well circumscribed, have a more variable architecture similar to the background thyroid, and do not compress adjacent tissue. Most benign thyroid tumors are multiple and occur in longstanding thyroid disease, where there is thought to have been a deficiency of thyroid hormones over a protracted period of time. Diffuse enlargement (hyperplasia) of the thyroid eventually develops into multinodular goiter. Such nodules are thought to result from over distension of some involuted follicles, persistence or enlargement of epithelial hyperplasia, and localized degenerative and reparative changes. Some nodules may be poorly demarcated, but others may be well circumscribed and may resemble adenomas. The term adenomatous goiter is often used to distinguish between glands that contain multiple tumors and the classic solitary adenoma, which may harbor a malignant neoplasm. In practice, distinction between nodule and adenoma is not always easy in a multinodular gland.

The majority of benign tumors do not take up greater amounts of radioiodine than the surrounding thyroid. However, some, termed “hot” nodules, are functionally more active in radioactive uptake than the surrounding tissue, and a proportion of these have been shown to have activating mutations in exon 10 of the TSH receptor (23).

In humans, it is possible to separate two types of differentiated thyroid carcinoma that derive from the follicular cell: follicular carcinoma and papillary carcinoma. It is thought that follicular carcinomas arise from preexisting follicular adenomas, whereas papillary carcinomas arise de novo from the follicular cell. The two types of carcinoma are distinct in respect to molecular biology, architecture, and clinical course. Follicular adenomas are composed microscopically of follicles ranging from small (microfollicular) to large (macrofollicular). Occasionally they may show a trabecular pattern rather than follicular architecture, but on closer examination tiny colloid spaces can usually be seen. Follicular carcinomas account for 10–40% of clinically evident thyroid carcinomas and, like nodules and adenomas, are more common in females. They are more common in areas of iodide deficiency (24,25) and microscopically are often generally similar to adenomas, being composed of follicles, trabeculae, or solid cell islands, but showing evidence of vascular and capsular invasion. An example of follicular carcinoma can be seen in Figure 1.

Figure 1 Section from a follicular carcinoma in humans stained with hematoxylin and eosin. The tumor is composed of sheets of follicular epithelium, and there is clear evidence of capsular invasion.

Papillary carcinoma is the most common type of thyroid carcinoma in humans, usually accounting for up to 75% of all cases diagnosed clinically. It is more common in the young, shows a greater frequency in females, occurs more frequently in areas of iodide excess (26), and has a clear association with irradiation of the neck (27–30) or exposure to radioiodine in fallout (31). Papillary carcinoma can rarely occur in families (32). It is so named because it was originally separated on the basis of its papillary architecture, but it is now recognized that its main characteristics are cytological. Microscopically, the diagnosis of papillary carcinoma depends on the presence of several features found together; when present individually these features may not be specific. The single most important feature is the presence of large, crowded, cuboidal epithelial cells with pale staining cytoplasm and a high nuclear cytoplasmic ratio. Nuclei are characteristically grooved, and have a pale, ground glass appearance (Fig. 2A). Psammoma bodies, which are calcified concentric structures, may also be present; they are thought to arise from dying papillae. Tumor papillae with fibroblastic cores are characteristic, although the degree of tumor papillarity varies. Some papillary carcinomas are composed entirely of follicles (confusingly referred to as the follicular variant of papillary carcinoma; Fig. 2B) but show the other features associated with papillary carcinomas; others may be solid. The latter are relatively more common in young children (33) and also show a greater relative frequency post-Chernobyl (34,35). Papillary carcinomas characteristically metastasize via the lymphatics. Minute papillary carcinomas (papillary microcarcinomas) are a relatively common finding at autopsy, occurring in as much as 30% of the population (36), but such lesions are not considered to be clinically relevant in humans.

Undifferentiated or anaplastic carcinoma is relatively rare in humans (fewer than 15% of all thyroid carcinomas) and extremely rare in animals. There is evidence that it arises by progression from preexisting papillary or follicular carcinomas.

In animals the majority of thyroid tumors, whether spontaneous or chemically induced, are benign. As in humans, two morphological types can be distinguished—nodule and adenoma—and separated using criteria similar to those in humans. It is also difficult to draw a definite line between the two in animals, as both tumors are most likely to be multiple and to be found in glands after prolonged hyperplasia. In addition, benign tumors in rodents often show areas of papillary architecture projecting into the lumen of distended cystic follicular structures (Fig. 3).

Distinction between nodule and adenoma has been made on grounds of structure and difference in clonal origin (36). Nodules are benign tumors composed of structures of similar cytology to the background thyroid,

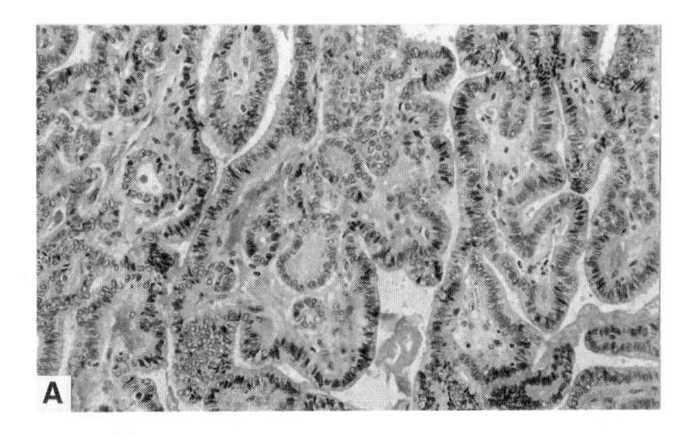

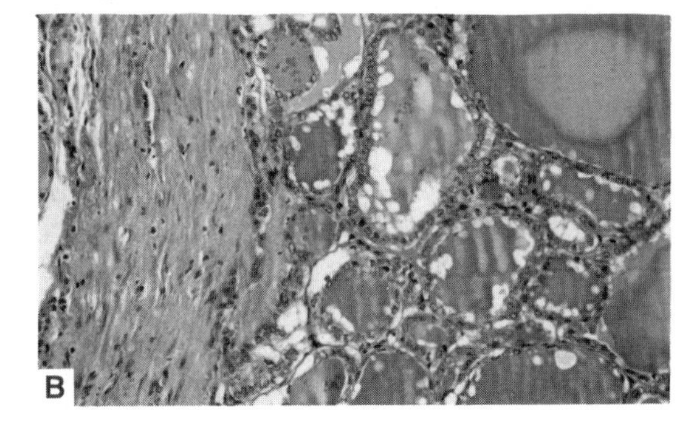

Figure 2 (A) Section from a human papillary carcinoma, stained with hematoxylin and eosin. The tumor is composed of papillae with fibroblastic cores. The overlying epithelial cells have crowded, grooved nuclei, which are strikingly different from those of follicular carcinoma. (B) Section from a human papillary carcinoma (follicular variant), stained with hematoxylin and eosin. In contrast to the tumor showin in Figure 2A, this variant of papillary carcinoma is composed of follicular structures containing some colloid. However, the epithelial cell nuclei are similar to those seen in Figure 2A and distinct from those in Figure 1.

but more variable architecture and containing excessive stroma; these were polyclonal in origin (Fig. 4 A,B). Adenomas are benign tumors composed of crowded uniform follicular cells with a follicular or trabecular architecture; these were monoclonal in origin (Fig. 4 C,D).

Carcinoma in rodents is relatively rare, and, unlike in humans, there is no clear distinction between papillary and follicular carcinoma, either on the basis of morphology or molecular biology. The one exception is that tumors showing some of the characteristic features of human papillary carcinoma, including grooved nuclei, have been observed in mice showing targeted transgenic expression of a translocation known to be associated with papillary carcinoma in humans (38,39).

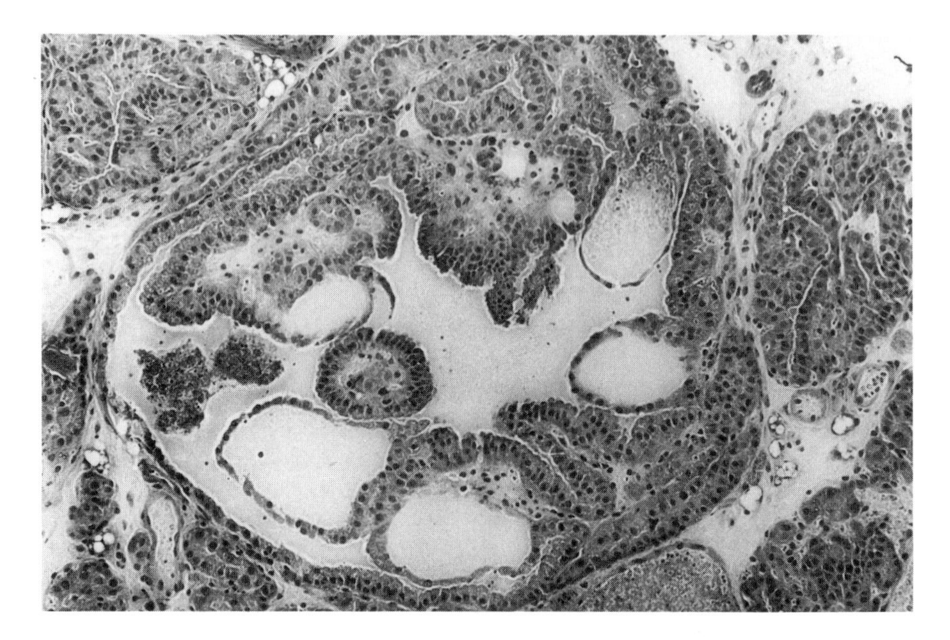

Figure 3 Small benign tumor induced in a mouse after a single dose of radiation and 46 weeks of goitrogen (aminotriazole) administration. The tumor shows papillary infoldings, but does not show the characteristic nuclei associated with papillary carcinoma in humans.

III. PRODUCTION OF THYROID HORMONES

Structure and function is intimately linked in the thyroid, and in order to understand the histogenesis of thyroid tumors, it is necessary to first have an understanding of thyroid physiology. Thyroid homeostasis utilizes a multitude of mechanisms, many of which have been shown to be affected by diet and xenobiotic agents. Synthesis of the thyroid hormones tri-iodothyronine (T3) and tetra-iodothyronine (thyroxine, T4) is dependent upon the dietary supply of iodine. Briefly, iodine is taken via a symporter into the thyroid follicular cell, where it is quickly oxidized by the thyroid specific enzyme thyroid peroxidase and bound at the apical membrane to tyrosyl residues on the thyroid specific protein thyroglobulin. The newly iodinated thyroglobulin lies next to the apical membrane, just within the follicular lumen; the iodide is in the form of monoiodotyrosine (MIT) and di-iodotyrosine (DIT). These two compounds are then coupled (again under the influence of thyroid peroxidase) to give T3 and T4, which remain bound to thyroglobulin. Thyroglobulin containing thyroid hormones is taken back up into the follicular cell by pinocytosis, and MIT, DIT, T3, and T4 are released. T3 and T4 are passed into the circulation where they are bound to plasma proteins, whereas MIT and DIT are enzymatically deiodinated to release iodine for more thyroid hormone production.

Thyroid-binding globulin (TBG) is the main plasma protein that binds thyroid hormones in humans. It has a greater affinity for T4 than T3 (40). Two other proteins are known to bind thyroid hormones in humans: thyroxine-binding prealbumin (TBPA), which is responsible for 15% of bound plasma thyroid hormone, and albumin. In the rat the majority of thyroid hormone in plasma is bound to albumin. Only low levels of a protein showing 70% homology to human TBG are present (41). However, the level of TBG in rat is upregulated by thyroidectomy, suggesting that it may play some physiological role (42). More T4 than T3 is released from the thyroid, but T4 is metabolized peripherally to give local production of T3, the more active hormone.

IV. METABOLISM OF THYROID HORMONES

Thyroid hormones are taken up into the liver, where most conversion to T3 takes place, by an active process (43) and are then metabolized by deiodination. There are three distinct deiodinases, which show a different tissue distribution. The enzyme 5′ deiodinase I is found in most tissues, including thyroid itself, but is most abundant in liver and kidney. It is capable of both outer ring deiodination to give T3 and inner ring deiodination to give rT3, which is not produced by the thyroid, and as yet has no known function. Deiodinase II occurs in the central nervous system, anterior pituitary, and brown adipose tissue. It deiodinates the outer ring only, and its preferred substrate is T4. Deiodinase III is found only in the brain, skin, and placenta and deiodinates only the inner ring. It shows a high affinity for both T3 and T4. The latter two deiodinases are thought to give rise only to local T3 and rT3 production.

Aside from local production of the more active hormone, deiodination is the major route of catabolism of the thyroid hormones in humans, resulting ultimately, after stepwise deiodination, in the production of thyronine. Thyroid hormones are also sulfated by the action of phenol sulphotransferases, and this process is believed to facilitate deiodination. In addition, the thyroid hormones are also subject to glucuronidation. This pathway is less important in humans, but in rats it is the major route for catabolism of thyroid hormones. This pathway may take on greater relevance in patients treated with phenytoin (44), carbemazepine (45), or phenobarbitone (46) or exposed to pollutants such as polychlorobenzenes (47)

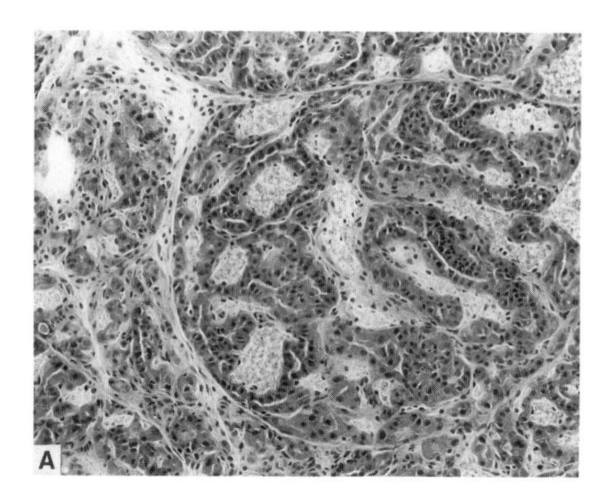

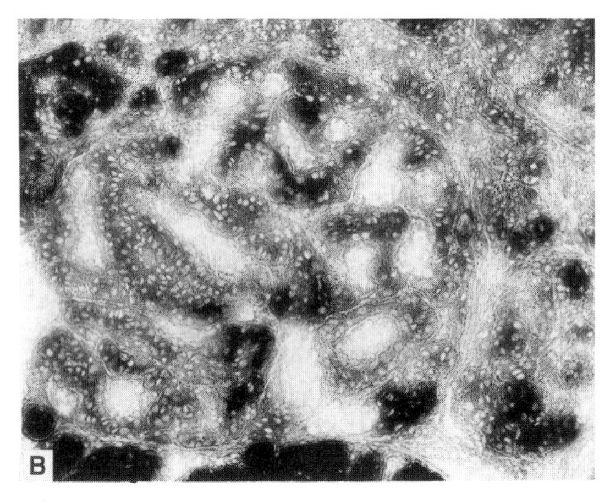

Figure 4 Clonality of the two morphological types of benign tumors induced in mice after prolonged goitrogen administration. The mice used were heterozygous for a histochemically demonstrable polymorphism of the X-linked enzyme glucose-6-phosphate dehydrogenase (G6PD). Nodule: (A) Frozen H- and E-stained section from a mouse maintained on 1% perchlorate in the drinking water for 60 weeks. The nodule is composed of follicles of mainly normal cytology and architecture, but there is an increased stromal content in the lesion. (B) Serial section stained with the histochemical technique for G6PD activity. Both enzyme phenotypes are clearly visible in both the nodule and the background hyperplastic thyroid, indicating that the epithelial component of the nodule is of polyclonal origin. Adenoma: (C) Frozen H- and E-stained section from a mouse given a single intraperitoneal injection of ^{131}I (3 μCi) at 3 weeks of age and subsequently aminotriazole (0.2%) in the drinking water for 60 weeks. There is a well-demarcated adenoma characterized by

and polybrominated biphenyls (48), as all of these compounds are known to affect the enzymes involved in glucuronidation.

The circulating T4 level is monitored by the thyrotrophs of the anterior pituitary, which are responsible for the production of the major trophic hormone involved in both thyroid function and growth—thyroid-stimulating hormone (TSH). T4 is metabolized to T3 by 5′ deiodinase II in the thyrotroph: T3 then binds to nuclear receptors in the cell. A decrease in T3 receptor occupancy results in an increased synthesis of TSH. A higher tier of control also exists between the hypothalamus, which secretes thyrotropin-releasing hormone (TRH), which also stimulates release of TSH from the thyrotrophs.

The complexity of both the production and the catabolism of thyroid hormones provides many mechanisms with which xenobiotics can interact. These agents can be broadly divided into five categories (Fig. 5). Class 1 comprises directly acting agents, such as those that interfere with thyroid hormone production either by inhibiting iodine uptake or by inhibiting the thyroid peroxidase-stimulated organification of iodide. Class 2 includes those compounds whose effect is predominantly on the liver. Class 2A compounds are those that induce hepatic microsomal enzymes leading to an increased biliary clearance of T4. Class 2B compounds affect the hepatic transport of thyroid hormones. Class 3 compounds comprise deiodinase inhibitors, which affect the deiodination of thyroid hormones. Class 4 consists of compounds that affect the plasma protein binding of thyroid hormones, and Class 5 contains certain neurotransmitters, including dopamine, which have been implicated in the control of TSH output via TRH. An exhaustive list of examples of these compounds is given by Atterwill et al. (49).

Figure 4 FACING PAGE
abnormal follicular/papillary architecture, with crowded basophilic epithelial cells and a scanty stromal component. (D) Serial section to 4A, stained with the enzyme histochemical technique for G6PD. In this heterozygous female, two cellular patterns of enzyme reaction can be seen in the normal background hyperplastic thyroid. Cells in which the mutant deficient G6PD gene is on the active X chromosome show low enzyme activity, whereas those in which the normal gene is on the active X chromosome show high levels of G6PD activity. The tumor is entirely composed of cells that express high G6PD activity, showing that it is monoclonal in origin. (From Ref. 143.)

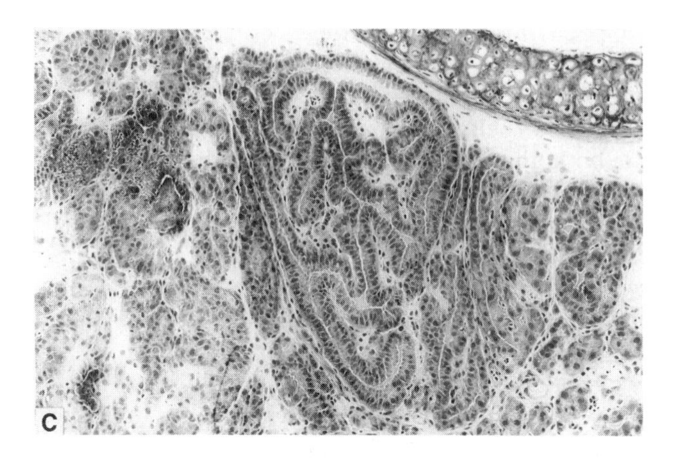

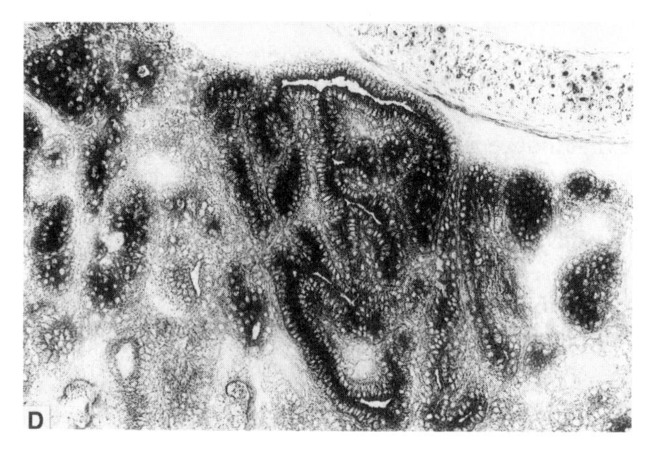

Figure 4 Continued.

V. INDUCTION OF THYROID TUMORS

The majority of agents that result in the production of thyroid tumors, usually after prolonged administration, interfere with thyroid hormone homeostasis (50). TSH is not only the major factor to control thyroid follicular cell function, but it is also the key factor in thyroid follicular cell growth. When the decrease in circulating T4 is small and short-lived, the deficiency can be redressed by an increase in the functional capacity of the gland. However, if the decrease in circulating T4 is profound and prolonged and cannot be redressed by increase in thyroid function alone, an increase in the units of thyroid hormone production is required. Then an increase in

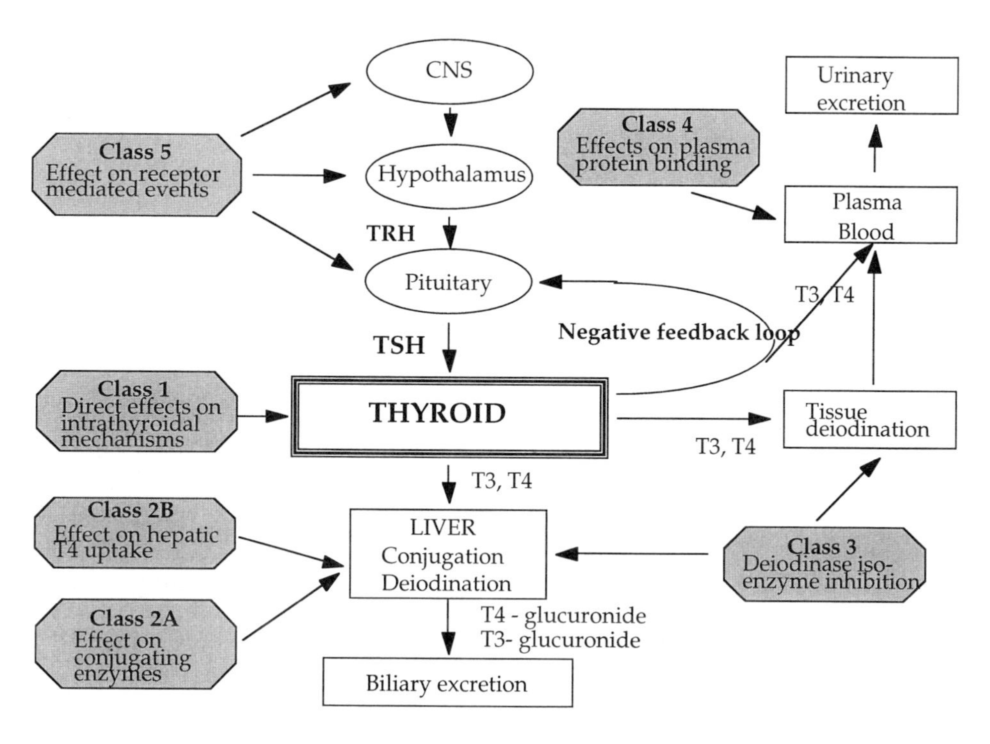

Examples of xenobtiotics for each class:
1: Agents which act directly on the thyroid: Perchlorate (iodide uptake); thinamides (thyroid peroxidase inhibitor)
2A: Agents affecting enymes which conjugate thyroid hormones in the liver: phenobarbital
2B: Agents which affect transport of thyroid hormones into hepatocytes: temelastine, lupitidine
3: Agents which inhibit deiodinases: propylthiourea (agents in this class may also inhibit thyroid peroxidase)
4: Agents affecting plasma protein binding: salicylates (in man)
5: Agents affecting certain neurotransmitters implicated in the control of TRH/TSH e.g. dopamine (bromocriptine)
For more detailed information see Atterwill et al (49)

Figure 5 Xenobiotic interaction with thyroid hormone synthesis and catabolism.

thyroid follicular cell number occurs, leading to the development of goiter (hence the term goitrogen for agents that provoke this reaction). However, the thyroid follicular cell has only a limited capacity for growth in the rat. After high-dose goitrogen administration there is initially a rapid increase in mitotic activity, but this falls after about 7–14 days, returning to relatively low levels. An increase in thyroid weight is observed, which reaches a plateau after about 3 months of goitrogen administration. If administration of the goitrogen continues long term, tumors of the follicular cell are observed after about 8 months in the rat (51). Similar observations have been made in the mouse (52). It is assumed that these lesions develop after loss of the mechanism that restricts the normal follicular cell's growth response to TSH.

Many compounds have been shown to produce benign lesions in mouse and rat after prolonged administration (49). The frequency of these lesions can be influenced by the iodine content of the diet and by administration of carcinogens. TSH is believed to be a key component in the genesis of these tumors as both goitrogenesis and tumorigenesis can be inhibited by either administration of T4 or by hypophysectomy (53). A variety of other experiments or naturally occurring conditions show the importance of TSH-induced growth in thyroid tumorigenesis. Congenital hypothyroidism as a result of an inherited defect in one of the key steps in thyroid hormonogenesis (usually in thyroid peroxidase or in thyroglobulin) results in goiter and tumorigenesis in both humans (54–57) and animals (58–60). Thyroid tumors are also more common in humans in areas of iodide deficiency (24,25) and can be induced by iodide deficiency in animals (61,62). Iodide deficiency can be exacerbated by the presence of goitrogens in the drinking water (63) or in the food (64).

The majority of tumors induced in the mouse and rat by prolonged goitrogen administration are benign, with carcinomas in the mouse being particularly rare. Benign murine tumors can be transplanted over several generations and may eventually invade and metastasize (65), suggesting that clonal evolution of thyroid tumors in the mouse is slow. The finding that follicular tumors produced by prolonged goitrogen administration regressed on withdrawal of goitrogen feeding led to the suggestion that these lesions were hyperplastic, rather than neoplastic (66). A similar conclusion had been drawn from previous studies in mice, using long-term administration of methimazole (a thyroid peroxidase inhibitor), when thyroid deposits were found in the lung (17). However, more recent studies showed that monoclonal lesions, which clearly showed capsular and vascular invasion, induced by a combination of mutagen and goitrogen also regressed on withdrawal of the TSH stimulus (67). The clonal nature of these lesions suggests that they are true tumors in that they arise by heritable alteration in growth control within a single clone. However, their regression on withdrawal of goitrogen feeding indicates that these thyroid lesions are more akin to the so-called "carcinogen-dependent lesions" found in rodent liver (68), rather than directly comparable with thyroid tumors in humans that have developed in the absence of a high TSH. Earlier work in the rat, using transplantation of malignant rat thyroid tumors into first euthyroid and then hypothyroid hosts, suggested that the malignant phenotype could be reinduced by elevated TSH (69). These results suggest that prolonged goitrogen treatment gives rise to lesions that show some of the features associated with human carcinomas, i.e., they are monoclonal and show capsular and vascular invasion, but that at least some rodent tumors retain dependence on elevated levels of TSH for growth. In the last aspect rodent tumors appear to be different from human tumors, where TSH dependence is often the subject of speculation but rarely, if ever, clearly demonstrated.

The mechanism by which goitrogen-induced tumors arise is far from clearly understood. Three key steps are likely to be common to most spontaneous thyroid carcinomas. First, the normal follicular cell possesses a growth-limiting mechanism; the continued growth of the tumor cells infers that any such mechanism must have been lost in neoplasia. Second, development of TSH-independent growth is also likely to be essential for spontaneous thyroid carcinogenesis. Third, a key step is the cellular ability to invade—an essential feature of malignancy.

The development of these successive defects that lead to neoplasia can be considered as a process of natural selection at a cellular level: manipulation of the environment in which selection takes place may alter the selection process. In experimental thyroid carcinogenesis, the induction of persistently high TSH levels creates an environment in which any cell that suffers a mutation leading to loss of its growth limiting mechanism will give rise to a clone of cells that will continue to grow and therefore continue to be at risk of further mutation in the progeny of that cell. When invasiveness develops in a cell that has not acquired TSH-independent growth, the resulting carcinomas retain TSH dependency and will regress when TSH stimulation is withdrawn.

VI. ROLE OF MUTAGENS IN THE GENERATION OF THYROID TUMORS

The majority of compounds that induce thyroid neoplasia in rodents do so via an effect on TSH secretion. Aromatic amines such as aminofluorene (70), azo dyes such as 4,4′-methylene-bis-(*N,N*-dimethyl)-benzamine (71), nitrosamines, e.g., diisopropanolnitrosamine (72), *N*-bis-(2-hydroxypropyl)-nitrosamine (73), and the nitrosoureas, e.g., methylnitrosourea (MNU) (74–76), have all induced a low frequency of thyroid tumors in rats after administration of multiple doses. However, in one of these studies (76) administration of the carcinogen alone (MNU) was shown to decrease circulating T4. In addition, the frequency of thyroid tumors can be greatly increased when these mutagens are given in combination with agents known to interfere with thyroid function (70–76). Agents that demethylate DNA, such as 5-azacytidine, can also potentiate thyroid carcinogenesis induced by goitrogens in mice (77), although such agents do not show the marked effect of the combination of mutagens with goitrogens. All these results suggest that nongenotoxic effects on thyroid homeostasis play a key role in the generation of thyroid tumors in rodents. There have been no confirmed reports of induction of thyroid tumors in humans as a result of exposure to chemical mutagens. However, one important mutagen, radiation, is linked to thyroid tumorigenesis in both humans and animals.

VII. RADIATION AND THYROID CANCER

The role of radiation in rodent thyroid carcinogenesis is twofold. Experimental evidence in rats suggests that radiation does not show a linear dose-response curve. The dose of ^{131}I that gives the highest frequency of thyroid tumors in adult rats (30 μCi) (78) has been shown to lie within a dose range that demonstrably interferes with thyroid function (79). In addition, Nichols et al. (80) showed that after exposure to X-irradiation with shielding of one of the two thyroid lobes, lesions developed in both thyroid lobes but were more common in the unshielded lobe. Both these experiments suggested that there was a second component to radiation-induced carcinogenesis. Observation that hypophysectomy (81) and concomitant T4 administration (82) inhibited radiation-induced thyroid carcinogenesis suggested that the mitotic stimulus provided by TSH was an important element in radiation-induced thyroid carcinogenesis. The bell-shaped dose-response curve to radiation with respect to thyroid carcinogenesis can be explained in this manner. Low doses of radiation cause little thyroid toxicity and result in a low frequency of mutations in the DNA of follicular cells. As the dose of radiation increases, the cytotoxicity of radiation increases, resulting in a loss of functioning thyroid follicular cells, as well as an increase in mutation frequency in those cells which survive. The increased cytotoxicity results in a decrease in T4 and increase in TSH, which stimulates the remaining thyroid follicular cells to grow. If, however, the dose of radiation is so great that the majority of the gland is rendered incapable of cell division, too few exposed and viable cells remain to give an increase in tumor frequency.

The role of TSH in the generation of thyroid tumors is also shown by the fact that radiation administration can also be combined with goitrogen administration to further increase the yield of thyroid tumors in rats and mice (70). Thyroid cancer induced in animals by radiation does not show a particular phenotype, and morphologically the lesions are similar to those induced by goitrogen alone. Combination of radiation and goitrogen leads to the development of relatively more carcinomas than either radiation or goitrogen alone.

Due to the April 1986 accident at the Chernobyl Nuclear Plant, the role of radiation exposure in the development of human thyroid cancer has recently assumed more importance. There was a considerable release not only of radioisotopes of iodine, but of other isotopes, such as ^{132}tellurium, which decays to the short-lived isotope of iodine ^{132}I (83). There has been a considerable increase in the incidence of thyroid cancer, particularly in those areas that received the highest level of radioiodine in fallout (19–21), and particularly in those who were children at the time of the accident (22). There is considerable published evidence that age at exposure to radiation (given for therapeutic purposes) is a modifying factor in human thyroid cancer. Individuals exposed in the very early years of life (0–5 years) have a much greater risk of thyroid cancer later in life (84–87). The risk declines with increasing age at exposure, being negligible after about 40 years of age. Other studies have shown that there is a risk associated with even low doses of radiation exposure in childhood—as low as 0.093 Gy in one study (86). There are few reports of age-related increases in susceptibility to thyroid cancer in animals after radiation exposure. However, in one study using CBA mice, irradiation at 18 days gestation (the onset of the ability to concentrate radioiodine in the mouse thyroid) led to a dose-dependent increase in the number of thyroid tumors observed after 2 years. Thirty-four percent of male mice and 25% of female mice irradiated at 18 days gestational age had thyroid tumors 2 years later, compared with only 9% of mice irradiated at 4 months of age (88,89).

Radioactive iodine is used therapeutically in treating thyrotoxicosis in patients with Graves' disease. There have been occasional anecdotal reports of cases of thyroid carcinoma following this treatment (90). However, large epidemiological follow-up studies of adults treated with radioiodine for Graves' disease have not found that this treatment confers any detectable risk of subsequent development of thyroid cancer (91,92). It is interesting to note, in the light of the shape of the dose-response curve in the rat, that the dose of radioiodine used therapeutically is in the cell-sterilizing range.

A seemingly excessive amount of thyroid cancer in the population of the Marshall Islands occurred after atomic bomb tests were conducted there (93,94). Here the main component of the thyroid dose resulted from short-lived isotopes of iodine: ^{132}I, ^{133}I, and ^{135}I. Other epidemiological studies of environmental radiation or fallout from nuclear tests have suggested that thyroid cancer rates were not elevated, but further studies will be needed in the light of more recent information on levels of exposure.

The increase in thyroid carcinoma in Belarus and Ukraine since the Chernobyl accident has been dramatic in children who were under age 15 at the time of the nuclear power plant accident. The incidence of thyroid cancer in individuals who were of less than 3 months of intrauterine age at the time of the accident falls sharply, which suggests that the isotope in fallout responsible for the induction of thyroid carcinogenesis is relatively short-lived. In addition, the fact that thyroid cancer is the only cancer that has yet shown a significant increase in the exposed population and that the thyroid is the only organ in the body to both take up and organically bind iodine suggests that ^{131}I and the shorter-lived isotopes of iodine are a likely cause of the increased thyroid carcinomas. Other tissues such as breast and salivary gland also contain an iodide pump, but do not bind iodide within the tissue.

The rise in thyroid cancer incidence after the Chernobyl incident is unlikely to be due to better screening. Only a few papillary microcarcinomas have so far been reported (95); the majority of cancers are relatively large and many show extensive invasion. In addition, it is unlikely to be due to iodide insufficiency in the areas around Chernobyl, as there is no correlation between those areas that show the highest frequency of goiter formation and those that show the highest frequency of thyroid cancer. In addition, virtually all of the thyroid cancers (>95%) are of the papillary subtype. No significant increase has yet been observed in follicular carcinoma post-Chernobyl. Further longer-term studies are needed to ascertain whether more follicular carcinomas will appear as the exposed cohort ages and to determine whether there is a link between the morphology, molecular biology, and etiology of thyroid cancer.

VIII. THE PATHOBIOLOGY OF THYROID CANCER

Few studies have compared the molecular biology of thyroid cancer in humans and animals. However, there are now several interesting transgenic mouse models which provide an insight into genes that may be involved in thyroid carcinogenesis as well as an increasing literature on the role of different genes in human thyroid carcinoma.

A. Involvement of Oncogenes

The two main types of follicular carcinoma in humans—papillary and follicular carcinoma—show distinct molecular biological profiles. Mutation of the *ras* genes are more frequent in the follicular type of carcinoma. Translocations of the *ret* (96) and *trk* (97) oncogenes are more common in the papillary type of carcinoma. Indeed, translocation of the *ret* oncogene occurs exclusively in papillary carcinoma (98). The *ret* oncogene, which encodes the tyrosine kinase–linked receptor for glial-derived neurotrophic factor (99), is thought to be normally silent in follicular cells. Activation of the tyrosine kinase segment of the *ret* gene occurs on translocation. Three distinct rearrangements have so far been described, two intrachromosomal and one interchromosomal [termed PTC1 (gene rearrangement caused by the fusion of 5′ H4 with 3′ ret), PTC2 and PTC3] (96,100,101), and there are reports of other, as yet uncharacterized rearrangements. The frequency of *ret* activation varies in different series (102–104); whether this is due to different etiology, genetic background, or the sensitivity of the technique used to detect the translocation is uncertain. There have been some recent suggestions that the frequency of *ret* activation is higher in children exposed to fallout from the Chernobyl accident (105–107) and even that the frequency of the different types of rearrangement may be related to radiation. However, these studies were carried out on small series of tumors and lack sex- and age-matched controls. Using much larger series of tumors, our own studies suggest that *ret* activation may be more common in females (108), that there is no relative increase in the frequency of *ret* activation in post-Chernobyl thyroid papillary carcinomas, and that PTC1 is the most common translocation observed, with a frequency similar to that observed in adults (95). No studies have been published on the frequency of *ret* activation in thyroid tumors induced in normal animals, but two independent groups have published results from mice transgenic for thyroid-specific expression of the PTC1 rearrangement of *ret* (38,39). Both report the induction of thyroid tumors with morphological similarity (particularly with respect to the characteristic nuclear changes) to papillary carcinoma in humans.

The role of *ras* genes in thyroid neoplasia has also been extensively studied. Early work using slot blot hybridization suggested that mutation of Ha *ras* was a relatively common finding in thyroid follicular adenomas and carcinomas (109), and that Ki *ras* may be preferentially mutated in follicular tumors after exposure to radiation (110). A lower frequency of *ras* mutation was found in papillary carcinoma (109). However, more recent studies have suggested that the frequency of *ras* mutation is lower in follicular adenomas than previously thought and is likely to be a late event rather than an initial event (111). Recent studies have also suggested that *ras* mutation is not associated with papillary carcinoma (111) even after radiation (95). The observations made in mice that are transgenic for thyroid-targeted expression of mutant Ki *ras* are generally in agreement with the suggestion that *ras* mutation is not an early event in thyroid carcinogenesis in that they show only a low frequency and a late onset of thyroid lesions (112,113). Interestingly in both *ret* and *ras* transgenics, the induced lesions are discrete, which suggests that alteration in other genes in addition to constitutive expression of the transgene is required in thyroid neoplasia.

Mutation of p53 is not thought to play a role early in tumor formation in the thyroid, but has been shown to be involved at the transition from differentiated to undifferentiated carcinoma (114).

Using activated oncogenes, which have not yet been implicated in human thyroid neoplasia, other transgenic mouse models have produced varying pathologies. A model of very aggressive thyroid carcinogenesis has been developed by thyroid-specific expression of the large T antigen of the SV40 virus (115). Large T antigen binds to two different antioncogenes: p53 and p105Rb (116). the mice are hypothyroid, and after an initial stage of hyperplasia, multiple poorly differentiated cancers develop, and the mice die quickly after birth from tracheal compression caused by the large tumors. Two other transgenic models now have been produced targeting E6 or E7 proteins from the human papilloma virus to the thyroid epithelium. These two proteins bind to p53 and p105Rb, respectively. The E6 transgenic model was asymptomatic but presented with a slightly enlarged thyroid gland at the age of 6 months. However, the E7 transgenic mice develop large colloid goiters and malignant, poorly differentiated tumors after about 1 year (117). Another transgenic mouse line, this time with constitutive activation of the pA2 receptor, shows diffuse thyroid hyperplasia, and the animals are hyperthyroid (118). Interestingly, breeding of the E7 mice with mice transgenic for constitutive activation of the pA2 receptor results in mice that show a rapid development of thyroid neoplasia with early and frequent metastases (119).

One important distinction should be made between tumorigenesis in transgenic animals and tumorigenesis in humans. The site and number of copies of the transgene can affect the phenotype of the animals, and in addition the transgene is expressed from early development (from about a gestational age of 17–18 days in the case of transgenes produced using a thyroglobulin promoter). In the human the assumption is that thyroid tumors arise as a result of the accumulation of somatic mutations in later life. Transgenic mice provide a useful tool for studying the effects of putative oncogenes of thyroid epithelial cells in vivo. However, it is still necessary to study the expression of such oncogenes in human tumors, as there are undoubtedly differences between species in the level of expression of genes and their role in carcinogenesis in different tissues of different species. For example, mice transgenic for c-*mos* expression show a tumor pattern similar to human MEN2, with pheochromocytomas and medullary thyroid carcinomas (120). However, there is no evidence for involvement of this oncogene in human MEN2 (121). In addition, mice that are heterozygous for loss of function activity of the adenopolyposis coli (APC) gene (Min mice) (122) show multiple adenomas in the small intestine, as well as tumors of the colon, whereas in humans the same congenital defect results in multiple tumors of the colon but only rarely in small intestinal tumors. It is interesting to note, however, that a specific morphological type of thyroid carcinoma is found in a few cases of patients heterozygous for truncation of the APC gene (123). Whether this is the result of loss of the normal APC gene in clones of thyroid epithelial cells is not yet known, but loss of APC gene function does not appear to play a role in thyroid tumorigenesis in general (124).

B. Growth Factors and Thyroid Neoplasia

Although TSH is the major growth factor for the growth of the thyroid epithelium, other factors are required for the growth response to TSH. Early studies showed that high levels of insulin were required to elicit a growth response to TSH in vitro (125,126), but later it was considered that the supraphysiological levels of insulin were likely to be acting via the insulinlike growth factor 1 (IGF1) receptor. IGF1 was later shown to be the critical permissive growth factor for the follicular cell's growth response to TSH in vitro both in rodent (125), dog (127), and human (128). The cell producing IGF1 in the thyroid in vivo was initially thought to be the fibroblast, but studies using in situ hybridization in the mouse showed that thyroid follicular cells synthesized IGF1 mRNA during developmental and goitrogen-induced growth (129). Further studies have shown that in humans the normal follicular epithelium shows variable levels of expression of IGF1 mRNA, whereas both thyroid follicular and papillary carcinomas show more homogeneous expression of IGF1 mRNA and usually show low levels of IGF1 receptor protein on immunocytochemistry (130). However, there is a specific morphological subset of papillary carcinoma with a widespread intimate plasma cell infiltrate, which shows high levels of expression of IGF1 mRNA. This subset of carcinomas is associated with overlying epithelial cells, which exhibit strong positivity for IGF1 receptor protein. This has led to the suggestion of a stromal-epithelial interaction in this tumor (130). Benign thyroid tumors in humans and animals do not show elevated levels of IGF1 mRNA, suggesting that overexpression of IGF1 may be a later event in thyroid carcinogenesis (G. A. Thomas et al., unpublished observations). However, there are a plethora of binding proteins that regulate the bioavailability of IGF1 to the cell, so these results should be interpreted with care. A recent study also suggested that insulin itself, in addition to IGF1, might be mitogenic for the dog thyrocyte (131).

Other growth factors and their receptors are also believed to be important in thyroid carcinogenesis in humans, but little work has been carried out in animals. Hepatocyte growth factor has been shown to be a potent mitogen for follicular cells in vitro (132), and its receptor, encoded by the oncogene *met*, shows increased expression in the majority of papillary carcinomas in humans by both Western blotting (133) and by immunocytochemistry (134,135). Interestingly, this oncogene, like *ret* and *trk*, does not appear to be involved in follicular carcinomas. It has also been suggested that papillary carcinomas exhibit an autocrine loop for transforming growth factor alpha (TGF-α) via the epidermal growth factor (EGF) receptor (136).

IX. INTERPRETATION OF THYROID CARCINOGENESIS IN RODENTS AND ITS RELEVANCE TO HUMANS

Two important mechanisms—genotoxic and nongenotoxic—are involved in thyroid carcinogenesis in both humans and animals. The nongenotoxic mechanism involves elevation of TSH through either inhibition of thyroid hormone synthesis or stimulation of thyroid hormone clearance, and virtually all xenobiotics that cause an increase in thyroid tumors after prolonged administration show an effect on thyroid hormone production or catabolism. The majority of these xenobiotics can be shown to lack positivity in tests for genotoxic agents.

The exact mechanism by which these tumors are induced is not yet determined, but it is clear that sustained high doses of goitrogen (presumably resulting in a complete or almost complete block of thyroid hormone synthesis) induces thyroid tumors in rats, whereas lower doses of goitrogen do not (137). It is also clear that there are species differences in response to the same dose of goitrogen. One possible explanation for this is the finding that thyroid peroxidase from monkeys in vitro shows a reduced sensitivity to the effects of phenyl thiourea compared to rat thyroid peroxidase (138). Monkeys would therefore have a lower block of thyroid hormone synthesis than rats, and presumably a lower frequency of tumor induction. Several studies indicate that monkeys are less sensitive to the goitrogenic effect of other peroxidase inhibitors, for example, sulfamethoxazole (139). Interestingly, there have been no reports of adverse effects on the thyroid after long-term clinical use of sulfonamides (140,141), suggesting that human thyroid peroxidase may be more like that in monkey than in rat.

Very few chemical mutagens, which show no effect on thyroid homeostasis, cause a significant increase in thyroid tumors in animal studies when administered on their own. However, used in combination with a xenobiotic, which results in an increase in TSH, the incidence of tumors is greatly increased, together with the relative frequency of carcinomas compared with administering either mutagen or goitrogen alone. In addition, the tumorigenic effect of radiation has been shown to be a combination of direct damage to the DNA, and an effect on TSH production as a result of the cytotoxic effect of irradiation. The role of thyroid growth in radiation-associated thyroid carcinogenesis is further demonstrated by the finding that tumorigenesis induced by radiation can be further potentiated either by administration of a TSH-elevating chemical after radiation or by administration of radiation to a child, where the thyroid is still undergoing a period of developmental growth. In addition to the arguments for a greater dose effect in the child's thyroid, there are two other factors that may contribute to the greater risk of thyroid neoplasia following exposure to radiation early in life. Carcinogenesis is a multistep phenomenon and requires the acquisition of multiple mutations in a clone of cells. Mutation is more likely to occur at S phase. The child's thyroid has a higher proportion of dividing cells compared with the adult and is therefore likely to be more susceptible to the mutational effects of radiation. In addition, the phase of developmental growth that occurs results in cells that have been exposed to radiation. These cells may harbor mutations deleterious to growth regulation, undergoing several more rounds of cell division, thus increasing the probability of acquiring additional deleterious mutations. It therefore follows that exposure to any mutagen during thyroid developmental growth carries the potential risk of development of thyroid neoplasia in later life. It is difficult to obtain data on chemical mutagen exposure in early life in humans. Very few animal studies address this point in general and in thyroid in particular.

The majority of tumors induced in animals after protracted administration by substances that interfere with thyroid homeostasis are benign and occur against a background of hyperplasia. Little is known about the interaction of two or more substances that interfere with thyroid homeostasis by different mechanisms, but in humans, evidence that goiter formation can be exacerbated in iodide-deficient areas by either goitrogens contained in food (64), or goitrogens in the drinking water (65), or deficiency of other elements known to play a role in thyroid homeostasis [e.g., selenium: (142)] suggests that additive effects may be observed. The evidence from radiation-induced carcinogenesis clearly suggests that age plays a role in the development of thyroid neoplasia after exposure to a mutagen, but little attention has been paid so far to the possible differential effects of administration of nongenotoxic thyroid carcinogens on the development of thyroid neoplasia to the young and mature thyroid. However, it is unlikely that early age at exposure to a nongenotoxic thyroid carcinogen carries a significant risk in humans.

In terms of risk assessment, an agent that induces thyroid follicular tumor formation after long-term administration but is not mutagenic in conventional genotoxic assays and can be demonstrated to interfere with thyroid hormone homeostasis is unlikely to pose significant risk of thyroid tumorigenesis in humans. Long-term low-dose exposure has been shown in rats to be inefficient with respect to tumorigenesis, and evidence suggests that humans are very much less sensitive to goitrogens than rats. Even in patients with dyshormonogenesis, where there is an almost complete block of thyroid hormone synthesis for the lifetime of the patient (and is therefore analogous to prolonged exposure to high doses of goitrogen in animals), benign tumors are common, but carcinomas are very rare.

There is still much to learn about the molecular biological mechanisms involved in thyroid tumorigenesis. It is likely that future research into genetic predisposition will throw light on other oncogenes that may be involved in human thyroid carcinogenesis. However, it is extremely important to be aware of differences in physiology and biochemistry that may lead to different tumorigenic mechanisms between species. Awareness of these species differences can help ensure that the toxicological tests we conduct ensure the safe exposure of humans, rather than mice, to environmental and pharmaceutical agents.

ACKNOWLEDGMENTS

I would like to thank the Cancer Research Campaign, Medical Research Council, and the European Commission, who have generously supported my work on thyroid tumors, and Professor Sir Dillwyn Williams for his help and advice over many years.

REFERENCES

1. Mulligan LM, Kwok JB, Healey CS, et al. Germline mutations of the ret proto-oncogene in multiple endocrine neoplasia type 2A. Nature 1993; 363: 458–460.
2. Hofstra RMW, Landsvater RM, Ceccherini I, et al. A mutation in the ret proto-oncogene associated with multiple endocrine neoplasia type 2B and sporadic medullary carcinoma. Nature 1994; 367:375–376.
3. Larsson LI. Differential changes in calcitonin, somatostatin and gastrin/cholecystokinin-like immunoreactivities in rat thyroid parafollicular cells during ontogeny. Histochemistry 1985; 82:121–130.
4. Thomas GA, Neonakis E, Davies HG, Wheeler MH, Williams ED. Synthesis and storage in rat thyroid C cells. J Histochem Cytochem 1994; 42:1055–1060.
5. DeLellis RA. Changes in structure and function of thyroid C cells. In: Mohr U, Dungworth DL, Capen CC, eds. Pathobiology of the Aging Rat. Washington, DC: ILSI, 1994:285–300.

6. Ward JM, Goodman DG, Squire RA, et al. Neoplastic and non neoplastic lesions in aging (C57BL/6NxC3H/HeN)F1 (B63C3F1) mice. J Natl Cancer Inst 1979; 63:849–854.
7. Vander JB, Gaston EA, Dawber TR. Significance of solitary non-toxic thyroid nodules. N Engl J Med 1954; 251:970–973.
8. Sokal JE. The problems of malignancy in nodular goitre: recapitulation and a challenge. JAMA 1959; 170:405–412.
9. Tunbridge WMGH, Evered DC, Hall R, et al. The spectrum of the thyroid disease in a community: the Wickham survey. Clin Endocrinol 1977; 7:481–493.
10. Robinson E, Horn Y, Hochmann A. Incidence of thyroid cancer in nodules. Surg Gynaecol Obstet 1966; 123:1024–1026.
11. Psarras A, Papadopoulos SH, Livadas D, Pharmakiotis AD, Koutras DA. The single thyroid nodule. Br J Surg 1972; 59:545–548.
12. Liechty RD, Stoffel PT, Zimmerman DE, Silverberg SG. Solitary thyroid nodules. Arch Surg 1977; 112:59–64.
13. Hoffmann GL, Thomson NW, Heffron C. The solitary thyroid nodule. Arch Surg 1972; 105:379–385.
14. Thomas GA, Williams ED. Changes in structure and function of thyroid follicular cells. In: Mohr U, Dungworth DL, Capen CC, eds. Pathobiology of the Aging Rat. Washington, DC: ILSI, 1994:269–284.
15. Littlefield NA, Gaylor DW. Chronic toxicity/carcinogenicity studies of sulfamethazine in B6C3F1 mice. Food Cosmet Toxicol 1989:455–463.
16. Steinhoff D, Webere H, Mohr U, Böhme K. Evaluation of amitrole (aminotriazole) for potential carcinogenicity in orally dosed rats, mice and golden hamsters. Toxicol Appl Pharmacol 1983; 69:161–169.
17. Jemec B. Studies on the goitrogenic and tumorigenic effect of two goitrogens. Cancer 1977; 40:2188–2202.
18. Prejean JD, Peckham JC, Casey AE, et al. Spontaneous tumours in Sprague-Dawley rats and Swiss mice. Cancer Res 1973; 33:2768–2773.
19. Kazakov VS, Demidchik EP, Astakhova LN. Thyroid cancers after Chernobyl. Nature 1992; 359:21.
20. Baverstock K, Egloff B, Pinchera A, Ruchti C, Williams D. Thyroid cancer after Chernobyl. Nature 1992; 359:21–22.
21. Likhtarev IA, Sobolev BG, Kairo IA, Tronko ND, Bogdanova TI, Oleinic VA, Ephstein EV, Beral V. Thyroid cancer in the Ukraine. Nature 1995; 375:365.
22. Williams ED, Cherstvoy E, Egloff B, Höfler H, Vecchio G, Bogdanova T, Bragarnik M, Tronka ND. Interaction of pathology and molecular characterisation of thyroid cancers. In: Karaoglou A, Desmet G, Kelly GN, Menzel HG, eds. The Radiological Consequences of the Chernobyl Accident. European Commission EUR 16544 EN 1996:785–789.
23. Parma J, Duprez L, Van Sande J, Cochaux P, Gervy C, Mockel J, Dumont J, Vassart G. Somatic mutations in the thyrotropin receptor gene cause hyperfunctioning thyroid adenomas. Nature 1993; 365:649–651.
24. Cuello C, Correa P, Eisenberg H. Geographic pathology of thyroid carcinoma. Cancer 1969; 23:230–239.

25. Hedinger C. Geographic pathology of thyroid diseases. Pathol Res Pract 1981; 171:285–292.
26. Williams ED, Doniach I, Bjarnason O, Michie W. Thyroid cancer in an iodide rich area. A histopathological study. Cancer 1977; 39:215–222.
27. Duffy BJ, Fitzgerald PJ. Cancer of the thyroid in children: a report of 28 cases. J Clin Endocrinol Metab 1950; 10:1296–1308.
28. Hemplemann LH. Risk of thyroid neoplasms after radiation in childhood. Science 1969; 160:159–163.
29. Modan B, Baidatz D, Mart H, Steinitz R, Levin SG. Radiation induced head and neck tumours. Lancet i:277–279.
30. Shore R. Issues and epidemiological evidence regarding radiation induced thyroid cancer. Radiat Res 1992; 131: 98–111.
31. Williams D, Pinchera A, Karaoglou A, Chadwick K. Thyroid cancer in children living near Chernobyl. EUR 15248 EN CEC, Luxembourg, 1993.
32. Lote K, Andersen K, Nordal E, Brenn-Hovd IO. Familial occurrence of papillary thyroid carcinoma. Cancer 1980; 46:1291–1297.
33. Harach HR, Williams ED. Childhood thyroid cancer in England and Wales. Br J Cancer 1995; 72:777–783.
34. Bogdanova T, Bragarnik M, Tronko ND, Harach HR, Thomas GA, Williams ED. The pathology of thyroid cancer in Ukraine post Chernobyl. In: Karaoglou A, Desmet G, Kelly GN, Menzel, HG, eds. The Radiological Consequences of the Chernobyl Accident. European Commission EUR 16544 EN, 1996:785–789.
35. Cherstvoy E, Pozcharskaya V, Harach HR, Thomas GA, Williams ED. The pathology of childhood thyroid carcinoma in Belarus. In: Karaoglou A, Desmet G, Kelly GN, Menzel HG, eds. The Radiological Consequences of the Chernobyl Accident. European Commission EUR 16544 EN, 1996:779–784.
36. Fransilla KO, Karach HR. Occult papillary carcinoma of the thyroid in children and young adults. Cancer 1986; 58:715–719.
37. Thomas GA, Williams D, Williams ED. The clonal origin of thyroid nodules and adenomas. Am J Pathol 1989; 134:141–147.
38. Santoro M, Chiappetta G, Cerrato A, Salvatore D, Zhang L, Manzo G, Picone A, Portella G, Santelli G, Vecchio G, Fusco A. Development of thyroid papillary carcinomas secondary to tissue-specific expression of the RET/PTC1 oncogene in transgenic mice. Oncogene 1996; 12:1821.
39. Jhiang SM, Sagartz JE, Tong Q, Parker TJ, Capen CC, Cho JY, Xing S, Ledent C. Targeted expression of the ret/PTC1 oncogene induces papillary thyroid carcinomas. Endocrinology 1996; 137:375–378.
40. Robbins J, Rall JE. The iodine containing hormones. In: Gray CH, James AHT, eds. Hormones in the Blood. Vol 1. London: Academic Press, 1979: 576–688.
41. Shuji I. Molecular cloning of the primary structure of rat thyroxine binding globulin. Biochem J 1991; 3:22–27.
42. Nanno M, Ohtska R, Kikuchi N, Oki Y, Ohgo S, Kurahachi H, Yoshimi T, Hamada S. Frontiers in thyroidology. In: Medeiros-Neto G, Gaitan E, eds. Vol 1. 1986:481–484.

43. Krenning EP, Docter R, Bernard B, Visser TJ, Henneman G. Active transport of triiodothyronine into isolated liver cells. FEBS Lett 1978; 91:113–116.
44. Rootwelt K, Ganes T, Johannessen SI. Effect of carbamezepine, phenytoin and phenobarbital on serum levels of thyroid hormones and thyrotropin in humans. Scand J Lab Invest 1978; 38:731–736.
45. Oppenheimer JH, Bernstein G, Surks ML. Increased thyroxine turnover and thyroidal function after stimulation of hepatocellular binding of thyroxine by phenobarbital. J Clin Invest 1968; 47:1399–1406.
46. Cavalieri RR, Gavin L, Bui F, McMahon F, Hammon M. Serum thyroxine, free thyroxine, triiodothyronine and reverse triiodothyronine in diphenylhydantoin-treated patients. Metab Clin Exp 1977; 28:1161–1165.
47. Kimborough RD. The toxicity of polychlorinated polycyclic compounds and related chemicals. Crit Rev Toxicol 1974; B:445–498.
48. Koopman-Esseboom C, Morse DC, Weisglas-Kuperus N, Lutkeschipolt IJ, van der Paauw CG, Tuinistra LG, Brouwer A, Sauer PJ. Effects of dioxins and polychlorinated biphenyls on thyroid hormone status of pregnant women and their infants. Pediatr Res 1994; 36:468–473.
49. Atterwill CK, Jones C, Brown CG. Thyroid gland II – mechanisms of species dependent thyroid toxicity, hyperplasia and neoplasia induced by xenobiotics. In: Atterwill CK, Flack JD, eds. Endocrine Toxicology. Cambridge: Cambridge University Press, 1992:137–182.
50. Hill RN, Edreich LS, Paynter OE, Roberst PA, Rosenthal SL, Wilkinson CF. Thyroid follicular cell carcinogenesis. Fundam Appl Toxicol 1989; 12: 629–697.
51. Wynford-Thomas D, Stringer BMJ, Williams ED. Dissociation of growth and function in the rat thyroid during prolonged goitrogen administration. Acta Endocrinol 1982; 101:210–216.
52. Peter HJ, Gerber H, Studer H, et al. Comparison of FRTL-5 cell growth in vitro with that of xenotransplanted cells in the thyroid of the recipient mouse. Endocrinology 1991; 128:211–219.
53. Jemec B. Studies on the goitrogenic and tumorigenic effect of two goitrogens in combination with hypophysectomy or thyroid hormone treatment. Cancer 1980; 45:2138–2202.
54. Stanbury JB, Chapman EM. Congenital hypothyroidism with goitre: absense of an iodide concentrating mechanism. Lancet 1960; i:1162–1165.
55. Stanbury JB, Hedge AN. A study of a family of goitrous cretins. J Clin Endocrinol 1950; 10:1741–1758.
56. Alexander NM, Burrow GN. Thyroxine biosynthesis in human goitrous cretinism. J Clin Endocrinol Metab 1970; 30:308–315.
57. Salvatore G, Stanbury JB, Rall JE. Inherited defects of thyroid hormone biosynthesis. In: De Visscher M, ed. Comprehensive Endocrinology: The Thyroid Gland. New York: Raven, 1980:443–487.
58. van Jaarsfeld P, van der Walt B, Theron CN. Afrikander cattle congenital goiter: purification and partial identification of the complex iodoprotein pattern. Endocrinology 1972; 91:470–482.

59. Falconer IR, Roitt IM, Seamark RF, Torrigiani G. Studies of the congenitally goitrous sheep. Iodoproteins of the goitre. Biochem J 1965; 117:417–424.
60. de Vijlder JM, van Voorthuizen WF, van Dijk JE, Rijnberk A, Telegaers WHH. Hereditary congenital goiter with thyroglobulin deficiency in a breed of goats. Endocrinology 1978; 102:1214–1222.
61. Axelrad AA, Leblond CP. Induction of thyroid tumors in rats by a low iodine diet. Proc Am Assoc Cancer Res 1955; 1:2.
62. Shaller RT, Stevenson JK. Development of carcinoma of the thyroid in iodine deficient rats. Cancer 1966; 19:1063–1080.
63. Gaitan E. Goitrogen. Balliere's Clin Endocrinol Metab 1988; 2:683–702.
64. Ekpechci OL, Dimitriadou A, Fraser R. Goitrogenic activity of cassava (a staple Nigerian food). Nature 1966; 210:1137–1138.
65. Morris HP, Dalton AJ, Green CD. Malignant thyroid tumors occurring in the mouse after prolonged hormonal imbalance during the ingestion of thiouracil. J Cell Endocrinol 1951; 11:1281–1295.
66. Todd GC. Induction and reversibility of thyroid proliferative changes in rats given an antithyroid compound. Vet Pathol 1986; 23:110–117.
67. Thomas GA, Williams D, Williams ED. The reversibility of the malignant phenotype in monoclonal thyroid tumours in the mouse. Br J Cancer 1991; 63:213–216.
68. Grasso P, Hinton RH. Evidence for the possible mechanisms of non genotoxic carcinogenesis in the rodent liver. Mutat Res 1991; 248:271–290.
69. Matovinovic J, Nishiyama RH, Poissant G. Transplantable thyroid tumours in the rat: development of normal appearing thyroid follicles in the differentiated tumours, and development of differentiated tumours from iodide deficient, thyroxine involuted goiters. Cancer Res 1970; 30:504–514.
70. Doniach I. The effect of radioactive iodine alone and in combination with methylthiourea and acetylaminofluroene upon tumour production in the rats' thyroid gland. Br J Cancer 1950; 4:223–234.
71. Murthy ASK. Morphology of the neoplasms of the thyroid gland in Fischer 344 rats treated with 4,4′-methylene-bis-(N,N′-dimethyl)-benzylamine. Toxicol Lett 1980; 6:391–397.
72. Mohr U, Reznik G, Pour P. Carcinogenic effect of diisopanolnitrosamine in Sprague-Dawley rats. JNCI 1977; 58:361–366.
73. Hiasa Y, Ohshima M, Kiathori Y, Yuasa T, Fujita T, Iwata C. Promoting effects of 3-amino-1,2,4-triazole on the development of thyroid tumours in rats treated with N-bis-(2-hydroxypropyl)-nitrosamine. Carcinogenesis 1982; 3:381–384.
74. Tsuda H, Fukunshima S, Imaida K, Kurata Y, Ito N. Organic specific promoting effect of phenobarbital and saccharin in induction of thyroid, liver and bladder tumours in rats after initiation with N-nitrosomethylurea. Cancer Res 43:3292–3296.
75. Schäffer R, Müller HA. On the development of metastasizing tumors of the thyroid gland after combined administration of nitrosomethylurea and methylthiouracil. J Cancer Res Clin Oncol 1980; 96:281–285.

76. Milmore JE, Chandraskaran V, Weisburger JH. Effects of hypothyroidism on development of nitrosomethylurea-induced tumors of the mammary gland, thyroid gland and other tissues. Proc Soc Exp Biol Med 1982; 169: 487–493.
77. Thomas GA, Williams ED. Production of thyroid tumours in mice by demethylating agents. Carcinogenesis 1992; 13:1039–1042.
78. Doniach I. The effect of radioactive iodine alone and in combination with methylthiouracil upon tumour production in the rats thyroid gland. Br J Cancer 1953; 7:181–202.
79. Maloof F, Dobyns B, Vickery AL. The effects of various doses of radioactive iodine on the function and structure of the thyroid of the rat. Endocrinology 1952; 50:612–638.
80. Nichols CW, Lindsay S, Sheline GE, Chaikoff IL. Induction of neoplasms in rat thyroid glands by X-irradiation of a single lobe. Arch Pathol 1965; 80: 177–183.
81. Nadler NJ, Mandavia M, Goldberg M. The effect of hypophysectomy on the experimental production of rat thyroid neoplasia. Cancer Res 1970; 30:1909–1911.
82. Doniach I. Carcinogenic effect of 100, 200, 250 and 500 rad X-rays on the rat thyroid gland. Br J Cancer 1974; 30:487–495.
83. Williams D, Pinchera A, Karaoglou A, Chadwick K. Thyroid cancer in children living near Chernobyl. EUR 15248 EN CEC Luxembourg, 1993.
84. Akiba S, Lubin J, Ezaki H, et al. Thyroid cancer incidence among atomic bomb survivors in Hiroshima and Nagasaki 1958–79. Technical report TR 5-91. Hiroshima, Japan: Radiation Effects Foundation, 1991.
85. Ron E, Kleinerman R, Boice J, LiVolsi V, Flannery J, Fraumeni J. A population based case-control study of thyroid cancer. JNCI 1987; 79:1–12.
86. Ron E, Modan B, Preston D, Alfandary E, Stovall M, Boice J. Thyroid neoplasia following low dose irradiation in childhood. Radiat Res 1989; 120: 516–531.
87. Schneider A, Shore-Freedman E, Weinstein R. Radiation induced thyroid and other head and neck tumors: occurrence of multiple tumors and analysis of risk factors. J Clin Endocrinol Metab 1986; 63:107–112.
88. Walinder G. Late effects of irradiation on the thyroid gland in mice. I Irradiation of adult mice. Radiol Ther Phys Biol 1972; 11:433–451.
89. Walinder G, Sjöden A-M. Late effects of irradiation on the thyroid gland in mice. II Irradiation of mouse fetuses. Acta Radiol Ther Phys Biol 1972; 11: 577–589.
90. McDougal R. Thyroid cancer after iodine 131 therapy. J Am Med Assoc 1974; 227:438.
91. Holm LE, Dahlquist I, Israelsson A, Lundell G. Malignant thyroid tumours after iodine 131 therapy, a retrospective cohort study. N Engl J Med 1980; 303:188–191.
92. Holm LE, Wiklund K, Lundell G, et al. Thyroid cancer after diagnostic doses of iodine 131. JNCI 1988; 1132–1138.

93. Robbins J, Adams W. Radiation effects in the Marshall Islands. In: Nagataki S, ed. Radiation and the Thyroid. Amsterdam: Excerpta Medica, 1989:11–24.
94. Lessard ET, Brill A, Adams W. Thyroid cancer in the Marshall Islands: relative risk of short-lived emitters and external radiation exposure. In: Schlafke-Stelson A, Watson E, eds. Fourth International Radiopharmaceutical Dosimetry Symposium. Springfield, VA: CONF-851113 National Technical Information Service, 1985:628–647.
95. Williams ED, Tronko ND. Molecular, cellular, biological characterisation of childhood thyroid cancer. EUR 16538 EN, EC, Luxembourg.
96. Fusco A, Grieco M, Santoro M, et al. A new oncogene in human thyroid papillary carcinomas and their lymph nodal metastases. Nature 1987; 328: 170–172.
97. Grieco A, Pierotti MA, Bongarzone I, Pagliardini J, Lanzi C, Della Porta G. Trk-T1 is a novel oncogene formed by the fusion of TPR and Trk genes in human papillary cancer. Oncogene 1992; 7:237–242.
98. Santoro M, Carlomango F, Hay ID. Ret oncogene activation in human thyroid neoplasms is restricted to the papillary cancer subtype. J Clin Invest 1992; 89:1517–1522.
99. Durbec P, Marcos GC, Kilkenny C, Grigorou M, Wartiwaara K, Suvanto P, Smith D, Ponder B, Constantini F, Saarma M, Sariola H, Pachnis V. GDNF signalling through the ret receptor kinase. Nature 1996; 381:789–793.
100. Bongarzone I, Monzini N, Borello MG, et al. Molecular characterisation of thyroid specific transforming sequence formed by the fusion of ret tyrosine kinase and the regulatory subunit of cyclic AMP dependent protein kinase C. Mol Cell Biol 1993; 13:358–366.
101. Santoro M, Dathan NA, Berlingieri MT, et al. Molecular characterisation of ret/PTC3: a novel rearranged version of the ret protooncogene in human papillary thyroid carcinoma. Oncogene 1994; 9:509–516.
102. Ishizaka Y, Kobayashi S, Ushijima T, Hirohashi S, Sugimura T, Nagao M. Detection of ret TPC/PTC transcripts in thyroid adenomas and adenomatous goiter by an RT-PCR method. Oncogene 1991; 6:1667–1672.
103. Jhiang SM, Caruso DR, Gilmore E, Ishizaka Y, Tahira Y, Nagao M. Detection of the PTC/retTPC oncogene in human thyroid cancers. Oncogene 1994; 7:1331–1337.
104. Zou M, Shi Y, Farid NR. Low rate of ret proto-oncogene activation (PTC/retTPC) in papillary thyroid carcinomas from Saudi Arabia. Cancer 1994; 73:176–180.
105. Ito T, Seyama T, Iwamoto KS, Mizuno T, Tronko ND, Komissarenko IV, Cherstvoy ED, Satow Y, Takiechi N, Dohi K, Akiyama M. Activated ret oncogene in thyroid cancers of children from areas contaminated by Chernobyl accident. Lancet 1994; 344:259.
106. Fuggazzola L, Pilotti S, Pinchera A, Vorontsova TV, Mondellini P, Bongarzone I, Greco A, Astakhova L, Butti MG, Demidchik EP, Pacini F, Pierotti MA. Oncogenic rearrangements of the ret proto-oncogene in papillary thy-

roid carcinomas from children exposed to the Chernobyl nuclear accident. Cancer Res 1995; 55:5617–5620.
107. Klugbauer S, Lengfelder E, Demidchik EP, Rabes HM. High prevalence of ret rearrangement in thyroid tumors of children from Belarus after the Chernobyl reactor accident. Oncogene 1995; 11:2459–2467.
108. Williams GH, Rooney S, Carrs A, Tronko ND, Bogdanova TI, Voscoboinik L, Thomas GA, Williams ED. The influence of gender, age and radiation exposure on the frequency of ret activation in papillary carcinoma of the thyroid. submitted.
109. Lemoine NR, Mayall ES, Wyllie FS, Williams ED, Goyns M, Stringer BMJ, Wynford-Thomas D. High frequency of ras oncogene activation in all stages of human thyroid tumourigenesis. Oncogene 1989; 2:159–164.
110. Wright PA, Williams ED, Lemoine NR, Wynford-Thomas D. Radiation associated and "spontaneous" human thyroid carcinomas show a different pattern of ras oncogene mutation. Oncogene 1991; 6:471–473.
111. Manenti G, Pilotti S, Re FC, Della Porta G, Pierotti MA. Selective activation of ras oncogenes in follicular and undifferentiated thyroid carcinomas. Eur J Cancer 1994; 30A:987–993.
112. Rochefort P, Caillou B, Michiels FM, Ledent C, Talbot M, Sclumberger M, Lavelle F, Monier R, Feunteun J. Thyroid pathologies in transgenic mice expressing a human activated ras gene driven by a thyroglobulin promoter. Oncogene 1996; 12:111–118.
113. Santelli G, de Franciscis V, Portella G, et al. Production of transgenic mice expressing the Ki ras oncogene under the control of a thyroglobulin promoter. Cancer Res 1993; 53:5523–5527.
114. Donghi R, Longoni A, Pilotti S, Michieli P, Della Porta G, Pierotti MA. p53 gene mutations are restricted to poorly differentiated and undifferentiated tumors of the thyroid gland. J Clin Invest 1993; 91:1753–1760.
115. Ledent C, Dumont JE, Vassart G, Parmentier M. Thyroid adenocarcinomas secondary to tissue-specific expression of simian virus-40 large T antigen in transgenic mice. Endocrinology 1991; 129:1391–1401.
116. Vousden K. Interactions of human papillomavirus transforming proteins with the products of tumour suppressor genes. FASEB J 1993; 7:972–979.
117. Ledent C, Dumont JE, Vassart G, Parmentier M. Thyroid expression of an A2 adenosine receptor transgene induces thyroid hypeplasia and hyperthyroidism. EMBO J 1992; 11:37–542.
118. Lendent C, Marcotte A, Dumont JE, Vassart G, Parmentier M. Differentiated carcinomas develop as a consequence of the thyroid specific expression of a thyroglobulin-human papillomavirus type 16 E7 transgene. Oncogene 1995; 10:1789–1797.
119. Ledent C, Coppee F, Parmentier M. Transgenic models of metastatic differentiated thyroid carcinomas. J Endocrinol Invest 1994; 17(suppl)1–6: 17A.
120. Schulz N, Propst, Rosenberg MM, Linnolia RI, Paules RS, Schulte D, van de Woude GF. Patterns of neoplasia in c-moc transgenic mice and their

relevance to human multiple endocrine neoplasia. Henry Ford Hosp Med J 1992; 40:307–311.
121. Eng C, Foster KA, Healey CS, Houghton C, Gayther SA, Mulligan LM, Ponder BA. Mutation analysis of the c-mos proto-oncogene and the endothelin B receptor gene in medullary thyroid carcinoma and phaeochromocytoma. Br J Cancer 1996; 74:339–341.
122. Moser AR, Luongo C, Gould KA, McNeley MK, Shoemaker AR, Dove WF.APC^{min}: a mouse model for intestinal and mammary tumourigenesis. Eur J Cancer 1995; 31A:1061–1064.
123. Harach HR, Williams GT, Williams ED. Familial adenomatous polyposis associated thyroid carcinoma: a distinct type of follicular cell neoplasm. Histopathology 1994; 25:549–561.
124. Colletta G, Sciacchitano S, Palmirotta R, Ranieri A, Zanella E, Cama A, Constantini RM, Battista P, Pontecorvi A. Analysis of adenomatous polyposis coli gene in thyroid tumours. Br J Cancer 1994; 70:1085–1088.
125. Smith P, Wynford-Thomas D, Stringer BMJ, Williams ED. Growth factor control of rat thyroid follicular cell proliferation. Endocrinology 1986; 119: 1439–1445.
126. Roger PP, Servais P, Dumont JE. Stimulation by thyrotropin and cyclic AMP of the proliferation of quiescent canine thyroid cells cultured in a defined medium containing insulin. FEBS Lett 1983; 157:323–329.
127. Roger PP, Dumont JE. Factors controlling proliferation and differentiation of canine thyroid cells cultured in reduced serum conditions: effects of thyrotropin, cyclic AMP and growth factors. Mol Cell Endocrinol 1984; 36:79–93.
128. Williams DW, Williams ED, Wynford-Thomas ED. Control of human thyroid follicular cell proliferation in suspension and monolayer culture. Mol Cell Endocrinol 1987; 51:33–40.
129. Thomas GA, Davies HG, Williams ED. Expression of IGF1 in the normal and short-term goitrogen treated mouse thyroid. J Pathol 1994; 173:355–360.
130. Takahashi MH, Thomas GA, Williams ED. Evidence for mutual interdependence of epithelium and stroma in a subset of papillary carcinomas. Br J Cancer 1995; 72:813–817.
131. Burikanov R, Coulonval K, Pirson I, Lamy F, Roger PP. TSH induces insulin receptor expression and insulin comitogenic stimulation in dog thyroid cells. J Endocrinol Invest 1996; 19(suppl 1–6):23A.
132. Dremier S, Taton M, Coulonval K, et al. Mitogenic, differentiating and scattering effects of hepatocyte growth factor on dog thyroid cells. Endocrinology 1994; 135:135–140.
133. Di Renzo MF, Olivero M, Ferro S, et al. Overexpression of the c-met/HGF receptor gene in human thyroid carcinomas. Oncogene 1992; 7:2549–2553.
134. Thomas GA, Takahashi MH, Davies HG, Hooton N, Williams ED. Cellular localisation of three tyrosine kinase receptors in thyroid tumours. J Endocrinol 1995; 144(suppl RC3).
135. Belfiore A, Gangemi P, Santonocito MG, La Rosa GL, Constantino A,

Fiumara A, Vigneri R. Prognostic value of c-met expression in papillary thyroid carcinomas. Thyroid 1995; 5(suppl 1):264A.

136. Haugen DRF, Akslen LA, Varhaug JE, Lillehaug JR. Demonstration of a TGFα- EGF receptor autocrine loop and c-myc overexpression in papillary thyroid cancers. Int J Cancer 1993; 55:37–43.
137. Paynter OE, Burin GJ, Jaeger RB, Gregorio CA. Goitrogens and thyroid follicular cell neoplasia: evidence for a threshold process. Regul Toxicol Pharmacol 1988; 8:102–109.
138. Takayama S, Aihara K, Onodera T, Akimoto T. Antithyroid effects of phenylthiourea and sulfamonomethoxine in rats and monkeys. Toxicol Appl Pharmacol 1986; 82:191–199.
139. Swarm RL, Roberts GK, Levy AC, Hines LR. Observations on the thyroid gland of rat following administration of sulfamethoxazole and trimethoprim. Toxicol Appl Pharmacol 1973; 24:351–363.
140. Kutscher AH, Lane SL, Segall R. The clinical toxicity of antibiotics and sulfonamides. J Allergy 1954; 25:135–150.
141. Koch-Weser J, Sidel VW, Dexter M, Parish C, Finer DC, Kanarek P. Adverse reactions of sulfisoxazole, sulfamethoxazole and nitrofurantoin. Arch Intern Med 1971; 128:399–404.
142. Larsen PR, Berry MJ. Nutritional and hormonal regulation of thyroid hormone diseases. Ann Rev Nutr 1995; 15:323–352.
143. Thomas GA, Williams D, Williams ED. Clonal origin of thyroid tumours. In: Wynford-Thomas D, Williams ED, eds. Thyroid Tumours: Molecular Basis of Pathogenesis. Edinburgh: Churchill Livingstone, 1989:38–56.

19
Lung Carcinogenesis

F. F. Hahn
Lovelace Respiratory Research Institute
Albuquerque, New Mexico

I. DESCRIPTION OF PULMONARY CARCINOGENESIS

A. Introduction

Lung cancer is an important public health problem. It is the most frequent cause of cancer fatalities in the United States, and the incidence of lung cancer has been rising for at least 50 years (Travis, 1995). Only recently has the incidence rate in men declined; however, this decline is more than matched by an increased rate in women. In addition, the survival rate of those affected has not changed appreciably in the past 20 years, suggesting that therapeutic advances have had little effect. The overwhelming role of tobacco smoking in the causation of lung cancer has been repeatedly demonstrated (Boyle, 1995). The current lung cancer rates reflect cigarette smoking habits of men and women in the past decades. Although cigarette smoking dominates in the causation of lung cancer, other potential causative agents are of importance because they may interact with cigarette smoke or be part of ambient or indoor air pollution or occupational exposure that may be preventable (Samet, 1993).

Inhalation is the most important route of exposure to agents causing lung cancers. Of the 57 human carcinogenic agents or factors identified by the International Agency for Research on Cancer (IARC) as of 1987, 21 induce cancer of the lung or pleura (Tomatis, 1989). And of these 21, essentially all enter the body by the inhalation route (Table 1). For example, none of the pharmaceuticals, which are not administered by inhalation, cause lung cancer. The IARC list does not include a number that are exposures associated with industrial processes, such as aluminum or coke pro-

Table 1 Human Carcinogens Affecting Lung[a]

Types	Total number	Number affecting lung/pleura
Industrial processes	12	6
Chemicals and chemical groups	17	13
Environmental agents and cultural risk factors	11	2
Pharmaceuticals	17	0
Total	57	21

[a]Based on IARC Monographs.
Source: Tomatis et al., 1989.

duction and iron or steel founding, or to specific chemicals with occupational exposure, such as asbestos, chromium compounds, and coal tars (Tomatis, 1989).

Compounds administered by routes other than inhalation can cause lung cancer. For example, 4-(*N*-methyl-*N*-nitrosamino)-1-(3-pyridyl)-butanone (NNK) administered to rats by intravenous injection is metabolized by Clara cells in the lung, resulting in cytotoxicity and ultimately lung cancer (Belinsky, 1986). In addition, several polycyclic hydrocarbons, nitrosamines, hydrazine, and aflatoxin B, when injected intravenously or in the peritoneal cavity, will increase the number of pulmonary tumors in strain A mice (Stoner, 1991). However, even in experimental studies, inhalation is the primary route for entry of carcinogens that cause lung cancer. For example, well over 60 organic chemicals (either in vapor or particulate form), inorganic compounds (such as metals, nonmetals, and radionuclides), and complex mixtures have been shown to be carcinogenic in rats or mice after inhalation exposure.

B. Human Pulmonary Carcinogenesis

General Morphological Features

The major histological types of human lung cancer are squamous cell carcinoma, adenocarcinoma, small-cell lung carcinoma (SCLC), and large-cell carcinoma (WHO, 1981). These four types comprise about 88% of all lung cancers, as shown in Table 2. Only in the last 10–15 years has the frequency of adenocarcinomas exceeded that of squamous cell carcinomas, probably due to changes in smoking habits over the past 30 years. The use of filtered cigarettes results in less inhalation of cigarette particulate material but a

Table 2 Major Histological Types of Human Lung Cancer[a]

Lung cancer type	Percent of total lung cancers
Adenocarcinoma	31.5
Squamous cell carcinoma	29.4
Small-cell carcinoma	17.8
Large-cell carcinoma	9.2
Other	12.1

[a]Based on 59,260 cases in the NCI SEER Program.
Source: Travis et al., 1995.

pattern of deposition deeper in the lung. The site of lung cancers may be central, in or around the bronchi, or peripheral, in the lung parenchyma (Travis, 1996). About two thirds of the squamous cell carcinomas and the SCLC are central. The adenocarcinomas and large-cell carcinomas tend more often to be peripheral.

For treatment purposes, human lung cancers are frequently classified merely as SCLC or non–small-cell lung cancer (NSCLC) because there are major differences in the therapeutic approach to patients with SCLC compared with all other major histological types of lung cancer. Another factor in such a simplified classification is the cellular heterogeneity in lung cancer, which can make classification by histological type problematic (Travis, 1996). At the cellular level, electron microscopic analyses of primary lung cancers demonstrate that many are composed of cells each of which contains ultrastructural characteristics of multiple histological types.

Molecular Features

Understanding of the molecular biology of lung cancer has expanded rapidly in the past few years. The focus of many investigations in human lung cancer has been the similarities and differences of the molecular changes in SCLC and NSCLC to effect better treatments of lung cancer. Some of the most consistent molecular events associated with lung cancer are shown in Table 3. The most widespread genetic abnormality observed is the loss of portions of the short arm of chromosome 3 (3p).This 3p abnormality is present in nearly 100% of SCLC and about 75% of NSCLC. A number of genes have been cloned from in or near the area of deletions on 3p, but none appears to be the specific tumor-suppressor gene (or genes) whose deletion or mutation is required for the development of lung cancer. Other

Table 3 Molecular Changes[a] Associated with Human Lung Cancer

Molecular Change	Small-cell lung cancer	Non–small-cell lung cancer	Squamous cell carcinoma	Adeno-carcinoma	Large-cell carcinoma
Oncogene expression[b]					
myc	18	8	–	–	–
K-ras	0	16	–		–
H-ras	0	2	–	[30]	–
N-ras	0	10	–		–
HER2/neu[c]	0	–	–	30	–
Suppressor genes[c] (mutation or loss)					
Retinoblastoma	60	10	–	–	–
p53	77	49	–	–	–
Chromosome 3p	~100	75	–	–	–
Growth factor expression[a]					
Bombesin/gastrin releasing peptide	31	6	–	–	–
Epidermal growth factor	0	–	86	50	55
Insulinlike growth factor	95	82	–	–	–

[a]Percent of total tumor type showing change.
[b]From Richardson and Johnson, 1993.
[c]From Gazdar, 1994.

common molecular abnormalities are the mutations and loss of p53 or RB1 tumor suppressor genes. p53 is mutated or lost in >75% of SCLC and ~50% of NSCLC. RB1 is mutated or lost in a majority of SCLC, but in few of NSCLC. A major portion of both SCLC and NSCLC produce the peptide hormone insulinlike growth factor-1, which may serve as an autocrine growth factor (Richardson and Johnson, 1993). Several molecular abnormalities serve as discriminators between SCLC and NSCLC. The ras family of oncogenes and the HER/neu oncogene are found almost exclusively in NSCLC. Among the different histological types of NSCLC, these ras oncogenes are usually associated with adenocarcinomas. Epidermal growth factor is also found almost exclusively in NSCLC and frequently in the squamous cell carcinomas. The oncogene myc and the growth factor, bombesin, are not found frequently in the SCLC, but these two factors are found with a much higher frequency than in NSCLC.

Some molecular changes have a correlation with carcinogen exposure. K-ras mutations are significantly less frequent in patients who have never smoked than in those with a smoking history. About 30% of smokers or former smokers have a K-ras mutation in their tumor, whereas less than 5% of adenocarcinomas of patients who never smoked contain such a mutation (Rodenhuis, 1996). The absence of germline RB1 mutations in all but a few cases of lung cancer indicates that cigarette smoking or other environmental carcinogens are the most important factors in causation of the disease (Carbone and Kratzke, 1996).

C. Animal Pulmonary Carcinogenesis

General Morphological Features

In mice, the spontaneous incidence of lung neoplasms is highly dependent on the strain of mouse (Table 4). In those strains frequently used in toxicology studies ($B6C3F_1$, Balb/c, C57B1), the lifetime incidence generally ranges from about 5% to 20%, although in any given study it may be as high as 50%. In strain A mice the incidence may be as high as 90% and begins to increase before 4 months of age. The incidence in males tends to be higher than in females.

Rats have a low spontaneous incidence of lung neoplasms. The range of lifetime incidence among different strains and colonies of rats is 0.3–2.3% (Hahn, 1993). Table 4 shows the incidence in the rat strains frequently used in bioassays. The incidence is higher among males than females by a factor of about 2.

The major histological types of rodent lung cancer are listed in Table 5. Although the lung tumors of mice and rats can be histologically classified using the same criteria, they are not really the same disease in the two species. In all mouse strains the vast majority of lung neoplasms are adenoma or adenocarcinomas. A clear progression has been shown from alveolar epithelial hyperplasia to adenoma to adenocarcinoma in unexposed mice and those responding to pulmonary carcinogens of several types (Foley et al., 1991). These neoplasms have been classified in a number of ways over the years, including pulmonary adenoma or carcinoma, alveolar-bronchiolar adenoma or carcinoma, broncho-alveolar adenoma or carcinoma, papillary adenoma or adenocarcinoma, and alveologenic nodules. The many names for the same neoplastic process results from the difficulty in clearly distinguishing proliferative from benign from malignant pulmonary lesions in mice and from controversy over the cellular origin of the neoplasms. Most recent data indicate that the majority of pulmonary neoplasms arise

Table 4 Lung Tumor Incidence in Laboratory Mice and Rats Commonly Used in Bioassays

		Lung tumor incidence[a]		
Strain	Sex	Benign	Malignant	Ref.
Mice				
B6C3F$_1$	Female	5.5 ± 3.6	2.0 ± 2.3[b]	Haseman et al., 1984a
B6C3F$_1$	Male	12 ± 6.7	5.1 ± 4.3	Haseman et al., 1984a
C$_{57}$/B1/1crf	Female	4.4	0.4	Rowlatt et al., 1976
C$_{57}$/B1/1crf	Male	6.0	0	Rowlatt et al., 1976
Balb/c	Female	(21)		Heath et al., 1982
A/J	Female	(75)		Shimkin et al., 1966
A/J	Male	(78)		Shimkin et al., 1966
Rats				
F344	Female	0.8 ± 1.4	0.4 ± 0.9	Haseman et al., 1984b
F344	Male	1.5 ± 2.1	0.9 ± 1.6	Haseman et al., 1984b
Sprague-Dawley	Female	0.1	0.3	Prejean et al., 1973; MacMartin et al., 1992
Sprague-Dawley	Male	0.3	0.4	Prejean et al., 1973; MacMartin et al., 1992
Wistar	Female	0.0	0.3 (0–2.0)[c]	Walsh and Poteracki, 1994
Wistar	Male	0.9 (0–5.7)	0.2 (0–0.8)	Walsh and Poteracki, 1994

[a]Incidence in animals observed long-term (22 months to life span).
[b]± SD.
[c]Range.

from pulmonary Type II cells, but that some also arise from Clara cells (Boorman and Dixon, 1991).

A progression of alveolar proliferative lesions from hyperplasia to benign to malignant in unexposed animals is not clearly recognizable in rats as it is in mice. This may be due to the low number of spontaneous lung neoplasms available for study in rats. The number of neoplasms in rats that are classified as benign is much lower than in mice. The fraction of spontaneous lung tumors in rats that are benign is usually less than 0.33.

Table 5 Major Histological Types of Rodent Lung Cancers

Lung cancer type	Mouse	Rat
Adenoma	+	+
Keratinizing cystic epithelioma	O	+
Adenocarcinoma (bronchioloalveolar carcinoma)	+	+
Squamous cell carcinoma	+	+
Adenosquamous carcinoma	O	+

+ = Present; O = not present in spontaneous or induced lung cancer.

Rats generally respond to pulmonary carcinogens with an increase in the number of carcinomas of the lung. The number of adenomas may increase but not as much as carcinomas, resulting in an even lower fraction of adenomas. Squamous cell carcinomas may also be increased, a situation frequently associated with higher doses of carcinogen and the resultant marked lesions of pulmonary injury and repair.

The cystic keratinizing epithelioma and its precursor, the keratinizing squamous cyst, appear to be proliferative lesions limited to rats and rarely found in other species (Boorman et al., 1996). This phenomenon is rarely found in unexposed rats and is most prominent in rats exposed to heavy burdens of particles in the lung, although it also occurs after inhalation of radioactive particles, a situation where the particle load is low. The lesions are part of a spectrum of lesions that appears to progress from squamous metaplasia to keratin cyst to keratinizing cystic epithelioma to squamous cell carcinoma. The distinction between cyst and epithelioma is still controversial (Carlton et al., 1994). However, because these lesions occur infrequently and late after inhalation exposure, there has been little opportunity to study their biological behavior in detail.

Molecular Features

Molecular biology techniques are being used to study the pathogenesis of lung cancer in animals so that, with better understanding of the process, the disease in humans can be prevented, detected, or treated. The focus of study is on early lesions, progression of lesions and correlation of molecular changes with types of carcinogens.

Mice are good species to use for the study of the progression of lung cancer because the timing of morphological changes has been well characterized in several strains. The sequence of some of the major molecular changes in the lungs of mice after carcinogen exposure is shown in Figure 1. The earliest changes detected are DNA methylation and adduct

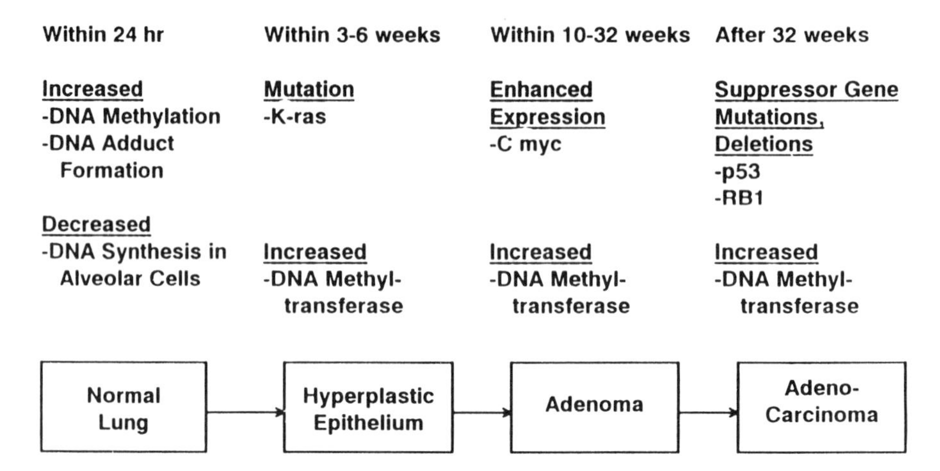

Figure 1 Sequence of molecular changes in mouse lung after carcinogen exposure. (From Malkinson and Belinsky, 1996.)

formation (Wilson et al., 1992). For example, urethane injection (a classic procedure for inducing lung tumors in susceptible mouse strains) causes a rise in DNA methylation detectable 6 hours after injection and lasting about one week. In resistant strains methylation briefly decreases after injection. The direct-acting chemical carcinogen NNK causes increased DNA methylation 7 days after exposure (Belinsky et al., 1996a). Subsequently, DNA methyltransferase gene expression is increased in hyperplastic alveolar epithelium, adenomas, and carcinomas that develop many months later. The induction and persistence of DNA adducts also occurs early after exposure and is correlated with increased lung tumor incidence.

Many lung tumors of mice have K-ras mutations. The frequency of these mutations is dependent on the strain of mouse and the nature of the initiating carcinogen. The high susceptibility of the strain A mouse to develop lung tumors has been directly linked to three "pulmonary susceptibility" genes, one of which correlates with the K-ras gene (Malkinson, 1991; You et al., 1992). Activating K-ras mutations are found in a high percentage (>90%) of lung tumors in the strain A mouse, regardless of how the tumors are induced. Other strains, with lower spontaneous incidence of lung tumors, have a lower frequency of K-ras mutations and respond to pulmonary carcinogens with tumors that have a low frequency of K-ras mutations. The spectrum of K-ras mutations is dependent on the nature of the initiating carcinogen, not on the strain of mouse (You et al., 1991; Devereux et al., 1993). Characteristic mutational spectra have been found in lung tumors induced by tetranitromethane (GGT → GAT codon 12), butadiene (GGC → CGC codon 13), and urethane (codon 61 mutations) (Stowers et al., 1987; Goodrow et al., 1990; Ohmori et al., 1992). These characteristic mutations are important as they may reveal mechanisms underlying tumor initiation by these carcinogens. Mutations or deletions in the p53 or RB1 tumor-suppressor genes have been detected in low frequencies (<10%) in mouse carcinomas induced by chemical carcinogens (Hegi et al., 1993; Malkinson et al., 1996).

The molecular changes of lung cancer in rats have been studied primarily because of the wide variety of carcinogens that will induce these cancers in rats. The sequence of some of the major cellular and molecular changes in the lungs of rats after carcinogen exposure is shown in Figure 2. The progression of the proliferative changes in the rat lung are not as stereotyped as they are in the mouse. Metaplasia of the alveolar epithelium with progression to squamous cell carcinoma is a pathway in rats that is not readily expressed in mice (Herbert et al., 1993). The sequence of morphological events and correlated molecular events in Figure 2 are based on the reaction to inhaled plutonium dioxide, a radioactive metal oxide. Timing may be different with other carcinogens. The initial event noted is a increase in cell turnover in pulmonary cells, most likely a reaction to cellular DNA damage. Epithelial hyperplasia and metaplasia are apparent by 6–9 months and neoplasms start to appear at 10–12 months after exposure. K-ras muta-

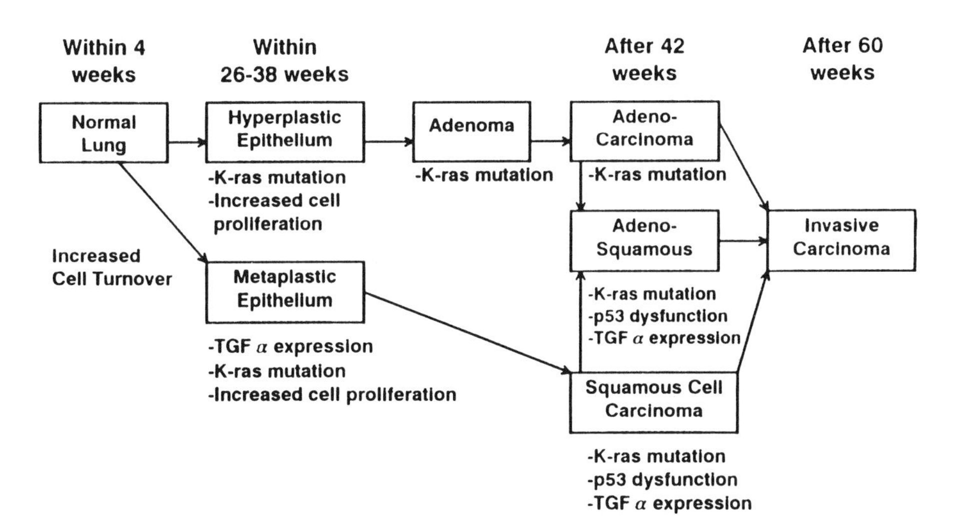

Figure 2 Sequence of molecular changes in lung of rats after carcinogen exposure. (From Herbert et al., 1993.)

Table 6 Carcinogen Dependence of Molecular Changes in Lung Tumors of Rats

Carcinogen	K-ras mutation (%)[a]	p53 dysfunction (%)	Estrogen receptor methylation (%)
Tetranitromethane[b]	100	ND	ND
Plutonium dioxide[c]	40	8 (of SCC only)	81
X-irradiation[d]	3	17 (of SCC only)	38
Beryllium metal[e]	10	0	ND
4-(N-methyl-N-nitrosamino)-1-(3-pyridyl)-butanone[f]	0	0	14
Diesel exhaust/carbon black[g]	6	54 (of SCC only)	ND

ND, Not done; SCC, squamous cell carcinoma.
[a]Tumors examined showing change.
[b]From Stowers et al., 1987.
[c]From Stegelmeier et al., 1991; Kelly et al., 1995; Issa et al., 1996.
[d]From Issa et al., 1996; Belinsky et al., 1996b.
[e]From Nickell-Brady, 1994; Belinsky et al., 1996b.
[f]From Issa et al., 1996.
[g]From Swafford et al., 1995.

tions are associated with 27% of the hyperplasias, 69% of the adenomas, and 41% of the carcinomas. p53 dysfunctions were found only in squamous cell carcinomas and in only 8% of this tumor type. Further studies of lung tumors induced by other carcinogens in rats show that the type and frequency of molecular changes varies widely depending on the inducing carcinogen (Table 6). For example, the frequency of K-ras mutations can vary from 0 to 100% and of p53 dysfunctions from 0 to 54%. Another important point is that p53 dysfunctions, if present at all, are found only in squamous cell carcinomas. The most striking feature of these studies is that neither K-ras nor p53, both important molecular changes in human lung cancer, are of much importance in pulmonary carcinogenesis.

II. PULMONARY CARCINOGENS IN HUMANS AND ANIMALS

A. Known Human Pulmonary Carcinogens

Smoking is estimated to be the cause of about 85% of lung cancer deaths in the United States (Schottenfeld, 1996). Early epidemiological studies in the 1950s showed the association between tobacco smoking and lung cancer, and accumulating evidence in the 1960s firmly established the causal relationship (Wu-Williams and Samet, 1994). The relative risk of death from lung cancer among males who smoke more than 25 cigarettes a day is about 25 compared to the risk for nonsmokers (Schottenfeld, 1996). The risk depends on various aspects of smoking behavior, such as age at starting smoking, number of cigarettes smoked per day, the products smoked and inhalation pattern, as well as individual susceptibility. Gender, ethnicity, and metabolic detoxifying capabilities are some of the factors determining individual susceptibility (Caporaso et al., 1990). Environmental tobacco smoke, or side stream smoke, is also associated with elevated relative risk, but at a much lower level, about 2.

The chemical complexity of cigarette smoke has made it difficult to identify the relative importance of the many putative carcinogenic agents present, which include polynuclear aromatic hydrocarbons, nitrosamines, aromatic amines, benzene, aldehydes, arsenic, and polonium 210. Some epidemiological studies have shown that smoking filter cigarettes that yield medium levels of tar is not as strongly related to lung cancer as the smoking of higher tar yield cigarettes (Wynder and Muscat, 1995). Thus, the carcinogenicity of cigarette smoke seems to be related to the tar yield. However, other factors, such as changes in smoke-inhalation patterns or smoking habits, may be involved; consequently continued studies are needed. Exposures to other environmental and occupational carcinogens may be interactive with cigarette smoking and may also influence trends of lung cancer incidence and mortality.

Exposure to radon and its alpha-particle–emitting decay products is a well-described situation for causing lung cancer, if exposures are sufficiently high (National Research Council, 1988). Radon is an inert gas that can migrate from soils and rocks and accumulate in enclosed areas, such as underground mines and tightly built homes. Studies of experimentally exposed laboratory animals and epidemiological studies of uranium, tin, and iron miners have clearly shown an excess risk of lung cancer with exposure. To date, however, epidemiological studies of residential radon exposure have not convincingly demonstrated an excess risk of lung cancer. Such studies are confounded by uncertain exposure estimation, low total exposure, and anticipated small risks. Consequently, estimates of the hazards of indoor radon have been derived primarily from studies of underground miners. A recent analysis of pooled epidemiological data of underground miners was used to estimate the risk for indoor radon exposure (Lubin et al., 1995). Based on this analysis, about 10% of all lung cancer deaths in the United States and 30% of the lung cancer deaths in those that never smoked might be due to indoor radon exposure. The authors warn that these estimates should be interpreted with caution, because concomitant exposures of miners to agents such as arsenic or diesel exhaust may modify the radon effect and, when considered together with other differences between homes and mines, might reduce the applicability of findings in miners.

Estimates vary on the proportion of lung cancers in the general population caused by occupational and other environmental agents (Coultas, 1994). For example, Doll and Peto (1981), in a review of the causes of cancer, estimated that 5% of lung cancers in women and 15% in men were related to occupational exposures. A population-based, case-control study in New Mexico empirically documented that about 10% of lung cancer cases in men could be attributed to occupation. A review of the literature cited risk estimates for occupation and lung cancer ranging from 4 to 40% (Vineis and Simonato, 1991). These estimates indicate that occupation is an important factor among the causes of cancer, although the total number of cases is small compared with those associated with cigarette smoking.

In specific occupations or exposure situations, the lung cancer risk can be significant. For example, occupations with significant exposure to asbestos, bis(chloromethyl)ether, chromium compounds, nickel and nickel compounds, polycyclic aromatic hydrocarbons, or vinyl chloride all have an significantly increased risk for lung cancer. In addition, epidemiological studies have associated a number of industrial processes with increased lung cancers. In most cases, a specific carcinogenic compound is suspected but cannot be identified with certainty. Table 7 shows the agents identified by the International Agency for Research on Cancer as known human lung carcinogens.

B. Known Rodent Pulmonary Carcinogens

A wide variety of vapors, dusts, and aerosols can induce lung cancer in laboratory rats and mice. These agents can be divided into three general classes: organic chemicals (Table 8), inorganic chemicals (Table 9), and complex mixtures (Table 10).

Organic chemicals include both vapors and particles. The inhaled organic chemicals that result in lung tumors are primarily in vapor form (Table 8). Several of these, such as bis(chloromethyl)ether and dimethyl sulfate, also cause nasal tumors. Others, such as vinyl chloride and butadiene, cause tumors in other organs as well as lung. Chemicals that cause neoplasms at multiple sites frequently require metabolism in the body to be carcinogenic. Often, the neoplasms in other organs have a higher incidence and are more serious health risks than the lung tumors. For example, many studies of vinyl chloride in rats have shown a high incidence of Zymbal's gland carcinoma and hepatic hemangiosarcoma, but few lung tumors. Bu-

Table 7 Agents or Processes Causally Associated with Human Lung Cancer

Agents or processes
Chemicals and groups of chemicals
Arsenic and arsenic compounds
Asbestos
Beryllium compounds
Bis(chloromethyl)ether and chloromethyl methyl ether
Cadmium compounds
Chromium compounds, hexavalent
Coal tars
Coal tar pitches
Mustard gas (sulfur mustard)
Nickel and nickel compounds
Soots
Talc containing asbestiform fibers
Vinyl chloride
Environmental agents and cultural risk factors
Tobacco smoke
Radon and its decay products
Industrial processes
Aluminum production
Coal gasification
Coke production
Hematite mining, underground with exposure to radon
Iron and steel founding
Painter, occupational exposure

Based on database of IARC Monographs Program.
Source: Tomatis, 1989.

tadiene is such a potent carcinogen in mice that the initial inhalation bioassay was terminated early because of high mortality and multiple fatal tumors involving the heart, ovary, and stomach, as well as lung (Huff, 1985).

Organic chemicals in a particulate form can induce lung tumors by the inhalation route. In this form, tumors are usually restricted to the respiratory tract. Benzo-*a*-pyrene causes a marginal increase in lung tumors in rats that inhale SO_2. On the other hand, Syrian hamsters developed lung tumors after instillation of benzo-*a*-pyrene, but carcinogenesis is not enhanced by inhaled sulfate.

Inorganic chemicals that are pulmonary carcinogens in rodents include metallic particles, nonmetallic particles, fibers, and radioactive particles and gases. The metallic compounds of antimony, beryllium, cadmium, nickel, and titanium in several different chemical forms result in pulmonary

Table 8 Animal Lung Carcinogens: Organic Chemicals

	Rat	Mouse	Ref.
Vapors			
Benzene	ND	+	Farris et al., 1993; Snyder et al., 1988
Bis(chloromethyl)ether	+	+	Leong et al., 1971
1,3-Butadiene	−	+	Melnick et al., 1990
Bromoethane (ethyl bromide)	−	+	NTP, 1989a
Chloroethane (ethyl chloride)	−	+	NTP, 1989b
Chloromethyl methyl ether	−	±	Laskin et al., 1975
Diazomethane	±	±	Pepelko, 1984
1,2-Dibromoethane (ethylene dibromide)	+	+	NTP, 1982a
1,2-Dibromo-3-chloropropane	+	−	NTP, 1982b
Dichloromethane (methylene chloride)	−	+	IARC, 1986
1,3-Dichloropropene	−	+	Lomax, 1989
1,2-Epoxybutane	+	−	Huff et al., 1991
Ethylene oxide	−	+	IARC, 1994
Hydrazine	ND	+	MacEwen, 1974
Mustard gas	ND	+	IARC, 1987
Nitrobenzene	−	+	Cattley et al., 1994
3-Nitro-3-hexene	+	+	Pepelko, 1984
Tetranitromethane	+	+	Bucher et al., 1991
Trichlorethylene	−	+	IARC, 1987
Urethane	ND	+	Leong et al., 1971
Vinyl chloride	+	+	IARC, 1987
Vinylidene chloride	−	+	IARC, 1986
Particles			
Polymeric methylene diphenyl diisocyanate	+	ND	Reuzel et al., 1994

ND = No data; + = positive; − = negative; ± = limited.

carcinoma (Table 9). The chemical form of the metal is a major factor and relates to the solubility of the compound. For example, inhaled nickel subsulfide, which is relatively insoluble, causes more lung tumors than nickel chloride, which is very soluble. However, nickel oxide, which is quite insoluble, causes fewer lung neoplasms than nickel subsulfide. Consequently, factors other than solubility must be involved. The physical size of the particle may play a role in the carcinogenicity. Finely divided forms of titanium dioxide are more toxic than larger particles (Oberdörster, 1995). In all cases of carcinogenicity, the differences probably relate to differences in the actual doses of chemical to the target cells and tissues.

A number of nonmetallic particles may cause lung tumors if inhaled

Table 9 Animal Lung Carcinogens: Inorganic Chemicals

	Rat	Mouse	Ref.
Metallic particles			
Antimony ore, trioxide, trisulfide	+	ND	Groth et al., 1986
Beryllium fluoride, hydrogen phosphate, sulfate	+	ND	Schepers, 1961, 1971
Beryllium metal	+	±	Finch et al., 1996a
Beryl ore	+	ND	Wagner et al., 1969
Cadmium chloride, sulfate, sulfide	+	–	Heinrich et al., 1989; Glaser et al., 1990
Cadmium oxide	+	+	Heinrich et al., 1989; Glaser et al., 1990
Calcium chromate	+	+	Nettesheim et al., 1971
Chromium dioxide	±	ND	Lee et al., 1988a
Nickel carbonyl	+	ND	Sunderman, 1965
Nickel metal, subsulfide	+	–	Heuper et al., 1958
Nickel oxide	+	±	Dunnick et al., 1995
Titanium dioxide	+	–	Lee et al., 1985
Titanium tetrachloride (hydrolysis products)	+	ND	Lee et al., 1986
Zinc manganese beryllium silicate	+	ND	Schepers, 1961
Nonmetallic particles			
Carbon black	+	–	Nikula et al., 1995
Coal dust	±	ND	Martin et al., 1977
Oil shale dust	+	ND	Holland et al., 1985
Silica (crystalline)	+	–	Saffioti et al., 1995; Wilson et al., 1986
Talc (asbestos-free)	+	–	NTP, 1993
Volcanic ash	+	ND	Wehner et al., 1986
Fibers			
Aramid	+	ND	Lee et al., 1988b
Asbestos—amosite	+	–	Reeves et al., 1971
Asbestos—anthophyllite	+	ND	IARC, 1977
Asbestos—chrysotile	+	±	Mast et al., 1995a
Asbestos—crocidolite	+	–	Reeves et al., 1971
Ceramic—kaolin	+	ND	Mast et al., 1995b
Ceramic—alumina zirconia silica	+	ND	Mast et al., 1995a
Potassium octatitanate	±	ND	Lee et al., 1981
Rockwool	±	ND	Wagner et al., 1984
Radionuclides			
Alpha-emitting particulate radionuclides	+	+	Lundgren et al., 1987, 1995
Beta-emitting particulate radionuclides	+	+	Lundgren et al., 1974, 1992
Radon and decay products	+	–	Monchaux et al., 1994a
Gases			
Ozone	–	±	NTP, 1994

+ = Positive; ND = no data; ± = limited; – = negative.

by rats (Table 9). None of these is a very potent carcinogen; they require high aerosol concentrations and long exposure times for much increase in lung tumor incidence. And none of these particles is mutagenic in in vitro assays. This group of compounds appears to induce neoplasms in the lungs of rats by epigenetic mechanisms related to particle overload, a phenomenon discussed in a later section.

Asbestos is the classic nonmetallic fiber that causes lung cancer in rodents after inhalation (Table 9). Mesotheliomas, tumors of the pleural lining of the lung and thoracic cavity, also result from inhaled asbestos. Many substitutes for the commercial uses of asbestos have been developed. In bioassays of these newly created fibers, asbestos is frequently used as a positive control. Two such fibers, ceramic aluminosilicate and aramid, have been found to be carcinogenic in rats. Quartz and silica are generally not considered carcinogenic in exposed human populations, however, several studies in rats have demonstrated the induction of lung cancers.

Many studies of inhaled radionuclides have been conducted (Table 9). For the purposes of inhalation hazards, radioactive materials can be divided into three general categories: particulate radioactive materials that emit alpha particles, particulate materials that emit beta particles, and radon gas, with its associated daughter radionuclides that emit alpha particles. When these radioactive materials in particulate form are inhaled and deposited in the lungs, a radiation dose is delivered primarily to the pulmonary parenchyma because many particles are delivered to the deep lung, and penetration of the emitted alpha or beta particles is low (micrometers to millimeters). Radon gas, on the other hand, delivers most of its radiation dose to the cells lining major airways with little gas reaching the alveolar parenchyma of the deep lung. All of these radiations can cause tumors in rats or mice.

Complex mixtures that are known rodent pulmonary carcinogens include atmospheres that contain particles with several, if not hundreds, of organic and inorganic chemicals and various gases that may be toxic or carcinogenic (Table 10). These atmospheres are usually generated directly from a source found in the environment, such as a cigarette or an automobile engine. Cigarette smoke has frequently been studied because of its importance to the human population. It is not, however, an impressive inducer of lung tumors in animals. Significant increases in lung tumors have been observed in long-term smoking studies with strain A (Essenberg, 1952), C57B1 (Harris et al., 1974), and albino mice strains (Otto, 1963). Rats experienced a significant increase in lung tumors in two studies of cigarette smoke, but the increases are not impressive (Dalby et al., 1980; Finch et al., 1995a). The reason that rodent species do not readily develop lung cancer after inhalation of cigarette smoke is not clear, but may relate

Table 10 Animal Lung Carcinogens: Complex Mixtures by Inhalation

	Rat	Mouse	Ref.
Artificial smog (ozonized gasoline vapor)	ND	+	Kotin et al., 1958
Benzo(*a*)pyrene + sulfur dioxide	+	−	Laskin et al., 1970
Cadmium oxide + zinc oxide	+	ND	Glaser et al., 1990
Coal smoke	+	+	Liang et al., 1988
Coal tar aerosols	+	+	Heinrich et al., 1994
Coal tar pitch + carbon black (CB)	+	ND	Heinrich et al., 1994
Coal tar pitch + CB + SO_2 + NO_2 + HCO	+	ND	Heinrich et al., 1994
Diesel exhaust	+	±	Mauderly et al., 1987
Plutonium-239 dioxide + beryllium metal	+	ND	Finch et al., 1996a
Plutonium-239 dioxide + tobacco smoke	+	ND	Finch et al. 1995
Radon + diesel exhaust	+	ND	Monchaux et al., 1994a
Radon + ozone	+	ND	Monchaux et al., 1994b
Radon + tobacco smoke	+	ND	Monchaux et al., 1994a
Tobacco smoke	+	±	Dalby et al., 1980
Urban air	±	−	Ito et al., 1989
Wood smoke	+	+	Liang et al., 1988

ND = No data; + = positive; ± = limited; − = negative.

to the carcinogenic dose to the pulmonary cells at risk. Animals are generally more susceptible to the acute toxic effects of cigarette smoke (due to nicotine and carbon monoxide) and cannot inhale as much smoke as humans. In addition, many animals will avoid inhaling smoke if possible, thus reducing the dose to the tissues. Several large, long-term studies in the United States, Europe, and Japan have shown a carcinogenic effect of diesel exhaust in rats (Ishinishi et al., 1986). More recent studies have attempted to determine the role of the organic and inorganic fractions of the exhaust in production of these tumors. Results indicate that the inorganic fraction alone can account for the entire carcinogenicity of diesel exhaust (Nikula et al., 1995).

III. RESPONSE OF ANIMALS TO KNOWN HUMAN PULMONARY CARCINOGENS

The pulmonary response of rats and mice to inhalation of known human lung carcinogens is shown in Table 11. Rats develop lung tumors after inhalation of human lung carcinogens in all instances where they were tested except for one. Hexavalent chromium compounds caused only an

Table 11 Pulmonary Tumor Response of Laboratory Rodents to Inhalation of Known Human Pulmonary Carcinogens[a]

	Human	Rat	Mouse
Chemicals and groups of chemicals			
Arsenic and arsenic compounds	+	ND	ND
Asbestos	+	+	±
Beryllium compounds	+	+	±
Bis(chloromethyl)ether and chloromethyl methyl ether	+	+	+
Cadmium compounds	+	+	±
Chromium compounds, hexavalent	+	±	ND
Coal tars	+	+	+
Coal tar pitches	+	ND	ND
Mustard gas (sulfur mustard)	+	ND	±
Nickel and nickel compounds	+	+	±
Soots	+	+	ND
Talc containing asbestiform fibers	+	+	−
Vinyl chloride	+	+	+
Environmental agents and cultural risk factors			
Tobacco smoke	+	+	±
Radon and its decay products	+	+	−

[a]Not included are industrial processes known to be associated with human lung cancer.
+ = Positive; ND = no data; ± = limited; − = negative.
Source: Based on database of IARC Monographs Program (Tomatis, 1989) and Refs. in Tables 8, 9, and 10.

equivocal increase in lung tumors in rats. On the other hand, both tobacco smoke and radon and its decay products caused lung tumors in rats. This finding is important because of the major role of these two compounds in causing human lung cancer.

In some studies of cigarette smoke in rats, statistically significant increases in lung tumor incidence are present; however, none of the responses seems indicative of a potent pulmonary carcinogen. One reason for this rather weak response may be the dose of tobacco smoke, which is usually lower in rats than in humans. Usually smokers intentionally inhale relatively large amounts of smoke with each puff—that is the reason they are smoking. Animals, on the other hand, don't really like to smoke and use any number of mechanisms to avoid inhalation. For example, rodents exposed nose-only in tubes will attempt to hold their breath when smoke is present. Those exposed whole-body in chambers will huddle and hide their noses. The net result is a smaller deposition of smoke particles in rodents than in humans.

Another factor may be our perception of how potent a pulmonary carcinogen tobacco smoke really is. Although cigarette smoking is the dominant cause of lung cancer, only about 15% of cigarette smokers develop lung cancer. In view of the heavy dosage cigarette smokers administer themselves relative to most studies in rats, we should not expect a very high incidence of lung tumors in cigarette smoke-exposed rats.

Mice responded with an increase in lung tumors in only three of the 11 human lung carcinogens (Table 11). Most often the results were equivocal or, in two cases, negative. The three carcinogens which were positive in mice were all vapors; none of the particles was positive.

IV. SCIENTIFIC INTERPRETATION OF COMPARISONS BETWEEN ANIMAL AND HUMAN PULMONARY CARCINOGENIC RESULTS

A. Problem of Mouse Strains

It is well known that the incidence of lung tumors in various strains of mice varies widely. However, this does not appear to be a factor in the relatively low response of mice to inhaled human lung carcinogens. Instructive is the response of the strain A mouse to inhaled potentially carcinogenic materials.

The strain A mouse has been used in short-term carcinogen bioassays because of its marked sensitivity to spontaneous and carcinogen-induced lung neoplasia (Shimkin and Stoner, 1975). In a typical assay, groups of weanling strain A mice are adminstered the test compound by intraperitoneal injection. Four to six months later, the mice are killed and the lung tumors counted. Urethane injection is used as a positive control and results in about 90% lung tumor incidence. The assay has been used to screen potential carcinogens; however, the test is not as straightforward as it might appear. A comparative evaluation has shown poor agreement of test results between laboratories and with test results from genotoxicity and 2-year carcinogenicity tests (Maronpot et al., 1986). This evaluation has been criticized, however, because a significant portion of the compounds compared were aromatic amines and aliphatic halides, chemicals known to be weakly carcinogenic in the rodent lung (Stoner, 1991). Assays with classes of compounds to which the assay is sensitive (e.g., polycyclic hydrocarbons, nitrosamines, carbamates, hydrazines) are more fruitful.

From the standpoint of inhalation carcinogenesis, it is important to note that few inhalation studies have been conducted with strain A mice. Urethane, the standard positive control compound, will induce increased tumor incidence by either intratracheal instillation (McNeill et al., 1990) or inhalation (Leong et al., 1971). Similar exposure to other, known pulmonary carcinogens has resulted in few positive results. Inhalation studies using cigarette smoke have yielded contradictory, but generally negative results. Early studies showed a marginal increase in the number of cigarette smoke-exposed, tumor-bearing A/J mice (Essenberg, 1955). Other studies have failed to find an increased incidence of lung tumors in strain A mice exposed to either mainstream (Lorenz et al., 1942; Finch et al., 1996b) or sidestream cigarette smoke (Witschi et al., 1995). Inhalation exposure to compounds known to be pulmonary carcinogens in rats have produced either no or marginal increases in incidence and number of tumors per mouse. For example, asbestos (Lynch et al., 1957) or bis(chloromethyl)ether (Leong et al., 1971) resulted in only marginal increases, and beryllium (Finch et al., 1996a) and nickel subsulfide (by intratracheal instillation) produced no increases in lung tumors.

Thus, it can be seen that even the strain A mouse yields no or only marginal increases in lung tumors after inhalation exposure to known human lung carcinogens, such as cigarette smoke, asbestos, and bis(choromethyl) ether. These findings are in spite of the strain A having a high background of lung tumors and high susceptibility to carcinogenesis from polycyclic hydrocarbons and nitrosamines given by other routes.

B. Problem of Particle Overload

Particle overload is a phenomenon that influences the interpretation of carcinogenicity bioassays of poorly soluble compounds inhaled by rats (Mauderly, 1996). The term refers to an impaired clearance of poorly soluble, weakly toxic particles from the lung by alveolar macrophages and is most frequently used in reference to chronic exposures in which the rate of pulmonary deposition exceeds the rate of pulmonary clearance. The overload of particles may result in marked pulmonary inflammation, proliferation, and neoplasia. The neoplasms are thought to result from the marked tissue damage and repair, a mechanism thought not to occur in humans.

Particle overload has been described most often in rats, but has also been noted in mice and Syrian hamsters. Limited data from dogs and monkeys chronically exposed to aerosols of UO_2 do not indicate that particle overload occurs in these species (Snipes, 1996). A study of coal miners using a magnetopneumographic method showed that pulmonary clearance was substantially prolonged (Freedman and Robinson, 1988). In addition, evaluation of lung particle burdens from coal miners exposed to heavy dust concentrations showed that more particles were present than predicted from mathematical models of lung clearance. These findings indicate that particle overload may occur in these populations (Mauderly, 1996).

Particle overload, as documented by increasing lung burdens of particles and a slowing of the clearance of tracer particles, has been shown in rodents exposed to diesel exhaust (Mauderly et al., 1990), volcanic ash (Wehner et al., 1986), polyvinyl chloride particles (Muhle et al., 1988), plastic copolymer toner particles (Muhle et al., 1988), and carbon black (Nikula et al., 1995). Altered particle clearance kinetics caused by depressed alveolar macrophage function is regularly seen at particulate burdens above about 1–3 mg in the lung for particles that have poor solubility and low cytotoxicity in rats. Experimental evidence, however, suggests that the volume, not the mass, of the particles phagocytized by alveolar macrophages is most critical for causing impaired clearance function (Morrow, 1992). Impairment becomes apparent once the retained lung particle burden reaches a level equivalent to a volume of about 1 μl/g of lung. The physical form of the particles may also be a factor; ultrafine particles in the nanometer size range and natural and man-made fibers are more toxic than their chemical characteristics might indicate.

More cytotoxic particles may affect macrophage-mediated particle clearance as well, but at much lower lung burdens (Oberdörster, 1995). Cytotoxic particles may kill macrophages or other cells, causing clearance impairments by affecting macrophages directly or causing inflammation that may secondarily impair clearance. Examples of such cytotoxic compounds are crystalline SiO_2, BeO, and Ni_3S_2.

Rats respond differently than other species to the overload of particles in the lung. Adverse effects observed in rats with particle overload include chronic inflammation, septal fibrosis, alveolar proteinosis, epithelial hyperplasia, squamous metaplasia, and benign and malignant lung neoplasms (Table 12). These effects have been documented in rats exposed to a number

Table 12 Pulmonary Effects in Different Species After Exposure to High Concentrations of Particles

	Rat	Mouse	Syrian Hamster	Human
Impaired particle clearance	+ +	+ +	+ +	Yes
Chronic inflammation	+ + +	+ +	+	Yes
Septal fibrosis	+ +	+	+	Yes
Squamous metaplasia	+ +	No	No	Yes
Epithelial hyperplasia	+ + +	+	+	Yes
Neoplasia	Yes	No	No	No

+ = Minimal; + + = mild; + + + = moderate.
Source: Oberdörster, 1995; Heinrich, 1996.

of poorly soluble, weakly toxic compounds including diesel exhaust, carbon black, coal dust, and plastic copolymer toner. Although impaired particle clearance also occurs in mice and Syrian hamsters, subsequent chronic effects are seen to a much lesser degree, and no increased incidence of pulmonary neoplasms has been seen in either species.

Rats readily develop lung tumors after inhalation of metallic and nonmetallic particles and fibers (Table 9). In some cases, the overload status of the rats during the study is known, but in others it is not. In mice exposed to similar particles and fibers, few positive test are recorded.

V. SUMMARY

Much is known about pulmonary carcinogenesis in humans from epidemiological studies of human populations. This is particularly true with respect to cigarette smoking and exposure to radon and its decay products. These two exposure situations are estimated to account for much of the lung cancer in the United States. Studies in animals are important for helping determine specific causative agents related, for example, to occupational exposures or to potential future exposures. Interpreting the pulmonary carcinogenicity of compounds in humans from studies in animals is not a straightforward task. Many factors must be considered—most importantly, the test species and its characteristic response to specific carcinogens. These responses in animals will become better understood as more experience is gained with inhaled carcinogens. In the future, more will become known about the cellular and molecular responses to pulmonary carcinogens, allowing a less empirical approach to bioassays in animals.

REFERENCES

Belinsky SA, White CM, Boucheron JA, Richardson FC, Swenberg JA, Anderson MW. Accumulation and persistence of DNA adducts in respiratory tissue of rats following multiple administrations of the tobacco specific carcinogen 4-(N-methyl-N-nitrosamino)-1-(3-pyridyl)-1-butanone. Cancer Res 1986; 46: 1280–4.

Belinsky SA, Nikula KJ, Baylin SB, Issa JJ. Increased cytosine DNA-methyltransferase activity is target-cell-specific and an early event in lung cancer. Proc Natl Acad Sci 1996a; 93:4045–50.

Belinsky SA, Middleton SK, Picksley SM, Hahn FF, Nikula KJ. Analysis of the K-*ras* and *p53* pathways in X-ray-induced lung tumors in the rat. Radiat Res 1996b; 145:449–56.

Boorman GA, Dixon D. Histogenesis of mouse lung tumors: an overview. Exp Lung Res 1991; 17:107–9.

Boorman GA, Brockmann M, Carlton WW, Davis JMG, Dungworth DL, Hahn FF, Mohr U, Reichhelm H-BR, Turusov VS, Wagner BM. Classification of cystic keratinizing squamous lesions of the rat lung: Report of a workshop. Toxicol Pathol 1996; 24(5):564–72.

Bucher JR, Huff JE, Jokinen MP, Haseman JK, Stedham M, Cholakia JM. Inhalation of tetranitromethane causes nasal passage irritation and pulmonary carcinogenesis in rodents. Cancer Lett 1991; 57:95–101.

Caporaso NE, Tucker MA, Hoover RN, Hayes RB, Pickle LW, Issaq HJ, Muschik GM, Green-Gallo L, Buivys D, Aisner S, Resau JH, Trump BF, Tollerud D, Weston A, Harris CC. Lung cancer and the debrisoquine metabolic phenotype. J Natl Cancer Inst 1990; 82:1264–72.

Carbone D, Kratzke R. *RB1* and *p53* genes. In: Pass HI, Mitchell JB, Johnson DH, Turrisi AT, eds. Lung Cancer: Principles and Practice. Philadelphia: Lippincott-Raven, 1996:107–22.

Carlton WW. "Proliferative keratin cyst," a lesion in the lungs of rats following chronic exposure to para-aramid fibrils. Fundam Appl Toxicol 1994; 23: 304–7.

Cattley RC, Everitt JJ, Gross EA, Moss OR, Hamm TE Jr, Popp JA. Carcinogenicity and toxicity of inhaled nitrobenzene in $B6C3F_1$ mice and F344 and CD rats. Fundam Appl Toxicol 1994; 22:328–40.

Coultas DB. Other occupational carcinogens. In: Samet JM, ed. Epidemiology of Lung Cancer. New York: Marcel Dekker, 1994:299–333.

Dalbey WE, Nettesheim P, Griesemer R, Caton JE, Guerin MR. Chronic inhalation of cigarette smoke by F344 rats. J Natl Cancer Inst 1980; 64:383–90.

Devereux TR, Belinsky SA, Maronpot RR, White CM, Hegi ME, Patel AC, Foley JF, Greenwell A, Anderson MW. Comparison of pulmonary O^6- methylguanine DNA adduct levels and Ki-*ras* activation in lung tumors from resistant and susceptible mouse strains. Mol Carcinog 1993; 8:177–85.

Doll R, Peto R. The causes of cancer. J Natl Cancer Inst 1981; 66:1191–309.

Dunnick JK, Elwell MR, Radovsky AE, Benson JM, Hahn FF, Nikula KJ, Barr EB, Hobbs CH. Comparative carcinogenic effects of nickel subsulfide, nickel oxide or nickel sulfate hexahydrate chronic exposures in the lung. Cancer Res 1995; 55:5251–6.

Essenberg JM. Cigarette smoke and the incidence of primary neoplasm of the lung in the albino mouse. Science 1952; 116:561–2.

Farris GM, Everitt JI, Irons RD, Popp JA. Carcinogenicity of inhaled benzene in CBA mice. Fundam Appl Toxicol 1993; 20:503–7.

Finch GL, Nikula KJ, Barr EB, Bechtold WE, Chen BT, Griffith WC, Hahn FF, Hobbs CH, Hoover MD, Lundgren DL, Mauderly JL. Effects of combined exposure of F344 rats to radiation and chronically inhaled cigarette smoke. In: Inhalation Toxicology Research Institute Annual Report, 1994–1995, ITRI-146. Albuquerque, NM: Inhalation Toxicology Research Institute, 1995: 77–9.

Finch GL, Hoover MD, Hahn FF, Nikula KJ, Belinsky SA, Haley PJ, Griffith WC. Animal models of beryllium-induced lung disease. Environ Health Perspect 1996a; 104(5):1–7.
Finch GL, Nikula KJ, Belinsky SA, Barr EB, Stoner GD, Lechner JF. Failure of cigarette smoke to induce or promote lung cancer in the A/J mouse. Cancer Lett 1996b; 99:161–7.
Foley JF, Anderson MW, Stoner GD, Gaul BW, Hardisty JF, Maronpot RR. Proliferative lesions of the mouse lung: progression studies in strain A mice. Exp Lung Res 1991; 17:157–68.
Freedman AP, Robinson SE. Noninvasive magnetopneumographic studies of lung dust retention and clearance in coal miners. In: Frantz RL, Ramani RV, eds. Respirable Dust in the Mineral Industries: Health Effects, Characterization and Control. University Park, PA: Penn State University Press, 1988: 181–6.
Gazdar AF. The molecular and cellular basis of human lung cancer. Anticancer Res 1994; 13:261–8.
Glaser U, Hochrainer D, Otto FJ, Oldiges H. Carcinogenicity and toxicity of four cadmium compounds inhaled by rats. Toxicol Environ Chem 1990; 27:153–62.
Goodrow T, Reynolds S, Maronpot R, Anderson M. Activation of K-*ras* by codon 13 mutations in C_{57}B1/6 × C3H F_1 mouse tumors induced by exposure to 1,3-butadiene. Cancer Res 1990; 50:4818–23.
Groth DH, Stettler LE, Burg JR, Busey WM, Grang GC, Wong L. Carcinogenic effects of antimony trioxide and antimony ore concentrate in rats. J Toxicol Environ Health 1986; 18:607–26.
Hahn FF. Chronic inhalation bioassays for respiratory tract carcinogenesis. In: Gardner DE, et al., eds. Toxicology of the Lung. 2nd ed. New York: Raven Press, 1993:435–59.
Hahn FF, Boorman GA. Neoplasia and preneoplasia of the lung. In: Bannasch P, Gössner W, eds. Pathology of Neoplasia and Preneoplasia in Rodents: EULEP Color Atlas. Vol. 2. New York: Schattauer, 1996:29–42.
Harris RJC, Negroni G, Ludgate S, Piek CR, Chesterman FC, Maidment BJ. The incidence of lung tumours of C57B1 mice exposed to cigarette smoke-air mixtures for prolonged periods. Int J Cancer 1974; 14:130–6.
Haseman JK, Huff J, Boorman GA. Use of historical control data in carcinogenicity studies in rodents. Toxicol Pathol 1984a; 12:126–35.
Haseman JK, Crawford DD, Huff JE, Boorman GE, McConnell EE. Results from 86 two-year carcinogenicity studies conducted by the National Toxicology Program. J Toxicol Environ Health 1984b; 14:621–39.
Heath JE, Frith CH, Wong PM. A morphologic classification and incidence of alveolar-bronchiolar neoplasms in BALB/c female mice. Lab Anim Sci 1982; 32:638–47.
Hegi ME, Söderkvist P, Foley JF, Schoonhoven R, Swenberg JA, Kari F, Maronpot R, Anderson MW, Wiseman RW. Characterization of *p53* mutations in methylene chloride-induced lung tumors from B6F3F_1 mice. Carcinogenesis 1993; 14:803–10.

Heinrich U. Comparative response to long-term particle exposure among rats, mice, and hamsters. In: Mauderly JL, McCunney RJ, eds. Particle Overload in the Rat Lung and Lung Cancer: Implications for Human Risk Assessment. Washington, DC: Taylor and Francis, 1996:51–71.

Heinrich U, Peters L, Ernst H, Rittinghausen S, Dasenbrock C, Konig H. Investigation on the carcinogenic effects of various cadmium compounds after inhalation exposure in hamsters and mice. Exp Pathol 1989; 37:253–8.

Heinrich U, Peters L, Creutzenberg O, Dasenbrock C, Hoymann H-G. Inhalation exposure of rats to tar/pitch condensation aerosol or carbon black alone or in combination with irritant gases. In: Mohr U, Dungworth DL, Mauderly JL, Oberdörster G, eds. Toxic and Carcinogenic Effects of Solid Particles in the Respiratory Tract. Washington, DC: LSI Press, 1994:433–42.

Herbert RA, Gillett NA, Rebar AH, Lundgren DL, Hoover MD, Chang IY, Carlton WW, Hahn FF. Sequential analysis of the pathogenesis of plutonium-induced pulmonary neoplasms in the rat: morphology, morphometry, and cytokinetics. Radiat Res 1993; 134:29–42.

Heuper WC. Experimental studies in metal cancerigenesis. IX. Pulmonary lesions in guinea pigs and rats exposed to prolonged inhalation of powdered metallic nickel. AMA Arch Pathol 1958; 65:600–7.

Holland LM, Wilson JS, Tillery MI, Smith DM. Lung cancer in rats exposed to fibrogenic dusts. In: Goldsmith DF, Winn DM, Shy CM, eds. Silica, Silicosis, and Cancer—Controversy in Occupational Medicine: Cancer Research Monographs. New York: Praeger, 1985:267–70.

Huff JE, Melnick RL, Solleveld HA, Hasmane JK, Powers M, Miller RA. Multiple organ carcinogenicity of 1,3-butadiene in B6C3F1 mice after 60 weeks of inhalation exposure. Science 1985; 227:548–9.

Huff J, Cirvello J, Haseman J, Bucher J. Chemicals associated with site-specific neoplasia in 1394 long-term carcinogenesis experiments in laboratory rodents. Environ Health Perspect 1991; 93:247–70.

IARC. Asbestos. Vol. 14. In: IARC Monographs on the Evaluation of Carcinogenic Risk of Chemicals to Man. Lyon, France: International Agency for Research on Cancer, 1977.

IARC. Some halogenated hydrocarbons and pesticide exposure. Vol. 41. In: IARC Monographs on the Evaluation of Carcinogenic Risks to Humans. Lyon, France: International Agency for Research on Cancer, 1986.

IARC. Overall Evaluations of Carcinogenicity: An Updating of IARC Monographs. Vols. 1 to 42, Supplement 7. Lyon, France: International Agency for Research on Cancer, 1987.

IARC. Some industrial chemicals. Vol. 60. In: IARC Monographs on the Evaluation of Carcinogenic Risks to Humans. Lyon, France: International Agency for Research on Cancer, 1994.

Ishinishi N, Kuwabara N, Nagase S, Suzuki T, Ishiwata S, Kohno T. Long-term inhalation studies on effects of exhaust from heavy and light duty diesel engines on F344 rats. In: Ishinishi N, Koizumi A, McClellan RO, Stöber W, eds. Carcinogenic and Mutagenic Effects of Diesel Engine Exhaust. Amsterdam: Elsevier, 1986:329–48.

Issa JJ, Baylin SB, Belinsky SA. Methylation of the estrogen receptor CpG island in lung tumors is related to the specific type of carcinogen exposure. Cancer Res 1996; 56:3655–8.

Ito T, Ikemi Y, Kitamura H, Ogawa T, Kanisawa M. Production of bronchial papilloma with calcitonin-like immunoreactivity in rats exposed to urban ambient air. Exp Pathol 1989; 36:89–96.

Kelly G, Stegelmeier BL, Hahn FF. *p53* alterations in plutonium-induced F344 rat lung tumors. Radiat Res 1995; 142:263–9.

Kotin P, Falk HL, McCammon CJ. The experimental induction of pulmonary tumors and changes in the respiratory epithelium in C_{57}B1 mice following their exposure to an atmosphere of ozonized gasoline. Cancer 1958; 11:473–81.

Laskin S, Kuschner M, Drew RT. Studies in pulmonary carcinogenesis. In: Hanna MG Jr, Nettesheim P, Gilbert JR, eds. Inhalation Carcinogenesis. Springfield, VA: U.S. Department of Commerce, 1970:321–47.

Laskin S, Drew RT, Cappiello V, Kuschner M, Nelson N. Inhalation carcinogenicity of alpha halo ethers. II. Chronic inhalation studies with chloromethyl methyl ether. Arch Environ Health 1975; 30:70–2.

Lee KP, Barras CE, Griffith FD, Waritz RS, Lapin CA. Comparative pulmonary responses to inhaled inorganic fibers with asbestos and fiberglass. Environ Res 1981; 24:167–91.

Lee KP, Trochimowicz HJ, Reinhard CF. Pulmonary response of rats exposed to titanium dioxide (TiO_2) by inhalation for two years. Toxicol Appl Pharmacol 1985; 79:179–92.

Lee KP, Kelly DP, Schneider PW, Trochimowicz HJ. Inhalation toxicity study on rats exposed to titanium tetrachloride atmospheric hydrolysis products for two years. Toxicol Appl Pharmacol 1986; 83:30–45.

Lee KP, Ulrich CE, Geil RG, Trochimowicz HJ. Effects of inhaled chromium dioxide dust on rats exposed for two years. Fundam Appl Toxicol 1988a; 10: 125–45.

Lee KP, Kelly DP, O'Neal TO, Stadler JC, Kennedy GL Jr. Lung response to ultrafine kevlar aramid synthetic fibrils following 2-year inhalation exposure in rats. Fundam Appl Toxicol 1988b; 11:1–20.

Leong BKJ, Macfarland HN, Reese WH Jr. Induction of lung adenomas by chronic inhalation of bis (chloromethyl) ether. Arch Environ Health 1971; 22:663–6.

Liang CK, Quan NY, Cao SR, He XZ, Ma F. Natural inhalation exposure to coal smoke and wool smoke induces lung cancer in mice and rats. Biomed Environ Sci 1988; 1:42–50.

Lomax LG, Stott WT, Johnson KA, Calhoun LL, Yano BL, Quast JF. The chronic toxicity and oncogenicity of inhaled technical-grade 1,3-dichloropropene in rats and mice. Fundam Appl Toxicol 1989; 12:418–31.

Lorenz E, Stewart HL, Daniel JH, Nelson CV. The effects of breathing tobacco smoke on strain A mice. Cancer Res 1942; 3:123–4.

Lubin JH, Boice JD Jr, Edling C, Hornung RW, Howe GR, Kunz E, Kusiak RA, Morrison HI, Radford EP, Samet JM, Tirmarche M, Woodward A, Yao SX,

Pierce DA. Lung cancer in radon-exposed miners and estimation of risk from indoor exposure. J Natl Cancer Inst 1995; 87:817–27.
Lundgren DL, McClellan RO, Thomas RL, Hahn FF, Sanchez A. Toxicity of inhaled $^{144}CeO_2$ in mice. Radiat Res 1974; 58:448–61.
Lundgren DL, McClellan RO, Thomas RL, Hahn FF, Griffith WC, McClellan RO. Effects of protraction of the alpha dose to the lungs of mice by repeated inhalation exposure to aerosols of $^{239}PuO_2$. Radiat Res 1987; 111:201–24.
Lundgren DL, Hahn FF, Diel JH. Repeated inhalation exposure of rats to aerosols of $^{144}CeO_2$. II. Effects on survival and lung, liver, and skeletal neoplasms. Radiat Res 1992; 132:325–33.
Lundgren DL, Haley PJ, Hahn FF, Diel JH, Griffith WC, Scott BR. Pulmonary carcinogenicity of repeated inhalation exposure of rats to aerosols of $^{239}PuO_2$. Radiat Res 1995; 142:39–53.
Lynch KM, McIver FA, Cain JR. Pulmonary tumors in mice exposed to asbestos dust. AMA Arch Ind Health 1957; 15:207–14.
Mabry M, Nelkin B, Baylin S. Evolutionary model of lung cancer. In: Pass HI, Mitchell JB, Johnson DH, Turrisi AT, eds. Lung Cancer: Principles and Practice. Philadelphia: Lippincott-Raven, 1996:133–42.
MacEwen JD, McConnell EE, Back KC. The effect of a six month chronic level inhalation exposure to hydrazine on animals. L-Medical Research Laboratory, Wright-Patterson Air Force Base, Ohio. AMRL-FR-74-125, Paper No. 16. Distributed by National Technical Information Service, Publication No. AD-A-011865. Springfield, VA: U.S. Department of Commerce, 1974.
Malkinson AM. Genetic studies on lung tumor susceptibility and histogenesis in mice. Environ Health Perspect 1991; 93:149–59.
Malkinson AM, Belinsky SA. The use of animal models in preclinical studies. In: Pass HI, Mitchell JB, Johnson DH, Turrisi AT, eds. Lung Cancer: Principles and Practice. Philadelphia: Lippincott-Raven, 1996:273–84.
Maronpot RR, Shimkin MB, Witschi HP, Smith LH, Cline JM. Strain A mouse pulmonary tumor test results for chemicals previously tested in the National Cancer Institute carcinogenicity tests. J Natl Cancer Inst 1986; 756:1101–12.
Martin JC, Daniel H, LeBouffant L. Short- and long-term experimental study of the toxicity of coal-mine dust and of some of its constituents. In: Walton WH, ed. Inhaled Particles IV. Oxford: Pergamon Press, 1977:361–70.
Mast RW, McConnell EE, Anderson R, Chevalier J, Kotin P, Bernstein DM, Thevenaz P, Glass LR, Miller WC, Hesterberg TW. Studies on the chronic toxicity (inhalation) of four types of refractory ceramic fiber in male Fischer 344 rats. Inhal Toxicol 1995a; 7:425–67.
Mast RW, McConnell EE, Hesterberg TW, Chevalier J, Kotin P, Thevenaz P, Bernstein DM, Glass LR, Miller WC, Anderson R. Multiple-dose chronic inhalation toxicity study of size-separated kaolin refractory ceramic fiber in male Fischer 344 rats. Inhal Toxicol 1995b; 7:469–502.
Mauderly JL. Lung overload: the dilemma and opportunities for resolution. In: Mauderly JL, McCunney RJ, eds. Particle Overload in the Rat Lung and

Lung Cancer: Implications for Human Risk Assessment. Washington, DC: Taylor and Francis, 1996:1–28.

Mauderly JL, Jones RK, Griffith WC, Henderson RF, McClellan RO. Diesel exhaust is a pulmonary carcinogen in rats exposed chronically. Fundam Appl Toxicol 1987; 9:208–11.

Mauderly JL, Cheng YS, Snipes MB. Particle overload in toxicological studies: friend or foe? J Aerosol Med 1990; 3(1):S169–S187.

McMartin DN, Sahorta PS, Gunson DE, Hsu HH, Spaet RH. Neoplasms and related proliferative lesions in control Sprague-Dawley rats from carcinogenicity studies. Historical data and diagnostic considerations. Toxicol Pathol 1992; 20:212–25.

McNeil DA, Chrisp CE, Fisher GL. Tumorigenicity of nickel subsulfide in strain A/J mice. Drug Chem Toxicol 1990; 13(1):71–86.

Melnick RL, Huff J, Chou BJ, Miller RA. Carcinogenicity of 1,3-butadiene in C_{57}Bl/6 × C3H F_1 mice at low exposure concentrations. Cancer Res 1990; 50:6592–99.

Monchaux G, Morlier JP, Morin M, Chameaud J, Lafuma J, Masse R. Carcinogenic and cocarcinogenic effects of radon and radon daughters in rats. Environ Health Perspect 1994a; 102(1):64–73.

Monchaux G, Morlier JP, Morin M, Fritsch P, Tredaniel J, Masse R. Carcinogenic and cocarcinogenic effects of ozone in rats: preliminary results. Pollut Atmos 1994b; 142:84–8.

Morrow PE. Contemporary issues in toxicology. Dust overloading on the lungs: update and appraisal. Toxicol Appl Pharmacol 1992; 113:1–12.

Muhle H, Bellman B, Heinrich U. Overloading of lung clearance during chronic exposure of experimental animals to particles. Ann Occup Hyg 1988; 32(suppl 1):141–6.

National Research Council. Report on the Committee on the Biological Effects of Ionizing Radiation. Health Effects of Radon and Other Internally Deposited Alpha Emitters (BEIR IV). Washington, DC: National Academy Press, 1988.

Nettesheim P, Hanna MG Jr, Doherty DG, Newell RF, Hellman A. Effect of calcium chromate dust, influenza virus, and 100 R whole-body X radiation on lung tumor incidence in mice. J Natl Cancer Inst 1971; 47:1129–44.

Nickell-Brady C, Hahn FF, Finch GL, Belinsky SA. Analysis of K-*ras*, *p53* and c-raf-1 mutations in beryllium-induced rat lung tumors. Carcinogenesis 1994; 15:257–62.

Nikula KJ, Snipes MB, Barr EB, Griffith WC, Henderson RF, Mauderly JL. Comparative pulmonary toxicities and carcinogenicities of chronically inhaled diesel exhaust and carbon black in F344 rats. Fundam Appl Toxicol 1995; 25: 80–94.

NTP. Carcinogenesis Bioassay of 1,2-Dibromoethane in F344 Rats and B6C3F_1 Mice. NTP-80-28, NIH Publication No. 82-1766. U.S. Department of Health and Human Services. Research Triangle Park, NC: National Toxicology Program, 1982a.

NTP. Carcinogenesis Bioassay of 1,2-Dibromoethane-3-Chloropropane in F344 Rats and B6C3F_1 Mice. NTP-81-21, NIH Publication No. 82-1762. U.S. Department of Health and Human Services. Research Triangle Park, NC: National Toxicology Program 1982b.

NTP. Toxicology and Carcinogenesis Studies of Bromoethane in F344 Rats and B6C3F$_1$ Mice. NTP Tech. Rep. 363, NIH Publication No. 90-2818. U.S. Department of Health and Human Services. Research Triangle Park, NC: National Toxicology Program, 1989a.

NTP. Toxicology and Carcinogenesis Studies of Chloroethane in F344/N Rats and B6C3F$_1$ Mice. NTP Tech. Rep. 346, NIH Publication No. 90-2801. U.S. Department of Health and Human Services. Research Triangle Park, NC: National Toxicology Program, 1989b.

NTP. Toxicology and Carcinogenesis Studies of Talc in F344/N Rats and B6C3F$_1$ Mice. NTP Tech. Rep. 421, NIH Publication No. 94-3152. U.S. Department of Health and Human Services. Washington, DC: National Toxicology Program, 1993.

NTP. Toxicology and Carcinogenesis Studies of Ozone and Ozone/NNK in F344/N Rats and B6C3F$_1$ Mice. NTP Tech. Rep. 440, NIH Publication No. 95-3371. U.S. Department of Health and Human Services. Research Triangle Park, NC: National Toxicology Program, 1994.

Obersdörster G. Lung particle overload: implications for occupational exposures to particles. Regul Toxicol Pharmacol 1995; 27:123–35.

Ohmori H, Abe T, Hirano H, Murakami T, Katoh T, Gotoh S, Kido M, Kuroiwa A, Nomura T, Higashi K. Comparison of K-*ras* gene mutation among simultaneously occurring multiple urethane-induced lung tumors in individual mice. Carcinogenesis 1992; 13:851–5.

Otto H. Experimentelle Untersuchungen an Mausen mit passiver Zigarettenrauchbeatmung. Frankf Z Pathol 1963; 73:10–23.

Pepelko WE. Experimental respiratory carcinogenesis in small laboratory animals. Environ Res 1984; 33:144–88.

Prejean JD, Peckham JC, Casey AE, Griswold DP, Weisburger EK, Weisberger JH. Spontaneous tumors in Sprague-Dawley rats and Swiss mice. Cancer Res 1973; 33:2768–73.

Reeves AL, Puro HE, Smith RG, Vorwald AJ. Experimental asbestos carcinogenesis. Environ Res 1971; 4:496–511.

Reuzel PGJ, Arts JHE, Lomax LG, Kuijpers MHM, Kuper CF, Gembardt C, Feron VJ, Loser E. Chronic inhalation toxicity and carcinogenicity study of respirable polymeric methylene diphenyl diissocyanate (polymeric MDI) aerosol in rats. Fundam Appl Toxicol 1994; 22:195–210.

Richardson GC, Johnson BE. The biology of lung cancer. Semin Oncol 1993; 20(2): 105–27.

Rodenhuis S. *RAS* oncogenes and human lung cancer. In: Pass HI, Mitchell JB, Johnson DH, Turrisi AT, eds. Lung Cancer: Principles and Practice. Philadelphia: Lippincott-Raven, 1996:73–82.

Rowlatt C, Chesterman FC, Sheriff MU. Lifespan, age changes, and tumour incidence in an ageing C57B1 mouse colony. Lab Anim 1976; 10:419–42.

Saffioti U, Williams AO, Lambert ND, Kaighn ME, Mao Y, Shi X. Carcinogenesis by crystalline silica: animal, cellular, and molecular studies. In: Castranova

V, Vallyathan V, Wallace WE, eds. Silica and Silica-Induced Lung Diseases. Boca Raton, FL: CRC Press, 1995:345–81.

Samet JM. The epidemiology of lung cancer. Chest 1993; 103(suppl):21S–29S.

Schepers GWH. Neoplasia experimentally induced by beryllium compounds. Prog Exp Tumor Res 1961; 2:203–44.

Schepers GWH. Lung tumors of primates and rodents. Ind Med 1971; 40(11):23–31.

Schottenfeld D. Epidemiology of lung cancer. In: Pass HI, Mitchell JB, Johnson DH, Turrisi AT, eds. Lung Cancer: Principles and Practice. Philadelphia: Lippincott-Raven Publishers, 1996:305–21.

Shimkin MB, Stoner GD. Lung tumors in mice: application to carcinogenesis bioassay. In: Keane G, Wernboise S, Haddow A, eds. Advances in Cancer Research New York: Academic Press, 1975:1–48.

Shimkin MB, Weisburger JH, Weisburger EK, Gubareff N, Suntzeff V. Bioassay of 29 alkylating chemicals by the pulmonary-tumor response in strain A mice. J Natl Cancer Inst 1966; 36:915–35.

Snipes MB. Current information on lung overload in nonrodent mammals: contrast with rats. In: Mauderly JL, McCunney RJ, eds. Particle Overload in the Rat Lung and Lung Cancer: Implications for Human Risk Assessment. Washington, DC: Taylor and Francis, 1996:91–110.

Stegelmeier BL, Gillett NA, Rebar AH, Kelly G. The molecular progression of plutonium-239-induced rat lung carcinogenesis: Ki-*ras* expression and activation. Mol Carcinog 1991; 4:43–51.

Stoner GD. Lung tumors in strain A mice as a bioassay for carcinogenicity of environmental chemicals. Exp Lung Res 1991; 17:405–23.

Stowers SJ, Glover PL, Reynolds SH, Boone LR, Maronpot RR, Anderson MW. Activation of the K-*ras* protooncogene in lung tumors from rats and mice chronically exposed to tetranitromethane. Cancer Res 1987; 47:3212–19.

Sunderman FW, Donnelly AJ. Studies of nickel carcinogenesis: metastasizing pulmonary tumors in rats induced by the inhalation of nickel carbonyl. Am J Pathol 1965; 46:1027–41.

Swafford DS, Nikula KJ, Mitchell CE, Belinsky SA. Low frequency of alterations in *p53*, K-*ras*, and *mdm2* in rat lung neoplasms induced by diesel exhaust or carbon black. Carcinogenesis 1995; 16(5):1215–21.

Snyder CA, Sellakumar AR, James DJ, Albert RE. The carcinogenicity of discontinuous inhaled benzene exposures in CD-1 and C_{57}B1/6 mice. Arch Toxicol 1988; 72:331–5.

Tomatis L, Aitio A, Wilbourn J, Shuker L. Human carcinogens so far identified. Jpn J Cancer Res 1989; 80:795–807.

Travis WD, Travis LB, Devesa SS. Lung cancer. Cancer 1995; 75(suppl):191–202.

Travis WD, Linder J, Mackay B. Classification, histology, cytology, and electron microscopy. In: Pass HI, Mitchell JB, Johnson DH, Turrisi AT, eds. Lung Cancer: Principles and Practice. Philadelphia: Lippincott-Raven, 1996:361–95.

Vineis P, Simonato L. Proportion of lung and bladder cancers in males resulting from occupation: a systematic approach. Arch Environ Health 1991; 46:6–15.

Wagner WD, Groth DH, Holtz JL, Madden GE, Stokinger HE. Comparative chronic inhalation toxicity of beryllium ores, bertrandite and beryl, with production of pulmonary tumors by beryl. Toxicol Appl Pharmacol 1969; 15:10–29.

Wagner JC, Berry GB, Hill RJ, Munday DE, Skidmore JW. Animal experiments with MMM(V)F-effects of inhalation and intrapleural inoculation in rats. In: Biological Effects of Man-Made Mineral Fibres (Proceedings of a WHO/IARC Conference). Vol. 2. Copenhagen: World Health Organization, 1984: 209–33.

Walsh KM, Poteracki J. Spontaneous neoplasms in control Wistar rats. Fundam Appl Toxicol 1994; 22:65–72.

Wehner AP, Dagle GE, Clark ML, Buschbom RL. Lung changes in rats following inhalation exposure to volcanic ash for two years. Environ Res 1986; 40:499–517.

WHO. Histological Typing of Lung Tumors. 2nd ed. World Health Organization: 1981.

Wilson T, Scheuchenzuber WJ, Eskew ML, Zarkower A. Comparative pathological aspects of chronic olivine and silica inhalation in mice. Environ Res 1986; 39:331–44.

Wilson VS, Wei Q, Miley FB, et al. 5-Methyldeoxycyidine patterns and lung cancer susceptibility in mice. Proc Am Assoc Cancer Res 1992; 33:102–3.

Witschi H, Oreffo VK, Pinkerton KE. Six month exposure of strain A/J mice to cigarette sidestream smoke: cell kinetics and lung tumor data. Fundam Appl Toxicol 1995; 26:32–40.

Wu-Williams AH, Samet JM. Lung cancer and cigarette smoking. In: Samet JM, ed. Epidemiology of Lung Cancer. New York: Marcel Dekker, 1994:71–99.

Wynder EL, Muscat JE. The changing epidemiology of smoking and lung cancer histology. Environ Health Perspect 1995; 103(suppl 8):143–8.

You M, Wang Y, Lineen A, Stoner GD, You L, Maronpot RR, Anderson MW. Activation of protooncogenes in mouse lung tumors. Exp Lung Res 1991; 17: 389–400.

You M, Wang Y, Stoner G, You L, Maronpot R, Reynolds SH, Anderson MW. Parental bias of K-*ras* oncogenes detected in lung tumors from mouse hybrids. Proc Natl Acad Sci USA 1992; 89:5804–8.

20
Mammary Cancer

Minako Nagao and Takashi Sugimura
National Cancer Center Research Institute, Tokyo, Japan

I. CHEMICALS WIDELY USED IN RODENT MAMMARY CARCINOGENESIS MODELS

Induction of mammary cancer after application of an aromatic carcinogen to a site remote from the breast was first observed by Maisin and Coolen in 1936 (1). They painted the skin of mice with 3-methylcholanthrene (3MC) and observed a high incidence of mammary tumors in addition to skin cancers. In 1939 Strong and Smith found that a single injection of 3MC into the breast of mice induced adenocarcinoma and squamous cell carcinomas of the mammary gland (2). Repeated instillation of 3MC into the mouth or stomach of mice or rats was also demonstrated to induce mammary tumors with high frequency (3). The first compound that was found to induce mammary cancer by the oral route was 2-acetylaminofluorene (2-AAF) (4). The potency of 2-AAF was not high, even with chronic dietary administration to a susceptible strain (5). The influence of age on induction of mammary gland tumors by a single instillation of 3MC dissolved in sesame oil into the stomach of Sprague-Dawley (SD) rats was studied. The tumor incidence was zero when 3MC was administered to 23-day-old rats, reached a maximum (100%) at 50–75 days old, and then decreased, with only a small proportion of animals developing tumors after treatment between 100 and 365 days of age (5). A single intragastric dose of 7,12-dimethylbenz[*a*]anthracene (DMBA) dissolved in sesame oil also proved efficient at inducing mammary tumors in SD rats, with dose dependence observed between 1 and 15 mg (5). DMBA gavage is now routinely used as the standard Huggins method to obtain mammary tumors of confirmed reproducibility. 1-Methyl-1-nitrosourea (MNU) has also been established to be an efficient mammary carcinogen in the SD rat (6) with a single intravenous injection of 50 mg/kg inducing a near 100% incidence after 6 months, each animal having two to seven tumors. Intravenous or subcutaneous administration of this carcinogen also specifically causes mammary carcinoma (7). Since in practical terms subcutaneous injection is by far the easiest, many studies have been carried out using this technique. The common routes for chemical carcinogen administration are intramuscular (i.m.), subcutaneous (s.c.), intragastric (i.g.), intraperitoneal (i.p.), and intravenous (i.v.). However, none of the widely used carcinogens—3MC, 2-AAF, MNU, or DMBA—is naturally occurring, and therefore the likelihood of human exposure to these chemicals is very low.

The heterocyclic amines (HCAs), which are ubiquitously present in cooked food, provide a clear contrast. 2-Amino-1-methyl-6-phenylimidazo [4,5-*b*]pyridine (PhIP), 2-amino-3,4-dimethylimidazo[4,5-*f*]quinoline (MeIQ) and 2-amino-3-methylimidazo[4,5-*f*]quinoline (IQ) (8–12,77) all induce mammary tumors in the Fischer 344 (F344) or SD strains of rats, as shown in Table 1. Further, 10 750 mg/kg i.g. doses of PhIP dissolved in sesame or corn oil within a 2-week initiation period followed by administration of a high-fat (23.5%) diet caused development of mammary tumors histologically very similar to those in humans including infiltrating duct carcinomas and tubular adenocarcinomas, with an incidence of 53% by 27 weeks (12). Since humans are exposed to these HCAs on a daily basis (13,14), mammary carcinogenesis by HCAs is an important and appropriate animal model system.

Applying the Huggins method, orally administered 400 μmol doses of benzo[*a*]pyrene (B[*a*]P) and PhIP were found to show similar carcinogenic potencies in the CD(SD) strain of rats (15), while 1-nitropyrene was much weaker. In a 41-week experimental period, adenocarcinoma-incidences were 73%, 77%, and 6% for B[*a*]P, PhIP, and 1-NP, respectively (15).

Under the auspices of the U.S. NTP, 301 chemicals were tested for their carcinogenicities in F344 rats and $B6C3F_1$ mice, and 157 proved positive in some organ. However, only 17 compounds were demonstrated to induce mammary tumors in both or either of these species (16). They are listed in Table 2 and their structures are depicted in Figure 1.

In a larger U.S. NTP data base of 450 chemical tested for mammary cancer in different strains of rats and mice, 34 chemicals (including the above 17 chemicals) were positive in rodents: 29 were positive in female rats and 5 were positive in female mice only (17). The rat strains that developed mammary cancer were the F344/N, Sprague-Dawley, and Osborne-Mendel. The $B6C3F_1$ mouse was the responding mouse strain. Mammary tumors occurred after dermal and inhalation exposure as well as by oral and injection routes of chemical administration (17). The additional 17 chemicals

Table 1 Carcinogenicity of Various Heterocyclic Amines in F344 and SD Rats

			Animal				% Tumor incidence			
HCA	Route of application	Amount	Strain	Sex	No.	Exp. period (weeks)	Mammary gland	Colon	Liver	Ref.
PhIP	Diet	400 ppm	F344	M	29	52	0	55	0	8
				F	30		47	7	0	
		100 ppm	F344	M	30	104	0	43	0	8
				F	30		47	13	0	
		25 ppm	F344	M	30	104	0	0	0	8
				F	30		7	0	0	
PhIP	i.g.	75 mg/kg × 10	SD	F	15	29	53	–	–	12
MeIQ	Diet	300 ppm	F344	M	20	40	0	35	5*	9
				F	20	40	5	25	0	
IQ	i.g.	0.22 mmol/kg × 43	SD	F	32	52	22	0	0	10
IQ	Diet	300 ppm	F344	M	40	55	0	63	68	75
				F	40	72	0	23	45	

*Not significant.
–, Not available.

Table 2 Mammary Carcinogens Detected by the U.S. NTP

	Rat (% TBA)								Mice (% TBA)							
	♂				♀				♂				♀			
	C	L	M	H	C	L	M	H	C	L	M	H	C	L	M	H
2,4-Diaminotoluene					5	76		82								
Hydrazobenzene					1	6		12								
5-Nitroacenaphthene					0	11		16								
o-Toluidine · HCl					35	40		71								
Nitrofurazone					16	72		72								
3,3′-Dimethoxybenzidine · 2HCl					2	4	19	33								
3,3′-Dimethylbenzidine · 2HCl					0	2	4	10								
2,4-Dinitrotoluene					18	24		46								
Furosemide													0	4		10
Benzene													0	4	10	20
Reserpine													0	14		15
1,2-Dibromoethane					8	58		48					4	28		16
1,2-Dichloroethane					1	2		36					0	18		15
Phenesterin					6	10	41	40								
Glycidol	6	16		14	28	68		74					4	12		30
Dichloromethane	0	4	10	8	10	22	26	46								
Sulfallate					0	14		23					0	48		24

TBA = Tumor-bearing animals; C = control; L = low dose; M = medium dose; H = high dose.
Source: Data from Ref. 16.

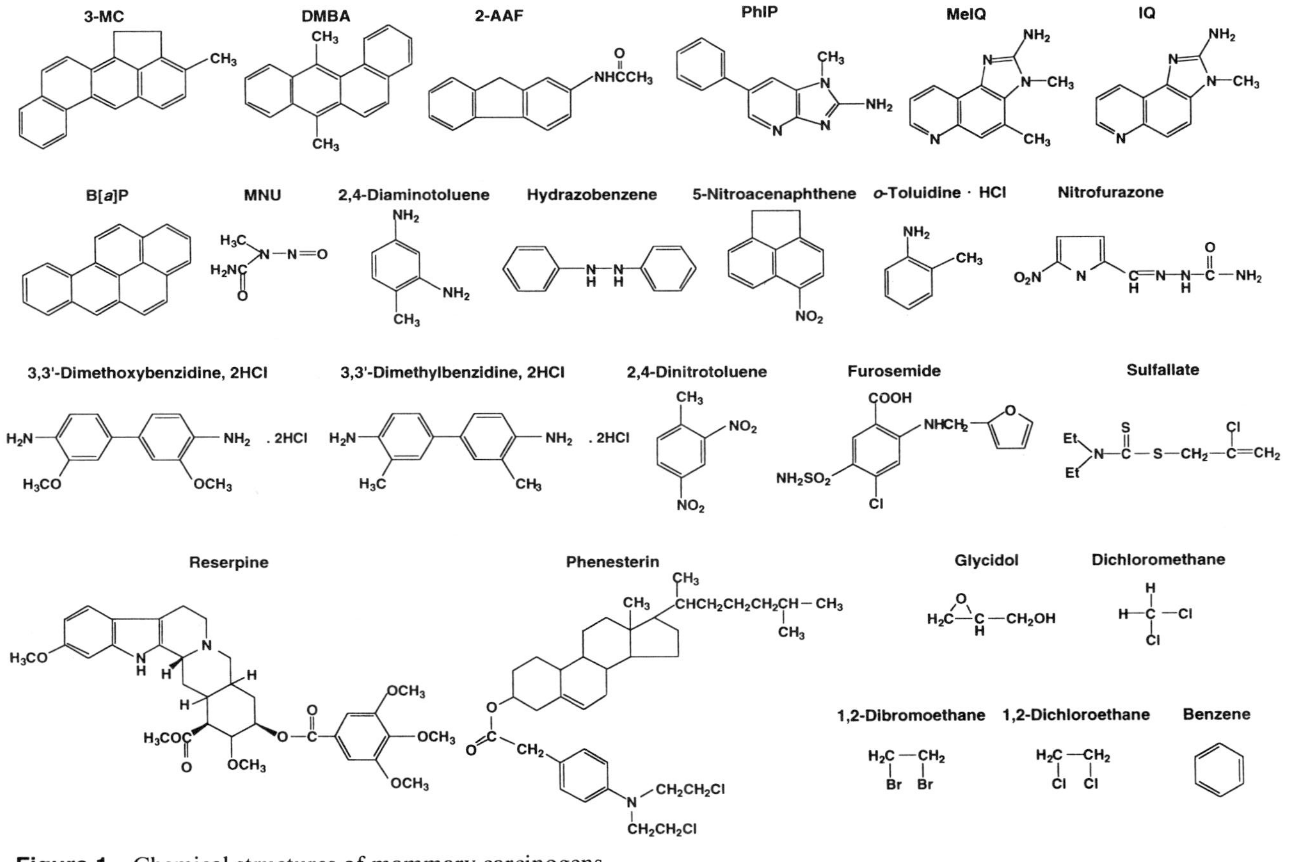

Figure 1 Chemical structures of mammary carcinogens.

found to cause mammary neoplasms are acronycine, 1,3-butadiene, 2-chloroacetophenone, clonitralid, cytembena, 1,2-dibromo-3-chloropropane, 2,3-dibromo-1-propanol, 1,1-dichloroethane, 1,2-dichloropropane, dichlorvos, ethylene oxide, isophosphamide, nithiazide, orchratoxin A, procarbazine hydrochloride, 2,4-2,6-toluene diisocyanate and 1,2,3-trichloropropane, although in some of these cases the evidence was only equivocal.

II. CHEMICAL CARCINOGEN SPECIFICITY FOR THE MAMMARY GLAND

One HCA, PhIP, induces colon cancer in males and mammary tumors in females of the F344 strain of rats (8). Although DNA-adduct levels do not necessarily correlate with carcinogenic potency (18), DNA-adduct formation is generally accepted as the first step in carcinogenesis of genotoxic compounds. Oral administration of PhIP is associated with high levels of DNA adducts in mammary epithelial cells of the SD female rat. It has been suggested that PhIP is metabolized to *N*-hydroxy-PhIP outside the mammary gland and then activated to an ultimate *N*-acetoxy-PhIP form in situ by *O*-acetyltransferase (19). However, HCAs including PhIP, MeIQ, and IQ were found to be metabolized to active forms producing DNA adducts in first-passage human breast epithelial cells in vitro (20).

Male and female F344 rats form similar levels of PhIP-DNA adducts in colon mucosa (18), but only males develop a significant incidence of colon cancers (8). Prostate cancers are induced with significantly high levels of PhIP-DNA adducts (21). In female Balb/c mice PhIP induces mammary tumors and tumors of the lymphatic system (unpublished results). Thus, while PhIP only causes tumors in mammary, colon, and the lymphatic system, the mammary gland is not the sole target.

At 300 ppm in the diet, MeIQ, another HCA, also induces mammary tumors in F344 female rats with an incidence of 25% (9). Other target organs included the large intestine, the Zymbal glands, the oral cavity of both males and females, and the skin of males. IQ is known to induce mammary gland tumors in female SD but not F344 rats (10,13,77), but again the mammary gland is not the sole target (10).

A single i.m. or s.c. injection of 3MC induces sarcomas in mice and rats (22). Skin painting of this compound causes skin cancers in addition to mammary tumors in mice (22). 3MC given i.p. induces lung adenoma and malignant lymphoma in the mouse. 3MC given i.g. results in mammary tumors in the rat and hamster, although only forestomach squamous cell carcinomas develop in mice (22).

DMBA causes both skin and mammary tumors in mice when painted onto the skin. Cheek pouch exposure to DMBA targets the gastrointestinal tract in the rat and hamster. However, i.v., i.p., or i.g. application of DMBA induces exclusively mammary tumors in rats.

MNU induces tumors in various organs such as the skin, lungs, brain, kidney, and gastrointestinal tract, as well as extrathymic lymphomas. Administration of MNU in the drinking water induces tumors in glandular stomach in the male rat (22,23).

Of the 450 compounds tested by the U.S. NTP, the eight site-specific mammary carcinogens were 2-chloroacetophenone, clonitalid, 1,1-dichloroethane, 1,2-dichloropropane, nithiazide, nitrofurazone, and phenesterin in F344, SD, or Osborne-Mendel female rats and furosemide in female B6C3F1 mice (17).

III. SPECIES AND STRAIN DIFFERENCES IN THE DEVELOPMENT OF MAMMARY CANCER

The mouse and rat are the most frequently used experimental models. In particular, mouse strains carrying milk-transmitted mouse mammary tumor virus (MMTV) demonstrate a high propensity for development of mammary cancers. Mouse strains most frequently used for studies on mammary tumorigenesis are as follows: (a) C57BL and O20, which are resistant to both MMTV infection and tumor induction, with no or a very low spontaneous mammary tumor incidence; (b) BALB/c, which is susceptible to tumor induction, but free of milk-transmitted MMTV, with a low spontaneous mammary tumor incidence; (c) C3H, RIII, A, DBA, GR, and DD, which are all susceptible, carrying milk-transmitted MMTV, with a high spontaneous mammary tumor incidence (24). Chemicals such as 3MC, B[*a*]P, dibenz[*a*]anthracene, and DMBA induce high yields of mammary tumors in the susceptible strains (24). PhIP also causes mammary tumors in BALB/c mice as stated above. Although some role for retroviruses in human carcinogenesis has been suggested (25), the histological features of mouse tumors are quite different from those of humans. More than 90% of human mammary cancers are considered to be of ductal origin. In contrast, most mouse mammary tumors arise in alveoli or ductules, developing mostly in the terminal structures most distal from the nipple (24). Thus, there is no exact established experimental system that duplicates human mammary cancer, although DMBA and 3MC have been reported to induce neoplasms of ductal origin in mice (26).

Rats—the SD strain in particular—develop mammary tumors spontaneously, demonstrating an incidence of 54% within the average life span of

760 ± 21 days. There is no evidence of virus involvement in rat mammary carcinogenesis (27).

DMBA in SD rats and MNU in F344 rats are the most widely used experimental models of mammary carcinogenesis. Among rat strains treated with DMBA, the Wistar-Furth (WF) and Lewis strains have a relatively high susceptibility to mammary cancer. F344 rats have less susceptibility, and ACI and Copenhagen (COP) rats are resistant to mammary carcinogenesis (28). Strain differences similar to MNU have been observed for WF (sensitive), F344 (moderate), and Wistar Kyoto (WKY, resistant) rats (29).

Histologically malignant rat mammary tumors have features in common with human intraductal and infiltrating ductal carcinomas. Comparisons of neoplastic and nonneoplastic changes in rats and humans have established that almost all lesions found in the rat have their counterparts in humans. However, in situ or invasive lobular carcinomas and Paget's disease of the nipple have not been reported in the rat mammary gland (28).

SD rats are sensitive to the food contaminant carcinogen, PhIP, and a wide variety of mammary tumors are induced when an initiating dose is followed by either a low-fat or a high-fat diet (19). The lesions include (a) hypertrophy, fibrocystic changes, and a papillomatosis like that considered benign in the human mammary gland, (b) intraductal hyperplasia and atypical intraductular hyperplasia, which are associated with a high risk for developing subsequent invasive cancer in humans, and (c) intraductal, tubular, and infiltrating duct carcinomas, which are often associated with human metastases.

IV. GENETIC BASIS OF MAMMARY CARCINOGENESIS IN RODENTS AND HUMANS

Only 5–10% of American patients with mammary cancers have a family history of the disease. Including reproductive parameters related to lifetime exposure to endogenous estrogen, known risk factors account for only about 30% of cases (30,31). Recent advances in DNA technology and molecular biology have demonstrated the presence of at least three distinct loci for familial breast cancer: BRCA-1, BRCA-2, and BRCA-3 (32–34). Androgen receptor mutations have been implicated in hereditary male breast cancer (35). The structures of the *BRCA-1* and *BRCA-2* genes and the amino acid sequences of the encoded proteins are known, and the presence of mutations in familial cases has been clarified (36–38). Li Fraumeni families develop mammary cancers with a rate of 25% along with various other tumors (39). However, the penetration of the mammary cancer phenotype among *BRCA-1* or *BRCA-2* mutant allele carriers is not necessarily 100%. Genetic background and environmental agents may cause a lowering of the penetration. Meiotic processes may also exert an influence (40).

Genetic factors also determine the rate of tumor induction by chemical carcinogens to a large extent. Strain differences in DMBA or MNU mammary tumorigenesis have been well studied in the rat (41). There is evidence from genetic crosses in various combinations that rat mammary carcinogenesis is regulated by multiple genes, with quantitative traits (42). A similar situation may also exist with regard to human cases, and it is possible that both environmental and hereditary factors are involved. Using the F_1 progeny of WF and COP rats, which are susceptible and resistant to mammary carcinogenesis, respectively, the resistant trait was demonstrated to be dominant (29,42). Thirty days after administration of MNU, Ha-*ras* mutations were induced at the same rate in the COP (resistant strain) as Buffalo/N (BUF/N) (susceptible strain) (43). However, in BUF/N rats the number of cells with the mutated Ha-*ras* allele increased 10- to 100-fold between days 30 and 60, whereas no increase was observed in the COP rats. This suggests that some factor operates to block clonal expansion of Ha-*ras* mutants in the COP strain resistant to mammary carcinogenesis. The Ha-*ras* mutation rate in mammary tumors developing in the COP strain after administration of MNU is very low (44,45).

Chromosomal mapping of a rat mammary carcinoma suppressor gene (*Mcs-1*) has been performed using 211 (WF × COP) F_1 × WF backcross progeny. Genotyping of these backcross animals with 113 microsatellite markers allowed a quantitative trait analysis to be made for mammary carcinoma multiplicity, and *Mcs-1* was mapped to chromosome 2 (42). Carcinogenicity studies with DMBA and MNU in the progeny of crosses among the WF, WKY, COP, and F344 strains, and further crosses of the F_1 generation with one of these strains suggested that WKY has the same resistance gene, *Mcs-1*, as COP rats. WKY was also suggested to have multiple copies of the susceptible gene(s), analogous to the susceptible WF, which carries three copies of a dominant susceptibility gene. The *Mcs-1* was proposed to be epistatic in regard to these susceptibility genes (29). For finer localization, a denser map is necessary. Although the number of rat polymorphic markers available at present is limited, the number of markers is increasing rapidly. Representational difference analysis (RDA) markers, which can be utilized as probes for dot-blot analysis, are also being rapidly developed (46–49). For the genotyping of large numbers of animals, this approach is most efficient. Further, genotypically directed RDA may provide useful markers for regions of interest (46). A partial cDNA clone of the rat *Brca-1* has in fact been isolated from testis and mapped to chromosome 10 by linkage analysis (50). Possible linkage of *Brca-1* to the carcinogen-resistant phenotype is now also being studied (50).

V. MOLECULAR BIOLOGY OF MAMMARY CANCER DEVELOPMENT

Analysis of genetic changes is essential to clarify the mechanisms involved in mammary carcinogenesis as well as providing clues for new prevention and therapeutic measures. Furthermore, if certain types of genes involved in tumorigenesis suffer mutations at high rates, their analysis will provide mutational fingerprints. Since these are chemical carcinogen specific, information on causative environmental carcinogens should be obtained. For example, *p53* mutational spectra are valuable data pointing to specific compound exposure (51,52).

Cumulative data on *p53* mutation spectra indicates that there are significant geographical/population differences between Japan and Europe/North America: G → T transversions are 2.1% in the former and 14.3% in the latter, and A → G transitions are 16.7% in the former and 5.7% in the latter (M. Hollstein, personal communication).

Estimation of the number of events required for mammary carcinogenesis has been attempted based on dose effects of DMBA and MNU and the effects of multiple treatments with suboptimal doses of DMBA in female SD rats. The results suggest that at least two genetic or epigenetic events are necessary for cancer development (53).

About 90% of MNU-induced rat mammary tumors demonstrate Ha-*ras* activation by a G → A transition in the 12th codon of the oncogene (54). DMBA induced tumors frequently feature A → T transversion mutations in codon 61 (54). The same Ha-*ras* oncogene activation could be detected in 70–80% of transplantable hyperplastic outgrowth lines established from mouse alveolar nodules (55). Ki-*ras* mutations have also been detected in 50% of MNU-induced rat mammary tumors (54). Studies on the role of Ha-*ras* mutations in rat mammary carcinogenesis demonstrated their appearance as early as 2 weeks after MNU treatment. However, another research group reported that preexisting G_{35} → A_{35}Ha-*ras* mutations in rat mammary epithelial cells are responsible for the findings and that MNU exposure merely selects for already existing mutated cells (56). The rate of Ha-*ras* activation found in the MNU-induced tumors appears to vary with the timing of carcinogen administration, being highest (85%) when MNU is applied during metestrus, and lowest (8%) when administered during estrus (54). However, activation of Ha-*ras* may not in itself be sufficient to complete tumorigenesis, because when activated Ha-*ras* was transfected into mouse mammary epithelial cells, lobular alveolar nodules were induced that are mortal and do not form tumors (54).

The role of activated Ha-*ras* seems to be small in rat mammary tumors resulting from PhIP feeding with a reported rate of only 18%, due to G → A transitions at the 2nd position of codon 12, and no Ki-*ras* mutations, as in the case of DMBA-induced rat mammary tumors and human breast cancers (57).

In PhIP-induced mammary tumors no *neu* (*erbB2*) activation by mutation in the transmembrane region has been found and no mutations or amplification of *neu* were detected in ENU-induced tumor lines or in DMBA-induced rat tumors (58). Introduction of an activated *neu* gene into rat mammary cells using a retrovirus vector has potent partial transforming potential (it alone may not be sufficient to transform the mammary cells) (59).

Only one of 10 PhIP-induced mammary tumors examined was found to have a mutation in the *p53* gene. So far no *p53* gene mutations have been reported in rat mammary tumors induced by other agents (57).

Loss of heterozygosity (LOH) or imbalance of the distal part of chromosome 1 has been observed in 50% of MNU-induced mammary tumors in BUF/NCr × ACI/Vsp F_1 rats, with no parental bias (60). However, no equivalent LOH could be detected in DMBA- or PhIP-induced tumors in SD × F344 F_1 rats where involvement of Ha-*ras* mutation is rare. The rate of LOH in MNU-induced lesions appears to be two times higher in the tumors with rather than without Ha-*ras* mutations (60). Cytologically, structural and numerical abnormalities such as partial duplications and trisomy are associated with advanced stages of malignant progression (60). Similar phenomena are also observed for mouse chromosome 7, showing synteny with rat chromosome 1, in mouse skin tumors induced by DMBA and 12-*O*-tetradecanoyl phorbol-13-acetate (TPA) in which Ha-*ras* activation is frequent (61). Rat chromosome 1 shows synteny with human chromosome 11 p15.5, and allelic loss affecting this region has been reported in 30% of human breast cancers (62,63). However, whether a tumor-suppressor gene exists in this region remains to be confirmed. In about 30% of PhIP- and DMBA-induced tumors, the distal part of chromosome 10 shows LOH. The region is syntenic with the distal part of human chromosome 17 (64). While any involvement of *BRCA-1* has yet to be demonstrated, LOH is observed in about 30% of human sporadic breast cancers, and the presence of another tumor-suppressor gene has also been suggested (65).

Microsatellite instability is a feature of about 30% of human breast cancers, associated with reduced survival and a poor prognosis (66). Rat mammary tumors induced by PhIP also show microsatellite mutations, although the underlying mechanisms remain to be clarified (64). Such alterations are rare in mammary tumors caused by DMBA (64). In MNU-induced primary tumors, 2- to 6-fold increases in expression levels of cyclin A and 10- to 15-fold increases in cyclin D1 are observed with a high incidence (67). Most also show high levels of CDK2 and CDK4 proteins. Thus, it has been proposed that aberrant expression of cyclins and other cyclin-related genes occurs frequently in the mammary gland in both rodents and humans (67).

VI. HORMONES AND MAMMARY CANCER

In 1932, Lacassagne discovered that administration of estrone could induce mammary tumors in mice of the RIII strain (68). Although DNA-damaging agents that have no hormonal action also induce mammary tumors, lesion growth is hormone dependent and the hormone status of the treated animals affects the tumor incidence. Mammary tumors occur in males after treatment with carcinogens but their frequency is much less serious than in female animals and the latent period required for tumor development is very long. Bielschowsky established early that the presence of ovaries is of great importance in facilitating the development of mammary cancers due to 2-AAF (69). The same is also true for MNU (6). Thus, initiated cells require the presence of estrogen for further progression towards tumors. The negligible response of ovariectomized animals can be returned to original levels by estrogen administration (70). Thus it is considered that initiated cells harboring activated Ha-*ras* can not grow autonomously until they became engaged in hormone-mediated differentiation. However, high levels of estrogen are also known to inhibit development and/or growth of 3MC- or DMBA-induced rat mammary carcinomas (71). Implantation of pituitary tissue increases mammary tumor incidences; prolactin is the pituitary hormone that appears to be most important (reviewed in Ref. 69). This hormone stimulates tumor growth in ovariectomized rats, in ovariectomized adrenalectomized rats, and in ovariectomized-adrenalectomized hypophysectomized rats. The detection of specific membrane prolactin receptors on the lesions points to a role for a signaling pathway. DMBA- or 3MC-induced rat mammary tumors also bear appreciable quantities of estrogen receptors. Furthermore, administration of an antiestrogenic compound, tamoxifen, to DMBA-treated rats significantly suppresses the development and/or growth of mammary carcinomas. Progesterone also modifies mammary tumorigenesis, but the mechanisms are complicated. An inhibitory action of androgens on the development and growth of polycyclic aromatic hydrocarbon–induced rat mammary carcinomas has been observed. The fact that pseudopregnancy or pregnancy may act as a promoting stimulus for mammary carcinogenesis could well be explained by the increased secretory rates of estrogen, progesterone, and/or prolactin under these conditions (71). The inhibition of mammary carcinoma growth during lactation, despite high blood levels of prolactin, may be due to a deficiency of estrogen and/or progesterone (72,73). It is known that cessation of nursing causes greater tumor regression (73,74). MNU-induced tumors are also influenced by hormones. Pregnancy without lactation reduces the mammary cancer incidence, and the mean latent period before tumor appearance is significantly reduced (75). This could be related to the fact that genetic changes in tumors induced by MNU are probably not homogeneous, so that the response to pregnancy could vary considerably. However, growth of MNU-induced tumors which express receptors for estrogen, progesterone, androgens, and prolactin is inhibited by tamoxifen (76). Conclusive evidence for prolactin involvement in human mammary carcinogenesis remains elusive. Hormones are known to be important not only in cancer of the mammary gland but also in carcinogenesis of the uterus/cervix (see Chapter 28) and of the prostate gland (see Chapter 25).

VII. CONCLUSION

There is a great deal of evidence that environmental factors play some role in neoplasia in the human mammary. The Huggins method, with gavage of a single or several doses of test chemicals in oil into a mammary cancer–susceptible strain of rats, followed by a high-fat diet, may be the most efficient and appropriate method for examination of mammary carcinogenicity. Since genetic alterations in mammary tumors differ depending upon the inducing carcinogen, it is to be recommended that compounds to which humans are exposed be used in studies to facilitate cancer prevention and therapy. Prevention of mammary tumors in humans inheriting cancer susceptibility genes is an important and urgent problem, whose solution should be a high research priority.

REFERENCES

1. Maisin J, Coolen M-L. Au sujet du pouvoir cancérigene du méthyl-cholanthrene. CR Soc Biol Paris 1936; 123:159–160.
2. Strong LC, Smith GM. Local induction of carcinoma of mammary gland by methylcholanthrene. Yale J Biol Med 1939; 11:589–592.
3. Proshaska JV, Brunschwig A, Wilson H. Oral administration of methylcholanthrene to mice. Arch Surg 1939; 38:328–333.

4. Wilson RH, DeEds F, Cox AJ Jr. The toxicity and carcinogenic activity of 2-acetaminofluorene. Cancer Res 1941; 1:595–608.
5. Huggins C, Grand LC, Brillantes FP. Mammary cancer induced by a single feeding of polynuclear hydrocarbons, and its suppression. Nature 1961; 189: 204–207.
6. Gullino PM, Pettigrew HM, Grantham FM. N-Nitroso-methylurea, a mammary gland carcinogen in rats. J Natl Cancer Inst 1975; 54:404–414.
7. Thompson HJ, Meeker LD. Induction of mammary gland carcinomas by the subcutaneous injection of 1-methyl-1-nitrosourea. Cancer Res 1983; 43:1628–1629.
8. Ito N, Hasegawa R, Sano S, Tamano S, Esumi H, Takayama S, et al. A new colon and mammary carcinogen in cooked food, 2-amino-1-methyl-6-phenylimidazo[4,5-b]pyridine (PhIP). Carcinogenesis 1991; 12:1503–1506.
9. Kato T, Migita H, Ohgaki H, Sato S, Takayama S, Sugimura T. Induction of tumors in the Zymbal gland, oral cavity, colon, skin and mammary gland of F344 rats by a mutagenic compound, 2-amino-3,4-dimethylimidazo[4,5-f]quinoline. Carcinogenesis 1989; 10:601–603.
10. Tanaka T, Barnes WS, Williams GM, Weisburger JH. Multipotential carcinogenicity of the fried food mutagen 2-amino-3-methylimidazo[4,5-f]quinoline in rats. Jpn J Cancer Res 1985; 76:570–576.
11. Shirai T, Tamano S, Sano M, Masui T, Hasegawa R, Ito N. Carcinogenicity of 2-amino-1-methyl-6-phenylimidazo[4,5-b]pyridine (PhIP) in rats: dose response studies. In: Adamson R, Gustafsson JA, Ito N, Nagao M, Sugimura T, Wakabayashi K, Yamazoe Y, eds. Heterocyclic amines in cooked foods: possible human carcinogens. Princeton, NJ: Princeton Scientific Publication Co., Inc., 1995:232–239.
12. Ghoshal A, Preisegger K-H, Takayama S, Thorgeirsson SS, Snyderwine EG. Induction of mammary tumors in female Sprague-Dawley rats by the food-derived carcinogen 2-amino-1-methyl-6-phenylimidazo[4,5-b]pyridine and effect of dietary fat. Carcinogenesis 1994; 15:2429–2433.
13. Nagao M, Ushijima T, Wakabayashi K, Ochiai M, Kushida H, Sugimura T, Hasegawa R, Shirai T, Ito N. Dietary carcinogens and mammary carcinogenesis. Cancer 1994; 74(suppl 3):1063–1069.
14. Layton D, Bogen K, Knize MG, Hatch FT, Johnson VM, Felton JS. Cancer risk of heterocyclic amines in cooked foods: an analysis and implications for research. Carcinogenesis 1995; 16:39–52.
15. El-Bayoumy K, Chae Y-H, Upadhyaya P, Rivenson A, Kurtzke C, Reddy B, Hecht SS. Comparative tumorigenicity of benzo[a]pyrene, 1-nitropyrene, and 2-amino-1-methyl-6-phenylimidazo[4,5-b]pyridine administered by gavage to female CD rats. Carcinogenesis 1995; 16:431–434.
16. Ashby J, Tennant RW. Definitive relationships among chemical structure, carcinogenicity and mutagenicity for 301 chemicals tested by the U.S. NTP. Mutat Res 1991; 257:229–306.
17. Dunnick JK, Elwell MR, Huff J, Barrett JC. Chemically induced mammary gland cancer in the National Toxicology Program's carcinogenesis bioassay. Carcinogenesis 1995; 16:173–179.

18. Ochiai M, Watanabe M, Kushida H, Wakabayashi K, Sugimura T, Nagao M. DNA adduct formation, cell proliferation and aberrant crypt focus formation induced by PhIP in male and female rat colon with relevance to carcinogenesis. Carcinogenesis 1996; 17:95–98.
19. Ghoshal A, Davis CD, Schut HJA, Snyderwine EG. Possible mechanisms for PhIP-DNA adduct formation in the mammary gland of female Sprague-Dawley rat. Carcinogenesis 1995; 16:2725–2731.
20. Carmichael PL, Stone EM, Grover PL, Gusterson BA, Phillips DH. Metabolic activation and DNA binding of food mutagens and other environmental carcinogens in human mammary epithelial cells. Carcinogenesis 1996; 17:1769–1772.
21. Shirai T, Sano M, Tamano S, Takahashi S, Hirose M, Futakuchi M, Hasegawa R, Imaida K, Wakabayashi K, Sugimura T, Ito N. The prostate is also a target for carcinogenicity of 2-amino-1-methyl-6-phenylimidazo[4,5-b]pyridine (PhIP) which derived from cooked foods. Cancer Res 1997; 57:195–198.
22. Survey of compounds which have been tested for carcinogenic activity. 1985–1986 Vol, Sect II. NIH Publication 87-2883. Bethesda: NCI, 1987:1518–1558.
23. Hirota N, Aonuma T, Yamada S, Kawai T, Saito K, Yokoyama T. Selective induction of glandular stomach carcinoma in F344 rats by N-methyl-N-nitrosourea. Jpn J Cancer Res 1987; 78:634–638.
24. Squaritini F, Pingitore R. Tumors of the mammary gland. In: Turosov V, Mohr U, eds. Pathology of Tumors in Laboratory Animals. Vol 2: Tumors of the Mouse. 2nd ed. Lyon, France: IARC Scientific Publications, 111, 1994: 47–100.
25. Wang Y, Holland JF, Bleiweiss IJ, Melana S, Liu X, Pelisson I, Cantarella A, Stellrecht K, Mani S, Pogo BGT. Detection of mammary tumor virus ENV gene-like sequences in human breast cancer. Cancer Res 1995; 55:5173–5179.
26. Medina D, Warner MR. Mammary tumorigenesis in chemical carcinogen-treated mice. IV. Induction of mammary ductal hyperplasia. J Natl Cancer Inst 1976; 57:331–337.
27. Durbin PW, Williams MH, Jeung N, Arnold JS. Development of spontaneous mammary tumors over the life-span of the female charles river (Sprague-Dawley) rat: the influence of ovariectomy, thyroidectomy, and adrenalectomy-ovariectomy. Cancer Res 1966; 26:400–411.
28. Russo J, Gusterson BA, Rogers AE, Russo IH, Wellings SR, van Zwieten MJ. Comparative study of human and rat mammary tumorigenesis. Laboratory Investigation 1990; 62:244–278.
29. Haag JD, Newton MA, Gould MN. Mammary carcinoma suppressor and susceptibility genes in the Wistar-Kyoto rat. Carcinogenesis 1992; 13:1933–1935.
30. Marshall E. The politics of breast cancer. Science 1993; 259:616–632.
31. Henderson JC. Risk factors for breast cancer development. Cancer 1990; 71: 2127–2140.
32. Hall JM, Lee MK, Newman B, Morrow JE, Anderson LA, Huey B, King MC. Linkage of early-onset familial breast cancer to chromosome 17q21. Science 1990; 250:1684–1689.

33. Wooster R, Neuhausen SL, Mangion J, Quirk Y, Ford D, Collins N, Nguyen K, Seal S, Tran T, Averill D, et al. Localization of a breast cancer susceptibility gene, BRCA2, to chromosome 13q12-13. Science 1994; 265:2088–2090.
34. Kerangueven F, Essioux L, Dib A, Noguchi T, Allione F, Geneix J, Longy M, Lidereau R, Eisinger F, Pébusque M-J, Jacquemier J, Bonaïti-Pellié C, Sobol H, Birnbaum D. Loss of heterozygosity and linkage analysis in breast carcinoma: indication for a putative third susceptibility gene on the short arm of chromosome 8. Oncogene 1995; 10:1023–1026.
35. Wooster R, Mangion J, Eeles R, Smith S, Dowsett M, Averill D, Lee PB, Easton DF, Ponder BAJ, Stratton MR. A germline mutation in the androgen receptor gene in two brothers with breast cancer and Reifenstein syndrome. Nat Genet 1992; 2:132–134.
36. Miki Y, Swensen J, Shattuck-Eidens D, Futreal PA, Harshman K, Tavtigian S, Liu Q, Cochran C, Bennett LM, Ding W, et al. A strong candidate for the breast and ovarian cancer susceptibility gene BRCA1. Science 1994; 266:66–71.
37. Inoue R, Fukutomi T, Ushijima T, Matsumoto Y, Sugimura T, Nagao M. Germline mutation of *BRCA1* in Japanese breast cancer families. Cancer Res 1995; 55:3521–3524.
38. Wooster R, Bignell G, Lancaster J, Swift S, Seal S, Mangion J, Collins N, Gregory S, Gumbs C, Micklem G, et al. Identification of the breast cancer susceptibility gene BRCA2 [see comments]. Nature 1995; 378:789–792.
39. Malkin D, Li FP, Strong LC, Fraumeni JF, Nelson CE, Kim DH, Kassel J, Gryka MA, Bischoff FZ, Tainsky MA, Friend SH. Germ line p53 mutations in a familial syndrome of breast cancer, sarcomas and other neoplasms. Science 1990; 250:1233–1238.
40. Krontiris TG. Minisatellites and human disease. Science 1995; 269:1682–1683.
41. Isaacs JT. Genetic control of resistance to chemically induced mammary adenocarcinogenesis in the rat. Cancer Res 1986; 46:3958–3963.
42. Hsu L-C, Kennan WS, Shepel LA, Jacob HJ, Szpirer C, Szpirer J, Lander ES, Gould MN. Genetic identification of Mcs-1, a rat mammary carcinoma suppressor gene. Cancer Res 1994; 54:2765–2770.
43. Richardson KK, Richardson FC, Crosby RM, Swenberg JA, Skopek TR. Proc Natl Acad Sci USA 1987; 84:344–378.
44. Lu S-J, Archer MC. Ha-ras oncogene activation in mammary glands of N-methyl-N-nitrosourea-treated rats genetically resistant to mammary adenocarcinogenesis. Proc Natl Acad Sci USA 1992; 89:1001–1005.
45. Lu S-J, Archer MC. ras oncogene activation in mammary carcinomas induced by N-methyl-N-nitrosourea in Copenhagen rats. Mol Carcinog 1992; 6:260–265.
46. Toyota M, Canzian F, Ushijima T, Hosoya Y, Kuramoto T, Serikawa T, Imai K, Sugimura T, Nagao M. A rat genetic map constructed by representational difference analysis markers with suitability for large-scale typing. Proc Natl Acad Sci USA 1996; 93:3914–3919.
47. Lisitsyn N, Lisitsyn N, Wigler M. Cloning the differences between two complex genomes. Science 1993; 259:946–951.

48. Yamada J, Kuramoto T, Serikawa T. A rat linkage map and comparative maps for mouse or human homologous rat genes. Mamm Genome 1994; 5:63–83.
49. Jacob HJ, Brown DM, Bunker RK, Daly MJ, Dzau VJ, Goodman A, Koike G, Kren V, Kurtz T, Lernmark A, Levan G, Mao YP, Pettersson A, Pravenec M, Simon JS, Szpirer C, Szpirer J, Trolliet MR, Winer ES, Lander ES. A genetic linkage map of the laboratory rat, Rattus norvegicus. Nat Genet 1995; 9:63–69.
50. Chen K-S, Shepel LA, Haag JD, Heil GM, Gould MN. Cloning, genetic mapping and expression studies of the rat Brca1 gene. Carcinogenesis 1996; 17:1561–1566.
51. Biggs PJ, Warren W, Venitt S, Stratton MR. Does a genotoxic carcinogen contribute to human breast cancer? The value of mutational spectra in unravelling the aetiology of cancer. Mutagenesis 1993; 8:275–283.
52. Sidransky D, Hollstein M. Clinical implications of the p53 gene. Annu Rev Med 1996; 47:285–301.
53. Isaacs JT. Determination of the number of events required for mammary carcinogenesis in the Sprague-Dawley female rat. Cancer Res 1985; 45:4827–4832.
54. Sukumar S, McKenzie K, Chen Y. Animal models for breast cancer. Mutat Res 1995; 333:37–44.
55. Dandekar S, Sukumar S, Zarbl H, Young LJT, Cardiff RD. Specific activation of cellular Harvey-ras oncogene in dimethylbenz(a)anthracene-induced mouse mammary tumors. Mol Cell Biol 1986; 6:4104–4108.
56. Cha RS, Thilly WG, Zarbl H. N-nitroso-N-methylurea-induced rat mammary tumors arise from cells with preexisting oncogenic Hras1 gene mutations. Proc Natl Acad Sci USA 1994; 91:3749–3753.
57. Ushijima T, Kakiuchi H, Makino H, Hasegawa R, Ishizaka Y, Hirai H, Yazaki Y, Ito N, Sugimura T, Nagao M. Infrequent mutation of Ha-ras and p53 in rat mammary carcinomas induced by 2-amino-1-methyl-6-phenylimidazo-[4,5-b]pyridine. Mol Carcinog 1994; 10:38–44.
58. Hal DG, Stoica G. Evaluation of potential oncogene alterations in the ENU1564 rat mammary tumor model. Anticancer Res 1994; 14(2A):481–487.
59. Wang B, Kennan WS, Yasukawa-Barnes J, Lindstrom MJ, Gould MN. Frequent induction of mammary carcinomas following neu oncogene transfer into in situ mammary epithelial cells of susceptible and resistant rat strains. Cancer Res 1991; 51:5649–5654.
60. Gollahon LS, Chen A, Aldaz CM. Loss of heterozygosity at chromosome 1q loci in rat mammary tumors. Mol Carcinog 1995; 12:7–13.
61. Aldaz CM, Trono D, Larcher F, Slaga TJ, Conti CJ. Sequential trisomization of chromosomes 6 and 7 in mouse skin premalignant lesions. Mol Carcinog 1989; 2:22–26.
62. Ali IU, Lidereau R, Theillet C, Callahan R. Reduction to homozygosity of genes on chromosome 11 in human breast neoplasia. Science 1987; 238:185–197.
63. Winqvist R, Mannermaa A, Alavaikko M, et al. Refinement of regional loss

of heterozygosity for chromosome 11p15.5 in human breast tumors. Cancer Res 1993; 53:4486–4488.

64. Toyota M, Ushijima T, Weisburger JH, Hosoya Y, Canzian F, Rivenson A, Imai K, Sugimura T, Nagao M. Microsatellite instability and loss of heterozygosity on chromosome 10 in rat mammary tumors induced by 2-amino-1-methyl-6-phenylimidazo[4,5-b]pyridine. Molecular Carcinogenesis 1996; 15: 176–182.
65. Cropp CS, Nevanlinna HA, Pyrhönen S, Stenman U-H, Salmikangas P, Aibertsen H. Evidence for involvement of BRCA1 in sporadic breast carcinomas. Cancer Res 1994; 54:2548–2551.
66. Paulson TG, Wright FA, Parker BA, Russack V, Wahl GM. Microsatellite instability correlates with reduced survival and poor disease prognosis in breast cancer. Cancer Res 1994; 56:4021–4026.
67. Sgambato A, Han EK, Zhang YJ, Moon RC, Santella RM, Weinstein IB. Deregulated expression of cyclin D1 and other cell cycle-related genes in carcinogen-induced rat mammary tumors. Carcinogenesis 1995; 16(9):2193–2198.
68. Lacassagne A. Apparition de cancers de la mamelle chez le souris mâle, soumise a des injections de folliculine. Compt Rend Acad 1932; 195:630–632.
69. Bielschowsky F. Distant tumours produced by 2-amino- and 2-acetyl-amino-fluorene. Br J Exp Path 1994; 25:1–4.
70. Kumar R, Sukumar S, Barbacid M. Activation of ras oncogenes preceding the onset of neoplasia. Science 1990; 248:1101–1104.
71. Welsch CW. Host factors affecting the growth of carcinogen-induced rat mammary carcinomas: a review and tribute to Charles Brenton Huggins. Cancer Res 1985; 45:3415–3443.
72. Sakamota S, Imamura Y, Sassa S, Okamoto R. DMBA-induced mammary tumor and hormone environment in the rat during pregnancy postpartum, and long term lactation. Toxicol Lett 1979; 4:237–240.
73. McCormick GM, Moon RC. Hormones influencing postpartum growth of 7, 12-dimethylbenzanthracene induced rat mammary tumors. Cancer Res 1967; 27:626–631.
74. McCormick GM. The effect of varying the length of the nursing period on the postpartum growth of chemically induced rat mammary tumors. Cancer Res 1972; 32:1574–1576.
75. Grubbs CJ, Hill DL, McDonough KC, Peckham JC. *N*-Nitroso-*N*-methylurea-induced mammary carcinogenesis: effect of pregnancy on preneoplastic cells. J Natl Cancer Inst 1983; 71:625–628.
76. Manni A, Wright C. Effect of tamoxifen and α-difluoromethylornithine on clones of nitrosomethylurea-induced rat mammary tumor cells grown in soft agar culture. Cancer Res 1983; 43:1084–1086.
77. Ohgaki H, Takayama S, Sugimura T. Carcinogenicities of heterocyclic amines and cooked food. Mutat Res 1991; 259:399–410.

21

Cancers of the Hematopoietic System in Rodents and Humans: Chemical Factors, Physical Factors, and Molecular Epidemiological Studies

Elizabeth W. Newcomb
New York University Medical Center, New York, New York

I. INTRODUCTION

From an historical point of view, the World Health Assembly established The International Agency for Research on Cancer (IARC) in 1965 to systematically conduct studies on the epidemiology of cancer worldwide (1). The research programs determine potential carcinogens in the environment and identify biological agents and any other factors that might contribute to the causes of cancer. Over the past three decades, more than 50 monographs have been published by IARC working groups, which have evaluated carcinogenic agents for their risk to human health. Agents that show increased risk for cancer in humans have been identified in in vitro and animal bioassays (2, 3). Exposure to these agents in both animals and humans is evaluated for cancer-inducing potential and used to divide exposure into four categories. Group 1 agents are *carcinogenic to humans* because a causal relationship has been established. Group 2A agents are *probably carcinogenic to humans* because an association with cancer has been observed in exposed humans and they are also carcinogenic in animal studies. Group 2B agents are *possibly carcinogenic to humans* because there is evidence for carcinogenic potential from animal studies but no clear associ-

Table 1 Chemicals Known to Induce Hematopoietic Tumors in Standardized Animal Bioassays

Chemical (IARC Group)[a]	Species	Route	Site	TD_{50}	$p<$	Ref.
Acetamide[b] (2B)	M cb6-m	eat	mln	2–6 g	0.002–0.0005	4,20
trans-5-Amino-3[2-(5-nitro-2-furyl)vinyl]-1,2,4-oxadiazole (2A)	M cdl-m,f	eat	thm	100–200 mg	0.005–0.0005	5
Benzene[b] (1)	M c56-m	inh	thm	460 mg	0.004–0.0005	6,7
	R sda-m	gav	leu	506 mg	0.008	
1,3-Butadiene[b] (2B)	M b6c-m,f	inh	lym	46–75 mg	0.0005	8–10,20
Chlorambucil[b] (1)	M swi-m,f	ipj	lys	400–700 mg	0.003–0.0005	11
	R cdr-m	ipj	lys	1.4–1.7 mg	0.004–0.0005	
Cyclophosphamide[b] (1)	M swi-m,f	ipj	lys	7–8 mg	0.003–0.004	11,20
Dacarbazine[b] (2B)	M swi-m,f	ipj	leu/lys	8–16 mg	0.05–0.0005	11
	R sda-f	eat	thm	4 mg	0.02	
3,3′-Dimethoxybenzidine · 2HCl (2A)	R f34-m,f	wat	leu	2.5–7 mg	0.003–0.0005	10
1,4-Dioxane (2B)	M b6c-m	wat	lym	4 gm	0.009	12
Furan (2A)	M b6c-m,f	gav	lym	2–4 mg	0.0005	10
	R f34-m,f	gav	lym	0.5–2 mg	0.0005	
Isophosphamide[b]	M b6c-f	ipj	lym	5 mg	0.0005	12
	R sda-m	ipj	leu/lym	3 mg	0.0005	
Melphalan[b] (1)	M swi-m,f	ipj	lys	0.3–0.7 mg	0.02–0.06	11
	R cdr-m,f	ipj	lys	0.4–1.5 mg	0.09–0.003	

Nitrosoureas (2B)						
1-amyl-1-	R don-m,f	wat	leu	9–20 mg	0.005	13
1,3-dibutyl	R don-f	wat	leu	32 mg	0.06	14
1-ethyl-1-	R f3d-f	wat	leu	150 μg	0.0005	15
1-(2-hydroxyethyl)-1-	M swi-m,f	wat	lym	0.6–1 mg	0.0005	16
	R mrw-m	wat	lym	2 mg	0.0005	17
N-Nitroso-*N*-methylurea (2A)	M cb6-m,f	ipj	lym	.09 mg	0.0005	10
Phenacetin (2A)	R sda-m,f	eat	leu	2.5 g	0.0005	8,12
Procarbazine · HCl[b] (2A)	M b6c-m,f	ipj	leu/lym	0.9 mg	0.0005	11
	M swi-f	ipj	leu/lys	3.7 mg	0.0005	
	R sda-m,f	ipj	lym	2.5–5.2 mg	0.0005	
	P cym-m,f	ipj	leu	15 mg	0.005	18
	P rhe-m,f	ipj	leu	4 mg	0.0005	19
Thiotepa[b] (1)	M b6c-m,f	ipj	leu/lym	0.2–0.3 mg	0.0005	12,20
	R sda-m	ipj	leu/lym	0.3–0.4 mg	0.0005	

[a]Chemical group: (1) = carcinogenic; (2A) = probably carcinogenic; (2B) = possibly carcinogenic (1).
[b]Chemical tested positive at some site in both mice and rats (3).
M = Mouse (strains, b6c = B6C3F1; cb6 = C57BL/6; cdl = Charles River CDl; c56 = C57BL/6J; swi = Swiss); R = rat (strains cdr = Charles River CD; don = Donyru; f3d = F344/DuCrj; f34 = Fischer 344; mrw = MRC-Wistar; sda = Sprague-Dawley) P = monkeys (strains, cyn = Cynomolgus [Macaca fascicularis]; rhe = Rhesus [Macaca mulatta] Route of administration: eat = chemical in food; gav = gavage; inh = inhalation; inj = intraperitoneal injection, wat = water. Site code: mln = mesenteric lymph node, thm = thymus; leu = leukemia, lym = lymphoma, lys = lymphosarcoma. TD_{50} = Tumorigenic dose rate (mg/kg body weight/day) for 50% of the test animals to get tumors for a given target site. P = significance value of tumorigenic potency determined from two-tailed test.
Source: Ref. 3.

ation with cancer has been established for exposed humans. Groups 3 and 4 agents are *probably not carcinogenic to humans* or not able to be classified. Table 1 lists chemicals and their IARC classification, if known, that induce hematopoietic tumors in standardized animal bioassays (4–20).

Cancer is a multigenic disease process. Normal cells accumulate genetic or epigenetic changes (DNA methylation) over an extended time interval, which contribute to their subsequent progression to malignancy. Some of the causative agents for human cancers, although not completely understood at the mechanistic level, include a large array of chemical carcinogens, radiation exposure, and certain viruses (1). Susceptibility to cancer-causing agents is influenced by many genes, including enzymes that control the response of the cell to carcinogens such as cytochrome P-450 (21,22) and glutathione S-transferase (23,24). Many other genes serve as targets for DNA damage induced by the interaction with the carcinogen such as oncogenes (25,26) and tumor-suppressor genes (27–30).

The development and application of molecular genetic techniques in the past decade now makes it possible to study cancer-susceptibility genes for any given cancer. In this chapter we will discuss the evidence, using molecular epidemiology, for the role of chemical and physical factors as causal agents for hematopoietic tumors in rodents and humans. In addition, the spectrum of DNA damage to critical target genes such as the *ras* oncogene (25,26), the *p53* (28,29), *p16* or *Rb* (30) tumor-suppressor genes will be summarized. Mutations in genes closely associated with cancer in humans such as the *ras* or *p53* genes may serve as useful biomarkers to identify etiological agents associated with the selective induction of cancers of the hematopoietic system.

II. ORGAN- AND SPECIES-SPECIFIC CARCINOGENESIS

A. Chemicals Known to Induce Hematopoietic Tumors

Table 1 lists 20 chemicals that have tested positive in animal bioassays for the induction of tumors of the hematopoietic system. Chemicals with an asterisk have tested positive in both rats and mice and may have also induced tumors at multiple organ sites (3,20). Five of the agents classified by IARC as *carcinogenic* (Group 1), including chlorambucil, cyclophosphamide, melphalan, and thiotepa, are chemotherapeutic agents commonly used to treat several human cancers. As discussed below, many secondary cancers arise in the hematopoietic system in patients who have been treated with a number of these agents. An additional six chemicals are classified as *probably carcinogenic* (Group 2A). Notable among them is the chemotherapeutic drug procarbazine · HCl, which has tested positive in three species including mice, rats, and monkeys.

B. Genetic Susceptibility and Carcinogenesis

The response of an organism to chemical carcinogen exposure depends on many factors, including enzyme-mediated metabolism, affinity of the compound for DNA, DNA repair, and the rate of cell proliferation, which may differ between cell types in different organs. The genetics of susceptibility in chemical carcinogenesis is an important consideration for both rodent and human studies of cancer incidence. It is well known that some chemicals selectively induce tumors in only certain animal species and some chemicals induce tumors in only a single tissue within a given species (31–36). This inherent genetically determined variation in induction of hematopoietic tumors by chemicals in animals makes extrapolation for potential risk to the human population more difficult to ascertain (37–39). To determine more meaningful relative risk for humans, several proposals have been made regarding evaluation of potential carcinogens in animal bioassays (37,40–42). Tennant (42) has suggested "stratification" of results from animal bioassays in order to identify those chemicals which are most likely to be *carcinogenic* in humans. Chemicals that induce tumors at common organ sites across sexes and species are less likely to be influenced by species-specific polygenic traits and are more likely to induce tumors *via* a common mechanism(s). The majority of the chemicals listed in Table 1 induce hematopoietic tumors at common sites in at least two species and generally in both sexes. Thus, these chemicals may form a subset that poses a greater risk for human health.

III. EXPERIMENTAL STUDIES IN ANIMALS

A. Identifying Cancer-Susceptibility Genes

The development of inbred strains of mice began in the early 1900s for the purpose of studying the genetic basis for cancer (43). As a result many strains of mice are available, which differ in their susceptibility to cancer. Spontaneous tumors of the hematopoietic system occur in most strains of mice, with their incidence strongly influenced by the inheritance of different genes, by age, and by the presence of sex hormones (44,45). Some, like the AKR strain, are high-incidence strains, while others, such as DBA/2, are low-incidence strains (43). Use of recombinant inbred strains of mice derived from a cross between mice with high and low incidences of hematopoietic tumors can help define cancer susceptibility genes (46,47).

Cancer is a polygenic disease. With the identification of simple sequence genetic differences in the DNAs between mouse strains, cloning complex trait loci is now a reality (48). Once chromosomal regions are identified in mice, the putative susceptibility genes may be located from known chromosomal genetic maps or new genes can be cloned by positional cloning techniques. Because chromosomal regions between human and mouse are highly conserved in evolution, identification of cancer susceptibility genes in one species allows for the identification of similar genes in humans. Therefore, animal models are increasingly important to understand the genetic basis of disease, particularly cancers induced by carcinogenic agents.

B. Experimental Leukemogenesis in Animal Models

Animal Models for Chemically Induced Tumors

Alkylnitroso compounds occur ubiquitously in the human environment and have been shown to be powerful carcinogenic agents in many species and in many organs (49). Table 2 summarizes the data from a large number of inbred strains of mice on the induction of thymic lymphomas by these agents. Early studies tested various nitroso compounds for their potential to induce tumors in different strains of mice and defined the different variables such as age, sex, and dosage (single *versus* multiple treatments) and route of injection that influenced tumor incidence (50–55). The results suggested that all mouse strains tested were susceptible to lymphoma development, particularly to the alkylating agent *N*-methylnitrosourea (MNU) (50,51,54,55). Since that time, MNU has been widely used in different animal models of carcinogenesis to dissect the genetic events underlying the development of thymic lymphomas (56–70).

The mode of action of MNU is well established as a direct-acting carcinogenic alkylating agent (71–76). It induces O^6-methylguanine adduct, which leads to mispairing at thymines resulting in a characteristic mutation spectrum, a G-to-A transition mutation (77). The observed tissue specificity for tumor induction is related to the capacity of a particular organ to repair MNU-induced DNA damage through the DNA repair enzyme O^6-alkylguanine-DNA alkyltransferase (70,78,79). Further studies utilizing transgenic animals expressing high levels of the human alkyltransferase have shown that MNU-induced thymic lymphomas are inhibited (70) due to the rapid removal of MNU-induced DNA adducts from thymus tissue (80). These animals can now be used to probe further the chemical actions of different *N*-nitroso compounds in carcinogenesis (81).

Other transgenic models have been useful in studying the cooperation of different genes which predispose to carcinogen-induced lymphomagenesis, such as the *pim-1* transgenic mice (68). The *pim-1* gene was identified due to its frequent activation by proviral insertion in murine leukemia virus (MuLV)-induced tumors (82). *Pim-1* transgenic mice express the *pim-1* proto-oncogene in lymphoid tissue and develop tumors in approximately 10% of the animals by 9 months of age (67). Treatment of *pim-1* mice with ENU increases the tumor incidence to 100% and accelerates the formation of tumors to 2 months (67). Because of the increased susceptibility of *pim-1* mice to develop T-cell tumors, they have been used in short-term animal bioassays to screen compounds for carcinogenic potency (83). This is only one of several animal models with potential use in carcinogen screening bioassays.

Another chemical, 3-methylcholanthrene (MCA), also induces hematopoietic tumors following percutaneous application in some strains of mice (84,85). Susceptibility to tumor induction is strongly influenced by the genes at the Ah locus, which control the expression of the aryl hydrocarbon (Ah) hydrolase (86,87). Mice homozygous for the recessive allele Ah^d, such as the RF/J strain, develop thymic lymphomas rather than fibrosarcomas at the site of MCA application like mice with the Ah^b allele (88). Recently, the aryl hydrocarbon receptor (AHR) was knocked out by homologous recombination in 129/SV ES cells (89). The Ah receptor is a transcription factor that functions in mice to mediate the harmful effects of many toxic chemicals. It is constitutively expressed in many tissues and regulates the induction of several classes of genes such as cytochrome P-450s and uridine diphosphate-glucuronosyltransferase (89). These AHR-deficient mice should also prove useful as animal models in the assessment of carcinogenic risk posed by a variety of environmental pollutants such as dioxin, benzo[a]pyrene, and polychlorinated biphenyls.

Animal Models for Radiation-Induced Tumors

Exposure to radiation induced hematopoietic tumors in survivors of the atomic bombs in Japan (90–92). Following that historical event, several groups in the 1950s developed animal models to test the variables inherent in the induction of tumors in mice following gamma or neutron radiation (93,94). Early studies defined the different variables such as age, sex, and dosage (single *versus* multiple treatments) that influenced tumor incidence (93,94). Table 3 summarizes the data from a number of inbred strains of mice on the induction of hematopoietic tumors by different radiation sources (94–97). The most widely used model for studying radiation-induced disease is that of Kaplan in C57BL mice (93).

The mode of action of gamma and neutron radiation in inducing

Table 2 Summary of Genetic Changes Induced in Hematopoietic Tumors by Chemical Compounds

Mouse strain	Sex	Compound (dosage)	Tumor incidence (%)	Gene alterations (frequency)	Ref.
Swiss	m,f	MNU 1X (75–125 mg/kg)	60–70	ND	50
	f	MNU 5X (30–50 mg/kg)	>90	ND	51
	m	MNUN 1X (20 mg/kg)	6	ND	50
	f	MNTS 1X (7 mg/kg)	4	ND	50
	f	MNNG 1X (40 mg/kg)	0	ND	52
CBA	m,f	NBU (0.04% water)	80	Trisomy 15	53
CBA/Ca	f	MNU 1X (50 mg/kg)	70	ND	54
NMRI	f	MNU 1X (50 mg/kg)	80–90	ND	54
DBA/2	f	MNU 1X (50 mg/kg)	80–90	ND	54
C57BL/6	f	MNU 1X (50 mg/kg)	80–90	ND	54
(C57BL/6J × DBA/2)F1	f	MNU 1X (50 mg/kg)	80–90	ND	54
C57BL/Cbi	f	MNU 1X (50–80 mg/kg)	60–90	ND	55
	f	ENU 1X (160–240 mg/kg)	13–38	ND	55
AKR/J	f	MNU 1X (50 mg/kg)	100	K-*ras* (75%) c-*myc* (0%) *Pim*-1 (0%)	56
	f	None	100	K-*ras* (0%) c-*myc* (7%) *Pim*-1 (17%)	56

AKR/J	f	MNU 1X (50 mg/kg)	100	K-*ras* (22%) c-*myc* (4%) *Pim*-1 (11%)	57
	f	MMS 1X (120 mg/kg)	100	K-*ras* (0%) c-*myc* (30%) *Pim*-1 (4%)	57
AKR/J	f	MNU 1X (75 mg/kg)	100	K-*ras* (19%)	58
		MNU 5X (30 mg/kg)	100	K-*ras* (58%)	
	m	MNU 1X (75 mg/kg)	90	K-*ras* (16%)	
		MNU 5X (30 mg/kg)	90	K-*ras* (50%)	
(AKR/J × RF/J)F1	f	MNU 5X (30 mg/kg)	100	N-*ras* (83%)	59,60
RF/J	f	MNU 5X (30 mg/kg)	>90	K-*ras* (50%) N-*ras* (50%) *p53* (3%) *MDM-2* (0%)	60–62
129/J	f	MNU 5X (30 mg/kg)	>90	K-*ras* (50%) N-*ras* (50%)	60,61
C57BL/6J	f	MNU 5X (30 mg/kg)	90	K-*ras* (80%) *p53* (13%) Trisomy 15 (78%)	63–65
C57BL/6J	f	MNU 1X (80 mg/kg)	46	K-*ras* (18%)	66
Transgenic strains and nitroso compounds					
(CBA × C57BL)F1-pim-1	f	ENU 1X (60 mg/kg)	100	K-*ras* (70%) N-*ras* (30%) c-*myc* (90%) N-*myc* (0%)	67,68

(*continued*)

Table 2 Summary of Genetic Changes Induced in Hematopoietic Tumors by Chemical Compounds

Mouse strain	Sex	Compound (dosage)	Tumor incidence (%)	Gene alterations (frequency)	Ref.
Nontransgenic controls	f	ENU 1X (60 mg/kg)	20	K-*ras* (100%)	
				c-*myc* (90%)	
				N-*myc* (0%)	
				K-*ras* (16%)	
				p53 (10%)	
(C57BL/6 × DBA/2)F1-N-ras	m	MNU 5X (30 mg/kg)	57	*MDM-2* (0%)	62,69
				K-*ras* (20%)	
				N-*ras* (5%)	
				p53 (5%)	
Nontransgenic controls	m	MNU 5X (30 mg/kg)	68	*MDM-2* (0%)	
(C57BL/6 × SJL)F1-MGMT	m,f	MNU 1X (80 mg/kg)	10	ND	70
Nontransgenic controls	m,f	MNU 1X (80 mg/kg)	90	ND	
Chemicals other than nitroso compounds					
RF/J	m,f	MCA 2X/wk (0.5%)	22	ND	71
(AKR × RF)F1	m,f	MCA 2X/wk (0.5%)	67	ND	71
RF/J	m,f	MCA 5X (1%)	75	K-*ras* (53%)	72

Compounds: DMN, dimethynitrosoamine; ENU, *N*-ethyl-*N*-nitrosourea; MNU, *N*-methylnitrosourea; MNNG, *N*-methyl-*N′*-nitro-nitrosoguanidine; MNUN, methylnitrourethan; MNTS, methylnitroso-*p*-tolysolfonamide; NBU, *N*-nitoso-*N*-butylurea; MMS, methyl methanesulfonate, MCA, methylcholanthracene.
Transgenic strains: *pim*-1, mouse proto-oncogene (67); N-*ras*, mouse proto-oncogene (69); MGMT, human O^6-methylguanine-DNA-methyltransferase (70).

Table 3 Summary of Genetic Changes Induced in Hematopoietic Tumors by Gamma or Neutron Radiation

Mouse strain	Sex	Radiation (dosage)[a]	Tumor incidence (%)	Gene alterations (frequency)	Ref.
C57BL/6J	m,f	gamma 4X (150 rads)	>90	Marker chrom	93
RF	m,f	gamma 3X (150 rads)	30–40	ND	94
129/J	f	gamma 4X (150 rads)	0	ND	61
(AKR/J × RF/J)F1	f	gamma 4X (150 rads)	>90	K-*ras* (100%)	59
RF/J	f	gamma 4X (150 rads)	>90	K-*ras* (88%) N-*ras* (12%) *p53* (0%) *MDM-2* (0%)	61,62
C57BL/6J	f	gamma 4X (175 rads)	>90	K-*ras* (17%) N-*ras* (42%) *p53* (13%) Trisomy 15 (55%) Marker chrom (40%)	63–65
(C57BL/6J × RF/J)F1	m,f	gamma 4X (175 rads)	>90	LOH *p16* (34%)	95
RFM	f	neutron 1X (47 rads)	30–35	ND	96
RF/J	f	neutron 1X (100 rads)	71	K-*ras* (75%) N-*ras* (25%) *p53* (0%) *MDM-2* (0%)	62,97

[a]Gamma rays are low linear energy transfer (LET) radiation; neutron rays are high LET radiation. A neutron dose of 47 rads is equivalent to 150 rads of gamma rays (94). ND, None detected.

DNA damage is different. Unlike MNU or similar nitroso compounds, which act directly on DNA to produce miscoding lesions unless repaired, gamma rays or low-LET (linear energy transfer) radiation interacts with water molecules in the cell to produce free radicals, which then chemically react with DNA (98,99). Neutrons are high-LET radiation that deposit large amounts of energy that react directly with DNA or free radical intermediates to produce DNA strand breaks, chromosomal rearrangements, and deletions (100,101).

Unlike nitroso compounds, which are potent carcinogens in all mouse strains that have been tested, some strains of mice such as 129/J are resistant to radiation-induced tumors (61,102). Host genes that control resistance or susceptibility to radiation-induced disease have been mapped, in some cases, using recombinant inbred strains of mice (46). However, identification of additional genes which predispose to radiation-induced disease is an area requiring future research efforts.

Human O^6-alkylguanine-DNA alkyltransferase activity in transgenic mice does not prevent radiation-induced lymphomas (70). This result is expected because of the differences in the mode of action of MNU *versus* gamma radiation in the induction of thymic lymphomas. Transgenic as well as nontransgenic littermates developed gamma radiation–induced lymphomas with equal frequency. As will be noted below, despite the differences in the mechanisms for DNA damage induced by MNU or radiation, the genetic alterations detected in thymic lymphomas induced in animals of the same inbred strain are remarkably similar (Tables 2 and 3).

Alterations in Cancer Susceptibility Genes in Carcinogen- and Radiation-Induced Hematopoietic Tumors

Cancer development is a complex process thought to involve several genetic events occurring over a relatively long time interval converting a normal cell into a fully malignant state (103–105). At least two families of genes have been strongly implicated in this process in human cancers. Genetic alterations, mostly missense mutations, have been observed in the *ras* oncogene (25,26) and the *p53* tumor-suppressor gene (29,106). Not until the mid-1980s, with the discovery of the *ras* and *p53* genes together with the development of molecular genetic assays to detect altered versions of both genes, did extensive analysis of *ras* and *p53* mutations in experimentally induced tumors in animals get underway (60,107–114).

In an attempt to understand some of the events involved in carcinogen-induced tumors in humans, we have studied tumor induction in several inbred strains of mice using the DNA alkylating agent MNU, ionizing radiation, and neutron radiation (59–66,69,95,97,111,112,115,116). Several questions concerning tumor initiation and progression have been addressed using these animal model systems. First, use of different but genetically defined strains of mice treated with the same carcinogenic agent has allowed one to determine the contribution of the host genetic background on the spectrum of alterations induced in oncogenes and tumor-suppressor genes. Second, treatment of the same strain of mice with different carcinogenic agents has allowed one to determine whether alterations induced in oncogenes and tumor-suppressor genes are carcinogen specific. Third, since the process of tumor development occurs stepwise, this allows one to determine the timing of genetic alterations in each cancer susceptibility gene in relationship to each other.

The inbred strains C57BL/6J, RF/J, 129/J, and the F_1 hybrids between (AKR/J × RF/J) and (C57BL/6J × DBA/2J) were treated with ionizing radiation according to the Kaplan protocol (4 weekly doses at 150–175 rads/dose at 4 weeks of age), with neutron radiation (one dose of 100 rads of 0.4 MeV neutrons at 4 weeks of age), or MNU according to the Frei protocol (5 weekly doses at 30 mg/kg or one dose at 80 mg/kg i.p. at 4 weeks of age). The results are summarized in Tables 2 and 3 and indicate the tumor incidence for each strain and inducing agent and the frequency of alteration in oncogenes (*ras*, c-*myc*, *pim-1*, *MDM-2*) or tumor-suppressor genes (*p53*, *p16*) and any other consistent chromosomal changes.

Activation of ras *Genes in Hematopoietic Tumors*

Several conclusions can be drawn from the large survey of activated *ras* genes detected in thymic lymphomas induced in different strains of mice by different carcinogenic agents. Although there are three members of the ras family—K-, N-, and H-*ras*—only K-*ras* and N-*ras* have been detected in lymphoid tumors (56–61,63,66,69,97). Overall, the K-*ras* gene is preferentially mutated in most strains of mice irrespective of the inducing agent (gamma radiation, neutron radiation, or MNU). However, there were some exceptions that suggest that host genes influence the frequency of *ras* gene activation as well as which *ras* gene becomes activated. MNU treatment of RF/J or 129/J mice produced activation of K- and N-*ras* genes with similar frequencies. In contrast, C57BL/6J or (AKR × RF)F_1 mice treated with MNU showed almost exclusive activation of K-*ras* or N-*ras* genes, respectively, whereas radiation treatment of the same strains produced almost exclusive activation of the other *ras* gene. One strain, SJL/J, was not susceptible to radiation-induced thymic lymphomas. The frequency of tumor induction and activation of *ras* genes was directly proportional to the dose of the carcinogen administered (58,63,66). This result implies that fractionated doses or continued exposure to a carcinogen has a greater chance of producing a mutation in a critical cancer susceptibility gene.

Only limited studies have been performed with other carcinogens such as ENU, MMS, or MCA (Table 2). The K-*ras* gene was preferentially activated in thymic lymphomas with ENU and MCA treatments, including transgenic mice harboring the *pim-1* and N-*ras* oncogenes (67–69). The methylating agent MMS did not induce mutations in the *ras* gene in AKR/J mice given a single dose, whereas a single dose of MNU produced a low frequency (22%) of K-*ras* mutations in the same strain.

Activation of Cooperating Oncogenes

It is clear from a number of studies that *ras* activation by itself is not sufficient to cause malignant transformation (117). Rather, the c-*myc* oncogene has been shown to be one of several genes that can cooperate with *ras* in bringing about cellular transformation. Similarly, the *Pim-1* oncogene co-operates with c-*myc* and N-*myc* genes to predispose to lymphomas (118). The c-*myc* as well as the *pim-1* genes can be activated by insertion of a provirus leading to gene overexpression at the mRNA level (56,57,119). Alternatively, elevated levels of c-*myc* transcripts have been detected in MNU-induced tumors in the absence of proviral integrations (68). The mechanism for the upregulation of c-*myc* transcription is currently unknown. However, the results in both cases lead to the possible interaction between *ras* and c-*myc* or between c-*myc* and *pim-1* proteins and an increased potential for malignant transformation.

The c-*myc* locus is localized to chromosome 15 of the mouse (120). Genetic rearrangements involving trisomy of chromosome 15 is the most consistent chromosomal change documented in mouse lymphomas irrespective of whether the inducing agent is viral, radiation or chemical (121,122). Trisomy of chromosome 15 was the most consistent chromosomal abnormality observed in both MNU- and gamma radiation–induced tumors (65–70%) (64). Deregulation of the *myc* gene is a common occurrence in both human and animal lymphoid malignancies (123,124). Trisomy of chromosome 15 may be an alternative mechanism for deregulation of the *myc* gene in mouse tumors. The frequency of trisomy 15 was examined in thymic lymphomas induced by murine leukemia virus (125,126). The results showed that trisomy 15 and c-*myc* rearrangements were mutually exclusive events. One can postulate that cooperation between c-*myc* and mutant *ras* genes or other oncogenes can give rise to a malignant T cell as they have been shown to complement each other in transgenic animal models (127).

Radiation frequently induces double strand breaks in DNA, which can result in the formation of rearranged chromosomes (98,101). We identified a specific marker chromosome (Table 3) consisting of a translocation between mouse chromosomes 1 and 5 (64). Since radiation would produce random damage, the finding of a marker chromosome in a significant number of radiation-induced tumors argues for some selective advantage for this particular translocation. Specific chromosome translocations are associated with the development of many human lymphoid malignancies and activation of oncogenes including the c-*myc* gene in Burkitt's lymphomas (128), the c-*abl* gene in chronic myelocytic leukemia (129), and the *bcl-2* gene in follicular lymphomas (130). Clearly one of the recurring themes in animal models of leukemogenesis is cooperation of oncogenes whether activation occurs via point mutation, chromosome rearrangement, or upregulation of genes by some mechanism such as hypomethylation of gene promoters (131–133).

Inactivation of Tumor-Suppressor Genes

The *p53* tumor-suppressor gene is the most frequently mutated gene in human cancer, differing in frequency depending on the tissue type involved in the malignancy (29). We and others have analyzed the frequency of *p53* gene mutations in thymic lymphomas induced in different strains of mice by different carcinogenic agents (Tables 2 and 3). The findings from these surveys show a low frequency of *p53* gene mutations (10–13%) in T-cell tumors induced by MNU or gamma radiation in C57BL/6J mice (65) or MNU-treated (C57BL/6J × DBA/2)F_1 mice (69). No mutations of the *p53* gene were detected in tumors arising in RF/J mice induced by either gamma or neutron radiation (62). Thus, *p53* gene mutations were detected at very low frequencies. Therefore, *MDM*-2 (murine double-minute-2 gene), which is known to bind to and inactivate the p53 protein (134), was also examined in a large series of tumors (62). No alterations in the *MDM*-2 gene were found.

The *p16* tumor-suppressor gene was recently identified and shown to be the second most altered gene in a wide variety of human tumors next to the *p53* gene (135,136). In many human cancers it is inactivated through gene deletion (136). Radiation-induced thymic tumors in (C57BL/6J × RF/J)F_1 hybrids were screened for loss of heterozygosity (LOH) for chromosome 4, the location of the mouse homolog of the *p16* gene (95). LOH for *p16* was detected in 34% of the tumors (95).

The overall conclusions from these studies indicate a low frequency of *p53* mutations, implying that other transformation-associated genes are involved in carcinogen-induced leukemogenesis. Few studies to date have examined the frequency of *p16* gene deletions since they require F_1 hybrids to follow LOH at the *p16* gene locus (95). Future studies will be needed to look at cell cycle control regulatory genes other than *p53* and *p16*, such as the *Rb* tumor-suppressor gene, which undergoes alterations in many human cancers (30).

C. An Overview of Genetic Alterations in MNU-Induced Tumors: Relevance to Human Studies

Tumor initiation and subsequent outgrowth of malignant clones of cells is believed to be accompanied by a stepwise acquisition of further genetic changes. Comparison of the frequency of K-*ras* mutations, trisomy of chromosome 15, and *p53* mutations in thymocytes from MNU-treated animals at different stages of disease can determine the timing of the genetic events with tumor development. The results of our studies have shown that *ras* mutations occur early in tumor development and can be detected in tissues from carcinogen-treated animals well in advance of the development of frank tumors using sensitive PCR-based methods (115). Although *ras* mutations are frequent and early initiating events in MNU-induced thymic lymphomas in C57BL/6J mice, clearly not every tumor has a mutation of the *ras* gene and not all the tumors with *ras* mutations have trisomy of chromosome 15. Therefore, there are several different pathways involved in the development of MNU-induced T-cell tumors. The preferred pathway in 90% of the MNU-induced thymic lymphomas in C57BL/6J mice involves mutation at the hotspot G^{35} of codon 12 of the K-*ras* gene together with trisomy of chromosome 15. Approximately 10% of the MNU-induced tumors presumably have other non-*ras* genes, as yet unidentified, as targets for MNU-induced mutations. Further studies will be needed to characterize them. Since some of the tumors lack *ras* mutations but have trisomy of chromosome 15, this would suggest that other oncogenes can substitute for mutations in K-*ras* and cooperate with the same chromosomal changes to produce thymic lymphomas.

For human cancers, the spectrum of mutations in the *p53* gene in different tissues has begun to yield clues about environmental factors that may influence characteristic or so-called "hot spots" for mutation (28). The frequency of *p53* gene mutations in MNU-induced thymic lymphomas of mice is low and a late event in tumor progression. Similarly, lymphoid tumors in humans have a low incidence of *p53* mutations (137). The *p53* mutations detected in MNU-induced tumors, unlike *ras* mutations, are unlikely to be due to the carcinogen treatment as they occurred at CpG dinucleotides. CpG dinucleotides are known hot spots for mutations in the *p53* gene in human cancers. These mutations result from spontaneous deamination of 5-methylcytosine (29). Taken together these findings may indicate that other genes within the *p53* regulatory pathway are targets for mutation. The *MDM-2* oncogene is one gene in this pathway (138). It has been shown to complex with p53 protein and inactivate its ability to suppress cell growth (139). The *MDM-2* gene has been shown to be amplified in some human tumors which lack *p53* gene mutations (134). A survey of 94 thymic lymphomas, 48 of which were induced by MNU, found no amplification of the *MDM-2* gene (62). These findings suggest that other oncogenes and tumor suppressor genes are likely to play a significant role in MNU-induced T-cell tumors.

Ras mutations in human T-cell malignancies have not been reported to date (26). The high frequency of *ras* mutations observed in MNU-induced murine thymic lymphomas *versus* no *ras* mutations observed in human T-cell tumors may reflect a basic difference in the biology of these tumors. Spontaneous T-cell tumor formation in C57BL/6J mice is nil.

These results suggest that *ras* oncogene activation plays a critical role in the development of murine T-cell malignancies and that other non-*ras* genes may be important in the formation of human T-cell tumors.

IV. CANCERS OF THE HEMATOPOIETIC ORGANS IN HUMANS

A. Classification of Hematopoietic Tumors

Tumors of the hematopoietic system are classified by hematological criteria using morphology of cells in blood and bone marrow smears. In addition to cell morphology, immunophenotyping for cell surface markers is used to characterize cells for cell lineage (B, T, myeloid) and stage of maturation (immature *versus* mature). Another valuable criteria for classification is the analysis of T-cell receptor (T cell) or immunoglobulin gene (B cell) gene rearrangements. In addition, *myc* and *bcl*-2 oncogene rearrangements are notable in some subtypes of B-cell tumors. Assessment of specific chromosome rearrangements such as the Philadelphia chromosome is used in classifying chronic myelogenous leukemia (CML).

Leukemias comprise 3% of the total world cancer incidence (1) and consist of acute lymphocytic (ALL), chronic lymphocytic (CLL), acute myeloid (AML), and chronic myeloid (CML) diseases. ALL of the B-cell type is more common and occurs most often in children. CLL of the B-cell type is the most common disease in the western world and is a disease of the elderly.

Lymphomas are generally grouped as Hodgkin's disease (HD) or non-Hodgkin's lymphoma (NHL). HD is relatively rare, occurs in young and old, and can be distinguished from NHL tumors by the presence of characteristic Reed-Sternberg cells. NHL is a group of several different tumor types generally classified according to characteristic morphology, such as diffuse large cell (DLC), follicular mixed (FM), follicular small cell (FSC), mantle cell (MCL), and anaplastic large cell (ALCL). However, the classification is not uniform, and there is still confusion in classifying some subtypes of NHL tumors.

There are three types of plasma cell tumors: multiple myeloma (MM) is associated with older age, plasma cell leukemia (PCL), and monoclonal gammopathies of undetermined significance (MGUS). The incidence of MM is increasing worldwide.

Three diseases are associated with viral infections: Burkitt's lymphoma (BL), adult T-cell leukemia (ATL), and human immunodeficiency virus (HIV). BL has two forms: endemic to Africa and nonendemic elsewhere in the world. Endemic BL is associated with infection with the Epstein-Barr virus (EBV). Sporadic BL is associated with activation of c-*myc* via translocations of chromosome 8 with other partner chromosomes. Other EBV lymphomas arise frequently in immunocompromised organ transplant patients or those with acquired immunodeficiency syndrome (AIDS). AIDS NHL are typically classified as B-immunoblastic (BIBL), large noncleaved cell (LNCLL), and small noncleaved (SNCLL). Human T-lymphotropic virus type 1 (HTLV-1) is associated with tumors of mature T cells. It occurs around middle age with a long latency period between exposure to the virus and development of the disease.

Thymomas in humans, unlike rodents, are rare, with the majority being benign tumors often associated with myasthenia gravis and cytopenias.

B. Alterations in Cancer-Susceptibility Genes in Hematopoietic Tumors

Genetic alterations in two critical classes of cancer-susceptibility genes, tumor suppressor genes and oncogenes, are frequent in human cancers and are strongly implicated in their development. Table 4 summarizes the frequency of alterations detected in three tumor suppressor genes—*p53*, *p16*, and *Rb*—and the oncogene *ras*.

Inactivation of Tumor Suppressor Genes

The *p53* Gene. The *p53* tumor-suppressor gene is the most studied tumor-suppressor gene in human cancer. More than 3800 primary hematopoietic tumors have now been screened for mutations in the *p53* tumor-suppressor genes (137,140–183). As shown in Table 4, low frequencies of 0–4% were observed in B-cell tumors PLL (151), HD (153–155), FSC (137,158,159,162,163), MGUS plasma cell tumors (168), and primary T-ALL T-cell tumors (137,140–142,145,170,171). Intermediate frequencies of 7–22% occurred in B-cell tumors ALL (137,140–147), CLL (137,143,148–152), DLC (137,141,155,157–162), FM (137,155,158,159,162,163), MCL (156,161,164), the plasma cell tumor MM (137,166–168), and all of the myeloid tumors AML (137,141–143,150,172–176), CML (137,141,142,177–180), and MDS (137,141,143,150,173,174,176,181–183). Higher frequencies of p53 gene mutations were observed in three diseases associated with viral infection. Thirty-six percent of BL (137,142,146,156,157), which is associated with EBV infection, had *p53* gene mutations. The AIDS-NHL B-cell tumor SNCLL had the highest frequency of 53% gene mutation (161,165). Likewise, ATL T-cell tumors associated with infection by HTLV-1 had 25% *p53* gene mutations (137,141,169). The higher frequencies in these diseases may be associated with increased lymphoproliferation due to viral infection where *p53* may play a role in tumor progression. Similar high frequencies of 25% were detected in the plasma cell tumor PCL (168) and relapsed cases of T-ALL (141,170,171).

The *p16* Gene. The *p16* tumor-suppressor gene was first identified in 1994 (135,136) and is considered to be the second most altered genetic lesion in human cancers. Over 2000 primary hematopoietic tumors have been screened for inactivating homozygous deletions of *p16* (152,184–195). In some cases the same tumors were simultaneously analyzed for deletion of the companion *p15* tumor-suppressor gene on chromosome 9 (188,189,191–193). ALL as a disease is most frequently targeted for *p16* gene deletions occurring in 21% of B-ALL (184–190) and 55% of T-ALL cases (184,186, 188,190,193–195) (Table 4). In general, tumors showing loss for the *p16* gene also showed loss for the *p15* gene (188,189,191–193). The only other disease with a significant frequency of *p16* deletions (21%) was the T-cell tumor ATL (184,190,191). Low frequencies of 5–10% were detected in CML (184,186,187,189,190) and DLC (185,186,190–182), respectively. The majority of B-cell, plasma cell, and myeloid tumors did not show deletion for the *p16* gene. However, it is now known that the *p16* gene can become inactivated by hypermethylation of the promoter region (184). Hematopoietic tumors without homozygous deletions of *p16* have not been systematically analyzed to date to see if the *p16* gene may be silenced *via* changes in methylation status (184).

The *Rb* Gene. Inactivation of the *Rb* tumor-suppressor gene has been implicated in many human cancers. It was first cloned and characterized in 1986 (27). Table 4 summarizes the data from approximately 600 hematopoietic tumors analyzed for alterations in the *Rb* locus, including gene deletions, point mutations, and loss of *Rb* expression by Northern and Western blot analysis (137,165,168,179,196–201). The B-cell tumor B-ALL (196) and plasma cell tumors MM (168) and PCL (168) showed the highest frequency for *Rb* gene inactivation, ranging from 30 to 36%. Both AIDS-associated B-cell tumors BIBL (165) and SNCLL (165) and the T-cell tumor ATL (137) lacked *Rb* alterations. This result may by due to the small sample size analyzed for these three diseases. Of the remaining six types of hematopoietic tumors analyzed (B-CLL, FM, T-ALL, AML, CML, MDS), on average, 15% of the cases showed inactivation of the *Rb* gene (179,196–201). It is known that the *Rb* gene can also be inactivated by hypermethylation (202) in its promoter sequence similar to that recently described for the *p16* gene (203,204). To date, no hematopoietic tumors have been examined with respect to the methylation status of the *Rb* gene and its promoter.

Loss of *Rb* and *p16* have been shown to be mutually exclusive events

Table 4 Summary of Genetic Alterations in Human Hematopoietic Tumors

Diagnosis	*p53* gene		*p16* gene		*Rb* gene		*ras* gene	
	Positive/ Tested (%)	Ref.	Positive/ Tested (%)	Ref.	Positive/ Tested (%)	Ref.	Positive/ Tested (%)	Ref.
B-cell tumors								
B-ALL	39/563 (7%)	137,140–147	104/494 (21%)	184–190	10/33 (30%)	196	11/157 (7%)	26,144,209,210, 215
B-CLL	59/521 (11%)	137,143,148–152	3/244 (1%)	152,184–186, 189,190	56/289 (19%)	196–201	0/89 (0%)	208,215
B-PLL	3/90 (3%)	151	0/7 (0%)	185				
HD	1/38 (3%)	153–155	0/8 (0%)	186			2/28 (7%)	26
NHL								
BL	33/91 (36%)	137,142,146,156, 157	1/21 (5%)	184,185,191			0/38 (0%)	208
DLC	38/197 (19%)	137,141,155, 157–162	17/164 (10%)	185,186,190–192			0/15 (0%)	208
FM	19/132 (14%)	137,155,158,159, 162,163	0/85 (0%)	185–187,191	2/15 (13%)	197	0/79 (0%)	26,208
FSC	7/158 (4%)	137,158,159,162, 163	0/85 (0%)	192				
MCL	10/46 (22%)	156,161,164	0/3 (0%)	191			0/15 (0%)	164
AIDS-NHL								
BIBL	9/72 (13%)	153,156–158,165	0/7 (0%)	191	0/5 (0%)	165	1/5 (29%)	165
SNCLL	10/19 (53%)	161,165			0/16 (0%)	165	3/16 (19%)	165

Plasma cell tumors								
MM	23/212 (11%)	137,166–168	0/22 (0%)	184,190	5/14 (36%)	168	46/190 (24%)	137,166,217, 222–224
MGUS	0/15 (0%)	168					0/30 (0%)	217
PCL	4/18 (25%)	168			3/9 (33%)	168	4/13 (31%)	217
T-Cell tumors								
ATL	22/87 (25%)	137,141,169	9/44 (20%)	184,190,191	0/10 (0%)	137	0/10 (0%)	137
T-ALL								
1^0	2/159 (1%)	137,140,141,142, 145,170,171	141/258 (55%)	184,186,188,190, 193–195	6/43 (14%)	196,197	1/20 (5%)	209
R	13/52 (25%)	141,170,171						
Myeloid tumors								
AML	51/419 (12%)	137,141–143,150,172–176	8/331 (2%)	184,186,187,189, 190	9/54 (17%)	196	154/618 (25%)	26,173,215,216, 219,220
CML	26/269 (10%)	137,141,142, 177–180	7/148 (5%)	184,186,187,189, 190	9/84 (11%)	179,196	11/299 (4%)	26,177,178,213, 214,221
MDS	56/653 (9%)	137,141,143,150, 173,174,176, 181–183	0/100 (0%)	184,186,189,190	3/18 (17%)	196	106/474 (22%)	26,173,211,212, 218

ALL = Acute lymphoblastic leukemia; AML = acute myelogenous leukemia; ATL = adult T-cell leukemia; B-ALL = B-cell acute lymphoblastic leukemia; B-CLL = B-cell chronic lymphocytic leukemia; BIBL = B-immunoblastic lymphoma; B-PLL = B-prolymphocytic leukemia; CML = chronic myelogenous leukemia; DLC = diffuse large-cell lymphoma; FM = follicular mixed lymphoma; FSC = follicular small cleaved lymphoma; HD = Hodgkin's disease; MCL = mantle cell lymphoma; MDS = myelodysplastic syndrome; MGUS = monoclonal gammopathy of undetermined significance; MM = multiple myeloma; NHL = non-Hodgkin's B-cell lymphoma; PCL = plasma cell leukemia; SNCLL = small noncleaved lymphoma; T-AA = T-cell acute leukemia (1^0 = primary, R = relapse).

in other cancers (205–207). As seen in Table 4, several different tumors that show no to low frequencies (1–5%) of *p16* gene deletions such as B-CLL, FM, MM, AML, CML, and MDS have 11–19% of cases with *Rb* alterations. Similarly, the T-cell tumor ATL shows 20% of tumors with *p16* gene deletions and no cases with alterations in *Rb*. However, unlike brain and lung cancers (205–207), where both *p16* and *Rb* genes have been assessed for alterations in the same tumors, few hematopoietic tumors have been studied for both *p16* and *Rb* genes. We have looked at alterations in multiple tumor suppressor genes in B-CLL including the simultaneous analysis of *p16* and Rb genes (152). We found that in the five cases of B-CLL with deletion of the *p16* gene, the *Rb* gene and, where studied, the Rb protein were retained. To date, loss of *Rb* and *p16* appear to be mutually exclusive events in hematopoietic cancers. Further studies of other hematopoietic tumors are required to extend the analysis of inactivation of these tumor-suppressor genes.

Activation of Cooperating Oncogenes

***ras*.** Members of the *ras* oncogene family (H-, K-, and N-*ras*) are found to be altered by point mutation in a wide variety of human cancers (25,26,208–225). For hematological malignancies, activation is limited to N-*ras* and K-*ras* genes (25,26,225). Table 4 summarizes the recent literature for the frequency of *ras* mutations analyzed in more than 2000 different tumors (137,144,164–166,173,177,178,208–225). *Ras* mutations are found most frequently associated with the AIDS-associated B-cell tumors BIBL (29%) and SNCLL (19%), the plasma cell tumors MM (24%) and PCL (31%), and the myeloid tumors AML (25%) and MDS (22%). The majority of all other B-cell tumors consisting of HD, NHL, T-cell tumors ATL and T-ALL, and the myeloid tumor CML show no to low frequencies (4–7%) of *ras* mutations.

The overwhelming majority of mutations occur in the N-*ras* gene. Of 339 reported *ras* mutations, less than 5% have been detected in the K-*ras* gene (26,216,219,224). Often MM patients, in which K-*ras* mutations have been detected, also show concomitant mutations in the N-*ras* gene (224). Activation of the *ras* gene appears to be a late event associated with the acquisition of a more transformed phenotype and poorer patient prognosis (209,212,214).

***MDM*-2.** The *MDM*-2 oncogene was discovered in 1991 (138) and is now known to play an important role in the p53 regulatory pathway. The p53 protein is a transcription factor that can bind to specific DNA sequences and control the expression of any gene which contains p53 DNA-binding elements. The *MDM*-2 gene contains a p53 DNA binding site and therefore its expression can be regulated by the level of expression of wild-type p53 protein (226). In addition, the MDM-2 protein can complex with the p53 protein and block its ability to act as a positive regulator of transcription (139). This forms an autoregulatory loop, which can regulate both *p53* and *MDM*-2 gene activity within the cell. Tumors that lack *p53* gene mutations are suspect for amplification of the *MDM*-2 gene or protein overexpression due to other translational or transcriptional regulatory mechanisms (134).

Several groups have assessed hematopoietic tumors for the expression of the *MDM*-2 gene at the mRNA and protein levels (147,227–230). In addition, some have studied the relationship between the presence or absence of *p53* gene mutations and expression levels of the *MDM*-2 gene. To date, fewer than 200 tumors have been assessed for levels of *MDM*-2 gene overexpression. A very high proportion of HD tumors (66/77, 86%) express the MDM-2 protein. Using immunohistochemistry it was shown that 58 of the tumors coexpressed both the MDM-2 and p53 proteins within the nuclei of Reed-Sternberg cells (228). As would be predicted, the rate of *p53* gene mutations is extremely low (3%, Table 4) in HD tumors. Among ALL, 29% (5/17) of the cases overexpress MDM-2 protein, and in all cases it was associated with the expression of the wild-type p53 protein (147,230). Approximately 27% (8/30) of NHL tumors (DLC, BIBL, SNCLL) express the MDM-2 protein (229). Immunohistochemical analysis for p53 in the same tumors showed that all MDM-2–positive tumors were also p53 positive (229). Since MDM-2 protein overexpression has been detected in the presence of low to normal mRNA levels, measurement of mRNA transcripts can only lead to an inaccurate picture of the actual levels of MDM-2 protein within the cell (147). An additional 63 tumors consisting of CLL, ALL, AML, CML, and MDS were screened using Northern blot analysis only for levels of *MDM*-2 gene expression (227). If one scores only those tumors showing high amounts of MDM-2 mRNA transcripts, then 33% of B-CLL (4/12), 25% of AML (8/32), and 43% of MDS (3/7) overexpressed MDM-2. The status of the *p53* gene, wild-type *versus* mutant, was not assessed in these tumors, however, all three diseases show a low frequency (9–11%) of *p53* gene mutations. This data suggests that MDM-2 overexpression disrupts the p53 regulatory pathway. In the future, many more studies will be needed to establish whether deregulated expression of MDM-2 contributes significantly to the pathogenesis of the majority of hematopoietic tumors showing low frequency for *p53* gene mutations.

Chromosomal Translocations and Activation of Oncogenes

A high frequency of specific chromosome translocations occur in both T- and B-cell malignancies, presumably due to mistakes during the normal process of T-cell receptor and immunoglobulin gene rearrangement. The major classes of genetic abnormalities detected in hematopoietic tumors have been extensively reviewed (231,232). The c-*myc* oncogene is rearranged to immunoglobulin gene loci in 100% of BL and approximately 10% of ALL (T and B) (231,232). The c-*abl* oncogene undergoes fusion to the *bcr* gene locus in greater than 95% of CML cases with the Philadelphia chromosome (232). The *bcl*-2 oncogene is found rearranged in greater than 20% of DLC and 75% of FM. Among ALL, a low frequency of several different translocations, ranging from 5 to 15%, has been detected. All of the translocations in T-ALL result in fusion of different genes to a T-cell-receptor gene (232). In B-ALL, the transcription factor E2A becomes fused with a homeobox gene, *PBX1*, or a gene *HLF*. Just as the c-*myc*, c-*abl*, and *bcl*-2 oncogenes were identified, many other putative oncogenes are being cloned from the nonrandom chromosome translocations identified in hematological malignancies.

The Spectrum of Alterations in Cancer Susceptibility Genes: A Comparison Between Human and Animal Models of Lymphoid Disease

Tables 2–4 present a comprehensive overview of some of the alterations detected in several of the known cancer susceptibility genes *p53*, *p16*, *Rb*, *ras*, *MDM*-2, and *myc* in hematopoietic tumors induced in animals *versus* those occurring in sporadic human tumors. In murine animal models, the major target organ for leukemia/lymphoma development is the thymus, regardless of the inducing agent (viral, chemical, radiation) (233). In contrast, hematopoietic tumors in humans arise mainly in the B-cell or myeloid compartments. Despite these differences in hematopoietic stem cell lineages between human and mouse, the frequency and spectrum of alterations in critical target cancer genes remain remarkably similar in tumors of both lymphoid systems.

Alterations in the tumor-suppressor gene *p53* and oncogenes of the *ras* family have been studied extensively in hematopoietic tumors arising in both species. The overall frequency of *p53* mutations is generally low in lymphoid malignancies occurring in 11% of human and 6% of murine tumors (Tables 2–4). The mechanism of inactivation of the *p53* gene is the same in both species and consists of inactivating point mutations in one allele with frequent loss of the remaining wild-type allele. In both species *p53* alterations occur late in disease progression in hematopoietic tumors. In humans, the frequency of *p53* gene mutations is similar in most hematopoietic tumors, with the exception of B-PLL, HD, NHL-FSC, MGUS, and primary T-ALL cases, where they are extremely low. In the case of HD, the MDM-2 regulatory protein is overexpressed in the majority of tumors. Since MDM-2 can block the function of p53, it provides an alternative pathway for malignant transformation in this disease. Similarly, in the mouse, the *MDM*-2 gene was examined only with respect to gene amplification and none was found (62). Studies are needed in the future to analyze the levels of *MDM*-2 gene expression in hematopoietic tumors of both species in order to determine the importance of the p53 regulatory pathway in the pathogenesis of lymphoid tumors.

For *p53* mutations, the spectrum of mutations detected in a given malignancy can yield clues as to possible etiological agents (28,29). We have analyzed the spectrum and patterns of *p53* mutations in 42 cases of B-CLL (234). Transversion mutations were noted at a significantly higher frequency in B-CLL compared with other lymphoid tumors. In addition, over 65% of the *p53* mutations detected in B-CLL showed a strand bias for *p53* mutations on the untranscribed DNA strand. This feature of strand bias is common in cancers of the lung, esophagus, and head and neck that result from high exposure to chemical carcinogens. Thus, the spectrum of *p53* mutations observed in B-CLL may implicate environmental carcinogen exposure associated with *p53* gene damage.

Ras genes play an important function in the control of cell growth and differentiation. Unlike the pattern of *p53* alterations, which show a broad distribution and similar frequency among tumors from all different cell lineages, *ras* alterations are closely associated with distinct tumor subtypes (Table 4). Although the overall frequency of *ras* mutations, specifically the N-*ras* gene, in human tumors is 16%, the majority occur in aggressive malignancies of B-cell and myeloid cell types. In certain situations, *ras* mutations may be directly related to chemotherapeutic treatments (discussed below). In mouse strains, the frequency of *ras* mutations is low to nil in the absence of some inducing agent (Table 2). Genetic background of the host inbred strain influences the incidence of *ras* mutations as well as the target gene, N-*ras* versus K-*ras*, in response to chemical or radiation treatments (Tables 2, 3). However, in the mouse, K-*ras* is the most frequently mutated in contrast to N-*ras* observed in human tumors. There may be a difference between the stem cell origin of the T- and B-cell tumors, respectively, in the two species.

Malignant transformation requires the cooperation of different genes: some contribute to immortilization like c-*myc*, while others contribute to tumorigenic potential like *ras* (117). Hematopoietic tumors of humans and mice often show activation of more than one oncogene. Activation occurs through point mutations, such as the *ras* gene, or by gene rearrangements induced in humans by chromosome translocations of c-*myc*, *bcl*-2, or c-*abl* and in mice by viral insertion into the c-*myc* or *pim*-1 loci. An alternative pathway in murine T-cell tumors for deregulating c-*myc* expression involves trisomy of chromosome 15.

In summary, human and mouse hematopoietic tumors undergo similar alterations in the same classes of cancer-susceptibility genes. Inactivation of the *p53* tumor-suppressor gene and activation of oncogenes occur by similar mechanisms (point mutation, gene amplification, chromosome translocation), although with somewhat different frequencies. Other key regulators of the cell cycle include *p16* and *Rb* and many cyclins and their inhibitors (30,235,236). Few alterations in cell cycle regulatory genes have been examined in hematopoietic tumors in animal models. However, the rapid development of transgenic models for studying cancer predisposition of known cancer-susceptibility genes should provide considerable information in the future. The fact that hematopoietic tumors of human and mouse share many of the same genetic alterations indicates that animal models, particularly some of the new transgenic animal models, may become useful as carcinogen screening systems (237).

C. Risk Factors Associated with Hematopoietic Tumors

Risks Linked with Radiation Exposure

Radiation can be emitted from natural sources, such as radon gas, or man-made sources, such as x-rays or atomic energy sources. Beginning 7 years after Roentgen discovered x-rays in 1895, higher incidences of leukemia were recorded in radiologists (1,238). The systematic epidemiological study of risks associated with acute exposure to ionizing radiation was initiated in 1948 with the creation of the Atomic Bomb Casualty Commission (238) and the United Nations Scientific Committee on the Effects of Atomic Radiation (90–93). Results of these studies show an increased risk for the development of AML and CML. However, the risks associated with chronic, low-level exposure to radon gas remains controversial (139,239). Radon exposure among mine workers predisposes for lung cancer (30,238). A study conducted in England showed no evidence for increased incidence of childhood cancers and radon exposure (239). Indeed, the entire question of cancer risk assessment to low levels of chronic radiation exposure *versus* acute levels is an important issue (240). Recently, a study conducted on the incidence of childhood leukemia in Greece following the Chernobyl accident showed that low levels of *in utero* exposure to radiation increased the risk for ALL approximately threefold (241). Use of molecular genetics may help define alterations associated with the different risks associated with different exposure conditions. Genetic alterations occurring in *p53*, *Rb*, and H-, K-, and N-*ras* genes will be systematically analyzed in archival tumors occurring in atomic bomb survivors and more recently in persons living in the vicinity of the Chernobyl nuclear reactor accident (242). The results of these studies, which are part of an investigation to characterize the molecular events important in human radiation carcinogenesis, in contrast to animal radiation studies, should lead to much valuable information.

Patients with nonmalignant conditions, such as ankylosing spondylitis, as well as patients with cancers are frequently given therapeutic x-ray treatment and are at increased risk for developing leukemia/lymphomas (1,243,244). Genetic alterations of the key cancer-susceptibility genes arising in radiation-induced second malignancies has not been specifically or systematically addressed to date. The development of sensitive detection methods for use on archival samples will allow molecular genetic studies of these tumors in the near future (242).

Risks Linked with Chemical Exposure

Development of second malignancies following treatment with a variety of chemotherapeutic agents such as the alkylating agents chlorambucil, cyclophosphamide, melphalan, thiotepa, dacarbazine, and procarbazine of the nitrosourea class are now well-documented (245–249). All of these chemical agents are recognized by IARC to be carcinogenic to humans (Table 1) (1). The most frequent therapy-related hematopoietic tumors include acute nonlymphocytic leukemia (ANLL), ALL, AML, CML, and MDS (247–249). Second malignancies often occur 3–9 years following initial therapy (245). Therapy-induced leukemias serve as an important source of information for identifying the genetic basis for these diseases. However, only a few studies are known that have assessed mutations in *p53* or *ras* genes in therapy-related tumors (250,251). Patients with HD or NHL were treated with chemotherapy, radiation therapy, or a combination of both (250). No patient had overt disease, but *ras* mutations were detected in the DNA of normal peripheral blood from 9 of 70 patients. In a second study, 50 pediatric patients in complete remission from ALL following cytotoxic therapy were assessed for mutations in the *ras* and *fms* genes (251). *Ras* mutations were detected in 6% of the patients and *fms* in one patient, who also had a *ras* mutation. These patients are under follow-up to determine whether presence of activating mutations in these genes predispose to the development of secondary leukemias (251).

Analysis of chromosomal translocations has shown distinct differences between de novo acute leukemias and therapy-related leukemias (252). Many leukemias show rearrangements at 11q23, the site of the *MLL* gene. Approximately 1–15% of cancer patients treated with topoisomerase II inhibitors develop therapy-related AML (252). Study of the translocations in de novo AML versus therapy-related AML showed that >75% of the drug-induced AML cases had breakpoints within the *MLL* gene that contain the topoisomerase II DNA-binding sites (252). Results of studies such as these give clues as to the mechanisms underlying chromosome translocations and gene activation, which play critical roles in specific hematopoietic tumors.

Risks Linked with DNA Repair Systems

DNA damage and repair processes are tightly linked with cancer development (253,254). Heritable genes can affect repair of DNA damage. As noted above, induction of secondary leukemias in patients is most likely due to treatment with one of several chemotherapeutic agents, which are methylating or chloroethylating compounds that induce DNA damage. Overexpression of the human O^6-alkylguanine-DNA alkyltransferase gene protected transgenic mice from MNU-induced disease due to rapid repair of the DNA adducts (80). Hematopoietic cells normally show low levels of DNA alkyltransferase and therefore are potentially at higher risk for DNA damage from alkylating agents. Recently, retroviral gene therapy was used to transfer the human O^6-alkylguanine-DNA alkyltransferase gene to long-term cultures of mouse bone marrow and protected them against repeated exposures to MNU (255). Similar gene therapy strategies in individuals who must undergo repeated exposures to cytotoxic DNA-damaging agents could afford protection to the bone marrow from therapy-related cancers.

Cancer susceptibility can also be related to the inheritance of enzymes that affect cancer risk. The glutathione S-transferase (*GST*) genes are one example of an enzyme system whose activity to metabolize genotoxic agents is dependent on inheritance of specific alleles (22). Two *GST* genes are known, *GST M1* and *GST theta 1* (*GSTT1*), and each has a null variant allele. Persons inheriting the null genes show absence of enzyme activity and therefore might be at greater risk for carcinogen-induced cancers due to a decreased ability to metabolize genotoxic agents. Using polymerase chain reaction (PCR)–based assays, the DNA from lymphocytes or bone marrow samples from patients can be easily assessed for the presence of absence of *GST* alleles. A recent study of 96 MDS patients and 200 case controls found an increased risk for MDS associated with inheritance of the defective *GSTT1* gene (256). The *GSTT1* null allele was associated with increased risk of developing carcinogen-induced cancers of the lung and bladder (256). An important area of future research will be genotyping individuals for inherited genetic polymorphisms in *GST* alleles as well as other genes such as the human cytochrome P-450 genes involved in activation of a number of procarcinogenic chemicals.

V. CONCLUSIONS AND FUTURE DIRECTIONS

This chapter has attempted to summarize the current findings in animal models of carcinogen- and radiation-induced hematopoietic disease. Advances in molecular genetics have allowed the detection of alterations in several key regulators of cell growth and differentiation such as the tumor-suppressor genes *p53*, *p16*, and *Rb* and the oncogenes *ras* and c-*myc*. The frequency of alterations in these genes is summarized from our work and the literature and compared with a tabulation of genetic alterations occurring in the same genes in human hematopoietic neoplasms. The pattern and spectrum of mutations in tumor-suppressor genes and oncogenes found in humans and mice are remarkably similar. The findings presented here support the use of animal models, including some of the new transgenic models predisposed to develop lymphoid cancers, as suitable test systems for evaluating the risk to humans of potential carcinogenic compounds (81,83,89, 237).

Cancer is a complex disease, and an important part of the equation is the inherent genetic variation in disease susceptibility in humans. The association between cancer and exposure to environmental carcinogens is related to many variables such as age, sex, length of exposure, dose (chronic *versus* acute), and inheritance of genes for proteins (GST, cytochrome P-450s) which metabolize genotoxic substrates. Molecular genetic studies are yielding clues on the DNA damage to some of the critical cancer-susceptibility genes. Future studies are needed on human tumors known to be induced by chemotherapy (248–251) or by radiation (241,242). The study of hematopoietic tumors arising in these patients should afford the most direct analysis of the genetic basis for cancers of the hematopoietic system in humans.

REFERENCES

1. Tomatis L. Cancer: Causes, Occurrence and Control. Vol. 100. Lyon: IARC, 1990.
2. Tennant RW, Margolin BH, Shelby MD, Zeiger E, Haseman JK, Spalding J, Caspary W, Resnick M, Stasiewicz S, Anderson B, Minor R. Prediction of chemical carcinogenicity in rodents from in vitro genetic toxicity assays. Science 1987; 236:933–941.
3. Gold LS. Combined Plot of the Carcinogenic Potency Database. Vol. 1–3. Berkeley: Carcinogenic Potency Project, 1996.
4. Fleischman RW, Baker JR, Hagopian M, Wade GG, Hayden DW, Smith ER, Weisburger JH, Weisburger EK. Carcinogenesis bioassay of acetamide, hexanamide, adipamide, urea and p-tolyurea in mice and rats. J Environ Pathol Toxicol 1980; 3:149–170.
5. Dunsford HA, Keysser CH, Dolan PM, Seed JL, Bueding E. Carcinogenicity of the antischistosomal nitrofuran trans-5 amino-3-[2-(5-nitro-2-furyl)vinyl]-1,2,4-oxadiazole. J Natl Cancer Inst 1984; 73:151–160.
6. Maltoni C, Conti B, Cotti G, Belpoggi F. Experimental studies on benzene

carcinogenicity at the Bologna Institute of Oncology: current results and ongoing research. Am J Ind Med 1985; 7:415–446.

7. Snyder CA, Goldstein BD, Sellakumar AR, Bromberg I, Laskin S, Albert RE. The inhalation toxicology of benzene: incidence of hematopoietic neoplasms and hematotoxicity in AKR/J and C57BL/6J mice. Toxicol Appl Pharmacol 1980; 54:323–331.
8. Gold LS, Slone TH, Backman GM, Magaw R, Da Costa M, Lopipero P, Blumenthal M, Ames BN. Second chronological supplement to the Carcinogenic Potency Database: standardized results of animal bioassays published through December 1984 and by the National Toxicology Program through May 1986. Environ Health Perspect 1987; 74:237–329.
9. Gold LS, Maley NB, Slone TH, Garfinkel GB, Rohrbach L, Ames BN. The fifth plot of the Carcinogenic Potency Database: results of animal bioassays published in the general literature through 1988 and by the National Toxicology Program through 1989. Environ Health Perspect 1993; 100:65–135.
10. Gold LS, Manley NB, Slone TH, Garfinkel GB, Ames BN, Rohrbach L, Stern BR, Chow K. Sixth plot of the Carcinogenic Potency Database: results of animal bioassays published in the general literature 1989–1990 and by the National Toxicology Program 1990–1993. Environ Health Perspect 1995; 103(suppl 8):3–126.
11. Skipper HE. Phase I studies on the carcinogenic activity of anticancer drugs in mice and rats: final report. Birmingham, AL: Southern Research Institute, 1976:Booklet 1.
12. Gold LS, Sawyer CB, Magaw R, Backman GM, de Veciana M, Levinson R, Hooper NK, Havender WR, Bernstein L, Peto R, Pike MC, Ames BN. A carcinogenic potency database of the standardized results of animal bioassays. Environ Health Perspect 1984; 58:9–319.
13. Fujii K, Nakadate M, Ogiu T, Odashima S. Induction of digestive tract tumors and leukemias in Donryu rats by administration of 1-amyl-1-nitrosourea in drinking water. Gann 1980; 71:464–470.
14. Ogiu T, Kajiwara T, Furuta K, Takeuchi M, Odashima S, Tada K. Mammary tumorigenic effect of a new nitrosourea, 1,3-dibutyl-1-nitrosourea (B-BNU), in female Donryu rats. Cancer Res Clin Oncol 1980; 96:35–41.
15. Maekawa A, Ogiu T, Matsuoka C, Onodera H, Furuta K, Kurokawa Y, Takahashi M, Kokubo T, Tanigawa H, Hayashi Y, Nakadate M, Tanimura A. Carcinogenicity of low dose of N-ethyl-N-nitrosourea in F344 rats; a dose-response study. Gann 1984; 75:117–125.
16. Mirvish SS, Weisenburger DD, Joshi SS, Nickols J. 2-Hydroxyethylnitrosourea induction of B cell lymphoma in female Swiss mice. Cancer Lett 1990; 54:101–106.
17. Bulay O, Mirvish SS, Garcia H, Pelfrene AF, Gold B, Eagen M. Carcinogenicity test of six nitrosamides and a nitrosocyanamide administered orally to rats. J Natl Cancer Inst 1979; 62:1523–1528.
18. Adamson RH, Sieber SM. Chemical carcinogenesis studies in nonhuman primates. In: Langenback R, Nesnow S, Rice JM, eds. Organ and Species

Specificity in Chemical Carcinogenesis. New York: Plenum Press, 1982:129-156.
19. Thorgeirsson UP, Dalgard DW, Reeves J, Adamson RH. Tumor incidence in a chemical carcinogenesis study in nonhuman primates. Reg Toxicol Pharmacol 1994; 19:130–151.
20. Gold LS, Slone TH, Stern BR, Bernstein L. Comparison of target organs of carcinogenicity for mutagenic and non-mutagenic chemicals. Mutat Res 1993; 286:75–100.
21. Weisburger EK. General principles of chemical carcinogenesis. In: Waalker MP, Ward JM, eds. Carcinogenesis. New York: Raven Press, Ltd., 1994:1–23.
22. Raunio H, Husgafvel-Pursiainen K, Anttila S, Hietanen E, Hirvonen A, Pelkonen O. Diagnosis of polymorphisms in carcinogen-activating and inactivating enzymes and cancer susceptibility—a review. Gene 1995; 159:113–121.
23. Mannervik B, Danielson UH. Glutathione S-transferases—structure and catalytic activity. Crit Rev Biochem 1988; 23:283–337.
24. Tsuchida S, Sato K. Glutathione transferases and cancer. Crit Rev Biochem Mol Biol 1992; 27:337–384.
25. Barbacid M. Ras genes. Annu Rev Biochem 1987; 56:779–827.
26. Bos JL. Ras oncogenes in human cancer: a review. Cancer Res 1989; 49: 4682–4689.
27. Weinberg RA. Tumor suppressor genes. Science 1991; 254:1138–1145.
28. Harris CC. p53: at the crossroads of molecular carcinogenesis and risk assessment. Science 1993; 262:1980–1981.
29. Greenblatt MS, Bennett WP, Hollstein M, Harris CC. Mutations in the p53 tumor suppressor gene: clues to cancer etiology and molecular pathogenesis. Cancer Res 1994; 54:4855–4878.
30. Cordon-Cardo C. Mutation of cell cycle regulators: biological and clinical implications for human neoplasia. Am J Pathol 1995; 147:545–560.
31. Diwan BA, Rice JM. Organ and species specificity in chemical carcinogenesis and tumor promotion. In: Sirica AE, ed. The Pathobiology of Neoplasia. New York: Plenum Press, 1989:149–171.
32. Barrett JC, Shelby MD. Mechanisms of human carcinogens. In: D'Amato R, Slaga TJ, Farland WH, Henry C, eds. Relevance of Animal Studies to the Evaluation of Human Cancer Risk. New York: Wiley-Liss, Inc., 1992:415-434.
33. Huff, JE. Chemicals and cancer in humans: first evidence in experimental animals. Environ Health Perspect 1993; 100:201–210.
34. Huff J. Chemicals causally associated with cancer in humans and in laboratory animals. A perfect concordance. In: Waalker MP, Ward JM, eds. Carcinogenesis. New York: Raven Press, Ltd., 1994:25–37.
35. Huff J, Cirvello J, Hasemman J. Bucher J. Chemicals associated with site specific neoplasia in 1394 long-term carcinogenesis experiments in laboratory rodents. Environ Health Perspect 1991; 93:247–270.
36. Gold LS, Slone TH, Manley NB, Bernstein L. Target organs in chronic bioassay of 533 chemical carcinogens. Environ Health Perspect 1991; 93:233-246.

37. Squire RA. Ranking animal carcinogens: a proposed regulatory approach. Science 1981; 214:877–880.
38. Lave L, Ennever FK, Rosenkranz HS, Ornen G. Information value of the rodent bioassay. Nature 1988; 336:631–633.
39. Ames BN, Gold LS. Chemical carcinogenesis: too many rodent carcinogens. Proc Natl Acad Sci 1990; 87:7772–7776.
40. Ashby J, Morrod RS. Detection of human carcinogens. Nature 1990; 352: 185–186.
41. Nesnow S. A multi-factor ranking scheme for comparing the carcinogenic activity of chemicals. Mutat Res 1990; 239:83–115.
42. Tennant RW. Stratification of rodent carcinogenicity bioassay results to reflect relative human hazard. Mutat Res 1993; 286:111–118.
43. Altman PL, Katz DD. Inbred and Genetically Defined Strains of Laboratory Animals. Part I: Mouse and Rat. Vol. 3. Bethesda, MD: Federation of American Societies for Experimental Biology, 1979.
44. Cole RK, Furth J. Experimental studies on the genetics of spontaneous leukemia in mice. Cancer Res 1941; 1:957–965.
45. McEndy DP, Boon MC, Furth J. On the role of thymus, spleen, and gonads in the development of leukemia in a high-leukemia stock of mice. Cancer Res 1944; 4:377–383.
46. Meruelo D, Offer M, Rossomando A. Induction of leukemia by both fractionated x-irradiation and radiation leukemia virus involves loci in the chromosome 2 segment H-30-A. Proc Natl Acad Sci 1983; 80:462–466.
47. Mucenski ML, Taylor BA, Jenkins NA, Copeland NG. AKXD recombinant inbred strains: models for studying the molecular genetic basis of murine lymphomas. Mol Cell Biol 1986; 6:4236–4243.
48. Frankel WN. Taking stock of complex trait genetics in mice. Trends Genet 1995; 111:471–477.
49. Magee PN, Barnes JM. Carcinogenic nitroso compounds. Adv Cancer Res 1967; 10:163–246.
50. Frei JV. Toxicity, tissue changes, and tumor induction in inbred Swiss mice by methylnitrosamine and -amide compounds. Cancer Res 1970; 30:11–17.
51. Joshi VV, Frei JV. Effects of dose and schedule of methylnitrosourea on incidence of malignant lymphoma in adult female mice. J Natl Cancer Inst 1970; 45:335–339.
52. Frei JV, Joshi VV. Lack of induction of thymomas and of pulmonary adenomas in inbred Swiss mice by N-methyl-N′-nitro-N-nitrosoguanidine. Chem Biol Interact 1974; 8:131–133.
53. Yoshida MA, Takagi N, Sasaki M. Influence of strain difference on the karyotypic changes in N-nitroso-N-butylurea-induced mouse lymphomas. Cancer Genet Cytogenet 1982; 7:19–31.
54. Seidel HJ, Kreja L. Leukemia induction by methylnitrosourea (MNU) in selected mouse strains. J Cancer Res Clin Oncol 1984; 108:214–220.
55. Frei JV, Lawley PD. Thymomas induced by simple alkylating agents in C57BL/Cbi mice: kinetics of the dose response. J Natl Cancer Res 1980; 64: 845–856.

56. Warren W, Lawley PD, Gardner E, Harris G, Ball JK, Cooper CS. Induction of thymomas by N-methyl-N-nitrosourea in AKR mice: interaction between the chemical carcinogen and endogenous murine leukaemia viruses. Carcinogenesis 1987; 8:163–172.
57. Warren W, Clark JP, Gardner E, Harris G, Cooper GS, Lawley PD. Chemical induction of thymomas in AKR mice: interaction of chemical carcinogens and endogenous murine leukemia viruses. Comparison of N-methyl-N-nitrosourea and methyl methanesulphonate. Mol Carcinog 1990; 3:126–133.
58. Romach E, Moore J, Rummel S, Richie E. Influence of sex and carcinogen treatment protocol on tumor latency and frequency of K-ras mutations in N-methyl-N-nitrosourea-induced lymphomas. Carcinogenesis 1994; 15:2275–2280.
59. Guerrero I, Calzada P, Mayer A, Pellicer A. A molecular approach to leukemogenesis: mouse lymphomas contain an activated c-ras oncogene. Proc Natl Acad Sci 1984; 81:202–205.
60. Newcomb EW, Diamond LE, Sloan SR, Corominas M, Guerrero I, Pellicer A. Radiation and chemical activation of ras oncogenes in different mouse strains. Environ Health Perspect 1989; 81:33–37.
61. Diamond LE, Guerrero I, Pellicer A. Concomitant K- and N-ras gene point mutations in clonal murine lymphoma. Mol Cell Biol 1988; 8:2233–2236.
62. Saez GT, Oliva MR, Mangues R, Pellicer A. Absence of MDM-2 gene amplification in experimentally induced tumors regardless of p53 status. Mol Carcinog 1994; 9:40–45.
63. Newcomb EW, Steinberg JJ, Pellicer A. ras Oncogenes and phenotypic staging in N-methylnitrosourea- and γ-irradiation-induced thymic lymphomas in C57BL/6J mice. Cancer Res 1988; 48:5514–5521.
64. McMorrow LE, Newcomb EW, Pellicer A. Identification of a specific marker chromosome early in tumor development in γ-irradiated C57BL/6J mice. Leukemia 1988; 2:115–119.
65. Brathwaite O, Bayona W, Newcomb EW. p53 mutations in C57BL/6J murine thymic lymphomas induced by γ-irradiation and N-methylnitrosourea. Cancer Res 1992; 52:3791–3795.
66. Corominas M, Perucho M, Newcomb EW, Pellicer A. Differential expression of the normal and mutated K-ras alleles in chemically induced thymic lymphomas. Cancer Res 1991; 51:5129–5133.
67. Breuer M, Slebos R, Verbeek S, van Lohuizen M, Wientjens E, Berns A. Very high frequency of lymphoma induction by a chemical carcinogen in pim-1 transgenic mice. Nature 1989; 340:61–63.
68. Breuer M, Wientjens E, Verbeek S, Slebos R, Berns A. Carcinogen-induced lymphomagenesis in pim-1 transgenic mice: dose dependence and involvement of myc and ras. Cancer Res 1991; 51:958–963.
69. Mangues R, Kahn JM, Seidman I, Pellicer A. An overexpressed N-ras protooncogene cooperates with N-methylnitrosourea in mouse mammary carcinogenesis. Cancer Res 1994; 54:6359–6401.
70. Dumenco LL, Allay E, Norton K, Gerson SL. The prevention of thymic lymphomas in transgenic mice by human O^6-alkylguanine-DNA alkyltransferase. Science 1993; 259:219–222.

71. Kleihues P, Magee PN. Alkylation of rat brain nucleic acids by N-methyl-N-nitrosourea and methyl methane-sulphonate. J Neurochem 1973; 20:595–606.
72. Kleihues P, Margison GP. Carcinogenicity of N-methyl-N-nitrosourea: possible role of excision repair of O^6-methylguanine from DNA. J Natl Cancer Inst 1974; 53:1839–1841.
73. Buecheler J, Kleihues P. Excision of O^6-methylguanine from DNA of various mouse tissues following a single injection of N-methyl-N-nitrosourea. Chem Biol Interact 1977; 16:326–333.
74. Loveless A. Possible relevance of O-6 alkylation of deoxyguanosine to the mutagenicity and carcinogenicity of nitrosamines and nitrosamides. Nature 1969; 223:206–207.
75. Frei JV, Swenson DH, Warren D, Lawley PD. Alkylation of deoxyribonucleic acid in vivo in various organs of C57BL mice by the carcinogens N-methyl-N-nitrosourea, N-ethyl-N-nitrosourea and ethyl methanesulphonate in relation to induction of thymic lymphoma. Biochem J 1978; 174:1031–1044.
76. Pegg AE. Methylation of the O^6 position of guanine in DNA is the most likely initiating event in carcinogenesis by methylating agents. Cancer Invest 1984; 2:223–231.
77. Toorchen D, Topal MD. Mechanisms of chemical mutagenesis and carcinogenesis: effects on DNA replication of methylation at the O^6-guanine position of dGTP. Carcinogenesis 1983; 4:1591–1597.
78. Montesano R. Alkylation of DNA and tissue specificity in nitrosamine carcinogenesis. J Supramolec Structure Cell Biochem 1981; 17:259–273.
79. Pegg AE. Mammalian O^6-alkylguanine-DNA alkyltransferase: regulation and importance in response to alkylating carcinogenic and therapeutic agents. Cancer Res 1990; 50:6119–6129.
80. Liu L, Allay E, Dumenco LL, Gerson SL. Rapid repair of O^6-methylguanine-DNA adducts protects transgenic mice from N-methylnitrosourea-induced thymic lymphomas. Cancer Res 1994; 54:4648–4652.
81. Gerson SL, Zaidi NH, Dumenco LL, Allay E, Fan CY, Liu L, O'Connor PJ. Alkyltransferase transgenic mice: probes of chemical carcinogenesis. Mutat Res 1994; 307:541–555.
82. Cuypers HT, Selten G, Quint W, Zijlstra M, Robanus-Maandag E, Boelens W, van Wezenbeek P, Melief C, Berns A. Murine leukemia virus-induced T-cell lymphomagenesis: integration of proviruses in a distinct chromosomal region. Cell 1984; 37:141–150.
83. Storer RD, Cartwright ME, Cook WO, Soper KA, Nichols WW. Short-term carcinogenesis bioassay of genotoxic procarcinogens in PIM transgenic mice. Carcinogenesis 1995; 16:285–293.
84. McEndy DP, Boon MC. Induction of leukemia in mice by methylcholanthrene and X-rays. J Natl Cancer Inst 1942; 3:227–247.
85. Eva A, Trimmer RW. High frequency of c-K-ras activation in 3-methylcholanthrene-induced mouse thymomas. Carcinogenesis 1986; 7:1931–1933.
86. Nebert DW, Atlas SA, Guenthner TM, Kouri RE. The Ah locus: genetic regulation of the enzymes which metabolize polycyclic hydrocarbons and the

risk for cancer. In: Gelboin HV, Tsao POP, eds. Polycyclic Hydrocarbons and Cancer. New York: Academic Press, 1978:345–390.
87. Kouri RE, Ratrie H, Whitmire CE. Evidence of a genetic relationship between susceptibility to 3-methylcholanthrene-induced subcutaneous tumors and inducibility of aryl hydrocarbon hydroxylase. J Natl Cancer Inst 1973; 51:197–200.
88. Duran-Reynolds ML, Lilly F, Bosch A, Blank KJ. The genetic basis of susceptibility to leukemia induction in mice by 3-methylcholanthrene applied percutaneously. J Exp Med 1978; 147:459–469.
89. Fernandez-Salguero P, Pineau T, Hilbert DM, McPhail T, Lee SST, Kimura S, Nebert DW, Rudikoff S, Ward JM, Gonzalez FJ. Immune system impairment and hepatic fibrosis in mice lacking the dioxin-binding Ah receptor. Science 1995; 268:722–726.
90. United Nations Scientific Committee on the Effects of Atomic Radiation. Sources and Effects of Ionizing Radiation. New York: United Nations, 1977.
91. United Nations Scientific Committee on the Effects of Atomic Radiation. Sources and Effects of Ionizing Radiation. New York: United Nations, 1988.
92. United Nations Scientific Committee on the Effects of Atomic Radiation. Sources and Effects of Ionizing Radiation. New York: United Nations, 1994.
93. Kaplan HS, Brown MB. A quantitative dose-response study of lymphoid-tumor development in irradiated C57 black mice. J Natl Cancer Inst 1952; 13:185–208.
94. Upton AC, Wolff FF, Furth J, Kimball AW. A comparison of the induction of myeloid and lymphoid leukemias in X-radiated RF mice. Cancer Res 1958; 18:842–848.
95. Santos J, Perez de Castro I, Herranz M, Pellicer A, Fernandez-Piqueras J. Allelic losses on chromosome 4 suggest the existence of a candidate tumor suppressor gene region of about 0.6 cM in γ-radiation-induced mouse primary thymic lymphomas. Oncogene 1996; 12:669–676.
96. Ullrich RL, Jernigan MC, Cosgrove GE, Satterfield LC, Bowles ND, Storer JB. The influence of dose and rate on the incidence of neoplastic disease in RFM mice after neutron irradiation. Radiat Res 1976; 68:115–131.
97. Sloan SR, Newcomb EW, Pellicer A. Neutron radiation can activate K-ras via a point mutation on codon 146 and induces a different spectrum of ras mutations than does gamma radiation. Mol Cell Biol 1990; 10:405–408.
98. Breimer LH. Ionizing radiation-induced mutagenesis. Br J Cancer 1988; 57: 6–18.
99. Hutchinson F. Chemical changes induced in DNA by ionizing radiation. Prog Nucleic Acids Res Mol Biol 1985; 32:115–154.
100. Ward JF. Biochemistry of DNA lesions. Radiat Res 1985; 104:103–105.
101. Tates AD, Broerse JJ, Neuteboom I, de Vogel N. Differential persistence of chromosomal damage induced in resting rat-liver cells by X-rays and 4.2-MeV neutrons. Mutat Res 1982; 92:275–290.
102. Mayer A, Dorsch-Hasler R. Endogenous MuLV infection does not contribute to onset of radiation- or carcinogen-induced murine thymomas. Nature 1980; 295:253–255.

103. Knudson AG. Hereditary cancer, oncogenes and antioncogenes. Cancer Res 1985; 45:1437–1443.
104. Peto R, Roe FJC, Lee PN, Levy L, Clark J. Cancer and ageing in mice and man. Br J Cancer 1975; 32:411–426.
105. Weinberg RA. Oncogenes, antioncogenes, and the molecular basis of multistep carcinogenesis. Cancer Res 1989; 49:3713–3721.
106. Hollstein M, Rice K, Greenblatt MS, Soussi T, Fuchs R, Sorie T, Hovig E, Smith-Sorensen B, Montesano R, Harris CC. Database of p53 gene somatic mutations in human tumors and cell lines. Nucleic Acids Res 1994; 22:3551–3555.
107. Guerrero I, Pellicer A. Mutational activation of oncogenes in animal model systems of carcinogenesis. Mutat Res 1987; 185:293–308.
108. Balmain A, Brown K. Oncogene activation in chemical carcinogenesis. Adv Cancer Res 1988; 51:147–182.
109. Anderson MW, Reynolds SH. Activation of oncogenes by chemical carcinogens. In: Sirica AE, ed. The Pathobiology of Neoplasia. New York: Plenum Press, 1989:291–304.
110. Sukumar S. ras oncogenes in chemical carcinogenesis. Curr Topics Microbiol Immunol 1989; 148:93–114.
111. Newcomb EW, Corominas M, Bayona W, Pellicer A. Multistage carcinogenesis in murine thymocytes: involvement of oncogenes, chromosomal imbalances, and T cell growth factor receptor. Anticancer Res 1989; 9:1407–1416.
112. Corominas M, Sloan SR, Leon J, Kamino H, Newcomb EW, Pellicer A. ras activation in human tumors and in animal model systems. Environ Health Perspect 1991; 93:19–25.
113. Mangues R, Pellicer A. ras activation in experimental carcinogenesis. Sem Cancer Biol 1992; 3:229–239.
114. Barrett JC, Wiseman RW. Molecular carcinogenesis in humans and rodents. In: Klein-Szanto AJP, Anderson MW, Barrett JC, Slaga TJ, eds. Comparative Molecular Carcinogenesis. New York: Wiley-Liss, Inc., 1992:1–30.
115. Newcomb EW, Bayona W, Pisharody S. N-Methylnitrosourea-induced Ki-ras codon 12 mutations: early events in mouse thymic lymphomas. Mol Carcinog 1995; 13:89–95.
116. Newcomb EW. Clonal evolution of N-methylnitrosourea-induced C57BL/6J thymic lymphomas by analysis of multiple genetic alterations. Leukemia Res 1997; 21:189–198.
117. Land H, Parada L, Weinberg RA. Tumorigenic conversion of primary embryo fibroblasts requires at least two co-operating oncogenes. Nature 1983; 304:596–600.
118. Van Lohuizen M, Verbeek S, Krimpenfort P, Domen J, Saris C, Radaszkiewicz T, Berns A. Predisposition to lymphomagenesis in pim-1 transgenic mice: cooperation with c-myc and N-myc in murine leukemia virus-induced tumors. Cell 1989; 56:673–682.
119. Selten G, Cuypers HT, Berns A. Proviral activation of the putative oncogene pim-1 in MuLV-induced T-cell lymphomas. EMBO J 1985; 4:1793–1798.

120. Harris LJ, Lang RB, Marcu KB. Non-immunoglobulin-associated DNA rearrangements in mouse plasmacytomas. Proc Natl Acad Sci 1982; 79:4175–4179.
121. Chang TD, Biedler JL, Stockert E, Old LJ. Trisomy of chromosome 15 in x-ray-induced mouse leukemia. Proc Am Assoc Cancer Res 1977; 18:225.
122. Chan FPH, Ball JK, Sergovich FR. Trisomy #15 in murine thymomas induced by chemical carcinogens, x-irradiation, and an endogenous murine leukemia virus. J Natl Cancer Inst 1975; 62:605–610.
123. Potter M. Neoplastic development in B-lymphocytes. Carcinogenesis 1990; 11:1–13.
124. Mushinski JF, Bauer SR, Potter N, Reddy EP. Increased expression of myc-related oncogene mRNA characterized most BALB/c plasmacytomas induced by pristane or Abeleson murine leukemia virus. Proc Natl Acad Sci 1983; 80:1073–1077.
125. O'Donnell PV, Fleissner E, Lonial H, Koehne CF, Reicin A. Early clonality and high-frequency proviral integration into the c-myc locus in AKR leukemias. J Virol 1985; 55:500–503.
126. O'Donnell PV, Woller R, Koehne C, Fotheringham S, Jahnwar S. Trisomy of chromosome 15 and insertional mutagenesis of c-myc appear to be unrelated events in murine T-cell leukemogenesis. In: RNA Tumor Viruses. Cold Spring Harbor, NY: Cold Spring Harbor Laboratory, 1986;172.
127. Sinn E, Muller W, Pattengale P, Tepler I, Wallace R, Leder P. Coexpression of MMTV/V-Ha-ras and MMTV/C-myc genes in transgenic mice: synergistic action of oncogenes in vivo. Cell 1987; 49:465–475.
128. Dalla-Favera R, Bregni M, Erikson J, Paterson D, Gall RC, Croce CM. Human c-myc oncogene is located on the region of chromosome 8 that is translocated in Burkitt lymphoma cells. Proc Natl Acad Sci 1982; 79:7824–7827.
129. Rowley JD. A new consistent chromosomal abnormality in chronic myelogenous leukemia identified by quinacrine and Giemsa staining. Nature 1973; 243:290–293.
130. Tsujimoto Y, Croce CM. Analysis of the structure, transcripts and protein product of bcl-2, the gene involved in human follicular lymphoma. Proc Natl Acad Sci 1986; 83:5214–5218.
131. Feinberg AP, Vogelstein B. Hypomethylation of ras oncogenes in primary human cancers. Biochem Biophys Res Commun 1983; 111:47–54.
132. Counts JL, Goodman JI. Hypomethylation of DNA: an epigenetic mechanism involved in tumor promotion. Mol Carcinog 1994; 11:185–188.
133. Mangues R, Schwartz S, Seidman I, Pellicer A. Promoter demethylation in MMTV/N-ras^{N} transgenic mice required for transgene expression and tumorigenesis. Mol Carcinog 1995; 14:94–102.
134. Oliner JD, Kinzler KW, Meltzer PS, George DL, Vogelstein B. Amplification of a gene encoding a p53-associated protein in human sarcomas. Nature 1992; 358:80–83.
135. Kamb A, Gruis NA, Weaver-Feldhaus J, Liu Q, Harshman K, Tavtigian

SV, Stockert E, Day RS, Johnson BE, Skolnick MH. A cell cycle regulator potentially involved in genesis of many tumor types. Science 1994; 264:436-440.
136. Nobori T, Miura K, Wu DJ, Lois A, Takabayashi K, Carson DA. Deletions of the cyclin-dependent kinase-4 inhibitor gene in multiple human cancers. Nature 1994; 368:753-756.
137. Newcomb EW. P53 gene mutations in lymphoid diseases and their possible relevance to drug resistance. Leuk Lymphoma 1995; 17:211-221.
138. Fakharzadeh SS, Trusko SP, George DL. Tumorigeneic potential associated with enhanced expression of a gene that is amplified in a mouse tumor cell line. EMBO J 1991; 10:1565-1569.
139. Oliner JD, Pietenpol JA, Thiagalingam S, Gyuris J, Kinzler KW, Vogelstein B. Oncoprotein MDM2 conceals the activation domain of tumour suppressor p53. Nature 1993; 362:857-860.
140. Felix CA, Nau MM, Takahashi T, Mitsudomi T, Chiba I, Poplack DG, Reaman GH, Cole DE, Letterio JJ, Whang-Peng J, Knutsen T, Minna JD. Hereditary and acquired p53 gene mutations in childhood acute lymphoblastic leukemia. J Clin Invest 1992; 89:640-647.
141. Mori N, Wada M, Yokota J, Terada M, Okada M, Teramura M, Masuda M, Hoshino S, Motoji T, Oshimi K, Mizoguchi H. Mutations of the p53 tumour suppressor gene in haematologic neoplasms. Br J Haematol 1992; 81:235-240.
142. Wada M, Bartram CR, Nakamura H, Hachiya M, Chen D-L, Borenstein J, Miller CW, Ludwig L, Hansen-Hagge TE, Ludwig W-D, Reiter A, Mizoguchi H, Koeffler HP. Analysis of p53 mutations in a large series of lymphoid hematologic malignancies of childhood. Blood 1993; 82:3163-3169.
143. Lepelley P, Preudhomme C, Vanrumbeke M, Quesnel B, Cosson A, Fenaux P. Detection of p53 mutations in hematological malignancies: comparison between immunocytochemistry and DNA analysis. Leukemia 1994; 8:1342-1349.
144. Kawamura M, Kikuchi A, Kobayashi S, Hanada R, Yamamoto K, Horibe K, Shikano T, Ueda K, Hayashi K, Sekiya T, Hayashi Y. Mutations of the p53 and ras genes in childhood t(1;19)-acute lymphoblastic leukemia. Blood 1995; 85:2546-2552.
145. Hamdy N, Bhatia H, Shaker H, Kamel A, El Mawla N-G, Abou-Enein M, Yassin D, El-Sharkawy N, Magrath I. Molecular epidemiology of acute lymphoblastic leukemia in Egypt. Leukemia 1995; 9:194-202.
146. Preudhomme C, Dervite I, Wattel E, Vanrumbeke M, Flactif M, Lai JL, Hecquet B, Coppin MC, Nelken B, Gosselin B, Fenaux P. Clinical significance of p53 mutations in newly diagnosed Burkitt's lymphoma and acute lymphoblastic leukemia: a report of 48 cases. J Clin Oncol 1995; 13:812-820.
147. Marks DI, Kurz BW, Link MP, Ng E, Shuster JJ, Lauer SJ, Brodsky I, Haines DS. High incidence of potential p53 inactivation in poor outcome childhood acute lymphoblastic leukemia at diagnosis. Blood 1996; 87:1155-1161.

148. Aguilar-Santelises M, Magnusson KP, Wiman KG, Mellstedt H, Jondal M. Progressive B-cell chronic lymphocytic leukaemia frequently exhibits aberrant p53 expression. Int J Cancer 1994; 58:474–479.
149. Gandini D, Aguiari GL, Cuneo A, Piva R, Castoldi GL, del Senno L. Novel small deletions of the p53 gene in late-stage B-cell chronic lymphocytic leukaemia. Br J Haematol 1994; 88:881–885.
150. Wattel E, Preudhomme C, Hecquet B, Vanrumbeke M, Quesnel B, Dervite I, Morel P, Fenaux P. p53 mutations are associated with resistance to chemotherapy and short survival in hematologic malignancies. Blood 1994; 84: 3148–3157.
151. Dohner H, Fischer K, Bentz M, Hansen K, Benner A, Cabot G, Diehl D, Schlenk R, Coy J, Stilgenbauer S, Volkmann M, Galle PR, Poustka A, Hunstein W, Lichter P. P53 gene deletion predicts for poor survival and non-response to therapy with purine analogs in chronic B-cell leukemias. Blood 1995; 85:1580–1589.
152. Newcomb EW, Rao LS, Giknavorian SS, Lee SY. Alterations of multiple tumor suppressor genes (p53 (17p13), $p16^{INK4}$ (9p21), and DBM (13q14)) in B-cell chronic lymphocytic leukemia. Mol Carcinog 1995; 14:141–146.
153. Nakamura H, Said JW, Miller CW, Koeffler HP. Mutation and protein expression of p53 in acquired immunodeficiency syndrome-related lymphomas. Blood 1993; 82:920–926.
154. Trumper LH, Brady G, Bagg A, Gray D, Loke SL, Griesser H, Wagman R, Braziel R, Gascoyne RD, Vicini S, Iscove NN, Cossman J, Mak TW. Single-cell analysis of Hodgkin and Reed-Sternberg cells: molecular heterogeneity of gene expression and p53 mutations. Blood 1993; 81:3097–3115.
155. Weintraub M, Lin AY, Franklin J, Tucker MA, Magrath I, Bhatia KG. Absence of germline p53 mutations in familial lymphoma. Oncogene 1996; 12:687–691.
156. Villuendas R, Piris MA, Algara P, Sanchez-Beato M, Sanchez-Verde L, Martinez JC, Orradre JL, Garcia P, Lopez C, Martinez P. The expression of p53 protein in non-Hodgkin's lymphomas is not always dependent on p53 gene mutations. Blood 1993; 82:3151–3156.
157. Edwards RH, Raab-Traub N. Alterations of the p53 gene in Epstein-Barr virus-associated immunodeficiency-related lymphomas. J Virol 1994; 68: 1309–1315.
158. Lo Coco F, Gaidano G, Louie DC, Offit K, Chaganti RSK, Dalla-Favera R. p53 mutations are associated with histologic transformation of follicular lymphoma. Blood 1993; 82:2289–2295.
159. Farrugia MM, Duan L-J, Reis MD, Ngan BY, Berinstein NL. Alterations of the p53 tumor suppressor gene in diffuse large cell lymphomas with translocation of the c-myc and bcl-2 proto-oncogenes. Blood 1994; 83:191–198.
160. Kocialkowski S, Pezzella F, Morrison H, Jones M, Laha S, Harris AL, Mason DY, Gatter KC. Mutations in the p53 gene are not limited to classic 'hot spots' and are not predictive of p53 protein expression in high-grade non-Hodgkin's lymphoma. Br J Haematol 1995; 89:55–60.

161. Louie DC, Offit K, Jaslow R, Parsa NZ, Murty VVVS, Schluger A, Chaganti RSK. p53 overexpression as a marker of poor prognosis in mantle cell lymphomas with t(11;14)(q13;q32). Blood 1995; 86:2892–2899.
162. Koduru PRK, Raju K, Tse W, Kolitz J, Broome JD. Mutations in p53 gene are frequent in diffuse large cell histologic type of B-cell non-Hodgkin's lymphoma. Ann Oncol 1996; 7:15.
163. Sander CA, Yano T, Clark HM, Harris C, Longo DL, Jaffe ES, Raffeld M. p53 mutation is associated with progression in follicular lymphomas. Blood 1993; 82:1994–2004.
164. Baldini L, Fracchiolla NS, Cro LM, Trecca D, Romitti L, Polli E, Maiolo AT, Neri A. Frequent p53 gene involvement in splenic B-cell leukemia/lymphomas of possible marginal zone origin. Blood 1994; 84:270–278.
165. Ballerini P, Gaidano G, Gong JZ, Tassi V, Saglio G, Knowles DM, Dalla-Favera R. Multiple genetic lesions in acquired immunodeficiency syndrome-related non-Hodgkin's lymphoma. Blood 1993; 81:166–176.
166. Portier M, Moles J-P, Mazars G-R, Jeanteur P, Bataille R, Klein B, Theillet C. p53 and ras gene mutations in multiple myeloma. Oncogene 1992; 7:2539–2543.
167. Preudhomme C, Facon T, Zandecki M, Vanrumbeke M, Lai JL, Nataf E, Loucheux-Lefebvre MH, Kerckaert JP, Fenaux P. Rare occurrence of p53 gene mutations in multiple myeloma. Br J Haematol 1992; 81:440–443.
168. Corradini P, Inghirami G, Astolfi M, Ladetto M, Voena C, Ballerini P, Gu W, Nilsson K, Knowles DM, Boccadoro M, Pileri A, Dalla-Favera R. Inactivation of tumor suppressor genes, p53 and Rb1, in plasma cell dyscrasias. Leukemia 1994; 8:758–767.
169. Nishimura S, Asou N, Suzushima H, Okubo T, Fujimoto T, Osato M, Yamasaki H, Lisha L, Takatsuki K. p53 gene mutation and loss of heterozygosity are associated with increased risk of disease progression in adult T cell leukemia. Leukemia 1995; 9:598–604.
170. Diccianni MB, Yu J, Hsiao M, Mukherjee S, Shao LE, Yu AL. Clinical significance of p53 mutations in relapsed T-cell acute lymphoblastic leukemia. Blood 1994; 84:3105–3112.
171. Hsiao MH, Yu AL, Yeargin J, Ku D, Haas M. Nonhereditary p53 mutations in T-cell acute lymphoblastic leukemia are associated with the relapse phase. Blood 1994; 83:2922–2930.
172. Hu G, Zhang W, Deisseroth AB. p53 gene mutations in acute myelogenous leukemia. Br J Haematol 1992; 81:489–494.
173. Buhler-Leclerc M, Gratwohl A, Senn H-P. Occurrence of point mutations in p53 gene is not increased in patients with acute myeloid leukemia carrying an activating N-ras mutation. Br J Haematol 1993; 84:443–450.
174. Preudhomme C, Lubin R, Lepelley P, Vanrumbeke M, Fenaux P. Detection of serum anti p53 antibodies and their correlation with p53 mutations in myelodysplastic syndromes and acute myeloid leukemia. Leukemia 1994; 8:1589–1591.
175. Trecca D, Longo L, Biondi A, Cro L, Calori R, Grignani F, Maiolo AT,

Pelicci PG, Neri A. Analysis of p53 gene mutations in acute myeloid leukemia. Am J Hematol 1994; 46:304–309.
176. Adamson DJA, Dawson AA, Bennet B, King DJ, Haites NE. p53 mutation in the myelodysplastic syndromes. Br J Haematol 1995; 89:61–66.
177. Gaidano G, Guerrasio A, Serra A, Carozzi F, Rege-Cambrin G, Petroni D, Saglio G. Mutations in the p53 and ras family genes are associated with tumor progression of BCR/ABL negative chronic myeloproliferative disorders. Leukemia 1993; 7:946–953.
178. Neubauer A, He M, Schmidt CA, Huhn D, Liu ET. Genetic alterations in the p53 gene in the blast crisis of chronic myelogeneous leukemia: analysis by polymerase chain reaction based techniques. Leukemia 1993; 7:593–600.
179. Gaidano G, Serra A, Guerrasio A, Rege-Cambrin G, Mazza U, Saglio G. Genetic analysis of p53 and RB1 tumor-suppressor genes in blast crisis of chronic myeloid leukemia. Ann Hematol 1994; 68:3–7.
180. Nakai H, Misawa S, Horiike S, Maekawa T, Kashima K, Ishizaki K. Hemizygous expression of the wild-type p53 allele may confer a selective growth advantage before complete inactivation of the p53 gene in the progression of chronic myelogenous leukaemia. Br J Haematol 1995; 90:147–155.
181. Ludwig L, Schulz AS, Janssen JWG, Grunewald K, Bartram CR. P53 mutations in myelodysplastic syndromes. Leukemia 1992; 6:1302–1304.
182. Sugimoto K, Hirano N, Toyoshima H, Chiba S, Mano H, Takaku F, Yazaki Y, Hirai H. Mutations of the p53 gene in myelodysplastic syndrome (MDS) and MDS-derived leukemia. Blood 1993; 81:3022–3026.
183. Kaneko H, Misawa S, Horiike S, Nakai H, Kashima K. TP53 mutations emerge at early phase of myelodysplastic syndrome and are associated with complex chromosomal abnormalities. Blood 1995; 85:2189–2193.
184. Hirama T, Koeffler HP. Role of the cyclin-dependent kinase inhibitors in the development of cancer. Blood 1995; 86:841–854.
185. Stranks G, Height SE, Mitchell P, Jadayel D, Yuille MAR, De Lord C, Clutterbuck RD, Treleaven JG, Powles RL, Nacheva E, Oscier DG, Karpas A, Lenoir GM, Smith SD, Millar JL, Catovsky D, Dyer MJS. Deletions and rearrangement of CDKN2 in lymphoid malignancy. Blood 1995; 85:893–901.
186. Fizzotti M, Cimino G, Pisegna S, Alimena G, Quartarone C, Mandelli F, Pelicci PG, Lo Coco F. Detection of homozygous deletions of the cyclin-dependent kinase 4 inhibitor (p16) gene in acute lymphoblastic leukemia and association with adverse prognostic features. Blood 1995; 85:2685–2690.
187. Sill H, Goldman JM, Cross NCP. Homozygous deletions of the p16 tumor-suppressor gene are associated with lymphoid transformation of chronic myeloid leukemia. Blood 1995; 85:2013–2016.
188. Rasool O, Heyman M, Brandter LB, Liu Y, Grander D, Soderhal S, Einhorn S. $p15^{ink4B}$ and $p16^{ink4A}$ gene inactivation in acute lymphocytic leukemia. Blood 1995; 85:3431–3436.
189. Haidar MA, Cao X-B, Manshouri T, Chan LL, Glassman A, Kantarjian HM, Keating MJ, Beran MS, Albitar M. $p16^{INK4A}$ and $p15^{INK4B}$ gene deletions in primary leukemias. Blood 1995; 86:311–315.

190. Ogawa S, Hangaishi A, Miyawaki S, Hirosawa S, Miura Y, Takeyama K, Kamada N, Ohtake S, Uike N, Shimazaki C, Toyama K, Hirano M, Mizoguchi H, Kobayashi Y, Furusawa S, Saito M, Emi N, Yazaki Y, Ueda R, Hirai H. Loss of the cyclin-dependent kinase 4 inhibitor (p16;MTS1) gene is frequent in and highly specific to lymphoid tumors in primary human hematopoietic malignancies. Blood 1995; 86:1548–1556.
191. Gombart AF, Morosetti R, Miller CW, Said JW, Koeffler HP. Deletions of the cyclin-dependent kinase inhibitor genes $p16^{INK4A}$ and $p15^{INK4B}$ in non-Hodgkin's lymphomas. Blood 1995; 86:1534–1539.
192. Koduru PRK, Zariwala M, Soni M, Gong JZ, Xiong Y, Broome JD. Deletion of cyclin-dependent kinase 4 inhibitor genes p15 and p16 in non-Hodgkin's lymphoma. Blood 1995; 86:2900–2905.
193. Okuda T, Shurtleff SA, Valentine MB, Raimondi SC, Head DR, Behm F, Curcio-Brint AM, Liu Q, Pui C-H, Sherr CJ, Beach D, Look AT, Downing JR. Frequent deletion of $p16^{INK4a}$/MTS1 and $p15^{INK4b}$/MTS2 in pediatric acute lymphoblastic leukemia. Blood 1995; 85:2321–2330.
194. Cayuela J-M, Madani A, Sanhes L, Stern M-H, Sigaux F. Multiple tumor-suppressor gene 1 inactivation is the most frequent genetic alteration in T-cell acute lymphoblastic leukemia. Blood 1996; 87:2180–2186.
195. Ohnishi H, Kawamura M, Ida K, Sheng XM, Hanada R, Nobori T, Yamamori S, Hayashi Y. Homozygous deletions of p16/MTS1 gene are frequent but mutations are infrequent in childhood T-cell acute lymphoblastic leukemia. Blood 1995; 86:1269–1275.
196. Ahuja HG, Jat PS, Foti A, Bar-Eli M, Cline MJ. Abnormalities of the retinoblastoma gene in the pathogenesis of acute leukemia. Blood 1991; 78: 3259–3268.
197. Ginsberg AM, Raffeld M, Cossman J. Inactivation of the retinoblastoma gene in human lymphoid malignancies. Blood 1991; 77:833–840.
198. Newcomb EW, Thomas A, Selkirk A, Lee SY, Potmesil M. Frequent homozygous deletions of D13S218 on 13q14 in B-cell chronic lymphocytic leukemia independent of disease stage and retinoblastoma gene inactivation. Cancer Res 1995; 55:2044–2047.
199. Neubauer A, De Kant E, Rochlitz C, Laser J, Zanetta AM, Gallardo J, Oertel J, Herrmann R, Huhn D. Altered expression of the retinoblastoma susceptibility gene in chronic lymphocytic leukaemia. Br J Haematol 1993; 85:498–503.
200. Kay NE, Suen R, Ranheim E, Peterson LC. Confirmation of Rb gene defects in B-CLL clones and evidence for variable predominance of the Rb defective cells within the CLL clone. Br J Haematol 1993; 84:257–264.
201. Kornblau SM, Chen N, del Giglio A, O'Brien S, Deisseroth AB. Retinoblastoma protein expression is frequently altered in chronic lymphocytic leukemia. Cancer Res 1994; 54:242–246.
202. Sakai T, Toguchida J, Ohtani N, Yandell DW, Rapaport JM, Dryja TP. Allele-specific hypermethylation of the retinoblastoma tumor-suppressor gene. Am J Hum Genet 1991; 48:880–888.

203. Herman JG, Merlo A, Mao L, Lapidus RG, Issa J-PI Davidson NE, Sidransky D, Baylin SB. Inactivation of the CDKN2/p16/MTS1 gene is frequently associated with aberrant DNA methylation in all common human cancers. Caner Res 1995; 55:4525–4530.
204. Gonzalez-Zulueta M, Bender CM, Yang AS, Nguyen TD, Beart RW, van Tornout JM, Jones PA. Methylation of the 5′ CpG island of the p16/CDKN2 tumor suppressor gene in normal and transformed human tissues correlates with gene silencing. Cancer Res 1995; 55:4531–4535.
205. Otterson GA, Kratzke RA, Coxon A, Kim YW, Kaye FJ. Absence of $p16^{INK4}$ protein is restricted to the subset of lung cancer lines that retains wildtype RB. Oncogene 1994; 9:3375–3378.
206. Shapiro GI, Edwards CD, Kobzik L, Godleski J, Richards W, Sugarbaker DJ, Rollins BJ. Reciprocal Rb inactivation and $p16^{INK4}$ expression in primary lung cancers and cell lines. Cancer Res 1995; 55:505–509.
207. Ueki K, Ono Y, Henson JW, Efird JT, von Deimling A, Louis DN. CDKN2/p16 or Rb alterations occur in the majority of glioblastomas and are inversely correlated. Cancer Res 1996; 56:150–153.
208. Neri A, Knowles DM, Greco A, McCormick F, Dalla-Favera R. Analysis of RAS oncogene mutations in human lymphoid malignancies. Proc Natl Acad Sci 1988; 85:9268–9272.
209. Lubbert M, Mirro J, Miller CW, Kahan J, Isaac G, Kitchingman G, Mertelsmann R, Herrmann F, McCormick F, Koeffler HP. N-ras gene point mutations in childhood acute lymphocytic leukemia correlate with a poor prognosis. Blood 1990; 75:1163–1169.
210. Terada N, Miyoshi J, Kawa-Ha K, Sasai H, Orita S, Yumura-Yagi K, Hara J, Fujinami A, Kakunaga T. Alteration of N-ras gene mutation after relapse in acute lymphoblastic leukemia. Blood 1990; 75:453–457.
211. Jacobs A. Gene mutations in myelodysplasia. Leukemia Res 1992; 16:47–50.
212. Horiike S, Misawa S, Nakai H, Kaneko H, Yokota S, Taniwaki M, Yamane Y, Inazawa J, Abe T, Kashima K. N-ras mutation and karyotypic evolution are closely associated with leukemic transformation in myelodysplastic syndrome. Leukemia 1994; 8:1331–1336.
213. Collins SJ, Howard M, Andrews DF, Agura E, Radich J. Rare occurrence of N-ras point mutations in Philadelphia chromosome positive chronic myeloid leukemia. Blood 1989; 73:1028–1032.
214. LeMaistre A, Lee M-S, Talpaz M, Kantarjian HM, Freireich EJ, Deisseroth AB, Trujillo JM, Stass SA. RAS oncogene mutations are rare late stage events in chronic myelogenous leukemia. Blood 1989; 73:889–891.
215. Browett PJ, Ganeshaguru K, Hoffbrand AV, Norton JD. Absence of Kirsten-ras oncogene activation in B-cell chronic lymphocytic leukemia. Leukemia Res 1988; 12:25–31.
216. Aurer I, Labar B, Nemet D, Ajdukovic R, Bogdanic V, Gale RP. High incidence of conservative RAS mutations in acute myeloid leukemia. Acta Haematol 1994; 92:123–125.
217. Corradini P, Ladetto M, Voena C, Palumbo A, Inghirami G, Knowles DM,

Boccadoro M, Pileri A. Mutational activation of N- and K-ras oncogenes in plasma cell dyscrasias. Blood 1993; 81:2708–2713.
218. Paquette RL, Landaw EM, Pierre RV, Kahan J, Lubbert M, Lazcano O, Isaac G, McCormick F, Koeffler HP. N-ras mutations are associated with poor prognosis and increased risk of leukemia in myelodysplastic syndrome. Blood 1993; 82:590–599.
219. Neubauer A, Dodge RK, George SL, Davey FR, Silver RT, Schiffer CA, Mayer RJ, Ball ED, Wurster-Hill D, Bloomfield CD. Prognostic importance of mutations in the ras proto-oncogenes in de novo acute myeloid leukemia. Blood 1994; 83:1603–1611.
220. Tien HF, Wang CH, Lin MT, Lee FY, Liu MC, Chuang SM, Chen YC, Shen MC, Lin KH, Lin DT. Correlation of cytogenetic results with immunophenotype, genotype, clinical features, and ras mutation in acute myeloid leukemia. A study of 235 Chinese patients in Taiwan. Cancer Genet Cytogenet 1995; 84:60–68.
221. Watzinger F, Gaiger A, Karlic H, Becher R, Pillwein K, Lion T. Absence of N-ras mutations in myeloid and lymphoid blast crisis of chronic myeloid leukemia. Cancer Res 1994; 54:3934–3938.
222. Tanaka K, Takechi M, Asaoku H, Dohy H, Kamada N. A high frequency of N-RAS oncogene mutations in multiple myeloma. Int J Hematol 1992; 56: 119–127.
223. Paquette RL, Berenson J, Lichtenstein A, McCormick F, Koeffler HP. Oncogenes in multiple myeloma: point mutation of N-ras. Oncogene 1990; 5: 1659–1663.
224. Neri A, Murphy JP, Cro L, Ferrero D, Tarella C, Baldini L, Dalla-Favera R. RAS oncogene mutation in multiple myeloma. J Exp Med 1989; 170:1715–1725.
225. Rodenhuis S. Ras and human tumors. Sem Cancer Biol 1992; 3:241–247.
226. Wu X, Bayle JH, Olson D, Levine AJ. The p53-mdm-2 autoregulatory feedback loop. Genes Dev 1993; 7:1126–1132.
227. Bueso-Ramos CE, Yang Y, deLeon E, McCown P, Stass SA, Albitar M. The human MDM-2 oncogene is overexpressed in leukemias. Blood 1993; 82: 2617–2623.
228. Chilosi M, Doglioni C, Menestrina F, Montagna L, Rigo A, Lestani M, Barbareschi M, Scarpa A, Mariuzzi GM, Pizzolo G. Abnormal expression of the p53-binding protein MDM2 in Hodgkin's disease. Blood 1994; 84:4295–4300.
229. Maestro R, Gloghini A, Doglioni C, Gasparotto D, Vukosavljevic T, De Re V, Laurino L, Carbone A, Boiocchi M. MDM2 overexpression does not account for stabilization of wild-type p53 protein in non-Hodgkin's lymphomas. Blood 1995; 85:3239–3246.
230. Zhou M, Yeager AM, Smith SD, Findley HW. Overexpression of the MDM2 gene by childhood acute lymphoblastic leukemia cells expressing the wild-type p53 gene. Blood 1995; 85:1608–1614.
231. Croce CM, Nowell PC. Molecular basis of human B cell neoplasia. Blood 1985; 65:1–7.

232. Cline MJ. The molecular basis of leukemia. N Engl J Med 1994; 330:328–336.
233. Harleman JH, Schuurman HJ, Kuper CF. Carcinogenesis of the hematopoietic system. In: Waalker MP, Ward JM, eds. Carcinogenesis. New York: Raven Press, Ltd., 1994:403–428.
234. Newcomb EW, El Rouby S, Thomas A. A unique spectrum of p53 mutations in B-cell chronic lymphocytic leukemia distinct from that of other lymphoid malignancies. Mol Carcinog 1995; 14:227–232.
235. Morgan DO. Principles of CDK regulation. Nature 1995; 374:131–134.
236. Grana X, Reddy P. Cell cycle control in mamalian cells: role of cyclins, cyclin dependent kinases (CDKs), growth suppressor genes and cyclin-dependent kinase inhibitors (CKIs). Oncogene 1995; 11:211–219.
237. Sullivan N, Gatehouse D, Tweats D. Mutation, cancer and transgenic models: relevance to the toxicology industry. Mutagenesis 1993; 8:167–174.
238. Doll R. Hazards of ionizing radiation: 100 years of observation on man. Br J Cancer 1995; 72:1339–1349.
239. Parker L, Craft AW. Radon and childhood cancers. Eur J Cancer 1996; 32A: 201–204.
240. Goldman M. Cancer risk of low-level exposure. Science 1996; 271:1821–1822.
241. Petridou E, Trichopoulos D, Dessypris N, Flytzani V, Haidas S, Kalmanti M, Koliouskas D, Kosmidis H, Piperopoulos F, Tzortzatou F. Infant leukaemia after in utero exposure to radiation from Chernobyl. Nature 1996; 382: 352–353.
242. Iwamoto KS, Mizuno T, Ito T, Akiyama M, Takeichi N, Mabuchi K, Seyama T. Feasibility of using decades-old archival tissues in molecular oncology/epidemiology. Am J Pathol 1996; 149:399–406.
243. Darby SC. Irradiation for non-malignant conditions. In: Coleman MP, ed. Cancer risk after medical treatment. New York: Oxford University Press, 1991:29–45.
244. Birch JM. Childhood malignancy. In: Coleman MP, ed. Cancer Risk After Medical Treatment. New York: Oxford University Press, 1991:101–121.
245. Kaldor JM, Lasset C. Cytotoxic chemotherapy for cancer. In: Coleman MP, ed. Cancer Risk After Medical Treatment. New York: Oxford University Press, 1991:50–65.
246. Ellis M, Lishner M. Second malignancies following treatment in non-Hodgkin's lymphoma. Leuk Lymphoma 1993; 9:337–342.
247. Ellis M, Ravid M, Lishner M. A comparative analysis of alkylating agent and epipodophyllotoxin-related leukemias. Leuk Lymphoma 1993; 11:9–13.
248. Gerson SL. Molecular epidemiology of therapy-related leukemias. Curr Opinion Oncol 1993; 5:136–144.
249. Travis LB, Curtis RE, Boice JD, Platz CE, Hankey BF, Fraumeni JF. Second malignant neoplasms among long-term survivors of ovarian cancer. Cancer Res 1996; 56:1564–1570.
250. Carter G, Hughes DC, Clark RE, McCormick F, Jacobs A, Whittaker JA, Padua RA. RAS mutations in patients following cytotoxic therapy for lymphoma. Oncogene 1990; 5:411–416.

251. Taylor C, McGlynn H, Carter G, Baker AH, Warren N, Ridge SA, Owen G, Thompson E, Thompson PW, Jacobs A, Padua RA. RAS and FMS mutations following cytotoxic therapy for childhood acute lymphoblastic leukemia. Leukemia 1995; 9:466–470.
252. Broeker PLS, Super HG, Thirman MJ, Pomykala H, Yonebayashi Y, Tanabe S, Zeleznik-Le N, Rowley JD. Distribution of 11q23 breakpoints within the MLL breakpoint cluster region in de novo acute leukemia and in treatment-related acute myeloid leukemia: correlation with scaffold attachment regions and topoisomerase II consensus binding sites. Blood 1996; 87: 1912–1922.
253. Holliday R. A new theory of carcinogenesis. Br J Cancer 1979; 40:513–522.
254. Jones PA. DNA methylation errors and cancer. Cancer Res 1996; 56:2463-2467.
255. Jelinek J, Fairbairn LJ, Dexter TM, Rafferty JA, Stocking C, Ostertag W, Margison GP. Long-term protection of hematopoiesis against the cytotoxic effects of multiple doses of nitrosourea by retrovirus-mediated expression of human O^6-alkylguanine-DNA-alkyl-transferase. Blood 1996; 87:1957–1961.
256. Chen H, Sandler DP, Taylor JA, Shore DL, Liu E, Bloomfield CD, Bell DA. Increased risk for myelodysplastic syndromes in individuals with glutathione transferase theta 1 (GSTT1) gene defect. Lancet 1996; 347:295–297.

22
Urinary Bladder Cancer

Shoji Fukushima, Hideki Wanibuchi, and Shinji Yamamoto
Osaka City University Medical School, Osaka, Japan

I. INTRODUCTION

Urinary bladder cancers have attracted major attention as one of the occupational cancers strongly linked to chemical exposure to several dyestuffs. However, in the majority of cases of urinary bladder cancers, there is no obvious explanation for the tumor development, although factors such as smoking, coffee drinking, and so on have been speculated as playing causal roles.

From work with experimental model systems, our knowledge of chemical carcinogenesis has certainly progressed over the past 20 years. This chapter describes the available evidence regarding etiological factors, the processes involved in carcinogenesis, and molecular events responsible for development of urinary bladder cancers.

II. ETIOLOGICAL FACTORS IN HUMANS AND EXPERIMENTAL ANIMALS

A. Occupational Exposure

Urinary bladder cancer was the first human malignant tumor thought to be related to industrial chemical exposure. The high incidence of urinary bladder cancer in dye industry workers led to the suggestion that aniline was the chemical agent causing tumors, but experimental studies failed to prove any carcinogenicity in animals. Eventually experiments with the arylamine 2-napthylamine induced urinary bladder cancers in the dog (1), and this chemical was confirmed to be a causal factor in aniline dye industry–linked neoplasia. Benzidine was also early implicated as a urinary bladder carcinogen in humans (2). In 1955 a close relation between 4-aminobiphenyl exposure in industry and an increased risk of bladder cancer was reported (3), and later carcinogenicity was demonstrated in the mouse, dog, and rabbit. In addition, *p*-cresidine, anisidine, 4,4′-methylene-bis(2-chloroaniline), and 3,3′-dichlorobenzidine have all been found to be urinary bladder carcinogens in dogs or rodents (4,5). Thus, the data indicate that many arylamines may have important roles in urinary bladder cancer development as etiological factors.

B. Cigarette Smoking

There are a number of epidemiological studies supporting a relationship between cigarette smoking and an increased risk of urinary bladder cancer in humans (6,7). The association has been observed for both males and females, and the risk correlates with numbers of cigarettes smoked, the duration of smoking, and the degree of inhalation of the smoke (8). However, no unequivocal experimental data with cigarette smoking have so far been published to support the epidemiological findings. While specific carcinogens have not been identified, cigarette smoke contains a number of carcinogenic nitroso compounds and arylamines (9), and recently it was reported that one component, acrolein, can initiate rat urinary bladder carcinogenesis (10). In addition, since cigarette smokers exhibit increased cell proliferation in the urinary bladder as evidenced by epithelial hyperplasia (11), cigarette smoke may exert nongenotoxic promotional effects on urinary bladder carcinogenesis as well as genotoxic effects.

C. Other Environmental Chemicals

N-*Nitroso Compounds*

N-Nitroso compounds produce cancers in various organs of various animal species. Experimentally, nitrosamines such as *N*-butyl-*N*(4-hydroxybutyl)-nitrosamine (BBN) (12–14), *N*-ethyl-*N*-(4-hydroxybutyl)nitrosamine (EHEN) (15), and *N*-methyl-*N*-nitrosourea (MNU) (16) are well known to induce urinary bladder carcinomas in rodents and dogs. There have been no epidemiological studies pointing to urinary bladder carcinogenicity due to *N*-nitroso compounds in humans, but the fact that human beings are continuously exposed to endogenous and exogenous *N*-nitroso compounds suggests that careful attention should be paid to the potential relationship of these known chemical carcinogens with urinary bladder cancer development. Endogenous nitrosation can be mediated by bacteria and macrophages in infected organs (17). *N*-Nitroso compounds have been found in infected urinary bladders, in particular in patients having the *Schistosoma hematobium*, which is significantly associated with neoplasia in the urinary bladder (18).

Heterocyclic Amines

Mutagenic and carcinogenic heterocyclic amines have been identified as pyrolysis products formed during cooking and broiling of meat and fish. One of the series of heterocyclic amines tested, 3-amino-1-methyl-5H-pyridol[4,3-b]indole, otherwise known as Trp-P-2, was found to be carcinogenic in rat urinary bladder and liver (19). Cigarette smoke contains carcinogenic heterocyclic amines, and Trp-P-2 was reported to be present at levels of 0.95 ng/cigarette. DNA adduct formation due to heterocyclic amines has been found in organs such as the liver, colon, and kidney. Humans are certainly exposed through lifestyle choices such as cigarette smoking. Overall, the data imply that heterocyclic amines may be causative factors for human urinary bladder neoplasia.

Medicines

There are also some medicines which may induce urinary bladder cancers in man. The analgesic agent, phenacetin, is used alone or in combination with aspirin and caffeine. A possible relationship between phenacetin abuse and renal pelvic cancers was first pointed out in Sweden (20). Long phenacetin exposure was also found to increase the risk of urinary bladder cancer development and finally phenacetin was found to be carcinogenic for the human urinary tract (21). Experimental studies revealed carcinogenic effects on the kidney of (C57BL/6 X C3H) F_1, mice, and the nasal cavity and urinary bladder of SD rats (22–24). In rats fed phenacetin, an increase of the labeling index of renal pelvic and renal papillary epithelium and a promoting effect on urinary tract carcinogenesis have been reported (25,26), However, it has not been satisfactorily proven that phenacetin alone can induce renal pelvic carcinomas in experimental animals as it appears to do in humans. Recently we examined this question in spontaneous hydronephrosis-bearing rats (Table 1) (27). Fifty-five SD/cShi male rats were fed 2% phenacetin-containing diet for 85 weeks, and a further 32 animals received basal diet as controls. Forty-three of 53 rats given the phenacetin developed renal pelvic carcinomas, with lung metastases in three of them. The mean induction time was 78 weeks. Ureteral and urinary bladder carcinomas were also observed in 2 and 6 treated animals, respectively, and no such lesions found in control animals. In addition, large amounts of phenacetin and its metabolites, *N*-hydroxyphenacetin and *N*-

Table 1 Incidences of Urinary Tract Lesions in SD/cShi Rats Given Phenacetin

	Incidence (%)	
Location and histology	Phenacetin-treated rats	Control
Total number of animals	53	30
Renal pelvis		
Carcinoma	43 (81)[a]	0
Papilloma	10 (19)	0
PN hyperplasia	42 (79)[a]	0
Ureter		
Carcinoma	2 (4)	0
Papilloma	6 (11)	
PN hyperplasia	7 (13)[c]	0
Urinary bladder		
Carcinoma	6 (11)	0
Papilloma	7 (13)[c]	0
PN hyperplasia	6 (11)	0

[a] $p < 0.001$.
[b] $p < 0.01$.
[c] $p < 0.05$, phenacetin-treated rats versus control.
SD/cShi rats spontaneously develop hydroenphrosis.

acetyl-*p*-aminophenol, were detected in the urine and plasma. This study thus provided experimental proof of phenacetin carcinogenicity toward the renal pelvis in an animal model.

Since the first report of urinary bladder cancers related to cyclophosphamide therapy in 1971 (28), many cases have been described. Patients undergoing this type of treatment have a ninefold increased risk of urinary bladder cancer development. IARC has classified cyclophosphamide as a human carcinogen (21). Schmahl and Habs (29) reported that oral administration of cyclophosphamide induced transitional cell carcinomas of the urinary bladder in rats, and Hicks et al. (30) found that cyclophosphamide acted as a cocarcinogen in rat urinary bladder. Acrolein, a cyclophosphamide metabolite, binds to DNA, and Cohen et al. (10) reported that it initiates urinary bladder carcinogenesis in the rat. Acrolein is also an important industrial chemical, and it is present in cigarette smoke (10–140 mg/cigarette). *N,N*-Bis(2-chloroethyl)-2-naphthylamine (chlornaphazine) has been used for therapy of polycythemia, but 13 urinary bladder carcinomas developed among 61 treated patients. IARC thus identified chlornaphazine as a human carcinogen (21), although there is no literature regarding any carcinogenicity of this chemical in rodents.

Tryptophan

Tryptophan and its metabolites have been speculated to be etiological factors in urinary bladder cancer because patients show increased levels of these agents in their urine (31). While there are no carcinogenicity data for tryptophan in rodents, it was demonstrated that its dietary administration can promote urinary bladder carcinogenesis in rats (32). As noted above, the heterocyclic amine Trp-P-2, which is a pyrolysis product of tryptophan, is known to induce urinary bladder carcinoma in rats by oral administration (19).

Arsenics

Epidemiological investigations have revealed that arsenics are carcinogenic to humans, especially affecting the skin and lung. In the blackfoot disease–endemic area of Taiwan (which had a high concentration of arsenics in the drinking water) an elevated mortality from internal cancers, especially those of the urinary bladder, kidney, liver, and lung, as well as skin cancer, has been reported (33). The tumor incidence was found to be dependent on the concentration of arsenics in the water. In experimental animals, arsenics have not exerted complete carcinogenicity. Using a multiorgan carcinogenesis bioasssay in rats, dimethylarsinic acid (DMA, a major metabolite of inorganic arsenic in most animals) promoted carcinogenesis in the urinary bladder, kidney, liver, and thyroid gland (34). In a single organ model system for rat urinary bladder carcinogenesis, the development of preneoplastic lesions and tumors (papillomas and carcinomas) was also enhanced in a dose-dependent manner from 10 ppm DMA upward (35). The promoting potential of DMA at this low dose in experimental animals thus provides support for the epidemiological data pointing to development of urinary bladder cancer due to arsenics. The age-dependent lethality of DMA to rats is very marked. Seven of 10 rats starting 200 ppm DMA treatment at 6 weeks of age were dead within 4 weeks. In contrast, DMA-induced lethality was only 1 of 10 and none of 10 rats aged 8 and 10 weeks, respectively (unpublished data). These data reveal that chronic DMA administration is more toxic for younger rats. Since it was reported that DMA induces chromosomal alterations and DNA damage, such as DNA single-strand breaks and DNA-protein cross-links (36), clastogenic effects might be important for arsenic-induced rat carcinogenesis. On the other hand, it was recently reported that sodium arsenite increases rat hepatic ODC activity and hepatic heme oxygenase activity, but causes no rat liver or lung DNA damage, indicating that it might be a promoter rather than an initiator of carcinogenesis (37).

D. Coffee and Tea Drinking

An elevated risk of urinary bladder cancer development has been reported in association with coffee and tea drinking (8). However, IARC stated in a recent review that there is no clear link between intake of coffee and tea and increased cancer risk (38). Experimentally caffeine administration was found not to promote urinary bladder carcinogenesis in rats (39).

E. Infection

In North Africa and the Middle East, where infection with *S. hematobium* is endemic, there is a very high incidence of urinary bladder neoplasia. The tumors are generally squamous cell carcinomas, which have a poor prognosis. The etiology of cancers associated with *S. hemalobium* involves many factors, including carcinogen exposure and mechanical irritation due to bilharzial eggs. Secondary bacterial infection is also suspected to be important. It is known that under such conditions, urinary nitrates can convert to nitrites and result in *N*-nitroso compound formation by reaction of nitrites with secondary amines (40). In fact, *N*-nitroso compounds have been found in the urine of bilharzial patients with or without urinary bladder cancers (18).

Urinary stone formation is commonly associated with urinary tract infection, and this has also been reported to increase human cancer risk (40). Experimentally, the presence of urinary stones is linked with urinary bladder cancer in rats and mice (41).

III. ETIOLOGICAL FACTORS KNOWN ONLY FROM EXPERIMENTAL STUDIES

Table 2 classifies the many known promoters of urinary bladder carcinogenesis into six groups.

A. Sodium or Potassium Salts

Sodium saccharin has been shown to act as a urinary bladder carcinogen in rats when given in large doses (42), but the interpretation of complete carcinogen status for sodium saccharin is controversial. Hicks et al. (30)

Table 2 Classification of Urinary Bladder Cancer Promoters in Experimental Animals

1. Sodium or potassium salts
Sodium saccharin
Sodium L-ascorbate
Sodium *o*-phenylphenate
Sodium bicarbonate
Sodium citrate
Sodium erythorbate
Sodium phenobarbital
Sodium barbital
Potassium carbonate with or without ascorbic acid
2. Urolithiasis-inducing chemicals
Uracil
Diphenyl
3. Antioxidants
Butylated hydroxyanisole
Butylated hydroxytoluene
Ethoxyquin
t-Butylhydroxyquinone
2-*t*-Butyl-4-methylphenol
4. Anticancer agents
Adriamycin
Mitomycin C
5. Amino acids
DL-Tryptophan
L-Leucine
L-Isoleucine
6. Others
Urinary components (fractions I and II)
Allopurinol (?)

demonstrated that prior instillation of a subcarcinogenic dose of MNU resulted in a high incidence of sodium saccharin-induced urinary bladder carcinomas. Subsequently, Cohen et al. (32) confirmed that sodium saccharin is a strong promoter in rats. Its effects are dose dependent (43) but are not shared by the parent acid (44). Sodium cyclamate has also been shown to act as a promoter of urinary bladder carcinogenesis in rats (30).

In light of these experimental findings for sodium saccharin, we examined the promoting effects of sodium L-ascorbate (Na-AsA) and found that oral administration of a 5% dietary dose clearly enhanced urinary bladder carcinogenesis (45), whereas L-ascorbic acid (AsA) did not. In a subsequent study, attention was focused on the roles played by urinary pH and Na ion concentration in such promotion (46). After initiation with BBN, male rats were given a diet containing no added chemical, 5% AsA, 3% $NaHCO_3$, 5% AsA + 3% $NaHCO_3$, 5% Na-AsA, 1% NH_4Cl, or 5% Na-AsA + 1% NH_4Cl. As summarized in Table 3, $NaHCO_3$ was found to significantly promote the urinary bladder carcinogenesis. Like Na-AsA, AsA + $NaHCO_3$ brought about marked enhancement of neoplastic lesion induction. Addition of NH_4Cl to Na-AsA reduced its promotion activity. Urinary analyses revealed differences in the concentration of AsA of the group treated with AsA + $NaHCO_3$ versus the $NaHCO_3$-treated group. No differences were found in the pH and Na ion concentration of these groups; these latter parameters were significantly elevated in both groups. As expected, no increase in urinary pH was observed after combined treatment with NH_4Cl and Na-AsA. The results thus suggested important roles for urinary Na ion concentration and pH in modulation of urinary bladder carcinogenesis. AsA was further found to act as a copromoter (an amplified) under conditions of increased urinary pH and Na ion concentration (47). The promoting activity of $NaHCO_3$ was confirmed by Lina and Woutersen (48). Further studies demonstrated that administration of sodium hippurate and 1% NaCl was not associated with any promotion of urinary bladder carcinogenesis; increased urinary Na ion concentration in these cases was not accompanied by an elevation of pH (49). It is also of interest that treatment with cetazolamide, which is a potent carbonic anhydrous inhibitor, did not enhance rat bladder carcinogenesis, although it did bring

Table 3 Relationship Between Promoting Activity of Urinary Bladder Carcinogenesis and Changes in Urinary Parameters

		Urine		
Test chemicals	Promoting activity in BQN-initiated rats	pH	Na^+	Total ascorbic acid concentration
AsA	−	↓	−	↑
$NaHCO_3$	+	↑	↑	−
AsA + $NaHCO_3$	+ + +	↑	↑	↑
Na-AsA	+ + or + + +	↑	↑	↑
NH_4CL	−	↓	↓	−
Na-AsA + NH_4CL	±	−	↑	↑

+ + +, Very strong; + +, moderate; +, slight; ±, very slight; −, no change; ↑, increase; ↓, decrease; −, no change; AsA, L-ascorbic acid; Na-AsA, sodium L-ascorbate.

about increased urinary pH. In this case increased Na ion concentration was absent. Therefore, combined elevation of both urinary pH and Na ion concentration appears to be important. Several other sodium compounds were found to share promoting activity (49), thus providing overwhelming evidence that the Na ion is of importance in the promotion of urinary bladder carcinogenesis.

Examination of ions other than Na revealed that dietary administration of AsA plus K_2CO_3 clearly promoted the development of bladder carcinomas while inducing the expected changes in urinary parameters: elevation of pH, increased K ion concentration, and increase of AsA (50). K_2CO_3 alone also exerted weak promoting activity. In this case increases of urinary pH and K ion concentration were observed without change in AsA. Other researchers (48) have recently stressed that the K ion is as potent as the Na ion in promoting urinary bladder cancer under conditions of elevated urinary pH. Thus, elevated K ion concentration together with elevated pH are also associated with promoting activity, especially when acting in concert with the co-promoter AsA. Although Ca ion or Mg ion concentrations were elevated in groups given $CaCO_3$ or $MgCO_3$ and an increase in pH was found for $CaCO_3$, these treatments did not exert any promoting activity. To summarize, the monovalent cations Na and K show promotional activity; the divalent cations Ca and Mg do not.

Case-control epidemiological studies on sodium saccharin have provided little evidence for increased risk of urinary bladder cancer from sodium saccharin. Cohen and Ellwein (51) reported little effect at even the highest level of human consumption.

B. Urolithiasis

The coexistence of urinary bladder carcinoma and calculi in rats and mice has been observed in many studies. For example, two rat strains, the Brown Norway and DA/Han, have high incidences of spontaneous urinary bladder tumors that are often associated with the presence of calculi. Insertion of foreign bodies into the urinary bladder can also result in the development of urinary bladder tumors (52). These findings indicate that tumor formation may be caused directly by stimulation of proliferation by calculi or foreign body–associated trauma.

Rats given a diet containing 3% uracil suffer urolithiasis accompanied by severe and extensive but reversible papillomatosis in the urinary bladder (53). Interestingly, the calculi induced by 3% uracil feeding exerted a strong promoting activity on urinary bladder carcinogenesis in rats initiated with BBN or MNU (54). Diphenyl administration is similarly associated with urinary crystals (consisting of the metabolite *p*-phenylphenol) and also enhances BBN bladder carcinogenesis in rats (49). The experimental results with uracil and diphenyl strongly suggest that tumor promotion is in some way caused by the continued increased urinary bladder cell proliferation associated with the irritant effects of calculi or severe crystal build-up. In this context it is interesting to note that treatment with uracil or diphenyl does not influence the urinary pH or Na ion concentration. Long-term chronic stimulation by calculi can result in development of urinary bladder carcinomas in rats and mice (41). For example, administration of 3% uracil to F344 rats for 104 weeks (41) induced tumors, and particularly transitional cell carcinomas, in the urinary bladders of 90% of the males and 19% of the females (Table 4). Squamous cell carcinomas also developed in 10% of the males, but not in females. Calculi were present in the urinary bladder of almost all males, but in only 30% of the females. Thus, tumor induction was directly related to the presence of calculi. This is in agreement with the epidemiological finding by Kantor et al. (40) that urinary bladder stones increase the risk of bladder cancer. In the induction of calculi there are biological threshold levels for chemicals. For example, feeding a diet containing 1% uracil caused neither urinary calculi nor urinary bladder carcinomas in rats (55). Therefore, the threshold levels for formation of calculi are of both biological and regulatory importance. Human risk assessment should consider the dose-response relationship for such threshold-type events.

C. Antioxidants

Antioxidants, such as butylated hydroxyanisole (BHA), butylated hydroxytoluene (BHT), α-tocopherol, and propyl gallate, have been widely used as

Table 4 Induction of Urinary Bladder Carcinomas in Rats Treated with 3% Uracil in the Diet for 104 Weeks

Sex	Uracil	No. of rats	No. with carcinoma (%) TCC	No. with carcinoma (%) SCC	No. with urinary calculi (%)
Male	+	30	27 (90)[a]	3 (10)	29 (97)[a]
	−	30	0	0	0
Female	+	27	5 (19)[b]	0	8 (30)[b]
	−	30	0	0	1 (3)

TCC, Transitional cell carcinoma; SCC, squamous cell carcinoma.
[a,b] $p < 0.01$, $p < 0.05$ (significantly different from the control by Fisher's exact probability test).

additives in various processed foods. Antioxidant use has generally been thought to be without human hazard. Antioxidants have actually been demonstrated to have anticarcinogenic activities when given before and/or together with carcinogens (56). However, several recent reports have demonstrated the enhancement of tumor formation in animals by antioxidants.

In our two-stage urinary bladder carcinogenesis model, various kinds of antioxidants were examined in rats after initiation with BBN for 4 weeks. BHA and BHT exerted strong, and ethoxyquin and tertiary butylhydroquinone (TBHQ) weak promoting activities (57,58). However, α-tocopherol and propyl gallate administration did not result in any enhancement of urinary bladder carcinogenesis (58). No specific relationship between urinary components and the promoting activity of BHA, BHT, ethoxyquin, or TBHQ could be established. The potency of antioxidant action does not correlate with the demonstrated enhancing effects on carcinogenesis. Recently it was suggested that toxicity of glutathione conjugates of TBHQ to the urinary bladder may contribute to the promoting activity of BHA and TBHQ (59).

D. Anticancer Agents

Intravesical instillation therapy with anticancer agents is widely accepted as effective for urinary bladder cancer patients. However, some data have introduced controversy into this question. For example, intravesical instillation of adriamycin and mitomycin C was found to be associated with enhancement of urinary bladder carcinogenesis in female rats initiated with BBN (60). Thus, when long follow-up periods are taken into account, patients who receive intravesical chemotherapy may suffer later from more frequent recurrence of urinary bladder cancer than those not receiving this therapy.

E. Amino Acids and Others

Cohen et al. (32) found that DL-tryptophan demonstrated promoting activity for bladder carcinogenesis in male rats initiated with FANFT, although the effects were less marked than with sodium saccharin. However, L-tryptophan, which is the biologically active amino acid, did not exert any significant promoting activity. Allopurinol can also promote the induction of bladder carcinomas by FANFT treatment (61). Allopurinol is widely used in the treatment of gout and is known to inhibit tryptophan oxygenase and thus may alter the pattern of urinary tryptophan metabolites. Further studies, however, are required to confirm this result, because allopurinol did not promote carcinogenesis after BBN initiation (49).

Other amino acids found to exert promoting activities on urinary bladder carcinogenesis of rats are L-isoleucine and L-leucine (62). This suggests a possible relation between the high incidence of urinary bladder cancers in western, developed countries where the diet is rich in protein.

Oyasu et al. (63) developed a heterotopically transplanted rat bladder (HTB) model to investigate the role of urinary factors in urinary bladder carcinogenesis. Exposure of the HTB to test fluids first revealed that normal rat urine exerts promoting activity after initiation with BBN or MNU (64). Urine fractions designated as fraction I (MW = 37,000) and II (MW = 4300). Fraction I strongly promoted the development of tumors. It is now thought that the main constituents of fractions I and II are epidermal growth factor and a related molecule (65).

F. Cell Proliferation and Promoters

Tumor promoters are themselves all able to induce epithelial cell proliferation in the urinary bladder without prior initiation (49), resulting in epithelial hyperplasia. Elevation of the activity of rat urinary bladder epithelial ornithine decarboxylase and spermidine/spermine *N′*-acetyltransferase, enzymes related to polyamine metabolism is caused by treatment with promoters (65,66). Increased levels of prostaglandins, especially prostaglandin E_2, have also been found in urinary bladder epithelium following application of tumor promoters (49) to rats.

Our study (46) showed that administration of a high dose of Na-AsA to rats increased AsA concentrations in bladder epithelial cells; no such result was evident with AsA administration itself. Since BHA treatment similarly increased the AsA content of bladder tissue, this experimental result may have significance for the promotion process, but its exact significance remains to be elucidated.

IV. GENETIC FACTORS

A. Familial Occurrence of Urinary Bladder Cancers

Fraumeni and Thomas (67) first reported familial occurrence of urinary bladder cancer in 1967, and subsequently a number of cases involving different family members have been described (68). However, such familial aggregation of urinary bladder cancers is rare.

B. Phenotypic Variation in *N*-Acetyltransferase Activity

N-Acetylation has been demonstrated to be involved in deactivation of some arylamines. Human populations show a genetically based polymorphism and phenotypic variation in the activity of *N*-acetyltransferase. These differences are of significance to urinary bladder cancers, particularly in workers exposed to aromatic amines. Individuals have one of two phenotypes and are either slow and fast acetylators. The slow acetylator has a lower detoxication capacity than the fast acetylator and is at an increased risk of urinary bladder cancers (69). Indeed, quantification of molecular adducts due to aromatic amines might be useful to test for genetic susceptibility to this type of carcinogenesis.

C. Sex, Strain, and Species Differences in Urinary Tract Carcinogenesis in Rodents

Sex, strain, and species differences in response to urinary bladder carcinogenesis are well known. Male rats and mice are more susceptible to cancer induction by BBN (70,71). The male-to-female ratio is 3–4 : 1 for tumors of the human urinary bladder and renal pelvis. Hormonal factors may thus play an important role. Examination of the influence of sex on renal pelvic carcinogenesis by BBN in NON/Shi mice (71) revealed a shortening of the cancer induction time in males, which appeared to influence the progression of renal pelvic carcinomas.

There have been many reports concerning strain differences with regard to urinary bladder carcinogenesis in rodents. In a study of BBN carcinogenesis in rats, analbuminemic rats were found to be most sensitive, followed in decreasing order by SD > ACI > Wistar > F344 (72). Interestingly the relative promoting effects of Na-AsA on two-stage urinary bladder carcinogenesis of rats were F344 > Lewis > Wistar/Shi, although Na-AsA treatment induced essentially the same increases of urinary pH and Na^+ concentration in all cases (73). It has been clearly shown that BALB/c mice are more sensitive to 2-acetylaminofluorene urinary bladder carcinogenesis than C57BL/c mice (74), with males developing more tumors than females.

Rats, mice, hamsters, and guinea pigs have been tested for different susceptibility to BBN, EHBN, and *N*-butyl-*N*-3(-carboxypropyl)nitrosamine. Rats proved to the most sensitive to all three compounds, and especially to EHBN (75). Mice were less so, but demonstrated more induction of invasive cancers. Hamsters were far less susceptible, and guinea pigs were the most resistant of the four species.

D. Influence of Aging on Urinary Tract Carcinogenesis

Spontaneous tumors appear mainly in older experimental animals and neoplastic lesions generally show age-associated development, this being a reflection of the long-term nature of the disease. However, there have also been many reports of an independent influence of animal age on the induction of neoplastic lesions by chemical carcinogens. Much evidence has indicated that young animals are more sensitive than old ones (76), although in some cases an age-related increase in susceptibility was reported (77). We found an age-related increase in the development of urinary bladder carcinomas in rats of both sexes treated with BBN for 20 weeks and established that this was not dependent on intake of carcinogen (Table 5) (78). Since indirect carcinogens such as BBN need metabolic activation to exert their carcinogenicity, age-related metabolic activation might play a role. In general, however, the activities of liver microsomal enzymes show age-related decreases, resulting in reduced metabolic activation of carcinogens. Differences in the susceptibility of young and old animals to BBN can therefore not simply be explained in terms of its metabolism, although this is clearly a complicating factor.

The effects of aging on the multiorgan carcinogenesis induced in rats by MNU, a direct carcinogen which does not need metabolic activation to exert carcinogenicity, have also been examined in male F344 rats at 6, 52, and 98 weeks of age (79). In young rats, malignant lymphomas, particularly thymic types, were observed at a significant incidence. In middle-aged rats a significantly higher incidence of adenocarcinomas in the small intestine than in young or old animals and elevated induction of proliferative and neoplastic lesions in the large intestine was observed. In addition, epithelial

Table 5 Age-Dependent Development of Urinary Bladder Carcinomas in Rats Treated with BBN

	Age (weeks) at				Carcinoma	
Group	Start[a]	Sacrifice	Sex	No. of rats	Incidence (%)	Area[b]/cm^2
1	6	26	Male	20	8 (40)	0.24 ± 0.14
			Female	20	7 (35)	0.20 ± 0.21
2	52	72	Male	19	8 (42)	1.21 ± 0.95
			Female	20	6 (30)	0.51 ± 0.66
3	98	118	Male	10	9 (90)[c]	2.08 ± 1.83[c]
			Female	10	10 (100)[c]	2.42 ± 2.71[c]

[a]F344 rats at 6, 52, or 98 weeks of age were given water containing 0.025% BBN for 20 weeks.
[b]Areas of carcinomas in histological specimens measured with a color video image processor.
[c]Significantly different from groups 1 and 2 at $p < 0.01$.

hyperplasia of the tongue, but not the forestomach, occurred at the highest incidence in middle-aged rats. However, there were no intergroup differences in the induction of epithelial lesions in the urinary bladder. The results demonstrated that while some target organs other than the urinary bladder demonstrate age-dependent variation in their response to MNU, levels of DNA synthesis and O^6-methylguanine DNA adduct formation in most cases cannot explain the observed differences.

Examination of the influence of aging on renal pelvic carcinogenesis induced by BBN in 6-, 15-, and 45-week-old male and female NON/Shi mice suffering from hereditary hydronephrosis revealed an age-related shortening of the time required for renal pelvic carcinomas to appear in males (71).

V. HISTOGENESIS OF EXPERIMENTAL URINARY BLADDER CARCINOMAS

The histopathological lesions observed in the urinary bladder epithelium of rats treated with BBN have been classified into four types: simple hyperplasia, papillary or nodular hyperplasia, papilloma, and carcinoma (80). Simple hyperplasia consists of diffuse or focal thickening of the epithelium with four to eight layers of transitional epithelial cells. In papillary or nodular hyperplasia, the epithelium is six to eight cells thick, and in most cases the changes are strictly localized. Cellular atypia and mitotic figures are only rarely observed in areas of hyperplasia. Areas of hyperplasia demonstrate either exophytic growth, with a delicate fibrovascular core and protrusion into the lumen of the urinary bladder, or endophytic growth. Papillomas are defined as benign epithelial tumors in which the transitional epithelium cells are arranged in branched fingerlike processes surrounding a delicate fibrovascular core. They are generally exophytic but may show an endophytic growth pattern. Cellular irregularity is slight, and few mitotic figures are present. Carcinomas have morphological characteristics of atypia, invade the muscularis, and demonstrate a high degree of mitotic activity.

Papillary or nodular hyperplasia of the urinary bladder in rats treated with 0.05% BBN develops before the induction of papillomas or carcinomas (13,80) and can be induced at high incidence by large doses of BBN within a short period. A dose-response relationship has been observed (80). The period of carcinogen exposure also plays a role. Moreover, a good correlation exists between the degree of development and the numbers of eventual carcinomas. Thus, papillary or nodular hyperplasia is considered a preneoplastic lesion in the rat urinary bladder.

VI. CHARACTERISTICS OF URINARY BLADDER CARCINOGENS

Carcinogens are generally classified into two categories: genotoxic and nongenotoxic. Genotoxic carcinogens are DNA reactive, with or without enzymatic activation in the liver or other organs, and cause genetic alterations and usually also cell proliferation. Nongenotoxic carcinogens lack reliable evidence of direct interaction with DNA, and their exertion of other kinds of biological effects appears to be the basis for their carcinogenicity. Many etiological agents belong to this nongenotoxic category. Common characteristics of nongenotoxic carcinogens are an ability to increase DNA synthesis, mitosis, and cell division, often resulting in a hyperplastic response in the early stage of carcinogenesis (81). Some of them, however, may induce DNA damage through indirect mechanisms, such as producing oxygen radicals (82). In general, high doses of these agents are required to exert carcinogenicity. Our strong promoter, uracil, is a nongenotoxic carcinogen.

Carcinogens are also classified into direct and indirect types, the latter needing enzymatic activation to cause tumor induction. The majority of genotoxic agents are indirect carcinogens, demonstrating strain, species, and sex variation in their activities. Direct carcinogens such as MNU interact with macromolecules in the target cells without metabolic activation.

VII. DEVELOPMENT OF CARCINOMAS IN THE RENAL PELVIS AND URETER

Attempts to induce tumors of the renal pelvis and ureter in rats and mice by carcinogens have met with little success. *N*-Nitroso-*bis*(2-oxopropyl)amine has been shown to have low carcinogenicity toward the renal pelvic epithelium in SD rats (83). MNU may also cause such lesions in Fischer-344 rats (85). However, artificially produced stagnation of urine flow in the renal pelvis has long been known to result in augmented induction of tumors in this site (84). Thus, the SD/cShi rat, which has nearly a 100% incidence of spontaneous bilateral hydronephrosis and hydroureter, shows a high incidence of carcinomas in the renal pelvis and ureter, when given 0.05% BBN for 12 weeks and then maintained without further treatment for 10 weeks (86). Long-term contact of target cells with carcinogen is presumably responsible for the induction of tumors in this case. BBN administration also causes development of renal pelvic carcinomas, which often metastasize to the lung in NON/Shi mice with spontaneous hydronephrosis (87).

VIII. GENETIC ALTERATION

It is generally accepted that tumor development occurs as the result of accumulation of genetic alterations. Putative candidate genes for urinary bladder carcinogenesis include the retinoblastoma (*Rb*) (88) and *p53* (89) tumor-suppressor genes and several oncogenes, H-*ras*, c-*myc* (90), and c-*erb* B-2 (91). Particular attention has been concentrated on the possible participation of *p53* gene alterations in urinary bladder carcinogenesis (89,92,93). It has been reported that *p53* mutations are common in invasive and/or high-grade urinary bladder carcinomas and roles for *p53* have been suggested in differentiation or tumor progression (89,92,93). Since wild-type *p53* serves as a critical regulator of a G1 cell cycle checkpoint following exposure of cells to DNA-damaging agents, defects in *p53* protein function allow cells containing heritable DNA template alterations to enter S phase without adequate check and repair. As a result, *p53* insufficiency might accelerate the accumulation of genetic alterations and has been regarded as one of the most important indicators of likely further progression and poor prognosis.

To elucidate differences and/or similarities in genetic events between human and animal models, we initially evaluated the presence of *p53* and H-*ras* gene alterations using rat and mouse urinary bladder tumors induced by BBN. In the rat lesions, characterized by papillary structures and rare metastasis, *p53* mutations were observed in 6 of 10 (60%) early neoplastic lesions (94). In contrast to rat lesions, mouse urinary bladder carcinomas induced by BBN are typically flat type and rapidly progress to invasion and metastasis. In NON/Shi mice treated with BBN, *p53* mutations were found in 14 of 18 (78%) primary urinary bladder carcinomas without metastasis (95) and 9 of 10 (90%) with metastasis (unpublished data). Significant participation of this tumor-suppressor gene in progression was also indicated by the good correlation with malignant potential observed with various animal urinary bladder carcinomas. The incidence of *p53* alterations in mouse primary carcinomas without metastasis was similar to those for human invasive carcinomas reported previously (89,92,93). In rats *p53* alterations in early neoplastic lesions were higher than the value for corresponding human papillary low-grade tumors.

Generally, the spectrum of mutations within the *p53* gene in human urinary bladder carcinomas has varied without any evident hot-spot. However, Shibata et al. (96) reported a distinct patter in *p53* lesions from the endemic area of blackfoot disease in Taiwan, which might be related to high arsenic levels in artesian well water. Such a phenomenon is indicative of preferential interaction of a carcinogen with certain bases of DNA and may be helpful in predicting the agent responsible for base-pair substitutions. In contrast to Taiwan's studies, no mutational hot spot was observed in any of our BBN studies, although mutations were relatively concentrated within *p53* gene exons 5 and 7.

Some reports concerning human urinary bladder carcinomas have stated that mutational defects in either *p53* allele are accompanied by the loss of remaining allele (89). Presence of loss of heterozygosity (LOH) on the *p53* allele in experimentally induced urinary bladder carcinomas was evaluated using F_1 hybrid mice and a microdissection procedure (unpublished data). Invasive urinary bladder carcinomas were induced after 12 weeks administration of BBN in male (NON/Shi × C3H/HeN/Shi) F_1 offspring within a 21-week total experimental period. Abnormal band-shifts suggesting *53* mutations and loss of either allele were observed in 57% and 29% of carcinomas, respectively. Of four carcinomas with allelic losses, three had *p53* mutations in the remaining allele.

Microsatellite instability (MSI), due to DNA replication errors in repetitive nucleotide sequences, has been reported to occur at a low rate in human urinary bladder carcinomas (97,98). We also evaluated the MSI frequencies for 20 loci in a rat urinary bladder carcinogenesis model using BBN (unpublished data). MSI were present in 4 of 12 (33%) urinary bladder carcinomas, which were also used for *p53* mutational analysis. The significance of this must be considered low, however, since 1 of the 4 cases harbored *p53* mutations.

Masui et al. (99) earlier described a lack of H-, K-, and N-*ras* mutations in rat urinary bladder carcinomas induced by BBN. In mouse lesions, H-*ras* mutations were only found in 2 of 18 (11%) carcinomas without metastasis and 2 of 10 (20%) with metastasis. These incidences are also similar to those reported for human urinary bladder cancer cases (100). *c-erb*B-2/*neu*, another possible gene responsible for urinary bladder carcinogenesis, was not reported to play a significant role in *N*-[4-(5-nitro-2-furyl)-2-thiazolyl]formamide-induction of rat urinary bladder carcinomas and 2-amino-4-(5-nitro-2-furyl)thiazole transformation of rat bladder epithelial cells (101).

IX. CONCLUSIONS

Urinary bladder cancer was the first neoplastic disease proven to be related to industrial chemical exposure, and a large number of environmental chemicals have now been evaluated to be carcinogens or promoters of the urinary bladder carcinogenesis in humans and rodents. In general, carcinogens and promoters exert their activities on target cells in the urinary tract via their presence in the urine. Etiological factors in human, in particular those due to occupational exposure, are mostly genotoxic. The causative agents in the environment, like arylamines, cigarette smoking, and arsenics, are potentially avoidable. Infection with *Schistosoma hematobium* can be eradicated by education of the exposed populations, improvements in social welfare, and drug therapy. Medicines such as cyclophosphamide and phenacetin are associated with a high risk of iatrogenic induction of urinary tract cancers; physicians should pay particular attention to their injurious as well as beneficial effects. A reasonably adequate data base of long-term carcinogenicity study results in rodents is required to assess the carcinogenic potential of new drugs. The present data suggest that coffee and tea drinking are unlikely to pose a risk for urinary bladder cancer induction.

Causative factors such as urinary bladder cancer promoters in rats are nongenotoxic. Quite high doses are needed for activity in urinary bladder carcinogenesis. As nongenotoxicity, a high dose requirement and considerable carcinogenic potency are all characteristics of urinary bladder promoters; further studies are required for the human risk assessment of these chemicals. Recently, it was concluded that sodium saccharin, the artificial sweetener, is unlikely to pose a major cancer risk to humans. Experimental results regarding cell proliferation and carcinogenesis in the urinary bladder show that cell proliferation is a significant factor in the mechanism of tumor induction by nongenotoxic agents.

Genetic factors exist as evidenced by familial aggregation of urinary bladder cancers and polymorphism in the activity of *N*-acetyltransferase in humans. In addition, sex, age, strain, and species differences have been demonstrated in rodents, further suggesting a genetic influence on urinary bladder carcinogenesis. Progress in molecular analysis should contribute to future risk assessment of urinary bladder carcinogenesis.

Multistep carcinogenesis involving multiple accumulating changes at the DNA level has been recently accepted as a general model of carcinogenesis. For the purpose of investigating genetic alterations in carcinogenesis, utilization of animal models features both benefits and disadvantages. For humans the variety of possible carcinogenic factors ensures a different level and interplay of complexity for animal models. The relationship between chemical exposure and host defense mechanisms cannot be assumed to be identical for humans and rodents. However, genetic evidence derived from animal models does add to our understanding of the complexity of carcinogenesis. The availability of animal models means that genetic analysis of the whole spectrum of changes involved in urinary bladder cancer by chemicals is experimentally possible.

REFERENCES

1. Lower GM Jr. Chemically induced human urinary bladder cancer. Cancer 1982; 49:1056–1066.
2. IARC Monographs Suppl 6, Genetic and Related Effects. Vol. 1–42. An Updating of Selected IARC Monographs. 1987: 113–115.
3. Melick WF, Escue HM, Naryka JJ, Mezera RA, Wheeler EP. The first reported cases of human bladder tumors due to a new carcinogen xenylamine. J Urol 1955; 74:760–766.
4. Bioassay of p-cresidine for possible carcinogenicity. Natl Cancer Inst Carcinog Tech Rep Ser 1979; 142:1–63.
5. Stula EF, Barnes JR, Sherman H, Reinhardt CF, Zapp JA Jr. Urinary bladder tumors in dogs from 4-4′-methylene-bis(2-chloroaniline)(MOCA). J Environ Pathol Toxicol 1978; 1:31–50.
6. Burch JD, Rohan TE, Howe GR, Risch HA, Hill GB, Steele R, Miller AB. Risk of bladder cancer by source and type of tobacco exposure. A case-control study. Int J Cancer 1989; 44:622–628.
7. Clavel J, Cordier S, Boccon-Gibod L, Hemon D. Tobacco and bladder cancer in males. Increased risk for inhalers and smokers of black tobacco. Int J Cancer 1989; 44:605–610.
8. Morrison AS, Buring JE, Verhoek WG, Aoki K, Leck I, Ohno Y, Obata K. An international study of smoking and bladder cancer. J Urol 1984; 131:650–654.
9. Patrianakos C, Hoffmann D. Chemical studies of tobacco smoke. LXIV. On the analysis of aromatic amines in cigarette smoke. J Anal Chem 1979; 3: 150–154.
10. Cohen SM, Garland EM, John M St, Okamura T, Smith RA. Acrolein initiates rat urinary bladder carcinogenesis. Cancer Res 1992; 52:3577–3581.
11. Auerbach O, Garfinkel L. Histologic changes in the urinary bladder in relation to cigarette smoking and use of artificial sweeteners. Cancer 1989; 64: 983–987.
12. Druckrey H, Preussman R, Ivankovic S, Schmidt CH, Mennel HD, Stahl KW. Selektive Erzeugung von Blasenkrebs an Ratten durch Dibutyl-und N-Butyl-N-butanol(4)nitrosamine. Z Krebsforsch 1964; 66:280–290.
13. Ito N, Hiasa Y, Tamai A, Okajima E, Kitamura H. Histogenesis of urinary bladder tumors induced by N-butyl-N-(4-hydroxybutyl)nitrosamine in rats. Gann 1969; 60:401–410.
14. Okajima E, Hiramatsu T, Hirao K, Ijuin M, Hirao Y, Babaya K, Ikuma S, Ohara S, Shiomi T, Hijioka T, Ohishi H. Urinary bladder tumors induced by N-butyl-N-(4-hydroxybutyl)-nitrosamine in dogs. Cancer Res 1981; 41:1958–1966.
15. Hashimoto Y, Iiyoshi M, Okada M. Rapid and selective induction of urinary bladder cancer in rats with N-ethyl-N-(4-hydroxybutyl)nitrosoamine and by its principal urinary metabolite. Gann 1974; 65:565–566.
16. Hicks RM, Wakefield JSTJ. Rapid induction of bladder cancer in rats with

N-methyl-N-nitrosourea I. Histology. Chem Biol Interactions 1972; 5:139-152.
17. Leaf CD, Wishnok JS, Tannenbaum SR. Mechanisms of endogenous nitrosation. Cancer Surv 1989; 8:323-334.
18. El-Merzabani MM, El-Aaser AA, Zakhary NI. A study on the etiological factors of bilharzial bladder cancer in Egypt, 1. Nitrosamines and their precursors in urine. Eur J Cancer 1979; 15:287-291.
19. Takahashi M, Toyoda K, Aze Y, Furuta K, Mitsumori K, Hayashi Y. The rat urinary bladder as a new target of heterocyclic amine carcinogenicity; tumor induction by 3-amino-1-methyl-5H-pyridol(4,3-6)indol acetate. Jpn J Cancer Res 1993; 84:852-858.
20. Hultengren N, Lagergren C, Ljungquist A. Carcinoma of the renal pelvis in renal papillary necrosis. Acta Chir Scand 1968; 130:314-320.
21. IARC Monographs on the evaluation of the carcinogenic risk of chemicals to humans. Overall Evaluations of Carcinogenicity: An Updating of IARC Monographs Vol. 1-24(suppl 7). Lyon, France: IARC,1987.
22. Isaka H, Yoshii H, Otuji A, Koike M, Nagai Y, Koura M, Sugiyasu K, Kanabayashi T. Tumors of Sprague-Dawley rats induced by long-term feeding of phenacetin. Gann 1979; 70:29-36.
23. Johansson SL. Carcinogenicity of analgesics: long-term treatment of Sprague-Dawley rats with phenacetin, phenazone, caffeine and paracetamol (acetamidophen). Int J Cancer 1981; 27:521-529.
24. Nakanishi K, Kurata Y, Oshima M, Fukushima S, Ito N. Carcinogenicity of phenacetin: long-term feeding study in $B6C3F_1$ mice. Int J Cancer 1982; 29: 439-444.
25. Johansson SL, Radio SJ, Saidi J, Sakata T. The effects of acetaminophen, antipyrine and phenacetin on rat urothelial cell proliferation. Carcinogenesis-(Lond.) 1989; 10:105-111.
26. Kunze E. Mohlmann R. Proliferation-stimulating effect of phenacetin on the urothelium and papillary epithelium in rats. Urol Int 1983; 38:223-228.
27. Murai T, Mori S, Machino S, Hosono M, Takeuchi Y, Ohara T, Makino S, Takeda R, Hayashi Y, Iwata H, Yamamoto S, Ito H, Fukushima S. Induction of renal pelvic carcinoma by phenacetin in hydronephrosis-bearing rats of the SD/cShi strain. Cancer Res 1993; 53:4218-4223.
28. Worth PHL. Cyclophosphamide and the bladder. BMJ 1971; 3:182.
29. Schmahl D, Habs MR. Prevention of cyclophosphamide-induced carcinogenesis in the urinary bladder of rats by administration of mensa. Cancer Treat Rev 1983; 10(suppl A):57-61.
30. Hicks RM, Wakefield J, Chowaniel J. Evaluation of a new model to detect bladder carcinogens or co-carcinogens; results obtained with saccharin, cyclamate, and cyclophosphamide. Chem Biol Interact 1975; 11:225-233.
31. Wolf H. Studies on the role of tryptophan metabolites in the genesis of bladder cancer. Acta Clin Scand 1973; 433(suppl):154-168.
32. Cohen SM, Arai M, Jacobs JB, Friedell GH. Promoting effect of saccharin and D,L-tryptophan in urinary bladder carcinogenesis. Cancer Res 1979; 39: 1207-1217.

33. Chen C-J, Chuang Y-C, Lin T-M, et al. Malignant neoplasms among residents of a blackfoot disease endemic area in Taiwan: high-arsenic artesian well water and cancers. Cancer Res 1985; 45:5895–5899.
34. Yamamoto S, Konishi Y, Matsuda T, Murai T, Shibata M-A, Matsui-Yuasa I, Otani S, Kuroda K, Endo G, Fukushima S. Cancer induction by an organic compound, dimethylarsinic acid (cacodylic acid), in F344/DuCrj rats after pretreatment with five carcinogens. Cancer Res 1995; 55:1271–1276.
35. Wanibuchi H, Yamamoto S, Chen H, Yoshida K, Endo G, Hori T, Shoji Fukushima. Promoting effects of dimethylarsinic acid on N-butyl-N-(4-hydroxybutyl)nitrosamine-induced urinary bladder carcinogenesis in rats. Carcinogenesis (in press).
36. Endo G, Kuroda K, Okamoto A, Horiguchi S. Dimethylarsinic acid induces tetraploidy in Chinese hamster cells. Bull Environ Contam Toxicol 1992; 48: 131–137.
37. Brown JL, Kitchin KT. Arsenite, but not cadmium, induces ornithine decarboxylase and heme oxygenase activity in rat liver: relevance to arsenic carcinogenesis. Cancer Lett 1996; 98:227–231.
38. IARC Monographs on the Evaluation of the Carcinogenic Risks of Chemicals to Humans: Coffee, Tea, Mate, Methylxanthines and Methylglyoxal. Vol 51. Lyon, France: IARC, 1991.
39. Nakanishi K, Fukushima S, Shibata M, Shirai T, Ogiso T, Ito N. Effects of phenacetin and caffeine on the urinary bladder of rats treated with N-butyl-N-(4-hydroxybutyl)nitrosamine. Gann 1978; 69:395–400.
40. Kantor AF, Hartge P, Hoover RN, Narayana AS, Sullivan JW, Fraumeini JF Jr. Urinary tract infection and risk of bladder cancer. Am J Epidemiol 1984; 119:510–515.
41. Fukushima S, Tanaka H, Asakawa E, Kagawa M, Yamamoto A, Shirai T. Carcinogenicity of uracil, a nongenotoxic chemical, in rats and mice and its rationale. Cancer Res 1992; 52:1675–1680.
42. IARC Monographs on the Evaluation of Carcinogenic Risk of Chemicals to Humans: Some Non-nutritive Sweetening Agents. Vol. 22. Lyon, France: IARC,1980.
43. West RW, Beranek DT, Kadlubar FF. Dose-dependent effects of dietary saccharin on promotion of urinary bladder carcinogenesis and on urothelial DNA adducts after initiation with N-methylnitrosourea. Adv Bladder Cancer Res 1983; 2:87–93.
44. Cohen SM, Ellwein LB, Okamura T, Masui T, Johansson SL, Smith RA, Wehner JM, Khachab M, Chappel CI, Schoenig GP, Emerson JL, Garland EM. Comparative bladder tumor promoting activity of sodium saccharin, sodium ascorbate related acids, and calcium salts in rats. Cancer Res 1991; 51:1766–1777.
45. Fukushima S, Imaida K, Sakata T, Okamura T, Shibata M, Ito N. Promoting effects of sodium L-ascorbate on two-stage urinary bladder carcinogenesis in rats. Cancer Res 1983; 43:4454–4457.
46. Fukushima S, Shibata M, Shirai T, Tamano S, Ito N. Roles of urinary

sodium ion concentration and pH in promotion by ascorbic acid of urinary bladder carcinogenesis in rats. Cancer Res 1986; 46:1623–1626.
47. Fukushima S, Imaida K, Shibata M-A, Tamano S, Kurata Y, Shirai T. L-Ascorbic acid amplification of second-stage bladder carcinogenesis promotion by $NaHCO_3$. Cancer Res 1988; 48:6317–6320.
48. Lina BAR, Woutersen RA. Effects of urinary potassium and sodium ion concentrations and pH on N-butyl-N-(4-hydroxybutyl)nitrosamine-induced urinary bladder carcinogenesis in rats. Carcinogenesis 1989; 10:1733–1736.
49. Ito N, Fukushima S. Promotion o urinary bladder carcinogenesis in experimental animals. Exp Pathol 1989; 36:1–15.
50. Fukushima S, Shibata M-A, Shirai T, Kurata Y, Tamano S, Imaida K. Promotion by L-ascorbic acid of urinary bladder carcinogenesis in rats under conditions of increased urinary K ion concentration and pH. Cancer Res 1987; 47:4821–4824.
51. Ellwein LB, Cohen SM. The health risks of saccharin revisited. Crit Rev Toxicol 1990; 20:311–324.
52. Toyoshima K, Leighton J. Bladder calculi and urothelial hyperplasia with papillomatosis in the rat following insertion of chalk powder in the bladder cavity with subsequent trauma of the bladder wall. Cancer Res 1975; 35: 3786–3791.
53. Shirai T, Ikawa E, Fukushima S, Masui T, Ito N. Uracil-induced urolithiasis and the development of reversible papillomatosis in the urinary bladder of F344 rats. Cancer Res 1986; 46:2062–2067.
54. Shirai T, Tagawa Y, Fukushima S, Imaida K, Ito N. Strong promoting activity of reversible uracil-induced urolithiasis on urinary bladder carcinogenesis in rats initiated with N-butyl-N-(4-hydroxybutyl)nitorsamine. Cancer Res 1987; 47:6726–6730.
55. Okumura M, Shirai T, Tamano S, Ito M, Yamada S, Fukushima S. Uracil-induced calculi and carcinogenesis in the urinary bladder of rats treated simultaneously with N-butyl-N-(4-hydroxybutyl)nitrosamine. Carcinogenesis (Lond.) 1991; 12:35–41.
56. Wattenberg LW. Inhibition of chemical carcinogenesis. J Natl Cancer Inst 1978; 60:11–18.
57. Imaida K, Fukushima S, Shirai T, Ohtani M, Nakanishi K, Ito N. Promoting activities of butylated hydroxyanisole and butylated hydroxytoluene on 2-stage urinary bladder carcinogenesis and inhibition of γ-glutamyltranspeptidase-positive foci development in the liver of rats. Carcinogenesis 1983; 4: 895–899.
58. Tamano S, Fukushima S, Shirai T, Hirose M, Ito N. Modification by α-tocopherol, propyl gallate and teritary butylhydroquinone of urinary bladder carcinogenesis in Fischer 344 rats pretreated with N-butyl-N-(4-hydroxybutyl)nitrosamine. Cancer Lett 1987; 35:39–46.
59. Peters MMCG, Rivera MI, Jones TW, Monks TJ, Lau SS. Glutathione conjugates of *tert*-butyl-hydroquinone, a metabolite of the urinary tract tu-

mor promoter 3-*tert*-butyl-hydroxyanisole, are toxic to kidney and bladder. Cancer Res 1996; 56:1006–1011.
60. Ohtani M, Fukushima S, Okamura T, Sakata T, Ito N, Koiso K, Niijima T. Effects of intravesical instillation of antitumor chemotherapeutic agents on bladder carcinogenesis in rats treated with N-butyl-N-(4-hydroxybutyl)nitrosamine. Cancer 1984; 54:1525–1529.
61. Wang CY, Hayashida S, Pamukcu AM, Bryan GT. Enhancing effect of allopurinal on the induction of bladder cancer in rats by N-[4-(5-nitro-2-furyl)-2-thiazolyl]formamide. Cancer Res 1976; 36:1551–1555.
62. Nishio Y, Kakizoe T, Ohtani M, Sato S, Sugimura T, Fukushima S. L-Isoleucine and L-leucine: tumor promoters of bladder cancer in rats. Science 1986; 231:843–845.
63. Oyasu R, Iwasaki T, Matsumoto M, Hirao Y, Tabuchi Y. Induction of tumors in heterotopic bladder by topical application of N-methyl-N-nitrosourea and N-butyl-N-(3-carboxypropyl)-nitrosamine. Cancer Res 1978; 38:3019–3025.
64. Oyasu R, Hirao Y, Izumo K. Enhancement by urine of urinary bladder carcinogenesis. Cancer Res 1981; 41:478–481.
65. Babaya K, Izumi K, Ozono S, Miyata Y, Morikawa A, Chmiel JS, Oyasu R. Capability of urinary components to enhance ornithine decarboxylase activity and promote urothelial tumorigenicity. Cancer Res 1983; 43:1774–1782.
66. Matsui-Yuasa I, Otani S, Yano Y, Takada N, Shibata M-A, Fukushima S. Spermidine/spermine N^1-acetyltransferase, a new biochemical marker for epithelial proliferation in rat bladder. Jpn J Cancer Res 1992; 83:1037–1040.
67. Fraumeni JF Jr, Thoman LB. Malignant bladder tumors in a man and his three sons. JAMA 1967; 201:507–509.
68. Schulte PA. The role of genetic factors in bladder cancer. Cancer Detect Prev 1988; 11:379–388.
69. Cartwright RA, Glashan RW, Rogers HJ, Ahmad RA, Barham-Hall D, Higgins B, Kahn MA. Role of N-acetyltransferase phenotypes in bladder carcinogenesis: A pharmacogenetic epidemiological approach to bladder cancer. Lancet 1982; 2:842–845.
70. Bertram JS, Craig AW. Specific induction of bladder cancer in mice by butyl-(4-hydroxybutyl)nitrosamine and the effects of hormonal modifications on the sex difference in response. Eur J Cancer 1972; 8:587–594.
71. Murai T, Mori S, Hosono M, Takeuchi Y, Ohhara T, Makino S, Hayashi Y, Takeda R, Fukushima S. Influences of aging and sex on renal pelvic carcinogenesis by N-butyl-N-(4-hydroxybutyl)nitrosamine in NON/Shi mice. Cancer Lett 1994; 76:147–153.
72. Nakanowatari J, Fukushima S, Imaida K, Ito N, Nagase S. Strain differences in N-butyl-N-(4-hydroxybutyl)nitrosamine bladder carcinogenesis in rats. Jpn J Cancer Res 1988; 79:453–459.
73. Mori S, Kurata Y, Takeuchi Y, Toyama M, Makino S, Fukushima S. Influences of strain and diet on the promoting effects of sodium L-ascorbate in two-stage urinary bladder carcinogenesis in rats. Cancer Res 1987; 47:3492–3495.

74. Littlefield NA, Cueto C, Davis AK, Medlock K. Chronic dose-response studies in mice fed 2-AAF. J Toxicol Environ Health 1975; 1:25–37.
75. Hirose M, Fukushima S, Hananouchi M, Shirai T, Ogiso T, Takahashi M, Ito N. Different susceptibilities of the urinary bladder epithelium of animal species to three nitroso compounds. Gann 1976; 67:175–189.
76. Ebbesen, P. Aging increases susceptibility of mouse skin to DMBA carcinogenesis independent of general immune status. Science 1974; 183:217–218.
77. Stenback F, Peto R, Shubik P. Initiation and promotion at different ages and doses in 2200 mice. III. Linear extrapolation from high doses may underestimate low-dose tumor risks. Br J Cancer 1981; 44:24–34.
78. Fukushima S, Shibata M-A, Tamano S, Ito N, Suzuki E, Okada M. Aging and urinary bladder carcinogenesis induced in rats by N-butyl-N-(4-hydroxybutyl)nitrosamine. J Natl Cancer Inst 1987; 79:263–267.
79. Mizoguchi M, Naito H, Kurata Y, Shibata M-A, Tsuda H, Wild CP, Montesano R, Fukushima S. Influence of aging on multi-organ carcinogenesis in rats induced by N-methyl-N-nitrosourea. Jpn J Cancer Res 1993; 84:139–146.
80. Fukushima S, Murasaki G, Hirose M, Nakanishi K, Hasegawa R, Ito N. Histopathological analysis of preneoplastic changes during N-butyl-N-(4-hydroxybutyl)nitrosamine-induced urinary bladder carcinogenesis in rats. Acta Pathol Jpn 1982; 32:243–250.
81. Cohen SM, Ellwein LB. Genetic errors, cell proliferation, and carcinogenesis. Cancer Res 1991; 51:6493–6505.
82. Floyd RA. The role of 8-hydroxyguanine in carcinogenesis. Carcinogenesis 1990; 11:1447–1450.
83. Ketkar MB, Preussman R, Mohr U. The carcinogenic effect of bis-(2-oxopropyl)nitrosamine on Sprague-Dawley rats. Cancer Lett 1984; 24:73–79.
84. Maekawa A, Mtuoka C, Onodera H, Tanigawa H, Furuta K, Ogiu T, Mitsumori K, Hayashi Y. Organ-specific carcinogenicity of N-methyl-N-nitrosourea in F344 and ACI/N rats. Cancer Res Clin Oncol 1985; 109:178–182.
85. Ito N, Makiura S, Yokota Y, Kamamoto Y, Hiasa Y, Sugihara S. Effect of unilateral ureter ligation on development of tumors in the urinary system of rats treated with N-butyl-N-(4-hydroxybutyl)nitrosamine. Gann 1971; 62: 359–365.
86. Mori S, Hosono M, Machino S, Nakai S, Makino S, Nakao H, Takeuchi Y, Takeda R, Murai T, Fukushima S. Induction of renal pelvic and ureteral carcinomas by N-butyl-N-(4-hydroxybutyl)-nitrosamine in SD/cShi rats with spontaneous hydronephrosis. Toxicol Pathol 1994; 22:373–380.
87. Murai T, Mori S, Hosono M, Takeuchi Y, Ohara T, Makono S, Takeda R, Hayashi Y, Fukushima S. Renal pelvic carcinoma which shows metastatic potential to distant organs, induced by N-butyl-N-(4-hydroxybutyl)nitrosamine in NON/Shi mice. Jpn J Cancer Res 1991; 82:1371–1377.
88. Presti JCJ, Reuter VE, Galan T, Fair WR, Cordon-Cardo C. Molecular genetic alterations in superficial and locally advanced human bladder cancer. Cancer Res 1991; 51:5405–5409.
89. Sidransky D, Von EA, Tsai YC, Jones P, Summerhayes I, Marshall F, Paul

M, Green P, Hamilton SR, Frost P, Vogelstein B. Identification of p53 gene mutations in bladder cancers and urine samples. Science 1991; 252:706–709.

90. Perucca D, Szepetowski P, Simon M-P, Gaudray P. Molecular genetics of human bladder carcinomas. Cancer Genet Cytogenet 1990; 49:143–156.
91. Underwood M, Bartlett J, Reeves J, Gardiner DS, Scott R, Cooke T. C-erbB-2 gene amplification: a molecular marker in recurrent bladder tumors? Cancer Res 1995; 55:2422–2430.
92. Fujimoto K, Yamada Y, Okajima E, Kakizoe T, Sasaki H, Sugimura T, Terada M. Frequent association of p53 gene mutation in invasive bladder cancer. Cancer Res 1992; 52:1393–1398.
93. Spruck III CH, Ohneseit PF, Gonzalez ZM, Esrig D, Miyao N, Tsai YC, Lerner SP, Schmutte C, Yang AS, Cote R, Dubeau L, Nichols PW, Hermann GG, Steven K, Horn T, Skinner DG, Jones PA. Two molecular pathways to transitional cell carcinoma of the bladder. Cancer Res 1994; 54:784–788.
94. Masui T, Don I, Takada N, Ogawa K, Shirai T, Fukushima S. p53 mutations in early neoplastic lesions of the urinary bladder in rats treated with N-butyl-N-(4-hydroxybutyl)nitrosamine. Carcinogenesis 1994; 15:2379–2381.
95. Yamamoto S, Masui T, Murai T, Mori S, Oohara T, Makino S, Fukushima S, Tatematsu M. Frequent mutations of the p53 gene and infrequent H- and K-ras mutations in urinary bladder carcinomas of NON/Shi mice treated with N-butyl-N-(4-hydroxybutyl)nitrosamine. Carcinogenesis 1995; 16:2363–2368.
96. Shibata A, Ohneseit PF, Tsai YC, Spruck III CH, Nichols PW, Chiang H-S, Lai M-K, Hones PA. Mutational spectrum in the p53 gene in bladder tumors from the endemic area of black foot disease in Taiwan. Carcinogenesis 1994; 15:1085–1087.
97. Gonzalez-Zulueta M, Ruppert JM, Tokino K, Tsai YC, Spruck CH, Miyao N, Nichols PW, Hermann GG, Horn T, Steven K, Summerhayes IC, Sidransky D, Jones PA. Microsatellite instability in bladder cancer. Cancer Res 1993; 53:5620–5623.
98. Rosin MP, Cairns P, Epstein JI, Schoenberg MP, Sidransky D. Partial allelotype of carcinoma in situ of the human bladder. Cancer Res 1995; 55:5213–5216.
99. Masui T, Mann AM, Macatee TL, Okamura T, Garland EM, Smith RA, Cohen SM. Absence of ras oncogene activation in rat urinary bladder carcinomas induced by N-methyl-N-nitrosourea or N-butyl-N-(4-hydroxybutyl)-nitrosamine. Carcinogenesis 1992; 13:2281–2285.
100. Bos JL. ras oncogenes in human cancer: a review. Cancer Res 1989; 49:4682–4689.
101. Mann AM, Asamoto M, Masui T, Macatee T, Eklund SH, Cohen SM. *Neu* is not involved in N-[4-(5-nitro-2-furyl)-2-thiazolyl]formamide-induced bladder carcinoma or 2-amino-4-(5-nitro-2-furyl)thiazole transformation of rat bladder epithelial cells. Cancer Lett 1994; 84:7–13.

23
Cancers of the Oral Cavity

Dhananjaya Saranath
Laboratory of Cancer Genes, Cancer Research Institute, Tata Memorial Centre, Mumbai, India

I. INTRODUCTION

Cancer of the oral cavity is one of the 10 most common cancers in the world, with age-adjusted incidence rates in males for oral cancer (ICD-9, ICD-140,141,143–145), ranging from 2.2 per 100,000 males in Japan and several industrialized countries to 15–30 cases per 100,000 in South Asia (1), with more than 500,000 new cases projected worldwide annually (2). The incidence rates of cancers of the oral cavity in India are some of the highest in the world. The malignancy is generally more prevalent in males, and females generally show half the cancer rates as males in industrialized countries, whereas in the high risk areas of southern Asia, female rates are equivalent to the male rates (3). Tobacco is the major carcinogen in cancers of the oral cavity (4). Ten to 40% of oral cancer patients develop second primary tumors, most commonly in the oral cavity itself, or in the respiratory and upper digestive tract, with a significant association between the use of tobacco and incidence of second primary tumors (5,6). Relative survival rates for oral cancer are among the lowest of all major cancers, with an overall relative 5-year survival rate of ~40% reported in the United States, Italy, France, and Switzerland, and these survival rates have not improved appreciatively during the last two decades (6,7).

II. ETIOLOGICAL FACTORS IN ORAL CANCER

A. Tobacco Use

Epidemiological data have unequivocally proven that tobacco usage, whether chewed, dipped, or smoked, is the major source of intraoral carcinogens on a global scale (4,8), with smokeless tobacco, in particular chewing tobacco, responsible for majority of oral cancers observed in southern Asia. Tobacco habits in low-risk areas generally consist of tobacco smoking (e.g., cigarettes, pipe smoking, or snuff dipping) in certain parts of Europe and the United States. Tobacco chewing along with smoking is primarily seen in the high-risk oral cancer regions of southern Asia (9). Oral use of smokeless tobacco (unburnt), particularly in India, takes the form of chewing, sucking, and applying tobacco preparations (called Masheri) to the teeth and gums (10). In recent times, there has been a revival in the use of smokeless tobacco in the United States, with 6% of adult males and 0.6% of females (approximately 5.0 million adults) using smokeless tobacco (11).

B. Alcohol Consumption

Alcohol use is an independent risk factor in oral cancers, as well as a potentiating factor for tobacco-related carcinogenesis (12,13). However, it is unlikely that pure ethyl alcohol would be directly carcinogenic to man, since animal experiments have invariably been negative. It has been suggested that the alcohol-cancer association may be related to nutritional deficiency, or that it may act as a solvent and enhance penetration of carcinogens into target tissues, or that not alcohol per se, but contaminants in alcohol may be carcinogenic. Thus, the mode of action of alcohol in the carcinogenic process is not very clear, and remains to be resolved.

C. Dietary Factors

Dietary factors may also play a part in development of oral cancers. A protective role of dietary carotenoids and an inverse association between the consumption of fruits and vegetables and the incidence of oral cancers has been indicated (14,15,16).

D. Role of Human Papilloma Viruses

Increasing clinical evidence suggests a role for human papillomavirus (HPV) in oral cancers, coupled with experimental evidence indicating immortalization of oral keratinocytes and transformation of epithelial cells by HPV16/18 (17). The presence of HPV has been demonstrated in 28–62% of oral cancer patients (18–20). In Kerala, India, which has one of the worlds highest oral cancer incidence rates, HPV DNA was detected in 74% of oral cancers, with 42% showing HPV-16, HPV-18 detected in 47% patient tissues, HPV-6 in 13%, HPV-11 in 20%, and 41% showing multiple HPV infection (21). Our data indicate presence of HPV-16 in 35% of oral cancer tissues, as well as in premalignant leukoplakias, with other HPVs, including HPV-18, and multiple HPV infection not detected in our 80 patients (D. Saranath et al., unpublished).

E. Sunlight and Other Factors

There is a strong association between lip cancer and exposure to sunlight for fair-skinned individuals (22). Immune deficiency and suppression is often associated with oral cancer and increases with progression of the cancer, often to complete anergy (23). However, this may be more an effect, rather than a cause, of the malignancy.

F. Genetic Factors

It is obvious that not all heavy tobacco chewers and smokers develop cancer, and in fact a small percentage of oral cavity cancers develop in nonusers of tobacco (24). The possibility of genetic factors in oral carcinogenesis, as well as genetic predisposition to oral cancers, is suggested by the sporadic occurrence of the cancer in young adults and in nonusers of tobacco and alcohol. It needs to be stressed, however, that exposure rather than genes, drives risk. Susceptibility becomes unimportant if exposure is not present. Thus, individual susceptibilities as well as exposure to carcinogens has a role in oral carcinogenesis.

III. TOBACCO CARCINOGENS

The effect of tobacco and its derivatives is responsible for ~80% of oral cancers in the human population, and tobacco-induced oral cancer has been demonstrated in animal models. Topical application in hamster, baboon, and monkey induces histological progression up to the stages of hyperplasia and dysplasia akin to tissue changes seen in human oral epithelium (25,26).

Tobacco contains several known carcinogens, including volatile aldehydes, *N*-nitrosamines, lactones, benzo(*a*)pyrene, nickel, cadmium, radioactive polonium-210, and uranium (27). The definitive evidence on the role of tobacco-specific nitrosamines (TSNA) as major carcinogens is supported by epidemiological data.

TSNA are quantitatively the most abundant carcinogens present in tobacco and are considered to be important risk factors for tobacco-related cancers (28). *N*-nitroso compounds in Indian tobacco products are carcinogenic in rats (29). Increased endogenous nitrosation leading to formation of carcinogenic nitrosamines in tobacco chewers may also result in onset of malignancy (29). Of the TSNAs, *N*-nitrosonornicotine (NNN) and 40-(*N*-

nitrosomethylamino)-1-butanone (NNK) are the most potent carcinogens, inducing respiratory tract tumors in 30% of single (1 mg) dose-treated Syrian golden hamsters (30). NNN and NNK applied to the oral cavity of rats induce significant numbers of oral tumors (31).

Nitrosation of areca nut alkaloids, such as arecoline, may also contribute to oral carcinogenicity (32). Although the precise nature of genetic lesions due to areca nut chewing is not known, areca nut extract is genotoxic in vitro, causing single-stranded breaks as well as DNA protein cross-links (32,33).

IV. BIOASSAYS FOR SMOKELESS TOBACCO EFFECTS

Voluminous data exist using animal models to study the mode of action and the precise mechanisms leading to formation of oral cancers. Animal models permit investigation of lifelong effects of tobacco in the oral cavity, as well as possible interacting factors of exogenous or endogenous nature.

Treatment of the oral cavity of rats with smokeless tobacco shows the nicotine effect inducing morphological alterations in the microvasculature of oral mucosa (34). Repeated exposure of oral mucosa of rats to snuff placed in a surgically created canal in the lower lip led to induction of benign and malignant tumors of the oral cavity (35), and the early effects of snuff have been reported in Syrian golden hamsters as well (36). Chemical analyses of the mixed saliva of snuff-treated rats revealed the presence of nicotine and TSNA.

Betel quid with tobacco was shown to be capable of inducing tumors in the oral cavity in mice and hamsters (33). Tobacco extract given by gavage was shown to induce tumors in 50% of Swiss mice (10/20 mice developed tumors) and 21% (6/29) of Wistar rats within 21–25 months (37).

Hamster cheek pouch has been a favored site for study of oral carcinogenesis. Hamster cheek pouch has also been used in the study of inhibition of tumor formation using compounds such as β-carotene, in an initiation-promotion hamster buccal pouch system, with 0.1% dimethylbenzanthracene as initiator and 40% benzoyl peroxide as promoter (38). β-Carotene inhibited both initiation and promotion in the hamster pouch epithelial cells. Overexpression of TGF-α, H-ras, and p53 mutation was observed in the cells with hypertrophy and dysplasia, showing involvement in the early stages of transformation, whereas ErbB-1/EGF-R and K-ras seemed to increase only in the later stages (39).

Tobacco and herpes simplex virus (HSV) synergism in the development of precancerous lesions was demonstrated in mice, producing epithelial dysplasis and other premalignant histomorphological changes in mice (40). The role of viruses in oral carcinogenesis was confirmed by development of oral cancer in rats (41) and microinvasive squamous cell carcinoma in hamster buccal pouch tissue (42), using both tobacco and HSV. HSV is not a carcinogen per se, but enhances the oncogenicity of tobacco and related carcinogens in the oral cavity of animals.

Thus, epidemiological, experimental, and clinical investigations demonstrate that the major etiological factor in oral cancer is tobacco in the form of chewing, smoking, application to the teeth or gum, or snuff-dipping. Alcohol acts as an independent risk factor, sometimes synergistically with tobacco, in the etiopathogenesis of oral cancers. A different clinical and molecular picture is observed in this malignancy, dependent on the different geographic regions associated with the lifestyle and socioeconomic status of the populations. The possibility and involvement of distinct mutagenic and carcinogenic processes in each of the different populations is indicated.

Based on comparisons between the highest and lowest observed oral cancer incidences in the world, it is estimated that substantial reductions in these incidences can be achieved by the elimination of tobacco (43). Thus, oral cancer is a self-induced malignancy and is preventable. It is imperative to understand the pathological processes involved.

V. BIOLOGY OF ORAL CANCER

The development of oral cancer has been suggested to occur as a result of field cancerization, the exposure of an entire field of tissue to repeated carcinogenic insult, which predisposes the field to the development of multiple cancers (44,45). Field cancerization in the oral cavity is a feasible proposition, due to the exposure to tobacco-specific carcinogenic agents throughout the oral cavity and perhaps the upper aerodigestive tract.

The progression of epithelial carcinoma occurs in successive stages from normal, hyperplastic, dysplastic, carcinoma in situ to invasive carcinoma. The majority of oral cancers are squamous cell carcinomas that range from poorly differentiated to well-differentiated tumors. The TNM (tumor, node, metastasis) staging system integrates clinically available information (46). Generally, stages T1–T3 indicate the increasing size of the primary lesion, and T4 indicates the involvement of adjacent structures; lymph node status N1–N3 indicates progressively increasing lymph node involvement; and M1 indicates distant metastasis.

VI. PUTATIVE PREMALIGNANT LESIONS IN ORAL CANCER

Oral cancer is often preceded by or associated with putative precancerous lesions, such as leukoplakia, erythroplakia, and submucous fibrosis (SMF), in the high-risk south Asian regions (47). Leukoplakia is a raised white patch of oral mucosa measuring 5 mm or more, which cannot be scraped off and cannot be attributed to any other diagnosable disease (8). The rate of transformation of leukoplakia into invasive cancer is often related to the degree of histological abnormality. Prevalence of leukoplakia in Indians is high, with similar intraoral location as that of oral cancer. Leukoplakia is strongly associated with tobacco chewing and smoking, with almost all leukoplakias in India occurring only in tobacco users (48). In an Indian subpopulation group of 225 leukoplakias studied over a 10-year period, cancer developed in 4% cases, 47% remained stationary, 42% regressed, and 7% recurred (49). Further, nodular leukoplakias showed the highest rate of malignant transformation of 16% per year; while transformation rates in SMF was 1.6%, in lichen planus the rate was 0.08%, and in tobacco users with no oral lesions the transformation rate was as low as 0.005% per year (49). Studies from other countries have found transformation rates from leukoplakia to cancer of 11–14% (50). In an American population with a mean follow-up of 7 years, the long-term transformation rate for dysplastic lesions was 36% (51). Oral SMF is a chronic oral mucosal condition marked by rigidity of the mucosa of varying intensity, predominantly observed in Asian communities. Areca nut chewing in any form is the primary etiological agent for this condition (52,53). Unlike other precancerous lesions, SMF does not regress, either spontaneously or with cessation of areca nut chewing.

VII. MOLECULAR GENETICS OF ORAL CANCER

The etiological role of tobacco is well accepted in oral cancers. Nevertheless, the incidence of oral cancers is relatively rare with respect to the large number of individuals exposed to the carcinogens, suggesting that genetic host factors must affect individual disease susceptibility. Knowledge of the putative genes involved in oral cancer may provide evidence for understanding the biological diversity in the cancer and lead to identification of accurate diagnostic and prognostic factors, as well as indicate therapeutic implications in patient management. Head and neck tumors including oral cancers arise following 6–10 independent genetic events (54–56). An important component of the multistep oral carcinogenesis process may be the accumulation of genetic damage reflecting the degree of carcinogen exposure, the inherent sensitivity of the individual, and the degree of tissue damage. Those cells with the right hits at the premalignant stages may be more likely to progress towards malignancy. At the genetic level, oncogenes and tumor-suppressor genes are the two interacting classes of genes forming the two sides of a coin, which play a critical role in oral cancer pathobiology (55,56).

VIII. ACTIVATION OF ONCOGENES IN ORAL CANCER

Oncogenes, present as proto-oncogenes in normal cells, are highly conserved in nature and are involved in the normal biochemical circuitry of cells in the various regulatory pathways controlling cellular proliferation, differentiation, and apoptosis (57). The proto-oncogenes are activated by various mechanisms such as amplification, point mutation, and gene rearrangement, leading to consequent preneoplastic and neoplastic alterations in the cells (58). Activation of oncogenes often results in increased transcription and protein translation, with gain of function activity.

A. Oncogene Amplification in Oral Cancers

Amplification of c-myc, N-myc, K-ras, N-ras, EGF-R, int-2, and bcl-1 has been demonstrated in primary oral tumor tissues from 7 to 52% of the cancer patients in various studies as reviewed by Saranath and coworkers (56,59–61). H-ras, TGF-α, c-mos, c-erbB-2 and c-erbA-2 were not amplified in oral cancer tissues (62). A higher percentage of Indian patients demonstrated oncogene amplification as compared to patient groups from the United States, United Kingdom, and Australia (55,56). Multiple oncogene amplification was a consistent feature in the Indian patients, implying a complex, multihit simultaneous or sequential insult to the genes. The amplified oncogenes are generally overexpressed with an increase in the mRNA transcript, as well as oncoprotein expression (63–67). Enhanced expression of ras oncoprotein p21 and EGFR was observed in 50–70% of oral carcinomas and leukoplakias, confined to basal and basaloid cells, suggesting their association with cellular proliferation during tumor progression from normal cells to malignant phenotype (66,67).

Amplification and/or overexpression of cyclin D1 (on chromosome 11q13) appears in oral cancers, recurrent tumors, as well as in some preinvasive lesions (68). Gene amplifications have been associated with invasive tumors and correlated with recurrence in squamous cell carcinomas of head and neck cancers, poor prognosis, and shortened survival of patients (68).

B. Point Mutation of Oncogenes in Oral Cancer

Point mutation of the H-ras oncogene primarily at codon 12 (substituting glycine for valine) and codon 61 (resulting in glutamine to arginine/histidine substitution) was reported in 35% of Indian oral cancer patients with long-term tobacco chewing habits (69). In contrast, low incidence of ras oncogene activation by point mutation has been reported from the western patient population (70–72).

C. Oncogene Activation by Gene Rearrangement

Rearrangement of oncogenes may involve alterations detectable cytogenetically at the chromosomal level or may be detected at the genome level using molecular biology techniques. Restriction fragment length polymorphism of L-myc and H-ras oncogenes has been examined in oral cancers, with the idea of associating a specific oncogene allele with increased susceptibility to the malignancy or tumor-associated aberrations in the alleles (65,73,74). The L-myc EcoR1 L- and S-alleles showed equidistribution of the two alleles in Indian oral cancer patients and normal healthy individuals, indicating no predisposition to oral cancers by presence of either of the alleles (73). A similar distribution of H-ras BamH1 alleles was also reported in the Indian oral cancer patients, indicating no association of specific H-ras allele with predisposition to oral cancer (75).

D. A Potent, Predominant Oncogene in Oral Cancers

A potent, predominant transforming oncogene was detected in 12 of 14 primary tumor tissues of chewing-tobacco induced oral cancer patients in a NIH-3T3 cotransfection assay, followed by foci formation, soft agar cloning assay, and nude mice tumor induction (76). The gene was cloned in EMBL-3 bacteriophage after preparing a genomic library. Isolating and fine restriction mapping of the cloned oncogene has been reported by the authors. Further, Southern hybridization of the restricted fragments did not show homology to the oral cancer–associated oncogenes c-myc, N-myc,

K-ras, N-ras, H-ras, EGF-R, and L-myc. Nucleotide sequencing for identification of the cloned gene is in progress.

IX. TUMOR-SUPPRESSOR GENES IN ORAL CANCERS

Allelic losses representing chromosomal deletions signal inactivation of critical tumor-suppressor genes contained within the deleted region. Tumor-suppressor genes contribute to tumorigenesis by loss of functional activity, and hence generally both the alleles are inactivated (77,78). In general this involves a two-step inactivation of both tumor-suppressor gene copies, with point mutation of one of the alleles and loss of the other allele resulting in total inactivation of the gene.

An allelotype of oral cancer using microsatellite markers has recently been compiled (79,80). In practice, allelic losses can be determined by using highly polymorphic markers able to distinguish the two alleles in normal DNA. This pattern can then be compared to tumor DNA, where a recombination or deletion is represented as a loss of the allele. The advent of microsatellite markers, which are small repetitive units, highly polymorphic and occurring throughout the human genome, has revolutionized allelic mapping studies.

Consensus deletion regions have been identified on chromosomes 9p, 3p, 5q, and 19, with frequently occurring overrepresentations on chromosomes 3q and 5p (79–84). The most commonly deleted region, occurring in two thirds of all head and neck tumors, is on the chromosome 9p21-22 region (81). The allelic losses at this region occur in early lesions such as dysplasias and carcinoma in situ, suggesting that chromosome 9p21 losses occur early in the progression of this neoplasm. Contained within this region is a cell cycle gene called p16 (MTS-1 or CDKN-2). This critical inhibitor of cyclin-CDK complexes appears as an excellent candidate for an oral cancer tumor-suppressor gene.

Chromosome 3p region deletion was reported in 17 of 21 oral cancer patients, indicating that this may be a fundamental event in oral carcinogenesis (84). However, the chromosome 3p region is complex because of overlap of three distinct tumor-suppressor regions juxtaposed to one another. Chromosome 3p loss also occurs early, but subsequent to inactivation of the 9p region.

Another important tumor-suppressor gene involved in oral cancers is the p53 gene, on chromosome 17p (55,56). The normal tumor suppressor function of the p53 gene is inactivated by several mechanisms. These include (a) loss of an allele along with point mutation of the second allele, (b) overexpression of dominant inhibitors of p53 such as mdm-2 cellular protein, or (c) inactivation by viral proteins such as HPV E6 or SV40 large T antigen (77,78). Approximately 55% of head and neck tumors show loss of chromosome 17p (79,80), wherein the p53 gene is located, and 50% contain p53 mutations in preinvasive lesions as well as invasive lesions (85–87). Loss of heterozygosity of p53 alleles has been reported in oral cancers and premalignant tissues by polymeras chain reaction (PCR) analysis (88–90). In several cases p53 mutations reflect insult to the gene as a direct consequence of carcinogen exposure (77,78). Maintenance of identical p53 mutations throughout progression of squamous cell carcinoma of the tongue indicates the importance of the gene in oral cancers (91).

Identification of p53 mutations in surgical margins and lymph nodes from patients after surgical resection is also feasible using the PCR technique, which is able to identify one cancer cell in 10,000 normal cells. Many of the margins and lymph nodes that appear normal by light microscopy were found to contain infiltrating tumor cells by molecular analysis. After 2 years of follow-up of patients that were negative by p53 molecular analysis, a significant increase in patient disease-free survival was reported (92).

p53 mutations were much higher in patients exposed to tobacco and/or alcohol than among those with no tobacco or alcohol habits. On the other hand, mutations in repeat sequences such as CpG sequences were rare among patients who smoked cigarettes, while they constituted all of the mutations found in nonsmokers and nondrinkers. The CpG site mutations are important because through methylation and deamination, these CpG sites are thought to be potential sites of endogenous mutations.

Besides p53 point mutations, elevated levels of p53 protein were reported in oral cancers (93–96), which may reflect a response to genetic damage, either as a prelude to cell cycle exit, apoptosis, or underlying genetic instability of transformed cells.

During development of the progression model for oral cancer, Sidransky has reported that losses of chromosomes 9p and 3p uniformly precedes losses of 17p during tumor progression (80). However, precancerous lesions including minimal dysplasias also contain p53 mutation and other genetic changes (85–87), whereas loss of 13q containing the Rb gene region was associated with invasion, with a small number of tumors showing 13q loss. Immunohistochemical staining in these tumors detected the Rb protein (97). This suggests inactivation of the Rb protein via hyperphosphorylation or the involvement of an additional suppressor gene in the vicinity of Rb gene. Inverse correlation of E-cadherin expression with tumor dedifferentiation and lymph node metastasis in squamous cell carcinoma of head and neck has been reported, indicating tumor-suppressor–like function of the gene (98).

Microsatellite alterations and allelic alterations represent clonal markers that can be used for cancer detection (79,80,99). These can be identified in body fluids and may have use for analysis of surgical margins and lymph nodes. The identification of new genes and other molecular markers will help in early detection of oral cancers and may provide useful prognostic information about tumor behavior. The discoveries should lead to improved surgical techniques, chemoprevention strategies, and potentially novel therapeutic approaches.

Phenotypic changes including alterations in differentiation antigens, cytokeratins, blood group antigens, and other glycoproteins generally occur in later stages of carcinogenesis (100,101). Enhanced tenascin expression has been observed in both leukoplakia and squamous cell carcinoma of the oral cavity (102). Immunological alterations appear to affect both cell-mediated and humoral immunity (23,103).

As presented in this section, malignant as well as several nonmalignant oral lesions show multiple genetic changes, suggesting that morphology may not be the best determinant of tumor behavior. Genes may be altered prior to histopathologically visible changes in the tissues as early events in the process of carcinogenesis.

X. BIOMARKERS

Appropriate biomarkers for early detection of oral cancer, identification of putative premalignant lesions, prognostication of the disease pattern, monitoring the efficacy of treatment, and prediction of recurrence of the disease or development of second primary cancer would provide yeoman service to both clinicians and patients. No single biomarker can provide all this desired information. However, the following section briefly mentions the biomarkers now used in oral cancers.

Determination of nicotine and its metabolite in saliva by gas chromatography or radioimmunoassay is useful in determining the tobacco habits of the patients, and hence their exposure to the carcinogens (104). Measurement of globin adducts of NNN and NNK (both TSNA) in the blood of smokeless tobacco users may be a useful indicator of the risk of developing oral cancer (105). Mutagen-induced chromosome fragility is an independent risk factor for oral cancer and correlates with the prospective development of second primary tumors (106).

An elevated bleomycin clastogenicity score may identify individuals with constitutional hypersensitivity towards certain genotoxicants and may show an increased cancer susceptibility to head and neck cancer (107). Micronuclei and carcinogen-DNA adducts as indicators of DNA damage have been used as intermediate endpoints in nutrient intervention trials of precancerous lesions in the oral cavity (108,109). Abnormal DNA content indicating aneuploidy can be studied by flow cytometry, static cytometry, or image cytometry. Flow cytometry is capable of detecting alterations in DNA content, aneuploidy, and abnormalities of epithelial proliferation associated with oral epithelial dysplasia and carcinoma. In oral cancers, aneuploidy indicates aggressive behavior of the cancer, as well as poor prognosis (110,111). Such studies are a useful adjunct to predict the oral leukoplakias with a higher risk to progress to carcinoma. DNA content is a good indicator of prognosis, with a rapid tumor growth associated with poor prognosis (112). Besides flow cytometry studies, DNA content analysis to detect proliferative cells uses immunohistochemical methods including increased in vitro BrdU labeling and increased expression of Ki-67 antigen, EGF-R, transferrin receptor, and proliferating cell nuclear antigen (PCNA) (113).

Further, loss of expression of blood group antigen H on cells at invasive margins of the tumors has been associated with metastases and poor prognosis in oral cancers (113,114). With the advent of molecular biology techniques, several of the molecular biomarkers may prove useful in transferring current information and technology to the bedside of patients.

XI. THERAPY

In the United States, United Kingdom, and several other countries, about 40% of the patients present with highly confined T1 and T2 lesions and 60% show local or regionally advanced lesions (3,6). In India, 70% of the patients present with advanced disease (8). Five-year disease-free survival for the oral cancer patients is in the range of 30–40%. Loco-regional recurrence is the main reason for 50–60% treatment failure, and generally 10% of the patients die of metastases (7,115). Standard therapy of oral cancer is focused on local and regional approaches, such as surgery, radiation therapy, or both.

Radiation therapy for small lesions includes use of an intraoral cone or brachytherapy, defined as the placement of radioactive sources next to a tumor. Radioimmunotherapy using monoclonal antibodies specific to squamous cell carcinoma antigens, and labeled with technitium-99m, is a challenging option for adjuvant therapy of oral cancer (116).

Metastatic or recurrent disease is the single commonly accepted indication for chemotherapy, with the goal being palliation of treatment. Although oral cancer is considered a chemotherapy-resistant disease, certain chemotherapy regimens have shown 50% response rates in previously untreated patients (117–119). Active single chemotherapy agents include methotrexate, cisplatin, carboplatin, infusional fluorouracil, hydroxyurea, doxorubicin, and bleomycin (118). Ifosfamide, edatrexate, and taxol also have been effective (119). Combination chemotherapy has resulted in better response rates, but not in improved survival. Integrating chemotherapy and radiotherapy, neoadjuvant or adjuvant treatments may result in organ preservation. Immunotherapy has been tried in oral cancer, with equivocal results.

The standard therapy for local dysplastic leukoplakia lesions is surgical removal or laser ablation. Photodynamic therapy after sensitization by infusion of Photofrin and illumination by timed and measured pumped-dye laser light of 628 nm selectively destroys dysplastic skin or mucosa without scarring and may be the ideal treatment for early tumors of the lips (120).

Human oral cancers may provide a good model system for gene therapy, with easy accessibility to the tumors. Using the nude mouse as a model, tumors generated by transcutaneous injection of human oral cancer cells into the floor of mouth were treated with replication defective adenovirus containing the HSV-thymidine kinase gene (121). Mice received ganciclovir injection, and three fourths of the treated group showed near total tumor regression, with 50% of the treated group free of tumor at 160 days post–adenovirus injection followed by ganciclovir treatment. All controls died or required sacrifice within 43 days.

Introduction of wild-type p53 in two squamous cell carcinomas of the head and neck region via a recombinant adenoviral vector-Ad5CMV-p53 resulted in the cells producing 10 times more exogenous p53 mRNA and protein, with consequent growth arrest. Cell morphological changes were consistent with apoptosis, and the tumor volumes were significantly reduced in the nude mice. Thus, Ad5CMV-p53 is a potential novel therapeutic agent in squamous cell carcinoma of the head and neck region (122). Use of antisense oligonucleotides-RNA/DNA also has potential in the management of oral cancer (123). Antisense RNA to myc, fos, myb, HSV-1, HTLV-1, and HPV-16 have been used in experimental models. HPV-E6 and E7 antisense oligonucleotides inhibit the growth of HPV-positive oral cancer cells, but not HPV-negative cells. However, therapy with oligonucleotides will require continuous applications of the molecules.

XII. CHEMOPREVENTION

Chemoprevention is defined as the use of specific natural or synthetic agents to reverse, suppress, or prevent carcinogenesis before the development of invasive cancer (124). Chemoprevention is also applicable to the prevention of second primary tumor development in patients after clinical cure of the primary disease. Advances in the understanding of the molecular biology of carcinogenesis have generated the potential development of effective interventions to reduce cancer risk in humans. Elegant observations at the molecular level have revealed specific genetic alterations contributing to oral cancer chemopreventive strategies.

Leukoplakia of the oral cavity can be monitored easily and relatively noninvasively. It serves as an ideal in vivo model for testing the potential of chemopreventive agents that have low levels of toxicity and the potential for long-term administration in patients.

In head and neck cancers, two major systems have been used to screen for active chemopreventive agents—the 7,12-dimethylbenz(*a*)-anthracene-induced hamster buccal pouch carcinogenesis model and, more recently, 4-nitroquinoline 1-oxide–induced oral carcinogenesis in rats (125). Retinoids are the best studied chemopreventive agents, extensively studied in preclinical and clinical chemopreventive settings in the reversal of oral carcinogenesis (124,126–130). High levels of p53 protein accumulation are directly associated with retinoid resistance (131). Retinoic acid receptor RAR-β expression is suppressed in a high percentage of oral premalignant lesions and is significantly upregulated by isotretinoin (93,132). Development of cancer as the endpoint requires a longer time frame and greater resources to conduct chemopreventive trials. Biomarkers may therefore prove useful as intermediate endpoints in chemoprevention studies, provided their ability to predict cancer occurrence can be validated.

A study by Hong et al. reported a placebo-controlled trial of high-dose (2 mg/kg/day) isotretinoin (13 *cis*-retinoic acid or 13cRa) to oral leukoplakia patients (133). A significant remission of leukoplakia was reported in the treated group compared to the placebo group (67% vs. 10%). However, a relapse rate of 50% occurred within 2–3 months after cessation of 13cRA therapy, and second, unexpectedly high toxicity level was noted in this study. The authors followed up this trial with a low-dose, less toxic therapy (124). The study design had two phases: the induction phase and the maintenance phase. During the induction phase, all leukoplakia patients were treated with an intermediate dose of 13cRa of 1.5 mg/kg of body weight/day for 3 months. In the maintenance phase, patients whose lesions responded to the treatment were treated either with low-dose 13cRa (0.5 mg/kg/day) or β-carotene (30 mg/day) for 9 months. In the induction phase, the response was 67%, and in the maintenance phase 92% of the patients receiving 13cRa continued to respond or improved further, whereas patients on β-carotene did not show a significant response. The authors concluded that low-dose isotretinoin has a high therapeutic index for inducing and maintaining a beneficial response and preventing disease progression in patients with oral premalignant lesions.

Chemoprevention provides a new perspective regarding tumor control, and the role of chemoprevention is likely to expand in future. Several other chemopreventive agents have been investigated in oral cancers. Regression of oral leukoplakias with α-tocopherol was shown in a community-based chemoprevention study (134). Naturally occurring and synthetic isothiocyanates are also indicated as potential chemopreventive agents with potent inhibitory effects against tumor development and little or no toxicity (135). Inhibition of nitrosamine tumorigenesis by isothiocyanates is related to inhibition of cytochrome P-450 enzymes, which metabolically activate these nitrosoamines. *N*-Acetylcysteine (NAC) is another chemopreventive agent, due to its antioxidative or detoxifying properties (136).

XIII. CONCLUSIONS AND FUTURE PERSPECTIVES

A conservative estimate of the cost of treatment of the three major tobacco-related diseases—cancers, heart disease, and bronchitis—in India (a high-risk area for oral cancer) is 806 million U.S. dollars annually. This is 208 million U.S. dollars more than the revenue and the foreign exchange provided by tobacco to the government (137)]. Thus, in India the cost of tobacco far outweighs the gains due to tobacco revenue and export. The economics in India alone require the development of preventive strategies for oral cancer. Primary and secondary prevention of oral cancer is a feasible proposition within the existing health care infrastructure of India. The protective role and tumor-promoting role of certain dietary factors cannot be ignored. A balanced diet adequate in protein and nutrient intake, especially fresh vegetables and fruits rich in minerals and vitamins, coupled with avoidance of high intake of fats and meat and strict surveillance of food quality, preservation, and storage practices will help reduce to the lowest possible level the risk of oral cancer in the Indian population (16). Another approach towards reducing the carcinogenic effects of smokeless tobacco involves product modification by the manufacturers. Such product modification could reduce cancer and other health hazard risks.

Coupled with cancer-preventive measures, our growing understanding of the mechanism of oral carcinogenesis could result in improved strategies and reduced oral cancer mortality. There is certainly a need to differentiate potentially malignant lesions that will truly progress to malignancy from the majority that will not with the use of newer molecular and biological markers. Bringing about changes in the lifestyles of populations of countries associated with high cancer risks is a difficult, though not an impossible task. The understanding of the biology of oral cancer, through better technology and information, will certainly help in combating this malignancy.

ACKNOWLEDGMENTS

I gratefully acknowledge Ms. Perin Notani, consultant-cancer epidemiologist, Tata Memorial Hospital (TMH), Bombay, for discussions on oral cancer epidemiology/prevention and a critical assessment of the chapter, and Dr. V. Sanghavi, surgeon, TMH, for evaluation of the oral cancer therapy section of this chapter.

REFERENCES

1. Parkin SM, Laara E, Muir CS. Estimates of the worldwide frequency of sixteen major cancers in 1980. Int J Cancer 1988; 41:84–197.
2. Boring CC, Squires TS, Tong T. Cancer statistics, 1992. Ca-A Cancer J Clin 1992; 42(1):19–38.
3. Johnson NW, Waarnakulasuriya KAAS. Epidemiology and etiology of oral cancer in the United Kingdom. Community Dental Health 1993; 10(suppl 1): 13–29.
4. International Agency for Research on Cancer. Tobacco smoking. IARC Monographs on the evaluation of carcinogenic risk of chemicals to humans. Lyon:IARC, 1986:38.
5. Jovanovic A, van der Tol IG, Kostense PJ, Schulten EA, de Vries N, Snow GB, van der Waal I. Second respiratory and upper digestive tract cancer following oral squamous cell carcinoma. Oral Oncol, Eur J Cancer 1994; 30B(4):225–229.
6. Day GL, Blot WJ, Shore RE, Schoenberg JB, Kohler BA, Greenberg RS, Liff JM, Preston-Martin S, Austin DF, Mclaughlin JK, Fraumeni JF Jr. Second cancers following oral and pharyngeal cancers: patients' characteristics and survival patterns. Oral Oncol, Eur J Cancer 1994; 30B(6):381–386.
7. Boffetta P, Merletti F, Magnani C, Terracini B. A population-based study of prognostic factors in oral and oropharyngeal cancer: Oral Oncol, Eur J Cancer 1994; 30B(6):369–373.
8. Daftary DK, Murti PR, Bhonsle RR, Gupta PC, Mehta FS, Pindborg JJ. Risk factors and risk markers for oral cancer in high incidence areas of the world. In: Johnson NW, ed. Oral Cancer. Vol. 2. Cambridge: University Press, 1991:29–63.
9. Johnson NW. Orofacial neoplasms: global epidemiology, risk factors and recommendations for research. Int Dent J 1991; 41:365–375.
10. Bhonsle RB, Murti PR, Gupta PC. Tobacco habits in India. In: Gupta PC, Hamner JE,III, Murti PR, eds. Control of Tobacco-Related Cancers and Other Diseases. Bombay: Oxford University Press, 1992:25–46.

11. Winn DM. Smokeless tobacco in the USA; usage patterns, health effects, and extent of morbidity and mortality. In: Gupta PC, Hamner JE,III, Murti PR, eds. Control of Tobacco-Related Cancers and Other Diseases. Bombay: Oxford University Press, 1992:65–76.
12. International Agency for Research on Cancer. Alcoholic beverages. IARC Monographs on the evaluation of carcinogenic risk of chemicals to humans. Lyon: IARC, 1988:153–259.
13. Notani PN. Role of alcohol in the cancers of the upper alimentary tract; use of models in risk assessment. J Epidemiol Commun Health 1988; 42:187–192.
14. Peto R, Doll R, Buckley JD, Sporn MB. Can dietary beta-carotene materially reduce human cancer rates? Nature (Lond.) 1981; 290:201–218.
15. McLaughlin JK, Gridley G, Block G, Winn DM, Preston-Martin S, Schoenberg JB, Greenberg RS, Stenhagen A, Austin DF, Ershow AG, Blot WJ, Fraumeni JF,Jr. Dietary factors in oral and pharyngeal cancer. J Natl Cancer Inst 1988; 80(15):1237–1243.
16. Notani PN. Role of diet and alcohol in tobacco-related cancer at sites in the upper aerodigestive tract in an Indian population. In: Gupta PC, Hamner JE,III, Murti PR, eds. Control of Tobacco-Related Cancers and Other Diseases. Bombay: Oxford University Press, 1992:149–155.
17. Syrjanen SM, Syrjanen KJ, Happonen RP. Human papilloma virus (HPV) DNA sequences in oral precancerous lesions and squamous cell carcinomas demonstrated by in situ hybridisation. J Oral Pathol 1988; 17:273–278.
18. Scully C. Oncogenes, tumor suppressors and viruses in oral squamous cell carcinoma. J Oral Pathol Med 1993; 22:337–347.
19. Syrjanen S. Viral infections in oral mucosa. Scand J Dent Res 1992; 100:17–31.
20. Ostwald C, Mueller P, Barten M, Rusatz K, Sonnenburg M, Milde-Langosch K, Loning T. Human papilloma virus DNA in oral squamous cell carcinomas and normal mucosa. J Oral Pathol Med 1994; 23:220–225.
21. Balaram P, Nalinakumari KR, Abraham E, Balan A, Harindran MK, Bernard HV, Chan SY. Human papilloma viruses in 91 oral cancers from Indian betel quid chewers—high prevalence and multiplicity of infections. Int J Cancer 1995; 61:450–454.
22. Lindquist C, Teppo L. Epidemiological evaluation of sunlight as risk factor of lip cancer. Br J Cancer 1978; 37:983–989.
23. Gaze MN, Wilson JA. Head and neck tumor immunology. Clin Otolaryngol 1988; 13:495–499.
24. Ng SKC, Kabat GC, Wynder EL. Oral cavity cancers in non-users of tobacco. J Natl Cancer Inst 1993; 85:743–745.
25. Cohen B, Poswillo DE, Woods DA. The effect of exposure to chewing tobacco on the oral mucosa of monkey and man. Ann Roy Coll Surg Eng 1971; 48:255–273.
26. Hammner JE. Betel quid inducement of epithelial atypia in the buccal mucosa of baboons. Cancer 1972; 30:1001–1005.
27. Hoffman D, Riverson A, Prokopczyk B, Brunnemann KD. Smokeless tobacco and betel quid. In: Gupta PC, Hamner JE,III, Murti PR, eds. Control

of Tobacco-Related Cancers and Other Diseases. Bombay: Oxford University Press, 1992:193–204.

28. Hecht SS, Hoffman D. Tobacco specific nitrosamines, an important group of carcinogens in tobacco and tobacco smoke. Carcinogenesis 1988; 9:875–884.
29. Nair J, Chakradeo P, Bhide SV. Endogenous formation of N-nitroso compounds in tobacco users. In: Bhide SV, Rao KVK, eds. N-Nitroso Compounds: Biology and Chemistry. New Delhi: Omega Scientific Publishers, 1990:27–32.
30. Hecht SS, Adams JD, Numoto S, Hoffmann D. Introduction of respiratory tract tumors in Syrian golden hamsters by a single dose of 4-(methylnitrosamino)-1-(3-pyridyl)-1-butanone(NNK) and the effect of smoke inhalation. Carcinogenesis 1983; 4:1287–1290.
31. Hecht SS, Rivenson A, Braley J, DiBello J, Adams JD, Hoffmann D. Introduction of oral cavity tumors in F344 rats by tobacco-specific nitrosamines and snuff. Cancer Res 1986; 46:4162–4166.
32. Sundqvist K, Liu Y, Erhardt P, Nair J, Bartsch H, Grafstrom RC. Toxicity of areca nut extract related N-nitroso compounds and their precursor alkaloids in cultured human buccal epithelial cells. In: Bhide SV, Rao KVK, eds. N-Nitroso Compounds: Biology and Chemistry. New Delhi: Omega Scientific Publishers, 1990:101–108.
33. International Agency for Research on Cancer. IARC monographs on the evaluation of the carcinogenic risk to humans. Tobacco Habits Other Than Smoking: Betel-quid and Areca Nut Chewing and Some Related Nitrosamine. Lyon: IARC, 1985:37.
34. Johnson GK, Fung YK, Squier CA. Effects of systemic administration of nicotine on capillaries in rat oral mucosa. J Oral Pathol Med 1989; 18:230–232.
35. Hirsch JM, Larsson P-A, Johansson SZ. The reversibility of the snuff induced lesion: an experimental study in the rat. J Oral Pathol 1986; 15:540–543.
36. Shklar G, Niukian K, Hassan M, Herbose EG. Effects of smokeless tobacco and snuff on oral mucosa of experimental animals. J Oral Maxillofac Surg 1985; 43:80–86.
37. Bhide SV, Padma PR, Ammigan N, Amonkar AJ, Nair J. Mutagenicity and carcinogenicity of tobacco extract, nitrosonornicotine and 4-(methyl nitrosamine)-1-(3-pyridyl)-1-Butanone and the suppression of its tumorigenicity by betel-leaf extracts. In: Bhide SV, Rao KVK, eds. N-Nitroso Compounds: Biology and Chemistry. New Delhi: Omega Scientific Publishers, 1990:81–90.
38. Suda D, Schwartz J, Shklar G. Inhibition of experimental oral carcinogenesis by topical beta carotene. Carcinogenesis 1986; 7(5):711–715.
39. Wong DT, Biswas DK. Expression of c-erb B oncogene during dimethylbenzanthracene-induced tumorigenesis in hamster cheek pouch. Oncogene 1988; 2:67–72.
40. Park NH, Niukian K, Shklar G. Combined effect of herpes simplex virus and

tobacco on the histopathologic changes in lips of mice. Oral Surg Oral Med Oral Pathol 1985; 59:154–158.
41. Hirsch JM, Johansson SL, Vahlne A. Effect of snuff and herpes simplex virus 1 on oral rat mucosa: possible associations with the development of squamous cell carcinoma. J Oral Pathol 1983; 12:187–198.
42. Park NH, Sapp JP, Herbosa EG. Oral cancer induced in hamsters with herpes simplex virus infection and simulated snuff dipping. Oral Surg Oral Med Oral Pathol 1986; 62:164–168.
43. Tomatis L, Aitio A, Day NE, et al. Cancer: Causes, Occurrence and Control. IARC Scientific Publ. No. 100. Lyon: International Agency for Cancer, 1990.
44. Farber E. The multistep nature of cancer development. Cancer Res 1984; 44: 4217–4223.
45. Lippman SM, Hong WK. Second malignant tumors in head and neck squamous cell carcinoma: the overshadowing threat for patients with early-stage disease. Int J Radiat Oncol Biol Phys 1989; 17:691–694.
46. Harmer MN, ed. TNM Classification of Malignant Tumors. Geneva: Union Internationale Contre Le Cancer, 1988.
47. Mehta FS, Hamner JE, eds. Tobacco-Related Oral Mucosal Lesions and Conditions. India: TIFR, 1993.
48. Gupta PC, Mehta FS, Daftary DK, et al. Incidence rates of oral cancer and natural history of oral precancerous lesions in a 10-year follow-up study of Indian villagers. Community Dent Oral Epidemiol 1980; 8:287–333.
49. Gutpa PC, Bhonsale RB, Murti PR, Daftary DK, Mehta FS, Pindborg JJ. An epidemiologic assessment of cancer risk in oral precancerous lesions in India with special reference to nodular leukoplakia. Cancer 1989; 63:2247–2252.
50. Lind PO. Malignant transformation in oral leukoplakia. Scand J Dent Res 1987; 95:449–455.
51. Silverman S, Gorsky M, Lozada F. Oral leukoplakia and malignant transformation: a follow up of 257 patients. Cancer (Phila) 1984; 53:563–568.
52. Caniff JP, Harvey W. The etiology of oral submucous fibrosis: the stimulation of collagen synthesis by extracts of areca nut. Int J Oral Surg 1981; 10(suppl 1):163–167.
53. Sinor PN, Gupta PC, Murti PR, Bhonsle RB, Daftary DK, Mehta FS, Pindborg JJ. A case-control study of oral submucous fibrosis with special reference to the etiologic role of areca nut. J Oral Pathol Med 1990; 19:94–98.
54. Cowan JM, Beckett MA, Ahmed-Swan S, Weichselbaum RR. Cytogenetic evidence of the multistep origin of head and neck squamous cell carcinomas. J Natl Cancer Inst 1992; 84:793–797.
55. Field JK. Oncogenes and tumor suppressor genes in squamous cell carcinoma of the head and neck. Oral Oncol Eur J Cancer 1992; 28B:67–76.
56. Saranath D, Bhoite LT, Deo MG. Molecular lesions in human oral cancer: the Indian scene. Oral Oncol Eur J Cancer 1993; 29B:107–112.
57. Bishop JM. Molecular themes in oncogenesis. Cell 1991; 64:235–248.
58. Klein G, Klein E. Evolution of tumors and the impact of molecular oncology. Nature (Lond.) 1985; 315:190–195.

59. Saranath D, Panchal RG, Nair R, Mehta AR, Sanghavi VD, Sumegi J, Klein G, Deo MG. Oncogene amplification in squamous cell carcinoma of the oral cavity. Jpn J Cancer Res 1989; 80:430–437.
60. Saranath D, Panchal RG, Nair R, Mehta AR, Sanghavi VD, Deo MG. Amplification and overexpression of epidermal growth factor receptor gene in human oropharyngeal cancer. Oral Oncol Eur J Cancer 1992; 28B(2):139–143.
61. Deo MG, Saranath D. Oncogenes in tobacco induced oral cancer. In: Smokeless Tobacco or Health, An International Perspective. Monograph 2, NIH Publication, 1992.
62. Leonard JH, Kearsley JH, Chenevix-Trench G, Hayward NK. Analysis of gene amplification in head and neck squamous cell carcinoma. Int J Cancer 1991; 48:511–515.
63. Field JK, Spandidos DA. Expression of oncogenes in human tumors with special reference to the head and neck region. J Oral Pathol 1987; 16:97–107.
64. Azuma M, Furumato N, Kawamata H, Yoshida H, Yanagawa T, Yura Y, Hayashi Y, Takegawa Y, Sato M. The relation of ras oncogene product p21 expression to clinicopathological status criteria and clinical outcome of squamous cell head and neck cancer. Cancer J 1987; 1:375–380.
65. Sheng ZM, Barrois N, Klijanenko J, Micheau C, Richard JM, Riou G. Analysis of the H-ras-1 gene for deletion, mutation, amplification and expression in lymphnode metastasis of human head and neck carcinoma. Br J Cancer 1990; 62:398–404.
66. Kannan S, Balaram P, Pillai MR, Chandran GJ, Nair MK. Immunohistochemical analysis of ras p21 expression in normal, premalignant and malignant oral mucosa. Int J Cancer 1993; 55:700–702.
67. Hoellering J, Shuller CF. Localisation of H-ras mRNA in oral squamous cell carcinoma. J Oral Pathol Med 1989; 18:74–78.
68. Michalides R, van-Veelen N, Hart A, Loftus B, Wientjens E, Balm A. Overexpression of CyclinD1 correlates with recurrence in a group of forty-seven operable squamous cell carcinomas of the head and neck. Cancer Res 1995; 55:975–978.
69. Saranath D, Chang SE, Bhoite LT, Panchal RG, Kerr IB, Mehta AR, Johnson NW, Deo MG. High frequency mutation in codons 12 and 61 of H-ras oncogene in chewing tobacco-related human oral carcinoma in India. Br J Cancer 1991; 63:573–578.
70. Rumsby G, Carter RL, Gusterson BA. Low evidence of ras oncogene activation in human squamous cell carcinoma. Br J Cancer 1990; 61:365–368.
71. Chang SE, Bhatia P, Johnson NW, Morgan PR, McCormick F, Young B, Hiorns L. Ras mutations in United Kingdom examples of oral malignancies are infrequent. Int J Cancer 1991; 48:409–412.
72. Hirano T, Sheele P, Gluckman MD. Low incidence of point mutation at codon 12 of K-ras protooncogene in squamous cell carcinoma of the upper aerodigestive tract. Ann Otol Rhinol Laryngol 1991; 100:597–599.
73. Saranath D, Panchal RG, Nair R, Mehta AR, Sanghavi VD, Deo MG. Re-

striction fragment length polymorphism of the L-myc gene in oral cancer patients. Br J Cancer 1990; 61:530–533.
74. Saranath D, Bhoite LT, Mehta AR, Sanghavi VD, Deo MG. Loss of allelic heterozygosity at the harvey ras locus in human oral carcinomas. J Cancer Res Clin Oncol 1991; 117:484–488.
75. Bhoite LT, Saranath D, Nair R, Deo MG, Sanghavi V, Mehta A. H-ras-1 restriction fragment length polymorphism in normal individuals and oral cancer patients in India. J Oral Pathol Med 1993; 22:298–302.
76. Saranath D, Bhoite LT, Deo MG, Tandle AT, D'Costa J, Kolhapure R, Govardhan MK, Banerjee K. Detection and cloning of potent transforming gene(s) from chewing tobacco-related human oral carcinomas. Oral Oncol Eur J Cancer 1994; 30B:268–277.
77. Marshall CJ. Tumor suppressor genes. Cell 1991; 64:313–326.
78. Levine AJ. Tumor suppressor genes. Annu Rev Biochem 1993; 62:623–651.
79. Ah-See KW, Cooke TG, Pickford IR, Sourtard D, Balmain A. Allelotype of squamous cell carcinoma of the head and neck using microsatellite markers. Cancer Res 1994; 54:1616–1621.
80. Nawroz H, van der Riet P, Hruban RH, Koch W, Sidransky D. Allelotype of head and neck squamous cell carcinoma. Cancer Res 1994; 54:1152–1155.
81. van der Riet P, Nawroz H, Hruban RH, Corio R, Tokino K, Koch W, Sidransky D. Frequent loss of chromosome 9p21-22 early in head and neck cancer progression. Cancer Res 1994; 54:1156–1158.
82. Loughran O, Edington KG, Berry IG, Clark LJ, Parkinson EK. Loss of heterozygosity of chromosome 9p21 is associated with the immortal phenotype of neoplastic human head and neck keratinocytes. Cancer Res 1994; 54: 5045–5049.
83. Speicher MR, Howe C, Crotty P, du Manoir S, Costa J, Ward DC. Comparative genomic hybridisation detects novel deletions and amplifications in head and neck squamous cell carcinomas. Cancer Res 1995; 55:1010–1013.
84. Partridge M, Kiguwa S, Langdon JD. Frequent deletion of chromosome 3p in oral squamous cell carcinoma. Oral Oncol Eur J Cancer 1994; 30B(4):248–251.
85. Boyle JO, Hakim J, Koch W, van der Riet P, Hruban RH, Roa RA, Corio R, Eby YJ, Ruppert JM, Sidransky D. The incidence of p53 mutations increases with progression of head and neck cancer. Cancer Res 1993; 53:4477–4480.
86. Brachman DG, Graves D, Vokes E, Beckett M, Haraf D, Montag A, Dunphy E, Mick R, Yandell D, Weichsebaum RR. Occurrence of p53 gene deletions and human papilloma virus infection in human head and neck cancer. Cancer Res 1992; 52:4832–4836.
87. Chung KY, Mukhopadhyay T, Kim J, Casson A, Ro JY, Goepfert H, Hong WK, Roth JA. Discordant p53 gene mutations in primary head and neck cancers and corresponding second primary cancers of the upper aerodigestive tract. Cancer Res 1993; 53:1676–1683.
88. Tandle AT, Saranth D, Deo MG. p53 tumor suppressor gene in oral cancers

and putative premalignant lesions. In: Proceedings of XVI International Cancer Congress. Italy, 1994:913–917.

89. Saranath D. Clinical implications of the molecular biology of oral cancer. In: Desai PB, ed. Head and Neck Cancer: A Multidisciplinary Approach for Its Control and Cure. Bombay: Tata Press Ltd., 1995:181–192.
90. Largey JS, Meltzer SJ, Yin J, Norris K, Sauk JJ, Archibald DW. Loss of heterozygosity of p53 in oral cancers demonstrated by the polymerase chain reaction. Cancer 1993; 71:1933–1937.
91. Burns JE, McFarlane R, Clark LJ, Mitchell R, Robertson G, Soutar D. Maintenance of identical p53 mutations throughout progression of squamous cell carcinomas of the tongue. Oral Oncol Eur J Cancer 1994; 30B(5):335–337.
92. Brennan JA, Mao L, Hruban RH, Boyle JO, Eby YJ, Koch WM, Goodman SN, Sidransky D. Molecular assessment of histopathologic staging in squamous cell carcinoma of the head and neck. N Engl J Med 1995; 332(7):429–435.
93. Shin D, Kim J, Ro J, Hittelman J. Activation of p53 gene expression in premalignant lesion during head and neck tumorigenesis. Cancer Res 1994; 54:321–326.
94. Waarnakulasuriya KAAS, Johnson NW. Expression of p53 mutant nuclear phosphoprotein in oral carcinoma and potentially malignant oral lesions. J Oral Pathol Med 1992; 21(9):404–408.
95. Kaur J, Shrivastava A, Ralhan R. Overexpression of p53 protein in betel and tobacco-related human oral dysplasia and squamous cell carcinoma in India. Int J Cancer 1994; 58:340–345.
96. Kuttan NAA, Rosin MP, Ambika K, Priddy RW, Bhakthan NMG, Zhang L. High prevalence of expression of p53 oncoprotein in oral carcinomas from India associated with betel and tobacco chewing. Oral Oncol Eur J Cancer 1995; 31B(3):169–173.
97. Yoo GH, Xu HJ, Brennan JA, Westra W, Hruban RH, Koch W, Benedict WF, Sidransky D. Infrequent inactivation of human Rb gene despite frequent loss of chromosome 13q in head and neck squamous cell carcinoma. Cancer Res 1994; 54:4603–4606.
98. Schipper JH, Frixen UH, Behrens J, Unger A, Jahnke K, Birchmeier W. E-Cadherin expression in squamous cell carcinoma of head and neck: inverse correlation with tumor dedifferentiation and lymphnode metastasis. Cancer Res 1991; 51:6328–6337.
99. Mao L, Hruban RH, Boyle JO, Tochman M, Sidransky D. Detection of oncogene mutations in sputum precedes diagnosis of lung cancer. Cancer Res 1994; 54:1634–1637.
100. Vaidya M, Borges AM, Pradhan SA, Rajpal RM, Bhisey AN. Altered keratin expression in buccal squamous cell carcinoma. J Oral Pathol Med 1989; 18: 282–286.
101. Kannan S, Balaram P, Chandran GJ, Pillai MR, Mathew B, Nalinakumari KR, Nair MK. Differential expression of cytokeratin during tumor progression in oral mucosa. Epith Cell Biol 1994; 3:61–69.

102. Shreshta P. Enhanced tenascin immunoreactivity in leukoplakia and squamous cell carcinoma of the oral cavity: an immunohistochemical study. Oral Oncol Eur J Cancer 1994; 30B(2):132–137.
103. Vlock DR. Immunobiologic aspects of head and neck cancer: clinical and laboratory correlates. Hematol Oncol Clin North Am 1991; 5:797–820.
104. Sipahimalani AT, Chadka MS, Bhide SV, Pratap AI, Nair J. Detection of N-nitrosamines in the saliva of habitual chewers of tobacco. Food Chem Toxicol 1984; 22:261–264.
105. Carmella SG, Kagan SS, Kagan M, et al. Mass-spectrometric analysis of tobacco-specific nitrosamine hemoglobin adducts in snuff dippers, smokers, and nonsmokers. Cancer Res 1990; 50:5438–5445.
106. Schantz SP, Spitz MR, Hsu TC. Mutagen sensitivity in head and neck cancer patients: a biologic marker for risk of multiple primary malignancies. J Natl Cancer Inst 1990; 82:1773–1775.
107. Cloos J, Steen I, Joenje H, Ko JY, de Vries N, van der Sterre MLT, Nauta JJP, Snow GB, Braakhuis BJM. Association between bleomycin toxicity and non-constitutional risk factors for head and neck cancer. Cancer Lett 1993; 74:161–165.
108. Prasad MPR, Mukundan MA, Krishnaswamy K. Micronuclei and carcinogen DNA adducts as intermediate end points in nutrient intervention trial of precancerous lesion in the oral cavity. Oral Oncol Eur J Cancer 1995; 31B(3): 155–159.
109. Stich HF, Stich W, Rosin MP. The micronucleus test on exfoliated human cells. Basic Life Sci 1985; 34:337–342.
110. Kearsley JH, Bryson G, Battistutla D, Collins RJ. Prognostic importance of cellular DNA contents in head and neck cancers. A comparison of retrospective and prospective series. Int J Cancer 1991; 47:31–37.
111. Hemmer J, Kridler J. Flowcytomatric DNA ploidy analysis of squamous cell carcinoma of the oral cavity. Comparison with clinical staging and histologic grading. Cancer 1990; 66:317–320.
112. Balsara BR, Borges AM, Pradhan SA, Rajpal RM, Bhisey AN. Flow cytometric DNA analysis of squamous cell carcinoma of the oral cavity: correlation with clinical and histopathological features. Oral Oncol Eur J Cancer 1994; 30B:98.
113. Bryne M. Prognostic value of various molecular and cellular features in oral squamous cell carcinomas—a review. J Oral Pathol Med 1991; 20:413–420.
114. Carey TE, Wolf GT, Hsu S, Poore J, Peterson K, McClatchey KD. Expression of A9 antigen and loss of blood group antigens as determinants of survival in patients with head and neck squamous carcinoma. Otolaryngol Head Neck Surg 1987; 96:221–230.
115. Maraveyas A, Stafford N, Rowlinson-Busza G, Stewart JSW, Epenetos AA. Pharmacokinetics, biodistribution and dosimetry of specific and control radiolabelled monoclonal antibodies in patients with primary head and neck squamous cell carcinoma. Cancer Res 1995; 55:1060–1069.
116. de-Bree R, Roos JC, Quak JJ, den-Hollander W, Snow GB, Van-Dongen GAMS. Radioimmunoscintigraphy and biodistribution of Technitium-99m-

labelled monoclonal antibody 436 in patients with head and neck cancer. Clin Cancer Res 1995; 1:591–598.

117. Merlano M, Vitale V, Rosso R, Benasso M, Corvo R, Cavallan M, Sanguineti G, Bacigalupo A, Badellino F, Margarino G, et al. Treatment of advanced squamous cell carcinoma of the head and neck with alternating chemotherapy and radiotherapy. N Engl J Med 1992; 327(16):1115–1121.
118. Vokes EE. Concomitant chemoradiotherapy as investigational therapy for locoregionally advanced head and neck cancer. Front Radiat Ther Oncol 1992; 26:55–63.
119. Hamasaki VK, Vokes EE. Chemotherapy in head and neck cancer. Curr Opin Oncol 1992; 4/3:504–511.
120. Gregory GF, Hopper C, Fan K, Grant WE, Bown SG, Speight PM. Photodynamic therapy and lip vermilion dysplasia: a pilot study. Oral Oncol Eur J Cancer 1995; 31B(5):346–347.
121. O'Malley BW Jr., Chen SH, Schwartz MR, Woo SLC. Adenovirus mediated gene therapy for human head and neck squamous cell cancer in a nude mouse model. Cancer Res 1995; 55:1080–1085.
122. Liu TJ, Zhang WW, Taylor DL, Roth JA, Goepfert H, Clayman GL. Growth suppression of human head and neck cancer cells by the introduction of a wild type p53 gene via a recombinant adenovirus. Cancer Res 1994; 54: 3662–3667.
123. Shillitoe EJ, Lapeyre JN, Adler-Storthz K. Gene therapy—its potential in the management of oral cancer. Oral Oncol Eur J Cancer 1994; 30B(3):143–154.
124. Lippman SM, Batsakis JG, Toth BB, Weber RS, Lee JJ, Martin JW, Hays JL, Goepfert H, Hong WK. Comparison of low dose isotretinoin with beta carotene to prevent oral carcinogenesis. N Engl J Med 1993; 328:15–20.
125. Tanaka T, Makita H, Ohnishi M, Hirose Y, Wang A, Mori H, Satoh K, Hara A, Ogawa H. Chemoprevention of 4-nitroquinolene 1-oxide-induced oral carcinogenesis by dietary protocatechuic acid during initiation and post initiation phases. Cancer Res 1994; 54:2359–2365.
126. Hong WK, Lippman SM, Hittleman WN, Lotan R. Retinoid chemoprevention of aerodigestive cancer: from basic research to the clinic. Clin Cancer Res 1995; 1:677–686.
127. Stich HF, Rosin MP, Hornby AP. Remission of oral leukoplakia and micronuclei in tobacco/betel nut quid chewers treated with beta carotene and beta carotene plus vitamin A. Int J Cancer 1988; 42:195–199.
128. Han J, Jiao L, Lu Y, Sun Z, Gu QM, Scanlon KJ. Evaluation of N-4 (hydroxycarbophenyl) retinamide as a cancer prevention agent and as a cancer chemotherapeutic agent. In Vivo 1990; 4:153–160.
129. Costa A, Formelli F, Chiesa F, Decensi A, De Palo G, Veronesi U. Prospects of chemoprevention of human cancers with the synthetic retinoid fenretinide. Cancer Res 1994; 54:2032s–2037s.
130. Benner SE, Winn RJ, Lippman SM, Poland J, Hansen KS, Luna MA, Hong WK. Regression of oral leukoplakia with alpha-tocopherol; a community clinical oncology program chemoprevention study. J Natl Cancer Inst 1993; 85:44–47.

131. Lippman SM, Shin DM, Lee JJ, Batsakis JG, Lotan R, Tainsky MA, Hittleman WN, Hong WK. p53 and retinoid chemoprevention of oral carcinogenesis. Cancer Res 1995; 55:16–19.
132. Lotan R, Sozzi G, Ro J, Lee JS, Pastorino U, Pilotti S, Kurie J, Hong WK, Xu XC. Selective suppression of retinoic acid receptor β (RAR-β) expression in squamous metaplasia and in non-small cell lung cancers(NSCLC) compared to normal bronchial epithelium. Proc Am Soc Clin Oncol 1995; 14: 165.
133. Hong WK, Endicott J, Itri LM, Doos W, Batsakis JG, Bell R, Fofonoff S, Byers R, Atkinson EN, Vaughan C, et al. 13-cis-retinoic acid in the treatment of oral leukoplakia. N Engl J Med 1986; 315:1501–1505.
134. Benner SE, Pajak TF, Lippman SM, Earley C, Hong WK. Prevention of second primary tumors with isotretinoin in patients with squamous cell carcinoma of the head and neck: long term follow-up. J Natl Cancer Inst 1994; 86(2):140–141.
135. Lin JM, Amin S, Trushin N, Hecht SS. Effects of isothiocyanates on tumorigenesis by benzo(a)pyrene in murine tumor models. Cancer Lett 1994; 74: 151–159.
136. de Vries N, Pastorino U, Van-Zandwijk N. Chemoprevention of second primary tumors in Europe: Euroscan. Oral Oncol Eur J Cancer 1994; 30B(6): 367–368.
137. Luthra UK, Shrinivas V, Menon G, Prabhakar AK, Chaudhary K. Tobacco control in India: problems and solutions. In: Gupta PC, Hamner JE III, Murti PR, eds. Control of Tobacco-Related Cancers and Other Diseases. Bombay: Oxford University Press, 1992:241–248.

24

Reactive Intermediates and Skin Cancer

Moushumi Lahiri, Hasan Mukhtar, and Rajesh Agarwal*
Case Western Reserve University, Cleveland, Ohio

I. INTRODUCTION

Skin, the largest organ in the body, serves as a protective barrier against the deleterious effects of environmental factors. A plethora of studies have shown that exposure of the skin to environmental agents, both physical such as solar ultraviolet (UV) radiation (1) and chemicals such as polycyclic aromatic hydrocarbons (PAHs), nitrosamines, etc. present in air, automobile exhaust, tobacco smoke, etc. (1,2) contribute significantly to skin-related disorders, specifically nonmelanoma human skin cancer such as basal cell carcinomas (BCCs) and squamous cell carcinomas (SCCs). Epidemiological studies provide clear evidence that there is a steady increase in the incidence of skin cancers worldwide; in the United States alone, an estimated 900,000–1,200,000 new cases of human skin BCCs and SCCs are diagnosed each year (3,4). As the cases of human skin cancers are increasing every year, so do the efforts of a number of laboratories to investigate and understand the causative factors and the cellular, biochemical, and molecular mechanism(s) involved in human skin cancers. Humans are not alone in being susceptible to skin cancer. After exposure to several chemicals, mice also get skin cancer fairly easily; rats however, are quite resistant to developing skin cancer.

There are several general theories of carcinogenesis. Modern somatic mutation theories hypothesize that cells serially acquire more and more transforming or oncogenic events (normally thought of as mutations in the DNA of the cell). However, each individual transforming event of multi-

**Current affiliation*: AMC Cancer Research Center, Denver, Colorado.

stage carcinogenesis has a very low probability (e.g., only about one occurrence per million cell divisions). Therefore, in order to have enough cells at risk to actually progress along a multistep carcinogenesis pathway, substantial time and many generations of cell divisions of initiated cells must occur before multiple, rare genetic mutations can accumulate within an individual initiated cell.

In addition to somatic mutation theories of carcinogenesis, there are receptor-mediated theories of carcinogenesis, cytotoxicity theories of carcinogenesis, and theories that center on cell proliferation or signal transduction. Recently the free radical theory of carcinogenesis has come into greater acceptance. The free radical theory of cancer is somewhat related to the earlier reactive intermediate or electrophile theory of carcinogenesis brought into prominence by the work of the Millers at the University of Wisconsin. Recent data have shown that chemically reactive intermediates are generated inside the skin due to its exposure to UV radiation and/or environmental pollutants and due to treatment of the skin with chemicals in several dermatological conditions.

Oxygen (O_2), a necessary element for the life of a cell, unfortunately is also the source of deleteriously active states, which can disrupt cell structure and alter cell function. Reactive oxygen species (ROS) contribute to the induction and/or development of many human diseases such as heart attack, stroke, inflammatory bowel disease, cancer, emphysema, and the hereditary diseases ataxia telangiectasia, Fanconi's anemia, and Bloom syndrome (5). Though the integrity of all tissues is threatened by ROS, skin is especially at risk to the hazardous effects of oxygen, as it is exposed both from the inside at the level provided by blood and from the outside at the higher levels that occur in the atmosphere. In addition, several light-absorbing compounds are also present in skin, which act as photosensitizers. These photosensitizing compounds can absorb sunlight, become electronically excited, and then transfer their excitation energy to oxygen, resulting in the generation of free oxygen radicals. As mentioned earlier, skin is also continuously exposed to environmental agents such as benzo(*a*)-pyrene (BP), or intentionally exposed to drugs like benzoyl peroxide (BPO), which either are themselves converted to reactive intermediates or generate ROS indirectly, ultimately leading to skin disorders including inflammation and cancer. In this chapter, we have provided a brief overview of the formation of several reactive species and their relevance to skin cancer.

II. REACTIVE INTERMEDIATES AND THEIR FORMATION

A. Reactive Oxygen Species

Oxygen, which constitutes the energetic basis for all aerobic life, allows for much greater generation of useful energy than is possible in anaerobic conditions in living systems. In all aerobic living systems, more than 95% of the total O_2 consumed during cellular respiration is converted to water through a four-electron transfer (6). However, since in its ground state O_2 contains two unpaired electrons, it also has the capacity to undergo univalent and bivalent reduction, leading to the formation of certain intermediates, which possess an either odd or even number of electrons. These intermediates are highly reactive and are collectively referred as ROS. ROS that possess an odd number of electrons are known as free radicals; these include the superoxide anion radical ($O_2 \cdot^-$), the hydroxyl radical ($HO\cdot$), the peroxyl radical, the alkoxyl radical, and the hydroperoxyl radical. However, ROS that possess an even number of electrons, such as hydrogen peroxide (H_2O_2), the hypohalous acids, and the N-chlorinated amines, though not free radicals, are highly reactive in forming free radicals (7,8).

ROS are routinely being formed in the living cells; however, there are simultaneous defense mechanisms in operation, which constantly strive to maintain the ROS levels within limits by striking an enzymatic or nonenzymatic balance between the generation of these reactive intermediates and their dismutation/destruction by scavenging processes. Conversely, there are certain instances where an upheaval in this balance leads to an excess production of ROS, a condition often referred to as oxidative stress. In such a situation, the primary ROS formed from molecular oxygen is $O_2 \cdot^-$. $O_2 \cdot^-$, which is formed in almost all aerobic cells during respiration, can also be generated when chelated Fe^{3+} is reduced and undergoes autoxidation (9,10). Another important source of $O_2 \cdot^-$ is the "respiratory burst" of phagocytic cells such as neutrophils and macrophages that occurs when these cells come in contact with foreign particles or immune complexes (11). This phenomenon ultimately leads to inflammation-induced, radical-derived tissue damage, which occurs in both cutaneous inflammatory diseases as well as carcinogenesis. $O_2 \cdot^-$, which acts as both an oxidant and a reductant, can either directly or indirectly, by the generation of more potent radicals, modify a variety of biologically important molecules. In addition to the generation of free radicals, whenever two molecules of $O_2 \cdot^-$ are present in aqueous solution, they rapidly and spontaneously dismutate to generate H_2O_2 along with oxygen. While H_2O_2 is neither a free radical nor an aggressive oxidant, it is still viewed as a dangerous molecule primarily because of its small size and lack of charge, which allow it to cross biological membranes and move considerable distances in biological systems. This property of H_2O_2 is of enormous significance due to the fact that it can also collaborate with $O_2 \cdot^-$ in the Haber-Weiss reaction to form the most reactive and most toxic ROS, the hydroxyl radical ($HO\cdot$). This reaction is catalyzed by transition metals such as iron/copper in a Fenton reaction (12):

$$Fe^{3+} + O_2 \cdot^- \rightarrow Fe^{2+} + O_2 \qquad (1)$$

$$Fe^{2+} + H_2O_2 \rightarrow Fe^{3+} + HO^- + HO\cdot \quad (2)$$

$HO\cdot$, which is the most reactive free radical formed in biological systems, is also a highly indiscriminate oxidant. Unlike H_2O_2, the high reactivity of $HO\cdot$ renders it incapable of diffusing away from its site of formation. Thus, extensive tissue and molecular damage occurs at the proximate site of $HO\cdot$ production. The damage is specifically more significant when $HO\cdot$ is generated by the reduction of metals bound to key macromolecules such as nucleotide-iron complexes, ferritin, lactoferrin, transferrin, hemoglobin, and myoglobin (13). This site-specific $HO\cdot$ formation leads to $HO\cdot$-induced protein cross-links, amino acid bond cleavage, altered enzyme activity, fragmentation of macromolecules, DNA strand breaks, base modifications, and point mutations (14). Many of these events have a strong association with cancer induction and its subsequent development. Besides reacting with proteins and nucleic acids, $HO\cdot$ radicals may also attack nearby polyunsaturated fatty acids (LH) to yield lipid radicals, which then react with molecular oxygen to form fatty acid peroxides and peroxyl radical (6). Studies done in our laboratory have shown that the epidermis is a major site for ROS-mediated lipid peroxidation in mouse skin (15). The peroxyl radicals, which are relatively stable and can diffuse to cellular loci that are distant from their site of generation, propagate a peroxidative chain reaction with peroxides that ultimately results in the extensive disruption of architecture and function of membranes. The fact that skin has high levels of LH adds to its susceptibility to free radical damage. This effect, particularly with respect to lysosomal membranes, has been observed in UV-irradiated skin (16). Furthermore, the evidence showing a depletion in natural antioxidant ubiquinone and vitamin E in UV-irradiated skin suggests that oxidative stress–antioxidant balance has an important role in UV-mediated skin damage and carcinogenesis (17,18). It is also important to note that the peroxyl radicals and the lipid peroxides generated by oxidative stress eventually break down into different molecules like aldehydes, predominantly 4-hydroxynonenal, which are also very toxic to the cell (19,20).

In addition to high levels of LH, skin also contains several chromophores such as urocanic acid and melanin (21). These chromophores have the ability to absorb in the near-UV or visible region of the light spectrum and then transfer their excitation energy to O_2, which ultimately results in the generation of another ROS, singlet oxygen. Singlet oxygen, though not a free radical, has an extremely short half-life and is a highly potent oxidant (22–25). It has the capacity to attack several cellular components such as the double bonds present in unsaturated lipids and DNA, which can result in DNA single strand breaks (24–27). In fact, this is the primary basis for many of the light-driven adverse biological processes that occur in skin.

As mentioned earlier most of the ROSs are constantly being formed in vivo. There is a vast array of enzymes that either in the course of their normal cell function or sometimes inadvertently generate ROS in vivo. One such enzyme is NADPH oxidase. This enzyme complex, which is generally localized within the peroxisomes of neutrophils, is quiescent in resting cells (28). However, on being stimulated by agents such as phorbol esters (PMA), heat-aggregated IgG (HAGG), or unsaturated fatty acids, this enzyme complex is activated and then rapidly catalyzes the reduction of molecular O_2 by NADPH to yield $NADP^+$, $O_2 \cdot^-$, and H^+ (28–30). The $O_2 \cdot^-$ then generates H_2O_2 and $HO \cdot$. It is this large burst in oxygen consumption by neutrophils, often referred to as the respiration burst, that accompanies the process of phagocytosis. This generation of ROS ultimately results in skin inflammation, which is an early response of all skin tumor promoters. For detailed literature on this enzyme, the readers are referred to a review by Henderson and Chappel (28). Besides NADPH oxidases, other enzymatic sources that generate ROS include aldehyde oxidase, dihydro-orotate dehydrogenase, and amine oxidases such as monoamine, diamine, and amino acid oxidases (8).

While H_2O_2 is generated by several of these intracellular oxidases present in the peroxisome, one of the best known extraperoxisomal enzymes involved in H_2O_2 generation is xanthine oxidase. Xanthine oxidase in vivo is present in the form of xanthine dehydrogenase, which is a harmless NAD^+ reducing cytosolic enzyme incapable of generating ROS. However, under certain conditions of oxidative stress such as sulfhydryl oxidation, this enzyme is converted to xanthine oxidase, which utilizes O_2 instead of NAD^+ as the final electron acceptor. As a result, xanthine oxidase yields more H_2O_2 by divalent reduction of O_2 than $O_2 \cdot^-$ by univalent reduction of O_2 (8,31,32). Thus, the conversion of xanthine dehydrogenase to xanthine oxidase plays a critical role in free radical–derived tissue damage (8,31).

Another important class of enzymes that contributes significantly to the oxidative stress in a living system is the hemoprotein peroxidases. These enzymes, which are found in certain leukocytes, tissues, exocrine secretions, and subcellular membranes, catalyze the oxidation of a wide variety of electron donors, with H_2O_2 serving as the acceptor (8,33,34). In the process, these enzymes generate some of the most toxic and reactive oxygen intermediates, namely, the hypohalous acids. For example, during the respiratory bursts of polymorphonuclear leukocytes, the myeloperoxidase enzyme is released by neutrophils and monocytes into the extracellular medium, where it interacts with H_2O_2 to form an enzyme-substrate complex that ultimately oxidizes Cl^- by two electrons to yield hypochlorous acid. This hypochlorous acid is a highly potent oxidant capable of oxidizing a variety of biological macromolecules. The hypochlorous acid formed also chlorinates amines to yield another type of reactive intermediate, the *N*-chloramines. During phagocytosis, eosinophils are also activated and secrete the hemoprotein eosinophil peroxidase, which preferentially catalyzes the formation of hypobromous acid and hypothiocyanous acid (8). All these acids formed via the peroxide-catalyzed reactions are powerful oxidizing and halogenating agents that react rapidly with sulfhydryls and aromatic amines, and then eventually contribute significantly to the cytotoxic effects (8,35,36). Thus, both the myeloperoxidase- and the esoinophil-derived peroxide–catalyzed reactions represent very significant pathways of ROS-mediated cell injury during inflammation and carcinogenesis.

Distinct from these peroxidase is the ubiquitously present membrane-bound peroxidase prostaglandin H (PGH) synthetase; in fact, this enzyme system possesses two enzyme activities, cyclooxygenase and peroxidase. When cell membrane–disruptive agents such as UV radiation facilitate the release of membrane-bound arachidonic acid by activating the PLA_2 enzyme, the PGH synthetase (because of its cyclooxygenase and peroxidase activity) oxidizes arachidonic acid to form bicyclic peroxides, which are subsequently converted to form prostaglandins, thromboxanes, and prostacyclin (37,38) as well as reactive intermediates such as hydroendoperoxide and ROS (39,40). Evidence for a role of arachidonic cyclooxygenase products in skin disorders including carcinogenesis is supported by the reduction of UV-induced erythema by selective inhibitors of this enzymatic pathway (41,42).

Besides being generated enzymatically, ROS can also be generated nonenzymatically by autoxidation reactions occurring under certain physiological or pathological conditions. Autoxidation of iron (Fe^{2+} to Fe^{3+}) in molecules such as hemoglobin, myoglobin, and ferritin may be responsible for the generation of the superoxide radical during the operation of the mitochondrial respiratory chain or when hemoproteins are released from erythrocytes (43,44). Traces of transition iron ions also facilitate the autoxidation of several cellular compounds such as catecholamines, thiols, and ascorbic acid (45–48). This phenomenon is especially prevalent during UV-induced erythema, an event that has been linked to skin cancer. Another nonenzymatic mechanism by which $O_2\cdot^-$ may be generated in vivo is through the autoxidation of semiquinones. Endogenous quinones, semiquinones, and hydroquinones in both the electron transport chain of the mitochondria and the microsomal mixed-function oxidase system can initiate redox cycling via enzymatic reduction and concomitant autoxidation to generate $O_2\cdot^-$ and its dismutation product H_2O_2, which may trigger cell damage, toxicity, and carcinogenesis (49).

B. Reactive Nitrogen Species

It is increasingly proposed that, along with ROS, another chemical class of reactive intermediates that may play a significant role in skin-related disorders is the reactive nitrogen species (RNS) derived from nitric oxide (NO). Endogenous NO, a fascinating multifaceted molecule, has varied cell functions; it acts as a vasorelaxant (50), a regulator of the stomach and intestine wall dilation/relaxation (51), a neurotransmitter in the central nervous system (52), a cytocidal agent when released by activated macrophages (53), and a mediator of the human keratinocyte inflammatory responses (54). However, recent evidence suggests that in spite of being an essential signaling molecule, excess production of NO and its derivatives such as the peroxynitrite radical ($ONOO^-$), the nitrogen dioxide radical ($NO_2\cdot$) and other oxides of nitrogen (collectively referred to as RNS) may be deleterious to cell integrity, causing cell cytotoxicity, inflammation, and mutagenicity (55,56). These three events are associated with carcinogenesis.

NO reacts rapidly with $O_2\cdot^-$ to generate the $ONOO^-$ radical (57) and with alkoxyl and peroxyl radicals to generate organic peroxides (58). While the organic peroxides are relatively stable, the $ONOO^-$ radical is highly reactive. It can oxidize sulfhydryl groups, deaminate or nitrate DNA bases, peroxidize lipids, and induce cytotoxicity. Some $ONOO^-$-induced cytotoxicity may also be attributed to it being decomposed to highly toxic and reactive free radicals identical to or closely resembling free radicals such as $NO_2\cdot$, NO_2+, and $HO\cdot$ (59–61). The $NO_2\cdot$, a more reactive radical than its precursor $NO\cdot$, has the ability to react with alkanes, alkenes, and hemoproteins, initiate lipid peroxidation, oxidize thiols and thioethers, and induce DNA damage (8,62). It can also abstract $H\cdot$ from polyunsaturated fatty acids to form nitrous acid, a nitrosating agent involved in the biosynthesis of carcinogenic nitrosamines (8,63). Also included in the group of RNS are a variety of nitrogen oxides such as NO_2, N_2O_3, and NO^-, which are generated when $NO\cdot$ spontaneously and rapidly reacts with molecular oxygen; these oxides are potent oxidizing and N-nitrosating agents (64).

NO is generated enzymatically in vivo by a family of enzymes termed nitric oxide synthase (NOS). NOS catalyzes a five-electron oxidation of terminal guanidino nitrogen of L-arginine in the presence of O_2, requisite cofactors, and NADPH to yield L-citrulline and NO (65–67). Interestingly, NOS can also produce $O_2\cdot^-$ and H_2O_2 when NADPH is present but L-arginine is absent. There are two distinct forms of NOS occurring in vivo: the Ca^{2+}/calmodulin-dependent form, which is present in the brain and the lipopolysaccharide (LPS), and the cytokine-inducible form (iNOS), which is present in macrophages and hepatocytes. Recent data have revealed that normal human keratinocytes express both forms of NOS (68,69). Interestingly, from a pharmaceutical point of view, iNOS mRNA, which is expressed in human psoriatic vulgaris keratinocytes, can be inhibited by interleukin-10 (70) and by the anti-inflammatory agents retinoids and 1,25-$(OH)_2$-vitamin D_3 (71).

C. Xenobiotic-Derived Reactive Intermediates

Since the pioneering work of Elizabeth and Jim Miller (72), it has become clear that a variety of carcinogens, especially those belonging to the PAH class, are chemically inert and must be metabolically activated primarily by specific enzyme-catalyzed oxidative reactions to exert their neoplastic effect (73–76). While the biologically inert xenobiotics are referred as the "procarcinogens, or proximate carcinogens," their metabolically derived active forms that are capable of binding to DNA and initiate carcinogenic process are known as the "ultimate carcinogens" (73–76). BP and 7,12-dimethylbenz(*a*)anthracene (DMBA) are two representative PAHs known to initiate murine skin carcinogenesis only after being metabolically activated in vivo (75–77). The metabolic activation of PAHs occurs via several pathways. In one pathway, catalyzed by cytochrome P-450 (CYP)–dependent monooxygenases and epoxide hydrolase, first a non–K-region *trans*-dihydrodiol (proximate carcinogens) is formed, which is then converted to *anti*-diol epoxides (ultimate carcinogens), which can alkylate DNA. For example, BP is metabolized to an ultimate carcinogen named *anti*-7,8-dihydroxy-9,10-epoxy-7,8,9,10-tetrahydrobenzopyrene (BPDE), which alkylates the N^2 group of guanosine (78) and is a potent mutagen and carcinogen (76). Similarly, DMBA also is metabolized to DMBA-3,4-diol 1,2-epoxide, which forms DNA adducts by binding to the N^6 position of deoxyadenosine (3,79). An alternative route of *trans*-dihydrodiol metabolism that is catalyzed by dihydrodiol dehydrogenase involves the formation of ketols, which enolize to catechols that undergo autoxidation to form the corresponding PAH-*ortho*-quinones (80). A deleterious consequence of the formation of PAH-*ortho*-quinones is the generation of ROS and *ortho*-semiquinone anion radicals; the latter enters futile redox cycles, which results in ROS amplification, changed redox state mutations, and cell death (80–82). A second pathway of the PAH metabolism involves the one-electron oxidation of the PAH ring system to form a cation radical, which, in the case of BP, is localized at C-6 (83). This electrophilic intermediate has been implicated in the mutagenicity and carcinogenicity of BP due to its reactivity with DNA (84,85). A third pathway, catalyzed by the CYPs, involves the methylation of unsubstituted PAH, hydroxylation of the resultant methyl group, and conjugation to form a benzylic sulfate ester, which ultimately yields a highly reactive benzylic carbonium ion that can react with DNA. Indeed, 6-methyl-BP has been ranked as one of the most potent procarcinogens known (86).

The primary enzyme complex catalyzing all these pathways is the NADPH-dependent CYP mixed-function oxidases. Among the CYP isozymes, the one most relevant to PAH metabolism is the CYP1A1. This enzyme, which lies embedded in the endoplasmic reticulum of the cells of the skin, is one of the few CYPs expressed in the skin (87). Although CYP1A1 is not constitutively expressed in the skin to an appreciable degree, it can be highly induced by certain xenobiotics. During its catalytic action CYP binds to its substrate, accepts two electrons in a two-step sequence from the flavoprotein NADPH-CYP reductase (the second electron may also be donated by cytochrome b_5), to ultimately release the oxidized substrate (88), which in the case of most PAHs are highly reactive electrophiles capable of binding to macromolecules. If this catalytic cycle is interrupted (as by autoxidation of the CYP ferric-dioxyanion complex) after the introduction of the first electron, oxygen is released as $O_2 \cdot^-$. If the catalytic cycle is interrupted after the introduction of the second electron, oxygen is released as H_2O_2 (89). Both these ROS represent a potential toxication pathway, particularly when effective means of detoxification of ROS is lacking. In fact, recent studies done by Park et al. (90) have shown a direct association of CYP1A1 induction with oxidative DNA damage.

In mammalian skin, although the CYP-catalyzed pathway represents a major PAH-metabolizing route, a growing body of evidence suggests that other multiple oxidative pathways are also operative. The PGH synthetase catalyzed arachidonic acid–dependent cooxidation of molecules such as PAH dihydrodiols, aflatoxin B1, and aromatic amines constitutes one such distinctive pathway (91–96). The PGH synthetase enzyme is a membrane-bound enzyme unrelated to CYP, and is distributed in a wide variety of tissues including skin (38). PGH synthetase, which has a peroxidase activity, can metabolize BP to 6-OH-BP, which by subsequent autoxidation and by one-electron oxidation can generate three diones, namely, 1,6-2.6-, and 3.6-BP dione. All these BP quinones facilitate one-electron redox reactions producing DNA-damaging and cytotoxic ROS (97,98). Another relevant role of PGH synthetase is in the prostanoid metabolic pathway, where it catalyzes the biosynthesis of unsaturated hydroendoperoxides from arachidonic acid (99). These hydroendoperoxides and hematin (the prosthetic group of PGH synthetase) catalyze BP-7,8-diol epoxidation to the ultimate carcinogenic metabolic BPDE (99).

The lipoxygenase-catalyzed metabolism of PAH is another non–CYP-mediated pathway which is closely linked to PGH synthetase. The lipoxygenase enzymes are nonheme iron dioxygenases present in platelets (100), polymorphonuclear leukocytes (PMNs) (101), and organs such as brain, lung, spleen (102), and skin (103). The lipoxygenase activity represents an important part of the unsaturated fatty acid (such as arachidonic acid or linoleic acid) cascade leading to the formation of the hydroperoxyeicosatetraenoic acid (HPETE), leukotrienes, and lipoxins (104). By a mechanism analogous to the PGH synthetase pathway, fatty acid hydroperoxides have been shown to be metabolized by lipoxygenase to alkoxyl radicals (105, 106), and during this process oxidation of chemical carcinogens has been demonstrated (107).

Although several enzymatic activation pathways of carcinogen metabolism are known, present data also implicate free radicals in the activation of carcinogens in mammalian skin. Free radicals are known to metabolically activate a diverse group of carcinogens such as PAHs, nitroso compounds, aromatic amines, and aminoazo compounds (108). It has also been shown that free radicals derived from quinone components are capable of activating PAHs. Similarly autoxidative derivatives of sulfite may have a role in PAH activation (109). Several laboratories have shown that many chemical carcinogens that require metabolic activation are cooxidized to reactive intermediates during PGH synthetase–mediated generation of unsaturated fatty acid hydroperoxides. These hydroperoxides are converted to alkoxyl radical in a one-electron reduction step by hematin, the prosthetic moiety of PGH synthetase. The alkoxyl radical then cyclizes to the adjacent double bond generating a carbon-centered radical. This radical then couples with O_2 to form a peroxyl radical, which epoxidizes the PAH dihydrodiol to the mutagenic and carcinogenic bay region *anti*-PAH diol epoxides (37). BP activation by this pathway is perhaps one of the best documented examples of a peroxyl radical–catalyzed PAH-activation reaction. BP is initially converted by CYP to BP 7,8-diol, which is then converted to BPDE either by CYP or by a radical-dependent pathway. In vitro studies suggest that the stereoselectivity of the product generated in both of these pathways are distinct; the CYP pathway generates the (+)-enantiomer of *syn*-BPDE, whereas the peroxyl free radicals generate the (−)-enantiomer of *anti*-BPDE (110–114). Indeed when studied in vivo, the metabolic activation of BP 7,8-diol, which is inhibited by antioxidants and free radical scavengers and is also independent from NADPH, resulted in the formation of (−)-*anti*-BPDE and (−)-*anti*-BPDE DNA adducts in the mouse epidermis (112,115). In contrast, after treatment of skin with β-naphthoflavone (an established inducer of CYP1A1), (+)-*syn*-BPDE was the exclusively formed product. As expected, this epoxidation to (+)-*syn*-BPDE is dependent on NADPH and cannot be inhibited by free radical scavengers. Aside from these mentioned pathways, carcinogen activation may also be facilitated by macrophage-derived myeloperoxidase as well as metal-oxo complexes of certain proteins (6).

III. EPIDERMAL DEFENSE AGAINST REACTIVE INTERMEDIATES

The preceding section on reactive intermediates leads inescapably to the conclusion that our body tissues, more specifically the skin, are continuously exposed to dangerously reactive free radicals and other intermediates. During the course of evolution, the epidermis has developed defense mechanisms against free radicals that can be broadly classified into enzymatic and quenching reactions, as detailed in Figure 1.

The epidermal thioredoxin system (116) is an epidermal membrane–associated free radical–scavenging system that reduces two moieties of $O_2\cdot^-$ to H_2O_2 and H_2O outside the cell, thereby decreasing the likelihood of H_2O_2 combining with $O_2\cdot^-$ to form the highly reactive HO· (117). This enzyme acts in the presence of NADPH, which is an electron donor. The location of this enzyme system at the outer membrane surface in human epidermal keratinocytes suggests that it may be a first line of defense against ROS-mediated cell damage (118).

Glutathione reductase (GSSG-R) is another flavoprotein enzyme, which has similar activity. Electrons flow from NADPH to a protein-bound FAD molecule reducing a second disulfide active site. Glutathione peroxidase (GSH-Px) is a selenium (Se)-dependent enzyme with a cysteine residue at its catalytic site that rapidly detoxifies H_2O_2 and simple alkyl or phospholipid hydroperoxides. GSH-Px isozymes may be soluble (occurring both within and outside the cell) or membrane bound. Detoxification of H_2O_2 in the cytoplasm is accomplished largely by coupled GSH-Px–GSSG-R (117).

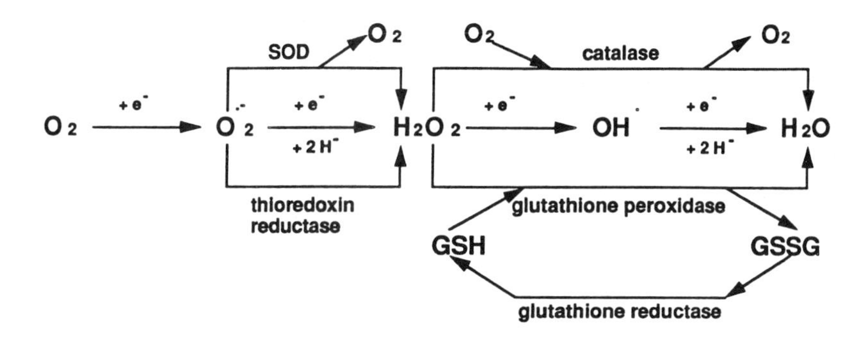

Intracellular antioxidants : **superoxide dismutase (SOD)**
catalase
peroxidase
thioredoxin reductase
transition metal chelators

Extracellular antioxidants : **extracellular SOD**
ceruloplasmin
ascorbate
vitamin E
thiols

Figure 1 Schematic representation of the sequential reduction of molecular oxygen and both intra- and extracellular defense mechanisms available in the skin.

Intracellular H_2O_2 is dismutated to form H_2O and O_2 by a hemoprotein, catalase, in the presence of glutathione. This enzyme is present in the skin, and is depleted by UV exposure and in several skin disorders including vitiligo (119).

Superoxide dismutase (SOD) is a family of metalloproteins that catalyzes the dismutation of $O_2\cdot^-$ to H_2O_2 and O_2. In human skin, both forms of SOD exist—one containing copper and zinc (Cu Zn–SOD) and the other containing manganese (Mn-SOD) at the active sites. Since the level of SOD is much lower in skin than in other tissues such as heart, liver, and kidney, these data suggest that other antioxidant defense systems like thioredoxin reductase and GSH-Px are more important in skin (120). Ironically, however, epidermal SOD activity is reported to decrease in various hyperproliferative keratinocytes including SCC, basal cell epithelioma, and benign hyperproliferative skin such as psoriatic epidermis (121). Similar to catalase, SOD also inhibits the formation of UVB-induced sunburn cells, which are a marker for UVB-induced epidermal injury and carcinogenesis (122).

In the skin the detoxification of carcinogens and other hydrophobic agents bearing electrophilic sites proceeds primarily via conjugation with glutathione catalyzed by glutathione-S-transferases (GSTs). The peroxidase component of GST is Se-independent. About 65–75% of the total GSH-Px activity found in the mouse skin is Se-dependent (123). For further details on this enzyme, readers are referred to several comprehensive reviews (124–127). Other skin enzymes that offer protection against toxic effects of reactive intermediates include indoleamine 2,3-dioxygenase, which utilizes $O_2\cdot^-$ as a substrate for the oxidation of L-tryptophan to *N*-formyl kynurenine and metalloenzymes (including metal complexes) containing transition metals in the higher oxidation state, which are capable of oxidation of $O_2\cdot^-$ to O_2 (128).

In addition to the antioxidant defense systems, skin also contains certain antioxidant molecules such as lipid-soluble membrane scavenger α-tocopherol (vitamin E), ascorbic acid (vitamin C), and vitamin D (17); melanin, which traps $O_2\cdot^-$ to produce reactive organoperoxy intermediates on the melanin polymer (128,129); urocanic acid, which also traps $O_2\cdot^-$; and finally glutathione, which (a) scavenges oxygen radicals (130), which helps to regenerate cellular antioxidants such as ascorbic acid to their reduced state (131), (b) detoxifies carcinogens (132), and (c) acts as a cofactor in prostaglandin and leukotrine synthesis (132). Exposure of skin to UV radiation and other oxidative stress–causing agents has been shown to deplete the levels of the antioxidants α-tocopherol, ubiquinols, and glutathione (133,134). Paradoxically, some of the cellular antioxidants such as ascorbic acid may under certain conditions actually contribute to the prooxidant state of a cell (135). Over and above these antioxidant defenses, there are several repair mechanisms that also participate in repairing reactive intermediate-induced cellular damages. These include (a) the macroxyproteinase enzyme, which hydrolyzes hydroxyl radical-modified proteins to constituent amino acids (136), (b) DNA glycolysases, which remove oxidized bases from DNA (137), and (c) poly(ADP-ribose) polymerase, which rejoins DNA strand breaks caused by ROS (138,139).

Regardless of the defense and repair mechanisms available in the skin, chronic and/or acute exposure to certain agents such as UV, skin carcinogens, skin tumor promoters, and wounding can overwhelm the protective and repair capacities of the skin, leading to dermatopathological conditions such as photodamage, photoaging, inflammation, and carcinogenesis (6,135,140–142).

IV. REACTIVE INTERMEDIATES AND SKIN CARCINOGENESIS

Skin cancer is a well-studied phenomenon in murine models consisting of three sequential steps: tumor initiation, promotion and progression (3,143–146). Although the significance of reactive intermediates in skin tumor promotion has been well established, data accumulated over the years also suggest a role of reactive species in tumor initiation and progression stages.

A. Reactive Intermediates and Skin Tumor Initiation

Tumor initiation is an irreversible and inheritable process characterized by the alteration/modification of DNA resulting in altered gene function and cell responsiveness (to its microenvironment) and a selective clonal expansion advantage (5,147,148). In view of this definition, one can appreciate the significance of the fact that DNA at different body sites in humans is constantly being attacked by endogenously generated oxidants at a rate of $\sim 10^4$ hits per cell per day (149). Indeed the pattern of damage to the four DNA bases that occurs in human cells and tissues suggests that $OH\cdot$ interacts with DNA to generate DNA lesions (55). Singlet oxygen attacks DNA selectively at the guanine base (150,151). Neither $O_2\cdot^-$ nor H_2O_2 reacts at all with DNA (152,153).

Some of the most common ROS-induced DNA base lesions that have been used as urinary markers for oxidative DNA damage are thymine radicals (154), primarily thymidine glycol, which is produced by the addition of $HO\cdot$ to the double bond at the 5,6 position of thymine (32,155,156), 5-hydroxymethyl uracil (32,156), 8-hydroxy-2′-deoxyguanosine, or 8-hydroxy guanine (157). Other ROS-induced lesions include 8-hydroxy-adenine, imidazole ring-opened guanine and adenine, cytosine glycol, and dihydroxycytosine (158,159). It is important to emphasize here that both endogenous or exogenous factors such as activated PMNs that occur during skin inflammation, UV irradiation, and exposure to xenobiotic carcinogens like BP are capable of generating ROS and thus these DNA lesions (160, 161).

Some of these lesions have been shown to block DNA replication or induce mutagenic and/or lethal events that may contribute to skin tumor initiation (162–164). The HO·-mediated 8-hydroxyguanosine (8-OHG) formation in the DNA template causes α-polymerase to miscode incorporation of nucleotides in the replicated strand and, not surprisingly, is mutagenic (165). In mammalian skin carcinogenesis, the data from several studies suggest that the G → T transversion in the codon 12 of Ki-ras and H-ras and at least some of the G → A transversions in p53 gene may be produced by misreplication of 8-OHG (162,166,167). These transversions in both the ras oncogene and p53 tumor suppressor gene are frequently observed in nonmelanoma human skin cancers (168). Significantly enough, recent studies have shown that UVB, an established etiological factor for nonmelanoma human skin cancers, and also BPO generate this DNA lesion in murine keratinocytes (169,170). In another study, UVA was also shown to generate this lesion in mouse skin treated with the carcinogen cyclic nitrosoalkylamine *N*-nitrosomorpholine (171), probably via the generation of a guanine cation radical. Besides directly binding DNA and causing mutations at polymerase-specific recognition sites (hot spots), it is also speculated that ROS may alter the conformation of the DNA template, thereby diminishing the accuracy of replication by DNA polymerases (52,172,173). This is because of the nonplanar conformation of many oxidized bases. The role of ROS in skin carcinogenesis is further exemplified by xeroderma pigmentosa patients, who lack the enzymes required to repair oxidative DNA damage and also exhibit an up to 1000-fold increased risk of skin cancers in sun-exposed body sites (174,139).

Although more information is needed about the tumor-initiating capacity, if any, of RNS, early reports suggest that some RNS may indeed damage DNA. Nitric oxide or, more likely, reactive products of NO, such as $NO_2\cdot$, $ONOO^-$, N_2O_3, and HNO_2, are mutagenic agents with the potential to produce nitration, nitrosation, and deamination reactions on DNA bases (56,175). NO, which is generated in human keratinocytes, has been shown to deaminate the 5-methylcytosine of DNA and cause DNA strand breaks (176,177). Since spontaneous deamination of 5-methylcytosine to thymine within coding regions has been proposed to be responsible for C → T transitions that occur in p53 tumor-suppressor gene (56,178,179), it is tempting to speculate that the deamination of 5-methylcytosine at CpGs by NO (176,177) may be responsible for the C → T transition mutations in the p53 gene in skin tumors. However, it has not yet been demonstrated that NO-induced DNA deamination can indeed cause point mutation in vivo. Additionally, besides the possibility of RNS directly contributing to skin tumor initiation, there are studies showing that $ONOO^-$ spontaneously decomposes to form HO·, which has certain serious implications in skin tumor initiation (60,180). Although not yet known to be relevant in skin tumorigenesis, it is interesting to note that during inflammation sufficient NO· is released by macrophages and granulocytes to cause nitrosation of endogenous amines, resulting in the formation of mutagenic and carcinogenic nitrosamines (56).

Most skin carcinogens belonging to the group of PAHs, such as BP and DMBA, are metabolically activated to form reactive intermediates (ultimate carcinogens) capable of binding to DNA and initiating skin carcinogenesis. Enzymes catalyzing the initial step in this complex metabolic pathway are CYP1A1 (181). PGH synthetase (cyclooxygenase), and lipoxygenase. The horseradish peroxidase/H_2O_2 model system has also been found to metabolize BP by one-electron oxidation (182). In a three-enzyme (monooxygenase, epoxide hydrolase, and monooxygenase)–catalyzed reaction, BP is converted initially to BP-7,8-diol (proximate carcinogen), which is then hydrolyzed to four stereoisomers of BP-7,8-diol-9,10-epoxide. However, it is only the *anti-trans*-BPDE (the ultimate carcinogen) that binds DNA and produces mutations, primarily because this metabolite contains an epoxy group adjacent to the bay region of the benzo ring (115). Indeed, the bay region diol epoxide of DMBA (DMBA-3,4-diol-1,2-epoxide) is the most potent ultimate carcinogenic isomer of the DMBA metabolites (183). On account of their electrophilic nature, BPDE, DMBA-3,4-diol-1,2-epoxide, and DMBA dihydrodiol epoxide (DDE) can all form covalent adducts with DNA. BPDE predominantly binds to the N^2 position of guanine, forming N^2-BPDE-deoxyguanosine adduct, and DMBA predominantly binds to the N^6 position of adenine, forming N^6-7,12-DMBA-diol-epoxide-deoxyadenosine adduct. Both these adducts generate characteristic base mutations (3,78,79). When the H-ras protooncogene mutation spectra of these carcinogen-DNA adducts were studied in several rodent tissues including mouse skin, it was observed that N^2-BPDE-deoxyguanosine produces G:C → T:A transversions and N^6-7,12-DMBA-diol-epoxide generates A:T → T:A transversions (78,79).

Besides forming the DNA-binding dihydrodiolepoxides, BP along with other PAHs can be converted to quinones by PGH synthetase, by ascorbate-dependent lipid peroxidation, and by the CYP-catalyzed one-electron oxidation pathway (80). The quinones facilitate one-electron redox reactions, producing ROS that may contribute to the DNA-damaging, mutagenic, and cytotoxic effects of BP (13,97,154). H_2O_2 appears to be one of the ROS generated during quinone-facilitated one-electron redox reactions (76), which may be indirectly responsible for the thymine glycols produced during BP metabolism in epithelial cells. It is also speculated that the one-electron oxidation represents the predominant CYP-dependent PAH metabolism pathway because several studies show that the major DMBA-DNA adducts (99%) are formed by one-electron oxidation and are lost from DNA by depurination, whereas DMBA-DNA adducts formed by the diol-epoxide pathway (184) are very minor.

Besides the generation of PAH quinones, the PGH synthetase pathway also generates peroxyl radicals that can convert PAH dihydrodiol to PAH dihydrodiol epoxide. Studies in untreated mouse skin using DMBA and BP indicate that the peroxyl radical–catalyzed PAH-diol epoxidation can occur in vivo even when the CYP is not induced, suggesting that the peroxyl radical–dependent epoxidation may be more important than CYP-dependent epoxidation (37).

B. Reactive Intermediates and Skin Tumor Promotion

Tumor promotion in mouse skin can be accomplished by repetitive treatment with tumor-promoting chemicals such as phorbol esters, benzoyl peroxide, etc. Tumor promotion results in proliferation and survival of the initiated cells to a greater extent than normal cells, and thus enhances the probability of additional genetic damage including endogenous mutations accumulating in the expanding population of initiated cells (185). The role of prooxidant states in skin tumor promotion is evident by the observations that (a) tumor promoters increase the generation and decrease the degradation of ROS and organic peroxides, (b) a free radical–generating system exhibits tumor-promoting activities and mimics/enhances some molecular events linked to tumor promotion, and (c) various antioxidants and free radical scavengers inhibit the biochemical and biological effects of the tumor promoters (17).

A number of free radical–generating compounds such as 12-*O*-tetradecanoyl-phorbol-13-acetate (TPA), BPO, etc., are skin irritants and tumor promoters. One of the foremost in vivo responses following TPA treatment is a remarkable increase in H_2O_2 production, especially in epidermal keratinocytes (186–188). Mezerein, BPO, 12-*O*-retinoyl-phorbol-13-acetate, UVA, UVB, etc., also produce substantial H_2O_2 in mouse skin in vivo (186,187). Interestingly, the increase in H_2O_2 production following tumor promoter treatment in vivo correlates with their activity as complete or stage II tumor promoters (6,156). It is also reasoned that the role of

PAHs as complete carcinogens may be due to their capacity to generate free radicals and thus be active in promotion as well as initiation.

As for H_2O_2, the role of $O_2\cdot^-$ in tumor promotion is also well substantiated by the observations that xanthine oxidase–generated $O_2\cdot^-$ mimics the effects of TPA on the promotion of transformation in JB6 cells and that the derivatives of tumor promoters such as anthralin and chrysarobin can autoxidize to generate peroxyl, semiquinone, anthronyl, and $O_2\cdot^-$ radicals (189,190). Studies by Fuchs and Packer (191) show that the anthrones can also generate the highly reactive OH·. Significantly, damage to the DNA in TPA-treated inflammatory cells was found to be characteristic of attack by the OH· radical (192). Increased 8-OHdG thymine glycol levels in the urine of TPA-treated mice and in the epidermis of nude mice exposed to near-UV is suggestive of a role of OH· in tumor promotion (161).

Studies from the laboratories of Marnett and Kensler (37) suggest that peroxyl radicals may also enhance skin tumor promotion. The fact that TPA promotes NO release and expression of iNOS in vitro suggests that RNS also play a role in tumor promotion (55,70). In addition to free radicals, a recent study by Guyton et al. (193) defines a role for electrophiles in skin tumor promotion. In this study, butylated hydroxytoluene quinone methide (BHTQM), the reactive electrophile derivative of the tumor promoter butylated hydroxytoluene hydroperoxide (BHTOOH), was shown to independently promote mouse skin tumors.

Early studies based on a correlation between promoting activity and the inflammatory activity of phorbol esters suggested a role for inflammation in tumor promotion (194,195). The TPA-induced inflammatory response is characterized by an influx of neutrophils and macrophages and an increased level of dermal myeloperoxidases in the mouse skin. In the oxidative burst, a large amount of $O_2\cdot^-$ is generated (195,196). In vitro studies show that the levels of $O_2\cdot^-$ released (as measured by its capacity to reduce ferricytochrome C to ferrocytochrome C) by neutrophils activated by different phorbol esters correlate with their tumor-promoting capability (197,198). Interestingly, phorbol myristate acetate, a metabolite of TPA in mouse skin that is equipotent to TPA in causing inflammation but is only one fifth as active as TPA as a tumor promoter, is also a weak stimulator of $O_2\cdot^-$ production (199). Additionally, inhibitors of promotion such as protease inhibitors and retinoids are also inhibitors of the TPA-induced oxygen burst (200,201). In TPA-treated animals, studies of inflammation also showed an increased $O_2\cdot^-$ release by macrophages as reflected by increased $O_2\cdot^-$-specific reduction of nitroblue tetrazolium. Also, murine epidermal cells and peripheral blood leukocytes isolated from skin treated with TPA in vivo oxidize the hydroperoxide-sensitive dye 2′,7′-dichlorofluorescin, an effect that can be inhibited by catalase. This suggests a role in tumor promotion for the hydroperoxide produced during TPA-induced inflammation.

Another characteristic feature of mouse skin tumor promotion is altered phospholipid metabolism. Administration of tumor promoters such as TPA or mezerein triggers significant lipid peroxidation in the mouse skin (37). The skin tumor–promotion capability of tumor-promoting chemicals can be linked to the production of hydroperoxide metabolites of arachidonic acid (202). Indeed, the hydroperoxide responses to thapsigargin and TPA are decreased by pretreatments that inhibit the phospholipase A_2–mediated release of arachidonic acid from membrane phospholipids (187,203). Most of the lipid peroxidation as well as inflammation and cell proliferation that occurs following the treatment with tumor promoters are attributed to the activation of the cyclooxygenase and lipoxygenase; the latter appears most critical if not essential to tumor promotion (193,203). The tumor promoters activate the PLA_2 enzyme, facilitating the release of membrane-bound arachidonic acid. Different lipoxygenases catalyze stereospecific incorporation of molecular oxygen at carbons 5, 8, 9, 11, 12, or 15 of arachidonic acid, resulting in the formation of corresponding HPETE (204,205). Studies using SENCAR mouse epidermal cells in vitro showed that nondihydroguaiaretic acid and other agents that inhibit both lipoxygenase and cyclooxygenase were effective in diminishing the TPA-induced chemiluminescence response, a marker for ROS production (206). However, when only cyclooxygenase inhibitors indomethacin and flurbiprofen were tested, no inhibitory effect was observed (206). These studies were further supported by the work from several other groups showing that both TPA- and thapsigargin- (a non TPA-type tumor promoter) induced production of hydroperoxide in mouse skin is inhibited by specific lipoxygenase inhibitors (187,203). Studies from our laboratory have shown that green tea polyphenols, which inhibit TPA-caused tumor promotion in mouse skin, also inhibit TPA-induced lipoxygenase and cyclooxygenase activities in mouse skin (207). Besides this enzyme complex, xanthine oxidase also appears to contribute to ROS production during tumor promotion since kinetic studies of H_2O_2 production and xanthine oxidase activity in TPA-treated mouse skin shows a positive correlation (189). Whereas the role of ROS in tumor promotion is well established, it is not clear whether they effect stage I tumor promotion or stage II tumor promotion. Tumor promotion in mouse skin is commonly subdivided into two stages: stage I and stage II (208). Studies show that different promoters produce single-strand breaks in DNA of epidermal cells in relation to their complete and stage II tumor-promoting activities, but they stimulate H_2O_2 formation and produce DNA base modification in relation to their potency as stage I tumor promoters (209,210). It is, therefore, suggested that the H_2O_2-induced damage may be responsible for the "memory" effect linked to stage

I tumor production (210). Indeed, H_2O_2 as well as other ROS are known to induce clastogenic effects, and it was therefore hypothesized that the convertogenic activity of TPA in stage I may be because of the clastogenic and mitogenic effects of H_2O_2 (211,212).

However, there is another line of thought that ROS may effect stage II tumor promotion based on the observations that inflammatory hyperplastic and stage I tumor-promoting agents fail to mimic the xanthine oxidase and hyperperoxide responses in skin treated with peroxides and other stage II tumor promoters (187,213–215). The protease inhibitor *N*-tosyl-L-phenylalanine chloromethyl ketone inhibits stage I promotion by TPA, but does not inhibit the stage II promotion by mezerien and the ODC, xanthine oxidase, hydroperoxide, or hyperplastic responses to TPA in mouse epidermis in vivo (144,203,215). Antioxidants such as α-tocopherol + glutathione or an SOD analog inhibit both ODC induction and complete tumor-promoting potential of TPA, but fail to inhibit skin tumor promotion when applied in combination with TPA in the first stage (216–218).

In addition to causing permanent genetic changes, oxy radicals activate cytoplasmic signal transduction pathways, which are related to cell growth, differentiation, and death. Indeed, oxy radicals in many ways mimic the action of polypeptide factors. For example, calcium ion mobilization from mitochondria and endoplasmic reticulum and influx from the extracellular space are some of the earliest events after exposure to oxy radicals (219). The consecutive changes in the activities of kinases (220) and phosphatases (221), which transduce the initial signal to a family of transcription factors, are only beginning to be examined. They form a part of a complete network that allows an immediate response to several forms of cellular stress. In some instances, oxidative stress results more directly in the alteration of the activity of transcription factors. For example, the nuclear transcription factor (NF) kappa-B is separated from its inhibitory subunit after exposure of cells to the oxy radicals TNF-α or IL-1 (222), and the oncoproteins Fos and Jun require intact, reduced cysteine residues in strategic positions for dimerization and full transcriptional activity (223). The transient activation of these and related transcription factors is quickly followed by their enhanced synthesis and the transactivation of subfamilies of effector genes, such as collagenase, metallothionein 11A, and certain viral long-terminal repeats (224). Activation of these circuits of gene expression by oxy radicals can participate in tumor promotion as well as progression.

C. Reactive Intermediates and Skin Tumor Progression

Tumor progression in skin involves the development of malignant skin carcinomas from preexisting papillomas and appears to be promoter independent (225,226). BPO is a classical example of the reactive intermediate class of agents which enhance tumor progression in mouse skin by generating peroxides (227). The inhibition of BPO-induced tumor progression in mouse skin by Cu(II)-(3,5-disopropylsalicylate)$_2$, a SOD mimic, substantiates the role of ROS in mouse skin tumor progression (228). Earlier studies from our laboratory showed that human skin carcinoma cells have the capacity to metabolize organic hydroperoxides into free radicals (229).

Tumor progression is proposed to involve a number of genetic alterations. Recent studies employing c-fos protooncogene–deficient mice indicate that c-fos is required for malignant progression of skin tumors (230). Other studies also substantiate this report (231,232). Several lines of evidence indicate the role of ROS in c-fos activation. In a study in Balb/3T3 cells, H_2O_2 and $O_2\cdot^-$ were shown to induce c-fos mRNA expression (232). In another study the antioxidant retinoic acid was found to inhibit c-fos expression in TPA-stimulated cells (233). In a recent study, the treatment of cells with the antioxidant dihydrolipoic acid resulted in suppression of TPA-activated c-fos expression in cells (231). Another characteristic feature of skin tumor progression is chronic inflammation (208), a condition where oxidative stress is imminent (35). Because benzoyl peroxide and H_2O_2 are more effective than TPA in stimulating malignant progression in skin tumors, the potency of various compounds in tumor progression may be linked to their ability to increase the epidermal levels of peroxidation and genetic lesions. In contrast to TPA, both BPO and DMBA generate DNA single-strand breaks, increase the carcinoma : papilloma ratio, and are resistant to retinoic acid (234). It has been proposed that BPO, on account of its capacity to generate free radicals, may induce extensive DNA strand breaks and exert cytotoxic effects in normal cell and indirectly favor the clonal expansion of initiated cells that are more resistant to ROS-induced genetic lesions (209,234).

V. CONCLUSION

As summarized in this chapter, convincing evidence shows that ROS, RNS, and xenobiotic-derived reactive species are involved in several cutaneous disorders. With regard to skin carcinogenesis, several studies employing experimental animal bioassay systems have demonstrated the involvement of oxidative stress in all three stages of carcinogenesis (initiation, promotion, and progression). However, because of the transient nature of reactive intermediates, limited information is available in human tissues. Once a suitable methodology to measure oxidative stress in human epidermal cells is established, it is believed that the role of oxidative stress in human skin diseases including cancer will be even better studied and established.

Reactive intermediates, generated by the multiple pathways of mammalian cellular systems, may be natural carcinogens, accelerate the formation of premalignant lesions, and help convert to malignancy. It is known that ROS released by inflammatory phagocytes can induce mutations and chromosomal aberrations in neighboring target cells. This genotoxicity may ultimately result in the activation of protooncogenes and tumor-suppressor genes. Such genetic alterations lead to uncontrolled cell growth, loss of cellular differentiation, and cancer induction.

The role of reactive intermediates in skin diseases including carcinogenesis is further established by the studies showing that antioxidants afford significant protection against skin disorders as well as carcinogenesis. The inhibition of free radical damage-induced pathophysiological processes by antioxidant therapy is now gaining considerable following. For example, SOD, an anti-inflammatory agent, has been used clinically for more than a decade. However, because some of the free radicals generated in our body have beneficial as well as deleterious effects, caution must be exercised in the use of antioxidant therapy. No global answer can be given about the role of individual antioxidants in human tumorigenesis. The outcome will depend on the interaction between the network of antioxidant components of the target tissue (skin) and the carcinogens that skin is exposed to.

ACKNOWLEDGMENTS

Our original experimental studies were supported by United States Public Health Service Grants CA51802, CA64514, PO1-CA48735, and P-30-AR-39750.

REFERENCES

1. Mukhtar H, et al. Photocarcinogenesis: mechanisms, models and human health implications. Photochem Photobiol 1996; 63:355–447.
2. Miller DL, Weinstock MA. Non melanoma skin cancer in the United States: incidence. J Am Acad Dermatol 1994; 30:774–778.
3. Agarwal R, Mukhtar H. Cutaneous chemical carcinogenesis. In: Mukhtar H, ed. Pharmacology of the Skin. Boca Raton, FL: CRC Press, 1991:371–387.
4. Mukhtar H, Mercurio MG, Agarwal R. Murine skin carcinogenesis: Relevance to humans. In: Mukhtar H, ed. Skin Cancer: Mechanisms and Human Relevance. Boca Raton, FL: CRC Press, 1995:3–8.

5. Cerutti PA. Prooxidant states and tumor promotion. Science 1985; 227:375–381.
6. Perchellet JP, Perchellet EM, Gali HU, Gao XM. Oxidant stress and multistage skin carcinogenesis. In: Mukhtar H, ed. Skin Cancer: Mechanisms and Human Relevance. Boca Raton, FL: CRC Press, 1995:145–180.
7. Moslen MT, Smith CV, eds. Free Radical Mechanisms of Tissue Injury. Boca Raton, FL: CRC Press, 1992.
8. Grisham MB, ed. Reactive Metabolites of Oxygen and Nitrogen in Biology and Medicine. Austin, TX: R. G. Landes, 1992.
9. Halliwell B. Superoxide-dependent formation of hydroxy radicals in the presence of iron salts. FEB Lett 1978; 96:238–242.
10. Liochev SI, Fridovich I. The role of $O_2 \cdot ^-$ in the production of $HO\cdot$ in vitro and in vivo. Free Radical Biol Med 1994; 16:29–33.
11. Grisham MB. Oxidants and free radicals in inflammatory bowel disease. Lancet 1994; 344:859–861.
12. Pyror WA. The role of free radical reactions in biological systems. In: Pyror WA, ed. Free Radicals in Biology. Vol. 1. London: Academic Press, 1976:1–43.
13. Hochstein P, Atallah AS. The nature of oxidants and antioxidant systems in the inhibition of mutation and cancer. Mutat Res 1988; 202:363–375.
14. Halliwell B, Gutteridge JMC, Blake DR. Metal irons and oxygen radical reactions in human inflammatory joint disease. Phil Trans R Soc London B 1985; 311:659–671.
15. Dixit R, Mukhtar H, Bickers DR. Studies on the role of reactive oxygen species in mediating lipid peroxide formation in epidermal microsomes of rat skin. J Invest Dermatol 1983; 81:369–375.
16. Johnson BE, Daniel F Jr. Lysosomes and reactions of skin to ultraviolet radiation. J Invest Dermatol 1969; 53:85–94.
17. Perchellet JP, Perchellet EM. Antioxidants and multistage carcinogens in mouse skin. Free Radical Biol Med 1989; 7:377–408.
18. Witt EH, Motchnik PA, Han P, Ames B, Packer L. Ultraviolet irradiation, destruction of lipophilic antioxidants, and formation of lipid hydroperoxides in skin of hairless mice (abstr). J Invest Dermatol 1991; 96:585A.
19. Barrera G, Brossa O, Fazio VM, Farace MG, Paradisi L, Gravela E, Dianzani MU. Effects of 4-hydroxynonenal on cell proliferation and ornithine decarboxylase activity. Free Radical Res Commun 1991; 14:81–89.
20. Glasgow WC, Afshari CA, Barrett JC, Eling TE. Modulation of the epidermal growth factor mitogenic response by metabolites of linoleic acid and arachidonic acid in Syrian hamster embryo fibroblasts. J Biol Chem 1992; 267:10771–10779.
21. Carbonare MD, Pathak MA. Skin photosensitizing agents and the role of reactive oxygen species in photoaging. J Photochem Photobiol 1992; 14:105–124.
22. Kanofsky JR. Singlet oxygen production by biological systems. Chem Biol Interact 1989; 70:1–28.

23. Epe B. Genotoxicity of singlet oxygen. Chem Biol Interact 1991; 80:239–260.
24. Piette J. Biological consequences associated with DNA oxidation mediated by singlet oxygen. J Photochem Photobiol 1991; 11:241–260.
25. Sies H. Damage to plasmid DNA by singlet oxygen and its protection. Mutat Res 1993; 209:183–191.
26. Schneider JE, Price S, Maidt L, Gutteridge JMC, Floyd RA. Methylene blue plus light mediates 8-hydroxy-2′-deoxyguanosine formation in DNA preferentially over strand breakage. Nucleic Acid Res 1990; 18:631–635.
27. Devasagayam TPA, Steenken S, Obendorf MSW, Schulz WA, Sies H. Formation of 8-hydroxy(deoxy)guanosine and generation of strand breaks at guanine residues in DNA by singlet oxygen. Biochemistry 1991; 30:6283–6289.
28. Henderson LM, Chappel JB. NADPH oxidase of neutrophils. Biochim Biophys Acta 1996; 1273:87–107.
29. Rossi F. The $O_2 \cdot^-$ forming NADPH oxidase of the phagocytes: nature, mechanisms of activation and function. Biochim Biophys Acta 1986; 853:65–89.
30. Morel F, Doussiere J, Vignais PV. The superoxide generating oxidase of phagocyte cells. Physiological molecular and pathological aspects. Eur J Biochem 1991; 201:523–546.
31. McCord JM. Oxygen-derived radicals: a link between reperfusion injury and inflammation. Fed Proc 1987; 46:2402–2406.
32. Cross CE, Halliwell B, Borish ET, Pryor WA, Ames BN, Saul RL, McCord JM, Harman D. Oxygen radicals and human disease. Ann Intern Med 1987; 107:526–545.
33. Ramos CL, Par S, Britigan BE, Cohen ME, Rosen GM. Spin trapping evidence for myeloperoxidase-dependent hydroxyl radical formation by human neutrophils and monocytes. J Biol Chem 1992; 267:8307–8312.
34. Geerdink JPM, Troost PW, Schalkwijk JS, Joosten LAB, Mier PD. The ‘metabolic burst’ in polymorphonuclear leucocytes from patients with quiescent psoriasis. Br J Dermatol 1985; 112:387–392.
35. Grisham MB. Role of reactive oxygen metabolites in inflammatory bowel disease. Curr Opin Gastroenterol 1993; 9:971–980.
36. Weitzman SA, Gordon LI. Blood: inflammation and cancer: role of phagocyte-generated oxidants in carcinogenesis. Blood 1990; 76:655–663.
37. Marnett LJ. Peroxyl free radicals: potential mediators of tumor initiation and promotion. Carcinogenesis 1987; 8:1365–1373.
38. Marnett LJ. Polycyclic aromatic hydrocarbon oxidation during prostaglandin biosynthesis. Life Sci 1981; 29:531–546.
39. Needleman P, Turk J, Jakshchik B, Morrison A, Letkowith JB. Arachidonic acid metabolism. Annu Rev Biochem 1986; 55:69–102.
40. Ohki S, Ogino N, Yamamoto S, Hayashi O. Prostaglandin hydroperoxidase, an integral part of prostaglandin endoperoxide synthetase from bovine vesicular gland microsomes. J Biol Chem 1979; 254:829–836.
41. Woodward DF, Raval P, Pipkin MA, Owen DA. Re-evaluation of the effect

of non-steroidal anti-inflammatory agents on uv-induced cutaneous inflammation. Agents Actions 1981; 11:711–717.
42. Miller WS, Ruderman FR, Smith JG Jr. Aspirin and ultraviolet light-induced erythema in man. Arch Dermatol 1967; 95:357–358.
43. Halliwell B, Gutteridge JMC. Oxygen toxicity, oxygen radicals, transition metals and disease. Biochem J 1984; 219:1–16.
44. Thomas CE, Morehouse LA, Aust SD. Ferritin and superoxide-dependent lipid peroxidation. J Biol Chem 1985; 260:3275–3280.
45. Scherer NM, Deamer DW. Oxidative stress impairs the functions of sarcoplasmic reticulum by oxidation of sulfhydryl groups in Ca^{2+}-ATPase. Arch Biochem Biophys 1986; 246:589–601.
46. Kukreja RC, Okabe E, Schrier GH, Hess ML. Oxygen radical-mediated lipid peroxidation and inhibition of Ca^{2+}-ATPase activity of cardiac sarcoplasmic reticulum. Arch Biochem Biophys 1988; 261:447–457.
47. Vile GF, Winterbourn CC. Thiol oxidation and inhibition of Ca-ATPase by adriamycin in rabbit heart microsomes. Biochem Pharmacol 1990; 39:769–774.
48. Vile GF, Tyrrell RM. UVA radiation-induced oxidative damage to lipids and proteins in vitro and in human skin fibroblasts is dependent on iron and singlet oxygen. Free Rad Biol Med 1995; 18:721–730.
49. Sies H. Oxidative stress: quinone redox cycline. 1S1 Atlas Sci Biochem 1988; 1:109.
50. Ignarro I. Biosynthesis and metabolism of endothelium derived nitric oxide. Ann Rev Pharmacol Toxicol 1990; 30:535–560.
51. Desai KM, Sessa WC, Vane JR. Involvement of nitric oxide in the reflex relaxation of the stomach to accommodate food or fluid. Nature 1991; 351: 477–479.
52. Synder SH. Nitric oxide: first in a new class of neurotransmitter? Science 1992; 257:494–496.
53. Mossalayi MD, Paul-Eugene PA, Ouaaz F, Arock M, Kolb JP, Kilchherr E, Debre P, Dugas B. Involvement of Fc epsilon RII/CD23 and L-arginine-dependent pathway in IgE-mediated stimulation of human monocyte functions. Int Immunol 1994; 6:931–934.
54. Becherel PA, Mossalayi MD, Ouaaz F, LeGoff L, Dugas B, Paul-Eugene N, Frances C, Chosidow O, Kilchherr E, Guillosson JJ, Debre' P, Arock M. Involvement of cyclic AMP and nitric oxide in immunoglobulin E-dependent activation of Fc epsilon RII/CD23+ normal human keratinocytes. J Clin Invest 1994; 93:2275–2279.
55. Wiseman H, Halliwell B. Damage to DNA by reactive oxygen and nitrogen species: role in inflammatory disease and progression to cancer. Biochem J 1996; 313:17–29.
56. Ohshima H, Bartsch H. Chronic infections and inflammatory processes as cancer risk factors: possible role of nitric oxide in carcinogenesis. Mutat Res 1994; 305:253–264.
57. Huie RE, Padmaja S. The reaction of NO with superoxide. Free Radicals Res Commun 1993; 18:195–199.

58. Padmaja S, Huie RE. The reaction of nitric oxide with organic peroxyl radicals. Biochem Biophys Res Commun 1993; 195:539–544.
59. Beckman JS, Chen J, Ischiropoulos H, Crow JP. Oxidative chemistry of peroxynitrite. Methods Enzymol 1994; 233:229–240.
60. Yermilov V, Rubio J, Becchi M, Friesen MD, Pignatelli B, Ohshima H. Formation of 8-nitroguanine by the reaction of guanine with peroxynitrite in vitro. Carcinogenesis 1985; 18:2045–2050.
61. Ischiropoulous H, Zhu L, Beckman JS. Peroxynitrite formation from macrophage-derived nitric oxide. Arch Biochem Biophys 1992; 298:446–451.
62. Bittrich H, Matik AK, Kraker J, Appel KE. NO_2-induced DNA single strand breaks are inhibited by antioxidative vitamins in V79 cells. Chem Biol Interact 1993; 86:199–211.
63. Ohshima H, Tsuda M, Adachi H, Ogura T, Sugimura T, Esumi H. L-Arginine-dependent formation of N-nitrosamines by the cytosol of macrophages activated with lipopolysaccharide and interferon-g. Carcinogenesis 1991; 12:1217–1220.
64. Heinzel B, John M, Klatt P, Bohme E, Mayer B. Ca^{2+}/calmodulin-dependent formation of hydrogen peroxide by brain nitric oxide synthase. Biochem J 1992; 281:627–630.
65. Pou S, Pou WS, Bredt DS, Snyder SH, Rosen GM. Generation of superoxide by purified brain nitric oxide synthase. J Biol Chem 1992; 267:24173–24176.
66. Klatt P, Schmidt K, Mayer B. Brain nitric oxide synthase is a haemoprotein. Biochem J 1992; 288:15–17.
67. White KA, Marletta MA. Nitric oxide synthase is a cytochrome P-450 type hemoprotein. Biochemistry 1992; 31:6627–6631.
68. Baudouin JE, Tachon P. Constitutive nitric oxide synthase is present in normal human keratinocytes. J Invest Dermatol 1996; 106:428–431.
69. Wang R, Ghahary A, Shen YJ, Scott PG, Tredget EE. Human dermal fibroblasts produce nitric oxide and express both constitutive and inducible nitric oxide synthase isoforms. J Invest Dermatol 1996; 106:419–427.
70. Becherel PA, LeGoff L, Ktorza S, Ouaaz F, Mencia-Huerta JM, Dugas B, Debre P, Mossalayi MD, Arock M. Interleukin-10 inhibits IgE-mediated nitric oxide synthase induction and cytokine synthesis in normal human keratinocytes. Eur J Immunol 1995; 25:2992–2995.
71. Becherel PA, Mossalayi MD, Le Goff L, Frances C, Chosidow O, Debre P, Arock M. Mechanism of anti-inflammatory action of retinoids on keratinocytes. Lancet 1994; 344:1570–1571.
72. Miller EC, Miller JA. The presence and significance of bound aminoazo dyes in the livers of rats fed p-dimethylaminoazobenzene. Cancer Res 1947; 7:468–480.
73. Levin W, Wood A, Chang R, Ryan D, Thomas P, Yagi H, Thakkar D, Vyas K, Boyd D, Chu SY, Conney AH, Jerina DM. Oxidative metabolism of polycyclic aromatic hydrocarbons to ultimate carcinogens. Drug Metab Rev 1982; 13:555–580.
74. Miller EC, Miller JA. The metabolism of chemical carcinogens to reactive electrophiles and their possible mechanisms of actions in carcinogenesis. In:

Searle CE, ed. Chemical Carcinogens. Washington, DC: American Chemical Society, 1976:737–762.
75. Conney AH. Induction of microsomal enzymes by foreign chemicals and carcinogenesis by polycyclic aromatic hydrocarbons: GHA Clowes memorial lecture. Cancer Res 1982; 42:4875–4917.
76. Mukhtar H, Agarwal R, Bickers DR. Cutaneous metabolism of xenobiotics and steroid hormones. In: Mukhtar H, ed. Pharmacology of the Skin. Boca Raton, FL: CRC Press, 1991:89–110.
77. Quintanilla M, Brown K, Ramsden M, Balmain A. Carcinogen-specific mutation and amplification of Ha-*ras* during mouse skin carcinogenesis. Nature 1986; 322:78–80.
78. Weinstein IB, Jeffrey AM, Jenette KW, Blobstein SH, Harvey RG, Harris CC, Autrup H, Kasai H, Nakanishi K. Benzo(a)pyrene diol epoxides as intermediates in nucleic acid binding *in vitro* and *in vivo*. Science 1976; 193:592–595.
79. Dipple A, Pigott M, Moschel RC, Constantino N. Evidence that binding of 7,12-dimethylbenz(a)anthracene to DNA in mouse embryo cell cultures results in extensive substitution of both adenine and quanine residues. Cancer Res 1983; 43:4132–4135.
80. Flowers-Geary L, Bleczinski W, Harvey RG, Penning TM. Cytotoxicity and mutagenicity of polycyclic aromatic hydrocarbon o-quinones produced by dihydrodiol dehydrogenase. Chem Biol Interact 1996; 99:55–72.
81. Flowers-Geary L, Harvey RG, Penning TM. Cytotoxicity of polycyclic aromatic hydrocarbon o-quinones in rat and human hepatoma cells. Chem Res Toxicol 1993; 6:252–260.
82. Flowers-Geary L, Harvey RG, Penning TM. Examination of polycyclic aromatic hydrocarbon o-quinones produced by dihydrodiol dehydrogenase as substrates for redox-cycling in rat liver. Biochem (Life Sci Adv) 1992; 11:49–58.
83. Cavalieri EL, Rogan EG. The approach to understanding aromatic hydrocarbon carcinogenesis. The central role of radical cations in metabolic activation. Pharmacol Ther 1992; 55:183–199.
84. Flesher JW, Myers SR, Stansbury KH. The site of substitution of the methyl group in the bioalkylation of benzo(a)pyrene. Carcinogenesis 1990; 11:493–496.
85. Surh YJ, Liem A, Miller EC, Miller JA. Metabolic activation of the carcinogen 6-hydroxymethylbenzo(a)pyrene: formation of an electrophilic sulfuric acid ester and benzylic DNA adducts in rat liver in vivo and in reaction in vitro. Carcinogenesis 1989; 10:1519–1528.
86. Flesher JW, Myers SR, Blake JW. Bioalkylation of polynuclear aromatic hydrocarbins in vivo: a predictor of carcinogenic activity. In: Cooke M, Dennis A, eds. Polynuclear Aromatic Hydrocarbons: A Decade of Progress. New York: Batelle Press, 1988:261–276.
87. Bickers DR, Dutta-Choudhury T, Mukhtar H. Epidermis: a site of drug metabolism in neonatal rat skin. Studies on cytochrome P450 content and

mixed function oxidase and epoxide hydrolase activity. Mol Pharmacol 1982; 21:239–247.
88. Guengerich FP, Martin MV. Purification of cytochrome P-450 NADPH-cytochrome P-450 reductase and epoxide hydratase from a single preparation of rat liver microsomes. Arch Biochem Biophys 1980; 205:365–379.
89. Ortiz de Montellano PR. Cytochrome P-450 catalysis: radical intermediates and dehydrogeneation reaction. Trends Pharmacol Sci 1989; 10:354–359.
90. Park JK, Shigenaga MK, Ames BN. Induction of cytochrome P450 1A1 by 2,3,7,8-tetrachlorodibenzo-p-dioxin or indolo (3,2-b) carbazole is associated with oxidative DNA damage. Proc Natl Acad Sci USA 1996; 93:2322–2327.
91. Needleman P, Turk J, Jakschik B, Morrison A, Lekowith JB. Arachidonic acid metabolism. Annu Rev Biochem 1986; 55:69–102.
92. Marnett LJ, Reed GA, Dennison DJ. Prostaglandin synthetase-dependent activation of 7,8-dihydroxy-7,8-dihydro-benzo(a)pyrene to mutagenic derivatives. Biochem Biophys Res Commun 1978; 82:210–216.
93. Guthrie J, Robertson IGC, Zeiger E, Boyd JA, Eling TE. Selective activation of dihydrodiol of several polycyclic aromatic hydrocarbons to mutagenic products by prostaglandin synthetase. Cancer Res 1982; 42:1620–1623.
94. Battista JR, Marnett LJ. Prostaglandin H synthase-dependent epoxidation of aflatoxin B_1. Carcinogenesis 1985; 6:1227–1229.
95. Robertson IGC, Sivarajah K, Eling TE, Zeiger E. Activation of some aromatic amines to mutagenic products by prostaglandin endoperoxide of syntetase. Cancer Res 1983; 43:476–480.
96. Marnett LJ, Dix TA, Sachs RJ, Siedlik PH. Oxidations by fatty acid hydroperoxides and prostaglandin synthetase. In: Samuelsson B, Ramwell P, Paoletti R, eds. Advances in Prostaglandin and Thromboxane Research. New York: Raven Press, 1983:79–86.
97. Cadenas E, Hochstein P, Ernster L. Pro and antioxidant functions of quinones and quinone reductases in mammalian cells. Adv Enzymol Related Areas Mol Biol 1992; 65:97–146.
98. Marnett LJ, Bienkowski MJ. Hydroperoxide dependent oxygenation of trans-7,8-dihydroxy-7,8-dihydrobenzo(a)pyrene by rat seminal vesicle microsomes. Source of the oxygen. Biochem Biophys Res Commun 1980; 96:639–647.
99. Dix TA, Marnett LJ. Free radical epoxidation of 7,8-dihydroxy-7,8-dihydrobenzo(a)pyrene by hematin and polyunsaturated fatty acid hydroperoxides. J Am Chem Soc 1981; 103:6744–6746.
100. Nugteren DJ. Arachidonate lipoxygenase in blood platelets. Biochim Biophys Acta 1975; 380:299–302.
101. Borgeat P, Hamberg M, Samuelsson B. Transformation of arachidonic acid and homo-gamma-linolenic acid by rabbit polymorphonuclear leukocytes. J Biol Chem 1976; 251:7816–7820.
102. Malle E, Leis HJ, Karadi I, Kostner GM. Lipooxygenases and hydroperoxy/hydroxy-eicosatetraenoic acid formation. Int J Biochem 1987; 19:1012–1022.
103. Henke D, Danilowicz R, Eling T. Arachidonic acid metabolism by isolated

epidermal basal and differential keratinocytes from the hairless mouse. Biochem Biophys Acta 1986; 876:271–279.
104. Marnett LJ, ed. Arachidonic Acid Metabolism and Tumor Initiation. Boston: Martinus Nijhoff Publishing, 1985.
105. deGroot JMC, Veldink GA, Vliegenthart JFG, Boldingh J, Wever R, van Gelder BF. Demonstration by EPR spectroscopy of the functional role of iron in soybean lipoxygenase-1. Biochem Biophys Acta 1975; 377:71–79.
106. Mansuy D, Cucurou C, Biatry B, Battioni JP. Soybean lipoxygenase-catalyzed oxidations by linoleic acid hydroperoxide: different reducing substrates and dehydrogenation of phenidone and BW 755C. Biochem Biophys Res Commun 1988; 151:339–346.
107. Eling TE, Thompson DC, Foureman GL, Curtis JF, Hughes MF. Prostaglandin H synthase and xenobiotic oxidation. Annu Rev Pharmacol Toxicol 1990; 30:1–45.
108. Sun Y. Free radicals, antioxidant enzymes and carcinogenesis. Free Radicals Biol Med 1990; 8:583–599.
109. Reed GA, Curtis JF, Mottley C, Eling TE, Mason RF. Epoxidation of (±)-7,8-dihydroxy-7,8-dihydrobenzo(a)pyrene during (bi)sulfate autoxidation: activation of a procarcinogen by a carcinogen. Proc Natl Acad Sci USA 1986; 83:7499–7502.
110. Fischer SM, Floyd RA, Copeland ES. Oxy radicals in carcinogenesis—A Chemical Pathology Study Section Workshop. Cancer Res 1988; 48:3882–3887.
111. Nishimura A, Ames BN. U.S.-Japan Meeting on "Oxygen Radicals in Cancer." Jpn J Cancer Res (Gann) 1986; 77:843–853.
112. Pruess-Schwartz D, Nimesheim A, Marnett LJ. Peroxyl radical and cytochrome-P450-dependent metabolic activation of (+)-7,8-dihydroxy-7,8-dihydrobenzo(a)pyrene in mouse skin *in vitro* and *in vivo*. Cancer Res 1989; 49: 1732–1737.
113. Melikian AA, Bagheri K, Hoffmann D. Oxidation and DNA binding of (+)-7,8-dihydroxy-7,8-dihydrobenzo(a)pyrene in mouse epidermis in vivo and effects of coadministration of catechol. Cancer Res 1990; 50:1795–1799.
114. Ji C, Marnett LJ. Oxygen radical-dependent epoxidation of (7S,8S)-dihydroxy-7,8-dihydrobenzo(a)pyrene in mouse skin in vivo. Stimulation by phorbol esters and inhibition by antiinflammatory steroids. J Biol Chem 1992; 267:17842–17848.
115. Reddy AP, Preuess Schwartz D, Ji CA, Gorycki P, Marnett LJ. ^{32}P-postlabeling analysis of DNA adduction in mouse skin following topical administration of (+)-7,8-dihydroxy-7,8-dihydrobenzo(a)pyrene. Chem Res Toxicol 1992; 5:25–33.
116. Schallreuter KU, Pittelkow MD, Wood JM. Free radical reduction by thioredoxin reductase at the surface of normal and vitiliginous human keratinocytes. J Invest Dermatol 1986; 87:728–732.
117. Trenam CW, Blake DR, Morris CJ. Skin inflammation: reactive oxygen species and the role of iron. J Invest Dermatol 1992; 99:675–682.

118. Schallreuter KU, Wood JM. The role of thioredoxin reductase in the reduction of free radicals at the surface of the epidermis. Biochem Biophys Res Commun 1986; 136:630–637.
119. Schallreuter KU, Wood JM, Berger J. Low catalase levels in the epidermis of patients with vitiligo. J Invest Dermatol 1991:1081–1085.
120. Carraro C, Pathak MA. Characterization of superoxide dismutase from mammalian skin epidermis. J Invest Dermatol 1988; 90:31–36.
121. Kobayashi T, Matsumoto M, Ilzuka H, Suzuki K, Taniguuchi N. Superoxide dismutase in psoriasis, squamous cell carcinoma and basal cell epithelioma: an immunohistochemical study. Br J Dermatol 1991; 124:555–559.
122. Miyachi Y, Horio T, Imamura S. Sunburn cell formation is prevented by scavenging oxygen intermediates. Clin Exp Dermatol 1983; 8:305–310.
123. Perchellet JP, Perchellet EM, Orten DK, Schneider BA. Inhibition of the effects of 12-O-tetradecanoylphorbol-13-acetate on mouse epidermal glutathione peroxidase and ornithine decarboxylase activities by glutathione level-raising agents and selenium-containing compounds. Cancer Lett 1985; 26: 283–293.
124. Chasseaud LF. The role of glutathione and glutathione S-transferases in the metabolism of chemical carcinogens and other electrophilic agents. Adv Cancer Res 1979; 29:175–274.
125. Mantle TJ, Pickett CB, Hayes JD, eds. Glutathione S-Transferases and Carcinogenesis. Philadelphia: Taylor and Francis, 1987.
126. Ketterer B. Protective role of glutathione and glutathione transferases in mutagenesis and carcinogenesis. Mutat Res 1988; 202:343–361.
127. Ketterer B, Meyer DJ. Glutathione transferases: a possible role in the detoxification and repair of DNA and lipid hydroperoxides. Mutat Res 1989; 214: 33–40.
128. Dogan R, Soyuer U, Tanrikulu G. Superoxide dismutase and myeloperoxidase activity in polymorphonuclear leukocytes and serum ceruloplasmin and copper levels in psoriasis. Br J Dermatol 1989; 120:239–244.
129. Hawk JLM, Parrish JA. Responses of normal skin to ultraviolet radiation. In: Parrish JA, Kripke ML, Morrison WL, eds. Photoimmunology. New York: Plenum Medical Books, 1983:219–260.
130. Connor MJ, Wheeler LA. Depletion of cutaneous glutathione by ultraviolet radiation. Photochem Photobiol 1987; 46:239–245.
131. Martensson J, Meister A, Martensson J. Glutathione deficiency decreases tissues ascorbate levels in newborn rats: ascorbate spares glutathione and protects. Proc Natl Acad Sci USA 1991; 88:4656–4660.
132. Meister A, Anderson ME. Glutathione. Annu Rev Biochem 1983; 52:711–760.
133. Fuchs J, Packer L. Ultraviolet irradiation and the skin antioxidant system. Photodermatol Photoimmunol Photomed 1990; 7:90–92.
134. Shindo Y, Witt E, Han D, Packer L. Dose-response effects of acute ultraviolet irradiation on antioxidants and molecular markers of oxidation in murine epidermis and dermis. J Invest Dermatol 1994; 102:470–475.

135. Black HS. Potential involvement of free radical reactions in ultraviolet light-mediated cutaneous damage. Photochem Photobiol 1987; 46:213–221.
136. Pacifici RE, Kono Y, Davies KJA. Hydrophobicity as the signal for selective degradation of hydroxyl radical-modified hemoglobin by the multicatalytic proteinase complex, proteosome. J Biol Chem 1993; 268:15405–15411.
137. Cathcart R, Schiviers E, Saul RL, Ames BN. Thymine glycol and thymidine glycol in human and rat urine: a possible assay for oxidative DNA damage. Proc Natl Acad Sci USA 1984; 81:5633–5637.
138. Satoh MS, Poirer GG, Lindahl T. NAD(+)-dependent repair of damaged DNA by human cell extracts. J Biol Chem 1993; 268:5480–5487.
139. Satoh M, Lindahl T. Enzymatic repair of oxidative DNA damage. Cancer Res 1994; 54:1899s–1901s.
140. Douglas D, Fridovich I. Free radicals in cutaneous biology. J Invest Dermatol 1994; 102:671–675.
141. Witz G. Active oxygen species as factors in multistage carcinogenesis. Proc Soc Exp Biol Med 1991; 198:675–682.
142. Hruza LL, Pentland AP. Mechanisms of UV-induced inflammation. J Invest Dermatol 1993; 100:35S–41S.
143. DiGiovanni J. Modification of multistage skin carcinogenesis in mice. In: Ito N, Sugano H, eds. Modification of Tumor Development in Rodents. Basel: Karger, 1991:192–199.
144. Slaga TJ. Multistage skin tumor promotion and specificity of inhibition. In: Slaga TJ, ed. Mechanisms of Tumor Promotion, Vol II: Tumor Promotion and Skin Carcinogenesis. Boca Raton, FL: CRC Press, 1984:189–196.
145. Boutwell RK. Some biological aspects of skin carcinogenesis. Prog Exp Tumor Res 1964; 4:207–249.
146. Yuspa SH. Cutaneous chemical carcinogenesis. J Amer Acad Dermatol 1986; 15:1031–1044.
147. Yuspa SH, Harris CC. Molecular and cellular basis of chemical carcinogenesis. In: Schottenfeld Fraumeni JF, eds. Cancer Epidemiology and Prevention. Philadelphia: WB Saunders Co., 1982:23–43.
148. Yuspa SH, Poirier MC. Chemical carcinogenesis: from animal models to molecular models in one decade. Adv Cancer Res 1988; 50:25–70.
149. Ames BN. Endogenous oxidative DNA damage, aging, and cancer. Free Rad Res Commun 1989; 7:121–128.
150. Epe B. Genotoxicity of singlet oxygen. Chem Biol Interact 1991; 80:239–260.
151. Van den Akker E, Lutgerink JT, Lafleur MV, Joenje H, Retel J. The formation of one-G deletions as a consequence of single-oxygen induced DNA damage. Mutat Res 1994; 309:45–52.
152. Halliwell B, Aruoma OI, eds. DNA and Free Radicals. Chichester: Ellis Horwood, 1993.
153. Halliwell B, Aruoma OI. DNA damage by oxygen-derived species. Its mechanism and measurement in mammalian systems. FEBS Lett 1991; 281:9–19.
154. Simic MG. Mechanisms of inhibition of free-radical processes in mutagenesis and carcinogenesis. Mutat Res 1988; 202:377–386.

155. Hegi ME, Sagelsdorff P, Lutz W. Detection of ^{32}-P postlabeling of thymidine glycol in g-irradiated DNA. Carcinogenesis 1989; 10:43–47.
156. Dizdarogln M. Measurement of radiation-induced damage to DNA at the molecular level. J Radiat Biol 1992; 61:175–183.
157. Shigenaga MK, Gimeno CJ, Ames BN. Urinary 8-hydroxy-2′-deoxyguanosine as a biological marker of in vivo oxidative DNA damage. Proc Natl Acad Sci USA 1989; 86:9697–9701.
158. Aruoma OI, Halliwell B, Dizdaroglu M. Iron ion-dependent modification of bases in DNA by superoxide radical-generating system hypoxanthine/xanthine oxidase. J Biol Chem 1989; 264:13024–13028.
159. Breimer LH. Molecular mechanisms of oxygen radical carcinogenesis and mutagenesis: the role of DNA base damage. Mol Carcinogenesis 1990; 3:188–197.
160. Wei H, Frenkel K. Suppression of tumor promoter-induced oxidative events and DNA damage in vivo by sarcophytol A: a possible mechanism of antipromotion. Cancer Res 1992; 52:2298–2303.
161. Hattori-Nakakuki Y, Nishigori C, Okamoto K, Imamura S, Hiai H, Toyokuni S. Formation of 8-hydroxy-2′-deoxyguanosine in epidermis of hairless mice exposed to near UV. Biochem Biophys Res Commun 1994; 201:1132–1139.
162. Cerutti PA. Oxy-radicals and cancer. Lancet 1994; 344:862–863.
163. Jackson JH. Potential molecular mechanisms of oxidant-induced carcinogenesis. Environ Health Perspect 1994; 102(suppl 10):155–157.
164. Brawn K, Fridovich I. DNA strand scission by enzymatically generated oxygen radicals. Arch Biochem Biophys 1981; 206:414–419.
165. Floyd RA. The role of 8-hydroxyguanosine in carcinogenesis. Carcinogenesis 1990; 11:1447–1450.
166. Daya-Grosjean L, Robert C, Drougard H, Suarez A, Sarasin A. High mutation frequency in *ras* genes of skin tumors isolated from DNA repair deficient xeroderma pigmentosum patients. Cancer Res 1993; 53:1625–1629.
167. Moriya M, Du C, Bodepudi V, Johnson F, Takeshita M, Grollman A. Site-specific mutagenesis using a gapped duplex vector: a study of translesion synthesis past 8-oxodeoxyguanosine in *E. coli*. Mutat Res 1991; 254:281–288.
168. Kanjilal S, Ananthswamy HN. The role of oncogenes and tumor suppressor genes in ultraviolet-induced carcinogenesis. In: Mukhtar H, ed. Skin Cancer: Mechanisms and Human Relevance. Boca Raton, FL: CRC Press, 1995:305–316.
169. Maccubbin AE, Przybyszewski J, Evans MS, Budzinski EE, Patrzyc HB, Kulesz-Martin M, Box HC. DNA damage in UVB-keratinocytes. Carcinogenesis 1995; 16:1659–1660.
170. King JK, Egner PA, Kensler TW. Generation of DNA base modification following treatment of cultured murine keratinocytes with benzoyl peroxide. Carcinogenesis 1996; 17:317–320.
171. Fujiwara M, Honda Y, Inoue H, Hayatsu H, Arimoto S. Mutations and oxidative DNA damage in phage M13mp2 exposed to N-nitrosomorpholine plus near-ultraviolet light. Carcinogenesis 1996; 17:213–218.

172. Feig DI, Loeb LA. Mechanisms of mutations by oxidative DNA damage reduced fidelity of mammalian DNA polymerase beta. Biochemistry 1993; 32:4466–4473.
173. Feig DI, Reid TM, Loeb LA. Reactive oxygen species in tumorigenesis. Cancer Res (suppl) 1994; 54:1890s–1894s.
174. Sands AT, Abuin A, Sanchez A, Conti CJ, Bradley A. High susceptibility to ultraviolet-induced carcinogenesis in mice lacking XPC. Nature 1995; 377: 162–165.
175. Routledge MN, Wink DA, Keefer LK, Dipple A. DNA sequence changes induced by two nitric oxide donor drugs in the supF assay. Chem Res Toxicol 1994; 7:628–632.
176. Wink DA, Kasprzak KS, Maragos CM, Elespuru RK, Misra M, Dunams TM, Koch WH, Andrews AW, Allen JS, Keefer LK. DNA deaminating ability and genotoxicity of nitric oxide and its progenitors. Science 1991; 254: 1001–1003.
177. Nyugen T, Brunson D, Crespi CL, Penman BW, Wishnok JS, Tannenbaum SR. DNA damage and mutation in human cells exposed to nitric oxide in vitro. Proc Natl Acad Sci USA 1992; 89:3030–3034.
178. Rideout WM III, Coetzee GA, Olumi AF, Jones PA. 5-Methylcytosine as an endogenous mutagen in the human LDL receptor and p53 genes. Science 1990; 249:1288–1290.
179. Ehlrich M, Norris KF, Wang RYF, Kuo K, Gehrke CW. DNA cytosine methylation and heat induced deamination. Biosci Rep 1990; 6:387–393.
180. Beckman JS, Beckman TW, Chen J, Marshall PA, Freeman BA. Apparent hydroxyl radical production by peroxynitrites: Implications for endothelial injury from nitric oxide and superoxide. Proc Natl Acad Sci USA 1990; 87: 1620–1624.
181. Gonzalez FJ, Crespi CL, Gelboin HV. DNA expressed human cytochrome P450s: a new age of molecular toxicology and human risk assessment. Mutat Res 1991; 247:113–127.
182. Cavalieri E, Rogan E. Metabolic activation by one-electron and two-electron oxidation in aromatic hydrocarbon carcinogenesis. In: Woo Y-T, Lai DY, Arcos JC, Argus MF, eds. Chemical Induction of Cancer. Vol. IIIB. New York: Academic Press, 1985:533–569.
183. Chouroulinkov I, Gentil A, Tierney B, Grover PL, Sims P. The initiation of tumors on mouse skin by dihydrodiols derived from 7,12-dimethylbenz(a)-anthracene and 3-methylcholanthrene. Int J Cancer 1979; 24:455–460.
184. Ramakrishna NV, Devanesan PD, Rogan EG, Cavalieri EL, Jeong H, Jankowiak R, Small GJ. Mechanism of metabolic activation of the potent carcinogen 7,12-dimethylbenz(a)anthracene. Chem Res Toxicol 1992; 5:220–226.
185. Harris CC. Chemical and physical carcinogenesis: Advances and perspectives for the 1990s. Cancer Res 1991; 51:5023s–5044s.
186. Perchellet EM, Jones D, Perchellet JP. Ability of the Ca 2+ ionophores A23187 and ionomycin to mimic some of the effects of the tumor promoter 12-O-tetradecanoylphorbol-13-acetate on the hydroperoxide production, or-

nithine decarboxylase activity, and DNA synthesis in mouse epidermis in vivo. Cancer Res 1990; 50:5806–5812.
187. Perchellet EM, Perchellet JP. Characterization of the hydroperoxide response observed in mouse skin treated with tumor promoters in vivo. Cancer Res 1989; 49:6193–6201.
188. Gali HU, Perchellet EM, Klish DS, Johnson JM, Perchellet JP. Hydrolyzable tannins: potent inhibitors of hydroperoxide production and tumor promotion in mouse skin treated with 12-O-tetradecanoylphorbol-13-acetate in vivo. Int J Cancer 1992; 51:425–432.
189. Nakamura Y, Gindhart TD, Winterstein D, Tomita I, Seed JL, Colburn NH. Early superoxide dismutase-sensitive event promotes neoplastic transformation in mouse epidermal JB6 cells. Carcinogenesis 1988; 9:203–207.
190. DiGiovanni J. Multistage carcinogenesis in mouse skin. Pharmacol Ther 1992; 54:63–128.
191. Fuchs J, Packer L. Investigations of anthralin free radicals in model systems and in skin of hairless mice. J Invest Dermatol 1989; 92:677–682.
192. Dizdaroglu M, Olinski R, Doroshow JH, Akman SA. Modification of DNA bases in chromatin of intact target human cells by activated human polymorphonuclear leukocytes. Cancer Res 1993; 53:1269–1272.
193. Guyton KZ, Bhan P, Kuppusamy P, Zweier JL, Trush M, Kensler TW. Free radical-derived quinone methide mediates skin tumor promotion by hydroxytoluene hydroperoxide: expanded role for electrophiles in multistage carcinogenesis. Proc Natl Acad Sci USA 1991; 88:946–950.
194. Trush MA, Seed JL, Kensler TW. Oxidant-dependent metabolic activation by polycyclic aromatic hydrocarbons by phorbol ester-stimulated human polymorphonuclear leukocytes: possible link between inflammation and cancer. Proc Acad Natl Sci USA 1985; 82:5194–5198.
195. Kensler TW, Egner PA, Moore KG, Taff BG, Twerdok LE, Trush MA. Role of inflammatory cells in the metabolic activation of polycyclic aromatic hydrocarbons in mouse skin. Toxicol Appl Pharmacol 1987; 90:337–346.
196. Hayashi O, Imamura S, Miyachi Y, eds. The Biological Role of Reactive Oxygen Species in Skin. New York: Elsevier Press, 1987.
197. Witz G. Active oxygen species as factors in multistage carcinogenesis. Proc Soc Exp Biol Med 1991; 198:675–682.
198. Witz G, Goldstein BD, Amoruso MA, Stone D, Troll W. Stimulation of human polymorphonuclear leukocyte superoxide anion radical ($O_2 \cdot -$) production by tumor promoters (Abstr). Proc Am Assoc Cancer Res 1980; 21:112.
199. Segal A, Van Duuren BL, Mate U, Solomon JJ, Seidman I, Smith A, Melchionne S. Tumor-promoting activity of 2,3-dihydrophorbol myristate acetate and phorbolol myristate acetate in mouse skin. Cancer Res 1978; 38:921–925.
200. Troll W, Klassen A, Janoff A. Tumorigenesis in mouse skin: Inhibition by synthetic inhibitors of proteases. Science 1970; 169:1211–1213.
201. Wattenberg LW. Inhibition of carcinogenic and toxic effects of polycyclic hydrocarbons by phenolic antioxidants and ethoxyquin. J Natl Cancer Inst 1972; 46:1425–1430.

202. Fischer SM. Arachidonic acid metabolism and tumor promotion. In: Fischer SM, Slaga TJ, eds. Arachidonic Acid Metabolism and Tumor Promotion. Boston: Martinus Nijhoff Publishing, 1985:21–47.
203. Perchellet EM, Gali HU, Gao XM, Perchellet JP. Ability of the non phorbol ester-type tumor promoter thapsigargin to mimic the stimulatory effects of 12-O-tetradecanoylphorbol-13-acetate on ornithine decarboxylase activity, hydroperoxide production, and macromolecule synthesis in mouse epidermis in vivo. Int J Cancer 1993; 55:1036–1043.
204. Fischer SM, Slaga TJ, eds. Arachidonic Acid Metabolism and Tumor Promotion. Boston: Martinus Nijhoff Publishing, 1985.
205. Needleman P, Turk J, Jakschik BA, Morrison AR, Lefkowith JB. Arachidonic acid metabolism. Annu Rev Biochem 1986; 55:69–102.
206. Fischer SM, Adams LM. Suppression of tumor promoter-induced chemiluminescence in mouse epidermal cells by several inhibitors of arachidonic acid metabolism. Cancer Res 1984; 45:3130–3136.
207. Katiyar SK, Agarwal R, Wood GS, Mukhtar H. Inhibition of 12-O-tetradecanoylphorbol-13-acetate-caused tumor promotion in 7,12-dimethylbenz(a)anthracene-initiated SENCAR mouse skin by a polyphenolic fraction isolated from green tea. Cancer Res 1992; 52:6890–6897.
208. Perchellet EM, Abney NL, Perchellet JP. Stimulation of hydroperoxide generation in mouse skin treated with tumor-promoting carcinogenic agents in vivo and in vitro. Cancer Lett 1988; 42:169–173.
209. Hartley JA, Gibson NW, Zwelling LA, Yuspa SH. The association of DNA strand breaks with accelerated terminal differentiation in mouse epidermal cells exposed to tumor promoters. Cancer Res 1985; 45:4864–4870.
210. Frenkel K, Chrzan K. Hydrogen peroxide formation and DNA base modification by tumor promoter-activated polymorphonuclear leukocytes. Carcinogenesis 1987; 8:455–460.
211. Farber B, Petrusevska RF, Fusenig NE, Kinzel V. Cytogenic effects caused by phorbol ester tumor promoters in HeLa cells: mechanistics aspects. Carcinogenesis 1989; 10:2345–2350.
212. Furstenberger G, Schurich B, Kaina B, Petrusevska RT, Fusenig NE, Marks F. Tumor induction in initiated mouse skin by phorbol esters and methyl methanesulfonate: correlation between chromosomal damage and conversion (stage I of tumor promotion) in vivo. Carcinogenesis 1989; 10:749–752.
213. Reiners JJ, Pence BC, Barcus MCS, Cantu AR. 12-O-Tetradecanoylphorbol-13-acetate-dependent induction of xanthine dehydrogenase and conversion to xanthine oxidase in murine epidermis. Cancer Res 1987; 47:1775–1779.
214. Perchellet EM, Maatta EA, Abney NL, Perchellet JP. Effects of diverse intracellular thiol delivery agents on glutathione peroxidase activity, the ratio of reduced/oxidized glutathione, ornithine decarboxylase induction in isolated mouse epidermal cells treated with 12-O-tetradecanoylphorbol-13-acetate. J Cell Physiol 1987; 131:64–73.
215. Pence BC, Reiners JJ. Murine epidermal xanthine oxidase activity: correlation with degree of hyperplasia induced by tumor promoters. Cancer Res 1987; 47:6388–6392.

216. Kensler TW, Taffe BG. Role of free radicals in tumor promotion and progression. Prog Clin Biol Res 1989; 298:233–248.
217. Schwarz M, Peres G, Kunz W, Furstenberger G, Kittstein W, Marks F. On the role of superoxide anion radicals in skin tumor promotion. Carcinogenesis 1984; 5:1663–1670.
218. Perchellet JP, Abney NL, Thomas RM, Guislain YL, Perchellet EM. Effects of combined treatments with selenium, glutathione, and vitamin E on glutathione peroxide activity, ornithine decarboxylase induction and complete and multistage carcinogenesis in mouse skin. Cancer Res 1987; 47:477–485.
219. Cerutti P, Trump B. Inflammation and oxidative stress in carcinogenesis. Cancer Cells 1991; 3:1–7.
220. Larsson R, Cerutti P. Oxidants induce phosphorylation of ribosomal protein s6. J Biol Chem 1988; 263:17452–17458.
221. Alessi C, Smythe C, Keyse S. The human CL100 gene encodes tyr/thr protein phosphatase which potently and specifically inactivates MAP kinase and suppresses its activation by oncogenic ras in Xenopus oocyte extracts. Oncogene 1993; 8:2015–2120.
222. Schreck R, Reiber P, Baeurle P. Reactive oxygen intermediates are apparently widely used messengers in the activation of NfkB transcription factor and HIV-1. EMBO J 1991; 10:2247–2258.
223. Abate C, Patel L, Rauscher F, Curran T. Redox regulation of fos and jun DNA binding activity in vitro. Science 1990; 249:1157–1161.
224. Stein B, Angel P, van Dam H, Ponta H, Herrlich P, van der Eb A, Rahmsdorf H. Ultraviolet-radiation induced c-jun gene transcription: two AP-1 like binding sites mediate the response. Photochem Photobiol 1992; 55:409–415.
225. Hennings H, Shores R, Wenk ML, Spangler EF, Tarone R, Yuspa SH. Malignant conversion of mouse skin tumors is increased by tumour initiators and unaffected by tumour promoters. Nature 1983; 304:67–69.
226. Hennings H, Shores R, Mitchell P, Spangler EF, Yuspa SH. Induction of papillomas with a high probability of conversion to malignancy. Carcinogenesis 1985; 6:1607–1610.
227. Slaga TJ, Klein-Szanto AJP, Triplett LL, Yotti LP, Trosko KE. Skin tumor-promoting activity of benzoyl peroxide, a widely used free radical generating compound. Science 1981; 213:1023–1025.
228. Duran HA, Lanfranchi H, Palmieri MA, deRey BM. Inhibition of benzoyl peroxide induced tumor promotion and progression by copper (II)(3,5-diisopropylsalicylate)2. Cancer Lett 1993; 69:167–172.
229. Athar M, Mukhtar H, Bickers DR, Khan IU, Kalyanaraman B. Evidence for metabolism of tumor promoter organic hydroperoxides into free radicals by human carcinoma skin keratinocytes: an ESR-spin trapping study. Carcinogenesis 1989; 10:1499–1503.
230. Saez E, Rutberg SE, Mueller E, Oppenheim H, Smoluk J, Yuspa SH, Spiegelman BM. C-fos is required for malignant progression of skin tumors. Cell 1995; 82:721–732.
231. Mizono M, Packer L. Suppression of protooncogene c-fos expression by antioxidant dihydrolipoic acid. Methods Enzymol 1995; 252:180–186.

232. Shibanuma M, Kuroki T, Nose K. Induction of DNA replication and expression of proto-oncogene c-myc and c-fos in quiescent Balbc/3T3 cells by xanthine/xanthine oxidase. Oncogene 1988; 3:17–22.
233. Busam KJ, Roberts AB, Spron MB. Inhibition of mitogen-induced c-fos expression in melanoma cells by retinoic acid. J Biol Chem 1992; 267:19971–19977.
234. Hartley JA, Gibson NW, Kilkenny A, Yuspa SH. Mouse keratinocytes derived from a initiated skin or papillomas are resistant to DNA strand breakage by benzoyl peroxide: a possible mechanism for tumor promotion mediated by benzoyl peroxide. Carcinogenesis 1987; 8:1827–1830.

25
Prostate Cancer

Maarten C. Bosland
Nelson Institute of Environmental Medicine, New York University Medical Center, New York, New York

Prostate cancer incidence and mortality rates have increased in the United States over the past two decades, and it is currently the most frequently diagnosed cancer and the second most frequent cause of death due to cancer in U.S. males (1). In contrast, spontaneously occurring prostate tumors are rare in most animal species (2,3), with the exception of the dog. It is not understood why prostate cancer is so common in men but very rare in almost all other species. The purpose of this chapter is to review known and suspected prostate carcinogens and to explore possible explanations for the enormous disparity in prostate cancer occurrence between humans and most other mammalian species. In addition, guidelines are proposed for interpreting carcinogenicity bioassay data that indicate prostate cancer induction.

I. EPIDEMIOLOGICAL STUDIES

The epidemiology of prostatic cancer has been reviewed in depth elsewhere (4–10). Environmental and hormonal factors, particularly androgens, are almost certainly involved in human prostatic carcinogenesis, as demonstrated in these reviews. However, with exception of a few previous reviews (4,5,10), no systematic attempts have been made to compare human and animal data on chemical exposures and prostate carcinogenesis.

A. Smoking

Tobacco smoke, which is a major cause of human cancer and contains many carcinogenic, co-carcinogenic, and tumor-promoting chemicals, has until recently never been found to be associated with prostate cancer risk (6,11). A few recent studies, however, have found a relationship between smoking and risk for prostate cancer (11–15). However, these studies have been criticized on the basis of their methods and interpretations (16–18). Furthermore, the vast majority of studies have demonstrated an absence of a relation between smoking and prostate cancer risk (6,11,18), including several recent case-control and cohort studies (e.g., 19–22, see also Ref. 18).

In summary, although some recent studies indicate a possible association between smoking, and perhaps also smokeless-tobacco use, and prostate cancer risk, the vast majority of studies found no such relationship, and in the few studies that did, only slightly elevated risks were found (18). However, a commonality in etiology between prostate cancer and smoking-related cancers is suggested by the 2 to 1.5 times higher risk of prostate cancer as a second primary cancer in men with lung cancer or other smoking-related cancers (pancreas, urinary bladder, upper aerodigestive tract), respectively (23). In addition, smoking appears to have effects on circulating levels of testosterone and other hormones that may be involved in prostate carcinogenesis (24,25). In conclusion, any association between tobacco use and prostate cancer risk is unlikely, but cannot be ruled out entirely (18).

B. Occupation and Occupational Exposures

An in-depth review of some of the current literature on occupational factors and prostate cancer risk by van der Gulden and coworkers (10) confirms some of the conclusions of earlier literature reviews on this subject (4,6). Although several studies have observed increased risk for a variety of occupations (see, e.g., some recent studies: 10,26–34), this is found with some consistency for only a few occupational categories, e.g., farmers/farm workers, armed services personnel (4,6), and auramine manufacture (35). Since excess risk has been found for most categories of armed service personnel, any specific chemical exposures are unlikely to be involved. An association between exposures in the rubber industry, coke production, and iron/steel founding and prostate cancer risk is limited to one or a few plants or one or a few cohorts (4,6,35–37). A possible elevated prostate cancer risk in mechanics, repairmen, and machine operators—a very diverse group of occupations with a large range of potential exposures to carcinogenic agents—is discussed by van der Gulden et al. (10).

Two studies found higher than expected rates of prostate cancer among workers exposed to acrylonitrile in textile fiber-production plants (38,39). Rates were highest in workers exposed to the highest levels and with the longest period between onset of exposure and cancer diagnosis. These studies were based on only a few cases (4–6), but exposure to dimethylformamide, a possible confounder in these workers, does not appear to be associated with prostate cancer risk (40). Recent studies about risk in farmers or farm workers and in cadmium-exposed populations is summarized in the following sections. In addition, the possible relationship between ionizing radiation exposure and prostate cancer risk is discussed.

Farming and Farm Work

A slightly increased risk for prostate cancer has been found for farmers and farm workers in many, but not all, studies (6,10,41,43). Van der Gulden et al. (10) evaluated a large number of incidence and mortality studies relevant to this subject. Prostate cancer risk was significantly higher than expected in 3 of 14 incidence studies (21%) and 6 of 17 mortality studies (35%). The measure of prostate cancer risk (odds ratio, etc.) was equal or higher than 1.10 in 9 of the incidence studies (64%) and 10 of the mortality studies (58%). Several of these studies were included in a meta-analysis on 22 mortality and incidence studies conducted by Blair et al. (41). A significantly elevated meta-relative risk of 1.08 was calculated, with a 95% confidence interval of 1.06–1.11 for a range of relative risks from 0.9 to 2.7. From recent epidemiological studies in farmers, van der Gulden et al. (33), Burmeister (43), and Fincham et al. (44) reported additional data indicating increased prostate cancer risk, whereas three other studies found no significantly elevated risk (43,46,47). From these studies and those summarized by Blair et al. (41) and van der Gulden et al. (10), it appears that there is inconsistent evidence for weak positive association between farming or farm work and prostate cancer risk.

Recently no association was found between prostate cancer risk and occupational exposure to insecticides or pesticides (48) or to herbicides (49). These data are relevant to the farmwork situation. Van der Gulden et al. (33), Morrison et al. (49), and Pearce and Reif (42) list a number of studies showing excess risk for prostate cancer in populations exposed to pesticides, herbicides, and/or fertilizers, as well as several studies in which no such association was found; this is consistent with the above-mentioned weak and inconsistent association between farming/farm work and prostate cancer risk.

Cadmium Exposure

The literature on cadmium exposure and prostate cancer risk, which has recently been summarized by the IARC (50), Waalkes and Oberdörster (51), and Waalkes and Rehm (52), indicates that there is no consistent evidence that cadmium exposure is associated with elevation of prostate cancer risk and if such an association existed it would be minimal at best (6). In none of some recent studies (30,53) were elevated risks for prostate cancer observed in (potentially) cadmium-exposed occupational cohorts. On the other hand, a high incidence of prostate cancer has been found to correlate with high cadmium levels in groundwater available for drinking purposes (54) or other potential environmental (55) and dietary sources of cadmium (6,19,26). This is in contrast with an earlier found absence of a correlation between prostate cancer risk and cadmium levels in drinking water (6). It is, however, in line with the aforementioned possible positive association with tobacco smoke, which is a major source of cadmium exposure (50).

Ionizing Radiation

A number of previously summarized (6) reports suggest that exposure to ionizing radiation increases risk for prostate cancer in workers in the nuclear industry. Several studies from the United Kingdom Atomic Energy Authority have demonstrated such an increased risk (56–58). An additional study of nuclear industry workers confirms this, but the elevation in risk was not significant (59), and another recent study did not find increased prostate cancer risk (60). Furthermore, studies of cancer risk among 76,000 survivors of the 1945 atomic bomb explosions in Japan, including revised estimates of radiation dose received, indicate that prostate cancer risk is not elevated in this cohort (61–63). Interestingly, an increased prostate cancer risk was found among airline pilots exposed to in-flight radiation (64). Also, a strong international correlation between prostate cancer incidence and indoor radon levels in 14 countries ($r = 0.72$; $p < 0.01$) has been reported (65), but inconclusive data also exist (66). Taken together, these studies provide some support for an association between prostate cancer risk and exposure to ionizing radiation (6), but the current evidence for this is equivocal.

C. Conclusions

Only a few risk factors for prostate cancer have consistently been found in epidemiological studies (4,6,9). Tables 1 and 2 summarize the current information on prostate cancer and occupation and occupational exposures. Although there is some evidence to suggest that farming or farm work, armed forces employment, and exposures to ionizing radiation are associated with a slight elevation in prostate cancer risk, there is to date no credible evidence that exposures to specific chemicals are risk factors for prostate cancer.

The single most important risk factor combination is to be of African descent and residing in the United States. Whereas genetic factors are unlikely to be involved, environmental exposures (in the broadest sense of the term) are most probably responsible for this, as discussed in depth previously (4,6). African Americans have a prostate cancer risk that is approximately twice that of white Americans. To date there are no definite reasons known for the black-white disparity in prostate cancer rates in the United States, but a relationship to similar racial disparities in exposure to potential carcinogens and high-risk dietary habits has been proposed (6). The strongest single risk factor is a western lifestyle, including western dietary habits. It is conceivable that this risk factor is mediated by a hormonal mechanism involving androgens (6). Familial aggregation of prostate cancer risk is very consistently observed and confers a considerable (3- to 4-fold or more) increase in risk. However, it explains only less than 10% of all cases.

In conclusion, with exception of exposure to a western lifestyle, there are no known specific chemical or other exposures associated with prostate cancer risk. This lack of known risk-enhancing exposures is remarkable in view of the high frequency of this malignancy in western countries. The following hypotheses to explain this have been suggested (4):

1. The factors responsible for high risk of clinically evident prostate cancer are very ubiquitously present in western countries, but not uncommon in other areas such as East Asia.
2. There is a large number of different prostate cancer-causing factors, probably present in different combinations throughout the world, that require an approximately conducive setting, i.e., a western lifestyle environment, to produce clinical prostate cancer.
3. The ubiquitously present risk factor *is* the western lifestyle environment factor (i.e., a combination of the above two hypotheses).

The last hypothesis is consistent with the observation that early prostate cancer that is not clinically evident (so-called latent cancer) occurs at similar rates throughout the world, whereas there are large geographic differences in the frequency of clinically evident prostate cancer (6,67). This implies that exposure to a western lifestyle causes or enhances progression from latent to clinically evident cancer by an unknown, perhaps hormonal, mechanism, which permits required additional genetic alterations to occur in the target cells (67). The existence of a large number of different prostate cancer–causing factors could explain why no specific exposures have been found to be related to prostate cancer risk and is consistent with the weak to very weak risks found for farming/farm work and armed services em-

Table 1 Association Between Exposure to Specific Chemicals and Prostate Cancer

	IARC general carcinogenicity[a]			
Chemical	human	animal	Human data on prostate cancer	Animal data on prostate cancer
Acrylonitrile	Yes	Yes	Two- to threefold excess frequency in 2 cohorts (small number of cases); no excess in several other studies (35,38,39); probably not a human prostatic carcinogen	No evidence of prostate cancer induction in several studies (35)
Cadmium	Yes	Yes	Overall no evidence for relation with prostate cancer (35,50–52); but slight excess risk found in a few studies with a low number of cases (10,26,50); possibly a very weak prostatic carcinogen, but most likely not a human prostatic carcinogen (50,53,146).	Induces non-invasive proliferative lesions of the rat ventral prostate, including adenomas, but invasive carcinoma is only found after intraprostatic cadmium injection (see text)
3,2′-Dimethyl-4-aminobiphenyl	N/A	Yes	No human exposure occurs	Induces rat ventral prostate adenoma; 32% maximum incidence; invasive cancer only after hormonal manipulation (see text)
7,12-Dimethylbenz[*a*]anthracene	N/A	Yes	No human exposure occurs	Induces invasive adenocarcinoma of rat dorsolateral prostate, but only after hormonal manipulation (see text)

N-Methyl-*N*-nitrosourea	N/A	Yes	No human exposure occurs	Induces invasive adenocarcinoma of rat dorsolateral prostate, but only after hormonal manipulation (see text)
N-Nitrosobis-(2-oxopropyl)amine and *N*-nitrosobis-(2-hydroxypropyl)amine	N/A	Yes	No human exposure occurs	Induces invasive squamous cell carcinoma of rat ventral prostate; adenocarcinoma of the dorsolateral prostate only after hormonal manipulation (see text)
Testosterone	Probable	Yes	Human exposure occurs as abuse of anabolic steroids and as androgen replacement in hypogonadic men or as substitution therapy in aging men; in addition, every man is exposed to endogenous testosterone	Invasive adenocarcinoma of dorsolateral rat prostate in 10–40% incidence (see text)
Diethylstibestrol (DES)	Yes	Yes	Prenatal exposure has occurred in a few million men between the mid 1940s and the early 1970s, but there is no information about prostate cancer risk in this cohort. DES is also used as prostate cancer therapy but this is unlikely to result in additional cancer	Induces a variety of proliferative lesions in rat and mouse accessory sex glands, including invasive cancer of prostate structures (see text)
17β-Estradiol	Yes	Yes	Every man is exposed to endogenous 17β-estradiol, but no exposure to 17β-exogenous estradiol occurs	When co-administered with testosterone induces a high incidence of prostatic adenocarcinomas in rats (see text)

N/A = Not applicable.
[a]Carcinogenicity for any site according to IARC (35,50).

Table 2 Association Between Exposure in Industrial Processes or Occupations and Prostate Cancer

Industrial process or occupation	IARC general carcinogenicity[a]		Human data on prostate cancer	Animal data on prostate cancer
	human	animal		
Armed forces employment	Not applicable		Slight excess risk has consistently been found in several studies of various design (6); unknown factor(s) related to armed forces employment are possibly carcinogenic for the prostate	No relevant data available
Auramine manufacture	Yes	Yes (auramine exposure)	Excess risk in two studies of one cohort, but no confirmation in other cohorts (35); probably carcinogenic for the prostate	No evidence of prostate cancer induction in a few studies (35)
Coke production	Yes	No data	1.6-fold excess mortality (58 cases) in one group of cohorts, but no evidence of excess prostate cancer risk in several other studies (35,37); possibly a very weak prostatic carcinogen, but most likely not a prostatic carcinogen	No relevant data available

Farming/Farm work	Not applicable		Slight excess risk has been found in a large number of studies of various design, but it is inconsistent (41,43, and see text); unknown farming-related factor(s) are probably carcinogenic for the prostate	No relevant data available
Iron/Steel founding	Yes	No data	Two-fold excess risk found in only one cohort (very small number of cases), but not in many other studies (35,37); probably not carcinogenic for the prostate	No relevant data available
Rubber industry	Yes	Probable	Excess risk has only been observed in one or a few U.S. plants, but not in several studies of other work sites (36)	Rats housed at sites in a rubber plant developed prostate carcinomas (6–13% incidence) (6,36)

[a]Carcinogenicity for any site according to IARC (35).

ployment and the inconsistencies in associations with chemical and radiation exposures.

II. EXPERIMENTAL STUDIES

As indicated earlier, for unknown reasons spontaneous prostate tumors are rare in most species, including rodents (2,3), with the exception of dogs and humans. The rodent prostate, unlike the human or canine prostate, consists of distinct paired lobes: the ventral, dorsal, lateral, and anterior lobes. The dorsal and lateral lobes are frequently referred to as the dorsolateral prostate, and the anterior lobe is more commonly termed the coagulating gland. In the human and canine prostate, these lobes have merged into one gland, in which different zones have been defined (68). A homolog of the rodent ventral lobe is not present in the human gland (69). The various lobes of the rat prostate differ in their propensity to develop prostate carcinomas, either spontaneously or induced by carcinogens or hormones (70). However, the exact lobe location of prostate tumors has not often been accurately reported in the literature.

The prostate is rich in certain drug-metabolizing enzymes, some of which are highly inducible. For example, the rat ventral prostate has a level of mRNA expression of the P450PCN gene that is similar to that in the liver (71). Although the activity of aryl hydrocarbon hydroxylase in the rat ventral prostate is 1000-fold lower than in the liver, it can be induced 1000-fold by β-naphthoflavone in the prostate but only 8-fold in the liver (72). Furthermore, there is a specific high-affinity receptor for phorbol esters in the rat ventral prostate (73). Also, two TCDD-binding proteins have been found in the rat ventral prostate, one of which has similarities with the TCDD receptor, and the other with the androgen-dependent prostatic binding/secretory protein (72,74). Selective metabolism of specific chemical carcinogens in the prostate may determine the sensitivity of this target tissue for these chemicals in rats and humans, as reviewed previously (6). Thus, there are multiple mechanisms whereby carcinogenic agents and tumor promoters can affect the rat and, perhaps, human prostate. However, there is hardly any information in this regard about the rat dorsolateral and anterior prostates and the human prostate.

A. Induction of Prostate Tumors by Carcinogenic Agents

No induction of prostatic neoplasms had been found in carcinogenicity bioassays involving hundreds of different chemicals (35,75–77). The only reports of induction of prostate tumors is from experimental studies on prostatic carcinogenesis animal models and the effects of cadmium administration and treatment with hormones.

Organic Chemical Carcinogens

Direct application of chemical carcinogens to prostate tissue in experimental animals can produce sarcomas or squamous cell carcinomas (3,78). Induction of prostate tumors by chemical carcinogens administered systemically or via the oral or inhalation routes is very rare. There are only two carcinogens that, without any additional concomitant or subsequent treatment, induce prostate tumors, both upon systemic administration, are the model carcinogens *N*-nitrosobis(oxopropyl)amine (BOP) and 3,2′-dimethyl-4-aminobiphenyl (DMAB). Pollard and coworkers (79,80) reported induction of prostate carcinomas in 2 of 20 (10%) and 4 of 20 (20%) of Lobund Wistar rats given a single i.v. injection of *N*-methyl-*N*-nitrosourea (MNU), but this was not confirmed by Hoover et al. (81) in a group of 45 rats of the same strain.

Weekly gavage administration of 10 mg/kg BOP to MRC rats resulted in a 33% incidence of squamous cell carcinomas in the ventral prostate with an average latency of 50 weeks (82). In this study, there was a high incidence of atypical glandular hyperplasia, which seemed to progress to a squamous differentiation. Administration for no longer than 20 weeks did not produce prostate tumors, and subcutaneous administration was less effective than intragastric treatment (82,83).

Katayama et al. (84) were the first to report induction of epithelial proliferative lesions of the F344 rat prostate by injection with 20–50 mg/kg DMAB for 20–37 weeks. This was subsequently confirmed by Shirai and coworkers (85). The induced lesions, referred to as carcinomas in situ, were noninvasive atypical epithelial proliferations confined to one or a few adjacent alveoli, and they occurred exclusively in the ventral prostate (84,85). The highest reported incidence following DMAB administration without any other treatment was 32% (84). One invasive adenocarcinoma apparently originating from the dorsolateral lobe was found in one of 293 DMAB-exposed rats (0.34%) (84). The age at which DMAB is administered has no effect on its prostatic carcinogenicity since similar in situ carcinoma responses were obtained in rats aged 5, 35, or 65 weeks (86). There are, however, important strain differences in the carcinogenic potential of DMAB for the rat prostate; the F344 and ACI/N strain are very susceptible and the LEW, Sprague-Dawley, and Wistar strains develop few or no in situ prostate carcinomas (87).

DMAB is metabolically activated by initial *N*-hydroxylation in the liver or prostate followed by O-acetylation via acetyl CoA. Administration of the *N*-hydroxy metabolite of DMAB at a weekly dose of 5 mg/kg for 20 weeks produced a 42% incidence of in situ carcinoma and atypical hyperplasia in 84% of the animals in F344 rats, but Wistar rats did not develop in situ carcinomas (88). In a subsequent experiment, the same treatment did not produce in situ carcinomas, and only 46% hyperplasia; a higher dose of 20 mg/kg, however, produced in situ carcinomas in 67% of animals and hyperplasia in 100% (89). Notwithstanding this inconsistency (perhaps due to instability of the carcinogen), these data indicate that *N*-hydroxy-DMAB is more carcinogenic for the rat prostate than DMAB itself, because DMAB at a dose of 25 mg/kg resulted in 25 and 50% incidences of in situ carcinomas and atypical hyperplasia, respectively (89).

O-acetylation occurs only in the ventral, but not the dorsolateral prostate in the rat, which may explain why DMAB selectively targets the ventral lobe (90). DMAB-DNA adduct formation studies using polyclonal antibodies have identified the ventral prostate as a major site of adduct formation, and the adducts persisted longer at that site than in any of the other tissues examined, including dorsal and lateral prostate, seminal vesicle, liver, and colon (87,91,92). However, when several rat strains that differed in susceptibility to DMAB-induced ventral prostate carcinoma in situ were compared, there was only little correlation with ventral prostate O-acetylation (90) and DMAB-DNA adduct formation (87). Thus, although there appears to be preferential O-acetylation of DMAB and DMAB-DNA adduct formation in the ventral prostate, the only prostate lobe target of DMAB, differences in these two processes do not explain differences among rat strains in sensitivity for prostate carcinoma induction by this compound. This suggests that there are additional, as yet unknown, determinants of the carcinogenicity of DMAB for the rat prostate.

Cadmium

Waalkes and coworkers (93,94) have demonstrated that a single injection of cadmium chloride in rats produces in situ carcinomas in the ventral prostate provided that cadmium-induced testicular toxicity is avoided by one of the following three methods: by lowering the cadmium dose to under 5 mg/kg, by administering cadmium intramuscularly rather than subcutaneously, or by antagonizing cadmium by simultaneous administration of a sufficient amount of zinc. The incidence of these noninvasive, intraalveolar lesions was between ~30 and 42%, while they occurred in ~10% of control rats. Oral cadmium exposure also resulted in such in situ lesions, but relatively few of these lesions (in less than 10% of the animals) qualified as adenoma (95), and no invasive carcinomas were found in any of these studies (93–95). In these experiments no proliferative lesions were induced in parts of the rat prostate other than the ventral lobe. These data indicate that only when testicular function, most likely testosterone production, is intact does cadmium induce proliferative lesions in the rat ventral prostate, but not in other accessory sex glands. Cadmium has been shown to be capable of producing in vitro transformation of rat ventral prostate epithelial cells (96), but these transformed cells developed into squamous cell carcinomas, but not adenocarcinomas, upon injection into syngeneic rats (97).

Metallothionein is a high-affinity metal-binding protein that is considered to play a major role in determining tissue sensitivity to the toxic and carcinogenic effects of cadmium. The rat prostate has been shown to lack metallothionein protein, but has some other cadmium-binding proteins that are distinct from metallothionein in their amino acid composition (98). The ventral prostate contained one iso-form, and the dorsolateral prostate five other separate forms of cadmium-binding proteins (98). Unlike in the liver, total cadmium-binding protein levels in both prostate lobes were not inducible by cadmium treatment, (98). Metallothionine gene mRNA expression is hardly detectable and is not inducible in the rat ventral prostate (52). Total cadmium-binding protein levels in the ventral prostate were approximately 30% of those in the dorsolateral prostate (98). The ventral prostate may be a selective target for the carcinogenic action of cadmium (52) because of this severe lack of cadmium-binding and the noninducibility of cadmium-binding proteins in the ventral prostate. These two factors are not present in the dorsolateral prostate and other tissues.

The IARC (50), Waalkes and Rehm (52), and Waalkes and Oberdörster (51) have extensively summarized these and other, older, experimental studies with cadmium. Although very little evidence of carcinogenicity of cadmium for the rodent prostate was found in the older studies, the results of the above summarized recent experiments of Waalkes and coworkers provide support for the hypothesis that cadmium is a human carcinogen. However, invasive prostate adenocarcinomas have only been described to occur occasionally following direct injection of cadmium into the prostate (50,52), and the rat ventral prostate may be unique in its lack of protective mechanisms against cadmium and may not be reflective of the human prostate.

Ionizing Radiation

Local exposure to x-rays of the pelvic area can induce prostate carcinomas in rodents. Eight exposures of 1000 rad at different intervals induced a 3.7% incidence of prostate carcinomas in ICR/JCL mice; at least some of these tumors occurred in the ventral prostate (99). Five exposures of

Sprague-Dawley rats to 1000 rad only produced prostate carcinomas, in a 33% incidence, when the animals were castrated and received androgen replacement prior to irradiation (100). Intact and castrated Sprague-Dawley rats did not develop prostate carcinomas following irradiation, indicating that testosterone treatment may be required for tumor development. Whole body x-irradiation at a single dose of 1000 rad of NEDH rats whose vascular system was surgically joined (parabiosis) to nonirradiated male partners produced prostate carcinomas in 2.2% of animals (101). The incidence of prostate carcinomas in nonirradiated partners and parabiosed controls was 0.2–0.3%. These studies indicate that x-irradiation can produce prostate cancer in rodents, providing support for the aforementioned, epidemiological indications for an association between exposure to ionizing radiation and prostate cancer risk.

B. Induction of Prostate Tumors by Stimulation of Cell Proliferation and Exposure to Chemical Carcinogens

Stimulation of cell proliferation in the prostate at the time of carcinogen administration is an effective way to increase tumor induction. Several chemical carcinogens that do not induce prostate carcinomas when administered alone have been shown to produce these tumors in a low incidence (5–25%) following stimulation of prostatic cell proliferation. This has been demonstrated for a single dose administration of the indirect-acting carcinogens DMAB and 9,12-dimethylbenz[*a*]anthracene and the direct-acting MNU (102–104). Furthermore, treatments that stimulate prostatic cell proliferation have been shown to markedly enhance the carcinogenic potential of DMAB (105,106). The site of carcinoma formation within the prostate was the ventral lobe for the DMAB studies of Shirai et al. (105) and Ito et al. (106), and the dorsolateral prostate and coagulating gland for the experiments with MNU, DMAB, and 9,12-dimethylbenz[*a*]anthracene conducted by Bosland et al. (102–104). However, other studies found only a very small or no enhancing effect of stimulation of prostatic cell proliferation on prostate carcinoma induction by MNU (107–109) and BOP (83). Differences between rat strains in their sensitivity to enhancement of prostatic carcinogenesis by stimulation of prostatic cell proliferation may explain these contrasting results (104). Although the method of stimulation of prostatic cell proliferation also differed among these studies, the levels of stimulation of prostatic cell proliferation obtained were of the same order of magnitude (83,110–112). Another possible source of variation between these studies are the carcinogens used, particularly those that are metabolized in the prostate itself, such as DMAB.

In summary, there may be multiple, as yet only partially known, factors that determine strain and lobe selectivity of prostate carcinoma induction by chemical carcinogens in rats, one of which is the rate of cell proliferation at the time of carcinogen exposure. This stimulation of cell proliferation is probably co-carcinogenic for prostate cancer induction by many chemical carcinogens. Stimulation of cell proliferation during carcinogen exposure increases the likelihood that promutagenic DNA damage, such as carcinogen-DNA adducts, gets "fixed" as permanent mutations. For example, a high percentage of rat prostate carcinomas induced by MNU plus chronic testosterone treatment contain activating G-to-A mutations in codon 12 of the K-*ras* genes (113). This suggests that base-mispairing due to formation of O^6-methylguanine is the first step in the process of multistage prostatic carcinogenesis by MNU.

C. Induction of Prostate Tumors by Hormones

Testosterone

Chronic administration of testosterone induces a low to moderate (5–56%) incidence of prostate carcinomas in several rat strains—Wistar WU, Wistar MRC, Lobund Wistar, and NBL (79,81,83,114–119)—but not in the F344 strain (112). These tumors developed from the dorsolateral prostate and/or coagulating gland, but not the ventral prostate (81,114–119), and they were adenocarcinomas in all studies but one, in which some squamous cell carcinomas were also observed (83). The prostate carcinoma incidence in most of these studies was low, i.e., between 5 and 20% (81,83,114,115). Only in the studies with the Lobund Wistar rat strain were sometimes higher carcinoma incidences found, but the reported incidences varied considerably (79–81,117,120). In conclusion, testosterone is a weak complete carcinogen for the rat prostate (121).

The mechanism of the carcinogenic and tumor-promoting effects of androgens on rat prostatic carcinogenesis is presently not known. It is quite conceivable that the normal androgen receptor–mediated mechanisms are involved. For example, it has been hypothesized (113) that prostate cells carrying critical genetic alterations may be selectively sensitive to the cell-proliferative actions, rather than to the cell-differentiating actions, of androgen. These cells would then have a selective growth advantage over normal cells, which do not respond to chronic testosterone treatment with sustained proliferation (109). The activation by a G-to-A transition in codon 12 of K-*ras* genes in a high percentage of rat prostate carcinomas induced by MNU plus chronic testosterone treatment may be one such critical genetic alteration (113).

Intact testicular androgen production is required for cadmium to produce proliferative lesions in the rat ventral prostate (93–95). This suggests that androgen may act as a tumor promoter in this system as well, but this hypothesis has not been tested specifically. Testosterone considerably increases cadmium disposition and retention in the rat ventral prostate (52,95); this pharmacokinetic effect may be relevant to cadmium carcinogenesis.

Estrogens

Noble (115), who first established that testosterone is carcinogenic for the rat prostate, also demonstrated that sequential treatment with testosterone and 17β-estradiol was even more effective, resulting in an approximately 50% prostate carcinoma incidence in the NBL rat strain that he developed. Combined long-term treatment of NBL rats with testosterone and 17β-estradiol leads to a 100% incidence of adenocarcinomas, which develop from the periurethral ducts of the dorsolateral and anterior prostate and are of microscopic size and therefore easily missed if the periurethral portion of the prostate is not examined histologically (70,122). Without testosterone the prostate atrophies in large part because the estrogen suppresses LH secretion and thereby testosterone production.

Similar long-term treatment of NBL rats with diethylstilbestrol (DES) plus testosterone produces a low carcinoma incidence in the dorsolateral prostate and some early-stage carcinomas (carcinoma in situ in the ventral lobe (122). When these treatments were given to Sprague-Dawley rats, carcinoma incidence was considerably lower (122). In conclusion, combined long-term treatment with testosterone and 17β-estradiol is strongly carcinogenic for the dorsolateral prostate in the NBL rat and weaker in other strains. The sensitivity for the carcinogenic effects of testosterone and 17β-estradiol exposure seems to be confined to the periurethral, proximal ducts of the dorsolateral and anterior prostate.

Estrogens have been shown to be capable of producing DNA damage in estrogen-carcinogenicity target tissues (123–125), independent of their interaction with the estrogen receptor (125). A direct DES-DNA adduct (126) and indirect (endogenous) DNA adducts of undetermined structure detectable by ^{32}P-postlabeling have been found in the kidney of male hamsters treated with DES (124). Sixteen-week treatment with testosterone plus 17β-estradiol enhances the formation of a chromatographically unique endogenous DNA adduct selectively in the periurethral region of the rat dorsolateral prostate, which is the site of the carcinogenic effect of this treatment (122,127,128). This enhancement of the adduct formation occurs selectively at the site of tumor formation and precedes carcinoma formation. This evidence suggests that the adduct is causally involved in the carcinogenic effect of testosterone plus 17β-estradiol treatment. In addition, there is evidence that these hormones generate oxidative DNA damage and lipid peroxidation in these target tissues (M. C. Bosland, unpublished observations). These observations suggest that these hormones act via a combination of genotoxic mechanism (most likely induced by estradiol) and other mechanisms including receptor mediation (tumor promotion by testosterone). However, the exact mechanisms whereby exposure to testosterone and estrogens leads to prostatic cancer in rats remain unknown.

Perinatal Estrogen Exposure

Carcinogenic effects of perinatal exposure to DES in male experimental animals have been described in mice, rats, and hamsters. McLachlan and coworkers (129,130) studied the effects in male offspring of CD-1 mice that had been treated with a daily dose of 100 μg/kg DES on days 9–16 of gestation. At an age of 9–10 months, 6 of 24 mice had nodular enlargements of the coagulating gland, ampullary glands, and colliculus seminalis. One animal had a hyperplastic and squamous metaplastic lesion in the area of the coagulating gland and colliculus seminalis that resembled early neoplasia (129). When eight prenatally DES-exposed male mice were allowed to live for 20–26 months, one animal had an adenocarcinoma of the coagulating gland, three animals had hyperplasia of coagulating gland, two had hyperplasia of ventral prostate, one had a carcinoma of the seminal vesicle, and two had squamous metaplasia of the seminal vesicle (130,131). No such lesions were found in control animals.

The long-term effects of neonatal exposure of Han:NMRI mice to DES or 17β-estradiol were studied by Pylkkänen and coworkers (132,133). DES at a dose of 2 μg/pup/day and 17β-estradiol at doses of 20–200 μg/pup/day for the first 3 days of life resulted after 12–18 months in a very high incidence of moderate to severe epithelial dysplasia of a region which included the periurethral glands, and the periurethral proximal parts of the dorsolateral prostate, coagulating glands, and seminal vesicles. Additional treatment with DHT and 17β-estradiol from 9 to 12 months of age increased severity of the dysplasia examined at 12 months.

Arai et al. (131,134,135) studied the effects in Wistar rats of neonatal exposure to DES. The rats were treated with DES from birth for 30 days at a dose of 1 μg/rat/day for the first 10 days, 2 μg/rat/day for the next 10 days, and 4 μg/rat/day for days 21–30. One group was neonatally castrated and the second group was left intact. Two of 11 castrated, DES-exposed rats developed squamous cell carcinomas in the area of the dorsolateral prostate, coagulating gland, and ejaculatory ducts, and all 11 animals had papillary hyperplastic and squamous metaplastic lesions of the coagulating gland and ejaculatory duct area. In eight noncastrated, DES-exposed rats, hyperplastic or neoplastic lesions were not observed.

In conclusion, perinatal estrogen exposure is carcinogenic for the male accessory sex glands in rodents. The sensitivity for the carcinogenic effects of perinatal estrogen exposure seems to be highest in the periurethral, proximal ducts of the dorsolateral and anterior prostate and seminal vesicle and the intraprostatic urethral epithelium. Interestingly, Driscoll and Taylor (136) reported hypertrophy and squamous metaplasia of the prostatic utricle and prostatic ducts in 55–71% of 31 human infants exposed to DES in utero that had died perinatally from unrelated causes. Squamous metaplastic changes were also found in human fetal prostatic tissue transplanted into DES-treated nude mice (137). The changes were confined to the prostatic utricle and urethra in tissue grown to a gestational age equivalent of 16.5 weeks or less, but included the prostatic ducts in tissue grown to a gestational age equivalent of 17 weeks or more.

Perinatal estrogen treatment may act via inducing permanent alterations in the secretion of pituitary hormones and testicular androgen and in the accessory sex gland androgen and prolactin receptors, resulting in impaired growth of these glands (138–141). LH and FSH plasma levels were elevated in neonatally estrogenized mice (141), while circulating testosterone levels were decreased (138,139) or unaltered (141). Nuclear androgen receptor levels in these mice were decreased in dorsal and ventral prostate but unchanged in the lateral lobe (138). However, there was an increase in the numbers of stromal cells with immunohistochemically detectable androgen receptors in all three lobes (138). The significance of these findings for the carcinogenic effects of perinatal estrogen exposure is not clear. No abnormalities in circulating estrogen and androgen levels were found in boys that had been exposed to DES in utero (133). Other mechanisms, including disturbed morphogenesis of the target tissues (129), and earlier mentioned genotoxic effects of the estrogen are possibly involved in the carcinogenic effects for the prostate of perinatal estrogen treatment of mice.

D. Induction of Prostate Tumors by Chemical Carcinogens and Androgens

Prostatic carcinogenesis is markedly enhanced by long-term administration of testosterone to rats initially treated with chemical carcinogens that target the prostate because of their tissue-specific metabolism (DMAB, BOP) and/or concurrent hormonal stimulation of prostatic cell proliferation, or because of the specific sensitivity of the rat strain (Lobund Wistar rat) to a particular carcinogen (MNU) (70,79–81,83,92,114,120). If not all requirements are adequately met, this enhancement may not occur (70,83). For example, when a single injection of BOP or MNU was given to F344 rats without concurrent stimulation of prostatic cell proliferation, testosterone treatment did not enhance prostatic carcinogenesis (109). Chronic testosterone administration was given following a single administration of MNU or BOP during stimulation of prostatic cell proliferation in Wistar WU or Wistar MRC rats, or during and after 10 repeated biweekly injections of DMAB in F344 rats; this treatment produced carcinomas of the dorsolateral and/or anterior prostate, but not the ventral prostate, in 66–83% of the animals (70,83,92,112,114). However, when the same treatments were given to Lobund Wistar rats, rather variable incidences of between 50 and 97% were reported by Pollard and coworkers (79,80,143,144) and only a 24% incidence was found by Hoover et al. (81).

It is remarkable that the enhancing effect of testosterone on prostate carcinogenesis is confined to the dorsolateral and anterior prostate and does not occur for the ventral prostate. In fact, long-term testosterone administration shifts the site of DMAB-and BOP-induced carcinoma occurrence from exclusively the ventral lobe to predominantly the dorsolateral and anterior lobes (83,92,112). The dose-response relationship between testosterone dose and prostate carcinoma yield is extremely steep; only very little (less than 1.5-fold) elevation of circulating testosterone levels is sufficient for a near-maximal increase in tumor response, and a 2-3-fold elevation is sufficient for a maximal response (114). Testosterone is thus a very powerful tumor promoter for the rat prostate at near-physiological plasma concentrations. The shape of the very steep relationship between testosterone dose and prostate carcinoma response also suggests the involvement of an androgen receptor–mediated mechanism (95).

Conclusions

Table 2 summarizes the literature on prostate cancer induction by specific chemical exposures. With the exception of cadmium, no chemicals for which human exposure occurs are related with known prostate cancer risk. The negative bioassay results with acrylonitrile support the notion that it is not a human carcinogen. The observation of some prostate tumors in rats housed in a rubber plant should lead to some doubt about the conclusion that exposures in the rubber industry are not carcinogenic for the human prostate.

Also mentioned in Table 2 are prostate cancer–inducing effects of hormonal exposures. Although, with the exception of some case reports of prostate cancer in men who had used anabolic steroids (4), there are no data about the potential prostate cancer risk that may result from the use of such drugs, this should be considered a serious possibility given the results of the animal studies with testosterone. Estrogen exposure to men poses more difficulties with interpretation. There are no current human exposures to estrogenic chemicals such as DES and estradiol with the possible exception of small groups of workers in pharmaceutical plants that produce these compounds. However, the cohort of men that have been exposed to DES in utero during the period that this drug was used in pregnant women may well experience an elevated risk for prostate cancer in view of the experimental animal data summarized above. In contrast, dietary exposure to estrogens of plant origin (phytoestrogens) is widespread, but seems to be associated with a low rather than a high prostate cancer risk; countries with a high consumption of phytoestrogen-rich foods such as soy generally have low prostate cancer rates (4,6). There are as yet no animal model data concerning the effects of phytoestrogens and foods such as soy on prostate cancer induction.

III. HUMAN PROSTATE CANCER RISK AND THE INTERPRETATION OF CARCINOGENESIS BIOASSAYS

In the first two sections of this chapter it was pointed out that prostate cancer is very rare in rodents used for carcinogenesis bioassays and very frequent in humans living in western societies. It was also demonstrated that laboratory rats can develop prostate cancer at high incidence provided that (a) prostatic cell proliferation is high during carcinogen exposure and/or (b) circulating androgen levels are slightly elevated, regardless of the specific carcinogen studied. It is possible that these two conditions are frequent in humans but rare in rats kept under laboratory conditions that are designated to minimize exposure to infectious and chemical agents and to standardize and optimize dietary variables. These chosen experimental conditions of the cancer bioassay could result in rats that have very low rates of bacterial prostatitis, which could be a major source of reactive cell proliferation, and rats that are fed diets with low fat and high dietary fiber content, and minimal contamination with chemical carcinogens of the rats' diet. The epidemiology of prostate cancer has identified a western lifestyle, particularly diet, as a major risk factor, but rats are not exposed to western lifestyle diets. It is, therefore, possible that the use for carcinogenicity testing of rodents kept under standard laboratory conditions is not appropriate for the purpose of detecting prostate carcinogens. On the other hand, it is possible that there are simply no strong prostate-specific human carcinogens, but only very weak ones that cannot be identified with the standard rodent bioassay because of their weak carcinogenic activity. The lack of identification of strong prostate carcinogens by epidemiologists despite many studies clearly supports the latter interpretation. In that case, use of animals which are co-treated with androgens as tumor promoters or cell proliferation stimulators to increase the power of the rodent bioassay to detect prostate carcinogens would lead to inappropriate false identification of "strong" prostate carcinogens that in reality are of no or only very minor concern as human prostate carcinogens. Nevertheless, the use of such prostate-specific carcinogen bioassays could be considered similar to the proposed use of mammary cancer-specific carcinogen bioassays (145).

Thus, the hypothesis emerges that there may be many chemical and physical agents that contribute to the development of human prostate cancer, but that factors that stimulate prostate cell proliferation or elevate circulating androgen levels and a western lifestyle and dietary habits are the major determinants of human prostate cancer risk. In this hypothesis, stimulation of cell proliferation is most likely to be involved in the early stages of prostate cancer development and may be a major factor leading to the formation of histological prostate cancer which is similar in rates of occurrence around the world. As tumor-promoting stimuli, factors leading to elevated circulating androgen levels and western lifestyle and diet may be the major determinants of the risk for clinically manifest prostate cancer. This hypothesis would not only predict that current carcinogenesis bioassay practices are adequate to identify strong human prostate carcinogens, but also indicate that chemical or physical agents that elevate circulating androgen levels or stimulate prostate cell proliferation are potential prostate cancer risk-enhancing factors. In that view, rodent bioassays should not only search for proliferative prostate lesions (2 for approach and criteria), but also attempt to detect these relevant noncancer effects. As an aside, it should be mentioned that adequate tissue preparation and histological examination of the male accessory sex glands is critical for experimental studies and carcinogenesis bioassays to be appropriate for detecting prostate cancer-inducing properties of exposures to chemicals and other agents (4).

Very sensitive yet simple endpoints that could be incorporated in standard bioassay protocols include weighing the various accessory sex glands using standardized procedures at the standard time points examined in a typical bioassay. Four weeks of repeated chemical treatment of weanling rats provides information on the effects on the accessory sex glands, which are very sensitive to androgenic stimuli during this time. Thirteen weeks of repeated treatment should be adequate to identify effects on circulating androgen levels. Treatment starting in utero is ideal to identify effects of perinatal exposure when animals are sacrificed before (weaning) or after puberty (8 or 17 weeks of age). Measurement of circulating testosterone levels (using stored serum) or prostatic cell proliferation indicators (on formalin-fixed, paraffin-embedded tissue) can be incorporated when organ weight measurements suggest relevant effects. This approach, which can be incorporated in the standard bioassay protocols at minimal cost and effort, is well worth considering in view of the very high prostate cancer frequency in our society.

Induced and spontaneously occurring proliferative lesions are more common in the ventral prostate and seminal vesicle than in the other accessory sex glands, including the dorsolateral and anterior prostate lobes (2); a few examples of induction of these changes by chemicals were mentioned earlier (4). However, a homolog of the rodent ventral prostate is not present in the human prostate gland (69), and seminal vesicle carcinomas are extremely rare in humans. This poses some problems for the proper extrapolation of the observation of proliferative lesions of the ventral prostate and seminal vesicle in rodent carcinogenesis bioassays. Nevertheless, both structures are androgen-sensitive tissues similar to the other rodent prostate lobes and the human prostate. Thus, the induction of hyperplastic lesions in the rodent ventral prostate or seminal vesicle by a chemical can be considered as evidence suggesting that the chemical may have the potential to produce rodent and human prostate cancer.

Induction of adenomas in the ventral prostate or seminal vesicle by a chemical treatment (using standard carcinogen bioassay interpretation criteria) is suggestive evidence that the chemical is an animal carcinogen and, thus, a possible human carcinogen; furthermore, this finding suggests that the compound is a possible animal prostate carcinogen. Chemically induced adenocarcinomas in the ventral prostate or seminal vesicle is conclusive evidence that the chemical is an animal carcinogen and, thus, a probable human carcinogen, and it suggests that the compound is a prostate-specific animal carcinogen. Chemical induction of adenocarcinomas in the dorsolateral and/or anterior prostate can be considered evidence that it is not only a probable human carcinogen, but also a possible human prostate carcinogen. Induction of adenomas or adenocarcinomas in the ventral prostate or seminal vesicle by a chemical treatment (using standard carcinogen bioassay interpretation criteria) is suggestive or conclusive evidence, respectively, that the chemical is an animal carcinogen and, thus, a probable or possible, respectively, human carcinogen, and it suggests that the compound is a possible prostate-specific carcinogen. The only clear example of this path of interpretation are the data of cadmium bioassays discussed earlier. Cadmium induces ventral prostate atypical hyperplasia and adenomas, which indicates that it is a probable animal carcinogen and a possible prostate carcinogen, but only suggests the possibility that it is a human prostate carcinogen. This interpretation is in line with the conclusion of the IARC (50) that there is limited evidence for the carcinogenicity of cadmium for animals, including its carcinogenicity for the rodent prostate. It also supports the conclusion of Waalkes and Rehm (52), Kazantis et al. (53), and Doll (147) that despite the predominant lack of evidence from epidemiological studies for cadmium being a human prostate carcinogen, this possibility cannot be entirely discounted because of these animal bioassay findings.

REFERENCES

1. American Cancer Society. Cancer Facts and Figures. Atlanta: American Cancer Society, 1996.
2. Bosland MC. Lesions in the male accessory sex glands and penis. In: Mohr U, Dungworth DL, Capen CC, eds. Pathobiology of the Aging Rat. Vol. 1. Berlin: Springer Verlag, 1987:252–260.
3. Bosland MC. Adenocarcinoma, prostate, rat. In: Jones TC, Mohr U, Hunt RD, eds. Genital System. Berlin: Springer Verlag, 1987:252–260.
4. Bosland MC. Male reproductive system. In: Waalkes MP, Ward JM, eds. Carcinogenesis. New York: Raven Press, 1994:339–402.
5. Bosland MC. Hormonal factors in carcinogenesis of the prostate and testis in humans and in animal models. In: Cellular and Hormonal Mechanisms of Hormonal Carcinogenesis: Environmental Influences. New York: Wiley-Liss, 1996:309–352.
6. Bosland MC. The etiopathogenesis of prostatic cancer with special reference to environmental factors. Adv Cancer Res 1988; 51:1–106.
7. Meikle AW, Smith Jr JA. Epidemiology of prostate cancer. Urol Clinics NA 1990; 17:709–718.
8. Muir CS, Nectoux J, Staszewski J. The epidemiology of prostatic cancer. Acta Oncol 1991; 30:133–139.
9. Nomura AMY, Kolonel LN. Prostate cancer: A current perspective. Am J Epidemiol 1991; 13:200–227.
10. van der Gulden JWJ, Kolk JJ, Verbeek ALM. Prostate cancer and work environment. J Occup Med 1992; 34:402–409.
11. Hsing AW, Mclaughlin JK, Hrubec Z, Blot WJ, Fraumeni JF. Tobacco use and prostate cancer: 26-year follow-up of US veterans. Am J Epidemiol 1991; 133:437–441.
12. Coughlin SS, Neaton JD, Sengupta. Cigarette smoking as a predictor of death from prostate cancer in 348,874 men screened for the Multiple Risk Factor Intervention Trial. Am J Epidemiol 1996; 309:901–911.
13. Hsing AW, McLaughlin JK, Schuman LM, Bjelke E, Gridley G, Wacholder S, Co Chien TH, Blot WJ. Diet, tobacco use, and fatal prostate cancer: results from the Lutheran Brotherhood cohort study. Cancer Res 1990; 50: 6836–6840.
14. Honda GD, Bernstein L, Ross RK, Greenland S, Gerkins V, Henderson BE.

Vasectomy, cigarette smoking, and age at first sexual intercourse as risk factors for prostate cancer in middle-aged men. Br J Cancer 1988; 57:326-331.

15. van der Gulden JWJ, Verbeek ALM, Kolk JJ. Smoking and drinking habits in relation to prostate cancer. Br J Urol 1994; 73:382-289.
16. Muscat JE, Taioli E. Re: A. W. Hsing et al., diet, tobacco use and fatal prostate cancer: results from the Lutheran brotherhood cohort study. Cancer Res 1991; 51:3067.
17. Mantel N. Re: Tobacco use and prostate cancer: 26 year follow-up of US veterans. Am J Epidemiol 1992; 135:326-327.
18. Lumey LH. Prostate cancer and smoking: a review of case-control and cohort studies. Prostate 1996; 29:249-260.
19. West DW, Slattery ML, Robison LM, French TIk Mahoney AW. Adult dietary intake and prostate cancer risk in Utah: a case-control study with special emphasis on aggressive tumors. Cancer Causes Control 1991; 2:85-94.
20. La Vecchia C, Negri E, D'Avanzo BD, Franceschi S, Boyle P. Dairy products and risk of prostatic cancer. Oncology 1991; 48:406-410.
21. Walker ARP, Walker BF, Tsotetsi NG, Sebitso C, Siwedi D, Walker AJ. Case-control study of prostate cancer in black patients in Soweto, South Africa. Br J Cancer 1992; 65:438-441.
22. Mills PK, Beeson WL. Re: Tobacco use and prostate cancer: 26-year follow-up of US veterans. Am J Epidemiol 1992; 135:326-327.
23. Schatzkin A, Baranovsky A, Kessler LG. Evidence from associations of multiple primary cancers in the SEER program. Cancer 1988; 62:1451-1457.
24. Handelsman DJ, Conway AJ, Boylan LM, Turtle JR. Testicular function in potential sperm donors: normal ranges and the effects of smoking and varicocele. Int J Androl 1984; 7:369-382.
25. Meikle AW, Bishop DT, Stringham JD, Ford MH, West DW. Relationship between body mass index, cigarette smoking, and plasma sex steroids in normal male twins. Genet Epidemiol 1989; 6:399-412.
26. Abd Elghany N, Schumacher MC, Slattery ML. Occupation, cadmium exposure, and prostate cancer. Epidemiology 1990; 1:107-115.
27. Brownson RC, Chang JC, Davis JR, Smith CA. Occupational risk of prostate cancer: A cancer registry-based study. J Occup Med 1988; 30:523-526.
28. Checkoway H, DiFerdinanco G, Hulka BS, Mickey DD. Medical life-style and occupational risk factors for prostate cancer. Prostate 1987; 10:79-88.
29. Fincham SM, Hill GB, Hanson J, Wijayasinghe C. Epidemiology of prostatic cancer: a case-control study. Prostate 1990; 17:189-206.
30. Le Marchand L, Laurence NK, Yoshizawa CN. Lifetime occupational physical activity and prostate cancer risk. Am J Epidemiol 1991; 133:103-111.
31. Oishi K, Okada K, Yoshida O, Yamabe H, Ohno Y, Hayes RB, Schoeder FH. Case-control study of prostatic cancer in Kyoto, Japan: demographic and some lifestyle risk factors. Prostate 1989; 14:117-122.
32. Pearce NE, Sheppard RA, Fraser J. Case-control study of occupation and

cancer of the prostate in New Zealand. J Epidemiol Comm Health 1987; 41: 130–132.

33. van der Gulden JWJ, Kolk JJ, Verbeek ALM. Work environment and prostate cancer risk. Prostate 1995; 27:250–257.
34. Yoshida O, Oishi K, Ohno Y, Schroeder FH. A comparative study in prostatic cancer in the Netherlands and Japan. In: Sasaki R, Aoki K, eds. Proceedings Monbusho 1989 International Symposium. Comparative Study of Etiology & Prevention of Cancer. Nagoya: Nagoya Press, 1990:73–84.
35. International Agency for Research on Cancer. IARC Monographs on the Evaluation of Carcinogenic Risks to Humans, Supplement 7, Overall Evaluations of Carcinogenicity: an updating of IARC Monographs, Vols 1–42. Lyon: IARC, 1987:139–142.
36. International Agency for Research on Cancer. IARC Monographs on the Evaluation of Carcinogenic Risks to Humans. Vol. 28, The Rubber Industry. Lyon: IARC, 1982.
37. International Agency for Research on Cancer. IARC Monographs on the Evaluation of Carcinogenic Risks to Humans. Vol. 28, Polynuclear Aromatic Compounds, Part 3: Industrial Exposures in Aluminum Production, Coal Gasification, Coke Production, and Iron and Steel Founding. Lyon: IARC, 1984.
38. O'Berg MT, Chen JL, Burke CA, Walrath J, Pell S. Epidemiologic study of workers exposed to acrylonitrile: an update. J Occup Med 1985; 27:835–840.
39. Chen JL, Walrath J, O'Berg MT, Burke CA, Pell S. Cancer incidence and mortality among workers exposed to acrylonitrile. Am J Indust Med 1987; 11:157–163.
40. Walrath J, Fayerweather WE, Gilby PG, Pell S. A case-control study of cancer among Du Pont employees with potential for exposure to dimethylformamide. J Occup Med 1989; 31:432–438.
41. Blair A, Zahm SH, Pearce NE, Heineman EF, Fraumeni JF. Clues to cancer etiology from studies of farmers. Scand J Work Environ Health 1992; 18: 209–215.
42. Pearce N, Reif JS. Epidemiologic studies of cancer in agricultural worker. Am J Indust Med 1990; 18:133–301.
43. Burmeister LF. Cancer in Iowa farmers: recent results. Am J Indust Med 1990; 18:295–301.
44. Fincham SM, Hanson J, Berkel J. Patterns and risks of farmers in Alberta. Cancer 1992; 69:1276–1285.
45. Stark AD, Chang H, Fitzgerald EF, Riccardi K, Stone RR. A retrospective cohort study of cancer incidence among New York State Farm Bureau members. Arch Environ Health 1990; 45:155–162.
46. Ronco G, Costa G, Lynge E. Cancer risk among Danish and Italian farmers. Br J Indust Med 1992; 49:220–225.
47. Gunnarsdottir H, Rafnsson V. Cancer incidence among Icelandic farmers. Scand J Social Med 1991; 19:170–173.
48. International Agency for Research on Cancer. IARC Monographs on the

Evaluation of Carcinogenic Risks to Humans. Vol. 53, Occupational Exposures in Insecticide Application, and Some Pesticides. Lyon: IARC, 1991.
49. Morrison HI, Wilkins K, Semenciw R, Mao Y, Wigle D. Herbicides and cancer. J Natl Cancer Inst 1992; 84:1866–1874.
50. International Agency for Research on Cancer. IARC Monographs on the Evaluation of Carcinogenic Risks to Humans. Vol. 58, Beryllium, Cadmium, Mercury, and Occupational Exposures in the Glass Manufacturing Industry. Lyon: IARC, 1993.
51. Waalkes MP, Oberdörster G. Biological effects of heavy metals. In: Foulkes ED, ed. Metal Carcinogenesis. Boca Raton, FL: CRC Press, 1990;129–158.
52. Waalkes MP, Rehm S. Cadmium and prostate cancer. J Toxicol Environ Health 1994; 32:251–269.
53. Kazantis G, Blanks RG, Sullivan KR. Is cadmium a human carcinogen? In: Nordberg GE, Herber RFM, Alessio L, eds. Cadmium in the Human Environment: Toxicity and Carcinogenicity. Lyon: IARC, 1992:435–446.
54. Garcia Sanchez A, Anotona JF, Urrutia M. Geochemical prospection of cadmium in a high incidence area of prostate cancer, Sierra de Gata, Salamanc, Spain. Sci Total Environ 1992; 116:243–251.
55. Bako G, Smith ESO, Hanson J, Dewar R. The geographical distribution of high cadmium concentrations in the environment and prostate cancer in Alberta. Can J Public Health 1982; 73:92–94.
56. Beral V, Inskip H, Fraser P, Booth M, Coleman D, Rose G. Mortality of employees of the United Kingdom Atomic Energy Authority, 1946–1979. Br Med J Clin Res 1985; 291:440–447.
57. Fraser P, Carpenter L, Maconochie N, Higgins C, Booth M, Beral V. Cancer mortality and morbidity in employees of the United Kingdom Atomic Energy Authority, 1946–1986. Br J Cancer 1993; 67:615–624.
58. Rooney C, Beral V, Maconochie N, Fraser P, Davies G. Case-control study of prostatic cancer in employees of the United Kingdom Atomic Energy Authority. Br Med J 1993; 307:1391–1397.
59. Checkoway H, Mathew RM, Shy CM, Watson JR, Tankersley WG, Wolfe SH, Smith JC, Fry SA. Radiation, work experience, and cause specific mortality among workers at an energy research laboratory. Br J Indust Med 1985; 42:525–533.
60. Cardis E, Gilbert ES, Carpenter L, Howe G, Kato I, Armstrong BK, Beral V, Cowper G, Douglas A, Fix J, Fry SA, Kaldor J, Lave C, Salmon L, Smith PG, Voelz GL, Wiggs LD. Effects of low doses and low dose rates of external ionizing radiation: cancer mortality among nuclear industry workers in three countries. Radiat Res 1995; 142:117–132.
61. Ron E, Preston DL, Mabuchi K, Thompson DE, Soda M. Cancer Incidence in atomic bomb survivors. Part IV: Comparison of cancer incidence and mortality. Radiat Res 1994; 137(suppl 2):S98–112.
62. Shimizu Y, Kata H, Schull WJ. Studies of the mortality of A-bomb survivors. Radiat Res 1990; 121:120–141.
63. Shimizu Y, Kato H, Schull WJ. Risk of cancer among atomic bomb survivors. J Radiat Res 1991; 2:54–63.

64. Band PR, Le ND, Fang R, Deschamps M, Coldman AJ, Gallagher RP, Moody J. Cohort study of Air Canada pilots: mortality, cancer incidence, and leukemia risk. Am J Epidemiol 1995; 143:137–143.
65. Eatough JP, Henshaw DL. Radon and prostate cancer. Lancet 1990; 335: 1292.
66. Axelson O, Fastiere F. Radon as a risk factor for extra-pulmonary tumours. Med Oncol Tumor Pharmacol 1993; 10:167–172.
67. Carter HB, Piantodosi S, Issacs JT. Clinical evidence for and implications of multistep development of prostate cancer. J Urol 1990; 143:742–746.
68. McNeal JE, Redwine EA, Freiha FS, Stamey TA. Zonal distribution of prostatic adenocarcinoma. Correlation with histologic pattern and direction of spread. Am J Surg Pathol 1988; 12:897–906.
69. Price D. Comparative aspects of development and structure in the prostate. In: Vollmer EP, Kauffmann G, eds. The Biology of the Prostate and Related Tissues. Natl Cancer Inst Monogr. Vol. 12. Washington, DC: NIH, 1962:1–28.
70. Bosland MC. Animals models for the study of prostate carcinogenesis. J Cell Biochem 1992; 16:89–98.
71. Simmons DL, Kasper C. Quantitation of mRNAs specific for the mixed-function oxidase system in rat liver and extrahepatic tissues during development. Arch Biochem Biophys 1989; 271:10–20.
72. Söderkvist P, Töftgård R, Gustafsson JA. Induction of cytochrome P-450 related metabolic activities in the rat ventral prostate. Toxicol Lett 1982; 10: 61–69.
73. Carmena MJ, Garcia-Paramio MP, Prieto JC. Receptors for tumor-promoting phorbol esters in rat ventral prostate. Cancer Lett 1993; 68:143–147.
74. Söderkvist P, Poellinger L, Gustafsson JA. Carcinogen-binding proteins in the rat ventral prostate: specific and nonspecific high-affinity binding sites for benzo(a)pyrene, 3-methylcholanthrene, and 2,3,7,8-tetrachlorodibenzo-p-dioxin. Cancer Res 1986; 46:651–657.
75. Huff J, Cirvello J, Haseman J, Bucher J. Chemicals associated with site-specific neoplasia in 1934 long-term carcinogenesis experiments in laboratory rodents. Environ Health Perspect 1991; 93:247–270.
76. Swirsky Gold L, Slone TH, Manley NB, Bernstein L. Target organs in chronic bioassay of 533 chemical carcinogens. Environ Health Perspect 1991; 93:233–246.
77. Marselos M, Vainio H. Carcinogenic properties of pharmaceutical agents evaluated in the IARC Monographs programme. Carcinogenesis 1991; 12: 1751–1766.
78. Rivenson A, Silverman J. The prostatic carcinoma in laboratory animals: a bibliographic survey from 1900 to 1977. Invest Urol 1979; 16:468–472.
79. Pollard M, Luckert PH. Production of autochthonous prostate cancer in Lobund-Wistar rats by treatments with N-nitroso-N-methylurea and testosterone. J Natl Cancer Inst 1986; 77:583–587.
80. Pollard M, Luckert PH, Snyder D. The promotional effect of testosterone

on induction of prostate-cancer in MNU sensitized L-W rats. Cancer Lett 1989; 45:209–212.
81. Hoover DM, Best KL, McKenney BK. Experimental induction of neoplasia in the accessory sex organs of male Lobund-Wistar rats. Cancer Res 1990; 50:142–146.
82. Pour PM. Prostate cancer induced in MRC rats by N-nitrosobis(2-oxopropyl)-amine and N-nitrosobis(2-hydroxypropyl)amine. Carcinogenesis 1983; 4:49–55.
83. Pour PM, Stepan K. Induction of prostatic carcinomas and lower urinary tract neoplasms by combined treatment of intact and castrated rats with testosterone propionate and N-nitrosobis(2-oxopropyl)amine. Cancer Res 1987; 47:5699–5706.
84. Katayama S, Fiala E, Reddy BS, Rivenson A, Silverman J, Williams GM, Weisburger JH. Prostate adenocarcinoma in rats: Induction by 3,2′-dimethyl-4-aminobiphenyl. J Natl Cancer Inst 1982; 68:867–873.
85. Shirai T, Sakata T, Fukushma S, Ikawa E, Ito N. Rat prostate as one of the target organs for 3,2′-dimethyl-4-aminobiphenyl-induced carcinogenesis: effects of dietary ethinyl estradiol and methyltestosterone. Jpn J Cancer Res 1985; 76:803–805.
86. Shirai T, Nakamura A, Fukushima S. Effects of age on multiple organ carcinogenesis induced by 3,2′-dimethyl-4-aminobiphenyl in rats, with particular reference to the prostate. Jpn J Cancer Res 1989; 80:312–316.
87. Shirai T, Nakamura A, Fukushima S. Different carcinogenic responses in a variety of organs, including the prostate, of five different rat strains given 3,2′-dimethyl-4-aminobiphenyl. Carcinogenesis 1990; 5:793–797.
88. Shirai T, Nakamura A, Fukushima S. Selective induction of prostate carcinomas in F344 rats treated with intraperitoneal injections of N-hydroxy-3,2′-dimethyl-4-aminobiphenyl. Jpn J Cancer Res 1990; 81:320–323.
89. Shirai T, Iwasaki S, Naito H, Masui T, Kato T, Imaida K. Dose dependence of N-hydroxy-3,2′-dimethyl-4-aminobiphenyl-induced rat prostate carcinogenesis. Jpn J Cancer Res 1992; 83:695–698.
90. Yamada H, Shirai T, Ito N. Species-and strain specific O-acetylation of N-hydroxy-3,2′-dimethyl-4-aminobiphenyl by liver and prostate cytosol. In: King CM, Schuetzle D, eds. Carcinogenic and Mutagenic Responses to Aromatic Amines and Nitroarenes. New York: Elsevier Science Publishing Co., 1988:223–227.
91. Shirai T, Nakamura A, Fukushima S, Tada M, Morita T, Ito N. Immunohistochemical demonstration of carcinogen-DNA adducts in target and non-target tissues of rats given a prostate carcinogen, 3,2′-dimethyl-4-aminobiphenyl. Carcinogenesis 1990; 11:653–657.
92. Shirai T, Imaida K, Iwasaki S, Mori T, Tada M, Ito N. Sequential observation of rat prostate lesion development induced by 3,2′-dimethyl-4-aminobiphenyl and testosterone. Jpn J Cancer Res 1993; 84:20–25.
93. Waalkes MP, Rehm S, Riggs CW, Bare RM, Devor DE, Poirier LA, Wenk ML, Henneman JR, Balaschak MS. Cadmium carcinogenesis in male Wistar

[CRL:(WI)BR] rats: dose-response analysis of tumor induction in the prostate and testes and at the injection site. Cancer Res 1988; 48:4656–4663.
94. Waalkes MP, Rehm S, Riggs C, Bare RM, Devor DE, Poirier LA, Wenk ML, Henneman JR. Cadmium carcinogenesis in male Wistar [CRL:(WI)BR] rats: dose-response analysis of effects of zinc on tumor induction in the prostate, in the testes and at the injection site. Cancer Res 1989; 49:4282–4288.
95. Waalkes MP, Rehm S. Carcinogenicity of oral cadmium in the male Wistar (WF/Ncr) rat; effect of chronic dietary zinc deficiency. Fundam Appl Toxicol 1992; 19:512–520.
96. Terracio L, Nachtigal M. Transformation of prostatic epithelial cells and fibroblasts with cadmium chlorode in vitro. Arch Toxicol 1986; 58:141–151.
97. Terracio L, Nachtigal M. Oncogenicity of rat prostate cells transformed in vitro with cadmium chloride. Arch Toxicol 1988; 61:450–456.
98. Waalkes MP, Perantoni A. Apparent deficiency of metalothionein in the Wistar rat prostate Toxicol Appl Pharmacol 1989; 101:83–94.
99. Hirose F, Takizawa S, Watanabe H, Takeichi N. Development of adenocarcinoma of the prostate in ICR mice locally irradiated X-rays. Jpn J Cancer Res 1976; 67:407–411.
100. Takizawa S, Hirose F. Role of testosterone in the development of radiation-induced prostate carcinoma in rats. Jpn J Cancer Res 1978; 69:723–736.
101. Brown CE, Warren S. Carcinoma of the prostate in irradiated parabiotic rats. Cancer Res 1978; 38:159–162.
102. Bosland MC, Prinsen MK, Kroes R. Adenocarcinomas of the prostate induced by N-nitroso-N-methylurea in rats pretreated with cyproterone acetate and testosterone. Cancer Lett 1983; 18:69–78.
103. Bosland MC, Prinsen MK. Induction of dorsolateral prostate adenocarcinomas and other accessory sex gland lesions in male Wistar rats by a single administration of N-methyl-N-nitrosourea, 7,12-dimethylbenz(a)anthracene, and 3,2′-dimethyl-4-aminobiophenyl after sequential treatment with cyproterone acetate and testosterone propionate. Cancer Res 1990; 50:691–699.
104. Bosland MC, Prinsen MK, Rivenson A, Silverman J, Fiala E, Williams GM, Kroes R, Weisburger JH. Induction of proliferative lesions of ventral prostate, seminal vesicle, and other accessory sex glands in rats by N-methyl-N-nitrosourea: effect of castration, pretreatment with cyproterone acetate and testosterone propionate, and rat strain. Prostate 1992; 20:339–353.
105. Shirai T, Fukushima S, Ikawa E, Tagawa T, Ito N. Induction of prostate carcinoma *in situ* at high incidence in F344 rats by combination of 3,2′-dimethyl-4-aminobiphenyl and ethinyl estradiol. Cancer Res 1986; 46:6423–6426.
106. Ito N, Shirai T, Tagawa Y, Nakamura A, Fukushima S. Variation in tumor yield in the prostate and other target organs of the rat in response to varied dosage and duration of administration of 3,2′-dimethyl-4-aminobiphenyl. Cancer Res 1988; 48:4629–4632.
107. Takai K, Kakizoe T, Tobisu K, Ohtani M, Kishi K, Sato S, Aso Y. Sequential

changes in the prostate of rats treated with chlormadinone acetate, testosterone and N-nitroso-N-methylurea. J Urol 1988; 139:1363–1366.
108. Shirai T, Ikawa E, Tagawa Y, Iwasaki S, Takahashi S, Ito N. Lesions of the prostate glands and seminal vesicles induced by N-methylnitrosourea in F344 rats pretreated with ethinyl estradiol. Cancer Lett 1987; 35:7–15.
109. Shirai T, Yamamoto A, Iwasaki S, Tamano S, Masui T. Induction of invasive carcinomas of the seminal vesicles and coagulating glands of F344 rats by administration of N-methyl-nitrosourea or N-nitrosobis(2-oxoprophl)amine and followed by testosterone propionate with or without high-fat diet. Carcinogenesis 1991; 12:2169–2173.
110. Bosland MC. Prostatic cell proliferation and induction of prostate carcinomas in rats by methylnitrosourea. Proc Am Assoc Cancer Res 1988; 29:254.
111. Shirai T, Ikawa E, Imaida K. Proliferative response of rat accessory sex organs to dietary sex hormones after castration or initial dietary administration of estrogen. J Urol 1987; 138:216–219.
112. Shirai T, Tamano S, Kato T, Iwasaki S, Takahashi S, Ito N. Induction of invasive carcinomas in the accessory sex organs other than the ventral prostate of rats given 3,2′-dimethyl-4-aminobiphenyl and testosterone propionate. Cancer Res 1991; 51:1264–1269.
113. Sukumar S, Armstrong B, Bruyntjes JP, Leav I, Bosland MC. Frequent activation of the Ki-ras oncogene at codon 12 in N-methyl-N-nitrosourea-induced rat prostate adenocarcinomas and neurogenic sarcomas. Mol Carcinogenesis 1991; 4:362–368.
114. Bosland MC, Dreef-Van Der Meulen H, Sukumar S, Ofner P, Leav I, Han X, Liehr JG. Multistage prostate carcinogenesis: the role of hormones. In: Harris CC, Hirohasi S, Ito N, Pitot HC, Sugimura T, Terada M, Yokota J, eds. Multistage Carcinogenesis. Boca Raton, FL: CRC Press, 1992:109–123.
115. Noble RL. Prostate carcinoma of the Nb rat in relation to hormones. Intl Rev Exp Pathol 1982; 23:113–159.
116. Bosland MC, Scherrenberg PM Ford H, Dreef-van der Muelen HC. Promotion by testosterone of N-methyl-N-nitrosourea-induced prostatic carcinogenesis in rats. Proc Am Assoc Cancer Res 1989; 30:272.
117. Pollard M, Luckert PH, Schmidt MA. Induction of prostate adenocarcinomas in Lobund Wistar rats by testosterone. Prostate 1982; 3:563–568.
118. Pollard M, Luckert PH. Promotional effects of testosterone and dietary fat on prostate carcinogenesis in genetically susceptible rats. Prostate 1985; 6: 1–5.
119. Pollard M, Luckert P. Promotional effects of testosterone and high fat diet on the development of autochthonous prostate cancer in rats. Cancer Lett 1986; 32:223–227.
120. Pollard M, Luckert PH. Autochthonous prostate adenocarcinomas in Lobund-Wistar rats: a model system. Prostate 1987; 11:219–227.
121. Bosland MC. Carcinogenic risk assessment of steroid hormone exposure in relation to prostate cancer. In: Li JJ, Nandi S, Li AA, eds. Hormonal Carcinogenesis. Berlin: Springer-Verlag, 1992:225–233.

122. Bosland MC, Ford H, Horton L. Induction of a high incidence of ductal prostate adenocarcinomas in NBL and Sprague Dawley rats treated with estradiol-17β or diethylstilbestrol in combination with testosterone. Carcinogenesis 1995; 16:1311–1317.
123. Banerjee SK, Banerjee S, Li SA, Li JJ. Cytogenetic changes in renal neoplasms and during estrogen-induced renal tumorigenesis in hamsters. In: Li JJ, Nandi S, Li AA, eds. Hormonal Carcinogenesis. Berlin: Springer-Verlag, 1992:247–253.
124. Liehr JG, Avitts TA, Randerath E. Estrogen-induced endogenous DNA adduction: possible mechanism of hormonal cancer. Proc Natl Acad Sci USA 1986; 83:5301–5305.
125. Liehr JG, Sirbasku DA, Jurka E, Randerath K, Randerath E. Inhibition of estrogen-induced renal carcinogenesis in male Syrian hamsters by tamoxifen without decrease in DNA adduct levels. Cancer Res 1988; 48:779–783.
126. Gladek A, Liehr JG. Mechanism of genotoxicity of diethylstilbestrol in vivo. J Biol Chem 1989; 264:16847–16852.
127. Han X, Liehr JG, Bosland MC. Induction of a DNA adduct detectable by ^{32}P-postlabeling in the dorsolateral prostate of NBL/Cr rats treated with estradiol-17β and testosterone. Carcinogenesis 1995; 16:951–954.
128. Bosland MC, Han X, Liehr JG. Enhancement of endogenous DNA adduct formation and induction of adenocarcinomas at the same site in the rat dorsolateral prostate by treatment with estradiol-17β and testosterone. Proc Am Assoc Cancer Res 1993; 34:241.
129. McLachlan JA, Newbold RR. Reproductive tract lesions in male mice exposed prenatally to diethylstilbestrol. Science 1975; 190:991–992.
130. McLachlan JA. Rodent models for perinatal exposure to diethylstilbestrol and their relation to human disease in the male. In: Herbst AL, Bern HA, eds. Developmental Effects of Diethylstilbestrol (DES) in Pregnancy. Stuttgart: Thieme Verlag, 1981:141–157.
131. Arai Y, Mori T, Suzuki Y, Bern HA. Long-term effects of perinatal exposure to sex steroids and diethylstilbestrol on the reproductive system of male mammals. Int Rev Cytol 1983; 84:234–268.
132. Pylkkänen L, Santti R, Newbold R, Mclachlan JA. Regional differences in the prostate of the neonatally estrogenized mouse. Prostate 1991; 18:117–129.
133. Pylkkänen L, Santti R, Mäentausta, Vihko R. Distribution of estradiol-17β hydroxysteroid oxidoreductase in the urogenital tract of control and neonatally estrogenized male mice: immunohistochemical, enzymehistochemical, and biochemical study. Prostate 1992; 20:59–72.
134. Arai Y, Chen CY, Nishizuka Y. Cancer development in male reproductive tract in rats given diethylstilbestrol at neonatal age. Gann 1978; 69:861–862.
135. Arai Y, Suzuki Y, Nishizuka Y. Hyperplastic and metaplastic lesions in the reproductive tract of male rats induced by neonatal treatment with diethylstilbestrol. Virchows Arch A Path Histol 1977; 376:21–28.
136. Driscoll SG, Taylor SH. Effects of prenatal maternal estrogen on the male urogenital system. Obstet Gynecol 1980; 56:537–542.

137. Sugimura Y, Cunha GR, Yonemura GU, Kawamura J. Temporal and spatial factors in diethylstilbestrol-induced squamous metaplasia of the developing human prostate. Hum Pathol 1988; 19:133–139.
138. Prins GS. Neonatal estrogen exposure induces lobe-specific alterations in adult rat prostate androgen receptor expression. Endocrinol 1992:2401–2412.
139. Jean C, Andre JM, Berger JM, De Turckheim M, Veyssiere G. Estimation of testosterone and androstenedione in the plasma and testes of cryptorchid off-spring of mice treated with oestradiol during pregnancy. J Reprod Fertil 1975; 44:235–247.
140. Edery M, Turner T, Dauder S. Influence of neonatal diethylstilbestrol treatment on prolactin receptor levels in the mouse male reproductive system. Proc Soc Exp Biol Med 1990; 194:289–292.
141. Dalterio S, Bartke A, Steger R. Neonatal exposure to DES in BALB/c male mice: effects on pituitary-gonadal function. Pharmacol Biochem Behav 1985; 22:1019–1024.
142. Ross RK, Gabeff P, Hill-Paganini A, Henderson BE. Effect of in utero exposure to diethylstilbestrol on age at onset of puberty and on postpubertal hormone levels in boys. Can Med Assoc J 1983; 128:1197–1198.
143. Pollard M, Luckert PH, Snyder D. Prevention of prostate cancer and liver tumors in L-W rat by moderate dietary restriction. Cancer 1989; 64:686–690.
144. Pollard M, Luckert PH, Snyder D. Prevention and treatment of experimental prostate cancer in Lobund-Wister rats. 1. Effects of estradiol, dihydrotestosterone, and castration. Prostate 1989; 15:95–103.
145. McCormick DL, Moon RC. Tumorigenesis of the rat mammary gland. In: Milman HA, Weisburger EK, eds. Handbook of Carcinogen Testing. Park Ridge, NJ: Noyes Publications, 1985:215–229.
146. Doll R. Cadmium in the human environment: closing remarks. In: Nordberg GF, Herber RFM, eds. Cadmium in the Human Environment: Toxicity and Carcinogenicity. Lyon: IARC, 1992:459–464.

26

Exocrine Pancreatic Cancer

H. B. Bueno-de-Mesquita and Henk J. van Kranen
National Institute of Public Health and the Environment, Bilthoven, The Netherlands

Marko Jan Appel and Ruud A. Woutersen
TNO Nutrition and Food Research Institute, Zeist, The Netherlands

I. INTRODUCTION

During the last decade in different areas of scientific endeavor new knowledge on the etiology of cancer of the exocrine pancreas has rapidly emerged. Collation of existing information is timely, but interpretation is fraught with uncertainties. First, we describe animal carcinogens and the two predominant experimental animal models of pancreatic cancer, including recently observed oncogenetic alterations. Next, we summarize the body of evidence obtained in human carcinogenesis, i.e., findings from molecular biology and observational epidemiology. Finally, we compare animal and human data for oncogenetic events, several lifestyle factors, and diabetes.

II. ANIMAL CARCINOGENESIS

A. Animal Carcinogens and Animal Models

A total of 15 carcinogens (1) have been reported to cause exocrine pancreatic tumors in several laboratory animal species, mainly rodents. Carcinogens of the exocrine pancreas include aflatoxin B_1, clofibrate, N^{δ}-(*N*-methyl-*N*-nitrosocarbamoyl)-L-ornithine (MNCO), nafenopin, nitrofen, dimethylbenz(*a*)anthracene (DMBA), *N*-methyl-*N*-nitrosourea (MNU), 4-

hydroxyaminoquinoline-1-oxide (4-HAQO), azaserine, and several propyl-nitrosamines. Most of the chemicals appear to be procarcinogens and need to be metabolized to become activated to a DNA-reactive form. The only directly acting agents are MNU and MNCO (Table 1). Two animal models are frequently used to study pancreatic carcinogenesis: the azaserine-treated rat leading to acinar adenocarcinomas (2) and the *N*-nitrosobis(2-oxo-propyl)amine (BOP)–treated hamster leading to ductular adenocarcinomas (3).

Azaserine (*O*-diazoacetyl-L-serine) is a naturally occurring compound, produced by *Streptomyces fragilis*. Azaserine is moderately toxic in mice and rats (LD_{50} of 150 and 170 mg/kg/day, respectively) and was initially used as an anticancer drug. A single i.p. dose of 30 mg azaserine per kg body weight is sufficient to induce putative preneoplastic atypical acinar cell foci (AACF) in the pancreas of all treated rats after 2–4 months. The mutagenicity and genotoxicity of azaserine are rather specific for pancreatic acinar cells in rats. L-Azaserine causes DNA adducts in acinar cells of rat pancreas that persist for at least 4 weeks. When isolated pancreatic acinar cells are incubated with azaserine, DNA damage has been identified by HPLC in the form of N^7-carboxymethylguanine (4). D-Azaserine causes no DNA damage, nor does it induce pancreatic AACF, suggesting an enzyme-

Table 1 Chemical Carcinogens of the Exocrine Pancreas of Animals

Chemical	Species
Aflatoxin B_1	Monkey
Azaserine[a]	Rat, mouse
Clofibrate	Rat
7,12-Dimethylbenz(*a*)anthracene (DMBA)	Rat
4-Hydroxyaminoquinoline-1-oxide (4-HAQO)	Rat
N-Methyl-*N*-Nitrosourea (MNU)	Guinea pig
N^L-(*N*-Methyl-*N*-nitrosocarbamoyl)-L-ornithine (MNCO)	Rat, hamster
Nafenopin ([2-methyl-2-(*p*-1,2,3,4-tetrahydro-1-naphtyl)-phenoxy]-proprionic acid)	Rat
Nitrofen (2,4-dichloro-1(4-nitrophenoxy)benzene)	Rat
N-Nitrosobis(2-hydroxypropyl)amine (BHP)	Hamster, rat
N-Nitrosobis(2-oxopropyl)amine[b] (BOP)	Hamster
N-Nitroso(2-hydroxypropyl)(2-oxopropyl)amine (HPOP)	Hamster, rat
N-Nitrosodimethylamine (DMN)	Rat
N-Nitroso-2,6-dimethylmorpholine (NDMM)	Hamster
N-Nitrosomethyl(2-oxopropyl)amine (MOP)	Hamster

[a]Azaserine does not cause pancreatic tumors in hamsters.
[b]BOP does not cause pancreatic tumors in mice and rats.

mediated stereospecific step in the uptake and/or activation of azaserine (5). Furthermore, Zurlo et al. (6) found less azaserine-induced DNA damage and AACF in pyridoxal-deficient rats in comparison with control rats. These data point to an activation of azaserine by formation of an azaserine-pyridoxal complex followed by an enzymatic cleavage of the ester bond by an α,β-elimination reaction, resulting in formation of diazoacetate, ammonia, pyruvate, and phosphate. Diazoacetate is highly reactive and can form adducts with DNA and other macromolecules (Fig. 1).

BOP is one of the most potent and specific of several propylnitrosamine derivatives, which are capable of inducing cancer of the pancreas in hamsters. The metabolism of BOP in vivo involves reduction to *N*-nitroso(2-hydroxypropyl)(2-oxopropyl)amine (HPOP) with subsequent reduction to *N*-nitrosobis(2-hydroxypropyl)amine (BHP) or conjugation with sulfate or glucuronic acid (7). After activation, BOP can cause DNA alkylation in both acinar and ductular pancreatic cells in rats and hamsters (8). The most abundant DNA adducts were N^7-methylguanine and O^6-methylguanine (9,10). HPOP also yields 7- and O^6-hydroxypropylguanines in addition to the methylguanine adducts and causes a fivefold increase in pancreatic DNA synthesis in hamster pancreas. No significant repair of O^6-hydroxypropylguanine was observed for at least 8 days. The persistence of the adducts and the mitogenic effect may contribute to the pancreatropic

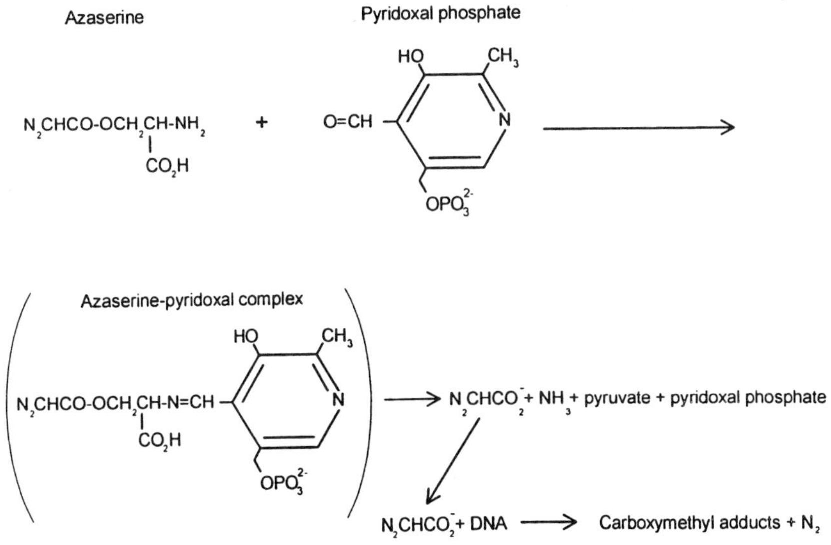

Figure 1 Metabolic pathways of azaserine.

carcinogenesis of BOP (11) (Fig. 2). Single or multiple s.c. injections of BOP result in proliferative cystic and ductular lesions in the pancreas of hamsters, which can be classified by grade of malignant phenotype. Adenocarcinomas in BOP-treated hamsters are invariably of the ductal/ductular type.

B. Oncogenetic Alterations

The (onco)genetic analysis of tumors induced in the most frequently used animal models, i.e., the azaserine-treated rat (2), the BOP-treated hamster (3), and available transgenic mice, are summarized in Table 2. At present our knowledge on tumors of acinar origin in the rat is limited to the K-*ras* oncogene and the involvement of the cholecystokinin (gastrin) and bombesin receptors. Several rat studies have concluded that activation of the K-*ras* oncogene is not detectable in tumors or its precursor lesions of acinar origin (12,13). A similar conclusion with respect to K-*ras* has been reached on

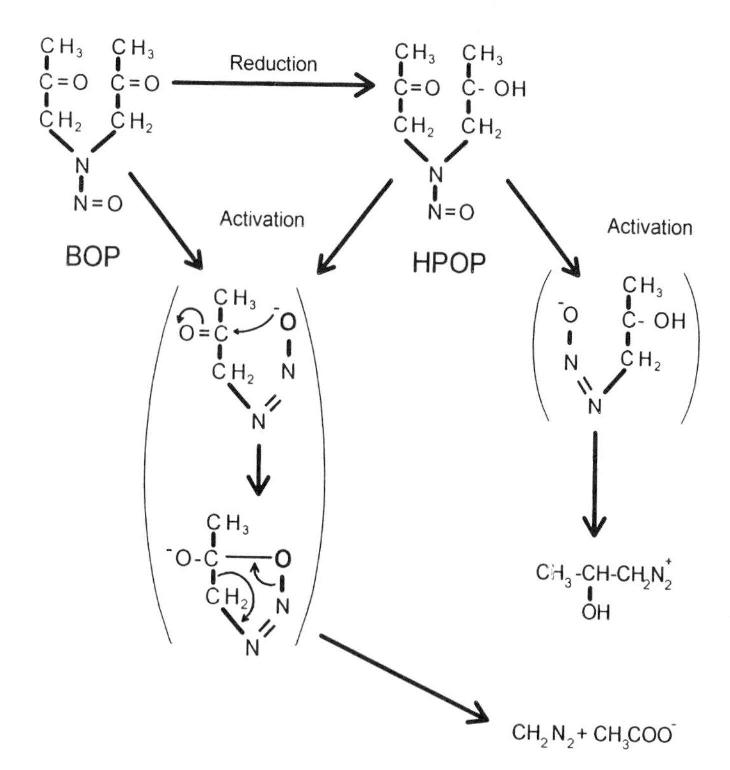

Figure 2 Metabolic pathways of *N*-nitrosobis(2-oxopropyl)amine (BOP).

Table 2 Frequency of Gene Alterations in Exocrine Pancreatic Cancer

Species-Tumor type	Gene Alteration (%)								
	K-*ras*	APC	DCC + DPC4	p53	mdm-2	MTS1	RB	LOH	Ref.
Human									
Adenocarcinoma	80–90	0–40[b]	88[c]	50–70	–*	38–85[g]	50[c]	chr5,22	81,147,149,150, 167,168
Ductular	0	–	–	–	–	–	–	–	
Acinar									
Hamster									
Syrian golden ductular (BOP-induced)	90	–	53[f]	0	26	–	42	–	12,13,16–18
Rat									
Acinar[(5)] (azaserine-induced)	0	–	–	–	–	–	–	–	12,13
Mouse									
(Mixed) acinar (transgenic, Ela-SV40, Ela-c-myc)	0[a]	+[d]	–	+[d]	–	–	–	–	14,15,152

– No information reported.

[a]In both acinar lesions from E1a-SV40 mice (6) as mixed acinar/ductal carcinomas from E1a-c-myc mice (14) only wild-type K-*ras* was detectable.

[b]No mutations detected in a well-described series from the United States (169), but 2.5–40% in two small series from Japan (150,167).

[c]Initially only allelic loss at or in the direct proximity of the DCC and RB loci, but no conformation yet at the DNA level was reported (147). Recently on chromosome 18 a second tumor-suppressor gene called DPC4 (deleted in pancreatic cancer locus 4) has been reported (82), excluding a role for the DCC gene as a tumor suppressor in human pancreatic cancer.

[d]Germline mutations in both *p53* and *APC* genes predisposes to acinar pancreatic neoplasia (152).

[e]Proliferation of acinar cells can be stimulated by cholecystokinin (CCK) through the CCK-B receptor.

[f]Loss or decreased expression of DCC as measured by RT-PCR (16).

[g]Allelic deletions of 9p21-22 in 85%, homozygous deletions in 41%, and mutations in 35% of cases analyzed (168).

tumors of acinar origin in mice (14,15). In Syrian golden hamsters, Scarpelli and coworkers (16) as well as other groups have demonstrated high incidences of mutations in the K-*ras* gene, overexpression of the mdm-2 gene, LOH in the direct proximity of the RB and DCC genes, and the absence of mutations in the *p53* gene (12,17,18). These rodent (but also human) data support the idea that different signal transduction pathways are involved with the (aberrant) growth of acinar and ductal cells.

C. The Role of Cell Proliferation

One of the main characteristics of tumor growth is increased cell proliferation. Cell proliferation correlates positively with the development of (pre)-neoplastic pancreatic lesions in both rats and hamsters (19,20). Carcinogen-induced hyperplastic acinar or ductular lesions in the exocrine pancreas, which are generally considered to be putative preneoplastic lesions, invariably exhibit a higher cell proliferation rate than both normal acinar and ductular tissue. In rat pancreas two types of AACF have been identified in hematoxylin and eosin (H&E)–stained tissue slides: basophilic and acidophilic AACF (21,22). The basophilic foci demonstrate a low growth potential, as measured by mitotic index, incorporation of ^{3}H-thymidine (21), or incorporation of BrdU (5-bromo-2′deoxyuridine) (19). The low proliferative capacity of the basophilic lesions suggests a minor involvement, if any, of these foci in the development of pancreatic tumors. Acidophilic AACF proliferate rapidly and show a high frequency of mitosis. The ^{3}H-thymidine labeling index (LI) was 2.3 in cells of acidophilic AACF in contrast to 0.1 in basophilic AACF (21). Studies with BrdU showed a 15- to 20-fold increase in LI in AACF in comparison with normal acinar cells. The LI in basophilic AACF resembled that of normal acinar tissue (19). Acidophilic AACF, adenomas, and carcinomas (in situ) all demonstrate ATPase activity in frozen tissue sections, whereas basophilic AACF do not (23). It is estimated that about 1% of the pancreatic proliferative lesions develop into acinar cell carcinomas 6–15 months after carcinogen treatment. In hamster pancreas, apart from the high proliferation of ductular lesions, BOP also induces a consistent and sustained proliferation of normal acinar cells. The labeling of acinar cells may in itself form a carcinogenic hazard (24), which may be ascribed to a regeneration of acinar tissue as a consequence of a cytotoxic insult.

D. Role of Growth Factors in Cell Proliferation

Visser et al. (25) demonstrated that increased levels of transforming growth factor-I (TGF-α) accompanied by decreased levels of epidermal growth factor (EGF) may form an early stimulus for increased cell proliferation in putative preneoplastic acinar cells in azaserine-treated rats. Expression of TGF-α, EGF, as well as of the EGF-receptor (EGFR) is reduced in AACF and absent in acinar adenocarcinomas. Thus, these growth factors may play a role in pancreatic acinar carcinogenesis during the transformation of histologically normal acinar cells into putative preneoplastic AACF. In BOP-treated hamsters, TGF-α and EGFR, but not EGF, are overexpressed in pancreatic ductular adenocarcinomas. This observation leads to the conclusion that TGF-α may have a paracrine or autocrine stimulatory effect on cell proliferation in neoplastic ductular lesions via binding to the EGF receptor (26).

E. Modulation of Cell Proliferation

Cell proliferation in putative preneoplastic lesions in rat pancreas can be modulated by dietary raw soya flour (27) and dietary fish oil (28). Dietary fish oil caused a decrease in cell proliferation accompanied by an increase in size of azaserine-induced pancreatic AACF in rats. These seemingly contradictory observations are difficult to explain. One possible explanation could be that the acinar cells of AACF become hypertrophic in fish oil–fed animals. Fatty acids from fish oil [mainly docosahexaenoic acid (DHA)] have been reported to cause cell swelling in vitro (29). Incorporation of ω-3 PUFA (polyunsaturated fatty acids) into acinar cell membranes may cause a similar effect. Another, more plausible explanation for the discrepancy between cell proliferation and size of AACF may be a difference in apoptotic rate in AACF from fish oil–fed animals. In fact, the apoptotic index in AACF from fish oil–fed rats is significantly lower than in rats maintained on a diet without fish oil (30). Cell proliferation in this study was measured as an accumulation of BrdU incorporation over 72 hours. It is estimated that the duration of the morphologically recognizable stage of apoptosis is approximately 3–4 hours in rat liver (31). Assuming that the duration of apoptosis is similar in rat pancreas, the resulting difference between decreased cell proliferation rate minus the decreased cell death rate may still result in an increase in cell numbers in AACF. This would explain the larger AACF in fish oil–fed rats than in rats maintained on a control diet. In other words, the decrease in apoptosis caused by fish oil may contribute more to the regulation of overall cell numbers in AACF than the decrease in cell proliferation caused by fish oil.

In addition to the traditional measurement of the number and size of acidophilic AACF in rat pancreas as a parameter for carcinogenesis, the quantitation of cell proliferation in (pre)neoplastic pancreatic tissue will add substantially to the information on the process (and modulation) of pancreatic tumor growth in experimental animals. However, it should be kept in mind that, although altered cell proliferation is an important characteristic of neoplastic development, the development of a tumor mass is dependent on both cell proliferation and cell death rates. Quantitation of cell turnover of AACF as a parameter of growth is most informative when it is expressed as the difference between cell proliferation and apoptosis. The difference in cell proliferation and cell death rates is well recognized as a major determinant of outcomes (both for cells and for multicellular organisms) in the cancer models developed by Moolgavkar and others.

F. Nongenotoxic Carcinogens

Unsaturated fat has been suspected of being a pancreatic nongenotoxic carcinogen based on the observation of a relatively high incidence of AACF and adenomas in male F344 rats that received corn oil by gavage as a control vehicle. Corn oil is especially rich in linoleic acid, the most important being the ω-6 fatty acid. A microscopic review of the pancreas of corn oil vehicle control and untreated F344/N male rats in 37 2-year carcinogenicity studies was conducted to determine the extent and the strength of the association of proliferative exocrine pancreas lesions with administration of corn oil by gavage (32). The overall incidence of acidophilic AACF and adenomas was about five times higher in male rats that received corn oil than in non–gavage-treated controls. In a 24-month study in which F344/N rats were treated with different doses of corn oil, an increase in AACF and acinar adenomas was observed with increasing doses of corn oil, leading to an overall incidence of these two lesions, which was approximately five times higher than in untreated controls (33).

Raw soybean flour and trypsin inhibitors (TI) have also been implicated as nongenotoxic pancreatic carcinogens. TI regulate the release of the pancreatropic gut hormone cholecystokinin (CCK). High trypsin activities downregulate the CCK release, and low trypsin activities stimulate CCK release. Inhibition of trypsin activity will result in increased release of CCK, followed by (over)stimulation of pancreatic secretion, leading to increased pancreatic growth (34). This feedback regulation has been demonstrated in rats, mice, and chickens, but not in several other species like dog, pig, or monkey (35). The initial effect of dietary TI on mice and rats is pancreatic growth, but when rats are fed raw soya flour for 2 years they eventually develop adenomas and carcinomas in the pancreas (36). CCK as well as the synthetic TI camostate also stimulate the development of azaserine-induced pancreatic carcinogenesis in rats (37). These effects contrast significantly with those observed in hamsters. Although BOP-treated hamsters receiving camostate showed increased pancreatic weights, they showed decreased numbers of pancreatic (pre)neoplastic ductular lesions (38). Treatment of hamsters with CCK resulted in a slight increase in pancreatic weight and in no effect on BOP-induced carcinogenesis (39). In other studies with BOP-treated hamsters, injections with CCK resulted either in enhancing effects (40) or inhibitory effects on pancreatic carcinogenesis (41), indicating that the role of CCK in pancreatic carcinogenesis in the hamster remains controversial. The relevance of protease inhibitors for human pancreatic function and pathology is not clear: high activity of duodenal proteases has been reported to suppress pancreatic exocrine secretion in humans (42), but inhibition of more than 95% of the intraduodenal trypsin activity had no effect on pancreatic secretion (43). An enhancing effect of TI on the frequency of pancreatic ductular tumors in humans, either in the presence or in the absence of predisposing environmental factors, seems highly unlikely, but cannot be excluded completely. Therefore, the role of TI in the development of pancreatic tumors in humans remains to be investigated.

It is beyond doubt that CCK is involved in azaserine-induced pancreatic carcinogenesis in rats. When CCK or its analog caerulein is injected subcutaneously, a dramatic increase in pancreatic weight and in number and size of AACF (37) and in number of pancreatic adenocarcinomas (25,44) has been observed. The enhancing effects of CCK on pancreatic carcinogenesis in azaserine-treated rats could largely be blocked by lorglumide, a specific CCK-receptor antagonist, whereas the effects of a high corn oil diet were not influenced by lorglumide (37,44). Furthermore, CCK, but not a high corn oil diet, causes an increase in pancreatic weight. From these results we concluded that CCK and dietary fat enhance pancreatic carcinogenesis via different mechanisms.

G. Modulation of Pancreatic Carcinogenesis

Apart from hormones and TI, several nutrients have been demonstrated to modulate pancreatic carcinogenesis in azaserine-treated rats or BOP-treated hamsters. Consistent promoting effects were observed with dietary saturated and unsaturated fat, whereas inconsistent results were obtained with ethanol, coffee, and several micronutrients like vitamins E and C, β-carotene, and selenium. Table 2 summarizes the effects seen with substances investigated in both rats and hamsters.

The results of a long-term study with azaserine-treated rats demonstrated that ethanol caused an increase in multiplicity but not in incidence of malignant pancreatic tumors, pointing to an enhancing effect on the development of acinar adenocarcinomas in carcinoma-bearing animals (45). Long-term ethanol ingestion in rats (14–53 weeks) produced changes in acinar, centroacinar, and ductular cells. Microscopic changes comprised degeneration and atrophic changes of acinar cells, fibrosis, and intraductal protein precipitates (46). The pseudoductular cysts are lined by cuboidal epithelium of ductal type, which may represent dedifferentiation of acinar structures or hyperplasia of centroacinar cells accompanied by atrophy of surrounding acini. A similar phenomenon has been observed in BOP-treated hamsters (model for ductular adenocarcinomas in exocrine pancreas) (47–53). In BOP-treated hamsters ethanol did not modulate pancreatic carcinogenesis (45,54). These findings are in agreement with those of Pour et al. (55), who found that ethanol given to outbred Syrian golden hamsters in drinking water at a 5% (w/v) concentration for life, beginning either before or after a single dose of BOP, had no effect on tumor induction. This observation was in contrast with the results of a previous study of this group (56) in which a higher concentration of ethanol (25% w/v) inhibited the development of BOP-induced pancreatic lesions. The observation of Pour et al. (55) that hamsters treated with BOP and maintained on ethanol for life exhibited a few atypical acinar cell foci might also point to the pancreatic acinar cell as the main target cell for ethanol and not the centro-acinar or ductular cell.

The results of two long-term studies performed at the Nutrition and Food Research Institute of the Netherlands Organization for Applied Scientific Research (TNO) with azaserine-treated rats (up to 15 months) and BOP-treated hamsters (up to 12 months) demonstrated that in rats rather than in hamsters chronic coffee consumption has an inhibitory effect on dietary fat–promoted pancreatic carcinogenesis (57). Nishikawa et al. (58) studied the modulating effects of caffeine instead of coffee on development of BOP-initiated pancreatic tumors and found a significantly higher multiplicity in animals receiving caffeine than in controls. This suggested discrepancy with our studies might be ascribed to the presence of such constituents as kahweol palmitate, cafestol palmitate, and chlorogenic acid in coffee, which have been shown to inhibit carcinogenesis in various organs.

In rodent studies, it appeared that natural ingredient diets containing mixtures of whole plant–derived products generally protect rodents against carcinogens in comparison with semi-purified diets containing only highly refined components of whole foods. We observed a significant inhibitory effect of a cereal-based laboratory chow versus an AIN^{76} semi-purified diet (i.e., a diet developed in 1976 by the American Institute of Nutrition) on pancreatic carcinogenesis in azaserine-treated rats (59). This interesting effect may be caused by the high fiber content of the lab chow, which contained 11.5% indigestible dietary fiber. This amount is 2.7 times higher than the percentage of dietary fiber (4.3%) in the purified diet.

The results of a long-term study with micronutrients demonstrated that selenium, β-carotene, and vitamin C, but not vitamin E, inhibited the development of acinar adenocarcinomas induced in rat pancreas by azaserine. The most pronounced inhibitory effects were observed with dietary supplementation of selenium and β-carotene, both of which reduced multiplicity and incidence of pancreatic tumors (60). Combination of β-carotene with vitamin C did not result in an additive or synergistic inhibitory effect on pancreatic carcinogenesis in rats as compared to these micronutrients alone. Selenium in combination with vitamin E, however, resulted in an enhanced inhibitory effect on pancreatic carcinogenesis in comparison with these components alone (60). The latter observation indicates synergy, since vitamin E alone did not have any effect. The combination of β-carotene, vitamins C and E, and selenium resulted in the strongest inhibition of pancreatic carcinogenesis. This effect was most likely due to the combination of β-carotene and selenium. Recently we investigated the effects of β-carotene and selenium either alone or in combination on the initiation and promotion phase of pancreatic carcinogenesis in azaserine-treated rats using cell proliferation and volumetric data of AACF as parameters. The most pronounced inhibitory effect on pancreatic carcinogenesis was observed when selenium was given either alone or in combination with β-carotene during the promotion phase (61).

In BOP-treated hamsters, dietary supplementation of vitamin C, alone as well as in combination with β-carotene, resulted in consistently lower numbers of advanced ductular lesions, but the differences with the controls did not reach the level of statistical significance (62). β-Carotene alone demonstrated no inhibitory effect on pancreatic carcinogenesis in hamsters. Neither vitamin E nor selenium nor a combination of both had any effect on the development of ductular lesions in BOP-treated hamsters (62).

III. HUMAN CARCINOGENESIS

A. Descriptive Epidemiology

Based on the Surveillance, Epidemiology, and End Results (SEER) Program data (23,116 patients with cancer of the pancreas during the period 1973–1987), the vast majority of all cancers of the exocrine pancreas were ductal adenocarcinomas, while less than 1% were recorded as acinar cell carcinomas (67). Presentation of pancreatic cancer is often late with established metastases. In the SEER dataset, 50% of newly diagnosed cases died within about 4 months, while the overall 5-year relative survival rate was as low as 2.5% (67). As survival of pancreatic cancer is so poor, mortality rates closely parallel incidence rates. In general, interpretation of differences in cancer-specific mortality rates are hampered due to differences in access to health services and in diagnostic and therapeutic procedures. Since there is no influence of therapeutic differences, patterns of incidence rates, i.e., the number of new cancers arising within a given population size in a defined period of time, better reflect differences in causal factors. However, the diagnosis of pancreatic cancer is notoriously difficult with relatively low histological verification rates greatly varying across cancer registries (68). Moreover, in the last decades diagnostic techniques have improved considerably, further compounding the interpretation of observed trends in incidence. Incidence rates of pancreatic cancer rise rapidly with age. Age-standardized rates are about 1.5- to 2-fold higher for men than for women. Worldwide there is a wide variation in (estimated) incidence rates of pancreatic cancer (69,70). During the period 1970–1985, highest truncated rates (30–74 years) of up to 30 per 100,000 in men and 15–20 per 100,000 in women have been seen in Hawaiians and New Zealand Maoris and in black populations in the United States (69). Asian and Caucasian populations appear to have lower rates generally—in the range of 10–20 for men and 3–10 for women (69). Among Asian populations, the Japanese again show relatively high rates. Lowest truncated rates in both genders, i.e., below 10 in men and below 5 in women, are seen in cancer registry areas in Israel (non-Jews), France, Spain, India, Hong Kong, and Middle and South America. As for most of the leading cancers, the 1990 estimated age-standardized incidence rates of pancreatic cancer are highest in the north and west of Europe, whereas the lowest incidence rates are seen in Mediterranean countries (70).

There is a general increase in incidence of pancreatic cancer over time (1965–1985) in many populations in Europe and Asia. However, in the early 1980s truncated rates in several (especially male) high-risk populations decelerated or even reversed [e.g., in Israel (Jews), Sweden, Norway, Denmark, the United Kingdom (Scotland and Birmingham), and Japan (Nagasaki) (69)]. In the European Community, incidence has been increasing in low-risk southern countries, narrowing the incidence difference with the north and west of Europe (70,71). Most rates in the United States and Oceania are stable, with some upward and downward trends depending on race and region. Several populations show continuous declines since the 1970s, notably among men in high-risk areas like Switzerland, New Zealand (Maoris), Hawaii (Hawaiians), the United States (Bay Area whites), and Canada (British Columbia) (69). The strongest decline in incidence in younger cohorts of 30–44 years in several registry areas with high incidence [e.g., in Nordic countries in Europe, Saarland (Germany), Israel (Jews), Nagasaki (Japan), South Australia, Canada, and the United States], may suggest a further fall in future incidence of pancreatic cancer (69). So, although in general geographical variations and secular changes in incidence rates are seen as reflecting differences in lifestyle factors rather than differences in genetic makeup, extreme caution is warranted in the case of pancreatic cancer. Most, but not all, of the observed temporal patterns in risk have been linked to changes in smoking habits (72,73).

B. Genetic Component

In a hospital-based case-control study in Italy, a family history of pancreatic cancer among first-degree relatives has been linked to risk of cancer of the pancreas (74). The association persisted even after adjustment for such known and suspected risk factors as smoking, alcohol, diet, and medical conditions. The authors calculated that about 3% of the newly diagnosed pancreatic cancers could be attributed to the familial component. Other studies have shown that as many as 7.8% of persons with pancreatic cancer have a family history of pancreatic cancer, compared with only 0.6% of controls (75). If cancer of the pancreas clusters in families, then identification of the responsible genetic defect becomes possible.

C. Oncogenetic Changes

Adenocarcinomas of the human exocrine pancreas are generally considered to be of ductal origin (76). However, next to putative ductal preneoplastic lesions (77), foci and nodules of acinar cell dysplasia are also frequently observed in pancreatic tissue, especially of elderly persons (78). A possible role of acinar cells in ductular human pancreatic carcinogenesis is still controversial but cannot be excluded (79,80). The best documented oncogenetic alteration in both preneoplastic ductal lesions and in ductal adenocarcinomas of the human pancreas is the frequently mutated K-*ras* gene. Initially there have been speculations about possible relations between exogenous agents and the nature or frequency of the K-*ras* mutation observed. An example of the latter is the suggested association between smoking and a higher rate of K-*ras* mutations, although at borderline statistical significance ($p = 0.046$) (81). As also summarized in Table 2, other major gene alterations identified so far in more recent years concern the *p53* gene, the MTS1 gene, probably the RB gene, and the very recently reported DPC4 gene (82). Analysis of the mutation spectra of the K-*ras* gene as well as the *p53* gene revealed the existence of several types of mutations, in the latter the majority consisting of G → A transitions (41%), A → G transitions (17%), and G → T transversions (13%) (83). The heterogeneity of base substitutions in codon 12 of the K-*ras* oncogene and the *p53* tumor-suppressor gene in pancreatic cancer provides little evidence to indicate the action of a single carcinogen (84). It is possible to speculate that the G → T transversions are caused by the mutagens present in tobacco smoke, but more data are needed to substantiate this hypothesis. More details on the state of the art of the oncogenetics of the human exocrine pancreas are presented in a recent review by Caldas and Kern (81).

D. Smoking

The majority of epidemiological investigations in different populations using different designs have implicated smoking in the etiology of cancer of the pancreas (85). Former and current smoking, usual amount smoked, age at start, duration, and time since quitting have been linked to risk. The dose-response relationship is generally weak. Only four case-control studies have examined the effect of type of cigarettes, suggesting that smoking of filter cigarettes may also be related to risk (86–89). The lower risks among former smokers and analyses by time period among current and former smokers indicate that smoking during the 10–20 years preceding diagnosis, rather than smoking longer previously, appears to be the critical factor (86,88–92).

E. Alcohol

In 1986, the evidence of an association between the drinking of alcohol and the development of cancer of the pancreas was judged insufficient (93). In 1988, the International Agency for Research on Cancer evaluated the carcinogenic risk of alcohol drinking and concluded that "taken together, the results of these (15 cohort and 14 case-control) studies suggest that consumption of alcoholic beverages is unlikely to be causally related to cancer of the pancreas" (94). Findings from seven out of eight more recently published population-based case-control (87,95–100) and three out of four cohort studies (101–103) argue against a role of alcohol in human pancreatic cancer. However, one cohort study among white men in the United States with 57 deaths due to pancreatic cancer reported that consumers of 10 or more drinks of alcohol per month had three times the risk of nondrinkers, but no clear dose-response trends among drinkers were seen (91). Inferences about the type of alcohol were limited, since most patients with pancreatic cancer who drank alcohol consumed both beer and hard liquor and increased risks were seen for both beer and hard liquor. Recent results of a large multicenter population-based case-control study in the United States on the basis of direct interviews only reported that alcohol drinking at the levels typically consumed by the general population is probably not a risk factor for pancreatic cancer (104). However, heavy drinking of hard liquor (more than 35 drinks per week in men and 8 or more per week for women) was associated with increased risk in blacks but not in whites, whereas heavy drinking of beer (more than 28 drinks of beer per week in men) was associated with increased risks in white men only. Estimates on the drinking of alcoholic beverages obtained from patients with cancer of the pancreas may be biased. This is not the case in cohort studies in which information on exposure is elicited prior to the occurrence of the disease of interest. The above findings have to be weighed against the results of the other three cohort studies. In the cohort study among members (60.4% were white) of a medical care program in the United States, after 6 years of follow-up 48 cases of pancreatic cancer with information about drinking habits were observed but no association with the drinking of alcohol (with a highest category of more than 1 drink a day), beer, or liquor was seen (101). In a cohort study among ethnically mixed subscribers to a medical care program in the United States, exposure information obtained at baseline from 405 of those who later developed pancreatic cancer was compared with information from a subcohort of 2678 controls (102). The drinking of alcohol even at levels as high as three or more drinks a day was not statistically significantly related to risk. Information on the drinking of beer or spirits was not available. In the third cohort study among elderly residents of a Californian retirement community who were almost entirely Caucasian, after 9 years of follow-up 63 incident cases of pancreatic cancer with information on the average number of drinks of types of alcohol were available for analysis (103). This study also showed no association with the drinking of alcohol even at the highest levels of more than two drinks per day. Risk estimates for the drinking of beer or spirits were also not provided. Studies among presumed heavy drinkers may yield further information. Although the incidence of pancreatic cancer was statistically significantly elevated among 6230 Swedish male brewery workers who received a daily ration of about 1 liter of beer, the possibility that the excess risk was due to smoking could not be excluded (105). A statistically significant 2.7-fold excess risk for cancer of the pancreas was also found among a cohort of 15,508 alcoholic women in Sweden, who mostly drank spirits (106). Confounding by smoking and dietary factors could, however, not be excluded. In brief, the possibility that heavy drinking of beer or spirits increases the risk of cancer of the pancreas remains of concern. There are also some suggestions that both drinking alcohol and smoking may increase the risk of pancreatic cancer more than smoking alone. On the basis of seven deaths, in the recent cohort study among white men in the United States, exposure to both smoking and alcohol was associated with a statistically significant sixfold increased risk (91). In the cohort study in Japan in which daily smoking increased the risk (RR 2.15) and daily alcohol drinking lowered the risk (RR 0.79), the relative risk associated with joint daily smoking and daily consumption of alcohol was much lower (1.81) and statistically nonsignificant (107). More research is needed to investigate the role of specific types of alcohol, especially at high levels of consumption, and the possibility of interaction with smoking.

F. Coffee and Tea

In 1991, after reviewing 6 cohort and 21 case-control studies, the International Agency for Research on Cancer concluded that "the data are suggestive of a weak relationship between high levels of coffee consumption and the occurrence of pancreatic cancer, but the possibility that this is due to bias or confounding is tenable" (108). However, none of the six cohort studies reviewed reported a significant association with increased consumption. More recently published cohort studies in the United States (91,102,103) and Norway (109) and seven out of eight case-control studies in such disparate populations as in Australia (96), Canada (87,97), the Netherlands (98), Greece (110), Poland (99), and Finland (111) further confirm the absence of a relationship with risk even at high levels of exposure of more than six cups of coffee per day (102,111). Only in one case-control study in Utah, in which all information was obtained from proxies for cases and controls, did risk increase with the lifetime amount of coffee drunk (112). Combined preliminary analyses of the effect of lifetime consumption of coffee assessed in a highly standardized manner in the five SEARCH case-control studies in Australia (96), Canada (87,97), the Netherlands (98), and Poland (99) did not show a relationship with risk. In several of the aforementioned case-control studies, however, exposure estimates were obtained from proxies and may have been misclassified (113). Yet, taken together, the weight of the epidemiological evidence is strongly against the drinking of coffee as influencing the development of cancer of the pancreas.

In 1991, after reviewing four cohort and six case-control studies on tea consumption and pancreatic cancer, the International Agency for Research on Cancer reported that one case-control study showed a positive association with risk, one cohort study documented a small protective effect, and the other studies reported no association, and concluded that there is inadequate evidence for the carcinogenicity in humans of tea drinking (108). In the cohort study, the protective effect of regular tea consumption at the time of a college physical examination was found after 16–50 years of follow-up of male former college students (114). However, there was no indication of a dose-response effect. One more recently published cohort study among residents of retirement community in the United States also found that risk of pancreatic cancer decreased with increasing tea consumption with a significant dose-response effect (103). In view of the many anticarcinogenic substances in tea, further study on the effect of tea should be directed to a possible protective rather than a risk-increasing effect.

G. Nutrition

The potential role of nutrition in the etiology of cancer of the pancreas was well summarized in a recent review (115). For cancer of the pancreas, it was concluded that "fairly consistent patterns of positive associations with the intake of meat, carbohydrates, and dietary cholesterol have been observed," whereas "consistent inverse relationships with the consumption of fruit and vegetable intakes and, in particular, with two markers of such foods, namely fiber and vitamin C, also have been noted." The review included the results of 17 case-control studies, of which 7 were conducted in Europe, 6 in the United States, 2 in Canada, and 1 each in Australia and Japan, and 4 cohort studies, of which 3 were conducted in the United States and 1 in Japan. The findings of 5 simultaneous population-based case-control studies in Adelaide (Australia), Montreal and Toronto (Canada), Utrecht (the Netherlands), and Opole (Poland), conducted under the auspices of the International Agency for Research on Cancer (IARC) SEARCH program, were considered to be of particular interest. The studies used the same protocol and were carried out with the objective to provide pooled estimates of risk. Interviewers collected comprehensive dietary information using very similar semi-quantitative food frequency questionnaires. In the combined analysis, more than 800 incident cases of pancreatic cancer and 2000 controls were included, representing the largest available dataset on nutrition and cancer of the pancreas to date. Positive associations were observed with the intake of carbohydrates and cholesterol and inverse associations with dietary fiber and vitamin C (116). Intake of total, saturated, mono-, and polyunsaturated fats and protein were not related to risk, suggesting that the nonenergy effect of carbohydrates may be the relevant factor (115). Pooled analysis on the level of foods was not performed. As dietary information obtained in case-control studies may be influenced by disease-related changes in diet and, in the case of pancreatic cancer, often dietary information obtained by proxies is used, the findings of cohort studies are of particular importance. Of the four cohort studies published to date, all reported analyses for foods/food groups (91,103, 107,117), while only one reported risk estimates for nutrients (103). The earliest and by far the largest cohort study has been conducted in Japan (107). In 1965, interviewers collected baseline information on frequency of consumption of seven food groups among 265,118 Japanese participants. After 17 years of follow-up, 679 deaths from cancer of the pancreas had occurred. Daily consumption of rice and wheat (the most important grains eaten in Japan) and of green-yellow vegetables, including carrots, spinach, green peppers, Italian broccoli, pumpkin, turnip leaves, green lettuce, chives, leeks (green), asparagus (green), chicory, and parsley, was not associated with male or female risk. In 1976, dietary habits of 34,000 Seventh-Day Adventists in the United States were measured by a food-frequency questionnaire containing 35 selected foods (117). After 7 years of follow-up, 17 men and 23 women died from cancer of the pancreas. Highly significant inverse associations with the consumption of vegetarian protein products, beans, lentils, and peas as well as dried fruit emerged, while no statistically significant effects were seen for the consumption of cooked green vegetables, green salad, tomatoes, or (non)citrus fruit. The influence of major sources of carbohydrates was not evaluated. In a cohort study among white men in the United States, after 20 years of follow-up 57 deaths from pancreatic cancer had occurred (91). In 1966, diet was assessed by asking the frequency of consumption of only 35 food items, and thus the study was unable to assess associations with dietary fat, protein, and carbohydrates. However, high consumption of bread and cereal, major sources of carbohydrates, was not found to be related to risk. Consumption of vegetables and fruits also showed no clear association with risk. The dietary portion of the questionnaire, used in 1982 in the prospective study among elderly residents of a Californian retirement community, was designed specifically to measure intake of foods rich in either vitamin A and its precursors or vitamin C (103). After 8 years of follow-up, 65 incident cases of pancreatic cancer were identified. It was concluded that higher intake of vegetables, fruits, dietary β-carotene, and vitamin C were each associated with a reduced risk, although none of these associations was statistically significant. Results on the effect of calorie-providing nutrients were not provided probably due to the limitations of the questionnaire and thus the independency of total energy of the association with vegetables and fruits could not be established. In summary, only two of the four cohort studies reported on intake of (sources of) carbohydrates, and both showed no association with risk (91,107). Two of the five SEARCH case-control studies reported on (sources of) carbohydrates and findings suggested that intake of simple rather than complex carbohydrates might be the relevant factor (96,1118). Four cohort studies reported on a possible link with the consumption of (markers of) vegetables and fruits, of which two showed no association with risk (91,107), one reported weak statistically nonsignificant inverse associations with the intake of vegetables, fruits, β-carotene, and vitamin C (103), while only one showed strongly inverse associations with specific types of vegetables and fruits, i.e., vegetarian protein products, specific types of pulses, and dried fruit (117).

It is clear that the evidence derived from cohort studies on the association between the consumption of vegetables and fruit and the intake or blood levels of related nutrients and the occurrence of cancer of the pancreas is less consistent than one would have expected on the basis of the results of case-control studies. A similar inference was drawn for other cancer sites (119). Although recall bias is considered one of the main drawbacks of case-control studies, cohort studies too may suffer from inherent limitations, explaining at least in part the observed inconsistency in results, e.g., measurement error in dietary exposure assessment when using single self-administered questionnaires and lack of heterogeneity of exposure. When reviewing the reported relative validity for ranking individuals according to intake of vegetables and fruit as assessed by self-administered food frequency questionnaires, relative validity for consumption estimates is poor for vegetables and moderate for fruit (120). Studies on biomarkers of the consumption of vegetables and fruits are therefore of particular importance. In a nested case-control study conducted in Maryland, prediagnostic (up to 9–12 years prior to diagnosis) serum levels of lycopene, but not β-carotene, were statistically significantly lower in 22 cases with pancreatic carcinoma compared to 44 matched controls (121). Such findings once again support a possible protective role of specific vegetables or fruits rich in lycopene but poor in β-carotene like tomatoes. In fact, the consumption of tomatoes has been linked to cancer of the pancreas (96,122) and other digestive tract cancers as well (123).

When summarizing the results of nine case-control and three cohort studies with respect to meat consumption, Howe concluded that three case-control studies found no evidence of an association, one case-control study reported an inverse association, while nine studies, including the three cohort studies, found positive associations with risk (115). It may be of interest that in the cohort study in Japan (107), among various combinations of lifestyle factors, the relative risk for the combination of daily smoking and daily meat intake was by far the highest (a relative risk of 3.07). One of the case-control studies reported a positive association with fried or grilled meat (124). Further support for the hypothesis that mutagens in (fried) meat may be involved in the development of cancer of the pancreas is provided by the findings of a cohort study on the intake of fried meat and cancer risk among 9990 Finnish men and women (125). A modified dietary history method was used to assess habitual consumption of fried meat from 1966 to 1972. After 24 years of follow-up, 29 incident cases of cancer of the pancreas were identified. After adjustment for total energy and other (dietary) variables, findings for cancer of the pancreas suggested nonsignificantly elevated risks among women with high intakes of fried meat.

H. Menstrual History

Several case-control studies have examined the association with menstrual variables in women (110,126–128). Three of four studies suggested that an earlier age at menarche was associated with an increased risk of pancreatic cancer (110,126,127). These findings of course have to be replicated, but they may indicate that hormonal factors account for part of the male-to-female difference in incidence of pancreas cancer. Dietary factors early in life may be involved as well. In a prospective study, early menarche was positively associated with consumption of meat, whereas aspects of a vegetarian diet, i.e., meat analogs, grains, nuts, beans, and other legumes, were associated with a late onset of menarche (129). A dietary pattern characterized by a high consumption of meat and/or a low consumption of vegetables not only may attribute to an early onset of menarche, but, if it is continued, may also predispose to cancer of the pancreas later in life.

I. Medical Conditions

Since early studies found that a history of chronic pancreatitis (130), diabetes (131) with early onset in women (132), cholecystectomy in women (132), or allergies (133) appears to be associated with risk, these medical conditions as well as others have remained a focus of interest. As such conditions are rare in patients with cancer of the pancreas, case-control studies are less suited to examine the association with risk. Therefore, following large cohorts of persons with such medical conditions is the method of choice to evaluate the relationship with risk of cancer of the pancreas.

Two large cohorts of subjects with chronic pancreatitis, one multicenter study in Europe and the United States (134) and a second in Sweden (135), were followed up for cancer of the pancreas, both showing a positive association with risk. In the multicenter cohort study over a 20-year period, risk of pancreatic cancer appeared to increase nearly linearly (134). A similar time-risk relationship was found in a case-control study in Italy (136). In the cohort study in Sweden, however, increased risks were restricted to the first 10 years after first discharge for pancreatitis (135). Although the latter type of association is generally assumed to be less supportive of a direct causal relationship and is more consistent with pancreatitis as a first sign of pancreatic cancer, it does not exclude the possibility that both diseases still have shared risk factors. Most cases of pancreas cancer, however, are not preceded by long-term clinically diagnosed chronic pancreatitis. So, even if smoking and/or alcohol consumption are causally related to pancreatitis and pancreatitis to cancer of the pancreas, pancreatitis may be an intermediate stage between normal pancreatic function in only some cases of pancreas cancer. In the absence of chronic pancreatitis, smoking and possibly the drinking of alcohol may act through other mechanisms.

Two recent large-scale case-control studies in Italy (137,138) reported declining risks with increasing time between diabetes and cancer of the pancreas. The associations became nonsignificant when excluding cases with onset of diabetes 3 or more years preceding diagnosis of cancer of the pancreas in one study (138) and 10 or more years in the other (137). These results suggest that diabetes probably is an early symptom of pancreatic cancer rather than an etiological factor. However, a meta-analysis of 20 case-control and cohort studies confirmed that persons with long-standing diabetes of at least 5 years duration are at increased risk of pancreatic cancer (139). Risks of cohort studies were found to be generally stronger than those found in case-control studies. The time–risk relationship in some cohort studies is also different. For instance, risks found in the cohort of subscribers to a medical care program remained virtually unchanged when excluding diabetes cases whose cancer developed within 5 years of the medical checkup, indicating that diabetes was indeed a predictor rather than a result of pancreatic cancer (102). However, other results were obtained in a cohort of about 7000 male working men in Paris, where 5992 normoglycemic and 312 known and newly diagnosed non–insulin-dependent diabetic subjects were followed for fatal pancreatic cancer (140). After exclusion of deaths due to pancreatic cancer occurring during the first 5 years of follow-up, the diabetic men experienced a fivefold increased risk for fatal pancreatic cancer. Such results clearly suggest that diabetes or some factor closely associated to diabetes may be a risk factor of cancer of the pancreas. In a nationwide cohort of 134,096 Swedish persons hospitalized for diabetes from 1965 through 1983, even 10 years after initial hospitalization, on the basis of 103 observed cases of cancer of the pancreas, a 1.7-fold statistically significant increased risk was seen (141). Non–insulin-dependent diabetes appears to be the most likely candidate (139,141). Further study is required on the type of diabetes and whether or not both diseases share common diet-related or genetic factors. For instance, the consumption of certain kinds of food (specifically sugar, fat, and refined carbohydrates) has been associated with an increased risk of diabetes. Such a finding is consistent with some epidemiological studies, suggesting that the positive association between the intake of carbohydrates and cancer of the pancreas might be explained by simple sugars rather than complex carbohydrates.

Although the possibility that cholecystectomy may act as a risk factor of cancer of the pancreas was supported by a recent cohort study among elderly residents of a Californian retirement community (103), these findings prompted the authors of another cohort study to reexamine their data set, and no association with risk was observed (142).

A number of case-control studies and at least two cohort studies (102,117) investigated the association with allergy-related conditions, and most (97,102,110,117,143–145), but not all (146), reported inverse associations with conditions such as asthma, eczema, hay fever, contact dermatitis, urticaria, or other allergies. In the preliminary common analysis of the five

SEARCH case-control studies, the protective effect was consistently present over type of allergy (asthma, eczema, hay fever, and other allergies), as well as proxy status and study center (Baghurst, personal communication). The results further support a possible protective effect of hyperactivity of the cell-mediated immune system.

IV. COMPARISON OF ANIMAL AND HUMAN CARCINOGENICITY

A. Oncogenetic Events

Although a role of acinar cells in the development of pancreatic cancer cannot be ruled out, it is now generally believed that human pancreatic adenocarcinomas arise from preneoplastic ductal lesions classified as pancreatic intraepithelial neoplasm (PIN) (147). Before addressing the similarities in oncogenetics between human pancreatic tumors and the corresponding adenocarcinomas induced in Syrian hamsters by BOP, the current knowledge on the (aberrant) proliferation of acinar cells will be briefly discussed.

The search for genetic alterations in (pre)neoplastic acinar lesions of the pancreas has not been very successful so far. As previously mentioned, it has been demonstrated that mutations in the K-*ras* gene are not detectable in azaserine-induced acinar adenocarcinomas in rats. Moreover, studies on comparable acinar lesions in the pancreas of mice and humans have led to the similar conclusion of no activating mutations present in the K-*ras* gene (148–150). From these data it can be concluded that activation of the K-*ras*–mediated signal transduction pathway is not crucial for the transformation of acinar cells. In transgenic mice three different oncogenes (activated H-*ras*, c-*myc*, and SV40 largeT), when targeted to acinar cells by the elastase promoter, produce various acinar lesions including adenocarcinomas (151). In more recent studies it has been demonstrated that cholecystokinin (CCK) through its CCK-B receptor is involved in acinar cell growth (80). Therefore, elucidation of the signal transduction route of the CCK-B receptor may eventually lead to the identification of more acinar specific transforming genes.

One of the best documented oncogenetic alterations in both preneoplastic ductal lesions and in ductal adenocarcinomas of the human pancreas is the frequently mutated K-*ras* gene. Results from several studies (12,17,18) have demonstrated that high incidences of K-*ras* mutations as well as its early appearance are also observed in ductal pancreatic tumors of the Syrian hamster. These observations provided support at the molecular level for

BOP-induced Syrian hamster pancreatic carcinogenesis as the most relevant model for human pancreatic cancer.

Over the past 3 years significant progress has been made in identifying other oncogenetic alterations in human pancreatic cancer. As summarized in Table 2, the major gene alterations identified so far concern the *p53* gene, the MTS1 gene, probably the RB gene, and, as very recently reported, the DPC4 gene (82). These observations support the following genetic model of pancreas tumorigenesis: ductal cells are initially and predominantly driven by a mutation of the K-*ras* oncogene, followed by further deregulation of RB mediated cell-cycle control through the frequent mutation of three genes involved in CDK (cyclic dependent kinases) inhibition: *p53*, MTS1, and DPC4. Scarpelli and coworkers have published a paper about several oncogenetic alterations in the Syrian hamster model (16). They reported the following genetic alterations in ductal adenocarcinomas: 90% for K-*ras*, 53% for DCC, 42% for RB, 26% for mdm-2, and an absence of *p53* mutations. While K-*ras* and RB alterations resemble those found in human pancreatic cancer, more data on DCC, DPC4, and the MTS-1 genes in both species are needed to make a more substantial comparison. The striking difference in *p53* alterations remains to be solved, but may in part be substituted for by the observed overexpression of mdm-2 in 26% of the hamster pancreatic tumors. Furthermore, this leaves room for alterations in other proteins in the *p53* pathway as well. Another important aspect with regard to *p53* dysfunction is the tissue specificity as demonstrated by Clarke and coworkers (152). They show that loss of function of one Apc allele in nullizygous *p53* mice predisposes to pancreatic neoplasia and not to the expected intestinal tumors. The authors suggest that the acinar origin of the observed lesions is largely due to the mutated Apc, although metaplasia cannot be excluded. Another explanation bears on different functions of *p53* in the pancreas and the intestine. It is speculated that *p53* may guard the genome of acinar cells specifically by causing cell-cycle arrest, whereas in the intestine failure of *p53*-dependent apoptosis contributes significantly to the progression of tumors. The latter is very similar to the recently suggested cooperation between *p53* deficiency and Wnt-1 in mouse mammary cancer and also to the progression of chemically induced skin cancer in a *p53* null genetic background (153,154).

Although considerably more information is available for ductal rather than acinar lesions, a full comparison between oncogenetic alterations in human exocrine pancreatic cancer and the relevant animal model systems is not possible because data on corresponding loci in the different species are still far from complete. Further studies with carefully selected mutant alleles of candidate pancreatic transforming genes in well-defined genetic backgrounds of mice are needed to clarify these issues. Besides the transgenic mice, the value of the Syrian hamster model for experimental pancreatic carcinogenesis is indicated by the multiple genetic lesions observed, many of which are identical to those found in human pancreatic cancer. However, the absence of *p53* mutations in BOP-induced tumors of the Syrian hamster are in sharp contrast to the *p53* analysis of human pancreatic carcinomas. This discrepancy is also observed in tumors of the intestine (151,152). Finally, our present knowledge of the oncogenetic alterations involved in pancreatic cancer is schematically presented in Figure 3.

B. Lifestyle

Factors modulating experimental pancreatic carcinogenesis in rats and hamsters and related findings from observational epidemiological studies in humans are shown in Table 3. Human dose-response data on the smoking of cigarettes are generally weak. However, the consistency of epidemiological findings across different designs, populations, and time periods, and the reduction in risk when quitting, leads credence to a causal role of tobacco-derived carcinogens in pancreatic cancer. Probably due to difficulties in establishing an appropriate experimental animal model, support from animal experiments is far from complete. Treatment of rats with tobacco-specific nitrosamines in drinking water did result in the induction of pancreatic acinar and ductal cell neoplasms (155). To our knowledge, controlled animal experiments on the effect of smoking of cigarettes on the incidence of cancer of the exocrine pancreas have not been published. Smoking may act through several mechanisms. For instance, nicotine and chemical oxidants may damage pancreatic cells, leading to impaired exocrine function and possibly CCK-mediated compensatory proliferation. Tobacco may also induce phase I enzymes that metabolize procarcinogens in

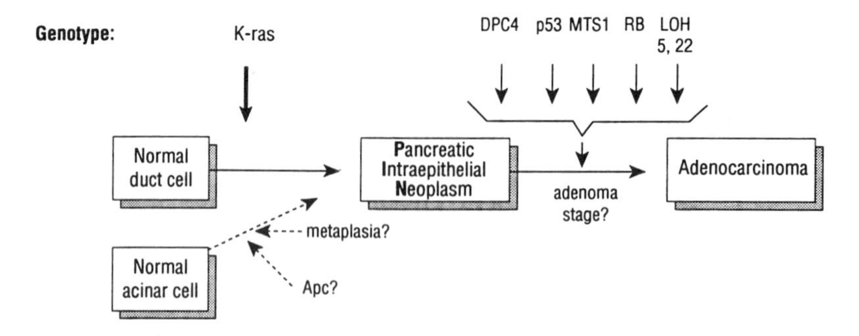

Figure 3 Multistage pancreatic carcinogenesis in humans.

Table 3 Factors Modulating Experimental Pancreatic Carcinogenesis in Rats and Hamsters and Related findings from Observational Epidemiological Studies in Humans

Modulating factor	Rats	Hamsters	Humans
Smoking	n.d.	n.d	+ [85]
Alcohol	+ [45]	0 [45]	0 [94], + [91,104]
Coffee	− [57]	0 [57]	0 [108]
Tea	n.d.	n.d.	0 [108], − [103]
Cholecystokinin (CCK)	+ [44]	+;0; − [39-41]	n.d.
Trypsin inhibitors (TI)	+ [41]	− [37]	n.d.
β-Carotene	− [60,61]	0 [62]	0 [116]
Vitamin E	0 [60,61]	0 [62]	0 [122]; − [96]
Vitamin C	− [60,61]	0 [62]	− [116]
Selenium	− [60,61]	+;0; − [62-64]	0 [96]
Vegetables and fruits	n.d.	n.d.	− [115]
Cereal-based laboratory chow	− [59]	n.d.	n.d.
Carbohydrates	n.d.	n.d.	− [116]
Dietary fiber	n.d.	n.d.	− [116]
Meat	n.d.	n.d.	+ [115]
Lard/Beef tallow (saturated fat)	+ [59]	+ [65]	0 [115]
Corn oil/Safflower oil (unsaturated fat; ω-6)	+ [45]	+ [45]	n.d.
Fish oil (unsaturated fat; ω-3)	−;+ [28,66]	0 [30]	n.d.
Dietary cholesterol	n.d.	+ [157]	+ [116]

0, No significant effect; +, significant increase; −, significant decrease in comparison with appropriate control groups; n.d., not determined.

tobacco smoke like polycyclic aromatic hydrocarbons and nitrosamines to chemical species that may directly damage DNA (85,156).

The results of epidemiological studies indicate that the drinking of alcohol at levels typically consumed by the general population is probably not a risk factor for cancer of the pancreas. However, recent findings do not exclude the possibility that heavy drinking may yet increase the risk. The BOP-induced hamster pancreatic tumors are morphologically closely similar to the pancreatic cancers occurring in humans, and the absence of an enhancing effect of ethanol on pancreatic carcinogenesis in hamsters may be more relevant to the human situation than the enhancing effects found in the rat studies.

Even at high levels of exposure, the weight of the epidemiological evidence is strongly against the drinking of coffee as increasing the risk of cancer of the pancreas. This is consistent with animal research indicating that long-term coffee consumption actually has an inhibitory effect on dietary fat–promoted pancreatic carcinogenesis in azaserine-treated rats and, though less pronounced, in BOP-treated hamsters. There is inadequate evidence for pancreatic carcinogenicity of the drinking of tea in humans. Recent cohort studies suggest the possibility that in fact the drinking of tea may protect against cancer of the pancreas. To our knowledge, no reports on the modulating effect of chronic tea consumption in carcinogen-treated animals have been published.

Nutritional epidemiological evidence points to fairly consistent patterns of positive associations with the intake of meat, carbohydrates, and dietary cholesterol. In contrast, consistent inverse relationships were obtained with the consumption of fruit and vegetables and in particular with two markers of such foods, namely, fiber and vitamin C. In both animal models consistent promoting effects with dietary saturated and unsaturated fat have been observed. This conclusion is in contrast with human research. The positive association with dietary cholesterol observed in humans is consistent with the promoting effect of a cholesterol-rich diet administered to Syrian hamsters following initiation with BOP (157). Although the BOP-treated hamster model is generally considered to be the best model for human pancreatic carcinogenesis, selenium, β-carotene, and vitamin C inhibited the development of acinar adenocarcinomas in the azaserine-treated rats, whereas no effect was seen in BOP-treated hamsters. A high-fiber cereal diet inhibited pancreatic carcinogenesis in the rat model, while the effect has not yet been examined in the hamster model. It is now increasingly recognized that for most cancers the protective effect of a high consumption of vegetables and fruits and fiber-rich cereals may be explained by a mix of naturally occurring anticarcinogens in such plant foods (158–160). Therefore, there is an urgent need to also test the modulating effects of whole vegetables and fruits in animal pancreatic carcinogenesis.

Since the heating/cooking of foods derived from muscle protein may give rise to the formation of heterocyclic amines, the positive associations in human epidemiological research with the consumption of (fried) meat might well be explained by the carcinogenic effect of formed heterocyclic amines (HAAs). Only one report on a relevant animal experiment with HAAs could be found showing that 2-amino-3-methylimidazo[4,5-f]-quinoline (IQ) induces atypical hyperplastic acinar cell lesions in rats (161). The morphology of these lesions resembles those of AACF. Foodborne HAAs can be metabolically activated by humans both through N-oxidation and O-oxidation to produce highly reactive metabolites that form DNA adducts (162). Administration of 2-amino-1-methyl-6-phenylimidazo-[4,5-b]pyridine (PhiP) to rats resulted in more adducts in the pancreas, lung, and heart than in the liver (163). Alternatively, the effect might be due to N-nitroso compounds or their precursors present in preserved meats.

C. Diabetes

Epidemiological evidence tends to support the view that persons with long-standing diabetes of at least 5 years duration are at increased risk of cancer of the pancreas. Findings point in the direction of non–insulin- rather than insulin-dependent diabetes. If that turns out to be the case, the exocrine pancreas will have been exposed for many years to compensatory high levels of insulin, because it receives a large proportion of its blood supply through the islets. The proliferative effect of insulin on pancreas cells may then explain the association with risk of cancer of the pancreas. Animal research, however, points to the importance of intact islet function, rather than insulin-mediated cell proliferation, for enhancing pancreatic carcinogenesis in hamsters. The modulating effect of streptozotocin (STZ)–induced diabetes (which acts by destroying islets function) in BOP-treated hamsters appears to depend on whether STZ is administered prior to, concurrent with, or following initiation with BOP. Enhancement of pancreatic carcinogenesis in hamsters is seen only when STZ and BOP are administered simultaneously (164), whereas STZ diabetes established after BOP tumor initiation plays no apparent role (165). Conversely, pretreatment of hamsters with STZ and 14 days later followed by BOP initiation inhibits pancreatic ductal/ductular cell carcinomas (166). The protective effect was potentiated by additional treatment with insulin. It was concluded that intact islets appear to be a prerequisite for exocrine pancreatic cancer induction by BOP and that the inhibitory action of insulin on ductal/ductular cancer induction by BOP seems to be related to suppression of ductal/ductular cell replication.

V. CONCLUSION

Smoking is still the only established risk factor of cancer of the exocrine pancreas. The potential role of our common diet to increase as well as lower the risk is increasingly acknowledged. Specific agents that cause or prevent cancer of the pancreas in humans have not yet been identified. As the consumption of vegetables and fruits appear to protect against most cancers—and cancer of the pancreas does not seem to be an exception—research on the possible protective effect of the intake of plant foods is of particular interest. Vegetables, fruits, and grains contain a large number of naturally occurring compounds for which cancer protective effects have been postulated. However, it is suspected that individual agents have only small effects. More emphasis on the study of the effects of whole foods is therefore justified and in line with the finding that, in contrast to single agents, the combination of β-carotene, vitamins C and E, and selenium resulted in the strongest inhibition of azaserine-induced pancreatic carcinogenesis in rats.

The multicausal etiology of cancer of the exocrine pancreas emerging from animal and human research is consistent with a number of initiating and promoting events explaining the sequential accumulation of genetic damage with intermittent clonal expansion of mutated cells. The genetic pathway of human pancreatic cancer more closely resembles the gene alterations observed in carcinogen-induced pancreatic cancers in the hamster than the rat. For several factors the modulating effects on pancreatic carcinogenesis in animals and humans are inconsistent or have not yet been investigated. Except enhancing effects for dietary cholesterol, the consistency of findings in both hamsters and humans mainly concerns factors for which no effect was seen. Somewhat surprisingly, findings on some potentially protective dietary factors in humans are more consistent with results in rats than in hamsters. Therefore, further research using both animal models is warranted, although the highest priority needs to be given to the hamster model. Since major elements of the genetic pathway leading to clinically overt cancer of the pancreas have been identified, the search is now on to try to link epidemiological risk factors to specific mutations. As cancer of the human pancreas is a relatively rare disease, large-scale cohort studies with banks of biological samples are of paramount importance. Due to the inherent limitations of questionnaire information, better exposure estimates may be obtained by using biomarkers of exposure to, for instance, smoking and the consumption of meat and plant foods. Susceptibility to cancer of the pancreas may not be equally distributed among humans. Banking of DNA would allow the study of the influence of genetic polymorphisms in carcinogen-metabolizing enzymes on the association between lifestyle factors (smoking and diet) and cancer of the pancreas. Further elucidation of mediating biological events will also require small-scale experimental (clinical) (feeding) studies. In brief, further progress in the etiology of cancer of the pancreas is best guaranteed by a multidisciplinary approach with a balanced interest in the role of carcinogenic and anticarcinogenic factors.

REFERENCES

1. Longnecker DS. Experimental models of exocrine pancreatic tumours. In: Go VLW, Garner JD, Brooks FP, Lebenthal E, DiMagno EP, Scheele GA, eds. The Exocrine Pancreas, Biology, Pathobiology and Diseases. New York: Raven Press, 1986:443–458.
2. Longnecker DS, Curphey TJ. Adenocarcinoma of the pancreas in azaserine-treated rats. Cancer Res 1975; 35:2249–2258.
3. Pour PM, Althoff J, Krüger FW, Mohr U. A potent pancreatic carcinogen in

Syrian hamsters: N-nitrosobis(2-oxopropyl)amine. J Natl Cancer Inst 1977; 58:1449–1453.
4. Zurlo J, Curphey TJ, Hiley R, Longnecker DS. Identification of 7-carboxymethylguanine in DNA from pancreatic acinar cells exposed to azaserine. Cancer Res 1982; 42:1286–1288.
5. Zurlo J, Longnecker DS, Cooney DA, Kuhlmann ET, Curphey TJ. Studies of pancreatic nodule induction and DNA damage by D-azaserine. Cancer Lett 1981; 12:75–80.
6. Zurlo J, Roebuck BD, Rutkowski JV, Curphey TS, Longnecker DS. Effect of pyridoxal deficiency on pancreatic DNA damage and nodule induction by azaserine. Carcinogenesis 1984; 5:555–558.
7. Kokkinakis DM, Scarpelli DG, Hollenberg PF. Metabolism and activation of nitrosamine pancreatic carcinogens. In: Scarpelli DG, Reddy JK, Longnecker DS, eds. Experimental Pancreatic Carcinogenesis. Boca Raton, FL: CRC Press, 1987:3–20.
8. Lawson T, Kolar C, Garrels R, Kirchmann E, Nagel D. The activation of ^{3}H-labeled N-nitrosobis(2-oxopropyl)amine by isolated hamster pancreas cells. J Cancer Res Clin Oncol 1989; 115:47–52.
9. Kokkinakis DM, Scarpelli DG. DNA-alkylation in the hamster induced by two pancreatic carcinogens. Cancer Res 1989; 49:3184–3189.
10. Bax J, Schippers-Gillissen C, Woutersen RA, Scherer E. Cell-specific DNA alkylation in target and non-target organs of N-nitrosobis(2-oxopropyl)amine induced carcinogenesis in hamster and rat. Carcinogenesis 1991; 12:583–590.
11. Kokkinakis DM, Subbarao V. The significance of DNA damage, its repair and cell proliferation during carcinogen treatment in the initiation of pancreatic cancer in the hamster model. Cancer Res 1993; 53:2790–2795.
12. van Kranen HJ, Vermeulen E, Schoren L, Bax J, Woutersen RA, van Iersel P, van Kreijl CF, Scherer E. Activation of c-K-ras is frequent in pancreatic carcinomas of Syrian hamsters, but is absent in pancreatic tumors of rats. Carcinogenesis 1991; 12:1477–1482.
13. Schaeffer BK, Zurlo J, Longnecker DS. Activation of c-Ki-ras not detectable in adenomas or adenocarcinomas arising in rat pancreas. Mol Carcinogen 1990; 3:165–170.
14. Schaeffer BK, Terhune PG, Longnecker DS. Pancreatic carcinomas of acinar and mixed acinar/ductal phenotypes in Ela-1-myc transgenic mice do not contain c-K-ras mutations. Am J Pathol 1994; 145:696–701.
15. Kuhlmann E, Terhune PG, Longnecker DS. Evaluation of c-K-ras in pancreatic carcinomas from Ela-1, SV40E offnsgenic mice. Carcinogenesis 1993; 14:2649–2651.
16. Chang KW, Laconi S, Mangold KA, Hubchak S, Scarpelli DG. Multiple genetic alterations in hamster pancreatic ductal adenocarcinomas. Cancer Res 1995; 55:2560–2568.
17. Fujii H, Egami H, Chaney W, Pour P, Pelling J. Pancreatic ductal adenocarcinomas induced in Syrian hamsters by N-nitrosobis(2-oxopropyl)amine contain a c-Ki-ras oncogene with a point-mutated codon 12. Mol Carcinogen 1990; 3:296–301.
18. Cerny WL, Mangold KA, Scarpelli DG. K-ras Mutation is an early event in

pancreatic duct carcinogenesis in the Syrian golden hamster. Cancer Res 1992; 52:4507–4513.

19. Appel MJ, Woutersen RA. Modulation of growth and cell turnover of preneoplastic lesions and of prostaglandin levels in rat pancreas by dietary fish oil. Carcinogenesis 1994; 15:2107–2112.
20. Appel MJ, Woutersen RA. Effects of dietary fish oil (MaxEPA) on BOP-induced pancreatic carcinogenesis in hamsters. Cancer Lett 1995; 94:179–189.
21. Rao MS, Upton MP, Subbarao V, Scarpelli DG. Two populations of cells with differing proliferative capacities in atypical acinar cell foci induced by 4-hydroxyaminoquinoline-1-oxide in rat pancreas. Lab Invest 1982; 46:527–534.
22. Roebuck BD, Baumgartner KJ, Thron CD. Characterization of two populations of pancreatic atypical acinar cell foci induced by azaserine in the rat. Lab Invest 1984; 50:141–146.
23. Bax J, Feringa AW, van Garderen-Hoetmer A, Woutersen RA, Scherer E. Adenosine triphosphate, a new marker for the differentiation of putative precancerous foci induced in rat pancreas by azaserine. Carcinogenesis 1986; 7:457–462.
24. Iversen OH. Role of cell proliferation in carcinogenesis: Is increased cell proliferation in itself a carcinogenic hazard? In: Iversen OH, ed. New Frontiers in Cancer Causation. Proc. 2nd Int. Conf. Theories of Carcinogenesis. Washington, DC: Taylor & Francis, 1992:97–108.
25. Visser CJT, Woutersen RA, Bruggink AH, van Garderen-Hoetmer A, Seifert-Bock I, Tilanus MGJ, de Weger RA. Transforming growth factor-α and epidermal growth factor expression in the exocrine pancreas of azaserine-treated rats: modulation by cholecystokinin or a low fat-high fibre (caloric restricted) diet. Carcinogenesis 1995; 16:2075–2082.
26. Visser CJT, Bruggink AH, Korc M, Kobrin MS, de Weger RA, Seifert-Bock I, van Blokland WTM, van Garderen-Hoetmer A, Woutersen RA. Overexpression of transforming growth factor-α and epidermal growth factor receptor, but not epidermal growth factor, in exocrine pancreatic tumours in hamsters. Carcinogenesis, in press.
27. Daly JM, Morgan RGH, Oates PS, Yeoh GCT, Tee LBG. Azaserine-induced pancreatic foci: detection, growth, labelling index and response to raw soya flour. Carcinogenesis 1992; 13:1519–1523.
28. Appel MJ, Woutersen RA. Dietary fish oil enhances pancreatic carcinogenesis in azaserine-treated rats. Br J Cancer 1996; 73:36–43.
29. Stillwell W, Ehringer W, Jenski LJ. Docosahexaenoic acid increases permeability of lipid vesicles and tumor cells. Lipids 1993; 28:103–108.
30. Appel MJ, Visser CJT, Woutersen RA. Cell proliferation and apoptosis in the exocrine pancreas of azaserine-treated rats and N-nitrosobis(2-oxopropyl)amine-treated hamsters: effects of dietary fish oil. Int J Oncol, in press.
31. Bursch W, Paffe S, Putz B, Barthel G, Schulte-Hermann R. Determination of the length of the histological stages of apoptosis in normal liver and in altered hepatic foci of rats. Carcinogenesis 1990; 11:847–853.

32. Haseman JK, Huff JE, Rao GN, Arnold JE, Boorman GA, McConnell EE. Neoplasms observed in untreated and corn oil gavage control groups of F344/N rats and (C57BL/6NxC3H/HeN)F1 (B6C3F1) mice. J Natl Cancer Inst 1985; 63:1291–1298.
33. Eustis SL, Boorman GA. Proliferative lesions of the exocrine pancreas: relationship to corn oil gavage in the National Toxicology Program. J Natl Cancer Inst 1985; 75:1067–1071.
34. Woutersen RA, van Garderen-Hoetmer A, Lamers CBHW, Scherer E. Early indicators of exocrine pancreas carcinogenesis produced by non-genotoxic agents. Mutat Res 1991; 248:291–302.
35. Göke B. A critical appraisal of studies of the pancreas. Animal models used in pancreas research: studies on feedback regulations of the pancreas. Int J Pancreatol 1990; 6:181–188.
36. McGuinness EE, Morgan RGH, Levison DA, Frape DL, Hopwood D, Wormsley KG. The effects of long-term feeding of soya flour on the rat pancreas. Scand J Gastroenterol 1980; 15:497–502.
37. Douglas BR, Woutersen RA, Jansen JBMJ, de Jong AJL, Rovati LC, Lamers CBHW. Modulation by CR-1409 (Lorglumide), a cholecystokinin receptor antagonist, of trypsin inhibitor-enhanced growth of azaserine-induced putative preneoplastic lesions in rat pancreas. Cancer Res 1989; 49:2438–2441.
38. Meijers M, van Garderen-Hoetmer A, Lamers CBHW, Rovati LC, Jansen JBMJ, Woutersen RA. Effects of the synthetic trypsin inhibitor camostate on the development of N-nitrosobis(2-oxopropyl)amine-induced pancreatic lesions in hamsters. Cancer Lett 1991; 60:205–211.
39. Meijers M, van Garderen-Hoetmer A, Lamers CBHW, Rovati LC, Jansen JBMJ, Woutersen RA. Role of cholecystokinin in the development of BOP-induced pancreatic lesions in hamsters. Carcinogenesis 1990; 11:2223–2226.
40. Howatson AG, Carter DG. Pancreatic carcinogenesis—enhancement by cholecystokinin in the hamster-nitrosamine model. Br J Cancer 1985; 51:107–114.
41. Pour PM, Lawson T, Helgeson S, Donnelly T, Stepan K. Effect of cholecystokinin on pancreas carcinogenesis in the hamster model. Carcinogenesis 1988; 9:597–601.
42. Chey WY. Hormonal control of pancreatic exocrine function. In: Go VLW, DiMagno EP, Garner JD, Lebenthal E, Reber HA, Scheele GA, eds. The Pancreas, Biology, Pathobiology and Disease. New York: Raven Press, 1993: 403–425.
43. Hotz J, Ho SB, Go VW, DiMagno EPD. Short term inhibition of duodenal tryptic activity does not affect human pancreatic, biliary or gastric function. J Lab Clin Med 1983; 101:488–495.
44. Appel MJ, Meijers M, van Garderen-Hoetmer A, Lamers CBHW, Rovati LC, Spreij-Mooij D, Jansen JBMJ, Woutersen RA. Role of cholecystokinin in dietary fat-promoted azaserine-induced pancreatic carcinogenesis in rats. Br J Cancer 1992; 66:46–50.
45. Woutersen RA, van Garderen-Hoetmer A, Bax J, Scherer E. Modulation of

dietary fat-promoted pancreatic carcinogenesis in rats and hamsters by chronic ethanol ingestion. Carcinogenesis 1989; 10:453–459.
46. Sarles H, Lebreuil G, Tasso F, Figarella C, Clemente F, Devaux MA, Fagonde B, Paean H. A comparison of alcoholic pancreatitis in rat and man. Gut 1971; 12:377.
47. Pour PM. Histogenesis of exocrine pancreatic cancer in the hamster model. Environ Health Perspect 1984; 56:229–243.
48. Pour PM. Mechanism of pseudoductular (tubular) formation during pancreatic carcinogenesis in the hamster model. An electron-microscopic and immunohistochemical study. Am J Pathol 1988; 130:335–344.
49. Meijers M, Bruijntjes JP, Hendriksen EGJ, Woutersen RA. Histogenesis of early preneoplastic lesions induced by N-nitrosobis(2-oxopropyl)amine in exocrine pancreas of hamsters. Int J Pancreatol 1989; 4:127–137.
50. Levitt MH, Harris CC, Squire R, Springer S, Wenk M, Mollelo C, Thomas D, Kingsbury E, Newkirk C. Experimental pancreatic carcinogenesis. Am J Pathol 1977; 88:5–28.
51. Takahashi M, Arai H, Kokubo T, Furukawa F, Kurata Y, Ito N. An ultrastructural study of precancerous and cancerous lesions of the pancreas in Syrian golden hamsters induced by BOP. Gann 1980; 71:825–831.
52. Flaks B, Moore MA, Flaks A. Ultrastructural analysis of pancreatic carcinogenesis III. Multifocal cystic lesions induced by N-nitroso-bis(2-hydroxypropyl)amine in the hamster exocrine pancreas. Carcinogenesis 1980; 1:693–706.
53. Flaks B, Moore MA, Flaks A. Ultrastructural analysis of pancreatic carcinogenesis IV. Pseudoductular transformation of acini in the hamster pancreas during N-nitroso-bis(2-hydroxypropyl)amine carcinogenesis. Carcinogenesis 1981; 2:1241–1253.
54. Woutersen RA, van Garderen-Hoetmer A, Bax J, Feringa AW, Scherer E. Modulation of putative preneoplastic foci in exocrine pancreas of rats and hamsters I. Interaction of dietary fat and ethanol. Carcinogenesis 1986; 7: 1587–1593.
55. Pour PM, Reber H, Stepan K. Modulation of pancreatic carcinogenesis in the hamster model. XII. Dose-related effect of ethanol. J Natl Cancer Inst 1983; 71:1085–1087.
56. Tweedie JH, Reber HA, Pour PM, Pounder DM. Protective effect of ethanol on the development of pancreatic cancer. Surg Forum 1981; 32:222.
57. Woutersen RA, van Garderen-Hoetmer A, Bax J, Scherer E. Modulation of dietary fat-promoted pancreatic carcinogenesis in rats and hamsters by chronic coffee ingestion. Carcinogenesis 1989; 10:311–316.
58. Nishikawa A, Furukawa F, Imazawa T, Yoshimura H, Mitsumori K, Takahashi M. Effects of caffeine, nicotine, ethanol and sodium selenite on pancreatic carcinogenesis in hamster after initiation with N-nitrosobis(2-oxopropyl)amine. Carcinogenesis 1992; 13:1379–1382.
59. Appel MJ, van Garderen-Hoetmer A, Woutersen RA. Azaserine-induced pancreatic carcinogenesis in rats: promotion by a diet rich in saturated fat and inhibition by a standard laboratory chow. Cancer Lett 1990; 55:239–248.

60. Appel MJ, Roverts G, Woutersen RA. Inhibitory effects of micronutrients on pancreatic carcinogenesis in azaserine-treated rats. Carcinogenesis 1991; 12:2157–2161.
61. Appel MJ, Woutersen RA. Effects of dietary β-carotene and selenium on initiation and promotion of pancreatic carcinogenesis in azaserine-treated rats. Carcinogenesis, in press.
62. Appel MJ, van Garderen-Hoetmer A, Woutersen RA. Lack of inhibitory effects of β-carotene, vitamin C, vitamin E and selenium on development of ductular adenocarcinomas in exocrine pancreas of hamsters. Cancer Lett, in press.
63. Birt DF, Julius AD, Runice CE, White LT, Lawson T, Pour PM. Enhancement of BOP-induced pancreatic carcinogenesis in selenium-fed Syrian golden hamsters under specific dietary conditions. Nutr Cancer 1988; 11:21–33.
64. Kise Y, Yamamura M, Kogata M, Uetsuji S, Takada H, Hioki K, Yamamoto M. Inhibitory effect of selenium on hamster pancreatic cancer induction by N′-nitrosobis(2-oxopropyl)amine. Int J Cancer 1990; 46:95–100.
65. Birt DF, Julius AJ, Dwork E, Hanna T, White LT, Pour PM. Comparison of the effects of dietary beef tallow and corn oil on pancreatic carcinogenesis in the hamster model. Carcinogenesis 1990; 11:745–748.
66. O'Connor TP, Roebuck BD, Peterson FJ, Lokesh B, Kinsella JE, Campbell TC. Effect of dietary omega-3 and omega-6 fatty acids on development of azaserine-induced preneoplastic lesions in rat pancreas. J Natl Cancer Inst 1989; 81:858–863.
67. Carriaga MT, Henson DE. Liver, gallbladder, extrahepatic bile ducts, and pancreas. Cancer 1995; 75:171–190.
68. IARC. Cancer Incidence in Five Continents. Vol. VI. IARC Sci. Pub. No. 120. Lyon: International Agency for Research on Cancer, 1992.
69. IRAC. Trends in Cancer Incidence and Mortality. IARC Sci. Pub. No. 121. Lyon: International Agency for Research on Cancer, 1993.
70. IARC. Facts and Figures of Cancer in the European Community. Lyon: International Agency for Research on Cancer, 1993.
71. Fernandez E, La Vecchia C, Porta M, Negri E, Lucchini F, Levi F. Trends in pancreatic cancer mortality in Europe, 1955–1989. Int J Cancer 1994; 57: 786–792.
72. Howe GR. Pancreatic cancer. In: Doll R, Muir CS, Fraumeni JF, eds. Cancer Surveys: Trends in Cancer Incidence and Mortality. Vol. 19/20, Cold Spring Harbor, NY: Cold Spring Harbor Press, 1994:139–158.
73. Zheng T, Holford TR, Ward BA, McKay L, Flannery J, Boyle P. Time trend in pancreatic cancer incidence in Connecticut. Int J Cancer 1995; 61:622–627.
74. Fernandez E, La-Vecchia C, D'Avanzo B, Negri E, Franceschi S. Family history and the risk of liver, gallbladder, and pancreatic cancer. CEBP 1994; 3:209–212.
75. Lumadue JA, Griffin CA, Osman M, Hruban RH. Familial pancreatic cancer and the genetics of pancreatic cancer. Surg Clin North Am 1995; 75:845–855.

76. Cubilla AL, Fitzgerald PJ. Classification of pancreatic cancer (nonendocrine). Mayo Clin Proc 1979; 54:449–458.
77. Pour PM, Sayed S, Sayed G. Hyperplastic, preneoplastic and neoplastic lesions found in 83 human pancreases. Am J Clin Pathol 1982; 77:136–152.
78. Longnecker DS, Hashida Y, Shinozuka H. Relationship of age to prevalence of focal acinar cell dysplasia in the human pancreas. J Natl Cancer Inst 1980; 65:63–66.
79. Scarpelli DG, Cerny WL, Mangold KA. Gene alterations in rodent and human tumors of the exocrine and endocrine pancreas. Prog Clin Biol Res 1992; 376:223–243.
80. Visser CJT. Hormones and growth factors in experimental exocrine pancreatic carcinogenesis. Thesis, 1995.
81. Caldas C, Kern SE. K-ras mutation and pancreatic adenocarcinoma—state of the art. Int J Pancreatol 1995; 18:1–6.
82. Hahn SA, Schutte M, Hoque ATMS, Moskaluk CA, daCosta LT, Rozenblum E, Weinstein CL, Fischer A, Yeo CJ, Hruban RH, Kern SE. DPC4, a candidate tumor suppressor gene at human chromosome 18q21.1. Science 1996; 271:350–353.
83. Greenblatt MS, Bennett WP, Hollstein M, Harris CC. Mutations in the p53 tumor suppressor gene: clues to cancer etiology and molecular pathogenesis. Cancer Res 1994; 54:4855–4878.
84. Barton CM, Staddon SL, Hughes CM, Hall PA, O'Sullivan C, Kloppel G, Theis B, Russell RC, Neoptolemos J, Williamson RC, et al. Abnormalities of the p53 tumor suppressor gene in human pancreatic cancer. Br J Cancer 1991; 64:1076–1082.
85. IARC. Tobacco Smoking. Lyon: International Agency for Research on Cancer, 1986.
86. Bueno-de-Mesquita HB, Maisonneuve P, Moerman CJ, Runia S, Boyle P. Life-time smoking and exocrine carcinoma of the pancreas: a population-based case-control study in the Netherlands. Int J Cancer 1991; 49:816–822.
87. Ghadirian P, Simard A, Baillargeon J. Tobacco, alcohol, and coffee and cancer of the pancreas. A population-based, case-control study in Quebec, Canada. Cancer 1991; 67:2664–2670.
88. Howe GR, Jain M, Burch JD, Miller AB. Cigarette smoking and cancer of the pancreas: evidence from a population-based case-control study in Toronto, Canada. Int J Cancer 1991; 47:323–328.
89. Silverman DT, Dunn JA, Hoover RN, Schiffman M, Lillemoe KD, Schoenberg JB, Brown LM, Greenberg RS, Hayes RB, Swanson GM, Wacholder S, Schwartz, Liff JM, Pottern LM. Cigarette smoking and pancreas cancer: a case-control study based on direct interviews. J Natl Cancer Inst 1994; 86: 1510–1516.
90. McLaughlin JK, Hrubec Z, Blot W, Fraumeni JF. Smoking and cancer mortality among U.S. veterans: a 26-year follow-up. Int J Cancer 1995; 60:190–193.
91. Zheng W, McLaughlin JK, Gridley G, Bjelke E, Schuman LM, Silverman

DT, Wacholder S, Co-Chien HT, Blot WJ, Fraumeni JF. A cohort study of smoking, alcohol consumption, and dietary factors for pancreatic cancer (United States). Cancer Causes Control 1993; 4:477–482.

92. Boyle P, Maisonneuve P, Bueno-de-Mesquita HB, Ghadirian P, Howe GR, Zatonski W, Baghurst P, Moerman CJ, Simard A, Miller AB, Przewoniak K, McMichael AJ, Hsieh CC, Walker AM. Cigarette smoking and pancreas cancer: a case-control study of the SEARCH program of the IARC. Int J Cancer, in press.
93. Velema JP, Walker AM, Gold EB. Alcohol and pancreatic cancer. Insufficient epidemiologic evidence for a causal relationship. Epidemiol Rev 1986; 8:28–41.
94. IARC. Alcohol Drinking. Lyon: International Agency for Research on Cancer, 1988.
95. Farrow DC, Davis S. Risk of pancreatic cancer in relation to medical history and the use of tobacco, alcohol and coffee. Int J Cancer 1990; 45:816–820.
96. Baghurst PA, McMichael AJ, Slavotinek AH, Baghurst KI, Boyle P, Walker AM. A case-control study of diet and cancer of the pancreas. Am J Epidemiol 1991; 134:167–179.
97. Jain M, Howe GR, Louis PS, Miller AB. Coffee and alcohol as determinants of risk of pancreas cancer: a case-control study from Toronto. Int J Cancer 1991; 47:384–389.
98. Bueno-de-Mesquita HB, Maisonneuve P, Moerman CJ, Runia S, Boyle P. Life-time consumption of alcoholic beverages, tea and coffee and exocrine carcinoma of the pancreas: a population-based case-control study in the Netherlands. Int J Cancer 1992; 50:514–522.
99. Zatonski WA, Boyle P, Przewozniak K, Maisonneuve P, Drosik K, Walker AM. Cigarette smoking, alcohol, tea and coffee consumption and pancreas cancer risk: a case-control study from Opole, Poland. Int J Cancer 1993; 53: 601–607.
100. Ji BT, Chow WH, Dai Q, McLaughlin JK, Benichou J, Hatch MC, Gao YT, Fraumeni JF. Cigarette smoking and alcohol consumption and the risk of pancreatic cancer: a case-control study in Shanghai, China. Cancer Causes Control 1995; 6:369–376.
101. Hiatt RA, Klatsky AL, Armstrong MA. Pancreatic cancer, blood glucose and beverage consumption. Int J Cancer 1988; 41:794–797.
102. Friedman GD, Van den Eeden SK. Risk factors for pancreatic cancer: an exploratory study. Int J Epidemiol 1993; 22:30–37.
103. Shibata A, Mack TM, Paganini-Hill A, Ross RK, Henderson BE. A prospective study of pancreatic cancer in the elderly. Int J Cancer 1994; 58:46–49.
104. Silverman DT, Brown LM, Hoover RN, Schiffman M, Lillemoe KD, Schoenberg JB, Swanson GM, Hayes RB, Greenberg RS, Benichou J, Schwartz AG, Liff JM, Pottern LM. Alcohol and pancreatic cancer in blacks and whites in the United States. Cancer Res 1995; 55:4899–4905.
105. Carstensen JM, Bygren LO, Hatschek T. Cancer incidence among Swedish brewery workers. Int J Cancer 1990; 45:393–396.

106. Sigvardsson S, Hardell L, Przybeck TR, Cloninger R. Increased cancer risk among Swedish female alcoholics. Epidemiology 1996; 7:140–143.
107. Hirayama T. Diet and mortality. In: Wahrendorf J, ed. Lifestyle and Mortality: A Large-Scale Census-Based Cohort Study in Japan. Basel: Karger, 1990: 73–95.
108. IARC. Coffee, Tea, Mate, Methylxanthines and Methylglyoxal. Lyon: International Agency for Research on Cancer, 1991:170–171, 261.
109. Stensvold I, Jacobsen BK. Coffee and cancer: a prospective study of 43,000 Norwegian men and women. Cancer Causes Control 1994; 5:401–408.
110. Kalapothaki V, Tzonou A, Hsieh CC, Toupadaki N, Karakatsani A, Trichopoulos D. Cancer Causes Control 1993; 4:375–382.
111. Partanen T, Hemminki K, Vainio H, Kauppinen T. Coffee consumption not associated with risk of pancreas cancer in Finland. Prev Med 1995; 24:213–216.
112. Lyon JL, Mahoney AW, French TK, Moser R. Coffee consumption and the risk of cancer of the exocrine pancreas: a case-control study in a low-risk population. Epidemiology 1992; 3:164–170.
113. Lyon JL, Egger MJ, Robinson LM, French TK, Gao R. Misclassification of exposure in a case-control study: the effects of different types of exposure and different proxy respondents in a study of pancreatic cancer. Epidemiology 1992; 3:223–231.
114. Whittemore AS, Paffenbarger RS, Anderson K, Halpern J. Early precursors of pancreatic cancer in college men and women. J Chron Dis 1983; 36:251–256.
115. Howe GR, Burch JD. Nutrition and pancreatic cancer. Cancer Causes Control 1996; 7:69–82.
116. Howe GR, Ghadirian P, Bueno-de-Mesquita HB, Zatonski WA, Baghurst PA, Miller AB, Simard A, Baillargeon J, De Waard F, Przewozniak K, McMichael AJ, Jain M, Hsieh CC, Maisonneuve P, Boyle P, Walker AM. A collaborative case-control study of nutrient intake and pancreatic cancer within the SEARCH programme. Int J Cancer 1992; 51:365–372.
117. Mills PK, Beeson L, Abbey DE, Fraser GE, Phillips RL. Dietary habits and past medical history as related to fatal pancreas cancer among Adventists. Cancer 1988; 61:2578–2585.
118. Bueno-de-Mesquita HB, Moerman CJ, Runia S, Maisonneuve P. Are energy and energy-providing nutrients related to exocrine carcinoma of the pancreas? Int J Cancer 1990; 46:435–444.
119. Bueno-de-Mesquita HB. Main hypotheses on diet and cancer investigated in the EPIC Study. Eur J Cancer Prev 1996;5(suppl 1):00–00.
120. Ocké MC, Bueno-de-Mesquita HB, Goddijn HE, Jansen A, Pols MA, Van Staveren WA, Kromhout D. The Dutch EPIC food frequency questionnaire. I. Description of the questionnaire, and relative validity and reproducibility for food groups. Int J Epidemiol, in press.
121. Burney PGJ, Comstock GW, Morris JS. Serologic precursors of cancer: serum micronutrients and the subsequent risk of pancreatic cancer. Am J Clin Nutr 1989; 49:895–900.

122. Bueno-de-Mesquita HB, Maisonneuve P, Runia S, Moerman CJ. Intake of foods and nutrients and cancer of the exocrine pancreas: a population-based case-control study in the Netherlands. Int J Cancer 1991; 48:540–549.
123. Franceschi S, Bidoli E, La Vecchia C, Talamini R, D'Avanzo, Negri E. Tomatoes and risk of digestive-tract cancers. Int J Cancer 1994; 59:181–184.
124. Norell SE, Ahlbom A, Erwald R, Jacobson G, Lindberg-Navier I, Olin R, Törnberg B, Wiechel KL. Diet and pancreatic cancer: a case-control study. Am J Epidemiol 1986; 124:894–902.
125. Knekt P, Steineck G, Järvinen R, Hakulinen T, Aromaa A. Intake of fried meat and risk of cancer: a follow-up study in Finland. Int J Cancer 1994; 59: 756–760.
126. Bueno-de-Mesquita HB, Maisonneuve P, Moerman CJ, Walker AM. Anthropometric and reproductive variables and exocrine carcinoma of the pancreas: a population-based case-control study in the Netherlands. Int J Cancer 1992; 52:24–29.
127. Fernandez E, La-Vecchia C, D'Avanzo B, Negri E, Franceschi S. Menstrual and reproductive factors and pancreatic cancer risk in women. Int J Cancer 1995; 62:11–14.
128. Ji BT, Hatch MC, Chow WH, McLaughlin JK, Dai Q, Howe GR, Gao YT, Fraumeni JF. Anthropometric and reproductive factors and the risk of pancreatic cancer: a case-control study in Shanghai, China. Int J Cancer 1996; 66:432–437.
129. Kissinger DG, Sanchez A. The association of dietary factors with the age of menarche. Nutrition Res 1987; 7:471–479.
130. Burch GE, Ansari A. Chronic alcoholism and carcinoma of the pancreas. Arch Intern Med 1968; 22:273–275.
131. Kessler II. Cancer mortality among diabetics. J Natl Cancer Inst 1970; 44: 673–686.
132. Wynder EL, Mabuchi K, Maruchi N, Fortner JG. A case-control study of cancer of the pancreas. Cancer 1973; 31:641–648.
133. Lin RS, Kessler II. A multifactorial model for pancreatic cancer in man. J Amer Med Ass 1981; 245:147–152.
134. Lowenfels AB, Maisonneuve P, Cavallini G, Ammann RW, Lankisch PG, Andersen JR, Dimagno EP, Andrén-Sandberg Å, Domellöf L, and the International Pancreatitis Study Group. N Engl J Med 1993; 328:1433–1437.
135. Ekbom A, McLaughlin JK, Karlsson BM, Nyrén O, Gridley G, Adami HO, Fraumeni JF. Pancreatitis and pancreatic cancer: a population-based study. J Natl Cancer Inst 1994; 86:625–627.
136. Fernandez E, La-Vecchia C, Porta M, Negri E, D'Avanzo B, Boyle P. Pancreatitis and the risk of pancreatic cancer. Pancreas 1995; 11:185–189.
137. La Vecchia C, Negri E, Franceschi S, D'Avanzo B, Boyle P. A case-control study of diabetes mellitus and cancer risk. Br J Cancer 1994; 70:950–953.
138. Gullo L, Pezzilli R, Morselli-Labate AM. Diabetes and the risk of pancreatic cancer. Italian Pancreatic Cancer Study Group. N Engl J Med 1994; 331:81–84.
139. Everhart J, Wright D. Diabetes mellitus as a risk factor for pancreatic cancer. A meta-analysis. JAMA 1995; 273:1650–1609.

140. Balkau B, Barrett-Connor E, Eschwege E, Richard JL, Claude JR, Ducimetiere P. Diabetes and pancreatic carcinoma. Diabet Metab 1993; 19:458–462.
141. Chow WH, Gridley G, Nyrén O, Linet MS, Ekbom A, Fraumeni JF, Adami HO. Risk of pancreatic cancer following diabetes mellitus: a nation-wide cohort study in Sweden. J Natl Cancer Inst 1995; 87:930–931.
142. Friedman GD. Cholecystectomy not confirmed as a risk factor for pancreatic cancer. Int J Cancer 1995; 61:745–746.
143. Mack TM, Yu MC, Hanisch R, Henderson BE. Pancreas cancer and smoking, beverage consumption, and past medical history. J Natl Cancer Inst 1986; 76:49–60.
144. Bueno-de-Mesquita HB, Maisonneuve P, Moerman CJ, Walker AM. Aspects of medical history and exocrine carcinoma of the pancreas: a population-based case-control study in the Netherlands. Int J Cancer 1992; 52:17–23.
145. Dai Q, Zheng W, Ji BT, Shu XO, Jin F, Zhu JL, Gao YT. Prior immunity-related medical conditions and pancreatic cancer risk in Shanghai. Int J Cancer 1995; 63:337-340.
146. Gold EB, Gordis L, Diener MD, Seltser R, Boitnott JK, Bynum TE, Hutcheon DF. Diet and other risk factors for cancer of the pancreas. Cancer 1985; 55:460–467.
147. Hahn SA, Kern SE. Molecular genetics of exocrine pancreatic neoplasm. Surg Clin North Am 1995; 75:857.
148. Vassar R, Hutton ME, Fuchs E. Transgenic overexpression of transforming growth factor alpha bypasses the need for c-Ha-ras mutations in mouse skin tumorigenesis. Mol Cell Biol 1992; 12:4643–4653.
149. Pellegata NS, Sessa F, Renault B, Bonato M, Leone BE, Solcia E, Ranzani GN. K-ras and p53 gene mutations in pancreatic cancer—ductal and nonductal tumors progress through different genetic lesions. Cancer Res 1994; 54: 1556–1560.
150. Yashima K, Nakamori S, Murakami Y, Yamaguchi A, Hayashi K, Ishikawa O, Konishi Y, Sekiya T. Mutations of the adenomatous polyposis coli gene in the mutation cluster region: comparison of human pancreatic and colorectal cancers. Int J Cancer 1994; 59:43–47.
151. Sandgren EP, Quaife CJ, Paulovich AG, Palmiter RD, Brinster RL. Pancreatic tumor pathogenesis reflects the causative genetic lesion. Proc Natl Acad Sci USA 1991; 88:93–97.
152. Clarke AR, Cummings MC, Harrison DJ. Interaction between murine germline mutations in p53 and APC predisposes to pancreatic neoplasia but not to increased intestinal malignancy. Oncogene 1995; 11:1913–1920.
153. Donehower LA, Godley LA, Aldaz CM, Pyle R, Shi YP, Pinkel D, Gray T, Bradley A, Medina D, Varmus HE. Deficiency of p53 accelerates mammary tumorigenesis in Wnt-1 transgenic mice and promotes chromosomal instability. Gene Develop 1995; 9:882–895.
154. Kemp CJ, Donehower LA, Bradley A, Balmain A. Reduction of p53 gene dosage does not increase initiation or promotion but enhances malignant progression of chemically induced skin tumors. Cell 1993; 74:813–822.
155. Rivenson A, Hoffmann D, Prokopczyk B, Amin S. Induction of lung and

exocrine pancreas tumors in F344 rats by tobacco-specific and areca-derived N-nitrosamines. Cancer Res 1988; 48:6912–6917.

156. Cuzick J, Routledge MN, Jenkins D, Garner RC. DNA adducts in different tissues of smokers and non-smokers. Int J Cancer 1990; 45:673–678.
157. Ogawa T, Makino T, Kosahara K, Koga A, Nakayama F. Promoting effects of both dietary cholesterol and cholestyramine on pancreatic carcinogenesis initiated by N-nitrosobis(2-oxopropyl)amine in Syrian golden hamsters. Carcinogenesis 1992; 13:2047–2052.
158. Steinmetz KA, Potter JD. Vegetables, fruit, and cancer. I. Epidemiology. Cancer Causes Control 1991; 2:325–357.
159. Steinmetz KA, Potter JD. Vegetables, fruit, and cancer. II. Mechanisms. Cancer Causes Control 1991; 2:427–442.
160. Block G, Patterson B, Subar A. Fruit, vegetables, and cancer prevention: a review of the epidemiological evidence. Nutr Cancer 1992; 18:1–29.
161. Tanaka T, Barnes WS, Williams GM, Weisburger JH. Multipotential carcinogenicity of the fried food mutagen 2-amino-3-methylimidazo[4,5-f]-quinoline in rats. Jpn J Cancer Res (Gann) 1985; 76:570–576.
162. Turesky RJ, Lang NP, Butler MA, Teitel CH, Kadlubar FF. Metabolic activation of carcinogenic heterocyclic aromatic amines by human liver and colon. Carcinogenesis 1991; 12:1839–1845.
163. Sugimura T, Wakabayashi K. Mutagens and carcinogens in food. In: Pariza MW, Felton JS, Aeschbacher HU, Sato S, eds. Mutagens and Carcinogens in the Diet. New York: Wiley-Liss, 1990:1–18.
164. Pour PM, Patil K. Modification of pancreatic carcinogenesis in the hamster model. X. Effect of streptozotocin. J Natl Cancer Inst 1983; 71:1059–1065.
165. Povoski SP, Fenoglio-Preiser CM, Sayers HJ, McCullough PJ, Zhou W, Bell RH. Effect of streptozotocin diabetes on development of nitrosamin-induced pancreatic carcinoma when diabetes induction occurs after nitrosamine exposure. Carcinogenesis 1993; 14:961–967.
166. Pour PM, Kazakoff K, Carlson K. Inhibition of streptozotocin-induced islet cell tumors and N-nitrosobis(2-oxopropyl)amine-induced pancreatic exocrine tumors in Syrian hamsters by exogenous insulin. Cancer Res 1990; 50:1634–1639.
167. Horii A, Nakatsuru S, Miyoshi Y, Ichii S, Nagase H, Ando H, Yanagisawa A, Tsuchiya E, Kato Y, Nakamura Y. Frequent somatic mutations of the APC gene in human pancreatic cancer. Cancer Res 1992; 52:6696–6698.
168. Caldas C, Hahn SA, Dacosta LT, Redston MS, Schutte M, Seymour AB, Weinstein CL, Hruban RH, Yeo CJ, Kern SE. Frequent somatic mutations and homozygous deletions of the p16 (MTS1) gene in pancreatic adenocarcinoma. Nat Genet 1994; 8:27–32.
169. Seymour AB, Hruban RH, Redston M, Caldas C, Powell SM, Kinzler KW, Yee CJ, Kern SE. Allelotype of pancreatic adenocarcinoma. Cancer Res 1994; 54:2761–2764.

27
Colorectal Cancer

Gabriel A. Kune and Luis Vitetta
University of Melbourne, Melbourne, Victoria, Australia

I. INTRODUCTION

Colorectal tumor etiology and carcinogenesis is best explained using a multicausal model. This model assumes the presence of several environmental exposures (as well as an inherited tendency in some), which lead to several pathophysiological changes in the environment of the colorectal epithelium, which over time, through the accumulation of several mutations, results in the development of colorectal cancer. This means that colorectal neoplasia is interpreted at four levels: causes, mechanisms of action, genetic changes, and morphological changes (Fig. 1).

Experimentally induced colorectal tumors in rodent models employ specific chemical agents such as azoxymethane and dimethylhydrazine, which have no relevance to the human situation. Specific chemical compounds that could be called "carcinogens" have not been identified with any certainty in human colorectal carcinogenesis. Moreover, the model depicted in Figure 1 implies a greater complexity of colorectal tumor development than the simple and somewhat misleading and dated hypothesis that a specific carcinogen will lead to the development of a cancer in a predictable manner. Interestingly, there is a far better understanding of specific and chemically identifiable "protective" factors, such as vitamin C, folate, methionine, butyrate, and β-carotene in human colorectal neoplasia, than there is of "risk" factors.

The important likely causes of colorectal tumors include an inherited predisposition, certain dietary factors, alcohol and particularly beer consumption, and for colorectal adenomas, smoking. The inherited aspects of

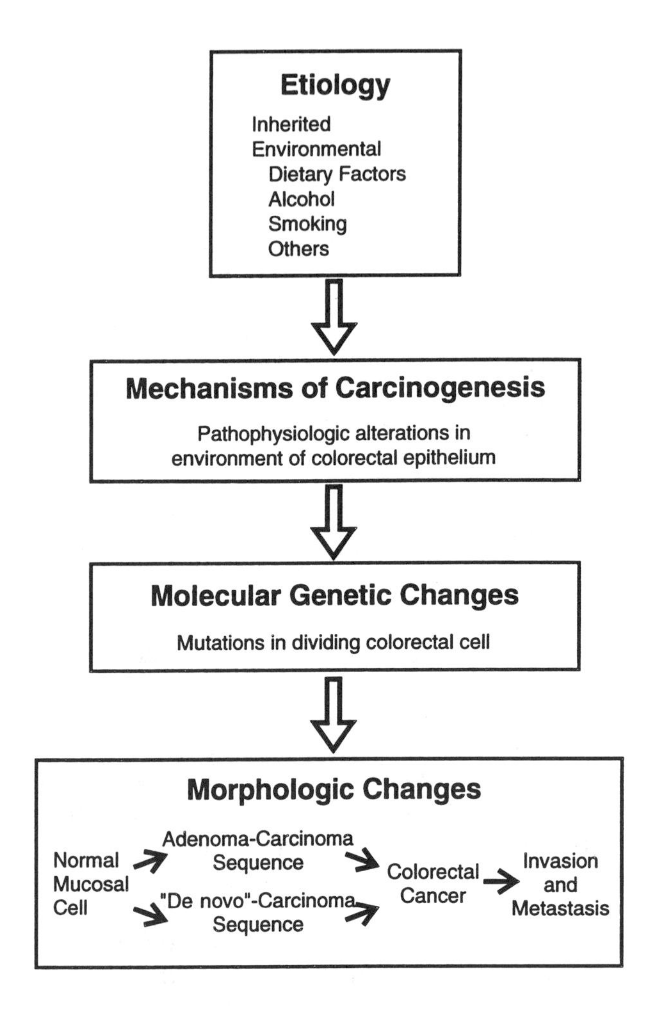

Figure 1 A multicausal model for the interpretation of colorectal tumor development. (Adapted from Ref. 1.)

colorectal tumor development will not be discussed further, and the reader is referred to recent publications on this subject (1).

This chapter will focus on the interpretation of carcinogenicity of so-called ordinary or sporadic colorectal tumors in humans (and representing 95% of incident cases) and in chemically induced lesions in rodent models in relation to dietary factors, alcohol consumption, and smoking. Occupational exposures appear unimportant for colorectal cancer and will be only briefly discussed. Colorectal cancer associated with hereditary syndromes and inflammatory bowel disease will not be discussed.

II. DIETARY FACTORS

Extensive human and experimental data over the past 30 years strongly suggest that dietary factors are relevant to the etiology of both colorectal cancer and its major precursor lesion, colorectal adenomas. There are two major dietary hypotheses, which, put in a simplified form, may be called the "meat/protein/fat/energy/fried/grilled food risk" and the "vegetable/fruit/cereal/fiber/starch/phytochemical food protection" hypotheses. Other and seemingly less important dietary factors are the protective effects of fish/fish oil and calcium, the risk associated with high salt and sugar consumption, and a high meal frequency.

III. MEAT/PROTEIN/FAT/ENERGY/FRIED/GRILLED FOOD RISK

A. Human Data

Evidence collected from almost 100 major epidemiological studies which investigated the relationship between diet factors and colorectal tumors indicates that a high intake of red meat, protein, animal fat, energy, and heavily fried or grilled meat is a risk for both colorectal adenomas and colorectal cancer (1–3). Protein is a risk, probably because in western diets meat is the main source of protein. A high energy intake is emerging as an important independent risk factor and therefore appears to be an important component of the fat risk. However, fats of vegetable origin do not appear to be a risk, and fish oils may be protective (1).

B. Experimental Data

Extensive experimental studies over the past 20 years in chemically induced colon tumors in rodents fairly consistently report a significantly increased tumor incidence in animals fed with a high fat or high meat or high protein diet (4–9). Fat of animal origin, as well as protein, was consistently associated with an increased tumor incidence in these studies, whereas the effect of unsaturated fat was inconsistent. However, fat from fish origin, administered as fish oil, had a protective effect for both colon tumors and for mucosal proliferative activity (1,10).

Energy restriction in experimental models has been consistently associated with a decrease in proliferative activity, a decrease in the incidence of preneoplastic lesions such as aberrant crypt foci, and a decrease in colorectal tumor incidence, confirming human data that a high energy consumption is an independent risk in colorectal neoplasia (1,11–12). In experimental models, energy restriction appears to be important early in the neoplastic process.

C. Mechanisms of Action

The mechanisms and compounds involved in the meat risk are uncertain; however, several modes of action have been proposed. Meat consumption may result in an increase in endogenously produced fecal nitrosamines, an increase in metabolites of tryptophan; however, recent data suggest that the most important mechanism is the result of frying or grilling of meat. Frying or grilling produces compounds such as heterocyclic amines and hydromethyl-formaldehyde, which are known to be carcinogenic in laboratory and animal studies (1). The carcinogenic process for these compounds is under genetic control involving acetylation and oxidation, and fast acetylators and fat oxidizers who also regularly eat fried and grilled meat are at an increased risk for colorectal cancer (1).

The mechanisms involved in the animal fat risk invoke the consequences of an increased bile acid production in the presence of a high fat diet (13). More fat is present in the large bowel as well as more bile acids, which are converted by fecal bacteria into secondary bile acids and other compounds found to be carcinogenic in the laboratory (1,13). The mechanism involved in relation to a high energy intake is unclear; however, it may be related to an increased activity and therefore an increased rate of cell division of the colorectal mucosa (1).

IV. VEGETABLE/FRUIT/CEREAL/FIBER/STARCH/PHYTOCHEMICAL PROTECTION

Although this volume focuses on carcinogenicity, dietary protective factors for colorectal cancer will also be briefly discussed because of their enormous potential importance in the prevention of colorectal tumors. These are essentially foods of plant origin. Apart from plant foods, a high consumption of fish appears to be protective, and a high calcium-containing diet may also provide modest protection for colorectal tumors (1).

A. Human Data

An analysis of almost 100 major epidemiological studies that have examined the association between dietary factors and colorectal tumors consistently found that plant foods have an important protective effect (1). Vegetables are probably the main protective factor, fruits are next in importance, and then cereals. A high consumption of starch is also emerging as a protective factor in colorectal tumors. Dietary fiber, from all sources, is an important protective factor, although up to now it is unknown to what extent the fiber itself is protective and to what extent the protection is due to anticancer compounds present in fiber-rich foods.

B. Experimental Data

In experimental models of chemically induced colon tumors and preneoplastic lesions, a high fiber diet has been found to decrease the incidence of these lesions with a moderately high consistency, and particularly so for so-called insoluble fiber, or fiber that is a rich source of butyrate (1).

C. Mechanisms of Action

Dietary fiber and starch appears to be protective to a large part because of its fermentation in the large bowel, with the production of short-chain fatty acids, particularly butyrate, which has been shown to have a protective effect on mucosal cell proliferation (1). Fiber also induces physical changes in the feces, such as lowering colonic pH, diluting intraluminal carcinogens, increasing fecal bulk and thereby decreasing transit time, which adds to its protective effect. Fiber also reduces conversion of primary to secondary bile acids (1).

The emerging study of naturally occurring compounds in foods of plant origin, "phytochemicals," which have anticancer properties, is most exciting. These compounds include carotenoids, vitamins C and E, folate, indoles, flavinoids, allylic sulfides, monoterpenes, linolenic acid, and others (1). These compounds exert their protective effects via several mechanisms, including antioxidant action, inhibition of nitrosamine formation, blocking hormone receptor sites, regulating prostaglandin production, and acting as cell differentiation agents. Much further research is required to understand how these compounds present in plant foods work; however, they appear to be important in protection against colorectal tumors over and above the fiber content of the food.

V. ALCOHOL CONSUMPTION

A. Human Data

Since 1957 over 90 epidemiological studies have examined the relationship between alcohol consumption and colorectal tumor risk (1,21). These studies indicate with moderate consistency that long-continued alcohol consumption presents about a twofold risk for both colorectal adenomas and colorectal cancer. Beer is the alcoholic beverage that poses the most important risk. The minimum consumption that results in an elevated risk is about three standard drinks per day continued over two decades or longer (1).

B. Experimental Data

The eight experimental studies that examined the relationship between alcohol consumption and colorectal tumors in chemically induced bowel tumors in rats were, overall, consistent with the human data (1,21). In four of the eight studies, a positive augmentation of tumors was found (22–25), in one there was a distal shift of tumor incidence (8), in two no augmentation was noted (26,27), and unexpectedly, inhibition of tumor development was found in one study (28).

C. Mechanisms of Action

The mechanisms of the alcohol effect in colorectal neoplasia are not well understood. Several mechanisms have been suggested. The first is a change in bile acid metabolism with alcohol ingestion, resulting in increased bile production and an increased production of secondary bile acids (1). The second is hypomethylation of DNA, an effect that probably occurs in conjunction with low dietary folate and methionine intakes. Hypomethylation of DNA in the long term can result in abnormal gene expression and can be an early event in neoplastic change (1). The third suggested mechanism is that alcohol and particularly beer consumption results in an increased endogenous synthesis of nitrosamines, which may be carcinogenic to the colorectal mucosa (1,21).

Alcohol ingestion in humans and in rat models has been shown to be associated with morphological changes, suggesting a stimulatory effect with a measurable increase in the rate of mucosal cell proliferation and regeneration (1,21,29). It is possible that ethanol ingestion results in an increased ornithine decarboxylase activity and in acetaldehyde production, which may trigger hyperregeneration.

VI. SMOKING

A. Human Data

The scientific basis for a causal link between colorectal tumors and smoking is recent and was first provided by a study of colorectal adenomas from Norway in 1987 (1). Up to the time of writing, 27 studies have examined the association between smoking and colorectal adenomas, and a highly consistent positive association in over 80% of these studies was found. This association was statistically significant in 86% of the positive studies (1). The effect of smoking seems to occur early in the colorectal neoplastic process, at the time of adenoma formation, or in the de novo pathway when a dysplastic epithelial cell is formed (Fig. 1). Established smokers have a twofold risk for developing a colorectal adenoma (1). There are about 20 colorectal adenomas for every colorectal cancer that develops. When lifelong smokers are compared to nonsmokers, the relative risk for rectal cancer is 1.5 (a 50% elevation) and for colon cancer is 1.2 (a 20% elevation) (1).

B. Mechanisms of Action

Tobacco may reach the colorectal epithelium either as ingested tobacco and/or as absorbed carcinogenic substances reaching the epithelium through the circulatory system. Currently it is unknown which of the almost 4000 substances present in tobacco and tobacco smoke are responsible for the neoplastic effect in colorectal tumors; however, nitrosamines, aromatic amines, and aromatic hydrocarbons are candidate compounds. In relation to molecular genetic changes, recent data suggest that smoking is associated with mutation of the *p53* gene, a mutation frequently found in colorectal neoplasia (1).

VII. OCCUPATIONAL EXPOSURES

A. Human Data

Sedentary occupations with their associated inactivity have been the most consistent occupational risk for colorectal cancer (1). On current evidence it appears unlikely that occupational exposure to asbestos fiber is causally linked to colorectal cancer, although epidemiological data are not sufficiently accurate to exclude the possibility that a few cases of colorectal cancer are contributed to by a heavy ingestion of asbestos fiber (1). Slightly elevated risks for colorectal cancer have been found in the chemical, textile, rubber, petroleum, automotive, woodworking, shoe, leather, and metal industries (1).

B. Experimental Data

With the exception of two positive studies (32,33), extensive studies of lifetime ingestion of various forms of asbestos fiber by rats and hamsters in the (U.S.) National Toxicology Program, as well as other experimental studies, were uniformly unsuccessful in increasing the number of bowel tumors over that expected spontaneously (1).

C. Mechanisms of Action

If asbestos is a carcinogen for colorectal tumors, which appears unlikely, the most likely route of action is via ingestion of asbestos fiber. To date, no mechanisms of action have been proposed for asbestos in the colorectal carcinogenesis area.

In the various occupational exposures in which elevated risks have been found for colorectal cancer, aromatic hydrocarbons, polypropylene fuel, dyes, solvents, metal and wood dust abrasives, as well as synthetic fibers, have been suggested as possible carcinogenic agents (1).

VIII. SUMMARY

Colorectal cancer development has been interpreted in terms of mechanisms of action, molecular genetic changes, and morphological changes (Fig. 1). The effects of certain dietary factors, alcohol consumption, smoking, and some occupational exposures are discussed. It is apparent that for colorectal cancer the broad causes are understood; however, the specific substances concerned are uncertain.

The principal dietary risk appears to be a high consumption of meat/protein/fat/energy/fried and grilled meat, and the principal protective factors are a high consumption of vegetables/fruit/cereals/fiber/starch and possibly fish and calcium-containing foods. These dietary factors appear to operate for both colorectal cancer and the main precursor lesion, colorectal adenoma.

For both colorectal adenomas and colorectal cancer, alcohol and particularly beer consumption of three or more drinks daily over two decades or longer poses about a twofold risk for colorectal tumors. For both the dietary factors and alcohol consumption, the human and experimental data are extensive and reasonably consistent.

Recent human data consistently indicate that established smokers have about a twofold risk for developing colorectal adenomas, as smoking appears to affect mainly or entirely the early part of the neoplastic process. As only about 5% of adenomas progress to cancer, the relative risk of smokers developing rectal cancer (1.5) or colon cancer (1.2) is considerably less than the relative risk of developing a colorectal adenoma (2.0), the preceding condition.

REFERENCES

1. Kune GA. Causes and Control of Colorectal Cancer. A Model for Cancer Prevention. Boston: Kluwer Academic Publishers, 1996.
2. Potter JD, Slattery ML, Bostick RM, Gapstur SM. Colon cancer: a review of the epidemiology. Epidemiol Rev 1993; 15:499–545.
3. Peipins LA, Sandler RS. Epidemiology of colorectal adenomas. Epidemiol Rev 1994; 16:273–297.
4. Reddy BS, Narisawa T, Vukusich D, Weisburger JH, Wynder EL. Effect of quality and quantity of dietary fat and dimethylhydrazine in colon carcinogenesis in rats. Proc Soc Exp Biol Med 1976; 151:237–239.
5. Reddy BS, Narisawa T, Weisburger JH. Effect of diet with high levels of protein and fat on colon carcinogenesis in F344 rats treated with 1,2-dimethylhydrazine. J Natl Cancer Inst 1976; 57:567–569.
6. Reddy BS, Ohmori T. Effect of intestinal microflora and dietary fat on 3,2-dimethyl-4-aminobiphenyl-induced colon carcinogenesis in F344 rats. Cancer Res 1981; 41:1363–1367.
7. El-Khatib S, Cora EM. Role of high fat diet in tumorigenesis in C57BL/1 mice. J Natl Cancer Inst 1981; 66:297-301.
8. Howarth AE, Pihl E. High fat diet promotes and causes distal shift of experimental rat colonic cancer—beer and alcohol do not. Nutr Cancer 1985; 6:229–235.
9. Pence BC, Butler MJ, Dunn DM, Miller MF, Zhao C, Landers M. Non promoting effects of lean beef in the rat colon carcinogenesis model. Carcinogenesis 1995; 16:1157–1160.
10. Lindner MA. A fish oil diet inhibits colon cancer in mice. Nutr Cancer 1991; 15:1–11.
11. Albanes D, Salbe AD, Levander OA, Taylor PR, Nixon DW, Winick M. The effect of early caloric restriction on colonic cellular growth in rats. Nutr Cancer 1990; 13:73–80.
12. Lasko CM, Bird RP. Modulation of aberrant crypt foci by dietary fat and caloric restriction: the effects of delayed intervention. Cancer Epidemiol Biomark Prev 1995; 4:49–55.
13. Cheah PY. Hypotheses for the etiology of colorectal cancer—an overview. Nutr Cancer 1990; 14:5–13.
14. Reddy BS, Mori H, Nicolais M. Effect of dietary wheat bran and dehydrated

citrus fiber on azoxymethane-induced intestinal carcinogenesis in F344 rats. J Natl Cancer Inst 1981; 66:553–557.
15. Reddy BS, Maeura Y, Wayman M. Effect of dietary corn bran and autohydrolysed lignin on 3,2-dimethyl-4-aminobiphenyl-induced intestinal carcinogenesis in female F344 rats. J Natl Cancer Inst 1983; 71:419–423.
16. Heitman DW, Hardman WE, Cameron IL. Dietary supplementation with pectin and guar gum on DMH induced colon carcinogesis in rats. Carcinogenesis 1992; 13:815–818.
17. McIntyre A, Gibson PR, Young GP. Butyrate production from dietary fibre and protection against bowel cancer in a rat model. Gut 1993; 34:386–391.
18. Thorup I, Meyer O, Kristiansen E. Effect of a dietary fiber (beet fiber) on DMH induced colon cancer in Wistar rats. Nutr Cancer 1992; 17:251–261.
19. Madar Z, Timar B, Nyska A, Zusman I. Effects of high-fiber diets on pathological changes in DMH-induced rat colon cancer. Nutr Cancer 1993; 20:87–96.
20. McIntosh GH, Jorgensen L, Royle P. The potential of an insoluble dietary fiber rich source from barley to protect from DMH-induced intestinal tumors in rats. Nutr Cancer 1993; 19:213–221.
21. Kune GA, Vitetta L. Alcohol consumption and the etiology of colorectal cancer: a review of the scientific evidence from 1957 to 1991. Nutr Cancer 1992; 18:97–111.
22. Seitz HK, Czygan P, Waldherr R, Veith S, Raedsch R. Enhancement of 1,2-dimethyl-carcinogenesis following chronic consumption in the rat. Gastroenterology 1984; 86:886–891.
23. Garzon FT, Seitz HK, Simanowski UA, Berger MA, Schmahl D. Enhancement of acetoxymethylmethylnitrosamine (AMMN) induced colorectal tumors following ethanol consumption in rats (abstr). Gastroenterology 1986; 90: 1424.
24. Hamilton SR, Hyland J, McAvinchey D, Boitnott JK. Effects of chronic dietary beer and ethanol consumption on experimental colonic carcinogenesis by azoxymethane in rats. Cancer Res 1987; 47:1551–1559.
25. Niwa K, Tanaka T, Sugie S, Shinoda T, Kato K. Enhancing effect of ethanol or sake on methylazoxymethanol acetate-initiated large bowel carcinogenesis in ACI/N rats. Nutr Cancer 1991; 15:229–237.
26. Nelson RL, Samelson SL. Neither dietary ethanol nor beer augments experimental colon carcinogenesis in rats. Dis Colon Rectum 1985; 28:460–462.
27. McGarrity TJ, Erwin B, Pegg AE, Colony PC. Polyamine levels in dimethylhydrazine (DMH) induced colorectal cancer: the effects of chronic alcohol (abstr). Gastroenterology 1986; 90:1543.
28. Hamilton SR, Sohn OS, Fiala ES. Inhibition by dietary ethanol of experimental colonic carcinogenesis by high-dose azoxymethane in F344 rats. Cancer Res 1988; 48:3313–3318.
29. Simanowski UA, Stickel F, Maier H, Gartner U, Seitz HK. Effect of alcohol on gastrointestinal cell regeneration as a possible mechanism in alcohol-associated carcinogenesis. Alcohol 1995; 12:111–115.
30. Hamilton SR, Luk GD. Induction of colonic mucosal ornithine decarboxylase

activity by chronic ethanol consumption in the rat. Gastroenterology 1987; 92: 1423.
31. Seitz HK, Simanowski UA, Garzon FT, Rideout JM, Peters TJ, Koch A, Berger MR, Einecke H, Mainwald M. Possible role of acetaldehyde in ethanol-related rectal cocarcinogenesis in the rat. Gastroenterology 1990; 98:406–413.
32. Donham KJ, Berg JW, Will LA. The effects of long-term ingestion of asbestos on the colon of F344 rats. Cancer 1980; 45:1073–1084.
33. Corpet DE, Pirot V, Goubet I. Asbestos induces aberrant crypt foci in the colon of rats. Cancer Lett 1993; 74:183–187.

28
Cervix Uteri and Uterus Cancer

Marios Marselos
Medical School, University of Ioannina, Ioannina, Greece

I. EPIDEMIOLOGY AND PATHOLOGY

Cancer of the cervix uteri and corpus uteri belongs to the neoplasms with a high incidence in humans. This is especially true for the cervix, which is the second most common site of cancer in women worldwide, after the mammary gland. The risk is highest in Central and South Africa and southeast Asia, where cancer of the cervix constitutes 20–30% of all cancers in women. A very high incidence has been described in Latin America. Particularly in northeastern Brazil, age-standardized rates of more than 80% have been reported. By contrast, cervical cancer accounts for only 4–6% of all female cancers in North America, Australia, and Europe (1).

In Argentina, a very strong positive correlation was traced between cervical cancer and various socioeconomic factors, such as illiteracy, lower socioeconomic occupations, and poor housing (2). The rates have been found to be generally twofold higher in urban than in rural populations, and they seem also to be related to marital status, being higher in married women than in single ones and higher in widowed or divorced women than in married ones (3). In addition, the rates are about four times higher in the wives of unskilled workers than in women married to professional men (4). The degree to which these variations can be explained by different exposures to a sexually transmitted agent has been pointed out and discussed extensively (4,5).

A wide variety of neoplasms may develop in the cervix of the uterus. However, generally cervical tumors are represented by three lesions: carcinomas, polyps, and papillomas.

Carcinoma of the cervix uteri has been widely studied and exemplifies a progressive change from dysplastic cells to fully atypical malignant cells (carcinoma in situ). These dysplastic changes appear as a continuum of the same basic change. Cervical adenocarcinoma has been associated with many different causes, such as hormonal and viral factors. However, the etiology of this lesion is still unclear. In some instances, malignancy is located in the cervix but has been developed on ectopic squamous epithelial cells similar to those normally encountered in the vagina. In very rare cases, adenocarcinoma of the cervix uteri may develop on benign neoplasms, such as polyps, condylomata acuminata, and papillomas.

The corpus uteri is a significantly less common site for cancer development, compared to the cervix. The frequency of uterine cancer is less than one third that of cervical cancer worldwide. The highest incidence reported (30 per 100,000) is in La Plata, Argentina, and high incidence rates were found also in U.S. whites, in Canada, and in Europe. Low rates of incidence (2 per 100,000) have been found in Asian populations (1).

Studies of migrants suggest that the risk increases in populations who move to areas with "westernized" life styles. For instance, Japanese living in the United States have incidence rates more than six times higher than those in Japan. However, differences in the risks of urban and rural residents are small and inconsistent, even when taking into account socioeconomic criteria.

The main site of uterine cancer is the endometrium. Unlike cervical cancer, endometrium does not seem to be affected by possible sexually transmitted factors. However, the importance of hormonal factors has been well established in the pathogenesis of both endometrial and cervical cancer (1,6). In the uterus there are three main types of neoplasia: endometrial polyps, leiomyomas, and adenocarcinomas (7,8).

Endometrial polyps are benign neoplasms, which can have a malignant change in very rare cases. Sometimes, frank carcinoma of the endometrium may be polypoid. Uterine leiomyomas are endocrine-dependent benign lesions, and estrogens play an important role in their development, although other mediators and cell growth factors are also implicated (9). Malignant transformation of leiomyomas is said to occur, but similarly to the polyps, it is very rare. Estrogens are implicated in the development of endometrial adenocarcinoma, especially when there is a prolonged estrogen stimulation.

The development of endometrial carcinoma appears to be similar to that of cervical carcinoma. There seems to be a progressive series of glandular atypicalities that proceed, for many years, to the development of localized carcinoma and eventually invasive cancer. The time interval between the atypical hyperplasia and cancer may vary from a few years to well over a decade.

Carcinoma of the human endometrium consists of three distinctive histological types. The majority (60–65%) are adenocarcinomas, but they may also be adenoacanthomas and adenosquamous carcinomas. Histologically, the adenocarcinoma is characterized by well-defined gland patterns lined by anaplastic epithelial cells, which may also have papillary formations. When epithelial cells have mucinous secretory activity, the tumor is characterized as "secretory carcinoma."

Adenoacanthoma is composed of glands and foci of squamous cells, which are the product of a gradual metaplasia in the columnar epithelial lining of the glands. They tend to be mature and benign in appearance.

Adenosquamous carcinoma of the endometrium consists of clearly malignant glandular and squamous elements. This tumor, in contrast to adenocarcinoma or adenoacanthoma, has a very poor prognosis.

Also of poor prognosis are some rare tumors, such as the leiomyosarcomas, the endometrial mixed mesodermal tumors, and the stromal tumors. Leiomyosarcomas almost always arise directly from the myometrium. They may develop on a leiomyoma, but this is the exception and not the rule. Mixed mesodermal and the stromal tumors are highly malignant, and they consist of stromal elements, with or without endometrial changes (7,8).

Chemical factors have been implicated in the mechanism of carcinogenesis in uterus and cervix uteri. Based on the available data, we can conclude that neoplasms in these organs are similar to mammary gland and prostate tumors in their apparent dependence on hormonal regulation. The effects of hormones that are relevant to carcinogenesis may vary according to the tissue, dose, presence of other endogenous or exogenous hormones, stage of carcinogenesis, and type of the target cell in terms of age. It generally takes many years for a clinically detectable cancer to arise from a normal cell. During the whole process, from its very beginning to the final stages, a hormone may influence the natural history of cancer development.

The role of hormones has been studied in several experimental protocols, where pharmacological doses of synthetic or naturally occurring hormonal compounds have been employed. In addition, several other chemicals have been implicated as possible carcinogens for these organs, by epidemiological studies, by experiments in animals, or by serendipitous findings. In this chapter, chemical carcinogenesis of the cervix uteri and uterus will be examined in connection with hormonal factors, other drugs, and environmental chemicals.

II. MECHANISTIC ASPECTS

A. Cervix Uteri

The data on the etiological factors for the cancer of cervix uteri suggest a synergism between various factors—idiosyncratic, infectious, chemical, and physical. Infectious and physical factors somewhat overlap, since the association of cancer of the uterine cervix with many sexual partners suggests the concept of increased contamination with an infectious factor. In this way, both microbial as well as viral factors have been implicated.

Current knowledge, though yet not conclusive, is inclined to implicate herpes simplex virus type 2 (HSV-2), especially the human papillomavirus (HPV) (10). It is believed that infection of the cervix by HPV may result in persistence of viral DNA. The persistent HPV-DNA may affect the genomic region responsible for cell-transforming properties through various oncoproteins and tumor-suppressor genes.

Other factors may be involved in this progression toward a precancerous or cancerous condition, such as various chemicals known to act as carcinogens or co-carcinogens. Under normal conditions, the cervix is influenced by estrogens and progestins. These hormonal effects may well contribute to the overall manifestation of the possible role of HPV in the etiology of cervical cancer. For example, progestins have been reported to enhance the transforming and transcriptional efficiencies of papillomaviruses (11,12).

Most studies have found that combined oral contraceptive use is associated with an increased risk for invasive cervical cancer, the risk being proportional to the duration of oral contraceptive use (12–14). The association was still statistically significant after adjusting for the women's sexual practices. The use of depot preparations of medroxyprogesterone acetate could also be associated with an increased risk for invasive cervical cancer (15), but inconsistency among studies has been noted (16). With the exception of diethylstilbestrol, little is known about a possible risk association between cervical cancer and exposure to hormones in utero. Also, any risk association between hormone replacement therapy and cervical cancer has not been established.

Apart from substances with apparent hormonal properties that could contribute by stimulating cell division, several environmental chemicals have been suspected or proven as carcinogens to cervix uteri. Although these chemicals may have a direct or indirect influence on hormonal parameters, their exact mechanism of action remains obscure in most cases.

B. Uterus

As mentioned already, hormonal factors are very important for the development of endometrial cancer. This is also true for the increased production of endogenous estrogens, e.g., from estrogen-secreting tumors. Patients with granulosa-theca cell tumors of the ovaries have a high incidence of uterine carcinoma, presumably due to the estrogens secreted by the ovarian tumor. On the other hand, women with early ovariectomy or ovarian agenesis do not, as a rule, develop endometrial carcinoma.

The paradoxical occurrence of endometrial cancer after menopause, when estrogen production is assumed to be decreasing, is explained by the presence of estrogens produced in body fat from adrenal and ovarian adrogen precursors. Although ovarian estradiol is indeed decreased, estrone synthesized outside the ovary from androgens is increased. It has been shown that endometrial hyperplasia occurs in patients with high conversion rates of estradiol to estrone in adipose tissue (17). This may also explain the association of increased risk of endometrial cancer with obesity (18,19).

Our understanding of hormonal effects and of their role in carcinogenesis is most clear for the endometrium. Estrogens increase the mitotic rate of endometrial cells. Sustained occupancy of estrogen receptors by estrogens is believed to be necessary for the estrogen-induced stimulation of DNA replication, a known early event in cell division. Progestins oppose this effect by decreasing the concentration of estrogen receptors (20) and by causing differentiation of the endometrial cells to a secretory state (21). The progestin medroxyprogesterone acetate causes either weak secretory changes or an atrophic endometrium (22). Progestins in general can inhibit the growth of endometrial cancer in organ culture (23). Progestins also increase the activity of the dehydrogenase, which oxidizes estradiol to the less potent estrone, but this enzyme does not oxidize synthetic estrogens (24). Rat strains with a high incidence of spontaneous pituitary adenomas have a remarkably increased mortality due to uterine adenocarcinomas (25). The progestin melengestrol acetate can lead to complete suppression of the occurrence of the endometrial neoplasms, presumably due to an antagonistic action on the excess of endogenous estrogens (26).

During natural menstrual cycles, the endometrial cell division rate rises quickly at the beginning of the follicular phase, reaching a plateau despite the continuing increase in estradiol concentrations, and then falls quickly to a very low level with the increase in progesterone during the luteal phase (27).

Increased production of estrogens has been associated with a persistent absence of ovulation and an increased risk of endometrial cancer. Two routes seem to be involved in this process. In the first, ovarian 17β-estradiol is secreted either continuously over long periods of time or intermittently, so that the endometrium is stimulated through the stages of normal proliferation to hyperplasia, reaching adenomatous hyperplasia, atypical hyperplasia, and eventually carcinoma in some cases. This would appear to be the most common route in premenopausal patients and may also be the route in those few postmenopausal women in whom progression through atypical hyperplasia to carcinoma takes place. The other route involves the production of estrone by extraglandular conversion of androstenedione. This second route seems to predominate in postmenopausal women. Under normal conditions, estrone does not have sufficient estrogenic activity to cause proliferative changes in the endometrium, but it does provide facultative support for the development of cancer (28).

III. CHEMICAL FACTORS IN CANCER OF CERVIX AND CORPUS UTERI

A. Steroid Hormones

Several hormones, of either natural or synthetic origin, have been used in medicine for various indications. During recent decades, combinations of estrogens and progestins have been introduced as oral contraceptives and have been taken regularly by many women worldwide. In addition, estrogens have a large spectrum of uses, such as for alleviation of climacteric symptoms, dysfunctional uterine bleeding, postpartum breast engorgement, female hypogonadism, mammary carcinoma, prostatic carcinoma, and vulvar dystrophies.

Epidemiological studies have shown unequivocally that estrogen replacement therapy increases the risk for endometrial cancer (29–34). This association has been observed for a variety of estrogens, including conjugated equine estrogens, estradiol valerate, and ethinylestradiol (Table 1). There is also evidence that increased risk correlates with increased strength of medication and increased duration of hormone use (30). The size of the observed risks varies somewhat among studies; typical relative risks (compared to nonusers) are 2.9 for a duration of use of 1–4 years, 5.6 for 5–9 years, and 10.0 for more than 10 years (35). A risk of 2.7 has been estimated for a usual dose of less than 0.625 mg of conjugated equine estrogens and 3.8 for more than 1.25 mg (36). The excess risk is evident within a few years after last use (32,35). These observations suggest that estrogen replacement therapy affects both late and early stages of carcinogenesis (30,31,37,38).

As previously mentioned, an enhanced estrogenic stimulus in the absence of progesterone favors the development of uterine tumors. The dramatic increase in uterine carcinoma in the late 1970s coincided with the use of "unopposed" estrogens for osteoporosis and other postmenopausal conditions (39,40). Following the first reports that estrogen replacement therapy increases the risk for endometrial cancer, the effect of progestins on the development of endometrial hyperplasia was studied. Studd et al. (41) reported that addition of a progestin to replacement therapy reduced the prevalence of hyperplasia from 56% in women who received unopposed

Table 1 Carcinogenicity in the Cervix and Corpus Uteri by Steroid Hormones and Drugs with Known Hormonal Action

Agent	Mice	Rats	Humans
Steroid hormones			
Estrogen replacement therapy	ND	ND	Uterus
17β-Estradiol and esters	Vagina, cervix, uterus	*	ND
Ethinylestradiol	Cervix, uterus	*	ND
Oral contraceptives			
Sequential	ND	ND	Uterus
Combined	ND	ND	Cervix, uterus
Norethynodrel and mestranol	Vagina, cervix	*	[Uterus]
Testosterone	Cervix, uterus	–	ND
Drugs with hormonal effects			
Diethylstilbestrol	Vagina, cervix, uterus	*	Vagina, cervix, [uterus]
Tamoxifen	*	*	Uterus
Bromocriptine	ND	Uterus	–

ND, No data available.
–, No effect detected.
*, Known carcinogenicity in other organs.
[ ], Inadequate or limited evidence of carcinogenicity.

estrogens for 2 months to 9, 3, and 0% in women who received norethisterone for 5, 10, and 13 days out of every 28 days, respectively. Few data are available on the effects of estrogen and progestin replacement therapy on cancer risk. Persson et al. (33) found a risk of 0.9 [95% confidence interval (CI), 0.4–2.0] for women given estrogen and progestin replacement therapy relative to untreated women, with no evidence of an increase in risk with increasing duration of use. In contrast, a relative risk of 2.2 (95% CI, 1.2–4.4) was found for estrogen replacement therapy, as conjugated estrogens, given for more than 4 years (33). In this study, women given estrogen and progestin replacement therapy usually received a progestin for 7–10 days of each treatment cycle. It seems that despite the demonstrated benefit of adding a progestin to replacement therapy, at least in these regimens, progestins did not prevent all endometrial cancer from developing.

Endometrial cancers caused by estrogen replacement therapy are, on average, well differentiated and have a better prognosis than other endometrial cancers (32). It is possible that hormone-responsive cancers grow quickly and reach a size that can be diagnosed rather early before accumulating further alterations associated with a poor prognosis.

Combined oral contraceptives are currently preferred in clinical practice. These drugs are strongly protective against endometrial cancer. Use for one year reduces the endometrial cancer risk by about 20%, and use for four years reduces the risk by more than 50%. The protective effect appears within 10 years of starting combined oral contraceptive use (42,43).

Sequential oral contraceptives were introduced in clinical use with the assumption that they mimic the normal alternate surge of hormones, having only estrogens during the first half of the cycle and only progestins during the second half. Nowadays, these drugs are considered obsolete, and they have a very limited application, if any. When in use, sequential oral contraceptives have been implicated in the development of endometrial cancer, presumably due to the unopposed action of estrogens (44,45).

Based on data from experimental animals, hormones can be evaluated separately. As presented in Table 1, some of them show a clear carcinogenic potential for uterus, cervix uteri, or both (Table 1).

17β-Estradiol and its esters (estradiol 3-benzoate, dipropionate, valerate, polyestradiol phosphate) have been used extensively in the past, but today these compounds are not very popular. 17β-Estradiol is a potent estrogenic agent of natural origin, and it is further metabolized to estrone. As early as 1941, Gardner et al. had shown that mice treated subcutaneously with 17β-estradiol and its esters developed carcinomas or invasive epithelial lesions in the uterine cervix (46–49).

These initial observations have been verified using subcutaneous implants of 17β-estradiol in mice, which developed squamous cell carcinomas of the cervix (50). Highman et al. (51) have also shown an increase in the incidence of uterine cancer of mice fed diets containing 17β-estradiol. Tumors were detected in uterus, cervix uteri, and vagina of tested mice, but not in rats or in hamsters (30).

Estrone, a metabolite of 17β-estradiol, has been shown in mice to possess carcinogenic properties for mammary gland, and in rats for pituitary and adrenals. However, it was not carcinogenic for cervix or uterus in the experimental animals tested (30,31,34).

There are no data available for cancer induction in humans after administration of either 17β-estradiol alone or estrone alone. Another natural estrogenic steroid, estriol, has not been shown to be carcinogenic for uterus and cervix in experimental animals or in humans.

Ethinylestradiol is a synthetic estrogenic steroid, which has been used extensively for the treatment of situations where estrogen replacement might help, i.e., with therapeutic indications similar to those of 17β-estradiol. Oral ethinylestradiol has been reported to produce malignant tumors of the uterine fundus and of the cervix in mice. However, this drug could not produce similar types of cancer in rats (30). Even when given in combination with progestins, ethinylestradiol retained its carcinogenicity for uterus and cervix in mice (30,52,53). No epidemiological data for ethinylestradiol alone are available for the evaluation of the carcinogenic potential of this drug in humans (31).

Mestranol is a synthetic estrogenic steroid, which has been used as a contraceptive in combination with progestins. Other clinical indications are endometriosis and amenorrhoea. When given alone, the drug could not be implicated in carcinogenicity of the uterus or cervix. In experiments in which mestranol was administered to female mice in combination with norethynodrel, vaginal and cervical squamous cell carcinomas were produced (31,54). No epidemiological data on mestranol alone are available. Considering, however, the inclusion of this drug in combination formulations with progestins, it can be stated that its possible carcinogenicity for humans should be estimated under the general paradigm of the use of oral contraceptives.

Norethynodrel is a synthetic compound that has been used primarily as a progestin in oral contraceptives. As mentioned above, when given to mice in combination with mestranol, norethynodrel could produce infiltrating carcinomas of the uterine cervix (54). When tested alone, it was associated with the development of tumors in organs other than the uterus or the cervix (30). Epidemiological data for possible carcinogenicity of norethynodrel in humans are not available.

Progesterone is a naturally occurring steroidal compound with distinct endocrinological properties. The drug has been tested for carcinogenicity in ovariectomized female animals (mice or rats), usually in rather complicated experimental protocols, including intrauterine or intravaginal implantation of an initiating polycyclic aromatic hydrocarbon (30). The results of such experiments are difficult to interpret in terms of possible carcinogenic effects, when progesterone is considered as a single agent. On the other hand, there are no available data on the effect of exogenous progesterone in humans. It should be added that the synthetic compound medroxyprogesterone acetate, which shares many of the properties of progesterone, was not found to be carcinogenic in mice (30). Similarly, there are no data implicating medroxyprogesterone as a human carcinogen, although it has been widely used as a depot contraceptive, after intramuscular injection. However, there is experimental and clinical evidence indicating that progesterone plays an important role in the development and growth of leiomyomas, together with estrogens and various cell growth factors (9,55,56).

Testosterone is the most active of the natural androgens. It is used in various endocrinic dysfunctions, in both males and females, usually as the propionate or enanthate. Mice implanted twice weekly with a pellet of testosterone propionate (1–2 mg) developed cervical and uterine tumors, which were infiltrating and in some instances metastasized to the lungs (57). There are no available epidemiological data on testosterone alone that could implicate this hormone as a human carcinogen.

B. Steroid Hormones in Two-Stage Carcinogenicity Studies

In search for an experimental model of uterine carcinogenesis, several protocols have been applied to the exogenous administration of estrogens (51,58–60). In general, the so-called two-stage model for the progress of chemical carcinogenesis is widely accepted, implicating at least one initiating agent and one or more promoters (61). In some experimental procedures, the estrogen is applied as a promoter after the administration of an initiator.

In mice, one protocol for the development of uterine sarcoma uses 1,2-dimethylhydrazine as the initiator and estradiol-dipropionate as the promoter (62). Ascorbic acid given simultaneously with estradiol-dipropionate could significantly decrease the incidence of uterine sarcoma, although sodium ascorbate did not exert an inhibiting effect. The dissimilarity of results between ascorbic acid and sodium ascorbate cannot be explained (63). In the same series of experiments, ascorbic acid diminished the increase of uterine weight after estradiol-dipropionate but did not affect the growth of transplantable uterine sarcoma (63). Also, in mice, high and rapid induction of uterine cancers could be achieved by using *N*-methyl-*N*-nitrosourea as the initiating agent and 17β-estradiol as the promoter (64).

Rats seem to be more resistant to initiator-promoter protocols. Fisher 344 rats did not develop uterine adenocarcinomas in a protocol applying *N*-methyl-*N*-nitrosourea as the initiator and estradiol dipropionate as the promoter (65). This strain of rats is well known to exhibit a very low incidence of spontaneous endometrial carcinoma (66,67). In contrast, Donryce rats were found to be rather sensitive to the coadministration of *N*-methyl-*N*-nitrosourea and estradiol dipropionate. In this strain, a rapid and high incidence of uterine adenocarcinomas was detected (65). However, as pointed out by the authors, these findings merely suggest an earlier appearance of uterine cancer in a rat strain known to develop spontaneous adenocarcinomas of the uterus with advancing age (65,68). As a matter of fact, 2-year-old Donryce rats have the same frequency of spontaneous adenocarcinomas of the uterus as *N*-methyl-*N*-nitrosourea– and estradiol dipropionate–treated rats at a much earlier age (65). It seems that for the time being there is not sufficient evidence that *N*-methyl-*N*-nitrosourea is carcinogenic for rat uterus.

The experimentally induced uterine sarcomas of CBA mice, after combined treatment with 1,2-dimethyl-hydrazine and estradiol diproprionate, bear light- and electron-microscopic resemblance to human endometrial stromal sarcomas (62). These tumors consist of cells similar to immature stromal cells and also of cells showing fibroblastic differentiation (69).

C. Nonsteroid Drugs with Hormonal Properties

1. *Diethylstilbestrol*

Of the nonsteroidal estrogens, there are data on carcinogenicity for diethylstilbestrol and its esters. Chemically, diethylstilbestrol is a stilbene congener (70). The esters of diethylstilbestrol have been preferred for estrogenic use in many instances, because they exhibit a sustained action. Among these, the dipropionate derivative has the longest activity (71). The indications for which diethylstilbestrol has been prescribed in the past are conditions related to threatened abortion and also postmenopausal estrogen replacement treatment. Even today, the drug is prescribed for carcinoma of the breast in postmenopausal women and for carcinoma of the prostate (30,72,73). The alkylated derivatives of stilbestrol, dienestrol and hexestrol, exhibit only minor structural and pharmacological differences from diethylstilbestrol. However, there are no data implicating these substances in possible carcinogenicity of the vagina, the cervix, or the uterus.

In contrast, diethylstilbestrol has been proven to be carcinogenic in human cervix and vagina. There are also some indications that this compound may be associated with human endometrial cancer (74). In animal studies, diethylstilbestrol could produce cancer of uterus and cervix uteri in mice and hamsters (75). This compound is a unique example of a carcinogen that affects the target organs transplacentally, during intrauterine life. Many epidemiological studies provide evidence that prenatal exposure to diethylstilbestrol before the 12th week of pregnancy is causally associated with vaginal and cervical clear-cell adenocarcinomas, a very rare type of cancer in the unexposed female population. Cancer usually appears after menarche at an age between 10 and 30 years. Up to the age of 24, the risk has been calculated to be 0.14–1.4/1,000. Cases after the age of 25 were very rare (76–78). It should be mentioned that the typical clear-cell carcinoma of the endometrium is an aggressive type of cancer that usually invades the myometrium. This neoplasm develops in postmenopausal women, and it is not related to diethylstilbestrol exposure during intrauterine life. However, the morphological features of endometrial clear-cell carcinoma are identical to those of clear-cell carcinoma of the ovary and vagina (8).

Adenocarcinomas of the vagina and the cervix have been found in several instances with no apparent relation to transplacental or other diethylstilbestrol exposure. A few cases of clear-cell adenocarcinomas were described in young women even before the introduction of this synthetic estrogen in therapeutics (79,80). However, the risk for the general unexposed population is considered to be several orders of magnitude lower, compared with the data obtained from women with a history of intrauterine exposure to diethylstilbestrol. Therefore, the evidence of a causal relationship between diethylstilbestrol and clear-cell adenocarcinoma of the vagina or cervix seems conclusive (81). Evaluating the available scientific data, the International Agency for the Research on Cancer (IARC/WHO) has estimated that there is sufficient evidence for diethylstilbestrol to be classified as a human carcinogen (30,31).

As mentioned already, administration of estrogens for the control of climacteric symptoms is known to be related to an increased incidence of endometrial cancer. Diethylstilbestrol most probably acts in a similar way, but the data are inadequate to allow any firm conclusion. On the other hand, there is no documentation on the development of endometrial cancer later in life in women treated with diethylstilbestrol during pregnancy (74).

Several early experimental studies have demonstrated the carcinogenicity of diethylstilbestrol either in adult animals or after intrauterine exposure. Among the organs affected are pituitary, mammary gland, liver, kidney, ovary, vagina, cervix and uterus (34). Epidermoid carcinomas of the vagina and the cervix developed in mice treated intravaginally with an oil solution or a pellet of diethylstilbestrol (49). Newborn female mice were injected subcutaneously with diethylstilbestrol as a single dose (2 mg). After several months, more than 30% of the animals developed cancer of the cervix and the vagina (54). In squirrel monkeys, pellets (60 mg) of diethylstilbestrol were implanted subcutaneously and were found to produce malignant uterine mesotheliomas in 70% of the animals after 5–14 months (82).

The transplacental carcinogenic effect of diethylstilbestrol has been shown in several animal species. However, carcinogenicity similar to that observed in humans was shown only in mice and hamsters. Mice were given the hormone at various dosage levels during days 9–16 of pregnancy. As a rule, offspring survival was inversely proportional to the dose level. Adenocarcinomas of the uterine endometrium and epidermoid tumors of the cervix and vagina were observed in the adult life of the surviving female offspring in all treated groups (83–86).

Also in studies with hamsters, administration of diethylstilbestrol during gestation produced neoplasms in the inner genitalia (87). In a series of experiments, this prenatal treatment was shown to lead to the development of endometrial adenocarcinoma and squamous cell papillomas of the cervix and vagina (88).

There is experimental evidence that prenatal treatment with diethylstilbestrol may increase the incidence of uterine carcinomas in second-generation descendants who were not exposed directly to the drug (89). These results, and the overall complexity of diethylstilbestrol carcinogenicity, have led to a theoretical model of a possibly multigenerational burden of tumor risk (90), a hypothesis advanced on the basis of experimental and epidemiological observations (91,92). Further experimental evidence for a transgenerational carcinogenic effect of diethylstilbestrol is provided by the increased incidence of uterine tumors in female mice obtained by mating males exposed prenatally to the hormone with unexposed females (93).

2. *Tamoxifen*

Tamoxifen as the citrate salt has a rather wide use in therapeutics. Chemically, it is (Z)-2-[4-(1,2-diphenyl-1-butenyl)phenoxy]-*N,N*-dimethylethanamine (70). The drug was shown to be an antifertility agent in rats (94), but it was soon proven to have the opposite effect in humans, in whom it induces ovulation. It is well documented that tamoxifen exhibits species-specific estrogenic or antiestrogenic effects.

One of the first clinical applications of tamoxifen concerned the palliative therapy for advanced breast cancer, particularly after it was shown to be equally efficacious as and less toxic than diethylstilbestrol (95–97). Later, tamoxifen was widely accepted as first-line endocrine therapy for metastatic disease in postmenopausal women. The drug has been an adjuvant therapy of choice for postmenopausal women with estrogen receptor–positive disease, and also in node-negative premenopausal women with an estrogen receptor–positive disease (98). It is estimated that more than 50% of all women who undergo surgery for breast cancer are currently receiving adjuvant tamoxifen for a rather long period of time. On the other hand, tamoxifen has been often used in combination with cytostatics or with another hormonal therapy, especially in postmenopausal node-positive women (99,100).

Because of its known inhibitory effect and because of the observed reduction in new contralateral breast cancers seen in many clinical trials (101), tamoxifen has been proposed as a preventive therapeutic agent in women at high risk, such as lobular carcinoma in situ or family history of breast cancer. However, this practice was criticized and never gained wide acceptance. Other indications suggested for the use of tamoxifen include male breast cancer (102), advanced endometrial cancer (103), hepatocellular carcinoma (104), gastric cancer (105), renal cell carcinoma (106), melanoma (107), carcinoma of the pancreas (108), carcinoma of the cervix uteri (109), and meningioma (110). Interest has also been focused on the positive effects of tamoxifen on cardiovascular lipid profiles and on cardiovascular diseases (111).

Tamoxifen has distinct hormonal effects, which are due to its ability to bind to estrogen receptors and act as agonist or antagonist. It is assumed that tamoxifen, as a partial agonist for these receptors, would be able to interact with estrogen-responsive elements on DNA and would then lead to enhanced mitotic activity and other biological changes, which may increase cancer risk. On the other hand, tamoxifen could produce a partially antagonistic effect by preventing estrogen access to the receptor in tissues where there are responsive elements for the tamoxifen-receptor complex. Because tamoxifen is not as potent as 17β-estradiol, there is a diminished response in estrogen-deficient tissues. This is more likely to occur in tissues that are rich in receptors and have estrogen levels high enough to elicit a response but low enough for tamoxifen to block estrogen access to a sufficient number of receptors, thereby decreasing normal estrogen responsiveness (112).

In tissues positive for estrogen receptors, but with low levels of estrogens, tamoxifen may act as an agonist because there is no tissue-specific estrogen response to diminish. Menstrual cycles remain regular, and secretory changes occur in the endometrium during tamoxifen treatment (22), suggesting that the cyclical process of endometrial proliferation and shedding is not suppressed and that tamoxifen is therefore not an effective antiestrogen in the human endometrium. As a matter of fact, tamoxifen increases the incidence of endometrial hyperplasia (113) and can stimulate the growth of human endometrial cancer in athymic mice (114).

These data are insufficient to allow firm conclusions to be drawn on the mechanism of action of tamoxifen, but suggest the possibility that tamoxifen may increase the risk of endometrial cancer by stimulating endometrial cell division. Estrogen receptor–mediated mitogenesis has been described in both uterine cells and hepatocytes. In both cell types, tamoxifen-enhanced proliferation may predispose to tumorigenesis, even moreso as tamoxifen is known to produce DNA damage by adduct formation (115). Although carcinogenic, in experimental animals tamoxifen has not been shown to cause de novo endometrial adenocarcinoma or to alter benign endometrial proliferating cells to initiate cancerous changes (116,117).

In mice, repeated administration of tamoxifen at doses up to 50 mg/kg/day for 13–15 months caused atrophy of the gonads and accessory sex organs with cystic endometrial hyperplasia, elongation of the vertebrae, and a marked increase in bone density with resorption and new bone formation in irregular patterns. These changes are consistent with the pharmacological action of tamoxifen in this species. In the liver, there were fatty changes and a swelling of the parenchymal cells (118).

Mice treated with tamoxifen orally (5 and 50 mg/kg/day for 3 months) developed tumors of the testes and the ovaries, but not of the uterus (118). Also in mice, tamoxifen (50 mg/kg 3 times/week for over one year) was found to reduce significantly cervical dysplasia and carcinoma, which were induced by local application of 3-methylcholanthrene in the canal of the uterine cervix (119).

In rats, administration of tamoxifen for 6–24 months at doses between 35 and 100 mg/kg/day caused atrophy of the gonads and accessory sex organs, while at lower doses (e.g., 2 mg/kg/day), the endometrium showed an absence of glands, flattening of the epithelium and occasional squamous metaplasia (116,118). Also, long-term administration of the drug has been associated with nodular hyperplasia of the liver (116).

Many carcinogenicity studies of tamoxifen have been performed in rats, with a wide range of doses and different routes of administration. In experiments with oral administration of tamoxifen (5–45 mg/kg/day for several weeks), the liver was the most usually affected organ. The drug could produce hepatocellular carcinoma in the rat strains Wistar (116), Sprague-Dawley (120–123), and Fisher 344 (124).

The uterus has been studied by Mäntylä et al. (125) as a possible target organ for the expression of tamoxifen-induced cancer. Female Sprague-Dawley rats were treated by gastric instillation of daily doses of tamoxifen citrate at 45 mg/kg for up to 52 weeks. In 104 rats given the drug for various periods, squamous metaplasia of the endometrium was found in 10 rats, dysplasia in 3 rats, with metaplasia and squamous-cell carcinoma in 2 rats. The carcinomas were found after 20 or 26 weeks of dosing. Among 109 control animals, no lesions of the uterus were found. This single report, however, provides only limited and inconclusive evidence for the classification of tamoxifen as a cause of rat uterine cancer.

Tamoxifen has also been coadministered to rats in various protocols along with known carcinogens. As a rule, it enhanced the yield of positive preneoplastic foci, as well as of tumors, in various protocols of hepatocarcinogenicity (126–132).

Paradoxically, in one study with Armenian hamsters (*Cricetulus migratorius*) tamoxifen greatly reduced the rate of liver carcinogenicity brought about by zeranol (133).

In rats, co-administration of tamoxifen with known carcinogens of the mammary gland significantly diminished the yield of tumors (134).

According to epidemiological studies in humans, tamoxifen produces a variety of changes in the endometrium. Although the most commonly reported change is endometrial atrophy, the drug may also produce hyperplasia. Since 1985 (135), several case reports have pointed out the connection between tamoxifen and the development of endometrial cancer in women receiving daily doses of 10 mg (136), 20 mg (135,137–140), 30 mg (141,142), 40 mg (143–146), or 60 mg (143). In these epidemiological studies, the duration of treatment was between one and five years.

Many case-control studies, cohort studies and randomized clinical trials have sufficiently established a positive association between endometrial cancer occurrence and tamoxifen treatment in patients with breast cancer (usually with a daily dose of 20 mg) (147–153).

On the other hand, it should be noted that tamoxifen was found to reduce significantly the risk of contralateral breast cancer, in patients who have already undergone surgery (150,153–156). In an overview of the data available till 1992 on the occurrence of contralateral breast cancer in women allocated to tamoxifen treatment (100), the original data of the clinical trials were meta-analyzed by using the method of life tables. It was estimated that women on tamoxifen had a 39% reduction in risk for contralateral breast cancer. There was a trend towards increased beneficial effect with longer duration of treatment. A reduction of risk of up to 53% was calculated for patients who had more than 2 years of tamoxifen treatment. Unlike experimental data with animals, tamoxifen administration to patients could not be associated with the development of primary hepatic cancer (157, 158).

3. *Bromocriptine*

In both humans and experimental animals, it is believed that hormones can induce tumors in various hormone-sensitive target-organs by other than strictly genotoxic mechanisms. However, experimental data are not always concordant with the human condition, as shown by the example of bromocriptine (159).

Bromocriptine is a dopamine receptor agonist, which is used chronically as a drug for parkinsonism. Due to its dopaminergic action, this drug leads also to hypoprolactinemia, which is a useful effect in situations such as galactorrhea due to adenoma of the hypophysis or normal postpartum galactorrhea that is present to an undesired degree. Despite its long clinical use, bromocriptine has not been implicated as a possible carcinogen to humans (34,160).

The situation is different for rodents, where prolactin has been found to act as a luteotrophic hormone (161). The menopause of rats is not caused by a loss of ovarian responsiveness, as in humans, but by a functional decline of hypothalamus and a deficit of dopamine. This leads to a relative hyperprolactinemia and consequently to corpus luteum persistence and a progesterone dominance.

The administration of bromocriptine to postmenopausal rats restores the dopaminergic functions and leads to luteolysis of the persistent corpora lutea, to the formation of new follicles, and eventually to estrogen dominance. The rats enter a state of permanent estrus after the previous state of diestrus due to the progesterone persistence. The overproduction of estrogens is also associated with endometrial stimulation, which may end up as carcinoma (162,163). As pointed out by Neumann (159), every drug that reduces prolactin increases the incidence of endometrial carcinoma in old rats.

Because prolactin is not luteotrophic in primates, it is obvious that dopaminergic drugs do not produce estrogen dominance in women. Therefore, bromocriptine or other dopamine agonists fail to be carcinogenic in human uterus.

4. *Other Drugs*

During the routine carcinogenicity assays for the development of new medicinal products, several substances have been implicated in the production of tumors in the inner genitalia of experimental animals. Table 2 presents some drugs that act as uterine carcinogens without an apparent hormonal mechanism of action. To this group belong the substances dacarbazine, glimepiride, isophosphamide, levobunolol, procarbazine hydrochloride, and thiophosphamide. The relevance of these experimental data to the clinical use of the drugs in humans is still inconclusive.

Dacarbazine is an antineoplastic drug acting as an alkylating agent

Table 2 Various Medicinal Drugs Found Carcinogenic in the Uterus

Agent	Mice	Rats	Humans
Dacarbazine	Uterus	*	ND
Glimepiride	–	[Uterus]	ND
Isophosphamide	*	Uterus	ND
Levobunolol	Uterus	ND	ND
Procarbazine HCl	Uterus	*	ND
Thiophosphamide	*	Uterus	*

ND, No data available.
–, No effect detected.
*, Known carcinogenicity in other organs.
[ ], Inadequate or limited evidence of carcinognicity.

after hepatic biotransformation. Clinically, it is applied in malignant melanoma, Hodgkin's disease, soft-tissue sarcoma, osteogenic sarcoma, and neuroblastoma (164,165). Dacarbazine has a rather limited use, and it is applied only in combination with other agents. These facts make confounding exposure a serious problem in interpreting any possible finding (166). In experimental models, the drug could produce cancer in both mice and rats at various organs. However, uterine tumors could be detected only in mice, when the drug was given with intraperitoneal injections (25–50 mg/kg bw 3 times a week for 6 months) (167,168). There are no epidemiological data implicating dacarbazine in the production of any type of cancer in humans (34).

Gilmepiride is a new hypoglycemic sulfonylurea, which has been found to cause a marginal increase of uterine adenocarcinomas in Wistar rats (169). This study provides limited evidence for possible carcinogenicity in rats, since tumor development was significant only at the highest tested dose level (5000 ppm) and after treatment for 31 months. On the other hand, glimepiride was ineffective in a carcinogenicity study in mice, and it was found to be devoid of mutagenic activity. Moreover, treatment of rats for 2 weeks at high dose levels (up to 5000 ppm) failed to affect the plasma levels of estrogens. The safety factors based on a comparison of systemic exposure (area under the curve values) of female rats and humans are high enough (about 20 times) to exclude a risk to patients at the proposed clinical doses (170).

Isophosphamide is an alkylating agent, with pharmacological properties similar to cyclophosphamide, of which it is a congener. It exerts antineoplastic and immunosuppressive actions and it has been applied in the treatment of oat cell tumors of the lung, ovarian cancer, breast cancer, and non-Hogkin's lymphomas (164,166,171). In Sprague-Dawley rats, the drug produced a high incidence of uterine leiomyosarcomas (31,166,172,173). Such an effect could not be shown in mice (173). On the other hand, there are no clinical data available on the possible carcinogenicity of this drug in humans (34).

Levobunolol is a nonselective β-adrenoreceptor antagonist, which has been found to cause uterine leiomyomas in mice treated at a rather high dose level (200 mg/kg/day for 80 weeks). Similarly high doses given to rats failed to produce neoplasia in any organ. Considering the rather low incidence of uterine leiomyomas in mice (4/50) and the high doses applied (more then 200 times the projected therapeutic dose), the clinical relevance of this findings is yet unknown (174). Moreover, the drug is used in humans only with local application in the eye for reducing intraocular pressure. Although some systemic absorption of levobunolol has been reported to occur (164), the risk of cancer in humans seems only theoretical.

Procarbazine hydrochloride is used as an antineoplastic agent, almost always in combination with other drugs, for the treatment of malignant melanoma, Hodgkin's disease, non-Hodgkin's lymphoma, and small-cell carcinoma of the lung (165). In experimental protocols, it has been shown to produce cancer at several sites in both mice and rats (167,168,173,175). When the uterus was examined as a target organ, the drug could produce uterine tumors in the mouse strain Swiss-Webster after prolonged intraperitoneal administration (12 or 25 mg/kg bw 3 times a week for 6 months) (167,168). With a similar experimental protocol, procarbazine hydrochloride could produce uterine tumors also in the mouse strain B6C3F1 (175). Procarbazine hydrochloride passes the blood-brain barrier readily, but it has not been established if the carcinogenic properties of the drug are secondary to a possible hormonal imbalance. Due to confounding exposure, procarbazine hydrochloride has not been able to be assessed as a single agent promoting carcinogenicity in humans (34).

Thiophosphamide (thiotepa) is a congener of phosphamide, with alkylating properties similar to the parent compound (165). The drug has been proven carcinogenic in both mice and rats, and there is also sufficient evidence that it can cause leukemia in humans (34,176). The development of uterine neoplasms has been shown only in rats treated intraperitoneally with thiotepa (0.7, 1.4, or 2.8 mg/kg 3 times a week for 52 weeks). Male rats had lymphocytic leukemia and lymphomas, whereas female rats developed uterine and mammary adenocarcinomas (34,176,177). There are no data implicating thiophosphamide for cancer in the cervix or uterus of women treated with the drug, although it is carcinogenic in humans (34).

D. Environmental Chemicals and Pollutants

1. Chemicals in the Work Environment

Several industrial substances have been implicated as possible carcinogens to the cervix uteri and uterus, either by epidemiological studies or by screening tests in animals (Table 3). In experiments performed in the National Toxicology Program, several of the examined substances have been proven carcinogenic in the rat uterus. To this group of chemicals belong 3-amino-9-ethylcarbazole, C.I. direct blue 15, 3,3′-dimethoxybenzidine, and 4,4′-thiodianiline (173).

3-Amino-9-ethylcarbazole is a poisonous solid used in the manufacture of dyes and also as an indicator for peroxidase activity. In National Toxicology Program studies this substance produced uterine tumors in the rat. However, such an effect could not be detected in mouse strains (173). There are no carcinogenicity data for this substance in humans. A similar

Table 3 Environmental Chemicals and Pollutants Implicated in Cancer of the Cervix and Corpus Uteri

Agent	Mice	Rats	Humans
3-Amino-9-ethylcarbazole	–	Uterus	ND
C.I. Direct blue 15	–	Uterus	ND
3,3′-Dimethoxybenzidine	–	Uterus	ND
4,4′-Thiodianiline	–	Uterus	ND
Bromoethane	Uterus	–	ND
Chloroethane	Uterus	*	ND
Ethylene oxide	Uterus	*	[*]
Perchloroethylene	ND	ND	[Cervix]
Tetrachloroethylene	ND	ND	[Cervix]
1,1,1-Trichloroethane	ND	ND	[Cervix]
Trichloroethylene	ND	ND	[Cervix]
1,2,3-Trichloropropane	Uterus	–	ND
2,3-Epoxy-1-propanol	Uterus	–	ND
Trimethylphosphate	Uterus	–	ND
Tobacco smoking	ND	ND	[Cervix]

ND, No data available.
–, No effect detected.
*, Known carcinogenicity in other organs.
[ ], Inadequate or limited evidence of carcinogenicity.

profile was found for the carcinogenic properties of the synthetic dye C.I. direct blue 15, which is used as a dye or in the manufacture of other dyes (173).

3,3′-Dimethoxybenzidine (dianisidine) is a synthetic substance used in the manufacture of azo dyes. It is also used as a dye for leather, paper, plastics, rubber, and textiles. Dianisidine has been found to produce cancer in the rat, at various sites, including the bladder, intestine and skin (31). It has been shown to produce also uterine cancer (173). There are no data for possible carcinogenicity of this substance in other species or experimental animals. Epidemiological data are also missing for a possible carcinogenic effect of dianisidine in humans.

4-4′-Thiodianiline is a synthetic substance, which is used exclusively as an intermediate in the preparation of various dyes. Carcinogenicity studies have been performed in mice and rats, with positive results in both species and in both sexes. Among the other neoplasms, uterine cancer has been demonstrated in female Fisher 344 rats, fed diets containing 1500 or 3000 mg/kg 4,4′-thiodianiline. Adenocarcinomas of the uterus were increased in a statistically significant manner at both dose levels (178,179).

There are no epidemiological data for possible carcinogenicity of this substance in humans.

Uterine carcinogenicity was shown to occur only in the mouse with some substances studied in the National Toxicology Program. To this group belong the industrial chemicals bromoethane, cloroethane, ethylene oxide, glycidol, 1,2,3-trichloropropane, and trimethylphosphate (173).

Bromoethane (ethyl bromide) has been detected as a natural compound in the air and in oceanic waters, presumably as a result of emission by various species of macroalgae. It also occurs in manufacturing units as an ethylating agent, as a refrigerant, and as an organic solvent. Unlike chloroethane, it has a very limited use as a local anesthetic (180). Chemically, it is a volatile liquid with ethereal odor and taste (70).

There are no data for possible carcinogenicity of bromoethane in humans. The compound has been tested for carcinogenicity in mice and rats (181). When given by inhalation to female mice at 100, 200, or 400 ppm of 98% pure bromoethane, uterine neoplasms developed. Histologically, female animals had adenomas and squamous cell carcinomas of the uterine endometrium. Overall, the incidence of neoplasms was significantly higher in exposed animals than in controls, and it was also dose-dependent (182). However, in an earlier study, intraperitoneal administration of bromoethane to mice failed to yield uterine neoplasms (183).

Rats treated with bromoethane did not develop tumors of the uterus, either when the compound was given subcutaneously or when it was given by inhalation (182,184). Bromoethane was reported to cause toxic effects in a subacute treatment (14 weeks) and at a rather high concentration (1600 ppm). Among the others, testicular atrophy was monitored in rats as was minimal atrophy of the endometrium in rats and mice (182).

Chloroethane (ethyl chloride) is a synthetic compound with a large variety of industrial uses. This compound has a wide application as a blowing agent in foamed plastics. At room temperature, it behaves as a gas, with a characteristic ethereal odor and taste. Exposure to this substance occurs mainly in the working environment. However, it can be considered as a common pollutant in the air and the water (181). It is also used as a local anesthetic because of its rapid cooling effect as it vaporizes (164).

Female mice exposed to chloroethane by inhalation (15000 ppm of 99.5% pure compound) developed uterine adenocarcinomas at a significantly higher rate than controls. In many of these animals, the cancer metastasized to a variety of organs (185). In rats, similar experimental conditions failed to show tumors of uterus or cervix, although carcinogenic effects were evident in other sites (185). There are no data for possible carcinogenicity of chloroethane in humans.

Mice exposed for 3 weeks to bromoethane and chloroethane, at concentrations similar to those known to produce uterine carcinoma, did not show any change in circulating sex hormones that would predispose to the development of endometrial cancer. In accordance, the animals maintained appropriate estrous cyclicity (186). Hormonal imbalance may ensue later during the chronic treatment with these two haloethanes. However, other mechanisms cannot be excluded, such as a direct action of these substances or their metabolites on the uterus.

Chloroethane and bromoethane are mutagenic in the Ames test, both with and without metabolic activation, but data for genotoxicity in other in vitro systems are rather weak (181). Experiments with in vivo tests showed no genotoxicity with chloroethane, and only scant information is available for bromoethane (182,185,187,188). Chloroethane and bromoethane have not been reported as carcinogenic in humans. However, cohort studies with occupational exposure to related halogenated hydrocarbons, such as the dry-cleaning solvents perchloroethylene (189), and trichloroethylene, tetrachloroethylene, or 1,1,1-trichloroethane, revealed an excess of cervical cancers (190). It is worth adding that 1,1,1-trichloroethane has been abandoned in most countries on account of its harmful environmental impact. In contrast, tetrachloroethylene has been used widely in dry cleaning facilities, in the metal industry, and in shoe factories. This chemical has been associated with reduced fertility in female workers (191), an effect that has not been examined yet in terms of possible hormonal disturbances.

Ethylene oxide is an industrial chemical used in manufacturing ethylene glycol, acrylonitrile, and nonionic surfactants. It also has applications as a fumigant for sterilizing surgical instruments, foodstuffs, and textiles. In agriculture, ethylene oxide is used as a fungicide (31,192).

There is sufficient evidence in both mice and rats for the carcinogenicity of ethylene oxide after administration by inhalation. However, only in mice did this substance produce adenocarcinoma of the uterus in a statistically significant way (193). Similarly, in humans there are epidemiological studies implicating ethylene oxide in the production of various neoplasms. The data are still inconclusive, because they refer to industry workers with confounding exposures. In any case, existing evidence for carcinogenicity in humans is in organs other than the uterus (31,192).

Experimental studies have shown that 1,2,3-trichloropropane, another polychlorinated hydrocarbon, can cause uterine cancer in mice (173). 1,2,3-Trichloropropane is a poisonous liquid that is used as a paint and varnish remover, as a solvent, as a decreasing agent, and as an extractant for resins, oils, fats, waxes, and chlorinated rubber. In organic synthesis, it is used for the manufacture of chemicals and as a copolymer. There are no epidemiological data for possible carcinogenicity of this substance to humans.

2,3-Epoxy-1-propanol (glycidol) is a liquid irritant to the skin used as a stabilizer for natural oils and as a demulsifier, especially in the manufacture of dyes. It has also a wide variety of applications in the chemical industry, such as the synthesis of glycerol, of glycidyl ethers, of esters and of amines. Glycidol is a carcinogen for the uterus of mice, but this substance was not found to exert a similar effect in rats (173). Moreover, no data are available for any possible carcinogenic effect of glycidol in humans.

Trimethylphosphate is a well-known experimental alkylating agent that has been found to interfere with the biosynthesis and biodynamics of steroid hormones in mice. In particular, trimethylphosphate seems to affect the biosynthesis and the function of testosterone, since it reduces the levels of the hormone in the plasma and the testes (194). Trimethylphosphate has been demonstrated to produce cancer in the uterus of mice (173).

2. *Tobacco Smoking*

The importance of estrogens in the pathogenesis of uterine cancer is also illustrated clearly in the case of tobacco smoking. It has been well established that smoking reduces the risk of endometrial cancer (195). This is an effect attributed to a reduction in the level of estrogens in the blood of smoking women. Similarly, smoking slightly lowers the age of menopause (196). These findings are probably related to the central actions of nicotine. It should be mentioned, however, that cigarette smoke contains polycyclic aromatic hydrocarbons, which have reportedly reduced the number of estrogen receptors in various organs (197).

Among substances inhaled with cigarette smoking, nicotine and cotinine have been identified in the cervical mucus of smokers, where they may exert a direct local action (198). Our knowledge of the possible mechanism of carcinogenicity of tobacco smoking is far from conclusive. It has been suggested that nicotine may favor abnormal cell proliferation by interfering with the function of cellular growth factors, as demonstrated in cultures of a cervical cell line (199). It should be mentioned that several cohort studies have shown higher rates of cancer of cervix uteri in smokers than in nonsmokers (200). However, the possibility of confounding with other factors exists, so no definite conclusion has been reached on this point.

IV. CONCLUSIONS

Various chemicals seem to be involved in the pathogenesis of cancer of the cervix uteri and uterus. In many instances, an assumed hormonal imbalance is the most probable causative factor.

Especially for cervical cancer, there is substantial evidence that contagious agents may be involved, such as herpes simplex virus and human papillomaviruses. Even so, hormones provide a facultative support in the progress of the neoplasm, as has been shown in the case of progestins and the human papillomavirus. Long-term use of combined oral contraceptives seems to be associated with the development of cervical cancer, even when confounding factors, such as age and sexual practices of the patients, are taken into account. An increased understanding of the etiology of cervical cancer may eventually lead us onto the right track toward the elimination of this neoplasia. If an infection indeed plays a role in the natural history of invasive cervical cancer, the development of a safe vaccine could prevent the spread of the disease.

In the case of endometrial cancer, estrogens undoubtedly play an important role, especially when their action is unopposed by the presence of progesterone or synthetic progestins. The epidemiological evidence is highly consistent in showing that administration of "unopposed estrogens" increases the risk for endometrial cancer. Under normal conditions, the direct or indirect actions of estrogens are eliminated by the presence of progesterone. Administration of estrogen replacement therapy to postmenopausal women, who have negligible production of endogenous progesterone, continuously stimulates endometrial cells. The combined oral contraceptive pill causes an atrophic endometrium with almost no cell division, due to the simultaneous administration of an estrogen and a progestin. Sequential oral contraceptives have been used in the past, but they were abandoned when it was found that they cause a considerable endometrial hyperplasia and possibly also endometrial cancer. Similarly, postmenopausal hormone replacement therapy with an estrogen alone or with an estrogen and a progestin caused endometrial cell division in proportion to the time during which estrogen is given without progestin opposition.

Two mechanisms have been proposed to explain the effect of "unopposed estrogens." The first hypothesis is that the risk for endometrial cancer is proportional to the mitotic rate of the endometrial cells and that the role of progestins is to shorten exposure to estrogens, so that risk will increase with all patterns of exposure to an estrogen unopposed by a progestin. The second proposed mechanism is that the observed protective effect of progestins (relative to the same estrogen exposure without a progestin) is at least partly due to prevention of endometrial hyperplasia and to shedding of the outer layer of the endometrium, which occurs on cessation of exposure to an estrogen and a progestin together and might cause loss of potentially cancerous cells.

For many carcinogenic chemical substances, no data are available that can give an adequate explanation for the underlying tumorigenic process.

On the other hand, the endocrinic dimension in the progress of these neoplasms may be very subtle and may involve species-specific mechanisms that are difficult to interpret. A typical example is the case of bromocriptine, which may cause uterine cancer only in old female rats of certain strains. Moreover, even when a hormonal effect is assumed as the most plausible mechanism in tumor induction, the real succession of events may be quite different. The transplacental effect of diethylstilbestrol in humans, which leads to the development of vaginal and cervical adenocarcinoma later on during puberty, illustrates the complexity of hormonal carcinogenesis.

The toxicological testing of drugs and industrial chemicals has revealed that many substances may have the cervix and corpus uteri as targets for carcinogenesis. These routine toxicological assays do not usually provide enough in-depth analysis of the underlying mechanism to understand the development of the neoplasms. It is apparent that a direct or indirect hormonal effect must be implicated only in a limited number of instances. Things become even more complicated when the assumed toxic effect is of only marginal magnitude or limited evidence, as is the case with the inconclusive epidemiological data on cervical cancer of women who smoke or who are exposed to organic solvents in their working environment.

The inconsistency of data on chemical carcinogenicity of cervix uteri and uterus between humans and experimental animals is a fact that imposes further difficulties in terms of interpreting toxic mechanisms, predicting relative risks and taking preventive measures for public health. Particular attention should be paid to the differentiation of the published results between species of experimental animals as well as between experimental animals and humans. Several substances induce cancer in the cervix or uterus of rats without affecting mice, and vice versa. As can be seen in Tables 1–3, this is the rule rather than the exception. Also, of the 30 substances or mixtures of substances described as carcinogenic in the literature, only 14 were found to be carcinogenic in mice, and only 7 in rats.

Taking into account the data on hormones or drugs with a hormonal action (Table 1), it is evident that mice match humans better than do rats. It is worth noticing that rats are even resistant to protocols using first a known carcinogen (an initiator) and then an estrogen as the promoter—the "two-stage model" of chemical carcinogenesis. Therefore, studies of carcinogenicity of the cervix and uterus are more reliable when done with mice, because these animals seem more sensitive to the development of this type of neoplasm than are rats. Thus, results obtained with mice are more readily extrapolated to humans. However, since mice and rats respond to carcinogens in a complementary way, it is still advisable that both species be used for experimental studies of uterine and cervical cancer.

REFERENCES

1. Tomatis L, Aitio A, Day NE, Heseltine E, Kaldor J, Miller AB, Parkin DM, Riboli E, eds. Cancer: Causes, Occurence and Control. IARC Scientific Publications No. 100. Lyon, IARC/WHO, 1990.
2. Poletto L, Morini JC. Cancer mortality and some socioeconomic correlates in Rosario, Argentina. Cancer Lett 1990; 49:201–206.
3. Leck I, Sibary K, Wakefield J. Incidence of cervical cancer by marital status. J Epidemiol Commun Health 1978; 32:108–110.
4. Beral V. Cancer of the cervix: a sexually transmitted infection? Lancet 1974; i:1037–1040.
5. Skegg DCG, Corwin PA, Paul C. Importance of the male factor in cancer of the cervix. Lancet 1982; ii:581–583.
6. Austin DF, Roe KM. The decreasing incidence of endometrial cancer: public health implications. Am J Public Health 1982; 72:65–68.
7. Robbins Sh, Cotran RS, Kumar V. Pathologic Basis of Disease. Female Genital Tract. 3rd ed. Philadelphia: Igaku-Shoin/Saunders Int Ed, 1984: 1109–1164.
8. Hendrickson MR, Kempson RL. Obstetrical and gynaecological pathology. In: Fox H, ed. Haines and Taylor-Obstetrical and Gynaecological Pathology. Churchill Livingstone, 1987:354–404.
9. Rein MS, Barbieri RL, Friedman AJ. Progesterone: a critical role in the pathogenesis of uterine myomas. Am J Obstet Gynecol 1995; 172:14–18.
10. Bornstein J, Rahat MA, Abramovici H. Etiology of Cervical Cancer: Current Conceps. Obstet Gynecol Survey 1995; 50:146–154.
11. Crook T, Storey A, Almond N, Osborn K, Crawford L. Human papillomavirus type 16 cooperates with activated ras and fos oncogenes in the hormone-dependent transformation of pmarymouse cells. Proc Natl Acad Sci USA 1990; 85:8820–8824.
12. Key TJA, Beral V. Sex hormones and cancer. In: Vainio H, Magee P, McGregor D, McMichaei AJ, eds. Mechanisms of Carcinogenesis in Risk Identification. IARC Scientific Publications No. 116. Lyon: IARC, 1992.
13. Beral V, Hannaford P, Kay C. Oral contraceptive use and malignancies of the genital tract. Results from the Royal College of general Practitioner's Oral Contraception Study. Lancet 1988; ii:1331–1335.
14. Brinton LA, Reeves WC, Brenes MM, Herrero R, de Britton RC, Gaitan E, Tenorio F, Garcia M, Rawls WE. Oral contraceptive use and risk of invasive cervical cancer. Int J Epidemiol 1990; 19:4–11.
15. Herrero R, Brinton LA, Reeves WC, Brenes MM, de Britton RC, Tenorio F, Gaitan E. Injectable contraceptives and risk of invasive cervical cancer: evidence for an association. Int J Cancer 1990; 46:5–7.
16. Thomas DB, Molina R, Cuevas HR, Ray Riotton G, Dabancens A, Benavides S, Martinez L, Salas O, Pallet JA, Lopez J. Monthly injectable steroid contraceptives and cervical carcinoma. Am J Epidemiol 1989; 130:237–247.
17. MacDonald PC, Edman CD, Hemsell DL, Porter JC, Siiteri PK. Effect of obesity on conversion of plasma and rostenedione to estrone in postmeno-

pausal women with and without endometrial cancer. Am J Obstet Gynecol 1978; 130:448–455.
18. Edman CD, MacDonald PC. Effect of obesity on conversion of plasma androstenedione to estrone in ovulatory and anovulatory young women. Am J Obstet Gynecol 1978; 130:456–461.
19. Weiss NS. Epidemiology of endometrial cancer: a review of hormonal and non-hormonal risk factors. In: Forastiere AA, ed. Gynecologic Cancer. New York: Churchill Livingstone, 1984:199–214.
20. Hsueh AJW, Peck EJ Jr, Clark JH. Progesterone antagonism of the oestrogen receptor and oestrogen-induced uterine growth. Nature 1975; 254:337–339.
21. Novak ER, Woodruff JD. Novak's Gynecological and Obstetric Pathology with Clinical and Endocrine Relations. Philadelphia: WB Saunders, 1979: 171–184.
22. Kokko E, Jänne O, Kauppila A, Vihko R. Effects of tamoxifen, medroxyprogesterone acetate, and their combination on human endometrial estrogen and progestin receptor concentrations. Am J Obstet Gynecol 1982; 143:382–388.
23. Kohorn EI. Gestagens and endometrial carcinoma. Gynecol Oncol 1976; 4: 398–411.
24. Tseng I, Gurpide E. Induction of human endometrial estradiol dehydrogenase by progestins. Endocrinology 1975; 97:825–833.
25. Deerberg F, Kaspareit J. Endometrial carcinoma in BDIIHan rats: model for a spontaneous hormone-dependent tumor. J Natl Cancer Inst 1987; 78:1245–1251.
26. Deerberg F, Pohlmeyer G, Lörcher K, Petrov V. Total suppression of spotaneous endometrial carcinoma in BDII/Han rats by melegestrol acetate. Oncology 1995; 52:319–325.
27. Key TJA, Pike MC. The dose-effect relationship between "unopposed" estrogens and endometrial mitotic rate: its central role in explaining and predicting endometrial cancer risk. Br J Cancer 1988; 57:205–212.
28. Siiteri PK, Schwarz BE, MacDonald PC. Estrogen receptors and the estrone hypothesis in relation to endometrial and breast cancer. Gynecol Oncol 1974; 2:228–238.
29. IARC. IARC Monographs on the Evaluation of the Carcinogenic Risks of Chemicals to Humans. Sex Hormones. Vol. 6. Lyon: WHO/IARC, 1974.
30. IARC. IARC Monographs on the Evaluation of the Carcinogenic Risks of Chemicals to Humans. Sex Hormones (II). Vol. 21. Lyon: WHO/IARC, 1979.
31. IARC. IARC Monographs on the Evaluation of the Carcinogenic Risks of Chemical to Humans. Overall Evaluation of Carcinogenicity: An Updating of IARC Monographs Volumes 1 to 42. Lyon: WHO/IARC, 1987.
32. Paganini-Hill A, Ross RK, Henderson BE. Endometrial cancer and patterns of use of oestrogen replacement therapy: a cohort study. Br J Cancer 1989; 59:445–447.
33. Persson I, Adami H-O, Bergkvist L, Lindgren A, Pettersson B, Hoover R,

Schairer C. Risk of endometrial cancer after treatment with oestrogens alone or in conjuction with progestogens: results of a prospective study. Br Med J 1989; 298:147–151.

34. Marselos M, Vainio H. Carcinogenic properties of pharmaceutical agents evaluated in the IARC monographs programme. Carcinogenesis 1991; 12: 1751–1766.
35. Shapiro S, Kelly JP, Rosenberg L, Kaufman DW, Helmrich SP, Rosenheim NB, Lewis JL, Jr, Knapp RC, Stolley PD, Schottenfeld D. Risk of localized and widespread endometrial cancer in relation to recent and discontinued use of conjugated estrogens. N Engl J Med 1985; 313:969–972.
36. Buring JE, Bain CJ, Ehrmann RL. Conjugated estrogen use and risk of endometrial cancer. Am J Epidemiol 1986; 124:434–441.
37. Horwitz RI, Feinstein AR. Estrogens and endometrial cancer. Responses to arguments and current status of an epidemiologic controversy. Am J Med 1986; 81:503–507.
38. IARC. Mechanisms of Carcinogenesis in Risk Identification. A Consensus Report of an IARC Monographs Working Group. IARC Internal Technical Report No 91/002. Lyon, June 11–18, 1991.
39. Henderson BE, Ross R, Bernstein L. Estrogens as a cause of human cancer. The Richard and Hinde Rosental foundation award lecture. Cancer Res 1988; 48:246–253.
40. Persky V, Davis F, Barrett R, Ruby E, Sailer C, Levy P. Recent time trends in uterine cancer. Am J Publ Health 1990; 80:935–939.
41. Studd JWW, Thom MH, Paterson MEL, Wade-Evans T. The prevention and treatment of endometrial pathology in post-menopausal women receiving exogenous estrogens. In: Pasetto N, Paoletti R, Ambrus JL, eds. The Menopause and Postmenopause. Lancaster: MTP Press, 1980:127–139.
42. Schlesselman JJ. Oral contraceptives in relation to the breast and reproductive tract—an epidemiological review. Br J Fam Plann 1989; 15:23–33.
43. Schlesselman JJ. Cancer of the breast and reproductive tract in relation to use of oral contraceptives. Contraception 1989; 40:1–38.
44. Weiss NS, Sayvetz TA. Incidence of endometrial cancer in relation to the use of oral contraceptives. N Engl J Med 1980; 302:551–554.
45. Henderson BE, Casagrande JT, Pike MC, Mack T, Rosario I, Duke A. The epidemiology of endometrial cancer in young women. Br J Cancer 1983; 47: 749–756.
46. Allen E, Gardner WU. Cancer of the cervix of the uterus in hybrid mice following long-continued edministration of estrogen. Cancer Res 1941; 1: 359–366.
47. Pan SC, Gardner WU. Carcinomas of the uterine cervix and vagina in estrogen- and androgen-treated hybrid mice. Cancer Res 1948; 8:337–341.
48. Gardner WU, Ferrigno M. Unusual neoplastic lesions of the uterine horns of estrogen-treated mice. J Natl Cancer Inst 1956; 17:601–613.
49. Gardner WU. Carcinoma of the uterine cervix and upper vagina: induction under experimental conditions in mice. Ann NY Acad Sci 1959; 75:543–564.

50. Munoz N. Effect of herpes virus type 2 and hormonal imbalance on the uterine cervix of the mouse. Cancer Res 1973; 33:1504–1508.
51. Highman B, Greenman DL, Norvell MJ, Farmer J, Shellenberger TE. Neoplastic and preneoplastic lesions induced in female C3H mice by diets containing diethylstilbestrol or 17beta-estradiol. J Environ Pathol Toxicol 1980; 4:81–95.
52. McKinney GR, Weikel JH Jr, Webb WK, Dick RG. Use of the life-table technique to estimate effects of certain steroids on probability of tumor formation in a long-term study in rats. Toxicol Appl Pharmacol 1968; 12:68–79.
53. Rudali G. Induction of tumors in mice with synthetic sex hormones. Gann Monogr 1975; 17:243–252.
54. Dunn TB, Green AW. Cysts of the epididymis, cancer of the cervix, glandular cell myoblastoma and other lesions after estrogen injection in newborn mice. J Natl Cancer Inst 1963; 31:425–455.
55. Koutsilieris M. Pathophysiology of uterine leiomyomas. Biochem Cell Biol 1992; 70:273–278.
56. Sadovsky Y, Kushner PJ, Roberts JM, Riemer RK. Restoration of estrogen-dependent progesterone expression in a uterine myocyte cell line. Endocrinology 1993; 132:1609–1613.
57. van Nie R, Benedetti EL, Mühlbock O. A carcinogenic action of testosterone, provoking uterine tumors in mice. Nature 1961; 192:1303.
58. Meissner WA, Sommers SC, Sherman G. Endometrial hyperplasia, endometrial carcinoma, and endometriosis produced experimentally by estrogen. Cancer 1957; 10:500–509.
59. Griffiths CT, Tomic M, Craig JM, Kistner RW. Effect of progestins, estrogens and castration on induced endometrial carcinoma in the rabbit. Surg Forum 1963; 14:399–401.
60. Newbold RR, Buclock BC, McLachlan JA. Uterine adenocarcinoma in mice following developmental treatment with estrogens: a model for hormonal carcinogenesis. Cancer Res 1990; 50:7677–7681.
61. Pitot HC. Fundamentals of Oncology. 3rd ed. New York: Marcel Dekker, 1986.
62. Turusov VS, Raikhlin NT, Smirnova EA, Trukhanova LS. Uterine sarcomas in CBA mice induced by combined treatment with 1,2-dimethylhydrazine and estradiol diproprionate—light and electron microscopy. Exp Toxicol Pathol 1993; 45:161–166.
63. Turusov VS, Trukhanova LS, Parfenov YuD. Modifying effect of ascorbic acid and sodium ascorbate on the promoting stage of uterine sarcomogenesis induced in CBA mice by 1,2-dimethylhydrazine and estradiol-dipropionate. Cancer Lett 1991; 56:29–35.
64. Niwa K, Tanaka T, Mori H, Yokoyama Y, Furui T, Mori H, Tamaya T. Rapid induction of endometrial carcinoma in ICR mice treated with N-methyl-N-nitrosourea and 17-beta-estradiol. Jpn J Cancer Res 1991; 82:1391–1396.

65. Nagaoka T, Takeuchi M, Onodera H, Mitsumori K, Lu J, Maekawa A. Experimental induction of uterine adenocarcinoma in rats by estrogen and N-methyl-N-nitrosourea. In Vivo 1993; 7:525–530.
66. Maekawa A. Neoplasms of the female reproductive organs. In: Stinson SF, Schuller HM, Reznik G, eds. Atlas of Tumor Pathology of the Fisher Rat. Boca Raton, FL: CRC Press, 1990:437–439.
67. Rao GN, Haseman JK, Grumbein S, Crawford DD, Eustis SL. Growth, body weight, survival and tumor trends in F344/N rats during an eleven-year period. Toxicol Pathol 1990; 18:61–70.
68. Nagaoka T, Onodera H, Matsuuushima Y, Todate A, Shibutani M, Ogasawara H, Maekawa A. Spontaneous uterine adenocarcinomas in aged rats and their relation to endocrine imbalance. J Cancer Res Clin Oncol 1990; 116: 623–628.
69. Akhtar M, Kim PY, Young I. Ultrastructure of endometrial stomal sarcoma. Cancer 1975; 35:406–412.
70. Budavari S, ed. The Merck Index. 12th ed. Rahway, NJ: Merck Co., 1995.
71. Dodds EC. Oestrus producing hormones. Br Med J 1934; 2:1187–1189.
72. Li JJ, Li SA. Estrogen-induced tumorigenesis in hamsters: roles for hormonal and carcinogenic activities. Arch Toxicol 1984; 55:110–118.
73. Liehr JG, Ballatore AM, Dague BB, Ulubelen AA. Carcinogenicity and metabolic activation of hexestrol. Chem-Biol Interact 1985; 55:157–176.
74. Marselos M, Tomatis L. Diethylstilbestrol: I, pharmacology, toxicology and carcinogenicity in humans. Eur J Cancer 1992; 28A:1182–1189.
75. Marselos M, Tomatis L. Diethylstilboestrol: II, Pharmacology, toxicology and carcinogenicity in experimental animals. Eur J Cancer 1993; 29A:149–155.
76. Herbst AL, Scully RE. Adenocarcinoma of the vagina in adolescence. A report of 7 cases including 6 clear-cell carcinomas (so-called mesonephromas). Cancer 1970; 25:745–757.
77. Herbst AL, Ulfelder H, Poskanzer DC. Adenocarcinoma of the vagina. Association of maternal stilbestrol therapy with tumor appearance in young women. N Engl J Med 1971; 284:878–885.
78. Greenwald P, Barlows JJ, Nasca PC, Burnett WS. Vaginal cancer after maternal treatment with synthetic estrogens N Engl J Med 1971; 285:390–392.
79. Fawcett KJ, Dockerty MB, Hunt AB. Mesonephric carcinoma of the cervix uteri: a clinical and pathologic study. Am J Obstet Gynecol 1966; 95:1068–1079.
80. Hammed K. Clear cell "mesonephric" carcinoma of uterine cervix. Obstet Gynecol 1966; 32:564–575.
81. Vessey MP. Epidemiological studies of the effects of diethylstilboestrol. In: Napalkov NP, Rice JM, Tomatis L, Yamasaki H, eds. Perinatal and Multigeneration Carcinogenesis. Lyon: Lyon International Agency for Research on Cancer, 1989:335–348.
82. McClure HM, Graham CE. Malignant uterine mesotheliomas in squirrel

monkeys following diethylstilbestrol administration. Lab Anim Sci 1973; 23: 493–498.
83. McLachlen JA. Prenatal exposure to diethylstilbestrol in mice. Toxicological studies. J Toxicol Environ Health 1977; 2:527–537.
84. McLachlan JA, Newbold RR, Bullock BC. Long-term effects on the female mouse genital tract associated with prenatal exposure to diethylstilbestrol. Cancer Res 1980; 40:3988–3999.
85. Newbold RR, McLachlan JA. Vaginal adenosis and adenocarcinoma in mice exposed prenatally or neonatally to diethylstilbestrol. Cancer Res 1982; 42: 2003–2011.
86. Walker BE. Uterine tumors in old female mice exposed prenatally to diethylstilbestrol. J Natl Cancer Inst 1983; 70:477–484.
87. Rustia M, Shubik P. Transplacental effects of diethylstilbestrol on the genital tract of hamster offspring. Cancer Lett 1976; 1:139–146.
88. Rustia M. Role of hormone imbalance in transplacental carcinogenesis induced in Syrian golden hamsters by sex hormones. Natl Cancer Inst Monogr 1979; 51:77–87.
89. Walker BE. Tumors of female offspring of mice exposed prenatally to diethylstilbestrol. J Natl Cancer Inst 1984; 73:133–140.
90. Walker BE. Animal models of prenatal exposure to diethylstilbestrol. In: Napalkov NP, Rice JM, Tomatis L, Yamasaki H, eds. Perinatal and Multigeneration Carcinogenesis. Lyon: IARC Scientific Publications, 1989:349–364.
91. Tomatis L, Narod S, Yamasaki H. Transgeneration transmission of carcinogenic risk. Carcinogenesis 1992; 13:145–151.
92. Walker BE, Kurth LA. Multi-generation carcinogenesis from DES investigated by blastocyst transfer in mice. Int J Cancer 1995; 61:249–252.
93. Turusov VS, Trukhanova LS, Parfenov YuD, Tomatis L. Occurence of tumours in the descendants of CBA male mice prenatally treated with diethylstilbestrol. Int J Cancer 1992; 49:131–135.
94. Harper MJK, Walpole AL. A new derivative of triphenylethylene effect on implantation and mode of action in rats. J Reprod Fertil 1967; 13:101–119.
95. Cole MP, Jones CT, Todd IDH. A new antioestrogenic agent in late breast cancer. An early clinical appraisal of ICI 46, 474. Br J Cancer 1971; 25:270–275.
96. O'Halloran MJ, Maddock PG. ICI 46, 474 in breast cancer. J Irish Med Assoc 1974; 67:38–39.
97. Ingle JN, Ahmann DL, Green SJ, Edmonson JH, Bisel HF, Kools LK, Nichols WC, Creagon ET, Hahn RG, Rubin J, Frytak S. Randomized clinical trial of diethylstibestrol versus tamoxifen in postmenopausal women with advanced breast cancer. N Engl J Med 1981; 304:16–21.
98. Glick JH, Gleber RD, Goldhirsch A, Senn HJ. Meeting highlights. Adjuvant therapy for primary breast cancer. J Natl Cancer Inst 84; 1992:1479–1485.
99. Early Breast Cancer Trialists' Collaborative Group. Effects of adjuvant tamoxifen and of cytotoxic therapy on mortality in early breast cancer: an

overview of 61 randomized trials among 28,896 women. N Engl J Med 1988; 319:1681-1692.
100. Early Breast Cancer Trialists' Collaborative Group. Systemic treatment of early breast cancer hormonal, cytotoxic or immune therapy: 133 randomized trials involving 31,000 recurrences and 24,000 death among 75,000 women. Lancet 1992; i:1-15, 71-85.
101. Consensus Conference (United States NIH). Consensus Conference, Adjuvant Chemotherapy for Breast Cancer. J Am Med Assoc 1985; 254:3461-3463.
102. Jaiyesimi IA, Buzdar AU, Sahin AA, Ross MA. Carcinoma of the male breast. Ann Intern Med 1992; 117:771-777.
103. Lentz SS. Advanced and recurrent endometrial carcinoma: hormonal therapy. Sem Oncol 1994; 21:100-1006.
104. Martinez-Cerezo FJ, Tomas A, Donoso L, Enriquez J, Guarnar C, Balanzo J, Nogueras AM, Vilardi F. Controlled trial of tamoxifen in patients with advanced hepatocellular carcinoma. J Hepatol 1994; 20:702-706.
105. Harrison JD, Morris DL, Ellis IO, Jones JA, Jackson I. The effect of tamoxifen and estrogen receptor status on survival in gastric carcinoma. Cancer 1989; 64:1007-1010.
106. Yagoda A, Abi-Rached B, Petrylak D. Chemotherapy for advanced renal-cell carcinoma: 1983-1993. Sem Oncol 1990; 22:42-60.
107. McClay EF, McClay MET. Tamoxifen: Is it useful in the treatment of patients with metastatic melanoma? J Clin Oncol 1994; 12:617-626.
108. Wong A, Chan A. Survival benefit of tamoxifen therapy in adenocarcinoma of pancreas. Cancer 1993; 71:2200-2203.
109. Vargas Rorg LM, Loft H, Olcese JE, Castro GL, Ciocca DR. Effects of short-term tamoxifen administration in patients with invasive cervical carcinoma. Anticancer Res 1993; 13:2457-2464.
110. Goodwin JW, Crowley J, Eyre HJ, Stafford B, Jaeckle KA, Townsend J. A phase II evaluation of tamoxifen in unresectable or refractory meningiomas: a Southwest Oncology Group Study. J Neurol Oncol 1993; 15:75-77.
111. Rutqvist LE, Mattson A. Cardiac and thromboembolic morbidity among postmenopausal women with early-stage breast cancer in a randomized trial of adjuvant tamoxifen. J Natl Cancer Inst 1993; 85:1398-1406.
112. Jordan VC, Allen KE, Dix CJ. Pharmacology of tamoxifen in laboratory animals. Cancer Treat Rep 1980; 64:745-759.
113. Neven P, de Muylder X, van Belle Y, Vanderick G, de Muylder E. Tamoxifen and the uterus and endometrium. Lancet 1989; i:375.
114. Gottardis MM, Robinson SP, Satyaswaroop PG, Jordan VC. Contrasting actions of tamoxifen on endometrial and breast tumor growth in the athymic mouse. Cancer Res 1988; 48:812-815.
115. Hemminki K, Widlak P, Hou SM. DNA adducts caused by tamoxifen and toremifene in human microsomal system and lymphocytes in vitro. Carcinogenesis 1995; 16:1661-1664.
116. Greaves P, Goonrtillebe R, Nunn G, Topham J, Orton T. Two year carcino-

genicity study of tamoxifen in Alderley Park Wistar derived rats. Cancer Res 1993; 53:3919–3924.
117. Jordan VC. Tamoxifen: toxicities and drug resistance during the treatment and prevention of breast cancer. Ann Rev Toxicol 1995; 35:195–211.
118. Tucker MJ, Adam HK, Patterson JS. Tamoxifen. In: Laurence DR, McLean AEM, Weatherall M, eds. Safety Testing of New Drugs, New York: Academic Press 1984:125–162.
119. Sengupta A, Dutta S, Mallick R. Modulation of cervical carcinogenesis by tamoxifen in a mouse model system. Oncology 1991; 48:258–261.
120. Hard GC, Iatropoulos MJ, Jordan K, Radi L, Kaltenberg OP, Imondi AR, Williams GM. Major difference in the hepatocarcinogenicity and DNA adduct forming ability between toremifene and tamoxifen in female Crl: CD(BR) rats. Cancer Res 1993; 53:4534–4541.
121. Hirsimäki P, Hirsimäki Y, Nieminen L, Payne BJ. Tamoxifen induces hepatocellular carcinoma in rat liver: a 1-year study with two antiestrogens. Arch Toxicol 1993; 67:49–54.
122. Williams GM, Iatropoulos MJ, Djordjevic MV, Kaltenberg OP. The triphenylethylene drug tamoxifen is a strong liver carcinogen in the rat. Carcinogenesis 1993; 14:315–217.
123. Ahotupa M, Hirsimäki P, Parssinen R, Mäntylä E. Alterations of drug metabolizing and antioxidant enzyme activities during tamoxifen-induced hepatocarcinogenesis in the rat. Carcinogenesis 15; 1994:863–868.
124. Carthew P, Rich KJ, Martin A, De Matteis F, Lim CK, Manson MM, Festing FW, White NH, Smith LL. DNA damage as assessed by ^{32}P-postlabelling in three rat strains exposed to dietary tamoxifen: the relationship between cell proliferation and liver tumour formation. Carcinogenesis 1995; 16:1299–1304.
125. Mäntylä ETE, Karlsson SH, Nieminen LS. Induction of endometrial cancer by tamoxifen in the rat. In: Li JJ, Li SA, Gustaffson J-A, Nandi S, Sekely LI, eds. Hormonal Carcinogenesis II. Proceedings of the Second International Symposium. New York: Springer-Verlag, 1991:442–445.
126. Mishkin SY, Farber E, Ho R, Mulay S, Mishkin S. Tamoxifen alone or in combination with estradiol-17b inhibits the growth and malignant transformation of hepatic hyperplastic nodules. Eur J Cancer Clin Oncol 1985; 21: 615–623.
127. Yager JD, Roebuck BD, Paluszcyk TL, Memoli VA. Effects of ethinyl estradiol and tamoxifen on liver DNA turnover and new synthesis and appearance of gamma glutamyl transpeptidase-positive foci in female rats. Carcinogenesis 1986; 7:2007–2014.
128. Ghia M, Mereto E. Induction and promotion of Á-glutamyltranspeptidase-positive foci in the liver of female rats treated with ethinyl estradiol, clomiphene, tamoxifen and their associations. Cancer Lett 1989; 46:195–202.
129. Dragan YP, Rizvi T, Xu YH, Hully JR, Bawa N, Campbell HA, Maronpot RR, Pitot HC. An initiation-promotion assay in rat liver as a potential com-

plement to the 2-year carcinogenesis bioassay. Fundam Appl Toxicol 1991; 16:525–547.
130. Dragan YP, Xu YD, Pitot HC. Tumor promotion as a target for estrogen/antiestrogen effects in rat hepatocarcinogenesis. Prev Med 1991; 20:15–26.
131. Servais P, Galand P. Increased yield in GST-P-positive liver preneoplastic foci induced by DENA or ENU in rats pre-treated with estradiol or tamoxifen. Int J Cancer 1993; 54:996–1001.
132. Dragan YP, Vaughan J, Jordan VC, Pitot HC. Comparison of the effects of tamoxifen and toremifene on liver and kidney tumor promotion in female rats. Carcinogenesis 1995; 16:2733–2741.
133. Coe JE, Ishak KG, Ward JM, Ross MJ. Tamoxifen prevents induction of hepatic neoplasia by zeranol, an estrogenic food contaminant. Proc Natl Acad Sci 1992; 89:1085–1089.
134. Zimniski SJ, Warren RC. Induction of tamoxifen-dependent rat mammary tumors. Cancer Res 1993; 53:2937–2939.
135. Killackey MA, Hakes TB, Pierce VK. Endometrial adenocarcinoma in breast cancer patients receiving antiestrogens. Cancer Treat Rep 1985; 69:237–238.
136. Segna RA, Dottino PR, Deligdisch L, Cohen CJ. Tamoxifen and endometrial cancer. Mt Sinai J Med 1992; 59:416–418.
137. Mathew A, Chabon AB, Kabakow B, Drucker M, Hirschman RJ. Endometrial carcinoma in five patients with breast cancer on tamoxifen therapy. NY State J Med 1990; 90:207–208.
138. Mignotte H, Sasco AJ, Lasset C, Saez S, Rivoire M, Bobin JY. Adjuvant therapy of breast cancer with tamoxifen and endometrial carcinoma. Bull Cancer 1992; 79:969–977.
139. Palacios A, Pertusa S, Montoya A, Martinez San Pedro R. Breast cancer, tamoxifen, and uterus [letter]. Med Clin Barc 1993; 100:479–479.
140. Jose R, Kekre AN, George SS, et al. Endometrial carcinoma in a tamoxifen treated breast cancer patient. Aust NZ J Obstet Gynaecol 1995; 53:201.
141. Rasmussen KI, Nielsen KM. Development of endometrial cancer during tamoxifen therapy. Ugeskr Laeger 153; 1991:2638–2638.
142. Dallenbach-Hellweg G, Hahn U. Mucinous and clear cell adenocarcinomas of the endometrium in patients receiving antiestrogens (tamoxifen) and gestagens. Int J Gynecol Pathol 1995; 14:7–15.
143. Atlante G, Pozzi M, Vincenzoni C, Vocaturo G. Four case reports presenting new acquisitions on the association between breast and endometrial carcinoma. Gynecol Oncol 1990; 37:378–380.
144. Rodier JF, Camus E, Janser JC, Renaud R, Rodier D. Tamoxifen and endometrial adenocarcinoma. Bull Cancer 1990; 77:1207–1210.
145. Spinelli G, Bardazzi N, Citernesi A, Fontanarosa M, Curiel P. Endometrial carcinoma in tamoxifen-treated breast cancer patients. J Chemother 1991; 3: 267–270.
146. Deprest J, Neven P, Ide P. An unusual type of endometrial cancer, related to tamoxifen? Eur J Obstet Gynecol Reprod Biol 1992; 46:147–150.
147. Andersson M, Storm HH, Mouridsen HT. Carcinogenic effect of adjuvant

tamoxifen treatment and radiotherapy for early breast cancer. Acta Oncol 1992; 31:259–263.
148. Nayfield SG, Karp JE, Ford LG, Dorr FA, Kramer BS. Potential role of tamoxifen in prevention of breast cancer. J Natl Cancer Inst 1991; 83:1450–1459.
149. Fornander T, Hellstrom A-C, Moberger B. Descriptive clinicopathologic study of 17 patients with endometrial cancer during or after adjuvant tamoxifen in early breast cancer. J Natl Cancer Inst 1993; 85:1850–1855.
150. Fisher B, Costantino JP, Redmond CK, Fisher ER, Wickerham DI, Cronin WM. Endometrial cancer in tamoxifen-treated breast cancer patients: findings from the national Surgical Adjuvant Breast and Bowel Project (NSABP)B-14. J Natl Cancer Inst 1994; 86:527–537.
151. Sasco AJ, Raffi F, Satge D, Godurdhun J, Fallouh B, Leduc B. Endometrial mullerian carcinosarcoma after cessation of tamoxifen therapy for breast cancer. Int J Gynecol Obstet 1995; 48:307–310.
152. Robinson DC, Bloss JD, Schiano MA. A retrospective study of tamoxifen and endometrial cancer in breast cancer patients. Gynecol Oncol 1995; 59: 186–190.
153. Rutqvist LE, Johansson H, Signomklao T, et al. Adjuvant tamoxifen therapy for early stage breast cancer and second primary malignancies. J Natl Cancer Inst 1995; 87:645–651.
154. Stewart HJ. The Scottish trial of adjuvant tamoxifen in node-negative breast cancer. Scottish Cancer Trials Breast Group. Monogr Natl Cancer Inst, 1992; 117–120.
155. Cummings FJ, Gray R, Tormey DC, Davis TE, Volk H, Harris J, Falkson G, Bennett JM. Adjuvant tamoxifen versus placebo in elderly women with nodepositive breast cancer: Long-term follow-up and causes of death. J Clin Oncol 1993; 11:29–35.
156. Cook LS, Weiss NS, Schwartz SM, White E, McKnight B, Moore DE, Daling JR. Population-based study of tamoxifen therapy and subsequent ovarian, endometrial, and breast cancers. J Natl Cancer Inst 1995; 87:1359–1364.
157. Andersson M, Storm HH, Mouridsen HT. Incidence of new primary cancers after adjuvant tamoxifen therapy and radiotherapy for early breast cancer. J Natl Cancer Inst 83; 1991:1013–1017.
158. Mühlemann K, Cook LS, Weiss NS, et al. The incidence of hepatocellular carcinoma in US white women with breast cancer after the introduction of tamoxifen in 1977. Breast Cancer Res Treat 1994; 30:201–204.
159. Neumann F. Early indicators for carcinogenesis in sex-hormone-sensitive organs. Mutat Res 1991; 248:341–356.
160. Besser GM, Thorner MO, Wass JAH, Doniach I, Canti G, Curling M, Grudziniskas JG, Setchell ME. Absence of uterine neoplasia in patients on bromocriptine. Br Med J 1977; 2:868.
161. Desclin L. Observations sur la structure des ovaires chez rats soumis a l' influence de la prolactine. Ann Endocr (Paris) 1949; 10:1–18.
162. Griffith RW. Bromocriptine and uterine neoplasia. Br Med J 1977; 6102–1605.

163. Griffith RW, Grauwiler J, Hodel Ch, Leist KH, Matter B. Ergot alkaloids and related compounds. In: Berde B, Schild HO, eds. Handbook of Experimental Pharmacology. Berlin: Springer, 1978:806–851.
164. Reynolds JEF, ed. Martindale—The Extra Pharmacopoeia. 30th ed. London: The Pharmaceutical Press, 1993.
165. Calabresi P, Chabner BA. Chemotherapy of neoplastic diseases. In: Hardman JG, Limbird LE, Molinoff PB, Ruddon RW, Goodman Gilman A, eds. Goodman and Gilman's—The Pharmacological Basis of Therapeutics. New York: Macmillan, 1996:1225–1287.
166. IARC. IARC Monographs on the Evaluation of Carcinogenic Risks of Chemicals to Humans. Some Antineoplastic and Immunosuppressive Agents. Vol. 26. Lyon: WHO/IARC, 1981.
167. Weisburger JH, Griswold DP, Prejean JD, Casey AE, Wood HB, Weisburger EK. The carcinogenic properties of some of the principal drugs used in clinical cancer chemotherapy. Recent Results Cancer Res 1975; 52:1–17.
168. Weisburger EK. Bioassays program for carcinogenic hazards of cancer chemotherapeutic agents. Cancer 1977; 40:1935–1949.
169. Hoechst Marion Roussel. File Toxicity Data on Amaryl (Glimepiride). Personal communication with A. Baader, 1996.
170. Summary of Product Characteristics for Amaryl (Glimepiride), as finalized by the Committee of Proprietary Pharmaceutical Products, at the European Agency for the Evaluation of Drugs (EMEA), London, April 1996.
171. Carter SK, Slavik M. Current investigational drugs of interest in the chemotherapy program of the National Cancer Institute. Natl Cancer Inst Monogr 1977; 45:116–117.
172. National Cancer Institute. Bioassay of isophosphamide for possible carcinogenicity. Washington, DC: U.S. Government Printing Office, 1977.
173. Griesemer RA, Eustis SL. Gender differences in animal bioassays for carcinogenicity. J Occup Med 1994; 36:855–859.
174. Rothwell CE, McGuire EJ, Martin RA, De La Iglesia FA. Chronic toxicity and carcinogenicity studies with the β-adrenoceptor antagonist levobunolol. Fundam Appl Toxicol 1992; 18:353–359.
175. National Cancer Institute. Bioassay of Procarbazine for Possible Carcinogenicity. Washington, DC: U.S. Government Printing Office, 1979.
176. IARC. IARC Monographs on the Evaluation of Carcinogenic Risks of Chemicals to Humans. Pharmaceutical Drugs. Thiotepa. Vol. 50. Lyon: WHO/IARC, 1990.
177. National Cancer Institute. Bioassay of Thiotepa for Possible Carcinogenicity. Washington, DC: U.S. Government Printing Office, 1978.
178. National Cancer Institute. Bioassay of 4,4′-Thiodianiline for Possible Carcinogenicity. Washington, DC: U.S. Government Printing Office, 1978.
179. Cueto C Jr, Chu KC. Carcinogenicity of dapsone and 4,4′-thiodianiline. In: Deichman WB, ed. Toxicology and Occupational Medicine. Vol. 4. Amsterdam: Elsevier, 1979:99–108.
180. Strobel K, Grummt T. Aliphatic and aromatic halocarbons as potential muta-

gens in drinking water. III. Halogenated ethanes and ethenes. Toxicol Environ Chem 1987; 15:101–128.
181. IARC. IARC Monographs on the Evaluation of Carcinogenic Risks of Chemicals to Humans. Chlorinated Drinking Water Chlorination By-Products; Some Other Halogenated Compounds; Cobalt and Cobalt Compounds. Vol. 52. Lyon: WHO/IARC, 1991.
182. National Toxicology Program. Toxicology and Carcinogenesis Studies of Bromoethane (Ethyl Bromide) (CAS No. 74-96-4) in F344/N Rats and B6C3F1 Mice (Inhalation Studies) (NTP Technical Report No. 363), Research Triangle Park, NC, 1989.
183. Poirier LA, Stoner GD, Shimkin MB. Bioassay of alkyl halides and nucleotide base analogs by pulmonary tumor response in strain A mice. Cancer Res 1975; 35:1411–1415.
184. Dipple A, Levy LS, Lawley PD. Comparative carcinogenicity of alkylating agents: comparisons of a series of alkyl and aralkyl bromides of differing chemical reactivities as inducers of sarcoma at the site of a single injection in the rat. Carcinogenesis 1981; 2:103–107.
185. National Toxicology Program. Toxicology and Carcinogenesis Studies of Chloroethane (Ethyl Chloride) (CAS No. 75-00-3) in F344/N Rats and B6C3F1 Mice (Inhalation studies) (NTP Technical Report No. 346), Research Triangle Park, NC, 1989.
186. Bucher JR, Morgan DL, Adkins B Jr, Travlos GS, Davis BJ, Morris R, Elwell MR. Early changes in sex hormones are not evident in mice exposed to the uterine carcinogens chloroethane or bromoethane. Toxicol Appl Pharmacol 1995; 130:169–173.
187. Hatch G, Anderson T, Elmore E, Nesnow S. Status of enhancement of DNA viral transformation for determination of mutagenic and carcinogenic potential of gaseous and volatile compounds. Abstract No Cd-26. Environ Mutagen 1983; 5:442.
188. Natarajan AT, Obe G. How do in vivo mammalian assays compare to in vitro assays in their ability to detect mutagens. Mutat Res 1986; 167:189–201.
189. Blair A, Srewart PA, Tolbert PE, Grauman D, Moran FX, Vaught J, Rayner J. Cancer and other causes of death among a cohort of dry cleaners. Br J Ind Med 1990; 47:162–168.
190. Anttila A, Pukkala E, Sallmen M, Hernberg S, Hemminki K. Cancer incidence among Finnish workers exposed to halogenated hydrocarbons. J Occup Environ Med 1995; 37:797–806.
191. Sallmén M, Lindbohm M-L. Reduced fertility among women exposed to organic solvents. Am J Ind Med 1995; 27:234–238.
192. IARC. IARC Monographs on the Evaluation of Carcinogenic Risks of Chemicals to Humans. Allyl Compounds, Aldehydes, Epoxides and Peroxides. Vol. 36. Lyon: WHO/IARC, 1981.
193. National Toxicology Program. Toxicology and Carcinogenesis Studies of Ethylene Oxide (CAS No 75-21-8) in B6C3F1 Mice (Inhalation Studies) (NTP TR 326; NIH Publ No 86-2582), Research Triangle Park, NC, 1986.

194. De Bruin A. Biochemical Toxicology of Environmental Agents. Amsterdam: Elsevier, 1976.
195. Doll R. Tobacco: an overview of health effects. In: Zaridze D, Peto R, eds. Tobacco—A Major International Helath Hazard. Lyon: World Health Organization/International Agency for Research on Cancer Scientific Publications, 1986:11–22.
196. Baron JA. Smoking and estrogen-related disease. Am J Epidemiol 1984; 119: 9–22.
197. Karageorgou M, Papadimitriou C, Marselos M. Sexual differentiation in the induction of the class 3 aldehyde dehydrogenase. In: Weiner H, Crabb D, Flynn G, eds. Advances in Experimental Medicine and Biology. New York: Plenum Press, 1993:123–129.
198. Sasson IM, Haley NJ, Hohhman D. Cigarette smoking and neoplasia of the uterine cervix: Smoke constituents in cervical mucus. N Engl J Med 1985; 312:315–319.
199. Rakowicz-Szulczynska EM, McIntosh DG, Smith M. Growth-factor mediated mechanisms of nicotine-dependent carcinogenesis. Carcinogenesis 1994; 15:1839–1846.
200. IARC. IARC Monographs on the Evaluation of Carcinogenic Risks of Chemicals to Humans. Vol. 38. Tobacco Smoking. Lyon: WHO/IARC, 1986.

29

Interspecies Differences in Response to Chemical Carcinogens

David B. Clayson*
CARP, Ontario, Canada

Kirk T. Kitchin
U.S. Environmental Protection Agency, Research Triangle Park, North Carolina

I. INTRODUCTION

It is clear from the preceding chapters that there are major differences in the quantitative response to individual carcinogens between different experimental animal species and humans. In some cases, a growing level of evidence indicates possible qualitative differences as well. Thus, substances can be carcinogenic in one or more species, but simultaneously there may exist substantive, usually mechanistic, evidence that they are not carcinogenic in other species. In other examples, the accumulated evidence suggests there may be no effect levels or thresholded responses. In this chapter we will first summarize what is currently known, or believed, about the overall way in which carcinogens exert their complex multistage effect, and then consider the evidence from preceding chapters that supports or argues against this overall view of carcinogenesis.

There are important differences between the conditions under which

---

This document has been reviewed in accordance with U.S. Environmental Protection Agency policy and approved for publication. Mention of trade names or commercial products does not constitute endorsement or recommendation for use.
*Retired, Health Protection Branch, Health Canada, Ottawa, Ontario, Canada.

the epidemiologist obtains results in human studies and those used by the experimentalist in demonstrating carcinogenicity in experimental animals. The relative genetic heterozygosity of the human compared to the inbred or closed-colony breeding of laboratory rodents has been emphasized on many previous occasions. The effects of dose exaggeration, resulting from the use of the maximum tolerated dose (MTD) in rodents as a predictor for much lower human exposures, are discussed in Chapter 3. Less interest has been displayed in the fact that, while the conditions under which animals are maintained in the bioassay (diet, breeding status, number of animals in a cage, temperature, humidity, light-dark cycle, and so on) are kept constant, the human population lives in a very variable environment in which different factors may differently affect the expression of cancer in individuals. Changing fashions in human behavior, such as the massive adoption of cigarette smoking by men during World War I and women during World War II, have led to immense increases in lung and other cancers. Seventy years ago lung cancer was considered to be a rare disease; now it represents about one third of the human cancer burden in the United States (1). Unknown factors have led to major decreases in gastric stomach cancer in Caucasians. Doll and Peto (1) found, in a survey of the United States' human cancer burden, that while 30% of the present total cancer burden is associated with tobacco use, a further 35% may be associated with diet and nutrition, an immensely complicated subject whose bearing on tumor induction is still in a most elemental state. There is evidence, mainly obtained from studies in animals (2), that the amount of food or calories eaten and the amount of exercise may have a major effect on the expression of both naturally occurring and chemically induced tumors (3–5). Similarly, Ames (6) has drawn attention to the presence of quite substantial levels of a few specific naturally occurring carcinogens in certain items in the human food supply; this area has not been adequately explored. Cooking-related mutagens and carcinogens represent a further problem (7), while the strictly controlled food additives and pesticides do not appear to pose a major cancer threat in view of their rigorous regulation and the relatively small quantities permitted to find their way into the food supply. However, we do not yet know the extent to which any of these dietary factors may affect tumor expression in individual men and women.

The only apparent way to resolve the many problems raised by different responses of different species and sexes of animals to individual carcinogens is through an understanding of the biological mechanisms by which they act. This chapter will first consider the known basic biological mechanisms of the induction of chemical carcinogenesis and then discuss on a tissue-by-tissue basis how the available information fits the presently available mechanistic knowledge.

II. MECHANISMS OF CARCINOGENESIS

Definitions of the terms carcinogen and cancer were elaborated before our present knowledge of carcinogenesis mechanisms was developed. Willis (8), for example, suggested that "a tumour is an abnormal mass of tissue, the growth of which exceeds and is uncoordinated with that of the normal tissues, and persists in the same excessive manner after the cessation of the stimuli which evoked the change." It is possible to define a carcinogen as "an agent that leads to a statistically significant increase in the number of benign and/or malignant tumors in a tissue" (9). The latter definition is unrelated to mechanistic considerations. Once multiple mechanisms of cancer induction by different chemicals are discovered and eventually appreciated, the interspecies differences in response to carcinogens that work by a variety of mechanisms will begin to make sense.

At this time we know of four major models or mechanisms (electrophile generation, oxidative damage, receptors, and cell proliferation) by which chemical carcinogens may exert their effects.

1. Electrophile generation through the metabolic activation or inherent breakdown of the chemical carcinogen proceeds to a highly reactive electron-deficient species possessing the ability to form covalent adducts with cellular macromolecules, including DNA (10).
2. More recently, attention has been drawn to the potential for oxidative damage to affect DNA in a similar fashion to electrophiles. This DNA damage may arise in a variety of ways through oxygen radical formation to affect DNA in a somewhat similar fashion to electrophiles. Nitrogen radical–induced damage has also been reported, although less frequently than oxygen radical damage.
3. Other chemicals such as estrogens or the peroxisome proliferators interact with specific receptor proteins to produce ligand-protein complexes that are capable of modifying the expression of critical regions of the genomal DNA.
4. Finally, there are agents that appear to exert none of these three effects on a significant scale but greatly increase cellular proliferation in one or more tissues. These agents have previously been identified as tumor-promoting agents (11–13). If they act on tissues that, for any reason, contain naturally occurring or background "initiated" cells that are aided by a cellular proliferation in developing into tumors, these agents will, under the present definitions, be identified as "complete" chemical carcinogens. Thus, these chemicals will be mechanistically promoting but operationally described as "complete" carcinogens.

It is necessary to consider briefly the evidence supporting these different ways in which cancer may be induced by chemicals and how much experimental work is required to identify whether a newly demonstrated carcinogen is effective through one or more of these mechanisms.

A. Electrophile Generation

James and Elizabeth Miller and their colleagues isolated a new metabolite of *N*,2-acetylaminofluorene (2-AAF), identified it as the *N*-hydroxy derivative, and showed it to be a rather more potent carcinogen in several rodent species than 2-AAF itself (14,15). As a result of a further 5 or so years investigation, DeBaun et al. (16) demonstrated in 1970 that in the rat liver *N*-hydroxy-2-AAF was converted by sulfotransferase to its sulfate ester, which, rather than being more readily excreted than 2-AAF, broke down into the positively charged electrophile *N*,2-acetyliminofluorene through the sulfate ion acting as a leaving group. *N*-Hydroxy-2-AAF sulfate ester was shown to be directly carcinogenic, causing subcutaneous sarcomas after injection into rats. It also formed adducts with cellular macromolecules. As a result of this sentinel observation, it has been shown that *N*-hydroxylation occurs with a wide variety of carcinogenic aromatic amines, although the conjugating group may differ between tissues or from chemical to chemical. Other important classes of carcinogens such as the nitrosamines, the polycyclic aromatic hydrocarbons, aflatoxins, and other similar chemicals are converted to electrophiles by metabolism. Substances generally known as biological alkylating or arylating agents, including the nitroamides, sulfur mustards, and nitrogen mustards, spontaneously break down in biological media into electrophiles and do not require metabolism to be carcinogenic (17).

Most attention has been paid to the formation of adducts with DNA by these electrophiles. Such DNA adducts are not of themselves genetic lesions. They may be faithfully repaired by the appropriate DNA repair enzymes present in the affected cell nuclei, or, if they affect the expression of critical functions within the cell, they may lead to cell death and ensuing cellular proliferation to restore the tissue mass and function to near normal levels. Recently, attention has been paid to the effects of such adducts on more complex cell processes such as apoptosis or programmed cell death (18,19). However, if cell replication occurs before the DNA adduct is removed by repair enzymes, the adduct may, for example, through base mispairing, lead to the formation of abnormal DNA sequences that are identifiable as mutational or possibly even clastogenic changes. If these changes occur in specific areas of oncogenes or tumor-suppressor genes (20,21), the affected cell may be set on the route to tumorigenicity. Mutational changes may be induced at sites other than the oncogenes and tumor-suppressor genes. If these changes permit the survival of mutated cells or have a positive effect on tumor development, they may persist into the final tumor, thus helping to explain the vast range of different properties associated with individual tumors, a subject of great fascination to the tumor pathologist.

Short-term in vitro and in vivo genotoxicity tests, such as the Ames *Salmonella typhimurium* assay, have proved singularly successful for identifying electrophile-generating carcinogens. For example, in the 1970s when most identified carcinogens were of the electrophile-generating type, concordances between the bioassay and the short-term tests were as high as 80–90% (22). The discovery of further chemical carcinogens, many of which operated through other mechanisms, reduced this level of agreement to 50–60% (23). These genotoxicity tests offer a qualitative approach to the identification of electrophile-generating carcinogens, largely because of the inclusion of mammalian microsomal fractions for metabolic activation. Because the isolation of the microsomal fraction upsets the natural balance of these organelles that exists within cells, generally these tests appear to have very little potential for quantitative discrimination between the effects of electrophile-generating carcinogens in humans and in rodents.

At this time, quantitative approaches to the rate of formation of DNA adducts with electrophiles appears much more promising than the use of genotoxicity tests for the quantization of this part of the carcinogenic process. A very sensitive accelerator mass spectroscopic process has been evolved (24,25). Its application to human rather than animal tissues presents ethical problems.

The problem posed by different species responses to an individual carcinogen is well illustrated by 2-naphthylamine, which induces bladder tumors in humans, dogs, and under special conditions, rats (27), liver tumors in mice, but relatively little in the other species tested over a substantial part of their life span (28). Physiologically based pharmacokinetics (PBPK) has proved useful in evaluating the distribution of systemically administered chemicals in rodents and in extrapolation of tissue burdens from rodents to humans (29). Because the deliberate exposure of humans to identified or suspected carcinogens is generally regarded as unethical in civilized communities, in vitro techniques that are not always satisfactory for the quantitative prediction of what occurs in vivo have to be relied on to gauge the level of conversion of a carcinogen to an electrophile. In addition, the efficiencies of DNA repair processes in different species and the longevity of individual species must be taken into account. Prediction of interspecies responses to carcinogens of this type is not yet at a stage where such predictions may be used with confidence.

B. Oxidation of DNA

The ability of certain substances to lead to the formation of oxidation products of DNA has recently begun to attract considerable attention (6,29,30). Fat metabolism provides a convenient approach to this topic. A number of reactive oxygen species are produced during fat metabolism, of which the hydroperoxide radicals ($RO^{\cdot}$, $ROO^{\cdot}$) are sufficiently stable to escape from the site of fat metabolism but reactive enough to interact with DNA. Hydrogen peroxide (H_2O_2) does not interact directly with DNA and is often degraded by enzymes such as catalase to harmless products. Excess hydrogen peroxide can be converted to the highly reactive hydroxyl radical ($HO^{\cdot}$) through interaction with transition metal ions including the essential ions of iron and copper. These radicals, if they reach the DNA, are capable of inducing one of about 20 different DNA lesions (31). These include, but are not limited to, single and double strand breaks and formation of mono- or dihydroxylated derivatives of the DNA bases. The most fully investigated of these derivatives is 8-hydroxyl-2-deoxyguanosine. It has been shown, using an acellular system, that this modified base leads to base mispairing at the 8-hydroxy-2-deoxyguanosine site or at an adjacent DNA base (32,33). Therefore, agents producing oxidative DNA damage mimic the electrophile-generating chemicals in this mutational respect.

These DNA oxidative reactions differ from electrophile generation and interaction with DNA in two important respects, that is, the frequency of lesions in untreated animals and the efficiency of DNA repair systems. First, as indicated by Ames and Gold (6) and subsequently by many others, DNA oxidative lesions are relatively common in human and animal cells. Thus Ames and Gold estimated that there may be as many as 10^6 oxidative lesions in the genome of a single liver cell derived from an untreated rat. Second, 66% partial hepatectomy in a living rat is followed by a wave of cellular proliferation that returns the liver mass and its functionality to near normal levels in about 48 hours (for additional references, see Ref. 30). This has two important implications. The ability of the liver cell to repair its oxygen radical–damaged DNA before the onset of its replication must, at least, be adequate to prevent disaster following necessary physiological processes such as fat metabolism that lead to oxidative DNA lesions. In contrast, it is also possible that the failure to repair every single lesion may account for the development of tumors and other diseases associated with aging as well as the failure of normal cells to exist beyond a certain number of replications in tissue culture (34,35). These observations imply that factors such as the consumption of excess fat or exposure to excessive levels of agents, such as butylated hydroxyanisole (BHA), which at excessive levels have the potential to lead to increased levels of hydrogen peroxide (36) that may overwhelm the catalase levels and repair systems and thereby increase the overall incidence of diseases of aging including tumors. The importance of the balance between oxygen radical damage and DNA repair holds for most tissues and is discussed for skin tumorigenesis in detail in Chapter 24.

Despite the fact that oxidative DNA lesions need to be regarded as pregenotoxic, these agents are not necessarily identified as genotoxic by the standard range of genotoxicity tests. This arises because in most cases the genotoxicity test systems do not contain the necessary enzyme functions to permit the detection of such agents. Attempts are now being made to remedy this deficiency (37). Until such methodology is readily available and validated, analytical measurements of tissue or urinary DNA-oxidation products must be relied on for the necessary information (6). It should therefore be apparent that a great deal of further investigation of the causes and effects of DNA oxidative damage is urgently needed for the understanding of the underlying factors leading to human and animal cancer burdens. In recent years, there has been increasing acceptance of the importance of free radical and oxidation-related damage of DNA as one mechanism of both radiation and chemical carcinogenesis.

Potentially oxidizing radicals may also arise in other ways. Ethanol, the cause of chronic alcoholism, has been shown under the conditions of chronic overconsumption to generate radicals that may be associated with both the formation of liver tumors as well as with the destruction of nerve cells (38,39).

In an article published in 1993, Pitot (40) emphasized the overall lack of knowledge concerning the precise genetic changes that have occurred in initiated cells, a concept reflected in Chapter 4, on mouse liver carcinogenesis. This lack of knowledge was attributed by Pitot to our present inability to isolate initiated cells from most tissues. Such a hypothesis is supported by the fact that the DNA-interacting species from either electrophile generation or oxygen radical formation are not limited to specific single-site targets on the DNA but may induce a very wide range of lesions that are not cancer specific.

One factor that needs to be considered is the fact that the activation of a gene, such as a proto-oncogene, may require interaction with a more precisely defined target on the DNA than does the inactivation of a tumor-suppressor gene, which might ensue from a variety of chemical-DNA interactions. Oxygen radicals appear to interact with DNA in multiple ways (31), whereas the totality of our present information suggests that electrophiles are rather more limited in their mode and site of attack. This could explain why in the most advanced colonic tumors in humans, in which four mutational events have occurred, three concern tumor-suppressor genes but only one a proto-oncogene (41). This, in turn, might be interpreted to support the importance of oxidative DNA damage particularly in background or apparently non–agent-related human tumorigenesis.

C. Receptor Protein Interactions

As discussed in Chapter 20, administration of estrogen has been shown to enhance the incidence of tumors of mammary, endocrine, and other endocrine-responsive tissues since the mid-1930s when crude extracts of estrogen first became available. Jensen (42) determined that the estrogens interacted physically with a receptor protein, which then reacted with the endocrine cell genome to modify the expression of a specific region of the DNA and, subsequently, to allow enhanced cellular proliferation at specific sites. A number of other receptor protein interactions concerned in cancer formation have subsequently been identified.

Scientists working for Imperial Chemical Industries Ltd. in England identified a receptor protein of the steroid hormone receptor superfamily involved in the mechanism of action of peroxisome-proliferating chemicals. Peroxisomes are subcellular organelles responsible for fat metabolism in rat liver, mouse liver, and other tissues (43,44). The discovery of this receptor protein was commercially important because it enabled those working with novel chemicals to screen out specific potential rodent liver carcinogens before enormous sums of money were invested in the development of these chemicals for marketing.

Poland et al. (45) suggested that 2,3,7,8-tetrachloro-dibenzo-*p*-dioxin (TCDD), the most potent rat liver carcinogen yet identified, as well as a potent inducer of toxicity in several other biosystems, interacts with a receptor protein to give a ligand-receptor complex that affects the expression of the Ah locus. Much further work, mainly at the molecular biological level, has followed this sentinel observation (46). Thus, other receptor proteins including the Arnt protein have been identified as playing a role in the effects of TCCD. There is limited, but not yet conclusive, evidence that humans elaborate a protein that is identical to the rat protein, forming ligands with TCDD and interacting with the Ah locus. This implies that humans will respond to TCCD, at least qualitatively, in a fashion similar to rodents.

Studies of tumorigenesis in rodents and other appropriate systems indicate that TCDD is probably nongenotoxic, that is, it does not form covalent DNA adducts, and therefore appears to increase tumor formation by a promotional action. There is limited evidence that TCCD enhances cell proliferation in tissues in which it promotes tumorigenesis (46). The demonstration of a possible mechanism at a biological rather than a molecular biological level may ultimately prove more useful in carcinogen risk assessment. Measurements of factors such as cellular proliferation in specific tissues may be obtained much more economically and routinely than the exploration of molecular biological mechanisms. Knowledge of the biomolecular mechanisms is, however, vital if we are to understand how such agents exert their effects.

It has been suggested, particularly in Europe, that a ligand-receptor protein interaction needs to attain a certain level of receptor occupancy before it will interact with specific regions of cellular DNA. This view has subsequently been broadened to include all nongenotoxic carcinogens. This suggestion is potentially misleading. Receptor proteins are more likely to have a specific and continuing physiological function in the cell than to be purposeless proteins waiting for specific test chemicals before exerting their effects on DNA. For example, the estrogen receptor is in continuous physiologically based use to modify the rate of cellular proliferation in the mammary gland during the menstrual cycle in humans or the estrous cycle in rodents. It has the potential to be modified by the addition of any amount of estrogen-mimetic substance, however small. Therefore, there should be specific regulatory concern about the effects of incremental doses in such cases. At this time there is evidence that some nongenotoxic carcinogens may demonstrate a threshold, but this needs to be carefully established on a case-by-case basis.

D. Cellular Proliferation

The best current approaches to the determination of cellular proliferation are both labor-intensive and costly (47). Careful experimental planning is needed if the maximum interpretive value is to be obtained. It is suggested that studies be limited, as far as possible, to those carcinogens that fail to interact covalently with DNA and to those tissues that give rise to cancer after adequate exposure to the agents of interest. Cell proliferation studies should be made on pathological slides of intact tissues. The following two examples should make it clear that this is necessary. In studies on the rat forestomach carcinogen BHA, it was found that the maximum level of cellular proliferation was confined to a relatively narrow strip of forestomach epithelium proximate to the gastric stomach (48,49), the area from which the majority of tumors subsequently arose. When the increase in cell proliferation was measured for the whole forestomach epithelium using cell suspensions and flow cytometry, the apparent increase was much smaller. This small increase could easily be dismissed as of limited biological significance (50) and would not be readily usable to demonstrate a possible threshold, which appears to exist if pathological slides are used. With the nongenotoxic liver carcinogen butylated hydroxytoluene (BHT) (51–53), there was only a minimum enhancement of cellular proliferation lasting 2–4 days from the start of BHT treatment, after which levels fell to below normal (54). However, carcinogenic levels of BHT enhanced cellular proliferation in preformed enzyme-altered foci in rat liver, suggesting that the major effect of BHT is on the later stages of tumorigenesis (55).

Methods for measuring cellular proliferation fall into two categories. The first measures the rate of DNA replication and includes the use of identifiable DNA-specific base substitutes such as [^{3}H]-thymidine, bromodeoxyuridine, or iododeoxyuridine. Recent methodological additions that include cells in DNA synthesis but are not limited only to the process of DNA replication include the immuno-identification of a protein or proteins elaborated only during cell replication, namely, the proliferative cell nuclear antigen (PCNA) (56), which may be used with suitably processed archival tissues, and the silver/nucleolar organizer region (AgNor) method (57–59).

The second approach is to identify, microscopically, cells during mitosis—the actual division of the cell. This requires even more labor and skilled personnel. In view of the association of mutational changes with DNA replication, the first methods mentioned, especially the use of DNA base substitutes, may possibly give results that are the most relevant to carcinogenesis. Perhaps the greatest need at present is to reduce the amount of labor associated with cell proliferation studies; this may soon be achieved by use of computer-based image analysis techniques that might appreciably increase the accuracy of the determinations as well as reduce the labor costs involved.

Enhanced cellular proliferation has been tentatively associated with the promotional stage of carcinogenesis for a considerable number of years. Little attention appears to have been given to the possibility that if such a promoting agent acts on a tissue that contains initiated cells occurring naturally or through accidental processes, such an agent will, using present definitions, be regarded as a "complete" carcinogen.

There is a need to identify tissues that are most likely to contain adequate numbers of initiated cells and to ensure that the same criteria apply in humans as in rodents. If it is assumed that in rodents naturally occurring tumors arise in a similar way to induced tumors that have been studied in depth, it is not unreasonable to further assume that the number of initiated cells will bear some relationship to the number of tumors occurring naturally in that tissue in a lifetime study. The rate of cellular proliferation naturally occurring in the affected tissue may be another important factor that needs to be considered, but this is unlikely to be generally true in view of the high expression of naturally occurring tumors in rodent liver combined with the usually very low rate of cellular proliferation in that tissue. Nevertheless, this may still imply that tissues such as the rodent liver, which are often susceptible to naturally occurring tumors, will be more susceptible to the effects of induced cell proliferation than tissues that are generally resistant to such tumors. The application of this concept to human tumorigenesis requires great care.

The most common naturally occurring tumors in rodents appear to depend on innate genetic factors or on factors such as the mammary tumor virus, which is passed from the dam to the offspring through the milk. In humans, on the other hand, the expression of the most common tumors is often linked to environmental factors. These environmental factors include cigarette smoking (lung or other cancers), dietary factors (mammary and colorectal cancer in Caucasians), aflatoxins and associated mold products with hepatitis virus infection (liver cancer) in those living in tropical and subtropical regions, as well as betel nut quid chewing (oral cavity cancer) in Asia, especially in parts of India. Great care will be needed to ensure that conclusions based on rodent studies do, in fact, apply to humans in such circumstances.

Another difficulty in using background tumors as an index against which the effects of cellular proliferation may be gauged is well illustrated by the $B6C3F_1$ mouse. Male mice of this type develop a very high and variable background incidence of liver tumors. Female mice produce fewer tumors. Vesselovinovitch and colleagues showed that the incidence of liver tumors was increased in female $B5C3F_1$ mice by surgically ablating endocrine glands (gonadectomy) associated with femininity (60). More recently, it has been shown that the liver tumor–enhancing effects of exposure to fractions of unleaded gasoline in female $B6C3F_1$ mice are the result of the antiestrogenic effects of these petroleum fractions (61–63).

E. Dose, Maximal Tolerated Dose, and Risk Assessment

Dose is a centrally important concept in both toxicology and carcinogenesis. As early as the sixteenth century the physician-alchemist Paracelsus wrote: "All substances are poisons; there is none which is not a poison. The right dose differentiates a poison from a remedy." These sixteenth-century writings in the area of toxicology were remarkably insightful and prescient. Twentieth-century laypersons, scientists, physicians, and other health professionals are insufficiently aware of the meaning, utility, and importance of Paracelsus's contributions. Dose can be considered as total dose to the organism, dose rate per unit time, delivered dose (to a target organ), biologically effective dose (e.g., adducted dose on cellular DNA), and so on. Much public hysteria, scientific controversy, and medical and public health confusion could be avoided or greatly diminished if we as individuals and as a society understood what Paracelsus knew and tried to teach us more than 400 years ago.

What is experimentally true of BHA at a 2.0% dietary concentration is not necessarily true at a concentration of 0.1%. At high doses, BHA is a toxic, pro-oxidant chemical; at low doses it is a safe, antioxidant useful in protecting our food supply. We tend to think in terms of exclusionary opposites (e.g., carcinogenic versus noncarcinogenic, toxic versus nontoxic, dangerous versus safe). In dose-response relationships there is often a continuous rather than a digital or quantal mathematical relationship between the dose of the chemical and the incidence and degree of severity of the biological effect.

The use of the MTD in animal carcinogenicity studies is required to adequately perform the first hazard identification step of risk assessment. Without use of the MTD in the first hazard identification step of risk assessment, we would commit too many false-negative errors.

Barnard (65) has estimated that the currently used prudent default assumptions that underlie present approaches to quantitative cancer risk assessment, as used by the regulatory agencies in the United States, may be 16–10,800 times too conservative in specific cases. Progress in meaningful quantitative risk assessment depends on elaboration of data that is less dependent on bioassay results using the MTD of the test chemical. The presently used data may be distorted either by a promotional or a toxic (inhibitory) action of such high doses, particularly on cell proliferation kinetics (66). The ability to measure low levels of DNA adduct formation (24,25), combined with computer-based image analysis for the lower-dose-level induction of cellular proliferation, promise the ability to derive quantitative data relevant to the major factors concerned in tumor formation at levels well below the MTD. This quantitative data could greatly improve the determination of the real shape of, at least, the mid to upper reaches of the chemical carcinogen dose-response curve and the possibility of escaping from the overt conservatism complained about by Barnard (65).

Modern risk assessment consists of four steps: hazard identification, exposure assessment, dose-response assessment, and risk characterization. In our scientific and technological society we perform the hazard identification and exposure assessment steps fairly well but often perform dose-response assessment poorly or not at all. In general we depend on default assumptions of a one-size-fits-all nature instead of actually performing an experimental dose-response assessment. For every step of the risk assessment process for which we do not have good data, we tend to assume the worst possible case or use a default assumption and carry forward with our risk assessment in spite of our acknowledged data gaps and model insufficiencies. There is a saying that a chain is as strong as its weakest link.

Contemporary risk assessment and risk characterization suffers in that we actually perform incomplete, rather than complete, risk assessments. What we actually do in most cases is to use default assumptions and assumed dose-response models in our risk-assessment process to make up for our deficiencies in both experimental data and extrapolation models. To describe the present-day state-of-the-art of risk assessment using the phrases "risk guessing," "risk guestimating," or "risk conjecturing" would not be too far wrong. In the dictionary, the word guestimate is defined as an estimate made without adequate information. This is exactly our present-day situation for almost all nongenotoxic chemicals.

In the second phase of the risk-assessment process, dose-response assessment should be performed either experimentally or as a second-best choice, assumed generically among members of a group of chemicals with a similar, demonstrated mode of action. For example, peroxisome-proliferating chemicals may act similarly in their dose-response relationship. This may also be true among those chemicals that act by necrosis and regenerative cell proliferating, by oxidation of DNA, and by some types of receptor-mediated mechanisms of carcinogenesis. A major fault of contemporary carcinogenicity risk assessment and characterization is that we have better studied the dose-response relationship for ionizing radiation and genotoxic chemical carcinogens, which act directly or indirectly though electrophile generation, DNA adduction, DNA damage and mutation, and then applied this theory and knowledge to different chemicals, often nongenotoxic chemicals, that may work predominantly or even exclusively through oxidation of DNA, receptor protein interactions, or cell proliferation. There are no present large-scale research endeavors to experimentally discover, as much as is possible within the limits of performable and interpretable animal experiments, what the dose-response relationship is for the various subcategories of nongenotoxic carcinogens. There is, however, reason for optimism. The second proposed U.S. EPA cancer risk-assessment guidelines allow the use of alternatives to the linear multistage model for risk assessment. This improvement will assist in experimental cancer work being performed and new extrapolation methods being developed and eventually accepted into broader use by members of the risk assessment community.

F. A Hypothesis Promising a Unified Approach to Cancer Risk

The preceding four sections should make it abundantly clear that it is most difficult, at this time, to decide how different rodent carcinogens exert their effects, let alone to determine the extent to which these effects will be exerted in humans. It may not be necessary, initially, to determine the mechanism by which every single rodent carcinogen exerts its effect before attempting to evaluate its likely effects in humans. Cohen and Ellwein (64) and others have suggested that cancer arises as a result of multiple mutational events interspersed with cellular proliferation to expand the number of mutated cells (Fig. 1). That is to say, if it is possible to define the likelihood that a specific agent induces either (a) effects on DNA that are capable of leading to mutations at critical genes, or (b) cell proliferation, or (c) both processes in rodents and in humans, it may be possible to judge the relevance of such rodent carcinogens to humans. For ethical reasons techniques by which these factors may be elaborated in human tissues either nondestructively or in vitro will be needed.

The experimental approach advocated is presented in Table 1. The most it can offer at this time would appear to be a qualitative judgement whether a given rodent carcinogen will also be a human carcinogen. Much further work on PBPK, the quantitative aspects of DNA lesion formation, specific DNA repair activities, and other scientific approaches, will be required before more meaningful and quantitative data may be adduced by this approach.

III. TISSUE-BY-TISSUE CONSIDERATIONS OF THE RESPONSE TO CARCINOGENS

In this book organs are classified into three groups: group A organs have a high animal and low human cancer rate (liver, kidney, glandular stomach, and thyroid gland), group B organs have a high animal and a high human rate (mammary, hematopoietic system, urinary bladder, oral cavity, and skin), and group C organs have a low animal and high human cancer rate (prostate, pancreas, colo/rectum, and cervix/uterus).

Organs were classified using two sets of criteria. First is the overall incidence of tumors demonstrated in the United States and other westernized countries. In experimental animals, cancer susceptibility depends a great deal on what animal and what genetic strain of the animal among the many available genetic strains is selected. Therefore, the susceptibility to chemically induced cancer of the commonly used strains of rats and mice was used as the second selection criterion. A compilation of the number of chemical carcinogens affecting 35 different organs in rats, in mice, and in both rats and mice was published by Ashby and Patton in the journal *Mutation Research* in 1993. This compilation of experimental animal chemical carcinogen data along with human cancer data was used to classify the organs covered in the Interpreting Carcinogenicity section of this book into group A, group B, and group C organs.

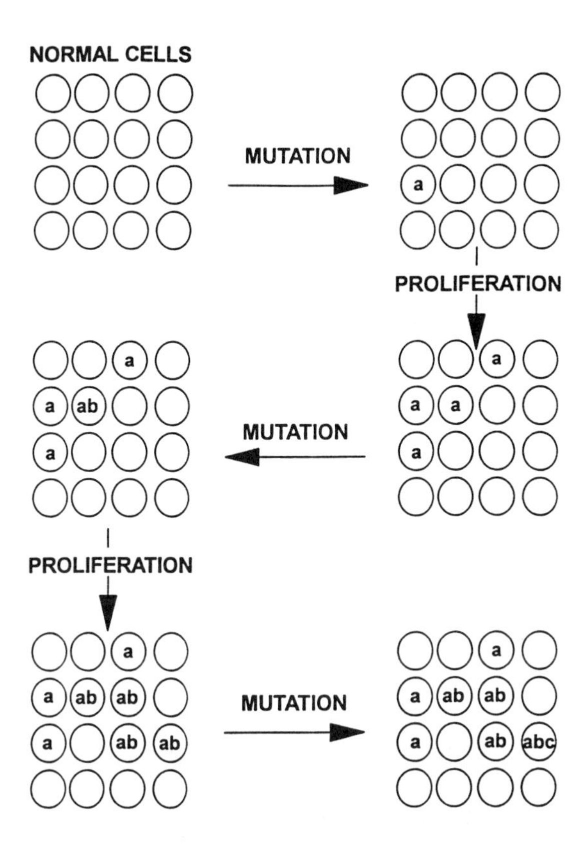

Figure 1 The mutation/cellular proliferation cascade model of carcinogenesis. Each specific mutation (a, b, c, etc.) affects either an oncogene or tumor-suppressor gene and leads to increasing malignant potential of the affected cells. A wide variety of other modifying factors may affect either the mutational process or cellular proliferation but are too numerous to be illustrated here. Cells containing one, two, or three letters have accumulated one, two, or three important mutations, respectively. This figure shows one possible order of a cell accumulating these three mutations. Many such orders of the mutational events are possible. Sequential mutation/cellular proliferation cascade models of carcinogenesis can include any number of mutational/proliferative steps. The particular model shown in this figure illustrates three mutations. O = Normal cells; a, b, and c are important mutational events. (From Ref. 64.)

Table 1 Evidence That May Help Determine the Relevance of Rodent Carcinogens to the Human Cancer Burden

	Relevance to	
	Rodents	Humans
Phase I		
Carcinogenicity Bioassay	+ + +	???
Phase II		
Genotoxicity tests—positive	+ +	+ +
DNA adduct formation[a]	+ +	+ + +
Both systems negative	−	???
Enhanced oxidative DNA damage	+ +	+ +
Phase III: Cell Proliferation Studies (nongenotoxic agents only)		
Enhanced proliferation[a]		
Acutely treated tissues	+ +	??
Enzyme-altered foci	+ +	??
Thresholded proliferation[a]	+ +	+

[a]Methodology has quantitative potential.
−, None; ?, doubtful; +, low; + +, medium; + + +, high.

An interesting question is why specific organ cancer rates are not the same or at least more similar for humans, rats, and mice. When we estimate the degree of similarity between humans, rats, and mice for the four mechanisms of carcinogenesis presented in this chapter, the answer becomes more apparent. Only for the carcinogenic mechanism of oxidation of DNA will the three species be roughly similar. For the three remaining carcinogenic mechanisms of electrophile generation, receptor protein interaction, and cell proliferation, vast differences are to be expected for the three species. Electrophile generation often depends on the amount and type of metabolic enzymes present for toxification and detoxification reactions. A vast literature in this area describes major qualitative and quantitative differences between human, mouse, and rat. The number and type of specific receptors present in a target organ will depend on the genetic makeup of an individual or species; again, major differences between species are expected. Three of the organs that do not match well in cancer rates between animals and humans have substantial involvement with receptor-mediated mechanisms of carcinogenesis—thyroid gland, prostate and cervix/uterus. Cell proliferation (both constitutive and chemically induced) is known to vary considerably between cell types, organs, and to a lesser extent species. For example, rats have a remarkable ability to regenerate liver cells and liver tissue following either chemically induced hepatic cell loss or surgical hepatectomy. Humans, unfortunately, do not often exhibit this remarkable hepatic regenerative ability. Interestingly, the hepatic cancer rate is high in experimental animals and low in Caucasian humans, matching the pattern of the capacity for chemically induced cell proliferation.

When experimental animals seem to be hyperresponders to chemically induced carcinogenesis, we should first think not of the cancer rate discrepancy between animals and humans as a possible false positive but rather as an opportunity to learn something about the carcinogenesis process itself. What is it about the chemically induced animal carcinogenesis process (e.g., in mouse liver, rat liver, thyroid, or forestomach) that may be valid and applicable to organs and exposure situations that are indeed germane to humans? In other words, we should think more about the similarities of the carcinogenic process and not about the differences between anatomical location of the cancer.

When experimental animals seem to be hyporesponders to chemically induced carcinogenesis, we should work hard to improve the available animal, transgenic animal, and nonanimal models systems available to detect possible causes of cancer in these organs. Candidate causes of carcinogenesis would include chemical carcinogens, lifestyle factors, dietary factors, infectious agents, and genetic susceptibilities. There are major human health problems in the areas of cancer of the prostate, pancreas, colo/rectum, and cervix/uterus. The quality of our presently available and utilized animal model systems (the rodent cancer bioassay) does not leave one with a confident, secure feeling that the health of American citizens is being well protected from chemicals or other factors that may cause cancer in these group C organs.

It is profoundly disappointing to have to admit that we are still lacking in knowledge of the genetic, chemical, lifestyle, dietary, or infectious factors responsible for the majority of human cancers other than cigarette smoking and lung cancer. In the United States, where people consume a very high-fat, high-calorie diet, colo/rectum, prostate, and pancreatic cancer are among the leading types of fatal cancers. It may be speculated that a high-fat diet increases the level of oxidative stress substantially in these tissues and that increased oxidative stress is responsible for the induction of these tumors. Acquiring experimental information that supports this speculative interpretation will require the development of nonintrusive methods that are acceptable for use in humans.

In the remainder of this chapter, the evidence presented in previous chapters on tumors arising in individual tissues will be briefly considered to determine the extent to which the evidence fits into the four-causal-element form utilized in Table 2. The previous chapters demonstrate that the amount

Table 2 Types of Carcinogenesis Associated with Different Tissues in Humans, Rats, and Mice

		DNA interactive		Nongenotoxic	
Tissue	Species	Electrophile-forming	Oxidative stress	Cell proliferation	Infection
Bladder	Human	+ + +			Schistosomiasis
	Rat	+ + +		+ + +	
	Mouse	+ + +		+ + +	
Cervix/uterus	Human	?		+ +	? virus
	Rat	+ +		+ +	
	Mouse	+ +		+ +	
Colorectum	Human	??	?/+ +		
	Rat	+ +			
	Mouse	+ +			
Forestomach	Human		Tissue does not exist		
	Rat	+ + +	?/+	+ + +	
	Mouse	+ + +			
Gastric stomach	Human	?			
	Rat	+ +			
	Mouse	+ +			
Hemopoietic system	Human	+	+ +		Epstein-Barr virus
	Mouse	+ + +	+ +		
Kidney					
Main organ	Human	No information			
	Rat	+ +		+ +	
	Mouse	+ +			
Renal pelvis	Human	+			
	Rat	+			
Liver	Human	+ +		? +	Hepatitis virus
	Rat	+ + +	+ +	+ +	
	Mouse	+ + +	+	+ + +	
Lung	Human	+ + +	+	+ + +	
	Rat	+ + +	+	+	
	Mouse	+ + +	+	+	
Mammary gland	Human	?	?	?	
	Rat	+ +		+ +	
	Mouse	+ +		+ +	Mammary tumor virus
Oral cavity	Human	+ +		?/+	
	Rat, mouse	+ + +			

Table 2 Continued

		DNA interactive		Nongenotoxic	
Tissue	Species	Electrophile-forming	Oxidative stress	Cell proliferation	Infection
Pancreas	Human	?	?		
(exocrine)	Hamster	+ +			
Prostate	Human	?	?	+ +	
	Rat	+ +	?	?	
Skin	Human	+ + +			
	Rat	+		+	
	Mouse	+ + +		+ + +	
Thyroid	Human	+ +		+ +	
	Rat	+ +		+ +	

?, Doubtful; +, limited examples; + +, some evidence; + + +, plentiful evidence.

of usable information differs considerably from tissue to tissue. Surprisingly, some tissues such as colo/rectum, prostate, and mammary gland, which are particularly prone to tumors in humans, are the least understood in so far as genotoxic chemical agents capable of inducing human tumors.

A. Liver Cancer

Liver cancer is a complex but important disease in humans, especially for those living in tropical and subtropical countries in Central Africa and Southeast Asia. Liver cancer is much rarer in humans living outside these areas. In general, the rank order of liver cancer is mice > rats > humans. Transspecies extrapolation of hepatocarcinogenesis from either rats or mice to humans exhibits special problems. In this book hepatocarcinogenesis has therefore been divided into three chapters. Dr. Choy discusses the human disease in Chapter 15, Drs. Williams and Enzmann the rat liver foci data in Chapter 13, and Dr. Goldsworthy and colleagues the mouse data in Chapter 4.

Dr. Choy discusses the two genotoxic chemicals for which there is substantial evidence of human liver cancer induction. The mold product aflatoxin B_1 is a common contaminant of foodstuffs in those communities with a high incidence of hepatic cell carcinoma. Vinylchloride monomer leads to the normally very rare liver angiosarcoma among heavily exposed workers in the chemical industry. The former use of colloidal thorotrast (radioactive thorium dioxide) as a radiocontrast medium in X-radiology was also associated with the induction of primary human hepatic cancer (67–69). The intravenous administration of thorotrast led to the colloid particles being removed from the blood stream into the liver sinusoids and to long-lasting irradiation of the surrounding liver tissues. The use of radio-opaque thorotrast to demonstrate lesions within sinuses and other bodily cavities led to local tumors if the removal of thorotrast was not complete. There is also limited evidence that a nongenotoxic substance induces benign liver tumors in women. Continuous use of the original high-estrogen oral contraceptives led to benign liver tumors, which were sometimes fatal due to bleeding. These tumors regressed when contraceptive use was stopped (70). This human observation needs to be considered when evaluating nongenotoxic rodent liver carcinogens, especially in view of the fact that these oral contraceptives had earlier been shown to be mouse liver carcinogens (71). Other chemicals may be involved in the induction of human liver cancer, but the requisite evidence has not yet been obtained.

The other factor strongly associated with human hepatic cell carcinoma is endemic infection with the hepatitis viruses, particularly hepatitis viruses B and C. Dr. Choy devotes considerable space to discussing the possibility that the hepatitis virus's DNA, or fragments thereof, may be incorporated into the liver cell DNA and thereby may transform it to a potentially malignant state. There is an alternative mechanism that, again, needs further research for its establishment. In healthy individuals, normal liver cells have a very low rate of cellular replication. The cirrhotic changes induced by hepatitis infection involve a substantial increase in the level of proliferation, and this greatly increases the opportunity for the development of liver tumors initiated by exposure to foodborne aflatoxin B_1 or other agents. Such an approach fits closely with the general initiation/cellular proliferation model of carcinogenesis proposed by Cohen and Ellwein (64).

The recent development of vaccines against certain hepatitis viruses promises an opportunity to control the high incidence of endemic hepatitis infection and thus liver cancer in the tropics. It can only be hoped that the proposed World Health Organization programs for the use of these vaccines in Africa will be successful and that eventually the liver cancer epidemic will be brought under control.

In Chapter 13, Drs. Williams and Enzmann discuss rat liver foci production as a predictive tool for hepatocarcinogens. It is suggested that such studies will be more economical than the routine 104-week animal bioassay, and, to some extent, they will enable the investigator to escape the consequences of very high dose levels (maximum tolerated dose, or MTD) in the cancer bioassay. The transplacental form of glutathione-S-transferase may be demonstrated by immunochemical staining and is used as the marker for the hepatic cell foci. In view of the fact, observed in other tissues, that different tumors induced by the same agent in the same tissue in the same type of animal demonstrated different ranges of altered oncogenes and tumor-suppressor genes, it might logically be asked whether one parameter is sufficient to detect all hepatocarcinogens. Using the methods described in Chapter 17 and more than 300 agents that had been tested in the complete bioassay, Dr. N. Ito and his colleagues in Nagoya, Japan, found a remarkable consistency between agents leading to rat hepatic foci and rat liver carcinogens. Only two rat liver foci inducers failed to induce tumors, while only a very few peroxisome proliferators (chemicals known to be carcinogenic) did not yield hepatic foci.

The major difficulty with the use of tissue-specific lesions, such as hepatic foci, for the rapid prediction of the carcinogenicity of chemicals is that different agents affect different tissues. Thus, to obtain coverage as adequate as provided by the whole animal bioassay, it would be necessary to develop and use some 30–40 specific tissue tests. This would be both an enormous amount of labor and very costly. Dr. N. Ito and colleagues have suggested that the initiation of animals with sequential doses of different tissue specific carcinogens followed by exposure to the test agent might, to some extent, circumvent this difficulty.

Drs. Williams and Enzmann devote considerable space to discussing liver foci formation in other species, including fish and birds (a rapid 24-day system is available using bird eggs). They suggest that use of these alternative species will help in predicting whether a particular agent is likely to induce liver cancer in humans. For our part, we would be much happier to base such transspecies extrapolation on a knowledge of the mechanism by which the agent induces cancer in the surrogate species and whether such a mechanism will work effectively in humans.

In Chapter 4, Dr. Goldsworthy and colleagues present a detailed account of the difficulties associated with the classification of mouse liver carcinogens and the use of molecular biological parameters such as oncogenes in resolving these problems. They begin with an account of current scientific perceptions of the degree of usefulness of tumors induced above a high naturally occurring incidence of the same lesion in the NTP's favored male $B6C3F_1$ mouse. They conclude that the only way of resolving differences in this difficult area is through an appreciation of the biological mechanisms associated with individual mouse liver carcinogens. We strongly agree with this opinion.

The possibility that biological markers, activated oncogenes, or deactivated tumor-suppressor genes may differ in naturally occurring and in induced tumors is explored. Despite a great deal of hard work during the past 10 years, the conclusions are far from clear-cut. In hindsight, this disappointing conclusion might have been expected since the tissues, especially the liver, are continuously exposed to oxygen radicals, produced, for example, by the metabolism of lipids. As already indicated, certain of these oxygen species lead to a wide variety of lesions in the cellular DNA and are thus able to induce, randomly, both mutations and clastogenic lesions. Dr. Goldsworthy's group addresses the biochemical changes associated with various oncogenes and tumor-suppressor genes and explains how these might influence tumorigenesis. He points out that not all mouse liver tumors induced by the same carcinogen treatment schedule possess the same complement of altered genes. Clearly, it is vitally necessary to determine the range of biological factors that may, if altered, cause a cell to move towards malignancy and then to determine how altered oncogenes and tumor-suppressor genes affect these processes.

The same detailed and meticulous approach is taken to the different types of stimulated cellular proliferation, which may involve a number of different mechanisms occurring at different periods of tumor induction. These include cytotoxicity followed by regeneration, growth factor or hormone induction of cellular replication, and other routes. The mouse liver chapter contains considerable material on apoptosis. Apoptosis, or programmed cell death, is a genetically controlled tissue process that permits cell renewal but controls the excessive multiplication of cells. If, for any reasons, such as genotoxicity, the controlling genes for apoptosis are prevented from functioning, such damaged cells will accumulate, ultimately leading to foci of relatively benign cells with a greater tendency to proliferate than the unaffected tissue cells. This will provide a more optimal environment for the development of cells that have been, or will later be, initiated and helps to explain the susceptibility of the mouse liver to carcinoma formation. It is unfortunately not yet possible to explain how these varied apoptosis parameters will help in interpreting the induction of tumors by specific individual nongenotoxic carcinogens, or how they can be applied to carcinogen risk assessment.

Perhaps the major conclusion to result from these three chapters on liver cancer is the relatively few adequately identified agents concerned in the genesis of human liver cancer compared to the plethora of agents that induce these tumors in rats and/or mice. Liver tumors are the most common tumor identified in bioassays in rats and mice. This might suggest that the rodent bioassay is strongly overpredictive for human (liver) cancer induction. Prudence dictates that without further information, largely based on mechanisms, such a heresy cannot be accepted unless there is adequate proof. The mechanisms set out in the present chapter may provide an initial approach.

B. Kidney Tumors

In Chapter 16, Dr. Hard points out that, like liver tumors, kidney tumors occur more frequently in chemically treated rodents, especially male rats, than in humans. The overall U.S. incidence of kidney tumors is about 2.5% of all cancers in men, while in women the level is about 1.5%. In Table 1 of Chapter 16, Dr. Hard lists all the rodent kidney carcinogens demonstrated in the NCI/NTP bioassay series. It is noteworthy first that only a fraction of these compounds are electrophile generating, such as aromatic amines and their derivatives or halogenated substances that might possibly be converted metabolically to an electrophile. Second, many of these tumors arise only in male rats and are histopathologically diagnosed as adenocarcinoma of the renal tubules. Although the most frequent renal tumor in rodents, its frequency in humans is only 12 per 100,000 Americans. These findings raise the question whether the causative agents are relevant to humans.

The male rat differs from the female rat and from humans and mice in that it elaborates appreciable quantities of a specific α_{u2}-globulin. This protein forms physically bound adducts with a variety of chemicals such as the food flavoring *d*-limonene. The protein, with its adducts, is filtered from the blood stream at a particular part of the renal tubules where the adducted chemical leads to the formation of hyaline droplets and eventually cancer. Dr. Hard supports the view that cell destruction followed by cell proliferation may be the driving force leading to cancer formation. It has been suggested that agents inducing kidney cancer in this way are more unlikely to be effective in humans as humans do not elaborate high levels of this globulin protein (72). There is an urgent need to decide on the requisite evidence to ensure that this surmise is correct.

Perhaps the major factor underlying renal carcinogenesis is the rate of blood flow through the kidney, which means that exposure to bloodborne carcinogens may be high. This combined with a relatively high capability to metabolize some carcinogens might suggest that this organ is particularly susceptible to carcinogenesis, which does not seem to be the case. Otherwise, information on renal carcinogenesis appears limited. As would be expected for a relatively rare human tumor, there is relatively little information on agents leading to human cancers of this organ. The induction of transitional cell tumors of the renal pelvis by a mixture of aspirin, phenacetin, and caffeine in both humans and male rats appears to be a result of the aromatic amine derivative phenacetin. People living in northern Sweden developed renal pelvic cancer as a result of abusing this former over-the-counter analgesic drug combination during and after World War I (73,74). It is also noteworthy that workers exposed to carcinogenic aromatic amines in the chemical and dyestuffs industries also develop transitional cell carcinomas of the renal pelvis on occasion.

As Dr. Hard makes clear in Chapter 16, there is little evidence that oxidative stress is a major factor in renal carcinogenesis. There is also very meager evidence on the oncogenes and tumor-suppressor genes that may be involved in renal carcinogenesis or in their utility as biological markers.

C. Gastric and Forestomach Carcinoma

In Chapter 17, Drs. Hirose and Ito point out that despite its recently declining incidence, especially in western countries such as the United States, gastric cancer is still an important human cancer that manifests different incidences in different parts of the world. In most cases it affects nearly twice as many men as women. The incidence in Japan is high, whereas that in the United States has declined steeply over the past 60 years.

In contrast to rodents, humans do not develop a forestomach. Consequently, humans cannot develop forestomach tumors. The forestomach is often a critically important site in rodent cancer studies, being affected by both genotoxic and nongenotoxic carcinogens. It is correctly pointed out by Drs. Hirose and Ito that genotoxic carcinogens that are active in the rodent forestomach may well affect other tissues in humans; nongenotoxic forestomach carcinogens require more research and especially more thought before relevance to other tissues in humans may be estimated.

In rodents, the forestomach demonstrates different shapes in different species. It appears to act as a collecting and storage vessel to hold excess food before it passes into the gastric stomach. The resulting increased contact time may account for the greater degree of response to carcinogens in the forestomach than in either the esophagus or the gastric stomach (75). Both the esophagus and forestomach are lined with squamous epithelium, which is separated from the glandular gastric epithelium by a slight thickening at the limiting ridge, the major point of attack for butylated hydroxyanisole, an apparently nongenotoxic antioxidant used as a food additive.

Perhaps the most interesting group of carcinogens to be discovered and discussed by Drs. Hirose and Ito is a series of phenolic agents that induce forestomach and, in some cases, glandular stomach tumors. There is evidence to support the concept that the quinonoid metabolites of these phenolic agents interact with certain redox enzymes to cyclically produce hydrogen peroxide. Hydrogen peroxide itself does not interact with DNA, and it is normally degraded by catalase to innocuous intermediates. If the level of hydrogen peroxide exceeds that which may be degraded efficiently by catalase, it may interact with transition metal ions, such as those derived from iron and copper, to give the extremely reactive hydroxyl radical. If it is generated sufficiently close to the DNA helix, the hydroxyl radical may produce one of about 20 different lesions and may affect oncogenes or tumor-suppressor genes. Despite their ability to induce massive DNA damage through the production of hydrogen peroxide, these phenolic agents are nongenotoxic, or at most only very slightly active, in the normal battery of tests for genotoxic agents. It is also interesting to note that agents such as butylated hydroxyanisole are pro-oxidant at high exposure levels but are effective antioxidants at lower levels. In rat forestomach the overall available dose-response evidence is consistent with thresholded tumor formation by butylated hydroxyanisole.

D. Thyroid Cancers

In Chapter 18, Dr. Thomas presents the structure of the thyroid and its tumors, its production of thyroid hormones, and the development of tumors in commendable detail. The thyroid contains two types of cells: the follicular cells that line the follicles and generate thyroid hormones and the C cells. Tumors derived from the C cells are rare in humans and have been suggested to be familial and associated with the occurrence of certain inherited gene mutations. In humans, follicular cell tumors are most common in the young (<29 years) and in the elderly (>60 years). This is an unusual pattern for human tumor appearance. One could speculate that those developing follicular cell tumors in childhood possess a mutated oncogene or tumor-suppressor gene either as a result of inheritance or through exposure to a mutagen early in their existence. It is noteworthy that radioactive fallout resulting from the Chernobyl nuclear accident has increased tumor frequency mainly in the younger age group.

Dr. Thomas discusses in detail the factors leading to follicular cell tumors in both humans and rodents. It is interesting to note how far the field has advanced since the pioneering observations of Bielschowsky (76,77). Perhaps the major conclusion to be drawn from this excellent chapter is the need to understand the physiological function of the tissue if the way in which it develops tumors is to become apparent.

E. Lung Cancer

In Chapter 19, Dr. Hahn spells out the nature of the agents that lead to lung cancer in humans, rats, and mice. This chapter discusses the morphological derivations of the various types of chemically induced pulmonary tumors and the genomic changes involved in inhibition of tumor suppressor gene function and the activation of oncogenes.

There is little, if any, discussion of the proliferative factors necessary for the expansion of clones of pretumor or tumor cells in lung carcinogenesis. One reason for a shortage of data in this area is the complexity and multicellularity of lung tissue compared to other tissues such as those of the liver or urinary bladder. This complexity is enough to deter all but the most determined investigators from embarking on cell proliferation studies in the lungs. Nevertheless, at least two series of observations in human lung carcinogenesis support the possible involvement of an important proliferative or promotional stage.

First, asbestos, and certain other mineral fibers, do not appear to be mutagenic in the standard battery of short-term genotoxicity tests such as those described in Chapter 6. These mineral fibers do, however, cause intensive irritation, which may persist if the geometry of the fibers make it difficult for them to escape from the lung alveoli and bronchioles. The consequent prolonged irritation, combined with possible minor exposure to trace concentrations of airborne carcinogens, represents an ideal situation for cancer development. Such chronic irritation may also produce oxidative stress.

The second example concerns cigarette smoking, the cause of the most prevalent preventable human cancers. Cigarette smoke contains a multitude of different chemicals, some of which are irritants. There are also trace levels of a wide range of genotoxic carcinogens including polycyclic aromatic hydrocarbons, aromatic amines, and N-nitroso compounds. The overall importance of the irritative components of cigarette smoke may be inferred from the fact that relatively heavy smokers who quit smoking at about the time when the first human lung tumors begin to appear in the heavy smoking population demonstrate a steady decline in their lung cancer risk. In a few years, the incidence of pulmonary cancer in ex-smokers approaches that of nonsmokers. This finding in the human data would not be anticipated if genotoxic carcinogens alone were responsible for these catastrophic tumors. If this scientific speculation is correct, the adverse effects of smoking-derived irritant chemicals acting on the lungs might be averted by providing addicted individuals with an alternative and controlled access to the addictive chemical(s). The addictive chemical(s) should be at a dosage level and in a chemical form in which the adverse cell proliferative effects of tobacco smoke on pulmonary and other tissues would be minimal. In this scenario, reducing the risks of tobacco smoking–incurred pulmonary cell proliferation could be used to lower the pulmonary cancer risk of tobacco smoke–addicted individuals.

Human cancer risk assessment uses experimental rat and mouse data and trusts that these two experimental species and humans have similar biological responses, chemical susceptibility, and dose-response relationships. In a societal sense, we use rodents and the cancer bioassays in two species for assessment of our own safety (66). In the caes of cigarette smoking-induced pulmonary cancer, Dr. Hahn, in Chapter 19, states that rats and mice respond to only a limited degree to even the maximal tolerated dose (MTD) of tobacco smoke or its concentrate. Thus, humans are by far the most sensitive of the three species to cigarette smoke-induced pulmonary cancer. In many other cases, rats and mice appear oversensitive to particular chemical carcinogens (66). With cigarette smoke, rats and mice are not sensitive enough to provide an adequate model for the human disease.

Dr. Hahn correctly draws attention to the high level of imprecision in using experimental animals with high incidences of background tumors as a model for human carcinogens. The use of strain A mice for the assay of lung carcinogens is a case in point. Such a bioassay gives no indication as to whether the test substance is an initiator or a promotor of carcinogenesis or whether increases in tumor incidence have occurred by chance. It is difficult to rule out the possible interpretation that the test substance has merely speeded up the normally high rate of pulmonary carcinogenesis in strain A mice.

F. Skin Cancers

In Chapter 24, Dr. Lahiri and coworkers draw attention to the increase in skin cancer incidence on a worldwide basis. This increase may be real, or it may be partly due to improved case reporting. One of the most common skin tumors, basal cell carcinoma, is relatively lacking in malignant potential. In the past, basal cell carcinoma was frequently surgically extirpated in the attending physician's office, and, as was often the case, if it did not recur with increasing malignancy, it was not reported in the appropriate manner. The advent of legally mandated cancer registries has greatly improved this situation. Thus, improved case reporting may have increased the apparent frequency of skin tumors.

Dr. Lahiri and colleagues devote the major part of Chapter 24 to discussing the role played by oxygen and nitrogen radicals and electrophiles in the generation of skin cancer. They focus their text on reactive intermediates, DNA oxidation, and tumorigenesis, discussing the formation and degradation of these radicals in considerable detail at the biochemical level. These free radicals and other factors are relevant to the genesis of cancer. Their account is of great value in understanding the complex interactions undergone by the tissue. The relevance of all this information to cancer formation may not be immediately apparent, as the type and frequency of skin DNA lesions is not discussed. Knowledge about the affected oncogenes and tumor-suppressor genes is potentially important. It is interesting to observe that while there is abundant information on reactive oxygen species, much less is known about the contribution made by reactive nitrogen species.

Some readers may be surprised by the emphasis placed by Dr. Lahiri and colleagues on the importance of reactive intermediates in skin tumor promotion. In the Cohen and Ellwein (64) model for cancer induction, appropriate reactive species may trigger both initiation, a mutational process, and promotion, induced by either agent cytotoxicity or by inactivation of apoptosis. Considerable knowledge and skill will be needed to decide whether a particular reactive intermediate will initiate or promote skin cancer. Sight should not be lost of the possibility that specific proteins such as epidermal growth factors and specific chemicals such as those that mimic hormones may play an important role in skin cancer promotion in humans.

G. Mammary Cancer

In Chapter 20, Drs. Nagao and Sugimura discuss in great detail the induction of mammary cancer in female rats and mice. The agents discussed are mainly polycyclic aromatic hydrocarbons such as 3-methylcholanthrene and aromatic amines including the heterocyclic cooking-related derivatives. As a result of these considerations, Drs. Nagao and Sugimura hypothesize that traces of environmental pollutants may be the underlying cause of human breast cancer.

In addition to these environmental considerations, other important factors include (a) the varying incidence of mammary cancer in different human populations, (b) the possible differences in diet and other characteristics of lifestyle on this incidence, and (c) more remotely, the influence of viruses. In Canada and the United States, mammary cancer has for many years been the most prevalent form of cancer among women. Recently the prevalence of cigarette smoking-related lung cancer in women is reaching a similar level. The incidence of mammary cancer in other countries, such as Japan, is considerably lower. Japanese migrating first to Hawaii and subsequently to the continental United States develop cancer incidences similar to those of Americans within one or two generations. This strongly indicates that it is lifestyle factors, perhaps diet, rather than heredity factors that underlie the higher American cancer rate.

The largest human exposure to chemicals is in the food we eat. The American diet differs considerably from the traditional Japanese diet in that it is much richer in fats and calories. A considerable amount of investigative work suggests that diets richer in fats and calories may influence the incidence of mammary cancer. For example, we have found in mice that excess calories enhanced cellular proliferation in the mammary gland ductules to a considerably greater extent than in six other tissues investigated (78). In a semisynthetic diet different fats induce different levels of cellular proliferation in the mammary gland (79). Much more work in this area needs to be considered. We need to know whether the excess calories (a) exert their effect directly on the mammary tissue, (b) interact with important hormonal factors, or (c) utilize both of these mechanisms.

Work on mammary cancer in mice up to about 1950 was complicated by the fact that many strains of mice carried the mammary tumor virus (MTV). This virus led to a substantial incidence of mammary tumors in older female mice. MTV thus complicated the interpretation of the earlier work, particularly that on hormonal effects. Bittner (80) found that this was not an infective virus in the normal sense of the term. MTV was transmitted in mother's milk from the mother mouse to the offspring. It was possible to clean up infected strains of mice by cross-suckling newborn mice to virus-free dams before they consumed any of their birth mother's milk. After about 1950 MTV was no longer a confounding problem in the interpretation of mouse mammary carcinogenesis.

H. Hemopoietic System Cancers

The hemopoietic system is one of the most complex of the bodily systems considered in this book. It consists of bone marrow and thymus as well as the contents of the arteries and veins. Many different types of lymphomas and leukemias, which are apparent to the pathologist, may arise in the hemopoietic series of cell types. In Chapter 21, Dr. Newcomb discusses her work and that of others on the nature of the genetic changes leading to tumorigenesis. She demonstrates that despite these different types of tumor, oncogene and tumor-suppressor gene alterations in this system showed considerable commonality within the same type of tumor, within the same species, and between mice and humans.

Dr. Newcomb first discusses agents capable of inducing hemopoietic tumors in animals, in particular, mice. There are three classes of such agents, each of which demonstrates a separate mechanism. First are the biological alkylating and arylating agents that form DNA adducts and may on DNA replication lead to base mispairing, that is to say, mutation. Evidence for the action of similar chemicals in humans is not as strong as in mice. It is largely confined to the induction of second primary tumors in patients treated for a different primary tumor with genotoxic chemotherapeutic treatments. Second, radiation is effective in inducing malignancies of the hemopoietic system in both animals and humans. The latter was established by epidemiological studies of atomic bomb survivors in Japan after World War II. Dr. Newcomb identifies the major action of high-energy radiation as interaction with water to produce reactive oxygen species (radicals). These have already been discussed in Section II.B of this chapter. They induce many different lesions in DNA (31). Third, neutron radiation leads to the direct induction of mainly clastogenic lesions in DNA. In humans, certain viral infections are associated with hemopoietic carcinogenesis. The best known example is the induction of Burkitt's lymphoma by the Epstein-Barr virus.

It is not intended to reiterate the wealth of information presented by Dr. Newcomb on the presence or absence of specific oncogene or tumor-suppressor gene alterations in each part of the hemopoietic system. Rather, it appears to us to be more useful to ask about the utility of this information in the detection and treatment of cancer. Dr. Newcomb mentions a mouse study in which a mutated oncogene could be detected in the animal's blood long before the tumor was apparent. She also refers to the possibility that specific gene mutations may lead to deficiencies in the functioning of specific enzymes important in DNA repair or in the deactivation of metabolically activated intermediates. From my perspective, the usefulness of the present vast efforts will only become fully apparent when the complete functions of each oncogene and tumor-suppressor gene is identified. Then, and only then, will it be possible to map out the changes that occur in cell initiation and to decide whether novel approaches to less traumatic therapy, or to the reversal of the carcinogenic process before tumors arise, will be possible. This applies to all tumor-bearing tissues. It is mentioned here in view of the enormous volume and quality of the work on the hemopoietic system described by Dr. Newcomb.

I. Urinary Bladder Tumors

In Chapter 22, Dr. Fukushima and coworkers present an account of the development of studies on urinary bladder cancer during the last 20 years. Bladder cancer has attracted a great deal of attention because its previous elevated frequency in parts of the chemical, textile-dyestuffs, and rubber-manufacturing industries made it easily recognizable and an excellent topic for epidemiological and experimental investigation (81). These studies together have, during the second half of this century, greatly helped in the definition and understanding of the principal factors involved in carcinogenesis. Some recent studies have helped confirm important earlier studies of the way in which bladder cancer arises.

Occupational bladder cancer due to certain aromatic amines (82) and the derived dyestuffs is rightly stressed as the major identified occupational cause of cancers of the human bladder. There is evidence that azo dyes, used in the textile dyestuffs industry, may be reduced by the chemically maintained intensely anaerobic conditions in the lower gut lumen (83). The aromatic amines liberated in this way are thus protected from degradation by the structure of the azo dye until they reach the part of the gastrointestinal tract from which they are absorbed as aromatic amines, thus increasing their carcinogenic efficiency. Epidemiological evidence strongly suggests that cigarette smoking is a major cause of bladder cancer in both men and women. Examples of drugs and other genotoxic substances leading to human bladder cancer are detailed in Chapter 22.

A wide variety of other, nongenotoxic, agents usually at high levels of exposure lead to bladder cancer in rats and mice. Dr. Fukushima and colleagues identify them collectively as bladder cancer–promoting agents because they react positively in their bladder cancer promoter model system. In many cases these agents would be more accurately classified as cocarcinogens because the agents alone are capable of inducing appreciable numbers of tumors. Perhaps the most interesting subgroup of these agents are sodium and potassium salts, such as sodium L-ascorbate alone or in the presence of sodium bicarbonate. Sodium saccharin is a member of this subgroup. It causes bladder tumors in male rats treated with 3% or more of sodium saccharin from conception or parturition (84) but apparently not when treated from an early adult age. The majority of the experiments designed to help understand why sodium saccharin, a commercially valuable artificial sweetener, induces bladder tumors in male rats were started at a young adult age, which makes them extremely difficult to interpret in terms of sodium saccharin's limited, time-dependent carcinogenicity (85).

The second major group of these nongenotoxic rodent carcinogens work through the induction of bladder stones. Rat and mouse bladder stones increase considerably the levels of cellular proliferation (86), which indicated the importance of cell proliferation in tumor promotion (85). Human bladder tumorigenesis in the presence of stones is more of a problem (87). Tumorigenesis in humans occurs at a very much lower frequency than in rodents. In humans, it is difficult to decide whether stones lead to tumors or, alternatively, tumors lead to stones.

Among other examples, attention is drawn to the antioxidants as bladder tumor promoters. Dietary butylated hydroxyanisole (BHA) at 2%, but not 0.5%, was shown to induce increased cellular proliferation in the rat urinary bladder after 3 months of feeding (88). BHA has not been shown to induce rodent bladder cancer when fed alone. BHA is thus established to be a bladder cancer–promoting agent that in all probability demonstrates an experimental threshold or no effect level in its promoting action on carcinogenesis. One of the major practical problems that now remains in this area is how to estimate the effects of mixtures of naturally occurring and man-made phenolic antioxidants so that the real exposures received by humans and rodents may be calculated and their probable effects estimated.

There is abundant evidence that the majority, if not all, of the identified human bladder carcinogens are electrophile generating (genotoxic). Rodent bladder carcinogens act either as electrophile generators or by considerably increasing the normally low levels of cellular proliferation in the urinary epithelium. It should be appreciated that electrophile generators are cytotoxic. Because the organisms need to ensure the integrity of the urothelium, the electrophile generators are also capable of increasing cellular proliferation. Dr. Fukushima and colleagues comment on aspects of these mechanistic principles including information on the effects on oncogene activation and tumor-suppressor gene inactivation. As far as enhanced cellular proliferation is concerned, the most interesting suggestion stemming from the work of Dr. Oyasu and colleagues on the heterotopically transplanted rat bladder is that an epidermal growth factor is able to escape into the urine through the renal tubules. If the surface membrane of the urothelium is no longer intact, the growth factor enters the epithelium and triggers cell division in all epithelial cell layers.

Many of the frequently occurring human bladder tumors in certain tropical and semi-tropical countries are associated with infection by the bigenetic nematode *Schistosomum haematobium*. This nematode lives for part of its life cycle in the human, but not the rodent, urothelium where it leads to intense inflammation and ultimately carcinoma of this tissue. In a recent review, Badawi (89) indicated that these tumors are often associated with mutation of the *p53* oncogene and speculates that the initiating factor may be a urinary nitrosamine. The rat bladder threadworm *Tricosomoides crassicauda* does not lead to bladder cancer in this species.

J. Oral Cavity Cancers

In Chapter 23, Dr. Saranath reminds the reader that cancer of the oral cavity is among the 10 most prevalent cancers worldwide. The majority of these cancers arise in people living in southern Asia, with India making a major contribution. Although it was originally believed that betel or areca nut chewing was the primary cause of these cancers in India, recent epidemiological investigations implicate tobacco chewing or, to a lesser extent, tobacco and betel nut chewing to be the major factors. Tobacco contains a variety of carcinogens capable of initiating carcinogenesis, such as tobacco-specific *N*-nitrosamines, aldehydes, benzo(*a*)pyrene, etc. Such carcinogens may be presumed to initiate the oral cancer incidence, although evidence for their metabolic activation in oral tissues is not available.

In India, both tobacco and betel nut are formed into a quid (a chewable wad) with (slaked) lime for insertion into the oral cavity. Relatively little attention appears to have been given to any possible contribution the lime may make in the genesis of these tumors. It might be expected that the lime would lead to irritation and inflammation in the oral cavity with a consequent increase in cellular proliferation that might serve to promote the tumor incidence. Irritation and inflammation are well known to be major factors in promotion in the mouse skin model system. Slaked lime-induced irritation and inflammation could possibly explain why the oral cancer incidence in India is so much worse than among tobacco chewers in the United States. The abundance of successful animal models of oral cavity cancer should provide at least one suitable experimental system to examine this hypothesis.

K. Prostatic Cancer

In the United States the frequency of prostatic cancer has increased during the past two decades to the most often diagnosed tumor in men and the second most frequent cause of cancer death. A well-referenced and detailed account of prostate cancer is presented in Chapter 25. At present there are no environmental chemical factors that are strongly associated with this form of human cancer. The present lack of environmental carcinogens of the human prostate is made more difficult by the fact that only men and dogs have an appreciable background yield of prostatic tumors. Rodents do not easily develop this tumor. Moreover, the prostatic structure is markedly different in men and dogs compared to rodents.

While several epidemiological studies suggest a relatively weak association between cigarette smoking and prostatic cancer, many others do not suggest such a correlation. Application of the principles of meta-analysis would seem to suggest a lack of meaningful correlation between smoking and prostate cancer. Dr. Bosland reviews a number of occupational settings, including (a) farming and farm work, (b) occupations and places of residence involving exposure to ionizing radiation, and (c) occupations involving exposure to cadmium. None of these factors gave an indication of being strongly associated with prostatic cancer.

It might be expected that dogs would be the primary species of choice for the experimental study of prostatic cancer in view of the similarity of this organ in humans and dogs both in background tumor yield and in structure of the normal gland. Dogs have not been heavily utilized because of the time and the expense involved. The prostate contains a reasonable complement of metabolizing enzymes and thereby leads to cancer when one of a variety of carcinogens is applied locally in rodents. In rats, prostatic cancer may be induced by the systematic administration of two chemicals: 3,2′-dimethyl-4-amino biphenyl or *N*-nitrobis(2-oxopropyl)amine. The latter is one of a class of agents leading to pancreatic cancer in hamsters. These two chemicals provide an opportunity to investigate the effects of secondary or promotional factors in prostatic carcinogenesis.

Dr. Bosland discusses the hormonal mechanisms that may be involved in the genesis of prostatic cancer. The continuous administration of testosterone induces prostatic cancer in some but not in all strains of rats. Coadministration of testosterone and estradiol considerably increases tumor frequency, but estradiol alone merely causes prostatic gland atrophy. The perinatal administration of estrogens to rodents leads to prostatic tumors. Dr. Bosland stresses the importance of cellular proliferation both in "locking in" DNA lesions and in tumor development.

The objective of the study of prostatic cancer is to explain why the human incidence is so high and is increasing, particularly in the United States, and ultimately to recommend how the disease may be averted. Dr. Bosland briefly notes two areas that we feel are most worthy of further research. Diethylstilbestrol was, over many years, given to pregnant women to "improve" pregnancy. While some female offspring developed vaginal cancer shortly after the menarche (90,91), no equivalent tumors have been described in male offspring. It would be worth an epidemiological study to determine if these male offspring are unusually liable to prostatic cancer. The second area concerns the average United States diet that is exceedingly rich in fat as well as in calories. As noted earlier, excess fats carry the potential to increase the level of oxidative DNA stress and thus to increase the level of mutational changes in a cell population; increased calories is an effective way of increasing the rate of cellular proliferation. Again, this needs direct epidemiological study.

L. Colorectal Cancer

Colorectal cancer is one of the most prevalent forms of cancer in both sexes of humans living in westernized communities. Despite this, as Drs. Kune and Vitetta emphasize in Chapter 27, we know little more about its etiology other than it appears to be multifactorial in origin. Animal studies have provided little help except in confirming the relevance of those factors that modify the expression of colorectal cancer in humans. The mechanistic considerations outlined in Chapter 27 suggest that the available evidence may be fit into a scheme that is capable of indicating those factors that appear to be most important and how research may proceed to clarify the present situation. It is important to be able to recognize initiating and promoting factors relevant to human colorectal cancer.

It is stressed in Chapter 27 that the traces of cooking-related (and smoking-related) heterocyclic aromatic amines as well as traces of nitrosamines may act as cancer initiators in the colorectum. Their level may not be high enough to account for more than a fraction of the induced tumors. The cells at the base of the intestinal crypts are normally proliferating very rapidly and, together with animal-derived evidence that dietary or energy restriction lowers both this rate of proliferation (78) as well as the tumor yield (3–5), supports the concept that this tissue may be self-promoting so far as tumor development is concerned. Thus, promoting agents may be less important than initiating agents in the colorectum.

The important evidence presented by Drs. Kune and Vitetta would appear to be the multiple epidemiological studies supporting the conclusion that human colorectal cancer is enhanced by diets rich in meat, fats, and similar constituents. Meats are rich in hemoglobin and iron as well as fats, each of which may lead to increased oxidative stress. Also, in the past, Japanese living in Japan had much lower incidences of colorectal cancer than Americans, an observation that may be linked with the fact that the Japanese diet contained only about 25% as much fat as that consumed in the United States. This may be interpreted to indicate that enhanced oxidative DNA stress is a major factor in the development of colorectal tumors in humans. This idea is further supported by epidemiological investigations showing that diets rich in fruit and vegetables and that contain appreciable levels of naturally occurring antioxidants are somewhat protective. The fact that high levels of ethanol consumption appear to increase the levels of colorectal cancer is supportive insofar as the carcinogenic properties of ethanol in other tissues may depend on the production of an ethanol-derived oxygen radical (38,39).

It would appear therefore that future investigations of the causes of human colorectal cancer should involve the levels of oxidative DNA stress in individuals consuming different diets. This will require either the less satisfactory estimation of degraded DNA oxidation products in urine or the development of novel techniques designed to limit their intrusiveness in human subjects. Further animal studies designed to support these conclusions will be of great value.

M. Tumors of the Exocrine Pancreas

In humans, exocrine pancreatic cancer is about twice as common in men as in women. Until the recent development of novel effective clinical techniques, it was exceptionally difficult to diagnose pancreatic cancer at an early stage. Consequently, the disease has had a very poor prognosis. About 50% of patients have died from the disease during the first 4 months after diagnosis. The 5-year survival rate was only 2.5%. Clearly, further research is needed to facilitate the early diagnosis of pancreatic cancer.

In Chapter 26, Dr. Bueno-de-Mesquita and coworkers carefully list a large number of human epidemiological and animal studies that address this subject. However, at present, it is not possible to reach many definite conclusions. It appears that in humans and in animals both mutational events and cell proliferation, due to various growth factors, are needed to generate pancreatic cancer. Each of the 17 chemicals that have led to pancreatic cancer in animals is genotoxic. The affected oncogenes show a degree of commonality between the species investigated. In humans, the great majority of exocrine pancreatic tumors are ductal adenocarcinomas. In rats, administration of the chemical asaserine led to acinar cell adenocarcinomas. In hamsters dosed with *N*-nitrosobis(2-oxopropyl)amine (BOP) or a series of structurally and metabolically related chemicals, the induced pancreatic tumors are ductal adenocarcinomas. Thus, treatment with BOP or its analogs matches the situation in humans. The oncogenes activated in hamsters appear to be marginally more similar to those activated oncogenes in humans than do those in rats. It is therefore to be expected that the experimentalist would use the hamster model, originally introduced by Dr. Pour and colleagues (92,93), rather than the rat model, because of the similarity of the resultant tumor type to those found in humans. The volume of work using the asaserine rat model suggests that this major point of human and animal tumor similarity has not always been fully considered.

The search for initiating, enhancing, or inhibiting factors concerned in the genesis of exocrine pancreatic cancer has not been very fruitful. Our present knowledge does not greatly help us in defining the mechanisms responsible for the disease. Epidemiology in humans suggests there may be an association with cigarette smoking, but this does not yet extend to the level of a dose-response relationship between the induction of pancreatic cancer and the number of cigarettes smoked. If, like lung cancer, the incidence of pancreatic cancer in women increases to reflect the fact that women acquired the cigarette-smoking habit considerably later in the twentieth century than men, there will be a greater degree of certainty about the smoking–pancreatic cancer association. Ethanol-containing beverages have been investigated on several occasions as factors in human pancreatic carcinogenesis, but overall the evidence is not sufficiently strong to be convincing. Many other, mainly dietary factors have been investigated, but again the results are far from convincing. Studies in animals have indicated that a group of chemicals such as selenium inhibit the onset of pancreatic cancer.

The terribly poor prognosis associated with human pancreatic cancer combined with the relative lack of knowledge about how these tumors arise in humans or animals makes it a prime target for further research. In planning further experiments, sight should not be lost of the fact that the best model for the induction of these tumors (BOP-treated hamsters) has only been known for about 20 years. Therefore, the greatest benefits may still to be derived from more classical studies designed to identify causative and modifying factors rather than from molecular biological observations.

N. Cervix/Uterus Carcinomas

Cancers of the cervix in women are, on a worldwide basis, second in frequency only to those of the breast. These tumors are much more common in women living in the less developed countries of Central and South Africa and Southeast Asia than in the more developed areas of Europe, North America, and Australia. This tumor distribution may suggest that levels of general hygiene and, in particular, sexual hygiene may be involved. It may also be asked whether the use of the recently developed uterine smear, the Pap test, has served to reduce the apparent incidence and mortality of this disease in western nations.

In Chapter 28, Dr. Marselos discusses the factors involved in the genesis of the two major cancers in this area of the female genital tract, uterine cervical cancer and endometrial cancer. The possible involvement of herpes simplex virus and human papillomavirus are mentioned, but the evidence that they might be able to effect cell transformation or tumor initiation appears, unfortunately, to be very limited. Major emphasis is placed on the role played by natural and synthetic hormones, particularly estrogens. In uterine carcinogenesis, different species react differently to the same chemicals. This is forcefully illustrated in the case of humans, mice, and rats (Chapter 28, Tables 1–3) where the available evidence indicates that only one of these three species may be affected by a particular estrogen.

Progress in this difficult area depends primarily on deciding the nature of the factor(s) that serves to "initiate" these tumors in humans or animals, or to demonstrate that both naturally occurring and synthetic estrogens may act in such tissues as "complete" carcinogens. The ability of estrogens to enhance cellular proliferation in the endometrial epithelium strongly suggests that these agents promote development of tumors of this tissue. This does not, however, help to identify the initiators that, by modifying oncogenes or tumor-suppressor genes, start up the process of carcinogenesis.

A tumor of another tissue in the female genital tract deserves mention. In the 1940s and 1950s many clinicians believed that the administration of quite large amounts of diethylstilbestrol (DES) to pregnant humans improved the outcome of the pregnancy. It has subsequently been found that

DES administration during the first trimester of pregnancy led in some female offspring to the appearance of vaginal adenocarcinoma shortly after the menarche (90,91). Information on how transplacental DES leads to these human tumors is still awaited. It does, however, emphasize the need to consider the whole life span in reaching conclusions about tumorigenesis by specific carcinogens. DES is a further rare example of an apparently nongenotoxic human carcinogen. Estrogen-induced benign liver tumors in women (Sec. III.A) are another example of a nongenotoxic human carcinogen.

IV. CONCLUSIONS

The wealth of information summarized in this book combined with our increasing knowledge of carcinogenesis mechanisms makes it abundantly clear that not all rodent carcinogens are necessarily carcinogenic in humans. An attempt to integrate mechanisms of cancer induction in various tissues with carcinogenic response in humans, rats, hamsters, and mice (Table 2) suggests that most identified human carcinogens react with DNA usually directly through electrophile generation or occasionally indirectly through oxidative stress. As pointed out repeatedly in the present chapter, the level of investigation of oxidative stress in the formation of tumors of many tissues needs detailed study. Two major exceptions to the concept that most human tumors are due to electrophile generators have been noted. First, the early oral (estrogen only) contraceptives cause benign liver tumors, and in utero diethylstilbestrol induces vaginal cancer following the menarche. Both chemicals have been used as human medicines in much greater doses than normally obtained from environmental contamination. Many nongenotoxic rodent carcinogens require very high exposure levels to be carcinogenic, and, under normal circumstances, humans do not receive such high exposures. Still it would be dangerous to assume without further evidence that nongenotoxic rodent carcinogens will be completely "safe" for humans. One approach to this problem would be to demonstrate firm evidence that the effect in rodents is thresholded and to apply a suitable but not excessive safety factor to the properly established data. Establishing such a threshold through animal bioassays is clearly impracticable because of the huge number of animals required and the consequential immense costs of such a study. What is required and is now becoming feasible in certain cases is the ability to identify one, or more, critical facets of the carcinogenesis mechanism and to demonstrate a threshold for these critical facets. This is illustrated by the fact that butylated hydroxyanisole is an antioxidant at lower exposure levels but pro-oxidative at higher exposure levels. At the change-over point between antioxidant and pro-oxidant properties, the induction of cellular proliferation and other pathological effects demonstrated a dose-response threshold. The assumption of a dose-response threshold without some direct experimental and/or mechanistic evidence is not scientifically or morally permissible for substances that may include serious human disease.

In summary, electrophile generators have a high probability of being carcinogenic in both humans and rodents. Chemicals leading to oxidative stress may well be carcinogenic in both humans and rodents, but a great deal more information, especially in humans, is required to determine the validity of this viewpoint. With nongenotoxic carcinogens (both receptor and cellular proliferation mediated), the parallelism between species is much less clear. If general mechanistic rules to determine which nongenotoxic rodent carcinogens pose a real threat to humans are to be established, an immense amount of research will be required. This point is very clearly demonstrated by Dr. Goldsworthy and colleagues in Chapter 4, in their discussion of mouse liver tumor induction.

REFERENCES

1. Doll R, Peto R. The Causes of Cancer: Quantitative Estimates of Avoidable Risk of Cancer in the United States Today. Oxford: Oxford University Press, 1981.
2. Committee on Diet, Nutrition and Cancer, Assembly of Life Sciences, National Research Council. Diet, Nutrition and Cancer. Washington, DC: National Academy Press, 1982.
3. Tannenbaum A, Silverstone H. Nutrition and the genesis of tumours. In: Raven RW, ed. Cancer. London: Butterworth, 1957:306–334.
4. Ross MH, Bras G. Influences of protein under-and overfeeding and overnutrition on spontaneous tumor prevalence in the rat. J Nutr 1973; 103:944–963.
5. Roe FJC, Lee PN, Conybeare G, Kelly D, Matter B, Prentice D, Tobin J. The Biosure Study: influence of composition of diet on longevity, degenerative diseases and neoplasia in Wistar rats studied for up to 30 months post weaning. Food Chem Toxicol 1995; 33(suppl 1):1S–100S.
6. Ames BN, Gold LS. Endogenous mutagens and the causes of aging and cancer. Mutat Res 1991; 250:3–16.
7. Sugimura T. Carcinogenicity of mutagenic heterocyclic aromatic amines formed during the cooking process. Mutat Res 1985; 150:33–41.
8. Willis RA. Pathology of Tumours. 3rd ed. London: Butterworth, 1960.
9. Clayson DB. Chemical Carcinogenesis. London: Churchill, 1962.
10. Miller JA. Chemical carcinogenesis—an overview: G.H.A. Clowes Memorial Lecture. Cancer Res 1970; 30:559–576.
11. Berenblum I, Shubik P. The role of croton oil applications, associated with a

single painting of a carcinogen, in tumour induction of the mouse's skin. Br J Cancer 1947; 1:379–382.
12. Berenblum I, Shubik P. A new, quantitative approach to the study of the stages of chemical carcinogenesis in the mouse's skin. Br J Cancer 1947; 1: 383–391.
13. Berenblum I, Shubik P. The persistance of latent tumour cells induced in the mouse's skin by a single application of 9,10-dimethyl-1,2-benzanthracene. Br J Cancer 1949; 3:384–386.
14. Miller EC, Miller JA, Hartmann HA. N-Hydroxy-2-acetylaminofluorene: a new metabolite of 2-acetylaminofluorene with increased carcinogenic activity in the rat. Cancer Res 1961; 21:815–824.
15. Miller EC, Miller JA, Enomoto M. The comparative carcinogenicity of 2-acetylaminofluorene and its N-hydroxy metabolite in mice, hamsters and guinea pigs. Cancer Res 1964; 24:2018–2031.
16. DeBaun JR, Miller EC, Miller JA. N-Hydroxy-2-acetylaminoflourene sulfotransferase: its probable role in carcinogenesis and in protein (methion-S-yl) binding in rat liver. Cancer Res 1970; 30:577–595.
17. Searle CE, ed. Chemical Carcinogens. 2nd ed. Washington, DC: American Chemical Society Press, 1984.
18. Brusch W, Kleine L, Tenniswood M. The biochemistry of cell death by apoptosis. Biochem Cell Biol 1990; 68:1071–1074.
19. Kimashima Y, Uehara T, Kishi K, Shiromizu K, Matsuzawa M, Tajayama S. Proliferative and apoptotic status in endometrial adenocarcinoma. Int J Gynacol Pathol 1995; 14:45–49.
20. Klein G. The approaching era of the tumor suppressor genes. Science 1987; 238:1539–1545.
21. Klein G, Klein E. Commentary: oncogene activation and tumor progression. Carcinogenesis 1984; 5:429–435.
22. McCann J, Ames BN. Detection of carcinogens in the Salmonella/microsome test: assay of 300 chemicals. Proc Nat Acad Sci USA 1975; 73:950–954.
23. Ashby J, Tennant RW. Chemical structure, Salmonella mutagenicity and the extent of carcinogenicity as indicators of genotoxic carcinogenesis among 222 chemicals tested in rodents by US NCI/NTP. Mutat Res 1988; 204:17–115.
24. Turteltaub KW, Felton JS, Gledhill BL, Vogel JS, Southon JR, Catfee MW, Finkel RC, Nelson DE, Proctor ID, Davis JC. Accelerator mass spectrometry in biomedical dosimetry: relationship between low-level exposure and covalent binding of heterocyclic carcinogens to DNA. Proc Nat Acad Sci USA 1990; 87;5288–5292.
25. Felton JS, Turteltaub KW. Accelerator mass spectrometry for measuring low-dose carcinogen binding to DNA. Environ Health Perspect 1994; 102:450–452.
26. Hicks RM, Wright R, Wakefield JSJ. The induction of rat bladder cancer by 2-naphthylamine. Br J Cancer 1982; 46:646–661.
27. Garner RC, Martin CN, Clayson DB. Carcinogenic aromatic amines and related compounds. In: Searle CE, ed. Chemical Carcinogens. Washington, DC: American Chemical Society, 1984:175–276.

28. Reitz RH. Distribution, persistence and elimination of toxic agents (pharmacokinetics). In: Clayson DB, Munro IC, Shubik P, Swenberg JA, eds. Progress in Predictive Toxicology. Amsterdam: Elsevier, 1990:79–90.
29. Halliwell B, Aruoma O. DNA damage by oxygen derived species: its mechanism and measurement in mammalian systems. FEBS Lett 1991; 281:9–19.
30. Clayson DB, Mehta R, Iverson F. Oxidative DNA damage: the effects of certain genotoxic and operationally nongenotoxic carcinogens. Mutat Res 1994; 317:23–42.
31. von Sonntag C. The Chemical Basis of Radiation Biology. London: Taylor and Francis, 1987.
32. Kuchino Y, Mori F, Masai H, Inoui H, Iwai S, Miura K, Ohisuka E, Nishimura S. Misreading of DNA templates containing 8-hydroxydeoxyguanosine as the modified base and at adjacent residues. Nature (London) 1987; 327:77–80.
33. Floyd RA. The role of 8-hydroxyguanosine in carcinogenesis. Carcinogenesis 1990; 11:1447–1450.
34. Hayflick L. The limited in vitro lifetime of human diploid cell strains. Exp Cell Res 1965; 37:614–636.
35. Hayflick L. Aging under glass. Mutat Res 1991; 356:69–80.
36. Cummings SW, Prough RA. Butylated hydroxyanisole-stimulated NADPH oxidase activity in rat liver microsomal fraction. J Biol Chem 1983; 258:12315–12319.
37. Gille JJP, van Berkel CGM, Joenje H. Mutagenicity of metabolic oxygen radicals in mammalian cultures. Carcinogenesis 1994; 15:2695–2699.
38. Sram RJ, Ginkova B, Gebhart JA. Impact of environmental mutagens on mental health. In: Bulyzenkov V, Christen Y, Philipko L, eds. Approaches to the Prevention of Mental Disorders. Berlin: Springer-Verlag, 1990:74–80.
39. Sram RJ, Topinka J, Binkova B, Kokisova J, Kubicek V, Gebhart JA. Genetic damage in peripheral lymphocytes of chronic alcoholics. In: Garner RC, Hradec J, eds. Biochemistry and Chemical Carcinogenesis. New York: Plenum Press, 1990:219–226.
40. Pitot HC. The molecular biology of carcinogenesis. Cancer 1993; 72:962–970.
41. Fearon ER, Vogelstein B. A genetic model for colorectal tumorigenesis. Cell 1990; 61:759–767.
42. Jensen EV. Estrogen-receptors in hormone-dependent breast cancer. Cancer Res 1975; 35:3362–3364.
43. Tugwood JD, Isseman I, Anderson RG, Bundell KR, McPheat WL, Green S. The mouse peroxisome proliferator activated receptor recognizes a response element in the 5′ flanking sequence of the rat acyl CoA oxidase gene. EMBO J 1992; 11:433–439.
44. Isseman I, Green S. Activation of a member of the steroid hormone receptor superfamily by peroxisome proliferators. Nature (London) 1990; 347:645–651.
45. Poland A, Glover AE, Kende AS. Stereospecific, high affinity of 2,3,7,8-tetrachlorodibenzo-p-dioxin by hepatic cytosol. J Biol Chem 1976; 251:4936–4938.
46. Whitlock JP. Mechanistic aspects of dioxin action. Chem Res Toxicol 1993; 6:754–763.

47. Alison MR. Assessing cellular proliferation: What's worth measuring? Human Exp Toxicol 1995; 14:935–944.
48. Nera EA, Lok E, Iverson F, Ormsby E, Karpinski KF, Clayson DB. Short-term pathological and proliferative effects of butylated hydroxyanisole and other phenolic antioxidants in the forestomach of Fischer 344 rats. Toxicology 1984; 32:197–213.
49. Iverson F, Lok E, Nera E, Karpinski K, Clayson DB. A 13-week feeding study of butylated hydroxyanisole: the subsequent regression of the induced lesions in the male Fischer 344 rat forestomach epithelium. Toxicology 1985; 35:1–11.
50. Verhagen H, Furnee C, Schutte B, Bosman FT, Blijnam GH, Henderson PT, ten Hoor F, Kleinjans JCS. Dose-dependent effects of short-term dietary administration of the food additive butylated hydroxyanisole on cell kinetic parameters in the rat gastro-intestinal tract. Carcinogenesis 1990; 11:1461–1468.
51. Bomhard EM, Bremmer JN, Herbold BA. Review of the mutagenicity/genotoxicity of butylated hydroxytoluene. Mutat Res 1992; 277:187–200.
52. Olsen P, Meyer O, Bille N, Wurtzen G. Carcinogenicity study on butylated hydroxytoluene (BHT) in Wistar rats exposed in utero. Food Chem Toxicol 1986; 24:1–12.
53. Inai K, Kobuke T, Nambu S, Takemoto T, Kou E, Nishina H, Fujijara M, Yonehara S, Suehiro S, Tsuya T, Horiuchi K, Tokuoka S. Hepatocellular tumorigenicity of butylated hydroxytoluene administered orally to B6C3F$_1$ mice. Jpn J Cancer Res 1988; 79:49–58.
54. Briggs D, Lok E, Nera EA, Karpinski K, Clayson DB. Short-term effects of butylated hydroxytoluene on the Wister rat liver, urinary bladder and thyroid gland. Cancer Lett 1990; 46:31–35.
55. Lok E, Mehta R, Jee P, Laver P, Nera EA, McMullen E, Clayson DB. The effects of butylated hydroxytoluene on the growth of enzyme-altered foci in male Fischer 344 rat liver tissue. Carcinogenesis 1995; 16:1071–1078.
56. Greenwell A, Foley JF, Maronpot RR. An enhancement method for immunohistochemical staining of proliferating cell nuclear antigen in archival rodent tissues. Cancer Lett 1989; 59:251–256.
57. Crocker J. Nucleolar organiser regions. Curr Top Pathol 1990; 82:91–149.
58. Derenzini M, Ploton D. Interphase nucleolar organiser regions in cancer cells. Int Rev Exp Pathol 1991; 32:149–192.
59. Mehta R. The potential for the use of cell proliferation and oncogene expression as intermediate markers during liver carcinogenesis. Cancer Lett 1995; 93: 85–102.
60. Vesselovinovitch SD, Itze L, Mihailovich N, Rao KVN. Modifying role of partial hepatectomy and gonadectomy on ethylnitrosourea-induced hepatocarcinogenesis. Cancer Res 1980; 40:1538–1542.
61. Standeven AM, Goldsworthy TL. Identification of hepatic mitogenic and cytochrome P 450-inducing fractions of unleaded gasoline in B6C3F1 mice. J Toxicol Environ Health 1995; 43:213–224.
62. Standeven AM, Wolf DC, Goldsworthy TL. Interactive effects of unleaded

gasoline and estrogen on liver tumor promotion in female B6C3F$_1$ mice. Cancer Res 1994; 54:1198–1204.
63. Standeven AM, Blazer DG, Goldsworthy TL. Investigation of antiestrogenic properties of unleaded gasoline in female mice. Toxicol Appl Pharmacol 1994; 127:233–240.
64. Cohen SM, Ellwein LB. Genetic errors, cell proliferation, and carcinogenesis. Cancer Res 1991; 51:6493–6505.
65. Barnard RC. Risk assessment: the default conservatism controversy. Reg Toxicol Pharmacol 1995; 21:431–438.
66. Clayson DB, Averson F. Carcinogenesis risk assessment at the crossroads: need for a biological turning? Reg Toxicol Pharmacol 1996; 24:45–59.
67. Dahlgren S. Thorotrast tumours. Acta Path Microbiol Scand 1961; 53:147–161.
68. Boyd JT, Longlands AO, MacCabe JJ. Long-term hazards of thorotrast. Br Med J 1968; ii:517–521.
69. Mori T, Sakai Y, Nozue Y, Okamoto T, Wada T, Tanuka T, Tsuya A. Malignancies and other injuries following thorotrast administration: follow-up study of 146 cases in Japan. Strahlentherapie 1967; 134:229–254.
70. Christopherson WM, Mays ET. Guest editorial—liver tumors and contraceptive steroids: experience with the first 100 registery cases. J Natl Cancer Inst 1977; 58:167–171.
71. Committee on Safety of Medicines. Carcinogenicity tests on oral contraceptives. A report of the Committee on Safety of Medicines. London: H.M. Stationery Office, 1963.
72. Flamm WG, Lehman-McKermann LD. The human relevance of the renal tumor-inducing potential of d-limonene in male rats: implications for risk assessment. Reg Toxicol Pharmacol 1991; 13:70–86.
73. Bengtsson U, Angerval L, Ekman H, Lehmann L. Transitional cell tumours of the renal pelvis in analgesic abusers. Scand J Urol Nephrol 1968; 2:145–150.
72. Rathert P, Melchoir H, Lutzeyer W. Phenacetin, a carcinogen for the urinary tract. J Urol 1975; 113:653–657.
75. Grice HC. Safety evaluation of butylated hydroxyanisole from the perspective of effects on forestomach and esophageal squamous epithelium. Food Chem Toxicol 1988; 26:717–723.
76. Bielschowsky F. Tumours of the thyroid produced by 2-acetylaminofluorene and allyl-thiourea. Br J Exp Pathol 1944; 25:90–95.
77. Bielschowsky F. Experimental nodular goitre. Br J Exp Pathol 1945; 26:270–275.
78. Lok E, Scott FW, Mongeau R, Nera EA, Malcolm S, Clayson DB. Caloric restriction and cellular proliferation in various tissues of the female Swiss Webster mouse. Cancer Lett 1990; 51:67–73.
79. Lok E, Ratnayake WMN, Scott FW, Mongeau R, Fernie S, Nera EA, Malcolm S, McMullen E, Clayson DB. The effect of varying the type of fat in a semipurified AIN-76A diet on cellular proliferation in the mammary gland

and intestinal crypts in female Swiss Webster mice. Carcinogenesis 1992; 13: 1735–1741.
80. Bittner JJ. "Incidences" of breast cancer development in mice. Publ Hlth Rep (Washington) 1938; 54:1590.
81. Clayson DB, Cooper EH. Cancer of the urinary tract. Advances Cancer Res 1970; 13:271–372.
82. Garner RC, Martin CM, Clayson DB. Carcinogenic aromatic amines and related compounds. In: Searle CE, ed. Chemical Carcinogens. 2nd ed. Washington, DC: American Chemical Society Press, 1984:175–276.
83. Childs JJ, Nakajima C, Clayson DB. The metabolism of 1-phenylazo-2-naphthol in the rat with reference to the action of the intestinal flora. Biochem Pharmacol 1967; 16:1555–1558.
84. Schoenig GP, Goldenthal EI, Geil RG, Frith CH, Richter WR, Calborg FW. Evaluation of the dose response and in utero exposure to saccharin in the rat. Food Chem Toxicol 1985; 23:475–490.
85. Clayson DB, Fishbein L, Cohen SM. Effects of stones and other physical factors on the induction of rodent bladder cancer. Food Chem Toxicol 1995; 33:771–784.
86. Clayson DB, Pringle JAS. The influence of a foreign body on the induction of tumours of the bladder epithelium of the mouse. Br J Cancer 1966; 20:564–568.
87. Burin BJ, Gibbs HJ, Hill RN. Human bladder cancer: evidence for a potential irritation-induced mechanism. Food Chem Toxicol 1995; 33:785–795.
88. Nera EA, Iverson F, Lok E, Armstrong CL, Karpinski K, Clayson DB. A carcinogenesis reversibility study of the effects of butylated hydroxyanisole on the forestomach and urinary bladder in male Fischer 344 rats. Toxicology 1988; 53:251–268.
89. Badawi AF. Molecular and genetic events in schistosomiasis associated human bladder cancer: role of oncogenes and tumor suppressor genes. Cancer Lett 1996; 105:123–138.
90. Herbst AL, Ulfekdwe H, Poskanzer DC. Adenocarcinoma of the vagina. Association of maternal stilbestrol therapy with tumor appearance in young women. N Engl J Med 1971; 284:878–881.
91. Poskanzer DC, Herbst AL. Epidemiology of vaginal adenosis and adenocarcinoma associated with exposure to stilbestrol in utero. Cancer 1977; 39:1892–1895.
92. Pour P, Althoff J, Gingell R, Koprer R, Kruger FW. N-Nitrosobis(2-hydroxypropyl)amine as a further pancreatic carcinogen in Syrian golden hamsters. Cancer Res 1976; 36:2877–2884.
93. Pour P, Althoff J, Kruger FW, Mohr U. A potential pancreatic carcinogen in Syrian golden hamsters: N-nitrosobis(2-oxopropyl)amine. J Natl Cancer Inst 1977; 58:1459–1463.

Index